ELEMENTS
of
STRUCTURAL ANALYSIS

ELEMENTS
of
STRUCTURAL ANALYSIS

N. C. Sinha
BE (Calcutta), MS (Stanford), PhD FIE (India)
Chartered Engineer
Formerly Professor and Head, Department of Civil Engineering
Bengal Engineering College (Bengal University of Science and Technology)

New Central Book Agency (P) Ltd
LONDON
HYDERABAD ERNAKULAM BHUBANESWAR
NEW DELHI KOLKATA PUNE GUWAHATI

NCBA

REGD OFFICE
8/1 Chintamoni Das Lane, Kolkata 700 009, India
email: ncbapvtltd@eth.net

OVERSEAS
NCBA (UK) Ltd, 149 Park Avenue North, Northampton, NN3 2HY, UK
email: ncbauk@yahoo.com

EXPORT
NCBA Exports Pvt. Ltd, 212 Shahpur Jat, New Delhi 110 049
email: ncbaexports@ncbapvtltd.com

BRANCHES

212 Shahpur Jat
New Delhi 110 049
email: ncbadel@ncbapvtltd.com

Shop Nos. 3 & 4, Vinayak Towers
681/B Budhwar Peth, Appa Balwant Chowk
Pune 411 002
email: ncbapune@ncbapvtltd.com

Shop No. 15, Bharati Towers, Ground Floor
Forest Park, **Bhubaneswar** 751 001
email: ncbabub@ncbapvtltd.com

House No. 3-1-315, 1st Floor
Nimboliadda, Kachiguda, **Hyderabad** 500 027
email: ncbahydb@ncbapvtltd.com

GSS Shopping Complex, 1st Floor
Opposite College Post Office, Convent Road
Ernakulam 682 035
email: ncbaernk@ncbapvtltd.com

Radhamayee Complex (Second Floor)
Opp. SBI, Panbazar, Jaswanta Road
Guwahati 781 001
email: ncbaguwahati@ncbapvtltd.com

ELEMENTS OF STRUCTURAL ANALYSIS • N. C. Sinha

First Published: February 2011
Reprinted: August 2013

PUBLISHER

New Central Book Agency (P) Ltd
8/1 Chintamoni Das Lane, Kolkata 700 009

TYPESETTER

Anin, Kolkata

PRINTER

New Central Book Agency (P) Ltd
Web-Offset Division, Dhulagarh, Sankrail, Howrah

COVER DESIGNER

Soumen Paul

COVER PRINTER

Biswanath Studio, Kolkata

TECHNICAL EDITOR

Dipan Roy

PROJECT TEAM

Prabhat Jas, Swapan Sarkar

ISBN 978 81 7381 638 3

Price: ₹ 635.00

Dedicated to the memory

of

my late Parents and late Wife

Contents

Preface

The book is the outcome of the author's long experience of teaching Structural Analysis to Undergraduate and Postgraduate students during his long teaching career—firstly in Jalpaiguri Government Engineering College and later in Bengal Engineering College. The book is based on the syllabi of Structural Analysis in various Technical Colleges and Universities. Even though the book is mainly meant for Undergraduate students who have to attend the courses on Structural Analysis mainly in the Fourth, Fifth, Sixth and Seventh Semesters, it contains some advanced topics, which, though mainly meant for Postgraduate students, have been included in the Undergraduate courses of some of the Technical Universities. Some such topics are the Matrix Methods of Analysis along with Finite Difference Method, Relaxation Technique, and Introduction to Structural Dynamics. In some parts of the book help have been taken from previous publications of the author in some Elementary and Advanced topics.

In writing the book a large number of Indian and foreign books on Structural Analysis have been consulted and taken help of. Some selected references in this respect have been listed at the end of the book.

An Appendix has been added to give ready reference to values for design in case of Statically Indeterminate Frames. This will be helpful to professionals taking up design and analysis of such structures.

A large number of examples have been worked out in the book so that the students can master the subjects. A large number of problems have also been included at the end of each chapter for practice of the students. It is quite possible that some computational and typographical errors are there and the author expresses regret for the same.

The author is grateful to New Central Book Agency (P) Ltd for taking up the publication of the book. Particular thanks are due to Sri R. M. Nath, Manager of New Central Book Agency (P) Ltd, for approaching the author and inducing him to write the book.

Rathayatra
13 July 2010
Kolkata

N. C. Sinha

Chapter 1

Concepts of Structural Engineering

1.1 Introduction

Structural Engineering recently has become a very vast subject. Basic job in Structural Engineering is to create a structure which will safely carry various imposed loads.

Structural analysis occupies a pivotal role in structural engineering as without a proper analysis it would not be possible to create a structure which can safely withstand various loads.

In case of structures, subjected to external forces, reactions and internal forces are induced in the members of the structure, along with deformations. Methods of calculations of these reactions and forces together with the deformations caused form the subject of structural analysis which requires the actual computing of the reactions and the internal forces, knowing which deformations produced in the structure can be readily evaluated.

For the analysis of structural problems perfectly rigid system completely restrained and subject to stationary load will be considered. Since no movement of any kind is involved, forces in the system will be considered, and the external forces acting in the system will be in a state of static equilibrium. Reactions, shears, moments and stresses in such systems may be obtained by application of the laws of statics.

Rapid development of Structural Engineering throughout the years have made it practically imperative that the engineers specializing in the design of modern structures are acquainted with the various methods of analysis available for the structure in which the application of the laws of statics is not sufficient. Laws of statics, however, are the fundamental tools of the structural engineer and have to be mastered thoroughly for proper understanding of the advanced methods of structural engineering.

1.2 Structural Forms

Important design decisions made by a structural engineer are the selection of suitable structural forms for the various requirements and objectives of a particular project. The most suitable structural form is the one which satisfies the functional, economic, aesthetic and other requirements to the highest degree.

From the point of view of efficiency, structural members may be classified into two broad categories :

(a) Uniform stress forms in which the stress is uniform over the depth of a member, or over the thickness of a panel as in Fig. 1.1.

(b) Varying stress forms in which the stress varies over the depth or thickness usually from a maximum tensile stress at one surface to a maximum compressive stress at the other as in Fig. 1.2.

Uniform stress forms utilise the available strength of the member in a more efficient manner than the varying stress forms. A structural engineer, therefore, tries to use a uniform stress form, when suitable from the point of view of functional requirements.

The simplest example of the uniform stress form is a member subjected to axial force in one direction (uniaxial) as in Fig. 1.1(a). Fig. 1.1(b) shows a plate of constant thickness subjected to uniform stress in two directions.

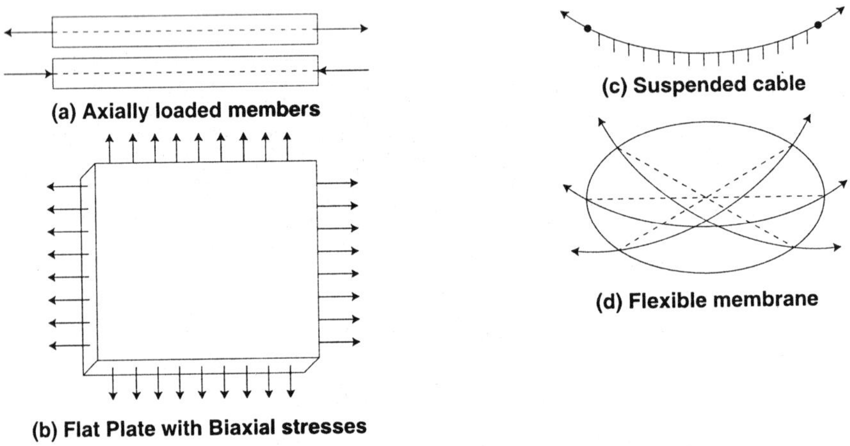

Fig. 1.1: Various forms of Structures (Uniform Stress)

Cable structure as in Fig. 1.1(c) illustrates a very efficient structural form. Cables can be considered perfectly flexible, and carry the loads by simple tension, even when subjected to lateral loads. They must adjust their shape to conform to the equilibrium polygon of loads applied to them. The magnitude of the tensile stress (uniaxial) in the cable varies along the length.

A flexible membrane as in Fig. 1.1(d) may be considered as a combination of large number of cables and it would be essentially this situation if the curved edges of the membrane were unsupported. The pattern of stress in this case is biaxial (called membrane stress) and the magnitude may vary from point to point.

Fig. 1.2(a) shows a slender member subjected to loads acting laterally, and subjected to bending moment and shear force is called a beam. The magnitude of the bending stress and shear stress may vary along the length of the beam, but will always vary along the depth. Fig. 1.2(b) showing a slab or plate supported and loaded laterally is subjected to bending moment, shear force and twisting moment. Stresses due to the effects vary along the width, length as well as the depth. The structural behaviour of a plate may be visualised as being similar to that of two layers of beam strips, each layer consisting of strips, which are cemented together side by side.

An arch in Fig. 1.2(c) is like an inverted cable, provided that its shape conforms to the equilibrium polygon for the loading imposed on it. Arch possesses flexural stiffness and strength as it would carry such a loading essentially with a uniform stress condition at each cross section. If, however, a completely different loading is applied to the arch, the arch tries to change in shape and being inflexible is subjected to bending moment and shear force over and above the direct force. Thus, a varying stress condition is developed in the arch, whereby the arch carries the load partly by flexural stress and partly by thrust.

An arch form may be an economical solution for cases, where there is one predominant loading condition. The equilibrium polygon of the predominant loading is then used to define the shape of the arch.

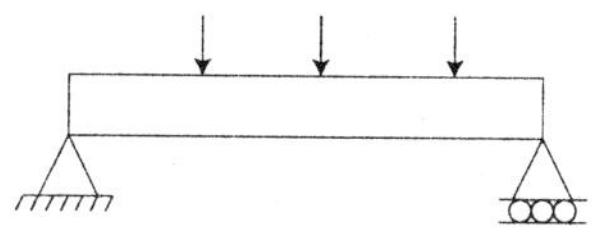

(a) Beam with lateral load

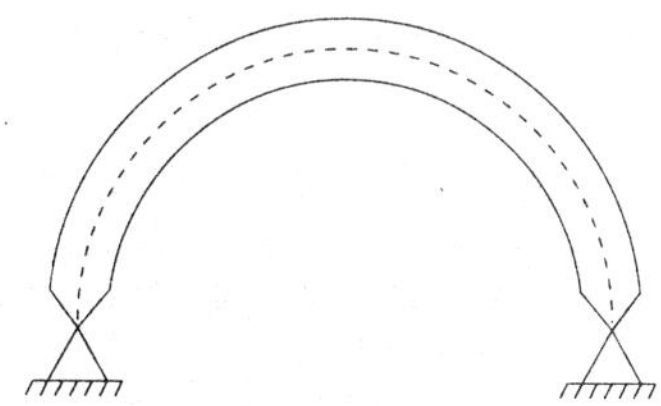

(c) Hinged Arch

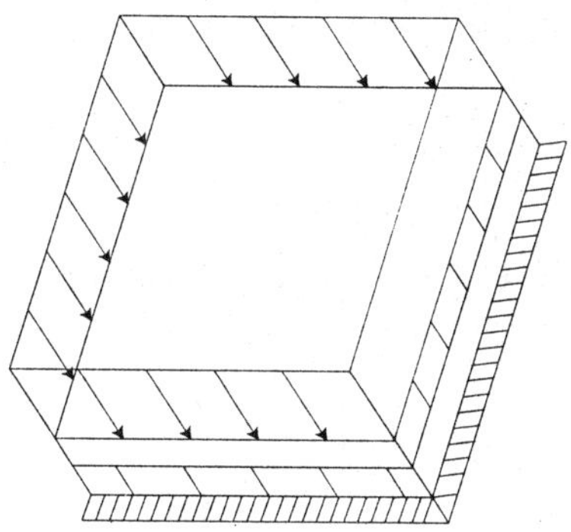

(b) Two-way slabs

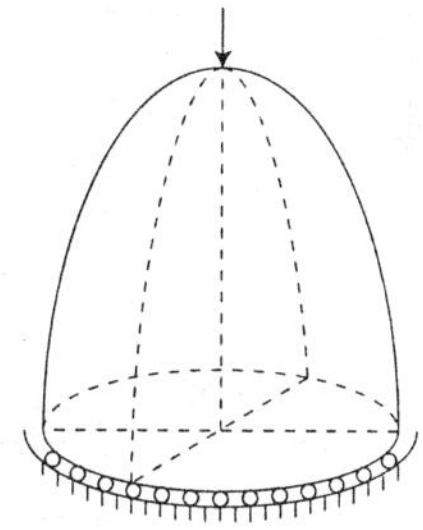

(d) Doubly curved shell

Fig. 1.2: Various forms of Structures (Varying Stress)

Fig. 1.2(d) shows a shell of revolution. If the edges are unrestrained and the loadings are symmetrical, only membrane stresses will exist. However, for restrained edges bending moment, shear force and membrane stresses will also develop.

Planer truss (or planer framework) is a very important uniform stress structural form. Fig. 1.3(a) shows pin connected trusses built up by combining several uniform stress elements. Even though such a structure as a whole tends to bend under the action of the loads, all elements of the structure will be uniformly stressed. Fig. 1.3(b) shows examples of a space truss in which even though members are uniformly stressed, the truss as a whole is subjected to bending and twisting under the loads.

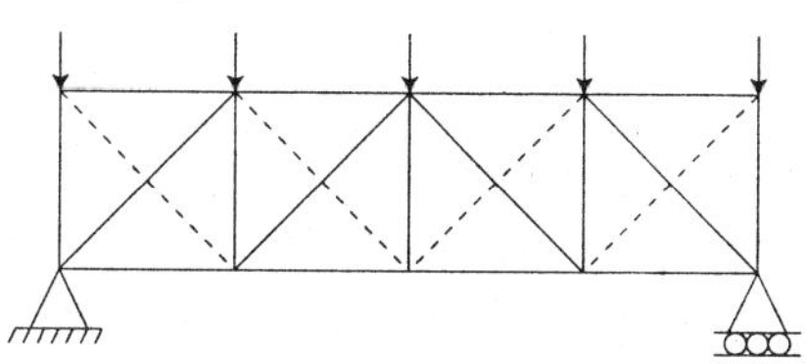

(a) Plane Truss

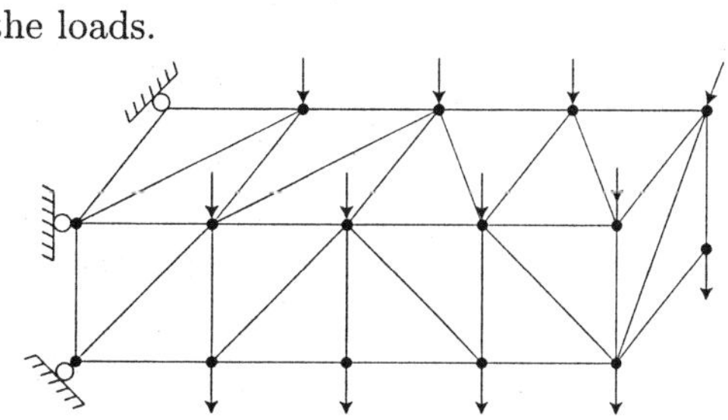

(b) Space Truss

Fig. 1.3: Types of Trusses

1.3 Idealisation of Practical Structure

A structure usually consists of assembly of intricate elements and connections having three dimensions. As analysis of three dimensional cases is rather complicated and the structure is wherever possible reduced to component two dimensional systems. There are of course structures in which the variation of stresses and strains in the three directions is important. In such cases simplification cannot be made and structures have to be treated as three dimensional problems.

Various three dimensional structures have uniform thickness in a particular direction or are long in the direction. In such cases, structures are normally subjected to uniform external forces along its thickness or length. All such structures can be reduced to plane stress problems.

Two dimensional structures are the most common systems encountered in practice and in such cases members are slender in the third direction. Such structures may be separated into several plane stress elements. A structure is again idealised and represented by its centre line dimensions. Examples of actual structures represented by centre line and their separation into elements for structural analysis are shown in Figs. 1.4, 1.5 and 1.6.

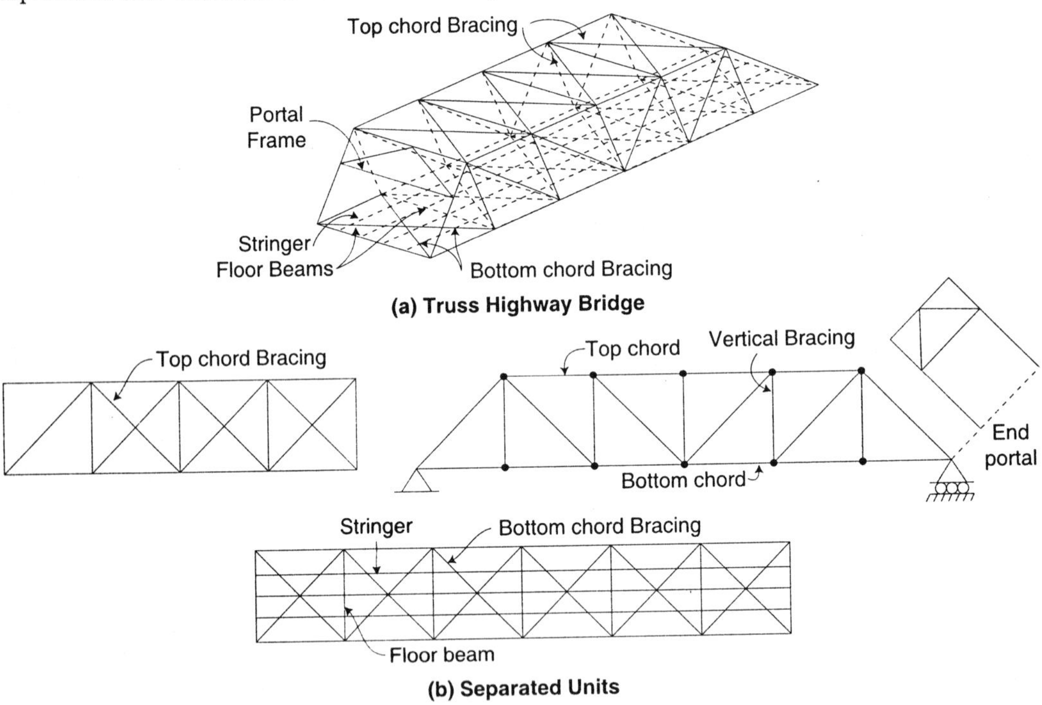

Fig. 1.4: Bridge Truss

Fig. 1.4(a) shows a truss highway bridge. The structure may be separated into several structural units, viz. the main truss, the lateral bracing system and the floor framing consisting of stringers and floor beams. The main trusses are designed to carry primarily vertical loads, transmitted to the trusses through the floor framing. The top and bottom chord bracings and the end portal frames which are part of the lateral bracing systems, are designed to resist the horizontal and lateral forces. The floor beams mainly transmit floor loads to the joints of the trusses and also act as part of the bottom chord bracing. Fig. 1.4(b) shows the separated units of the bridge truss used for simplified analysis.

Fig. 1.5 shows the framing plan of a highway plate girder bridge and its separation into various elements for analysis.

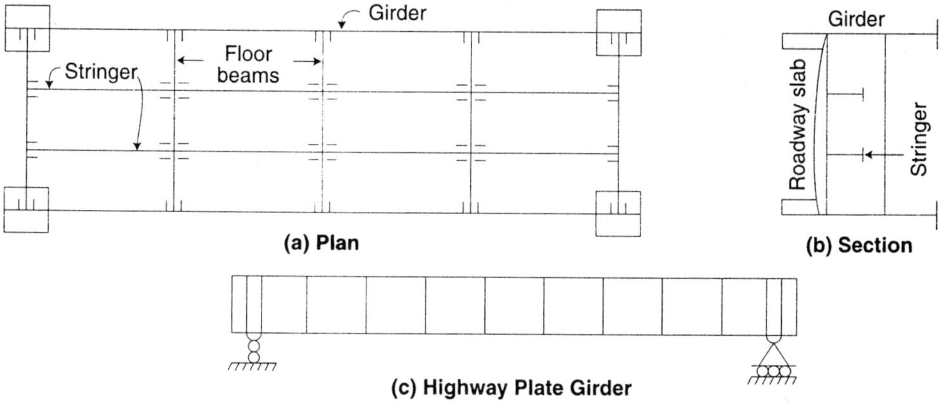

Fig. 1.5: Highway Bridges

Fig. 1.6 shows an actual structure and its idealised structure, for the purpose of practical analysis. In this case, the footing has been considered to be fixed and the joint has been assumed to be rigid. The effective height of the column has been taken as the distance from top of the foundation to the central line of the girder, the effective span of which is measured from the centre line of the column.

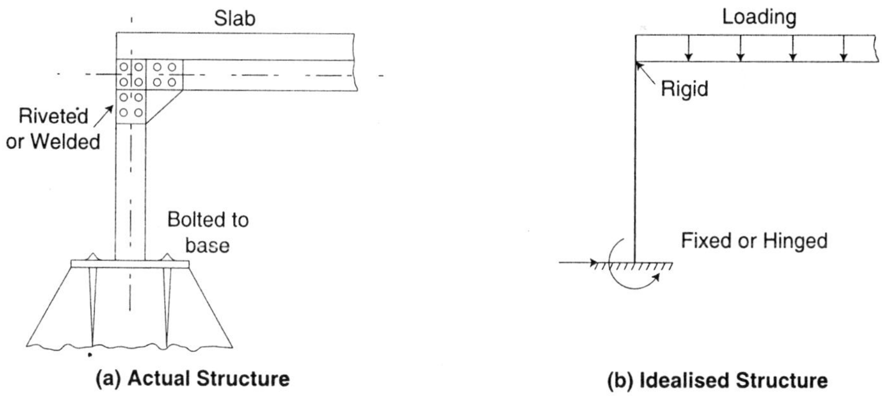

Fig. 1.6: Idealisation of Structure

It is necessary to idealise a structure for its practical analysis. Experience and judgement are necessary to determine the idealised structure. Where doubt exists, it is desirable to analyse on the basis of more than one type of idealised form.

1.4 Symbols and Convention

For the ideal structures, it will be convenient to make use of certain symbols and conventions for indicating the manner of support and construction of the structure under consideration.

Commonly used symbols are shown in Fig. 1.7. Fig. 1.7(a) represents roller supports in which movement normal to the surface on which rollers are supported is prevented, but in which rotation on the point of support and movement parallel to the surface of support are unimpeded. Fig. 1.7(b) represents hinged supports for beam or columns, in which movement in any direction at the point of support at A is prevented, but in which the beam or column is free to rotate on the point of support A. Fig. 1.7(c) represents fixed or restrained supports, where movement or rotation in any direction is absolutely prevented. Fig. 1.7(d) shows two members meeting at the corner A and represents a rigid joint there, so that the angle between the members at A cannot change. Fig. 1.7(e) represents a hinge joint at A, where angle between the members can change, although the members cannot separate.

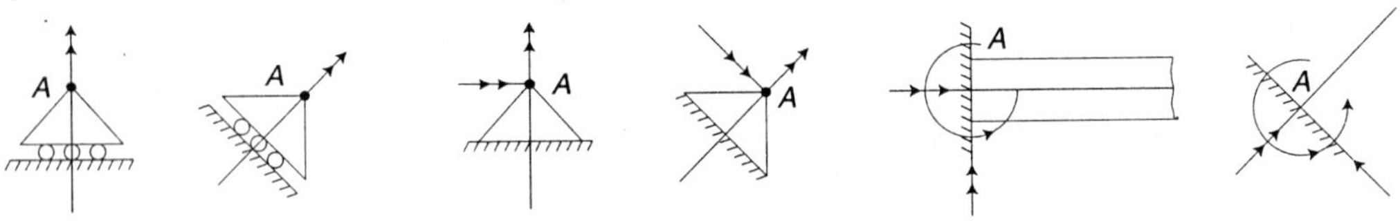

(a) Roller Support (Reaction Normal to Support) **(b) Hinged Support (Reactions in Two directions)** **(c) Fixed Support showing reactions**

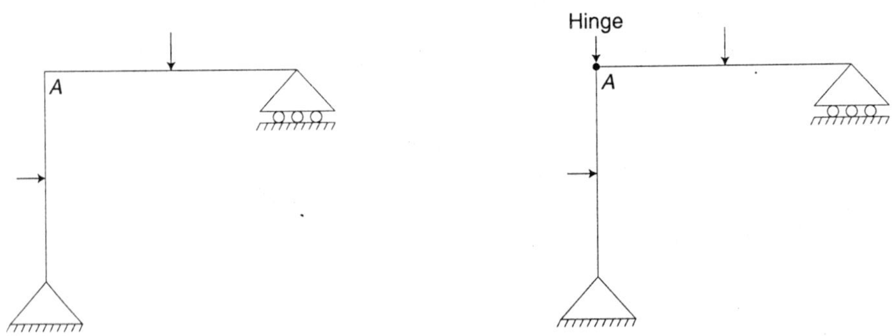

(d) Portal Frame with Rigid Joint at *A* **(e) Portal Frame with Hinge at *A***

Fig. 1.7: Symbols for supports with directions of Reactions (Reactions shown as ↑ → → ↻)

1.5 Force System

For a structural system acted upon by applied forces, rigid body movement is prevented by some restraining members in a state of equilibrium. Certain problems of practical importance such as determination of the number of restraining members required, the forces developed in the restraining members and the forces and stresses developed within the system itself will then arise. Buildings, bridges, machine, missiles, and others, are examples which may involve such problems.

Force system is in equilibrium if its resultant vanishes, and the resultant movement about any point is zero. Expressing these conditions for a space structure in a rectangular cartesian coordinate shown in Fig. 1.8. the equations for condition of equilibrium are

$$\sum F_x = 0; \quad \sum F_y = 0; \quad \sum F_z = 0;$$

i.e., the algebraic sum of all forces acting on a body in the equilibrium in the directions X, Y and Z must equal zero, and

$$\sum M_x = 0; \ \sum M_y = 0; \ \sum M_z = 0;$$

i.e., the algebraic sum of the moments of all the forces acting on body in equilibrium about any point on the X, Y and Z axes and on the plane of the loads must equal zero.

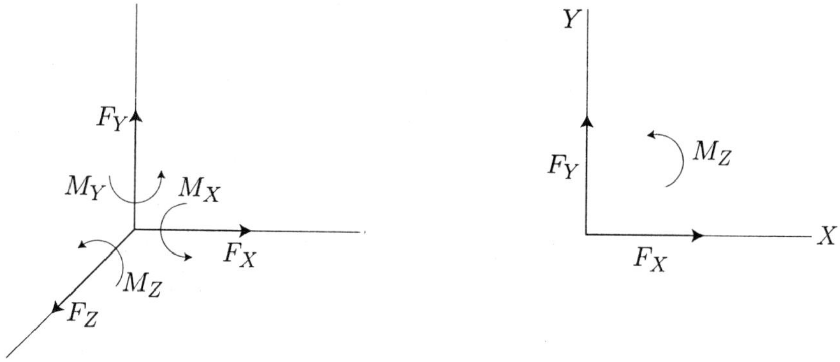

(a) Force System in a Space Structure **(b) Force System in a Plane Structure**

Fig. 1.8: Force System

For the case of a plane structure lying in the XY plane of the coordinate system, the set of equations reduces to

$$\sum F_x = 0; \ \sum F_y = 0 \quad \text{and} \quad \sum M_z = 0,$$

which for X- and Y- axes taken along the horizontal and vertical directions respectively, can be written in the form,

$$\sum H = 0; \ \sum V = 0 \quad \text{and} \quad \sum M = 0.$$

From the conditions of equilibrium for the plane structure several conclusions may be drawn. If a structural member is in equilibrium under the action of two forces, they must be equal, opposite and collinear as shown in Fig. 1.9(a). In such a case, one equation may be used. If a member is in equilibrium due to the application of three forces, they must be coplanar and their lines of actions must be concurrent or parallel as shown in Figs. 1.9(b) and 1.9(c) respectively. In such cases, two equations, either a force and a moment or two moment equations may be used. For concurrent or parallel system in space three equations involving force and moments may be used.

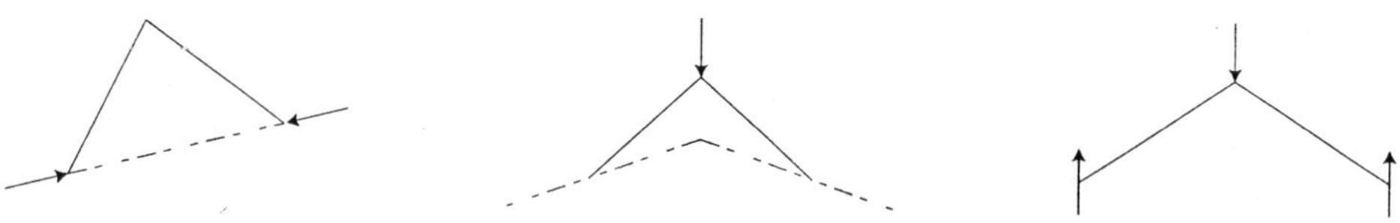

(a) Collinear Forces **(b) Coplanar Concurrent Forces** **(c) Coplanar Parallel Forces (Non-concurrent Forces)**

Fig. 1.9: Forces on Systems

1.6 Structures—Determinate and Indeterminate

A structural system has to be properly restrained for being under a state of equilibrium and the system satisfying this requirement is said to be in a position of stable equilibrium. If a system does not have the sufficient number of internal or external restraints, it may undergo a rigid body movement under actions of load. Such a system is said to be statically unstable, and is referred to as a mechanism. Sometimes, however, a structural system may be under stable equilibrium, under a particular loading, but may be unstable under another.

For a system in stable equilibrium, the reactions may be formulated in terms of external loads and the conditions of equilibrium. If the reactions in the system can be solved by the equations of equilibrium alone, the system is said to be a statically determinate structure. When the number of unknown reactions exceeds the number of conditions of equilibrium, they cannot be solved by statics alone and the system is known as a statically indeterminate structure. In such a system for solution of the reactions, equations involving the elastic behaviour of the structure will be necessary.

Structural system may be classified as externally redundant or internally redundant. Externally redundant structures are those which are indeterminate with respect to the reactions. Internally redundant structures are the ones which are indeterminate with respect to the unknown forces in one or more of the members.

Fig. 1.10 illustrates rigid systems under external restraints. Beam in Fig. 1.10(a) inadequately restrained is statically unstable under the load shown, even though, it will become stable when the inclined load is changed to a vertical one. Beam in Fig. 1.10(b) is adequately restrained and is statically determinate whereas beam in Fig. 1.10(c) is over-restrained and is externally statically indeterminate.

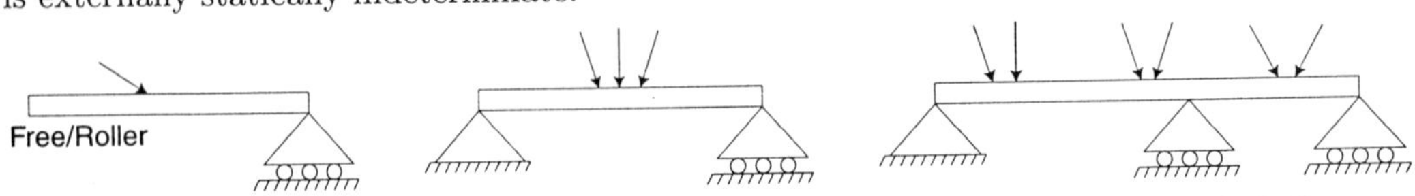

(a) Unstable Beam **(b) Statically Determinate Beam** **(c) Statically Indeterminate Beam**

Fig. 1.10: Beams

In Fig. 1.11, trusses in which members are subjected to axial forces are illustrated from the point of view of stability and determinancy. Fig. 1.11(a) shows a truss which is stable and is statically determinate externally but internally unstable whereas Fig. 1.11(b) shows a truss which is externally and internally statically determinate, Fig. 1.11(c) shows a statically indeterminate Truss (Internally).

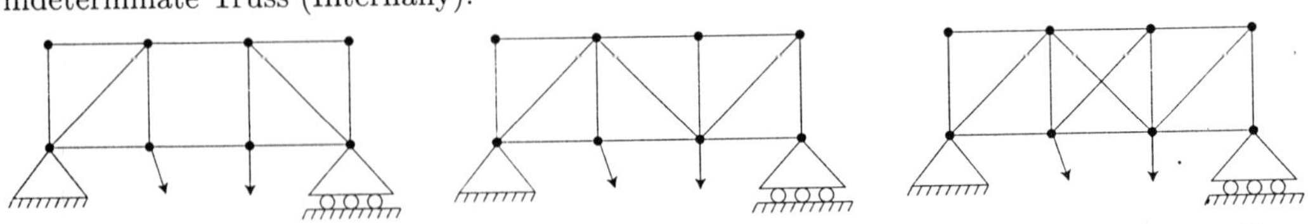

(a) Unstable Truss (Internally) **(b) Statically Determinate Truss** **(c) Statically Indeterminate Truss (Internally)**

Fig. 1.11: Trusses

1.7 General Principles of Analysis

The usual function of a structural system is to transmit external loads to a number of specified points, e.g., the loads on a bridge or a truss have to be carried to the abutment, piers or to some other type of supports. Since the external loading may vary, the supporting points must be capable of exerting reactive forces, which will statically balance any such loading.

In case of a body subjected to forces acting in a plane, three conditions of equilibrium must be satisfied :

$$\sum H = 0, \quad \sum V = 0 \quad \text{and} \quad \sum M = 0.$$

These conditions require three forces for their satisfaction and the supports must be such as to provide these forces. There are three things which must be known about each force, in order to have it fully determined, its magnitude, direction and point of application or line of action. If knowledge of all these in connection with one reaction is lacking, then the magnitude, direction, point of application or line of action of other forces acting on the structure must be known.

In normal cases, when loads are not parallel, there are two supports. One support provides a reaction along one of the axes only, while the other can provide reaction along both axes. The unknown quantities then are three, e.g., the magnitudes of three reactions or magnitudes of two reactions in a particular direction and the direction of one, and can be solved from conditions of statics.

Such supports may be provided by a frictionless roller and a hinge, and these supply the statically essential reactive forces. Such arrangement is illustrated in Fig. 1.12(a) to 1.12(c), where truss and beams are shown supported at A on a hinge, which can exert a reaction in any direction and at B on a frictionless roller, which can only exert a reaction normal to the bearing.

If both the supports are, however, hinged, the reactions are inclined and give components about the two chosen axes, thus introducing four forces. If, however, direction or line of action of one of the reactions can be fixed in such cases, the reactions can be solved by the condition of statics. Such a case of a truss in which reactions can be obtained by condition of statics is illustrated in Fig. 1.12(d).

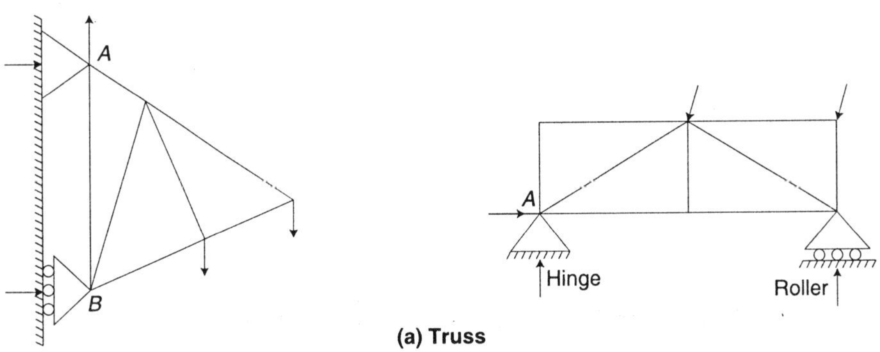

(a) Truss

Fig. 1.12: Various Forms of Supports

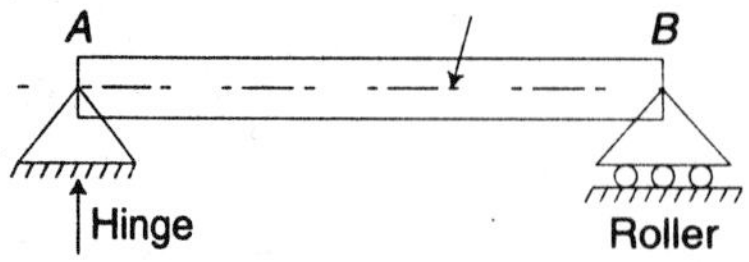

(b) Beam

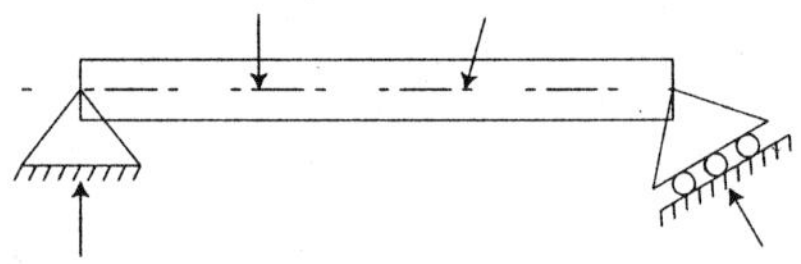

(c) Beam

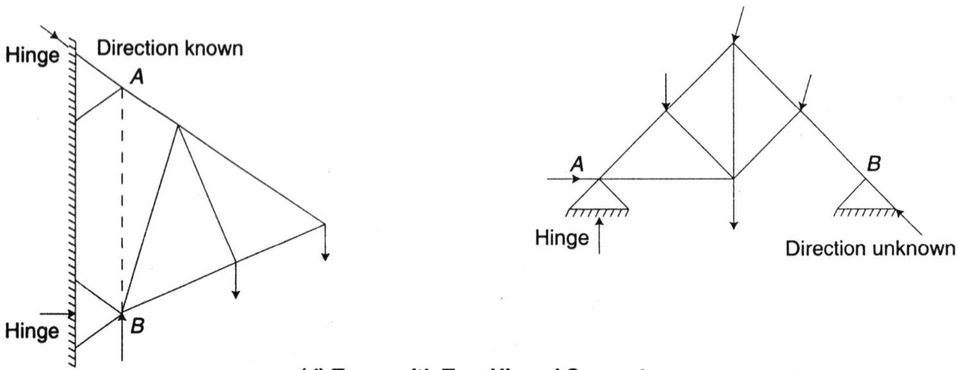

(d) Truss with Two Hinged Supports

Fig. 1.12: Various Forms of Supports

In case of space structural system, from conditions of statics six unknowns may be calculated. The essential reactions of a space system must, therefore, provide six-component

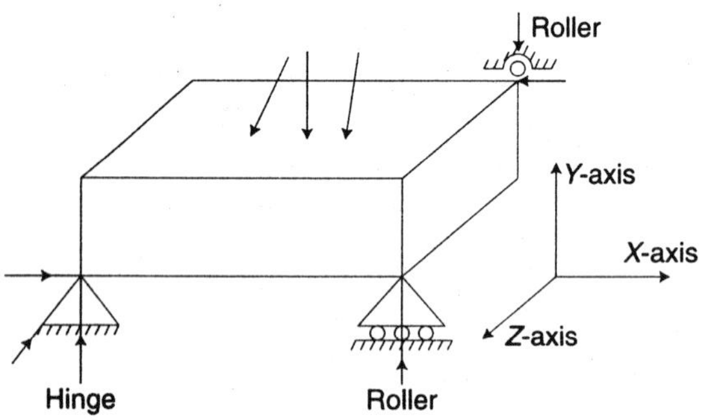

(a) Hinged and Roller Supports

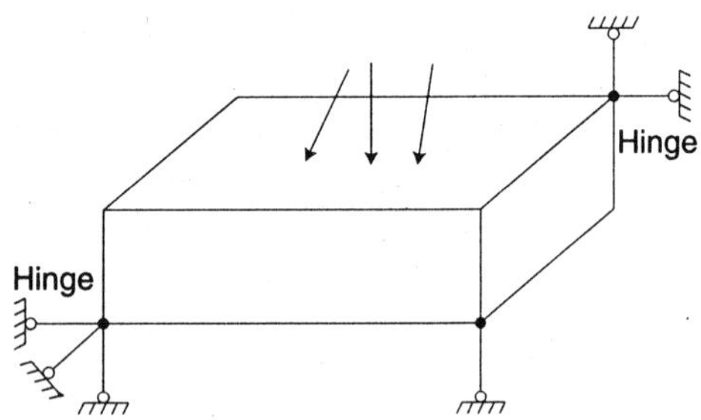

(b) Links as Supports

Fig. 1.13: Reactions and Space Structural System

forces suitably disposed. If one support consists of a type of support capable of providing reactions in the three directions, the other supports must provide reactions of which only three characteristics will be unknown, e.g., if there are two other supports one of them will provide restraint in two directions and the other will have one restraint providing reaction in one direction only. An arrangement of reactions in case of space frame for determination from conditions of statics is illustrated in Fig. 1.13(a). The components of reaction in the case may also be represented by means of links as shown in Fig. 1.13(b).

Chapter 2

Shear Forces, Bending Moments, Axial Forces and Twisting Moments for Fixed Loads

2.1 Introduction

Analysis of structural systems begins with the determination of reactions. After the reactions are calculated, the internal forces and moments at desired locations are obtained by taking sections through the locations, cutting the structure into two parts, choosing one of them and applying the laws of statics. The variations of the internal forces and moments specially their maximum values are of importance for design purposes.

In such an analysis three functions of the external loads usually are considered. Considering a section passing through a structural member normal to the axis of the member as in Fig. 2.1(a), the stress resultants acting at the section may be represented by an axial force N, a traverse force V (shear force) and a Moment M (Bending Moment) as in Fig. 2.1(b).

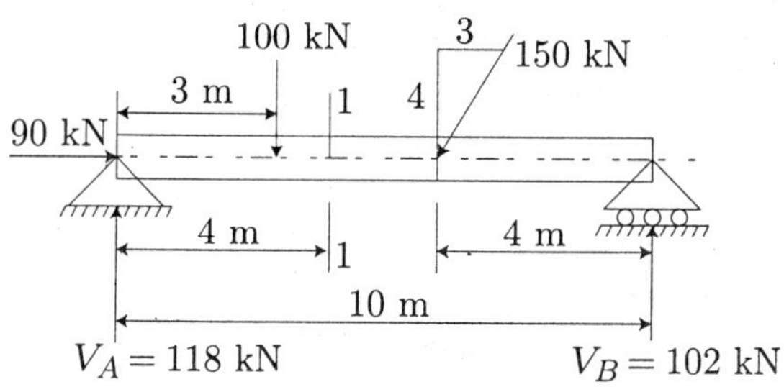

(a) Beam

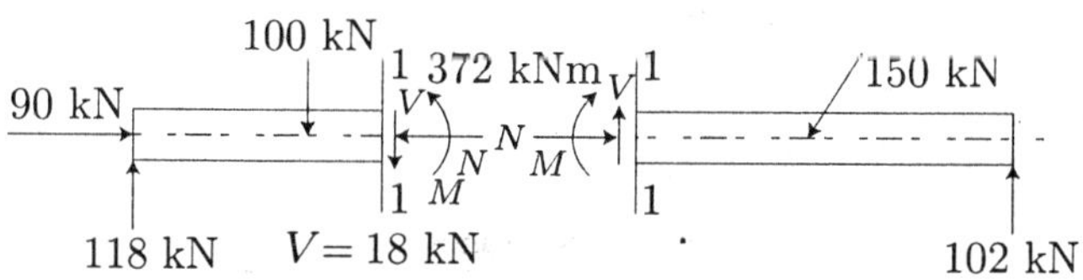

(b) Free Body Diagram

Fig. 2.1: B.M. and S.F. in Beam

From equilibrium of horizontal forces at the left of the section, $N = 90$ kN in the direction shown. Adding up the external vertical forces at the left of the section, the resultant upward force is 18 kN and this is the shear force.

To satisfy the requirement of $\sum V = 0$, there must be a vertical internal force of 18 kN acting down on the section as in Fig. 2.1(b).

Moments of the external force to the left of the section about a point at the centre of gravity of the section, a clockwise moment of 372 kNm results and this is the bending moment at the section. In order to satisfy one of the conditions of statics, i.e., $\sum M = 0$, an internal moment of 372 kNm acting on section 1-1, in a counter clockwise direction as shown in Fig. 2.1(b) is required. This internal moment is called the resisting moment, since it resists the effort of the external forces to bend the beam.

Since the uncut member is in equilibrium these forces and moment found by passing an imaginary plane at the section 1-1 actually exist within the structure on this section and be exerted on the part of the beam to the left of the section by that part to the right. The same values of the forces and moments will be obtained by considering part of the member to the right of the section 1-1.

2.2 Definitions

(a) **Shear Force (S.F.)** (Shear) at any section is defined as the algebraic sum of all the external forces acting parallel to the section on either side of the section. Such sections are taken normal to the axis of the member.

(b) **Bending Moment (B.M.)** (Moment) is defined as the algebraic sum of the moments of all the external forces including the reactions acting on right or left side of the section taken about the centre of gravity of the section.

As forces on either side of the section may be considered in calculating shear force and bending moment at a section in making the calculation, the side on which fewer external forces are acting will be preferred.

2.3 Convention of Signs

Selection of a particular sign convention is a matter of convenience. Fig. 2.2 shows the sign convention that is usually adopted in structural engineering. As in Fig. 2.2(a) axial force is considered positive when it produces tension in a member and is negative when it causes compression in the member. In Fig. 2.2(b), the pair of shear forces is designated as positive, if it tends to push the left side of the member upward and is negative if it pushes the left side of the member downward. In Fig. 2.2(c), the pair of moments applied on beam is considered

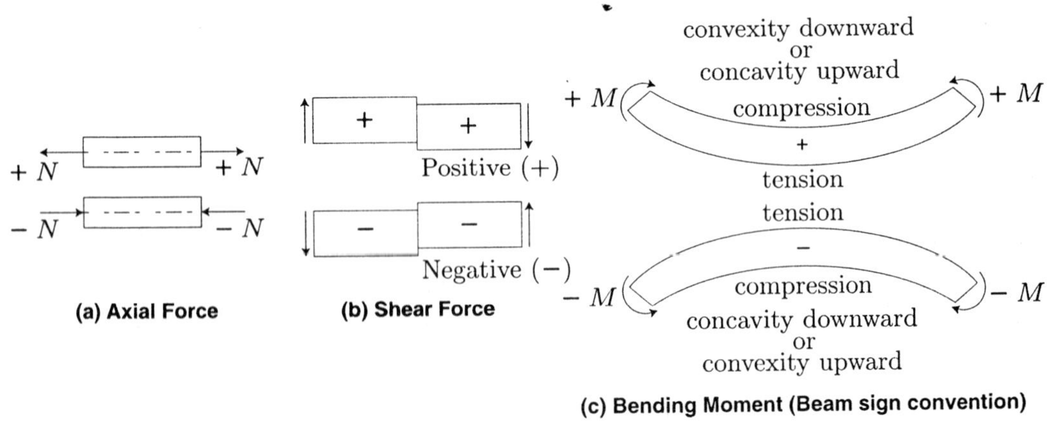

Fig. 2.2: Sign Conventions in Beams

positive, if it tends to bend a member concave upward, causing tension in the bottom fibre and is negative, if it causes the member to bend concave downward with tension at the top fibre.

The same conventions may be used for vertical members such as AB and CD in Fig. 2.3 by viewing from inside the structure.

When, however, it is not possible to differentiate as to which side of the vertical mark is the inside of two structure, as in case of member CD in Fig. 2.3(b), i.e., in case of multispan or multistorey rigid frames, the sheet may be turned so that the right hand side becomes the bottom. In case of Fig. 2.3(b), end C is then the right hand end. If this convention is used, the shear force and bending moment at section 1-1 of the rigid frame in Fig. 2.3(b) are both positive.

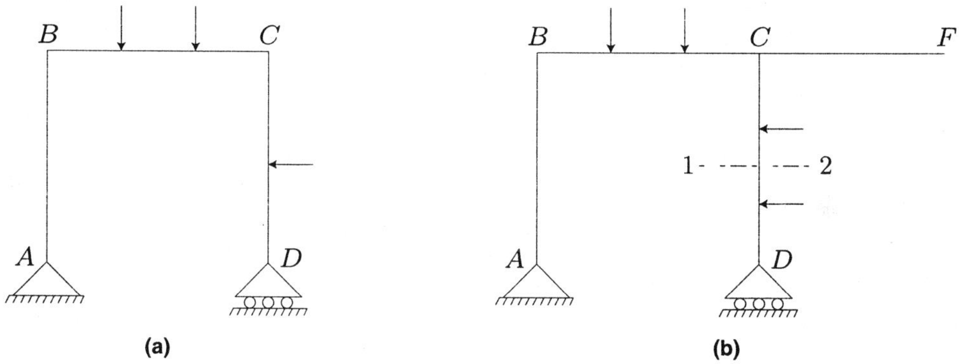

Fig. 2.3: Sign Conventions in Frames

2.4 Shear Force and Bending Moment Diagrams

Shear force and bending moment diagrams are graphical representation of variation of the shear forces and bending moments occurring simultaneously on all sections of member for a given loading system. Such diagrams are plotted on a base perpendicular to the direction of the cutting section. Positive shear forces and positive bending moments are normally plotted above the reference line. Reference lines for vertical member should be vertical, and positive shear forces and bending moments plotted above the line considering the sign convention used for their cases.

2.5 Examples of B.M. and S.F. Diagrams

(A) Simply Supported Beams

(a) Simply Supported Beam with a Concentrated load at centre (Fig. 2.4)

$$V_A = V_B = \frac{P}{2} \quad \text{and} \quad \text{S.F. at } A = +\frac{P}{2} \uparrow \downarrow$$

$$\text{S.F. at } x \text{ from } A = +\frac{P}{2} \text{ for } x < \frac{L}{2}$$

For $x > \frac{L}{2}$, S.F. at x from $A = -\frac{P}{2} \uparrow\downarrow$

S.F. at $B = -\frac{P}{2} \uparrow\downarrow$

B.M. at a distance x from $A = +\frac{P}{2}(x)$

Variation of B.M. is linear as in Fig. 2.4(c),

Maximum B.M. at $C = +\frac{P}{2}\left(\frac{L}{2}\right) = +\frac{PL}{4}$

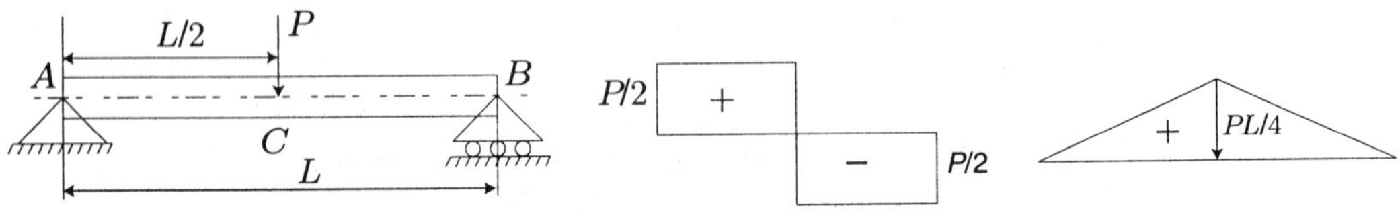

(a) Loading (b) S.F.D. (c) B.M.D.

Fig. 2.4: Simply Supported Beam with concentrated load at centre

(b) Simply Supported Beam with uniformly distributed load (u.d.1.) (Fig. 2.5).

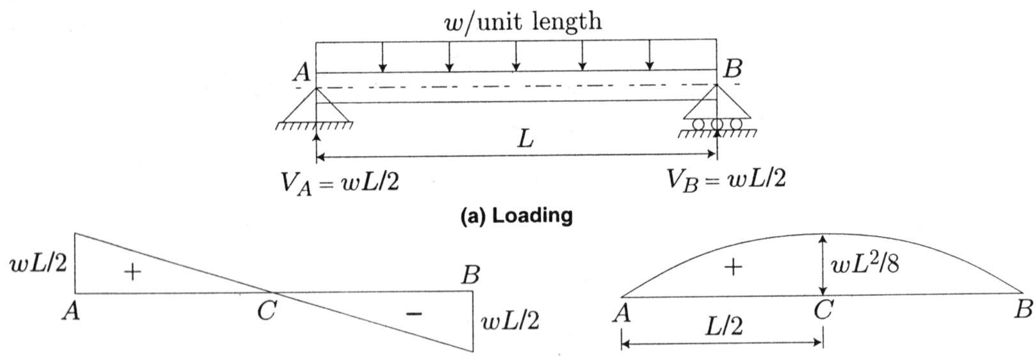

(a) Loading

(b) S.F.D. (c) B.M.D.

Fig. 2.5: Simply Supported Beam with unlformly distributed load

$V_A = V_B = \frac{wL}{2}$ and S.F. at $A = +\frac{wL}{2} \uparrow\downarrow$

S.F. at a dist. x from $A = \frac{wL}{2} - wx$.

Variation is linear as in Fig. 2.5(b).

S.F. at $B = -\frac{wL}{2} \uparrow\downarrow$

S.F. at centre $C = 0$.

B.M. at a dist. x from $A = V_A(x) - w(x)\left(\frac{x}{2}\right) = \frac{wL}{2}x - \frac{wx^2}{2} = \frac{wx}{2}(L - x)$.

Variation is parabolic as in Fig. 2.5(c)

Maxm. B.M. at $x = \frac{L}{2} = \frac{wl^2}{8}$.

(c) Simply Supported Beam with concentrated loan on the beam (Fig. 2.6)

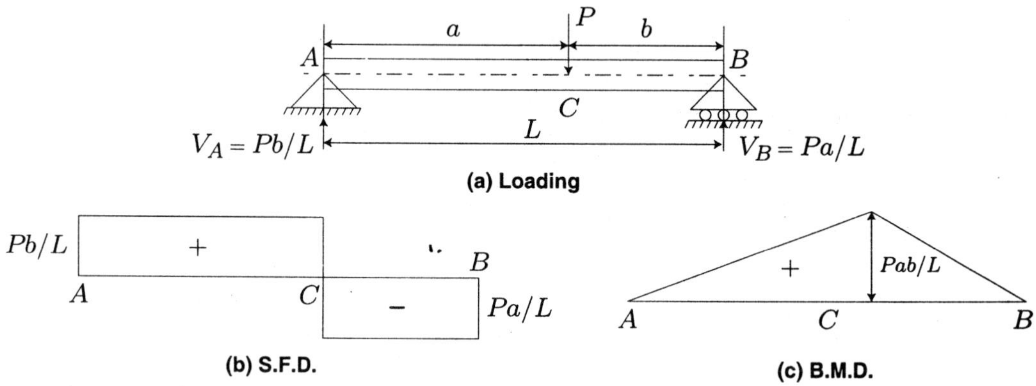

Fig. 2.6: Simply Supported Beam with concentrated load

(d) Simply Supported beam with triangular loading (Fig. 2.7)

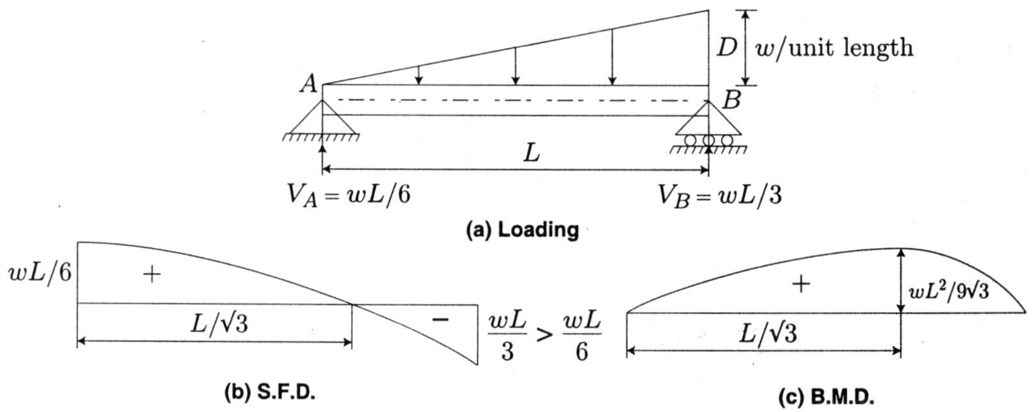

Fig. 2.7: Simply Supported Beam with triangular loading

$$V_A = \frac{wL}{6} \quad \text{and} \quad V_B = \frac{wL}{3}$$

$$\text{S.F. at } A = +\frac{wL}{6} \uparrow\downarrow$$

$$\text{S.F. at a distance } x \text{ from } A = \frac{wL}{6} - \frac{1}{2}\frac{wx}{L}(x) = \frac{w}{6L}(L^2 - 3x^2).$$

Variation of S.F. is parabolic as in Fig. 2.7(b).

Shear is zero at $L^2 - 3x^2 = 0$, i.e., at $x = \dfrac{L}{\sqrt{3}}$

$$\text{S.F. at } B = -\frac{wL}{3} \downarrow\uparrow$$

B.M. at a dist. x from $A = \dfrac{wL}{6}(x) - \dfrac{1}{2}\dfrac{wx^2}{L} \cdot \dfrac{x}{3} = \dfrac{wx}{6L}(L^2 - x^2)$.

Variation of B.M. is parabolic (cubical).

Maximum B.M. occurs where S.F. is zero, i.e., $\dfrac{dM}{dx} = 0$, whence $L^2 - 3x^2 = 0$ and $x = \frac{L}{\sqrt{3}}$.

Maximum B.M. $= \frac{w}{6L} \cdot \frac{L}{\sqrt{3}}\left(L^2 - \frac{L^2}{3}\right) = \frac{wL^2}{9\sqrt{3}}$.

(e) Simply Supported Beam with a Couple at any section (Fig. 2.8)

$$V_A = \frac{M}{L} \downarrow \quad \text{and} \quad V_B = \frac{M}{L} \uparrow$$

S.F. at $A = -\frac{M}{L} \downarrow \uparrow =$ S.F. at B

B.M. between A and C $= -\frac{M}{L}(x)$

and variation as in Fig. 2.8(c).

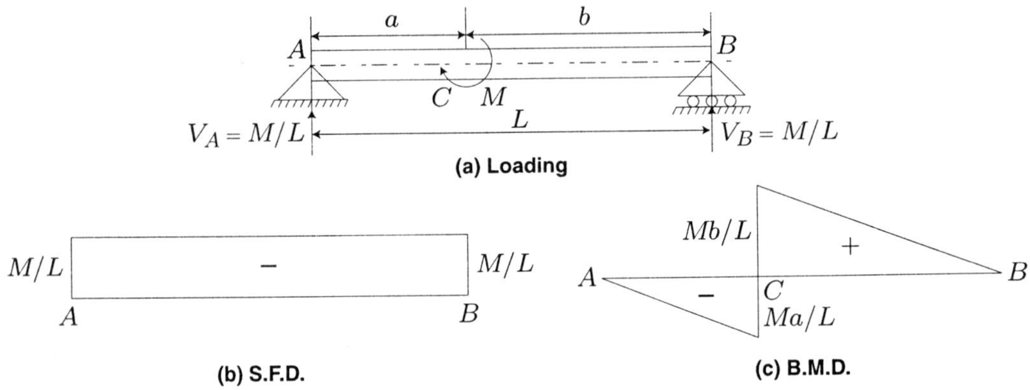

Fig. 2.8: Simply Supported Beam with a Couple

(f) Simply Supported Beam with several concentrated loads (Fig. 2.9)

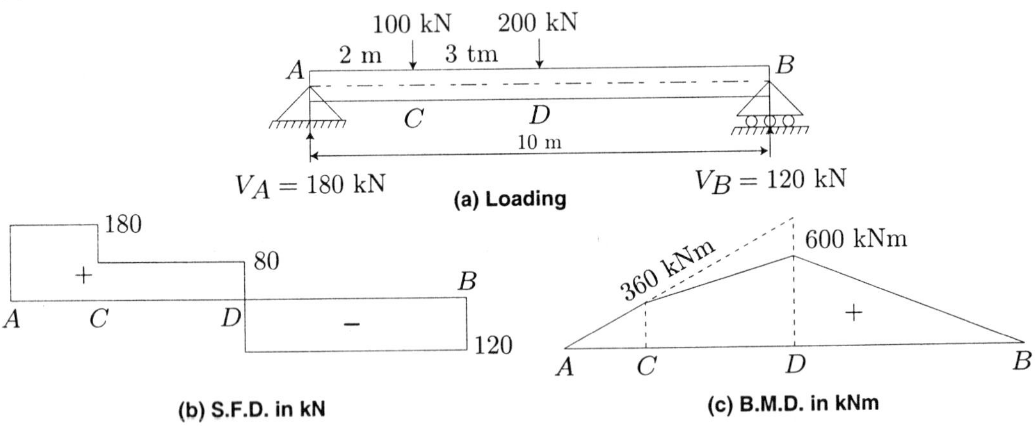

Fig. 2.9: Simply Supported Beam with Two Concentrated loads

Shear force curves in this case will be a series of straight lines having constant value between the concentrated loads and a change in ordinate under each concentrated load. The bending moment curve will be a series of straight lines, the equations changing at each concentrated load.

(g) Simply Supported Beam with combination of u.d.l. and Concentrated loads (Fig. 2.10)

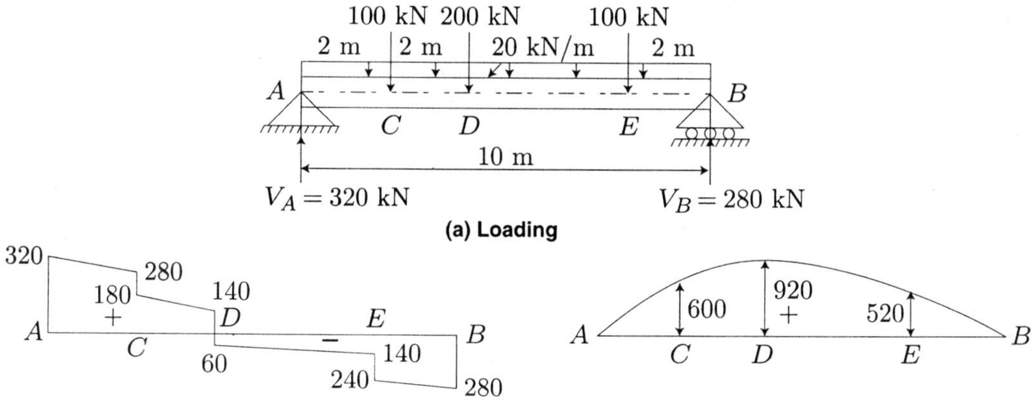

Fig. 2.10: Simply Supported Beam with u.d. l. and concentrated loads

Shear force curve in this case will be a series of straight lines having a slope equal to the magnitude of the uniformly distributed load and change in ordinate under each concentrated load equal to the load. The bending moment curve will be a series of parabolic arcs, the equations changing at each concentrated load.

(B) Cantilever Beams

(a) Cantilever Beam with uniformly distributed load (Fig. 2.11)

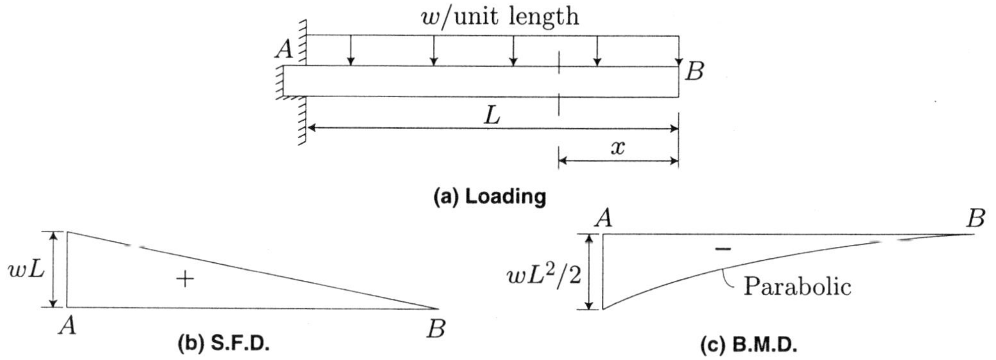

Fig. 2.11: Cantilever with uniformly distributed load

S.F. at x from the free end $B = +wx \uparrow\downarrow$

Variation of S.F. is linear as in Fig. 2.11(b).

Maxm. value at $A = +wL$.

B.M. at x from $B = -\dfrac{wx^2}{2}$.

Variation of B.M. is parabolic as in Fig. 2.11.

Maximum value at $A = -\dfrac{wL^2}{2}$.

(b) Cantilever Beam with several Concentrated loads (Fig. 2.12)

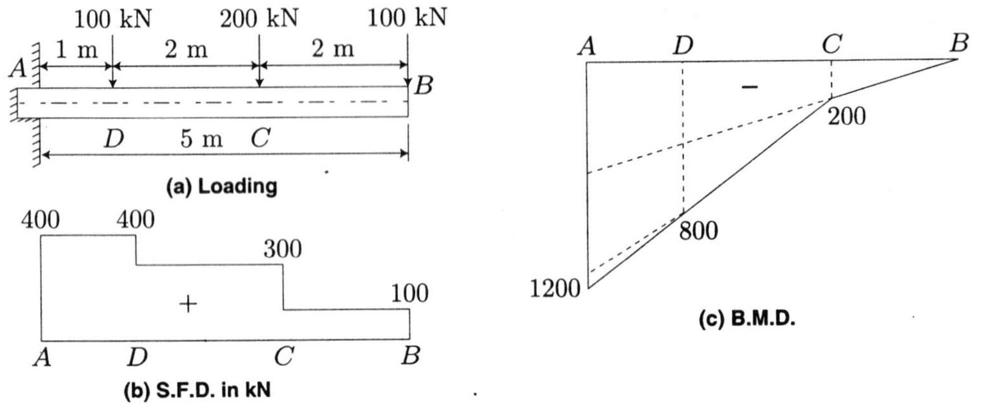

Fig. 2.12: Cantilever with several Concentrated loads

(C) Overhanging Beams and Beam with Internal Hinge

(a) Overhanging Beams with u.d.l and Concentrated Load (Fig. 2.13)

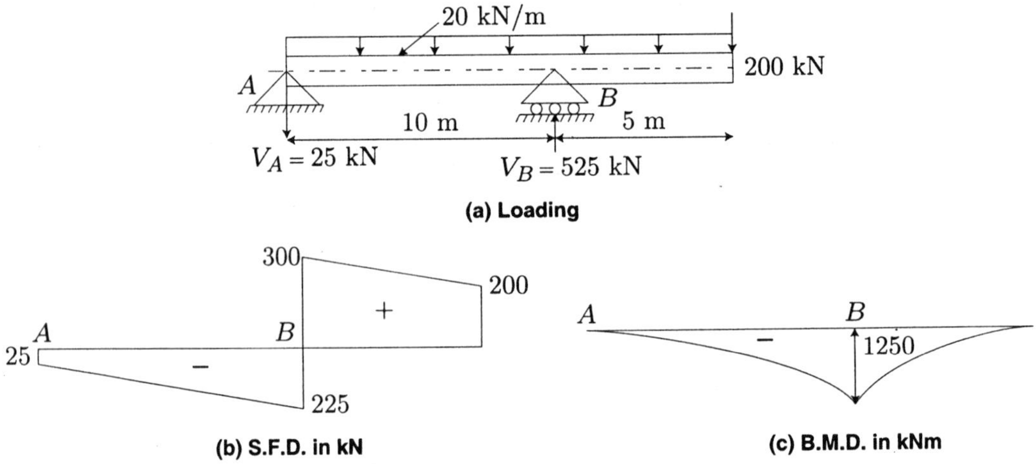

Fig. 2.13: Overhanging Beam with u.d.l. and Concentrated Load

(b) Overhanging Beam with u.d.l. only (Fig. 2.14)

S.F. and B.M. diagrams for BC will be the same as a cantilever beam. S.F. will be zero at C and $+\frac{wL}{2}$ at B.B.M. will vary in a parabolic form from zero at C to $-\frac{wL^2}{8}$ at B.

For a section at a distance x from A in AB.

$$\text{S.F. at } x = V_A - wx = \frac{3}{8}wL - wx.$$

$$\text{B.M. at } x = V_A(x) - \frac{wx^2}{2} = \frac{3}{8}wLx - \frac{wx^2}{2}.$$

The S.F. will vary linearly from $+\frac{3}{8}wL$ at A to $-\frac{5}{8}wL$ at B. The complete S.F. diagram is shown in Fig. 2.14(b). The **S.F. is zero at $\frac{3}{8}$ at L from A.**

The B.M. curve will be parabolic. The maximum positive B.M. will occur at the location where S.F. is zero, i.e., at $x = \frac{3}{8}$ L from A.

Maximum positive B.M. $= \dfrac{9}{128} wL^2$.

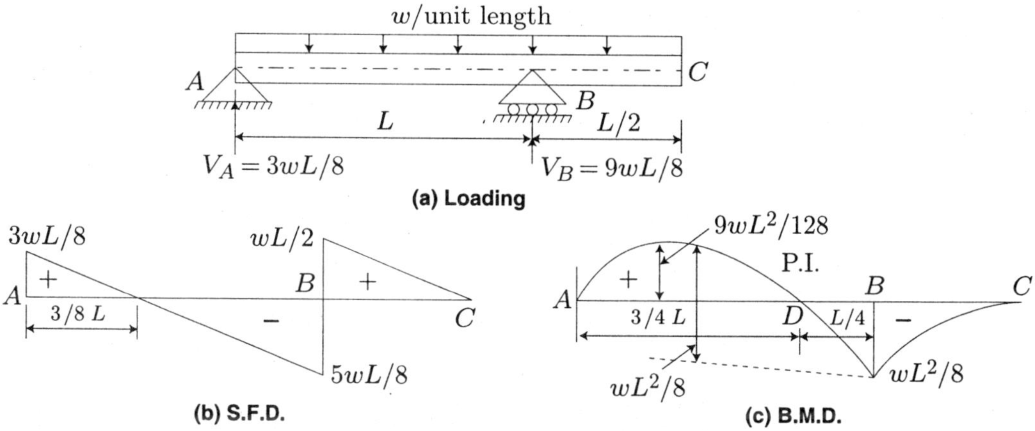

Fig. 2.14: Overhanging Beam with u.d.l. only

In span AB, the B.M. changes sign. The B.M. is zero at point D, a distance x from A, given by,

$$\frac{3}{8}wLx - \frac{wx^2}{2} = 0, \quad \text{where} \quad x = \frac{3}{4}L.$$

The B.M. is positive between A and this point D and negative between D and B. This point where the B.M. is zero and is of opposite sign, on either side of D is called the point of inflection (or point of contraflexure) P.I. in Fig. 2.14(c).

(c) Overhanging beam with Concentrated loads (Fig. 2.15)

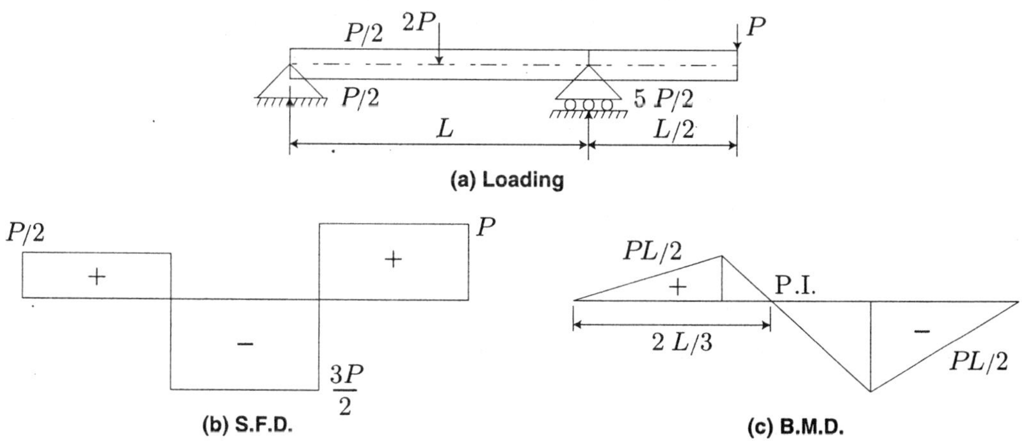

Fig. 2.15: Overhanging Beam with Concentrated Loads

(d) Beam with Internal Hinge (Fig. 2.16)

Taking moments about hinge C

$M_c = 0,$

whence $V_B = 60$ kN ↑ and $V_A = 270 - 60 = 210$ kN ↑

$$M_A = 60(9) - 30(4)(7) - 50(4) - 100(2) = -700 \text{ kN}$$

S.F. at $A = 210$ kN(+ve)

S.F. at $D = 210 - 100 = 110$ kN(+ve)

S.F. at $E = 110 - 50 = 60$ kN(+ve)

S.F. at $C = 60$ kN(+ve)

S.F. at $B = 60 - 120 = -60$ kN(−ve).

S.F.D. is drawn as shown in Fig. 2.16(b).

$$M_A = 70 \text{ kNm}$$

$$M_D = 210(2) - 700 = -280 \text{ kNm}$$

$$M_E = 210(4) - 100(2) - 700 = -60 \text{ kNm}$$

$$M_c = 0$$

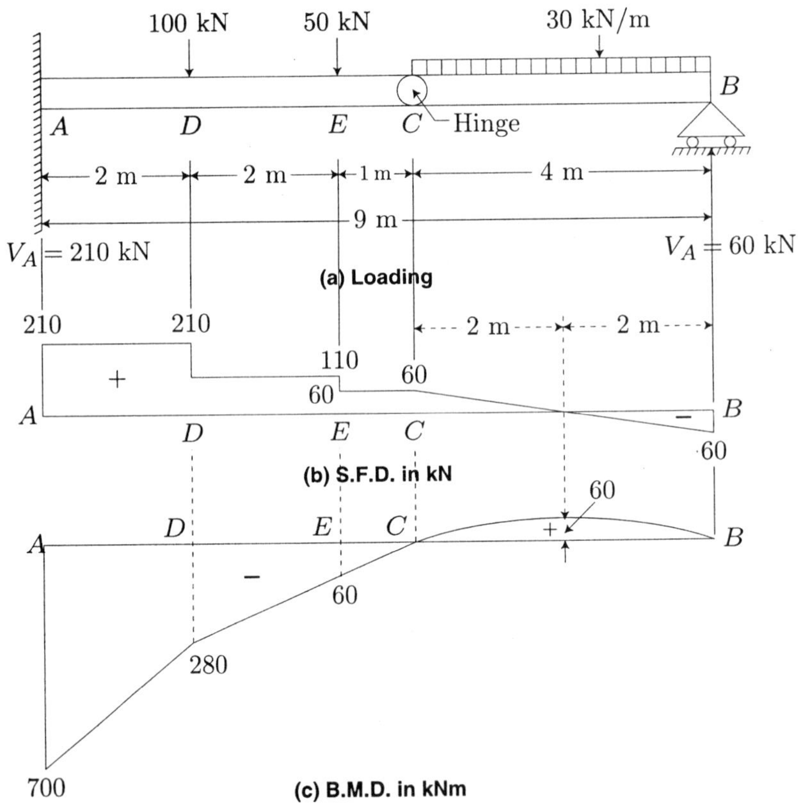

Fig. 2.16: Beam with Internal Hinge

Maximum +ve B.M. shall occur at mid-point of BC.

+ve maximum $= 60(2) - 30(2 \times 1) = 60$ kNm.

B.M.D. is drawn as shown in Fig. 2.16(c).

(e) Inclined Beam AB loaded as in Fig. 2.17

When an inclined beam is acted upon by a uniformly distributed vertical load per unit axial length such as its own weight, the load intensity may be resolved parallel and perpendicular to the beam axis. The component perpendicular to the axis contributes to shear and bending moment.

In this case the load intensity perpendicular to the beam axis is

$$w = 50 \cos 30° = 43.30 \text{ kN/m}$$

$$V_A = 43.30(5)\left(\frac{1}{2}\right) = 108.25 \text{ kN.}$$

Maxm. S.F. = 108.25 kN.

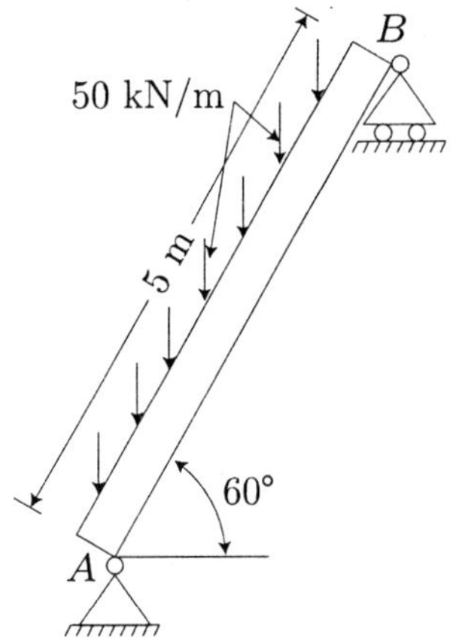

Fig. 2.17: Inclined Beam

$$\text{Maxm. B.M. } \frac{wl^2}{8} = \frac{43.30(5)^2}{8} = 135.32 \text{ kNm}$$

2.6 Examples of B.M., S.F. and Axial Force Diagrams

(a) Simply Supported Beam with inclined loading (Fig. 2.18)

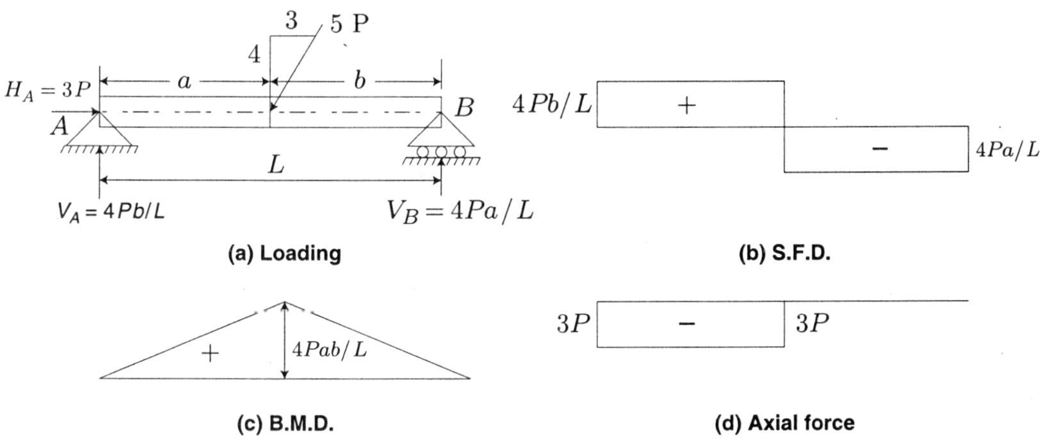

Fig. 2.18: Simply Supported Beam with inclined loading

(b) Portal Frame with Concentrated Loads (Fig. 2.19)

$\sum M_D = 0$, whence $V_A = 120$ kN ↑

$\sum V - 0$, whence $V_B = 80$ kN ↑

$\sum H = 0$, whence $H_A = 100$ kN →

S.F. in AB = 100 kN(−ve)

B.M. in AB, at a distance x from $A = -100(x)$.

Variation is linear with maximum at B, given by −400 kNm.

S.F. in BC, between B and E is 120 kN and between E and C is 80 kN.

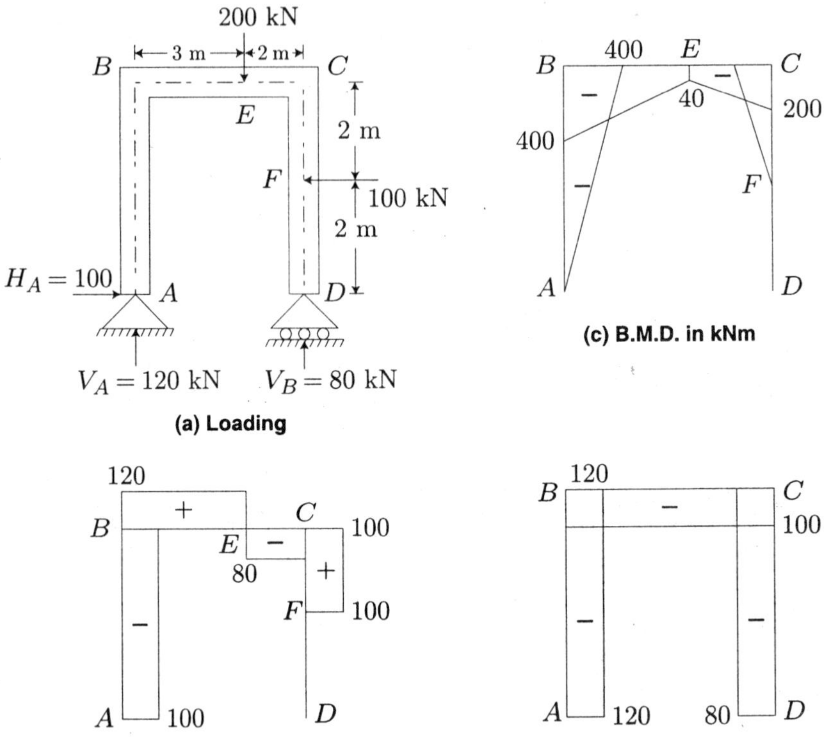

(a) Loading

(b) S.F.D. in kN

(c) B.M.D. in kNm

(d) Axial force in kN

Fig. 2.19: Portal Frame with Concentrated Load

B.M. in BC, between B and E, at a distance x from B is given by $(120x - 400)$.

Variation is linear, magnitude at E being 40 kNm.

Between E and C, at a distance x from B, B.M. is given by, $120x - 400 - 200(x - 3) = 200 - 80x$.

Variation is linear, magnitude at C being -200 kNm.

S.F. in CD, between F and C is 100, there being no S.F. between D and F.

B.M. in CD, between F and C at a distance x from F is given by $-100x$.

Variation is linear, maximum magnitude occurring at C being -200 kNm.

Axial forces on member AB is due to V_A, the nature of force being compressive.

In member BC, axial force is due to H_A, the nature of force being compressive.

In member CD, axial force is due to V_B.

It may be mentioned here that in designing such structure, effect of all the three joints are to be considered.

(c) Portal Frame with Concentrated Loading (Fig. 2.20)

The reactions of the frame are shown in Fig. 2.20(a) and the SF., B.M. and axial force diagrams are illustrated in Figs. 2.20(b), 2.20(c) and 2.20(d) respectively.

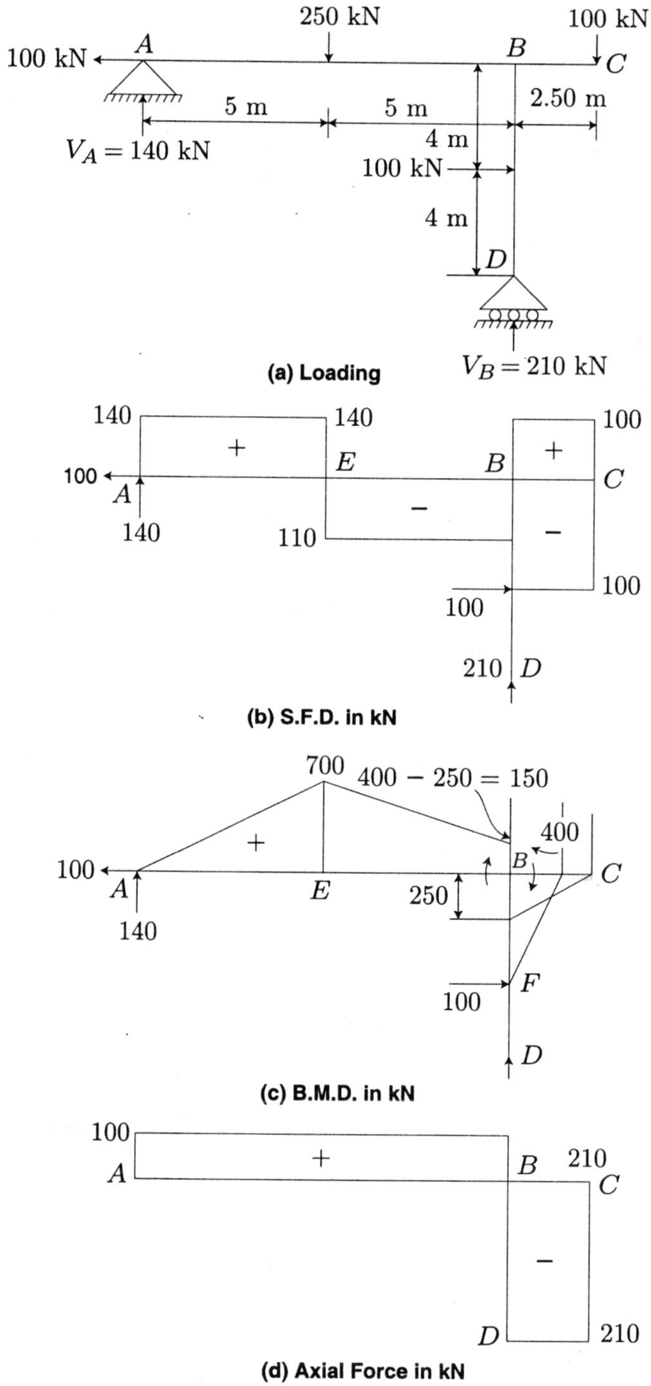

(a) Loading

(b) S.F.D. in kN

(c) B.M.D. in kN

(d) Axial Force in kN

Fig. 2.20: Portal Frame with Concentrated Loads

Vertical reactions V_A and V_B along with the external forces at C and E cause shear force in the member ABC. For member BD, the shear force is due to H_A and the external force at F.

In case of B.M., the diagram has been plotted on the basis of sign conventions. The directions of B.M. at joint B in members BA, BC and BD are shown in Fig. 2.20(c), which indicate that summation of moments at B is zero, i.e, the joint B is in equilibrium.

Axial force diagrams in Fig. 2.20(d) are obtained from reactions H_A on member AB and reaction V_D in member BD.

(d) Gable Frame loaded as in Fig. 2.21(a)

Vertical and horizontal reactions at the supports are shown in Fig. 2.21(a).

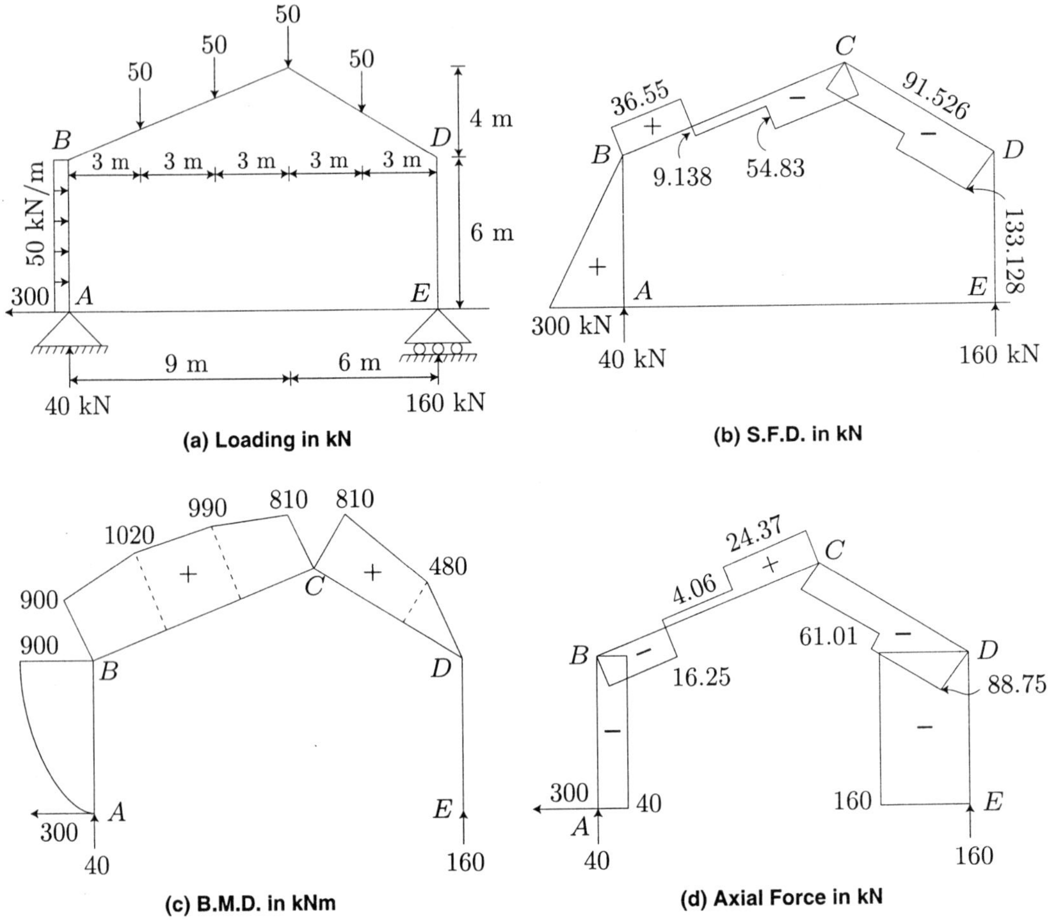

Fig. 2.21: Gable Frame with u.d.l. and Concentrated Load

S.F. in vertical member AB at a distance x is given by $(300 - 50x)$. Variation is linear with maximum magnitude 300 kN at A and minimum value zero at B.

S.F. in inclined member BC is due to the components at right angle to member of the vertical reaction at A and of the vertical loads on member BD. Thus S.F. at B on BC is given by

$$40 \frac{9}{\sqrt{9^2 + 4^2}} = 36.55 \text{ kN}$$

Values similarly obtained in BC are shown in Fig. 2.21(b).

S.F. in vertical member ED is zero. S.F. in inclined member DC is obtained similarly as in case of member BC and at D is given by $160\frac{6}{\sqrt{6^2+4^2}} = 133.12$ kN.

B.M. in member AB at a distance x from A is given by $\left(300x - \frac{50x^2}{2}\right)$, maximum value at B equal to 900 kNm.

B.M. at the first 50 kN load from B is given by $900 + 40(3) = 1020$ kNm.

Value at $C = 900 + 40(2) - 50(6) - 50(3) = 810$ kNm.

B.M. at D is zero.

Complete B.M. diagram is shown in Fig. 2.21(c).

Axial force in member AB is due to vertical reaction at A and is 40 kN (compression).

Axial force in other portions of the frame obtained similarly are shown in Fig. 2.21(d).

Axial force in inclined member BC at B is due to the component of vertical reaction at A along the member BC. This is given by $40\frac{4}{\sqrt{9^2+4^2}}$ and equals 16.25 kN (compression).

(e) Portal as in Fig. 2.22.

As D is on rollers, reaction at D shall be perpendicular to the bedding plane of the rollers. Since support at D is inclined at an angle of 45°, horizontal and vertical components of the reaction at D shall be equal in magnitude.

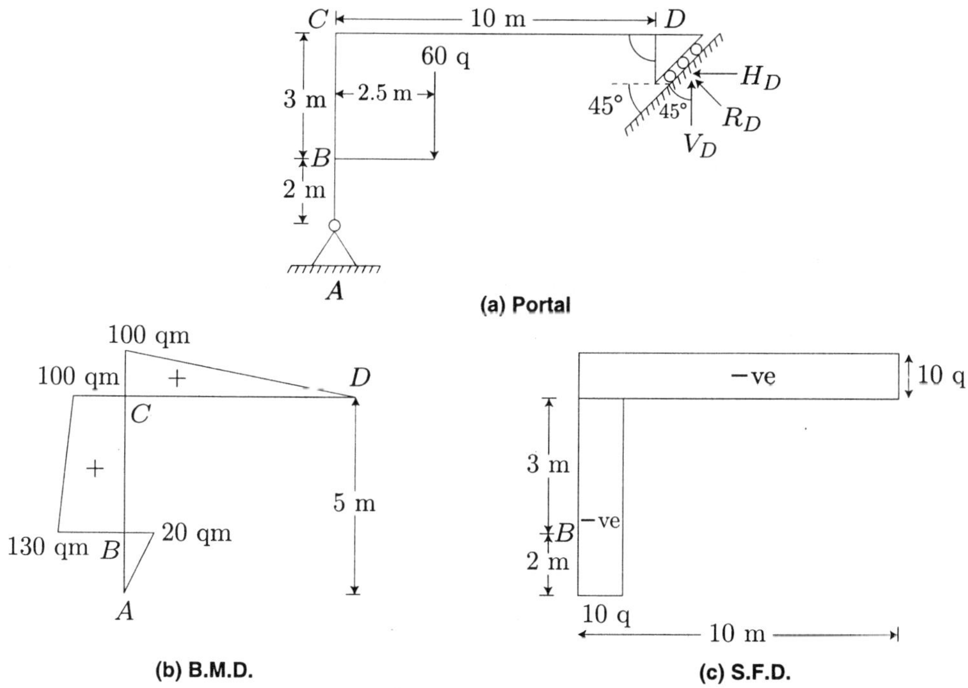

Fig. 2.22: Portal Frame

Taking moments about A,

$$\sum M_A = 0$$

$$V_D(10) + H_D(5) - 60(2.5) = 0$$

But $V_D = H_D$ and $15V_D = 150q$ and $V_D = 10q.$ ↑

Horizontal reaction at A, $H_A = 10q$.

Vertical reaction at A, $V_A = 60q - 10q = 50q$ ↑

Shearing force in ABC is constant throughout $= 10q \cdot (-\text{ve}) =$ S.F. in CD

B.M. at $C = 100qm(+\text{ve})$.

2.7 Relationship between B.M. and S.F.

Fig. 2.23 shows a beam with some sort of distributed loading. The free body diagram of a typical element of length dx of the beam is shown in Fig. 2.23(b). The element is subjected to bending moment and shear force as shown.

From equilibrium of the element in the vertical direction,

$$V + \frac{dV}{dx}(dx) - wdx - V = 0,$$

whence
$$\frac{dV}{dx} = w; \tag{2.1}$$

i.e., the rate of change of shear force at any section is the load per unit length at that section. The plus sign in front of w indicates that the shear force is increasing as is happening in the case. In case of a section under the effect of a concentrated load, the rate of change of shear force is infinite which means that the slope of the shear force diagram at a concentrated load becomes vertical under a concentrated load.

Considering that sum of moments on the element be zero,

$$\left(M + \frac{dM}{dx}dx\right) - M + wdx\frac{dx}{2} - \left(V + \frac{dv}{dx}\right)dx = 0.$$

Cancelling terms and neglecting small order terms,

$$\frac{dM}{dx} = V; \tag{2.2}$$

i.e., the rate of change of bending moment at any section is the shear force at that section. It also follows that the bending moment passes through a maximum or minimum value, wherever the shear force passes through zero.

If a section is considered at a concentrated load, there is a sudden change of shear force and, therefore, a similar change in the rate of change of bending moment. There must, therefore, be an abrupt change in the direction of the bending moment curve at a concentrated load.

From Eqs. (2.1) and (2.2), it follows that,

$$\frac{d^2M}{dx^2} = \frac{dV}{dx} = w,$$

whence
$$\frac{d^2M}{dx^2} = w. \tag{2.3}$$

Conversely, if the loading function w is known, Eqs. (2.1) and (2.2) can be integrated and expressed as

$$V = \int w dx + V_o \tag{2.4}$$

$$M = \int V dx + M_o = \int \left(\int w dx \right) dx + V_o x + M_o. \tag{2.5}$$

Eq. (2.4) indicates that the shear force at any section of a beam is an integral of the applied loads or the area under the loading curve. Eq. (2.5) means that the bending moment at any section normal to the beam is an integral of the shear forces or the area under the shear force curve. The constants of integration, e.g., V_o and M_o in the equations can be evaluated from the condition of shear force and bending moment at initial point O.

Eqs. (2.4) and (2.5) can be expressed in an alternate form, if definite integrals are used. Thus,

$$V_2 - V_1 = \int_{x_1}^{x_2} w dx \tag{2.6}$$

and
$$M_2 - M_1 = \int_{x_1}^{x_2} V dx = V_1(x_2 - x_1) + \int_{x_1}^{x_2} \int_{x_1}^{x_2} w dx\, dx.$$

$$= V_1(x_2 - x_1) + \int_{x_1}^{x_2} (x_2 - x)\, w dx. \tag{2.7}$$

Eq. (2.6) means that the difference in shear force between any two points is given by the area of the loading curve between the two points. According to Eq. (2.7), the difference in bending moments between any two points is given by the area of the shear force curve between the two points. This difference is related to the moment of shear force at one point about the other plus the double integration of the loading between the two points or the moment of the loading between the two points.

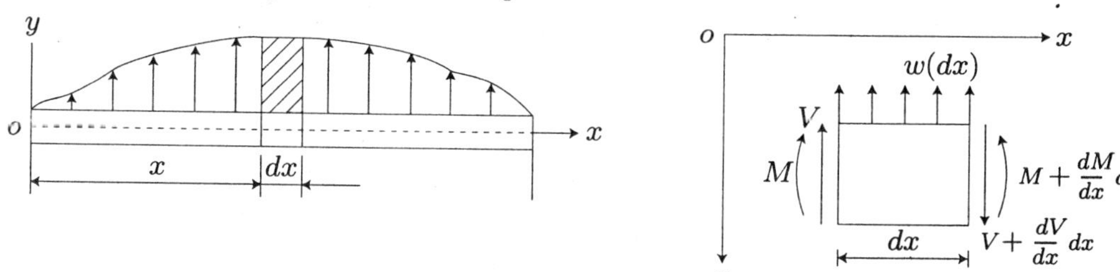

(a) Beam (b) Element

Fig. 2.23: Relationship between B.M. and S.F.

2.8 Characteristics of Shear Force and Bending Moment Diagrams

From the study of shear force and bending moment diagrams, the following broad conclusions may be made regarding their characteristics.

(a) The slope of the shear force diagram at any point is equal to the intensity of the distributed load at that point. Where loading is uniformly distributed, the change in shear force is constant and shear force diagram will be linear. At points of application of concentrated loads, abrupt change in the magnitude of the shear force will occur.

(b) The slope of the bending moment curve at any point is equal to the ordinate of the shear force diagram at that point. The bending moment is maximum or minimum at points where shear force is zero. Where the shear force diagram is horizontal, the change in bending moment is constant and the bending moment diagram will be a straight line, and have a positive or negative slope according to whether the shear force is positive or negative. Where the positive value of the shear force is decreasing or the negative value increasing (summing from the left), the bending moment diagram will be a curve and will be concave downward. At points of application of concentrated loads, abrupt change in the slope of the bending moment diagram occurs.

(c) The difference in the magnitudes of the shear force between any two points is equal to the total load applied to the beam between these two points.

(d) The difference in the magnitudes of the bending moment between any two points is equal to the area under the shear force diagram between the two points.

(e) At a section where a beam is subjected to an external couple, the bending moment will have two values, differing in magnitude by the value of the external couple.

2.9 Girders Supporting Floor Beams (Panelled Beams)

In case of many structures, the superimposed loads on floor system are not applied to the beam or girders directly, but to a secondary framing system, which is supported by the beam or girder. As an example, a girder supporting floor beams is shown in Figs. 2.24(a) and 2.24(b), in which the load from the floor comes to the floor beams through the stringers. In general, the span lengths of the stringers need not be equal and the floor beams may be irregularly spaced. The locations of the floor beams are called panel points, whereas the distances between them are considered as panels of the girders. The reactions of the stringers are the loads which are applied on the girders and may be computed as shown in Fig. 2.24(c).

The shear force and bending moment diagrams for the girder in Fig. 2.24(a) are indicated by the solid lines in Fig. 2.24(c) and 2.24(d) respectively. If the uniformly distributed loads were applied directly to an identical girder without a floor system, the resulting shear force and bending moment diagrams would be as shown by the dashed lines in the corresponding figures. It may be noted that the shear force at each panel point of the girder between floor beams is constant and is called the panel shear. The shear force at a point just to the left of a panel point is different from that just to the right of the same panel point. The bending moments at the panel points of the girder and the reactions at the ends are not affected by the use of the floor system.

Sometimes, however, as in case of buildings, the floor beams supported by the girders may be directly carrying the floor. In this case also the load is applied to the girders at certain fixed points. Although the entire floor panel may be covered with a uniformly distributed load, such girders are subjected to a series of concentrated loads acting at the points where the floor beams rest.

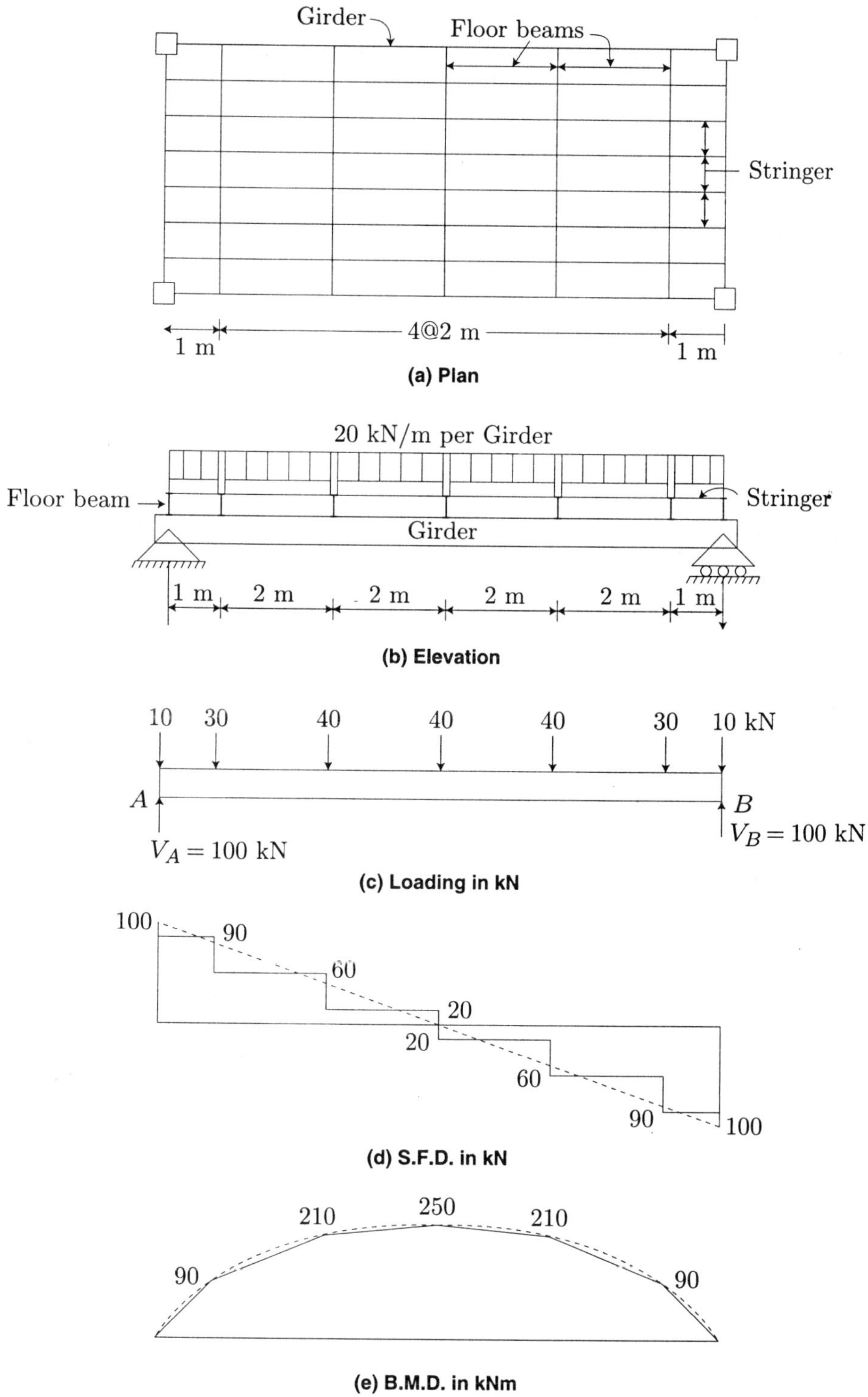

Fig. 2.24: Girder with Floor Beam

2.10 Three Hinged Arches

Three hinged arches are subjected to shear force, bending moment and axial force at a particular section. Analysis of such structures involves their calculation. Variation of these along the centroidal axis of the arches will depend on the nature of the curve representing the arch axis.

In drawing the shear force and bending moment diagrams for a curved member, however, certain difficulties not present in case of straight members are encountered, as the direction of axis of the member changes along its length. Thus, the axial force and shear force at any section depend on the angle between a horizontal line and the tangent to the curve at that point. Moreover, the base line for shear force and bending moment diagrams is a curve instead of a straight line, even though as a matter of convenience, shear force and bending moment diagrams on the horizontal projection of the curved members are also sometimes used.

Example 2.1 *Fig. 2.25 shows a three hinged arch of second degree parabolic shape. Find the bending moment diagram and the shear force (radial shear) and axial force (normal thrust) at* 10 m *from A of the arch.*

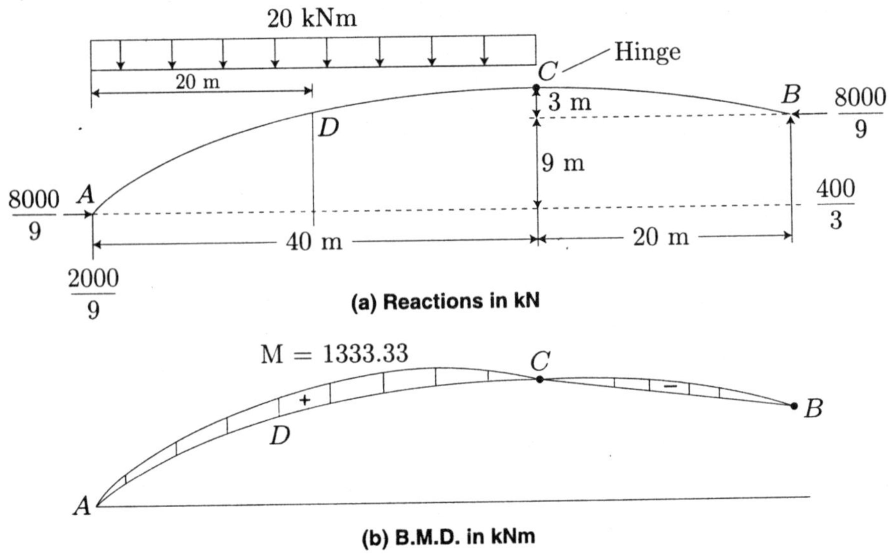

Fig. 2.25: For Example 2.1

Solution : Equation of a parabola of span L and central rise 12 m is given by

$$y = \frac{4(12)}{L^2}x(L - x).$$

As at central hinge tangent is horizontal,

$$L = 2(40) = 80\text{m},$$

whence $y = \dfrac{48}{(80)^2}x(80 - x) = \dfrac{3}{400}x(80 - x).$

The reactions of the arch are shown in Fig. 2.25(a). B.M. at any point between A and C will be given by

$$M = \frac{2000}{3}(x) - \frac{8000}{9}(y) - \frac{20x^2}{2} = \frac{2000}{3}(x) - \frac{20}{3}x(80 - x) - 10x^2.$$

Maxm. B.M. between A and C will occur at a place where $\frac{dM}{dx} = 0$.

i.e., $$\frac{2000}{3} - \frac{20}{3}(80 - 2x) - 20x = 0.$$

whence $x = 20$ m.

i.e., maximum positive B.M. will occur at D, where $y_D = 9$ m.

$$\text{B.M. at } D = \frac{2000}{3}(20) - \frac{8000}{9}(9) - \frac{20(20)^2}{2} = \frac{4000}{3} = 1333.33 \text{ kNm}.$$

B.M. at C is zero and hence considering CB, the resultant of vertical and horizontal reactions at B should pass through C. The pattern of the B.M. diagram is shown in Fig. 2.25(b). Maximum negative B.M. occurs midway between B and C, i.e., at a distance 10 m from B.

S.F. at a section 10 m from A:

Slope of the curve is given by $\frac{dy}{dx} = \frac{3}{400}(80 - 2x)$.

At $x = 10$ m,

$$\frac{dy}{dx} = \frac{3}{400}(80 - 20) = \frac{9}{20} = \tan\alpha,$$

from which $\sin\alpha = \frac{9}{\sqrt{481}}$ and $\cos\alpha = \frac{20}{\sqrt{481}}$.

$$\text{S.F. at 10 m from } A = \left[\frac{2000}{3} - 20(10)\right]\cos\alpha - \frac{8000}{9}\sin\alpha$$

$$= 425.56 - 364.76 = 60.80 \text{ kN} \uparrow$$

$$\text{Normal thrust at 10 m from } A = \left[\frac{2000}{3} - 200\right]\sin\alpha + \frac{8000}{9}\cos\alpha.$$

$$= 191.50 + 810.60 = 1002.10 \text{ kN}.$$

Example 2.2 *Fig. 2.26 shows a three hinged second degree parabolic arch with a uniform load of w/m throughout the entire span. Find the B.M. diagram for the case.*

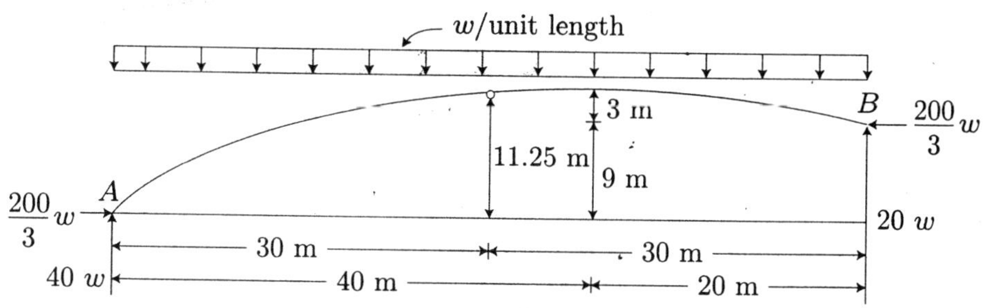

Fig. 2.26 For Example 2.2

Solution : In this case the arch is the same as in previous Example. Reactions of the arch are shown in Fig. 2.26.

B.M. at a distance x from A is given by

$$M = 40w(x) - \frac{w}{2}(x)^2 - \frac{200}{3}w(y) = 40wx - \frac{wx^2}{2} - \frac{200}{3}w\frac{3}{400}x(80-x)$$

$$= 40wx - \frac{wx^2}{2} - 40wx + \frac{wx^2}{2} = 0.$$

Thus the B.M. at any point on the arch is zero.

The above points to a very important feature of arches. The vertical reaction from the vertical loads on an arch tends to produce a positive bending moment at a section, while the horizontal reaction tends to reduce the magnitude of the moment. The moment at any section of an arch is much smaller than the bending moment at the same section of a beam with equal span subjected to the same vertical loads. For an arch span loaded on the entire span with uniform vertical load, the bending moment curve resulting from vertical forces alone has the shape of a second degree parabola. If the arch also has a second degree parabolic shape, the negative bending moment at a section caused by the horizontal reaction eliminates exactly the positive bending moment at that section caused by the vertical forces.

The reduction of bending moment along the arch axis is a very desirable characteristic. This is due to what is known as arch action, which refers to the tendency to develop horizontal thrust at the supports of the structure, even when subjected to vertical loads only. On the other hand, large horizontal reactions and axial forces are introduced in the structure.

Whatever be the shape of the arch, the centre line will always represent the bending moment due to horizontal thrust H, to an appropriate scale when subjected to vertical loads only. If the shape of the bending moment diagram in a beam of the same span, under the same loading follows the centre line axis of the arch, the resulting bending moment in the arch will be zero. Thus for arch subjected to a single concentrated force, the shape of the arch for no bending moment will be triangular. For several numbers of concentrated loads the shape of the arch axis for zero bending moment will consist of a combination of several lines.

2.11 Simple Structure in Space

A plane rigid frame or curved member, subjected to loads, which do not lie in the plane of the structure, are subjected to six stress resultants at a section normal to the axis of the member. In case of a beam, curved in the horizontal plane and subjected to inclined load, in a plane normal to the curved axis, the six stress resultants at a section are an axial force normal to the plane of the section, shear forces in the vertical and horizontal directions, bending moments in the horizontal and vertical planes about an axis, and torque about the longitudinal axis. Since six equations of equilibrium can be established for a free body in space, the six stress resultants at a section can be determined from the conditions of equilibrium.

If the inclined loads be resolved into vertical and horizontal components, the effect of the component in the plane of the structure will be the same as in case of arches or rigid frames. The effect of the component at right angles to the plane of the structure will only be considered here.

A curved or bent member loaded at right angles to the plane of the structure is called a bow girder, which will be subjected to bending moment, twisting moment and shear force.

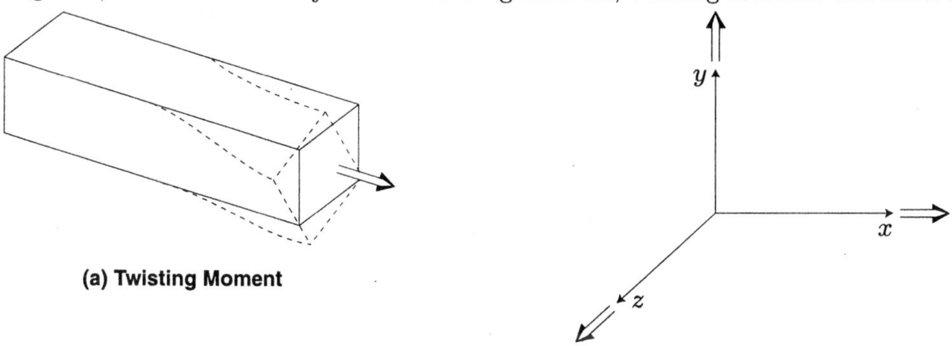

(a) Twisting Moment

(b) General Co-ordinate System

Fig. 2.27: Space Structure

Fig. 2.27(a) shows the direction of the twisting moment when it is considered positive. The convention follows from the vectorial representation of the twisting moment according to which an arrow pointed in the direction of movement of a right-hand screw, when turned in the direction of the twisting moment, represents the twisting moment vectorially. The twist vector will be distinguished from other force vector by double lines as shown.

For analysis of space structure, it is convenient to adopt a sign convention for forces in terms of framework coordinate system. For an XYZ coordinate system as shown in Fig. 2.27(b) the forces and moments acting on the structure can be considered positive when they act in a direction coincident with the coordinate axis, with the moments positive according to the right-hand screw rule. In this case, the moments are represented by vectors with double lines.

Example 2.3 *Obtain expressions for B.M., Twisting moment and S.F. for the circlar bow girder subjected to a concentrated load P at right angles to the plane of the structure as shown in Fig. 2.28.*

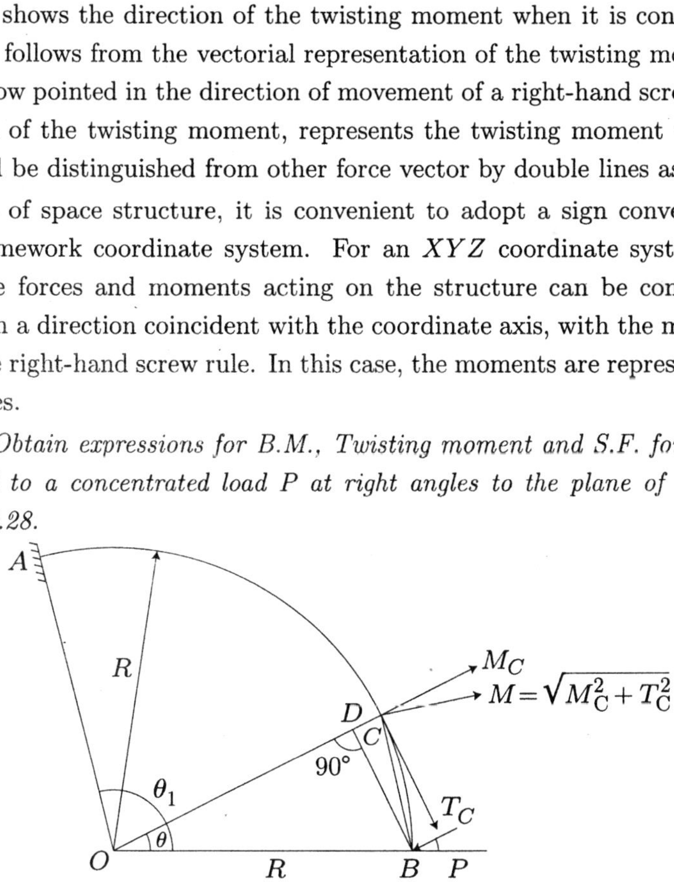

Fig. 2.28: For Example 2.3

Solution : B.M. at $C = P(BD) = PR \sin\theta$, shown vectorially as M_c.
Twisting moment at $C = P(CD) = PR(1 - \cos\theta)$, shown vectorially as T_c.
Resulting moment $M = \sqrt{M_c^2 + T_c^2}$.

$$\text{S.F. at } C = P.$$

Example 2.4 *Obtain expressions for B.M., Twisting moment and S.F. for the circular bow girder in Fig. 2.28 when subjected to uniformly distributed load of magnitude w per unit length over its entire length.*

Solution : B.M. at $C = wR^2 \int_0^\theta \sin\alpha \, d\alpha = wR^2[1 - \cos\theta]$.

Twisting moment at $C = wR^2 \int_0^\theta [1 - \cos\alpha] \, d\alpha = wR^2[\theta - \sin\theta]$.

S.F. at $C = wR \int_0^\theta d\alpha = wR\theta$.

Example 2.5 *Obtain B.M., Twisting moment and S.F. diagrams for the rigid right-angled bent shown in Fig. 2.29(a) when subjected to (i) a concentrated load P at the free end A and (ii) to a uniformly distributed load w per unit length on AB, loads being applied at right angles to the plane of the structure.*

Solution : The solution is illustrated in Figs. 2.29(b) and 2.29(c).

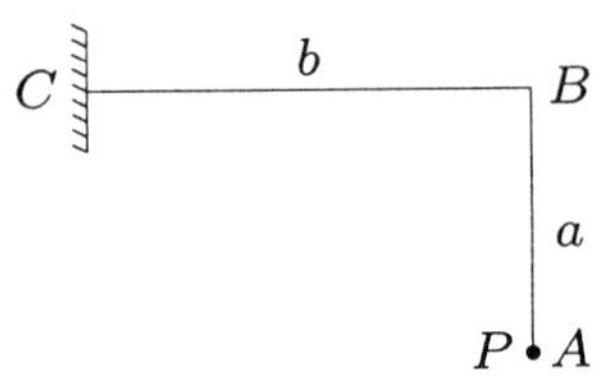

(a) *P* at right angles to the Plane of the Structure

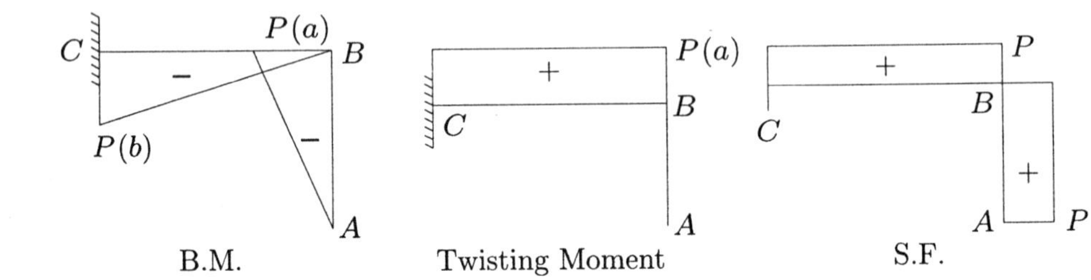

(b) For case (i) Load *P* at *A*

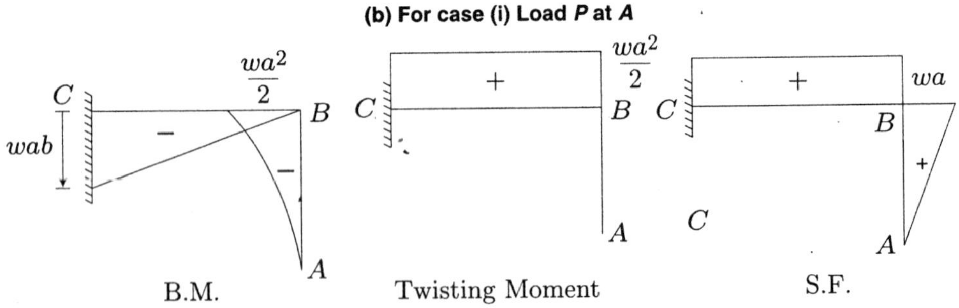

(c) For case (ii) U.D.L. on *AB* at right angles to the plane of the structure

Fig. 2.29: For Example 2.5

2.12 Graphical Determination of Bending Moments and Shearing Forces

The moment of a system of forces may be obtained graphically from the consideration that the algebraic sum of the moments of all forces about a point is equal to the moments of all forces about a point. This principle will also give the bending moment at a section, when forces or resultant of the forces on either side of the section are considered.

Thus, in Fig. 2.30(a), if R represents the resultant of the system of forces P_1, P_2, P_3 and P_4, the moment of the forces about a point A, will be given by

$$M_A = R(h_1),$$

where h_1 is the perpendicular distance from A to the line of action of R in the space diagram.

This method of determining moments graphically is universally applicable and generally as easy as any other, when the forces to be considered are not parallel. For forces which are parallel or nearly parallel, a method introduced by a Swiss Engineer, Professor Carl Culmaun of Zurich is very convenient.

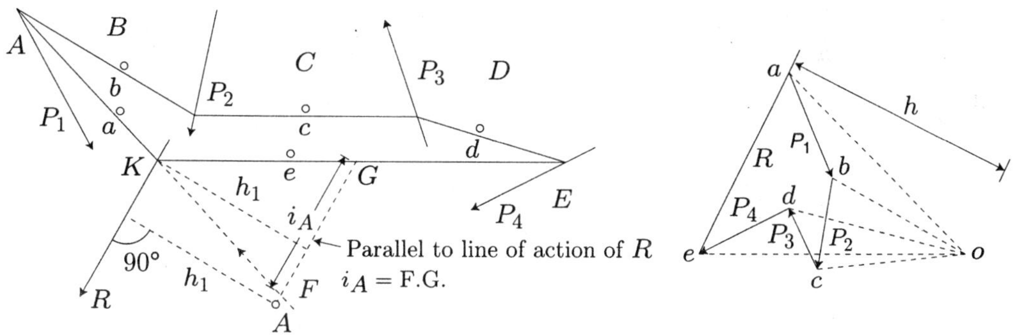

(a) Space diagram **(b) Force polygon**

Fig. 2.30: Graphical determination of B.M. and S.F.

In space diagram of Fig. 2.30(a), a line parallel to the line of action of resultant R is drawn through A. If this line intersects the rays oa and oe at F and G and if K is the point of intersection of rays oa and oe, a triangle KFG is formed, which is similar to triangle oae in the force polygon. From relationship of similar triangles,

$$\frac{h_1}{FG} = \frac{h}{ae} = \frac{h}{R},$$

where h is the perpendicular distance from pole O to side ae of the triangle oae and is called the pole distance.

Also $FG(h) = R(h_1) =$ moment of the force system about point A.

The distance FG denoted by i_A in Fig. 2.30(a) is called the intercept and the altitude h is referred to as the pole distance.

This means that the moment of any system of forces about any point A, is equal to the product of the pole distance of the resultant of the forces and the intercept which the strings or forces of the equilibrium polygon determining the line of action of the resultant cut from a line drawn through A parallel to the resultant. The pole distance is a force, the magnitude of which is fixed by the scale of the force polygon and the intercept, a distance measured to the scale of the space diagram.

Culman's method of determining moments of a force system graphically can be applied to obtain bending moments in beams. Since the shear force at a section of a beam equals the vertical component of the resultant of the forces applied on either to the left or to the right of the section, it can easily be found from the force polygon. The bending moment at a section is equal to the moment of the resultant of the forces applied either to the left or right of the section and may be obtained by the previous procedure. If the applied loads on the beam consists of distributed loads, the length of the beam must be divided into small segments, so that the distributed loads can be approximated by concentrated loads.

In Fig. 2.31(a), beam supporting four loads, P_1, P_2, P_3 and P_4 is shown, together with the force polygon and space diagram for the determination of the reactions R_A and R_B. The bending moment at D is the moment of resultant of R_A and P_1 about D. The line of

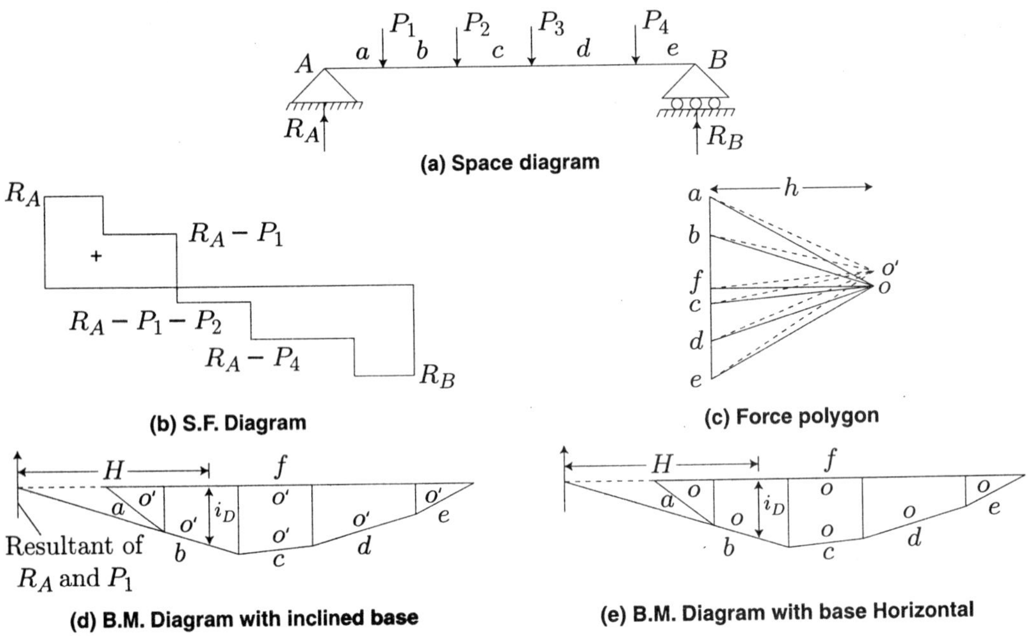

Fig. 2.31: Graphical determination of B.M. and S.F.

action of resultant R_A and P_1 is shown in Fig. 2.31(d), and is at the intersection of ray ob and the closing ray of. In accordance with the principles described above, a vertical line through D determines the intercept i_D and the bending moment at D then is

$$M_D = (R_A - P_1)h_1 = h(i_D), \text{ with } H = h_1$$

The bending moment thus obtained in Fig. 2.31(d) is shown with a horizontal base in Fig. 2.31(e), which can be obtained by choosing a new pole O', with $O'f$ horizontal as shown in Fig. 2.31(d). The shear force diagram is obtained by projecting the force polygon as illustrated in Fig. 2.31(b).

The bending moment may be scaled directly from the funicular polygon representing the bending moment diagram. If the space diagram be drawn so that 10 mm $= L$ and if the force scale is such that 10 mm represents P force, then the scale of bending moment diagram is PhL per 10 mm of intercept, the dimension of bending moment depending on the units of P and L.

Bending moment and shear forces in case of beams, supporting a distributed load may be constructed graphically using the method described above, by dividing the load into short lengths, applying the total load over such length at the centre of gravity of the load for that section. The graphical solution for reactions, shear force and bending moment is then carried out as usual considering the distributed load to be replaced by this series of concentrated loads. The values of reactions found in this manner are not in error, but the values of shear force and bending moment are exact only at the ends of the various short sections into which the distributed load is divided. The ordinates of shear force and bending moment curves at intermediate points are not appreciably in error, however, provided the lengths of the sections are reasonably small.

Fig. 2.32 illustrates the procedure. The beam shown in this case supports a uniformly distributed load which is divided into ten equal sections.

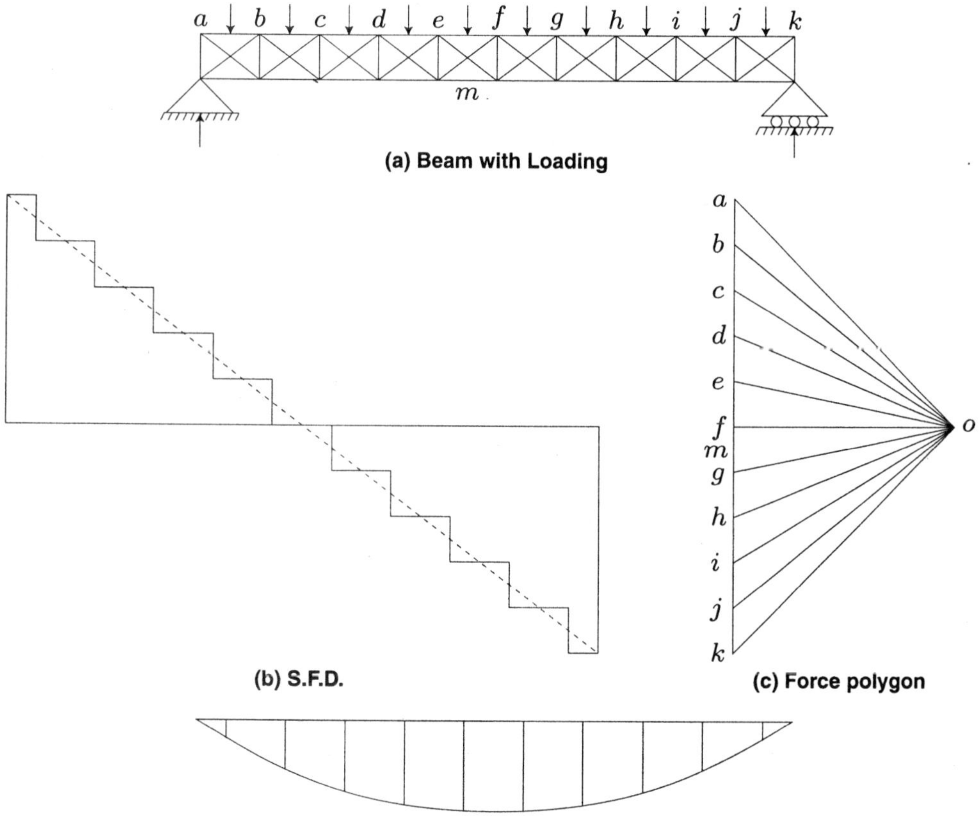

(a) Beam with Loading

(b) S.F.D.

(c) Force polygon

(d) B.M.D.

Fig. 2.32: Graphical determination of B.M. and S.F. for u.d.l.

Example 2.6 *Determine graphically the B.M. and S.F. diagram in the cantilever beam in* Fig. 2.33(a).

Solution : Fig. 2.33 illustrates the graphical solution.

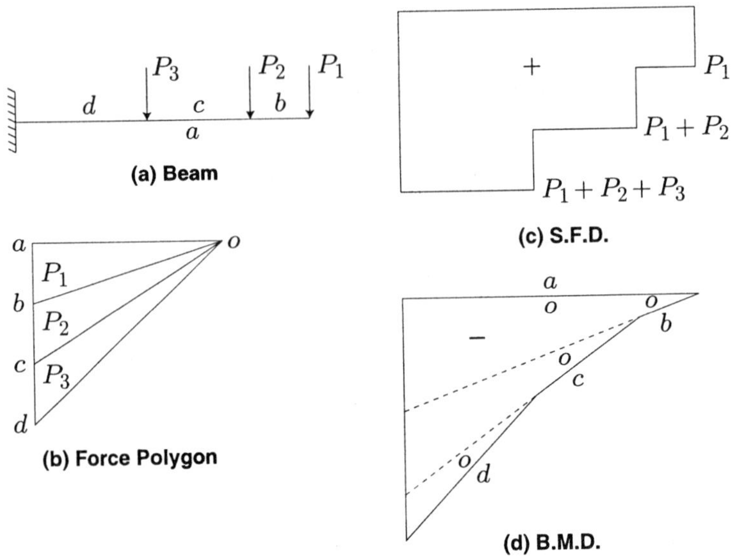

Fig. 2.33: For Example 2.6

Example 2.7 *Determine graphically the B.M. and S.F. diagram in the beam in* Fig. 2.34.

Solution Fig. 2.34 illustrates the graphical system.

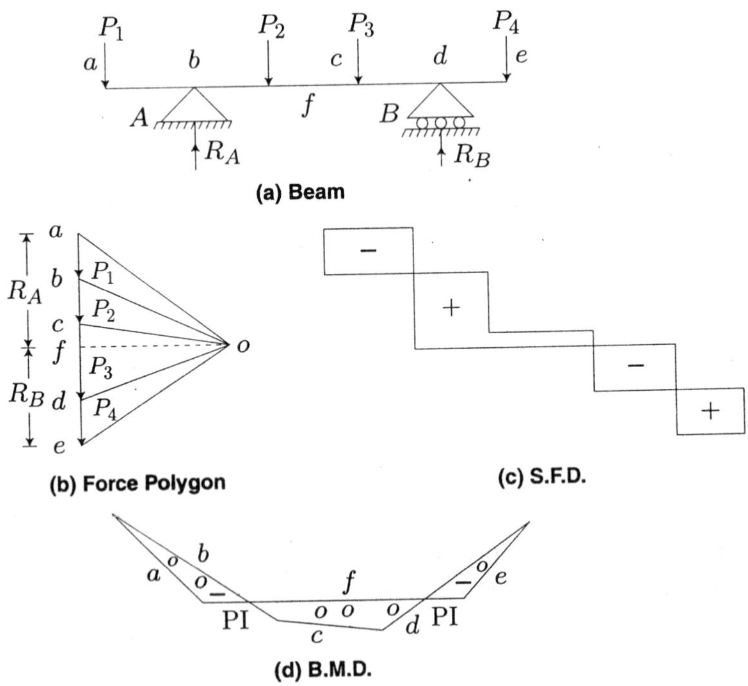

Fig. 2.34: For Example 2.7

Problems

2.1 Draw the shear force and bonding moment diagrams for cantilever loaded as shown in Fig. 3.1. Take $q = 1$ kN

2.2 A beam of 12 metres span carries a concentrated load of 1000 N placed at a distance of 3 metres from one end and another load of 6000 N placed 4 metres from the other end. Draw the shear force and bonding moment diagrams.

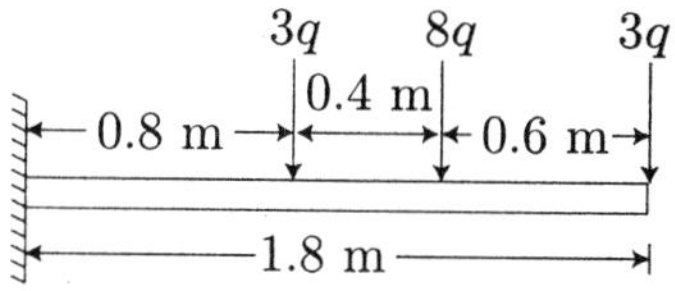

Fig. P-2.1

2.3 A beam is loaded as shown in Fig. P-2.2. Draw the diagrams of bending moment and shear force.

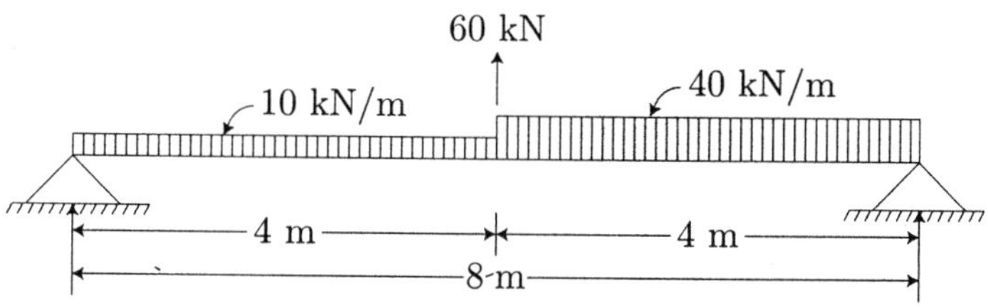

Fig. P-2.2

2.4 A beam 15 m long is supported at the left end and at a point 3 m from the other end. It carries a uniformly distributed load of 15 KN per metre over its whole length. Additional loads of 30 kN each are placed at 4 metres, 9 metres and 15 metres respectively from the left end. Draw the shearing force and bending moment diagrams.

2.5 An overhanging beam is loaded as shown in Fig. P-2.3. Find the magnitude of the load P so that the bending moment at C is zero.

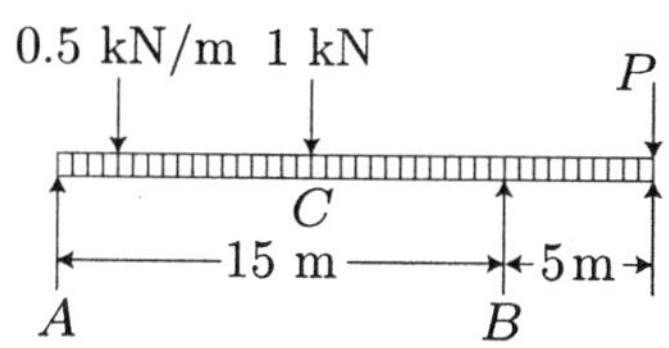

Fig. P-2.3

2.6 A cantilever beam projects 5m from a wall. It carries a uniformly distributed load of 20 kN/m on the entire projection and concentrated loads of magnitudes 50 kN and 100 kN at the free end and at 3 m from the wall. Draw the shear force and bending moment diagrams and state the maximum values of the shear force and bending moments.

2.7 A horizontal beam is simply supported at ends and carries a uniformly distributed load of 20 kN/m between supports 10 m apart. Counter-clockwise moments of 50

kNm and 100 kNm respectively are applied at the two ends of the beam at supports. Draw the shearing force and bending moment diagrams for the beam.

2.8 Draw the shearing force, bending moment and axial force diagrams for the structures in Figs. P-2.4 to P-2.7.

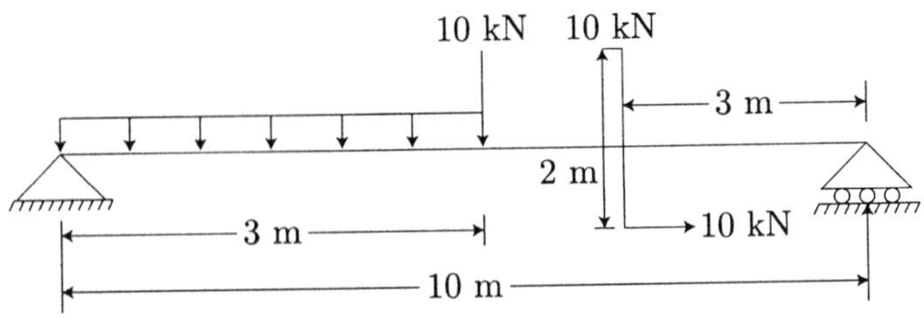

Fig. P-2.4

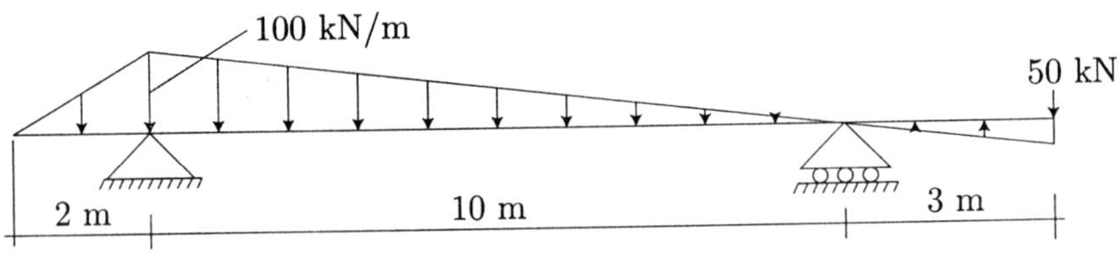

Fig. P-2.5

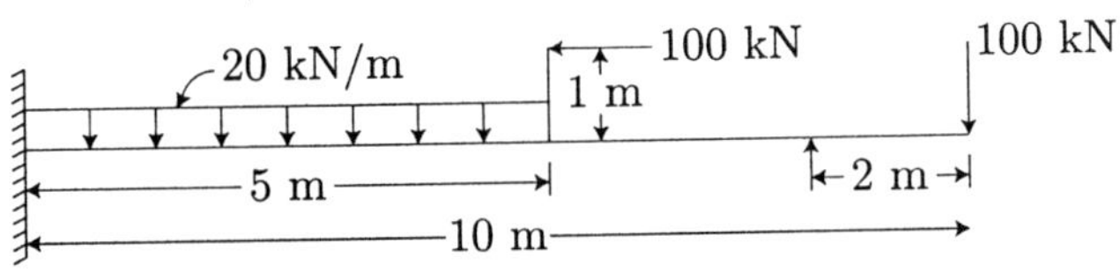

Fig. P-2.6

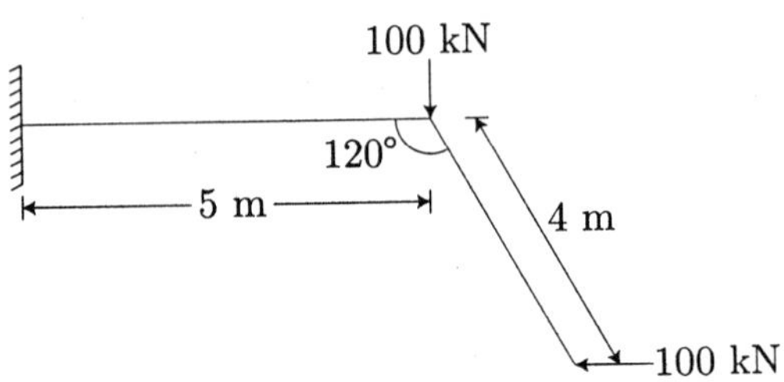

Fig. P-2.7

2.9 A uniform beam of weight W and length L rests on two supports A and B, A being at one end. Find the position of B, which gives the least value of the maximum bending moment. Draw the bending moment and shear force diagrams for the case.

2.10 A beam of length L carrying uniformly distributed load of w per unit length is simply supported on two intermediate supports movable along the length with equal overhang on either side. The supports are so adjusted that the maximum B.M. is the minimum possible. Determine the position of the supports and draw the B.M. and S.F. diagrams. Use this result to determine the most economic length for a railway sleeper if the rail centres are 1.60 m apart.

2.11 Compute the shear force and bending moments at various panel points and plot the diagram for each of the beams loaded as shown in Figs. P-2.8(a), P-2.8(b) and P-2.8(c). Also compute the corresponding values.

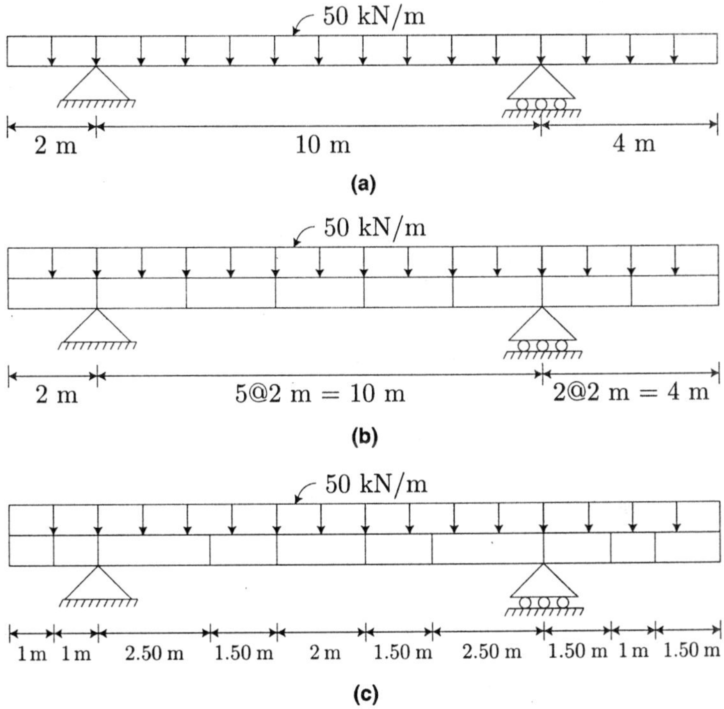

Fig. P-2.8

2.12 A beam AB is loaded as shown in Fig. P-2.9. Draw the shearing force and bending moment diagrams. Where does the maximum bending moment occur?

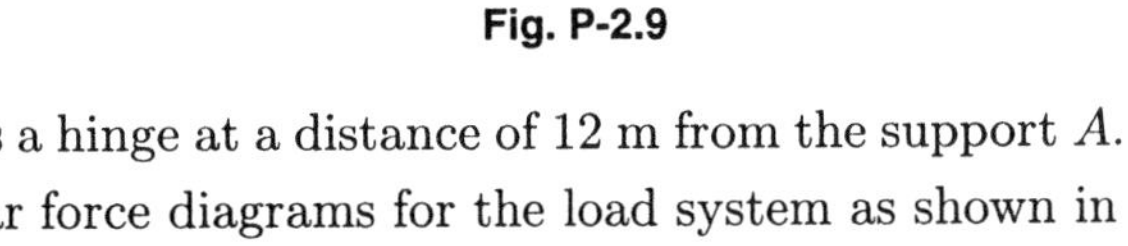

Fig. P-2.9

2.13 A beam ABC has a hinge at a distance of 12 m from the support A. Draw the bending moment and shear force diagrams for the load system as shown in Fig. P-2.10.

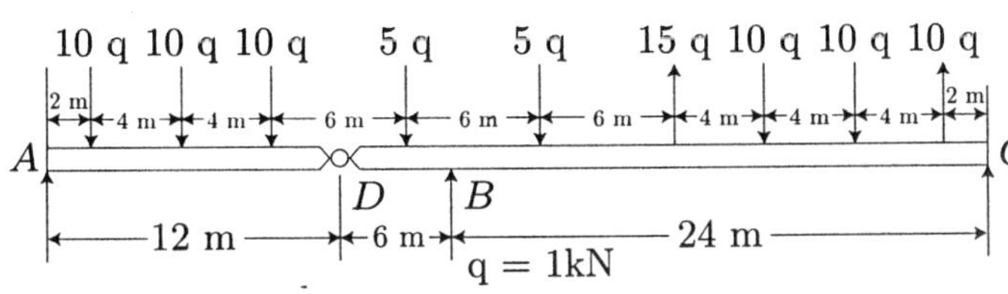

Fig. P-2.10

2.14 Draw the bending moment and shearing force diagrams for the beam *AB* (Fig. P-2.11)

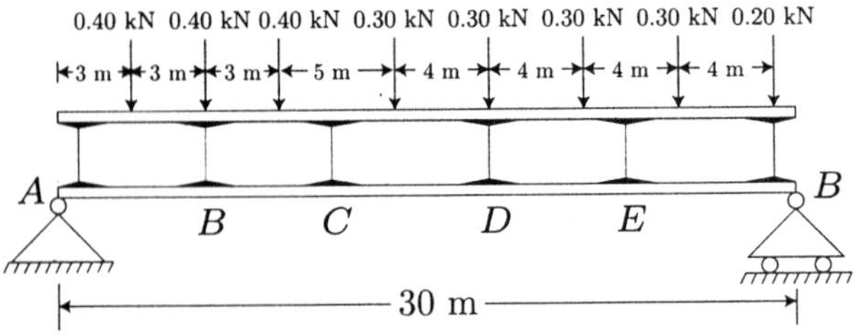

Fig. P-2.11

2.15 Evaluate the maximum bending moment and maximum shear force that can occur in a cantilever beam sloping as shown in Fig. P-2.12.

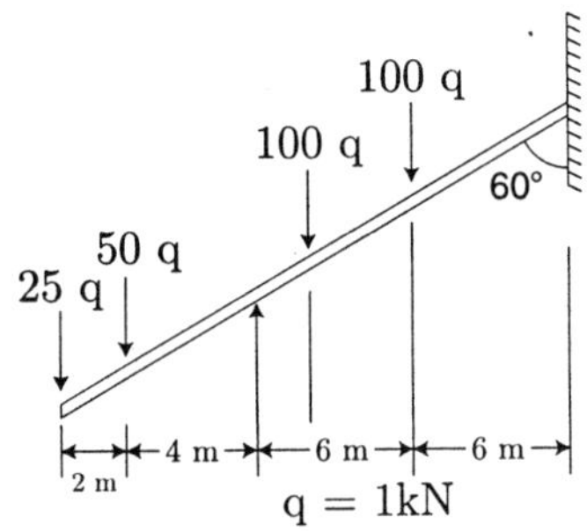

Fig. P-2.12

2.16 A beam is supported on roller at one end and hinged at the other. Determine the maximum bending moment due to the loads as shown in Fig. P-2.13.

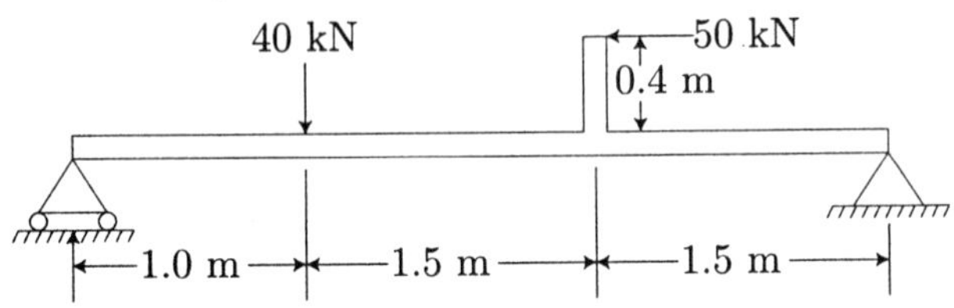

Fig. P-2.13

2.17 Evaluate the bending moments and shear forces at various sections of the structures drawn in Figs. P-2.14 to P-2.17.

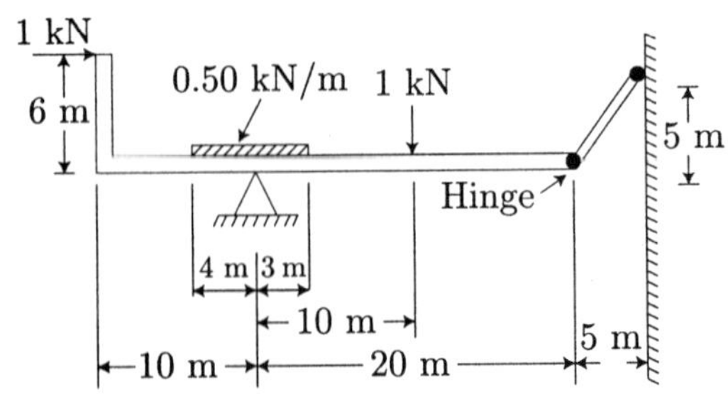

Fig. P-2.14

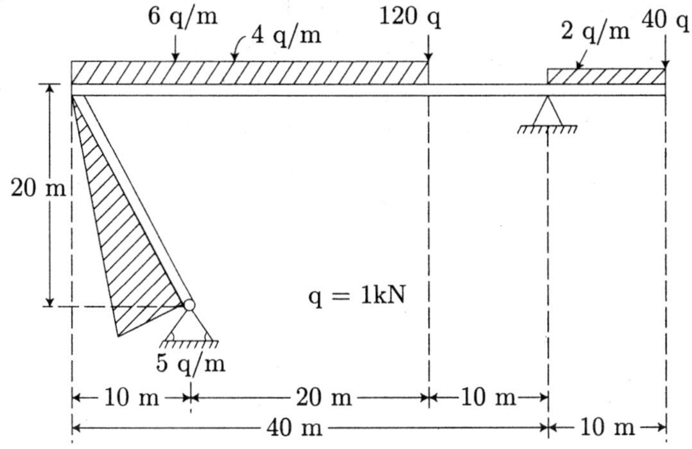

Fig. P-2.15

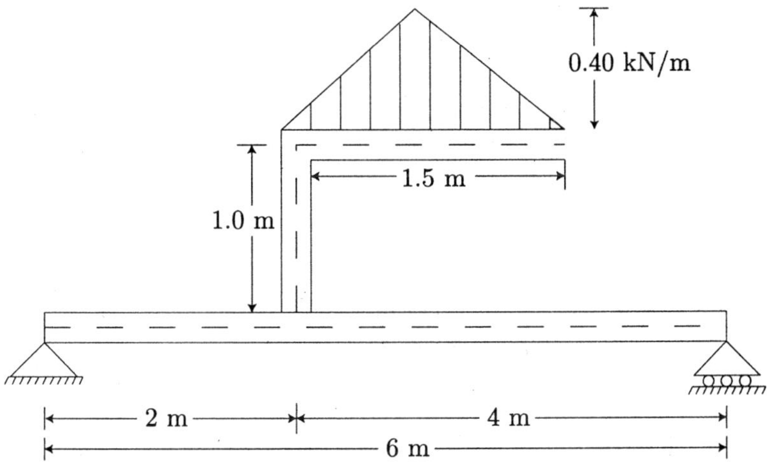

Fig. P-2.16

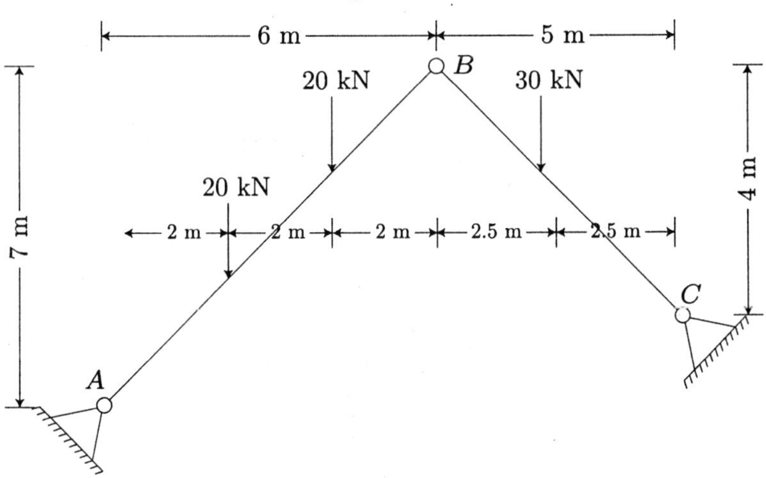

Fig. P-2.17

2.18 A spread footing supports two 1000 kN loads as shown in Fig. P-2.18. Construct shear force and bending moment diagrams for the footing. Assume that the resulting soil pressure on the base of the footing is uniform and the loads are distributed uniformly across the 800 mm walls.

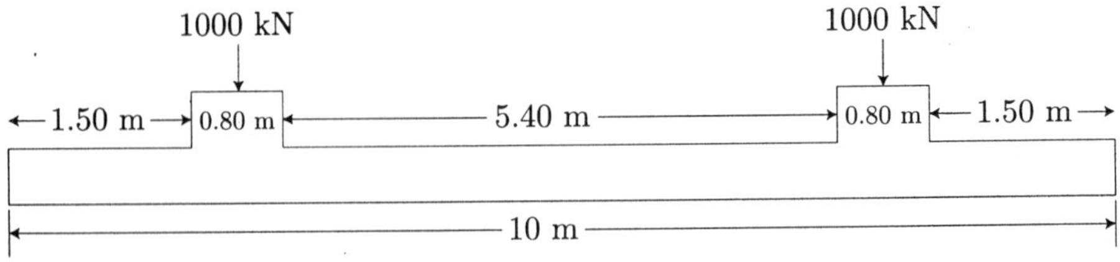

Fig. P-2.18

2.19 The beam shown in Fig. P-2.19 is floating in water. Determine the distance 'a' for making the upward reaction of water uniform across the beam. Also obtain the B.M. and S.F. diagrams.

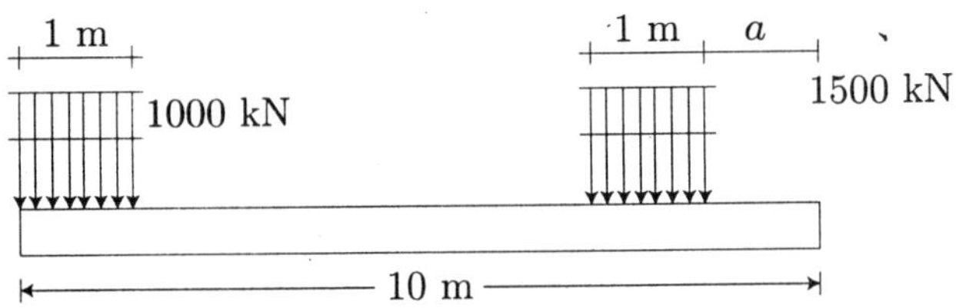

Fig. P-2.19

2.20 A three hinged symmetrical parabolic (2nd degree) arch of span L and central rise of y_c carries a uniformly distributed load of w per unit length over the left half of the span. Obtain the bending moment diagram and the magnitudes of the maximum positive and negative bending moments. Also calculate the normal thrust and radial shear at quarter span from left-hand support.

2.21 A three hinged parabolic arch of span 30 m has its crown 9 m above the left-hand support and 4 m above the right-hand support. The crown of the arch is at a distance 12 m from the left support. Obtain the bending moment diagram and calculate the maximum values when a concentrated horizontal loads of magnitude 100 kN placed 5 m above the left-hand support.

2.22 A symmetrical three-hinged parabolic arch has a span of 15 m and a rise of 3 m and is hinged at the springings and at the crown. It carries concentrated vertical loads of 50 kN each, one at 2 m from the left support and another at the crown. Find the bending moment diagrams and calculate the normal thrust and radial shear at quarter points.

2.23 symmetrical three-hinged circular arch has a span of 40 m and a rise of 10 m. The arch is loaded with 50 kN/m, over the entire span. Find the B.M. diagram and calculate the maximum value. Also calculate the normal thrust and radial shear at quarter points.

2.24 A three-hinged arch is formed as shown in Fig. P-2.20 from two segments. Determine the B.M. diagram and the maximum value when a load W acts as shown.

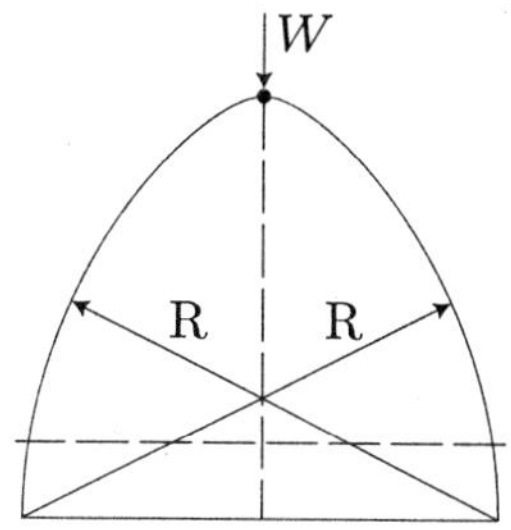

Fig. P-2.20

2.25 The deck of a symmetrical three-hinged parabolic arch is carried on spandrel columns. The span of the arch is 60 m, the rise at the crown is 10 m and the spandrel columns are at 10 m centres. Determine the bending moment diagram and the maximum values when a concentrated load of 60 kN is placed at quarter span from the left-hand support. Also calculate the normal thrust and radial shear at the section.

2.26 Obtain the vertical shear, bending moment and twisting moment diagrams for the rigid bents in horizontal plane shown in Fig. P-2.21 and loaded with vertical concentrated load as shown.

2.27 Obtain the shear force, bending moment and twisting moment diagrams for the bents in Fig. P-2.21 when the part AB of the bents is loaded with uniformly distributed load of magnitude w per unit length. Consider load at right angles to the plane of the structure.

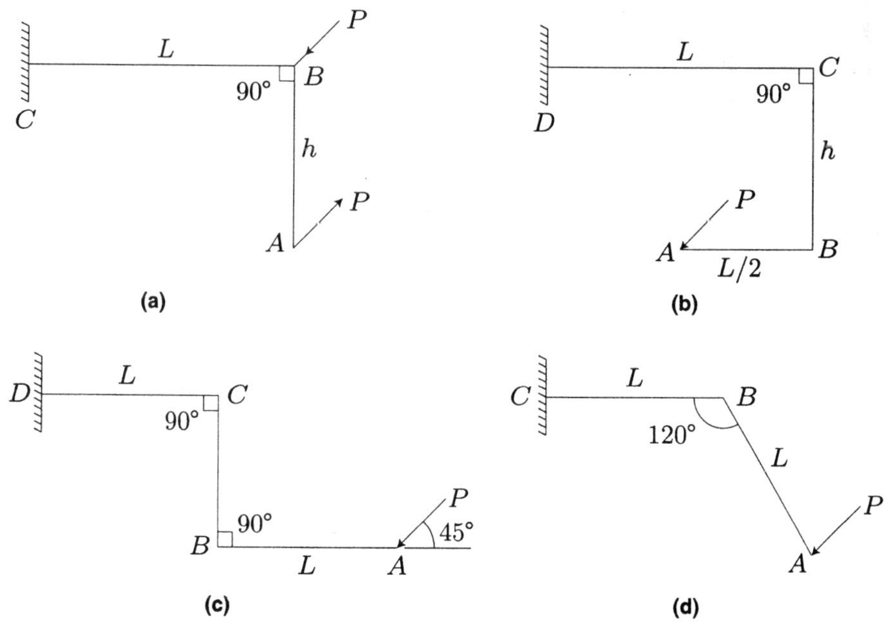

Fig. P-2.21: Load at right angles to the plane of the structure

Chapter 3

Framed Trusses and Structures

3.1 Introduction

A truss is any framework of bars placed together to form triangles. The members may be riveted or pin-jointed together. Ties are the members of a truss which are in tension, struts being those members which are in compression. When a frame has sufficient members to keep it in equilibrium, under all conditions of loading, and the forces in the members can be determined by the equations of statics directly, it is termed as a statically determinate structure. Whenever it has more members than those which can be analysed by the laws of statics, they are called statically indeterminate frames.

Members which form the outline or perimeter of a truss are generally called chord members or simply chords. The top members are called top cords and the bottom members are called bottom chords. The interior members which connect the joints of the chords are called web members or diagonals. Members which are in a vertical position in a truss are called as verticals.

3.2 Necessary Number of Members

The usual shape of a truss is a series of triangles, as triangle is the only figure which cannot be changed in shape without changing the length of one or more of its sides. If a force is

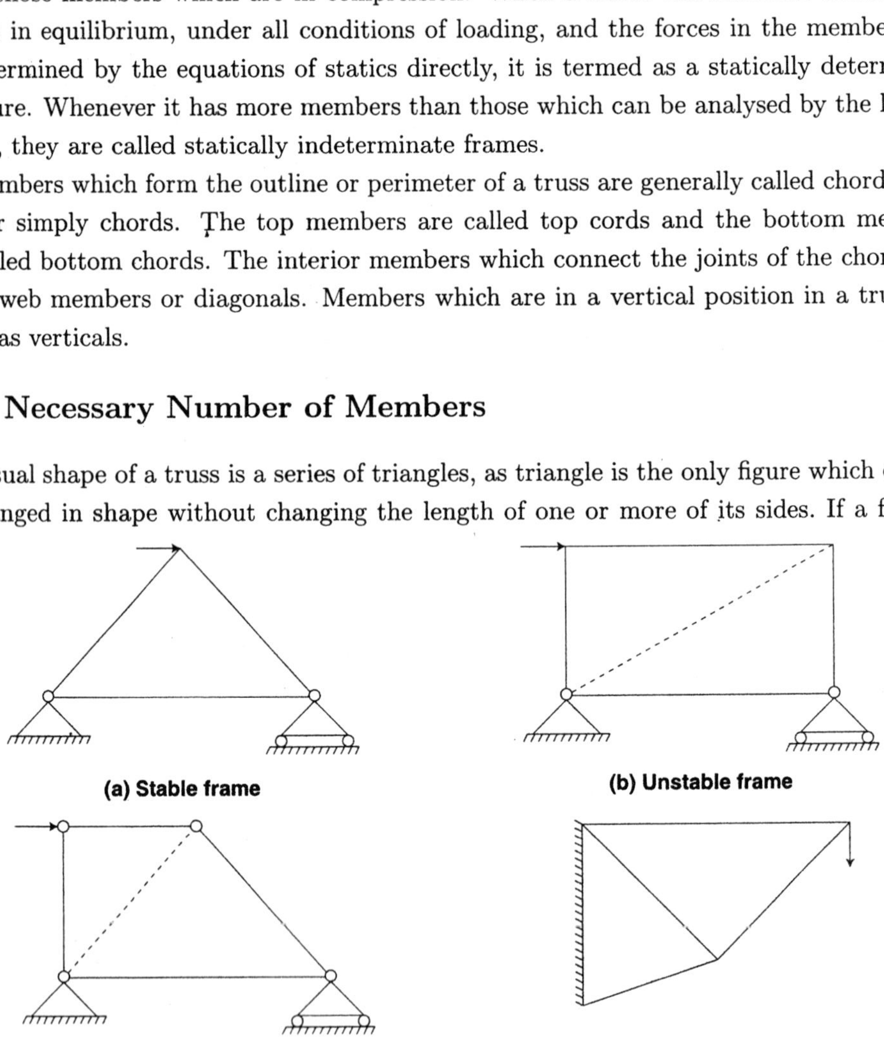

(a) Stable frame

(b) Unstable frame

(c) Unstable frame

(d) Stable Truss

Fig. 3.1: Frames

applied to a triangular frame, the members of which are connected by frictionless joints as in Fig. 3.1(a), it cannot distort in shape, whereas, if a similar force is applied to frames of a different shape, e.g., frames shown in Figs. 3.1(b) and 3.1(c), they will immediately change the shape of the truss unless prevented by the addition of members.

Since the forces at a joint form a group of concurrence forces each joint provides two equations of statics $\sum H = 0$ and $\sum V = 0$.

Hence two unknown forces can be determined at each joint provided not more than two unknown forces exist at that joint, which have not been determined at preceding joint. Considering the truss shown in Fig. 3.2, the forces in AB and AD can be determined from joint A. Once these are known, forces in DE and DB can be determined from the consideration of the equilibrium of joint D. Alternatively for this truss also considering the equilibrium of joint C forces in CE and CD can be determined.

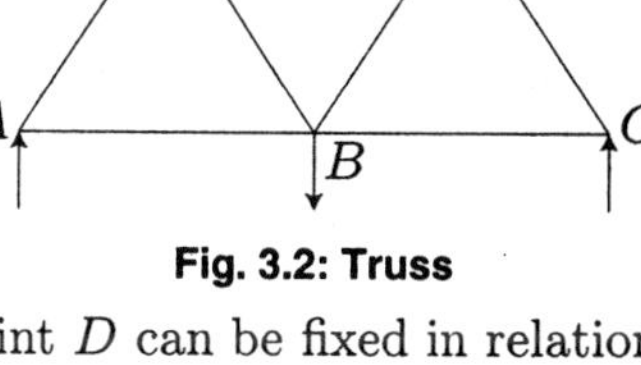

Fig. 3.2: Truss

If the bar AB is considered as fixed, the position of the joint D can be fixed in relation to AB with the help of two bars AD and DB. Similarly the position of joint E can be determined with the aid of bars BE and DE. Hence to establish a fixed relation between a series of joints, one bar and its end joints as a beginning and two bars for each additional joint are needed. Algebraically, if n = number of members and j = number of joints.

$$n = 1 + 2(j - 2) = 2j - 3 \tag{3.1}$$

Thus, the number of bar forces in a truss that can be found by statics is twice the number of joints minus three.

In case a simple truss hanging from the wall as in Fig. 3.1(d) and Fig. 3.3(a), the original triangle contains one joint and two members and the relationship becomes, $n = 2j$, (3.2) where the points of attachment to the walls are not considered as joints.

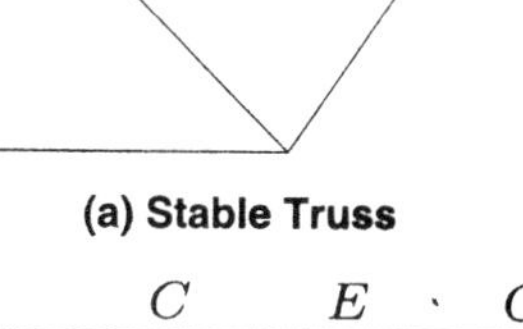

(a) Stable Truss

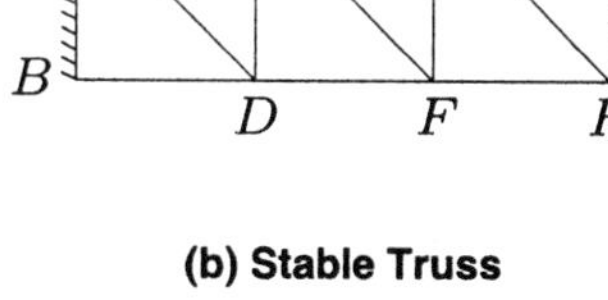

(b) Stable Truss

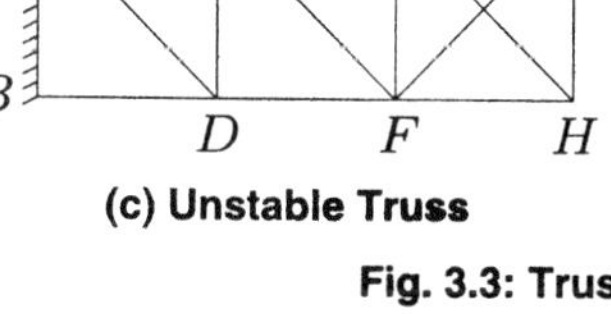

(c) Unstable Truss

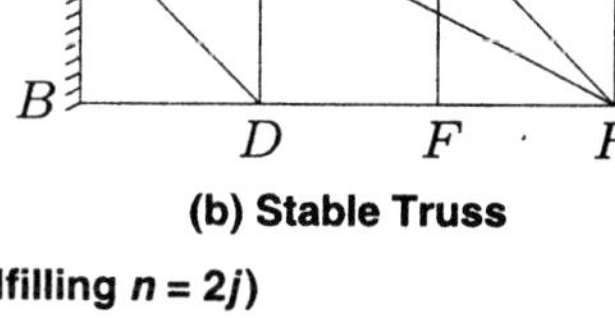

(b) Stable Truss

Fig. 3.3: Trusses (fulfilling $n = 2j$)

Simple truss of the first type requires three additional reactions in the form of bars or their equivalent, in order that the constraint in one plane is complete, whereas in the second type such constraints are already complete. Thus in either case, it may be said that for the complete constraint of j numbers of joints in a plane, they must be connected among themselves and the foundations by $n = 2j$ bars or equivalent constraints, i.e., it may be expressed in general as

$$n + r = 2j,$$

where r = no. of constraints or reactions.

Fulfilment of the relation $n = 2j$ or $n + r = 2j$ alone is not the complete criteria of the rigidity of a truss. As an example, considering the simple truss in Fig. 3.3(b), the rigidity may be destroyed by removing bar CF. If, however, this is replaced by bar FG, as in Fig. 3.3(c) the relation $n = 2j$ is fulfilled, but the truss loses its rigidity. Replacing bar CF by bar CH as in Fig. 3.3(d), however, does not affect the rigidity of the truss.

The criteria for rigidity, therefore, cannot be checked by the relationship alone. For rigidity it is necessary that the bars be properly arranged. However, even if instability of a truss is not always obvious, it can be detected if the results of the computed reactions or bar forces in a system turn out to be infinite or inconsistent.

3.3 Types of Plane Trusses

From the point of view of analysis, trusses may be statically determinate or statically indeterminate depending on whether the reactions and/or the axial forces in members (called bar forces or sometimes stresses) can be obtained from the conditions of statics alone. Normally rigid (i.e., stable) trusses fulfilling the relationship of Eqs. (3.1) and (3.2) are statically determinate truss. If for a truss $n > (2j - 3)$ or $n + r > 2j$, it is a case of a statically indeterminate truss, the degree of indeterminacy being given by $n - (2j - 3)$ or $(n + r) - 2j$.

Statically determinate truss may be classified into three categories, simple, compound and complex. A simple truss is defined as a system of bars assembled by first joining together three bars at their ends in the form of a triangle and then adding to this configuration two more bars for each additional joint. A compound truss is defined as a rigid system formed by interconnecting two or more simple trusses, the rigidity being provided by a sufficient number of independent components of constraints at the interconnection between the simple trusses. If the arrangement of bars in a truss fails to satisfy either the definition of a simple truss or that of a compound truss, even though the criteria for statical determinacy holds true, the system is said to be a complex truss.

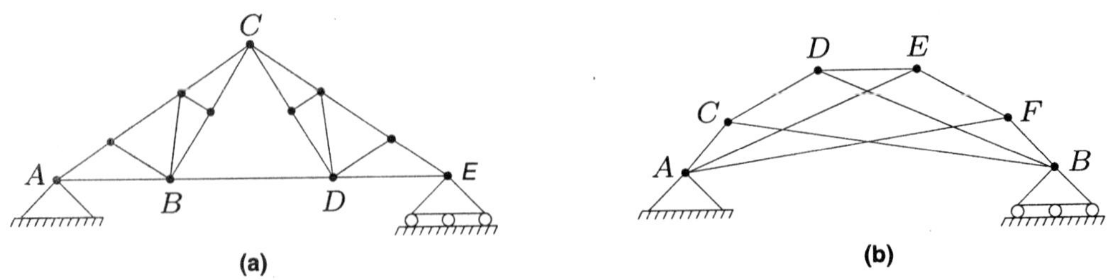

Fig. 3.4: Compound Trusses

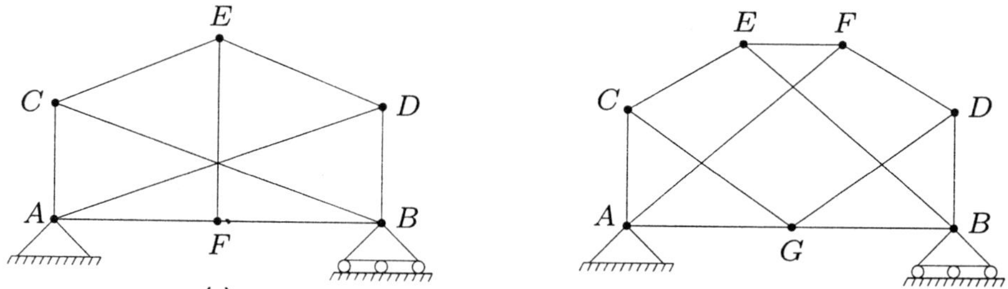

(a) (b)

Fig. 3.5: Complex Trusses

The compound and complex types of trusses are illustrated in Figs. 3.4 and 3.5.

3.4 Common Types of Plane Trusses

(i) **Roof Trusses :** The trusses shown in Fig. 3.6 are common types of plane trusses used for roofs of buildings.

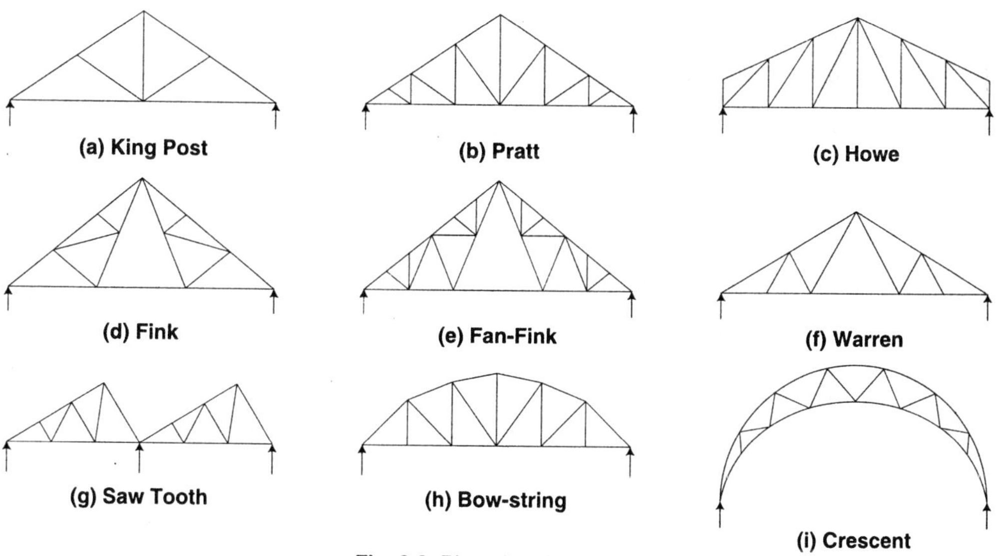

(a) King Post **(b) Pratt** **(c) Howe**

(d) Fink **(e) Fan-Fink** **(f) Warren**

(g) Saw Tooth **(h) Bow-string** **(i) Crescent**

Fig. 3.6: Plane Roof Trusses

(ii) **Bridge Trusses :** The trusses shown in Fig. 3.7 are common types of trusses used in Bridges.

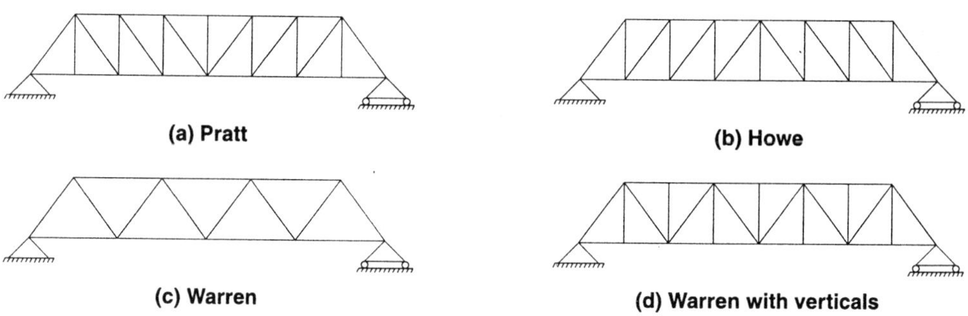

(a) Pratt **(b) Howe**

(c) Warren **(d) Warren with verticals**

Fig. 3.7: Plane Bridge Trusses (Contd.)

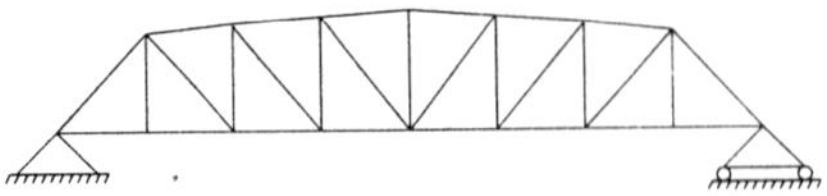

(e) Parker

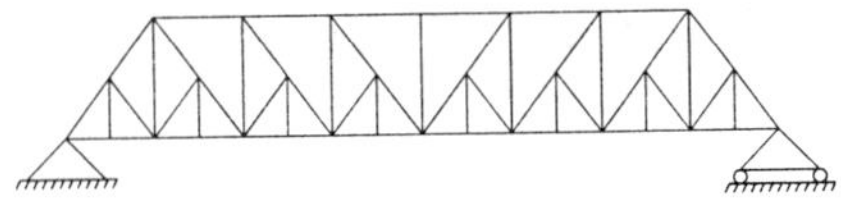

(f) Baltimore

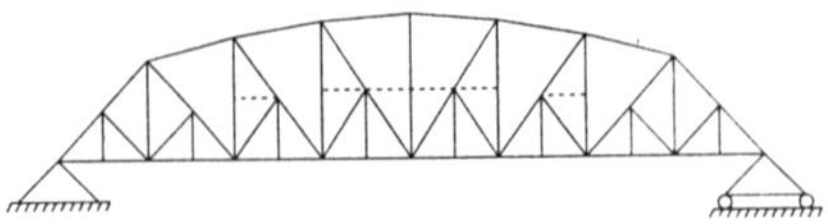

(g) Pettit or Pennsylvania

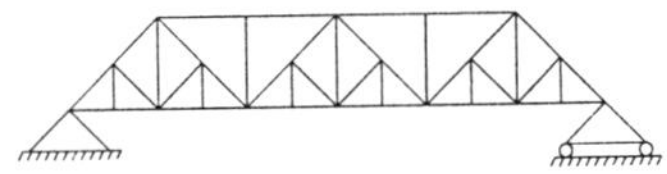

(h) Sub-divided Warren

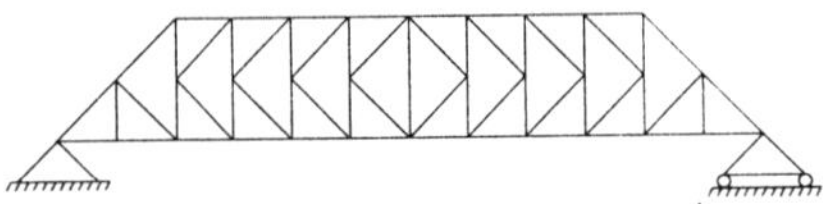

(i) K-Truss

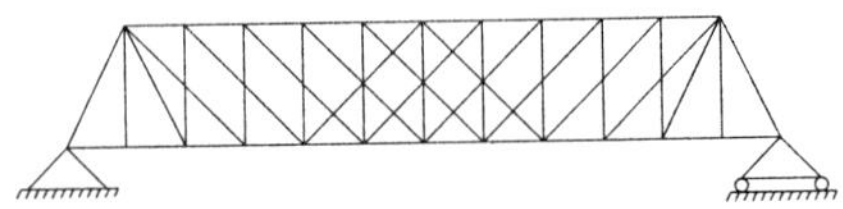

(j) Whippole

Fig. 3.7: Plane Bridge Trusses

3.5 Determination of Forces in Members of Frames

Fundamental Assumptions are

(a) The lines through the centre of gravity of the section of all members meet at a single point (at a joint).

(b) The joints of the truss are made with frictionless pins or hinges.

(c) It is assumed that there is no bending in any member of the truss.

(d) There is no change in the length of the individual members of the truss.

3.6 Method of Analysis

The following principles are to be kept in view while analysing forces in a framed structure :

(a) The frame as a whole acts as a rigid body and the external forces, i.e., the loads and the reactions acting on the frame form by themselves a system of forces in equilibrium.

(b) The pulls or thrusts produced in the several members meeting at any joint of the frame and the external load applied at that joint form a concurrent system of forces in equilibrium.

(c) Two equal and opposite forces acting at the ends of a member can replace the member.

(d) Any portion of the frame may be taken as a rigid body held in equilibrium by the external forces acting at the joints together with the forces exerted upon it through the members by the remainder of the structure.

The main methods of analysis are : (i) Method of joints. (ii) Method of sections or moments. (iii) Graphical method.

3.7 Method of Joints

This method involves the following steps.

(a) Determine the external reactions by taking the whole truss as a free body.

(b) Consider each joint and apply the two equations of statics : $\sum H = 0$ and $\sum V = 0$.

(c) It is advisable to start with the joint at the left and proceed joint by joint to the middle of the structure. Then start from the joint at the right hand side end and proceed to the middle. This will make it possible to have a check.

(d) The final forces are recorded on the line diagram of the truss. It may be helpful to write horizontal and vertical components of each force so that a check can be made.

(e) The components of forces may be found in horizontal and vertical directions by taking forces in proportion to the lengths of members.

The horizontal component of the force F in diagonal AC (Fig. 3.8)

$$F_h = F\cos\theta = F\frac{BC}{AC} = \text{Force}\left(\frac{\text{Component of length in that direction}}{\text{Length of the member}}\right)$$

Similarly, the vertical component

$$F_v = F\sin\theta = F\left(\frac{AB}{AC}\right).$$

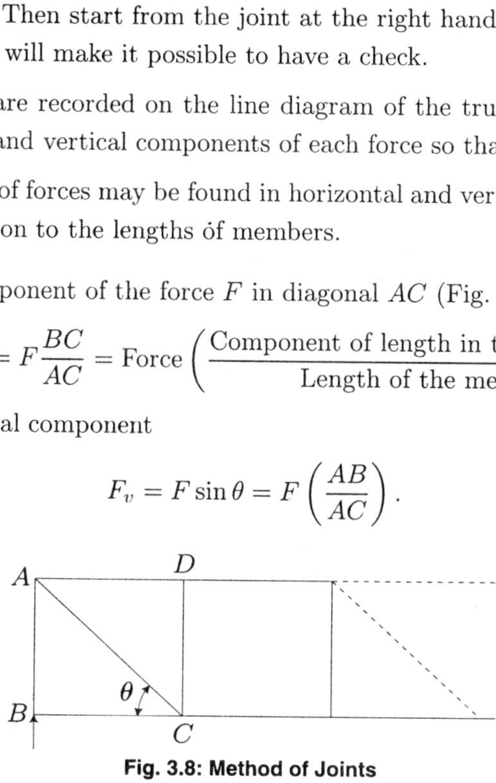

Fig. 3.8: Method of Joints

Example 3.1 *Analyse the Pratt truss given in Fig. 3.9 by method of joints.*

Solution : Reactions are as shown in Fig. 3.9(a).

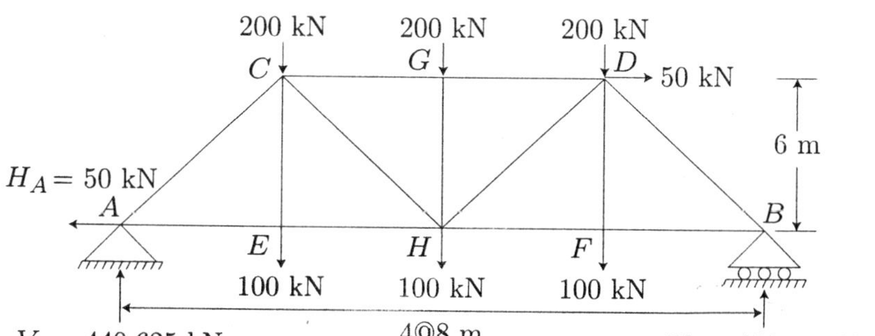

(a) Truss with Loading

Fig. 3.9: Truss for Example 3.1 (Contd.)

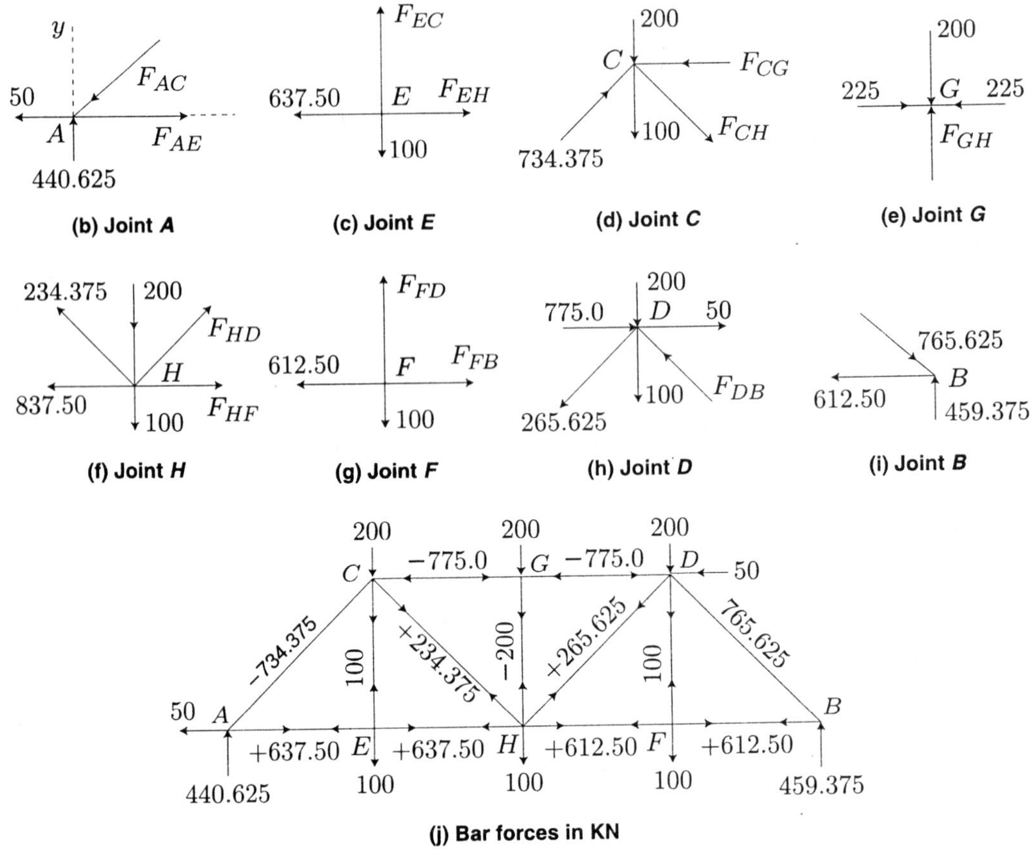

Fig. 3.9: Truss for Example 3.1

From $\sum M_B = 0$, $V_A = \dfrac{(300)(24) + (300)(16) + (300)(8) - 50(6)}{32} = 440.625$ kN $\uparrow$

From $\sum V = 0$, $V_B = 459.375$ kN $\uparrow$

From $\sum H = 0$, $H_A = 50$ kN $\leftarrow$

Bar Forces

Since there are only two unknown bar forces at both joints A and B, one of them may be chosen as the starting point. Free body diagram of joint A can be represented as shown in Fig. 3.8(b). F_{AC} and F_{AE} are the only unknown forces. The senses of direction of F_{AC} and F_{AE} may be arbitrarily assumed even though in many cases it will be possible to visualise the same.

In Fig. 3.9(b), F_{AC} has been assumed compressive and F_{AE} as tensile.

From $\sum F_y = 0$, i.e., $\sum$ Forces in the vertical direction,

$$\downarrow F_{AC}\left(\frac{6}{10}\right) - 440.625 = 0,$$

whence $F_{AC} = 734.375$ kN (compression)

Also from $\sum F_x = 0$,

$$\overrightarrow{F}_{AE} - \overleftarrow{50} - F_{AC}\left(\frac{8}{10}\right) = 0,$$

whence $F_{AE} = 637.50$ kN (tension).

The positive results confirm that the assumed senses of direction for both the members are correct.

Next proceeding to joint E, at which there are only two unknown bar forces, as shown in Fig. 3.9(c) from

$$\sum F_y = 0, F_{EH} = 637.50 \text{ kN (tension)}$$

and from $\sum F_x = 0$, $F_{EC} = 100$ kN (tension). Similarly for joint C, shown in Fig. 3.9(d).

$$\sum F_y = 0, \text{yields,}$$

$$\frac{6}{10}(F_{CH}) \downarrow +100 \downarrow +200 \downarrow -\frac{6}{10}(734.375) = 0,$$

whence $F_{CH} = 234.375$ kN (tension).

$$\sum F_c = 0 \quad \text{gives}$$

$$\overleftarrow{F}_{CG} - \frac{8}{10}(\overrightarrow{234.375}) - \frac{8}{10}(\overrightarrow{734.375}) = 0$$

whence $F_{CG} = 775$ kN (compression).

The bar forces in the remaining members may be obtained in a similar manner by the consideration of joints G, H, F and D as free bodies as shown in Figs. 3.9(e), 3.9(f), 3.9(g), 3.9(h) respectively. At point D, there is only one unknown bar force, which requires only one equation of equilibrium for its solution, the second equation of equilibrium being used as a check. The free body diagram of joint B, however, is not needed for solution of bar forces and is included for checking purpose only.

The results of analysis of the entire truss are summarised in Fig. 3.9(j) in which bar forces are shown as positive for tension and negative for compression. The components of the force in a bar in the vertical (i.e., y-axis) and horizontal (i.e., x-axis) directions may be obtained by multiplying the bar forces by the proportion of the length of the members in the two directions respectively with respect to their inclined lengths.

Example 3.2 *Determine the bar forces in all members of the truss shown in* Fig. 3.10.

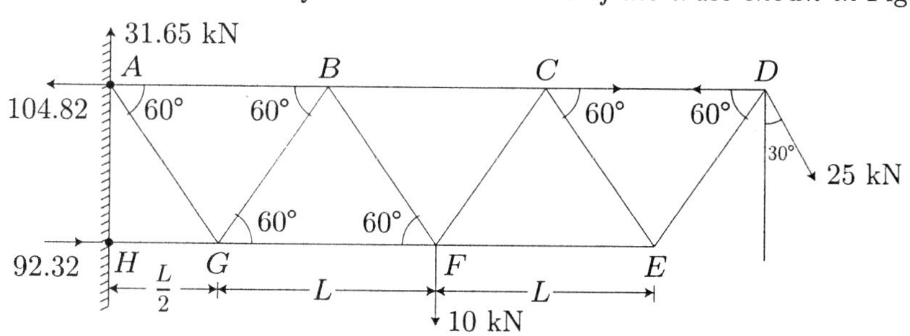

(a) Truss with Loading

Fig. 3.10: Truss for Example 3.2 (Contd.)

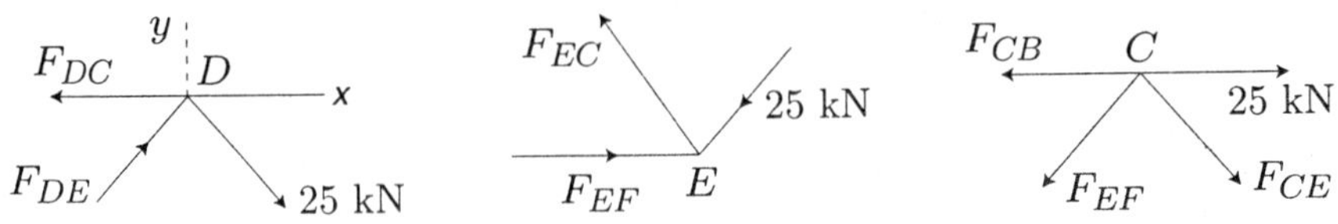

(b) Joint *D* (c) Joint *E* (d) Joint *C*

Fig. 3.10: Truss for Example 3.2

Solution : Bar forces can be obtained starting from joint D, Joints E, C, F, B and G have to be considered in the sequence mentioned in order to find bar forces in the other members. Calculation of reactions at A and H will serve as a check when equilibrium of these joints are considered with the bar forces thus obtained with method of joints.

Sample calculation for joint D as shown in Fig. 3.10b will yield

From $\sum F_y = 0$, i.e., $F_{DE} \sin 60° - 25 \cos 30° = 0$,

whence $F_{DE} = 25$ kN (compression)

$\sum F_x = 0$, i.e., $F_{DC} - F_{DE} \cos 60° - 25 \cos 60° = 0$,

whence $F_{DC} = 25$ kN (tension).

3.8 Method of Sections

This method can be used for finding bar force in a particular member of a truss, without first having to determine many other member forces by considerations of equilibrium of forces at various joints. The method consists essentially of isolating a portion of the truss by an imaginary section passing through the members for which bar forces are desired, and obtaining the unknown bar forces from consideration of equilibrium of the free body. In general, three unknown bar forces at a section can be determined from the solution of three simultaneous equations resulting from the consideration of equilibrium of free body isolated by the section. However, the numerical computation may be greatly simplified, if by the proper choice of centre of moments only one unknown appears in each of these equations. The critical aspect of this method is the choice of proper free body diagram so that forces in a particular member may be obtained without the necessity of calculating the bar forces in the other members cut by the section.

Example 3.3 *Determine the bar forces in the members BH, GH, BC and GB of the truss shown in* Fig. 3.11(a).

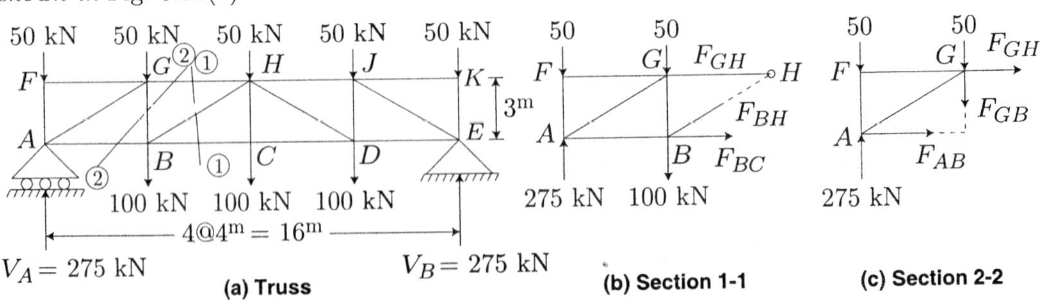

(a) Truss (b) Section 1-1 (c) Section 2-2

Fig. 3.11: Truss for Example 3.3

Solution : The free body diagram in Fig. 3.11(b) is obtained by cutting a section 1-1. in Fig. 3.11(a). Considering equilibrium of forces in the vertical direction, i.e., from $\sum F_y = 0$, $V_{BH} + 275 - (50 + 50 + 100) = 0$,

where V_{BH} = vertical component of bar force F_{BH}

$$V_{BH} = -75 \text{ kN} \quad \text{and} \quad F_{BH} = (-75)\frac{5}{3} = -125 \text{ kN (compression)}.$$

From $\sum M_B = 0$,

$$275(4) - 50(4) + F_{GH}(3) = 0 \text{ and } F_{GH} = -300 \text{ kN (compression)}.$$

From $\sum M_H = 0,\ 275(8) - 50(8) - 50(4) - 100(4) - F_{BC}(3) = 0$,

whence $F_{BC} = 400$ kN (tension).

For bar force F_{GB}, considering free body diagram, as in Fig. 3.11(c) for the section 2-2 from $\sum F_y = 0$, $F_{GB} - 275 + 50 + 50 = 0$ and $F_{GB} = 175$ kN (tension).

The same results would have been obtained if free body diagrams of the right half of the section were considered.

Example 3.4 *Calculate the bar forces in members CJ, CD, HJ, KF and FE of the truss shown in* Fig. 3.12(a).

Solution : The free body diagram in Fig. 3.12(b) for section 1-1, will be considered for bar forces F_{CJ}, F_{CD} and F_{HJ}.

From $\sum M_L = 0,\ F_{CJ}(6\sqrt{2}) + 383.33(4) - 500(8) - 500(12) = 0$,

whence $F_{CJ} \approx 997.80$ kN (Tension).

From $\sum M_J = 0,\ F_{CD}(4) + 500(4) + 500(8) - 383.33(12) = 0$,

whence $F_{CD} = -350$ kN (compression).

From $\sum M_C = 0$, $F_{HJ}(cc') - 500(4) + 383.33(8) = 0$

whence $F_{HJ} = -\dfrac{1066.64}{CC'} = -366.49$ kN (compression)

as $$CC' = CJ \sin\alpha = 4\sqrt{2}\frac{6\sqrt{2}}{\sqrt{(6\sqrt{2})^2 + (10\sqrt{2})^2}} = \frac{48}{16.492}.$$

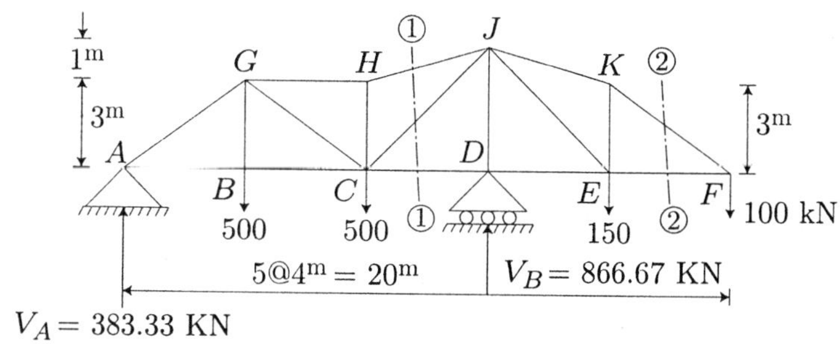

(a) Truss with Loading

Fig. 3.12: Truss for Example 3.4 (Contd.)

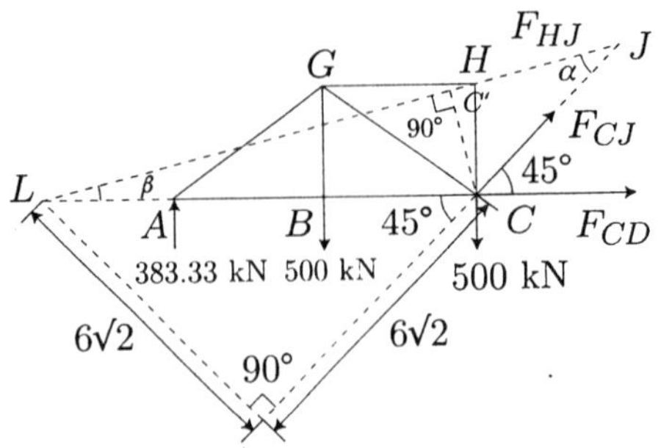

(b) Section 1-1

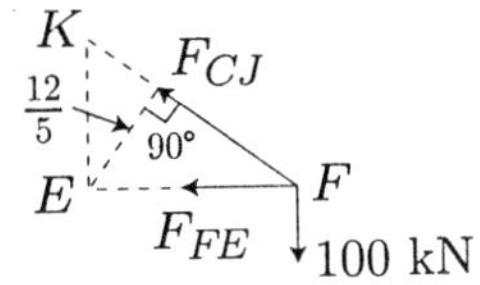

(c) Section 2-2

Fig. 3.12: Truss for Example 3.4

From free body diagram in Fig. 3.12(c) for section 2-2, considering $\sum M_E = 0$,

$$F_{FK} = \frac{100(4)}{3} = 166.67 \text{ kN (tension)}.$$

From $\sum F_x = 0,\ F_{FE} + F_{FK}\left(\frac{4}{5}\right) = 0,$

whence $F_{FE} = -133.34$ kN (compression).

Check Section 1-1,

$$\sum F_x = F_{HJ}\overrightarrow{\cos\beta} + F_{CJ}\cos 45^\circ + F_{CD}$$

$$= -366.49\frac{12}{\sqrt{(12)^2 + (3)^2}} + 997.80(0.707) - 350 \approx 0, \text{ O.K.}$$

$$\sum F_Y = F_{HJ}\sin\beta + F_{CJ}\sin 45^\circ + 383.33 - 500 - 500 \approx 0; \text{ O.K.}$$

The small errors if any involved in the checks are due to rationalisation or approximation of the magnitude of the bar forces.

Section 2-2.

$$\sum F_Y = F_{FK}\left(\frac{3}{5}\right) - 100 \approx 0, \text{ O.K.}$$

Example 3.5 *Compute the bar forces in the members M_4U_6, M_4L_5, U_4U_5, L_4L_5, U_4M_4 and M_4L_4 of the K truss shown in* Fig. 3.13.

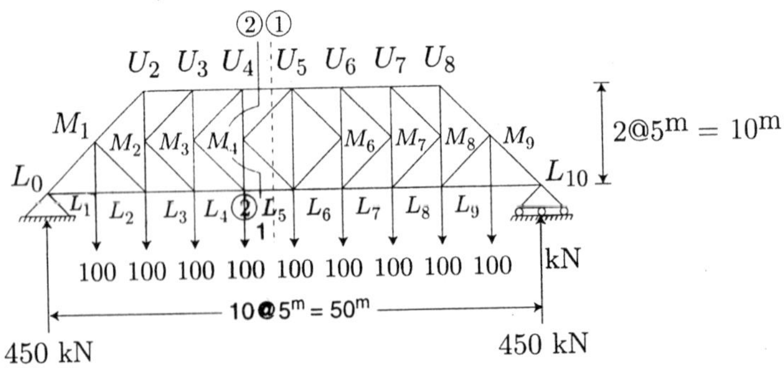

(a) Truss with Loading

Fig. 3.13: Truss for Example 3.5 (Contd.)

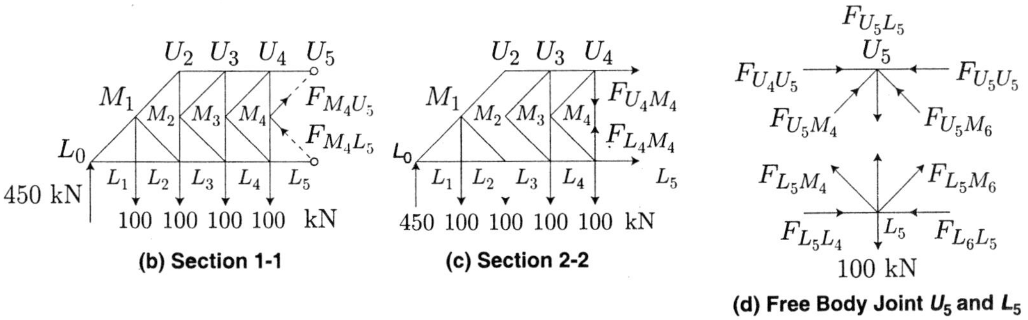

Fig. 3.13: Truss for Example 3.5

Solution : If a part of the structure is isolated by section 1-1 as in Fig. 3.13(b), there are four unknown bar forces in the free body. Since only three equations of equilibrium are available, the fourth equation are to be furnished by an additional requirement from the consideration of joint M_4 in equilibrium.

From $\sum F_x = 0$, of joint M_4, $F_{M_4U_5} = -F_{M_4L_5}$.

Then, from $\sum F_y = 0$

$$\uparrow F_{M_4U_5} \sin 45° - F_{M_4L_5} \sin 45° + 450 - 400 = 0,$$

whence $F_{M_4U_5} = -\dfrac{50}{2\sin 45°} = -25\sqrt{2}$ kN(compression)

and $F_{M_4L_5} = +25\sqrt{2}$ kN(tension).

From $\sum M_{L_4} = 0$,

$$F_{U_4U_5}(10) + 450(20) - 100(15) - 100(10) - 100(5) = 0,$$

(as moments of $F_{M_4U_5}$ and $F_{M_4L_5}$ about point L_4 cancel each other),

whence $F_{U_4U_5} = -600$ kN (compression),

From $\sum M_{U_4} = 0$,

$$F_{L_4L_5}(10) - 450(20) + 100(15) + 100(10) + 100(5) = 0,$$

whence $F_{L_4L_5} - 600$ kN (tension).

Bar forces in members U_4U_5 and L_4L_5 may also be obtained by considering section 2-2 and from

$$\sum M_{L_4} = 0, \text{and} \sum M_{U_4} = 0 \text{ repectively.}$$

Bar force in member U_5L_5 may be obtained from consideration of $\sum F_y = 0$ for joint U_5 or L_5.

Thus from Fig. 3.13(d).

$$\downarrow F_{U_5L_5} - F_{U_5M_4} \sin 45° - F_{U_5M_6} \sin 45° = 0,$$

whence $F_{U_5L_5} = 50$ kN (tension).

The same result will be obtained from the consideration of equilibrium of joint L_5.

Considerations of $\sum F_Y = 0$, for joints U_4 and L_4 will similarly give the bar forces in members U_4M_4 and L_4M_4 respectively.

3.9 Method of Sections for Parallel Chord Trusses

For trusses having parallel chords, the calculation of bar forces can be made in a very simple manner by considering them as beams and obtaining the resultant bending moment and shear force diagrams.

For example, for the simple truss with parallel top and bottom chords loaded as in Fig. 3.14(a), if a section 1-1 is considered, the equation of equilibrium for the portion of the truss to the left of the section gives

$$\left.\begin{aligned} F_{HJ} &= -\frac{V_A(4a) - P_1(2a)}{h} = -\frac{M_C}{h} \\ &= -\frac{\text{B.M. at } C \text{ when treated as a beam}}{h} \\ F_{BC} &= \frac{V_A(3a) - P_1(a)}{h} = \frac{M_H}{h} = \frac{\text{B.M. at } H}{h} \\ \text{and} \quad F_{HC} &= \frac{V_A - P_1}{\sin\alpha} = \frac{\text{S.F. at section 1-1}}{\sin\alpha}. \end{aligned}\right] \tag{3.3}$$

Eq. (3.3) means that in the case, the corresponding bar forces in various members of the truss are proportional to bending moment and shear force values at various joints and sections respectively. Similar conclusions may be reached for any section other than 1-1 and it may, therefore, be concluded that for the type of truss, bending moment and shear force diagrams are only needed for complete analysis of the bar forces. In case of truss, one of the important assumptions being that the loads are to be considered at joints only, the shear force will be constant between the joints, the bending moment varying linearly. Such methods of obtaining bar forces is also sometimes called the method of shears and moments.

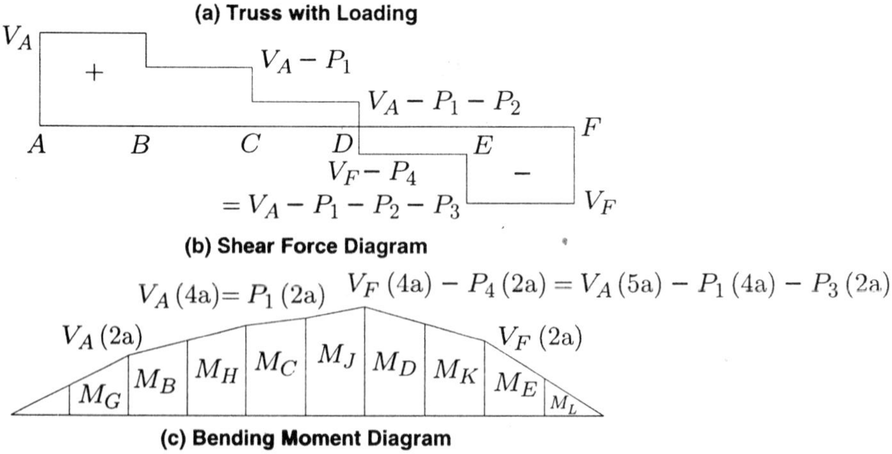

Fig. 3.14: Parallel Chord Trusses

3.10 Compound Trusses

Compound trusses can be formed by interconnecting two or more simple trusses in accordance with the requirements for complete constraint of a rigid body in one plane. In such cases it frequently happens that such trusses cannot be completely analysed by the method of joints alone.

Compound Trusses may be classified according to their pattern and method of solution. Classification may be considered as given below:

(a) Compound Truss as shown in Fig. 3.15(a) which consists of two simple trusses connected together at E. After calculation of reactions, bar forces in members AM, AF, FM, FG, MG and MB may be obtained using method of joint and considering joint A first and then joints F and M successively. Further calculations of bar forces cannot

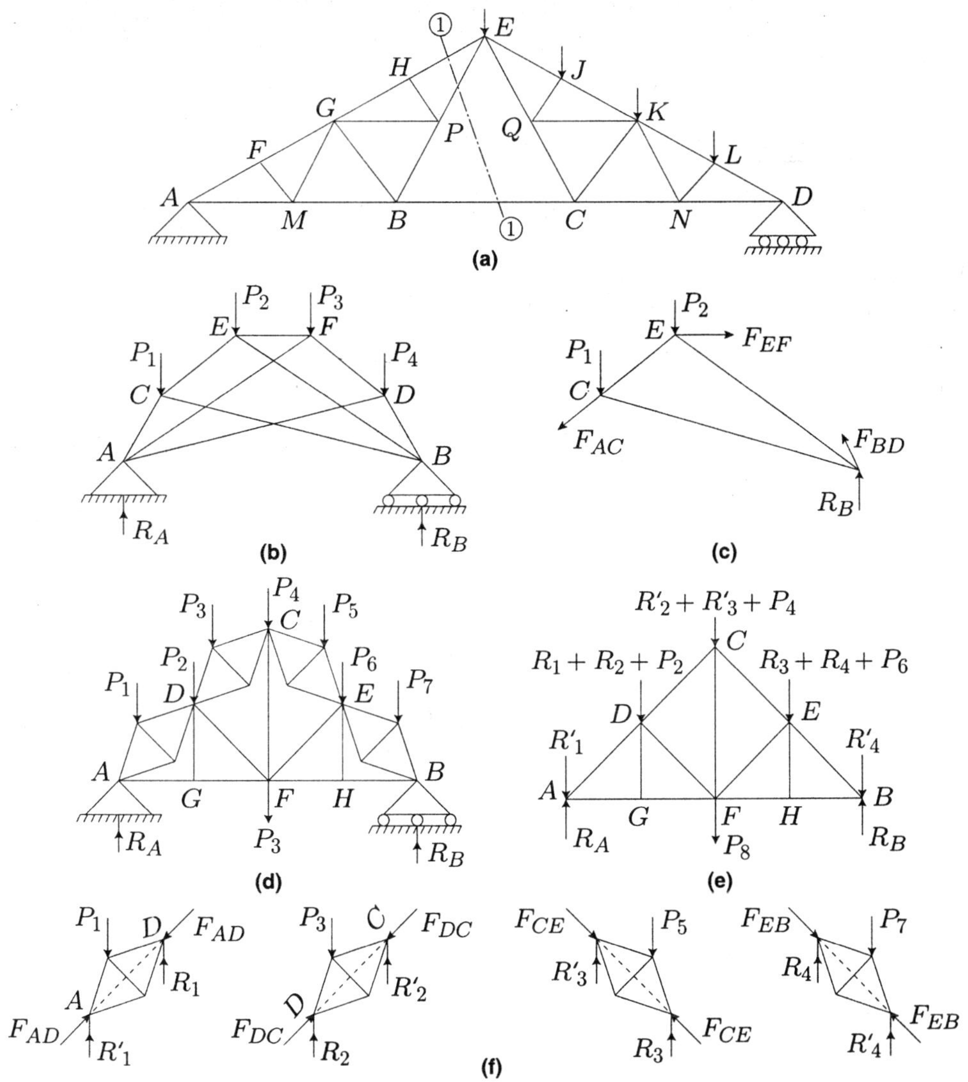

Fig. 3.15: Compound Trusses

be made by application of method of joint, as there is no other joint having two bar forces only as unknown. In order to analyse the truss, then, method of sections has to be used by taking section 1-1 and calculating bar force in member BC or HE from $\sum M_E = 0$ or from $\sum M_B = 0$. As soon as any of them are obtained, method of joints may be used.

(b) Compound truss as shown in Fig. 3.15(b) in which the method of joints fails at the very beginning, as there is no joint where less than three bars are attached. There is also no possibility of using the method of sections. The truss consists essentially of two simple trusses ADF and BCE, interconnected by three bars, which are neither parallel nor intersect at a point. After reactions are calculated one of the simple trusses, say BCE, may be isolated as shown in Fig. 3.15(b) and magnitudes of the three unknown bar forces F_{AC}, F_{EF} and F_{BD} may be obtained without difficulty. As soon as these bar forces are known the remainder of the analysis can be made by the method of joints.

(c) Compound Truss as shown in Fig. 3.15(d) are formed by combining several small size simple trusses on the members of a simple truss. The basic framework on which the smaller simple trusses are superimposed is considered as the primary system, whereas the smaller trusses are regarded as the secondary system. Such a truss system will involve special method of analysis. The primary and secondary systems have to be subdivided as shown in Figs. 3.15(e) and 3.15(f) and are analysed separately, the final results being obtained by superimposing one over the other.

3.11 Complex Trusses

Complex trusses do not satisfy the definition of a simple truss or a compound truss. Method of joints, method of sections or the method of subdivision usually used for their solutions fails to provide a direct solution of this type of truss. For a statically determinate complex truss $n+r = 2j$ or the sum of the number of unknown bar forces and the number of unknown components of reactions must be equal to twice the number of joints in the truss. Since two equations of equilibrium can be written for the force acting at a joint, the total number of equations available for the entire truss is $2j$. Solving the $2j$ numbers of simultaneous equations, bar forces and reactions may be obtained. This method, however, is impracticable due to large numbers of calculations involved.

A workable method developed by Henneberg can, however, be suitably used for solution of such complex trusses. The method called Hennebergs's method, involves replacing one or more members in a complex truss so as to convert it to a simple truss, which can be solved directly by the method of joints or the method of sections. A correction may then be applied to remove the effect of the substitute member or members. For the compound truss in Fig. 3.16(a) for example, member BF may be replaced by member CF resulting in simple truss as in Fig. 3.16(b) and subjected to the same loading, as in case of the original truss. The same simple truss is then subjected to unit load, and load X along the bar replaced, i.e., BF in this case, as shown in Fig. 3.16(c) and 3.16(d).

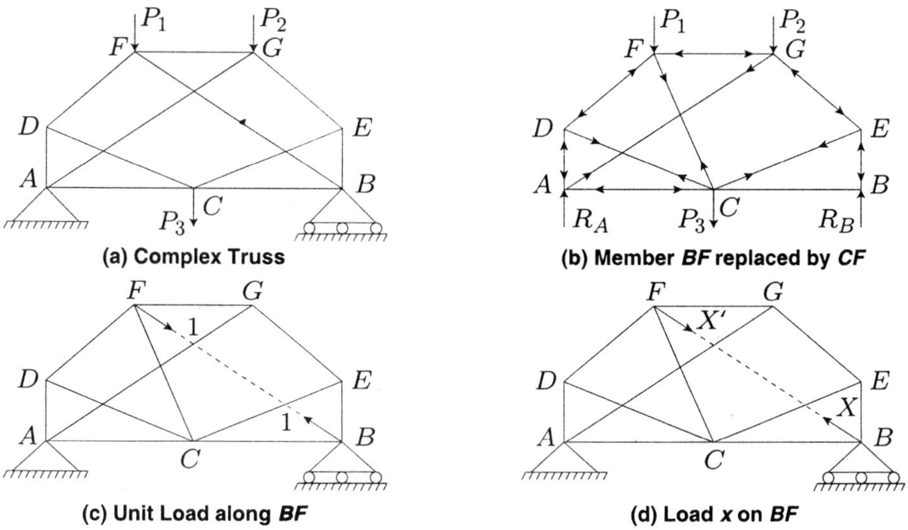

(a) Complex Truss

(b) Member ***BF*** replaced by ***CF***

(c) Unit Load along ***BF***

(d) Load ***x*** on ***BF***

Fig. 3.16: Complex Truss requiring replacement of one member

If F'_{CF} be the force in bar CF in simple truss in Fig. 3.16(b) and f_{CF} the force in bar CF in simple truss in Fig. 3.16(c), then X f_{CF} is the force in bar CF in simple truss in Fig. 3.16(d).

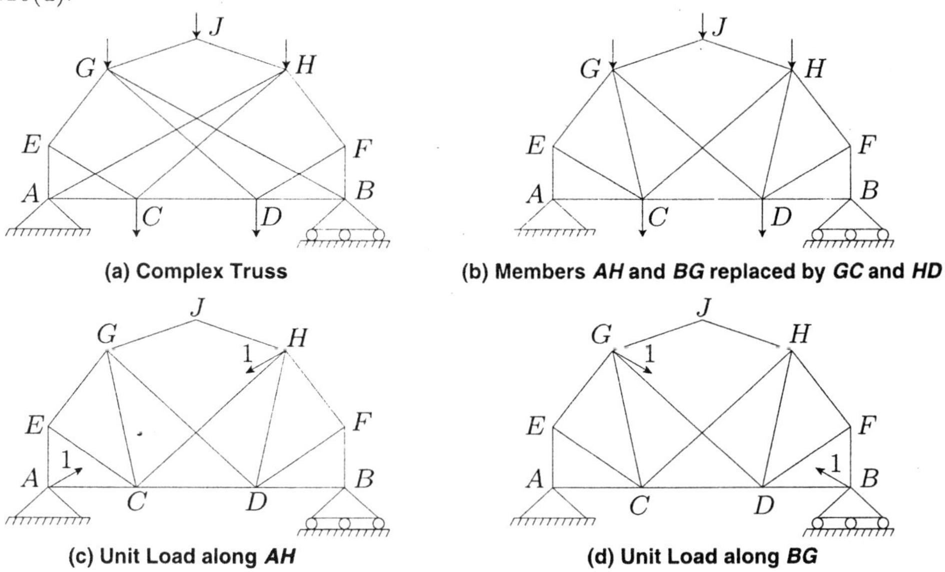

(a) Complex Truss

(b) Members ***AH*** and ***BG*** replaced by ***GC*** and ***HD***

(c) Unit Load along ***AH***

(d) Unit Load along ***BG***

Fig. 3.17: Complex Truss requiring replacement of one member

In order to make the substitute member (CF in this case) inactive when Fig. 3.16(d) is superimposed on Fig. 3.16(b), it is necessary that, $F_{CF} = F'_{CF} + X f_{CF} = 0$, whence

$$X = -\frac{F'_{CF}}{f_{CF}} = \text{force in bar } BF. \text{ i.e., } F_{BF} \tag{3.4}$$

The remaining bar forces obtained from the superposition of Figs. 3.16(b) and 3.16(d) will be given by,

$$F_I = F'_I + X(f_I),$$

where F_I = bar force in member I in complex truss of Fig. 3.16(a) F'_I = bar force in member I in simple truss of Fig. 3.16(b) and f_I = bar force in member I in simple truss of Fig. 3.16(c).

A similar procedure may be used in the analysis of a complex truss as in Fig. 3.17(a), where to arrive at a simple truss, it is necessary to replace two bars AH and BG by fictitious bars GC and HD respectively as shown in Fig. 3.17(b). Considering the three conditions of loading of this simple truss as shown in Figs. 3.17(b), 3.17(c) and 3.17(d), the corresponding bar forces F'_I, f_{I1} and f_{I2} in each bar of the truss may be obtained without difficulty by the method of joints. Denoting by X and Y, the unknown forces in the bars AH and BG respectively, of the given complex truss and from superposition, it is necessary that,

$$F_{GC} = F'GC + Xf_{GC1} + Yf_{GC2} = 0$$

and $F_{HD} = F'_{HD} + Xf_{HD1} + Yf_{HD2} = 0.$

Solving these two simultaneous equations,

$$\left.\begin{aligned} X &= \frac{(f_{HD2}(F'_{GC}) - (f_{GC2})(F'_{HD})}{(f_{GC2})(f_{HD1}) - (f_{HD2})(f_{GC1})} \\ \text{and} \quad Y &= \frac{(f_{HD1})(F'_{GC}) - (f_{GC1})(F'_{HD})}{(f_{HD2})(f_{GC1}) - (f_{HD1})(f_{GC2})} \end{aligned}\right] \tag{3.5}$$

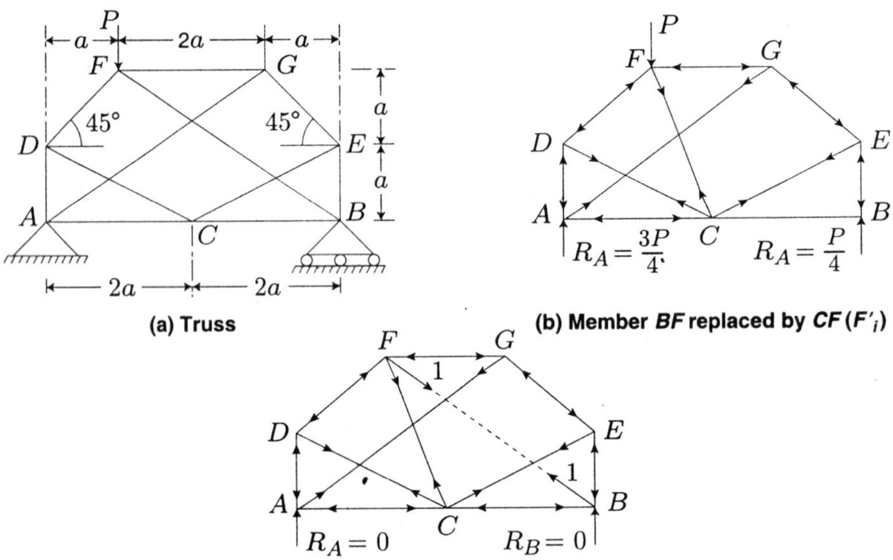

(a) Truss

(b) Member *BF* replaced by *CF* (F'_i)

(c) Unit Load along *BF*

Fig. 3.18: For Example 3.6

After obtaining values of X and Y, the bar forces in any member may be calculated from

$$F_I = F'_I + X(f_{I1}) + Y(f_{I2}).$$

In case of a complex truss of critical form the denominator of the expressions for X and Y in Eqs. (3.4) and (3.5) becomes zero and the truss in such a case is statically indeterminate.

Example 3.6 *Determine the bar forces in all members of the complex truss shown in* Fig. 3.18(a) *using Henneber'g method.*

Solution : Detail of calculations are presented in a tabular form according to the previous procedure.

Table 3.1 (Example 3.6)

Tension+; Compression−

Member	Forces due to external load P, i.e., F_I' (Fig. 3.18b)	Forces due to unit load along BF, i.e., f_I (Fig. 3.18c)	Forces due to load X along BF, i.e., Xf_I	Bar force, $F_I = F_I' + X$ $(f_I)(2) + (4)$
(1)	(2)	(3)	(4)	(5)
BE	$-0.25P$	-0.55	$+0.682P$	$+0.432P$
BC	$-0.00P$	-0.832	$+1.032P$	$+1.032P$
EC	$+0.186P$	$+0.413$	$-0.512P$	$-0.326P$
EG	$-0.234P$	-0.521	$+0.646P$	$+0.412P$
GF	$-0.416P$	-0.920	$+1.141P$	$+0.725P$
GA	$+0.300P$	$+0.664$	$-0.823P$	$-0.523P$
FD	$-0.313P$	-0.347	$+0.430P$	$+0.117P$
FC	$-0.434P$	-0.350	$+0.434P$	$0.000P$
DC	$+0.685P$	$+0.274$	$-0.340P$	$+0.345P$
DA	$-0.919P$	-0.369	$+0.458P$	$-0.461P$
AC	$-0.250P$	-0.552	$+0.685P$	$+0.435P$
BG	$-0.00P$	$+1.00$	$-1.240P$	$-1.240P$

$$X = -\frac{0.434P}{0.350} = -1.240P.$$

Complex truss may often be arranged so as to be geometrically unstable. It is, however, not always possible to see a critical form by inspection. Truss in Fig. 3.19 has a critical form, if $a = c$. Otherwise it has a stable form. In such cases detection is based on the principle that if the analysis for the truss yields a unique solution, then the truss is stable and determinate. If the analysis fails to yield a unique solution, the truss has a critical form.

The above principle suggests a convenient way of detecting critical form, known as Zero Load Test. For unloaded truss, one possible solution satisfying equilibrium is that all members of the truss have zero stress. If with no loads on a truss, a set of solutions for the member stresses other than zero can be found, which will satisfy the condition of equilibrium the truss has a critical form.

Considering complex truss (in Fig. 3.20) if tension in member C be taken as T, for joint B, $F_{AB} = T$ and $F_{BE} = -T$.

From joint C, $F_{CD} = T$ and $F_{CF} = -T$, whence from identity in each joint, $F_{AB} = F_{BC} = F_{CD} = F_{EF} = F_{FA} = T$ and $F_{AD} = F_{BE} = F_{CF} = -T$, which will satisfy condition of equilibrium at all joints. Since T may be assigned any value different from zero, this indicates that the truss has a critical form. In such cases no matter how the truss is actually loaded, the truss will have an infinite member of solutions satisfying the equilibrium requirements of statics. The truss is, therefore, statically indeterminate.

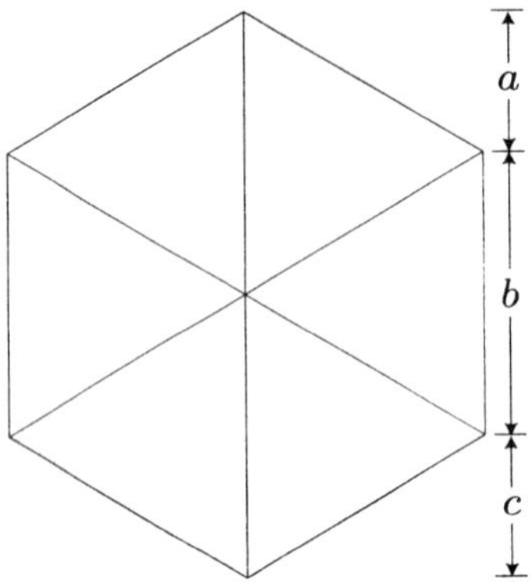

Fig. 3.19: Complex Truss

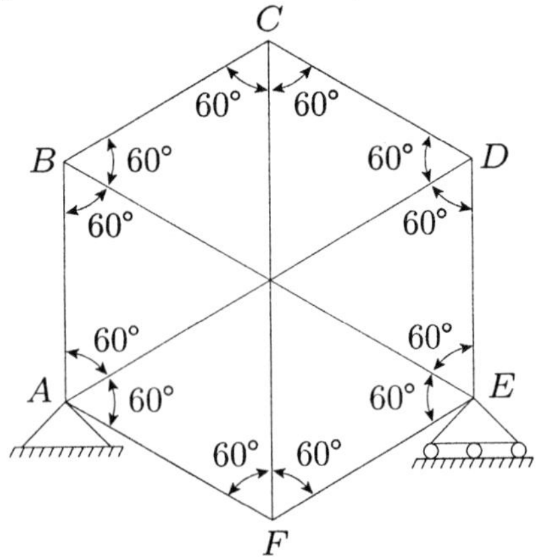

Fig. 3.20: Complex Truss

Example 3.7 *Determine the forces in all the members of the truss shown in* Fig. 3.21(a).

Reactions $V_A = V_B = \dfrac{3W}{2} \uparrow$ acting vertically.

The force in each of the verticals shall be equal to W each, at C, D and E.

$$F_{CF} = F_{DC} = F_{EH} = W(Compression)$$

The forces in inclined members at F, G and H can be determined by the method of joints.

For joint F, from $\sum H = 0$ (Fig. 3.21(b)),

$$F_{FA}\left(\frac{1}{\sqrt{2}}\right) - F_{FB}\left(\frac{3}{\sqrt{10}}\right) = 0,$$

whence $F_{FA} = \dfrac{3\sqrt{2}}{\sqrt{10}} F_{FB} = 1.342 F_{FB}$.

$$\sum V = 0$$

$$F_{FA}\left(\frac{1}{\sqrt{2}}\right) + F_{FB}\left(\frac{1}{\sqrt{10}}\right) - W = 0,$$

whence $F_{FB} = 0.7906W$(Tension) and $F_{FA} = 1.06W$(Tension).

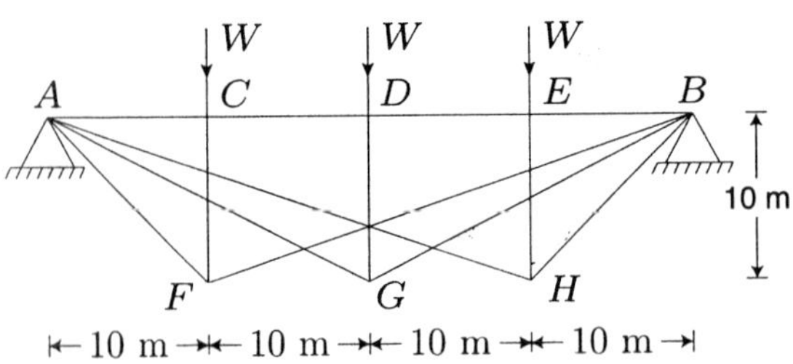

(a) Truss

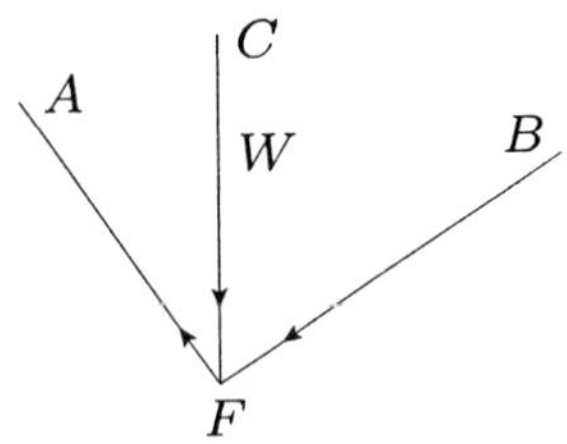

(b) Joint *F*

Fig. 3.21: Forces for Example 3.7 (Contd.)

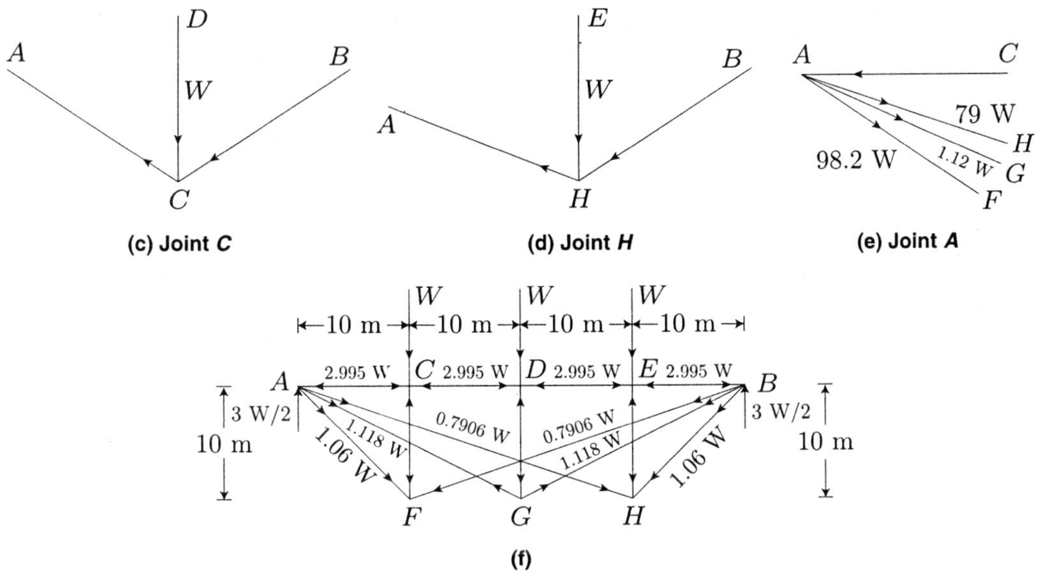

Fig. 3.21: Forces for Example 3.7

For Joint G, from $\sum H = 0$ [Fig. 3.21(c)]

$$F_{GA}\frac{2}{\sqrt{5}} - F_{GB}\frac{2}{\sqrt{5}} = 0,$$

whence $F_{GA} = F_{GB}$

$$\sum V = 0$$

$$F_{GA} \cdot \frac{1}{\sqrt{5}} + F_{GB} \cdot \frac{1}{\sqrt{5}} - W = 0$$

$$(F_{GA})\frac{2}{\sqrt{5}} = W$$

$$F_{GA} = 1.118W \text{ (tensile)}$$

$$F_{GB} = 1.118W \text{ (tensile)}$$

For Joint H, [Fig. 3.21(d)] $\sum H = 0$

$$F_{HA}\frac{3}{\sqrt{10}} - F_{HB}\frac{1}{\sqrt{2}} = 0$$

$$F_{HB} = 1.3416F_{HA}$$

$$\sum V = 0$$

$$F_{HA}\left(\frac{1}{\sqrt{10}}\right) + F_{HB}\left(\frac{1}{\sqrt{2}}\right) - W = 0$$

$$\frac{F_{HA}}{\sqrt{10}} + \frac{1.3416F_{HA}}{\sqrt{2}} = W,$$

whence $F_{HA} = 0.7906W$.

$$F_{HB} = 1.3416(0.7906) = 1.06W \text{ (tensile)}$$

For Joint A, [Fig. 3.21(e)] from $\sum H = 0$

$$F_{AC} - F_{AH}\frac{3}{\sqrt{10}} - F_{AG}\frac{2}{\sqrt{5}} - \frac{F_{AF}(1)}{\sqrt{2}} = 0,$$

whence $F_{AC} = 07906W\left(\frac{3}{\sqrt{10}}\right) + 1.118W \cdot \frac{2}{\sqrt{5}} + 1.06W \cdot \frac{1}{\sqrt{20}} = 2.995W.$

The forces in various members are shown in Fig. 3.21(f).

3.12 Simple Truss in Space

The simple element forming a space truss consists of three bars, which do not lie in one plane (as in Fig. 3.22) and the bars are connected by smooth ball and socket joints, each of which provides three components of constraints in three orthogonal directions. This also forms the basic element for a statically determinate space truss and additional joint may be introduced by adding three more bars, which do not lie in one plane and by connecting them to either previously established joints or the foundations as in Fig. 3.23.

As in case of plane truss, for space truss also a similar criteria for statical determinacy, can be developed,

$$n - 3 = 3(j - 3),$$

whence $n = 3j - 6.$ (3.6)

If the constraint or reactions involved are considered, the general relationship similar to the case of a plane truss comes to

$$n + r = 3j \tag{3.6a}$$

For a three dimensional framework, six independent equations of statics may be written regarding the equilibrium of the external loads and reactions on the entire structure. Considering the three axes as X, Y and Z, these equations are $\sum F_x = 0, \sum F_Y = 0, \sum F_Z = 0$, $\sum M_X = 0, \sum M_Y = 0$ and $\sum M_Z = 0$. The necessary condition for statical determination of a three dimensional framework with respect to its outer force is that the total number of independent reaction components shall equal to six.

For each joint three equations considering $\sum F_X = 0, \sum F_Y = 0$ and $\sum F_Z = 0$ can be written. If, therefore, in a space framework not more than three bars and not lying in a plane converge, bar forces on them may be obtained from consideration of statics. In using method of joints in a space framework such a joint may be taken as the starting point.

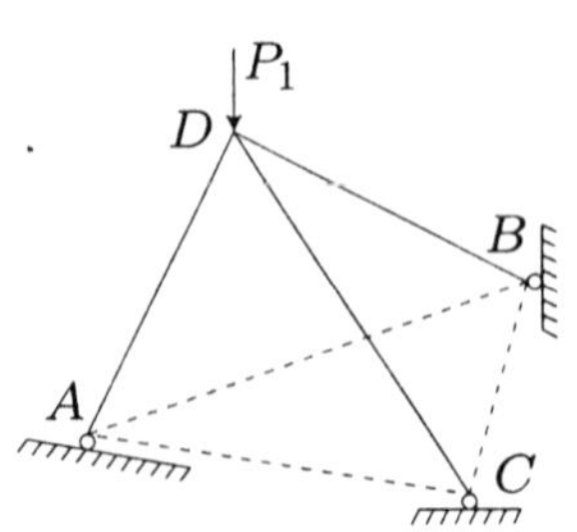

Fig. 3.22: Simplest Space Truss

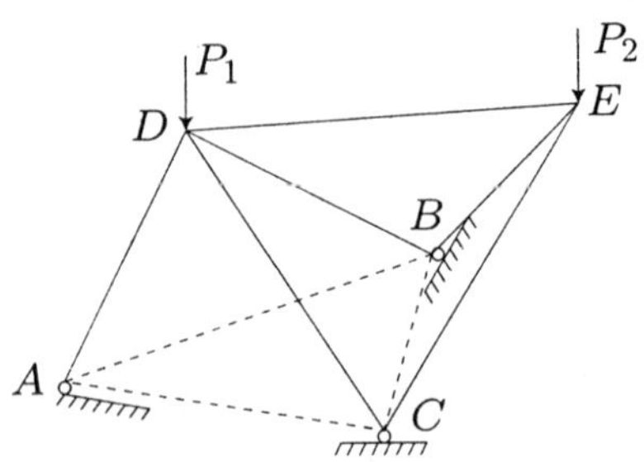

Fig. 3.23: Space Frame

3.13 Method of Tension Coefficient

The method of joints is always applicable in the analysis of simple trusses in space. However, the method can be considerably simplified for systematic treatment, if the ratios of the bar force to the length of the members are used instead of the bar forces themselves. Such a simplified procedure is called the method of Tension Coefficient and has been developed by Professor R.V. Southwell.[1] It is equally, applicable to plane trusses. The bar forces according to this method will be assumed to be tension. A positive answer will indicate that the bar force is tensile and a negative result will mean that the bar force is compressive.

If AB in Fig. 3.24 represents a bar of a space truss, having coordinates x_A, y_A and z_A for point A, and x_B, y_B, and z_B for point B, the tension T_{AB} in the member AB may be considered as being represented by

$$T_{AB} = L_{AB}(t_{AB}),$$

where L_{AB} = Length of the bar AB = $\sqrt{(x_B - x_A)^2 + (y_B - y_A)^2 + (z_B - z_A)^2}$

and $t_{AB} = \dfrac{T_{AB}}{L_{AB}}$ = Tension Coefficient for bar AB.

The components of T_{AB} in the directions of axes x, y and z will be given by

$$T_{AB}\frac{x_B - x_A}{L_{AB}} = t_{AB}(x_B - x_A)$$

$$T_{AB}\frac{y_B - y_A}{L_{AB}} = t_{AB}(y_B - y_A)$$

and $T_{AB}\dfrac{z_B - z_A}{L_{AB}} = (z_B - z_A)$ respectively.

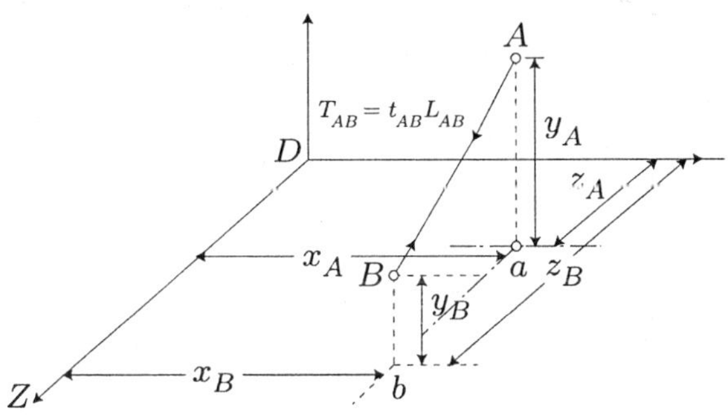

Fig. 3.24: Tension Coefficient method

At B, the other end of the bar, the components, along the three axes will be:

$$t_{AB}(x_A - x_B), t_{AB}(y_A - y_B) \text{ and } t_{AB}(z_A - z_B).$$

If at the joint A, several bars $AB, AC, \cdots, AN$ are connected together and Px_A, Py_A and Pz_A be the external forces at joint A, in the directions x, y and z respectively, from

[1]Southwell, R.V.,"Primary Stress determination in space frames"."Engineering"-February 6th, 1920, p. 165.

This method was also used by Muller-Bresleu in "Die Neueren Methoden der Festigkirtslehre" (Leipzig, 1924,and earlier editions).

consideration of equilibrium of joint, i.e., $\sum F_X = 0, \sum F_Y = 0$ and $\sum F_Z = 0$, equations obtained are as under:

$$\left.\begin{aligned} &t_{AB}(x_B - x_A) + t_{AC}(x_C - x_A) + \cdots + t_{AN}(x_N - x_A) + P_{XA} = 0 \\ &t_{AB}(y_B - y_A) + t_{AC}(y_C - y_A) + \cdots + t_{AN}(y_N - y_A) + P_{YA} = 0 \\ \text{and} \quad &t_{AB}(z_B - z_A) + t_{AC}(z_C - y_A) + \cdots + t_{AN}(z_N - z_A) + P_{ZA} = 0 \end{aligned}\right\} \quad (3.7)$$

On the basis of Eqs. (3.7), relationships can be formed for each joint in the truss involving $(x_B - x_A)$, $(y_B - y_A)$, $(z_B - z_A), \cdots$ etc., which can be obtained from the geometry of the structure and the unknown values of tension coefficients like t_{AB}, t_{AC}, etc.

From the set of simultaneous equations, the values of tension coefficients can be calculated and the tension in the bar obtained from

$$T_{AB} = t_{AB}\sqrt{(x_B - x_A)^2 + (y_B - y_A)^2 + (z_B - z_A)^2}.$$

A negative value of the tension coefficient will signify that the bar force is compressive.

As the reactive forces are calculated before the equations involving tension coefficients are formed, these equations do not all have to be solved simultaneously and the work is neither difficult nor involved.

Example 3.8 *Compute the bar forces in all members of the tripod shown in* Fig. 3.25.

Solution :

Table 3.2 (Example 3.8)

Joint	Axis	Equations
	x	$-3.50t_{AB} + 3t_{AC} + 1.5t_{AD} = 0$
A	y	$-2t_{AB} - 3.50t_{AC} - 3t_{AD} - 100 = 0$
	z	$-2t_{AB} - 3t_{AC} + 3t_{AD} = 0$

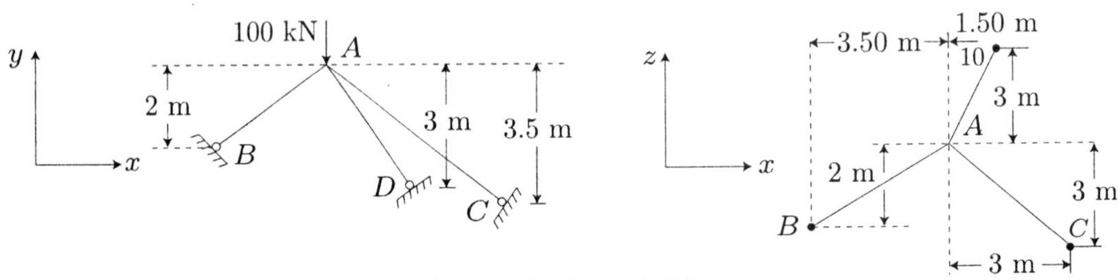

Fig. 3.25: For Exmaple 3.8

Solving equations, $t_{AB} = -\dfrac{1800}{137}$, $t_{AC} = -\dfrac{1000}{137}$ and $t_{AD} = -\dfrac{2200}{137}$

and forces in bars are:

$$F_{AB} = -\frac{1800}{137}\sqrt{(2)^2 + (2)^2 + (3.50)^2} = -59.12 \text{ kN(compression)}$$

$$F_{AC} = -\frac{1000}{137}\sqrt{(3)^2 + (3)^2 + (3.50)^2} = -40.146 \text{ kN(compression)}$$

$$\text{and} \quad F_{AD} = -\frac{2200}{137}\sqrt{(1.50)^2 + (3)^2 + (3)^2} = -72.263 \text{ kN(compression)}.$$

Example 3.9 *Determine the bar forces in all members of the space truss shown in* Fig. 3.26.

Solution : On the basis of tension coefficient method Table 3.3 shows the equations.

Bar forces for members obtained from solution of these tension coefficient equations are given in tabular form in Table 3.4.

Table 3.3 (Example 3.9)

Joint	Axis	Equations
	x	$-3t_{FE} - 3t_{FB} - 3t_{FH} = 0$
F	y	$1.56t_{FC} + 1.56t_{FJ} + 2.08t_{FH} + 2.08t_{FB} - 100 = 0$
	z	$0.90t_{FJ} - 0.90t_{FC} + 1.20t_{FH} - 1.20t_{FB} = 0$
	x	$-3t_{CH} - 3t_{CB} = 0$
C	y	$-1.56t_{CF} + 0.52t_{CB} + 0.52t_{CH} - 60 = 0$
	z	$1.80t_{CJ} + 2.10t_{CH} - 0.30t_{CB} + 0.90t_{CF} = 0$
	x	$-3t_{JH} = 0$
J	y	$0.52t_{JH} - 1.56t_{JF} - 30 = 0$
	z	$-1.80t_{JC} - 0.90t_{JF} + 0.30t_{JH} = 0$
	x	$3t_{EF} - 3t_{ED} - 3t_{EA} - 3t_{EG} = 0$
E	y	$2.08t_{EB} + 2.60t_{EA} + 2.08t_{EH} + 2.60t_{EG} = 0$
	z	$1.20t_{EH} - 1.20t_{EB} + 1.50t_{EG} - 1.50t_{EA} = 0$
	x	$3t_{BC} - 3t_{BA} + 3t_{BF} - 3t_{BG} = 0$
B	y	$-2.08t_{BE} - 2.08t_{BF} - 0.52t_{BC} + 0.52t_{BA} + 0.52t_{BG} = 0$
	z	$2.40t_{BH} + 2.70t_{BG} + 1.20t_{BE} + 1.20t_{BF} - 0.30t_{BA} + 0.30t_{BC} = 0$
	x	$3t_{HJ} + 3t_{HF} + 3t_{HC} - 3t_{HG} = 0$
H	y	$-0.52t_{HJ} - 2.08t_{HF} - 0.52t_{HC} + 0.52t_{HG} - 2.08t_{HE} = 0$
	z	$-2.40t_{HB} - 2.10t_{HC} - 1.20t_{HF} - 0.30t_{HJ} - 1.20t_{HE} + 0.30t_{HG} = 0$

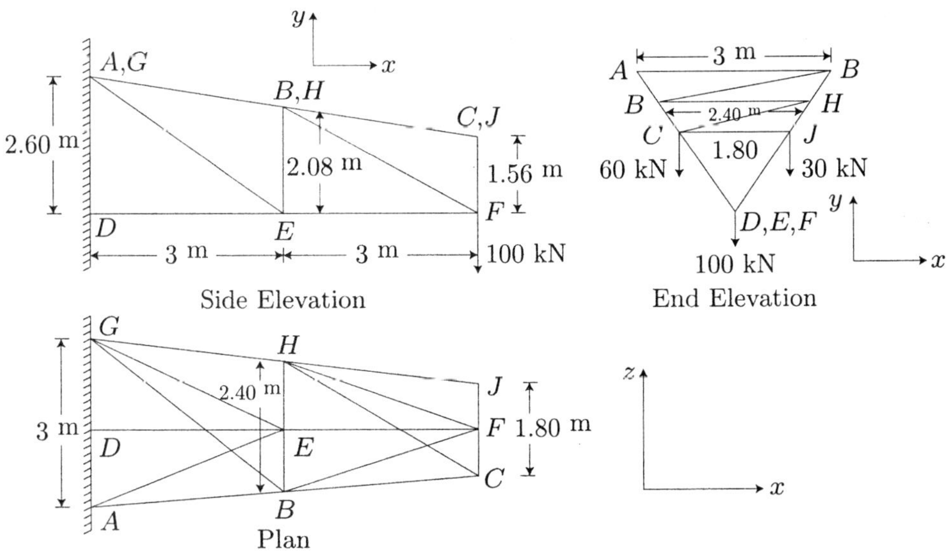

Fig. 3.26: For Exmaple 3.9

Table 3.4 (Example 3.9)

Tension+Ve; Compression–Ve

Member	Tension coefficient t in kN	Length L(m)	Bar force in kN
JH	0	3.06	0
JF	−19.20	1.80	−34.56
JC	9.60	1.80	17.28
FC	−38.33	1.80	−68.994
FB	53.0	3.84	203.52
FE	−91.667	3.00	−275.001
FH	38.667	3.84	148.48
CB	−7.167	3.06	−21.931
CH	7.167	3.69	26.446
EB	−39.80	2.40	−95.52
ED	−146.667	3.00	−440.001
EA	32.033	4.242	135.885
EG	22.80	4.242	96.7176
EH	−28.667	2.40	−68.80
BA	42.667	3.06	130.56
BG	4.433	4.068	18.033
BH	−5.533	2.40	−13.28
HG	45.833	3.06	140.249

3.14 Determination of Reactions for Three Dimensional Frame

For three dimensional framework statically determinate with respect to the external forces and supported at only three points, reactions can be readily determined by applying the equations of statics to the structure as a whole. When there are more than three points of supports, it is usually necessary to determine some or all of the bar forces before the reactions can be evaluated.

In particular structures, bar forces can be computed more easily if the reactions are calculated first. Referring to Fig. 3.27(a), there are six unknown reactions as shown in Figs. 3.27(b) and 3.27(c). As they can be calculated by means of statics in case of such a structure, reactions shall be computed first.

For the reactions, their values shall be calculated in terms of their components along the three axes. In forming the equations where possible it is preferable to form them in terms of one unknown component only.

Thus in case of truss in Fig. 3.27(a), moments of forces about X-axes, passing through points B and C give

$$-100(6) + 10R_{Dy} = 0,$$

whence $R_{Dy} = 60$ kN.

In taking the summation of moments, right-hand screw rule is used with positive direction of moment vector in the direction of the axes. Taking summation of moments about Z-axes passing through point B and using value of R_{Dy} determined above equation comes to

$$200(12) - 100(4) + 60(4) + 8R_{Cy} = 0, \quad \text{whence} \quad R_{Cy} = -280 \text{ kN}.$$

From $\sum F_y = 0,\ R_{By} = 320$ kN.

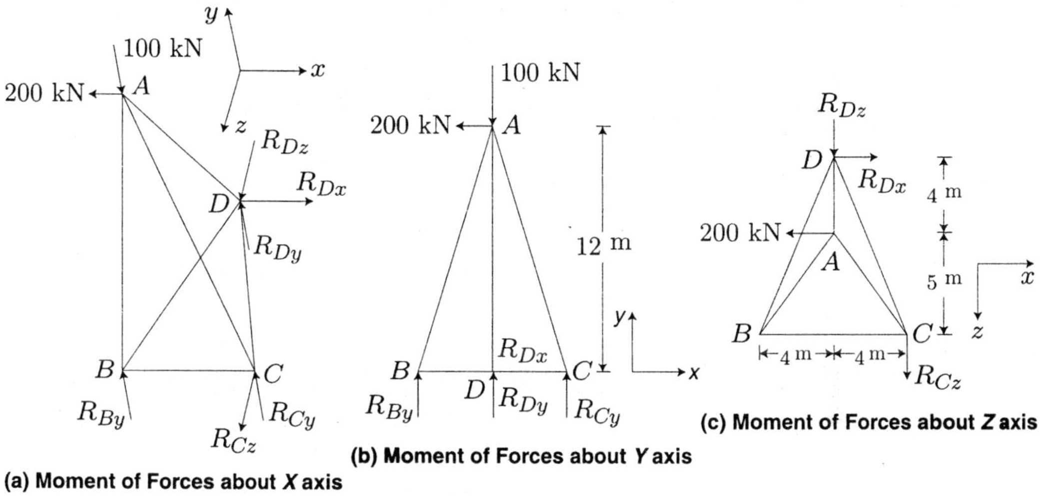

Fig. 3.27: Determination of Reactions

Value of R_{Cz} can be obtained by taking the summation of moments about a Y-axis passing through point A which gives

$$200(4) - 4R_{Cz} = 0, \quad \text{whence} \quad R_{Cz} = -200 \text{ kN}.$$

From $\sum F_Z = 0,\ R_{Dz} = 200$ kN

and from $\sum F_X = 0,\ R_{Dx} = 200$ kN.

The bar forces in the members may be obtained from equilibrium consideration of joints. A point having only three unknowns shall be taken as the starting point.

3.15 Graphical determination of Axial Forces in Trusses

A graphical method can be applied to determine the axial forces in the members of the trusses. For a given truss the forces meeting at a joint are known by their line of action, and all but two must be known in magnitude, since only two of the laws of statics are available. For any joint, where this condition exists, a section around the joint may be passed and by means of a force polygon, the unknown axial forces in the two members may be obtained. Such force polygons may be drawn separately for each joint or may be combined together into a single force polygon diagram, when it is called the Maxwell-Cremona diagram.

In the application of Maxwell Cremona diagrams for obtaining the axial forces in members of the truss, the procedure can be simplified by use of Bow's notation for representation of the forces. According to this, the spaces between the lines of actions of internal and external forces acting at a joint are denoted by a letter as shown in Fig. 3.28(a), where external force will be represented by bc, reaction V_A by ca, reaction V_B by ab, axial force in member AD by da and so on. Reactions V_A and V_B can be obtained analytically or graphically.

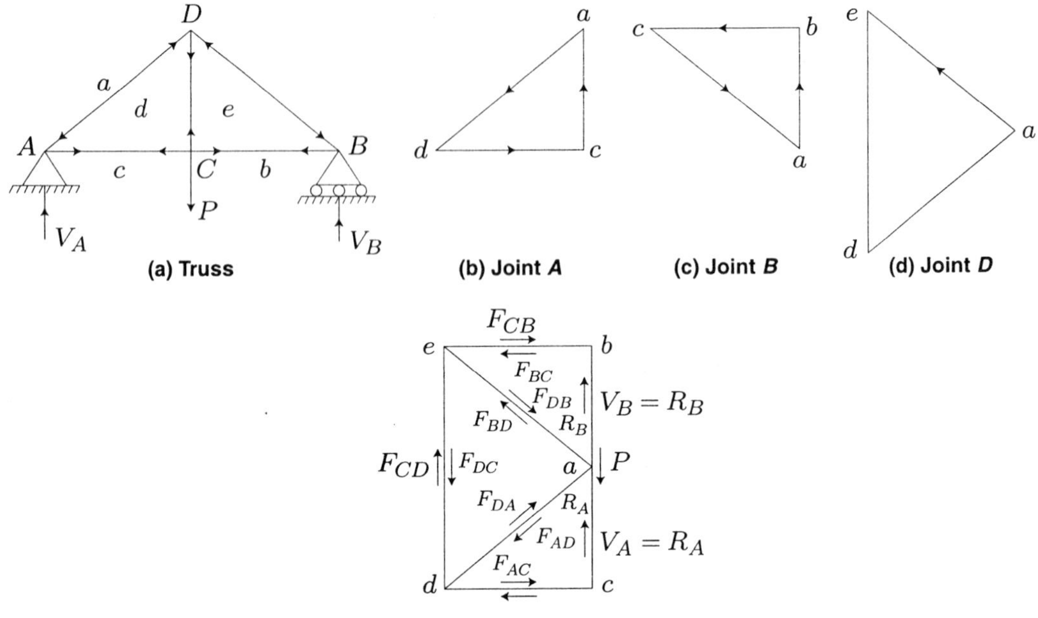

(e) Maxwell-Cremona Diagram

Fig. 3.28: Graphical method for Truss

In drawing the force polygons, each joint may be treated separately and axial forces in the unknown members of the truss at the joint can be obtained. Such force polygons are shown in Fig. 3.28(b) to 3.28(d) for the truss in Fig. 3.28(a). By combining these force polygons, Maxwell-Cremona diagram as in Fig. 3.28(e) can be obtained. In such a combined diagram, it is necessary to pass around each joint in the same direction and to draw the forces in the polygon in the order in which they are met in passing around the joint. In Fig. 3.28(e), each joint was passed around in a clockwise direction and the forces drawn in order. Whether the direction chosen is clockwise or anticlockwise is not important, but having started in one direction, it must be followed for the entire diagram.

The nature of the axial force in a member, i.e., whether it is tension or compression can be quickly determined by observing the direction in which the force acts around the force polygon for either joint to which the member connects and noting whether this direction is away from the joint (tension) or towards the joint (compression). For example, in case of the member AD (ad according to Bow's notation) of the truss in Fig. 3.28(a), as the force polygon was drawn clockwise for joint A, the direction of the force F_{AD} will be towards the joint which indicates that the axial force is compressive. The same result will also be obtained when joint D is referred to as in this case the member will be considered as 1a and the direction of the force is towards joint D (i.e., compression).

Example 3.10 *Determine the axial forces in the members of the Simple truss shown in* Fig. 3.29(a).

Fig. 3.29(b) shows the graphical solution for axial forces in the members of the truss.

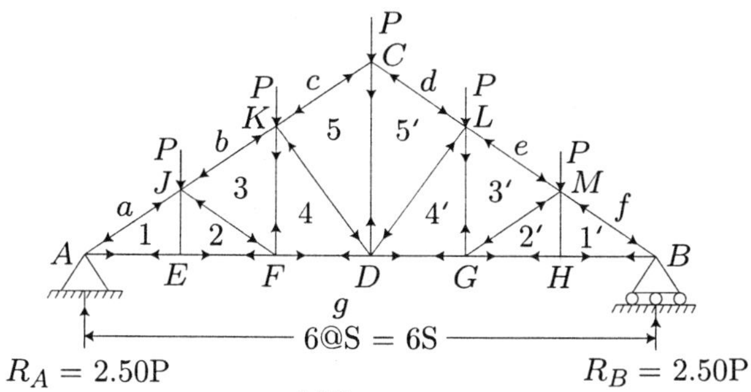

(a) Truss

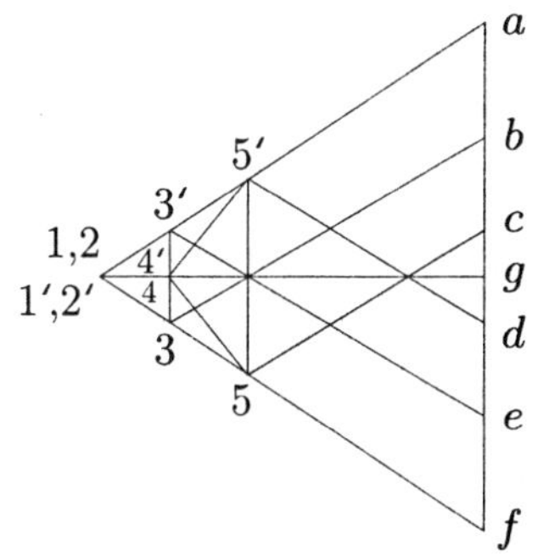

(b) Maxwell Diagram

Fig. 3.29: For Example 3.10

Example 3.11 *Determine the axial force in the members of the Compound truss (Fink truss) shown in* Fig. 3.30.

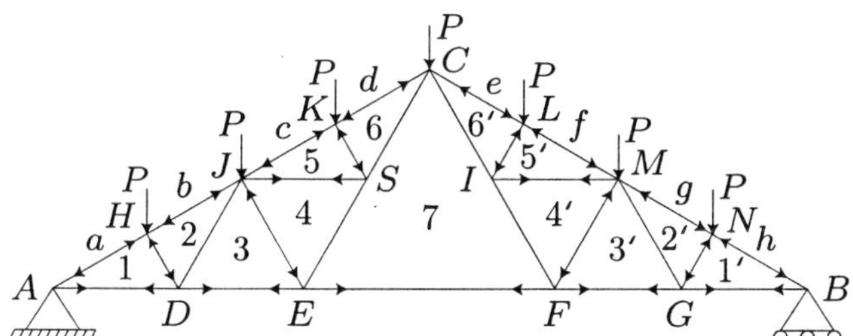

(a) Fink Truss

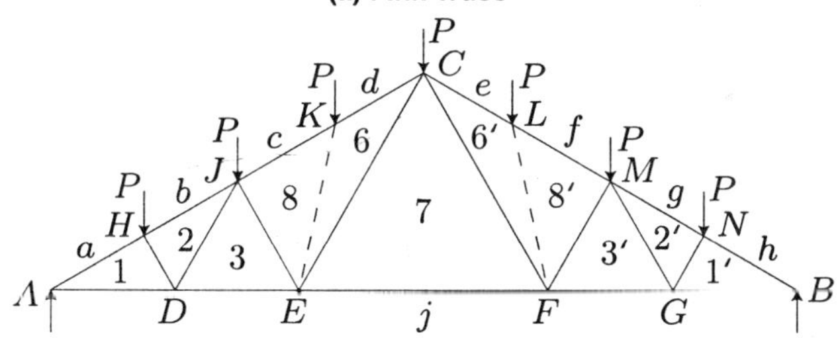

(b) Truss with Substitute Member

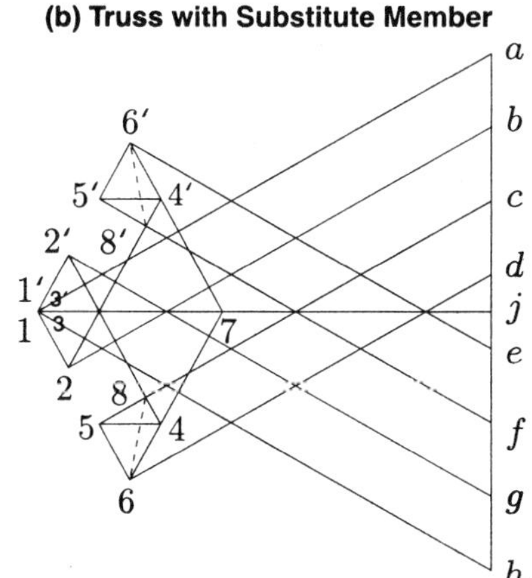

(c) Maxwell Diagram

Fig. 3.30: For Example 3.11

The graphical solution of the problem is similar to that for simple trusses. The construction of the force polygon can be started by considering first the forces at joint A, then at joints H and D. However, after this point is reached, force polygon for other joints cannot be drawn as there are more than two unknown at either joint J or joint E. Hence a substitute member EK is introduced to replace members JS and SK temporarily. It is noted that the axial forces in the members of the original truss except those bounded within the area $JESK$ will not be affected. Then the axial force in the substitute member EK is sought by locating points 8 and 6 in the Maxwell diagram. Referring back to the original truss in Fig. 3.28(a), joint K may now be considered, since the axial force in member CK is now known. The axial forces at joints J and E can be solved accordingly.

Problems

3.1 Determine whether the trusses shown in Fig. P-3.1 are stable or unstable and whether they are statically determinate or indeterminate. If they are indeterminate, state the degree of indeterminacy.

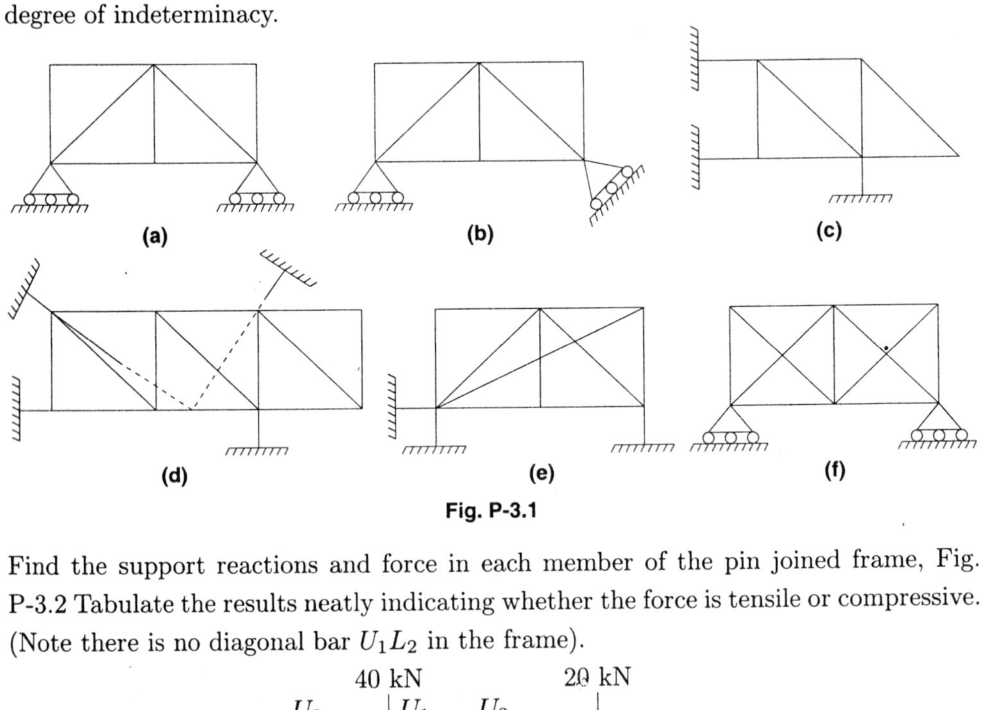

Fig. P-3.1

3.2 Find the support reactions and force in each member of the pin joined frame, Fig. P-3.2 Tabulate the results neatly indicating whether the force is tensile or compressive. (Note there is no diagonal bar U_1L_2 in the frame).

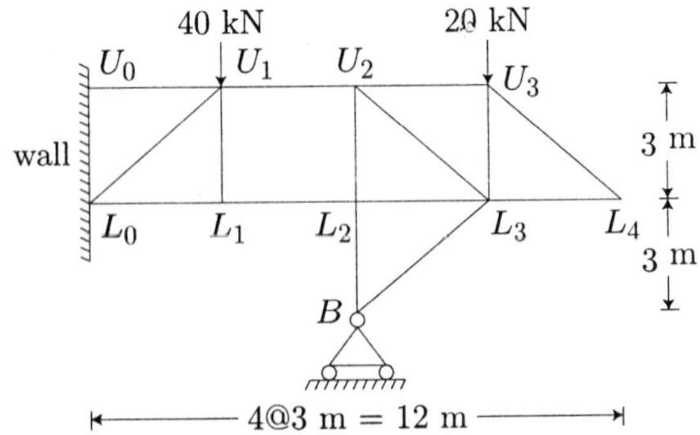

Fig. P-3.2

3.3 Determine the forces in the members of the frame shown in Fig. P-3.3. Use the method of joints.

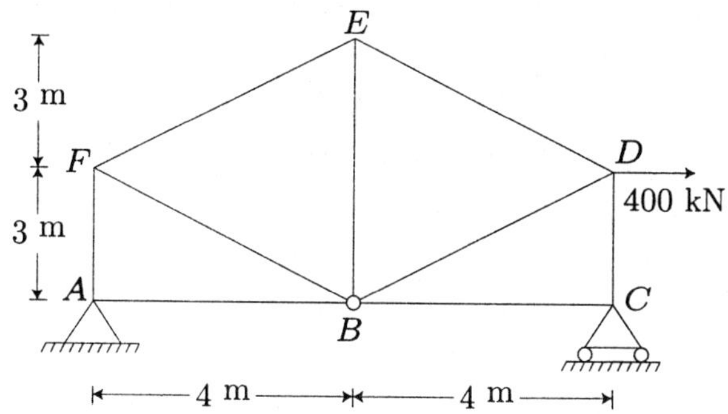

Fig. P-3.3

3.4 Fig. P-3.4 shows the outlines of a roof truss of span 10 m and height 2.5 m. Find the forces in each member.

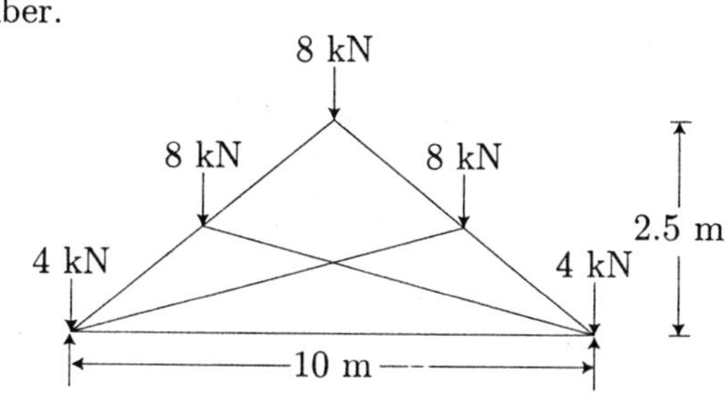

Fig. P-3.4

3.5 Determine the forces in all members of the trusses shown in Fig. P-3.5 by method of joints.

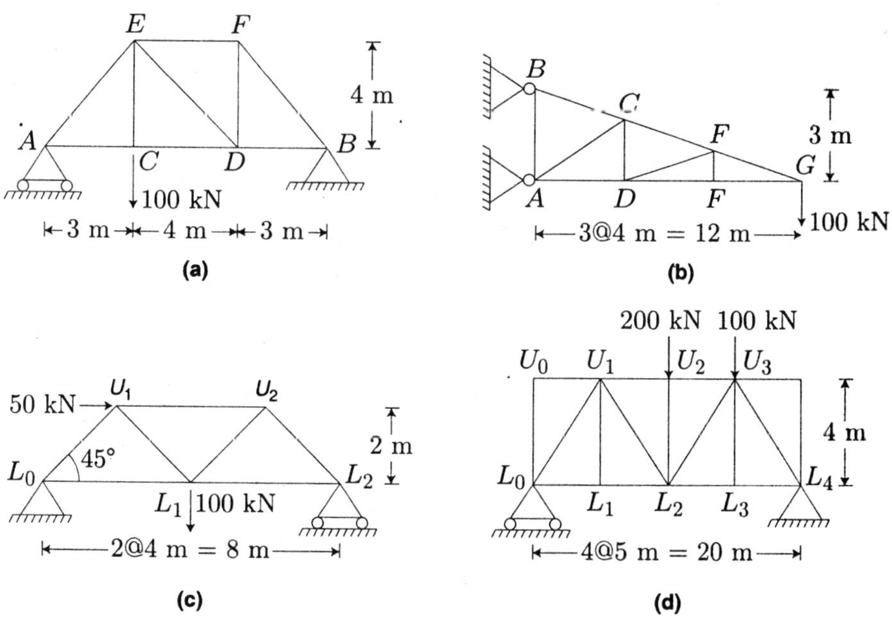

Fig. P-3.5

3.6 Calculate force in the member DE of truss shown in Fig. P-3.5(a) by the method of sections.

3.7 Calculate the force in the members U_1L_2, L_2U_3 and L_2L_3 of the truss shown in Fig. P-3.5(d) by the method of sections.

3.8 Calculate the force in the members a, b and c of the truss shown in Fig. P-3.6 by the method of sections.

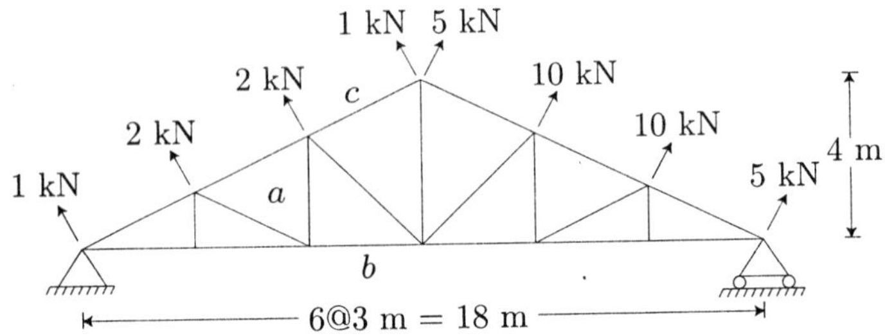

Fig. P-3.6

3.9 Calculate the forces in all the members in Kingpost truss shown in Fig. P-3.7.

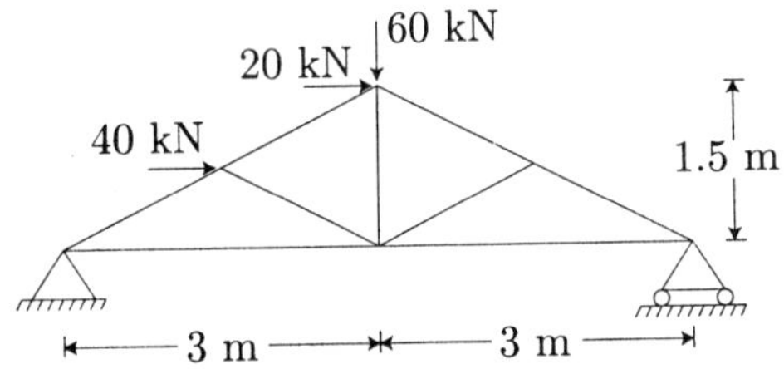

Fig. P-3.7

3.10 Calculate the force in the members a, b and c of the truss shown in Fig. P-3.8. by method of sections.

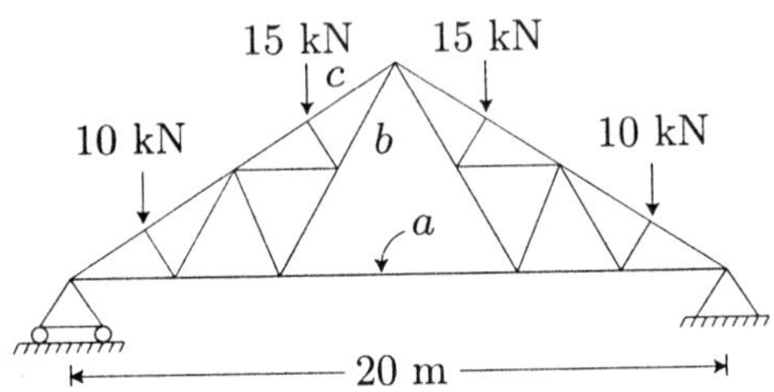

Fig. P-3.8

3.11 Determine the forces in the members a, b, c and d by the method of sections of the truss shown in Fig. P-3.9.

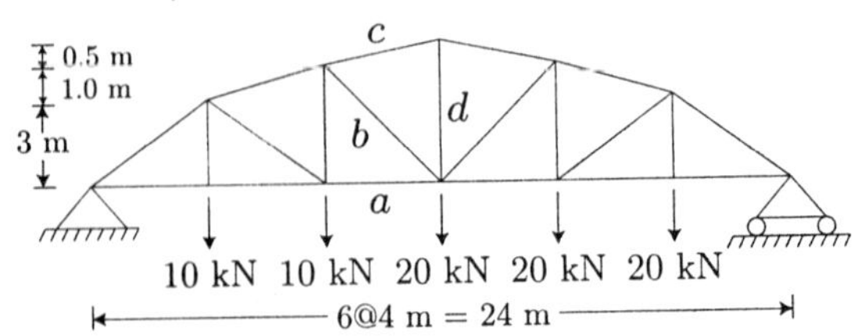

Fig. P-3.9

3.12 Determine the forces in all the members of the truss shown in Fig. P-3.6 by graphical method.

3.13 Determine the forces in all the members of the truss shown in Fig. P-3.8 by graphical method.

3.14 Determine the forces in all the members of the truss shown in Fig. P-3.5(d) by graphical method.

3.15 Calculate the forces in the members of the Pratt truss in Fig. P-3.10.

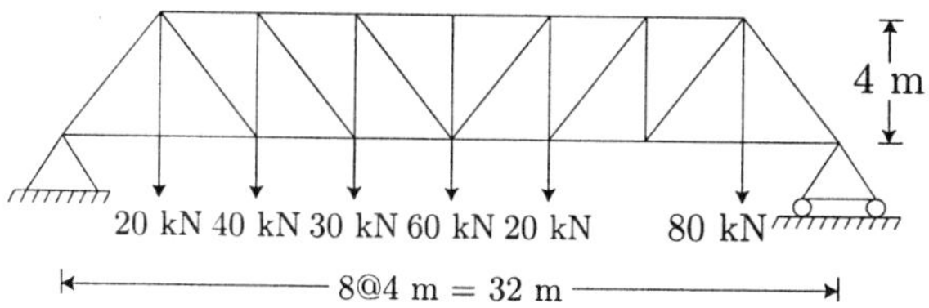

Fig. P-3.10

3.16 Calculate the forces in the members of the Warren truss shown in Fig. P-3.11.

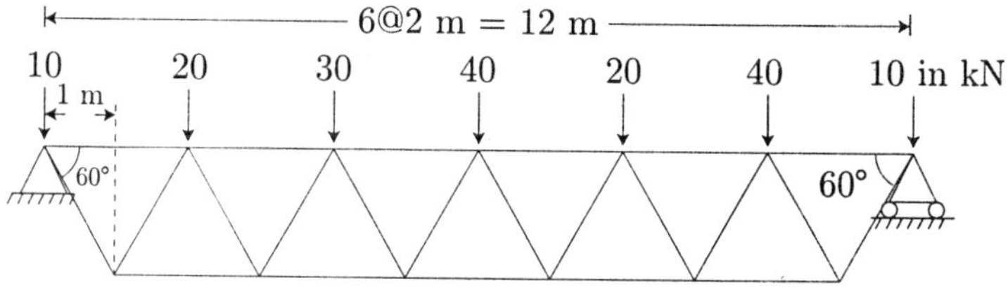

Fig. P-3.11

3.17 Determine the forces in the various members of Howe Truss shown in Fig. P-3.12.

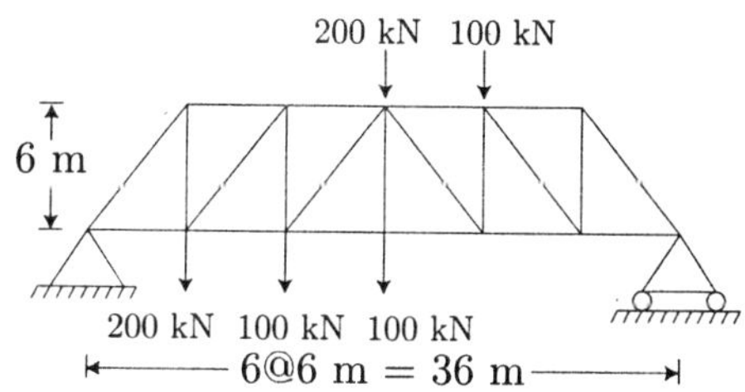

Fig. P-3.12

3.18 Determine the forces in all members of the truss shown in Fig. P-3.13.

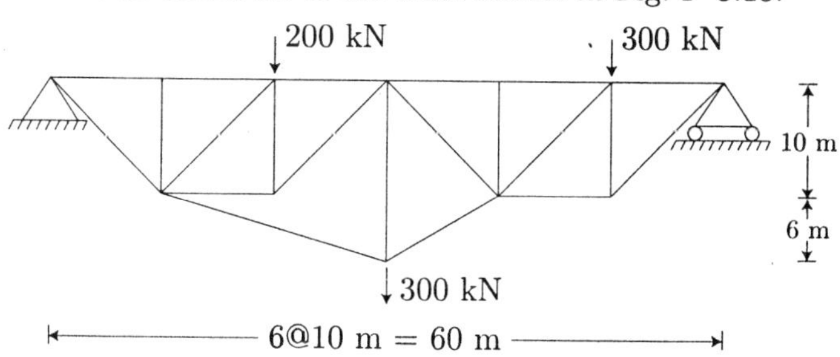

Fig. P-3.13

3.19 Determine the forces in members a, b and c of the truss shown in Fig. P-3.14.

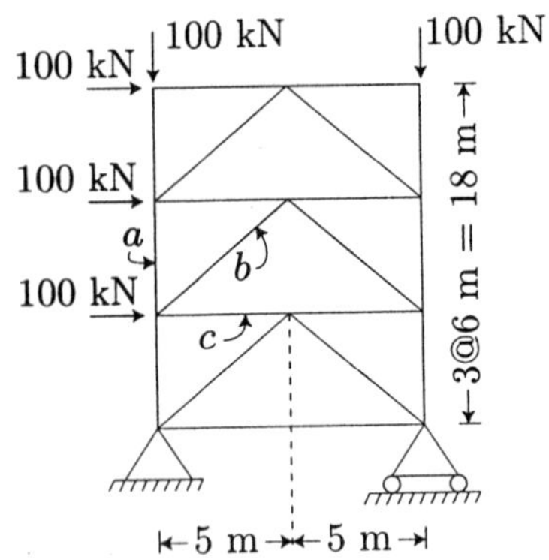

Fig. P-3.14

3.20 Compute the bar forces in the members a, b and c of truss shown in Fig. P-3.15.

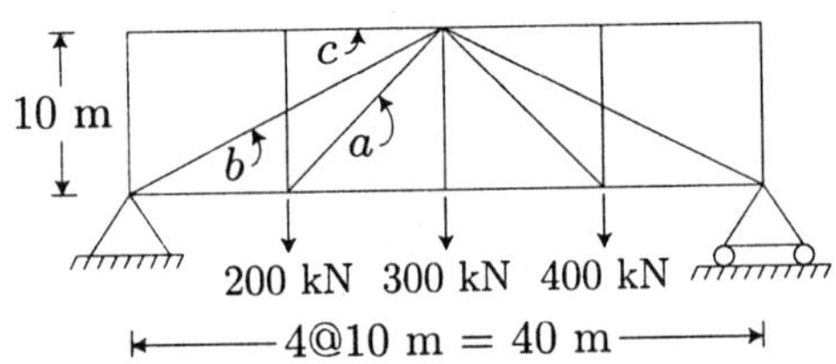

Fig. P-3.15

3.21 Compute the forces in bars a, b and c of the truss shown in Fig. P-3.16.

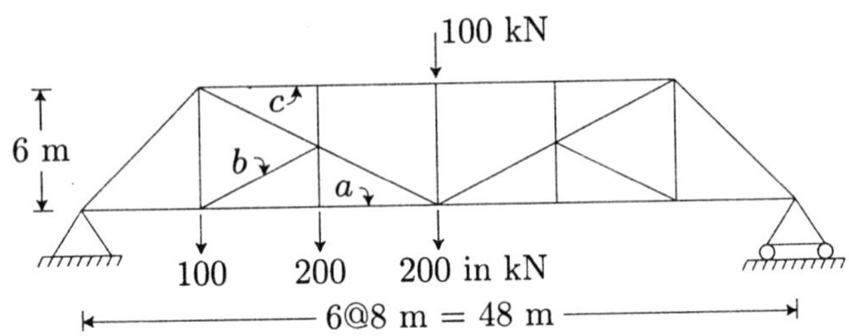

Fig. P-3.16

3.22 Draw the force diagram for the given roof truss shown in Fig. P-3.17 and determine the forces in the members.

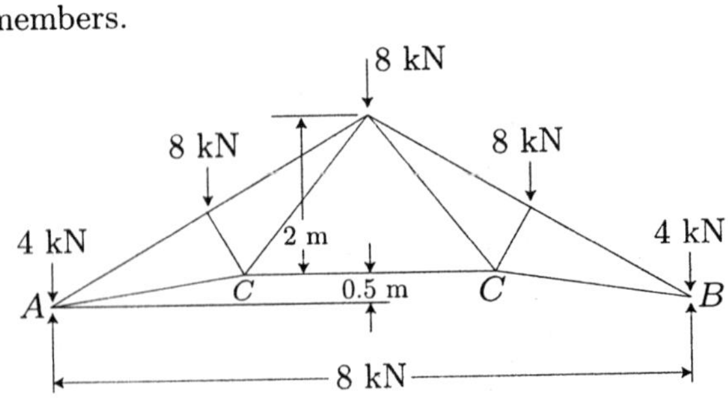

Fig. P-3.17

3.23 Determine the bar forces in all members of the trusses shown in Figs. P-3.18 to P-3.25.

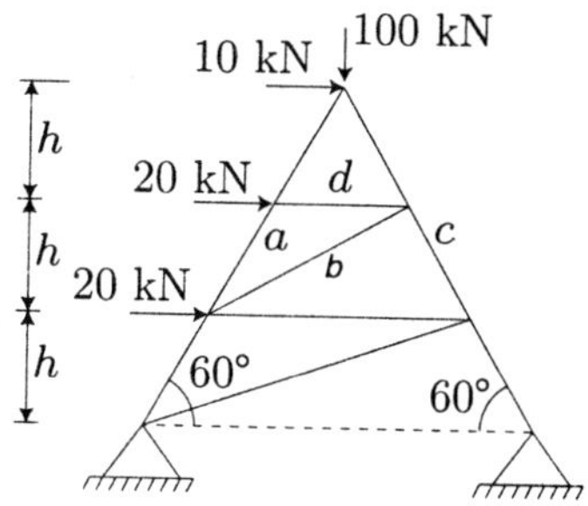

Fig. P-3.18

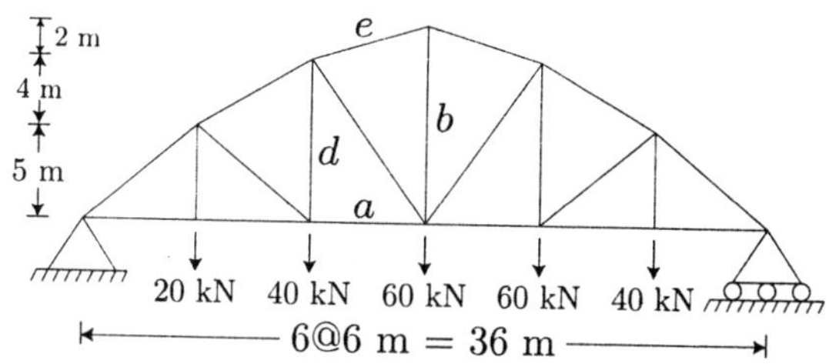

Fig. P-3.19

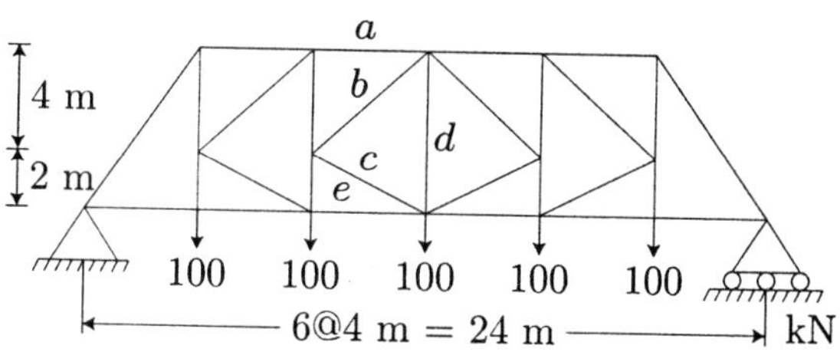

Fig. P-3.20

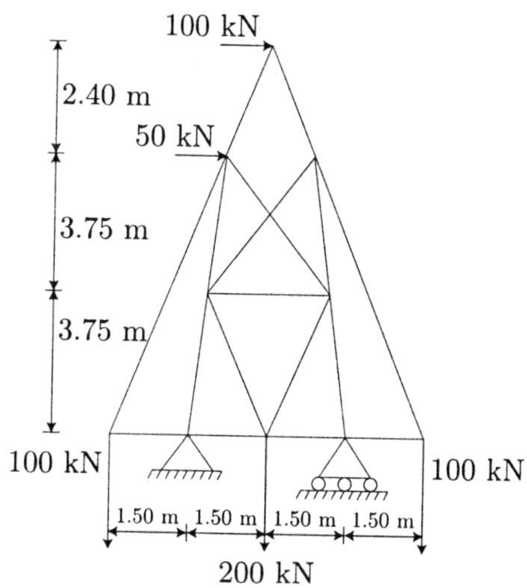

Fig. P-3.21

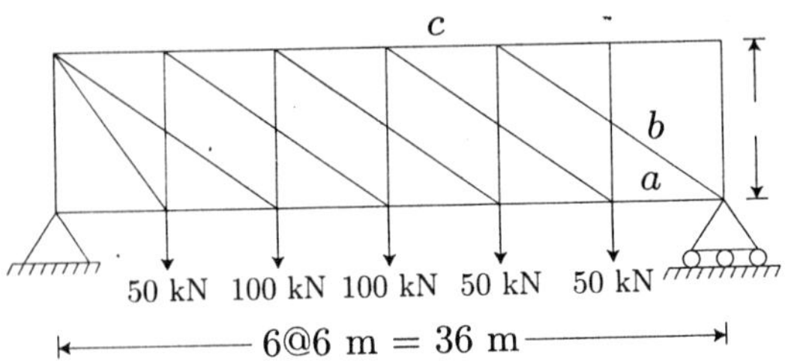

Fig. P-3.22

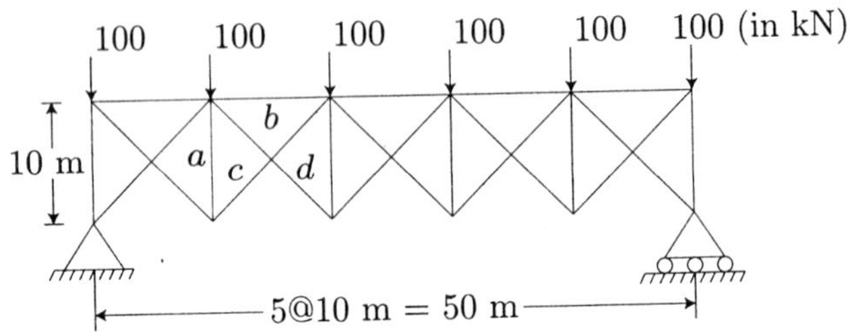

Fig. P-3.23

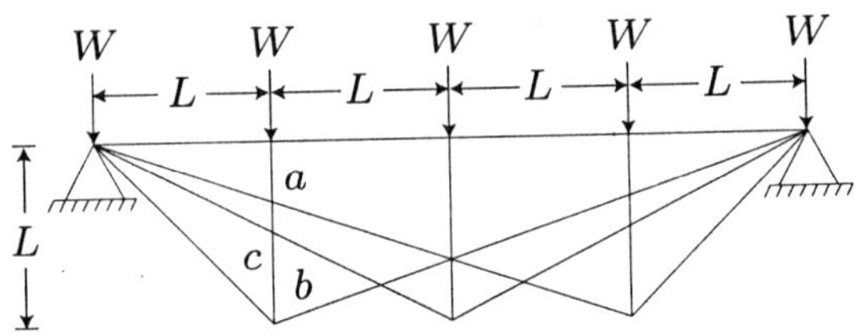

Fig. P-3.24

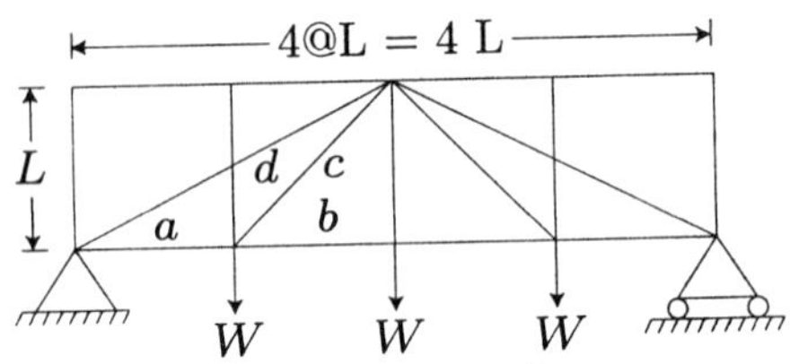

Fig. P-3.25

3.24 Calculate the bar forces in members marked by letters a, b, c, d and e of the trusses in Figs. P-3.18 to P-3.25 by the method of sections.

3.25 Determine the bar forces in all members of the trusses shown in Fig. P-3.26 and P-3.27 by the method of shears and moments.

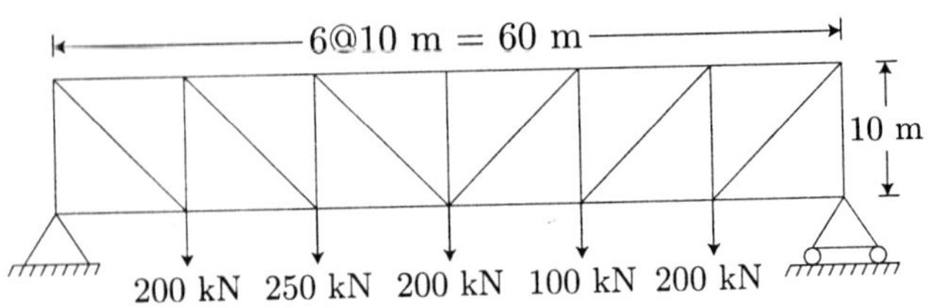

Fig. P-3.26 Load in kN

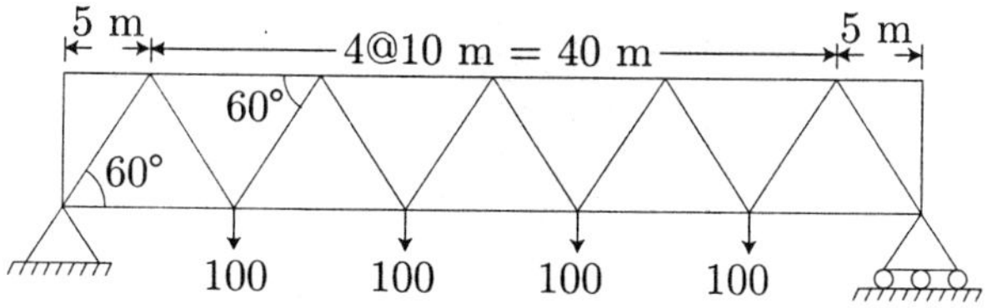

Fig. P-3.27 Load in KN

3.26 Calculate the bar forces in all members of the compound trusses shown in Figs. P-3.28 to P-3.31.

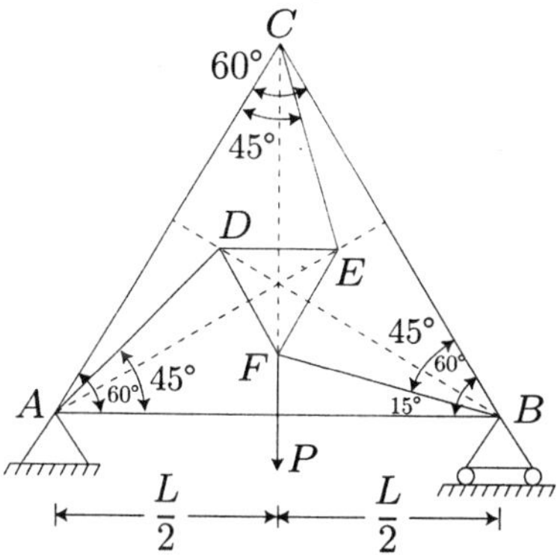

ABC and *DEF* are Equilateral Triangles

Fig. P-3.28

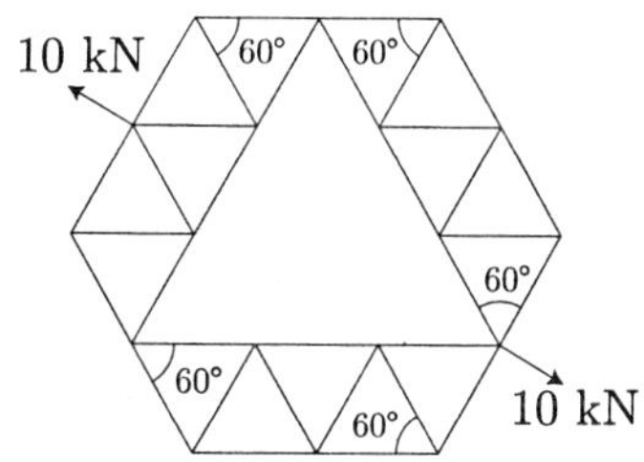

Fig. P-3.29

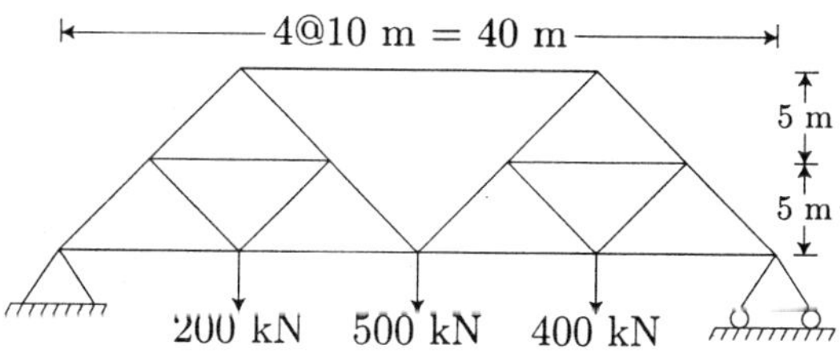

Fig. P-3.30

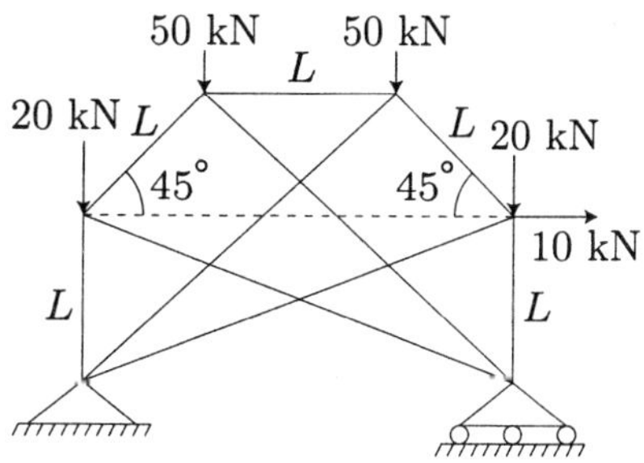

Fig. P-3.31

3.27 Calculate the bar forces in all members of the complex trusses shown in Figs. P-3.32 and P-3.33.

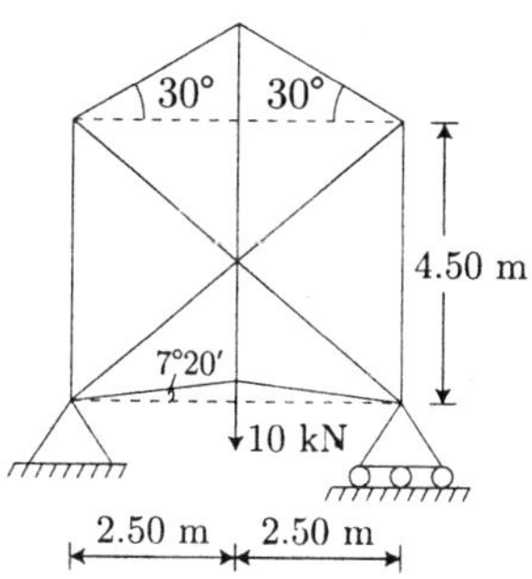

Fig. P-3.32

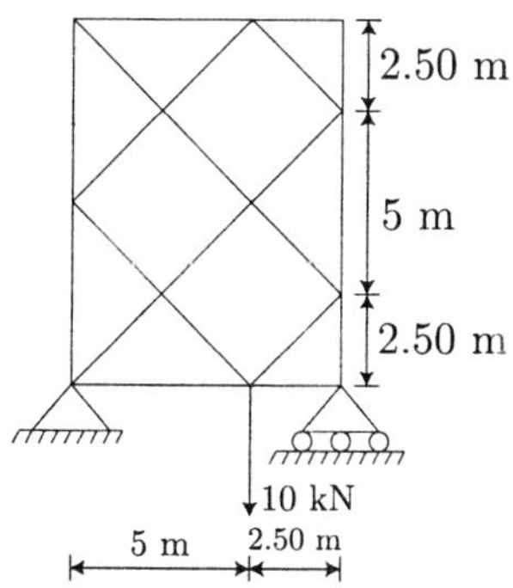

Fig. P-3.33

3.28 Using tension coefficients method, calculate the bar forces in all members of the space trusses shown in Figs. P-3.34 and P-3.35.

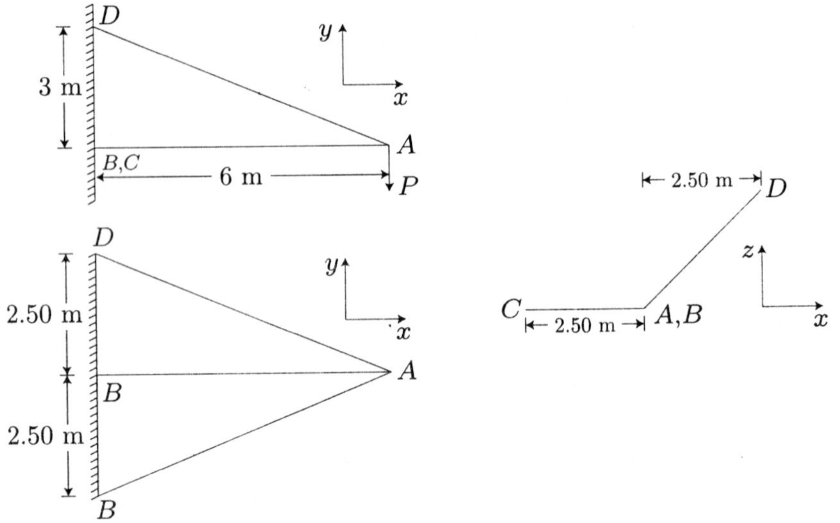

Fig. P-3.34

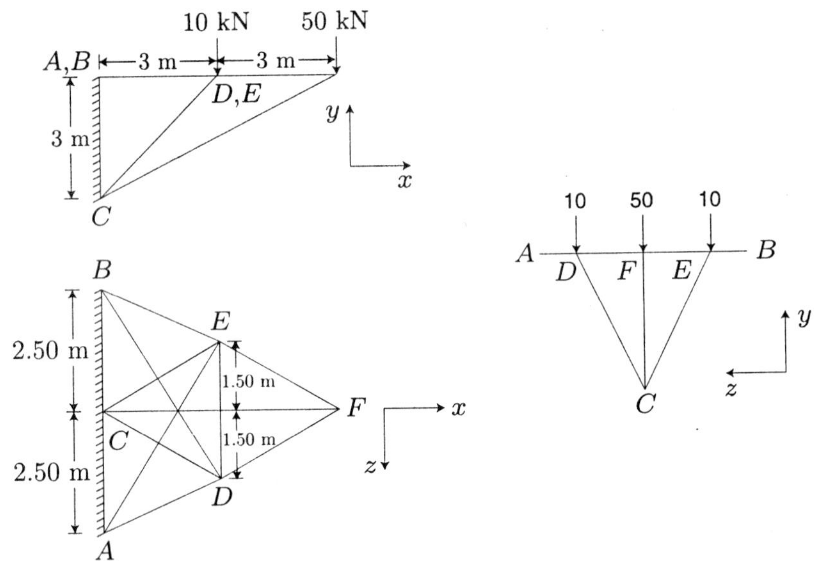

Fig. P-3.35

Chapter 4

Influence Lines and Moving Loads (Travelling Loads)

4.1 Introduction

When a live load, either moving or moveable changes position on a structure, it causes varying effects, depending on the location of the part being considered, the functions being reaction, shear, bending moment, axial force, stresses, etc. in members. It will be necessary to determine the position of the adopted system of loading, which will produce the maximum and minimum values of such functions.

In studying the effect of a system of loads moving across a structure, the use of diagrams called influence lines (denoted by I.L.) is often helpful. An influence line is a graphical representation showing the variation of functions at a given point for a single moving load occupying various positions on the structure. The effect may be the reaction, the shear, bending moment, the deflection, or the stress at a point of the structure. A shear or a moment influence line represents the shear or moment at a certain point for any position of the moving load, while a shear or a moment diagram indicates the shear or the moment at all points of the structure for a single position of the loads. Influence lines constructed for unit loads are called unit influence lines.

An unit influence line will be referred to in the following discussions. Influence lines are constructed by first calculating ordinates which represent the value of the desired function due to unit load as it moves across the structure. These ordinates are then plotted under the load at each critical position of the load and the line connecting the extremities of these ordinates is the influence line for the desired function.

4.2 Influence Lines for Reactions on a Beam

In order to draw influence lines for reactions, consider a beam AB of span L. Influence liens for left-hand reaction R_A and right-hand reaction R_B are to be drawn.

Consider a unit load travelling from A to B. When this load is at a distance 'x' from left support,

$$R_A = \frac{L-x}{L} \text{ and } R_B = \frac{x}{L}.$$

For the influence line of left reaction, the value of

$$R_A = \frac{L-x}{L},$$

is plotted directly under the unit load.

Since 'I.L.' diagram is a linear function of x, and its expression is valid for values of x between 0 and L influence line ordinate values are as under:

For $x = 0$, $R_A = \dfrac{L-x}{L} = \dfrac{L-0}{L} = 1.$

For $x = L$, $R_A = \dfrac{L-x}{L} = \dfrac{L-L}{L} = \text{zero}.$

The influence line for left-hand reaction R_A is obtained by splitting ordinates equal to one and zero at A and B respectively and connecting these by straight line as shown in Fig. 4.1(b).

Similarly, the influence line for right-hand reaction, R_B, is plotted in Fig. 4.1(c).

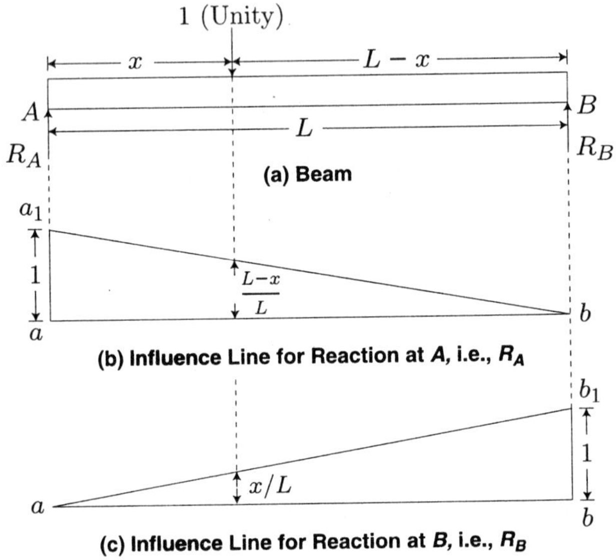

Fig. 4.1: Influence Lines for Reactions

The influence lines thus drawn can be used for calculating reaction R_A or R_B for various systems of loads.

4.3 Influence Lines for Shear in a Beam

Considering unit load travelling from A to B on a beam AB of span 'L', it is required to draw influence line for shear at a section 'C' distance 'a' from left support A.

When a unit load is between A and C, i.e., for value of x between zero to 'a' at section 'C'.

$$\text{S.F.} = R_B = -\frac{1(x)}{L} = -\frac{x}{L}$$

This shear force is negative as the force on the right-hand side of the section 'C' is acting upwards.

When $x = 0$, $R_B = 0$.

For $x = a$, S.F. $= R_A = \dfrac{L-a}{L}$.

Influence line ordinates for S.F. at C are the same as for the right-hand reaction between A and C and are plotted below the base line ab, since these ordinates are negative.

When the unit load is between C and B, at section C,

$$\text{S.F.} = R_A = \left(\frac{L-x}{L}\right),$$

which is valid for values of x between a to L

For $x = a$, S.F. at $C = \dfrac{L-a}{L}$.

For $x = L$, S.F. at $C = \dfrac{L-L}{L} = 0$.

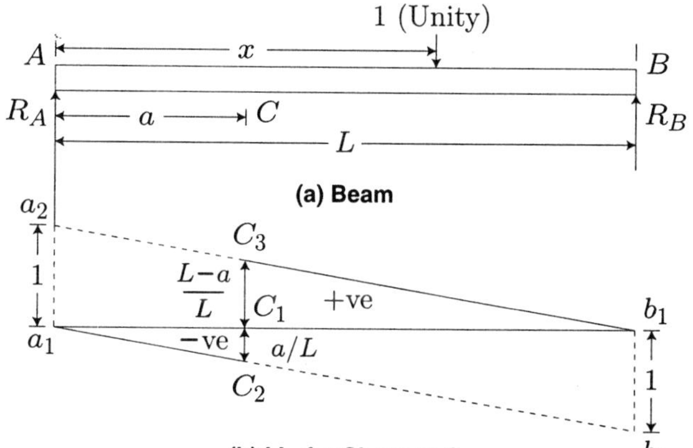

(b) I.L. for Shear at C

Fig. 4.2: I.L. for Shear

This S.F. is positive since the net force to left of the section 'C' is acting upwards.

For the portion of the beam between C and B the left-hand reaction influence line is also the shear influence line for C.

The diagram $a_1c_2c_1c_3b_1$ is the shear influence line for section C. The ordinate c_1c_3 is +ve S.F. at C when the unit load is placed at an infinitesimal distance to the right of C and the ordinate c_1c_2 is the −ve S.F. at C when the unit load is applied at an infinitesimal distance to the left of C.

In order to construct S.F. influence line at a section of a simply supported beam following should be remembered :

(i) *Draw influence lines for left and right reactions, with a_1b_1 as span and a_1a_2 and b_1b_2 as unity.*

(ii) *Project the section C on these two influence lines by a vertical line.*

(iii) $a_1c_2c_3b_1$ *is the required influence line.*

4.4 Influence Lines for Bending Moment in a Beam

The bending moment influence line, Fig. 4.3(b) gives the B.M. at the section C at a distance 'a' from the left support for a unit load moving from A to B.

When the load is between A and C, i.e., between $x = 0$ and $x = a$,

B.M. at $C = R_B(L-a) = \dfrac{x}{L}(L-a)$

This is a linear function of x.

When $x = 0$, B.M. at $C = \dfrac{(L-a)x}{L} = 0$.

For $x = a$, B.M. at $C = \dfrac{(L-a)a}{L}$.

c_1c_2 equals $\left[\dfrac{(L-a)}{L}a\right]$ on a_1b_1 as base, c_1 being the projection of point C of beam AB.

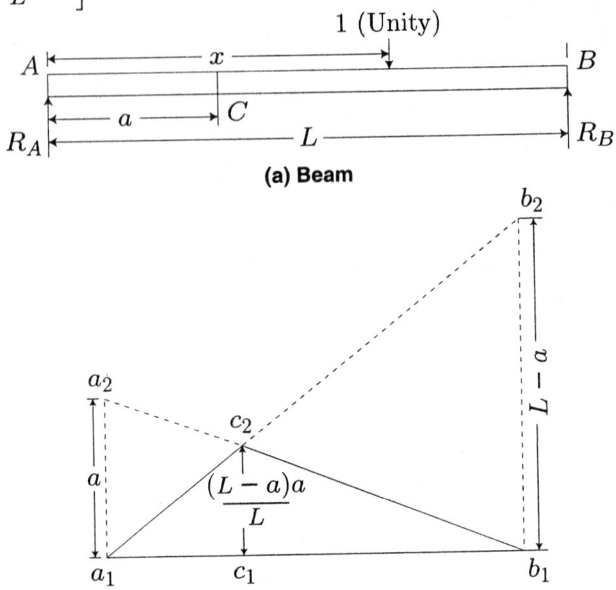

(a) Beam

(b) Influence Line for Bending Moment at *C*

Fig. 4.3: I.L. for Bending Moment

When the unit load is between C and B,

B.M. at $C = R_A(a) = \dfrac{(L-x)a}{L} = \left(\dfrac{a}{L}\right)(L-x)$

This is a linear function of x.

For $x = a$, B.M. at $C = \left(\dfrac{a}{L}\right)(L-x) = \dfrac{(L-a)a}{L}$.

For $x = L$, B.M. at $C = \left(\dfrac{a}{L}\right)(L-x) = \left(\dfrac{a}{L}\right)(L-L) = \text{zero}$.

Complete influence line for B.M. is given by the triangle $a_1c_2b_1$.

4.5 Use of Influence Line in Computation

Since the ordinate of an influence line equals the value of a particular function due to unit load acting at a point where the ordinate is measured, the following characteristics will hold :

(a) The value of a function due to the action of a single concentrated line load equals the product of magnitude of load and the ordinate of the influence line for that function measured at the point of application of load. Thus, referring to Figs. 4.1 to 4.3, when a concentrated load of magnitude W is placed at x from B, the reaction at A, i.e., R_A becomes $W\left(\frac{x}{L}\right) = \frac{Wx}{L}$, $\frac{x}{L}$ being dimensionless. The bending moment at C due to this load becomes

$$W\left(\frac{ax}{L}\right) = \frac{Wax}{L}.$$

Appropriate dimension of length and W are to be used.

From the consideration of the principle of superposition, the total value of a function due to more than one concentrated load can be obtained by superposing effect of each concentrated load determined separately.

(b) To obtain the maximum value of a function due to a concentrated live load, the load should be placed at the point where the ordinate of the influence line for that function is maximum. Thus, for the beam in Fig. 4.1 the maximum value of R_A due to a single concentrated load W occurs when the load is placed at A and equals $W(1)$, i.e., W. The maximum bending moment at C due to the single concentrated load W will similarly be given when W is at C and will be $\frac{Wa(L-a)}{L}$.

For cases when at a section, values may be either positive or negative depending on the position of load, the maximum positive and negative values will be determined by placing the concentrated load at the locations of maximum positive and maximum negative ordinates respectively. Thus maximum negative shear force at C of the beam in Fig. 4.2 will occur when the load W is just on the left of the section, the value being given by $W\frac{a}{L}$. The maximum positive shear force due to the load will be obtained, when the load W is just on the right of the section, the value being $\frac{W(L-a)}{L}$.

(c) The value of a function due to any position of a uniformly distributed load is equal to the product of the intensity of the loading and the net area under that position of the influence line, for the function under consideration.

Fig. 4.4 shows the influence line for a particular function on a position AB of a structure in which uniformly distributed load of magnitude w per unit length and of length 'a' is acting between two points C and D. Considering a short length dx of the loaded position of the structure, the effect of the uniformly distributed load may be treated as a concentrated load equal to wdx. If the ordinate of the influence line where dx is taken by y, the effect of the load is $wydx$ and the total effect due to the uniformly distributed load is given by

$$\int_0^a wydx = w\int_0^a ydx$$

$$= w\ [\text{area of influence line diagram between limits of load } w]$$

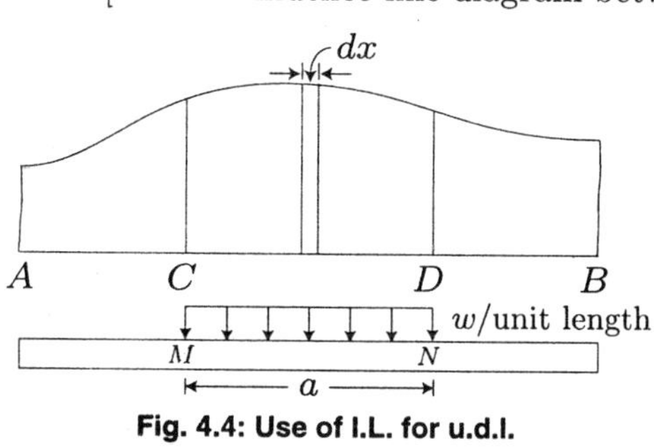

Fig. 4.4: Use of I.L. for u.d.l.

For example, if a uniformly distributed load of magnitude w per unit length is placed on the beam in Fig. 4.2 so as to extend from C to B, then

$$\text{Shear force at } C = +w\left[\frac{1}{2}\left(\frac{L-a}{L}+0\right)\right](L-a) = \frac{w}{2L}(L-a)^2$$

$$\text{and} \quad BM \text{ at } C = w\left[\frac{1}{2}\frac{a(L-a)}{L}\right](L-a) = \frac{w}{2L}a(L-a)^2.$$

(d) For the maximum value of a function due to a uniformly distributed live load larger than the span, the load shall be placed over all those portions of the structure, for which the ordinates to the influence line for that function have the sign of the character of the function desired. For example, referring to Figs. 4.1, 4.2 and 4.3 values of R_A, R_B and $B.M.$ at C will be maximum when the full span is loaded. For maximum positive shear force at C, referring to Fig. 4.2, only the portion CB of the beam shall be loaded.

(e) For the case when live load consists of uniform load, accompanied by a single concentrated load which may be placed at any point so as to produce the maximum effect under consideration, the maximum values of the functions may be obtained as per characteristics mentioned in (b) to (d).

(f) When the moving load consists of a series of concentrated load or uniformly distributed load of small length, the position of the loading for producing maximum values of the functions can be located by inspection or trial and error or by the criteria to be developed later.

Example 4.1 *Construct the influence lines for the reactions at supports, the shear forces at E (in between AB) and F on AC, the shear force just on the right and left of support A, the bending moment at E and the bending moment at A of the beam in Fig. 4.5.*

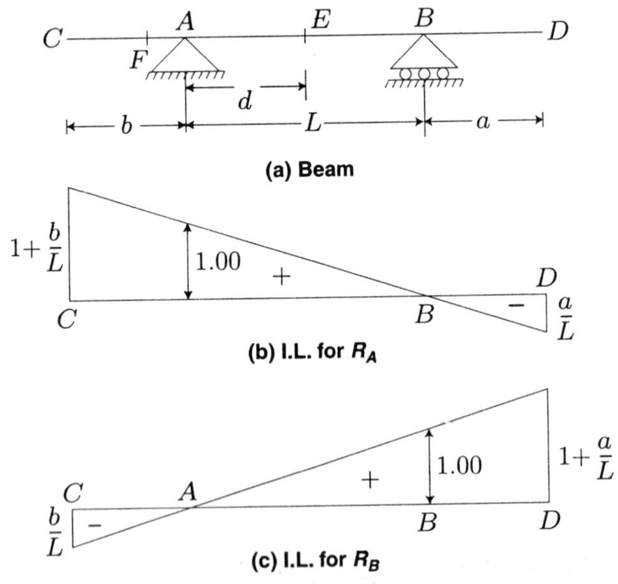

Fig. 4.5: For Example 4.1 (Contd.)

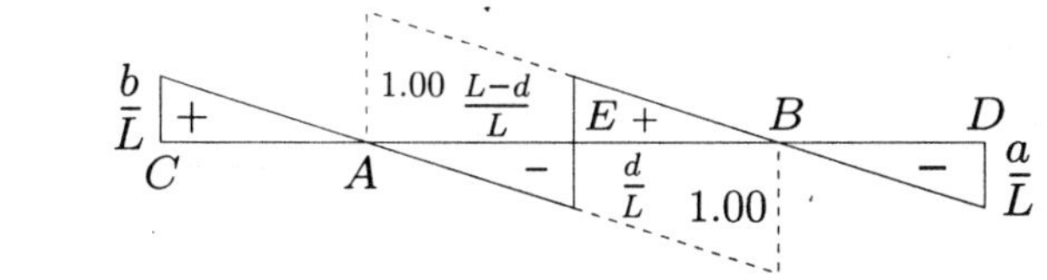

(d) I.L. for SF at *E*

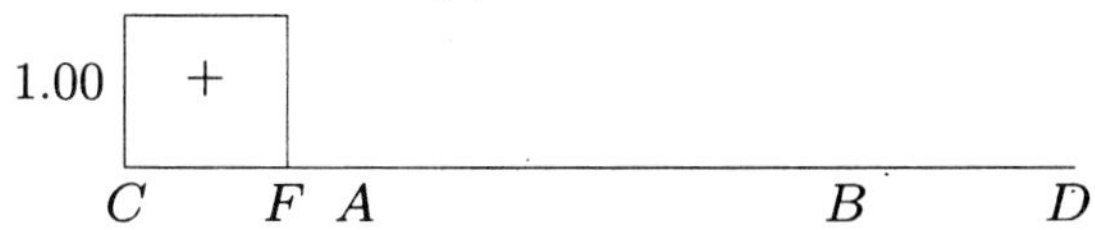

(e) I.L. for SF at *F*

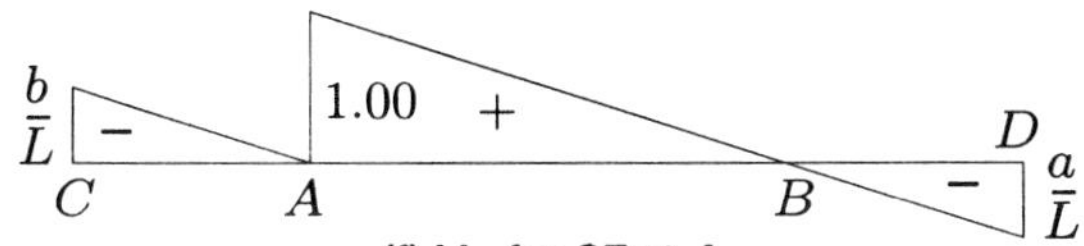

(f) I.L. for SF at *A*

Just on Right of *A*

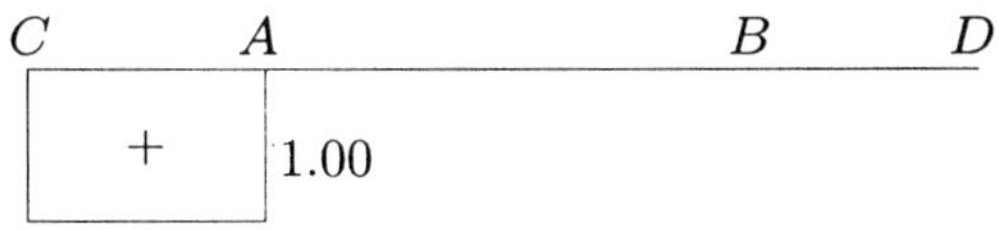

(g) I.L. for SF at *A*

Just on Left of *A*

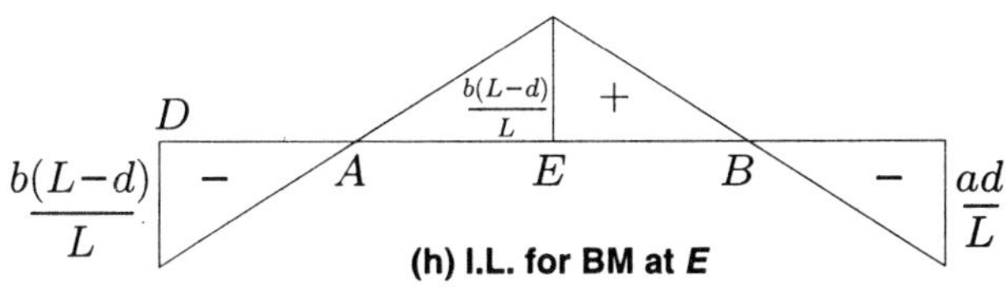

(h) I.L. for BM at *E*

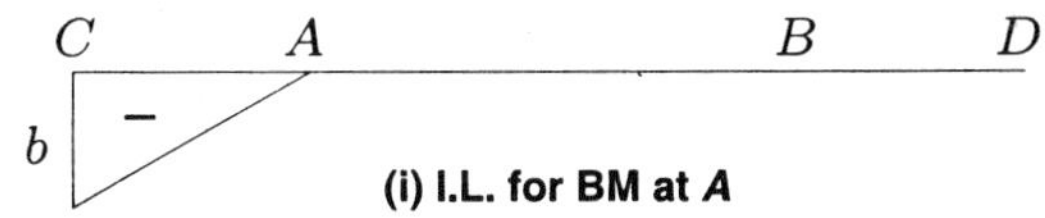

(i) I.L. for BM at *A*

Fig. 4.5: For Example 4.1

Solution : The influence lines for various functions in case of the beam in Fig. 4.5(a) may be obtained by calculating the control points. For unit load placed at C, the values of the reactions R_A and R_B are $\left(1+\frac{b}{L}\right)$ and $-\frac{b}{L}$ respectively. For unit load at A, $R_A = 1$ and $R_B = 0$, and for unit load at B, $R_A = 0$ and $R_B = 1$. When unit load is at D, $R_A = -\frac{a}{L}$ and $R_B = \left(1+\frac{a}{L}\right)$. The influence lines for R_A and R_B thus obtained are linear and shown in Figs. 4.5(b) and 4.5(c) respectively.

When a unit load is placed at C, the shear force at E becomes equal to R_B and has a magnitude of $+b/L$. As the unit load moves from C to A, the reaction at B and hence the shear force at E decreases to zero. As the unit load moves from A to a point just to the left of E, the shear force at E which equals R_B, increases from zero to $-d/L$. The variation of influence line for S.F. at E is thus linear for load between C and E as shown in Fig. 4.5(d).

When the unit load is just on the right of E, the shear force at E equals R_A and magnitude is given by $\frac{L-d}{L}$. The variation of influence line diagram for other position of the load varies linearly as shown in Fig. 4.5(d).

The influence lines for shear force for other sections on the beam may be obtained similarly as illustrated in Figs. 4.5(e) to 4.5(g).

To construct the influence line for the bending moment at E, the unit load may be considered being placed at C. The bending moment at E then becomes $R_B(L-d)$, i.e., $-\frac{b(L-d)}{L}$. As the load goes from C to A, the bending moment decreases linearly to zero. As the unit load goes from A to E, the reaction R_B increases linearly from zero to $\frac{d}{L}$ and bending moment at E, computed from R_B increases linearly from zero to $+\frac{d(L-d)}{L}$. As the unit load moves from E to B, the reaction at A, i.e., R_A decreases linearly from $\frac{L-d}{L}$ to zero. The bending moment at E thus decreases linearly from $\frac{d(L-d)}{L}$ to zero. For position of unit load between B and D, the bending moment at E may be similarly obtained from R_A. Variation of influence line for bending moment at E thus obtained is shown in Fig. 4.5(h).

Influence line for bending moment at A is shown in Fig. 4.5(i). There will be no bending moment at A for any position of load between A and D.

4.6 Influence Lines for Panelled Girder

Consider a panelled girder of 'n' panels having a panel length 'p' of each panel. It is required to draw influence lines for left and right reactions.

(a) Influence lines for Reactions. The influence lines for reaction of a girder with end floor beams are different from those for a girder without end floor beams.

When end floor beams are used, all the loads on the floor system are transferred to the abutments or piers through the end pedestals and the reaction at the end pedestals is unity when a pair of loads are placed symmetrically. When a unit load moves away from A to the intermediate panel points, the end reaction R_A decreases linearly from unity to zero. The required influence line for R_A is as shown in Fig. 4.6(b).

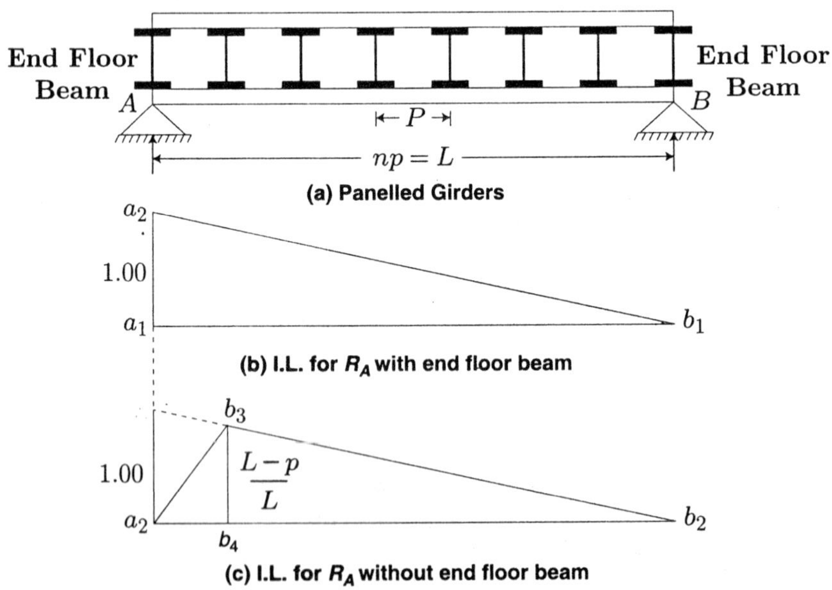

Fig. 4.6: Girder Supporting Floor Beams

When end floor beams are not used the exterior end of the stringers rests directly on abutments or piers. In this case when the loads are in the first panel, the stringers deliver part of the loads directly into the elevated portion of the abutment and not on the main bearing.

When the unit load is in the first panel at a distance x from A, x being less than p, the load on the panel point $b = \frac{x}{p}$ and R_A due to this panel load $\frac{x}{p}$ comes to

$$R_A = \frac{x}{p}\frac{(L-p)}{L} = \frac{x}{p}\left(1 - \frac{p}{L}\right) = \frac{x}{p} - \frac{x}{L}.$$

When $x = 0$, the unit load acts at stringer and the reaction goes directly to the abutment, and $R_A = 0$.

When $x = p$,

$$R_A = 1 - \frac{p}{L} = \frac{(L-p)}{L}.$$

The influence line for R_A is a triangle $a_2 b_3 b_2$ where the ordinate $b_3 b_4 = \frac{L-p}{L}$ as shown in Fig. 4.6(c).

(b) Influence Lines for Shear in a Panel

It is required to draw I.L. for shear in panel bc of a panelled girder of span l and having a number of panels. The panel length of each panel is 'p'.

When a unit load is between A and b. [as in **Fig. 4.7(a)**]

S.F. in the panel = right-hand reaction R_B and the part $a_1 b_2$ of I.L. for R_B plotted may be used for shear in the panel bc.

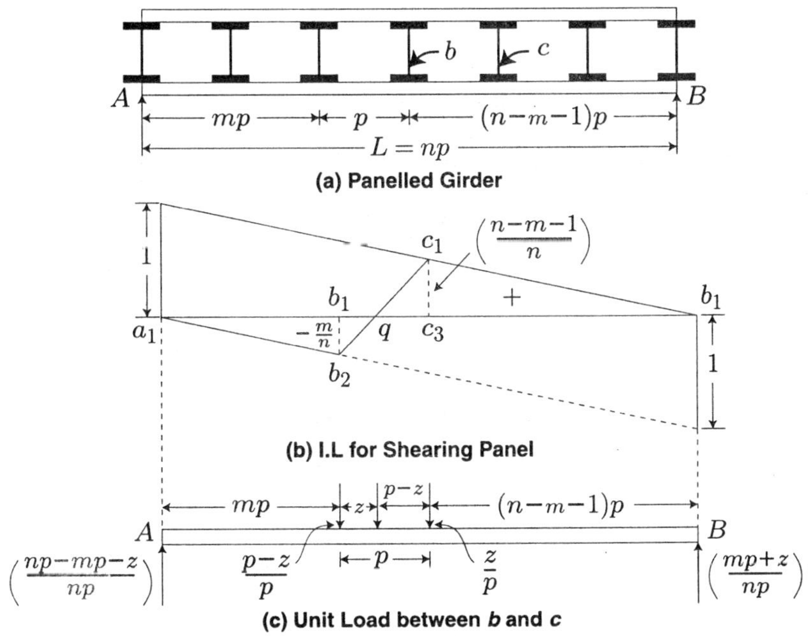

Fig. 4.7: I.L. for Shear

The ordinate $b_2 b_3 = \dfrac{mp}{np} = \dfrac{m}{n}$ and is −Ve.

When the unit load is between c and b. [as in **Fig. 4.7(a)**]

S.F. in panel bc = left-hand reaction R_A and the part b_1c_2 of I.L. for R_A, plotted as shown, can be used for shear in the panel bc.

The ordinate $c_2c_3 = \dfrac{(n-m-1)p}{np} = \dfrac{(n-m-1)}{n}$.

When the unit load is between b and c, the main girder receives its load at two points, i.e., at panel points b and c.

If the unit load be at a distance z from panel point b,

R_b = load at panel point $b = \dfrac{p-z}{p}$.

R_c = load at panel point $c = \dfrac{z}{p}$.

Reaction $R_A = \dfrac{(np-mp-p+p+z)}{np} = \dfrac{np-mp-z}{np}$.

Reaction $R_B = \dfrac{mp+z}{np}$.

$$\text{Shear at any point between } b \text{ and } c = R_A - \left(\frac{p-z}{p}\right) = \frac{np-mp-z}{np} - \frac{(p-z)}{p}$$
$$= \frac{[(n-1)z-mp]}{np}$$

Since z is the only variable, this expression represents a straight line. For $z = 0$, i.e., when load is at b,

S.F. in panel $bc = -\dfrac{mp}{np} = -\dfrac{m}{n}$.

For $z = p$, i.e., when load is at c,

S.F. in panel $bc = \left[\dfrac{(n-1)p-mp}{np}\right] = \dfrac{n-m-1}{n}$.

Therefore, complete influence line for S.F. in panel bc is given by $a_1b_2c_2b_1$, as in Fig. 4.7(b).

The point q where I.L. crosses the base a_1b_1, is called Neutral or Load Divide Point.

As triangles b_2b_3q and c_2c_3q are similar, $\dfrac{b_3q}{c_3q} = \dfrac{b_2b_3}{c_2c_3}$,

$$\text{whence}\quad b_3q = b_3c_3 \cdot \frac{b_2b_3}{b_2b_3+c_2c_3} = p\frac{\frac{m}{n}}{\frac{m}{n}+\frac{n-m-l}{n}} = \left(\frac{m}{n-1}\right)p.$$

The neutral point divides the panel in the same ratio as it divides the span.

(c) Influence Line for the Bending Moment at a Section in a Panelled Girder. [as in Fig. 4.8]

It is required to draw I.L. for bending moment at a section in the panel a_2a_3 of a panelled girder of span $np = L$.

Let the section C be at a distance of 'αp' from panel point a_2 and distance Aa_2 be mp.

When the unit load is between A and a_2 [Fig. 4.8(b)]

B.M. at section C,

$$M_C = R_B[(n-m-1+1-\alpha)p] = \frac{x}{np}(n-m-\alpha)p = \frac{x(n-m-\alpha)}{n}.$$

In this expression only x is variable, and it is a linear expression.

For $x = 0$, $M_C = 0$

and for $x = mp$, i.e., load at a_2 $M_C = \dfrac{(n - m - \alpha)}{n} \cdot mp$.

The part of the I.L. between A and a_2 is given by line ad.

When the load is between the panel points $a_2 - a_3$ and is at a distance z from a_2 the unit load gives a reaction of $\frac{p-z}{p}$ at a_2 and $\frac{z}{p}$ at a_3.

$$\text{B.M. at } C\text{, i.e., } M_C = R_A(mp + \alpha p) - \left(\frac{p - z}{p}\right) \alpha p.$$

$$= \left(\frac{np - mp - z}{np}\right)(mp + \alpha p) - \left(\frac{p - z}{p}\right) \alpha p.$$

In this expression only z is variable; therefore, this expression varies linearly.

When $z = 0$,

$$M_C = \left(\frac{np - mp}{np}\right)(m + \alpha)p - \alpha p = \frac{(np - mp)(m + \alpha)}{n} - \alpha p.$$

For $z = p$,

$$M_C = \frac{(np - mp - p)(mp - \alpha p)}{np} = \frac{p(n - m - 1)(m + \alpha)}{n}.$$

The variation of B.M. at C is shown by 'de' in Fig. 4.8(b).

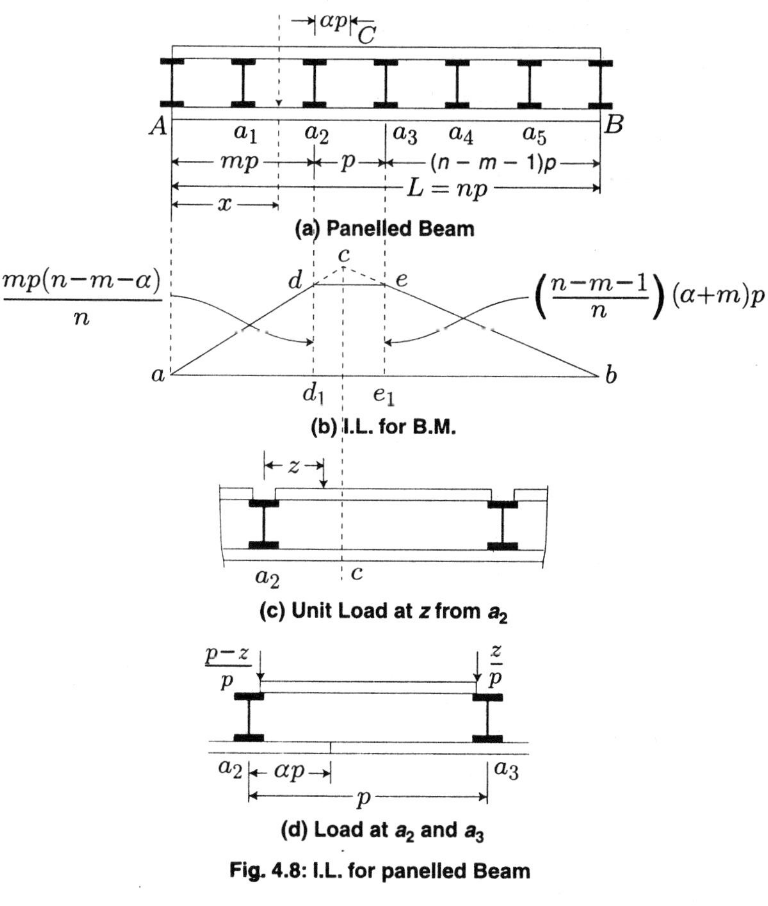

Fig. 4.8: I.L. for panelled Beam

When the load is between a_3 and B and load is at a distance x from A.

B.M. at section $C = M_C = R_A(mp + \alpha p) = \dfrac{(np - x)}{np}(m + \alpha)p;$

it is a linear expression.

When $x = mp + p$,

$$M_C = \frac{(np - mp - p)(m + \alpha)p}{np} = \frac{(n - m - 1)(\alpha + m)p}{n}.$$

When $x = np$,

$$M_C = 0.$$

I.L. for M_C between a_3 and B is shown by e_b in Fig. 4.8(b).

I.L. for M_C is shown in Fig. 4.8(b).

The simplest method to construct this I.L. is to draw the I.L. for M_C as if the load were taken directly by the main beam and to join the points d and e where the verticals from the panel points or cross girders meet the lines ac and bc.

4.7 Compound Beams

The procedure for constructing influence lines for various effects on simple beams can easily be applied in the case of compound beams. A unit load may be placed at a number of points along the beam where an abrupt change of effect is expected, the influence line between two computed ordinates being represented by a straight line.

The influence line for R_A of the compound beam in Fig. 4.9(a) is shown in Fig. 4.9(b). When the unit load is at A, $R_A = 1$. For unit load at B, $R_A = 0$ and when unit load is at D, $R_A = -\frac{b}{L_1}$. For unit load between D and C, R_A varies linearly and for unit load at C, $R_A = 0$. The influence line for R_A is then obtained by connecting the ordinates of influence lines for R_A thus obtained.

The influence lines for R_B and R_C may be similarly obtained as shown in Figs. 4.9(c) and 4.9(d).

The construction of influence lines for shear force at various points are illustrated in Figs. 4.9(e) to 4.9(l).

The influence lines for B.M. at A, B, F and G may be obtained in a similar manner and illustrated in Figs. 4.9(j) to 4.9(m). For example, in case of influence line for M_F, the considerations made are as below:

$$M_F = 0, \text{ for unit loads at } A, B \text{ and } C,$$

$$M_F = +\frac{e(L_1 - e)}{L_1}, \text{ for unit load at } F,$$

$$M_F = -\frac{a(L_1 - e)}{L_1}, \text{ for unit load at } E$$

$$M_F = -\frac{be}{L_1} \text{ for unit load at } D.$$

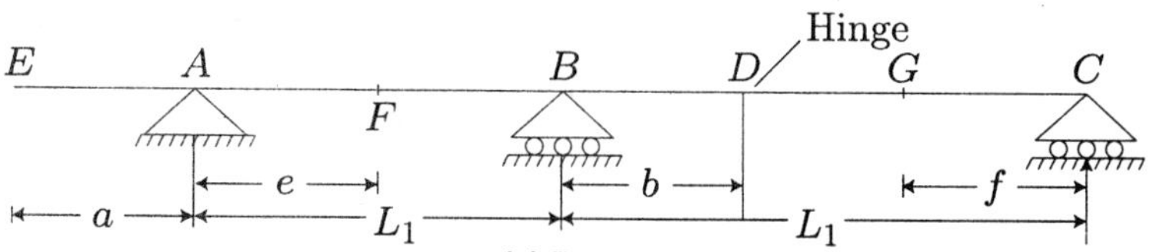

(a) Beam

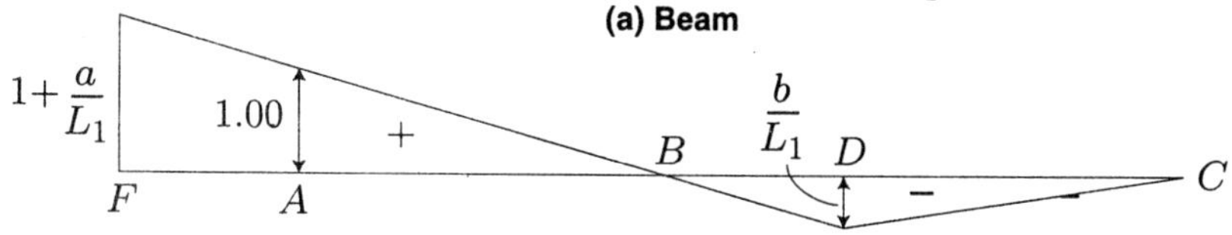

(b) I.L. for R_A

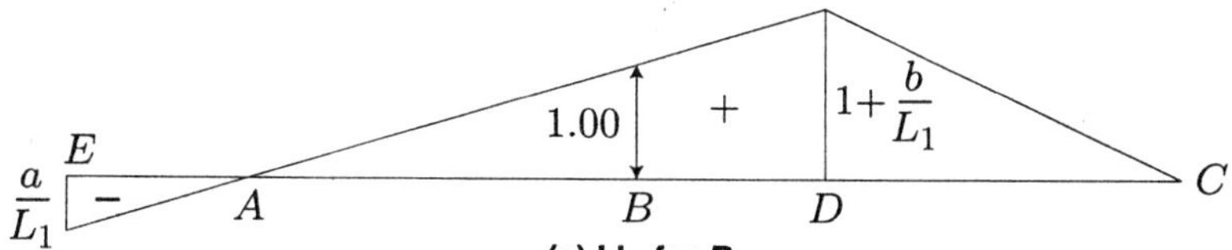

(c) I.L. for R_B

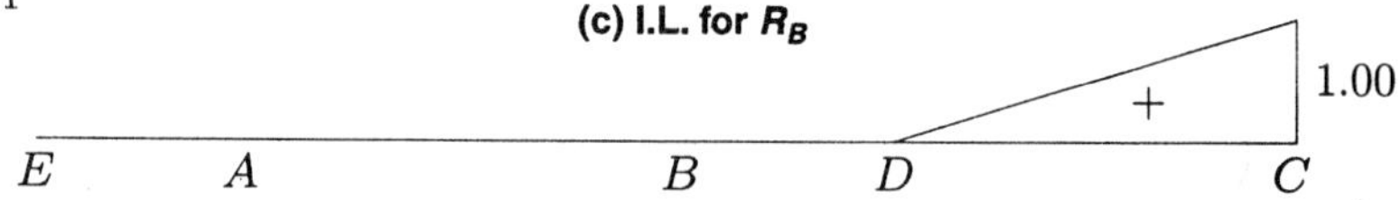

(d) I.L. for R_C

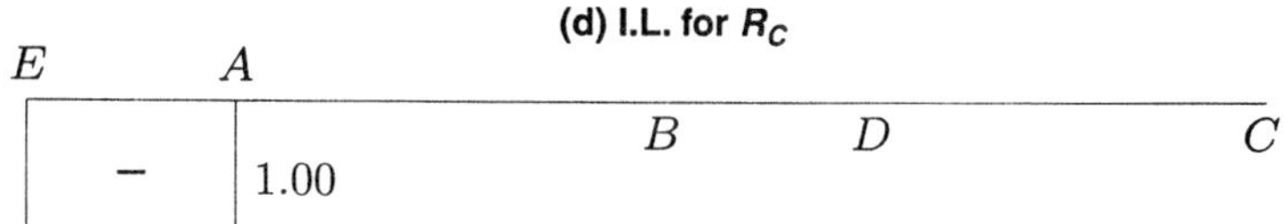

(e) I.L. for SF at *A* [Just on Left of A]

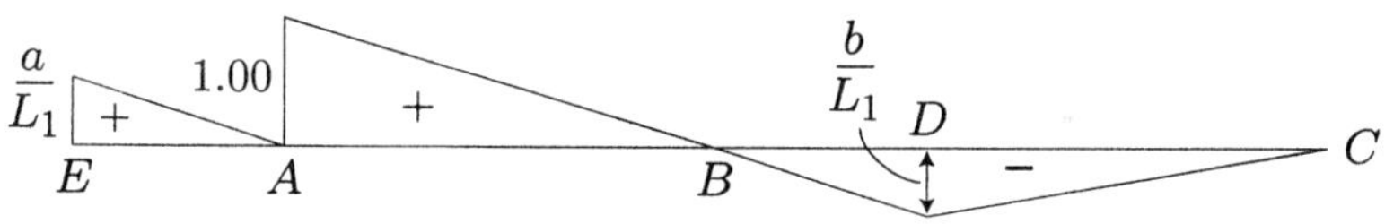

(f) I.L. for SF [Just on Right of A]

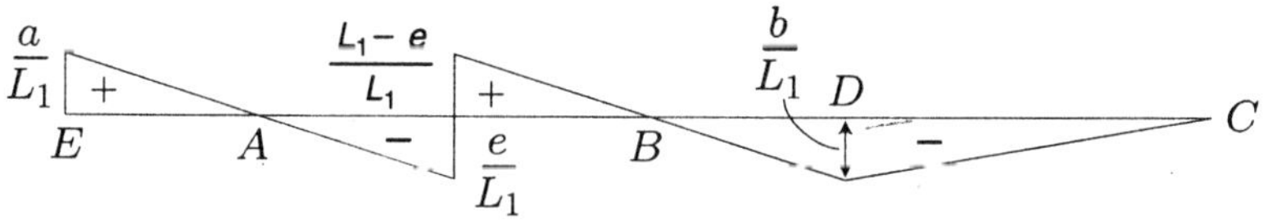

(g) I.L. for SF at *E*

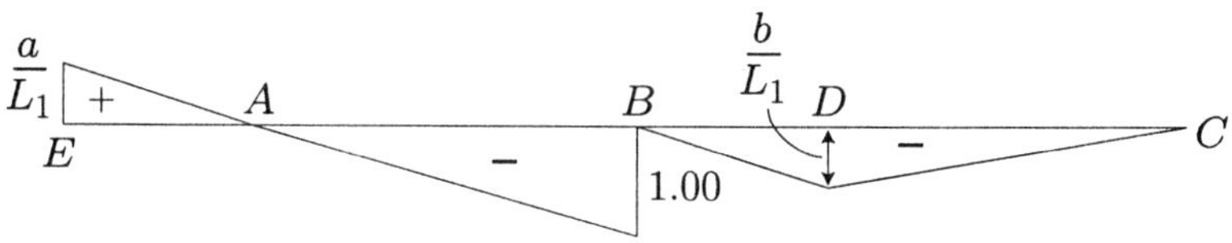

(h) I.L. for S.F. [Just on Left of B]

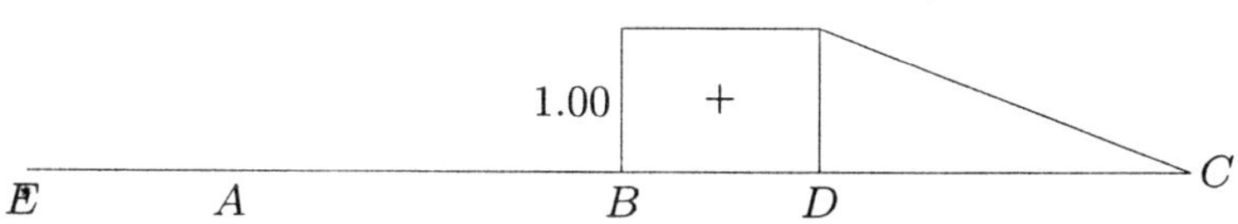

(i) I.L. for S.F. [Just on Right of B]

Fig. 4.9: I.L. for Compound Beam (Contd.)

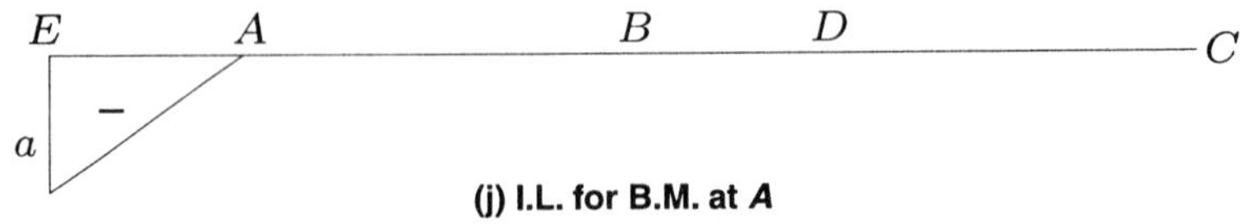

(j) I.L. for B.M. at *A*

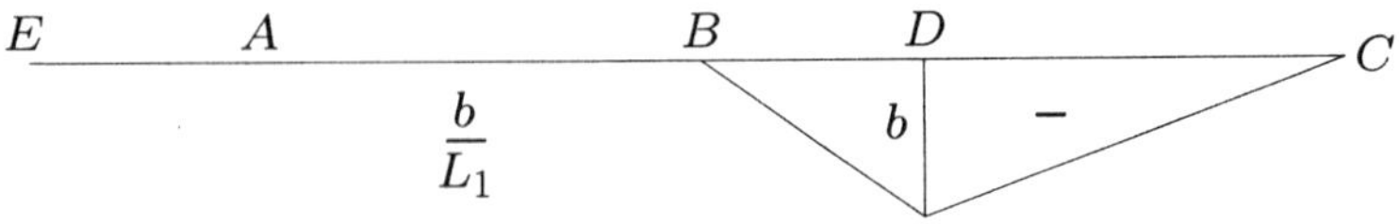

(k) I.L. for B.M. at *B*

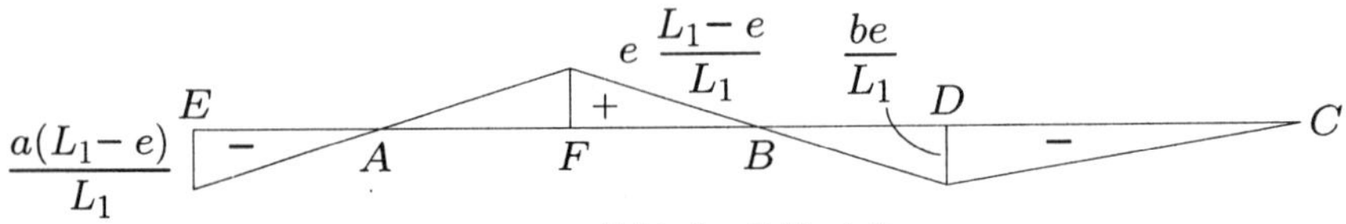

(l) I.L. for B.M. at *F*

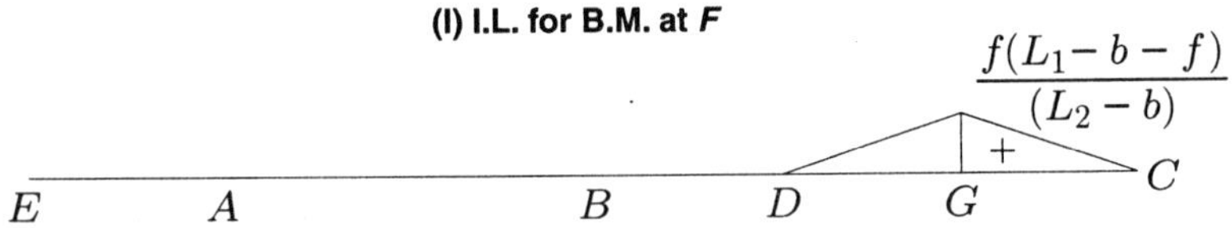

(m) I.L. for BM at *G*

Fig. 4.9: I.L. for Compound Beam

4.8 Three Hinged Arches

For such structures, the influence lines for axial force (Normal force), shear force (Radial shear) and bending moment at a section are influenced by the horizontal reaction H. It is necessary to obtain first the influence line for the horizontal thrust H in the analysis of arch rib.

For the three hinged arch shown in Fig. 4.10(a), when a moving unit load is applied directly to the arch rib as indicated, from the equilibrium of entire structure, $\sum F_X = 0$ gives $H_A = H_B = H$.

From $\sum M_B = 0$,

$$V_A = \frac{x}{L} + \frac{H_c}{L} = \frac{x}{L} + H \tan\alpha.$$

From $\sum M_A = 0$,

$$V_B = \frac{L-x}{L} - \frac{H_c}{L} = \frac{L-x}{L} - H \tan\alpha.$$

Considering the left half of the structure as a free body, $\sum M_C = 0$ gives $V_A(a) - H(h+f) = 0$, i.e.,

$$\left(\frac{x}{L} + \frac{H_c}{L}\right)(a) - H\left(h + \frac{ca}{L}\right) = \frac{ax}{L} - Hh = 0, \quad \text{whence } H = \frac{ax}{hL}.$$

If a fictitious beam having the span L as the arch be subjected to unit load as in the arch, reactions will be

$$R_A = \frac{x}{L} \quad \text{and} \quad R_B = \frac{L-x}{L}.$$

The moment at point B of the beam will be

$$M_B = R_A(a) = \frac{ax}{L}, \quad \text{and} \quad H = \frac{ax}{hL} = \frac{M_B}{h}.$$

Influence line for the horizontal thrust H of the three hinged arch is proportional to the moment at point B and may be obtained by dividing M_B by the factor h. Such influence line is shown in Fig. 4.10(b).

Influence lines for vertical reactions V_A and V_B of the arch, may be obtained from

$$V_A = \frac{x}{L} + H \tan\alpha \quad \text{and} \quad V_B = \frac{L-x}{L} - H \tan\alpha.$$

These influence line diagrams are shown in Figs. 4.10(c) and 4.10(d) and represent the influence line diagram for the corresponding vertical reactions of the fictitious beam in which the influence diagram representing $H \tan\alpha$ has been superimposed.

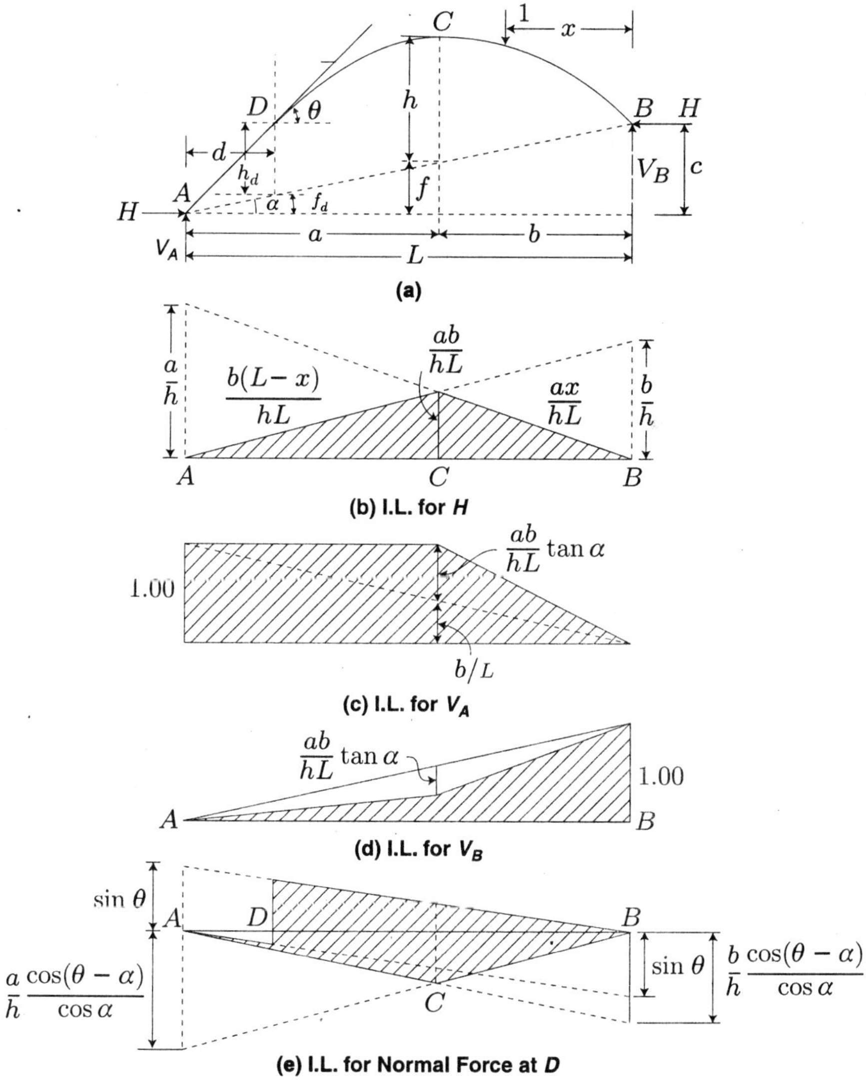

Fig. 4.10: I.L. for Three Hinged Arch (Contd.)

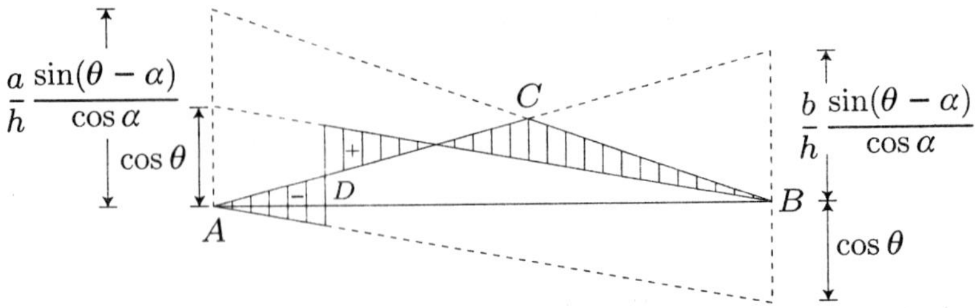

(f) I.L. for Radial Shear at *D*

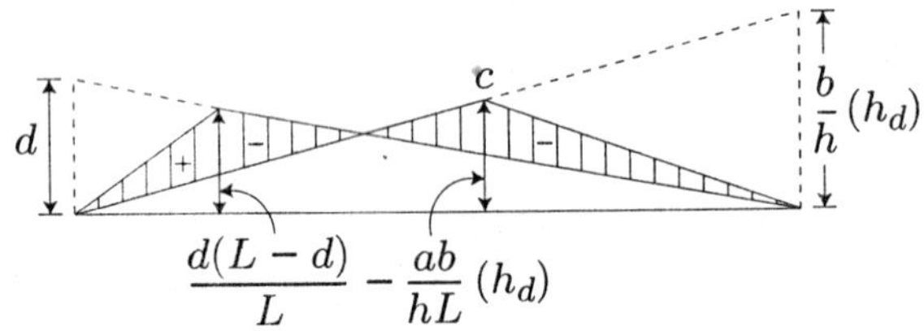

(g) I.L. for B.M. at *D*

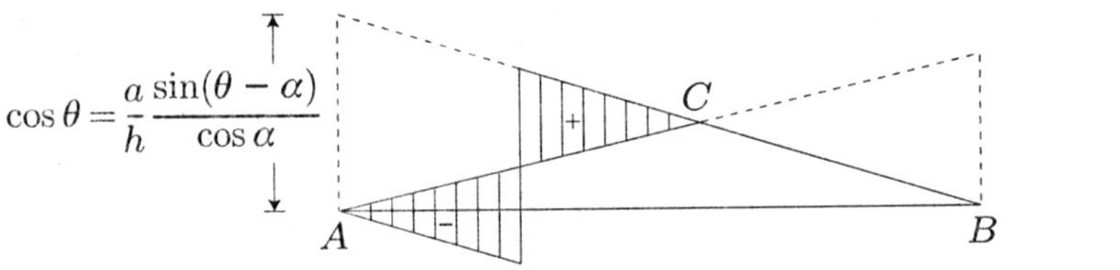

(h) I.L. for Radial Shear at Location where $\cos\theta = \frac{a}{h}\frac{\sin(\theta-\alpha)}{\cos\alpha}$

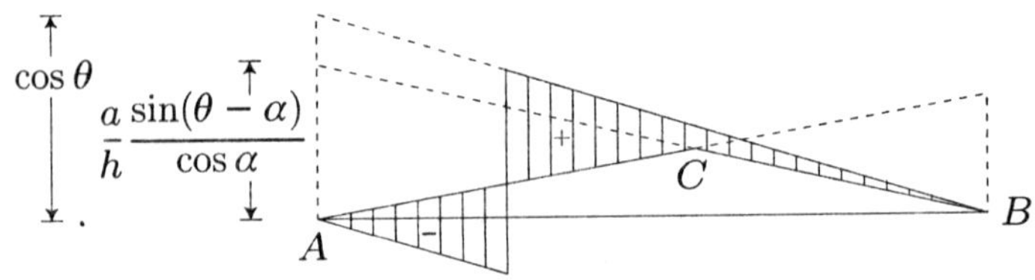

(i) I.L. for Radial Shear at Location where $\cos\theta > \frac{a}{h}\frac{\sin(\theta-\alpha)}{\cos\alpha}$

Fig. 4.10: I.L. for Three Hinged Arch

The magnitude of the normal force at any point D of the arch for unit load on the right of D may be expressed as

$$N_D = V_A\sin\theta + H\cos\theta = (R_A + H\tan\alpha)\sin\theta + H\cos\theta$$

$$= R_A\sin\theta + H\frac{\cos(\theta-\alpha)}{\cos\alpha}.$$

For the unit load placed at any point on the left of D, similarly,

$$N_D = \frac{H\cos(\theta-\alpha)}{\cos\alpha} - R_B\sin\theta.$$

The influence line diagram for N_D as obtained by the combination of the above two values is represented in Fig. 4.10(e).

For S.F. at D, the magnitude for unit load on the right of D may be expressed similarly as

$$\text{S.F. at } D = R_A \cos\theta - H\,\frac{\sin(\theta - \alpha)}{\cos\alpha}.$$

For unit load on the left of point D,

$$\text{S.F. at } D = -R_B \cos\theta - H\frac{\sin(\theta - \alpha)}{\cos\alpha}.$$

The influence line diagrams for S.F. at D thus obtained is represented in Fig. 4.10(f). Figs. 4.10(h) and 4.10(i) illustrate the variations of the influence line diagrams for S.F. for particular position of the sections.

For value of B.M. at D, i.e., M_D when unit load is placed on the right of D,

$$M_D = V_A(d) - H(h_d + f_d) = \left(R_A + \frac{H_C}{L}\right) d - H\left(h_d + \frac{dc}{L}\right) = R_A(d) - H(h_d).$$

For unit load on the left of D, similarly,

$$M_D = R_B(L - d) - H(h_d).$$

Combining these, the influence diagram as obtained is represented in Fig. 4.10(g).

In plotting the influence line diagrams for normal force, radial shear and bending moments at particular section of the arch, the angle θ, which the tangent to the arch axis at the section makes with the horizontal has to be computed from the shape of the arch rib. If desired, the ordinates of the influence line may also be plotted with a horizontal base.

4.9 Influence Line by Virtual Displacement Diagrams (from Deflected Shapes)

Influence lines at a section for any function can be developed on the basis of the principle of virtual displacements. This method of constructing influence lines is particularly useful for complicated structures and is based on the Muller-Breslau principle.

Muller-Breslau Principle. The influence line for any force quantity in linear elastic structure is represented by the deflected shape of the structure resulting from moving the force quantity under consideration through a small displacement (Virtual displacement).

To verify this principle, the simple beam of Fig. 4.11(a) for which influence lines have already been developed in Fig. 4.1 may be considered with a unit load considered being placed at an arbitrary point D of the beam. The vertical constraint at A is removed and replaced by reaction R_A, which is moved through a small virtual displacement by giving a small virtual rotation $\delta\phi$ about B. For small angles of $\delta\phi$, the movement of A along a circular arc can be approximated by the vertical distance δa as in Fig. 4.11(b). If the equilibrium of the nonrigid system is to be maintained, the virtual work δw for all forces acting on the system must be zero. Thus

$$\delta w = R_A(\delta a) - 1(\delta d) = 0 \quad \text{and} \quad R_A = \frac{\delta_d}{\delta_a}(1).$$

If the arbitrary displacement δ_a be made unity, the value of R_A due to unit load at D is given by δ_d of the deflected shape. The deflected shape thus obtained gives the influence line diagram for R_A.

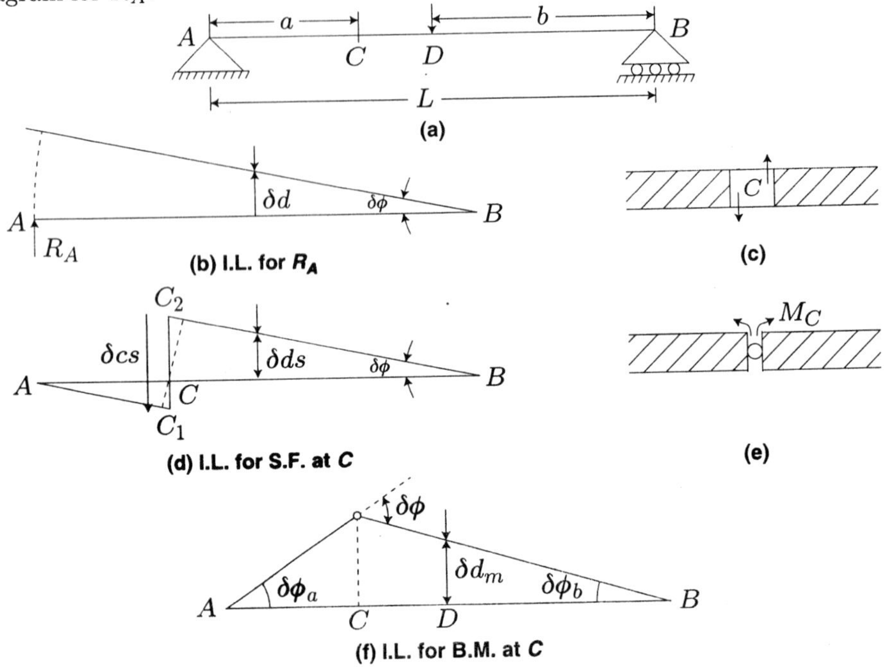

Fig. 4.11: I.L. by Virtual Work

To obtain the influence line diagram for shear force at point C of the beam in Fig. 4.11(a), the shearing constraint at C is removed and replaced by a pinned bar connection subjected to a pair of shear for V_C as shown in Fig. 4.11(c). This may be thought of as introducing an infinitesimal element at point C, so that no shear force can be transmitted through the element. If the element is given a small displacement causing an angular distortion in the element, it will produce rotations in beam segments AC_1 and BC_2. Since both must have the same angle of rotation, AC_1 and BC_2 are parallel as shown in Fig. 4.11(d). For an infinitesimal element, linear displacement $\delta_{CS} = C_1C_2$ may be considered as a vertical line. The virtual work done gives

$$V_C(\delta_{cs}) = 1(\delta_{ds}) \quad \text{and} \quad V_C = \frac{\delta ds}{\delta cs}(1)$$

If the virtual displacement δ_{cs} be made unity, the influence line for shear force at C is represented by the deflected shape, and ordinate $CC_1 = \frac{a}{L}$ and $CC_2 = \frac{L-a}{L}$.

Influence lines for bending moment at C can be obtained by introducing a pinned connection in the structure at the point and replacing the constraint by a moment M_C on each side of the hinge as shown in Fig. 4.11(e). If a small rotation $\delta\phi$ is applied at hinge C, the system will displace as shown in Fig. 4.11(f). The virtual work done by all forces will give

$$M_C(\delta\phi) - l(\delta dm) = 0,$$

whence $M_C = \dfrac{\delta dm}{\delta\phi}(1).$

Assuming a unit value of virtual rotation at C, i.e., $\delta\phi = 1$, the influence line for bending moment is represented directly by the deflected shape. The ordinates of the influence line at C can be obtained from the relations of similar triangles.

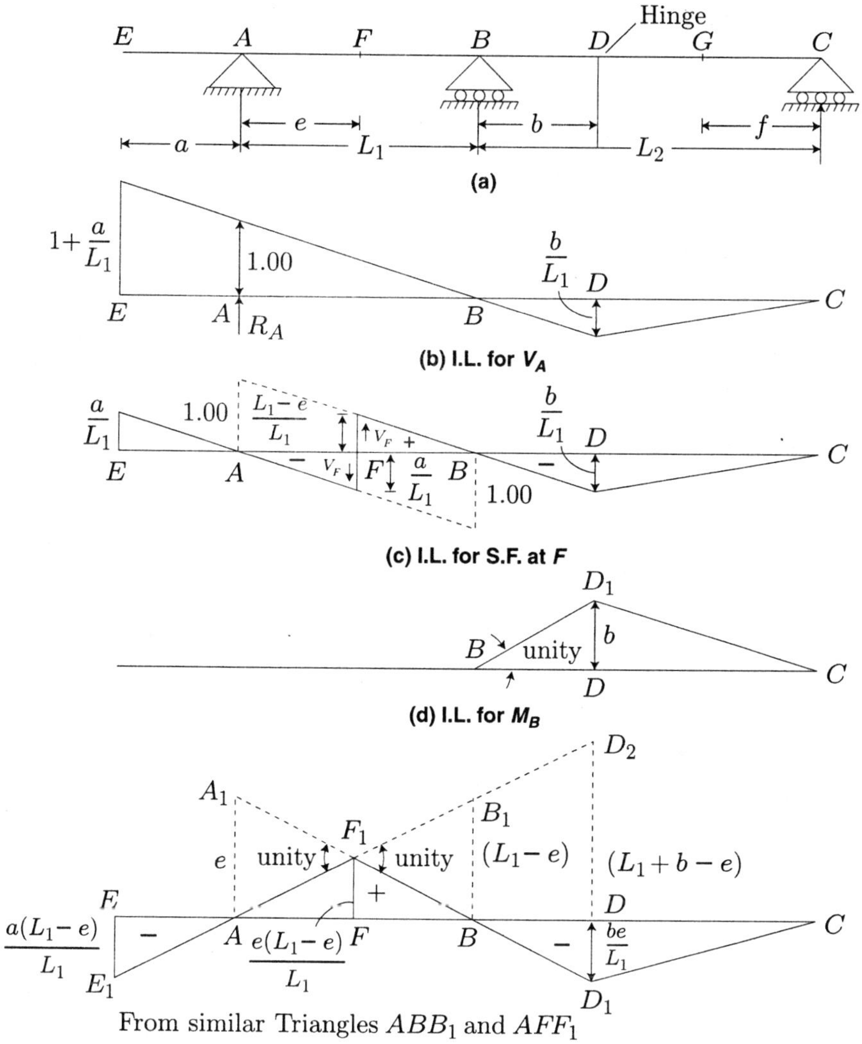

Fig. 4.12: I.L. for Compound Beam with Internal Hinge

For small angle $\delta\phi$,

$$CC' = \frac{a}{L}(L-a)\delta\phi = \frac{a}{L}(L-a), \quad \text{for } \delta\phi = \text{unity}.$$

Use of deflected shapes for determining influence lines becomes even more helpful in complicated structures. It is obvious that for statically determinate structures, the influence line consists only of straight line segments. Once the shape of an influence line is known, the ordinates may readily be determined from geometrical relations.

The use of the method for obtaining influence lines for R_A, S.F. at F, B.M. at B and B.M. at F in case of a compound beam with internal hinge are illustrated in Fig. 4.12.

4.10 Effects of Moving (or Travelling) Loads on Beams

Beams which support structures like bridges, travelling gantries, etc. are subjected to loads which change their positions. Magnitudes of shear force and bending moments also change with the change in position of the moving loads. In designing such beams, it is necessary to find the : magnitudes of the maximum shear force and bending moments at a given section as the set of loads moves over the beam and also the absolute maximum shear force and bending moment in the beam. The absolute maximum values of the functions will occur at particular sections depending on the span of the beam, and the pattern of loads moving over it.

The maximum values of the shear force and bending moment obtained may be plotted with the given span as the base and will indicate not only the absolute maximum values but also value at various sections. Such maximum value diagrams are very essential for design of beams.

4.11 Single Concentrated Load Moving on a Simply Supported Beam

Fig. 4.13(a) shows a beam AB of span L on which a single concentrated load is moving.

The influence line for shear force in Fig. 4.2 shows that the maximum positive S.F. occurs at a section C, when the load is just on the right of the section, value being

$$\text{S.F. (+Ve)} = \frac{W(L-a)}{L}.$$

Similarly, the maximum negative S.F. at C occurs when the load is just on the left of C, and is given by

$$\text{S.F. (−Ve)} = -\frac{Wa}{L}.$$

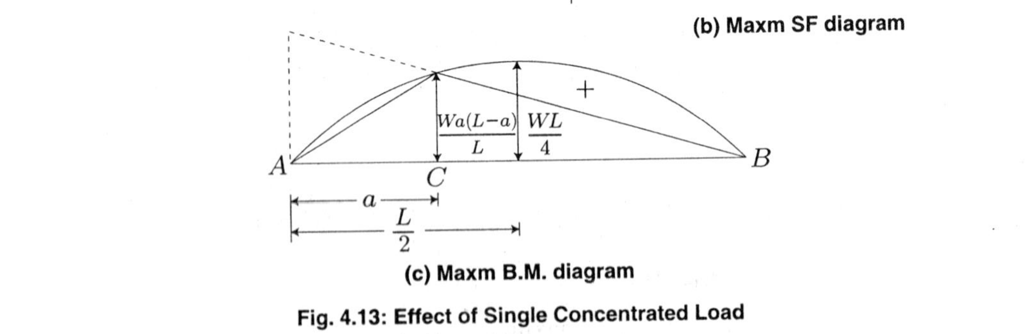

(a) Concentrated Load **W**

(b) Maxm SF diagram

(c) Maxm B.M. diagram

Fig. 4.13: Effect of Single Concentrated Load

Both of these expressions represent straight lines and are true for all values of a, between O and L. The diagram for maximum positive and negative shear forces at all points on the beam is given in Fig. 4.13(b). Two values of S.F. occur at any point of the beam, under different load conditions. At C, the shear force, just when load reaches C from left, is represented by CD. As the load crosses the section C, it changes from CD to CE, which is the maximum positive value at C and occurs when the load acts just to the right of the section.

From influence line for B.M. at a section in Fig. 4.13(c), the maximum bending moment at a section C as a single concentrated load W moves from A to B, occurs when it reaches C,

$$\text{Maxm. B.M. at } C = \frac{Wa(L-a)}{L}.$$

As 'a' can have any value between O to L, the curve for maximum bending moment for any section of the beam is given by the expression, which represents a parabola. Maximum ordinate of this parabola occurs at centre of span, i.e., at $a = \frac{L}{2}$ and is $\frac{WL}{4}$. The maximum bending moment diagram is shown in Fig. 4.13(c).

4.12 Uniformly Distributed Load Travelling on a Simply Supported Beam

Two cases considered are,

(i) When the length of the load is larger than the span of the beam and (ii) when the length of the load is smaller than the span of the beam.

Case I. U.D. Load length larger than the span. (w per unit length)

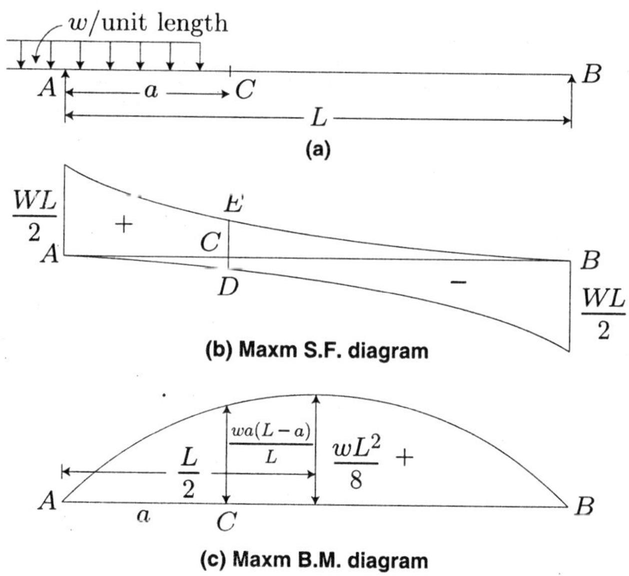

Fig. 4.14: Effect of moving u.d.l. larger than span

From Fig. 4.2(b) it is clear that the maximum negative and positive shear force at C will occur when AC and CB respectively are covered by the load, since, the negative and positive areas are then the greatest possible.

At section C,

$$\text{Maxm. } -\text{Ve S.F.} = -\frac{wa}{2}\frac{a}{L} = -\frac{wa^2}{2L}$$

and $$\text{maxm. } +\text{Ve S.F.} = +\frac{w}{2}(L-a)\frac{(L-a)}{2} = \frac{w}{2L}(L-a)^2.$$

Both these relationships hold for all values of 'a' between 0 and L and are parabolic with maximum ordinate of $-\frac{wL}{2}$ and $+\frac{wL}{2}$ at $a = L$ and $a = 0$ respectively. The maximum shear force diagram is shown in Fig. 4.14(b).

Referring to Fig. 4.3(b), the bending moment at any section C is represented by the area of the influence line diagram on the portion of the span covered by the load. The maximum bending moment will, therefore, occur, when the whole span is covered, as the area of the influence line diagram is then the greatest possible. The value of maximum B.M. at a section is given by

$$\text{maxm. at } C = \frac{wL}{2}\frac{a(L-a)}{L} = \frac{wa}{2}(L-a).$$

This represents the value of the maximum bending moment for any value of 'a' between 0 and L and represents a parabola with a maximum ordinate $\frac{wL^2}{8}$ at the centre as shown in Fig. 4.14(c).

Case II. U.D. Load length smaller than the span (w/unit length)

Fig. 4.15(a) shows a moving load of uniform intensity w/unit length covering a distance c which is less than L.

From the influence line for S.F. at C in Fig. 4.15(b), the maximum negative shear force will occur at any section C, when the load is on the part AC of the beam and the right hand of it is at C.

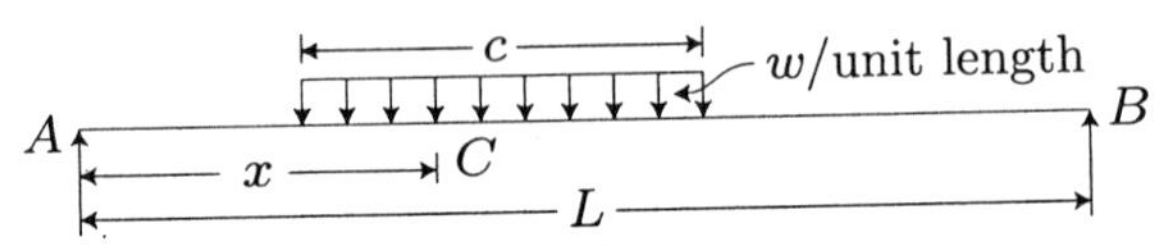

(a)

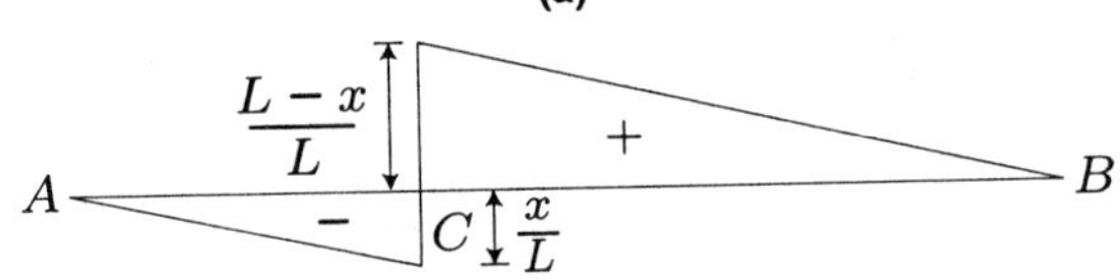

(b) I.L. for S.F. at C

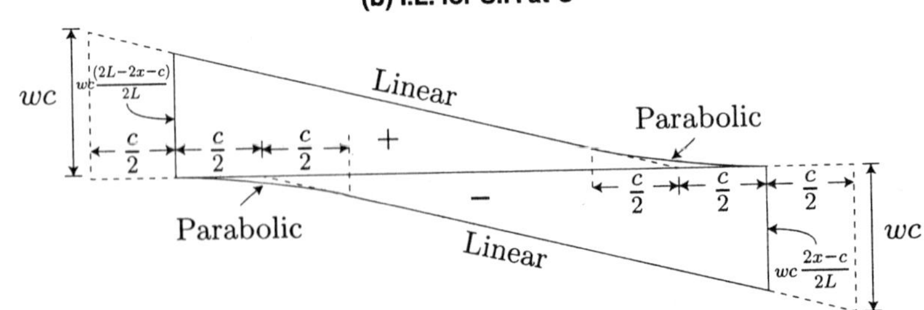

(c) Maxm. S.F. diagram

Fig. 4.15: Effect of moving U.D.L. smaller than span (Contd.)

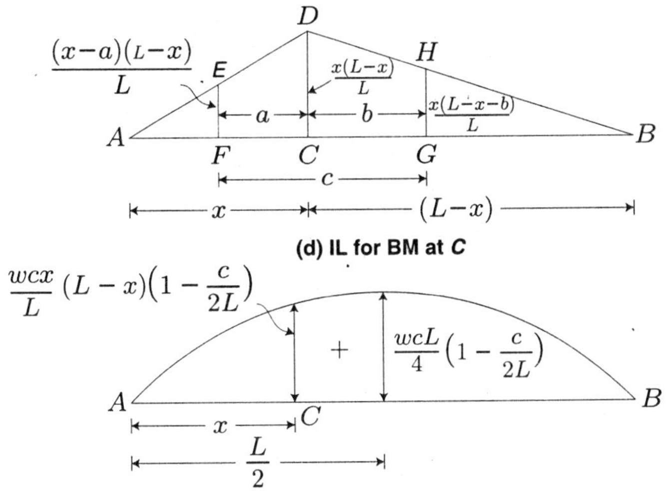

(e) Maxm. B.M. diagram

Fig. 4.15: Effect of moving U.D.L. smaller than span

The ordinate at the mid-point of the load is then

$$\frac{x - \frac{c}{2}}{x}\frac{x}{L} = \frac{2x - c}{2L}$$

and the maximum −Ve S.F. at $C = w(c)\dfrac{2x - c}{2L}$.

The maximum positive shear force at C will occur when the load is on the part CD with the left of the load at C.

Magnitude of the maximum positive S.F. then will be

$$\frac{wc}{2}\left[\frac{(2L - 2x - c)}{L}\right].$$

These represent two straight lines. For negative shear force, zero value occurs at

$$x = \frac{c}{2},$$

but as the equation is obtained on the basis of a full load width c, on span, the curve is a straight line up to $x = c$. For values of x less than c, the curve is parabolic, i.e., as in case of load larger than the span. Maximum negative shear force occurs at $x = L$, its magnitude being $\frac{wc}{2L}(2L - c)$.

Similar variation occurs for the case of maximum positive shear force. The complete maximum shear force diagram for the case is shown in Fig. 4.15(c).

From the influence line for bending moment at section C, shown in Fig. 4.15(d), it is clear that the maximum bending moment at C will occur, when the load covers a length FG, part of which lies to the left and part to the right of C, as in Fig. 4.15(d).

Assuming $CF = a$, $CG = b$, so that $FG = a + b = c$, from Fig. 4.15(d), then mean ordinate between F and C is

$$\frac{(2x - a)(L - x)}{2L}$$

and that between G and C is
$$\frac{[2(L-x)-b]}{2L}x.$$

$$\begin{aligned}\text{Area } DEFGH &= \text{area } DEFC + \text{area } DHGC\\ &= \frac{1}{2L}[(2x-a)(L-x)a + (2L-2x-b)bx]\\ &= \frac{1}{2L}[(2xL-2x^2)(c) - a^2(L-x) - (c-a)^2x] \text{ as } a+b=c,\end{aligned}$$

For this to be maximum, differentiating $-2a(L-x)+2(c-a)x = 0$,
$$\frac{x}{L-x} = \frac{a}{c-a} \quad \text{and} \quad \frac{x}{L} = \frac{a}{c}.$$

This indicates that the maximum bending moment occurs at any section C, when the point C divides the load and the span in the same ratio.

Maximum B.M. at C then is $= w$ (Area) $= \frac{wcx}{L}(L-x)\left(1-\frac{c}{2L}\right)$.

This represents a parabola having a maximum ordinate of $\frac{wcL}{4}\left(1-\frac{c}{2L}\right)$ at the centre. The maximum bending moment diagram is shown in Fig. 4.15(e).

4.13 Two Concentrated Loads on a Simply Supported Beam

Fig. 4.16(a) shows two concentrated loads of magnitude W_1 and W_2 separated by a distance 'a' moving from left to right across a simply supported beam AB of span L. The resultant of the loads is W acting at a distance 'c' from W_1 and 'b' from W_2, so that
$$W = W_1 + W_2 \quad \text{and} \quad a = b + c.$$

From the influence line for shear force in Fig. 4.2(b), it is evident that maximum negative shear force at a section will occur under any one of the following two conditions.

(i) Load W_2 on the section and load W_1 on left-hand portion of the section.

(ii) Load W_1 on the section and load W_2 on right-hand portion of the section.

For a section at a distance $x \le a$ from A, maximum negative S.F. will be due to W_2 at the section, magnitude being $W_2\frac{x}{L}$.

For sections beyond a distance 'a' from support A, from the first case
$$-\text{Ve S.F}_2 \text{ at } C = -\left[W_2\frac{x}{L} + W_1\frac{(x-a)}{L}\right].$$

In the second case when load W_1 is at the section, and W_2 is also on span,
$$\text{S.F}_1 \text{ at } C = -\left[W_1\frac{x}{L} - W_2\frac{(L-x-a)}{L}\right].$$

Maximum negative shear force at the section is the greater of the two values, depending on the magnitudes of W_1, W_2, a, L and x. If S.F$_2$ has to be greater than S.F$_1$, then
$$W_2\frac{x}{L} + W_1\frac{x-a}{L} > W_1\frac{x}{L} - W_2\frac{L-x-a}{L},$$
whence $a < \frac{W_2L}{W_1+W_2}$.

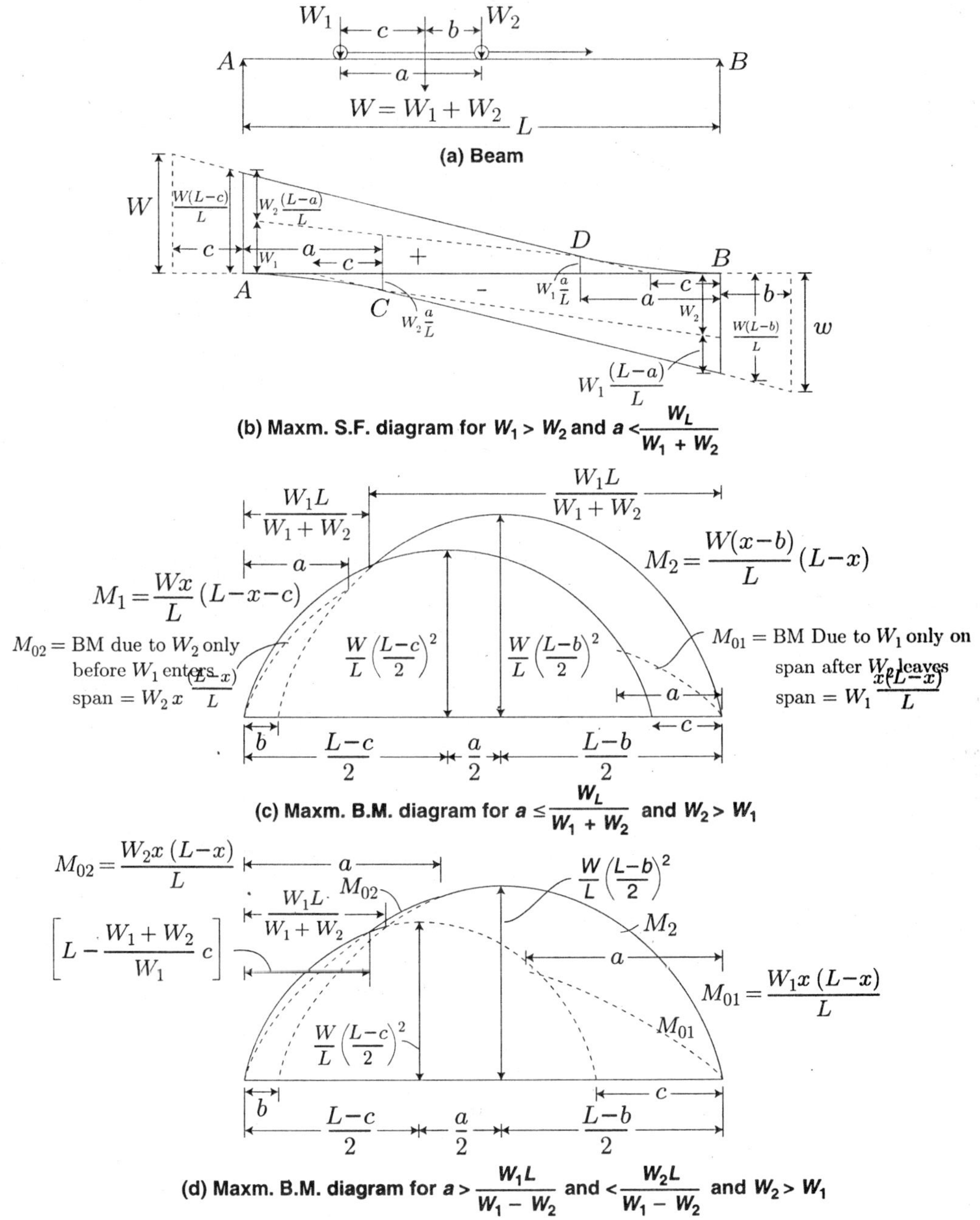

(a) Beam

(b) Maxm. S.F. diagram for $W_1 > W_2$ and $a < \frac{W_L}{W_1 + W_2}$

(c) Maxm. B.M. diagram for $a \le \frac{W_L}{W_1 + W_2}$ and $W_2 > W_1$

(d) Maxm. B.M. diagram for $a > \frac{W_1L}{W_1 - W_2}$ and $< \frac{W_2L}{W_1 - W_2}$ and $W_2 > W_1$

Fig. 4.16: Effect of Two Concentrated Loads

This is normally the case and maximum negative S.F. diagram has been plotted in Fig. 4.16(b).

For $x \le a$, S.F. $= -W_2\frac{x}{L}$.

For $x > a$, S.F. $= -\left[W_2\frac{x}{L} + W_1\frac{x-a}{L}\right]$.

Maximum positive shear force at a section will occur similarly either when load W_2 or W_1 is on the section. Similarly as above, for $x > a$,

$$\text{S.F}_2 \text{ at } C = +\left[W_2\frac{L-x}{L} - W_1\frac{x-a}{L}\right] \text{ with } W_2 \text{ at section}$$

and $$\text{S.F}_1 \text{ at } C = +\left[W_1\frac{L-x}{L} + W_2\frac{L-x-a}{L}\right] \text{ with } W_1 \text{ at section.}$$

Maximum positive S.F. is the greater of the two values and for $x > a$, S.F_1 will be larger than S.F_2, when

$$a < \frac{W_1 L}{W_1 + W_2}.$$

For a section close to A, i.e., for $x < a$, S.F_1 may be larger than S.F_2, even when the criteria is not fulfilled.

For a section close to B, where $(L - x) \leq a$, the maximum positive S.F. will occur when load W_1 is at the section, the magnitude being given by $W_1\frac{L-x}{L}$.

The limits of 'a' obtained above are usually for standard cases and the maximum S.F. diagrams are plotted in Fig. 4.16(b) on the basis of above discussion and for the case $W_2 > W_1$, i.e., when

$$\frac{W_1 L}{W_1 + W_2} < \frac{W_2 L}{W_1 + W_2}$$

From a consideration of the influence line diagram for bending moment in Fig. 4.16(e), it is evident that the maximum bending moment at any section will occur when one of the loads is on the section.

Assuming W_2 at a section, distance x from A, x being larger than 'a', when it is concluded that both the loads are on the span, the bending moment at the section becomes

$$M_2 = W_2\frac{x(L-x)}{L} + W_1\frac{(x-a)(L-x)}{L} = \frac{L-x}{L}[W_2 x + W_1(x-a)] = \frac{W}{L}(x-b)(L-x).$$

Similarly, when W_1 is at the section, such that $(L - x) > a$,

$$M_1 = W_1\frac{(L-x)x}{L} + W_2\frac{x(L-x-a)}{L} = W\frac{x}{L}(L-x-c).$$

The maximum B.M. at a given section is greater of the two values M_2 and M_1 depending on the magnitude of W_1, W_2, a, L and x,

Value of M_2 will be numerically greater than M_1, if

$$(L-x)(x-b) > x(L-x-c), \text{ i.e., if } -bL > -(b+c)x,$$

whence $\dfrac{x}{L} > \dfrac{b}{a}$, i.e., $\dfrac{x}{L} > \dfrac{W_1}{W_1 + W_2}$.

Intersection of the bending moment curves M_1 and M_2, therefore, occurs at

$$x = \frac{W_1 L}{W_1 + W_2}.$$

For $\dfrac{x}{L} < \dfrac{W_1}{W_1 + W_2}$, M_1 will be larger than M_2.

To plot the maximum bending moment diagrams, it is convenient to draw the curves for M_1 and M_2 separately and then use the appropriate one.

Thus $M_2 = \frac{W}{L}(L-x)(x-b)$ represents a parabola having the maximum ordinate at the point, where $L - x = x - b$, i.e. at $x = \frac{L+b}{2}$, from A.

The value of this maximum ordinate is $\frac{W}{L}\left(\frac{L-b}{2}\right)^2$.

Value of M_2 is zero at $x = L$ and at $x = b$.

Again, for $M_1 = \frac{Wx}{L}(L - x - c)$, which represents another parabola, with its maximum ordinate at $x = \frac{L-c}{2}$, the value of the maximum ordinate is given by $\frac{W}{L}\left(\frac{L-c}{2}\right)^2$ at $\frac{L-c}{2}$ from A.

Zero values of M_1 occurs at $x = 0$ and at $x = (L - c)$.

The maximum bending moment curves for the case $W_2 > W_1$ has been plotted on Fig. 4.16(c), in which

$$a < \frac{W_1 L}{W_1 + W_2}.$$

Bending moment diagrams for $x < a$, when only W_2 will be effective, if it is on the section, is denoted by M_{02} and shown by dotted line in the diagram. Similarly, M_{01} representing bending moment curve when W_1 is at the section (W_2 being outside the span) for $x > (L-a)$ is also indicated by dotted line. The maximum bending moment curve thus obtained is shown by full line.

Fig. 4.16(d) shows maximum bending moment diagram for the case where

$$W_2 > W_1 \quad \text{and} \quad a > \frac{W_1 L}{W_1 + W_2}, \quad \text{but} \quad < \frac{W_2 L}{W_1 + W_2}.$$

In such a case, part of the maximum bending moment curve will consist of curve for M_{02} also as shown. The M_{02} curve will intersect with M_2 curve at $x = a$ and with M_1 curve at

$$x = \left[L - \frac{W_1 + W_2}{W_1}(c)\right], \text{ as shown.}$$

The maximum positive shear force diagram for the case illustrated in Fig. 4.16(b) has to be modified as 'a' being larger than $\frac{W_1 L}{W_1+W_2}$. S.F$_1$ due to W_1 at section having a distance $x > a$ will no longer be larger than S.F$_2$ with W_2 at the section. In such a case, the equations for maximum positive shear force for a section at a distance $x > a$ has to be obtained with W_2 on the section. For a section $x < a$, equations for S.F. with W_1 and W_2 at section have to be investigated.

When the spacing between the wheel loads W_1 and W_2 is large and is larger than the bending moment, M_{02} will give the maximum value of bending moment in place of M_2.

4.14 Loads Moving in Reversed Order

For design of beams or girders, subjected to moving loads crossing a bridge in either direction, it is necessary to obtain the maximum shear force and bending moment diagrams by considering movement of loads from either direction. In case of simply supported beams, however, such maximum diagrams may be constructed by considering movement in one direction only.

In the case of the pair of loads W_1 and W_2 moving over the span, the maximum shear force diagram considering movement in reversible order may be obtained from Fig. 4.16(b) by taking the greater of the two ordinates positive or negative at equal distances on either side of the centre or from ends and using this in both positive and negative shear force diagrams. Such a maximum shear force diagram is shown in Fig. 4.17(b).

The maximum bending moment diagram for the case may be constructed by setting on either side of the centre of the span ordinates at the same distance from the centre in Fig. 4.17(c). Such a maximum bending moment diagram is illustrated in Fig. 4.17(c).

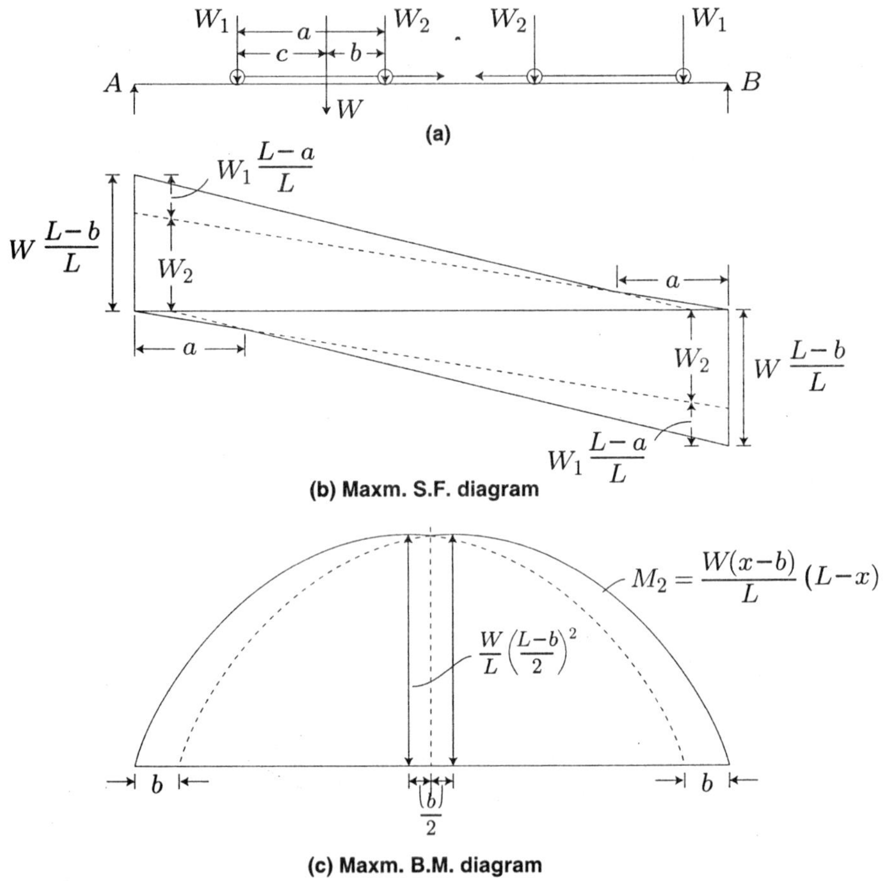

Fig. 4.17: Reversal of Loads

4.15 Series of Concentrated Moving (Travelling) Loads

The application of influence lines for determining the maximum effects of moving concentrated loads on beams can be illustrated by the development criteria of maximum reactions, shears and moments at specified locations of a simply supported beam. The criteria for the maximum of each of these effects can be established by considering the change in each case caused by the change in position of loads.

(a) Reactions

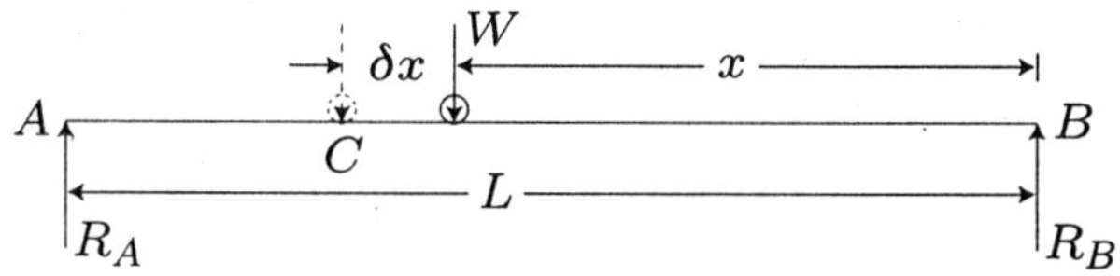

Fig. 4.18: Moving Concentrated Load

As a single concentrated load W moves across the span of a simple beam shown in Fig. 4.18 at a distance of x from support B, the magnitude of the reaction at A is $\frac{Wx}{L}$. If the load W moves to the left by a distance δx, the magnitude of the reaction at support A will be $\frac{W(x+\delta x)}{L}$, thus indicating a change of $W\frac{\delta x}{L}$ in the magnitude of the reaction at support A. When one of the series of concentrated loads on the beam shown in Fig. 4.19 travels to the left by the distance δx, each of the loads moves to the left by the same amount δx. If $\sum W$ denotes the sum of the loads on the beam, the change in reaction at support A will be $\sum W\frac{\delta x}{L}$, if no wheel on the span has left the span and no other wheel has entered the span.

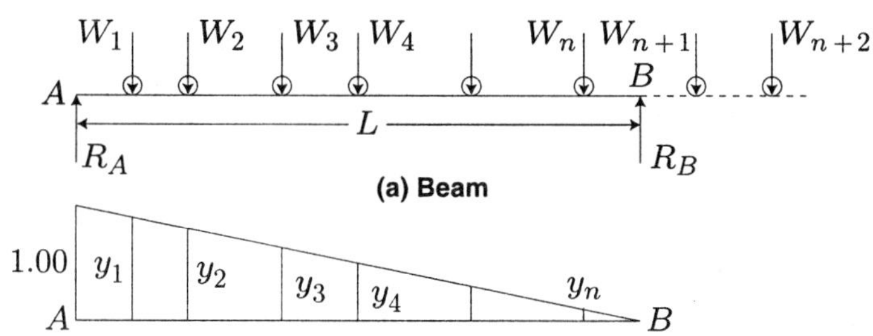

(a) Beam

(b) I.L. for R_A

Fig. 4.19: Series of moving Concentrated Loads

Fig. 4.20 shows a series of concentrated loads moving over the span AB from right to left. From the influence line diagram for reaction at A in Fig. 4.20(b) the change in reaction A as a result of the movement of loads to the left represents an increase until wheel W_1 reaches support A. When wheel load W_1 leaves the span, there will be a sudden decrease

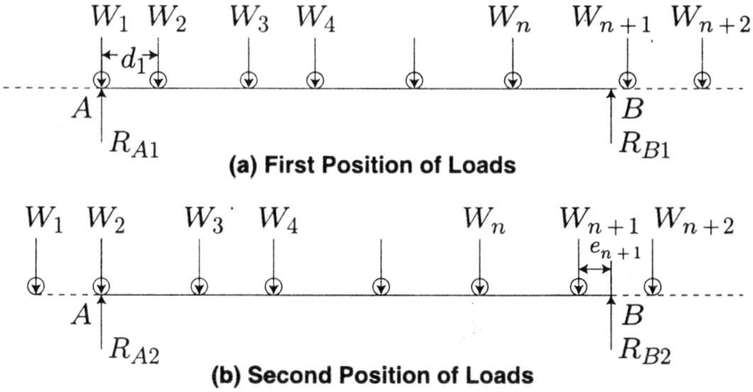

(a) First Position of Loads

(b) Second Position of Loads

Fig. 4.20: Criteria for Maximum Reactions

in the reaction at A. However, if the movement of the loads continues, the reaction at A will again increase gradually until wheel load W_2 reaches the support A. The same cycle will be repeated as each wheel load approaches support A. In course of each movement, it is possible that a wheel load may enter the span from the right. Although, any wheel load entering the span from the right will cause an increase in the reaction at A, the influence of the new load is relatively small, since the entering wheel is close to support B.

Since the reaction at A increases each time a wheel load approaches support A, and decreases suddenly as the load passes support A and leaves the span, it is necessary to compare only the values of the reaction when the wheel loads successively move over the support A. A general expression may be written for the change of the reaction at A by considering the loads moving off and on the span as well as those staying on the span during the movement. In this case, considering the movement of the wheel loads from the position in Fig. 4.20(a) to 4.20(b), if $\sum W$ be the sum of all the wheel loads that stay on the span during the movement of wheel loads by d_1, which is the distance between wheel loads W_1 and W_2, if W_{n+1} be a wheel that enters the span during the movement by a distance e_{n+1}, then the change of the reaction at support A due to movement is

$$R_{A2} - R_{A1} = \Delta R_A = \sum W \frac{d_1}{L} + W_{n+1} \frac{e_{n+1}}{L} - W_1. \tag{4.1}$$

A positive value of ΔR_A indicates that the movement causes an increase in the value of the reaction R_A, whereas negative value indicates a decrease. Values of ΔR_A between various positions are calculated and from the changes, position of loads for maximum value of R_A may be obtained.

(b) Shear force

When a single concentrated load W moves across the span of a simple beam as shown in Fig. 4.18, the shear force at a section passing through point C of the beam can be determined from the free body to the left of the section. As long as the load W is on segment BC of the beam, the shear force at C is positive and equal to the reaction at A. If the load moves to the left by a small distance δx within segment BC, the change in shear force at C and in reaction at A will be the same. As soon as the load moves to segment CA of the beam, the shear force at C is equal to the reaction at A minus the load. In that case, the change in shear force at C due to further movement of the load will be the same as the change in reaction at A minus the load.

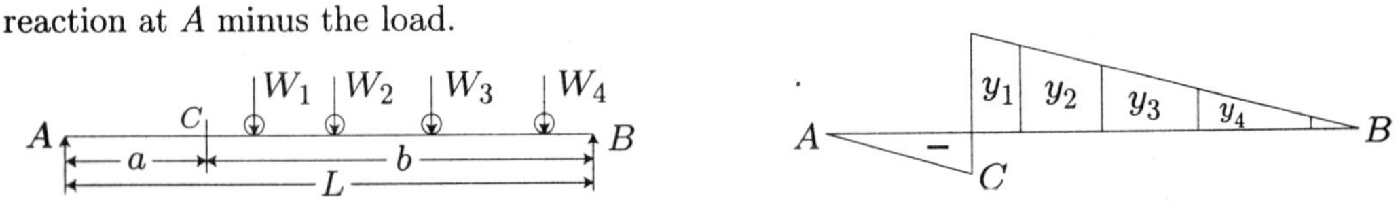

(a) Beam with Moving Load **(b) I.L. for S.F. at *C***

Fig. 4.21: Series of Concentrated Load

For a series of concentrated load moving onto the beam from the right as in Fig. 4.21(a), from the shape of the influence line for shear force at a section C, as in Fig. 4.21(b), it is clear that the shear force at C will increase when the loads move to the left until wheel load

W_1 reaches point C.. Any further movement will make wheel load W_1 pass to the left of point C and will cause a sudden decrease in the shear force at C. However, if the movement of the loads continues, the shear force at C will again increase gradually until the wheel load W_2 reaches point C. During each movement, it may be that a new wheel load will enter the span from the right and thus increase the shear force at C. A wheel may also leave the span from the left during the same movement and the removal of such a load from the span will reduce the negative shear force and, thus, increase the positive shear force.

As the shear force at C increases each time a wheel approaches point C and decreases suddenly as the load passes point C, it is necessary to compare only the values of the shear force at C when the wheel loads move successively over point C. The change in shear force resulting from the movement, as one wheel load succeeds another at the section, is equal to the change in the left reaction minus the wheel load that has just passed the section provided that no wheel has left the span during the movement. If a load that was to the left of the section leaves the span during the movement, the negative shear force that is caused at the section before leaving should be added to the change in shear force. A general expression for the change in shear force at C may be written, by considering the movement of all loads that are involved. Considering the movement of the loads from the position in Figs. 4.22(a) to 4.22(b), the change in shear force at C, due to the movement is given by,

$$\Delta V_c = \sum W \frac{d_i}{L} - W_i + W_{i-2} \frac{e_{i-2}}{L} + W_n \frac{e_n}{L}, \tag{4.2}$$

where $\sum W$ = sum of all the wheel loads that stay on the span during the movement, W_i = wheel load which has just passed the section, d_i = distance between wheel load W_i and the one immediately following, W_n = wheel load entering the span, e_n = distance by which the wheel load W_n travels on the span after entering, W_{i-2} = wheel load that has just left the span, and e_{i-2} = distance by which W_{i-2} travelled on the span before moving off.

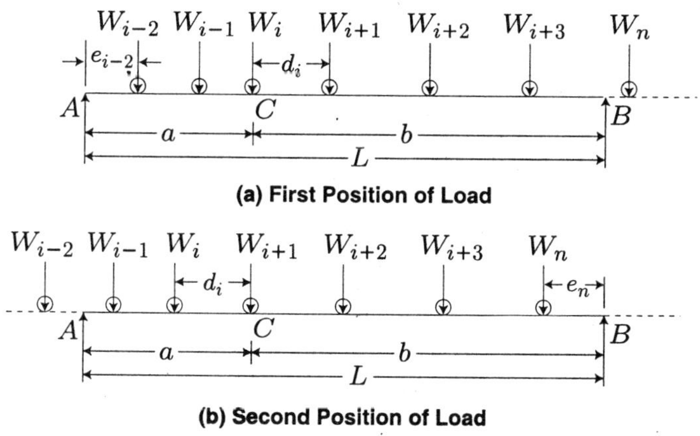

Fig. 4.22: Criteria for Maxm. S.F.

A positive value of ΔV_c indicates an increase in the shear force. and a negative value indicates a decrease. Values of ΔV_c between various positions of wheel loads are calculated and from the changes, position of load for maximum value of positive shear force at C can be obtained.

For calculation of maximum negative shear force at C, positions of load giving this may be similarly obtained considering wheel loads moving from left to right.

If a series of concentrated loads represented by $W_1, W_2, \cdots W_{i-1}, W_i, W_{i+1}, \cdots, W_n$ on a span AB the distance between them being $d_1, d_2, \cdots, d_i, \cdots, d_{n-1}$, moving from right to left, if the wheel load W_1 reaches the section C, the wheel loads W_i, $W_{i+1}, \cdots, W_n$ contribute positive shear forces and the loads $W_{i-1}, \cdots, W_1$ contribute negative shear forces to the shear force at C, distance x from end A of a beam AB of length L.

$$\text{Then, } V_{ci} = \frac{1}{L}\Bigg[(L-x)\sum_1^n W_i - \{W_{i+1}d_i + W_{i+2}(d_i + d_{i+1}) + \cdots + W_n(d_i + d_{i+1} + \cdots + d_{n-1})\} - \{W_{i-1}(L - d_{i-1}) + \cdots + W_1(L - d_{i-1} - \cdots d_1)\}\Bigg]$$

If the series of loads move so that W_{i+1} is at section C, the shear force at C then is given by,

$$Vc_{i+1} = \frac{1}{L}\Bigg[(L-x)\sum_1^n W_i - \{W_{i+2}d_{i+1} + \cdots + W_n(d_{i+1} + \cdots + \cdots d_{n-1})\} - \{W_i(L - d_i) + \cdots + W_1(L - d_i - d_{i-1} - \cdots - d_i)\}\Bigg].$$

Vc_i will be greater than Vc_{i+1}, if

$$Vc_i - Vc_{i+1} > 0, \text{ whence } Vc_i > Vc_{i+1}.$$

From the above expressions, this is fulfilled, if

$$W_i L - d_i \sum_1^n W_i > 0.$$

$$\text{i.e., if } \frac{W_i}{d_i} > \frac{\sum_1^n W_i}{L}.$$

In this analysis the wheel load W_i is supported to have just reached section C and contribute a positive shear force to the total at section C.

When considering positive shear force, the load system is to be taken as moving from right to left and when considering negative shear force, the load system is to move from left to right.

Thus, if $\frac{W_i}{d_i}$ or $\frac{W_i}{d_{i+1}}$ be represented by w_i and if w be the intensity of loading on the whole span, i.e., $\frac{\sum_1^n W_i}{L}$, then if $w_i > w$, the shear force, when the first wheel W_i is on the section, is larger than the shear force when the next wheel load W_{i+1} is on the section. If $w_i = w$, it signifies that the shear force at the section will be the same whether the leading or the following wheel load is on the section.

If the section of the beam under consideration is so close to the support that some of the loads are not on the span, w must be calculated from the total load on the span.

(c) Maximum Bending Moment at a section

When a single concentrated load W moves over the span of a simply supported beam as in Fig. 4.23(a), the change in bending moment at a point C can be studied from the influence line diagram, for bending moment at that point. Referring to Fig. 4.23(b), the change in bending moment when a load W in segment AC moves by a small distance Δx, so that the load remains on the segment, is given by

$$\Delta M_c = W y_d' - W y_d = W(\Delta y_d) = W \frac{y_c}{a}(\Delta x).$$

which indicates an increase.

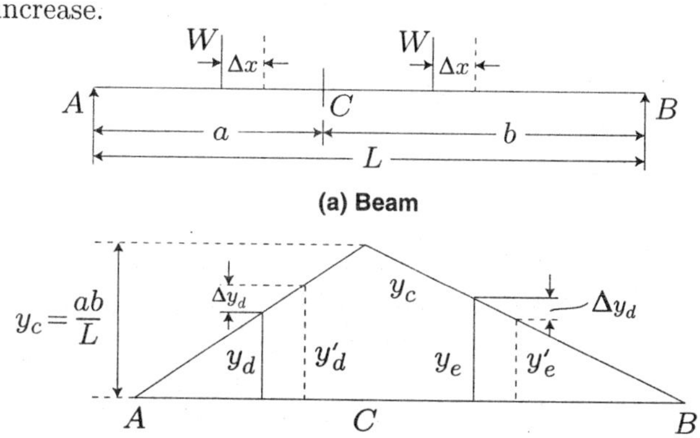

(a) Beam

(b) I.L. for B.M.

Fig. 4.23: Moving Load

Similarly, if the load W is in segment BC, for the same change in position in the same direction.

$$\Delta M_c = W y_e' - W y_e = -W(\Delta y_e) = -W \frac{y_e}{b}(\Delta x),$$

which indicates a decrease.

As series of concentrated loads moves across the span of a simply supported beam as in Fig. 4.24(a), all loads on the span will cause moment at C. If the loads on the span move towards the right by a small distance Δx, so that loads on the segments AC and BC remain on the segment, there will be an increase in the bending moment at C, due to the change in position of the wheel loads in segment AC, and a decrease in the bending moment at C, due to a similar change in position of the wheel loads in segment BC, the total change being,

$$M_c = W_1 \frac{y_c}{a}(\Delta x) - W_2 \frac{y_c}{b}(\Delta x),$$

where W_1 and W_2 are the resultant of the loads on the left and on the right of the section.

For the bending moment M_c to be maximum for a particular position, such a change in bending moment must become equal to zero, i.e., for a maximum,

$$\Delta M_c = 0, \quad \text{and} \quad W_1\left(\frac{y_c}{a}\right)(\Delta x) = W_2\left(\frac{y_c}{b}\right)(\Delta x)$$

i.e., $\dfrac{W_1}{a} = \dfrac{W_2}{b}$.

This expression means that for a maximum bending moment at any section C, the average load to the left of the section must equal the average load on the right.

From $\frac{W_1}{a} = \frac{W_2}{b}, \frac{W_1}{a} = \frac{W_1 + W_2}{a+b} = \frac{\sum W}{L} = \frac{W_2}{b},$ (4.3)

whence $\sum W$ = total load on the span and L = length of the span.

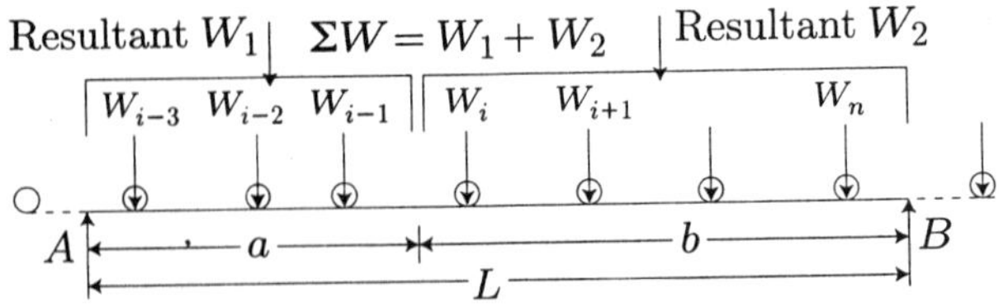

(a) Beam

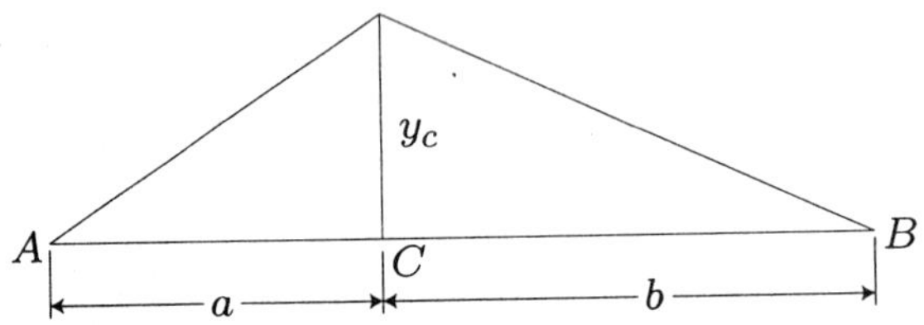

(b) I.L. for B.M.

Fig. 4.24: Criteria for Maxm. B.M.

The bending moment is maximum when the average load on the left and right of the section is equal to the average load on the span.

Maximum bending moment at a section due to a series of concentrated loads occurs when a concentrated load is on the section. The criteria should be that a particular load will satisfy the condition, if the average load on the span (or on the right of the section) is greater than the average load on the left of the section with this load at an infinitesimal distance to the right of the section and less than the average load on the left of the section with the same load at an infinitesimal distance to the left of the section.

Example 4.2 *A simply supported beam of clear span* 30 m *is subjected to series of concentrated loads as shown* in Fig. 4.25. *Determine position of loads for maximum values of (a) reaction at A, (b) shear force at* 10 m *from left-hand support and (c) bending moment at* 10 m *from the left-hand support.*

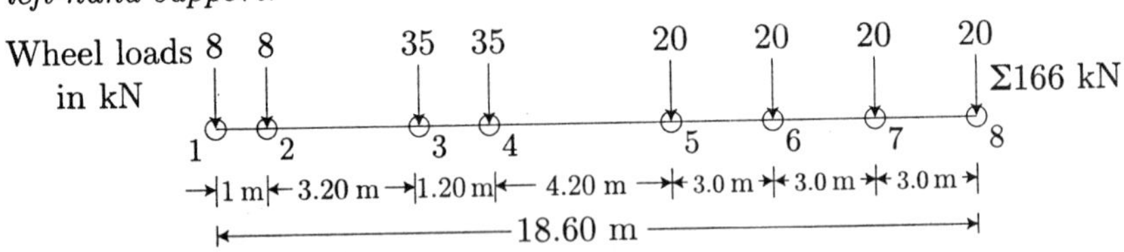

Fig. 4.25: For Example 4.2

Solution : *(a) For Maximum R_A:*

Consider loads moving from right to left on spend AB (30^m length)

Comparing position 1 when load 1 is on A and position 2 when load 2 moves over A, due to movement of all wheel loads by 1 m, change in reaction, i.e.,

$$\Delta R_A = \frac{158}{30}(1) - 8 = -2.73, \text{ i.e., a decrease.}$$

Considering position 3, when wheel load 3 moves over A, change in reaction between position 2 and 3 of wheel loads, i.e.,

$$\Delta R_A = \frac{150}{30}(3.20) - 8 = +8, \text{ i.e., an increase.}$$

Change in reaction between positions 1 and 3 = 8.00 − 2.73 = +5.27, an increase.

Change in reaction between position 3 and position 4, when wheel load 4 moves over A,

$$\Delta R_A = \frac{115}{30}(1.20) - 35 = -30.40, \text{ i.e., a decrease.}$$

Hence the maximum value of R_A is obtained when wheel load 3 is over A.

(b) Shear Force at section C which is 10 m *from left-hand support.*

Maximum shear force at the section will be positive.

Position 1: Wheel load 1 at C.

Position 2: Wheel load 2 at C.

Change in shear force between positions 1 and 2, i.e.,

$$\Delta V_A = \frac{166}{30}(1) - 8 = -2.47, \text{ a decrease.}$$

Position 3: Wheel load 3 at C.

Change in shear force between positions 2 and 3, i.e.,

$$\Delta V_A = \frac{166}{30}(3.20) - 8 = +9.71, \text{ an increase.}$$

Between positions 1 and 3,

$$\Delta V_A = +9.71 - 2.47 = +7.24, \text{ an increase.}$$

Position 4: Wheel load 4 at C.

Between position 3 and position 4,

$$\Delta V_A = \frac{166}{30}(1.20) - 35 = -28.36, \text{ a decrease.}$$

Maximum positive shear force at C occurs when wheel load 3 is at C.

Alternative Method

Average load on span, i.e.,

$$w = \frac{166}{30} = 5.53$$

$$w_1 = \frac{8}{1} = 8 > w$$

$$w_2 = \frac{8}{3.20} = 2.50 < w$$

$$w_3 = \frac{35}{1.20} = 29.17 > w$$

$$w_4 = \frac{35}{4.20} = 8.33 > w$$

$$w_5 = \frac{20}{3} = 6.67 > w$$

$$w_6 = w_7 = 6.67 > w.$$

The maximum positive shear force at C occurs when either wheel load 1 or 3 is at section C. From calculation of the shear forces placing the loads 1 and 3 separately at section C, it will be found that shear force is maximum when wheel load 3 is at the section.

(c) Maximum bending moment at section C, 10 m *from left-hand support.*

Average load on full span $= \dfrac{166}{30} = 5.53.$

Total load on the left-hand side of section C, has to be more than 5.53(10), i.e., 55.30, when wheel load at section considered on the left-hand side.

Wheel load 4 has to be at C.

With wheel load 4 at section C,

considered on left, $\dfrac{86}{10} > 5.53,$

considered on right, $\dfrac{115}{20} > 5.53.$

Wheel load 4 fulfills the criteria.

With other loads at section, the criteria is not fulfilled.

4.16 Graphical Method for Position of Loads for Maximum Bending Moment at a Section

Considering that for the case the concentrated loads on the span will be W_1, W_2, W_3, W_4, W_5 and W_6, BE is plotted at right angles to span AB drawn on a suitable scale to represent the length L as shown in Fig. 4.26. On length BE, wheel loads $W_1, W_2, \cdots, W_6$ are plotted on any suitable scale. Points A and E are joined and CF is drawn at right angles to AB at C, so as to intersect line AE at F. From F, FG is plotted perpendicular to BE. Then the load intercepted by FG has to be placed at C to get the maximum bending moment at C.

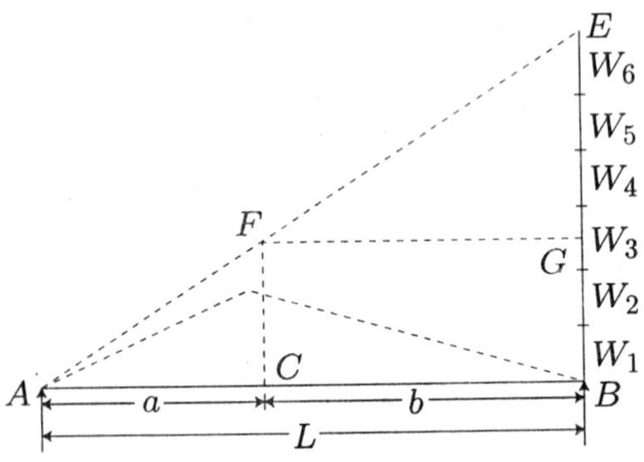

Fig. 4.26: Graphical Construction for Maxm. B.M.

Proof: Considering similar triangles AFC and EFG,

$$\frac{EG}{FG} = \frac{FC}{AC} = \frac{BG}{AC},$$

whence $\dfrac{EG}{b} = \dfrac{BG}{a}$ and $\dfrac{W_6 + W_5 + W_4 + nW_3}{b} = \dfrac{W_1 + W_2 + (n-1)W_3}{a},$

which fulfills the criteria for maximum bending moment at C.

4.17 Absolute Maximum Shear Force and Bending Moment in Beams

To determine the maximum shear force and bending moment at particular sections of a beam, it is often necessary to determine the magnitude and location of the absolute maximum shear force and bending moment. For the purpose of illustration, case of concentrated loads moving over a simply supported beam will be considered.

(a) Absolute maximum shear force

The absolute maximum shear force in a simply supported beam subjected to moving concentrated loads requires little discussion. It will occur at a point next to one of the reactions and, therefore, nearly entails positioning the loads such that the maximum value of the reaction is obtained.

(b) Absolute maximum bending moment

Analytical Method. In designing a beam or a girder, the designer is interested in the greatest possible B.M. that can occur anywhere in the beam. The maximum moment at the centre is generally not the absolute maximum B.M. and in a short beam the difference is appreciable.

Maximum B.M. under a given wheel. If W be the sum of all the loads on the span and x the distance from the centre of gravity of these loads to the right-hand support B, P be of one of the loads on the beam at a distance 'a' from the C.G. of all the loads and W_1 be the sum of all the loads to the left of P and b the distance from the C.G. of these loads to P as shown in Fig. 4.27, it is required to determine the position of the loads which will cause maximum moment under given wheel P.

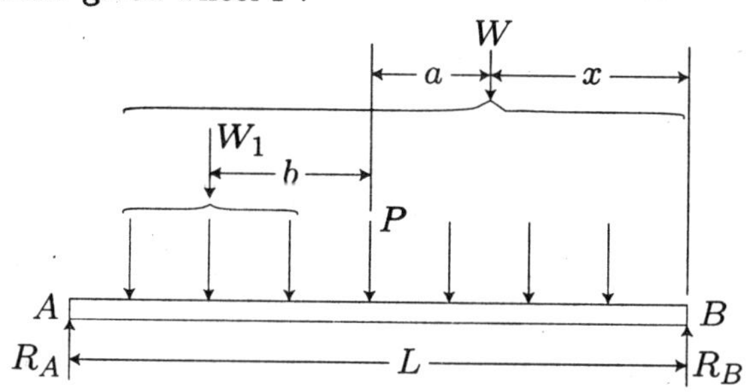

Fig. 4.27: Maxm. B.M. under a given load

The general expression for B.M. under the load P is :

$$M = R_A(L - a - x) - W_1 b, \quad \text{where } R_A = \frac{W \cdot x}{L}.$$

and $$M = \frac{W \cdot x}{L}(L - a - x) - W_1 b = \frac{W}{L}(xl - ax - x^2) - W_1 b.$$

Differentiating this with respect to x and equating to zero.

$$\frac{dM}{dx} = \frac{W}{L}(L - a - 2x) = 0,$$

whence $x = \dfrac{L}{2} - \dfrac{a}{2}.$

Thus the maximum moment under any load will occur when that load and the centre of gravity of all the loads on the span are equidistant from the centre of the beam.

The maximum possible B.M. will usually occur under the load nearest to the centre of gravity of all the loads. This may not be correct if this load is much smaller than the other loads which are near the centre of gravity.

If the criterion developed for maximum B.M. at a section be used in conjunction with the above criterion, it generally will not be necessary to try more than this load and the adjacent one.

It should be kept in mind while selecting the wheel which will cause absolute maximum B.M., that after the wheel is placed in accordance with the above criterion it must also satisfy the criterion for maximum moment at a point *(i.e., the average load to the left should equal the average load on the span)*. If it does not, it can be dropped and adjacent wheel can be tried.

Example 4.3 *Determine the absolute maximum bending moment in a simply supported beam of span* 10 m, *if the system of loads in* Fig. 4.28(a) *rolls over the span.*

Solution : Distance between resultant $\sum W$ and the 2nd 50 kN load = 0.54 m. Maxm. B.M. = 260.95 kNm. [as per Load Position in Fig. 4.28(b)].

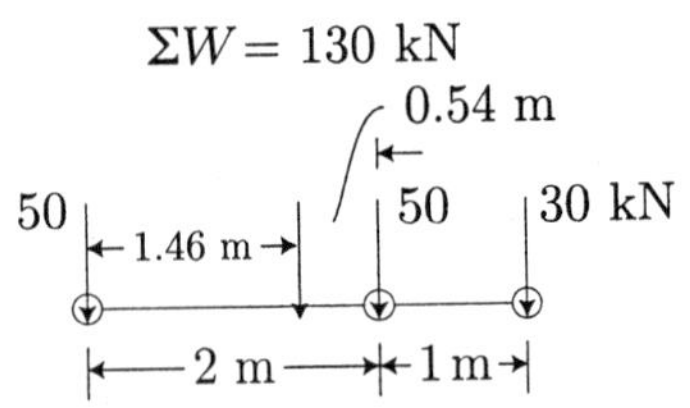

(a) Loads

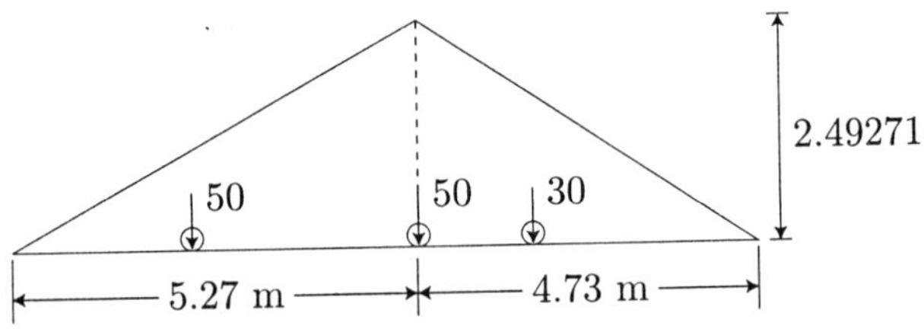

(b) I.L. for B.M.

Fig. 4.28: For Example 4.3

(ii) **Graphical Method (Tracing Paper Method).** The absolute maximum B.M. in a beam due to train of loads can be determined easily by a graphical procedure generally termed as *Tracing Paper Method.* This procedure also gives absolute maximum shear. The procedure is illustrated in Figs. 4.29 and 4.30.

Consider a beam AB, span l, across which travels, from left to right, a train of loads $W_1, W_2, W_3, \cdots$, etc., at a fixed distance apart.

The lines of action of the loads are set on a drawing sheet, with the assumption that these loads rest on a beam of unlimited span. These lines are extended above and below the beam as in Fig. 4.29(a). The spaces on either side of the lines of action as A, B, C, D, E, and F, etc. according to Bow's notation.

The magnitudes of the loads are set off on the load line $abcdef$ and complete the force polygon with pole O as shown in Fig. 4.29(b).

The stepped figures 1, 2, 3, 4, 5, 6, 7 are drawn as a part of the S.F.D. projecting horizontally the points a, b, c, $\cdots$, etc. as shown in Fig. 4.29(c).

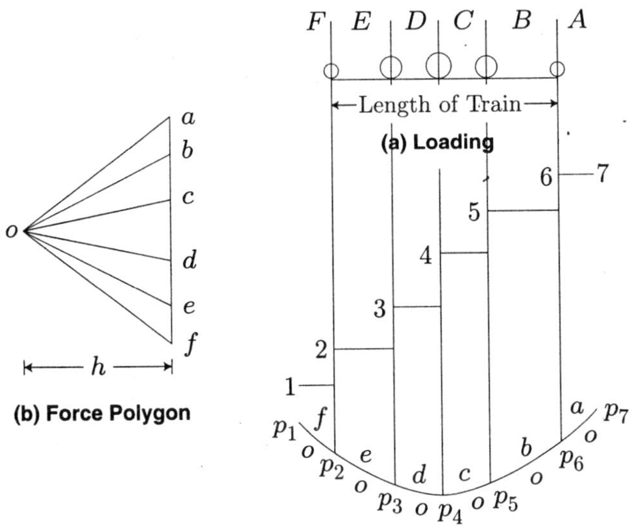

(a) Loading

(b) Force Polygon

(c) Link Polygon

Fig. 4.29: Tracing Paper Method

The link polygon $p_1\ p_2 \cdot p_3\ p_4\ p_5\ p_6\ p_7$ forming part of the B.M.D. are drawn by drawing lines, p_1p_2, p_2p_3, $\cdots$ parallel to rays of, oa, ob, $\cdots$ respectively.

On a sheet of tracing paper, the length scale used is drawn on the drawing paper. The horizontal line A_1B_1 represents the actual span L and perpendiculars A_1a_1 and B_1b_1 are drawn at the ends of the span. These lines should be prolonged upward and downwards as shown in Fig. 4.30.

Horizontal lines $a'b'$ and $a''b''$ are drawn as in Fig. 4.30(c) and 4.30(d), these lines being the base for maximum shearing force and bending moment diagrams.

The tracing paper is superimposed on the drawing paper, so that the combined appearance is as shown in Fig. 4.30(a). The span position is shown as A_1B_1, A_2B_2.

The perpendiculars at the end of span A_1B_1, cut the link polygon at points a_1 and b_1 in Fig. 4.30(a).

Line a_1b_1 is drawn on tracing paper. This will be the closing line of bending moment diagram for the position of loads relative to A_1B_1.

Line og is drawn parallel to a_1b_1, in the force polygon in Fig. 4.30(b). Project across, and draw 8–9, the closing line of S.F.D. *No other lines should be drawn on tracing paper below* A_1B_1. The vertical lines of A_1B_1 at the span cuts the shear force diagram at points 10, 8, 9 and 11. Then $8 - 9 - 11 - 4 - 3 - 2 - 10$ is the S.F.D. for the span A_1B_1 with the loads in this particular position.

This diagram is transferred to the base line $a'b'$ on the tracing paper. It appears as a'. 10. 11. b' in Fig. 4.30(c).

The bending moment diagram $a_1p_2p_3p_4$, b_1 is transferred to the base line $a''b''$, on the tracing paper appearing as a'', 1, 1, 1, b''.

The tracing paper is shifted until the span occupies another position A_2B_2 and the above procedure is repeated. Then a_2b_2 is the closing line to the B.M.D. in the link polygon. Line oh is drawn parallel to a_2b_2 and 12–13 is projected through h.

These new diagrams are transferred to $a''b''$ and $a'b'$ as before on the tracing paper. To avoid confusion, movements of base lines on tracing paper has not been shown.

The above operation is repeated for a sufficient number of times. Thus a series of S.F.D. and B.M.D. for number of positions of the travelling load are obtained on base lines $a'b'$ and $a''b''$.

The enveloping parabolas is drawn for maximum S.F.D. and maximum B.M.D. These enveloping curves are drawn as in Fig. 4.30(c) and 4.30(d).

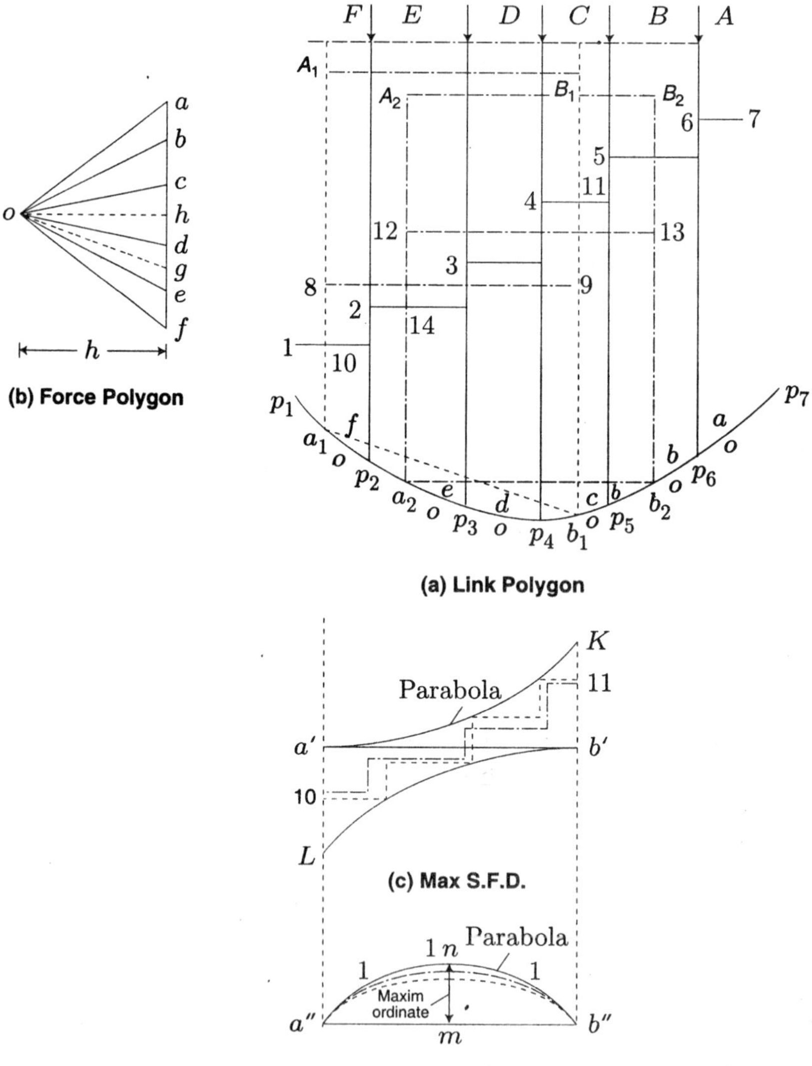

(b) Force Polygon

(a) Link Polygon

(c) Max S.F.D.

(d) Max B.M.D.

Fig. 4.30: Tracing Paper Method

The maximum ordinates of these curves give absolute, maximum, S.F. and B.M. respectively.

Scale. A unit mm of the ordinate of maximum B.M.D. represents :

$$BM = l.\text{w.h.},$$

where l = scale of length of span, say 10 mm = 5 m

w = load scale, say 10 mm = 100 kN (for force polygon)

h = distance of polar point O from load line $abcdef$ in mm, say 20 mm.

10 mm ordinate of B.M.D. = l.w.h. = 5(100)(20) = 10000 kNm.

The ordinate of S.F.D. will be on the basis of load scale used for force polygon.

4.18 Equivalent Uniformly Distributed Load

The maximum ordinate of enveloping parabola for the shear force diagram is given by $b'k$.

If w be the equivalent uniform load per unit length, which would produce the same maximum S.F. at every cross section, then

$$b'k = \frac{wL}{2}, \text{ whence } w = \frac{2}{l}(b'k).$$

The maximum ordinate of the enveloping parabola of the maximum bending moment diagram is mn and if w' be the equivalent uniform load per unit of length which would produce same maximum B.M. at that section, then maximum $M_C = \frac{w'l^2}{8}$ = mn (measured to B.M. scale (10 mm = lwh).

Also $w' = mn\frac{8}{l^2}$.

Example 4.4 *A system of wheel loads, shown in* Fig. 4.31(a) *rolls over a plate girder of* 16 m *span. Determine the absolute maximum bending moment.*

Solution : In this case

$$\sum W = 850 \text{ kN}$$

Resultant load act at a distance x from 180 kN load, given by 3.8 m.

According to the criterion for absolute maximum B.M. ***the loads must be so placed such that particular wheel load and C.G. of all the loads is equidistant from the centre. Also with the loads so arranged, load to left of the section under the particular wheel load be equal to average load on the span.***

Therefore, the loads are so arranged that wheel 3 of 160 kN and c.g. of the loads are equidistant from the centre of the beam.

The arrangement is shown in Fig. 4.31. The average loading condition is verified.

$$\text{Criterion : } \frac{W_1}{a} = \frac{W_1}{7.85}.$$

Position of wheel	*Average load on $A-3$* $\dfrac{W_1}{7.85}$	*Average load on span* $\dfrac{W}{L} = \dfrac{W}{16}$	*Criterion satisfied*
Wheel 3 at a distance of $8 - 0.15 = 7.85$ m from A Fig. 4.31(b)	Wheel 3 just to right of $C = \dfrac{340}{7.85} = 43.312$ Wheel 3 just to left of $C = \dfrac{180+160+160}{7.85} = \dfrac{500}{7.85} = 63.694$	$\dfrac{850}{16} = 53.125$	Yes

Because the average loading on AC changes from a decreasing to an increasing function with the passage of wheel 3, hence the arrangement is satisfactory and absolute maximum B.M. occurs under wheel 3 placed at 7.85 m from A.

$$R_A = \frac{850(7.85)}{16} = 417.031 \text{ kN}.$$

Absolute Max. B.M. under 160 kN load.

$$M_{\max} = R_A(7.85) - 180(3.50) - 160(1.50) = 2403.695 \text{ kNm}$$

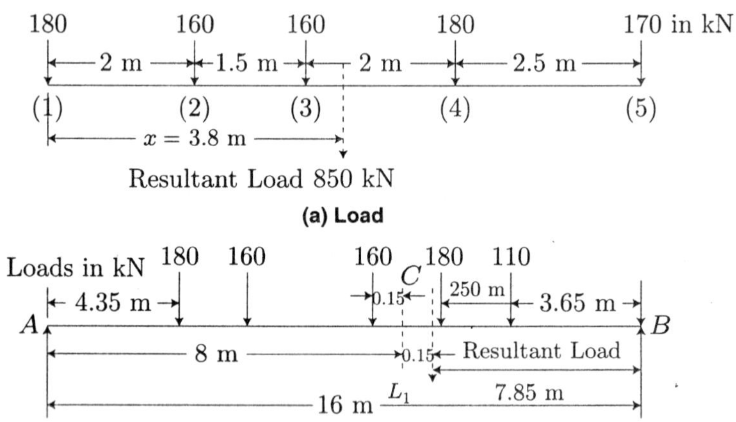

(a) Load

(b) Load position for Maxm. B.M.

Fig. 4.31: For Example 4.4

4.19 Influence Lines for Forces in Members of Trusses

Forces in the members of various trusses can be calculated by the method of sections easily. From the method of sections the external forces acting on one side of the section 11 of the truss in Fig. 4.32 and the internal forces in the members U_2U_3, U_2L_3 and L_2L_3 cut by the section must be in equilibrium.

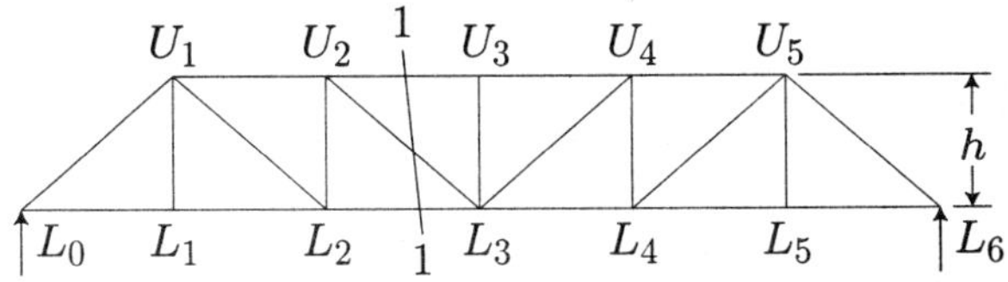

Fig. 4.32: Truss

From the principle of statics ($\sum M = 0$), the moment of external forces about point U_2 must be equal and opposite to the moment of the force in the bottom chord member L_2L_3 about U_2.

If the moment of the external forces about U_2 be M_{U2},

Force in $L_2L_3 = \frac{M_{U2}}{h}$, where '$h$' is the height of the truss which is constant for all chord members of parallel chord trusses. *Therefore, the force in the chord members of such trusses are proportional to the bending moment about the opposite node point.*

From $\sum V = 0$, the sum of vertical components of all external forces on one side of the section 11 must be equal and opposite to the vertical component of the force in the web member U_2L_3. Since the former sum represents the S.F. at 11 and there are no loads between L_2 and L_3, the S.F. between L_2 and L_3 is constant. Hence the force in the inclined web member of parallel chord trusses is equal to the S.F. in the panel L_2L_3 multiplied by cosec θ and the force in inclined members is proportional to the S.F. in the panel. It follows that :

(i) Force in a chord member of a parallel chord truss due to Travelling loads shall be given by the influence line of B.M. for the node point opposite to the members divided by the height of the truss.

(ii) Force in a web member of a parallel chord truss due to Travelling loads shall be given by the I.L. of S.F. for the panel containing the member.

Example 4.5 *Obtain the influence lines for the bar forces in members U_1L_1, L_0U_1, U_1L_2, U_1U_2 and L_1L_2 of the truss shown in Fig. 4.33(a).*

For members U_1L_1, the bar force can be directly determined from the consideration of equilibrium of joint L_1. When the unit load is placed at L_1, the bar force is equal to unity in tension. When the unit load is placed at any other joint, the bar force becomes zero. The influence line for the bar force in U_1L_1 is shown in Fig. 4.33(b).

For member L_0U_1, the bar force may be calculated from consideration of equilibrium of joint L_0 in the vertical direction, and the vertical component of the bar force in L_0U_1 balancces the summation of all vertical forces at joint L_0. The influence line thus obtained is shown in Fig. 4.33(c).

For member U_1L_2, the bar force may be calculated by multiplying the shear force in panel L_1L_2 by $\frac{\sqrt{a^2+h^2}}{h}$. Thus the ordinates of the influence line at joint L_1 and L_2 are given by $-\frac{1}{6h}\sqrt{a^2+h^2}$ and $\frac{2}{3h}\sqrt{a^2+h^2}$ respectively. The influence line for the bar force in U_1L_2 is shown in Fig. 4.33(d).

Bar force in member U_1U_2 can be obtained by considering any side of the section 1–1 as a free body and by taking moments about L_2. The bar force in member L_1L_2 is obtained by taking moments about joint U_1 of the free body. The influence line for bar forces in U_1U_2 and L_1L_2 obtained from the considerations are shown in Figs. 4.33(e) and 4.33(f) respectively.

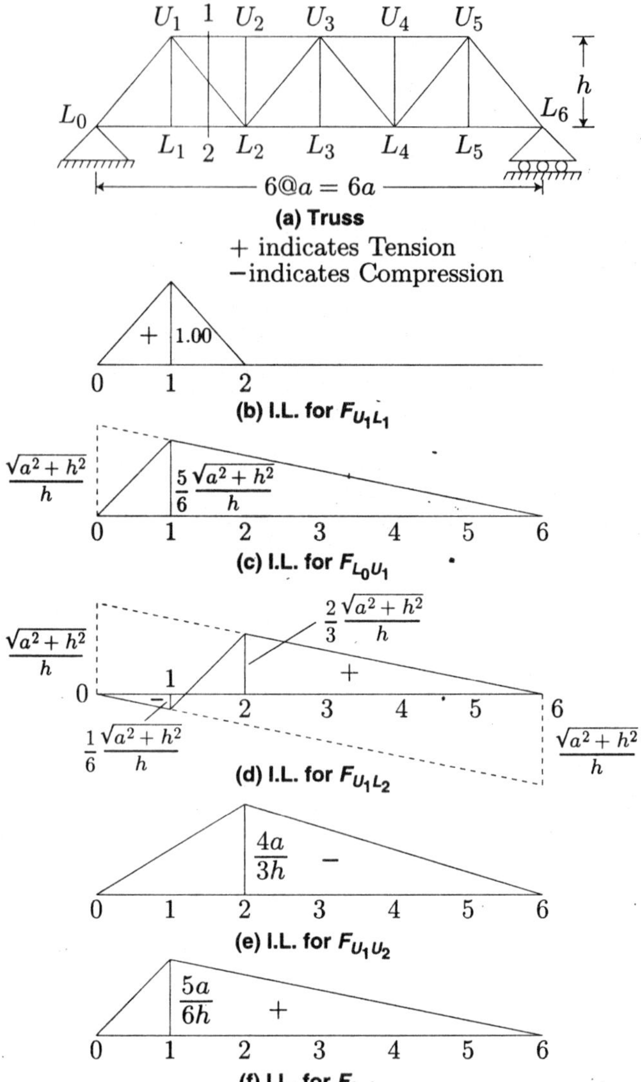

Fig. 4.33: For Example 4.5

Example 4.6 *Obtain the influence lines for bar forces in members* L_0U_0, L_0U_1, U_1L_2 *and* U_1U_3 *of the truss shown in* Fig. 4.34(a).

Solution : For member L_0U_0, the bar force can be directly determined from the consideration of equilibrium of joint U_0. For loads moving along bottom of truss, there is no bar force in member L_0U_0. For loads moving along top of truss, the influence line for the bar force in L_0U_0 is shown in Fig. 4.34(b) by dashed line.

For member L_0U_1, the bar force may be calculated by multiplying the shear force in panel L_0L_2 (for loads moving along the bottom of the truss) or in panel U_0U_1 (for loads moving along the top of the truss) by a factor $\frac{\sqrt{a^2+4h^2}}{2h}$. Influence lines thus obtained are shown in Fig. 4.34(c), the firm line showing the influence line for loads moving along the bottom, and the dashed line for loads moving along the top.

For member U_1L_2, the bar force may similarly be obtained from shear force in panel L_0L_2 or in U_1U_3 as in Fig. 4.34(d).

For member U_1U_3, the bar force may be obtained by considering free body diagram of any side of a section passing through the member and taking moments about L_1. The influence line for the bar forces in U_1U_3 is shown in Fig. 4.34(e), the dashed line showing the influence line diagram for loads moving over the top of the truss.

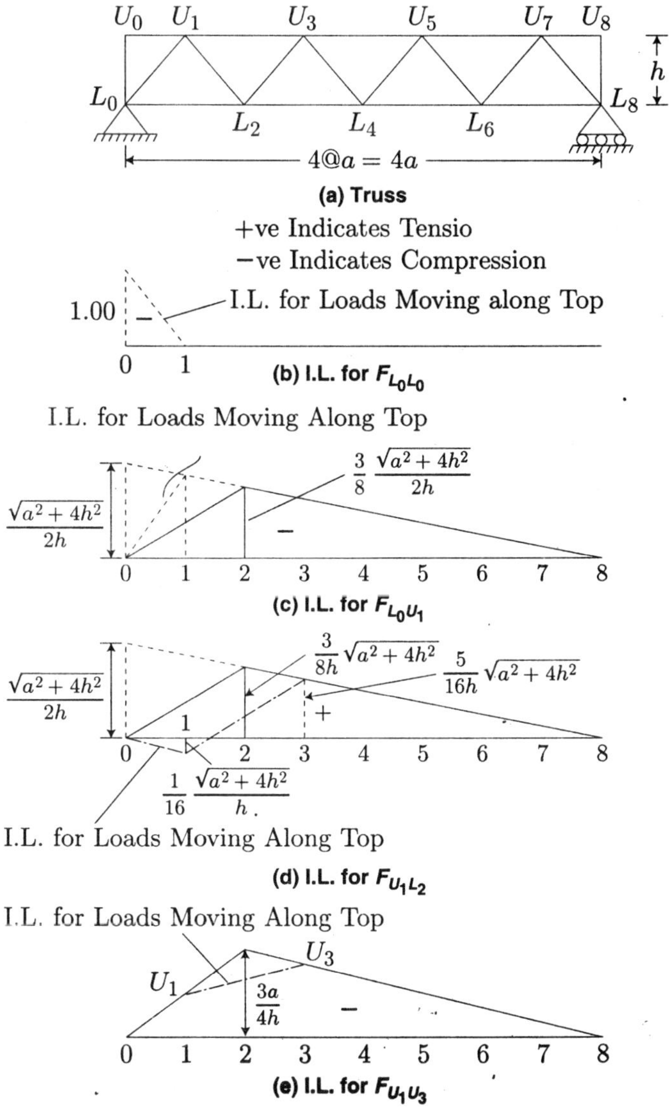

Fig. 4.34: For Example 4.6

Example 4.7 *Obtain the influence line diagrams for bar forces in members U_1U_2, U_2L_2, U_3L_3, U_2L_3 and U_3L_3 of the truss shown in* Fig. 4.35(a).

For member U_1U_2, the bar force may be obtained by considering free body diagram of any side of a section 1–1 and taking moments about L_2. Fig. 4.35(b) shows the influence line diagram for bar force in member U_1U_2.

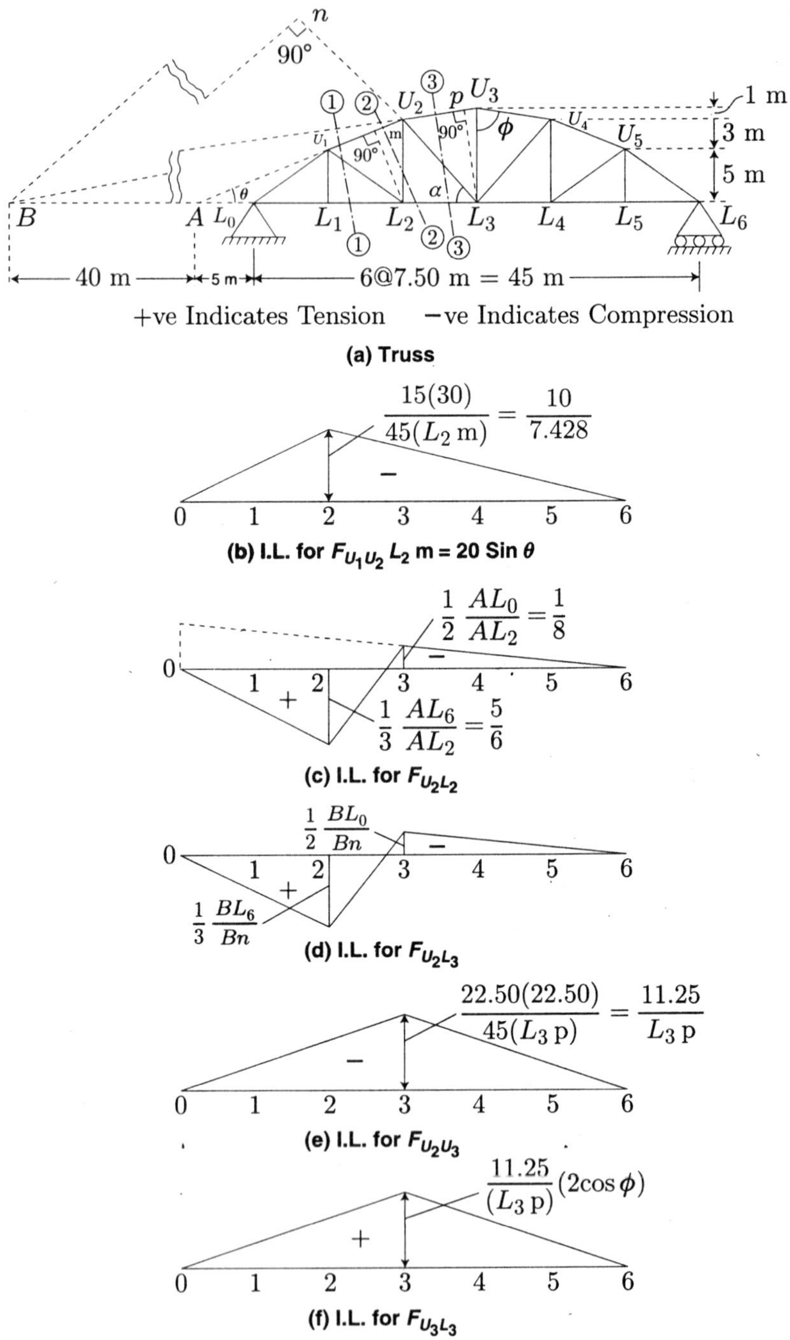

Fig. 4.35: For Example 4.7

For members U_2L_2 and U_2L_3, bar forces are obtained by taking moments about A and B respectively for free body diagrams considered for section through 2–2 and 3–3. The influence line diagrams thus obtained are shown in Figs. 4.35(c) and 4.35(d).

Fig. 4.35(e) shows the influence line diagram for bar force in member U_2U_3, this being obtained by taking section through 3–3 and taking moments about L_3.

For member U_3L_3, the bar force may be obtained by considering equilibrium of joint U_3, whence the influence line diagram in Fig. 4.35(f) results.

Example 4.8 *Obtain the influence lines for bar forces in members U_2U_4, $L_3L_4(L_2L_3)$, M_3L_4, U_2M_3, U_2L_2, M_3L_3, and M_3L_2 of the compound truss shown in* Fig. 4.36(a).

Bar force in member U_2U_4 can be obtained by taking moments about joint of the free body to the left of the section 1–1 and is not affected by the presence of the secondary systems. The influence line diagram thus obtained is shown in Fig. 4.36(b).

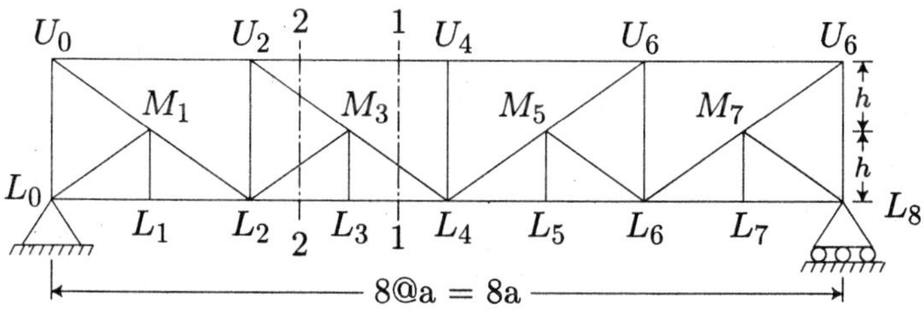

+ve Indicates Tension −ve Indicates Compression

(a) Truss

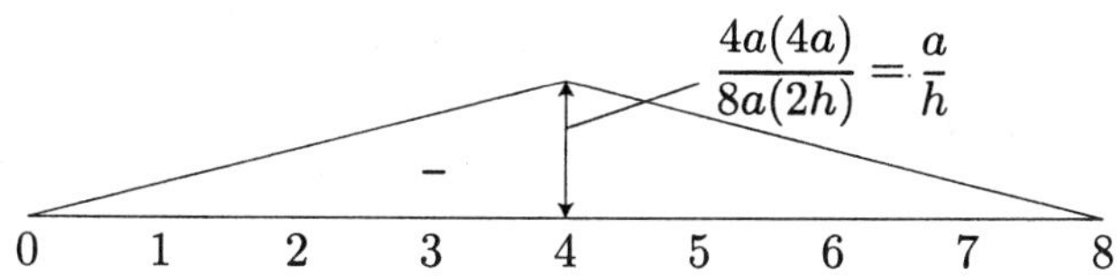

(b) I.L. for U_2U_4

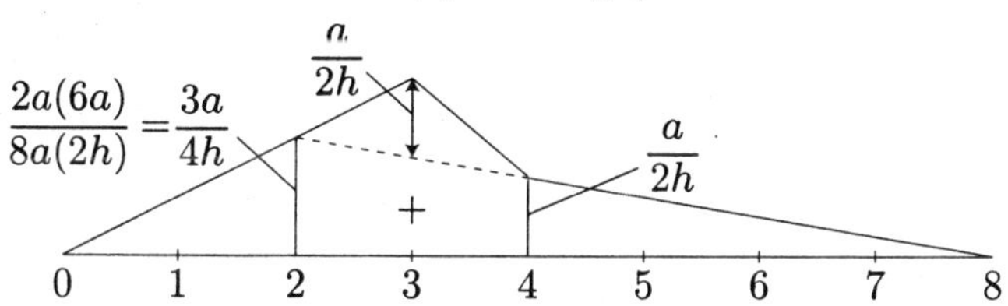

(c) I.L. for $L_3L_{4j}(L_2L_3)$

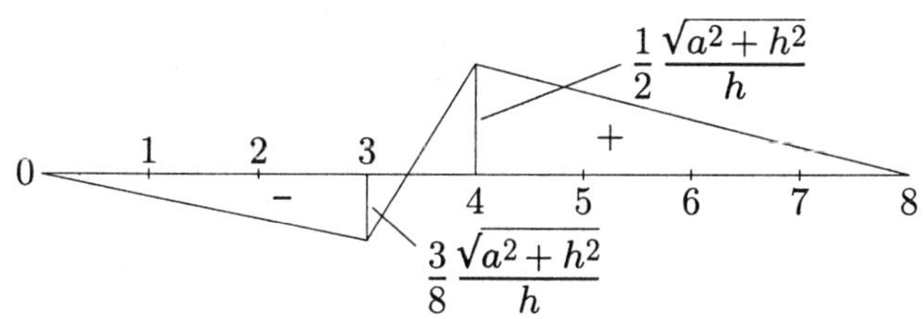

(d) I.L. for M_3L_4

Fig. 4.36: For Example 4.8 (Contd.)

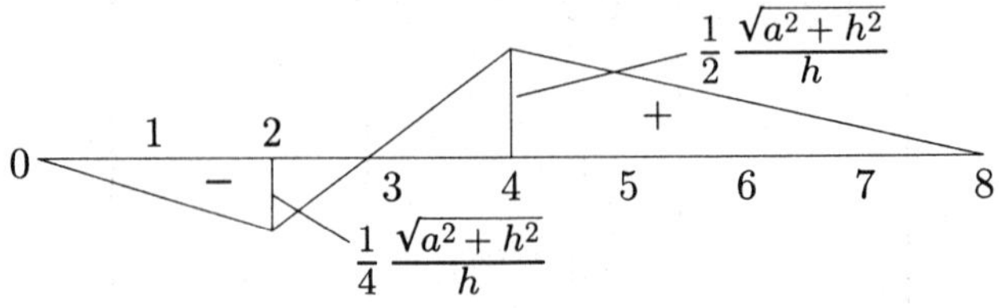

(e) I.L. for U_2M_3

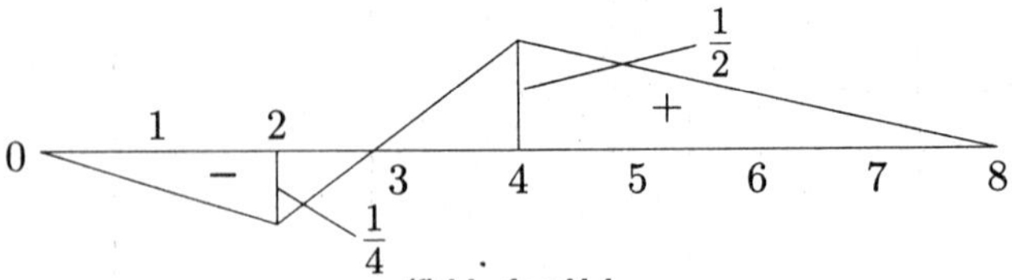

(f) I.L. for U_2L_2

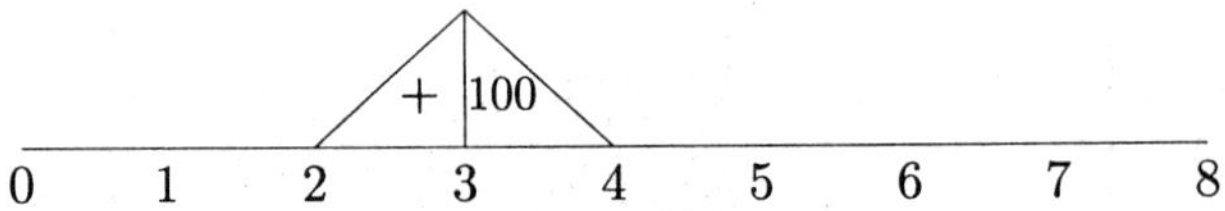

(g) I.L. for M_3L_3

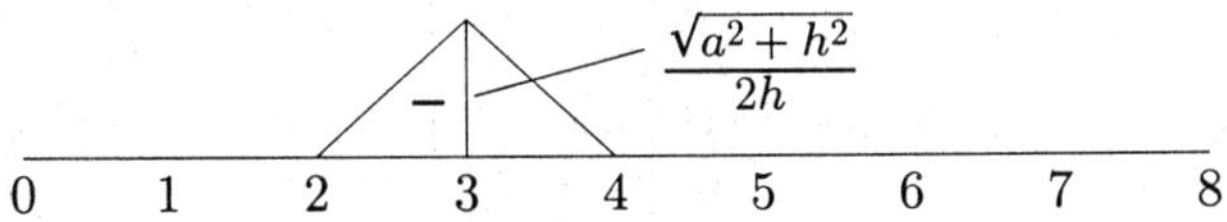

(h) I.L. for M_3L_2

Fig. 4.36: For Example 4.8

Bar force in member L_3L_4 (or L_2L_3) is affected by the small triangular secondary system $L_2L_4M_3$ only when the unit load is placed between L_2 and L_4, although it can be computed by taking moments about joint U_2 of the free body to the left of section 2-2. The influence line for the bar force except in panel L_2L_4 will have the pattern obtained from the influence line for bending moment at U_2 divided by $2h$. The effect varies linearly between L_2 and L_3 and between L_3 and L_4 due to transmission of applied load through the floor system. For unit load at L_3, the ordinate over the above diagram will be

$$\frac{(a)(a)}{(2a)(h)} = \frac{a}{2h}.$$

The combined effect is shown in Fig. 4.36(c).

Bar force in member M_3L_4 is shown in Fig. 4.36(d) in which the ordinates are calculated by considering the shear force in panel L_3L_4 for section through 1–1.

Bar force in member U_2M_3 can be related to the shear force in panel L_2L_4 as shown in Fig. 4.36(e), since the member occurs in the primary truss and is not affected by the secondary system.

Fig. 4.36(f) shows the influence line diagram for the bar force on member U_2L_2, which is not affected by the secondary system and is obtained by considering shear force in panel L_2L_4.

The influence lines for the bar forces in members M_3L_3 and M_3L_2 can be constructed in a straight forward manner as in case of simple trusses and are shown in Figs. 4.36(g) and 4.36(h) respectively.

4.20 Criteria for Maximum Bar Forces on Trusses

Bar forces in trusses may be obtained either by taking moments at a certain point or by considering the shear force at a certain panel or by considering equilibrium at a joint of the truss. The criteria for maximum bar forces in a truss due to a moving load are identical to those for maximum moments or maximum shear forces at appropriate locations of the truss. If the loaded chord of a truss be considered as a girder supporting floor beams, and criteria for shear forces in various panels and for moments at various panel points of such a girder be developed, the criteria for maximum bar forces in the truss itself can generally be obtained without difficulty.

(a) Maximum Shear Force in a Panel

As a series of concentrated load moves across the span from right to left, the maximum positive shear force in a panel is produced, when there are as many loads as possible, placed on the position of span, for which the area in the influence diagram is positive and no loads on the portion of the span for which the area in the influence diagram is negative, unless the moment of loads to the left beyond the neutral point causes an increase in shear force in the panel. The maximum negative shear force in a panel is similarly produced, when the series of loads moves from right to left and leaves this span from the left or when the series enters the span from the left, depending on the combination of loads.

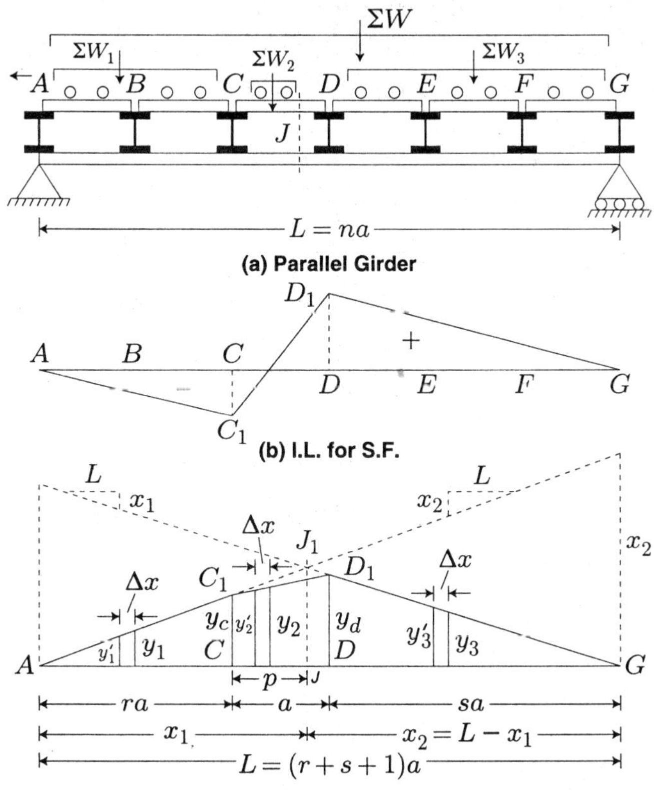

Fig. 4.37: Criteria for Maxm. Effects on Trusses

Fig. 4.37(a), shows a girder supporting floor beams, the influence line for shear force in the panel CD of the case being shown in Fig. 4.37(b). Fig. 4.37(a) shows position of a series of concentrated loads moving from right to left over the span, $\sum W_1$ being the sum of all wheel loads on the left of $C, \sum W_2$ the sum of all loads on panel CD and $\sum W$ the sum of all loads on the span. If all the loads move a small distance Δx to the left so that loads on a particular panel do not move beyond it, while the other loads do not move out of the supports, the change in shear force Fig. 4.37(b) will be

$$\Delta V = \sum W \frac{\Delta x}{L} - \sum W_2 \frac{\Delta x}{a},$$

where L = span length = na and a = panel length.

Term $\sum W_1$ does not appear in the above equation, as the sum of the reactions at A, B and C of the stringer AB and BC remains the same during the movement, although the reaction at B increases and that at C decreases at the same time.

When such a movement to the left causes an increase in shear force in the panel, the increase must continue until a load passes from the right into the panel, and, the shear force in the panel cannot change from an increasing to a decreasing function unless $\sum W_2$ in the above expression becomes larger or $\sum W$ becomes smaller. For the shear force to change from an increasing to a decreasing function, a load must pass from the right into the panel CD. The shear force passes through a maximum value as the change in shear force becomes zero, i.e., as

$$\sum W \frac{\Delta x}{L} = \sum W_2 \frac{\Delta x}{a},$$

whence
$$\frac{\sum W}{L} = \frac{\sum W_2}{a}$$

i.e.,
$$\frac{\sum W}{n} = \sum W_2 \qquad (4.4)$$

This expression states that the shear force in a panel is a maximum when the average load in the panel is equal to the average load on the span.

Maximum value of shear force in the panel CD will occur when a particular wheel load W_i is at the point D. The criteria for maximum shear force in the interior panel CD, as given by Eq. (4.4) may be stated as follows :

$$\left.\begin{aligned} &\text{For } W_i \text{ to the right of } D, \frac{\sum W}{n} > \sum W_2 \\ &\text{For } W_i \text{ to the left of } D, \frac{\sum W}{n} < \sum W_2 \end{aligned}\right] \qquad (4.5)$$

There can be more than one wheel load, which satisfy the criteria in Eq. (4.5). Under such circumstances, the maximum shear force may be obtained by trial.

As the shape of influence line diagram for shear force in the end panel is a triangle as in the case of influence line diagram for bending moment, the criteria as given by Eq. (4.5) may be adopted for determining the maximum shear force in the end panel. In such a case, therefore,

$$\frac{\sum W}{L} = \frac{\sum W_2}{a},$$

where $\sum W_2$ = sum of wheel loads on panel AB.

(b) Maximum Bending Moment in a Panel

The influence line for bending moment at a panel point of a girder supporting floor beams is identical to that of a beam of the same span without floor beams and the criteria for maximum bending moment in Eq. (4.3) are directly applicable to the computation of maximum values in such a case. For maximum bending moments at points in between panel points, suitable criteria have to be developed.

Fig. 4.37(c) shows the influence line diagram for moment at point J which is between panel points C and D of the girder in Fig. 4.37(a). Referring to Fig. 4.37(c),

$$y_c = \text{bending moment at } J \text{ for unit load at } C$$
$$= R_G(x_2) = \frac{r}{n}x_2$$

and

$$y_d = \text{bending moment at } J \text{ for unit load at } D$$
$$= R_A(x_1) = \frac{s}{n}x_1.$$

The slope of the segment AC_1 same as that of AJ_1 is given by

$$\frac{y_c}{AC} = \frac{x_1(L-x_1)}{L(x_1)} = \frac{x_2}{L}.$$

The slope of the segment GD_1 is similarly given by

$$\frac{y_d}{GD} = \frac{x_1}{L}.$$

The slope of the line segment C_1D_1 is given by,

$$\frac{y_d - y_c}{CD} = \frac{1}{a}\left[\frac{x_1}{L}s.a. - \frac{x_2}{L}ya\right] = \frac{sx_1 - rx_2}{L}.$$

Taking $p = x_1 - ra$, in Fig. 4.37(c),

$$\frac{sx_1 - rx_2}{L} = \frac{sx_1 + rx_1 + x_1 - rx_1 - x_1 - rx_2}{L}$$
$$= \frac{x_1(r+s+1) - (x_1+x_2)r - x_1}{L}$$
$$= \frac{x_1 - ra}{a} - \frac{x_1}{L} = \frac{p}{a} - \frac{x_1}{L}.$$

If the resultants of the loads on segment AC, panel CD, and segment DG, of the beam are denoted by $\sum W_1$, $\sum W_2$, $\sum W_3$ respectively and the ordinates of the influence line corresponding to the lines of action of these resultants are denoted by y_1, y_2, and y_3 respectively, then the bending moment at J is given by

$$M_J = \sum W_1 y_1 + \sum W_2 y_2 + \sum W_3 y_3.$$

If the series of loads moves a small distance Δx to the left, the bending moment at J becomes

$$M'_J = \sum W_1 y'_1 + \sum W_2 y'_2 + \sum W_3 y'_3;$$

the ordinates y'_1, y'_2 and y'_3 refer to those for the new locations of the resultants.

The change in bending moment is equal to

$$\Delta M_J = M'_J - M_J = \sum W_1[y'_1 - y_1] + \sum W_2[y'_2 - y_2] + \sum W_3[y'_3 - y_3].$$

From known slopes of the respective lines,

$$\Delta M_J = \left[-\sum W_1 \frac{x_2}{L} - \sum W_2\left(\frac{p}{a} - \frac{x_1}{L}\right) + \sum W_3 \frac{x_1}{L}\right]\Delta x,$$

whence
$$\frac{\Delta M_J}{\Delta x} = -\sum W_1 \frac{x_2}{L} - \sum W_2\left(\frac{p}{a} - \frac{x_1}{L}\right) + \left(\sum W - \sum W_1 - \sum W_2\right)\frac{x_1}{L}$$
$$= -\sum W \frac{x_1}{L} - \left[\sum W_1 + \sum W_2 \frac{p}{a}\right].$$

The maximum bending moment at J is reached when ΔM_J changes from positive to negative. This will occur when a wheel load is placed at either panel point C or point D. When a load W_i is placed at C or D, the criteria for maximum bending moment at J are satisfied if, for W_i to right of C or D,

$$\sum W \frac{x_1}{L} > \left[\sum W_1 + \sum W_2 \frac{p}{a}\right]$$

and for W'_i to the left of C or D, (4.6)

$$\sum W \frac{x_1}{L} < \left[\sum W_1 + \sum W_2 \frac{p}{a}\right].$$

Example 4.9 *A series of concentrated loads as in* Fig. 4.38(a) *moves across the top chord of the truss shown in* Fig. 4.38(b). *Determine the maximum forces in the members* U_4L_3 *and* U_2U_4 *of the truss.*

The influence line diagrams for the forces in members U_4L_3 and U_2U_4 are shown in Figs. 4.38(c) and 4.38(f) respectively.

(i) Maximum Compressive Force in member U_4L_3

Maximum value occurs when the wheel loads move from right towards left. Fig. 4.38(d) shows the movement of the series of concentrated loads over point U_4 in stages. The calculations are carried out in Table (a).

In this case, the length of the load is larger than the span, and various pattern of loads are on the span in various cases. The maximum compressive force in member U_4L_3 has to be obtained by calculation of the bar forces in the cases fulfilling the criteria. In this case, the maximum bar force will be obtained when W_3 is at U_4, the magnitude being 418.578 kN [compression] from Fig. 4.38(c).

Table (a) For Example 4.9

W_i at U_4	$\sum W_2$ W_i at left	$\dfrac{\sum W}{n}$	$\sum W_2$ W_i at right	Criteria Satisfied
W_1	200	$\dfrac{1100}{4} = 275.0$	0	No
W_2	400	$\dfrac{1450}{4} = 362.50$	200	Yes
W_3	750	$\dfrac{1750}{4} = 437.50$	400	Yes
W_4	700	$\dfrac{2050}{4} = 512.50$	350	Yes
W_5	700	$\dfrac{2150}{4} = 537.50$	350	Yes
W_6	650	$\dfrac{2250}{4} = 562.50$	350	Yes
W_7	600	$\dfrac{1900}{4} = 475.0$	300	Yes
W_8	600	$\dfrac{1550}{4} = 387.50$	300	Yes
W_9	600	$\dfrac{1200}{4} = 300.00$	300	No

(ii) Maximum Tensile Force in member U_4L_3

Maximum value will occur, when the loads cross panel point U_2. Fig. 4.38(f) shows movement of the series of concentrated loads over panel point U_2 in stages. The calculations have been carried out in Table (b).

Table (b) For Example 4.9

W_i at U_2	$\sum W_2$ W_i at left	$\dfrac{\sum W}{n}$	$\sum W_2$ W_i at right	Criteria Satisfied
W_8	300	$\dfrac{900}{4} = 225.0$	600	No
W_9	0	$\dfrac{600}{4} = 150.0$	300	Yes

This indicates that maximum tensile bar force in member U_4L_3 is obtained, when W_9 is at panel point U_2.

The magnitude of the tensile bar force is given by 115.47 kN

(iii) Maximum bar force in member U_2U_4 :

For the case, $p = 3$ m and $\dfrac{p}{a} = \dfrac{1}{2}$ and $x_1 = 9$ m, $L = 24$ m and $\dfrac{x_1}{L} = \dfrac{3}{8}$.

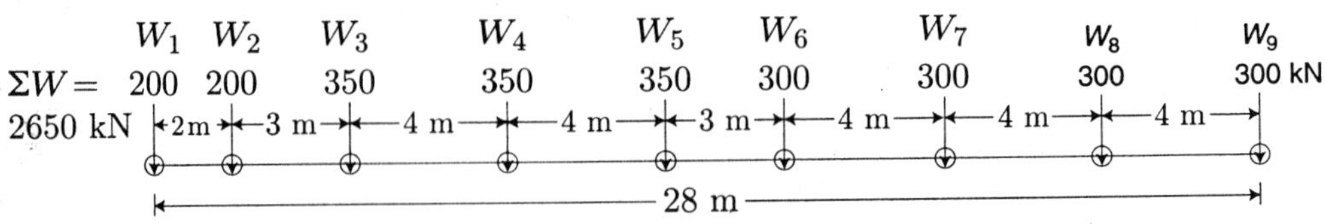

(a) Load

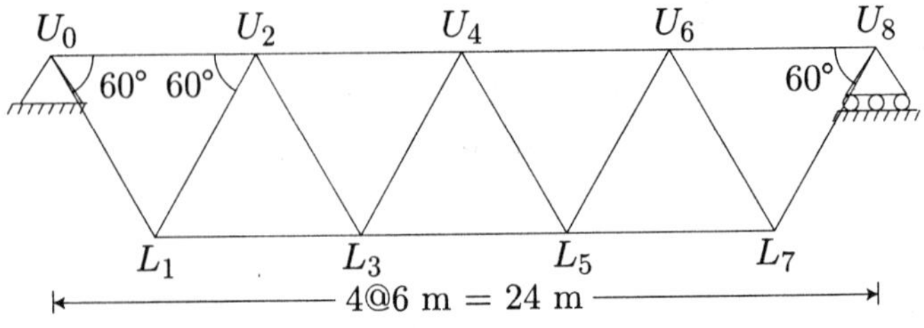

(b) Truss

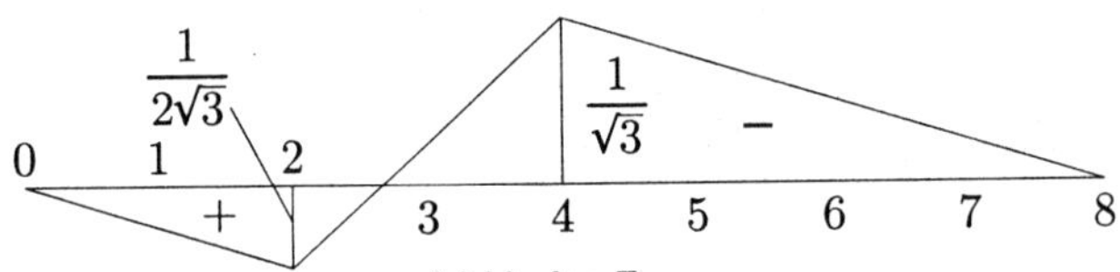

(c) I.L. for $F_{U_4L_3}$

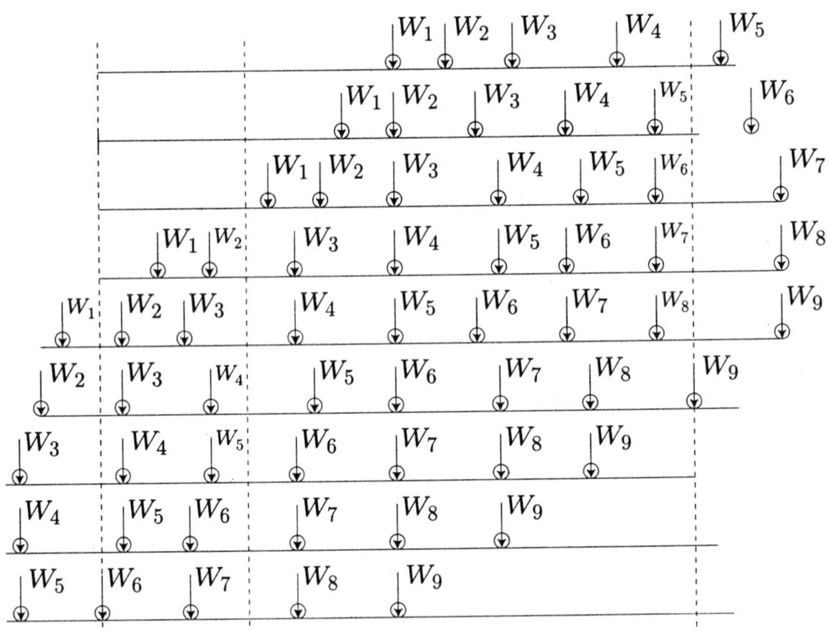

(d) W_i at U_4 (For $F_{U_4L_3}$)

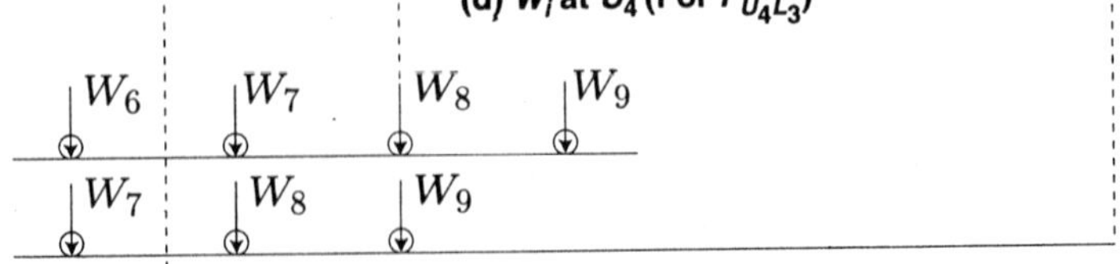

(e) W_i at U_2 (For $F_{U_4L_3}$)

Fig. 4.38: For Example 4.9 (Contd.)

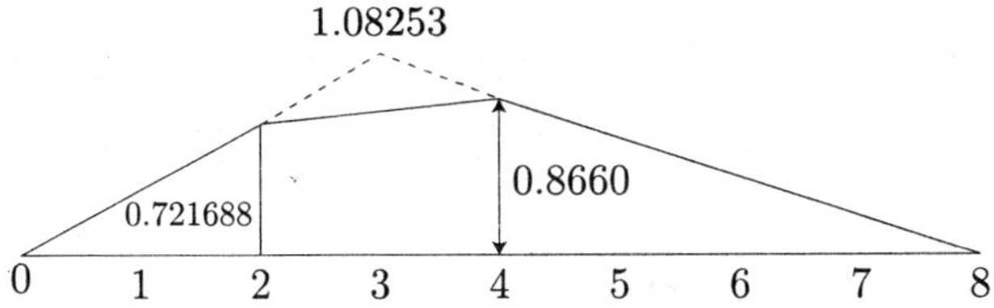

(f) I.L. for $F_{U_2U_4}$

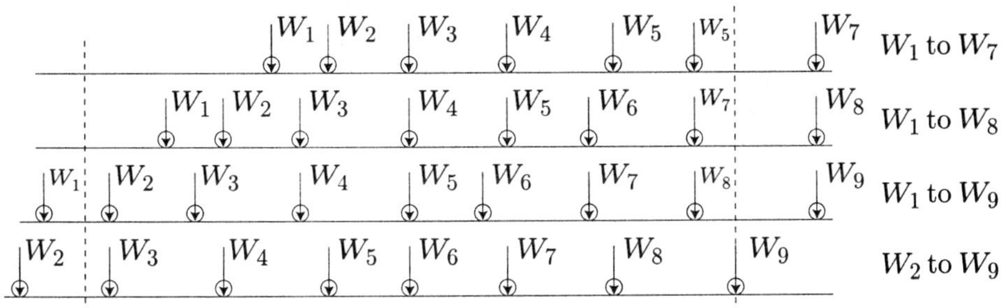

(g) W_i at U_4 for ($F_{U_2U_4}$)

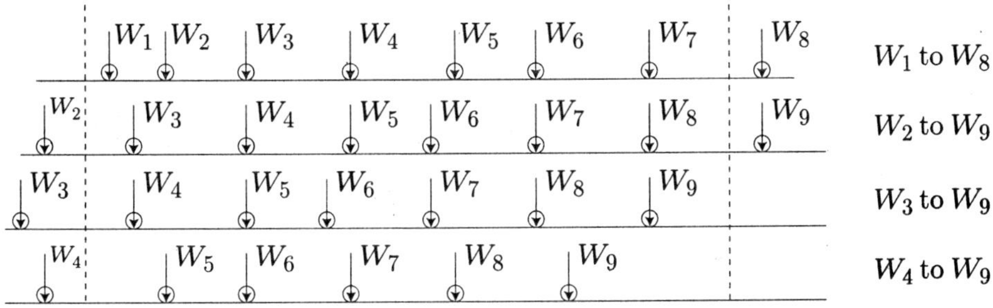

(h) W_i at U_2 for ($F_{U_2U_4}$)

Fig. 4.38: For Example 4.9

Tables (c) and (d) show the calculations involved for various loads at positions U_4 and U_2 respectively. It will be evident from Fig. 4.38(f) showing the I.L. for $F_{U_2U_4}$ that maximum bar forces on the member may occur when a concentrated load is at U_4 or U_2 or in both simultaneously.

Table (c): For Example 4.9

W_i at U_4	$\sum W_1 + \sum W_2 \frac{p}{a}$ W_i to right	$\sum W \frac{x_1}{L}$	$\sum W_1 + \sum W_2 \frac{p}{a}$ W_i to left	Criteria Satisfied
W_3	$0 + \frac{1}{2}(400) = 200$	$\frac{3}{8}(1750) = 656.25$	$0 + \frac{1}{2}(750) = 375$	No
W_4	$400 + \frac{1}{2}(350) = 575$	$\frac{3}{8}(2050) = 768.75$	$400 + \frac{1}{2}(700) = 750$	No
W_5	$550 + \frac{1}{2}(350) = 725.0$	$\frac{3}{8}(2150) = 806.25$	$550 + \frac{1}{2}(700) = 900$	Yes
W_6	$700 + \frac{1}{2}(350) = 875$	$\frac{3}{8}(2250) = 843.75$	$700 + \frac{1}{2}(650) = 1025$	No

Table (d): For Example 4.9

W_i at U_2	$\sum W_1 + \sum W_2 \frac{p}{a}$ W_i to right	$\sum W \frac{x_1}{L}$	$\sum W_1 + \sum W_2 \frac{p}{a}$ W_i to left	Criteria Satisfied
W_3	$400 + \frac{1}{2}(700) = 750$	$\frac{3}{8}(2050) = 768.75$	$750 + \frac{1}{2}(350) = 925$	Yes
W_4	$350 + \frac{1}{2}(700) = 700$	$\frac{3}{8}(1950) = 731.15$	$700 + \frac{1}{2}(350) = 875$	Yes
W_5	$350 + \frac{1}{2}(650) = 675$	$\frac{3}{8}(1900) = 712.50$	$700 + \frac{1}{2}(300) = 850$	Yes
W_6	$350 + \frac{1}{2}(600) = 650$	$\frac{3}{8}(1550) = 581.25$	$650 + \frac{1}{2}(300) = 800$	No

The maximum bar force in member U_2U_4 will be obtained when wheel load W_5 is at U_4 or wheel loads W_3, or W_4 or W_5 are at U_2.

From calculations of the bar forces for these positions of wheel loads, it will be found that the maximum bar force (compression) occurs when W_5 is at U_4. Magnitude = 1089.727 kN.

This indicates that in case of certain members there can be two maximum values of bar forces, one maximum compression and the other maximum tension. In designing such members, it is necessary to know both the maximum values. Such a change in the bar forces causing change of stress in a member from tension to compression or vice versa as a series of load moves over the span is called reversal of stress. Members in which reversal of stress takes place or in which tension or compression may exist for different positions of the load, even though the opposite signs do not occur in a single position of a series of load, are said to be subject to alternate stress. Actual reversal of stress can only occur when dead load stress which is always of one kind, is exceeded by live load stresses of opposite kind,

4.21 Uniform Load with Moving Concentrated Load

For highway bridge design, it is usual to specify for the structure a moving uniform load of constant magnitude per unit length accompanied by a concentrated load, which may occupy any position on the span, independent of the uniform load. Such a concentrated load is sometimes called *Knife edge load.*

Such a pattern of loading is entirely artificial and is intended as a convenient and reasonably representative substitute for the miscellaneous types of loading to which, a highway bridge is actually subjected to. The calculation of forces or stresses using a load of this kind is rapid and easy.

The loaded length is that part of a structure loaded with the uniform live loading to produce the maximum effect. This loaded length may be determined from the influence line for the member under consideration. For example, for the case shown in Fig. 4.33, the loaded length for maximum compression in member L_0U_1 [Fig. 4.33(c)] is the full length $6a$ of the truss with the Knife edge loading at panel point 1. For member U_1L_2 again, the loaded length for maximum tension, is up to the neutral point of the influence line diagram in Fig. 4.33(d) between panel points 1 and 2, and the right-hand support with the Knife edge load at panel point 2. For maximum compression again loaded length is between the neutral point and the left-hand support with the knife edge load at panel point 1.

4.22 Influence Lines for Bar forces in members of Truss from Deflected Shapes

Influence lines for bar forces in members of trusses can also be obtained from deflected shapes obtained from Muller Breslau principles. Considering the truss of Fig. 4.39(a) the influence line for the bar force in diagonal member U_1L_2 may be obtained if piston and cylinder connection as in Fig. 4.39(b) at some point on the member be introduced. If the force in the member is moved through a small displacement with the displacement represented by a contraction of the piston and cylinder connection, the deflected shape of truss will be as shown by dashed line in Fig. 4.39(a). Consideration of the virtual work involved will show that for a unit value of displacement in the member, the influence line for the bar force in U_1L_2 is represented by the deflected shape of the lower chord. For calculating the values of ordinates of the influence line diagrams, it is necessary to compute the ordinates at the two panel points 1 and 2 only, since the bottom chord would change slope at these panel points.

The influence line diagram for bar force in member L_1L_2 of the truss may be similarly obtained by introducing a piston and cylinder connection as in case of member U_1L_2. The deflected shape of the truss then obtained is shown by dashed line in Fig. 4.39(c).

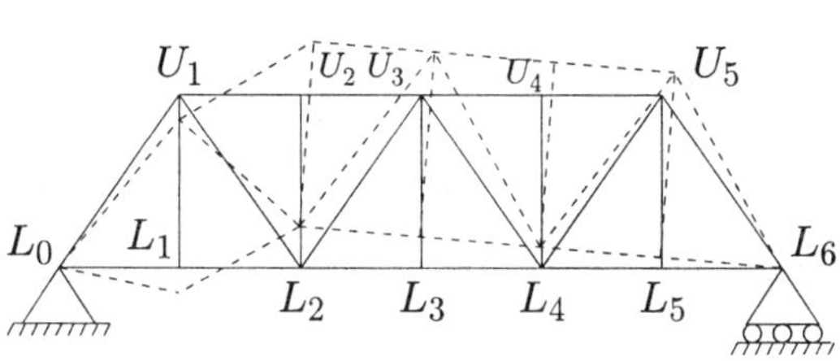

(a) I.L. for $F_{U_2U_4}$

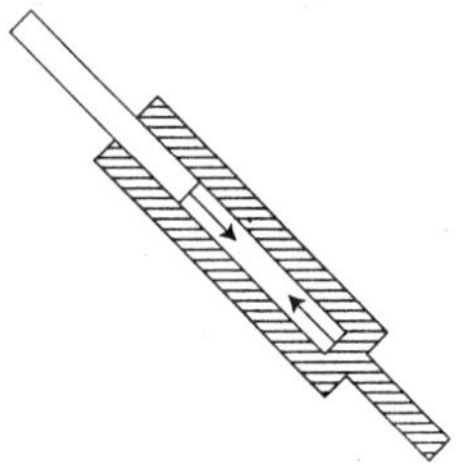

(b) Turnbuckle

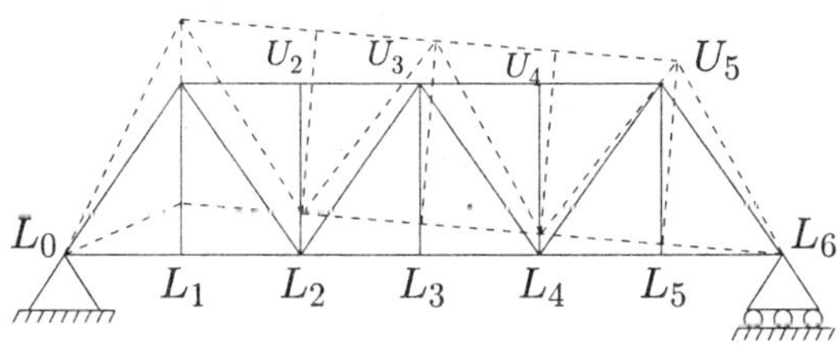

(c) I.L. for $F_{L_1L_6}$

Fig. 4.39: Influence Line from Deflected shape

Problems

4.1 A cantilever beam AB is fixed at A and free at B, length AB being 10 m. Obtain the influence lines for the following effects (a) Reaction (S.F.) at support A. (b) B.M. at A (c) S.F. and B.M. at a point C, 6 m from B.

4.2 A horizontal beam AB of length 6 m is hinged to the support at A and is supported at B by a wire, inclined at an angle of 30° to the horizontal. Obtain the influence line for the force in the wire. How will the influence line be modified if the wire having same inclination supports the beam at 2 m from B.

4.3 Obtain the influence lines for the following effects if a unit load moves from A to B in the structure shown in Fig. P-4.1. (a) Force in the tie rod (b) B.M. in the vertical mast at B.

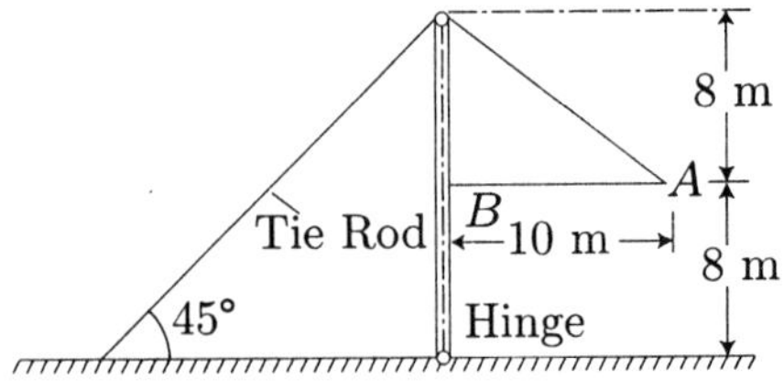

Fig. P-4.1

4.4 For the girders shown in Fig. P-4.2 obtain influence lines for the following effects :

(a) Reactions at supports A and B.

(b) S.F. and B.M. at E and F.

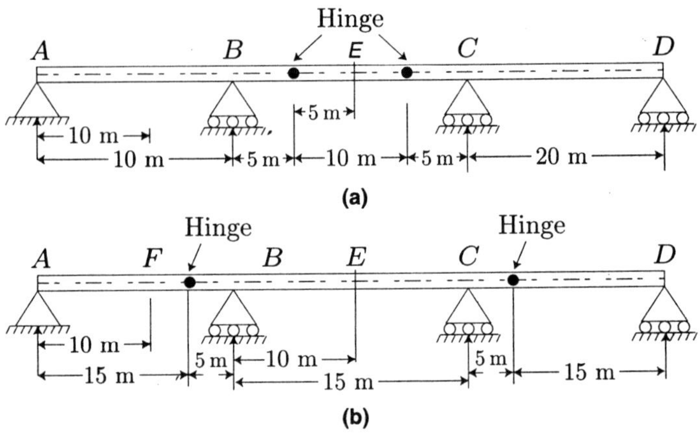

Fig. P-4.2

4.5 Obtain influence lines for vertical left-hand reaction and S.F. at quarter span of a beam 10 metres in span. Determine the values of reaction and shear force from the above influence lines when loads of 100 kN, 150 kN, 80 kN and 20 kN are placed at a distance of 1 m, 3 m, 4 m and 6 m, respectively from left-hand support.

4.6 Obtain influence line for shear force at a section 2 m from right-hand support of a simply supported beam of 10 m span. Determine the value of S.F. when a u.d.l. 5 m long is placed with its one end 2 m from left-hand support.

4.7 Obtain influence lines for B.M. at sections 2 m, 3.5 m, and 5.00 m, from right-hand support of a beam 15 m long. Evaluate the B.M. at these sections, when loads of 200 kN, 400 kN, and 600 kN are placed at 2 m, 3.5 m, 5.00 m and 10 m from right-hand support.

4.8 Obtain influence lines for B.M. and S.F. at a quarter and mid-span of 25 m long girder and determine values of S.F. and B.M. at the above section due to a u.d.l. of 200 KN/metre 8 metres long. The tail of the load is 4 m from left-hand supports.

4.9 Draw influence lines for forces in the members of U_1L_1, U_2L_3 and L_2L_3 of truss shown in Fig. P4.3.

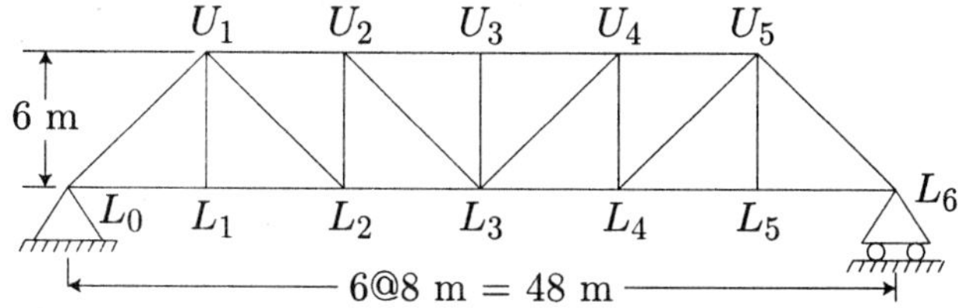

Fig. P-4.3

4.10 Obtain influence lines for forces in the members U_1U_2. L_1L_2 and U_1L_2 of the truss shown in Fig. P-4.4. The panel length is 6 m and height of the truss is 4 m.

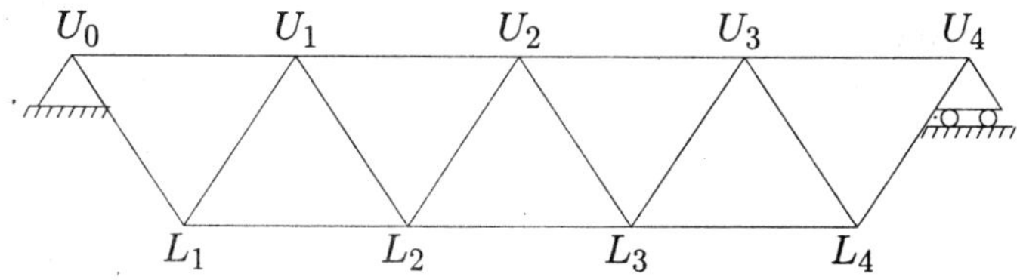

Fig. P-4.4

4.11 Draw influence lines for forces in the members U_2L_2, U_1L_1, L_1L_2 and U_2U_3 of the truss shown in Fig. P-4.5,

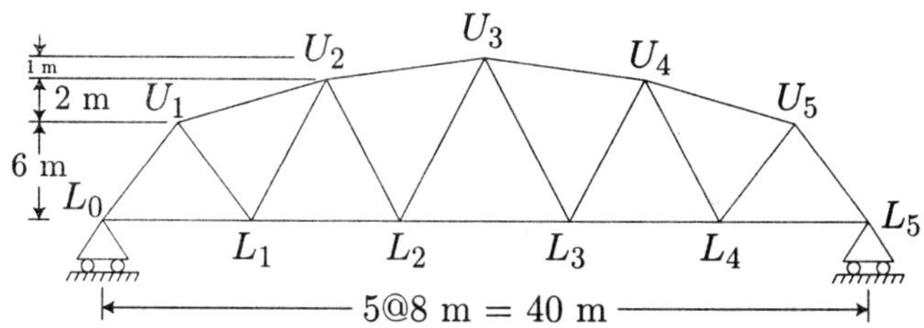

Fig. P-4.5

4.12 Draw influence lines vertical reaction and B.M. at B and B.M. and S.F. at a section 5 m from C of the overhanging beam as shown in Fig. P-4.6.

Fig. P-4.6

4.13 Draw influence lines for forces in the members L_4U_4, U_4M_5, M_5L_5 and U_4U_5 of the truss shown in Fig. P-4.7.

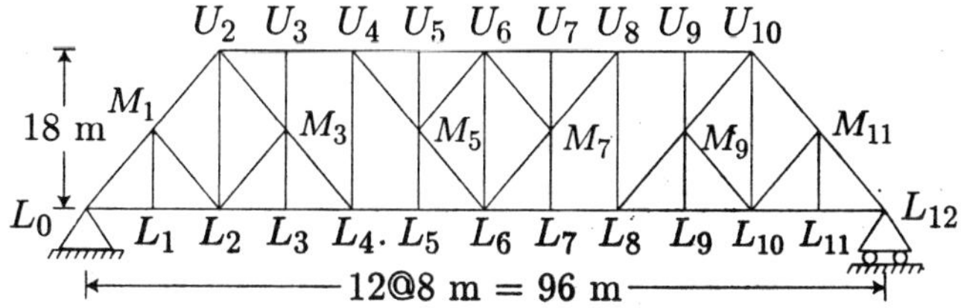

Fig. P-4.7

4.14 A through type four panel warren girder (Fig. P-4.8) is formed with equilateral triangles.

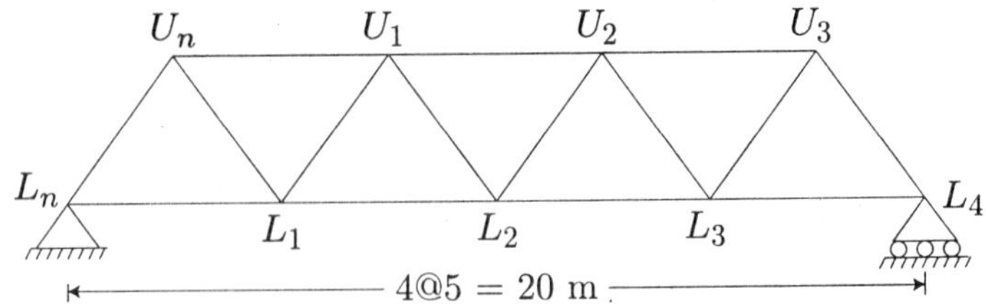

Fig. P-4.8

(a) Draw the influence line for the force in member U_2L_2 and U_1U_2.

(b) Determine the maximum tensile and compressive forces in the diagonal U_2L_2 and number U_1U_2 when a live load whose length is greater than the span of intensity 80 kN per metre per truss, crosses the girder dead weight being 20 kN per metre of the span.

4.15 What are influence lines? Where and how are they used?

4.16 Draw influence lines for forces in members U_1L_2, U_1U_2, U_2L_2 and L_1L_3 of the truss shown in Fig. P-4.9. The load is applied to truss only through the joints of the lower chord.

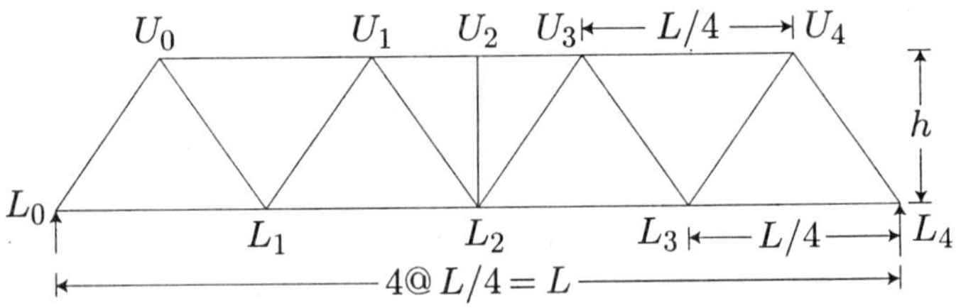

Fig. P-4.9

Hence evaluate the forces in the members mentioned above due to a moving load which is of uniformly distributed load of intensity w and longer than the span.

4.17 Draw the influence lines for the forces in the members x and y of the girder shown in Fig. P-4.10. If the girder is subjected to a uniform dead load of 30 KN/m and u.d.l. of 60 kN/m which can take up any position on the span, determine the maximum and minimum forces in the members x and y.

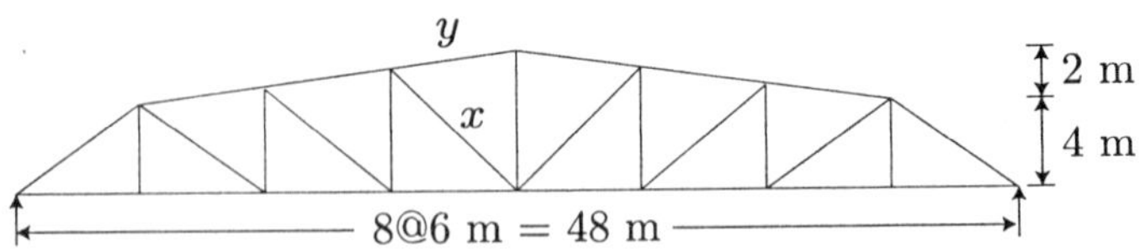

Fig. P-4.10

4.18 An N girder 48 m span is loaded on the top boom and also supported at the ends of the top boom. It has panels each 6 m long in the top and bottom boom. There is a dead load of 15 kN/m and a travelling load longer than the span of 30 kN/m runs. Determine the maximum forces in the three members cut by a vertical section 15 m from the left hand support.

4.19 Draw influence lines for the following functions if a unit load moves from B to D in the structures shown in Fig. P-4.11 (a) Reaction at A (b) S.F. and B.M. at E and F and (c) B.M. at B.

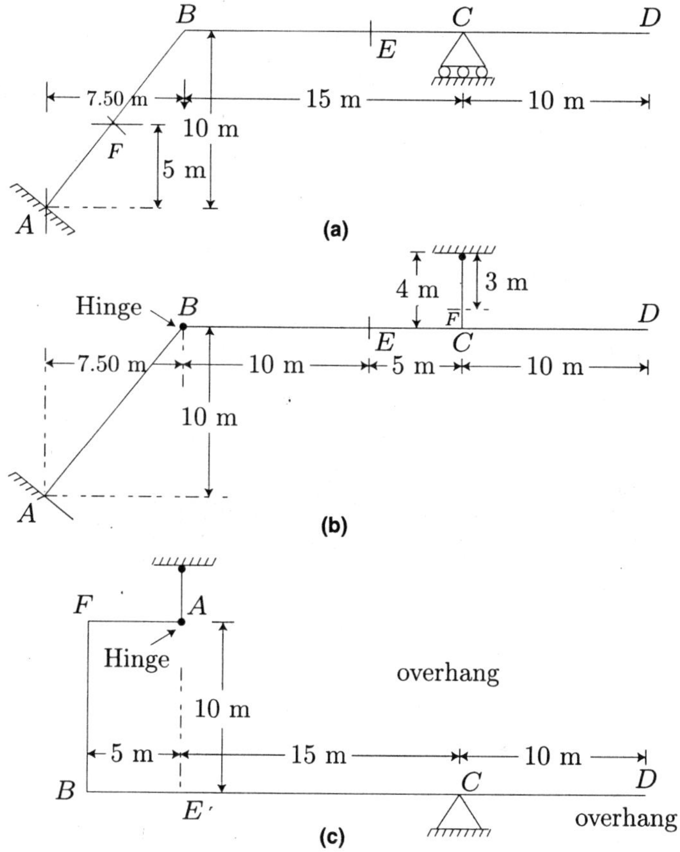

Fig. P-4.11

4.20 For the beams shown in Fig. P-4.12 draw influence lines for the following effects : (a) Reactions at supports A and B. (b) B.M. and S.F. at C.

Consider Load moving over the top beam.

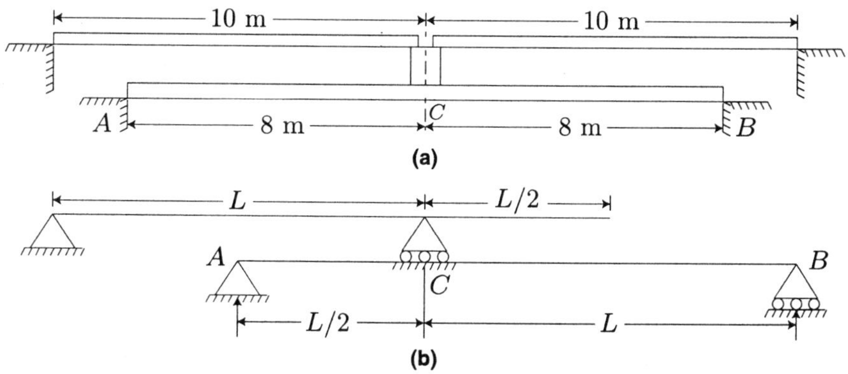

Fig. P-4.12

4.21 For each of the beams shown in Fig. P-4.13, draw influence lines for the following : (a) Reactions at supports (b) B.M. at points F, D and E (c) S.F. in panel CD.

Consider load moving over the top beam.

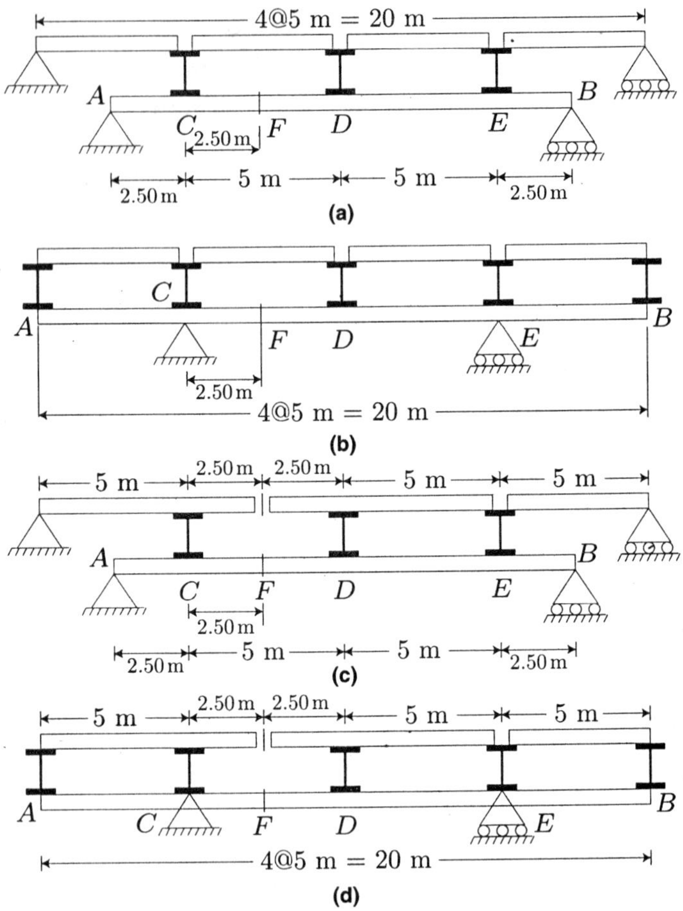

Fig. P-4.13

4.22 A symmetrical three hinged parabolic arch has a span of 60 m, the rise being 6 m. Draw the influence lines for horizontal thrust and normal force, radial shear and B.M. at quarter span.

4.23 For the arch in Fig. P-4.14, draw the influence line for the horizontal thrust at the supports B and C.

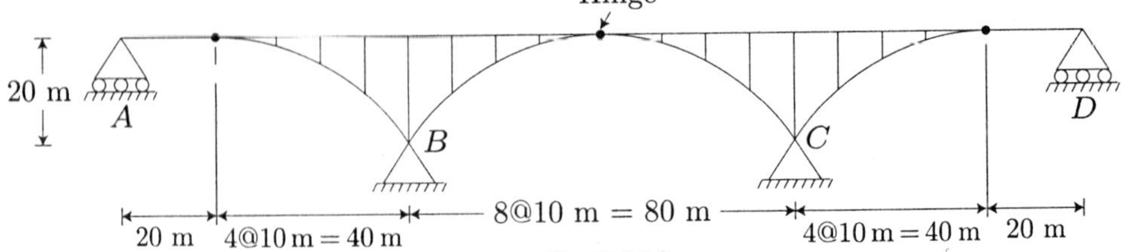

Fig. P-4.14

4.24 A load of 200 kN crosses a simply supported beam of 20 m span. Find the values of maximum positive and negative S.F. and B.M. at a section 8 m from the left-hand support. Also draw the maximum S.F. and B.M. diagrams.

4.25 A simply supported beam of span L is crossed by a uniformly distributed load of length 'a' and of total weight W. If $L > a$, obtain from first principle an expression for the maximum bending moment at any point, distance x from one support. Also show that a single point load of magnitude $W\left[1 - \frac{a}{2L}\right]$ travelling across the span will give the same maximum bending moment everywhere along the beam as the above uniformly distributed load.

4.26 Two point loads 100 kN and 50 kN, 4 m apart cross a simply supported girder, the smaller load leading from left to right. Construct the maximum S.F. and B.M. diagrams stating the position and values of maximum bending moments, if

(a) the span of the girder be 15 m.

(b) the span of the girder be 10 m.

(c) the span of the girder be 6 m.

Also calculate the equivalent uniformly distributed load from point of view of B.M. and S.F. for the cases, if the load system be reversible.

4.27 A system of load consisting of three point loads 30 kN, 50 kN and 50 kN spacing between 30 kN and 50 kN being 1 m and that between 50 kN and 50 kN being 2 m crosses a simply supported girder of span 10 m, the 30 kN load leading from left to right. Find (a) the maximum bending moment in the span, (b) the maximum reaction and (c) the maximum S.F. and B.M. at 4 m from left-hand support.

Also construct the maximum B.M. and S.F. diagrams in the beam. Find the equivalent uniformly distributed load, if the load system be reversible.

4.28 A beam AB, 13 m long is simply supported at left-hand A and at another point B, 3 m from the other end. The system of load as in Problem 4.27 moves over the beam. Determine

(a) Maximum positive B.M.

(b) Maximum negative B.M.

(c) Maximum S.F. at B and

(d) Maximum reactions at A and B.

4.29 A simply supported beam AB spans 16 m. A reversible load comprising of four loads of 50 kN, 200 kN, 100 kN and 100 kN all at 3 m centres cross the span. Obtain the magnitude of the maximum B.M. and its location and the maximum shear force at supports and at quarter point. Determine the equivalent uniformly distributed load from the point of view of B.M. and S.F.

4.30 Solve Problem 4.28 if the moving load consists of a uniformly distributed load of 10 kN/m and a Knife edge load of 50 kN.

4.31 A road bridge consists of two cantilevers AB and CD, each 6 m long and there is a suspended span BC, whose length is 8 m. Draw influence line for (a) B.M. and S.F. at A (b) S.F. at B and (c) B.M. at the middle of BC. Find the maximum B.M. and S.F. at A when a train of loads in kN of 40, 80, 140, 80 and 20 separated by distances of 3 m, 2 m, 2 m and 2 m crosses the span. Consider the load reversible.

4.32 A simply supported beam of span 50 m is traversed by loads moving from left to right as shown in Fig. P-4.15. Find for each case the maximum B.M. and S.F. at 10 m and 20 m from left support. Also calculate the magnitude of the maximum B.M. and S.F. in the beam.

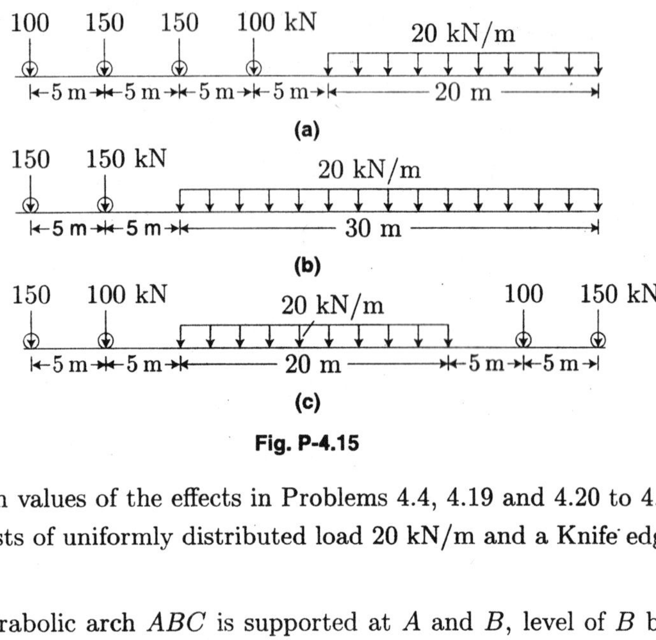

Fig. P-4.15

4.33 Find the maximum values of the effects in Problems 4.4, 4.19 and 4.20 to 4.23, when the live load consists of uniformly distributed load 20 kN/m and a Knife edge load of 200 kN.

4.34 A three-hinged parabolic arch ABC is supported at A and B, level of B being 5 m above A. The span of the arch is 20 m and the third hinge is at the crown at C, which is 12 m from A, its position being 9 m above A. Construct influence lines for horizontal thrust, and reactions and normal force, radial shear and B.M. at 6 m from support A. Find their maximum values when a live load consisting of a uniformly distributed load of 20 kN/m and a Knife edge load of 200 kN moves over the arch.

4.35 A simply supported beam of 20 m span is subjected to a uniform dead load of 10 kN/m. Obtain the combined B.M. and S.F. diagrams and the focal length of the beam for the following cases :

(a) Live load uniformly distributed of magnitude 20 kN/m larger than the span.

(b) Live load uniformly distributed of magnitude 20 kN/m and of length 10 m.

4.36 A train of loads consisting of 50 kN, 100 kN, 100 kN, 75 kN, 75 kN and 75 kN separated by distances of 3 m, 2 m, 2 m, 2 m and 3.50 m moves on a simply supported beam of span 10 m. Construct graphically the maximum B.M. and S.F. diagrams. Also compare the maximum values with the ones obtained analytically. Consider the load moving from left to right with 50 kN load leading.

4.37 A simply supported beam 10 m span carries a moving uniform load of 20 kN/m and two moving concentrated load of 50 kN each spaced 3 m on centres. Calculate the maximum B.M. and S.F. in the beam.

4.38 A portal frame $ABCD$ is hinged at the supports A and D at the same level and at a point E. AB and DC are vertical and of height 5 m. Length BC which is horizontal is 20 m and point E is midway between B and C. Construct influence lines for normal force, radial shear and B.M. at quarter span. Also find the maximum values when a live load consisting of uniformly distributed load 20 kN/m and a Knife edge load 200 kN moves over BC.

4.39 Determine the maximum B.M. induced in the beam AB for the system of beams and connected loads of 20 and 50 kN shown in Fig. P-4.16.

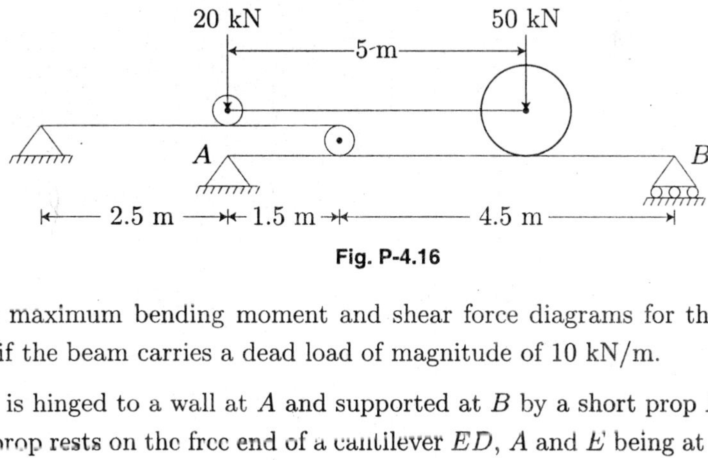

Fig. P-4.16

4.40 Construct the maximum bending moment and shear force diagrams for the case in Problem 4.26 if the beam carries a dead load of magnitude of 10 kN/m.

4.41 A beam ABC is hinged to a wall at A and supported at B by a short prop BD. The end D os the prop rests on the free end of a cantilever ED, A and E being at the same vertical plane. Length $AB = ED = 10$ m and $BC = 2$ m. Draw influence lines for B.M. and S.F. at E and at a section midway between E and D, if the load moves on AC. From the influence lines, determine B.M. and S.F. at the sections, if a uniform load of 20 kN/m is on AC and a concentrated load 40 kN is at C.

4.42 A symmetrical truss, simply supported at ends and of span 24 m, has horizontal top and bottom chords connected by verticals and diagonals in each panel. Height of the truss is 3 m. Bottom chord has 6 equal panels 4 m each and the top chord has 4 equal panels 4 m each, top and bottom chords being connected by means of diagonals. The other diagonal members are such that they will be under tension due to the effect of self weight of the truss.

Obtain I.L. diagrams for the following members of the truss, considering loads moving over the bottom chord :

(a) Central vertical member

(b) First vertical member

(c) First top chord and third bottom chord members

(d) Inclined member in the third panel.

Assuming D.L. as 5 kN/m uniformly distributed and L.L. as 10 kN/m uniformly distributed together with a Knife edge load of 200 kN, find the maximum forces in the members.

4.43 Determine the influence lines for bar forces in members marked a, b and c of the truss in Fig. P-4.17.

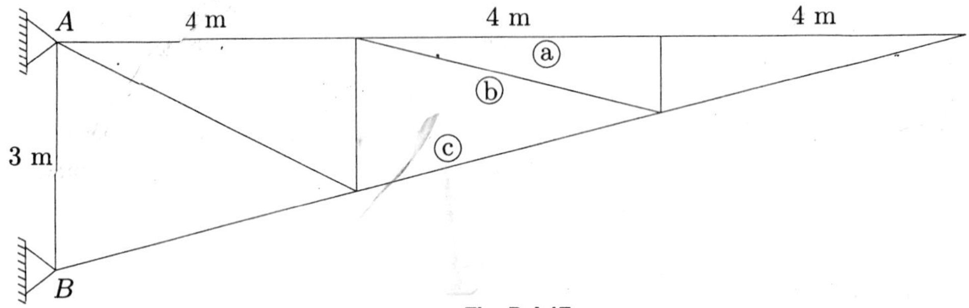

Fig. P-4.17

4.44 A bridge truss of N type has a simply supported span consisting of 10 panels each of length 5 m, the height of the truss being 6 m. The dead load is 20 kN/m and the L.L. is 40 kN/m longer than the span. Determine the maximum and minimum forces in the diagonal and the chord members of the fourth and fifth panels from the left-hand support. Also state if the panels may need counter bracing.

4.45 Determine the influence lines for bar forces in members marked a, b, c and d of the trusses in Figs. P-4.18 and P-4.19.

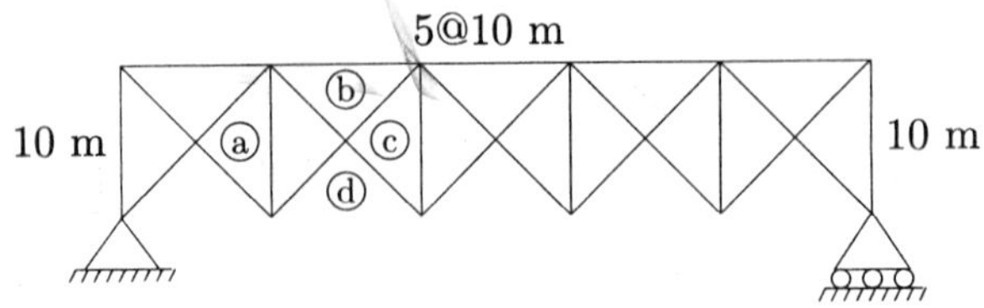

Fig. P-4.18

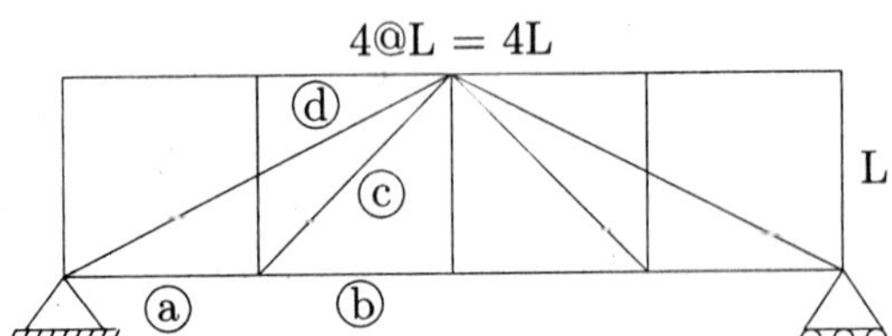

Fig. P-4.19

4.46 Construct influence lines for bar forces in members U_4U_5, L_4L_5, M_4U_5, M_4L_5, M_4U_4 and M_4L_4 of the K truss shown in Fig. 3.13 of chapter 3.

4.47 A warren girder has 5 panels 6 m each, the height being 4 m. Construct the influence line diagrams for the third bottom chord member, the second top chord member and the diagonal in the panel with the above two chord members.

Also determine the maximum values of the bar forces when a series of concentrated load consisting of 50, 60, 100, 100, 100 and 100 (all in kN) separated by 1 m, 2 m, 3 m, 3 m, and 3 m moves over the bottom chord of the truss from right to left with the 50 kN load leading.

4.48 Find the maximum bar forces in Problems 4.9 to 4.[illegible] of chapter 4 when the moving load is uniformly distributed of magnitude 20 kN/m and of length 12 m.

4.49 Derive the condition for maximum being moment on a span of 'L' when two equal concentrated loads of intensity 'W' each and distance 'd' apart cross the span. Determine the maximum bending moment on the span.

4.50 A gantry girder has a simply supported span of 10 m. Two wheel loads 120 kN each distant 2 m apart cross the span. Determine the maximum bending moment on the span.

4.51 A uniformly distributed load of length 1·2 m and intensity 90 kN per metre crosses a span of 30 metres. Determine the maximum bending moment and shearing force at a section 8 m from the left-hand support.

4.52 Determine the absolute maximum bending moment induced in a simply supported girder of 20 m span due to two concentrated loads of 300 kN and 400 kN moving at a fixed distance 2 m apart.

4.53 A load of 100 kN, uniformly distributed over a length of 3 m travels across a simply supported beam of span 12 m. Evaluate the maximum values of bending moment and shear at quarter span. Also find eq.u.d.l. values for the above cases.

4.54 (a) Derive the condition for absolute maximum B.M. on the span to occur when two concentrated loads W_1 and W_2 at a fixed distance d apart, are travelling over a sample span.

(b) Draw the maximum bending moment diagram and maximum shear force diagram for a simple beam of 25 m span when two concentrated loads of 50 kN and 100 kN 4 m apart move on the beam.

Chapter 5

Deflection of Structures, Strain Energy and Unit Load Methods

5.1 Introduction

The maximum load that can be applied to a beam sometimes may be limited by an elastic deflection, beyond which the beam would not function satisfactorily under the given service conditions. This means that the stiffness, rather than the strength, of the beam is the governing condition that the beam must satisfy. It is important that relationship between the loads acting on a beam and the elastic deformation or deflection caused by the loads is to be found. This relationship will include the stiffness of the material of which the beam is made (as indicated by the modulus of elasticity) and the dimensions of the beam. The general form of this relationship is expressed by the elastic-curve equation for the beam.

The main purpose is to derive the equation of the elastic curve for a beam. It is necessary to use the equation for determining the deflection at any point of the elastic curve (especially the maximum elastic deflection) of beams of various types.

The elastic curve of an originally straight beam is the curve of the longitudinal centroidal axis of the stressed beam, provided that the maximum stress in the beam does not exceed the proportional limit of the material. For beams of usual proportions and materials, this means that the deflections are small. The deflections in the figures are exaggerated in order to indicate the form of the elastic curve. Since the elastic curve lies in the neutral surface of the beam, it does not change its length as the beam is bent. Since the elastic deformations are small a differential length dL along the elastic curve of an originally horizontal straight beam usually may be assumed to be equal to the horizontal differential length dx, as in Fig. 5.1. The change in slope at two points on the curve the distance dL apart may be assumed to be equal to the change in the angle $d\theta$ itself ($\tan\theta = \theta = dy/dx$). The deflection y of any point of the elastic curve and especially the maximum deflection $y_{\max}$ will have to be found.

5.2 General Features of Elastic Behavior of Beams

The general shape of the elastic curve of a definite beam can be determined from the bending-moment diagram. Fig. 5.1(a) represents a horizontal simply supported beam that overhangs on its right-hand support. The bending-moment diagram is shown in Fig. 5.1(b). The elastic curve of the beam is represented in Fig. 5.1(a). The curve from A to I is concave upward since the bending moment is positive at all sections of the beam from A to I, whereas the curve is concave downward from I to C since the bending moment at all sections from I to

C is negative. The point of the elastic curve at which the curvature changes from concave upward to concave downward is called the *inflection point,* and it occurs at the section of the beam for which the bending moment is zero, as indicated in Fig. 5.1(b).

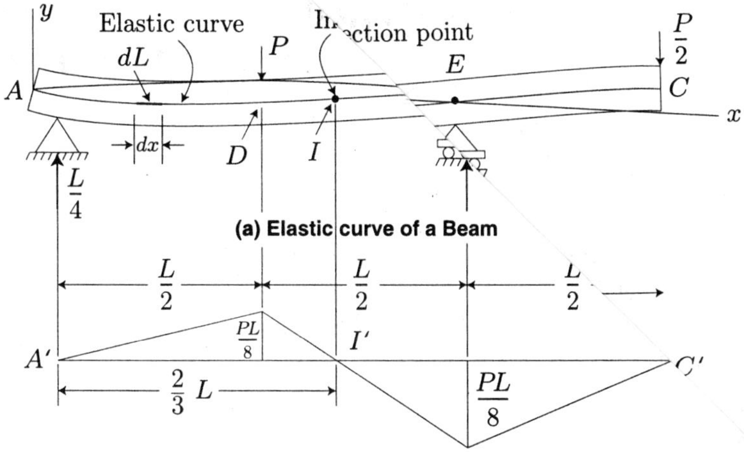

(a) Elastic curve of a Beam

(b) Bending moment diagram

Fig. 5.1: General characteristics of elastic curve of beam

In the solution of the main problem, it will be convenient to consider the beam to be made up of elements consisting of thin slices of length $dL = dx$. Such an element is shown as $GHFE$ in Fig. 5.2. The bending moments M and $M + dM$ acting on the vertical faces of any element cause one face of the element to rotate with respect to the other face through an angle $d\theta$. The value of M, of course, varies with the position of the element in the beam. The angle change $d\theta$ is equal to the change of slope of the elastic curve in the length dL and the change of slope $\Delta\theta$ over a finite length of the curve, as from A to B, is the summation of the angle changes in all the elements from A to B. Thus,

$$\Delta\theta_{AB} = \int_A^B d\theta. \tag{5.1}$$

To determine the elastic curve of the beam it is necessary to find an expression for $d\theta$ in terms of $M, E, I,$ and dx, so that the integral in Eq. (5.1) can be evaluated.

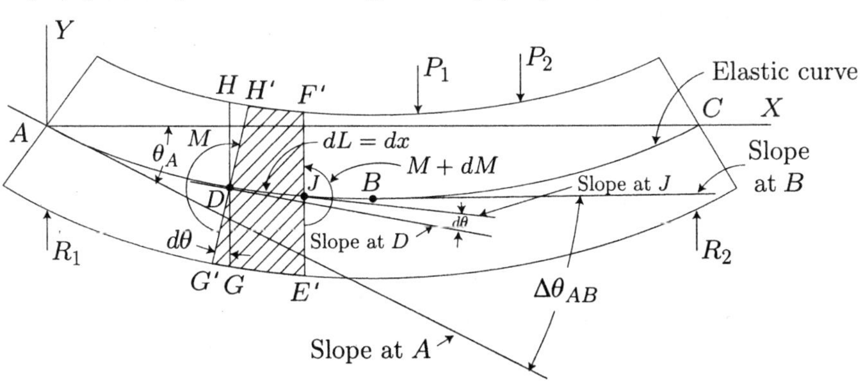

Fig. 5.2: Change of slope to elastic curve of beam

5.3 Equation of Elastic Curve in Terms of Angle Change

Fig. 5.2(a) shows a deflected beam similar to that in Fig. 5.2. The line HG is drawn parallel to $F'E'$ to show the angle chage $d\theta$ through which $H'G'$ has rotated relative to $E'F'$. In the triangle DGG' if the elongation GG' of the extreme fibre GE' be denoted by de, since GG' is small in comparison with the distance $c(DG)$ from the neutral axis to the extreme fibre, the angle GDG' may be obtained from the equation

$$\text{Angle } GDG' = d\theta = de/c. \tag{5.2}$$

But the elongation de of the extreme fibre is equal to ϵdL in which ϵ is the strain in the fibre and dL the fibre length. Furthermore, as previously shown $\epsilon = \sigma/E$ and $\sigma = Mc/I$, where σ is the stress in the extreme fibre. By making these substitutions in Eq. (5.2),

$$d\theta = \frac{\epsilon dL}{c} = \frac{\sigma dL}{Ec} = \frac{\frac{Mc}{I}dL}{Ec} = \frac{MdL}{EI}, \tag{5.3}$$

in which M is the bending moment (the change dM is neglected) in the element of length dL for which the angle change is $d\theta$, I, the moment of inertia of the cross section, and E the modulus of elasticity of the material. Eq. (5.3) is the elastic-curve equation for the beam expressed in terms of the change in slope $d\theta$ between two tangents drawn at the ends of the element of length dL of the curve in terms of M, E, I, and dL. The expression for $d\theta$ from Eq. (5.3) is substituted in Eq. (5.1), which gives

$$\Delta\theta = \int \frac{M}{EI} dL. \tag{5.4}$$

This equation can be used to great advantage in determining the deflection of a beam.

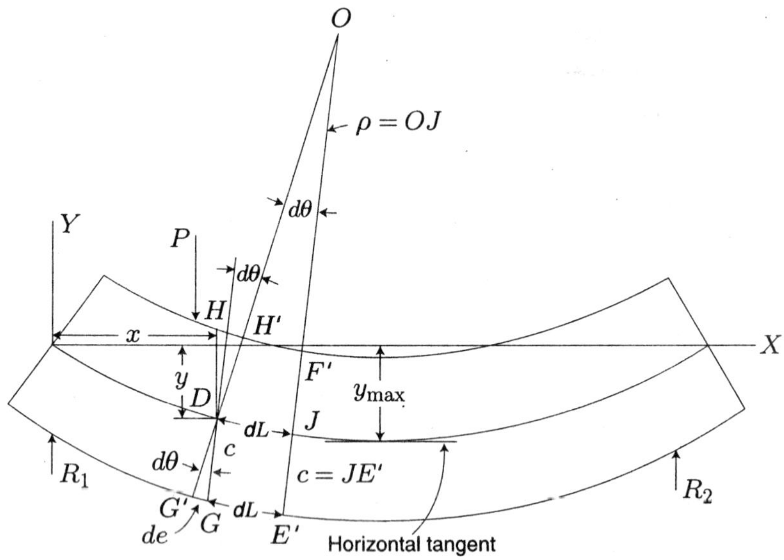

Fig. 5.2a: Analysis of strains in beam for determining deflection of beam

Example 5.1. *A cantilever beam (Fig. 5.3) of length L is fixed at its right end and supports a concentrated load P at its free end. If E be the modulus of elasticity of the material and*

I the moment of inertia of the cross section, compute the change in slope $\Delta\theta$ between the fixed end and the free end in terms of P, L, E and I.

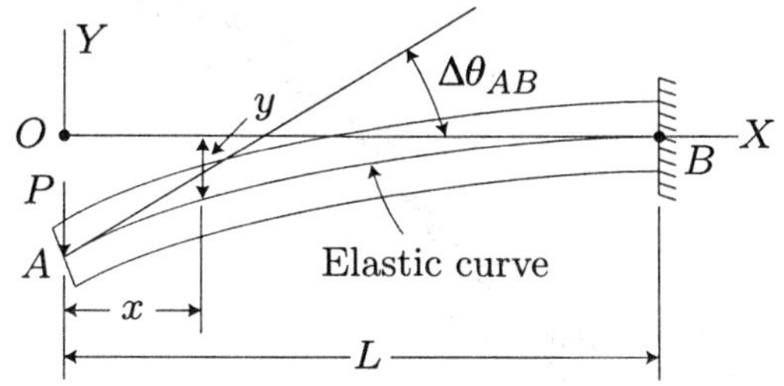

Fig. 5.3: For Example 5.1

Solution : From Eq. (5.4),

$$\Delta\theta_{AB} = \int_A^B \frac{M}{EI} dL = \int_0^L \frac{M}{EI} dx.$$

Since the changes in slope are assumed to be small (greatly exaggerated in Fig. 5.3.) $dL = dx$, and the bending moment $M = -Px$. Hence,

$$\Delta\theta_{AB} = \int_0^L -\frac{Pxdx}{EI} = -\left[\frac{Px^2}{2EI}\right]_0^L = -\frac{PL^2}{2EI}.$$

It is to be noted that the slope of the beam at its fixed end does not change when the beam is loaded. For this reason the value of $\Delta\theta_{AB}$ in the foregoing equation is the actual slope at the free end of the beam.

Example 5.2. *A beam of constant cross section is made of an aluminum alloy and is supported and loaded as shown in Fig. 5.4. The cross section of the beam is 50 mm wide and 200 mm deep. By making use of sketch of the elastic curve of the beam and the bending moment diagram, compute the following values : (a) maximum bending stress in the beam; (b) change in slope from A to B; (c) change in slope from C to B; (d) change in slope from A to C.*

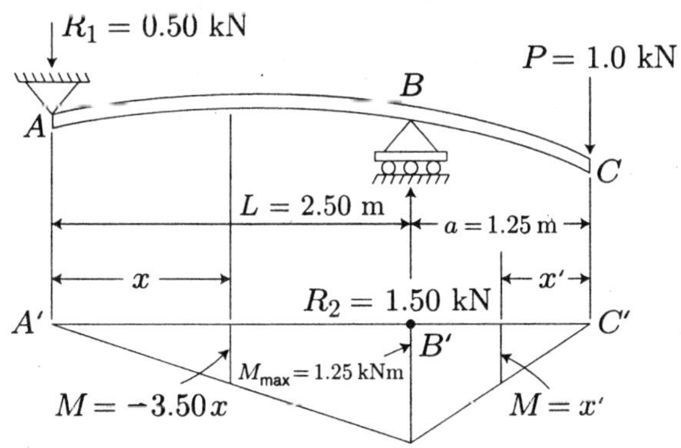

Fig. 5.4: For Example 5.2

Solution : (a) $I = \frac{1}{12}(50)(200)^3 = 33333333.33 \text{ mm}^4$

and $\sigma_{max} = \frac{M_{max}c}{I} = \frac{1.25(10)^6(100)}{I} = 3.75 \text{ N/mm}^2.$

This stress is less than the proportional limit.

(b) $\Delta\theta_{AB} = \int_A^B \frac{M}{EI}dx = \int_0^{2.50} -\frac{0.50x}{EI}dx = \left[-\frac{0.25x^2}{EI}\right]_0^{2.50} = \frac{0.046875}{E}.$

(c) $\Delta\theta_{CB} = \int_C^B \frac{M\,dx'}{EI} = \int_0^{1.25} \frac{1(x')}{EI}dx = -\frac{0.0234375}{E}.$

(d) $\Delta\theta_{AC} = \Delta\theta_{AB} + \Delta\theta_{CB} = -\frac{0.0703125}{E}$ radian.

5.4 Moment-Area Method

This is a semi-graphical method in which the bending moment diagram of a member is used for determination of slope and deflection.

Fig. 5.5(a) shows two points A and B, placed at a distance apart on the elastic line of a beam, ds being the length of a small element of the elastic line at a distance x horizontally from B. The angle $d\theta$ between the tangents or normals drawn at the two extremities of the element ds is given by

$$d\theta = \frac{ds}{R}, \tag{5.5a}$$

where R is the radius of curvature of the elastic line. As curvature $\frac{1}{R}$ equals $\frac{M}{EI}$, where M is the bending moment at x, I the moment of inertia of beam section about the neutral axis and E the modulus of elasticity, Eq(5.5a) reduces to

$$d\theta = \frac{M}{EI}ds. \tag{5.5b}$$

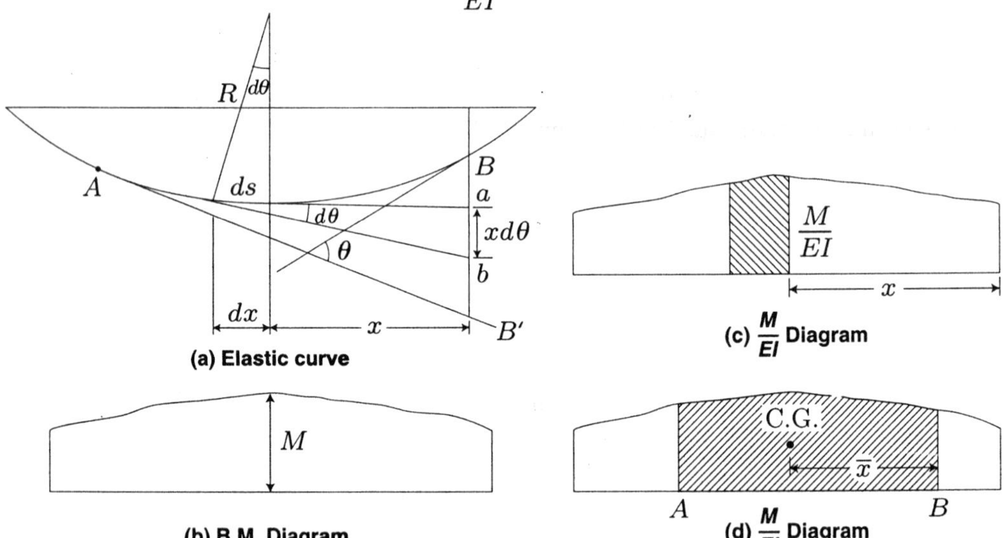

Fig. 5.5: Moment-area method

For small deflections, the elastic line is a very flat curve for which

$$ds \approx dx,$$

whence Eq. (5.5b) becomes

$$d\theta = \frac{M}{EI}dx. \tag{5.5c}$$

The angle θ between the tangents to the elastic curve at A and B can be obtained by integrating,

Thus
$$\theta = \int_0^L \frac{M}{EI} \cdot dx. \tag{5.5d}$$

Fig. 5.5(b) shows the corresponding B.M. diagram for the beam and Fig. 5.5(c) the $\frac{M}{EI}$ diagram obtained by dividing B.M. ordinates by EI values at the corresponding sections. Since dx is infinitely small, $\frac{M}{EI} \cdot dx$ indicates the area of the elemental strip of M/EI diagram of length dx. Hence the integral $\int_0^L \frac{M}{EI} dx$ on the right-hand side of Eq. (5.5d) represents the area of $\frac{M}{EI}$ diagram between the points A and B. Thus, the first theorem of moment-area method may be stated as follows :

Theorem I. The angle in radians between the tangents to the elastic line of a straight member at any two points is numerically equal to the area of $\frac{M}{EI}$ diagram between the said two points.

The intercept ab made by the tangents at the ends of the element ds on the line drawn through B perpendicular to the original axis of the beam is given by

$$ab = x d\theta = \frac{Mx}{EI} dx.$$

The intercept BB' made by the tangents at A and B on the same line through B is then obtained by integration as

$$BB' = \int_0^L \frac{Mx}{EI} dx = A\overline{X}, \tag{5.6}$$

where A is area of M/EI diagram between A and B and $\overline{X}$ is the distance of centroid of area from the point B. Since $\frac{Mxdx}{EI}$ represents moment of the elemental strip [Fig. 5.5(d)] about B, the integral $\int_0^L \frac{Mx}{EI} \cdot dx$ represents moment of the area of $\frac{M}{EI}$ diagram between A and B, shown shaded in Fig. 5.5(d) about B. Eq. (5.6) is the mathematical expression of the second theorem of moment-area method which may be stated as follows :

Theorem II. The deflection of a point B of a beam from the tangent to the elastic line at another point, A, measured perpendicular to the original axis of the beam is numerically equal to the moment of the area of $\frac{M}{EI}$ diagram between the points A and B about the point B.

5.5 Sign Convention of Moment-Area Method

With sagging bending moment as positive and hogging bending moment as negative, if the net area of $\frac{M}{EI}$ diagram between any two points is positive, the tangent to the elastic curve at the right-hand side point makes an anti-clockwise rotation with respect to the tangent at the left-hand side point and if the net area of $\frac{M}{EI}$ diagram is negative, the said rotation is clockwise.

With the sign convention for bending moment as above, if the net moment of $\frac{M}{EI}$ diagram between any two points A and B for a member about the point B is positive, then the deflection of the point B is such that it lies above the tangent to the elastic curve at A and if the net moment of $\frac{M}{EI}$ diagram is negative, B lies below the tangent at A.

5.6 Slope and Deflection by Moment-Area Method

Theorem I and Theorem II of moment-area method do not give directly the actual slope and deflection of a member except in case of a cantilever. Fig. 5.6(a) shows a simple beam with a given system of loads for which $\frac{M}{EI}$ diagram is as shown in Fig. 5.6(b) and deflected shape of the axis as shown in Fig. 5.6(c). At the point C at a distance x from A, the actual

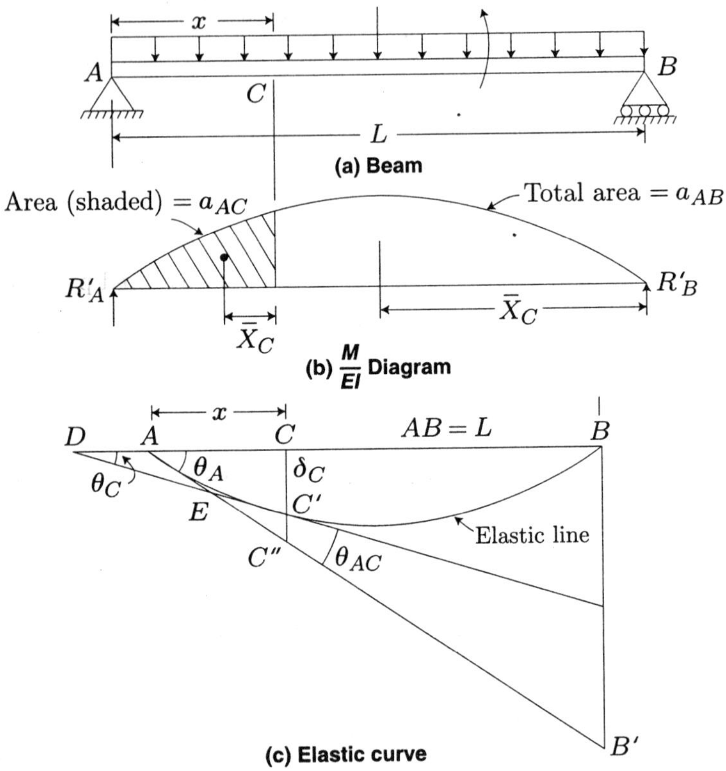

Fig. 5.6: Use of moment-area method

slope θ_C is given by the angle CDC', between the tangent to the elastic line at C', the displaced position of C, with the original axis AB. The actual deflection of C, i.e., δ_C is the distance CC' between the original and displaced positions of the point. But neither θ_C nor δ_C can be directly obtained from the two theorems of moment-area method. These can, however, be determined with the help of the moment-area method.

In Fig. 5.6(c), AB' drawn as tangent to the elastic curve at A intersects the lines through B and C drawn perpendicular to AB at B' and C''. Line DC' drawn as tangent to the elastic curve at C', intersect line AB' at E. Since the elastic curve is very flat

$$\theta_A = \frac{BB'}{L}.$$

According to theorem II of moment-area method,

$$BB' = \text{moment of } \frac{M}{EI} \text{ diagram between } A \text{ and } B \text{ about } B' = a_{AB} \cdot \overline{X}_B,$$

and $$\theta_A = \frac{a_{AE} \cdot \overline{X}_B}{L}.$$

From Fig. 5.6(c) again,

$$\theta_c = \angle ADE = \angle CAE - \angle AED = \angle CAE - \angle C'EC'',$$

whence $\theta_c = \theta_A - \theta_{AC}$.

In this expression θ_{AC} is the angle between the tangents to the elastic curve at A and C. θ_{AC} is obtained from theorem I of moment-area method as

$$\theta_{AC} = \text{area of } \frac{M}{EI} \text{ diagram between } A \text{ and } C = a_{AC}.$$

Hence $$\theta_C = \theta_A - \theta_{AC} = \frac{a_{AB} \cdot \overline{X}B}{L} - a_{AC}. \qquad (5.7)$$

The deflection σ_C is obtained from

$$\delta_C = CC' = CC'' - C'C'' = \theta_A \cdot x - C'C'',$$

where $C'C''$ is the deflection of C from the tangent at A measured perpendicularly to AB and is obtained from theorem II of moment-area method as

$$C'C'' = \text{ moment of } \frac{M}{EI} \text{ diagram between } A \text{ and } C \text{ about } C$$
$$= a_{AC} \cdot \overline{X}_C,$$

whence $$\delta_C = \theta_A \cdot x - a_{AC} \cdot \overline{X}_C = \frac{a_{AB}.\overline{X}_B}{L} \cdot x - a_{AC}.\overline{X}_C. \qquad (5.8)$$

For a cantilever as shown in Fig. 5.7, since the slope at A is zero, the slope θ_C at the point C is given by the angle between the tangents to the elastic curve at A and C and is directly obtained from Theorem I. The deflection δ_C at C is the deflection of point C from the tangent at A and is obtained directly from Theorem II as the moment of $\frac{M}{EI}$ diagram between A and C about C.

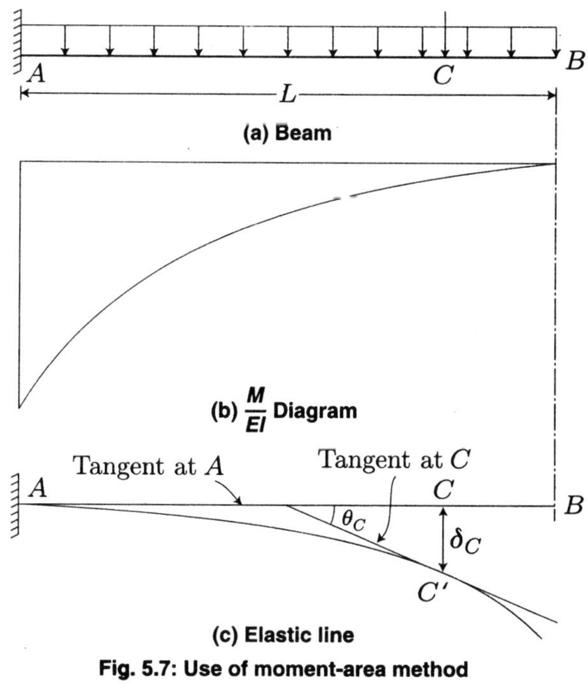

Fig. 5.7: Use of moment-area method

Example 5.3. *A beam ABC having depth 100 mm and width 48 mm is simply supported at A and B with BC overhanging. Length* $AB = 4$ m *and* $BC = 2$ m. *The beam carries a uniformly distributed load of 2 kN/m and a concentrated load of 3 kN at the free end C as shown in Fig. 5.8(a). Find slope at B and deflection at C, if* $E = 2(10)^5$ N/mm^2.

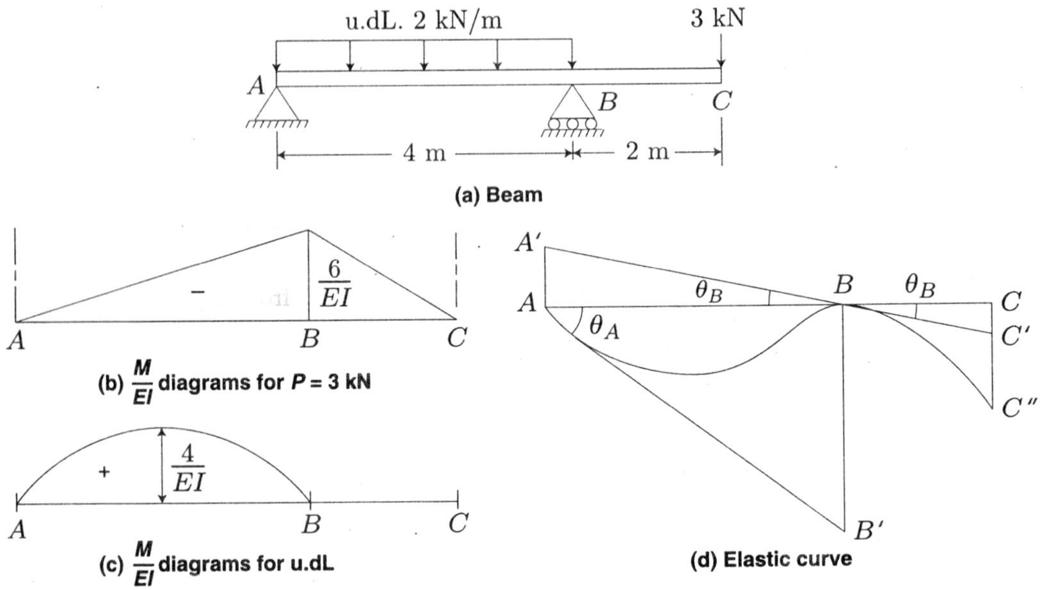

(a) Beam

(b) $\frac{M}{EI}$ diagrams for $P = 3$ kN

(c) $\frac{M}{EI}$ diagrams for u.dL

(d) Elastic curve

Fig. 5.8: For Example 5.3

Solution : Figs. 5.8(b) and 5.8(c) show $\frac{M}{EI}$ diagrams for the beam due to the concentrated and uniformly distributed loads respectively. In Fig. 5.8(d) ABC'' is the assumed deflected shape of the beam axis, ABC and $A'BC'$ being original axis and tangent at B to the elastic curve.

For the present case,

$$I = \frac{1}{12}(48)(100)^3 = (4)(10)^6 \text{ mm}^4.$$

and $$EI = (4)(10)^6(2)(10)^5 \text{ Nmm}^2 = 800 \text{ kNm}^2.$$

Also AA' = deflection of A from the tangent at B

$$= \text{moment of } \frac{M}{EI} \text{ diagram between } A \text{ and } B \text{ about } A$$

$$= \frac{2}{3}\left(\frac{4}{EI}\right)(4)(2) - \frac{1}{2}\left(\frac{6}{EI}\right)(4)\left(\frac{2}{3}\right)(4) = -\frac{32}{3EI},$$

in which negative sign indicates that A lies below the tangent at B as shown in the figure and slope θ_B is clockwise with respect to the original axis.

$$\theta_B = \frac{AA'}{AB} = \frac{(32)}{3EI}\left(\frac{1}{4}\right) = \frac{8}{3EI} = \frac{8}{(3)(800)} = 0.0033 \text{ radian.}$$

The deflection δ_C at C is $CC' + C'C''$

where $CC' = \theta_B \cdot BC = (0.0033)(2000) = 6.6$ mm

and $\quad C'C'' =$ deflection of C from the tangent at B

$= \text{moment of } \dfrac{M}{EI} \text{ diagram between } B \text{ and } C \text{ about } C$

$= \dfrac{1}{2}\left(\dfrac{6}{EI}\right)(2)\left(\dfrac{2}{3}\right)(2) = \dfrac{8}{EI} = \dfrac{8}{800} \text{ m} = 10 \text{ mm},$

whence $\delta_C \quad = 6.6 + 10 = 16.6$ mm

5.7 Conjugate Beam Method

This method is actually an alternate version of the moment-area method and can be easily derived:

Referring to Fig. 5.6 the slope and deflection at C are obtained from Eqs. (5.8) and (5.9) as

$$\theta_C = \theta_A - \theta_{AC} = \frac{a_{AB} \cdot \overline{X}_B}{L} - a_{AC} \tag{5.7}$$

and $$\delta_C = \theta_A \cdot x - a_{AC} \cdot \overline{X}_C = \frac{a_{AB} \cdot \overline{X}_B}{L} \cdot x - a_{AC} \cdot \overline{X}_C. \tag{5.8}$$

If a fictitious beam having the same length as the actual beam is considered to be subjected to $\frac{M}{EI}$ diagram of actual beam as load as shown in Fig. 5.6(b), the reactions of the fictitious beam are given by

$$R'_A = \frac{a_{AB} \cdot \overline{X}_B}{L} \tag{5.9}$$

and $$R'_B = \frac{a_{AB}(L - \overline{X}_B)}{L}. \tag{5.10}$$

The shear force V'_C and bending moment M'_C at point C of the fictitious beam can be obtained as

$$V'_C = R'_A - a_{AC} = \frac{a_{AB} \cdot \overline{X}_B}{L} - a_{ac} \tag{5.11}$$

and $$M'_C = R'_A \cdot x - a_{AC} \cdot \overline{X}_C = \frac{a_{AB} \cdot \overline{X}_B}{L}(x) - a_{AC} \cdot \overline{X}_C \tag{5.12}$$

Comparing Eq. (5.7) with Eq. (5.11) and Eq. (5.8) with Eq. (5.12), it is evident that

$$\theta_C = V'_C$$

and $$\delta_C = M'_C.$$

The two equations above can be expressed in the form of two theorems considering the fictitious beam loaded with $\frac{M}{EI}$ diagram as the conjugate beam having edge conditions to suit the geometric conditions of the actual beam.

Theorem I. The slope at a point of an actual beam is numerically equal to the shear force of the corresponding conjugate beam at corresponding point.

Theorem II. The deflection at a point of an actual beam is numerically equal to the bending moment of the corresponding conjugate beam at the corresponding point.

5.8 Sign Conventions for Conjugate Beam Method

If the bending moment of actual beam be sagging, the corresponding load on the conjugate beam will be taken as acting downward. Then the sagging bending moment of the conjugate beam will be taken as downward deflection of the actual beam and the positive shear force(↑↓) of conjugate beam as clockwise rotation of the tangent to the elastic curve.

5.9 Supports for Conjugate Beams

For the simply supported beam considered in Section 5.7, the supports of the conjugate beam are simple supports similar to that of the actual beam. This may not be true for other support conditions of actual beam. The support conditions of the conjugate beam have to be determined so that the relations enunciated in the theorems suit the geometrical conditions of the supports. As bending moment and shear force of the conjugate beam give respectively the deflection and slope of the actual beam, values of bending moment and shear force at the supports of the conjugate beam must be the same as that of deflection and slope at the supports of the actual beam. At the fixed end of an actual beam, the deflection and slope are both zero and, therefore, the bending moment and shear force at the corresponding end of the corresponding conjugate beam must be both zero as in the case of a free end. This means that the fixed end of an actual beam corresponds to the free end of the conjugate beam. Similarly, the free end of the actual beam corresponds to the fixed end of the conjugate beam. At an unsupported hinge point of the actual beam, there will

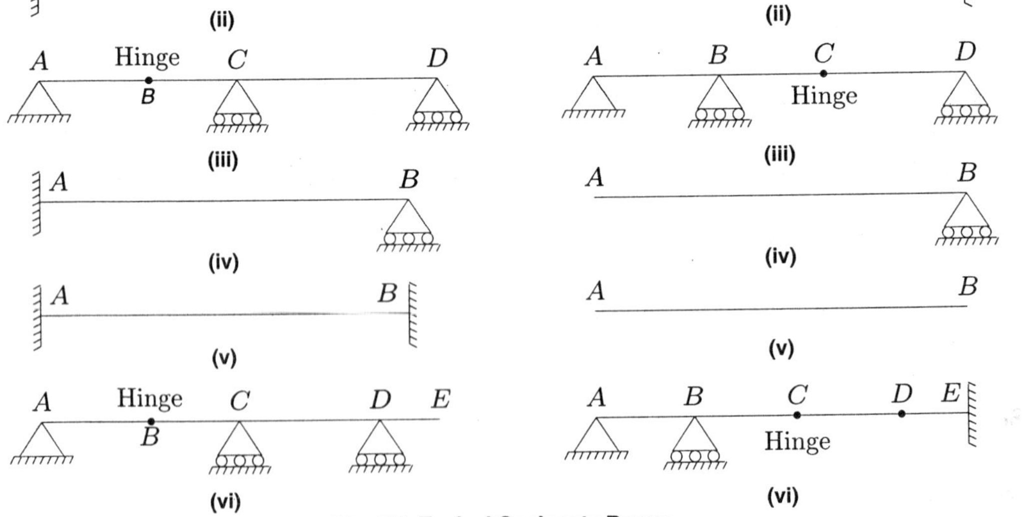

Fig. 5.9: Typical Conjugate Beams

be some deflection and variations of the slopes between the two sides of the hinge. In the corresponding point of the conjugate beam, there must be a bending moment and change of shear forces as in the case of an intermediate support. Similarly, the condition of zero deflection and non-zero slope at an intermediate support of the actual beam corresponds to the condition of zero bending moment and non-zero shear force at the corresponding hinge point of the conjugate beam. Fig. 5.9 shows the support conditions of some typical actual beams and that of the corresponding conjugate beams.

The relations between the supports of actual beam and those of corresponding conjugate beam can be expressed in the form of reversible equations as below :

(i) Simple end supports $\rightleftharpoons$ Simple end supports.

(ii) Fixed end supports $\rightleftharpoons$ free end.

(iii) Intermediate hinge $\rightleftharpoons$ Intermediate simple support.

Example 5.4. *A beam ABC (Fig. 5.10a) having depth 80 mm and width 40 mm is simply supported at A and B with BC overhanging and loaded as shown. Find the slope at A and deflection at C, if* $E = 2(10)^5$ N/mm^2.

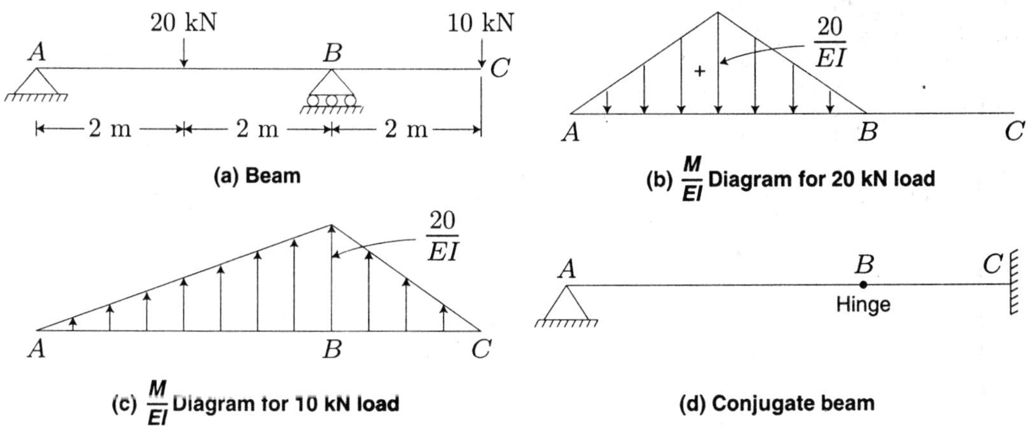

Fig. 5.10: For Example 5.4

Solution : Figs. 5.10(b) and 5.10(c) show the $\frac{M}{EI}$ diagrams due to 20 kN and 10 kN loads and Fig. 5.10(d) shows the corresponding conjugate beam. $\frac{M}{EI}$ diagrams of Fig. 5.10(b) and 5.10(c) will be considered as downward and upward loads on the conjugate beam as the bending moments are sagging and hogging respectively.

The reaction R'_A of the conjugate beam [Fig. 5.10(d)] is obtained by taking moment about the hinge at B and equating it to zero,

$$R'_A = \frac{1}{2}\left[\left(\frac{1}{2}\right)(4)\left(\frac{20}{EI}\right)(2) - \frac{1}{2}(4)\left(\frac{20}{EI}\right)\left(\frac{1}{3}\right)(4)\right]$$

$$= \frac{40}{3EI}, \text{ which is upward .}$$

Hence S.F. at A of the conjugate beam $= +\dfrac{40}{3EI}$.

The slope at A of the actual beam, i.e., rotation of the tangent to the elastic curve at A is $\dfrac{40}{3EI}$ clockwise.

Again, bending moment at C of conjugate beam

$$= \frac{40}{3EI}(6) + \frac{1}{2}(4)\left(\frac{20}{EI}\right)\left(2+\frac{4}{3}\right) + \frac{1}{2}(2)\left(\frac{20}{EI}\right)\left(\frac{2}{3}\right)(2) - \frac{1}{2}(4)\left(\frac{20}{EI}\right)(4)$$

$$= \frac{80}{EI}, \text{ which is sagging .}$$

Hence the deflection of the actual beam at $C = \dfrac{80}{EI}$ downward .

With $EI = 2(10)^8\left(\dfrac{1}{12}\right)(40)(80)^3(10)^{-12}\text{kN-m}^2 = \dfrac{1024}{3}$ kNm^2,

Slope at $A = \dfrac{40}{3EI} = \dfrac{40}{1024}$ radian $= 0.03906$ radian

and deflection at $C = \dfrac{80}{EI} = \dfrac{(80)(3)}{1024}$ m $= 234.38$ mm.

5.10 Strain Energy Method

Strain Energy methods provide a highly versatile means for computing deflections. Although other methods can be used for analysing complicated structures, the energy methods are often preferable, because of the manner in which the solution can be formulated.

A structure may be considered as a conservative system if no energy is added or taken away from it, when it is isolated from its environment. As a structure deforms under the action of loads applied, two types of work are done, viz., external work by external forces on the system and internal work against the deformation in the system by internal forces during the deformation. Internal work must be negative, as deformations are imposed on the system and is resisted by the internal forces.

The principle of conservation of energy is a basic law of Physics. According to this principle the total energy in an isolated structural system remains constant. It follows from the principle that the algebraic sum of the external work W_E and internal work W_i in the system must be zero. In general, however, equality of the external work and internal work is mentioned, which implies that only the absolute values of these quantities are considered. Thus in a structural system in which deformations are consistent with constraint, the displacements of he system can be related to the external forces by means of energy relations.

5.11 Strain Energy and Complementary Energy

In the analysis of displacements by means of energy approach, two energy quantities, viz., the strain energy and the complimentary energy are of particular interest. The first one, i.e., the strain energy which is the energy of deformation is related to the concept of real work, whereas the second one, i.e., the complementary energy which is stress energy is related to the concept of complementary work.

Taking the internal work as negative and strain energy U in a structural system, which is a measure of the internal work as a positive quantity,

$$U = -W_I, \text{ so that } W_E = U.$$

Fig. 5.11 shows a deformable system subjected to gradually applied external loads P_1, $P_2, \ldots, P_3$, which cause displacement $\Delta_1, \Delta_2, \ldots, \Delta_n$ in the directions of the respective loads. Due to the application of loads the various elements of the system are subjected to internal forces $S_1, S_2, \ldots, S_m$ and deformations $e_1, e_2, \ldots, e_m$ in the directions of the internal forces. The internal forces will also increase from zero to their full values, as the external loads are applied gradually to the system, increasing from zero to their full values.

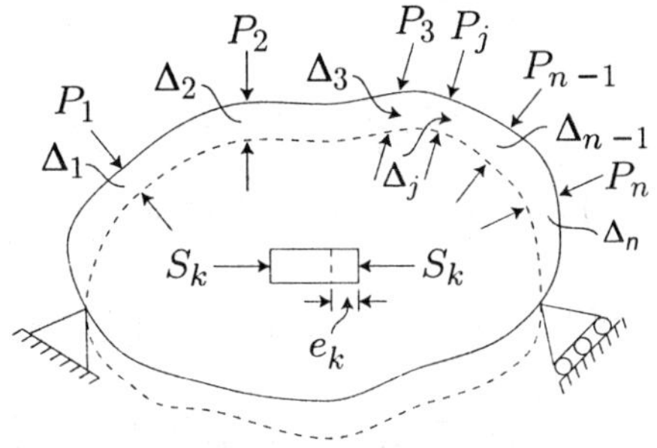

Fig. 5.11: Deformable System

Fig. 5.12(a) representing the force deformation characteristics of all elements of the system in Fig. 5.11, the strain energy of an element resulting from the deformation e_k in the direction of internal force S_k is given by $\int_0^{e_k} S_k \delta_{ek}$ or the area under the force deformation curve. Hence, the total strain energy of the system is equal to

$$U = \sum_1^m \int_0^{e_k} S_k \delta_{ek}. \tag{5.13}$$

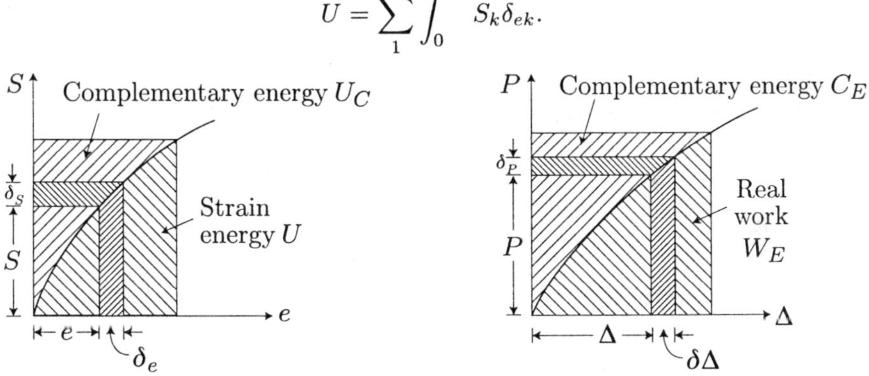

(a) Internal force deformation curve **(b) External force deformation curve**

Fig. 5.12: Force Deformation Curve

External work done on the same system by a force P_j for a displacement Δ_j in the direction of P_j is given by $\int P_j \delta\Delta_j$ or the area under the force deformation curve in Fig. 5.12(b). The external work done on the system then equals

$$W_E = \sum_1^n \int_0^{\Delta_j} P_j \delta\Delta_j. \tag{5.14}$$

From the principle of conservation of energy, i.e., from $W_E = U$,

$$\sum_1^n \int_0^{\Delta_j} P_j \delta\Delta_j = \sum_1^m \int_0^{e_k} S_k \delta e_k. \tag{5.15}$$

Complementary energy resulting from internal force S and deformation e is given by $\int e\delta S$ or the area above the force deformation curve in Fig. 5.12(a), whereas the external complementary work resulting from external force P for a displacement Δ in the direction of P is given by $\int \Delta \delta P$ or the area above the curve in Fig. 5.12(b). Total complementary energy U_C and the total complementary work C_E will be given by

$$U_C = \sum_1^m \int_0^{S_k} e_k \delta S_k \tag{5.16}$$

and $$C_E = \sum_1^n \int_0^{P_j} \Delta_j \delta P_j. \tag{5.17}$$

From principle of conservation of energy as in the case of strain energy and external work, $C_E = U_C$,

$$\text{i.e. } \sum_1^n \int_0^{P_j} \Delta_j \delta P_j = \sum_1^m \int_0^{S_k} e_k \delta S_k. \tag{5.18}$$

For the case of linearly elastic material, the areas below and above the curve are equal. The external complementary work on a linearly elastic system is identical to the real work on the system and the complementary and strain energies of the system are the same. In such a case, the integral representing the areas under or above the load deformation curve are areas of triangle. Eqs. (5.13), (5.16) and (5.14), (5.17) then become

$$U = \frac{1}{2} \sum_1^m S_k e_k = U_C \tag{5.19}$$

and $$W_E = \frac{1}{2} \sum_1^n P_j \Delta_j = C_E. \tag{5.20}$$

For the relative simplicity in the evaluations of these expressions, linear elasticity is frequently assumed in structural analysis, if the material characteristics agrees reasonably well with the assumption.

5.12 Strain Energy in Linear Elastic Systems

As a structure is subjected to external loading and external work is done, internal work or strain energy is generated and stored in the individual elements of the structure. The internal work can be expressed in terms of the stresses and strains developed in the individual members. The forms of stresses and strains commonly considered are those due to axial forces, bending moments, twisting moments and shear forces.

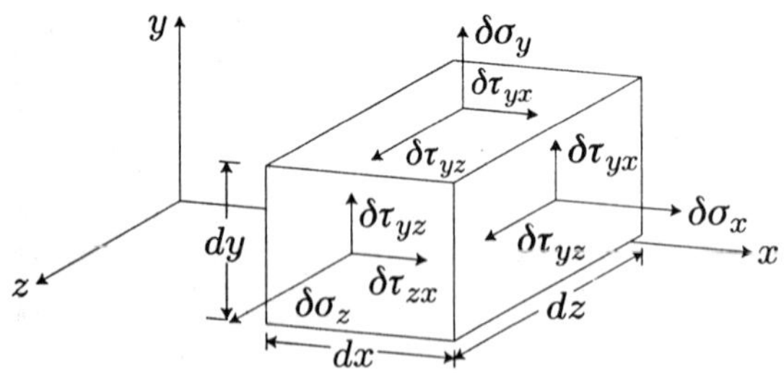

Fig. 5.13: Three dimensional element

Fig. 5.13 shows an element of sides dx, dy and dz of a member, with the possible stresses on the differential element as the structure is loaded.

Associated with a typical stress σ_x, there will be a strain ϵ_z. For a linearly elastic material the internal work done on the differential element only due to stress σ_x is given by

$$dU = dW_1 = \int_{\sigma_x=0}^{\sigma_x=\sigma_x} \frac{\sigma_x}{E} \cdot \delta\sigma_x \cdot dxdydz = \frac{\sigma_x^2}{2E} dxdydz = \frac{\sigma_x \epsilon_x dxdydz}{2}$$

$$= \frac{\sigma_x \epsilon_x dV}{2}, \quad \text{where } E = \frac{\sigma_x}{\epsilon_x}$$

and dV = differential volume of the element = $dxdydz$.

The internal work done on the differential element is given by

$$dU = \frac{1}{2}[\sigma_x\epsilon_x + \sigma_y\epsilon_y + \sigma_z\epsilon_z + \tau_{xy}\gamma_{xy} + \tau_{yz}\gamma_{yz} + \tau_{zx}\gamma_{zx}]dxdydz, \tag{5.21}$$

where γ represents the shearing strain and

$$\epsilon_x = \frac{1}{E}[\sigma_x - \mu(\sigma_y + \sigma_z)]; \ \gamma_{xy} = \frac{2(1+\mu)}{E}\tau_{xy},$$

$$\epsilon_y = \frac{1}{E}[\sigma_y - \mu(\sigma_z + \sigma_x)]; \ \gamma_{yz} = \frac{2(1+\mu)}{E}\tau_{yz},$$

$$\epsilon_z = \frac{1}{E}[\sigma_z - \mu(\sigma_x + \sigma_y)]; \ \gamma_{zx} = \frac{2(1+\mu)}{E}\tau_{zx},$$

where E = modulus of elasticity of the material μ = Poisson's ratio and $\dfrac{E}{2(1+\mu)} = G =$ shear modulus.

Eq. (5.21) can be integrated for finding the total strain energy stored in a linearly elastic material. Thus,

$$U = \frac{1}{2}\int\int\int(\sigma_x\epsilon_x + \sigma_y\epsilon_y + \sigma_z\epsilon_z + \tau_{xy}\gamma_{xy} + \tau_{yz}\gamma_{yz} + \tau_{zx}\gamma_{zx})dxdydz$$

$$= \int\int\int \frac{1}{2E}(\sigma_x^2 + \sigma_y^2 + \sigma_z^2) - \frac{\mu}{E}(\sigma_x\sigma_y + \sigma_y\sigma_z + \sigma_z\sigma_x)$$

$$+ \frac{1}{2G}(\tau_{xy}^2 + \tau_{yz}^2 + \tau_{zx}^2)dxdydz. \tag{5.22}$$

(a) Strain energy due to Axial Forces : For a structural member subjected to an axial force of magnitude S only, stress $\sigma = \frac{S}{A}$ and strain $\epsilon = \frac{e}{L}$, where A is the constant cross-sectional area, e the axial deformation and L the length of the member in the direction of the axial force. The total strain energy in the member then is

$$U_\alpha = \int\int\int \frac{\sigma^2}{2E}dxdydz = \int_0^L \frac{S^2}{2AE}dx = \frac{S^2L}{2AE}. \tag{5.23}$$

(b) Strain energy due to Flexure : For a member subjected to bending moment, the bending stresses also act in a direction normal to the cross section of the member, but are no longer uniformly distributed over the cross-sectional area. The stress σ_x due to bending moment may be expressed as

$$\sigma_x = \frac{M_x}{I}y,$$

where I = moment of inertia about the z-axis for stress in the direction σ_x, at a distance y from the neutral axis.

Total strain energy then becomes

$$U_f = \int\int\int \frac{1}{2E}\left(\frac{M}{I}y\right)^2 dxdydz = \int_0^L \frac{M^2}{2EI^2}\left[\int\int_A y^2dydz\right]dx = \int_0^L \frac{M^2dx}{2EI} \quad (5.24)$$

as $\int\int\int y^2dydz = I.$

An alternative expression for the total strain energy U_f due to flexure may be obtained from Eq. (5.24),

$$U_f = \int_0^L \frac{M^2dx}{2EI} = \int_0^L \frac{\left[EI\frac{d^2y}{dx^2}\right]^2}{2EI}dx = \int_0^L \frac{EI}{2}\left(\frac{d^2y}{dx^2}\right)^2 dx. \quad (5.25)$$

(c) Strain energy due to Twisting Moment : When an element of volume $dx\ dy\ dz$ is subjected to shear stress τ_{xy} and τ_{yz}, from the point of view of static equilibrium,

$$(\tau_{xy}dxdz)dy = (\tau_{yx}dydz)dx, \quad \text{and} \quad \tau_{xy} = \tau_{yx}.$$

Denoting the shear stress by $\tau_{xy} = \tau_{yx} = \tau$ and corresponding shear strain by γ, total strain energy in a structural member resulting from shear stresses in the xy-plane equals

$$U_t = \int\int\int \frac{\tau^2}{2G}dxdydz.$$

When a circular member of length L and constant cross section A is subjected to a constant twisting moment T, a state of pure shear exists. In such a case, shear stress at a radius R from the centre of the circle is given by

$$\tau = \frac{TR}{J},$$

where $J =$ polar moment of intertia $= \int\int R^2dydz.$
Total strain energy then,

$$U_t = \int\int\int \frac{1}{2G}\left(\frac{TR}{J}\right)^2 dxdydz = \int_0^L \frac{T^2dx}{2GJ} = \frac{T^2L}{2GJ}. \quad (5.26)$$

Since the angle of twist per unit length of a circular member is given by $\theta = \frac{T}{GJ}$, so that $T = GJ\theta$, total strain energy can also be expressed as

$$U_t = \frac{(GJ\theta)^2}{2GJ}L = \frac{GJ\theta^2 L}{2}. \quad (5.26a)$$

(d) Strain energy for Shear in Beams : In case of beam subjected to lateral loads, shear force as well as bending moment occurs in the beam. Shear stress resulting from the shear force is varying along the cross section and depends on the nature of the cross section. If V be the shear force at a particular section of cross-sectional area A, the shear stress at any point of the cross section can be expressed as

$$\tau = k\frac{V}{A},$$

where $k =$ variable factor depending on the shape of the cross section.

Total strain energy U_s in a member then equals

$$U_S = \int\int\int \frac{1}{2G}\left(k\frac{V}{A}\right)^2 dxdydz = \int_0^L \frac{V^2}{2GA^2}\left[\int\int_A k^2 dydz\right]dx$$
$$= \int_0^L \frac{\alpha V^2 dx}{2GA} = \alpha\int_0^L \frac{V^2 dx}{2GA}. \quad (5.26b)$$

where $\int\int k^2 dydz = \alpha A$, α being a rational constant called the *form factor.*

For a rectangular section of width b and depth d, the distribution of shear stress is given by

$$\tau = \frac{V}{Ib}\int_0^L\left[\int_0^b\int_{-\frac{d}{2}}^{\frac{d}{2}} \frac{V^2}{8GI^2}\left(\frac{d^2}{4}-y^2\right)^2 dy\, dz\right]dx = \int_0^L \frac{V^2 b}{8GI^2}\int_{-\frac{d}{2}}^{\frac{d}{2}}\left(\frac{d^2}{4}-y^2\right)^2 dy\bigg]\, dx.$$

As $\int_{\frac{d}{2}}^{\frac{d}{2}}\left(\frac{d^2}{4}-y^2\right)^2 dy = \frac{d^5}{30}$ and $I = \frac{bd^3}{12}$ and $A = bd$. It follows that,

$$U_s = 1.20\int_0^L \frac{V^2}{2GA}dx, \quad 1.20 \text{ being the value of } \alpha.$$

For an I section, value of α is close to unity.

5.13 Deflection by Real Work

Evaluation of deflection by real works follows from the consideration $W_E = U$. The internal work done can be calculated on the basis of the forces in the members. The types of force effects to be considered in evaluating the internal work of a particular structure depend on the types of structure under consideration. In evaluating the deflections of a two dimensional framed structure, it is usually necessary to consider only the bending moment effects, since the effects of shear force and axial force are so small in comparison with the bending moment effects that they can be neglected.

Evaluation of deflections by real work is limited. Only deflection associated with an applied load can be determined. If more than one load is applied to the structure, more than one unknown value of deflection will appear in the expression for real work and the resulting equation cannot be solved.

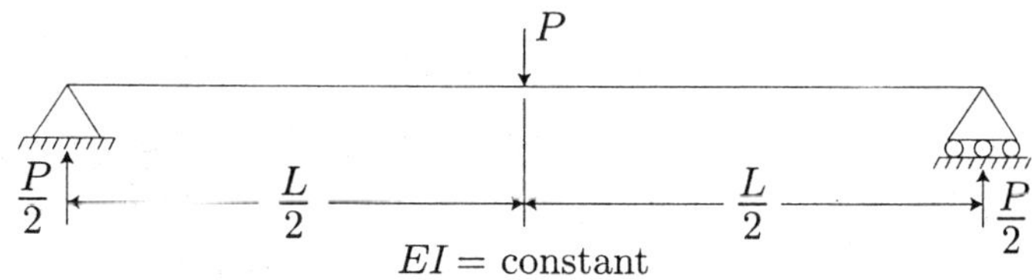

Fig. 5.14: Use of Real Work method

Referring to Fig. 5.14, which shows a simply supported beam subjected to concentrated load P at centre, considering strain energy due to flexure and shear force, and equating external and internal work,

$$\frac{1}{2}P\Delta = 2\int_0^{\frac{L}{2}} \frac{M^2}{2EI}dx + 2\int_0^{\frac{L}{2}} \alpha\frac{V^2}{2GA}dx = 2\int_0^{\frac{L}{2}} \frac{\left(\frac{P}{2}x\right)^2}{2EI}dx + 2\alpha\int_0^{\frac{L}{2}} \frac{\left(\frac{P}{2}\right)^2}{2GA}dx$$

$$= \frac{PL^3}{96EI} + \alpha\frac{P^2L}{8GA}$$

and $\Delta = \frac{PL^2}{48EI}\left[1 + 1.2\alpha\frac{EI}{GAL^2}\right].$

From the value of deflection, it is evident that for relatively long beams, the effect of shear force is quite small. Because of this, the effects of shearing strains are usually neglected in evaluating beam deflections except for short, deep beams.

5.14 Castigliano's Theorem

Fig. 5.15(a) shows an elastic body subjected to a system of forces $P_1, P_2, P_3, \ldots, P_n$ which are applied gradually giving rise to deflections $\Delta_1, \Delta_2, \Delta_3, \ldots, \Delta_n$ respectively in the directions of the applied forces. Then the strain energy of the body may be given by

$$U_1 = \frac{1}{2}P_1\Delta_1 + \frac{1}{2}P_2\Delta_2 + \cdots + \frac{1}{2}P_n\Delta_n = \frac{1}{2}\sum P_1\Delta_1.$$

If P_n is now increased by infinitely small amount dP_n, the additional strain energy is

$$U_2 = (\text{rate of change of } U \text{ with respect to} P_n)(dP_n) = \frac{\partial U}{\partial P_n} \cdot dP_n.$$

Hence the total strain-energy due to forces $P_1, P_2, \ldots, (P_n + dP_n)$ is

$$U = U_1 + U_2 = \frac{1}{2}\sum P_1\Delta_t + \frac{\partial U}{\partial P_n} \cdot dP_n. \tag{5.27a}$$

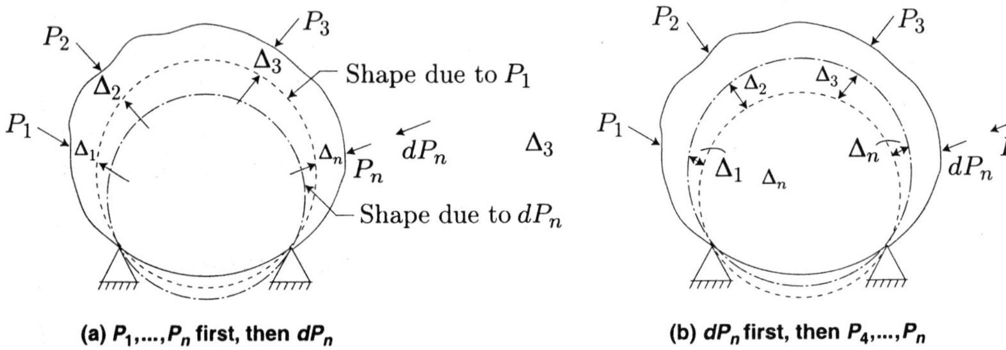

Fig. 5.15: Castigliano's Theorem

Fig. 5.15(b) shows the same elastic body which is first subjected to the small force dP_n and then to the system of forces $P_1, P_2, \ldots, P_n$ applied gradually causing additional deflections $\Delta_1, \Delta_2, \ldots, \Delta_n$ respectively. The strain-energy due to the force dP_n alone before the application of other forces is very negligible and can be ignored. The additional strain-energy developed subsequently due to application of forces is then the total strain-energy which is given by

$$U = \frac{1}{2}P_1\Delta_1 + \frac{1}{2}P_2\Delta_2 + \cdots + \frac{1}{2}P_n\Delta_n + dP_n \cdot \Delta_n = \frac{1}{2}\sum P_i\Delta_i + dP_n \cdot \Delta_n \qquad (5.27b)$$

The last term accounts for the work done by force dP_n which exists in full magnitude throughout the entire deflection Δ_n.

The cases represented by Figs. 5.15a and 5.15b are ultimately the same even though the sequence of loading has only changed. Since for a linear structure, the order of loading does not affect the final values of stress, strain, deflection or strain-energy, Eqs. (5.27a) and (5.27b) are identical. Hence

$$\frac{1}{2}\sum P_i\Delta_i + \frac{\partial U}{\partial P_n} \cdot dP_n = \frac{1}{2}\sum P_i\Delta_i + dP_n.\Delta_n,$$

whence $\dfrac{\partial U}{\partial P_n} = \Delta_n.$ (5.27)

Eq. (5.27) is actually the mathematical expression of Castigliano's Second Theorem for deflection, which may be stated as follows :

If the strain-energy of a linearly elastic structure be expressed in terms of the applied forces, then the partial derivative of the total strain-energy with respect to any chosen force gives the deflection of the body at the point of application of the force in the direction of the force.

Castigliano's first theorem can be derived in a similar manner and it gives

$$P_i = \frac{\partial U}{\partial \Delta_i}$$

The first theorem is an expression for loads in terms of the partial derivative of the strain-energy with respect to the displacement.

The first theorem of Castigliano imposes the requirement of equilibrium, while the second theorem expresses the requirement of compatibility. The first theorem has almost been forgotten, but the second theorem has been used extensively for the computation of deflections of statically determinate structures and for the requirement of compatibility for statically indeterminate structures.

5.15 Application of Castigliano's Theorem to Various Cases of Loading

Eq. (5.27) is the basic formula for finding out deformation using strain-energy. Expressions may be derived for different cases from the same basic formula.

(a) *Deflection and Rotation of Beams due to Flexure.* The strain-energy U_f, due to flexure is given by

$$U_f = \sum \frac{M^2 ds}{2EI}, \qquad (5.28a)$$

where M is the Bending Moment at any modulus of elasticity, I the moment of inertia about relative axis, ds is a very small length along the longitudinal axis of the beam and summation includes all sections over the entire length of the beam. The deflection under any given load P_i is given by

$$\Delta_f = \frac{\partial U_f}{\partial P_i} = \sum \frac{M}{EI} \cdot \frac{\partial M}{\partial P_i} \cdot ds. \tag{5.28}$$

Similarly, rotation θ_f at a section where the applied moment M_i acts is given by

$$\theta_f = \frac{\partial U_f}{\partial M_i} = \sum \frac{M}{EI} \cdot \frac{\partial M}{\partial M_i} \cdot ds. \tag{5.29}$$

(b) *Deflection and Rotation of Beams due to Flexural Shear Force.* The strain-energy U_s, of a beam due to flexural shear force is given by

$$U_s = \alpha \sum \frac{V^2 ds}{2AG}, \tag{5.30a}$$

where V is the shear force at a section, A is the area of the section, G is the shear modulus of the beam material and α is a factor depending on the shape of the section. The deflection under any given load P_i due to flexural shear force alone is given by

$$\Delta_s = \frac{\partial U_s}{\partial P_i} = \alpha \sum \frac{V}{AG} \cdot \frac{\partial V}{\partial P_i} \cdot ds \tag{5.30}$$

and the rotation θ_s at a section where moment M_i is applied is given by

$$\theta_s = \frac{\partial U_s}{\partial M_i} = \alpha \sum \frac{V}{AG} \cdot \frac{\partial V}{\partial M_i} \cdot ds. \tag{5.31}$$

(c) *Deflection, Slope and Angle of Twist due to Torsion.* The strain-energy of a member due to torsion is given by

$$U_t = \sum \frac{T^2 ds}{2GJ}, \tag{5.32a}$$

where T is the torque at any section, G is the shear modulus and J is the torsion constant. The linear deflection under a given force P_i is given by

$$\Delta_t = \frac{\partial U_t}{\partial P_i} = \sum \frac{T}{GJ} \cdot \frac{\partial T}{\partial P_i} \cdot ds. \tag{5.32}$$

Slope at a section where bending moment M_i is applied is given by

$$\theta_t = \frac{\partial U_t}{\partial M_i} \cdot ds \tag{5.33}$$

and the angle of twist ϕ_t at a section where torque T_i is applied is given by

$$\phi_t = \frac{\partial U_t}{\partial T_i} = \sum \frac{T}{GJ} \cdot \frac{\partial T}{\partial T_i} \cdot ds. \tag{5.34}$$

(d) *Deflection of Pin-jointed Trusses.* The strain-energy of a pin-jointed truss subjected to joint loads only is given by

$$U_a = \sum \frac{S^2 L}{2AE}, \tag{5.35a}$$

where S, L and A are respectively the force, length and sectional area of any member and E is the modulus of elasticity of the material. The deflection of any joint under any load P_i is given by

$$\Delta = \frac{\partial U_a}{\partial P_i} = \sum \frac{SL}{AE} \cdot \frac{\partial S}{\partial P_i}. \tag{5.35}$$

Example 5.5. *A cantilever beam AB [Fig. 5.16(a)] of length L and constant EI is subjected to a uniformly distributed load w per unit length and a concentrated load P at the free end. Determine (i) vertical deflection at the middle of the beam and (ii) slope at the free end by Castigliano's theorem.*

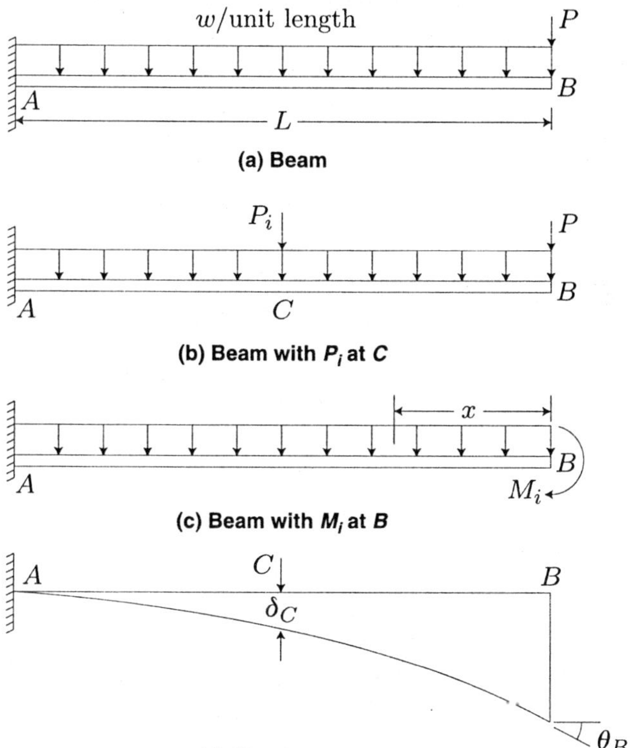

Fig. 5.16: For Example 5.5

Solution : (i) Since there is no concentrated load acting in the vertical direction at the mid-point C of the beam, a fictitious load P_i is assumed to be applied there as shown in Fig. 5.16(b). Expressions for bending moment and its partial derivatives for this loading conditions are given in the Table below :

Portion of the beam	Origin	Limits	M	$\frac{\partial M}{\partial P_i}$
BC	B	0 to $\frac{L}{2}$	$-\left[P \cdot x + \frac{wx^2}{2}\right]$	0
CA	C	0 to $\frac{L}{2}$	$-\left[P \cdot \left(\frac{L}{2}+x\right) + \frac{w}{2}\left(\frac{L}{2}+x\right)^2 + P_i x\right]$	$-x$

The central deflection Δ_c is given by

$$\Delta_c = \frac{\partial U_f}{\partial P_i} = \sum \frac{M}{EI} \cdot \frac{\partial M}{\partial P_i} \cdot dx$$

$$= \frac{1}{EI} \int_0^{\frac{L}{2}} \left(-P \cdot x - \frac{wx^2}{2}\right)(0) dx$$

$$+ \frac{1}{EI} \int_0^{\frac{L}{2}} \left[-P\left(\frac{L}{2}+x\right) - \frac{w}{2}\left(\frac{L}{2}+x\right)^2 - P_i x\right](-x) dx$$

$$= \frac{1}{EI} \int_0^{\frac{L}{2}} \left[P\left(\frac{L}{2}+x\right) + \frac{w}{2}\left(\frac{L}{2}+x\right)^2 + P_i \cdot x\right] x dx.$$

Since the force P_i does not actually exist, its value should be taken as zero for the above expression which then becomes

$$\Delta_c = \frac{1}{EI} \int_0^{\frac{L}{2}} \left[P\left(\frac{L}{2}x + x^2\right) + \frac{w}{2}\left(\frac{L^2}{4}x + Lx^2 + x^3\right)\right] dx = \frac{5}{48} \cdot \frac{PL^3}{EI} + \frac{17}{384} \cdot \frac{wL^4}{EI}.$$

The positive sign of the right-hand side of the above expression indicates that the deflection at C is in the direction of the assumed load P_i, i.e., downward.

(ii) Since there is no couple acting at B where the slope is required to be determined, a fictitious clockwise couple M_i is assumed to be acting at B as shown in Fig. 2.9(c). Taking B as origin, the bending moment at x from B for this loading condition is given by

$$M = -Px - \frac{wx^2}{2} - M_1$$

and hence, $\dfrac{\partial M}{\partial M_i} = -1.$

Then the slope θ_B is obtained as

$$\theta_B = \frac{\partial U}{\partial M_i} = \sum \frac{M}{EI} \cdot \frac{\partial M}{\partial M_i} \cdot dx = \frac{1}{EI} \int_0^L \left[-P \cdot x - \frac{wx^2}{2} - M_i\right](-1) dx$$

$$= \frac{1}{EI} \int_0^L \left(P \cdot x + \frac{wx^2}{2} + M_i\right) dx.$$

Since the couple M_i does not actually exist, its value should be taken as zero for the above expression which then becomes

$$\theta_B = \frac{1}{EI} \int_0^L \left(P \cdot x + \frac{wx^2}{2}\right) dx = \frac{PL^2}{2EI} + \frac{wL^3}{6EI}.$$

The positive sign of the right-hand side of the above expression indicates that the rotation of the tangent to the elastic line at B is in the same direction as that of the fictitious couple M_i, i.e., clockwise.

Example 5.6. *A semi-circular ring lying in a horizontal plane is fixed at the end A and free at the other end B. A vertical downward load P acts at the free end as shown in Fig. 2.10(a). Determine (i) the deflection under the load P, (ii) rotation of the tangent to the elastic line at B and (iii) angle of twist at B.*

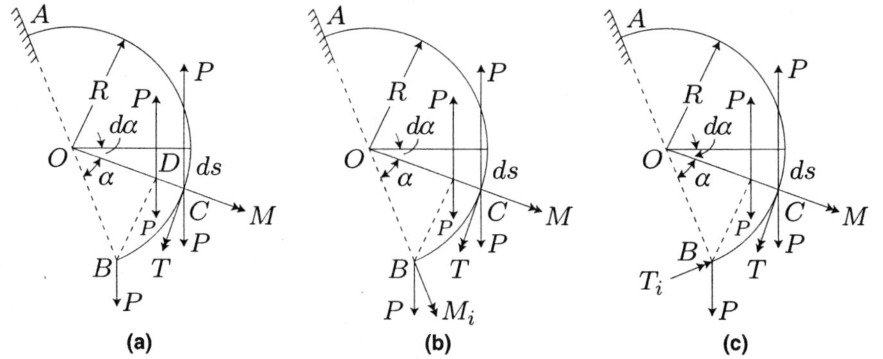

Fig. 5.17: For Example 5.6

Solution : (i) A section of the ring is, in general, subjected to bending moment, twisting moment and flexural shear. Neglecting the effect of flexural shear, the deflection under P is given by

$$\Delta = \frac{\partial U}{\partial P} = \frac{\partial}{\partial P}\left[\sum \frac{M^2 ds}{2EI} + \sum \frac{T^2 ds}{2GJ}\right] = \sum \frac{M}{EI} \cdot \frac{\partial M}{\partial P} ds + \sum \frac{T}{GJ}\frac{\partial T}{\partial P} \cdot ds.$$

Element of length ds of the ring at C is at an angular distance α from the free end and subtending an angle d_α at the centre of the ring. To obtain bending moment, twisting moment, and flexural shear force at C, a perpendicular BD is drawn to OC from B and equal and opposite forces P are assumed to act at D and C without affecting the loading system. The downward force P at B and upward force P at D produce twisting moment T at C, whereas the downward force P at C produces flexural shear V at C. Thus,

$$M = PR\sin\alpha, \quad T = PR(1 - \cos\alpha) \quad \text{and} \quad V = P.$$

The bending moment and twisting moment can also be represented by vectors drawn along the radius OC and tangent at C respectively as indicated in Fig. 5.17(a) with the double-arrow indicating the direction of movement of a right-hand screw when turned in the direction of the given couple.

Partially differentiating M and T with respect to P,

$$\frac{\partial M}{\partial P} = R\sin\alpha \quad \text{and} \quad \frac{\partial T}{\partial P} = R(1 - \cos\alpha).$$

Since $ds = Rd\alpha$,

$$\Delta = \frac{1}{EI}\int_0^{\pi}(P \cdot R\sin\alpha)(R\sin\alpha)(Rd\alpha) + \frac{1}{GJ}\int_0^{\pi} PR(1-\cos\alpha) \cdot R(1-\cos\alpha)(Rdx)$$

$$= \frac{\pi PR^3}{2}\left[\frac{1}{EI} + \frac{3}{GJ}\right].$$

(ii) Since there is no bending moment acting at the free end B where the slope of the axis is required, a fictitious bending moment M_i is assumed to act at B as shown in Fig. 5.17(b). The slope is given by,

$$\theta = \frac{\partial U}{\partial M_i} = \sum \frac{M}{EI} \cdot \frac{\partial M}{\partial M_i} ds + \sum \frac{T}{GJ} \cdot \frac{\partial T}{\partial M_i} \cdot ds.$$

Considering the force P and moment M_i acting at B, the bending and twisting moments at C are

$$M = PR\sin\alpha + M_i\cos\alpha \quad \text{and} \quad T = PR(1-\cos\alpha) + M_i\sin\alpha,$$

whence $\dfrac{\partial M}{\partial M_i} = \cos\alpha$ and $\dfrac{\partial T}{\partial M_i} = \sin\alpha$

$$\theta = \frac{1}{EI}\int_0^\pi (PR\sin\alpha + M_i\cos\alpha)\cos\alpha \cdot Rd\alpha + \frac{1}{GJ}\int_0^\pi \{PR(1-\cos\alpha) + M_i\sin\alpha\}\sin\alpha \cdot Rd\alpha.$$

Since M_i does not actually exist, putting zero for M_i, the above expression becomes

$$\theta = \frac{1}{EI}\int_0^\pi PR^2\sin\alpha\cdot\cos\alpha\cdot d\alpha + \frac{1}{GJ}\int_0^\pi PR^2(1-\cos\alpha)\sin\alpha\cdot d\alpha = \frac{2PR^2}{GJ},$$

which is in the direction of assumed bending moment M_i.

(iii) Since there is no twisting moment acting at the free end B where the angle of twist is required, a fictitious twisting moment T_i is assumed to act at B as shown in Fig. 5.17(c).

The angle of twist ϕ is given by

$$\phi = \frac{\partial U}{\partial T_i} = \sum \frac{M}{EI}\cdot\frac{\partial M}{\partial T_i}ds + \sum \frac{T}{GJ}\cdot\frac{\partial T}{\partial T_i}\cdot ds.$$

Considering the force P and torque T_i acting at B, the bending moment M and twisting moment T at C are

$$M = PR\sin\alpha + T_i\sin\alpha \quad \text{and} \quad T = PR(1-\cos\alpha) - T_i\cos\alpha.$$

Since $\dfrac{\partial M}{\partial T_i} = \sin\alpha$ and $\dfrac{\partial T}{\partial T_i} = \cos\alpha$

$$\phi = \frac{1}{EI}\int_0^\pi (PR\sin\alpha + T_i\sin\alpha)\cdot Rd\alpha + \frac{1}{GJ}\int_0^\pi [PR(1-\cos\alpha) - T_i\cos\alpha](-\cos\alpha)\cdot Rd\alpha.$$

Since T_i does not actually exist, putting zero for T_i in the above expression,

$$\phi = \frac{1}{EI}\int_0^\pi PR^2\sin^2\alpha\cdot d\alpha + \frac{1}{GJ}\int_0^\pi PR^2(\cos^2\alpha - \cos\alpha)d\alpha = \frac{\pi PR^2}{2}\left(\frac{1}{EI} + \frac{1}{GJ}\right),$$

which is in the direction of assumed torque T_i.

5.16 Unit Load Method

This method is also called Dummy Load Method or Virtual Work Method. It was developed by J.C. Maxwell in 1864 and independently by O. Mohr in 1874 and so often called Maxwell-Mohr method.

Fig. 5.18(a) shows an elastic body subjected to a system of loads $P_1, P_2, P_3, \ldots, P_n$ applied gradually causing deflection $\Delta_1, \Delta_2, \Delta_3, \ldots, \Delta_n$ respectively in the direction of the loads and deflection Δ at a point C in the direction x-x.

For this case,

$$\text{external work done} = \frac{1}{2}P_1\Delta_1 + \frac{1}{2}P_2\Delta_2 + \cdots + \frac{1}{2}P_n\Delta_n = \frac{1}{2}\sum_1^n P_i\Delta_i$$

and internal energy stored $= \dfrac{1}{2}\sum SdL,$

where S is the internal forces developed on an element of length dx and dL is the corresponding deformation of the element. Assuming that no energy is lost, law of conservation of energy gives

$$\frac{1}{2}\sum_1^n P_i\Delta_i = \frac{1}{2}\sum SdL. \tag{5.36a}$$

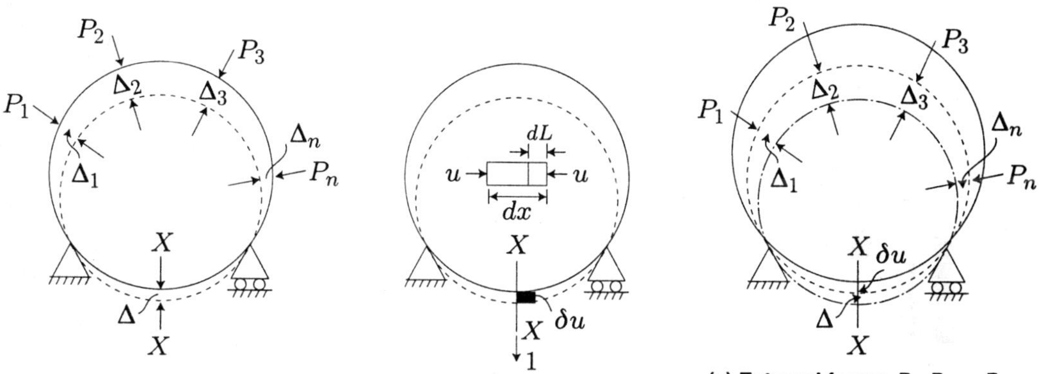

(a) External forces P_1, P_1,..., P_n **(b) Unit load at X** **(c) External forces P_1, P_1,..., P_n gradually applied on (b)**

Fig. 5.18: Unit load method

In Fig. 5.18(a) the same elastic body is subjected to a unit load at C in the direction x-x causing a deflection δ_u along x-x and internal force u on the element of length dx with a deformation of dL. Assuming no energy is lost, according to law of conservation of energy, internal energy is equal to the external work done, i.e.,

$$\frac{1}{2}(1)(\delta_u) = \frac{1}{2}\sum udL. \tag{5.36b}$$

The forces $P_1, P_2, \ldots, P_n$ are then applied gradually on the elastic body while the unit load is still acting on it, as shown in Fig. 5.18(c), As the structure is assumed to be linearly elastic, the additional deformation due to gradual application of loads $P_1, P_2, \ldots, P_n$, are the same as that shown in Fig. 5.18(a). Then the

$$\begin{aligned}\text{additional external work done} &= \frac{1}{2}P_1\Delta_1 + \frac{1}{2}P_2\Delta_2 + \cdots \frac{1}{2}P_n\Delta_n + (1)(\Delta)\\ &= \frac{1}{2}\sum_1^n P_i\Delta_i + (1)(\Delta).\end{aligned}$$

$$\text{Additional internal energy stored} = \frac{1}{2}\sum SdL + \sum udL.$$

Applying the law of conservation of energy to the additional external and internal work,

$$\frac{1}{2}\sum_1^n P_i\Delta_i + (1)(\Delta) = \frac{1}{2}\sum SdL + \sum udL. \tag{5.36c}$$

Subtracting Eq. (5.36a) from Eq. (5.36c),

$$\Delta = \sum udL. \tag{5.36}$$

which is the basic Eq. of Unit Load Method.

5.17 Application of Unit Load Method

From the basic equation of Unit Load Method, different formulae can be derived for different loading actions as follows:

(a) Deflection and Rotation of Beam due to Flexure : If M be the moments due to applied loads at a section C of the beam in Fig. 5.19(a) and m be that due to unit load acting at the point D where deflection is required, as in Fig. 5.19(b), then the force u on an element of length dx and sectional area dA at a distance y from the neutral axis due to unit load is given by

$$u = \frac{m}{I} \cdot ydA.$$

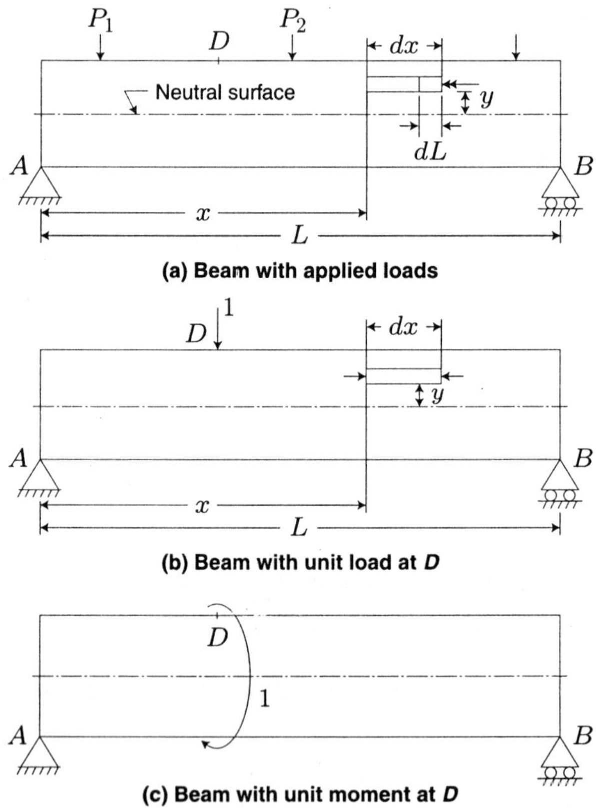

Fig. 5.19: Deflection and rotation due to flexure

The deformation dL of the same element due to the applied loads is

$$dL = \frac{M}{EI} \cdot ydx.$$

The deflection Δ_f at D due to flexure is

$$\Delta_f = \sum udL = \sum_0^L \sum_0^A \frac{m}{I} \cdot ydA \cdot \frac{M}{EI} \cdot ydx = \sum_0^L \frac{Mmdx}{EI^2} \sum_0^A y^2 dA = \sum_0^L \frac{Mmdx}{EI^2} \cdot I$$

$$= \sum_0^L \frac{Mmdx}{EI}. \tag{5.37a}$$

Similarly, the slope θ_f at D due to flexure alone is

$$\theta_f = \sum_0^L \frac{Mm'dx}{EI}, \tag{5.37b}$$

where m' is the moment at any section such as C due to a unit couple at the point D (Fig. 5.19(c)).

(b) Deflection and Rotation of Beams due to Flexural Shear : At any section C of the beam in Fig. 5.20(a), if the shear force due to applied load be V, the displacement along V for a small element of length dx, in Fig. 5.20(c), is given by

$$dL = \gamma \cdot dx = \frac{\alpha V}{AG} \cdot dx$$

where γ = shear strain, A = cross-sectional area of the beam, G = shear modulus and α = shape factor depending on the shape of the cross section.

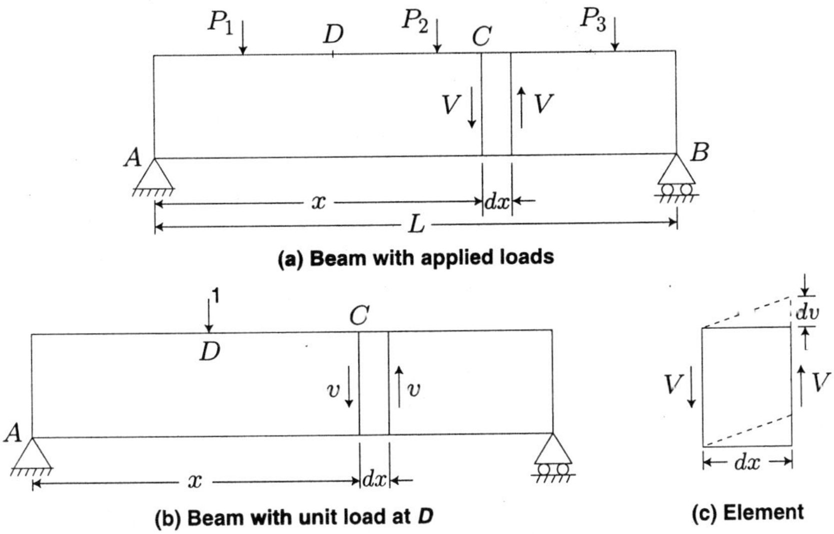

Fig. 5.20: Deflection due to shear force

If v be the shear force at the same section C due to a unit load at D, as in Fig. 5.20(b), then $u = v$ so that deflection Δ_s at D due to shear force only is

$$\Delta_s = \sum udL = \sum \frac{\alpha V v dx}{AG}. \tag{5.37c}$$

Similarly, if v' be the shear force at any section due to a unit couple acting at D, then the rotation θ_s at D due to shear only is

$$\theta_s = \sum \frac{\alpha V v' dx}{AG}. \tag{5.37d}$$

(c) Deflection and Angle of Twist due to Torsion of a Circular Shaft : If T be the torque on an element of length dx due to applied loads as in Fig. 5.21(a), and $d\phi$, the angle of twist over length dx, then

$$d\phi = \frac{Tdx}{GJ},$$

where J is the polar moment of inertia of the cross section.

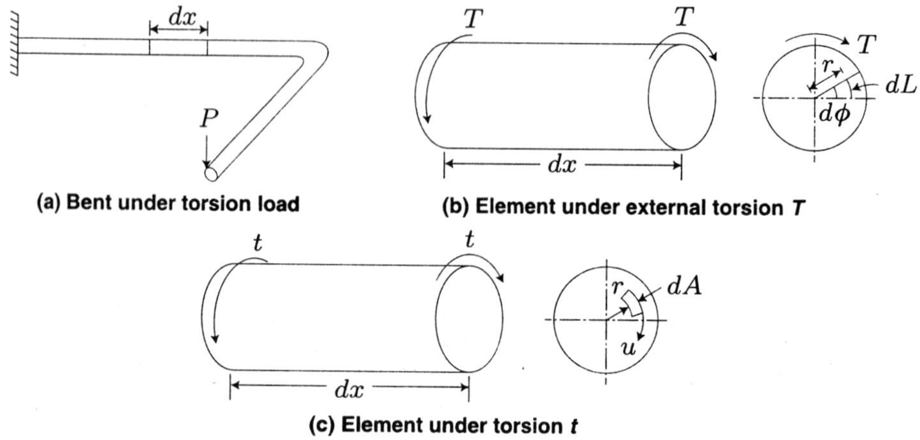

Fig. 5.21: Deflection and angle of twist due to torsion

The displacement dL of an elemental area dA located at a distance r from the centre of the section as in Fig. 5.21(b), is given by

$$dL = rd\phi = r \cdot \frac{Tdx}{GJ}.$$

If t be the torque on the same element due to unit load at the desired point where deflection is wanted as shown in Fig. 5.21(c), the tangential force acting on the elemental area dA is $\frac{t}{J}rdA$, so that,

$$u = \frac{t}{J} \cdot rdA.$$

Hence, deflection is given by

$$\Delta_t = \sum udL = \sum_0^L \sum_0^A \frac{Ttr^2}{GJ^2} dA \cdot dx = \sum_0^L \frac{Tt}{GJ^2} \cdot dx \sum_0^A r^2 dA = \sum_0^L \frac{Ttdx}{GJ^2} \cdot J$$

$$= \sum_0^L \frac{Ttdx}{GJ}. \tag{5.37e}$$

Similarly, if t' be the torque at any section due to a unit torque at the point where angle of twist is wanted, then the angle of twist ϕ_t is given by

$$\phi_t = \sum_0^L \frac{Tt'dx}{GJ}. \tag{5.37f}$$

(d) Deflection of Pin-jointed Trusses : If S be the axial force of any member of the truss due to applied loads, as in Fig. 5.22(a), then the deformation dL of the member due to applied loads is given by

$$dL = \frac{SL}{AE},$$

where L = length of the member, A = cross-sectional area of the member and E = modulus of elasticity.

If u be the axial force in the same member due to a unit load acting at the joint D where deflection is required as shown in Fig. 5.22(b), the deflection at the joint is given by

$$\Delta = \sum udL = \sum \frac{Sul}{AE}. \tag{5.37g}$$

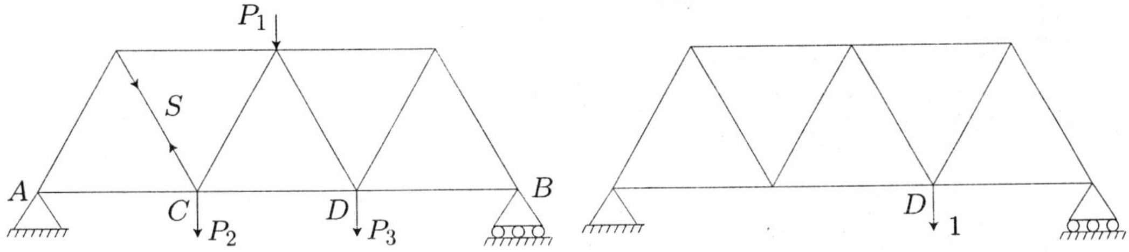

(a) Truss under external load **(b) Truss under unit load at *D***

Fig. 5.22: Deflection of pin-jointed truss

Figs 5.23 and 5.24 show how to select suitable unit load system for use in the computations of various types of deflection components, that may be required. It is necessary to select the unit load system in such a way that the desired deflection is the only unknown appearing on the left-hand side of Eq. (5.37). It may be noted that the line of action of the unit load is to be along the direction of deflection desired.

Deflection component	Unit load system	Basic equations
1. Vertical component of deflection of joint Δ_V	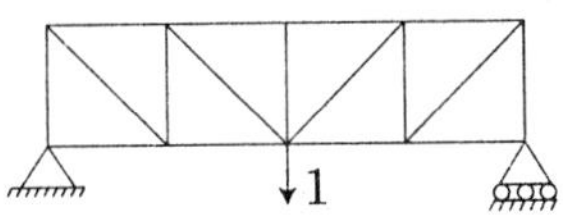	$1(\Delta_V\downarrow) = \sum udL = \sum \frac{SuL}{AE}$
2. Horizontal component of deflection of joint Δ_H	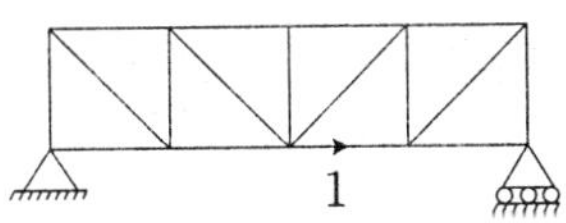	$1(\overrightarrow{\Delta_H}) = \sum udL = \sum \frac{SuL}{AE}$
3. Any component of deflection of joint Δ_α	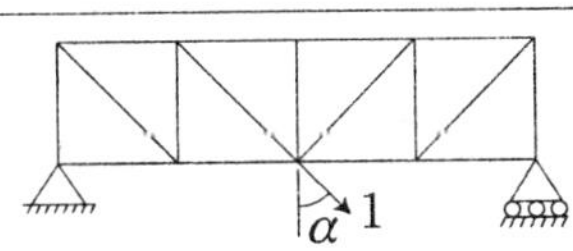	$1(\underset{\searrow}{\Delta_\alpha}) = \sum udL = \sum \frac{SuL}{AE}$
4. Relative deflection of two joints along line joining them Δ_{A-B}	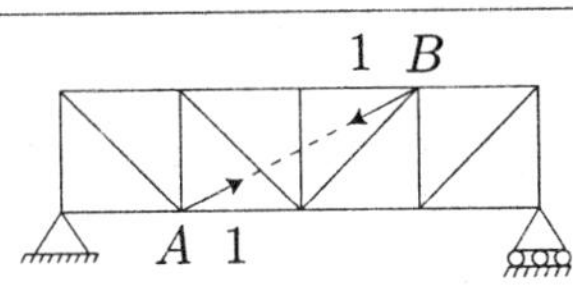	$1(\overset{\nearrow}{\Delta_A}) + 1(\overset{\swarrow}{\Delta_B}) = \sum udL$ $1(\Delta_{A-B}) = \sum udL = \sum \frac{SuL}{AE}$
5. Rotation of truss members change in angle CBD θ	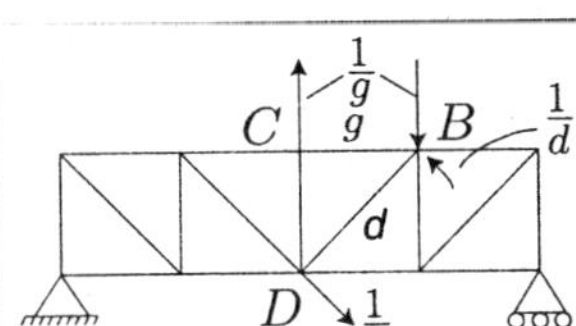	$1(\theta) = \sum udL = \sum \frac{SuL}{AE}$

Fig. 5.23: Typical unit load system for truss deflection problems

In using the method, it is particularly important to use proper sign convention. The external and internal works done are considered positive. This implies that Δ is to be considered positive when it is in the same direction as its corresponding unit force. Furthermore, $\sum udL$ is positive when both u and dL are of the same sense. If u is considered positive when tension, dL is plus when elongation. As far as the unit of the virtual load is concerned, it may be concerned dimensionless or of any suitable force unit.

Deflection component	Unit load system	Basic equations
1. Deflection in any direction Δ (unit load along deflection desired)	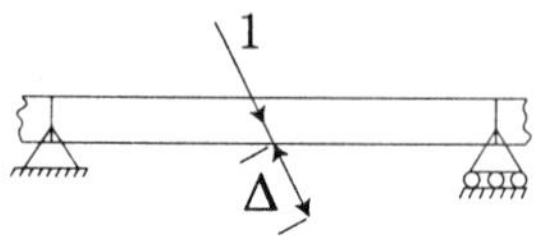	$1(\Delta_V) = \sum udL = \int \frac{Mmdx}{EI} + \alpha \int \frac{Vvdx}{AO} + \int \frac{Ttdx}{GJ}$
2. Slope at any point of the beam θ (unit moment at the point)	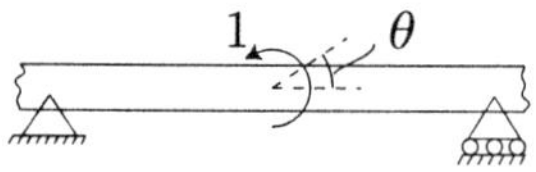	$1(\theta) = \sum udL = \int \frac{Mmdx}{EI} + \alpha \int \frac{Vvdx}{AO} + \int \frac{Ttdx}{GJ}$
3. Rotation due to twisting moment θ_1 (unit torque at the point)	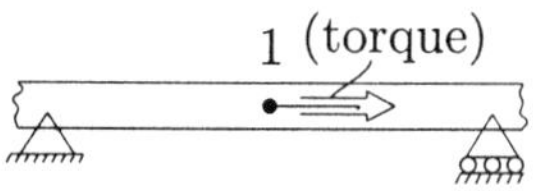	$1(\theta_1) = \sum udL = \int \frac{Mmdx}{EI} + \alpha \int \frac{Vvdx}{AO} + \int \frac{Ttdx}{GJ}$

Fig. 5.24: Typical unit load system for Beam Deflections

Example 5.7. *Compute the maximum deflection and rotation at the free end of the cantilever beam in Fig. 5.25(a). The bending stiffness EI of the beam is constant.*

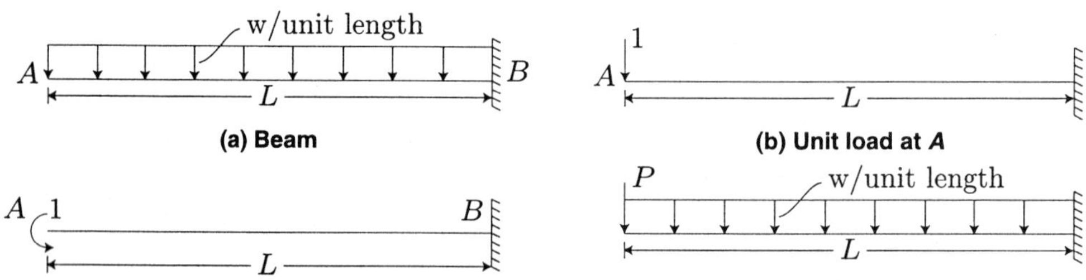

(a) Beam **(b) Unit load at A** **(c) Unit moment at A** **(d) Loading for Castigliano's Theorem**

Fig. 5.25: For Example 5.7

Solution : (i) *By Unit load method* : For deflection Δ at free end A, a downward unit load is considered acting at A as in Fig. 5.25(b). At any section of the beam,

$$\left.\begin{aligned} M &= -\frac{wx^2}{2} \\ V &= -wx \end{aligned}\right] \quad \text{and} \quad \left.\begin{aligned} m &= -1(x) \\ v &= -1 \end{aligned}\right\} \text{Fig. 5.25(b)}$$

Combining, from $\Delta = \int_0^L \frac{Mmdx}{EI} + \alpha \int_0^L \frac{Vvdx}{AG}$,

$$\Delta A = \int_0^L \frac{-\frac{wx^2}{2}(-x)dx}{EI} + \alpha \int_0^L \frac{-wx(-1)dx}{AG} = \frac{wL^4}{8EI} + \alpha \frac{wL^2}{2AG}.$$

The plus sign means that the deflection is in the direction of unit load. Neglecting effect of shear deflection which is rather small with respect to deflection due to moment,

$$\Delta A = \frac{wL^4}{8EI}.$$

Similarly combining Figs. 5.25(a) and 5.25(c),

$$\theta_A = \int_0^L \frac{-\frac{wx^2}{2}(-1)}{EI}dx = \frac{wL^3}{6EI} \text{ anticlockwise.}$$

(ii) By Castigliano's Theorem : For computing deflection at free end by Castigliano's theorem, a load P has to be assumed at free end as shown in Fig. 5.25(d)

As $M = -Px - \frac{wx^2}{2}, \; \frac{\partial M}{\partial P} = (-x).$

Neglecting effect of shear and as no load as P exists so that putting $P = 0$,

$$\Delta_A = \int_0^L \frac{M}{EI}\frac{\partial M}{\partial P}dx = \int_0^L \frac{-\frac{wx^2}{2}(-x)}{EL}dx = \frac{wL^4}{8EI}.$$

Example 5.8. *Compute the deflection at the third point and the rotation at support A of the beam of constant EI shown in Fig. 5.26(a).*

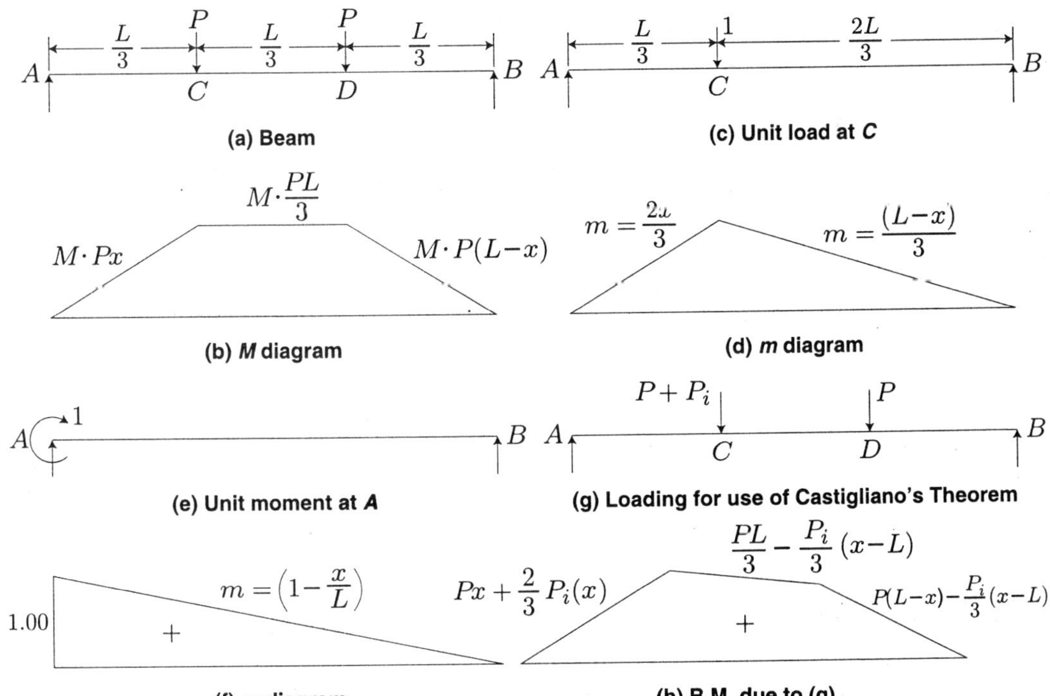

Fig. 5.26: For Example 5.8

Solution : (i) *By Unit Load Method* : Neglecting effect of shear, from Figs. 5.26(b) and 5.26(d),

$$\Delta = \int_0^L \frac{Mmdx}{EI} = \frac{1}{EI}\left[\int_0^{L/3} (Px)\left(\frac{2x}{3}\right)dx + \int_{L/3}^{2L/3}\left(\frac{PL}{3}\right)\left(\frac{L-x}{3}\right)dx + \int_{2L/3}^{L} P(L-x)(\frac{L-x}{3})dx\right]$$

$$= \frac{5}{162}\frac{PL^3}{EI}.$$

Rotation at A is given by combining Figs. 5.26(b) and 5.26(f)

and $$\theta_A = \int_0^L \frac{Mmdx}{EI} = \frac{1}{EI}\left[\int_0^{L/3}(Px)\left(1-\frac{x}{L}\right)dx + \int_L^{2L/3}\frac{PL}{3}\left(1-\frac{x}{3}\right)dx + \int_{2L/3}^{L}P(L-x)\left(1-\frac{x}{L}\right)dx\right]$$

$$= \frac{PL^2}{9EI}, \text{ clockwise.}$$

(ii) *By Castigliano's Theorem*: From Figs. 5.26(g) and 5.26(h),

$$M = Px + \frac{2}{3}(P_i)(x), \qquad \text{for } 0 \le x \le \frac{L}{3}$$

$$= \frac{PL}{3} - \left(\frac{P_i}{3}\right)(x-L), \qquad \text{for } \frac{L}{3} \le x \le \frac{2L}{3}$$

$$= P(L-x) - \left(\frac{P_i}{3}\right)(x-L), \quad \text{for} \frac{2}{3}L \le x \le L$$

whence $$\frac{\partial M}{\partial P_i} = \frac{2}{3}x, \qquad \text{for } 0 \le x \le \frac{L}{3}$$

$$= -\frac{x-L}{3} \quad \text{for } \frac{L}{3} \le x \le L.$$

As P_i does not exist, putting $P_i = 0$,

$$\Delta = \int \frac{M}{EI}\frac{\partial M}{\partial P_i}dx = \frac{1}{EI}\int_0^{L/3}(Px)\left(\frac{2x}{3}\right)dx + \int_{L/3}^{2L/3}\left(\frac{PL}{3}\right)\left(-\frac{x-L}{3}\right)dx + \int_{2L/3}^{L}P(L-x)\left(-\frac{x-L}{3}\right)dx\Bigg]$$

$$= \frac{5}{162}\frac{PL^3}{EI}.$$

Example 5.9. *Compute by unit load method the deflection at centre and the rotation at support A of the beam shown in Fig. 5.27(a).*

Solution : Neglecting effect of shear, deflection at C is given by [referring to Figs. 5.27(a) and 5.27(b)],

$$\Delta_C = \int_0^L \frac{Mmdx}{EI} = 2\int_0^{L/4}\frac{\left(\frac{P}{2}x\right)\left(\frac{x}{2}\right)dx}{EI} + 2\int_{\frac{L}{4}}^{\frac{L}{2}}\frac{\frac{P}{2}(x)\left(\frac{x}{2}\right)dx}{2EI} = \frac{3PL^3}{256EI}\downarrow$$

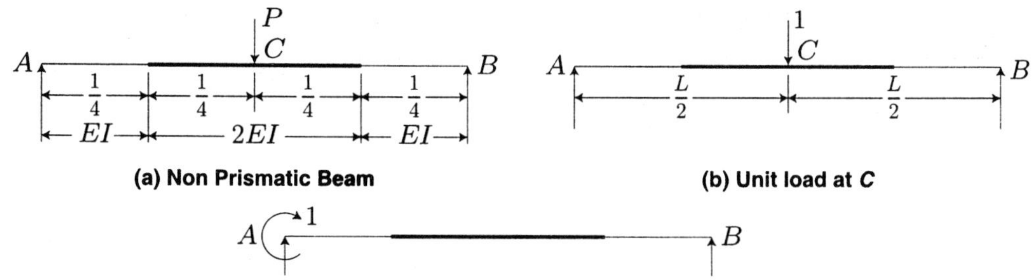

(a) Non Prismatic Beam **(b) Unit load at *C***

(c) Unit moment at *A*

Fig. 5.27: For Example 5.9

Referring to Figs. 5.27(a) and 5.27(c), slope at A is given by

$$\theta_A = \int_0^L \frac{Mmdx}{EI} = \int_0^{L/4} \frac{\left(\frac{Px}{2}\right)\left(1-\frac{x}{L}\right)dx}{EI} + \int_{\frac{L}{4}}^{\frac{L}{2}} \frac{\left(\frac{Px}{2}\right)\left(1-\frac{x}{L}\right)dx}{2EI}$$

$$+ \int_{\frac{L}{4}}^{\frac{L}{2}} \frac{\left(\frac{Px}{2}\right)\left(\frac{x}{L}\right)dx}{2EI} + \int_0^{\frac{L}{4}} \frac{\left(\frac{Px}{2}\right)\left(\frac{x}{L}\right)dx}{EI}$$

$$= \frac{5PL^2}{128EI} \text{ clockwise.}$$

Example 5.10. *Compute by unit load method the vertical deflection, the horizontal deflection and the rotation at free end D of the rigid frame shown in Fig. 5.28(a).*

Solution : Neglecting effect of shear and axial forces, vertical deflection, horizontal deflection and rotation are given by

$$\Delta_{DV} = \int_0^L \frac{Mm_V dx}{EI}; \quad \Delta_{DH} = \int_0^L \frac{Mm_H dx}{EI}; \quad \theta_D = \int_0^L \frac{Mm_\theta dx}{EI}.$$

Referring to Fig. 5.28, values of M, m_V, m_H and m_θ for various portions of the frame are given below :

Position of frame	AF	FB	BE	EC	CD
Origin	A	B	E	C	D
Limits	0 to 3 m	0 to 2 m	0 to 2 m	0 to 2 m	0 to 3 m
MI	I	I	$2I$	$2I$	I
M	$-420+60x$	-240	$-120x$	0	0
m_V	$+4$	$+4$	$(2+x)$	$+x$	0
m_H	$(2-x)$	$-(3+x)$	-3	-3	$-x$
m_θ	$+1$	$+1$	$+1$	$+1$	$+1$

$$\text{Hence, } \Delta_{DV} = \int_0^3 \frac{(-420+60x)(4)dx}{EI} + \int_0^2 \frac{(-240)(4)}{EI}dx + \int_0^2 \frac{(-120x)(2+x)}{2EI}dx$$

$$= -\frac{6280}{EI} \downarrow \text{ downward.}$$

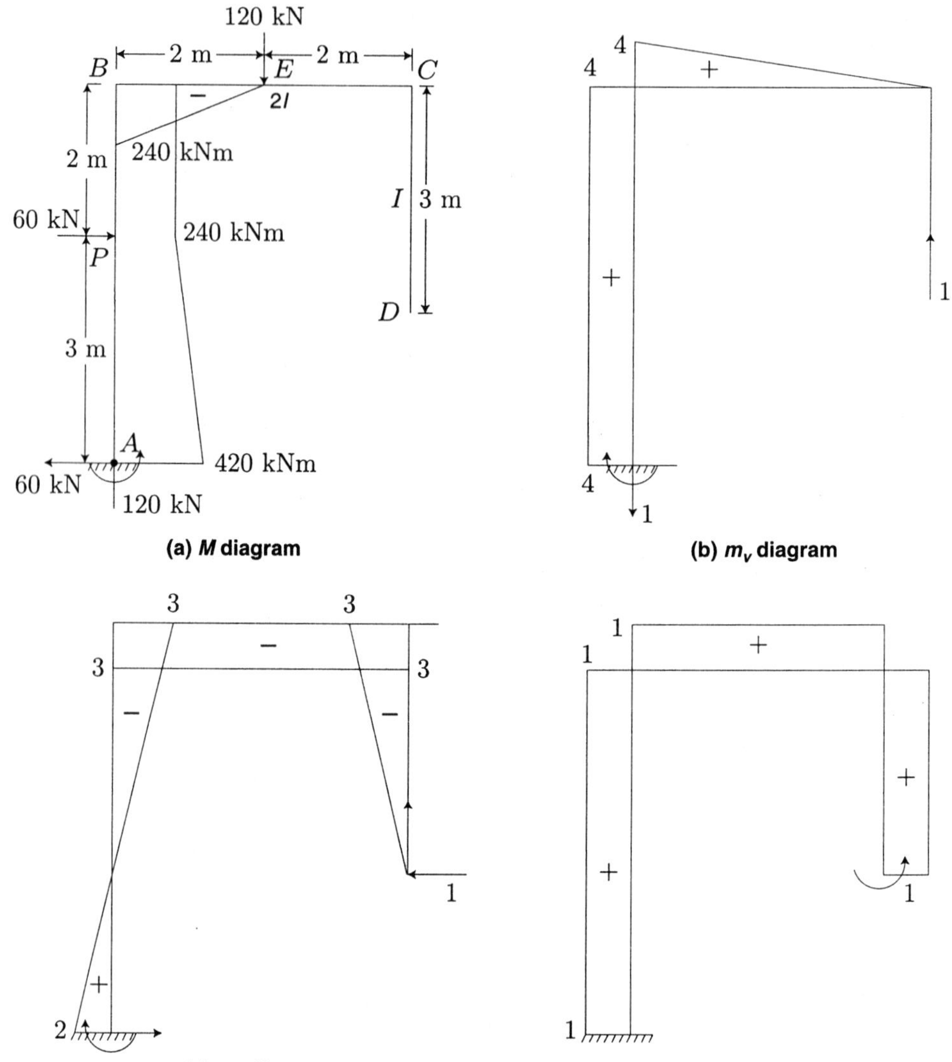

(a) *M* diagram (b) m_v diagram

(c) m_H diagram (d) m_θ diagram

Fig. 5.28: For Example 5.10

Similarly $\Delta_{DH} = \int_0^3 \frac{(-420+60x)(2-x)}{EI}dx + \int_0^2 \frac{(-240-3+x)}{EI}dx$

$$+\int_0^2 \frac{(-120x)(-3)}{2EI}dx = +\frac{690}{EI} \longleftarrow \text{ to the left.}$$

Also $\theta_D = \int_0^3 \frac{(-420+60x)(1)}{EI}dx + \int_0^2 \frac{(-240)(1)}{EI}dx + \int_0^2 \frac{(-120x)(1)}{2EI}dx$

$$= -\frac{1590}{EI} \text{ clockwise.}$$

Resulting deflection of D,

$$\Delta = \sqrt{(\Delta_{DV})^2 + (\Delta_{DH})^2} = \frac{6478.16}{EI}.$$

Example 5.11. *Compute by unit load method the horizontal deflection, the vertical deflection and the rotation at free end of the semicircular bent shown in Fig. 5.29(a).*

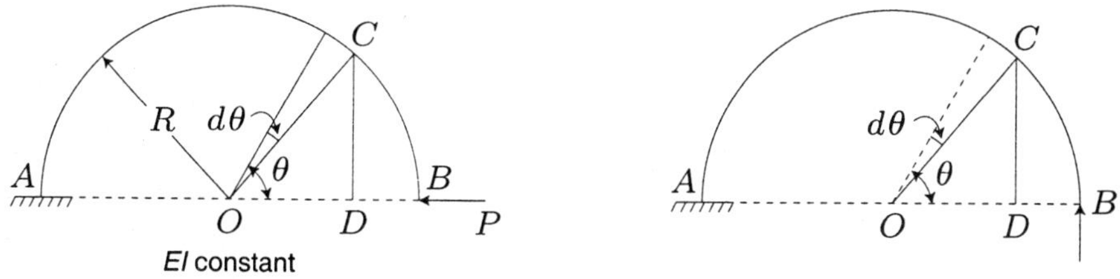

(a) Semicircular Bent **(b) Unit Load at *B***

Fig. 5.29: For Example 5.11

Solution : From Fig. 5.29(a), B.M. at C, $M_c = P(CD) = PR\sin\theta$.

For Δ_{BH} applying a unit horizontal load towards left at B, $m_H = R\sin\theta$.

In this case, integration has to be carried on along the longitudinal axis of the bent. Replacing dx by ds from Fig. 5.29(a), $ds = Rd\theta$.

$$\text{Then } \Delta_{BH} = \int_0^{\pi} \frac{(PR\sin\theta)(R\sin\theta)}{EI}(Rd\theta) = \frac{1}{EI}\int_0^{\pi} PR^3\sin^2\theta d\theta = \frac{Pr^3}{2EI}\pi \leftarrow$$

For vertical deflection at B, applying a unit vertical load at B as in Fig. 5.29(b),

$$m_V = -1(BD) = -R(1-\cos\theta).$$

$$\text{Then } \Delta_{BV} = \int_0^{\pi} \frac{(PR\sin\theta)[-R(1-\cos\theta)]}{EI}Rd\theta = \frac{PR^3}{EI}\int_0^{\pi}\left(\sin\theta - \frac{\sin 2\theta}{2}\right)d\theta$$

$$= -\frac{2PR^3}{EI} \downarrow \text{downward}$$

For rotation, applying an anticlockwise moment at B,

$$\theta_B = \int_0^{\pi} \frac{(PR\sin\theta)(1)}{EI}Rd\theta = \frac{2PR^2}{EI} \curvearrowright.$$

Example 5.12. *Compute the vertical deflection and rotation due to moment at the free end of the right-angled horizontal bent in Fig. 5.30(a) when vertical loads each of magnitude P acts at A and B of the bent. Assume EI as constant and $G = 0.40E$.*

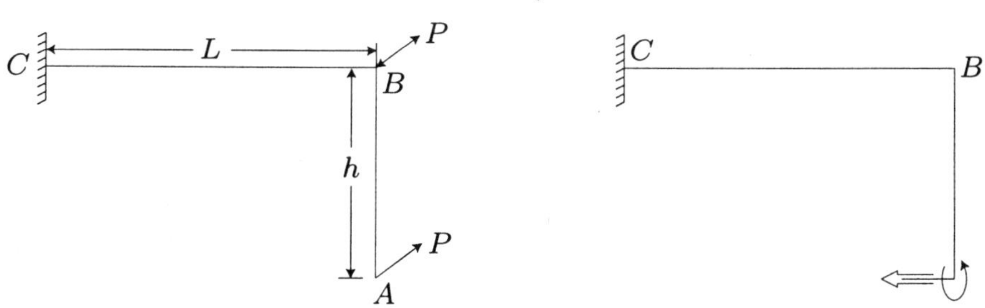

(a) Bent **(b) Unit moment at *A***

Fig. 5.30: For Example 5.12

Solution : From Fig. 5.30(a),

$$\begin{aligned} M &= +Px \text{ between } A \text{ and } B \\ &= 0 \text{ between } B \text{ and } C \\ T &= 0 \text{ between } A \text{ and } B \\ &= +Ph \text{ between } B \text{ and } C. \end{aligned}$$

Due to unit vertical upward load at A,

$$\begin{aligned} m &= +x \text{ between } A \text{ and } B \\ &= +x \text{ between } B \text{ and } C \\ t &= 0 \text{ between } A \text{ and } B \\ &= +h \text{ between } B \text{ and } C. \end{aligned}$$

Then, considering effect of moment and torque only,

$$\Delta_{AV} = \int_0^h \frac{(Px)(x)}{EI}dx + \int_0^L \frac{(Ph)(h)}{GJ}dx = \frac{Ph^3}{3EI} + \frac{Ph^2L}{GJ}.$$

Similarly rotation due to moment at A is given by combining M and T diagram from Fig. 5.30(a) with m and t diagram from Fig. 5.30(b). Thus,

$$\theta_A = \int_0^h \frac{(Px)(1)}{EI}dx + \int_0^L \frac{(Ph)(1)dx}{GJ} = \frac{Ph^2}{2EI} + \frac{PhL}{GJ}.$$

Example 5.13. *Determine the horizontal displacement and rotation of the roller support D of the simply supported portal frame shown in Fig. 5.31(a). The moments of inertia are as shown.*

Solution : $\Delta_D = \dfrac{Mmdx}{EI} \qquad \theta_D = \dfrac{Mm'dx}{EI}$

The expressions for bending moment M due to applied loads in Fig. 5.31(a), bending moment m due to unit load at D in Fig. 5.31(c) and bending moment m' due to unit couple at D in Fig. 5.31(e) for various portions of the frame are obtained with the help of free body diagrams shown in Figs. 5.31(b), 5.31(d) and 5.31(f) respectively, and tabulated for systematic computation.

Member portion	Origin	Limits m	M kN-m	m kN-m	m' kN-m	EI kN-m^2
DC	D	0 to 5	0	$(1)(x)$	1	EI_1
CE	C	0 to 4	$(5)(x)$	$\left(5 - \frac{x}{6}\right)$	$\left(1 - \frac{x}{6}\right)$	$3EI_1$
EB	E	0 to 2	$(20 - 5x)$	$\left(4\frac{1}{3} - \frac{x}{6}\right)$	$\left(\frac{1}{3} - \frac{x}{6}\right)$	$3EI_1$
BF	B	0 to 2	10	$(4 - x)$	0	$2EI_1$
FA	A	0 to 2	$5x$	$(1)(x)$	0	$2EI_1$

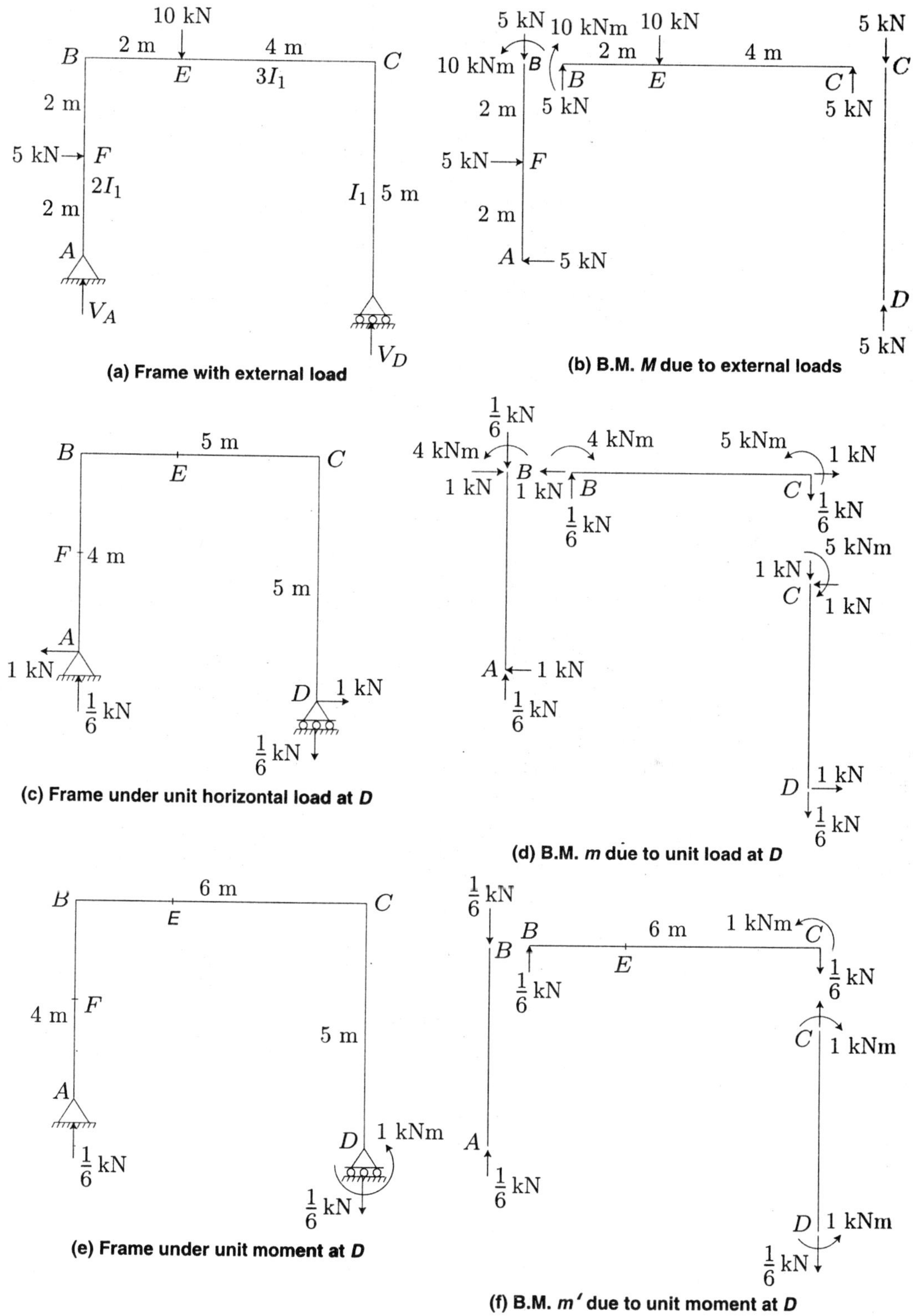

Fig. 5.31: For Example 5.13

$$\Delta_D = \int_0^5 \frac{(0)(x)dx}{EI_1} + \int_0^4 \frac{5x\left(5 - \frac{x}{6}\right)dx}{3EI_1} + \int_0^2 \frac{(20-5x)\left(4\frac{1}{3} - \frac{x}{6}\right)dx}{3EI_1} + \int_0^2 \frac{(10)(4-x)dx}{2EI_1} + \int_0^2 \frac{5x(x)dx}{2EI_1}$$

$$= \frac{139.22 \text{ kNm}^3}{EI_1}.$$

Since the calculated value of Δ_D is positive, it indicates that the deflection Δ_D is in the direction of the unit load supplied, i.e., to the right.

$$\theta_D = \int_0^4 \frac{5x\left(1 - \frac{x}{6}\right)dx}{3EI_1} + \int_0^2 \frac{(20-5x)\left(\frac{1}{3} - \frac{x}{6}\right)dx}{3EI_1} = \frac{9.26 \text{ kNm}^2}{EI_1}.$$

Since the calculated value of θ_D is positive, it indicates that rotation θ_D is in the same direction as the unit couple applied, i.e., counter-clockwise.

Example 5.14. *Determine the vertical and horizontal deflections of the point L_2 of the truss in Fig. 5.32(a). Assume bars of same cross-sectional area of* 3000 mm^2 *and* $E = 2(10)^2$ N/mm^2.

Solution : The example will be solved by unit load method for vertical and horizontal components of deflection of point L_2. For comparison of the method based on Castigliano's theorem with the unit load method vertical deflection of L_2 will be obtained by Castigliano's theorem. Results are given in tabular form.

Vertical and horizontal deflections at L_2, i.e., ΔL_{2V} and ΔL_{2H}.

By Unit Load Method. [Refer to Figs. 5.32(a), 5.32(b) and 5.32(c).]

Member	L(m)	A(mm^2)	S(kN)	u_V	u_H	$\frac{Su_VL}{A}$	$\frac{Su_VL}{A}$ (L in mm)
L_0U_1	5	3000	−43.75	−0.625	0	+45.58	0
L_0L_2	6	3000	+26.25	+0.375	+1.00	+19.69	52.50
U_1L_2	5	3000	−6.25	+0.625	0	−6.51	0
U_1U_3	6	3000	−22.50	−0.750	0	+33.75	0
U_3L_2	5	3000	+6.25	+0.625	0	+6.51	0
L_2L_4	6	3000	+18.75	+0.375	0	+14.06	0
U_3L_4	5	3000	−31.25	−0.625	0	+32.55	0
					$\sum$	145.63	+52.50

$$\text{Deflection } \Delta L_{2V} = \frac{145.63}{2(10)^2} = 0.72815 \text{ mm} \downarrow$$

$$\text{Deflection } \Delta L_{2H} = \frac{52.50}{2(10)^2} = 0.2625 \text{ mm} \rightarrow$$

Vertical deflection at L_2, i.e., ΔL_{2V}

By Castigliano's Theorem [Fig. 5.32(d)]

A load P ia assumed at L_2 as shown. Bar forces are obtained considering the external loadings and P. From the values of bar forces $\frac{\partial S}{\partial P}$ is obtained. Finally, value of P is put as zero before calculating $S\frac{\partial S}{\partial P}\frac{L}{A}$ as actually no force P is at L_2.

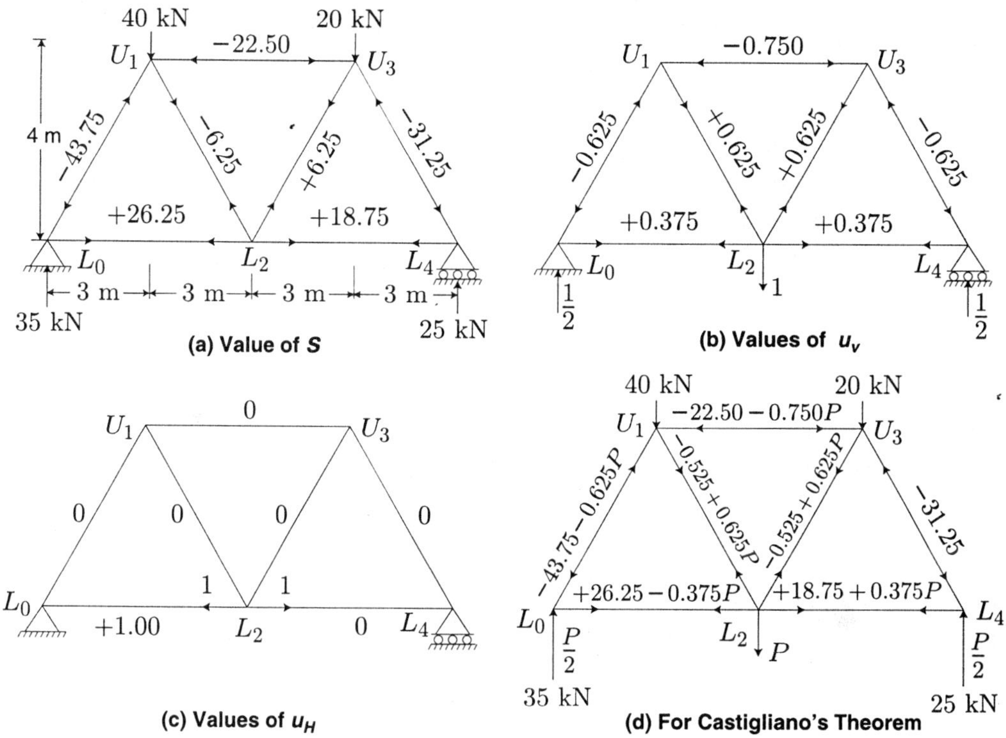

Fig. 5.32: For Example 5.14

If deflection at any point is desired where an external load in the direction of deflection has been applied bar forces have to be calculated taking an additional load P_i. After calculating $\frac{\partial S}{\partial P_i}$, for bar forces to be obtained from $S\frac{\partial S}{\partial P_i}\frac{L}{AE}$, P_i has to be taken as zero.

Member	L(m)	A(mm^2)	S	$\frac{\partial S}{\partial P}$	$\frac{S\frac{\partial S}{\partial P}L}{A}$ (L in mm)
L_0L_1	5	3000	-43.75 $-0.625P$	-0.625	$+45.58$
L_0L_2	6	3000	$+26.25$ $+0.375P$	$+0.375$	$+19.69$
U_1L_2	5	3000	-6.25 $+0.625P$	$+0.625$	-6.51
U_1U_3	6	3000	-22.50 $-0.750P$	-0.750	$+33.75$
U_3L_2	5	3000	$+6.25$ $+0.625P$	$+0.625$	$+6.51$
L_2L_4	6	3000	$+18.75$ $+0.375P$	$+0.375$	$+14.06$
U_3L_4	5	3000	-31.25 $-0.625P$	-0.625	$+32.55$
				$\sum$	145.63 Same as in case of unit load method.

Example 5.15. *Compute the change in distance L_1U_2 of the truss in Fig. 5.33(a), if cross-sectional areas of vertical and horizontal members be* 2000 mm^2 *and that of the inclined members* 3000 mm^2. *Assume* $E = 2(10)^5$ N/mm^2.

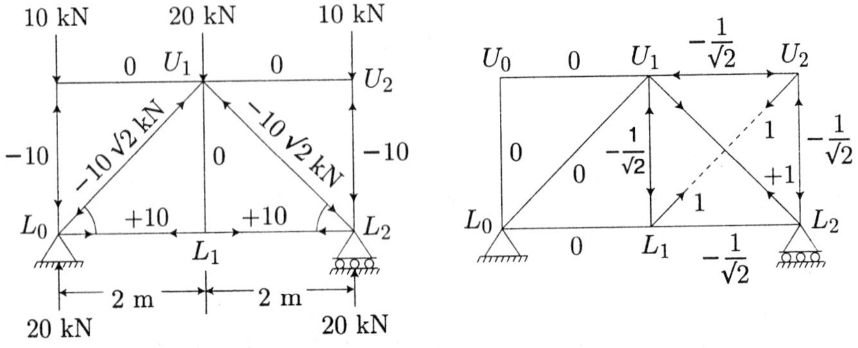

(a) Value of *S* in kN (b) Value of *u*

Fig. 5.33: For Example 5.15

Solution : The example will be solved by the application of the unit load method. Referring to Figs. 5.33(a) and 5.33(b), the calculations are shown in tabular form.

Table for Example 5.15.

Member	L(m)	A(mm)^2	S(kN)	u	$\frac{SuL}{AE}$ (L in mm)
U_0L_0	2	2000	-10	0	0
U_0U_1	2	2000	0	0	0
L_0U_1	$2\sqrt{2}$	3000	$-10\sqrt{2}$	0	0
L_0L_1	2	2000	$+10$	0	0
L_1U_1	2	2000	0	$-\frac{1}{\sqrt{2}}$	0
L_1L_2	2	2000	$+10$	$-\frac{1}{\sqrt{2}}$	$-\frac{10}{\sqrt{2}}$
U_1L_2	$2\sqrt{2}$	3000	$-10\sqrt{2}$	$+1$	$-\frac{40}{3}$
U_1U_2	2	2000	0	$-\frac{1}{\sqrt{2}}$	0
U_2L_2	2	2000	-10	$-\frac{1}{\sqrt{2}}$	$+\frac{10}{\sqrt{2}}$
					$\Sigma -\frac{40}{3}$

$$\text{Change in distance } L_1U_2 = -\frac{40}{(3)(2)(10)^2} = -\frac{1}{15} \text{ mm.}$$

Distance increases.

Example 5.16. *Fig. 5.34(a) indicates the* $\frac{S}{A}$ *values in* kN/mm^2 *positive indicating tension and negative compression, for the members of the two central points of a truss. Compute E time the change of angle* $L_1L_2U_2$.

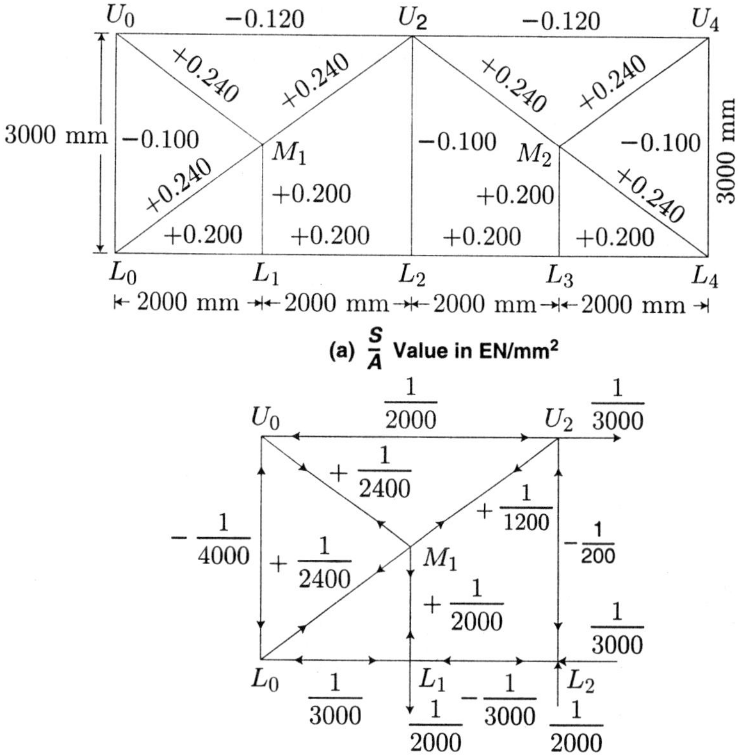

(a) $\frac{S}{A}$ Value in EN/mm²

(b) *u* Value

Fig. 5.34: For Example 5.16

Solution : Calculations are given in Tabular form.

Member	L(mm)	$\frac{S}{A}$ kN/mm^2	u	$\frac{SuL}{A}$
U_0U_2	4000	−0.12	$-\frac{1}{3000}$	+0.16
U_2U_4	4000	−0.12	0	0
U_0M_1	2500	+0.240	$+\frac{1}{2400}$	+0.25
U_0L_0	3000	−0.10	$-\frac{1}{4000}$	+0.075
L_0M_1	2500	+0.24	$+\frac{1}{2400}$	+0.25
L_0L_1	2000	+0.20	$-\frac{1}{3000}$	−0.133
L_1L_2	2000	+0.20	$-\frac{1}{3000}$	−0.133
L_1M_1	1500	+0.20	$+\frac{1}{2000}$	+0.150
U_2M_1	2500	+0.24	$+\frac{1}{1200}$	+0.50
L_2U_2	3000	−0.10	$-\frac{1}{2000}$	+0.15
For all other members $u = 0$			0	0
			$\sum$	1.269

Change in angle $L_1L_2U_2 = \dfrac{1.269}{E}$ radians.

Positive sign means that the angle increases.

Example 5.17. *Fig. 5.35(a) shows a pin-jointed truss in which members U_2U_4, U_0U_2 and U_4L_4 are coiled spring with a constant of 40,000 (N/mm.) Member U_2L_2 is essentially a hydraulic jack with a piston working inside a cylinder. All other members are of the same material with E as $2(10)^5$ N/mm² and other vertical and horizontal members and inclined members have cross-sectional areas of 4000 mm² and 2500 mm² respectively. Compute changes in distance between M_1 and M_3 that would be produced by the following factors operating simultaneonsly :*

(*a*) *Joint L_4 is subjected to a 240 kN load as shown.*

(*b*) *The length of the member U_2L_2 is increased by 5 mm by movement of the piston inside the cylinder.*

(*c*) *The temperature of the lower chord L_0L_4 is increased by 30 °C, while that of the remainder structure remains unchanged. Assume coefficient of thermal expansion as $2(10)^{-5}$per °C.*

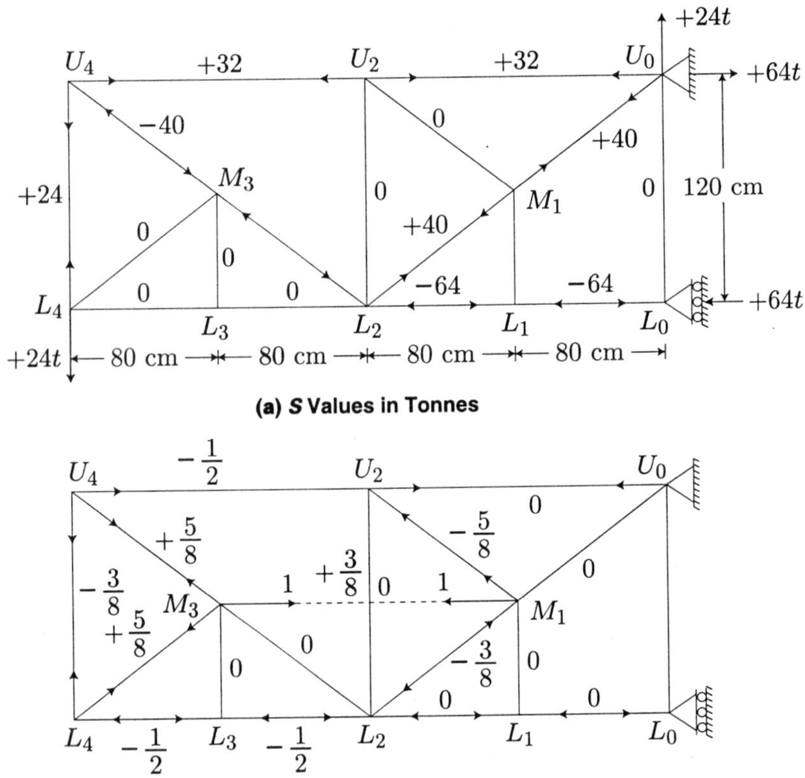

(a) *S* Values in Tonnes

(b) *u* Values

Fig. 5.35: For Example 5.17

Solution : In this example basic equation of unit load method, i.e., $\sum udL$ will be used, dL being the total deformation of a member due to external causes. In this case $dL = dL_1 + dL_2 + dL_3$ where dL_1 is due to load 240 kN acting at L_4, dL_2 due to movement of member U_2L_2 and dL_3 due to temperature change.

Calculations are given in Tabular form.

Table for Example 5.17.

Member	L(mm)	A(mm²)	E (kN/mm)²	S (kN)	Due to Loading $dL_1 = \frac{SL}{AE}$	Due to movement of member dL_2	Due to temp change dL_3	$dL = dL_1 + dL_2 + dL_3$	u	udL
L_0U_0	1200	4000	$2(10)^2$	0	0	0	0	0	0	0
L_0L_1	800	4000	do	−640	−0.640	0	+0.480	−0.160	0	0
U_0U_2	1600	–	40,000 N/mm	+320	+8.00	0	0	+8.00	0	0
U_0M_1	1000	2500	$2(10)^2$ kN/mm²	+400	+0.800	0	0	+0.800	0	0
L_1M_1	600	4000	do	0	0	0	0	0	0	0
U_2M_1	1000	2500	do	0	0	0	0	0	$-\frac{5}{8}$	0
L_1L_2	800	4000	do	−640	−0.640	0	+0.480	−0.160	0	0
L_2U_2	1200	–	–	0	0	+5.00	0	+5.00	$+\frac{3}{8}$	+1.875
L_2L_3	800	4000	$2(10)^2$	0	0	0	+0.480	+0.480	$-\frac{1}{2}$	−0.240
L_2M_1	1000	2500	do	+400	+0.800	0	0	+0.800	$-\frac{5}{8}$	−0.500
L_2M_3	1000	2500	$2(10)^2$	−400	−0.800	0	0	−0.800	0	0
U_2U_4	1600	–	40.000 N/mm	320	+8.00	0	0	+8.00	$-\frac{1}{2}$	−0.400
L_3L_4	800	4000	$2(10)^2$	0	0	0	+0.480	+0.480	$-\frac{1}{2}$	−0.240
L_3M_3	600	4000	do	0	0	0	0	0	0	0
L_4M_3	1000	2500	do	0	0	0	0	0	$+\frac{5}{8}$	0
L_4U_4	1200	–	40.000	+240	+6.00	0	0	+6.00	$-\frac{3}{8}$	−2.250
U_4M_3	1000	2500	$2(10)^2$	−400	−0.800	0	0	−0.800	$-\frac{5}{8}$	−0.500

Change in distance between M_1 and $M_3 = \sum udL = -5.855$ mm (increase in distance)

Problems

5.1 Calculate the slope at B and deflection at C of the cantilever beams shown in Figs. P-5.1 by (i) moment-area method and (ii) conjugate beam method.

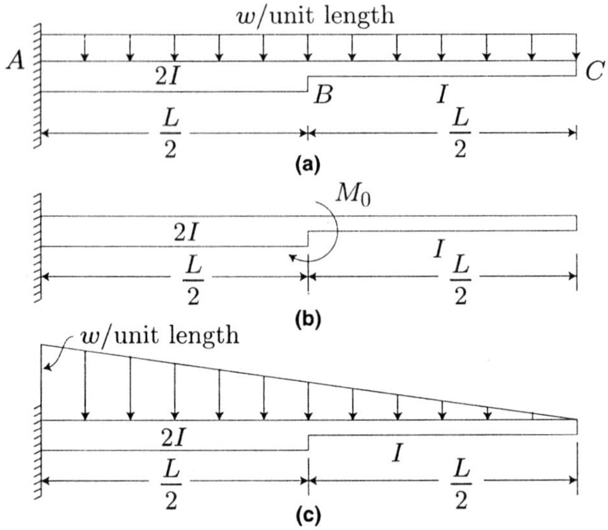

Fig. P-5.1

5.2 Calculate slopes at A and B and deflection at C of the bemas shown in Figs. P-2.2 by (i) moment-area method and (ii) Conjugate beam method.

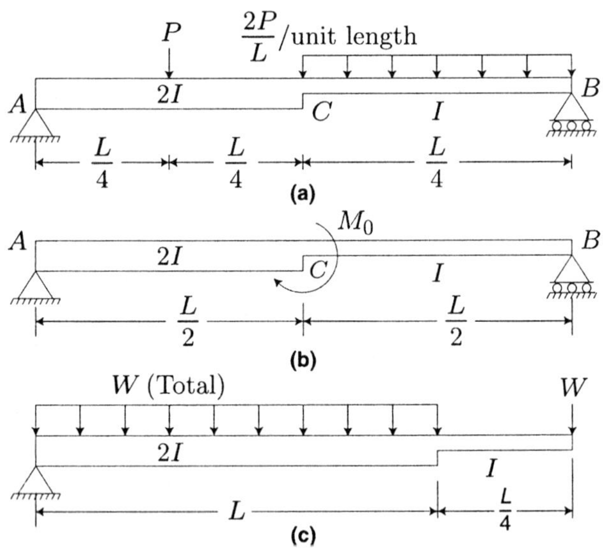

Fig. P-5.2

5.3 Solve Probs. 5.1 and 5.2 by unit load method.

5.4 Solve Probs. 5.1 and 5.2 by Castigliano's theorem.

5.5 A beam ABC is simply supported at A and B with BC overhanging. The depth of the beam varies uniformly from d at A to $2d$ at B to $d/2$ at C, while the width is constant.

The beam is subjected to a uniformly distributed load of 10 kN/m. $AB = 6$ m and $BC = 2$ m. Calculate the slopes at A and B and deflection at C by (i) unit load method and (ii) Castigliano's theorem.

5.6 Calculate the horizontal and vertical displacements and rotation at D of the frame shown in Fig. P-5.3 by (i) unit load method and (ii) Castigliano's theorem.

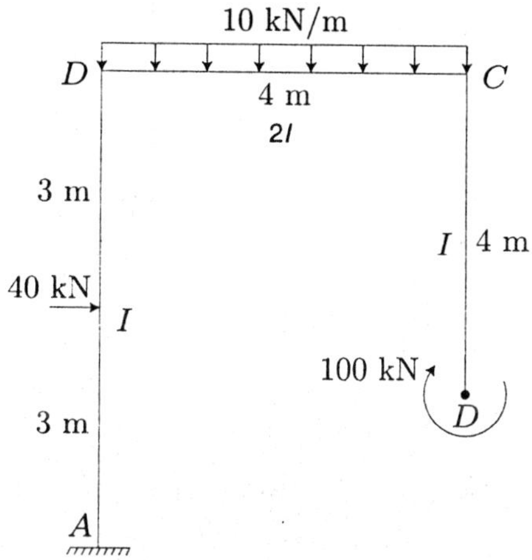

Fig. P-5.3

5.7 Calculate the horizontal displacement and rotation at D of the frame shown in Fig. P-5.4 by (i) unit load method and (ii) Castigliano's theorem.

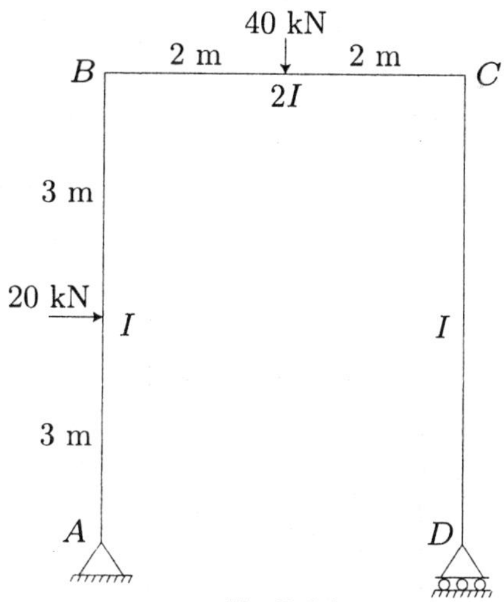

Fig. P-5.4

5.8 Calculate the vertical deflection and angle of twist at the end B of a semi-circular ring AB of mean radius R having centre line in the horizontal plane and end A fixed, if a vertical force P acts at the middle of the ring. Use (i) Unit load method and (ii) Castigliano's theorem.

5.9 Calculate by unit load method the vertical and horizontal deflection at E of the pin-jointed truss shown in Fig. P-5.5 due to the applied loads and a temperature rise of

40 °F, if the cross-sectional area be 200 mm^2 for each of the horizontal and vertical members and 300 mm^2 for each of the diagonal members. Take $E = 2(10)^5$ N/mm^2 and coefficient of thermal expansion $= 6(10)^{-6}$ °F. Calculate also what additional deflection will occur if the member BC is too short originally by 5 mm.

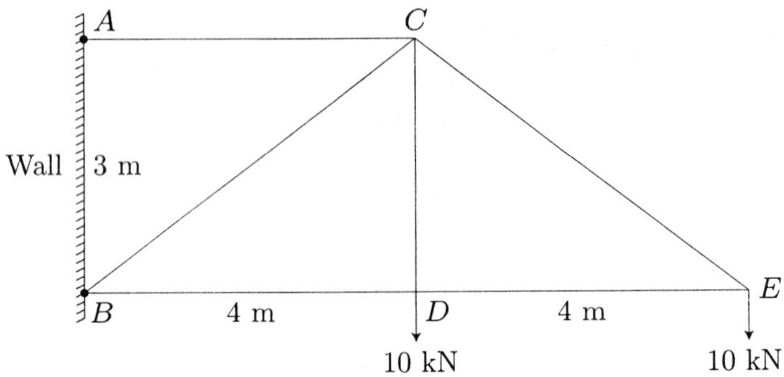

Fig. P-5.5

5.10 A uniform horizontal cantilever beam of length L is propped at a distance L/4 from the free end. If a clockwise couple M_0 acts at a distance $\frac{L}{2}$ from the fixed support, obtain the reaction of the prop and draw B.M. and S.F. diagrams for the beam.

5.11 A beam of length $2L$ and uniform EI is fixed at one end and supported on a prop at the other end. If the prop settles by k for each unit value of the reaction, calculate the load on the prop when a uniformly distributed load of w/unit length acts on the beam.

5.12 A cantilever beam ABC of length L is fixed at A and held by a wire BD of the same material as shown in Fig. P-5.6. Determine the deflection at C due to a load P acting there if 'a' and 'I' be respectively the sectional area of the wire and moment of inertia of the beam.

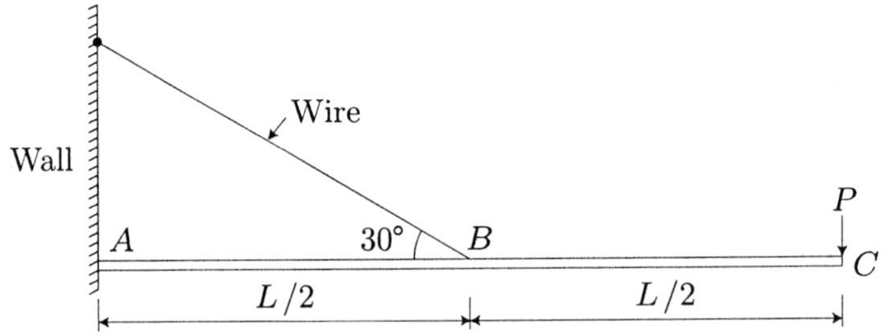

Fig. P-5.6

5.13 A cantilever beam carries a uniformly distributed load throughout its length and is propped at the free end. What would be the reaction and settlement of the prop so that the bending stress in the beam is least possible? If the prop be at the same level as the fixed end, where will it be placed to achieve the least possible bending stress?

5.14 A rectangular cantilever beam of length L has a constant breadth b and a varying depth given by (a) $k\sqrt{x}$, (b) kx and (c) kx^2, where k is a constant and x is the distance measured from the free end. The cantilever carries a uniformly distributed load of magnitude w/unit length. Find the deflection at the free end if d is the depth at fixed end.

5.15 A cantilever beam of rectangular cross section has a width b and depth varying uniformly from d at a support to $\frac{d}{2}$ at free end. Find the slope and deflection at the free end when a concentrated load P is placed there. M.I. at the fixed end is I.

5.16 A beam rests on supports at its ends and carries a load P midway between them. M.I. of the beam is $2I$ at centre and I at ends, variation being linear. Find the deflection at the centre.

What will be the deflection when uniformly distributed lead w per unit length is place over the beam.

5.17 A cantilever 4 m long is 400 mm deep throughout, but is tapered in such a way that when a load is placed at the free end, the stress is the same at all sections. If this stress be 100 N/mm^2, calculate the deflection under the load, if $E = 2(10)^5$ N/mm^2.

5.18 A uniform beam of length L and depth d is to carry a total load W kN uniformly distributed. If the maximum deflection is not to exceed $\frac{L}{400}$ when the maximum bending stress is 150 N/mm^2, obtain the maximum value of $\frac{L}{d}$, if

(a) the beam is a cantilever (b) the beam is simply supported.

Assume $E = 2(10)^5$ N/mm^2.

5.19 A cantilever beam has a flexural stiffness varying linearly from EI at the free end to $2\,EI$ at the fixed end. Calculate the deflection at the free end when a vertical load P is applied there.

5.20 Calculate the horizontal and vertical displacements and rotation of the joint D of the frames shown in Figs. P-5.7 and P-5.8 using

(a) Moment area method (b) Unit load method.

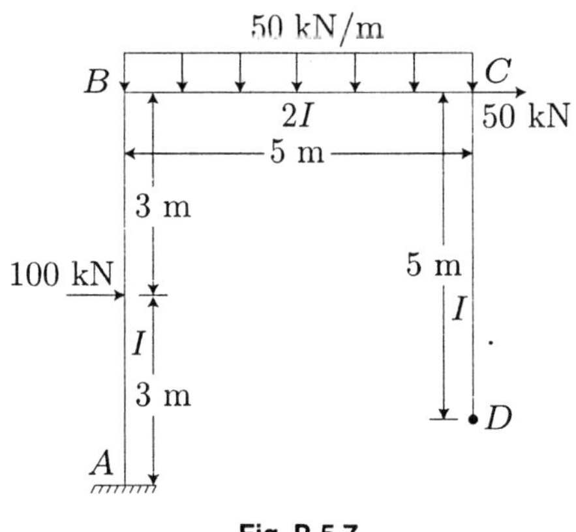

Fig. P-5.7

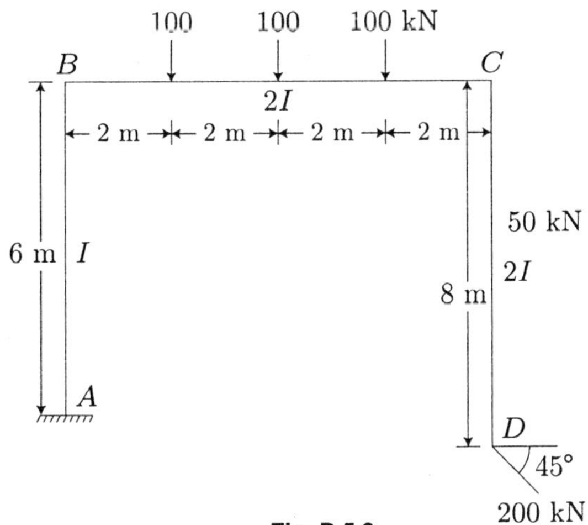

Fig. P-5.8

5.21 Calculate the horizontal displacement and rotation of the joint D for the frames shown in Figs. P-5.9, P-5.10 and P-5.11.

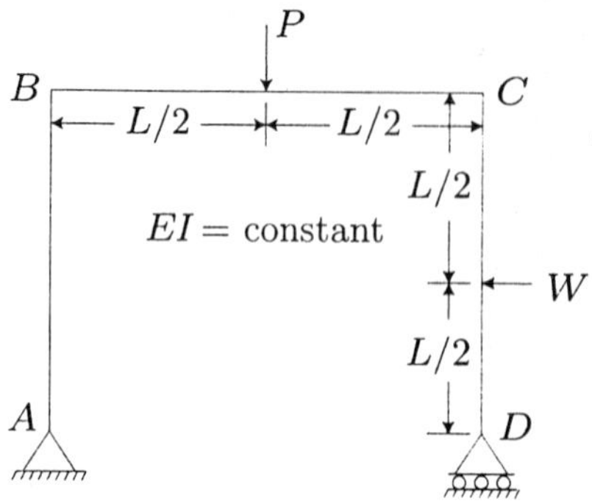

Fig. P-5.9

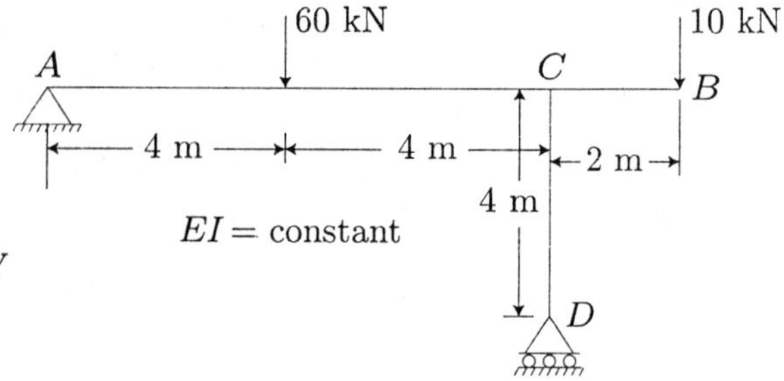

Fig. P-5.10

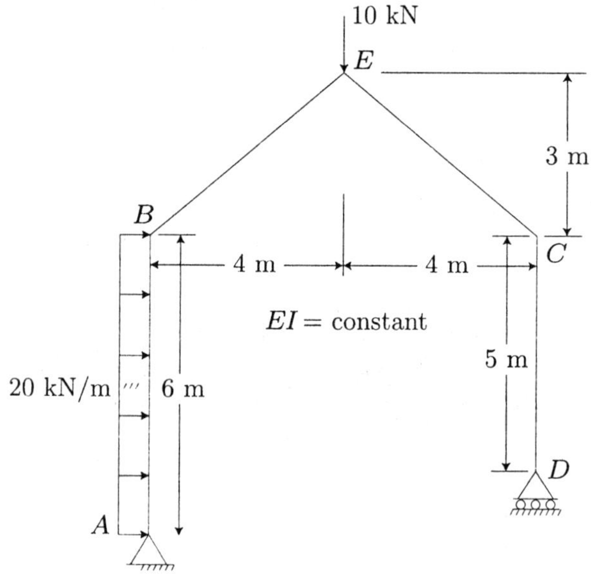

Fig. P-5.11

5.22 For the truss and beam structure of Fig. P-5.12, determine the vertical deflections at B and C and the slope at B. Assume E for each member as $2(10)^5$ N/mm^2.

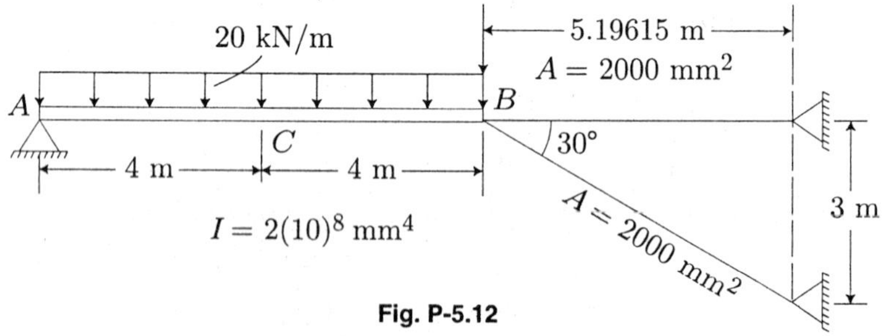

Fig. P-5.12

5.23 Calculate the vertical deflection at D and the rotation due to flexure and twisting moment at D of the horizontal frame consisting of circular bar as shown in Fig. P-5.13, when a vertical load of magnitude 10 kN acts at D. Assume $I = 5(10)^6\ \text{mm}^4, E = 2(10)^5\ \text{N/mm}^2$ and $G = 8(10)^4\ \text{N/mm}^2$.

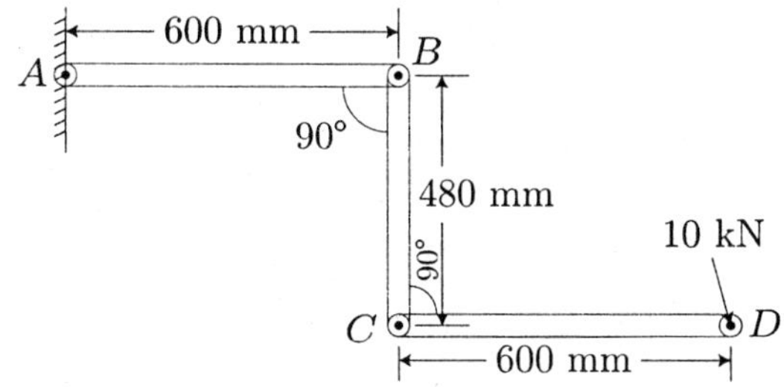

Fig. P-5.13

5.24 Calculate the deflection at end D of the frame shown in Fig. P-5.14, in the direction X, Y and Z. For each member assume $E = 2(10)^5\ \text{N/mm}^2, G = 8(10)^4\ \text{N/mm}^2, I = 2(10)^7\ \text{mm}^4$ and $J = 4000(10)^4\ \text{mm}^4$.

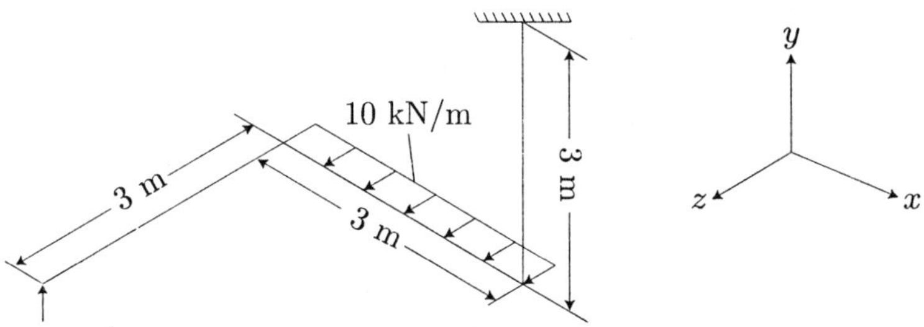

Fig. P-5.14

5.25 The truss shown in Fig. P-5.15 has the value of $\frac{L}{A}$ of members as 100 for top and bottom chord members, 150 for vertical members and 200 for inclined members :-

(a) Find the vertical and horizontal deflection of the joints L_3 and U_2 due to the loading shown.

(b) Compute the changes in distance L_3U_4 of the truss.

(c) Compute change in angle $L_2U_2U_3$.

(d) If the whole truss suffers a change of temperature of 40 °C, determine the vertical and horizontal deflection of joint L_3. Assume $\alpha = 12(10)^-6$ per °C.

(e) What will be the vertical and horizontal deflections of L_3 when only bottom chord suffers a change of temperature of 40 °C?

Assume E for all members as $2(10)^5\ \text{N/mm}^2$.

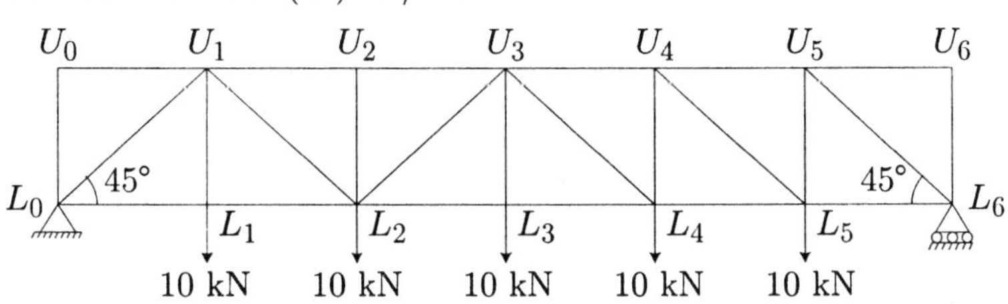

Fig. P-5.15

5.26 A pin-jointed frame shown in Fig. P-5.16 has members all of the same material and cross-sectional area A except for member L_1U_2, which has an area kA. Find the value of k, so that when a vertical load is applied at U_2, the resultant deflection is in a direction inclined at 30° to the vertical.

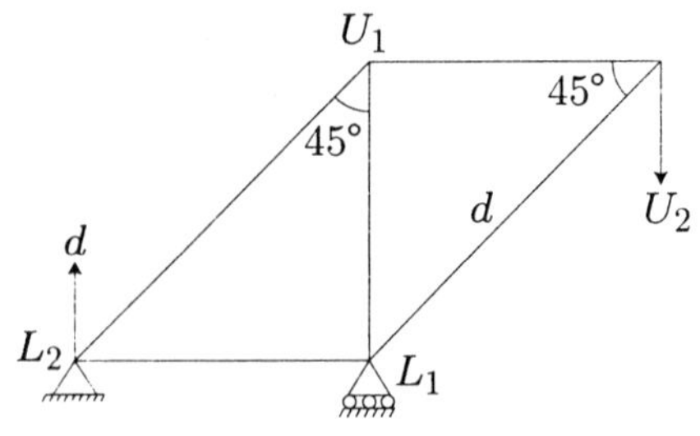

Fig. P-5.16

5.27 The cross-sectional areas of the truss shown in Fig. P-5.17 are such that due to loading stresses in tension and compression members are 150 N/mm^2 and 100 N/mm^2 respectively. Determine the vertical and horizontal deflections of joint L_4. Assume E for all members as $2(10)^5$ N/mm^2.

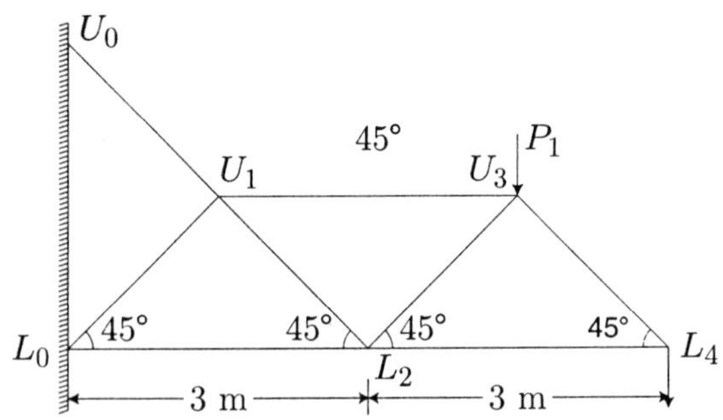

Fig. P-5.17

Chapter 6

Redundant Structures

6.1 Introduction

For the last eighty years statically indeterminate structures are being used extensively due to their economy and increased rigidity under moving loads. A knowledge of the analysis of indeterminate structures has, therefore, become increasingly important. A structure may be statically indeterminate either externally or/and internally. Examples of external indeterminacies are propped cantilever, fixed ended beam, continuous beams and trusses, one bay rigid frames, arches, etc. Rigid frames, building frames trusses with redundant members are examples of internal indeterminacies.

Statically indeterminate structures differ from statically determinate ones in the following respects.

(a) Stress analysis involves not only the geometry of the structure, but also its modulus of elasticity, cross-sectional areas and moment of inertia. To arrive at the final design of a statically indeterminate structure, it is necessary to assume preliminary sizes for the members, making a stress analysis of this design, redesigning the members for these stresses making a new stress analysis of the revised design, redesigning again and reanalysing until convergence on the final design is obtained.

(b) Stress in indeterminate structures are developed not only by loads but also on account of temperature changes, settlement of supports, etc.

Various methods of analysis for statically indeterminate structures are there, some applicable to the general problem and others particularly adopted to certain types. In this Chapter a general method of analysis of indeterminate structures will be discussed. This method can be called method of consistent deformation based on the application of the method of superposition and is applicable in the case of beams as well as trusses. The other method is based on the use of Castigliano's theorem and is also known as the method of least work. These methods, however, are based on the known deformation of the structures.

6.2 Methods of Analysis

(a) Method of Consistent Deformation (based on the Method of Superposition)

This method depends mainly on making the statically indeterminate structure statically determinate and stable and calculating the forces or moments or both necessary to restore the structure to its original condition—the moments or forces or both thus calculated, in combination with the laws of statics, make possible a complete analysis of the structure.

The method of consistent deformation is the most general method of analysing statically indeterminate structure. In using this method, it is first necessary to understand to what degree a structure is statically indeterminate. The restraining actions of the redundant forces in the structure are first removed.

The basic determinate structure is assumed to be subjected to the applied loads as well as the redundant reactions, so that the resultant deformations corresponding to the redundant reactions due to the applied loads and the redundant reactions are the same as the known values of the deformations. These conditions give consistent deformation equations which are equal in number to that of the redundant reactions. The fixed ended beam of Fig. 6.1(a) is subjected to vertical loads only and is indeterminate to the second degree. Choosing the vertical reactions F_1 and moment F_2 at fixed support B as the redundants, the basic determinate structure becomes a cantilever beam as shown in Fig. 6.1(b). The actually fixed beam of Fig. 6.1(a) is equivalent to the cantilever beam of Fig. 6.1(b) with end A fixed and end B free subjected to the applied loads and redundant reactions F_1 and F_2.

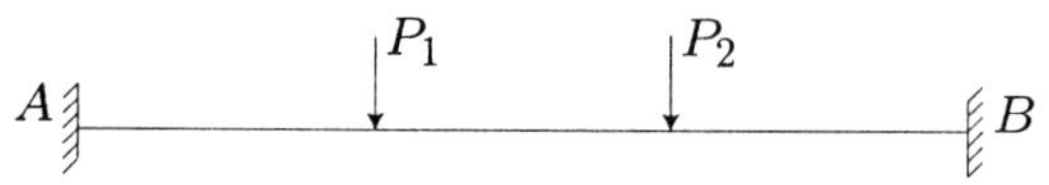

(a) Fixed end beam with external load

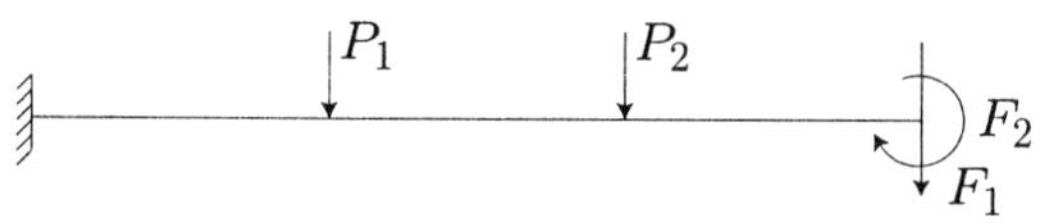

(b) Basic determinate structure for (a) with unknown forces

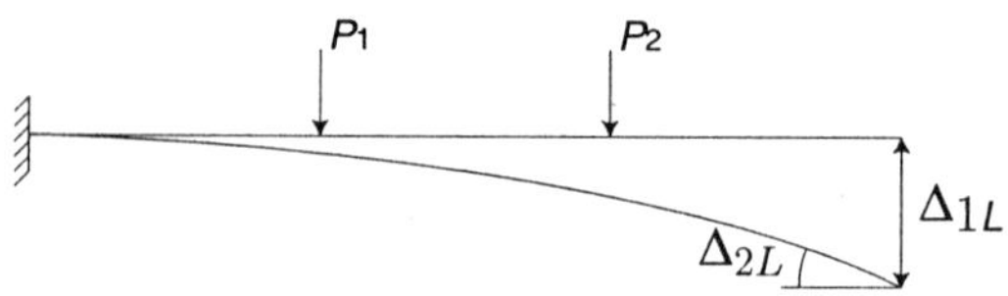

(c) Basic determinate structure under external loads

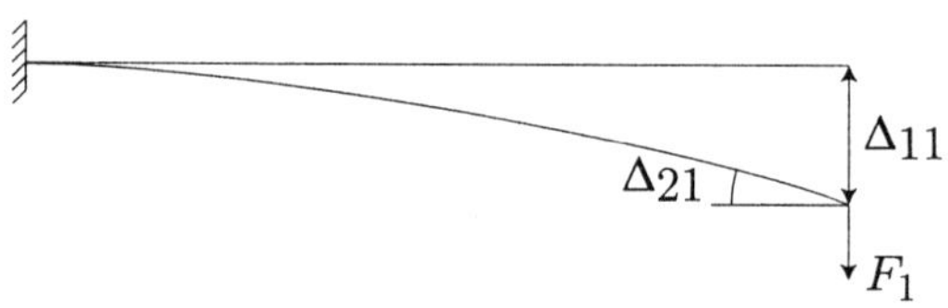

(d) Basic determinate structure under load F_1

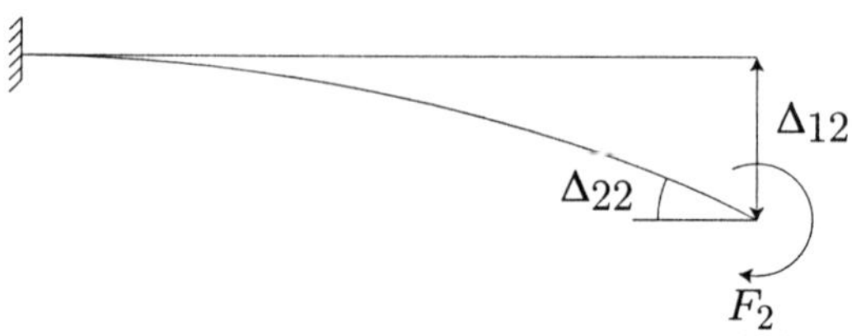

(e) Basic determinate structure under load F_2

Fig. 6.1: Consistent deformation equations

The applied loads, the reaction F_1 and the reactions F_2 are shown acting separately on the basic determinate structure in Figs. 6.1(c), 6.1(d) and 6.1(e) respectively. For each of these three loading cases, the deformations corresponding to the reactions F_1 and F_2 can be obtained.

If Δ_{1L} = vertical displacement at B due to applied loads in the direction of F_1,

Δ_{2L} = rotation at B due to applied loads in the direction of F_2,

Δ_{11} = vertical displacement at B due to F_1,

Δ_{21} = rotation at B due to F_1,

Δ_{12} = vertical displacement at B due to F_2, and

Δ_{22} = rotation at B due to F_2, then the resultant deformations

Δ_1 and Δ_2 corresponding to F_1 and F_2 are respectively given by

$$\left.\begin{aligned} \Delta_1 &= \Delta_{1L} + \Delta_{11} + \Delta_{12} \\ \text{and} \quad \Delta_2 &= \Delta_{2L} + \Delta_{21} + \Delta_{22} \end{aligned}\right] \tag{6.1a}$$

For this particular problem, the vertical displacement and rotation at B are known to be zero, so that,

$$\Delta_1 = 0 \quad \text{and} \quad \Delta_2 = 0.$$

Eq. (6.1a) then becomes,

$$\left.\begin{aligned} \Delta_{1L} + \Delta_{11} + \Delta_{12} &= 0 \\ \text{and} \quad \Delta_{2L} + \Delta_{21} + \Delta_{22} &= 0 \end{aligned}\right] \tag{6.1b}$$

It may be noted that suffixes of the deformations Δ_{11}, Δ_{12}, Δ_{21} and Δ_{22}, follow a definite order. The first figure indicates the reaction in the direction of which the deformation occurs and the second figure indicates the load causing the deformation. Thus, Δ_{12} is the deformation in the direction of F_1 caused by F_2 and Δ_{11} is the deformation in the direction of F_1 caused by F_1, Δ_{1L} and Δ_{2L} being the deformations in the directions due to applied loads.

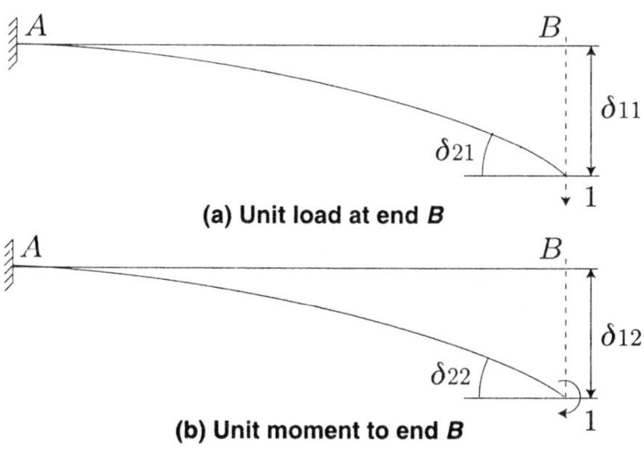

(a) Unit load at end *B*

(b) Unit moment to end *B*

Fig. 6.2: Deformations due to unit loads

If δ_{11} and δ_{21} be the deformations in the directions of F_1 and F_2 respectively caused by a unit value of F_1, and δ_{22} be the deformations in the directions of F_1 and F_2 respectively caused by a unit value of F_2, as shown in Figs. 6.2(a) and 6.2(b), then,

$$\Delta_{11} = \delta_{11} F_1$$

$$\Delta_{21} = \delta_{21} F_1$$

$$\Delta_{12} = \delta_{12} F_2$$

$$\Delta_{22} = \delta_{22} F_2.$$

Eq. (6.1a) then can be written as,

$$\left.\begin{aligned} \Delta_{1L} + \delta_{11}F_1 + \delta_{12}F_2 = 0 \\ \Delta_{2L} + \delta_{21}F_1 + \delta_{22}F_2 = 0 \end{aligned}\right] \tag{6.1c}$$

In the above example, the final deformations at B corresponding to the F_1 and F_2 are taken to be zero. There may, however, be some deformations due to yielding of support as shown in the Fig. 6.3.

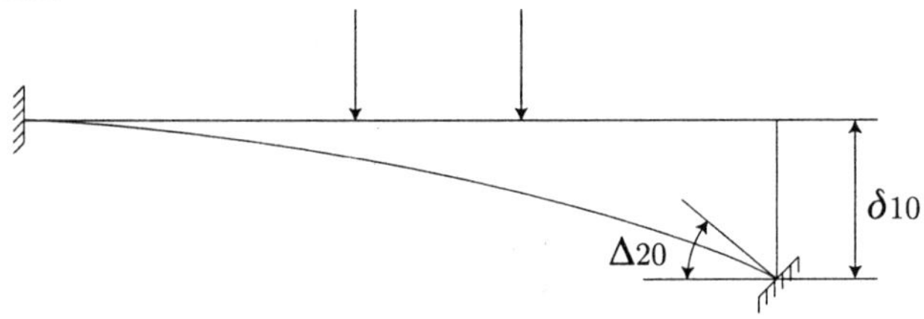

Fig. 6.3: Deformations due to yielding of support

For such a case, the deformation equations are :

$$\Delta_{1L} + \delta_{11}F_1 + \delta_{12}F_2 = \Delta_{10}$$

$$\Delta_{2L} + \delta_{21}F_1 + \Delta_{22}F_2 = \Delta_{20},$$

which after rearrangement becomes

$$\left.\begin{aligned} \delta_{11}F_1 + \delta_{12}F_2 + (\Delta_{1L} - \Delta_{10}) = 0 \\ \delta_{21}F_1 + \delta_{22}F_2 + (\Delta_{2L} - \Delta_{20}) = 0 \end{aligned}\right] \tag{6.1d}$$

If there are n redundants the deformation equations will be

$$\left.\begin{aligned} \delta_{11}F_1 + \delta_{12}F_2 + \delta_{13}F_3 + \cdots + \delta_{1n}F_n + (\Delta_{1L} - \Delta_{10}) &= 0 \\ \delta_{21}F_1 + \delta_{22}F_2 + \delta_{23}F_3 + \cdots + \delta_{2n}F_n + (\Delta_{2L} - \Delta_{20}) &= 0 \\ \delta_{31}F_1 + \delta_{32}F_2 + \delta_{33}F_3 + \cdots + \delta_{3n}F_n + (\Delta_{3L} - \Delta_{30}) &= 0 \\ \delta_{n1}F_1 + \delta_{n2}F_2 + \delta_{n3}F_3 + \cdots + \delta_{nn}F_n + (\Delta_{nL} - \Delta_{n0}) &= 0 \end{aligned}\right] \tag{6.1}$$

It may be noted here that some of the deformation coefficients in the above equations can be shown identical from Maxwell's theorem of reciprocal deflections, such as

$$\delta_{12} = \delta_{21}; \delta_{23} = \delta_{32}; \delta_{31} \text{ and so on.}$$

Example 6.1. *Determine the support moments and reactions of a fixed-ended beam AB, Fig. 6.4(a), of span L and uniform flexural rigidity EI, subjected to a concentrated load P at a distance a from AB and b from B.*

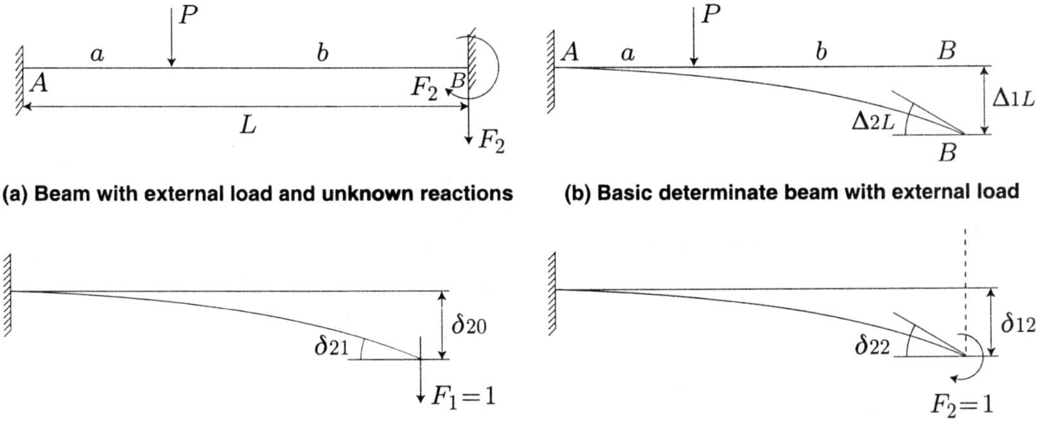

(a) Beam with external load and unknown reactions **(b) Basic determinate beam with external load**

(c) Deformation due to unit load F_1=1 **(d) Deformation due to unit load F_2=1**

Fig. 6.4: For Example 6.1

Solution : The vertical reaction and support moment at the fixed support B are assumed to be the two redundant reactions F_1 and F_2 respectively.

After removing the constraints corresponding to F_1 and F_2, the basic determinate structure becomes a cantilever as shown in Fig. 6.4(b). Figs. 6.4(c) and 6.4(d) show the deformation due to unit values of F_1 and F_2 respectively. Using standard methods for loadings shown in Figs. 6.4(b), 6.4(c) and 6.4(d).

$$\Delta_{1L} = \frac{Pa^3}{3EI} + \frac{Pa^2b}{2EI}$$

$$\Delta_{2L} = \frac{Pa^2}{2EI}$$

$$\text{Also } \delta_{11} = \frac{L^3}{3EI}; \; \delta_{21} = \frac{L^2}{2EI}$$

$$\text{and } \delta_{12} = \frac{L^2}{2EI}; \; \delta_{22} = \frac{L}{EI}.$$

In the case, $\Delta_{10} = 0$ and $\Delta_{20} = 0$.

The consistent deformation Equations for this case are

$$\delta_{11}F_1 + \delta_{12}F_2 + \Delta_{1L} = 0$$

$$\delta_{21}F_1 + \delta_{22}F_2 + \Delta_{2L} = 0.$$

Putting the known values these equations become

$$\frac{L^3}{3EI}F_1 + \frac{L^2}{2EI}F_2 + \frac{Pa^3}{3EI} + \frac{Pa^2b}{2EI} = 0$$

$$\frac{L^2}{2EI} \cdot F_1 + \frac{L}{EI}F_2 + \frac{Pa^2}{2EI} = 0.$$

Solving, $F_2 = \dfrac{Pa^2b}{L^2}$

$$F_1 = -\left[\frac{2Pa^2b}{L^3} + \frac{Pa^2}{L^2}\right].$$

The fixed end moment at $B = \dfrac{Pa^2b}{L^2}$ and the vertical reaction at $B = \dfrac{2Pa^2b}{L^3} + \dfrac{Pa^2}{L^2}$ acting upward.

The fixed end moment and vertical reaction at A can now be obtained from consideration of equilibrium.

The fixed end moment at $A = \dfrac{Pab^2}{L^2}$ and the vertical reaction at $B = \dfrac{2Pab^2}{L^3} + \dfrac{Pb^2}{L^2}$ acting upward.

Example 6.2. *A two-span continuous beam ABC of uniform flexural rigidity* $EI = 4,000$ kN-m^2 *is subjected to a concentrated load* 40 kN *on span BC as shown in Fig. 6.5(a). Solve the beam by consistent deformation method and draw B.M. and S.F. diagrams.*

Solution : Considering the support reaction at B as the redundant F_1, the basic determinate structure becomes a simply supported beam as shown in Fig. 6.5b.

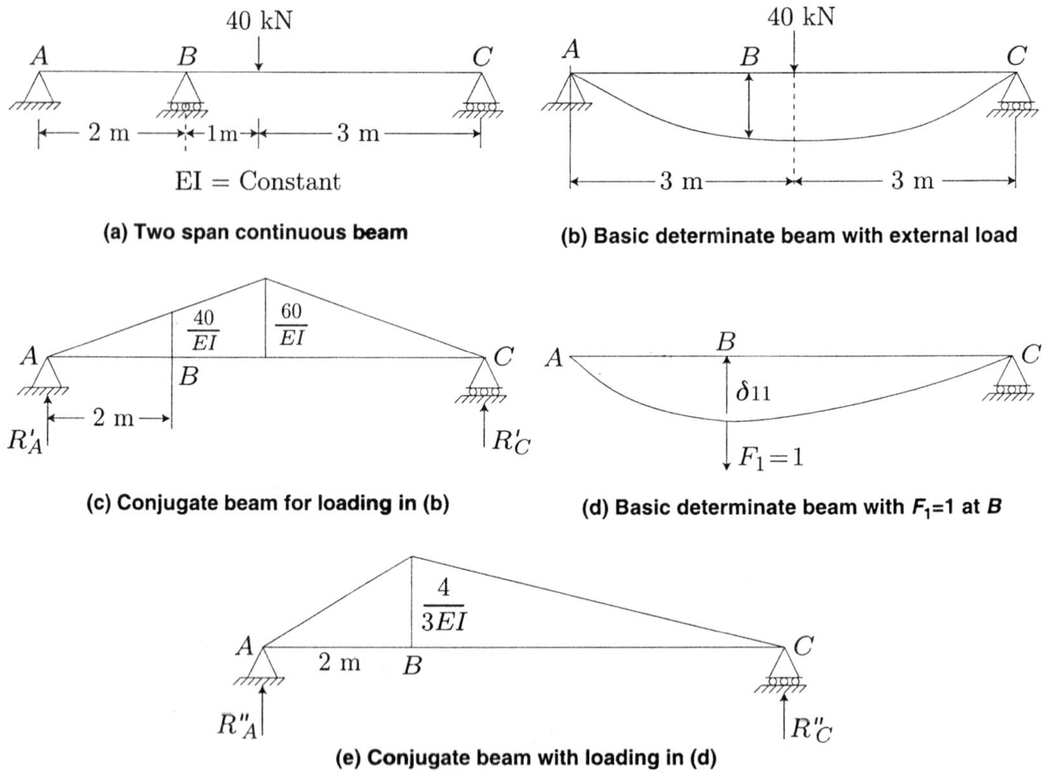

Fig. 6.5: For Example 6.2

Fig. 6.5(c) shows $\frac{M}{EI}$ diagram of the basic determinate beam for loading shown in Fig. 6.5(b). Fig. 6.5(d) shows the deflection for a unit value of F_1 for which the $\frac{M}{EI}$ diagram is shown in Fig. 6.5(c). The deflection Δ_{1L} in Fig. 6.5(b) and δ_{11} in Fig. 6.5(d) can be obtained by the conjugate beam method. Referring to Fig. 6.5(d), the conjugate beam reaction R'_A is

$$R'_A = \left(\frac{1}{2}\right)\left(\frac{1}{2}\right)\left(\frac{60}{EI}\right)(6) = \frac{90}{EI} = \frac{90}{4000} = 0.0225$$

and the bending moment of the conjugate beam at B is

$$M'_B = (R'_A)(2) - \frac{1}{2}\left(\frac{40}{EI}\right)(2)\left(\frac{1}{3}\right)(2) = \frac{460}{3EI} = \frac{460}{(3)(4000)} \text{ m}$$

$$= \frac{23}{600} \text{ m}$$

and $\quad \Delta_{1L} = M'_B = \dfrac{23}{600}$ m.

From Fig. 6.5(c) the conjugate beam reaction R''_A is

$$R''_A = \frac{1}{6}\left[\frac{1}{2}\left(\frac{4}{3EI}\right)(4)\left(\frac{2}{3}\right)(4) + \frac{1}{2}\left(\frac{4}{3EI}\right)(2)\left(4+\frac{2}{3}\right)\right] = \frac{20}{9EI} = \frac{1}{1800}$$

and bending moment of the conjugate beam about B is

$$M''_B = (R''_A)(2) - \frac{1}{2}\left(\frac{4}{3EI}\right)(2)\left(\frac{1}{3}\right)(2) = \frac{1}{1125} \text{ m},$$

whence $\delta_{11} = M''_B = \dfrac{1}{1125}$ m.

The consistent deformation equation is then

$$F_1\delta_{11} + \Delta_{1L} = 0$$

i.e., $\quad F_1\left(\dfrac{1}{1125}\right) + \dfrac{23}{600} = 0,$

whence $\quad F_1 = -\left(\dfrac{23}{600}\right)(1125) = -43.125$ kN.

Hence, the reaction R_B is 43.125 kN acting upward.

Fig. 6.6(a) shows the B.M. diagram which is obtained by superimposing the B.M. diagrams of basic determinate beam due to the applied load of 40 kN and the support reaction at B.

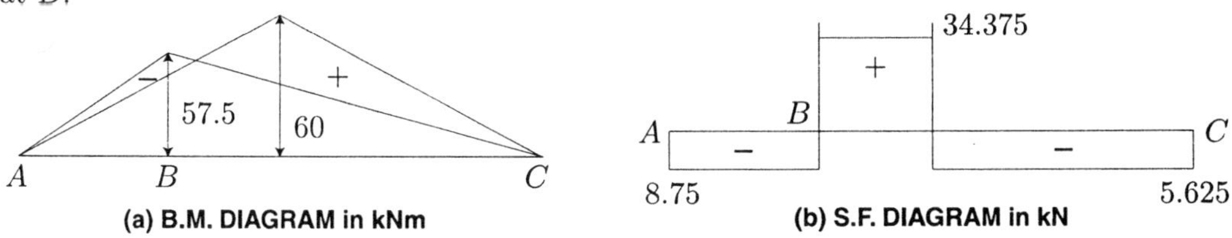

Fig. 6.6: B.M. and S.F. diagrams for Example 6.2

The reactions are

$$R_A = \frac{(40)(3) - (43.125)(4)}{6} = -8.75 \text{ kN (acting downward)}$$

$$R_C = \frac{(40)(3) - (43.125)(2)}{6} = 5.625 \text{ kN (acting upward)}$$

The S.F. diagram is shown in Fig. 6.6(b).

(b) Applications of Castigliano's Theorem

Castigliano's second theorem gives a force method of analysis and may be used to express the conditions of compatibility.

Considering the case of a linearly elastic system with n nos. of indeterminacies, i.e., with 'n' redundant forces, say X_A, $X_B, \cdots, X_N$ the deflection at the point of application of any redundant, say X_M is given by

$$\frac{\partial U}{\partial X_M} = \Delta_M, \tag{6.2}$$

where U denotes the total strain energy in the system including those in the redundants.

If points of application of the redundants do not deflect,

$$\frac{\partial U}{\partial X_M} = 0; \ (M = 1, 2, \cdots, N). \tag{6.2a}$$

This gives a set of simultaneous equations corresponding to the conditions of compatibility as in case of the previous method and reduces to the statement that the first partial derivative of the strain energy with respect to the redundants is equal to zero. This is equivalent to stating that the values of the redundants must be such as to minimise the strain energy. This special case of Castigliano's second theorem is often called the Theorem of Least Work and may be stated as follows :-

In a statically indeterminate system if there are no movement of support of all possible sets of values the redundant forces in the system may assume, the correct set of values is that which makes the strain energy U a minimum.

Castigliano further extended the Theorem of Least Work to system which are self-strained. In that case, the point of application of a redundant force X_M may be displaced by an amount Δ_M in the direction of the force, whence

$$\frac{\partial U}{\partial X_M} = \Delta_M; \ (M = 1, 2, \cdots, N). \tag{6.2b}$$

The self-straining may be due to settlement of the support of a redundant reaction or by the initial lack of fit of a member. It may be stated that the partial derivative of the total strain energy of a linearly elastic system with respect to a redundant force is equal to the initial lack of fit of the redundant.

The computations involved in Castigliano's method are essentially the same as involved in the method of consistent deformation. The only difference in the two approaches lies in the somewhat more automatic manner of setting up the solution when Castigliano's approach is used.

In Castigliano's method, the redundants carry their own units throughout a solution unlike in the case of method of Consistent deformation.

6.3 Reciprocity of Static Deflections. Maxwell's Theorem of Reciprocal Deflections and Betti's Law

Most of the work in the solution of statically indeterminate structures by the method of consistent deformation involves the computations of deflections or rotations. The computation work can be reduced if a very useful relation indicating reciprocity of deflections at different locations be observed.

Maxwell's law of reciprocal deflections states that in any elastic structure obeying Hooke's law, having unyielding supports and no change in temperature, the deflection Δ_{12} at any point 1 due to a load P at any other point 2 is equal to the deflection Δ_{21} at the point 2 caused by the same load at point 1.

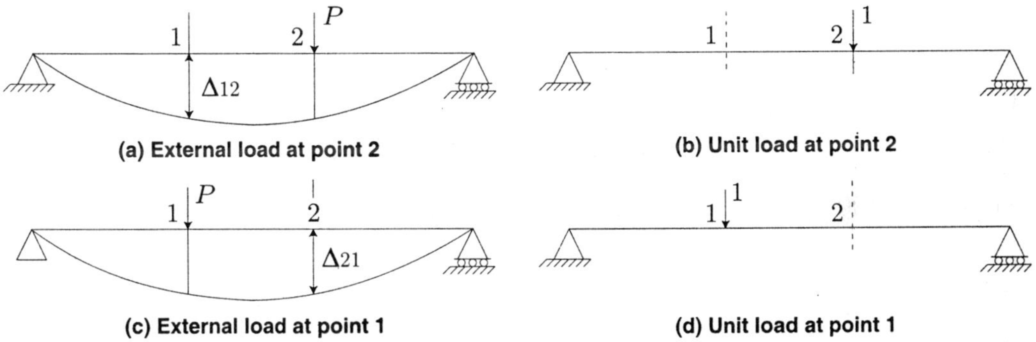

Fig. 6.7: Reciprocal deflections

This theorem can be easily proved with the help of Unit Load Method.

If M = bending moment at any section due to load P at point 2, [Fig. 6.7(a)]

m = bending moment at any section due to a unit load at point 2, [Fig. 6.7(b)]

M' = bending moment at any section due to load P at point 1, [Fig. 6.7(c)]

m' = bending moment at any section due to unit load at point 1, [Fig. 6.7(d)].

By Unit Load Method,

$$\Delta_{12} = \sum \frac{Mm'dx}{EI} \quad \text{and} \quad \Delta_{21} = \sum \frac{M'mdx}{EI}.$$

Since $M = P.m$ and $M' = P.m'$, the above expressions become

$$\Delta_{12} = \sum \frac{P\,mm'dx}{EI}$$

$$\Delta_{21} = \sum \frac{P\,mm'dx}{EI},$$

whence $\Delta_{12} = \Delta_{21}$. (6.3)

A further generalisation of the theorem has been made by Betti, and the modified form is usually called the Maxwell-Betti theorem (or Betti's theorem)*. According to this in

*The reciprocity theorem for static deflection is attributed to J.C. Maxwell, who discovered the relationship for two forces and to E. Betti, who proved it for the general case. Lord Rayleigh extended the principle to dynamical situations.

any elastic structure following Hooke's law, in which the supports are unyielding and the temperature is constant, the external virtual work done by a system of forces P_m during the deformation caused by a system of forces P_n is equal to the external virtual work done by the P_n system during the deformation caused by the P_m system.

If Δ_{mn} = displacement of the point of application of force P_m in the direction of this force caused by the application of P_n, and Δ_{nm} = displacement of the point of application of force P_n in the direction of this force caused by the application of P_m.

External work done by system of P_m forces,

i.e., $$W_{E1} = \sum_{l}^{m}\sum_{l}^{n} P_m \Delta_{mn} \quad \text{and} \quad W_{E2} = \sum_{l}^{n}\sum_{l}^{m} P_n \Delta_{nm}.$$

Now $\Delta_{mn} = kP_n$ and $\Delta_{nm} = kP_m$,

whence it is seen that,

$$\sum_{l}^{m}\sum_{l}^{n} P_m \Delta_{mn} = \sum_{l}^{n}\sum_{l}^{m} P_n \Delta_{nm}. \tag{6.4}$$

If all forces of the first set except P_m and all forces of the second set except P_n be zero, from Eq. (6.4), $P_m\Delta_{mn} = P_n\Delta_{nm}$.

When $P_m = P_n = P$, the expression is reduced to Maxwell's theorem of reciprocal displacement, i.e., $\Delta_{mn} = \Delta_{nm}$.

Example 6.3. *Determine the reactions and obtain shear force and bending moment diagrams for the propped cantilever beam of Fig. 6.8(a), if the supports are unyielding.*

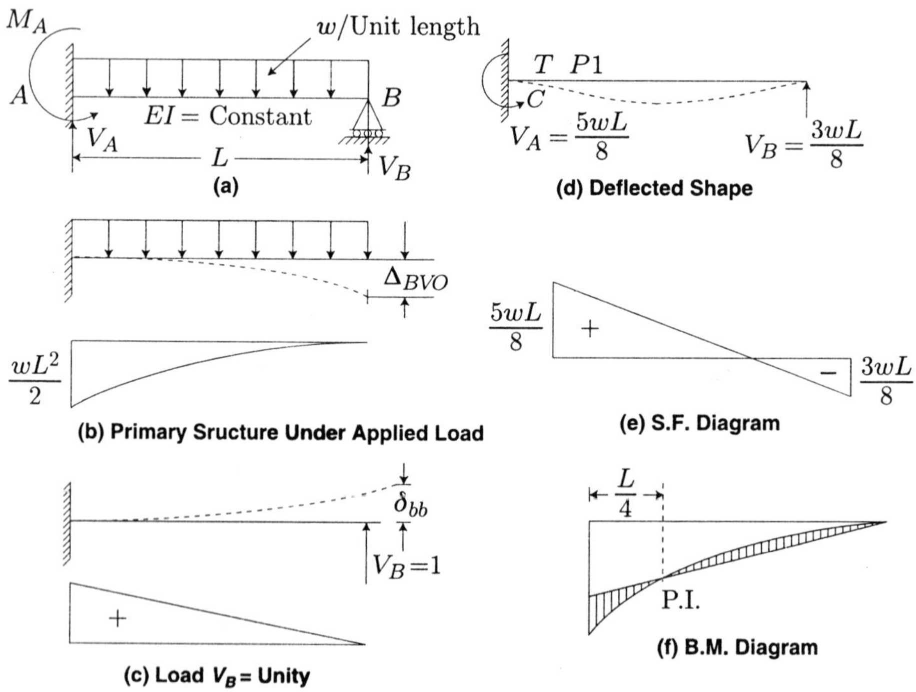

Fig. 6.8: For Example 6.3

(a) By the Method of Consistent Deformation

The beam is statically indeterminate to the first degree as it has three components of reactions, V_A, V_B and M_A. Taking a primary structure as in Fig. 6.8(b) by moment-area theorem,

$$\Delta_{BVO} = \frac{1}{3}\frac{wL^2}{2EI}(L)\left(\frac{3}{4}L\right) = \frac{wL^4}{8EI} \downarrow.$$

Again from Fig. 6.8(c)

$$\delta_{bb} = -\frac{1}{2}\frac{L}{EI}(L)\left(\frac{2}{3}L\right) = -\frac{L^3}{3EI} \uparrow.$$

For no settlement of support,

$$\Delta_{BVO} + V_B\delta_{bb} = \Delta_{BV} = 0,$$

whence $V_B = \dfrac{3}{8}wL \uparrow.$

(b) By Castigliano's Theorem

Using effect of flexure only and with V_B as the redundant,

$$\frac{\partial U}{\partial V_B} = \int \frac{M\partial M}{EI\partial V_B}dx = 0.$$

In this case, $M = -\dfrac{wx^2}{2} + V_B(x)$

and $\dfrac{\partial M}{\partial V_B} = x,$

so that $\dfrac{\partial U}{\partial V_B} = \displaystyle\int_0^L \frac{\left[-\frac{wx^2}{2} + V_B(x)\right]x}{EI}dx = \frac{1}{EI}\left(-\frac{wL^3}{8} + V_B\frac{L^3}{3}\right) = 0,$

whence $V_B = \dfrac{3wL}{8}.$

Example 6.4. *Determine the reactions for the case of the beam in Example 6.3 where*

(a) the support at B settles by Δ.

(b) the support at B is in the form of a spring having stiffness k.

Solution : (a) Using method of consistent deformation and referring to previous example,

$$\frac{wL^4}{8EI} \downarrow - V_B\frac{L^3}{3EI} \uparrow = \Delta_B \downarrow,$$

whence $V_B = \dfrac{\frac{wL^4}{8EI} - \Delta}{\frac{L^3}{3EI}}.$

Similar solution can be obtained using Castigliano's method considering self-straining.

(b) From method of consistent deformation,

$$\frac{wL^4}{8EI} \downarrow - V_B\frac{L^3}{3EI} \uparrow = \frac{V_B}{k} \downarrow,$$

whence $V_B = \dfrac{\frac{wL^4}{8EI}}{\frac{L^3}{3EI} + \frac{1}{k}}.$

Example 6.5. *Determine the reactions and draw shear force and bending moment diagrams for the fixed ended beam in Fig. 6.9(a). Also draw the deflected shape.*

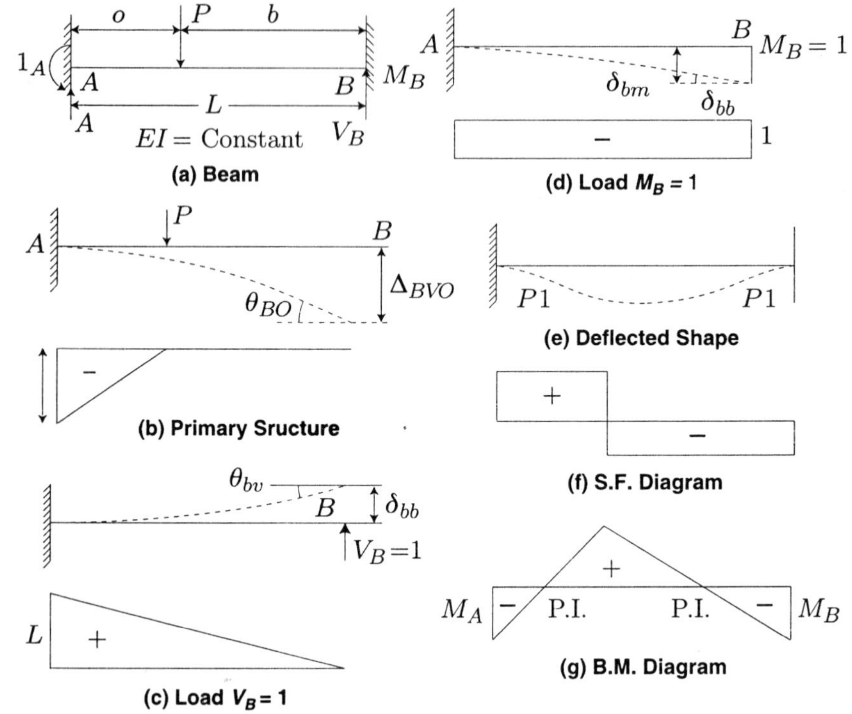

Fig. 6.9: For Example 6.5

Solution : By the method of consistent deformation.

The beam is statically indeterminate to the second degree as it has four components of reactions, viz., V_A, M_A, V_B and M_B. A primary structure as in Fig. 6.9(b) by removing redundants V_B and M_B is considered in this case.

From moment-area theorem [Fig. 6.9(b)],

$$\theta_{BO} = \frac{1}{2}\frac{P(a)(a)}{EI} = \frac{Pa^2}{2EI} \curvearrowright$$

and $$\Delta_{BVO} = \frac{1}{2}\frac{Pa(a)}{EI}\left(L - \frac{a}{3}\right) = \frac{Pa^2}{EI}\left(L - \frac{a}{3}\right) \downarrow$$

From Fig. 6.9(c),

$$\theta_{bv} = -\frac{1}{2}\frac{L(L)}{EI} = -\frac{L^2}{2EI} \quad \text{and} \quad \delta_{bb} = -\frac{L^2}{2EI}\left(\frac{2}{3}L\right) = -\frac{L^3}{3EI} \uparrow.$$

From Fig. 6.9(d),

$$\theta_{bb} = \frac{1(L)}{EI} = \frac{L}{EI} \quad \text{and} \quad \delta_{bm} = \frac{L}{EI}\left(\frac{L}{2}\right) = \frac{L^2}{2EI} \downarrow.$$

For no yielding of supports at B,

$$\frac{Pa^2}{2EI} = V_B\frac{L^2}{2EI} + M_B\frac{L}{EI} = 0 \qquad \text{(i)}$$

and $$\frac{Pa^2}{2EI}\left(L - \frac{a}{3}\right) - V_B\frac{L^3}{3EI} + M_B\frac{L^2}{2EI} = 0. \qquad \text{(ii)}$$

Solving simultaneously, $M_B = \dfrac{Pa^2b}{L^2}$ and V_B is also obtained,

Eqs. (i) and (ii) may be obtained from application of moment-area theorems in the original beam.

Both at A and B, slope is zero so that applying moment-area theorem,

$$\sum_A^B \frac{M}{EI}dx = 0, \qquad \text{(iii)}$$

which in this case will give the same equation as (i). Again, from the consideration that deflection at B or A beyond slope at A or B is zero, from moment-area theorem,

$\sum$ moment of $\dfrac{M}{EI}$ diagram between A and B about B or A,

whence $$\sum_A^B \frac{M}{EI}xdx = 0. \qquad \text{(iv)}$$

This will give the same equation as in (ii).

It is more advantageous to form equations with respect to considerations in (iii) and (iv) for fixed end beams.

(c) By Castigliano's theorem.

Considering V_B and M_B as redundants,

$$M = V_B(x) - M_B, \quad \text{for } 0 \le x \le b.$$
$$= V_B(x) - M_B - P[x - b], \quad \text{for } b \le x \le L,$$

whence $$\frac{\partial M}{\partial M_B} = -1 \quad \text{and} \quad \frac{\partial M}{\partial V_B} = (x).$$

Applying Method of least work, i.e., $\dfrac{\partial U}{\partial M_B} = 0$ and $\dfrac{\partial U}{\partial V_B} = 0$,

$$\int_0^b \frac{(V_Bx - M_B)}{EI}(-1)dx + \int_b^L \frac{[V_Bx - M_B - P(x-b)]}{EI}(-1)dx = 0$$

and $$\int_0^b \frac{(V_Bx - M_B)}{EI}(x)dx + \int_b^L \frac{[V_Bx - M_B - P(x-b)]}{EI}(x)dx = 0,$$

whence equations obtained will be the same as in case of method of consistent deformation.

S.F. and B.M. diagrams are shown in Figs. 6.9(f) and 6.9(g).

Example 6.6. *Determine the B.M. and S.F. diagrams of fixed end beam of constant EI and of span L, having a concentrated load P at center.*

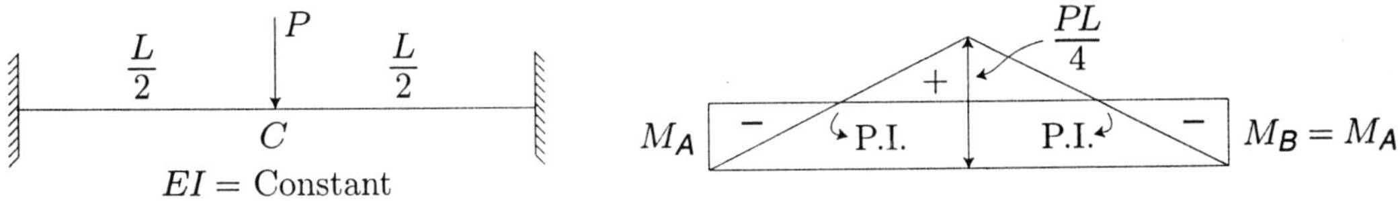

(a) Beam **(b) B.M. Diagram**

Fig. 6.10: For Example 6.6

Solution : The beam is statically indeterminate to the second degree. But from symmetry,

$$M_A = M_B \quad \text{and} \quad V_A = V_B = \frac{P}{2}.$$

Using the conditions, obtained from considerations of moment-area theorem as in previous example, from $\sum_A^B \frac{M}{EI}dx = 0$,

$$\text{Area of +ve } \frac{M}{EI} \text{ diagram} = \text{area of −ve } \frac{M}{EI} \text{ diagram},$$

whence $\frac{1}{2}\frac{PL}{4EI}(L) = M_A(L)$, which gives $M_A = \frac{PL}{8}$.

Maximum deflection occurs at centre C and $y_{\max} = y_{\text{centre}}$ = moment of $\frac{M}{EI}$ diagram between A and C about $C = \frac{PL^3}{192EI}$, i.e., $\frac{1}{4}$th as that in case of a simply supported beam. The effects of the restraining couples at the supports are to reduce the B.M. throughout, the maximum in this case being half the value of a simply supported beam. Consequently, the slopes and deflections are also reduced considerably. The obvious effect of fixation on a beam of a given section is, therefore, to make it stronger as well as stiffer.

Example 6.7. *A built-in beam of 30 m span and of constant EI carries a uniformly distributed load of* 5 kN *per metre run over the left-hand half of the span. Calculate the support moments and the reactions.*

Solution : This is the case of a beam statically indeterminate to the second degree, the resulting B.M. being obtained combining B.M. diagrams in Figs. 6.11(b) and 6.11(c).

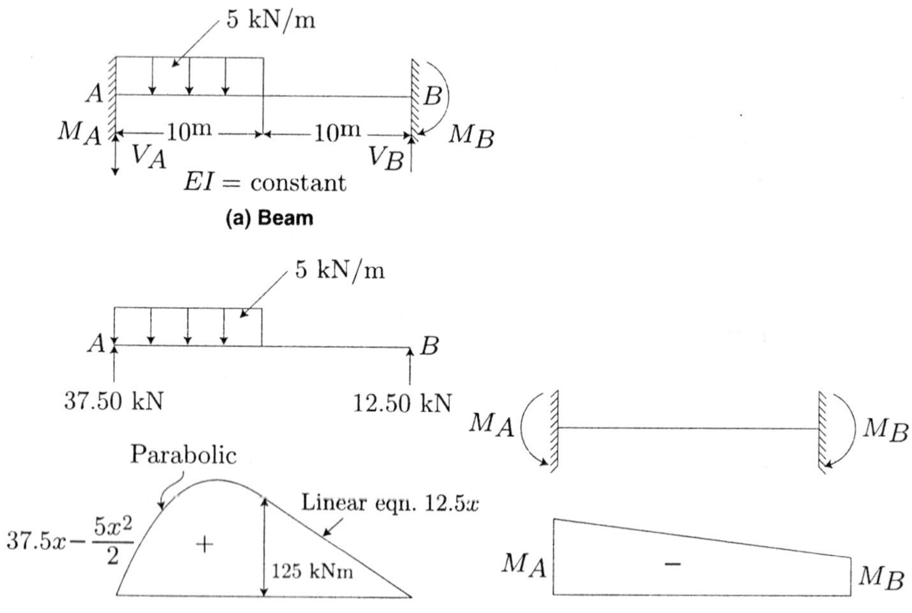

Fig. 6.11: For Example 6.7

The conditions obtained from moment-area method give the following equations:

From $\sum_A^B \frac{M}{EI} dx = 0,$

$$\frac{1}{EI}\left[\int_0^{10}\left(37.5x - \frac{5x^2}{2}\right)dx + \frac{1}{2}(125)(10) - \frac{M_A + M_B}{2}(20)\right] = 0,$$

whence $M_A + M_B = 166.67$ (i)

From $\sum_A^B \frac{M}{EI} x dx = 0,$

$$\frac{1}{EI}\left[\int_0^{10}\left(37.5x - \frac{5x^2}{2}\right)xdx + \frac{1}{2}(12.5)(10)\left(10 + \frac{10}{3}\right) - \frac{M_A}{2}(20)\left(\frac{20}{3}\right) - \frac{M_B}{2}(20)\left(\frac{40}{3}\right)\right] = 0,$$

whence $M_A + 2M_B = 218.75.$ (ii)

Solving (i) and (ii), $M_A = 114.59$ kNm

and $M_B = 52.08$ kNm.

Reactions at supports,

$$V_A = 37.50 + \frac{M_A - M_B}{L} = 40.626 \text{ kN}$$

$$V_B = 12.50 - \frac{M_A - M_B}{L} = 9.374 \text{ kN.}$$

Example 6.8. *Determine the B.M. and S.F. diagrams for the continuous beam shown in Fig. 6.12(a). The beam has a constant value of EI and the supports do not settle.*

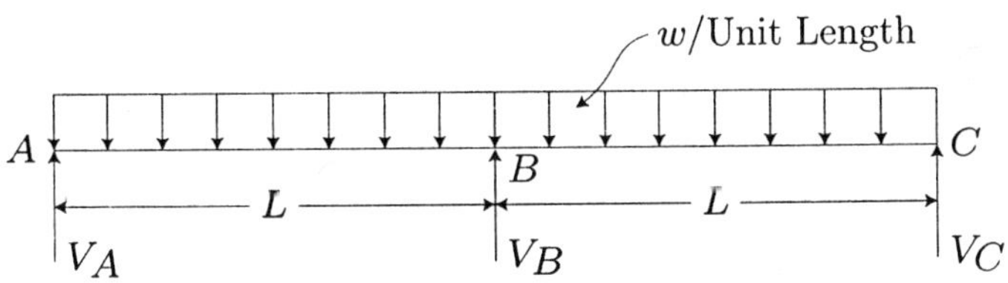

(a) Beam

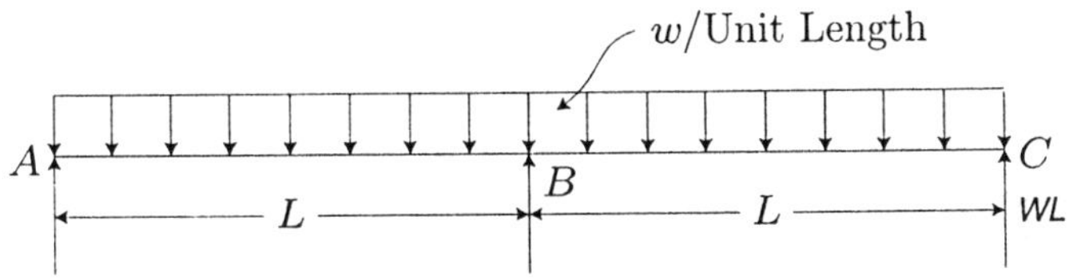

(b) Primary Structure with Loading

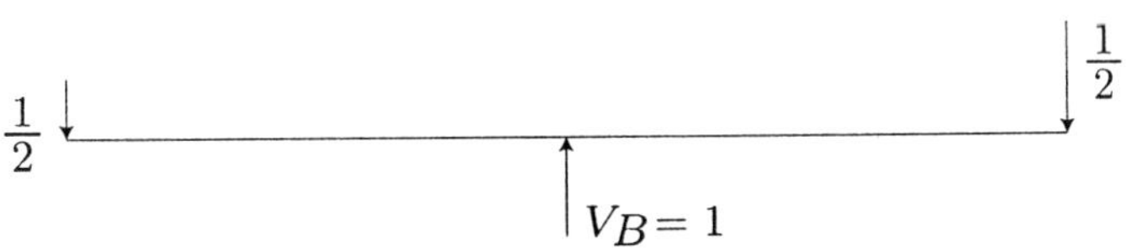

(c) Primary Structure with $V_B = 1$

Fig. 6.12: For Example 6.8 (Contd.)

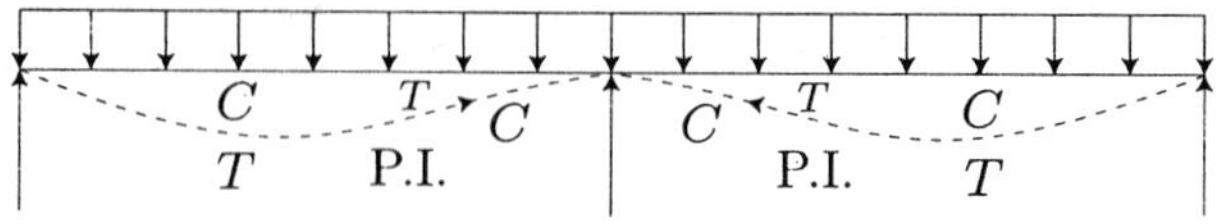

(d) Deflected Shape

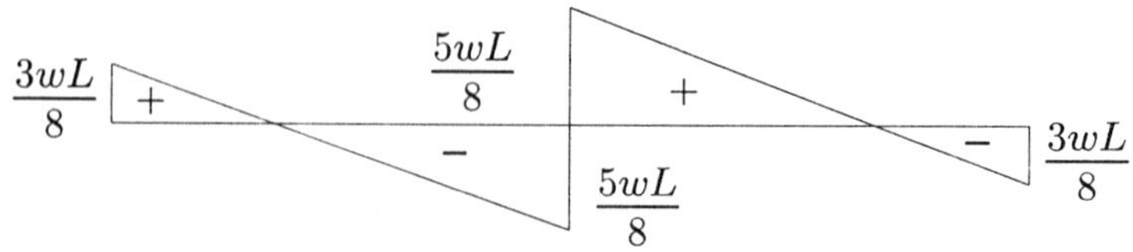

(e) S.F. Diagram

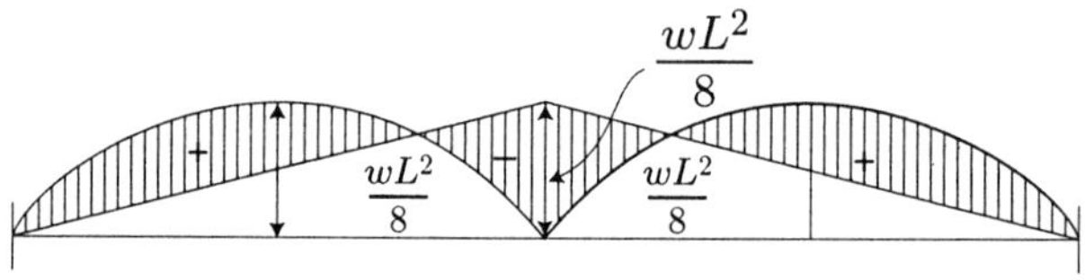

(f) B.M. Diagram

Fig. 6.12: For Example 6.8

Solution : In this case the number of indeterminacy is one. Considering the primary structure in Fig. 6.12(b) from moment-area theorem,

$$\Delta_{BVO} = \frac{2}{3}\frac{w(2L)^2}{8EI}(L)\left(\frac{5}{8} - L\right) = \frac{5w(2L)^4}{384EI} \downarrow.$$

From Fig. 6.12(c) for $V_B = 1$,

$$\delta_{bb} = \frac{1}{2}\frac{L(L)}{2EI}\left(\frac{2}{3}L\right) = \frac{(2L)^3}{484EI} \uparrow.$$

Equating $V_B = \dfrac{\Delta_{BVO}}{\delta_{bb}} = \dfrac{5wL}{4}$ and $V_A = V_C = \dfrac{3wL}{8}$.

S.F. and B.M. diagrams are shown in Figs. 6.12e and 6.12f.

Example 6.9. *Obtain the bending moment diagram of the closed ring shown in Fig. 6.13(a), if the radius of curvature is large with respect to the thickness of the ring.*

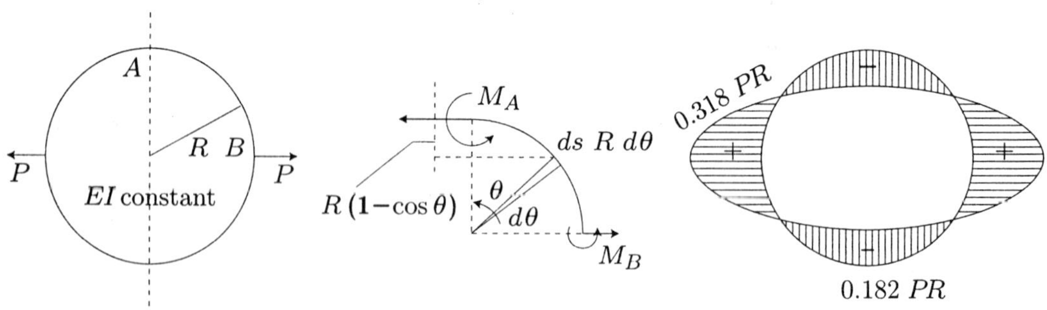

(a) Closed Ring **(b) Free body diagram** **(c) B.M. diagram**

Fig. 6.13: For Example 6.9

Solution : The circular ring is subjected to bending moment, axial force and shear force. If the ring is cut at any section A, the cut may be replaced by three types of forces, viz., M_A, H_A and V_A. Treating these as the redundant internal actions, from Castigliano's theorem,

$$\frac{\partial U}{\partial M_A} = 0, \frac{\partial U}{\partial H_A} = 0 \quad \text{and} \quad \frac{\partial U}{\partial V_A} = 0.$$

These will give three simultaneous equations, involving the three redundants, solving which the redundant forces may be obtained.

If the ring is symmetrically loaded as in the case, by suitable choice of the cut section, the number of redundant forces can be reduced. In this case choosing A as the cut section, the only unknown M_A may be obtained from only one condition, i.e., $\frac{\partial U}{\partial M_A} = 0$.

In the case of ring, only effect of bending moment of deformation will be considered as effects of the other actions are rather small.

From Fig. 6.13(b),

$$M = M_A + \frac{PR}{2}(1 - \cos\theta) \text{ and } ds = Rd\theta$$

$$\frac{\partial M}{\partial M_A} = 1$$

and $$\frac{\partial U}{\partial M_A} = \int \frac{M}{EI}\frac{\partial M}{\partial M_A} ds = \int \frac{M}{EI}\frac{\partial M}{\partial M_A}(Rd\theta)$$

$$= 4\int_0^{\frac{\pi}{2}} \left[M_A + \frac{\text{PR}}{2}(1 - \cos\theta)\right] Rd\theta = 0,$$

whence $$M_A = \frac{\text{PR}}{2}\left[\frac{2}{\pi} - 1\right] = -0.182 \text{ PR}$$

and $$M_B = \frac{\text{Pr}}{2}\left[\frac{2}{\pi} - 1\right] + \frac{\text{PR}}{2} = 0.318 \text{ PR}.$$

The B.M. diagram is shown in Fig. 6.13c.

6.4 Analysis of Statically Indeterminate Trusses

Trusses can have indeterminacies from the following points of view:

(i) External redundancy, where the indeterminacy is due to provision of more reactions than can be solved by means of statics.

(ii) Internal redundancy, where the indeterminacy is due to providing internal redundant members, as a result of which the forces in the members cannot be obtained by statics.

(iii) Combinations of external and internal redundancies.

For all cases, the analysis can be done by application of Castigliano's theorem or by method of consistent deformation. The redundant forces in the form of external reactions or in the form of internal bar forces are selected. From known deformation of the truss along the redundant forces their magnitudes can be obtained. In calculating the deformations effects of axial forces are considered in case of trusses.

(a) Application of Castigliano's Theorem

If U be the total strain energy of the truss, including the effects of redundant forces X_A, X_B, etc. from Castigliano's theorem,

$$\frac{\partial U}{\partial X_A} = \Delta_A = \lambda_A \tag{6.5a}$$

$$\frac{\partial U}{\partial X_B} = \Delta_B = \lambda_B, \text{ etc.,} \tag{6.5b}$$

where λ_A, λ_B, etc. represent the initial lack of fit of the redundant members. Values of redundants may be calculated from the simultaneous equations thus obtained.

(b) Method of Consistent Deformation

In this case if S represents the bar forces in the primary truss (basic determinate truss after removal of redundants) and u_a, u_b, etc. the axial forces due to application of unit load in the direction of redundants X_A, X_B, etc. From unit load method, the movement along directions of redundants are given by

$$\sum \frac{Su_aL}{AE}, \sum \frac{Su_bL}{AE} \text{ and so on.} \tag{6.6}$$

Also $$X_A \sum \frac{u_a^2 L}{AE}, X_A \sum \frac{u_a u_b L}{AE}, \tag{6.7a}$$

$$X_B \sum \frac{u_a u_b L}{AE}, X_B \sum \frac{u_b^2}{AE} \text{ and so on} \tag{6.7b}$$

represent the movement in the directions of the redundant forces due to various effects. From known deformation conditions,

$$\sum \frac{Su_aL}{AE} + X_A \sum \frac{u_a^2 L}{AE} + X_B \sum \frac{u_a u_b L}{AE} + \cdots = \text{actual deformation} \tag{6.8a}$$

and $$\sum \frac{Su_bL}{AE} + X_A \sum \frac{u_a u_b L}{AE} + X_B \sum \frac{u_b^2 L}{AE} + \cdots = \text{actual deformation.} \tag{6.8b}$$

The term on the right-hand side includes the change in length of the redundant member plus any initial lack of fit if it is the case of internal redundancy and only settlement, if any, in case of external redundancy. As for the change in length of the member, the terms which will be $X_A \frac{u_a^2 L}{AE}$ and $X_B \frac{u_b^2 L}{AE}$, etc. can be included in the left-hand side of the equation as well.

Example 6.10. *All members of the pin-jointed truss in Fig. 6.14(a) are of the same material and cross-sectional area. The truss is subjected to a vertical load W at B.*

Solution : The truss is externally redundant to the first degree. The vertical reaction at support C, i.e., V_C in this case can be considered as the redundant force. The solution of the problem is given in tabular form.

(a) By Castigliano's Theorem

From $\frac{\partial U}{\partial V_C} = 0$, it follows that $\sum \frac{S}{AE} \frac{\partial S}{\partial V_C} L = 0$, which will yield value of redundant reaction V_C.

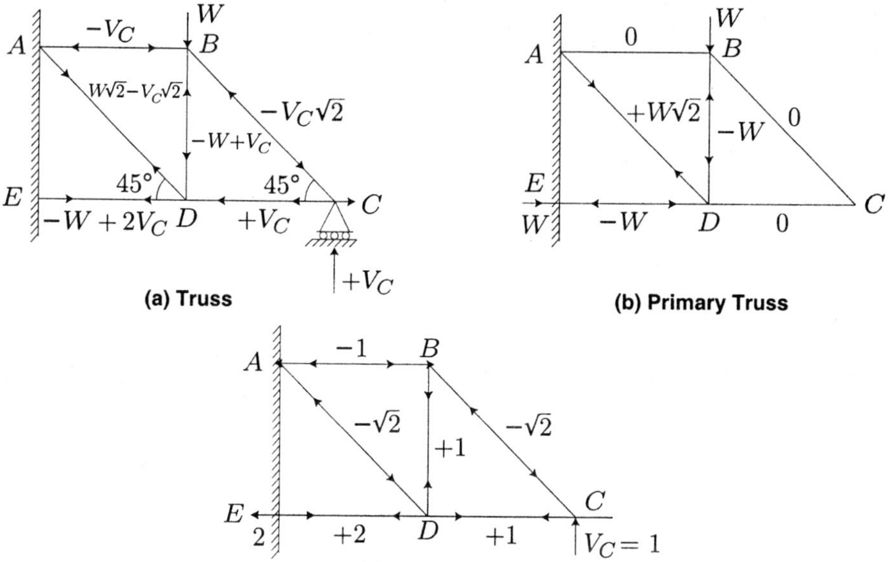

(a) Truss (b) Primary Truss (c) Load $V_C = 1$

Fig. 6.14: For Example 6.10

(a) By Castigliano's Theorem. Example 6.1(a)

Member	Length L	AE	S [From Fig. 6.14(a)]	$\frac{\partial S}{\partial V_C}$	$S\frac{\partial S}{\partial V_C}\frac{L}{AE}$
AB	L	AE	V_C	-1	$\frac{V_C L}{AE}$
BC	$L\sqrt{2}$	AE	$\sqrt{2}V_C$	$-\sqrt{2}$	$\frac{2\sqrt{2}V_C L}{AE}$
CD	L	AE	$+V_C$	$+1$	$\frac{V_C L}{AE}$
DA	$L\sqrt{2}$	AE	$W\sqrt{2}-V_C\sqrt{2}$	$-\sqrt{2}$	$\frac{-2\sqrt{2}WL+2\sqrt{2}V_C L}{AE}$
DB	L	AE	$-W+V_C$	$+1$	$\frac{-WL+V_C L}{AE}$
DE	L	AE	$-W+2V_C$	$+2$	$\frac{-2WL+4V_C L}{AE}$
			$\sum \frac{S}{AE}\frac{\partial S}{\partial V_C}L =$		$-\frac{2\sqrt{2}+3}{AE}WL + \frac{7+4\sqrt{2}}{AE}V_C L$

From $\sum \frac{S}{AE}\frac{\partial S}{\partial V_C}L = 0$, it follows that,

$$V_C = \frac{2\sqrt{2}+3}{7+4\sqrt{2}} \text{ W}^\circ = 0.460 \text{ W} \uparrow.$$

(b) By Method of Consistent Deformation (Example 6.10)

Member	Length (L)	AE	S (From Fig. 6.14b)	u (From Fig. 6.14c)	$\frac{SuL}{AE}$	$\frac{u^2L}{AE}$
AB	L	AE	0	-1	0	$\frac{L}{AE}$
BC	$L\sqrt{2}$	AE	0	$-\sqrt{2}$	0	$\frac{2\sqrt{2}L}{AE}$
CD	L	AE	0	$+1$	0	$\frac{L}{AE}$
DA	$L\sqrt{2}$	AE	$+W\sqrt{2}$	$-\sqrt{2}$	$\frac{-2\sqrt{2}WL}{AE}$	$\frac{2\sqrt{2}L}{AE}$
DB	L	AE	$-W$	$+1$	$-\frac{WL}{AE}$	$\frac{L}{AE}$
DE	L	AE	$-W$	$+2$	$-\frac{2WL}{AE}$	$\frac{4L}{AE}$
				Σ	$-\frac{2\sqrt{2}+3}{AE}WL$	$\frac{7+4\sqrt{2}}{AE}$

For no settlement of support c,

$$-\frac{2\sqrt{2}+3}{AE}WL + V_C\frac{7+4\sqrt{2}}{AE}L = 0,$$

whence $V_C = 0.460W \uparrow$

Example 6.11. *Find the force in the member BD of the pin-jointed truss shown in Fig. 6.15(a). Vertical and horizontal members have cross-sectional areas a, while the diagonal members are of area 2a. All members are of the same material.*

Solution : The truss is internally redundant to the first degree, and member BD may be considered as the redundant with a force R.

(a) By Castigliano's Theorem [Refer Table (a)]

From $$\frac{\partial U}{\partial R} = 0, \text{ i.e., from } \sum S\frac{\partial S}{\partial R}\frac{L}{AE} = 0,$$

$$R = -\frac{2+2\sqrt{2}}{3+2\sqrt{2}}\text{ W} = -0.828 \text{ W} \quad \text{(Compression).}$$

(b) By Method of Consistent Deformation [Refer Table (b)]

From compatibility of deformation,

$$\frac{(2+2\sqrt{2})WL}{aE} + R\frac{(3+2\sqrt{2})L}{aE} = 0,$$

whence $R = -0.828$ W (compression).

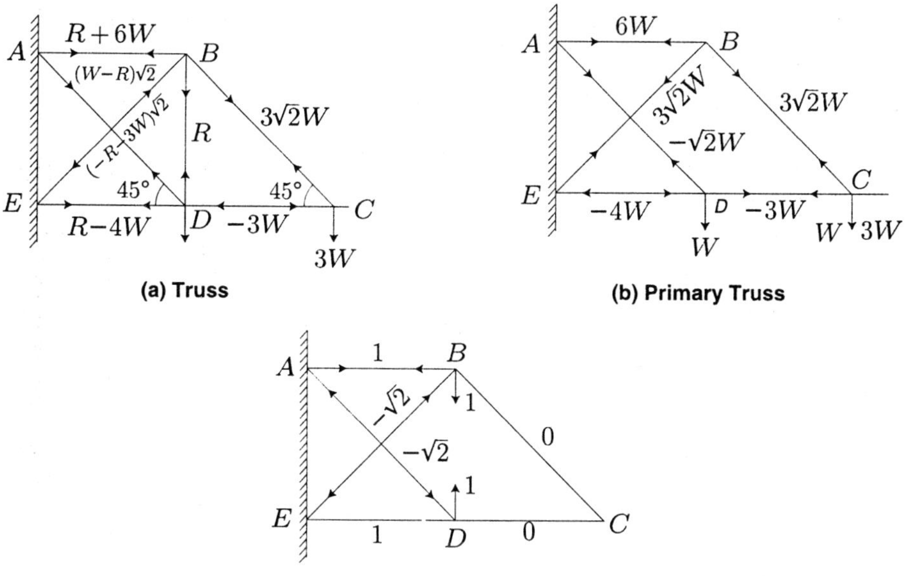

(a) Truss

(b) Primary Truss

(c) Unit Load on *BD*, i.e., *R* = 1

Fig. 6.15: For Example 6.11

(a) By Castigliano's Method (Example 6.11)

Member	Length L	AE	S [From Fig. 6.15(a)]	$\frac{\partial S}{\partial R}$	$S\frac{\partial S}{\partial R}\frac{L}{AE}$
AB	L	aE	$R+6W$	1	$\frac{R+6W}{aE}L$
BC	$L\sqrt{2}$	$2aE$	$3\sqrt{2W}$	0	0
CD	L	aE	$-3W$	0	0
DA	$L\sqrt{2}$	$2aE$	$(W-R)\sqrt{2}$	$-\sqrt{2}$	$\frac{R-W}{aE}L\sqrt{2}$
DB	L	aE	R	1	$\frac{RL}{aE}$
DE	L	aE	$R-4W$	1	$\frac{R-4W}{aE}L$
BE	$L\sqrt{2}$	$2aE$	$(-R-3W)\sqrt{2}$	$-\sqrt{2}$	$\frac{R+3W}{aE}L\sqrt{2}$
				$\sum\frac{\partial S}{\partial R}\frac{L}{AE}=$	$\frac{(3+2\sqrt{2})RL}{aE}+\frac{(2+2\sqrt{2})WL}{aE}$

(b) By method of Consistent Deformation (Example 6.11)

Member	Length (L)	AE	S (From Fig. 6.15b)	u (From Fig. 6.15c)	$\frac{SuL}{AE}$	$\frac{u^2L}{AE}$
AB	L	aE	$+6W$	$+1$	$\frac{6WL}{aE}$	$\frac{L}{aE}$
BC	$L\sqrt{2}$	$2aE$	$+3\sqrt{2W}$	0	0	0
CD	L	aE	$-3W$	0	0	0
DA	$L\sqrt{2}$	$2aE$	$+W\sqrt{2}$	$-\sqrt{2}$	$-\frac{\sqrt{2}WL}{aE}$	$\frac{L\sqrt{2}}{aE}$
DB	L	aE	0	$+1$	0	$\frac{L}{aE}$
DE	L	aE	$-4W$	$+1$	$-\frac{4WL}{aE}$	$\frac{L}{aE}$
BE	$L\sqrt{2}$	$2aE$	$-3\sqrt{2W}$	$-\sqrt{2}$	$\frac{3\sqrt{2}WL}{aE}$	$\frac{L\sqrt{2}}{aE}$
				Σ	$\frac{(2+2\sqrt{2})WL}{aE}$	$\frac{(3+2\sqrt{2})L}{aE}$

Example 6.12. *Find the force in the member AD of the pin-jointed frame in Fig. 6.16(a), if (i) the member AD has no initial lack of fit and (ii) if the member AD is larger by* 2 mm *than required. Assume* $E = 2(10)^2$ kN/mm^2. *Areas of members AD, BD and CD are* 3000 mm^2, 2000 mm^2 *and* 1000 mm^2 *respectively. Also calculate the force in member AD for the initial lack of fit only.*

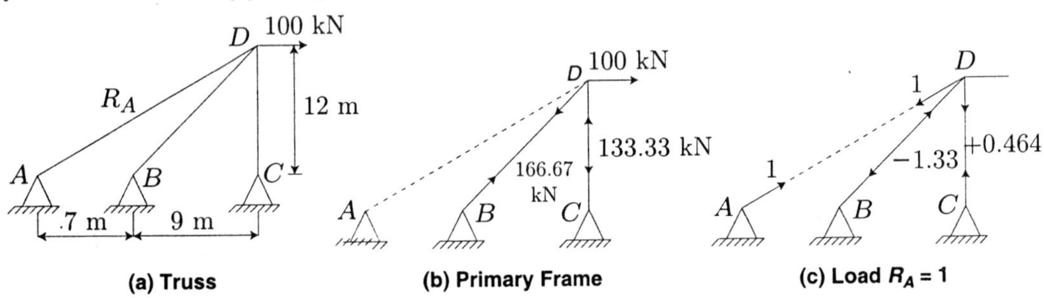

Fig. 6.16: For Example 6.12

Solution : In this case the frame is internally redundant to the first degree. Analysis will be made by the method of consistent deformation by considering AD as the redundant member.

Refer to Table on the following page.

(i) For no initial lack of fit of member AD,

$$-\frac{2404.9}{E} + \frac{22.517}{E}R_A = 0,$$

whence $R_A = 106.80$ kN (tension as assumed).

(ii) When member AD is 2 mm longer than required.

$$-\frac{2404.9}{E}\nearrow + \swarrow \frac{22.517}{E}R_A = -\frac{2}{10(10)^2}\nearrow,$$

whence $R_A = 89.04$ kN (tension as assumed).

This indicates that for initial lack of fit only, the bar force in member AD is -17.76 kN (compression).

Table By the Method of Consistent Deformation (Example 6.12)

Member	Length L(m)	AE (A in mm^2)	S (From Fig. 6.16b)	u (From Fig. 6.16c)	$\frac{SuL}{AE}$	$\frac{u^2L}{AE}$ (L in mm)
AD	20	3000 E	0	+1	0	$+\frac{6.667}{E}$
BD	15	2000 E	+166.67	−1.33	$-\frac{1,662.53}{E}$	$+\frac{13.267}{E}$
CD	12	1000 E	−133.33	+0.464	$-\frac{742.38}{E}$	$+\frac{2.583}{E}$
				Σ	$\nearrow -\frac{2404.91}{E}$	$+\frac{22.517}{E}\swarrow$

Example 6.13. *Compute the bar force in the member AC of the square panel pin-jointed frame shown in Fig. 6.17. All other members except AC are of the same material and vertical and horizontal cross-sectional areas are* 1000 mm^2, *and for the two diagonal members cross-sectional areas are* 500 mm^2. *Assume* $E_{AC} = 1(10)^2$ kN/mm^2 *and* $E_{\text{others}} = 2(10)^2$ kN/mm^2. *Consider the whole frame subjected to an increase of temperature of* 20°C *and assume* $\alpha_{AC} = 2(10)^{-5}$ *per* °C *and* $\alpha_{\text{others}} = (10)^{-5}$ *per* °C.

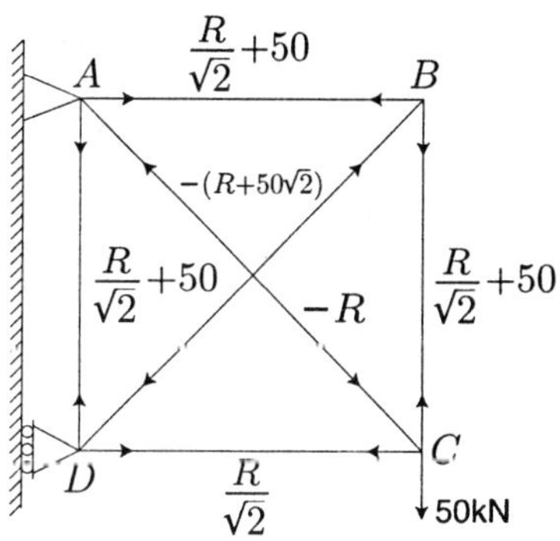

Fig. 6.17: For Example 6.13

Solution : The frame is internally redundant to the first degree and member AC may be considered as the redundant. In case of framed structure in all of which the members are made of the same material, whether it is statically determinate or not, a uniform change in temperature only causes proportional alteration in length, no stresses are induced and the structure is geometrically similar to the original. If, on the other hand, the material for the members differs then stress is induced by a uniform change in temperature. If again one or some of the members only be subjected to change in temperature stress is also induced.

Solution has been done by Castigliano's theorem. The bar forces on the redundant member AC is considered as R. (Refer to Table)

From Castigliano's theorem $\frac{\partial U}{\partial R} = \lambda$,

$$\frac{RL}{2(10)^5}(2+6\sqrt{2}) + \frac{50L}{(10)^5}\left(\frac{3}{2\sqrt{2}}+2\right) = L\sqrt{2}(1)(10)^{-5}(20),$$

whence $R = 23.795$ kN (Tensile).

6.5 Composite Structure

Most of the structures fall into one of the three classifications, viz., beams, frames or trusses. A beam is subjected to bending stresses only and members of a truss with smooth hinged joints are subjected to only direct stresses. The members of a frame with rigid joints are usually subjected to both bending and direct stresses, but the effect of the direct stresses in the members on deflections is in most cases small compared with that due to the bending stresses. In analysis of statically indeterminate structure of the various types, only effects of the predominant actions were considered.

By Castigliano's Theorem (Example 6.13)

Member	Length L	AE (A in mm^2)	S (From Fig. 6.17)	$\frac{\partial S}{\partial R}$	$S\frac{\partial S}{\partial R}\frac{L}{AE}$
AB	L	$2(10)^5$	$\frac{R}{\sqrt{2}}+50$	$\frac{1}{\sqrt{2}}$	$\frac{L}{2(10)^5}\left(\frac{R}{2}+\frac{50}{\sqrt{2}}\right)$
BC	L	$2(10)^5$	$\frac{R}{\sqrt{2}}+50$	$\frac{1}{\sqrt{2}}$	$\frac{L}{2(10)^5}\left(\frac{R}{2}+\frac{50}{\sqrt{2}}\right)$
CD	L	$2(10)^5$	$\frac{R}{\sqrt{2}}$	$\frac{1}{\sqrt{2}}$	$\frac{RL}{4(10)^5}$
DA	L	$2(10)^5$	$\frac{R}{\sqrt{2}}+50$	$\frac{1}{\sqrt{2}}$	$\frac{L}{2(10)^5}\left(\frac{R}{2}+\frac{50}{\sqrt{2}}\right)$
BD	$L\sqrt{2}$	$1(10)^5$	$-(R+50\sqrt{2})$	-1	$\frac{L}{(10)^5}(R\sqrt{2}+100)$
AC	$L\sqrt{2}$	$5(10)^4$	$-R$	-1	$\frac{RL\sqrt{2}}{5(10)^4}$
				$\sum S\frac{\partial S}{\partial R}\frac{L}{AE}$	$\frac{RL}{2(10)^5}(2+6\sqrt{2}) + \frac{50L}{(10)^5}\left(\frac{3}{2\sqrt{2}}+2\right)$

There are structures however, in which some members are primarily subjected to more than one type of actions. "Beam trusses" or "Truss beams" are examples of such structure, in which the predominant actions are due to bending stresses on some members and direct stresses on others. Such structures are known as *composite structures* and in analysis of statically indeterminate structures of the type, effects of more than one type of actions have to be considered. The following example will illustrate the basic ideas involved in the analysis of such structures.

Example 6.14. *Analyse the king post truss shown in Fig. 6.18.*

Solution : The structure will be analysed by the use of Castigliano's theorem. In this case beam action in member AB and truss actions in the frame in Fig. 6.18(b) will be taken into consideration.

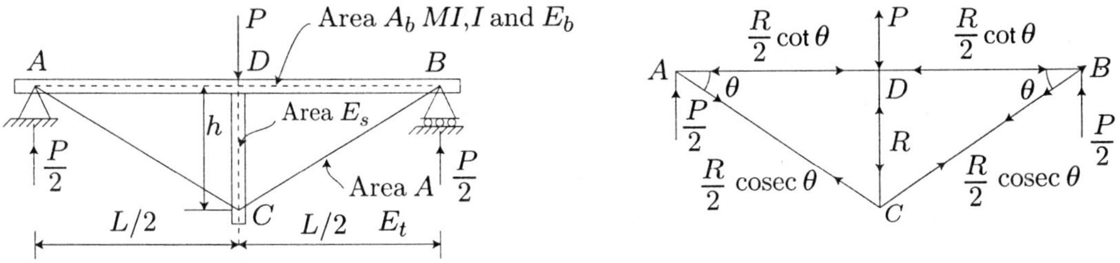

(a) King Post Truss **(b) Truss action**

Fig. 6.18: For Example 6.14

From Castigliano's theorem considering direct force in member CD, i.e., R as the redundant,

$$\frac{\partial U}{\partial R} = 0, \quad \text{whence} \quad \int \frac{M}{EI}\frac{\partial M}{\partial R}dx + \sum \frac{S}{AE}\frac{\partial S}{\partial R}L = 0.$$

In member AB,

$$M = \frac{P-R}{2}x, \quad \text{for} \quad 0 \le x \le \frac{L}{2},$$

whence $$\frac{\partial M}{\partial R} = -\frac{x}{2}.$$

Values of S in various members are shown in Fig. 6.18(b), from which $\frac{\partial S}{\partial R}$ may be obtained.

$$\frac{2}{E_b I}\int_0^{\frac{L}{2}} \left(\frac{P-R}{2}x\right)\left(\frac{x}{2}\right)dx + \left[2\left(\frac{R}{2}\cot\theta\right)\left(\frac{1}{2}\cot\theta\right)\frac{L}{2A_bE_b} + (S)(1)\frac{h}{aE_s}\right.$$
$$\left. + 2\left(\frac{S}{2}\operatorname{cosec}\theta\right)\left(\frac{1}{2}\operatorname{cosec}\theta\right)\frac{L}{2}\sec\theta\right] = 0.$$

As $S = R$ and $h = \frac{L}{2}\tan\theta$,

$$R = \frac{\dfrac{PL^2}{48E_bI}}{\dfrac{L^2}{48E_bI} + \dfrac{\cot^2\theta}{4A_bE_b} + \dfrac{\tan\theta}{2aE_s} + \dfrac{\operatorname{cosec}^2\theta\sec\theta}{4AE_t}}.$$

Neglecting effect of direct stresses on AB,

$$R = \frac{\dfrac{PL^2}{48E_bI}}{\dfrac{L^2}{48E_bI} + \dfrac{\tan\theta}{2aE_s} + \dfrac{\operatorname{cosec}^2\theta\sec\theta}{4AE_t}}.$$

Example 6.15. *Analyse the sructure shown in Fig. 6.19.*

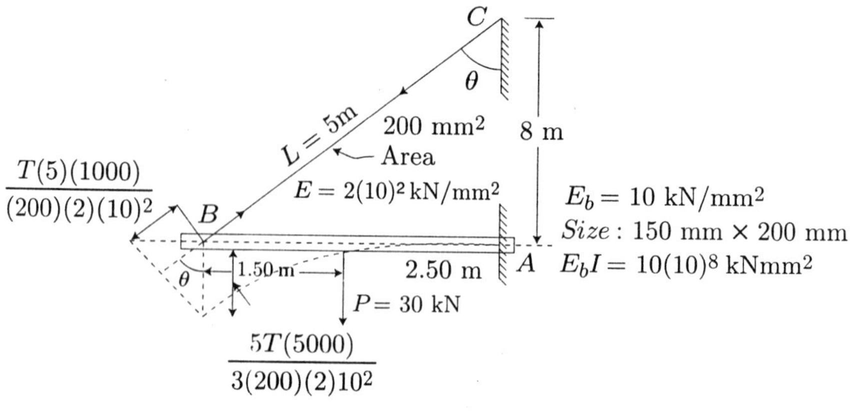

Fig. 6.19: For Example 6.15

Solution : This problem will be solved by the method of consistent deformation. Considering beam AB, vertical deflection at B, i.e.,

$$\Delta_{BV} = \frac{1}{(10)(10)^8}\left[\frac{1}{2}(30)(2.50)(1000)(2500)\left(1.50+\frac{5.00}{3}\right)(1000) - \frac{1}{2}(0.60T)(4000)(4000)\left(\frac{8000}{3}\right)\right]$$

$$= 296.875 - 12.80T$$

From the method of consistent deformation,

$$\Delta_{BV} = \frac{5}{3}\frac{T(5000)}{(200)(2)(10)^2} \text{ (from Fig. 6.19).}$$

whence $T = 22.822$ kN.

6.6 Continuous Beams

A continuous beam is supported at a number of points along its length. Such a beam may be of the same section throughout or may vary from span to span. The supports may be rigid, in which case the beam does not deflect at the supported points when loaded or they may be elastic thus allowing settlement to occur. The beam may again be pinned to the supporting points or be so attached that the joints between the supports and the beam are rigid. There are various methods of analysis of such continuous beams. The methods frequently used are,

(a) Wilson's method based on the use of Double Integration method.

(b) Method of Consistent Deformation or Castigliano's theorem.

(c) Theorem of Three Moments (known as Clapeyron's theorem).

(a) Wilson's Method

If a continuous beam is of constant section throughout its length, a method of analysis due to Dr. George Wilson is often convenient. In this method the unknown reactions are treated as loads and a generalised equation for bending moment written following Macaulay's method.

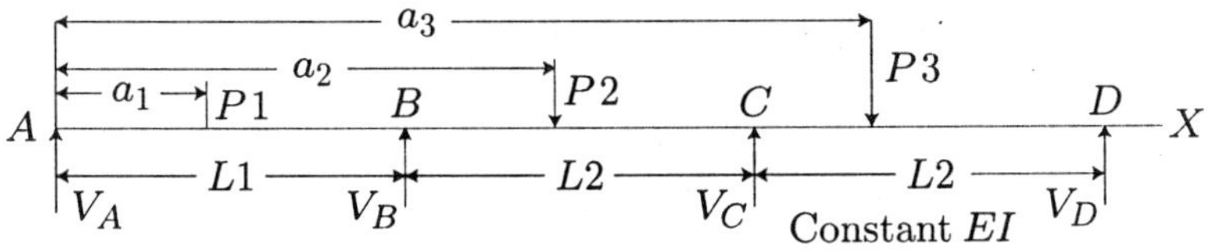

Fig. 6.20: Wilson's Method

Considering a continuous beam $ABCD$ of uniform section loaded as shown in Fig. 6.20, the bending moment at any section x from A is given by

$$-EI\frac{d^2y}{dx^2} = V_A(x) - P_1[x-a_1] + V_B[x-L_1] - P_2[x-a_2] + V_C[x-L_1-L_2] - P_3[x-a_3]. \qquad (6.9a)$$

Integrating,

$$-EI\frac{dy}{dx} = V_A\frac{x^2}{2} - \frac{P[x-a_1]^2}{2} + V_B\frac{[x-L_1]^2}{2} - P_2\frac{[x-a_2]^2}{2} + V_C\frac{[x-L_1-L_2]^2}{2} - P_3\frac{[x-a_3]^2}{2} + C_1. \qquad (6.9b)$$

Integrating

$$EIy = V_A\frac{x^3}{6} - \frac{P[x-a_1]^3}{6} + V_B\frac{[x-L_1]^3}{6} - P_2\frac{[x-a_2]^3}{6} + V_C\frac{[x-L_1-L_2]^3}{6} - P_3\frac{[x-a_3]^3}{6} + C_1x + C_2. \qquad (6.9c)$$

In Eqs. (6.9a), (6.9b) and (6.9c), V_A, V_B, V_C represent the reactions of supports at A, B and C, and C_1, and C_2 the constants of integration. There are six unknowns for solutions of which six conditions are required. Four conditions are obtained from the known deformations at supports. If there is no settlement of any one of the supports, these four conditions are as under:

At $x = 0,\ y = 0$

At $x = L_1,\ y = 0$

At $x = L_1 + L_2,\ y = 0$

At $x = L_1 + L_2 + L_3,\ y = 0.$

If supports settle, values of such settlements at supports are to be used in place of zero settlement.

The other two conditions are obtained from conditions of equilibrium. Thus from $\sum F_Y = 0, V_A + V_B + V_C + V_D = P_1 + P_2 + P_3$. From $\sum M = 0$ at A, $V_B(L_1) + V_C(L_1 + L_2) + V_D(L_1 + L_1 + L_3) = P_1a_1 + P_2a_2 + P_3a_3$.

From these six conditions, the four reactions can be obtained and bending moment and shear force at any section can then be calculated.

(b) Method of Consistent Deformation

In this method, the primary determinate beam is made by removing the redundants. Considering the beam in Fig. 6.20, primary structure may be obtained by removing supports B and C. Deflection Δ_{BVO} and Δ_{CVO} at points B and C of the primary structure subjected to applied load may be obtained. The primary structure is then subjected to $V_B = 1$ and $V_C = 1$, separately, and deflections δ_{bb}, δ_{cb} due to $V_A = 1$, and δ_{bc}, δ_{cc} due to $V_C = 1$ at points B and C calculated. From compatibility of deformations at B and C, equations can be formed as

$$\left.\begin{aligned} &\Delta_{BV} + \Delta_{BVO} + V_B\delta_{bb} + V_c\delta_{bc} = 0 \\ \text{and}\quad &\Delta_{CV} + \Delta_{CVO} + V_B\delta_{cb} + V_c\delta_{cc} = 0 \end{aligned}\right]$$

where Δ_{BV} and Δ_{CV} represent the known settlements of the supports at B and C respectively.

Solution of the simultaneous equations thus obtained will yield values of V_B and V_C. V_A and V_D may then be obtained from conditions of statics.

Equations as obtained by the method of consistent deformation may also be obtained by use of Castigliano's theorem from the considerations,

$$\frac{\partial U}{\partial V_B} = \Delta_{BV} \quad \text{and} \quad \frac{\partial U}{\partial V_c} = \Delta_{CV}.$$

(c) Theorem of Three Moments (Clapeyron's Theorem)

This theorem was first presented by the French Engineer Clapeyron. It expresses the relation between the bending moments at three consequitive support points of a continuous beam to the loadings on the intermediate span. The relation is derived on the basis of the continuity of the elastic curve over the middle support, which implies that the slope of the tangent to the right end of the left span must be equal to the slope of the tangent to the left end of the right span.

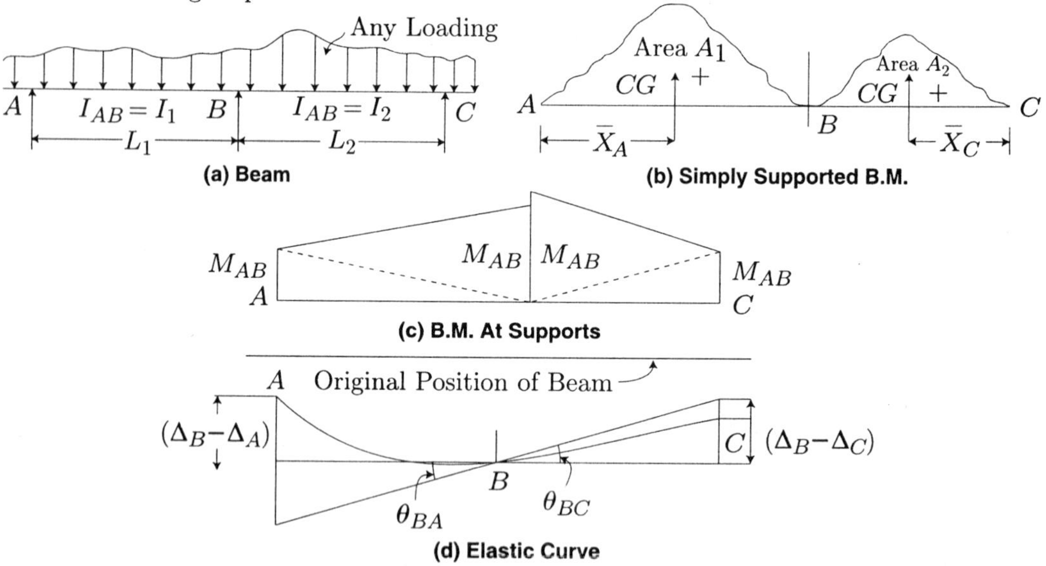

Fig. 6.21

Fig. 6.21(a) shows two spans ABC of a continuous bean subjected to any pattern of loading, the characteristics of the span AB and BC being as shown. The bending moment diagram in the spans AB and BC may be obtained by combining the bending moment diagrams shown in Figs. 6.21(b) and 6.21(c).

From moment-area theorem referring to the elastic curve shown in Fig. 6.21(c) from span AB, deflection of A beyond tangent at B,

$$\frac{1}{EI_{AB}}\left[A_1\overline{X}_A - \frac{M_{AB}}{2}(L_1)\left(\frac{L_1}{3}\right) - \frac{M_{BA}}{2}(L_1)\left(\frac{2L_1}{3}\right)\right],$$

whence $$\theta_{BA} = \frac{1}{EI_{AB}L_1}\left[A_1\overline{X}_A - \frac{M_{AB}L_1^2}{6} - \frac{M_{BA}L_1^2}{3}\right] - \frac{\Delta_B - \Delta_A}{L_1}.$$

Similarly, considering span BC,

$$\theta_{BC} = \frac{1}{EI_{BC}L_2}\left[A_2\overline{X}_C - \frac{M_{CB}L_2^2}{6} - \frac{M_{BC}L_2^2}{3}\right] - \frac{\Delta_B - \Delta_C}{L_2}.$$

As $\theta_{BA} = -\theta_{BC}$, it follows that,

$$M_{AB}\frac{L_1}{I_{AB}} + 2M_{BA}\frac{L_1}{I_{AB}} + 2M_{BC}\frac{L_2}{I_{BC}} + M_{CB}\frac{L_2}{I_{BC}}$$
$$= \frac{6A_1\overline{X}_A}{I_{AB}L_1} + \frac{6A_2\overline{X}_C}{I_{BC}L_2} - \frac{6E(\Delta_B - \Delta_A)}{L_1} - \frac{6E(\Delta_B - \Delta_C)}{L_2}. \quad (6.10a)$$

This is the generalised form of the theorem of three moments for the case where the continuous beam is prismatic in the span. The terms $\overline{x}_A$, $\overline{x}_C$ indicate the distances of C.G. of the positive B.M. diagram from the two ends of the spans being considered.

A consideration of the beam at support B, shows that if an external couple M_O be applied there,

$$M_{BA} = M_{BC} + M_O.$$

If no external couple is applied at B,

$$M_{BA} = M_{BC} = M_B.$$

In such a case, the three moments equation reduces to

$$M_A\frac{L_1}{I_{AB}} + 2M_B\left[\frac{L_1}{I_{AB}} + \frac{L_2}{I_{BC}}\right] + M_C\frac{L_2}{I_{BC}}$$
$$= \frac{6A_1\overline{X}_A}{L_1I_{AB}} + \frac{6A_2\overline{X}_C}{L_2I_{BC}} - \frac{6E(\Delta_B - \Delta_A)}{L_1} - \frac{6E(\Delta_B - \Delta_C)}{L_2}, \quad (6.10b)$$

taking $M_{AB} = M_A$, and $M_{CB} = M_C$.

For the case when $I_{AB} = I_{BC} = I$, equation reduces to

$$M_AL_1 + 2M_B(L_1 + L_2) + M_CL_2 = \frac{6A_1\overline{X}_A}{L_1} + \frac{6A_2\overline{X}_C}{L_2}$$
$$- 6EI\frac{(\Delta_B - \Delta_A)}{L_1} - 6EI\frac{(\Delta_B - \Delta_C)}{L_2} \quad (6.10c)$$

For a continuous beam of n spans, there will be $(n - 2)$ equations of the type of Eq. (6.10). The remaining two equations necessary for the complete determination of the support moments are obtained from a consideration of the conditions at the end supports. For example, if the left-hand point A of the beam is simply supported, $M_{AB} = M_A = 0$.

If the continuous beam has a fixed end, the bending moment at the fixed support is one of the unknowns. In such a case an imaginary span of any length L_O having an infinitely large moment of inertia (alternatively $L_O = 0$ for any values of I) may be assumed on the outer side of the end span. Thus, if the end A be fixed, a span A_OA of length L_O and of which I_O is infinitely large be assumed. As moments of A_1 and A_2 diagrams have to be taken about points A_O and B respectively when spans A_OA and AB are considered, obviously $\frac{L_O}{I_O} = 0$, and the three moments equation becomes

$$M_{AO}\left(\frac{L_O}{I_O}\right) + 2M_A\left(\frac{L_O}{I_O} + \frac{L_1}{I_{AB}}\right) + M_B\left(\frac{L_1}{I_{AB}}\right) = \frac{6A_1\overline{X}_B}{I_{AB}L_1} - \frac{6E(\Delta_A - \Delta_B)}{L_1},$$

whence $$2M_A\frac{L_1}{I_{AB}} + M_B\frac{L_1}{I_{AB}} = \frac{6A_1\overline{X}_B}{I_{AB}L_1} - 6E\frac{(\Delta_A - \Delta_B)}{L_1},$$

where $\overline{X}_B$ = distance of C.G. of positive B.M. on span AB from B.

An alternative scheme for fixed end A is to consider spans on left of A similar as on the right of A, so that the three moments equation reduces to

$$M_B\frac{L_1}{I_{AB}} + 2M_A\left(\frac{L_1}{I_{AB}} + \frac{L_1}{I_{AB}}\right) + M_B\frac{L_2}{I_{AB}}$$
$$= \frac{6A_1\overline{X}_B}{I_{AB}L_1} + \frac{6A_1\overline{X}_B}{I_{AB}L_1} - 6E\frac{(\Delta_A - \Delta_B)}{L_1} - 6E\frac{(\Delta_A - \Delta_B)}{L_1}.$$

This gives the same equation as before.

In using the equations derived, the following sign conventions are to be used :

(i) Simply supported bending moments are considered positive and support moments negative.

(ii) $(\Delta_B - \Delta_A)$, $(\Delta_B - \Delta_C)$ considered positive when B is at a lower level of A and C.

Example 6.16. *Analyse the continuous beam shown in Fig. 6.22(a), using the three moments equation, if*

(a) the supports do not settle (b) the interior support at B settles by 10 mm.

Assume for the beam,

$$E = 2(10)^2 \text{ kN/mm}^2 \quad \text{and} \quad I = (10)^8 \text{ mm}^4$$

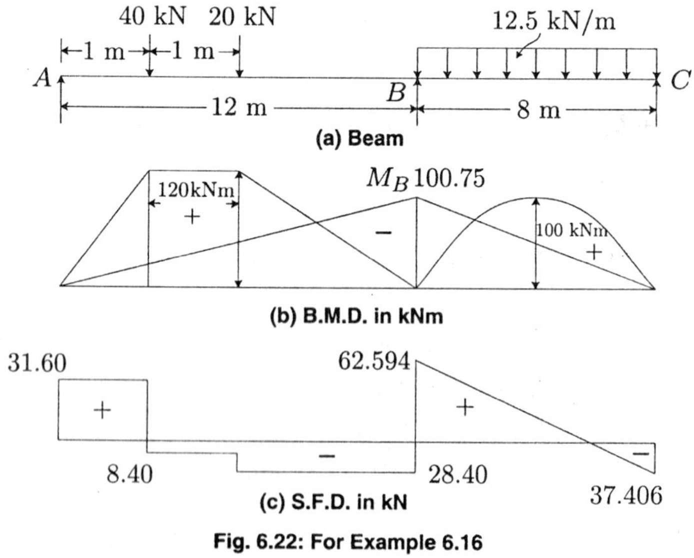

Fig. 6.22: For Example 6.16

Solution : (a) *No settlement at supports*

In this case $M_A = M_C = 0$.

The three moments equation becomes

$$2M_B(L_1 + L_2) = \frac{6A_1\overline{X}_A}{L_1} + \frac{6A_2\overline{X}_C}{L_2}.$$

Here $$\frac{6A_1\overline{X}_A}{L_1} = \frac{6}{12}\left[\frac{1}{2}120(3)(2) + (120)(3)(4.5) + \frac{1}{2}(120)(6)(8)\right]$$

$$= 2430$$

and $$\frac{6A_2\overline{X}_C}{L_2} = \frac{6}{8}\left[\frac{2}{3}(100)(8)(4)\right] = 1600,$$

whence $40M_B = 2430 + 1600 = 4030$

and $M_B = 100.75$ kNm.

Positive sign of M_B means that the support moment is actually negative.

Reactions—considering span AB:

$$V_A = \frac{1}{12}[40(9) + 20(6) - 100.75] = 31.604 \text{ kN}$$

$$V_{BA} = \frac{1}{12}[(40)(3) + 20(6) + 100.75] = 28.396 \text{ kN}$$

$$V_{BC} = \frac{1}{8}[(12.5)(8)(4) + 100.75] = 62.594 \text{ kN}$$

so that $V_B = V_{BA} + V_{BC} = 90.999$ kN

$$V_C = (40 + 20 + 100) - (V_A + V_B) = 37.406 \text{ kN}.$$

B.M. and S.F. diagrams are shown in Figs. 6.22(b) and 6.22(c).

(b) *When support B settles by* 10 mm.

In this case the three moments equation becomes

$$2M_B(L_1 + L_2) = \frac{6A_1\overline{X}_A}{L_1} + \frac{6A_2X_C}{L_2} - 6EI\left[\frac{\Delta_B - \Delta_A}{L_1} + \frac{\Delta_B - \Delta_C}{L_2}\right],$$

whence $$40M_B = 4030 - \frac{6(2)(10)^2(10)^8}{(10)^6}\left[\frac{1}{1200} + \frac{1}{800}\right]$$

$$= 4030 - 250 = 3780,$$

whence $M_B = 94.50$ kNm

Example 6.17. *A continuous beam ABCD of uniform section* 30 m *long rests on supports A, B, C and D all at the same level. Length AB* = 12 m, *BC* = 10 m *and CD* = 8 m. *It carries two concentrated load* 90 kN *and* 80 kN *at* 4 m *and* 16 m *from A and a uniformly distributed load of* 15 kN *per metre run, over CD. Find the moments and reactions at the supports.*

Solution : Considering spans AB and BC and as $M_A = 0$,

$$44M_B + 10M_C = \frac{6A_1\overline{X}A}{12} + \frac{6A_2\overline{X}_C}{10}$$

$$= \frac{6}{12}\left[\frac{1}{2}(240)(4)\left(\frac{2}{3}\right)(4) + \frac{1}{2}(240)(8)\left(4+\frac{8}{3}\right)\right] + \frac{6}{10}\left[\frac{1}{2}(192)(4)\left(6+\frac{4}{3}\right) + \frac{1}{2}(192)(6)\left(\frac{2}{3}\right)(6)\right],$$

whence $44M_B + 10M_C = 6912.0.$ (i)

Similarly, for span BC and CD,

$$10M_B + 36M_C = 4608.0. \quad \text{(ii)}$$

Solving (i) and (ii), $M_B = 136.625$ kNm and $M_C = 90.05$ kNm.

Reactions at supports.

$$V_A = \frac{1}{12}[90(8) - 136.625] = 48.615 \text{ kN} \uparrow$$

$$V_{BA} = \frac{1}{12}[(90)(4) + 136.625] = 41.385 \text{ kN}$$

$$V_{BC} = \frac{1}{10}[(80)(6) + 136.625 - 90.05] = 52.658 \text{ kN} \uparrow$$

so that $V_B = V_{BA} + V_{BC} = 94.043 \text{ kN} \uparrow$

$$V_{CB} = \frac{1}{10}[(80)(4) - 136.625 + 90.05] = 27.343 \text{ kN}$$

$$V_{CD} = \frac{1}{8}[(15)(8)(4) + 90.05] = 71.256 \text{ kN}$$

$$V_C = V_{CB} + V_{CD} = \uparrow 98.599 \text{ kN}$$

$$V_D = \frac{1}{8}[(15)(8)(4) - 90.05] = \uparrow 48.747 \text{ kN}.$$

B.M. and S.F. diagrams are shown in Figs. 6.23(b) and 6.23(c).

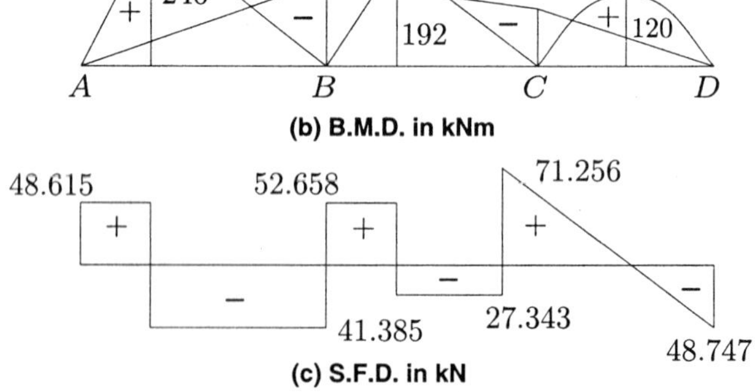

Fig. 6.23: For Example 6.17

Example 6.18. *Analyse the continuous beam shown in Fig. 6.24, if the support at B settles by* 10 mm. *Beam fixed at end A.*

Assume $E = 2(10)^2$ kN/mm^2 *and* $I = 40000000$ mm$^4 = 4(10)^{-5}$ m^4.

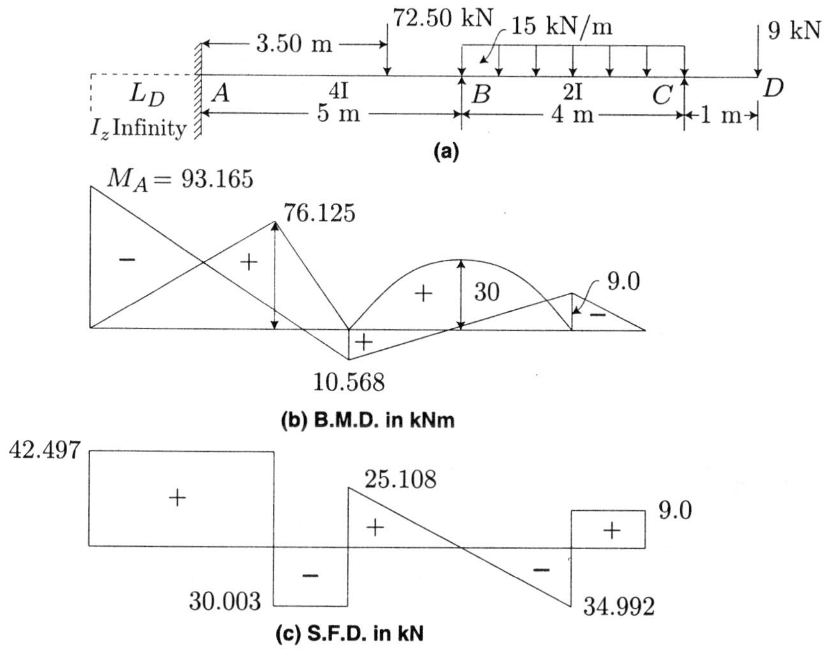

Fig. 6.24: For Example 6.18

Solution : In this case M_C is statically determinate and is given by $M_C = 9.0(1) = 9$ kNm and is negative.

As end A is fixed, considering an imaginary beam A_OA of length L_O and of infinite stiffness, I_O, for spans A_OA and AB,

$$M_{AO}\left(\frac{L_O}{I_O}\right) + 2M_A\left(\frac{L_O}{I_O} + \frac{5}{41}\right) + M_B\left(\frac{5}{41}\right)$$
$$= \frac{6}{5(41)}\left[\frac{1}{2}(76.125)(3.50)\left(1.50 + \frac{3.50}{3}\right) + \frac{1}{2}(76.125)(1.50)(1)\right] - \frac{6E}{5}\left(-\frac{10}{1000}\right).$$

As $\dfrac{L_O}{I_O} = 0, 2.50M_A + 1.25M_B = 219.703.$ (i)

From spans AB and BC,

$$M_A\left(\frac{5}{41}\right) + 2M_B\left(\frac{5}{41} + \frac{4}{21}\right) + M_C\left(\frac{4}{21}\right)$$
$$= \frac{6}{(5)(41)}\left[\frac{1}{2}(76.125)(3.50)\left(\frac{7}{3}\right) + \frac{1}{2}(76.125)(1.50)(4)\right]$$
$$+ \frac{6}{(4)(21)}\left[\frac{2}{3}(30)(4)(2)\right] - 6E\left[\frac{10}{5000} + \frac{10}{4000}\right].$$

As $M_C = 9$ kNm,

$$1.25M_A + 6.50M_B = 47.765. \quad \text{(ii)}$$

Solving, $M_B = -10.568$ and $M_A = 93.165$.

Negative value of M_B means that the B.M. at B is positive.

Reactions :

$$V_A = \frac{1}{5}[72.50(1.50) + 93.165 + 10.568] = 42.497 \text{ kN}$$

$$V_B = V_{BA} + V_{BC} = \frac{1}{5}[(72.5)(3.50) - 93.165 - (10.568)] + \frac{1}{4}[(15)(4)(2) - 9.0 - (10.568)]$$

$$= 30.003 + 25.108 = 55.111 \text{ kN}$$

$$V_C = 60 - (25.108) + 9.0 = 43.892 \text{ kN}$$

B.M. and S.F. diagrams are shown in Figs. 6.24(b) and 6.24(c).

Problems

6.1 A uniform horizontal cantilever beam of length L is propped at a distance $\frac{L}{4}$ from the free end. If a uniformly distributed load of magnitude w/unit length be placed on the beam, obtain the reaction of the prop and draw the B.M. and S.F. diagrams. Also calculate the deflection at the free end.

6.2 A cantilever of length L is propped at $\frac{L}{4}$ from the free end to the level of the fixed end. It carries a load P at the free end. Determine the ratio of the deflections at the free end for the prop in position and the prop removed.

6.3 A beam of length $2L$ and uniform EI is fixed at one end and supported on a prop at L from the fixed end. If the prop settles by K times the load on the support, calculate the load on the prop when a uniformly distributed load w/unit length is placed on the beam.

6.4 A vertical post of height H is built into the ground. The upper end is connected to a point in the ground by a bracing wire of length $2H$ and cross-sectional area 'a'. The M.I. of the post is 1 and its cross-sectional area is A. Both the columns and the wire are of the same material. If a horizontal force P is applied to the top of the column, estimate the tension in the bracing wire.

6.5 A cantilever beam carries a uniformly distributed load throughout its length. a) If the beam be propped at the free end, what fraction of the load should the prop carry if the intensity of the bending stress in the cantilever is to be the least possible? What will be the resulting settlement of the prop?

(b) At what fraction of its length from the free end should the cantilever be propped at the level of the fixed end so that the intensity of the bending stress shall be the least possible. What proportion of the load is then carried by the prop?

6.6 A beam AB of constant EI and of length 10 m has its ends constrained in such a way that the ends A and B rotate through unit radian, due to a moment of 500 kNm and 1000 kNm respectively. Draw the B.M. and S.F. diagrams when a concentrated load of 50 kN is placed at the centre of the span. Assume $E = 2(10)^2$ kN/mm^2 and $I = 4(10)^{-4}$ m^4. Also calculate the deflection at the centre of the beam.

6.7 The ends of a beam of length L which carries a concentrated load P at $\frac{1}{3}$rd of the span, are so constrained that they assume slopes one half of those which would occur if the beam were simply supported. Find B.M. and S.F. diagrams. Assume EI as constant.

6.8 Calculate the reactions and obtain the B.M. and S.F. diagrams for the components of the arrangement shown in Figs. P-6.1 to P-6.4.

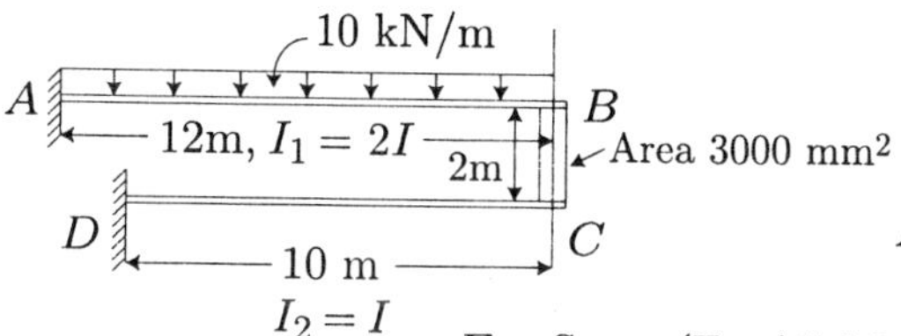

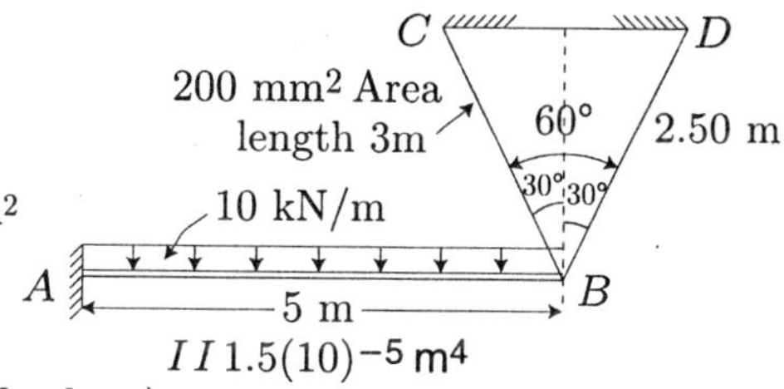

E = Same (For All Members)

Fig. P-6.1

Fig. P-6.2

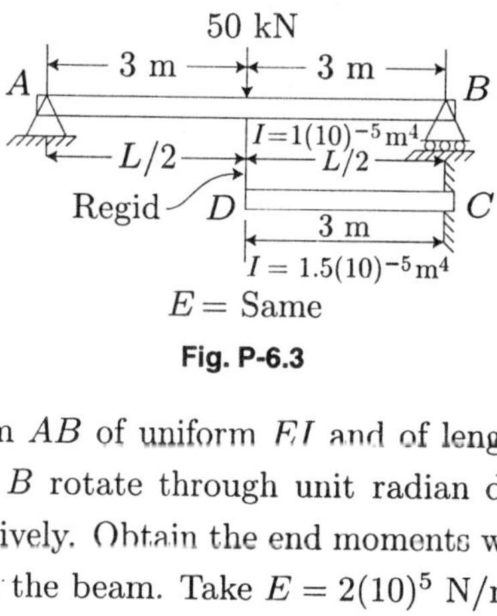

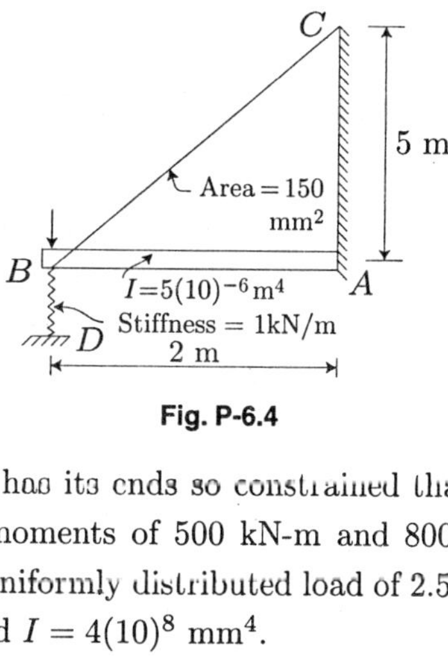

Fig. P-6.3

Fig. P-6.4

6.9 A beam AB of uniform EI and of length 6 m has its ends so constrained that ends A and B rotate through unit radian due to moments of 500 kN-m and 800 kN-m respectively. Obtain the end moments when a uniformly distributed load of 2.5 kN/m acts on the beam. Take $E = 2(10)^5$ N/mm^2 and $I = 4(10)^8$ mm^4.

Also calculate the deflection at the centre of the beam.

6.10 The end B of a fixed-end beam AB of constant EI and of length L settles by Δ with respect to end A. Calculate the fixed-end moments and draw B.M. and S.F. diagrams for the beam.

6.11 The end A of a fixed-end beam AB of constant EI and of length L rotates clock-wise by θ radians. Determine the end moments and draw B.M. and S.F. diagrams for the beam.

6.12 A fixed beam of uniform flexural rigidity and of length L is subjected to two downward loads each of magnitude P at $\frac{L}{3}$ and $\frac{2L}{3}$ from an end and an upward load R at $\frac{L}{2}$ from and end. Calculate the value of R in terms of P so that the greatest B.M. in the beam will be a minimum. Calculate also the central deflection and draw B.M. and S.F. diagrams.

6.13 A prismatic beam of length L with ends fixed is subjected to two vertical loads P acting in opposite directions at distance 'a' from each end. For what value 'a' will the bending moments at end supports be the greatest?

6.14 Determine the support reactions for the following rigid-jointed frame shown in Fig. P-6.5 and draw B.M., S.F. and axial force diagrams.

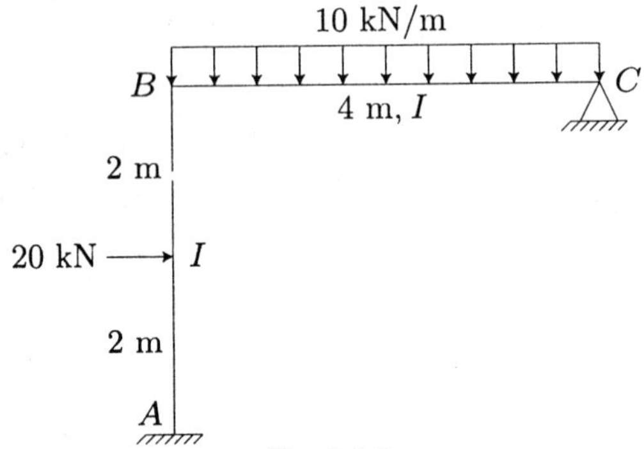

Fig. P-6.5

6.15 A uniform beam of stiffness EI and span L carries central load P. It is so constrained at the ends that when the end slope is θ, the restraining moment at the support is $K\theta$. Show that the magnitude of the moment at each end is $\frac{KPL^2}{8(KL+2EI)}$ and the magnitude of the central deflection is $\frac{PL^3}{192EI}\frac{KL+8EI}{KL+2EI}$.

6.16 Analyse the fixed end beams shown in Figs. P-6.6 to P-6.11 and obtain the bending moment and S.F. diagram. Also calculate the maximum deflection.

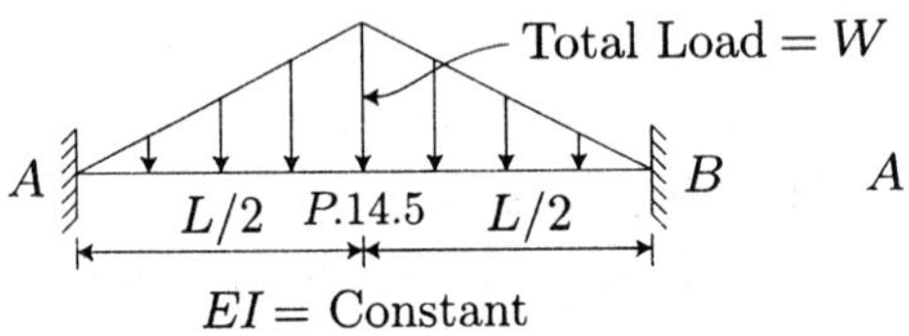

Fig. P-6.6

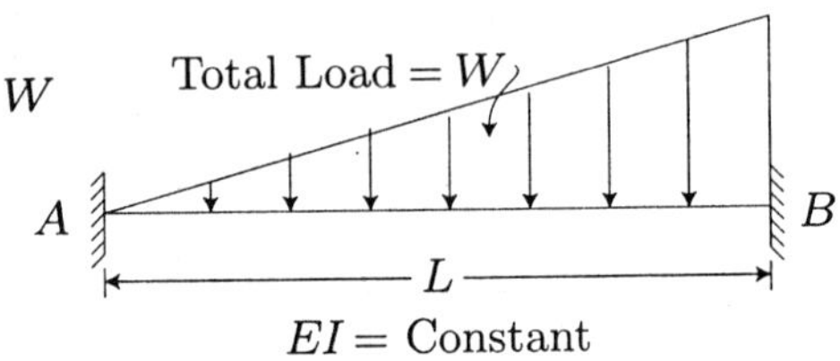

Fig. P-6.7

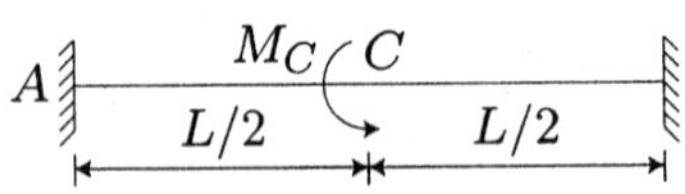

Fig. P-6.8

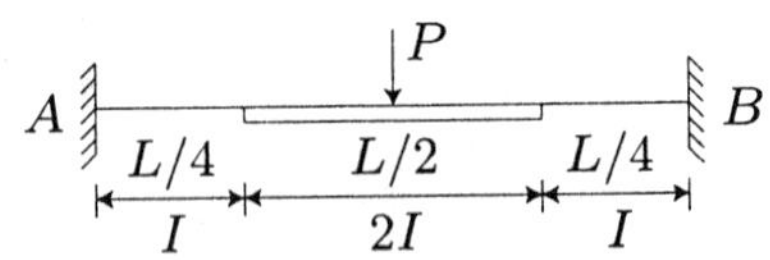

Fig. P-6.9

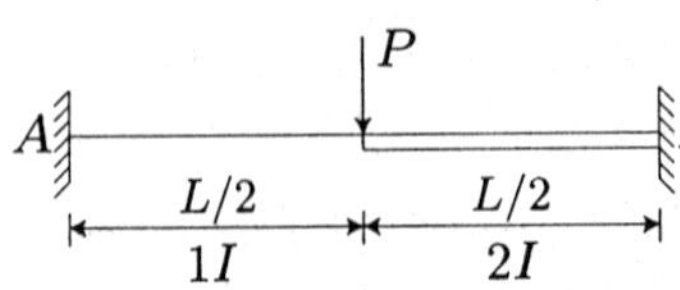

Fig. P-6.10

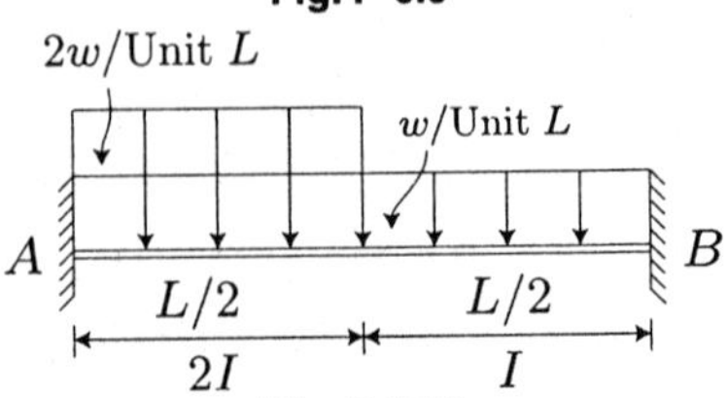

Fig. P-6.11

6.17 A beam AB of constant EI and of length L is fixed at both ends A and B. The support at B settles by Δ without rotation. Show that the magnitude of the bending moment at ends is given by $\frac{6EI\Delta}{L^2}$. Draw the B.M. and S.F. diagrams.

6.18 A beam AB of constant EI and length L is placed on a simple support at A and fixed at B. If a couple M be applied at A and if it causes a rotation θ, find the magnitude of the moment. Draw the B.M. and S.F. diagrams.

How will the B.M. and S.F. diagrams be modified when end B is also placed on a simple support.

6.19 A horizontal uniform beam of span L is fixed at both ends and two equal loads of magnitude P each are placed at equal distances 'a' from either end. Show that the bending moments at the centre and at the ends of the beam are $\frac{Pa^2}{L}$ and $\frac{Pa(L-a)}{L}$ respectively.

6.20 A beam AB of 6 m span is fixed at both ends and carries a load of 30 kN at C two metres from A. The M.I. of the portion $AC = 1(10)^{-5}\text{m}^4$ and for the portion BC is $2(10)^{-5}\text{m}^4$. Calculate the fixed end moments and the central deflection. Assume E as $2(10)^2$ kN/mm^2.

6.21 A beam of uniform section and span L is fixed at the ends and carries two loads each of magnitude P symmetrically at $\frac{L}{3}$ from ends. The beam is propped at mid-span by a force of such magnitude that the greatest B.M. to which the beam is subjected to is as small as possible.

Determine the magnitude of the supporting force and sketch the B.M. and S.F. diagrams. Also calculate the resulting deflection at mid-span in terms of EI.

6.22 A fixed beam of length L and of uniform flexural rigidity is subjected to the action of two vertical point loads P acting in opposite direction at distance 'a' from each end. Find the value of 'a' giving the largest bending moments at the end supports.

6.23 A horizontal cantilever beam AB of length 5 m is propped at the free end B. The beam is subjected to a vertical uniformly distributed load of magnitude 50 kN/m. Obtain the B.M. and S.F. diagrams for the cases stated below:

(a) When the support at B goes down by 10 mm

(b) When the support at B does not settle

(c) When the support at B goes up by 10 mm

(d) When the support at B goes down by 10 mm for a load of 50 kN

(e) When the end B is supported by a vertical steel wire having length 1 m and cross-sectional area 300 mm^2. Assume E for the beam and wire as $2(10)^2$ kN/mm^2.

6.24 A roof truss of 9 m span as shown in Fig. P-6.12 carries a load of magnitude P at apex. All the bars are pin-jointed, have the same cross-sectional area and same Young's modulus. If the end supports are connected by a bracing wire, estimate the tension in the wire to bring the apex of the frame to its original position.

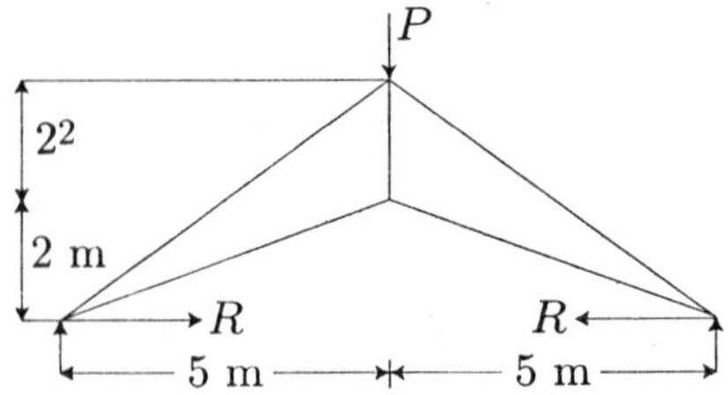

Fig. P-6.12

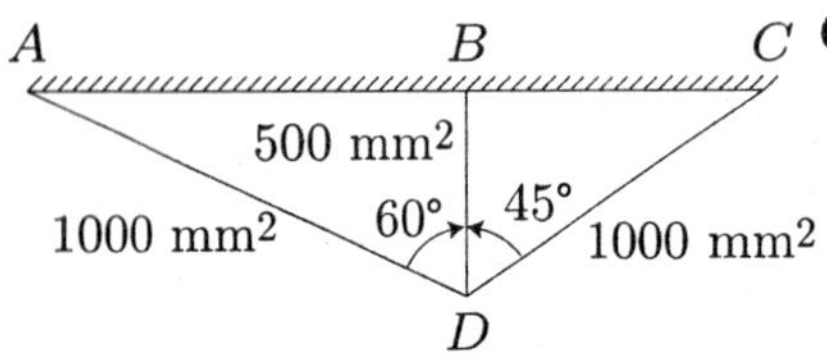

Fig. P-6.13

6.25 A frame consists of three hinged bars of cross-sectional areas shown in Fig. P-6.13. If an initial compressive stress of 20 N/mm^2 exists in BD, calculate the magnitude of the vertical load at D, if maximum stress in the members is not to exceed 200 N/mm^2.

6.26 A pin-joined frame is shown in Fig. P-6.14. If the cross-sectional area and material of all of the members are the same, show that the force produced in DE is given by $\frac{3+6\sqrt{2}}{7+12\sqrt{2}}P$.

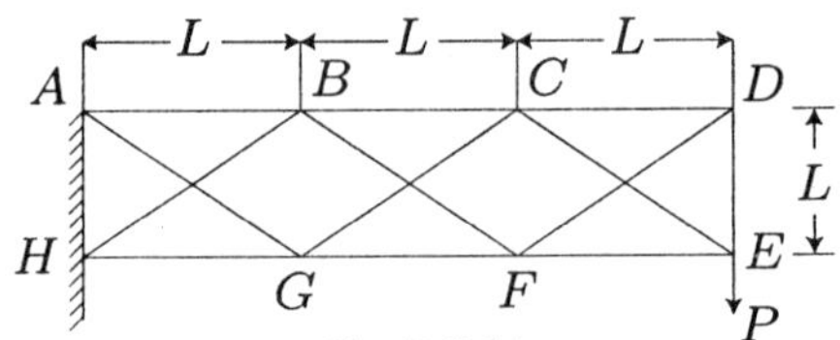

Fig. P-6.14

6.27 Find the forces on the various members of the frame shown in Example 6.10 if an additional diagonal member is provided along BE.

6.28 A square pin-jointed frame $ABCD$, provided with two diagonal members along AC and BD is hung at A from a wall, with the diagonal AC vertical. If a vertical downward load of magnitude 50 kN is placed at C, obtain the forces in various members. Assume cross-sectional areas of all the bars as same.

6.29 A pin-jointed frame is shown in Fig. P-6.15. If the cross-sectional areas of the chord members be two times that of the diagonals, determine the forces in all the members of the truss. Assume all bars are of the same material.

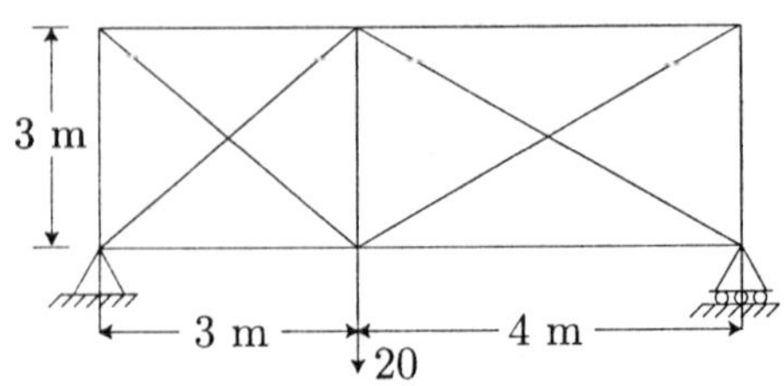

Fig. P-6.15

6.30 Analyse the king post truss of Example 6.14, when uniformly distributed load of magnitude w per unit length is placed on the truss.

6.31 A continuous beam of uniform cross section consists of two spans, each of equal length L and carries a uniform load of w per unit length. Find the settlement at the middle support for which the reactions at all supports will be the same.

6.32 Analyse the continuous beams shown in Figs. P-6.16 to P-6.18, for the conditions given. Draw the B.M. and S.F. diagrams and sketch the deflected shape

(a) when the supports do not settle.

(b) when the support B settles by 10 mm.

(c) when the support A settles by 10 mm.

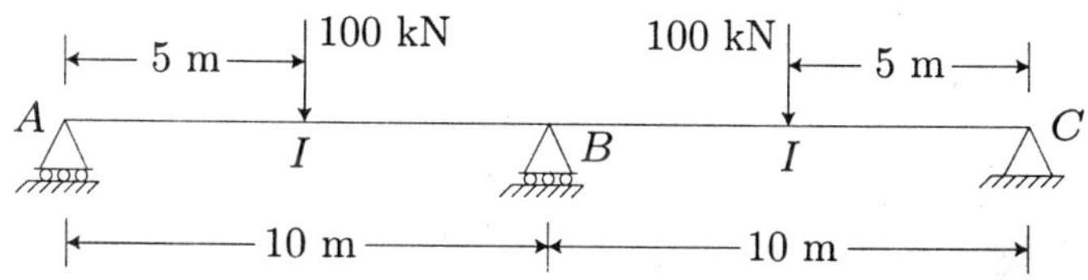

$E = 2(10)^5$ N/mm^2 $I = 4(10)^{-4}$ m^4

Fig. P-6.16

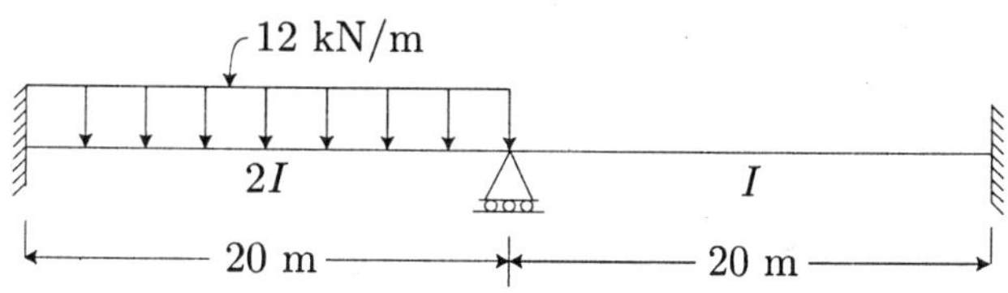

Fig. P-6.17

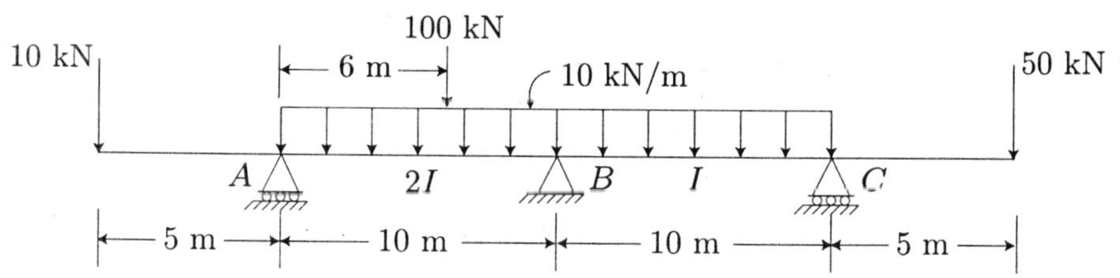

$E = 2(10)^5$ N/mm^2 and $I = 4(10)^{-4}$ m^4

Fig. P-6.18

6.33 Continuous girders ABC and CDE hinged at C form a pontoon bridge, the supports at A and E forming the end abutments. The intermediate supports at B, C and D are pontoons which sink 50 mm for every 10 kN load. If the girders carry u.d.l. of 5 kN/m, obtain the B.M. and S.F. diagrams. Assume $AB = BC = CD = DE = 10$ m and EI = constant.

6.34 A straight elastic beam $ABCD$ of uniform section rests on four similar floats at A, B, C and D which are spaced L metres apart. The buoyancy of each float is such that every additional kN of load increases its immersion by Δ. Initially all the floats are equally immersed. If two loads each P kN are placed on the girder at B and C, show that the proportion of the load carried by the central float is $\frac{P\left[1+\frac{6EI}{5L^3}\right]}{1+\frac{12EIA}{5L^3}}$, where EI is the flexural rigidity of the beam.

6.35 A beam ABC 10 m long and of uniform section rests freely on three supports and at the same level, placed so that $AB = 6$ m and $BC = 4$ m. The dead load is 20 kN/m and the live load is 10 kN/m uniformly distributed. The live load may act separately on each span or on the whole span.

Obtain the maximum B.M. and S.F. diagrams and state the maximum values.

6.36 A beam of constant flexural rigidity EI and length $2L$ carries a uniformly distributed load w per unit length and is supported on three springs one at its centre and one at each end. When the load is applied to the beam, the end springs are found to deflect $\frac{L}{100}$ and the centre spring to deflect $\frac{3L}{100}$. Find the deflection constants of the spring in terms of EI, L and w.

6.37 A beam ABC of constant flexural rigidity EI and length $2L$ carries a uniformly distributed load w per unit length and simply supported on rigid supports at ends A and C and on a spring at centre B, the stiffness of the spring being K. Draw the B.M. and S.F. diagrams and sketch the deflected shape.

6.38 A beam ABC of constant flexural stiffness EI and length $2L$ carries a uniformly distributed load w per unit length and simply supported on rigid support at the centre and to springs at A and C, the stiffness of the springs being K_a and K_c respectively. Obtain the B.M. and S.F. diagrams of the beam. Sketch the deflected shape.

6.39 Analyse a circular ring as in Example 6.9, if it be subjected to the following loads.

(a) Load of magnitude P acting outwards along the horizontal diameter and $2P$ acting onward along the vertical diameter.

(b) Load w per unit length (u.d.l) along the horizontal acting downwards from top and upward from bottom.

Chapter 7

Portal Frames and Arches

7.1 Introduction

Two hinged and fixed portal frames and arches are statically indeterminate to the first and third degrees respectively. In this Chapter method of solution of these structures based on principle of Consistent Deformation and Castigliano's Theorem will be evolved. There are other efficient methods for solution of these structures and those methods will be discussed in the subsequent Chapters.

Both portal frames and arches are subjected to axial force, shear and moment at any section. The axial force acts in a direction perpendicular to the section at its centroid. The shear is the total force acting in a direction parallel to the section, the moment being the total moment of forces about the centroidal axis in the cross section.

The axial force, shear and moment at any section of such structures can be obtained readily by the laws of statics, if all the reaction components at the supports are known.

7.2 Two-Hinged Portal Frame

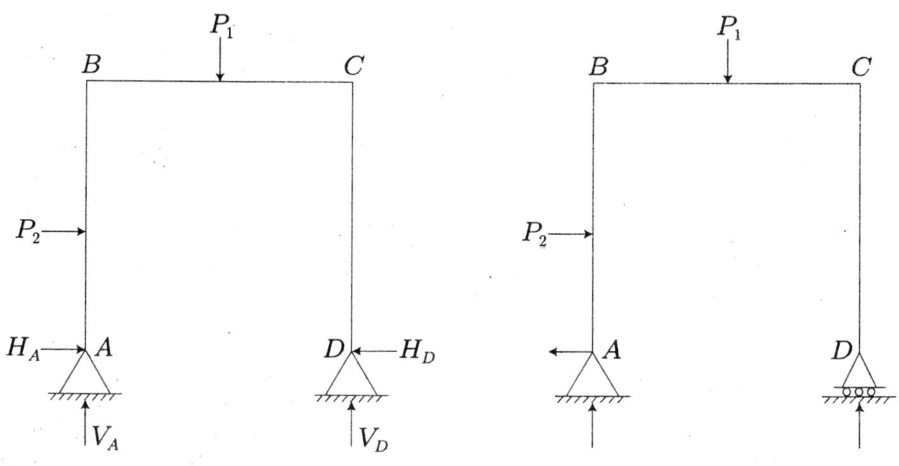

(a) Frame

(b) Reduced determinate structure

Fig. 7.1: Two-hinged portal frame

Fig. 7.1(a) shows a two-hinged portal frame, in which there are four unknown support reactions viz., H_A, V_A, H_D and V_D against three equations of equilibrium. Any of these may be considered as the redundant. Considering H_D as the redundant reaction, the reduced determinate structure becomes a simply supported portal frame as shown in Fig. 7.1(b) having freedom of lateral displacement at D. The governing condition for solution of this case is that the total horizontal displacement at D due to the applied loads and the redundant reaction as shown in Fig. 7.2(a) and 7.2(b), is zero, i.e.,

$$\Delta + \Delta' = 0, \tag{a}$$

where Δ = horizontal displacement at D due to applied loads and Δ' = horizontal displacement at D due to H_D. Neglecting the effect of shear and axial force and considering flexure alone, both Δ and Δ' can be determined by unit load method as under :

$$\Delta = \sum \frac{M_s m ds}{EI},$$

and $$\Delta' = \sum \frac{(mH_D)(m)ds}{EI} = H_D \sum \frac{m^2 ds}{EI}, \tag{b}$$

where M_s = Bending moment at any point of the reduced determinate structure due to applied loads,

m = bending moment at any point of the reduced determinate structure due to unit value of H_D.

For evaluation of Δ', the bending moment at any point due to H_D is taken as mH_D as m is the bending moment for $H_D = 1$. Substitution of the expressions in Eq. (b) in Eq. (a) gives,

$$\sum \frac{M_s m ds}{EI} + H_D \sum \frac{m^2 ds}{EI} = 0,$$

whence $$H_D = -\frac{\sum \frac{M_s m ds}{EI}}{\sum \frac{m^2 ds}{EI}}. \tag{7.1}$$

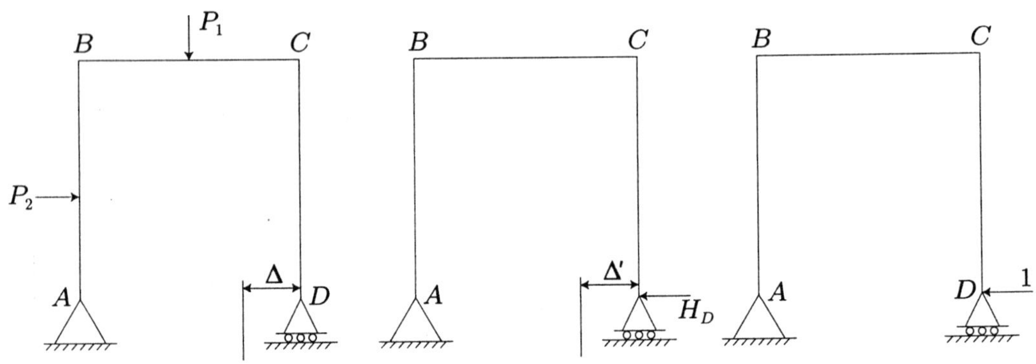

(a) Reduced frame (b) Reduced frame with H_D (c) Reduced frame with $H_D = 1$

Fig. 7.2: Reduced frame with loading

The same expression can also be obtained by strain-energy method according to which

$$\frac{\partial U}{\partial H_D} = 0, \quad \text{whence} \quad \sum \frac{M}{EI} \cdot \frac{\partial M}{\partial H_D} \cdot ds = 0, \tag{c}$$

where M = net bending moment at any point of the actual (indeterminate) structure.

Following the above notations, M can be expressed as

$$M = M_s + mH_D, \quad \text{whence} \quad \frac{\partial M}{\partial H_D} = m$$

and Eq. (c) reduces to, $\sum \dfrac{(M_s + mH_D)(m)ds}{EI} = 0,$

whence $\sum \dfrac{M_s m ds}{EI} + H_D \sum \dfrac{m^2 ds}{EI} = 0,$

from which H_D is obtained as in Eq. (7.1).

The deductions made above do not take into consideration the deformation due to axial force and shear, such effects being rather small with respect to the deformation due to flexure. Equations, however, can be obtained taking into consideration such effects.

Example 7.1 *Analyse the two-hinged portal frame shown in Fig. 7.3(a) for the loads and values of moments of intertia shown.*

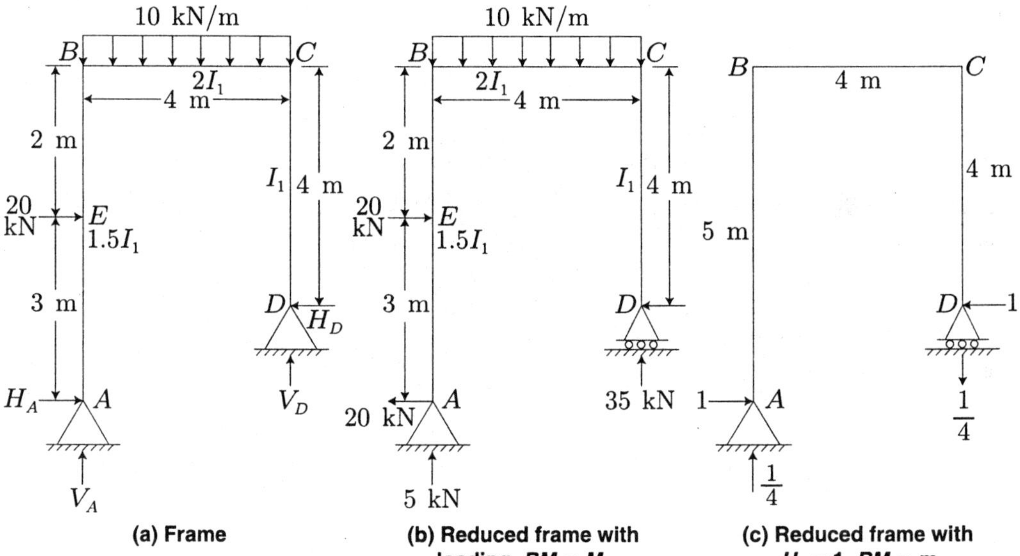

Fig. 7.3: For Example 7.1

Solution : Considering the horizontal reaction H_D at D as the redundant reaction, the reduced determinate structure with the applied loads is as shown in Fig. 7.3(b). Fig. 7.3(c) shows the reduced determinate structure subjected to unit value of H_D. Hence Figs. 7.3(b) and 7.3(c) show the loading conditions for the bending moments M_s and m of Eqn. (7.1). The general expressions for M_s and m for various portions of the frame may be tabulated as below for a systematic computation. The bending moment causing tension inside is taken as positive.

Portion of the structure	Origin	Limits (m)	M_s kN-m	m	ds	EI
DC	D	0 to 4	0	$-x$	dx	EI_1
CB	C	0 to 4	$35x - 5x^2$	$-\frac{1}{4}x - 4$	dx	$2EI_1$
BE	B	0 to 2	60	$x - 5$	dx	$1.5EI_1$
EA	A	0 to 3	$20x$	$-x$	dx	$1.5EI_1$

Summing up,

$$\sum \frac{M_s m ds}{EI} = \frac{1}{EI_1}\int_0^4 (0)(-x)dx + \frac{1}{2EI_1}\int_0^4 (35x - 5x^2)\left(-\frac{x}{4} - 4\right)dx + \frac{1}{1.5EI_1}\int_0^2 (60)(x-5)dx + \frac{1}{1.5EI_1}\int_0^3 (20x)(-x)dx$$

$$= -\frac{920}{EI_1}$$

$$\sum \frac{m^2 ds}{EI} = \frac{1}{EI_1}\int_0^4 x^2 dx + \frac{1}{2EI_1}\int_0^4 \left(\frac{x}{4} + 4\right)^2 dx + \frac{1}{1.5EI_1}\int_0^2 (x-5)^2 dx + \frac{1}{1.5EI_1}\int_0^3 x^2 dx$$

$$= \frac{90}{EI_1}$$

and $$H_D = -\frac{\sum \frac{M_s m' ds}{EI}}{\sum \frac{m^2 ds}{EI}} = -\frac{\frac{-920}{EI_1}}{\frac{90}{EI_1}} = 10\frac{2}{9} \text{ kN.}$$

The positive value of H_D indicates that the direction of H_D is the same as that assumed, i.e., H_D acts towards left.

From $\sum H = 0,$

$$H_A - H_D + 20 = 0$$

and $$H_A = -20 + 10\frac{2}{9} = -9\frac{7}{9} \text{ kN (acting towards left).}$$

Taking moment about A,

$$V_D(4) + \left(10\frac{2}{9}\right)(1) - \frac{(10)(4)^2}{2} - (20)(3) = 0,$$

whence $V_D = 32.44$ kN.

From $\sum V = 0$

$$V_A = (10)(4) - 32.44 = 7.56 \text{ kN.}$$

Fig. 7.4 shows free body diagrams helpful for drawing BM, SF and axial force diagrams, which are shown in Figs. 7.5, 7.6 and 7.7 respectively.

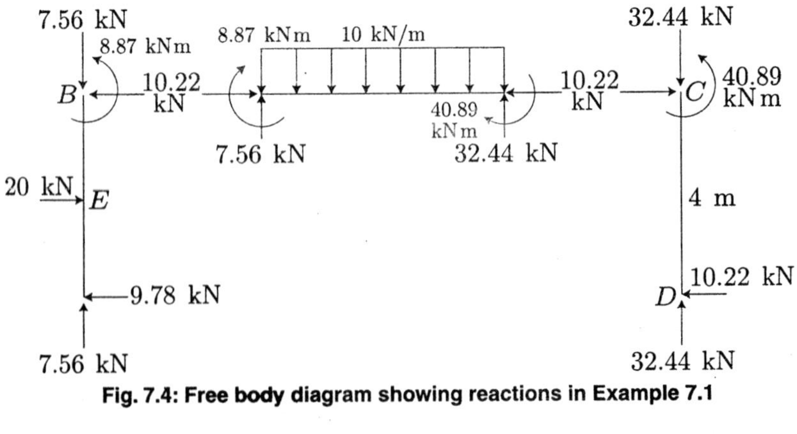

Fig. 7.4: Free body diagram showing reactions in Example 7.1

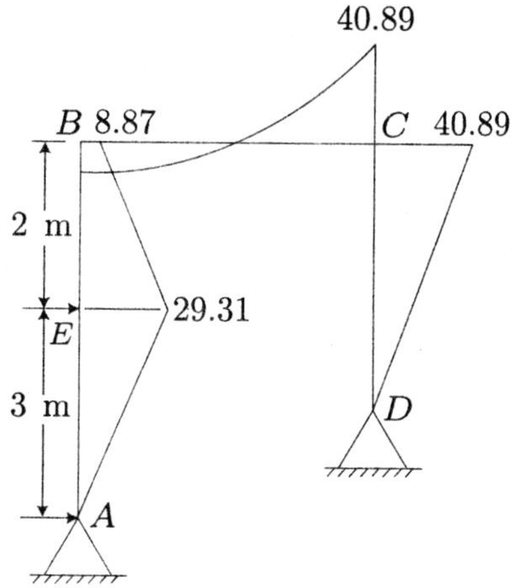

Fig. 7.5: B.M. in Example 7.1

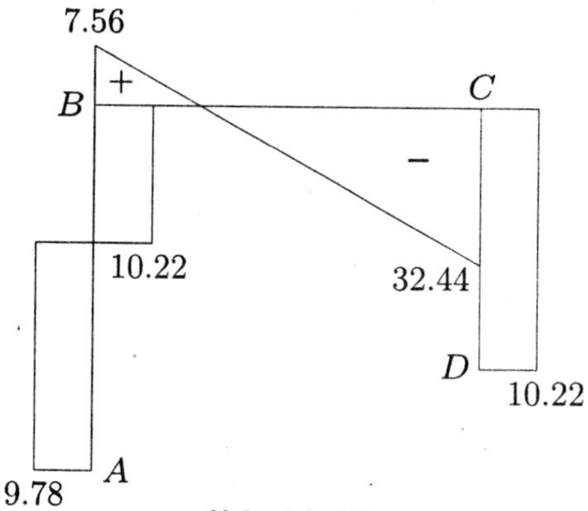

Fig. 7.6: S.F. diagram for Example 7.1

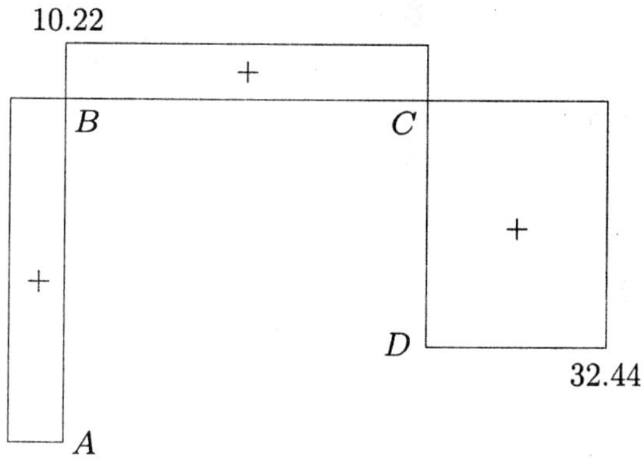

Fig. 7.7: Axial force diagram for Example 7.1 (+ve indicates compression)

As mentioned earlier, the analysis does not take into consideration deformation due to axial force and shear, but considers only flexural deformations. This does not affect the resulting values of the redundant H_D considerably. When all the effects are included in analysing the structure, it is considered that composite effect has been considered.

7.3 Two-Hinged Arch

The analysis of a two-hinged arch is similar to that of a two-hinged portal frame. For the two-hinged arch AB shown in Fig. 7.8(a), considering the horizontal thrust H_B at support B as redundant, the actual arch becomes equivalent to sum of the two cases shown in Fig. 7.8(b) and 7.8(c) so that

$$\Delta + \Delta' = 0.$$

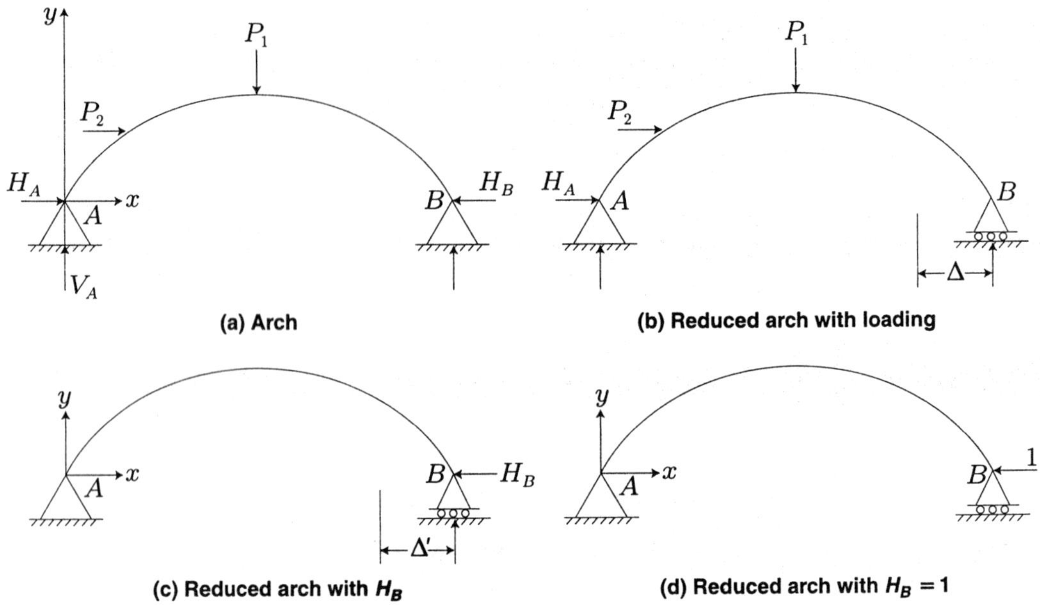

Fig. 7.8: Two-hinged arch

Following the same procedure as in Section 7.2 and considering flexure only, H_B can be obtained as under :

$$H_B = -\frac{\sum \frac{M_s m ds}{EI}}{\sum \frac{m^2 ds}{EI}},$$

where M_s and m are bending moments at any section for loading conditions shown in Figs. 7.8(b) and 7.8(d) respectively. From Fig. 7.8(d) with the positive directions of co-ordinate axes as shown and taking moment causing tension inside as positive,

$$m = -y,$$

which reduces the above expression for H_B to

$$H_B = +\frac{\sum \frac{M_s y ds}{EI}}{\sum \frac{y^2 ds}{EI}}. \tag{7.2}$$

Eq. (7.2) can be modified further, if the shape of the arch is known. In the following Section some special forms of arches are considered.

7.4 Two-Hinged Circular Arch

Fig. 7.9 shows a circular arch of span $2L$, which subtends an angle $2a$ at the centre. It is subjected to any given system of loads and has uniform cross section so that flexural rigidity EI is constant. Length ds is a small element of the arch at an angular distance θ from the support B and subtends an angle $d\theta$ at the centre. The co-ordinate y of the element measured from the chord AB is given by

$$y = R(\cos\theta - \cos\alpha),$$

where R is the radius of the arch.

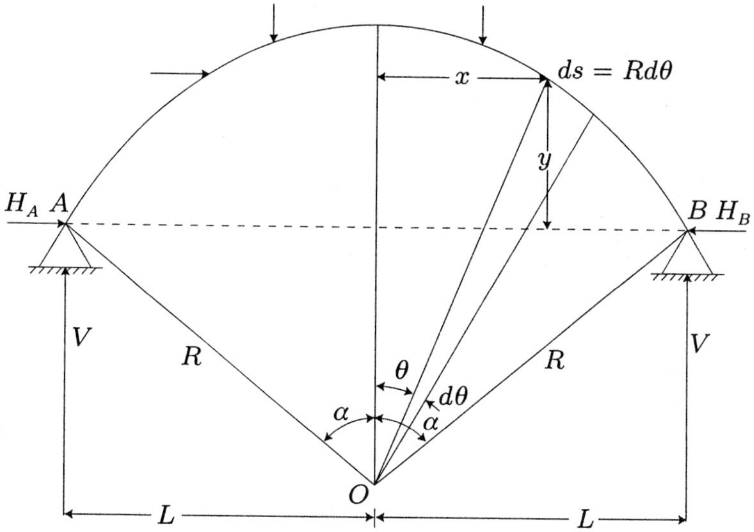

Fig. 7.9: Circular two-hinged arch

As $ds = Rd\theta$ and EI = constant, Eq. (7.2) may be written as

$$H_B = \frac{\sum M_s y ds}{\sum y^2 ds} = \frac{\sum M_s(\cos\theta - \cos\alpha)d\theta}{R\sum(\cos\theta - \cos\alpha)^2 d\theta}. \tag{7.3}$$

The expression for M_s depends on actual loading on arch.

Example 7.2. *A two-hinged circular arch of uniform section having span* 60 m *and central rise of* 10.50 m *is subjected to a concentrated load of* 10 kN *at the crown as shown in Fig. 7.10. The relevant moment of inertia of the arch section is* $16.65(10)^8$ mm^4 *and modulus of elasticity* $E = 2(10)^5$ N/mm^2. *Determine the horizontal thrust at each support and the bending moment at the crown.*

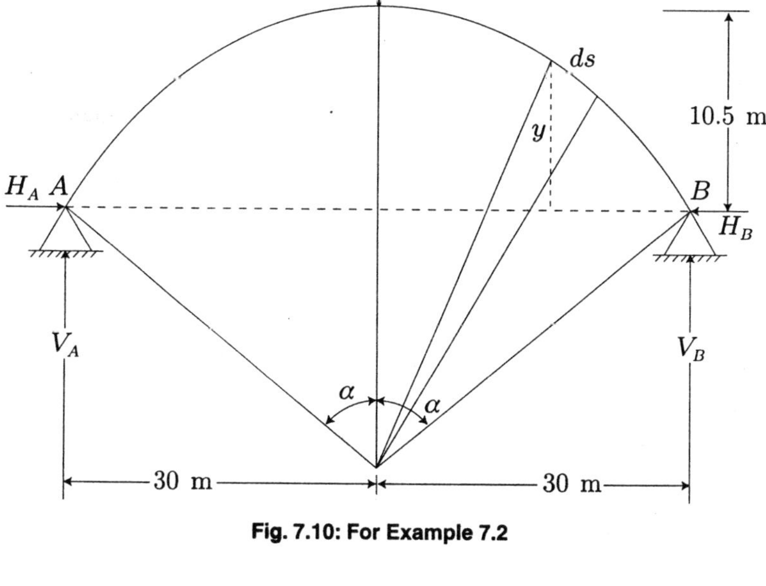

Fig. 7.10: For Example 7.2

Solution : From $\sum H = 0$, as there is no applied horizontal force,

$$H_A = -H_B = H.$$

Since $EI = \text{constant}$,

$$H = \frac{\sum M_s y ds}{\sum y^2 ds}.$$

Considering an element ds of the arch at an angular distance θ from the crown,

$$y = R(\cos\theta - \cos\alpha) \quad \text{and} \quad ds = Rd\theta.$$

Considering symmetry,

$$H = \frac{2\int_0^\alpha M_s R(\cos\theta - \cos\alpha).Rd\theta}{2\int_0^\alpha R^2(\cos\theta - \cos\alpha)^2.Rd\theta} = \frac{\int_0^\alpha M_s(\cos\theta - \cos\alpha)d\theta}{R\int_0^\alpha(\cos\theta - \cos\alpha)^2 d\theta}.$$

Radius R and angle α may be obtained from given data.

$$\text{Span} = 2R\sin\alpha = 60 \text{ m} \qquad \text{Rise} = R(1-\cos\alpha) = 10.5 \text{ m}.$$

Solving, $R = 48.11$ m and $\alpha = 38.53° = 0.673348$ radian

$$\sin\alpha = \sin 38.58° = 0.623607 \qquad \cos\alpha = \cos 38.58° = 0.781738.$$

Considering symmetry,

$$V_A = V_B = \frac{10}{2} = 5 \text{ kN}.$$

The bending moment M_s at the element ds for the simply supported condition as shown in Fig. 7.11 is

$$M_s = (5)(x) = (5)(30 - R\sin\theta).$$

Also $$\int_0^\alpha M_s(\cos\theta - \cos\alpha)d\theta = \int_0^\alpha 5(30 - R\sin\theta)(\cos\theta - \cos\alpha)d\theta$$

$$= \int_0^\alpha 5\left(30\cos\theta - \frac{1}{2}R\sin 2\theta - 3\theta\cos\alpha + R\cos\alpha\sin\theta\right)d\theta$$

$$= 5\left[30\sin\alpha + \frac{R}{4}(\cos 2\alpha - 1) - 30(\alpha)\cos\alpha - R\cos\alpha(\cos\alpha - 1)\right] = 8.854,$$

for $\alpha = 0.673348$ radian, and $R = 48.11$ m.

Also $$\int_0^\alpha (\cos\theta - \cos\alpha)^2 d\theta = \int_0^\alpha (\cos^2\theta - 2\cos\alpha\cos\theta + \cos^2\alpha)d\theta$$

$$= \frac{\sin 2\alpha}{4} + \frac{\alpha}{2} - 2\cos\alpha\sin\alpha + (\alpha)\cos^2\alpha = 0.016921.$$

$$H = \frac{\int_0^\alpha M_s(\cos\theta - \cos\alpha)d\theta}{R\int_0^\alpha(\cos\theta - \cos\alpha)^2 d\theta} = \frac{8.854}{(48.11)(0.016921)} = 10.876 \text{ kN}.$$

The bending moment at crown $= (5)(30) - (10.876)(10.5) = 35.802$ kNm.

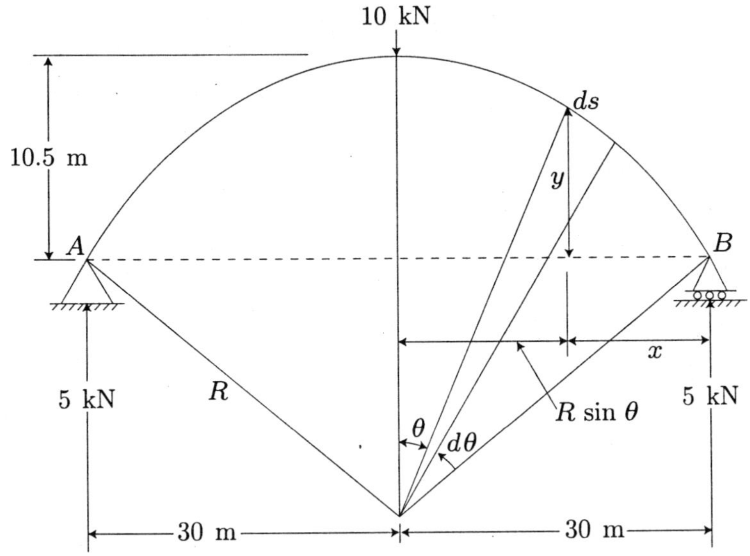

Fig. 7.11: For Example 7.2

Example 7.3. *Determine the horizontal thrust at support of the two-hinged circular arch of Example 7.2 if the concentrated load acts at left quarter-span point.*

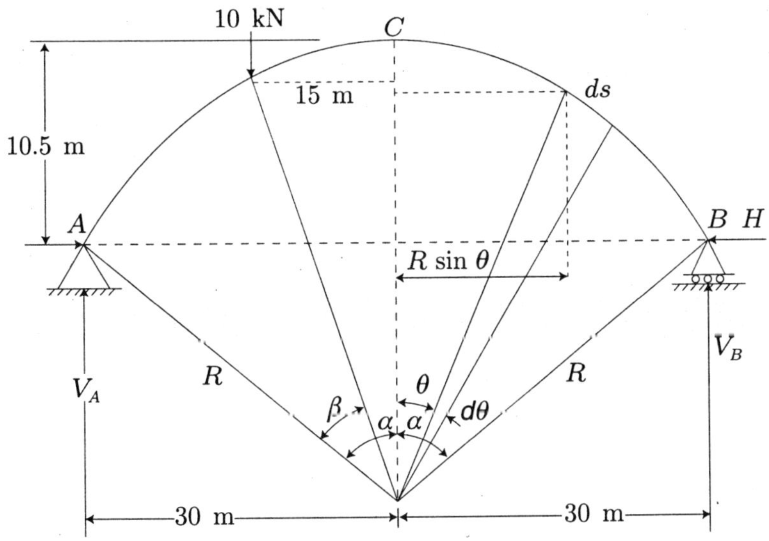

Fig. 7.12: For Example 7.3

Solution : Fig. 7.12 shows the determinate structure with the applied load. From previous Example, $R = 48.11$ m; $\alpha = 38.58° = 0.673348$ radian.

If the angular distance of the load from the crown be β, then

$$R \sin \beta = 15,$$

whence $\beta = \sin^{-1} \dfrac{15}{48.11} = 18.167° = 0.31707$ radian

$$V_A = \frac{(10)(45)}{60} = 7.5 \text{ kN} \quad \text{and} \quad V_B = 10 - 7.5 = 2.5 \text{ kN}.$$

The bending moment at a point in portion CB at an angle θ from C is

$$M_s = V_B[R\sin\alpha - R\sin\theta] = 2.5R(\sin\alpha - \sin\theta),$$

which is valid for $\theta = 0$ to α.

Similarly, the bending moment at a point in portion AD at an angle θ from C is

$$M_s = V_A[R\sin\alpha - R\sin\theta] = 7.5R(\sin\alpha - \sin\theta),$$

which is valid for $\theta = \beta$ to α.

The bending moment at a point in portion CD at an angle θ from C is

$$M_s = V_A R(\sin\alpha - \sin\theta) - 10(R\sin\alpha - R\sin\theta - 15) = +150 - 2.5R(\sin\alpha - \sin\theta),$$

which is valid for $\theta = 0$ to β.

Taking $y = R(\cos\theta - \cos\alpha)$ and $ds = Rd\theta$,

$$\begin{aligned}
\sum M_s y ds &= 2.5R^3 \int_0^\alpha (\sin\alpha - \sin\theta)(\cos\theta - \cos\alpha)d\theta \\
&\quad + 7.5R^3 \int_0^\alpha (\sin\alpha - \sin\theta)(\cos\theta - \cos\alpha)d\theta \\
&\quad + \int_0^\beta \{150 - 2.5R(\sin\alpha - \sin\theta)\}\, R(\cos\theta - \cos\alpha)\cdot Rd\theta \\
&= 2.5R^3 \left[\sin\alpha\sin\theta + \frac{\cos 2\theta}{4} - (\theta)\sin\alpha\cos\alpha - \cos\alpha\cos\theta\right]_\beta^\alpha \\
&\quad + 7.5R^3 \left[\sin\alpha\sin\theta + \frac{\cos 2\theta}{4} - (\theta)\sin\alpha\cos\alpha - \cos\alpha\cos\theta\right]_\beta^\alpha \\
&\quad - 2.5R^3 \left[\sin\alpha\sin\theta + \frac{\cos 2\theta}{4} - (\theta)\sin\alpha\cos\alpha - \cos\alpha\cos\theta\right]_0^\beta \\
&\quad + 150R^2\left[\sin\theta - (\theta)\cos\alpha\right]_0^\beta \\
&= 10R^3\left[\sin^2\alpha + \frac{\cos 2\alpha}{4} - (\alpha)\sin\alpha\cos\alpha - \cos^2\alpha\right] \\
&\quad - 10R^3\left[\sin\alpha\sin\beta + \frac{\cos 2\beta}{4} - (\beta)\sin\alpha\cos\alpha - \cos\alpha\cos\beta\right] \\
&\quad + 150R^2[\sin\beta - (\beta)\cos\alpha] \\
&= 10R^3[-0.4949] - 10R^3[-0.5015] + 150R^2[0.0639] \\
&= 0.066(48.11)^3 + (150)(48.11)^2(0.0639) = 29534.5.
\end{aligned}$$

$$\begin{aligned}
\sum y^2 ds &= 2R^3 \int_0^\alpha (\cos\theta - \cos\alpha)^2 d\theta \\
&= 2R^3(0.016921) = (2)(48.11)^3(0.016921) = 3768.4.
\end{aligned}$$

$$H_B = \frac{\sum M_s y ds}{\sum y^2 ds} = \frac{29534.5}{3768.4} = 7.837 \text{ kN}.$$

7.5 Two-Hinged Parabolic Arch

For the parabolic arch shown in Fig. 7.13(a) the equation of the arch axis with origin at a hinge support is

$$y = \frac{4y_c}{L^2} \cdot x(L-x), \tag{7.4}$$

where y_c = value of y co-ordinate of the crown C and L = span of the arch.

The cross section of a parabolic arch usually varies and in the subsequent analysis, the variation is taken to be such that the moment of inertia I at any point is given by

$$I = I_c \sec\alpha,$$

where I_c = moment of inertia of the section at crown, α = angle made by the tangent to the arch axis at the point with the horizontal.

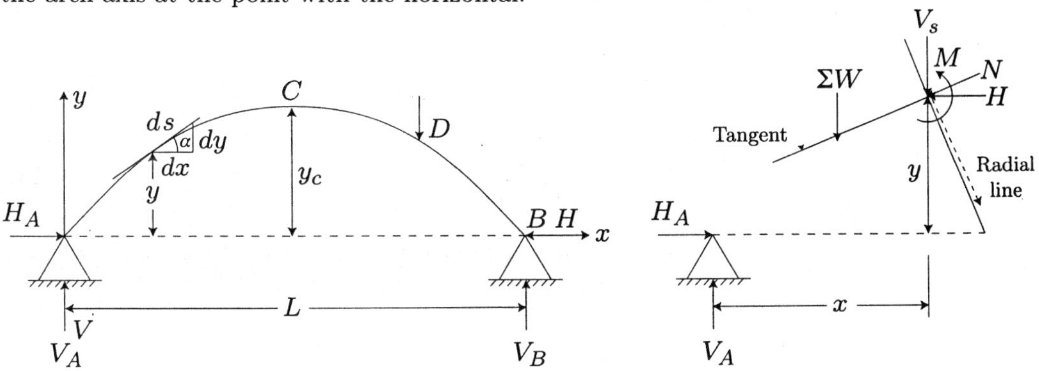

(a) Arch (b) Free body diagram showing forces

Fig. 7.13: Two-hinged parabolic arch

Since $\sec\alpha = \sqrt{1+\tan^2\alpha} = \sqrt{1+\left(\frac{dy}{dx}\right)^2} = \sqrt{\frac{dx^2+dy^2}{dx^2}} = \frac{ds}{dx}$ and $I = I_c\left(\frac{ds}{dx}\right)$

Hence, $\sum\frac{M_s y ds}{EI} = \sum\frac{M_s y ds}{EI_c\cdot\frac{ds}{dx}} = \sum\frac{M_s y dx}{EI_c}$ and $\sum\frac{y^2 ds}{EI} = \sum\frac{y^2 dx}{EI_c}$

It follows that,

$$H = \frac{\sum\frac{M_s y ds}{EI}}{\sum\frac{y^2 ds}{EI}} = \frac{\sum\frac{M_s y dx}{EI_c}}{\sum\frac{y^2 dx}{EI_c}} = \frac{\sum M_s y dx}{\sum y^2 dx}. \tag{7.5}$$

With reference to Fig. 7.13(b), the bending moment (M), shear force (S.F.) and normal thrust (M) at a horizontal distance x are given by

$$M = M_s - Hy \tag{7.6}$$

$$\text{S.F.} = V_s\cos\alpha - H\sin\alpha \tag{7.7}$$

$$N = V_s\sin\alpha + H\cos\alpha, \tag{7.8}$$

where M_s and V_s are bending moment and shear force at a horizontal distance x from origin (a support) as simply supported beam, first term M_s of Eq. (7.6) representing the simple beam bending moment and the second term represents contribution due to arch action. For a parabolic arch, term Hy is represented by a parabola as H is constant and y is related parabolically with x. For a concentrated load W acting at the crown as in Fig. 7.14(a), the bending moment at x from support is

$$M = \frac{W}{2} \cdot x - Hy.$$

Figs. 7.14(b) and 7.14(c) show B.M. diagrams due to simple beam action and arch action respectively. The total B.M. diagram is obtained by superimposing the above two diagrams as shown in Fig. 7.14(d).

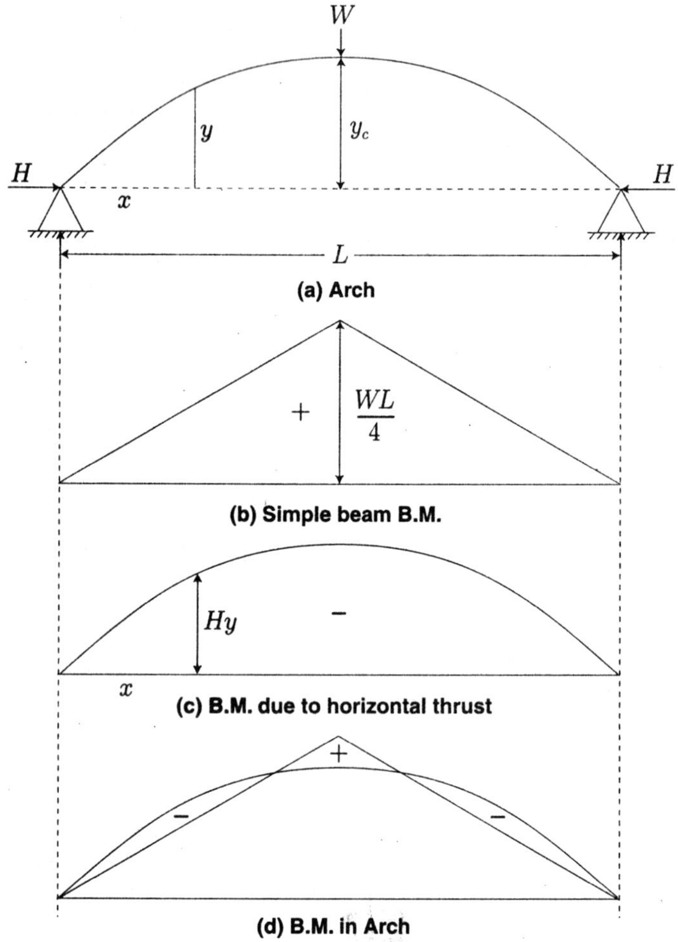

Fig. 7.14: Two-hinged parabolic arch

The shear force at x from the support is

S.F. $= V_A \cdot \cos\alpha - H \sin\alpha =$ (Simple beam shear) $\cos\alpha -$ (Horizontal thrust) $\sin\alpha$.

For a concentrated load acting away from the crown as in Fig. 7.15(a), the S.F. diagram is obtained by superimposing the diagrams due to simple beam shear and horizontal thrust as shown in Figs. 7.15(b) and 7.15(c) respectively. Fig. 7.15(d) shows a typical total *S.F.* diagram.

The normal thrust at x from support is

$N = V_A \sin\alpha + H \cos\alpha =$ (Simple beam shear) $\sin\alpha +$ (Horizontal thrust) $\cos\alpha$.

The total moment thrust diagram is shown in Fig. 7.15(g) which is obtained by superimposing the S.F. diagrams due to simple beam shear and horizontal thrust respectively as shown in Figs. 7.15(e) and 7.15(f).

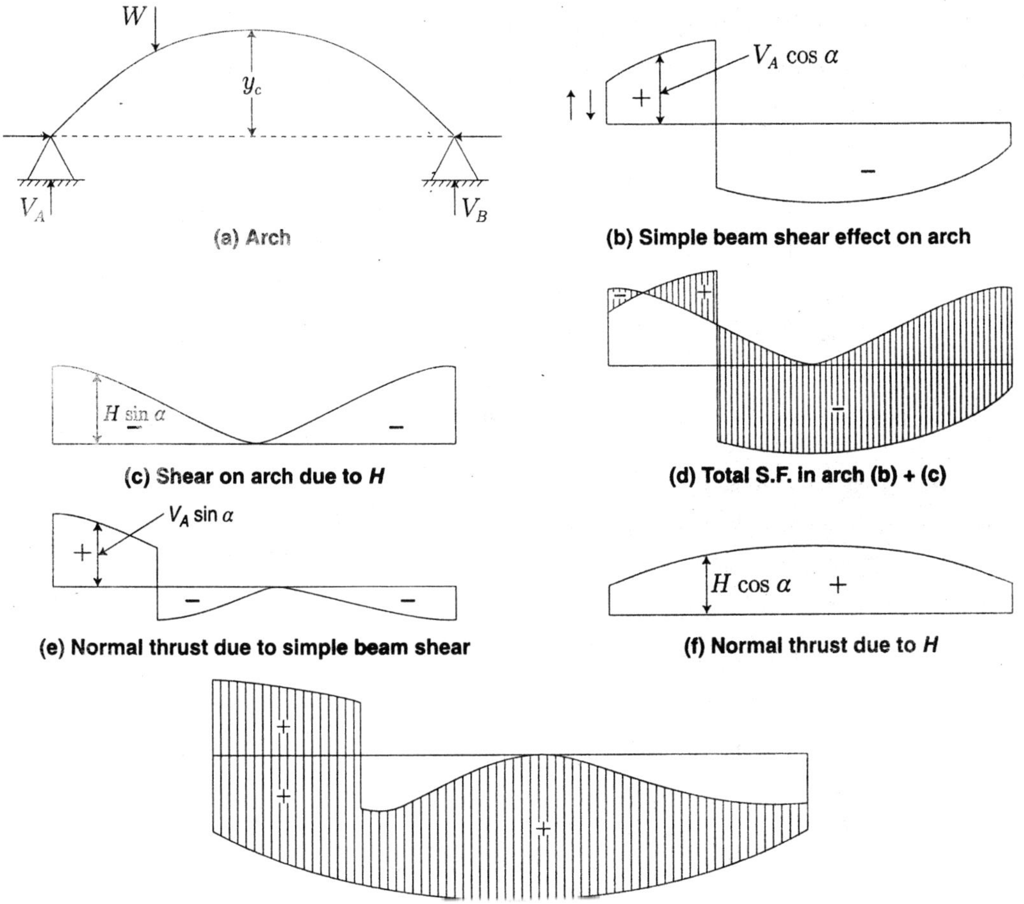

(a) Arch (b) Simple beam shear effect on arch

(c) Shear on arch due to H (d) Total S.F. in arch (b) + (c)

(e) Normal thrust due to simple beam shear (f) Normal thrust due to H

(g) Total normal thrust in arch

Fig. 7.15: Two-hinged parabolic arch

Example 7.4. *A two-hinged parabolic arch of span L and central rise y_c is loaded with a concentrated load W at the crown. Determine the horizontal thrust at the supports and B.M. diagram for the arch. Given $I = I_c \sec \alpha$.*

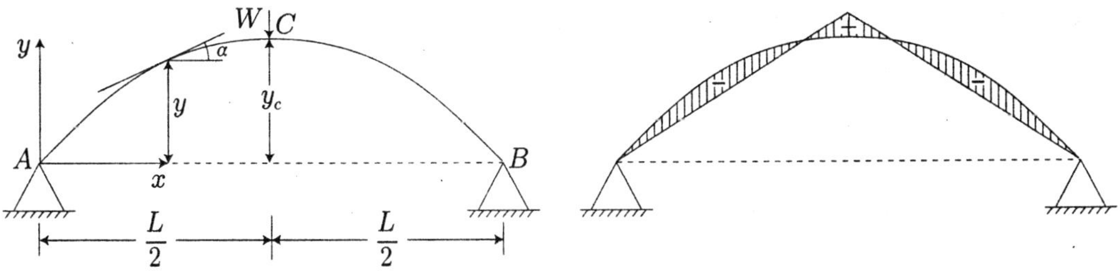

(a) Arch with $I = I_c \sec \alpha$ (b) B.M. Diagram

Fig. 7.16: Two-hinged parabolic arch

Solution : With reference to Fig. 7.16, considering the left support as origin and positive directions of axes as shown, at a distance x from origin,

$$y = \frac{4y_c}{L^2}x(L-x)$$

$$M_s = \frac{W}{2}(x)$$

and $$\sum M_s y dx = 2\int_0^{\frac{L}{2}} \frac{W}{2}x \cdot \frac{4y_c}{L^2}x(L-x)dx = \frac{4Wy_c}{L^2}\int_0^{\frac{L}{2}} x^2(L-x)dx = \frac{Wy_cL^2}{8}.$$

$$\sum y^2 dx = 2\int_0^{\frac{L}{2}} \left[\frac{4y_c}{L^2}x(L-x)\right]^2 dx = \frac{8}{15}y_c^2 L.$$

$$H = \frac{\sum M_s y dx}{\sum y^2 dx} = \frac{\frac{Wy_cL^2}{8}}{\frac{8}{15}y_c^2 L} = \frac{15}{64}W \cdot \frac{L}{y_c}.$$

Fig. 7.16(b) shows the B.M. diagram for the arch.

Example 7.5. *A two-hinged symmetrical parabolic arch having $I = I_c \sec\alpha$, is subjected to a concentrated load W at a horizontal distance $x = kL$, where L is the span of the arch and k is a numerical factor as shown in Fig. 7.17(a). Determine the value of the horizontal thrust at the supports and draw B.M. diagram for the arch.*

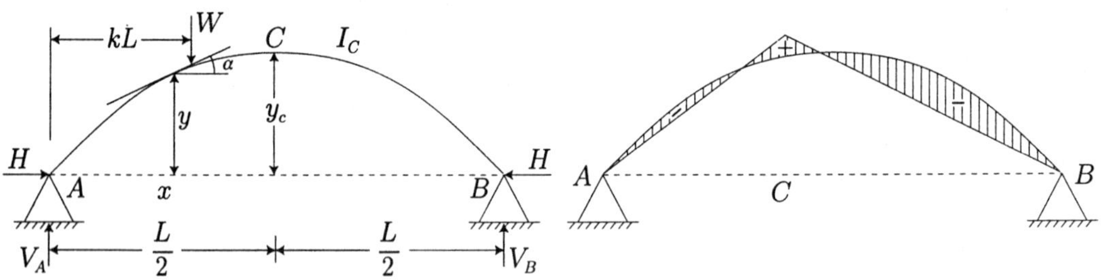

(a) Arch with $I = I_C \sec\alpha$ **(b) B.M. in Arch**

Fig. 7.17: For Example 7.5

Solution : With reference to Fig. 7.17(a)

$$y = \frac{4y_c}{L^2} \cdot x(L-x)$$

$$V_A = \frac{W(L-kL)}{L} = W(1-k)$$

$$V_B = kW.$$

$$\sum M_s y dx = \int_0^{kL} W(1-k)x \cdot \frac{4y_c}{L^2} \cdot x(L-x)dx + \int_0^{(1-k)L} kWx \cdot \frac{4y_c}{L^2} \cdot x(L-x)dx$$

$$= \frac{4Wy_c(1-k)}{L^2}\int_0^{kL}(x^2L - x^3)dx + \frac{4Wy_ck}{L^2}\int_0^{(1-k)L}(x^2L-x^3)dx$$

$$= \frac{Wy_ckL^2}{3}(1-2k^2+k^3)$$

$$\sum y^2 dx = \int_0^L \left[\frac{4y_c}{L^2} \cdot x(L-x)\right]^2 dx = \frac{8}{15}y_c^2 L$$

and $$H = \frac{\sum M_s y dx}{\sum y^2 dx} = \frac{\frac{1}{3} W y_c k L^2 (1 - 2k^2 + k^3)}{\frac{8}{15} \cdot y_c^2 L},$$

whence $$H = \frac{5}{8} \cdot \frac{WkL}{y_c}(1 - 2k^2 + k^3).$$

Fig. 7.17(b) shows B.M. diagram for the arch.

Example 7.6. *A two-hinged symmetrical parabolic arch having $I = I_c \sec \alpha$ and length L is subjected to a uniformly distributed load of w/unit length as in Fig. 7.18. It is required to determine the horizontal thrust at supports and to draw bending moment and shear force diagrams.*

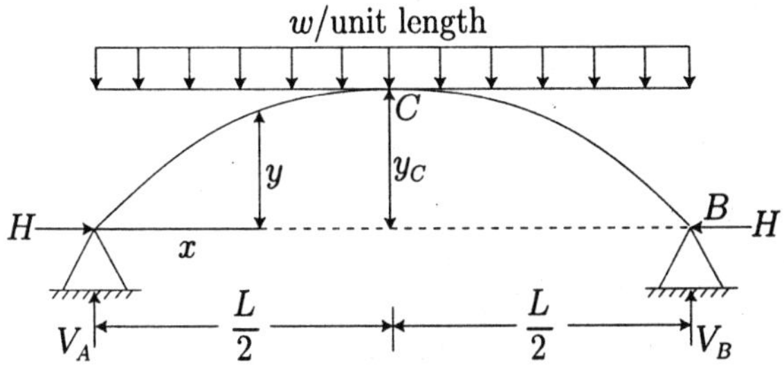

Fig. 7.18: For Example 7.6

Solution : $$H = \frac{\sum M_s y ds}{\sum y^2 dx}$$

In this case, $$y = \frac{4y_c}{L^2}[x(L - x)]$$

$$V_A = V_B = \frac{wL}{2}$$

$$M_s = \frac{wL}{2} \cdot x = \frac{wx^2}{2}$$

$$\sum M_s y dx = \int_0^L \left(\frac{wL}{2} \cdot x - \frac{wx^2}{2}\right) \cdot \frac{4y_c}{L^2} \cdot x(L - x) dx$$

$$= \frac{2wy_c}{L^2} \int_0^L (L^2x^2 - 2Lx^3 + x^4) dx = \frac{wL^3}{15} y_c.$$

$$y^2 dx = \int_0^L \left[\frac{4y_c}{L^2} \cdot x(L - x)\right]^2 dx = \frac{8}{15} \cdot y_c^2 L.$$

$$H = \frac{\sum M_s y dx}{\sum y^2 dx} = \frac{\frac{wL^3}{15} \cdot y_c}{\frac{8}{15} \cdot y_c^2 L} = \frac{wL^2}{8y_c},$$

i.e., $$H = \frac{\text{Simple beam bending moment at centre}}{y_c}.$$

The bending moment at any point at a horizontal distance x from a support is

$$M = \frac{wL}{2} \cdot x - \frac{wx^2}{2} - H(y) = \frac{wL}{2} x - \frac{wx^2}{2} - \frac{wL^2}{8y_c} \cdot \frac{4y_c}{L^2} \cdot x(L - x) = 0.$$

Thus the bending moment at any point of a parabolic arch due to uniformly distributed load is zero.

The shear force at any section at a distance x from a support A is

$$\text{S.F.} = V_A \cos\alpha - H \sin\alpha$$

Now, $\tan\alpha = \dfrac{dy}{dx} = \dfrac{4y_c}{L^2}(L - 2x)$

and $\sin\alpha = \tan\alpha \cdot \cos\alpha = \dfrac{4y_c}{L^2}(L - 2x)\cos\alpha.$

Then $\text{S.F.} = V_A \cos\alpha - H \sin\alpha = \left(\dfrac{wL}{2} - wx\right)\cos\alpha - \dfrac{wL^2}{8y_c} \cdot \dfrac{4y_c}{L^2}(L - 2x)\cos\alpha = 0.$

Then S.F. at any section due to u.d.l. is zero.

7.6 Yielding of Supports of Two-Hinged Arch

The effect of lateral yielding of a support of a two-hinged arch as shown in Fig. 7.19 can be easily determined using Castigliano's theorem.

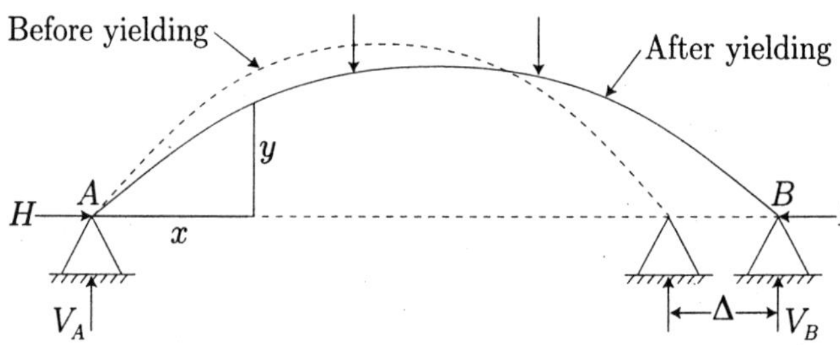

Fig. 7.19: Yielding of support in two-hinged arch

The lateral displacement Δ against the direction of the horizontal thrust H is given by

$$\frac{\partial U}{\partial H} = -\Delta, \quad \text{whence } \sum \frac{M}{EI} = \frac{\partial M}{\partial H} \cdot ds = -\Delta, \tag{a}$$

where U is the total strain-energy of the arch, M is the bending moment at any section and EI is the flexural rigidity of the arch at the section.

Since $M = M_s - H \cdot y$

and $\dfrac{\partial M}{\partial H} = -y,$

where M_s is the bending moment at a distance x from a support, the expression (a) can be written as

$$\frac{(M_s - H \cdot y)(-y)ds}{EI} = -\Delta,$$

whence $$H = \frac{\sum \frac{M_s y ds}{EI} - \Delta}{\sum \frac{y^2 ds}{EI}}. \tag{7.9}$$

If the support is elastic, so that Δ is proportional to H, i.e., $= kH$, where k is a constant, then

$$H = \frac{\sum \frac{M_s y ds}{EI}}{\sum \frac{y^2 ds}{EI} + k}. \tag{7.10}$$

7.7 Effect of Rib-shortening in Two-Hinged Arch

As in Fig. 7.13(b), an arch section is subjected to bending moment, shear force and normal thrust. The normal thrust in an arch causes shortening of its axial length. With reference to Fig. 7.20, the normal thrust N and bending moment M in a section at a horizontal distance x from origin (support A) is given by

$$N = V_s \sin\alpha + H\cos\alpha$$

and $\quad M = M_s - Hy, \qquad$ (a)

where M_s and V_s are the bending moment and shear force as in simply supported beam. Considering both flexure and normal thrust, the strain-energy is given by

$$U = \sum \frac{M^2 ds}{2EI} + \sum \frac{N^2 ds}{2AE},$$

where A is the sectional area of the arch.

Since $\dfrac{\partial U}{\partial H} = 0$, therefore,

$$\sum \frac{M}{EI} \cdot \frac{\partial M}{\partial H} \cdot ds + \sum \frac{N}{AE} \cdot \frac{\partial N}{\partial H} \cdot ds = 0.$$

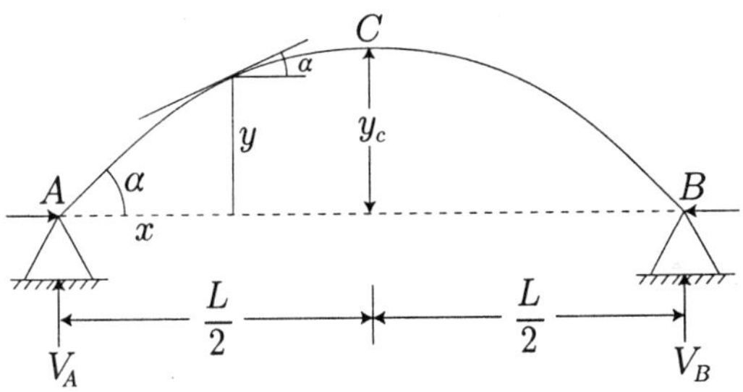

Fig. 7.20: Rib-shortening in two-hinged arch

Substituting for M and N from Eq. (a) and putting $\frac{\partial M}{\partial H} = -y$ and $\frac{\partial N}{\partial H} = \cos\alpha$, the above expression reduces to

$$\sum \frac{(M_s - Hy)(-y)ds}{EI} + \sum \frac{(V_s \sin\alpha + H\cos\alpha)\cos\alpha . ds}{AE} = 0,$$

whence $H = \dfrac{\sum \frac{M_s y ds}{EI} - \sum \frac{V_s \sin\alpha \cos\alpha ds}{AE}}{\sum \frac{y^2 ds}{EI} + \sum \frac{\cos^2 \alpha ds}{AE}}. \qquad$ (7.11a)

Neglecting the effect of V_s, which is comparatively small, horizontal thrust H is given by

$$H = \frac{\sum \frac{M_s y ds}{EI}}{\sum \frac{y^2 ds}{EI} + \sum \frac{\cos^2 \alpha ds}{AE}}. \qquad (7.11)$$

The sectional area A of an arch-rib usually varies from minimum value at the crown to some maximum value at the springings (supports) and the value of angle α which the tangent to the arch-rib makes with the horizontal decreases from some definite value at the springings to zero at the crown. Hence, the value of $\frac{A}{\cos\alpha}$ increases from a minimum at the crown to a maximum at the springings. Taking mean value of $\frac{A}{\cos\alpha}$ as A_m and assuming it to be constant throughout the entire length of the arch and putting $ds\cos\alpha = dx$,

$$\sum \frac{\cos^2 \alpha ds}{AE} = \int_0^L \frac{dx}{A_m E} = \frac{L}{A_m E},$$

where L is the span of the arch. With this, Eq. (7.11) reduces to

$$H = \frac{\sum \frac{M_s y ds}{EI}}{\sum \frac{y^2 ds}{EI} + \frac{L}{A_m E}}. \tag{7.12}$$

Eq. (7.12) above includes the effect of flexure and rib-shortening due to applied loads.

7.8 Effect of Temperature Change in Two-Hinged Arch

For the purpose of analysis, the hinged support B of the two-hinged arch of Fig. 7.21(a) is assumed to be roller-supported and the arch is allowed to extend freely due to a temperature rise of $t°$ as shown in Fig. 7.21(b).

The support B moves laterally towards right through a distance $L\alpha t$, which is the free lateral extension of the arch due to increase of temperature, α being the coefficient of thermal expansion. A horizontal force H' is now applied at B to bring the support to its original position as in the actual support condition when support B is hinged, no lateral movement of the support occurs. The magnitude of H' will then be such that it can produce a displacement $L\alpha t$ in its own direction. Then from Castigliano's theorem,

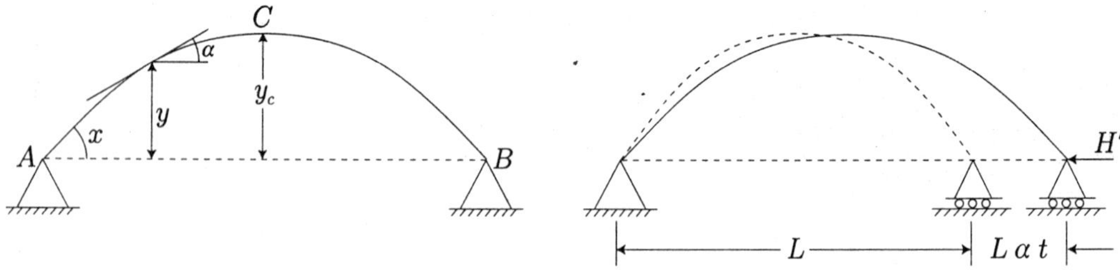

(a) Arch **(b) Effect of Temperature**

Fig. 7.21: Effect of temperature change in two-hinged arch

$$\frac{\partial U}{\partial H'} = L\alpha t. \tag{a}$$

In this case, bending moment and normal thrust at a horizontal distance x from origin (support A) are

$$M = H'y$$
$$N = H'\cos\alpha.$$

Considering the effect of flexure and normal thrust from Eq. (a),

$$\frac{\partial U}{\partial H'} = \sum \frac{M}{EI}\cdot\frac{\partial M}{\partial H'}ds + \sum \frac{N}{AE}\cdot\frac{\partial N}{\partial H'}ds = L\alpha t,$$

i.e., $$\sum \frac{(-H'.y)(-y)ds}{EI} + \sum \frac{(H'\cos\alpha)(\cos\alpha)ds}{AE} = L\alpha t,$$

whence $$H' = \frac{L\alpha t}{\sum \frac{y^2 ds}{EI} + \sum \frac{\cos^2 \alpha ds}{AE}}. \tag{7.13}$$

Using mean value of $\frac{A}{\cos \alpha}$ as A_m and taking it to be constant as an approximation, the horizontal thrust H' may be given by

$$H' = \frac{L\alpha t}{\sum \frac{y^2 ds}{EI} + \frac{L}{A_m E}}. \tag{7.14}$$

Eq. (7.14) gives horizontal thrust due to temperature rise only. In case of a temperature drop, the direction of H' will be reversed and it will be a pull rather than a thrust.

7.9 Arch with Tie-Bar

Fig. 7.22(a) shows a simply supported arch but provided with a tie-bar connecting the supports. Such a truss is externally determinate, but internally indeterminate. The analysis of such an arch is similar to that of a two-hinged arch with support yielding. In such a case, the roller support is displaced laterally through a distance equal to the extension of the bar as shown in Fig. 7.22(b). If H be the tension in the tie bar, its extension equals $\frac{HL}{A_t E_t}$, where A_t and E_t are the sectional area and modulus of elasticity of the tie-bar. This case is similar to a two-hinged arch with lateral support yielding of $\frac{HL}{A_t E_t}$. Considering lateral support yielding and effect of flexure and normal thrust, from Eqs. (7.10) and (7.12),

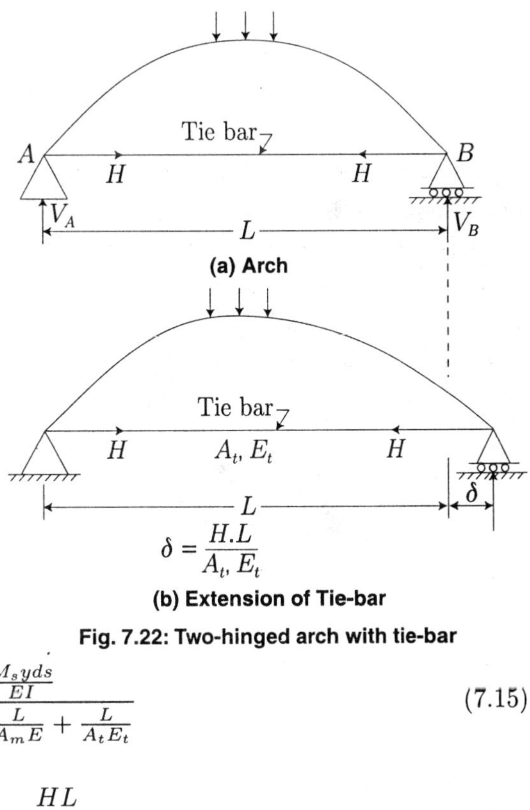

Fig. 7.22: Two-hinged arch with tie-bar

$$H = \frac{\sum \frac{M_s y ds}{EI}}{\sum \frac{y^2 ds}{EI} + \frac{L}{A_m E} + \frac{L}{A_t E_t}} \tag{7.15}$$

which can also be derived from

$$\frac{\partial U}{\partial H} = -\frac{HL}{A_t E_t}.$$

Example 7.7. *A two-hinged parabolic arch having I-section, with moment of inertia increasing in proportion to secant of slope from crown to springings has span* 36 m *and central rise* 8 m *and carries a uniformly distributed load of* 3 kN/m *on the left half span as shown in Fig. 7.23(a) and 7.23(b). It is required to determine the magnitude of the horizontal thrust at the springings considering flexure due to load only and to draw B.M. diagram for the arch. The bending moment, shear force and normal thrust at crown also are to be determined.*

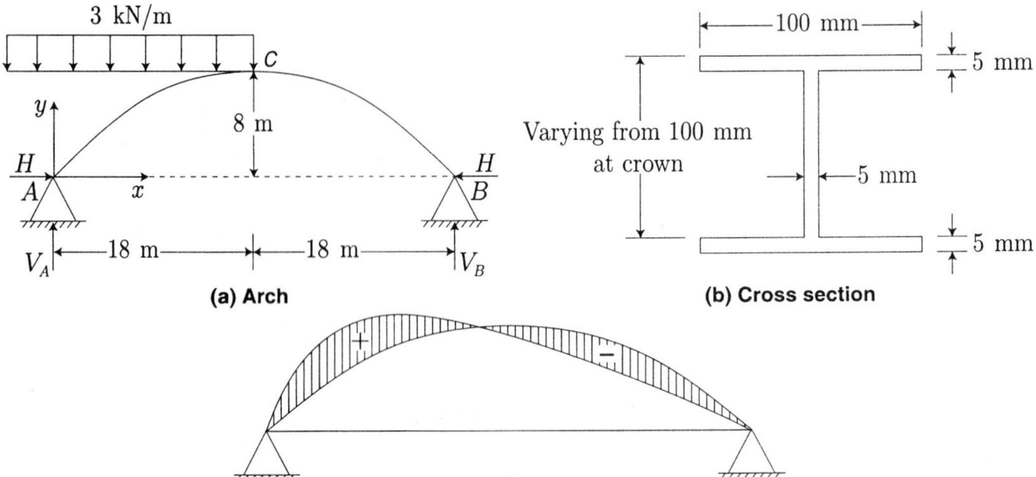

(a) Arch (b) Cross section

(c) B.M. Diagram

Fig. 7.23: For Example 7.7

How will the horizontal thrust be modified, if (i) rib-shortening is taken into account, (ii) temperature rises by 30 °C, and (iii) each support yields by 10 mm laterally?

Given, $E = 2(10)^5$ N/mm^2, and $\alpha = 12(10)^{-6}/°$C.

Portion of arch	Origin	Limits in m	M_s	y	dx
AC	A	0 to 18	$40.5x - 1.5x^2$	$\frac{2}{81}(36x - x^2)$	dx
BC	B	0 to 18	$13.5x$	$\frac{2}{81}(36x - x^2)$	dx

Solution : Since $I = I_c \sec\alpha$,

$$H = \frac{\sum M_s y dx}{\sum y^2 dx}$$

For the present case,

$$y = \frac{4y_c}{L^2} \cdot x(L - x) = \frac{(4)(8)}{(36)^2} x(36 - x) = \frac{2}{81}(36x - x^2)$$

$$V_A = \frac{(54)(27)}{36} = 40.5 \text{ kN}$$

$$V_B = 54 - 40.5 = 13.5 \text{ kN.}$$

$$\sum M_s y dx = \int_0^{18} (40.5x - 1.5x^2)\left(\frac{2}{81}\right)(36x - x^2)dx + \int_0^{18} 13.5x\left(\frac{2}{81}\right)(36x - x^2)dx$$

$$= 37324.8 \text{ kN-m}^3.$$

$$\sum y^2 dx = 2\int_0^{18} \left[\left(\frac{2}{81}\right)(36x - x^2)\right]^2 dx = 1228.8 \text{ m}^3.$$

$$H = \frac{\sum M_s y dx}{\sum y^2 dx} = \frac{37324.8}{1228.8} = 30.375 \text{ kN.}$$

The bending moment at crown $= M_s - H.y = (13.5)(18) - (30.375)(8) = 0.$

The shear force at crown

$$= V_s \cos\alpha - H\sin\alpha = (-13.5)\cos 0° - (30.375)\sin 0° = -13.5 \text{ kN}.$$

The normal thrust at crown

$$= V_s \sin\alpha + H\cos\alpha = (-13.5)\sin 0° + (30.375)\cos 0° = 30.375 \text{ kN}.$$

B.M. diagram : For portion AC, taking A as origin,

$$M = M_s - Hy = (40.5x - 1.5x^2) - (30.375)\left(\frac{2}{81}\right)(36x - x^2) = 13.5x - 0.75x^2,$$

which is a parabolic equation.

For portion BC, taking origin at B,

$$M = M_s - Hy = 13.5x - (30.375)\left(\frac{2}{81}\right)(36x - x^2) = 0.75x^2 - 13.5x.$$

Fig. 7.23(c) shows the B.M. diagram for the arch.

(i) Considering rib-shortening

$$H = \frac{\sum \frac{M_s y ds}{EI}}{\sum \frac{y^2 ds}{EI} + \frac{L}{A_m E}} = \frac{\frac{1}{EI_c}\sum M_s y dx}{\frac{1}{EI_c}\sum y^2 dx + \frac{L}{A_m E}}$$

$$I_c = \frac{1}{12}(100)(110)^3 - (95)(100)^3 = 3175(10)^3 \text{ mm}^4$$

$$EI_c = 3175(10)^3(2)(10)^5 = 635(10)^3 \text{ Nmm}^2 = 635 \text{ kNm}^2$$

$$y = \frac{4y_c}{L^2}\cdot x(L - x) = \frac{2}{81}(36x - x^2)$$

$$\tan\alpha = \frac{dy}{dx} = \frac{2}{81}(36 - 2x).$$

At the springing for which $x = 0$,

$$\tan\alpha = \frac{2}{81}(36) = \frac{8}{9},$$

whence $\alpha = 41.6335°$, $\cos\alpha = 0.74741$ and $\sin\alpha = 0.66436$.

Moment of inertia at the springing $= I_c \sec\alpha = \dfrac{3175(10)^3}{0.74741} = 4248(10)^3 \text{ mm}^4.$

If the depth of web be 'd' at the springing,

$$= \frac{1}{12}[(100)(d + 10)^3 - (95)(d)^3] = 4248(10)^3,$$

whence $d = 115$ mm.

The sectional area at the springing

$$= (2)(100)(5) + (115)(5) = 1575 \text{ mm}^2,$$

and that at crown $= (2)(100)(5) + (100)(5) = 1500 \text{ mm}^2.$

At crown, $\dfrac{A}{\cos\alpha} = \dfrac{1500}{\cos 0°} = 1500 \text{ mm}^2$

and at springing, $\dfrac{A}{\cos\alpha} = \dfrac{1575}{0.74741} = 2107 \text{ mm}^2.$

Average value of

$$\frac{A}{\cos\alpha} = A_m = \frac{2107 + 1500}{2} = 1803.5 \text{ mm}^2.$$

$$\frac{L}{A_m E} = \frac{36}{(1803.5)(10)^{-6}(2)(10)^8} = 0.99806(10)^{-4} \text{ m/kN}.$$

$$H = \frac{\frac{37324.8}{635}}{\frac{1228.8}{635} + 0.99806(10^{-4})}.$$

If rib-shortening effect be neglected $H = 30.375$ kN.

(ii) The horizontal thrust due to temperature rise alone is

$$H' = \frac{L\alpha t}{\sum \frac{y^2 ds}{EI} + \frac{L}{A_m E}} = \frac{(36)(30)(12)(10)^{-6}}{\frac{1228.8}{635} + 0.99806(10)^{-4}} = 6.7(10)^{-3} \text{ kN}.$$

Total horizontal thrust considering effect of temperature rise including that of rib-shortening is

$$H = 30.373 + 6.7(10)^{-3} = 30.3797 \text{ kN}.$$

(iii) Considering support yielding along with rib-shortening, the horizontal thrust is given by

$$H = \frac{\sum \frac{M_s y ds}{EI} - \Delta}{\sum \frac{y^2 ds}{EI} + \frac{L}{A_m E}}$$

i.e., $$H = \frac{\frac{37324.8}{635} - 0.02}{\frac{1228.8}{635} + 0.99806(10)^{-4}} = 30.363 \text{ kN}.$$

If temperature effect is included $H = 30.363 + 6.7(10)^{-3} = 30.3698$ kN.

Example 7.8. *For the linear arch shown in Fig. 7.24(a) determine*

(a) horizontal thrust and draw B.M. diagram neglecting the effect of rib-shortening and shear,

(b) horizontal thrust considering effect of rib-shortening and

(c) horizontal thrust for a temperature rise of 30 °C.

Assume the cross-sectional area A of the arch and moment of inertia about relevant axis as $2(10)^4$ mm^2 *and* $1(10)^8$ mm^4 *respectively,* $E = 2(10)^5$ N/mm^2 *and* $\alpha = 1(10)^{-5}/°$C.

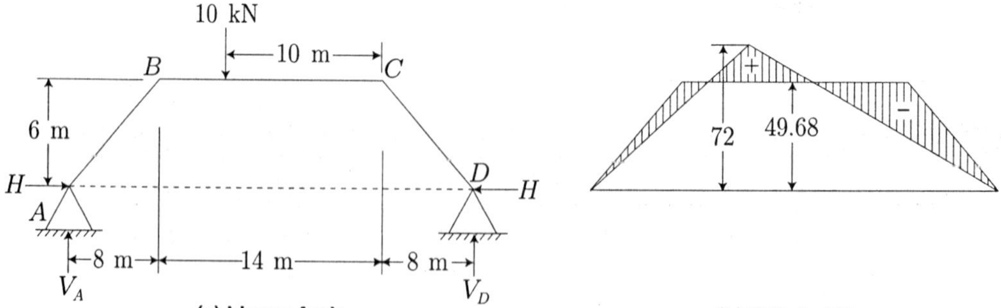

(a) Linear Arch

(b) B.M. in kNm

Fig. 7.24: For Example 7.8

Solution : $V_A = \dfrac{(10)(18)}{30} = 6 \text{ kN}$

$$V_D = 10 - 6 = 4 \text{ kN}$$

$$\tan\alpha = \frac{6}{8} = \frac{3}{4}$$

$$\sin\alpha = \frac{3}{5} \text{ and } \cos\alpha = \frac{4}{5}.$$

(a) Since EI is constant in this case,

$$H = \frac{\sum M_s y ds}{\sum y^2 ds}.$$

The equations for M_s and y for different portions of the arch are tabulated below.

Portion	Origin	Limit in m	M_s	y in m	ds
AB	A	0 to 8	$6x$	$x\tan\alpha = \frac{3}{4}x$	$\frac{dx}{\cos\alpha} = \frac{5}{4}dx$
BE	B	0 to 4	$6(x+8)$	6	dx
EC	C	0 to 10	$4(x+8)$	6	dx
CD	D	0 to 8	$4x$	$\frac{3}{4}x$	$\frac{5}{4}dx$

$$\sum M_s y ds = \int_0^8 (6x)\left(\frac{3}{4}x\right)\left(\frac{5}{4}dx\right) + \int_0^4 6(x+8)(6)dx$$
$$+ \int_0^{10} 4(x+8)(6)dx + \int_0^8 4x\left(\frac{3}{4}x\right)\left(\frac{5}{4}dx\right) = 6160.$$

$$\sum y^2 ds = \int_0^8 \left(\frac{3}{4}x\right)^2\left(\frac{5}{4}dx\right) + \int_0^4 (6)^2 dx + \int_0^{10} (6)^2 dx$$
$$+ \int_0^8 \left(\frac{3}{4}x\right)^2\left(\frac{5}{4}dx\right) = 744.$$

$$H = \frac{\sum M_s y ds}{\sum y^2 ds} = \frac{6160}{744} = 8.28 \text{ kN}.$$

B.M. diagram is shown in Fig. 7.24(b)

(b) Considering rib-shortening

$$H = \frac{\sum \frac{M_s y ds}{EI} - \sum \frac{V_s \sin\alpha\cos\alpha ds}{AE}}{\sum \frac{y^2 ds}{EI} + \sum \frac{\cos^2\alpha ds}{AE}},$$

where V_s is the simple beam shear force.

Portion	Origin	Limit	V_s	$\sin\alpha$	$\cos\alpha$	ds
AB	A	0 to 8	+6	$\frac{3}{5}$	$\frac{4}{5}$	$\frac{5}{4}dx$
BE	A	8 to 12	+6	0	1	dx
IC	A	12 to 22	−4	0	1	dx
CD	A	22 to 30	−4	$-\frac{3}{5}$	$\frac{4}{5}$	$\frac{5}{4}dx$

$$\sum \frac{V_s \sin\alpha\cos\alpha ds}{AE} = \frac{1}{AE}\left[\int_0^8 (6)\left(\frac{3}{5}\right)\left(\frac{4}{5}\right)\left(\frac{5}{4}\right)dx + \int_8^{12}(6)(0)(1)dx + \int_{12}^{22}(-4)(0)(1)dx + \int_{22}^{30}(-4)\left(-\frac{3}{5}\right)\left(\frac{4}{5}\right)\left(\frac{5}{4}\right)dx\right]$$

$$= \frac{48}{AE} = \frac{48}{(200)(10)^{-4}E} = \frac{2400}{E}.$$

$$\sum \frac{\cos^2\alpha ds}{AE} = \frac{1}{AE}\left[\int_0^8 \left(\frac{4}{5}\right)^2\left(\frac{5}{4}dx\right) + \int_8^{12}(1)^2dx + \int_{12}^{22}(1)^2dx + \int_{22}^{30}\left(\frac{4}{5}\right)^2\left(\frac{4}{5}\right)dx\right]$$

$$= \frac{26.8}{AE} = \frac{26.8}{(200)(10)^{-4}E} = \frac{1340}{E}.$$

$$\sum \frac{M_s y ds}{EI} = \frac{6160}{EI}\frac{6160}{(10^8)(10)^{-12}E} = \frac{(6160)(10)^4}{E}$$

$$\sum \frac{y^2 ds}{EI} = \frac{744}{EI} = \frac{744}{(10)^8(10)^{-12}E} = \frac{(744)(10)^4}{E}$$

and $$H = \frac{\frac{(6160)(10)^4}{E} - \frac{2400}{E}}{\frac{(744)(10)^4}{E} + \frac{1340}{E}} = 8.276 \text{ kN}.$$

Neglecting the effect of V_s,

$$H = \frac{(6160)(10)^4}{(744)(10)^4 + 1340} = 8.277 \text{ kN}.$$

(c) Additional thrust due to temperature rise alone is

$$H' = \frac{L\alpha t}{\sum \frac{y^2 ds}{EI} + \sum \frac{\cos^2\alpha ds}{AE}}$$

$$L\alpha t = (30)(1)(10)^{-5}(30) = (9)(10)^{-3} \text{ m}.$$

$$\sum \frac{y^2 ds}{EI} = \frac{(744)(10)^4}{E} = \frac{(744)(10)^4}{(2)(10)^8} = (372)(10)^{-4}$$

$$\sum \frac{\cos^2\alpha ds}{AE} = \frac{1340}{E} = \frac{1340}{(2)(10)^8} = (670)(10)^{-8}$$

$$H' = \frac{(9)(10)^{-3}}{(372)(10)^{-4} + (670)(10)^{-8}} = 0.242 \text{ kN}.$$

Example 7.9. *If the linear arch shown in Fig. 7.24(a) be supported on a roller at D and connected by a tie-rod of the same material and having sectional area of* 1000 mm^2 *between A and D, what will be the value of the horizontal thrust?*

Solution : For this case, the horizontal thrust is given by

$$H = \frac{\sum \frac{M_s y ds}{EI}}{\sum \frac{y^2 ds}{EI} + \sum \frac{\cos^2\alpha ds}{AE} + \frac{L}{A_t E_t}},$$

where A_t and E_t are the sectional area and modulus of elasticity of the tie-rod.

From Example 7.8,

$$\sum \frac{M_s y ds}{EI} = \frac{(6160)(10)^4}{E}$$

$$\sum \frac{y^2 ds}{EI} = \frac{(744)(10)^4}{E}$$

$$\sum \frac{\cos^2 \alpha ds}{AE} = \frac{1340}{E}$$

$$\frac{L}{A_t E_t} = \frac{30}{(1000)(10^{-6})E} = \frac{(3)(10)^4}{E}$$

$$H = \frac{(6160)(10)^4}{(744)(10)^4 + (1340) + (3)(10)^4} = 8.25 \text{ kN}.$$

7.10 Numerical Method for Solution of Two-Hinged Arch

Determination of horizontal thrust involves evaluation of integrals $\int \frac{M_s y ds}{EI}$ and $\int \frac{y^2 ds}{EI}$. The ease with which these integrals can be evaluated depends on the shape of the arch and the manner in which the section of the arch varies from the crown to springings. If has been found that in case of a parabolic arch with the section varying such that moment of inertia I at any section is given by $I_c \sec \alpha$, the evaluation of the integrals becomes relatively easy. But for any arbitrary nature of variation of the arch axis and cross section, direct evaluation

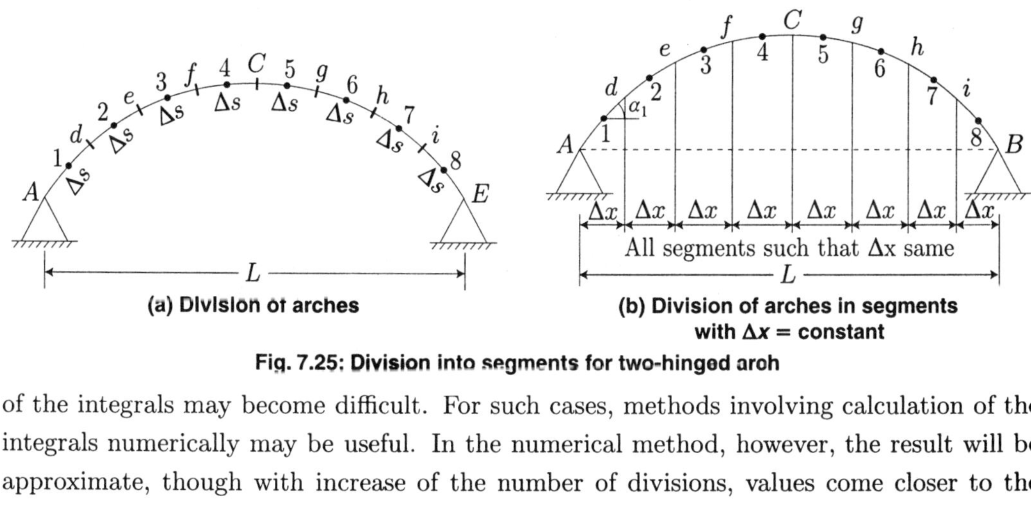

(a) Division of arches **(b) Division of arches in segments with Δx = constant**

Fig. 7.25: Division into segments for two-hinged arch

of the integrals may become difficult. For such cases, methods involving calculation of the integrals numerically may be useful. In the numerical method, however, the result will be approximate, though with increase of the number of divisions, values come closer to the exact ones. For numerical method, the arch is divided into a number of equal divisions Δs as shown in Fig. 7.25(a), and values of M_s, y and I are calculated at the centroids marked 1, 2, 3, etc. in Fig. 7.25(a) of these segments. Considering these values to be constant in the respective segments, the horizontal thrust H can be obtained from

$$H = \frac{\int \frac{M_s y ds}{EI}}{\int \frac{y^2 ds}{EI}} = \frac{\sum \frac{M_s y \Delta s}{EI}}{\sum \frac{y^2 \Delta s}{EI}} = \frac{\frac{\Delta s}{E} \sum \frac{M_s y}{I}}{\frac{\Delta s}{E} \sum \frac{y^2}{I}},$$

whence $$H = \frac{\sum \frac{M_s y}{I}}{\sum \frac{y^2}{I}}. \tag{7.16}$$

In Eq. (7.16) $\frac{M_s y}{I}$ and $\frac{y^2}{I}$ indicate summation of quantities $\frac{M_s y}{I}$ and $\frac{y^2}{I}$ at the centroids of all segments.

With $ds = \dfrac{dx}{\cos\alpha}$, the horizontal thrust may be expressed as,

$$H = \frac{\int \frac{M_s y ds}{EI}}{\int \frac{y^2 ds}{EI}} = \frac{\int \frac{M_s y dx}{EI\cos\alpha}}{\int \frac{y^2 dx}{EI\cos\alpha}}.$$

Dividing the arch into several segments such as Ad, de, ef, etc. as in Fig. 7.25(b) so that their horizontal projections are each equal to Δx, expression for H reduces to

$$H = \frac{\sum \frac{M_s y \Delta x}{EI\cos\alpha}}{\sum \frac{y^2 \Delta x}{EI\cos\alpha}} = \frac{\sum \frac{M_s y}{I\cos\alpha}}{\sum \frac{y^2}{I\cos\alpha}}. \tag{7.17}$$

The bending moment M_s at the centroid of any segment may be obtained by adding algebraically the change in bending moment due to change in x by Δx to that at the centre of the previous segment. Considering the loads acting at the centroids 1, 2, 3, etc. of the segments as shown in Fig. 7.26, the bending moment at point 2 is thus given by

$$M_2 = R_A \cdot x - P_1\Delta x.$$

At point 3, $M_3 = R_A(x + \Delta x) - P_2\Delta x - P_1(2\Delta x)$.

The increment in bending moment from point 2 to point 3 is thus given by

$$\begin{aligned}\Delta M &= M_3 - M_2 = R_A \cdot \Delta x - (P_1 + P_2)\Delta x \\ &= \Delta x(R_A - P_1 - P_2)\end{aligned}$$

i.e., $\Delta M = (\Delta x)$ (Shear on the left of the point).

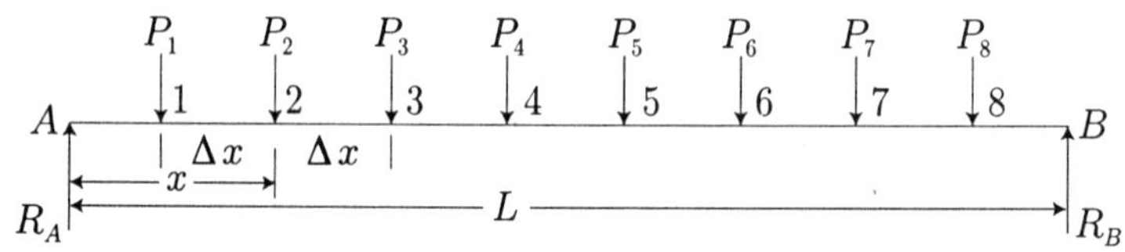

Fig. 7.26: Loading at centroids of segments

For a circular arch, it is convenient to divide the arch into equal segments and the co-ordinates of the centroids with respect to the left springing can be easily determined from the angle which the point makes with the vertical line through the centre of the arch.

The co-ordinates x and y of any point D on the circular arch in Fig. 7.27 with respect to the left springing A is given by

$$x_D = R(\sin\alpha - \sin\theta)$$

and $\quad y_D = R(\cos\theta - \cos\alpha)$,

where $\quad R =$ radius of the circular arch

$\alpha =$ half of the angle subtended by the arch at the centre

$\theta =$ angle made by the point D with the vertical line through the centre of arch.

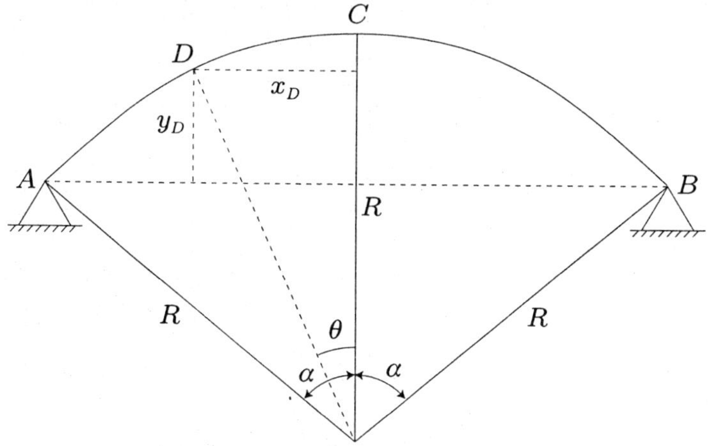

Fig. 7.27: Circular two-hinged arch

Example 7.10. *A symmetrical two-hinged circular solid web welded arch has an I-section with overall depth varying linearly from* 610 mm *at springing to* 410 mm *at crown, flange width of* 400 mm, *flange thickness of* 20 mm, *web thickness of* 20 mm *and has a span of* 60 m *between centres of hinge supports. The central rise of the arch measured from the centre of the springings to the centroid of the arch at the crown is* 10.50 m. *Such arch ribs are placed at* 10.50 m *centres and carry twenty purlins equally spaced around the rib, with the end purlins placed at half of the spacing from the end supports. Loads due to roof deck, purlins and arch ribs is* 1220 N/m² *on roof surface and live load may be taken as* 1950 N/m² *of the roof surface. Calculate the horizontal thrust.*

Solution : Fig. 7.28(a) shows the centre line of the circular arch rib with positions of the purlins marked as 1, 2, ..., 20. If the arch rib be divided into 20 equal segments, the concentrated loads through the purlins can be considered as applied at the centre of the segments. The radius R and the half central angle α of the arch are obtained from $R \sin \alpha = 30$ and $R(1 - \cos \alpha) = 10.5$, whence $R = 48.107$ m and $\alpha = 38.58° = 0.673348$ radian.

Length of the arch $= 2Rx = (2)(48.107)(0.673348) = 64.786$ m.

Length of arch segment $= \dfrac{64.786}{20} = 3.2393$ m.

Central angle subtended by each segment $= \dfrac{38.58}{10} = 3.858°$.

Total load $= 1220 + 1950 = 3170$ N/m² of surface.

Load per segment of the arch $= (3.2393)(10.5)(3170) = 107{,}820$ N $= 107.82$ kN.

Fig. 7.28(b) shows the loadings for simple beam condition which is considered as acting at the centroid of each segment.

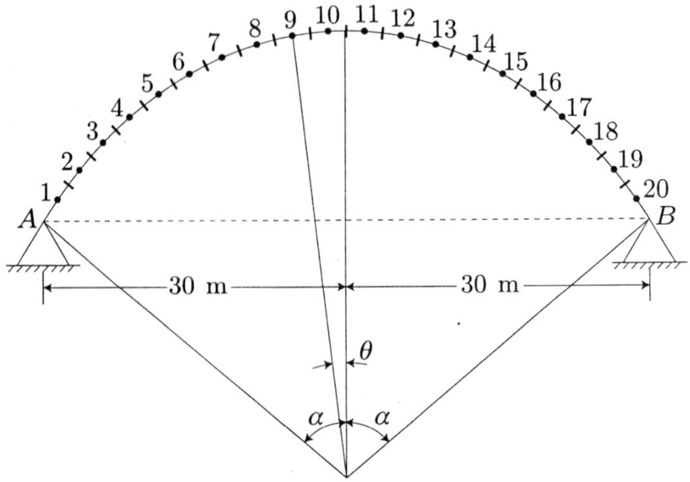

(a) Arch

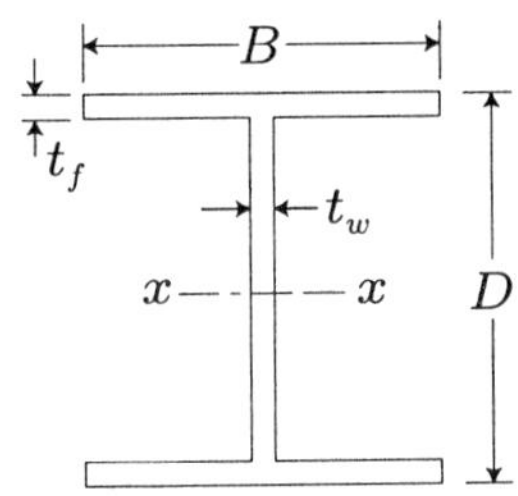

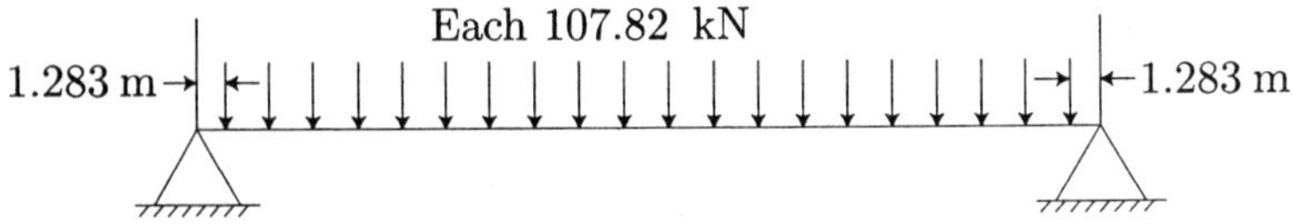

(b) Loadings for simple beam conditions

(c) Arch rib

Fig. 7.28: For Example 7.10

The co-ordinates x, y of the centroids of segments are calculated as shown in Table 7.1.

Table 7.1

Segment centroid marked	θ Degrees	$R\sin\theta$ (m)	$R\cos\theta$ (m)	x(m)	y(m)	Δx(m)
1	36.651	28.717	38.596	1.283	0.996	1.283
2	32.793	26.055	40.440	3.945	2.840	2.662
3	28.935	23.275	42.102	6.725	4.502	2.780
4	25.077	20.390	43.572	9.610	5.972	2.885
5	21.219	17.411	44.845	12.589	7.245	2.979
6	17.361	14.355	45.915	15.645	8.315	3.056
7	13.503	11.233	46.777	18.767	9.177	3.122
8	9.645	8.060	47.427	21.940	9.827	3.173
9	5.787	4.851	47.862	25.149	10.262	3.209
10	1.929	1.619	48.080	28.381	10.480	3.232

Table 7.2

Segment centroid marked	x (m)	y (m)	Δx (m)	Shear just on the left (kN)	B.M. incre-ments (kN-m)	M_s (kN-m)	$M_s y$ (kN-m^2)	y^2 (m^2)	D (m)	$I(10^4)$ (m^4)	$\frac{M_s y}{I}$ (10^4 kN/m^2)	$\frac{y^2}{I}$ (10^2 m^{-2})
1	1.283	0.996	1.283	1078.20	1383.33	1383.3	1378	0.992	0.60	16.388	84.07	6.05
2	3.945	2.840	2.662	970.38	2583.15	3966.5	11265	8.066	0.58	15.174	742.38	53.16
3	6.725	4.502	2.780	862.56	2397.92	6364.4	28653	20.268	0.56	14.013	2044.71	144.64
4	9.610	5.972	2.885	754.74	2177.42	8541.8	51012	35.665	0.54	12.905	3952.87	276.37
5	12.589	7.245	2.979	646.92	1927.17	10469.0	75848	52.490	0.52	11.849	6401.20	442.99
6	15.645	8.315	3.056	539.1	1647.49	12116.5	100749	69.139	0.50	10.844	9290.72	637.58
7	18.767	9.177	3.122	431.28	1346.46	13462.9	123549	84.217	0.48	9.889	12493.62	851.62
8	21.940	9.827	3.173	323.46	1026.34	14489.3	142386	96.570	0.46	8.984	15848.86	1074.91
9	25.149	10.262	3.209	215.64	691.99	15181.3	155790	105.309	0.44	8.128	19167.100	1295.63
10	28.381	10.480	3.232	107.82	348.47	15529.7	162752	109.830	0.42	7.320	22233.84	1500.41
										$\sum$	92259.37	6283.36

The arch rib has an I-section as shown in Fig. 7.28(c) with $B = 400$ mm, $t_f = 20$ mm, $t_w = 20$ mm and D varying from 610 mm at springings to 410 mm at the crown. The moment of inertia varies from point to point and at any point it is given by

$$I = (2)\left(\frac{1}{12}\right)(0.4)(0.02)^3 + (2)(0.4)(0.02)\frac{(D-0.02)^3}{2} + \frac{1}{12}(0.02)[D-0.04]^3$$
$$= (5.33)(10)^{-7} + (0.004)(D-0.02)^2 + (0.00167)(D-0.04)^3,$$

where D is in metre.

If S be arch distance of any point from the crown, then the depth D of the section at that point is

$$D = 0.410 + \frac{(0.610 - 0.410)}{\frac{64.786}{2}}(S) = 0.410 + \frac{0.400}{64.786}(S),$$

where D and S are in metre.

The calculation of D, I, $\frac{M_s y}{I}$ and $\frac{y^2}{I}$ for the centroids of the segments is shown in Table 7.2.

From Table 7.2,

$$\sum \frac{M_s y}{I} = 92259.37(10)^4 \text{ kN/m}^2$$

and $$\sum \frac{y^2}{I} = 6283.36(10)^2 \text{ m}^{-2},$$

whence $$H = \frac{(92259.37)(10)^4}{(6283.36)(10)^2} = 1468.3 \text{ kN}.$$

7.11 Influence Lines for Two-Hinged Arches

It has been shown in Example 3.5 that for a parabolic arch with moment of inertia $I = I_c \sec\alpha$, where I_c is the moment of inertia at crown and α, the angle made by the tangent at the point to the horizontal, the horizontal thrust H due to a concentrated load W at kL from the left springing, L being the span of the arch and k a numerical factor is given by

$$H = \frac{5}{8}\frac{WkL}{y_c}[1 - 2k^2 + k^3].$$

If $W = 1$, as shown in Fig. 7.29(a),

$$H = \frac{5kL}{8y_c} - [1 - 2k^2 + k^3]. \tag{7.18}$$

The values of H for $k = 0.1L$, $0.2L$, etc. can be calculated as shown in Table 7.3.

Table 7.3

Values of k	Values of H
$0.1L$	$0.0612L/y_c$
$0.2L$	$0.116L/y_c$
$0.3L$	$0.159L/y_c$
$0.4L$	$0.187L/y_c$
$0.5L$	$0.195L/y_c$

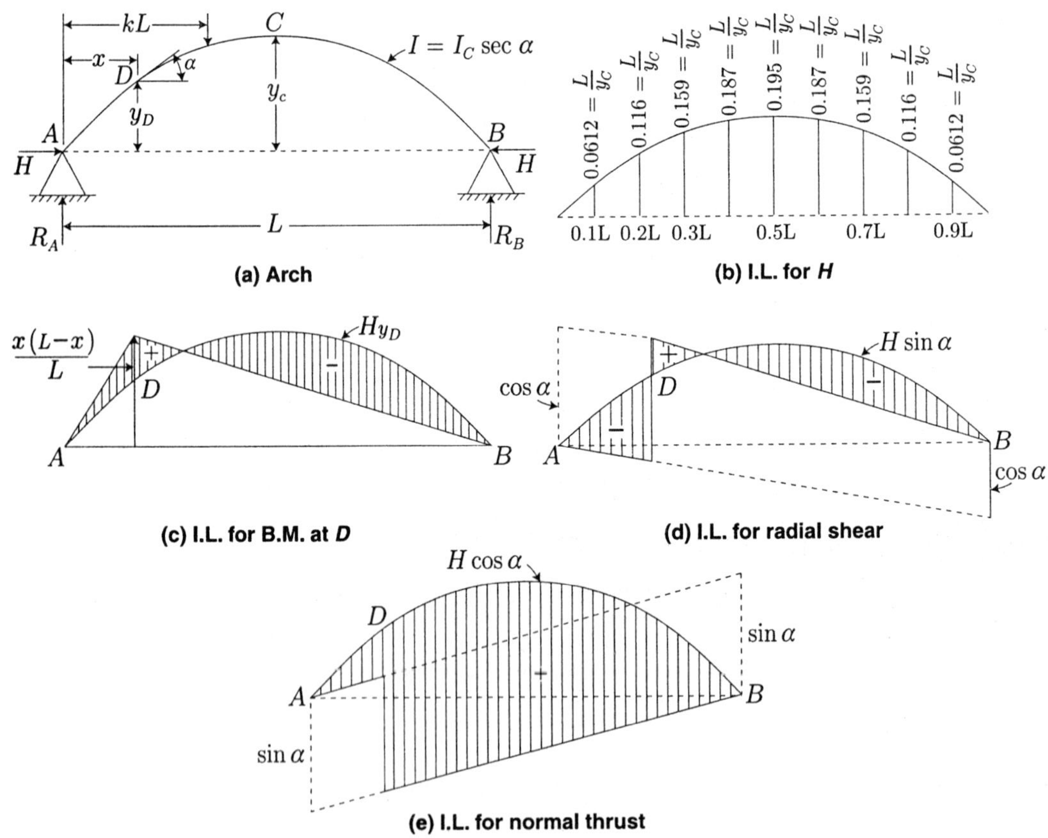

Fig. 7.29: I.L. for two-hinged arch

Plotting these values of H, the influence lines for H can be obtained as shown in Fig. 7.29(b).

The bending moment at any point D having co-ordinates (x, y) is given by

$$M = (\text{Simple beam bending moment}) - H(y).$$

The influence line for bending moment at D is obtained by superimposing the I.L. for B.M. at D due to simple beam condition over the I.L. for the horizontal thrust. The I.L. for B.M. at D due to horizontal thrust is obtained by multiplying the I.L. ordinates for H by the co-ordinate y_D at D, and is similar to I.L. for H, as y_D is constant for any given point. I.L. for B.M. at D is shown in Fig. 7.29(c).

The radial shear at D, where the slope of the tangent to the centre line of arch is α is given by $V = V_s \cos\alpha - H \sin\alpha$, where V_s is the simple beam shear force at D. The influence line for shear force at D is, therefore, obtained by multiplying the ordinates of influence lines for V_s and H by $\cos\alpha$ and $\sin\alpha$ respectively and then superimposing with appropriate sign as shown in Fig. 7.29(d).

The normal thrust at D is given by $N = V_s \sin\alpha + H \cos\alpha$.

Influence line for normal thrust is obtained by multiplying the ordinates of influence lines for V_s and H by factors $\sin\alpha$ and $\cos\alpha$ respectively and then superimposing for addition as shown in Fig. 7.29(c).

7.12 Two-Hinged Spandrel Braced Arches

A two-hinged spandrel braced arch as shown in Fig. 7.30(a) is solved as a statically indeterminate pin-jointed truss having one degree of indeterminacy. If the hinge support at B is replaced by a roller support as shown in Fig. 7.30(b), it is reduced to a determinate truss and in that case there will be horizontal displacement at B due to the external forces acting on the truss with no horizontal thrust at B. In actual case, i.e., with hinge supports at A and B, there is, however, no displacement at B. This implies that the magnitude of horizontal thrust occurring with the hinge support is such that it can neutralise the horizontal displacement caused by the external forces in the determinate condition. The horizontal displacements at B due to external forces only and that due to horizontal thrust H only in the determinate condition can be obtained by unit load method.

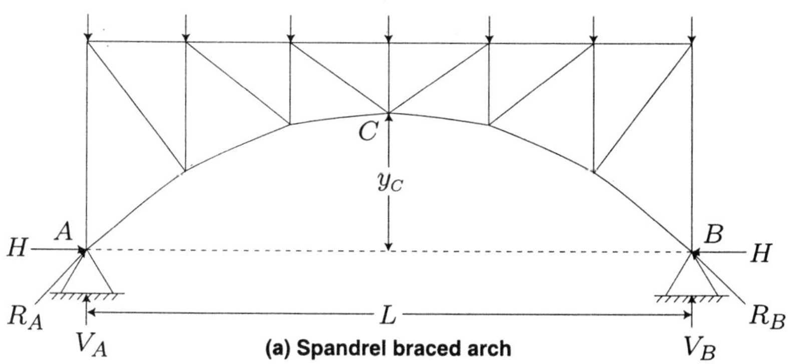

(a) Spandrel braced arch

Fig. 7.30 (Contd.)

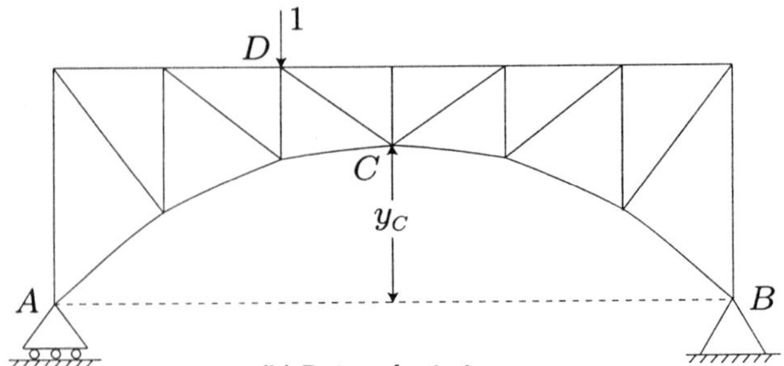

(b) Determinate truss

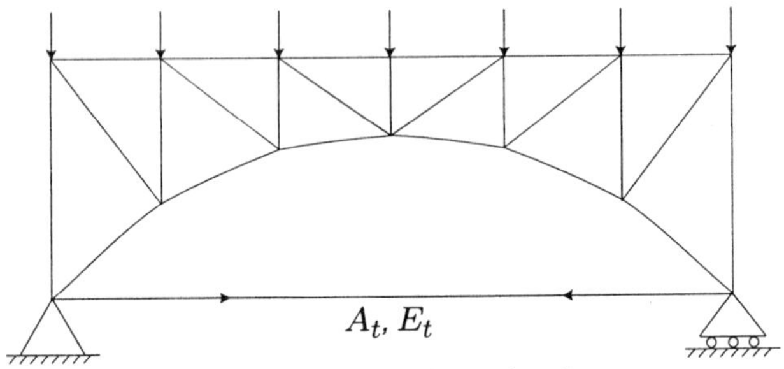

(c) Tied spandrel braced arch

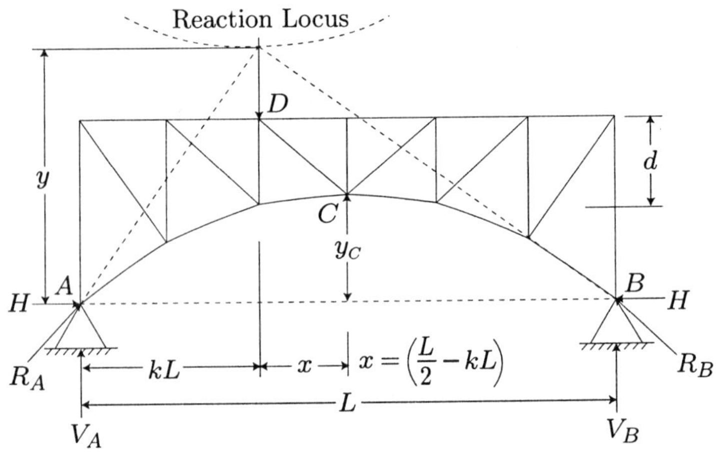

(d) Reaction locus for spandrel braced arch

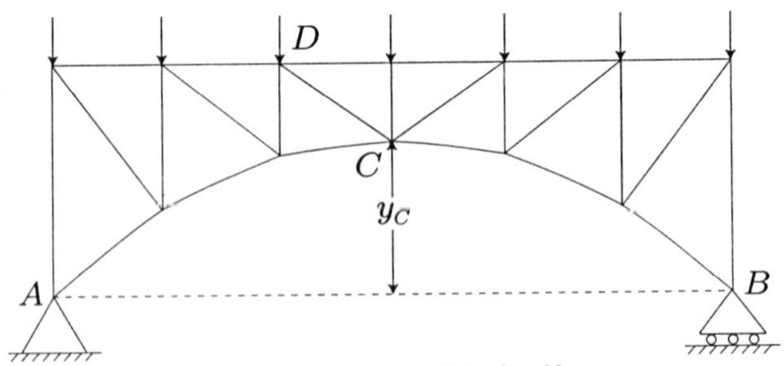

(e) Calculation of I.L. for *H*

Fig. 7.30: Two-hinged spandrel braced arch

If in determinate condition, S be the force in any member due to external forces only and u be the force in the member for a unit horizontal force at B in the direction of H, then the displacement at B due to external forces only

$$= \sum \frac{SuL}{AE},$$

and that due to $H = \sum \frac{(Hu)uL}{AE} = H \sum \frac{u^2 L}{AE}$.

Since the net displacement at B is zero,

$$\sum \frac{SuL}{AE} + H \sum \frac{u^2 L}{AE} = 0,$$

whence $H = -\dfrac{\sum \frac{Sul}{AE}}{\sum \frac{u^2 L}{AE}}$. (7.19)

The effect of temperature also can be determined from the consideration that the magnitude of the horizontal thrust is such that it produces a displacement equal and opposite to that due to temperature change in determinate condition, so that the governing condition of no displacement at B is satisfied. For a temperature rise of $t°$, the displacement of the roller support B to the right $= L\alpha t$, where α is the coefficient of thermal expression, and the displacement at B due to H to the left is given by $H \sum \frac{u^2 L}{AE}$.

As $H \sum \frac{u^2 L}{AE} - L\alpha t = 0$, it follows that,

$$H = \frac{L\alpha t}{\sum \frac{u^2 L}{AE}}. \tag{7.20}$$

For a drop of temperature, the direction of H will be opposite.

If the hinged support B is replaced by a roller support which is connected to hinge support A by a tie-bar as shown in Fig. 7.30(c), then the horizontal thrust on the arch is same as the tension in the tie-bar. The displacement at B in this case equals the elongation of the tie-bar, and the displacement equation is given by

$$\sum \frac{SuL}{AE} + H \sum \frac{u^2 L}{AE} = -\frac{HL}{A_t E_t}$$

or, $H = -\dfrac{\sum \frac{Sul}{AE}}{\sum \frac{u^2 L}{AE} + \frac{L}{A_t E_t}}$, (7.21)

where A_t and E_t are cross-sectional area and modulus of elasticity of the tie-bar.

If k be the rate of yielding of one support relative to the other, then from

$$\sum \frac{SuL}{AE} + H \sum \frac{u^2 L}{AE} = -kH,$$

$$H = -\frac{\sum \frac{SuL}{AE}}{\sum \frac{u^2 L}{AE} + k}. \tag{7.22}$$

As a general expression considering all effects together,

$$H = \frac{\sum \frac{SuL}{AE}}{\sum \frac{u^2L}{AE} + \frac{L}{A_tE_t} + k}. \tag{7.23}$$

If R_A and R_B are the reactions at the supports due to a unit load placed at a distance kL from support A as in Fig. 7.30(d) these two reactions and the unit load will be in equilibrium and their lines of action must meet at a point. The curve of the intersection point of R_A, R_B and unit load can be considered as the "Reaction locus". If the reaction locus is known, values of R_A and R_B can be determined and influence line for H can be drawn. Empirical formulae have been suggested for the reaction locus.

Formulae suggested by Johnson for a symmetrical case represent a parabolic curve as locus given by

$$\begin{aligned} y &= 2.5\frac{(x-d)x^2}{L^2} + y_c + 2.2d \\ &= \frac{5}{8}(y_c - d)(1 - 4k + 4k^2) + y_c + 2.2d, \end{aligned} \tag{7.24a}$$

where the dimensions are as in Fig. 7.30(d).

Freeman suggests a formula which represents a sine locus given by

$$\left.\begin{aligned} y &= 1.3(y_c + d) - 0.2(y_c + d)\sin\left(90^\circ - \frac{2x}{L}180^\circ\right) \\ &= (y_c + d)\{1.3 - 0.2\sin(k360^\circ - 90^\circ)\}. \end{aligned}\right] \tag{7.24b}$$

Influence line for horizontal thrust H for a unit load at a distance $kL = \left(\frac{L}{2} - x\right)$ from A is given by

$$\frac{H}{V_A} = \frac{kL}{y} \tag{7.25a}$$

whence $H = k(1-k)\dfrac{L}{y}$, as $V_A = 1 - k$.

Value of y from Eqs. (7.24a) and (7.24b) can be substituted in Eq. (7.25a). Thus substituting Eq. (7.24b),

$$H = \frac{kL(1-k)}{(y_c + d)\{1.3 - 0.2\sin(k360^\circ - 90^\circ)\}}. \tag{7.25}$$

For various values of k, values of H are tabulated in Table 7.4.

Table 7.4. Values of H

k	0	0.10	0.20	0.30	0.40	0.50
$H\frac{(y_c+d)}{L}$	0	0.062	0.117	0.170	0.211	0.227

The influence line ordinates for H can also be determined from the general expression in Eq. (7.19) for H, after evaluating the values of S and u and calculating the required terms $\sum \frac{SuL}{AE}$ and $\sum \frac{u^2L}{AL}$. This is a very laborious procedure. A quicker and easier method can be obtained by using Maxwell's Reciprocal Theorem.

Replacing the hinged support at B by a roller support as shown in Fig. 7.30(e), deflections δ_{DB} and δ_{BB}, the vertical and horizontal deflection at D and B due to unit horizontal load at B in the direction of the horizontal thrust H in the determinate arch, are obtained. From reciprocal theorem δ_{DB}, the vertical deflection at D due to unit horizontal load at A equals δ_{BD}, the horizontal deflection at A due to unit vertical load at D, i.e., $\delta_{DB} = \delta_{BD}$, whence from $H(\delta_{BB}) = 1(\delta_{DB})$,

$$H = \frac{\delta_{DB}}{\delta_{BB}}.$$

With the help of the influence line for horizontal thrust H, thus obtained, forces in various members can be calculated. Some of the influence lines thus obtained are illustrated in Fig. 7.31.

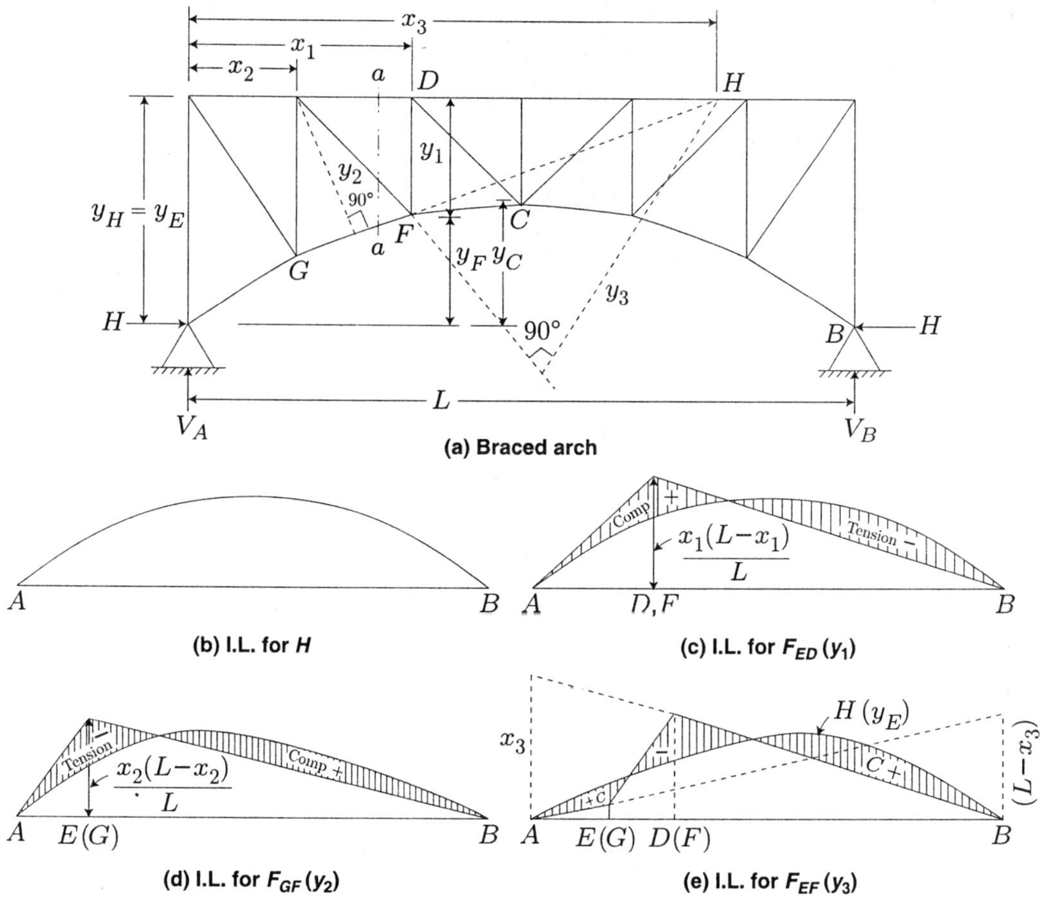

Fig. 7.31: I.L. for two-hinged spandrel arch

Fig. 7.32 shows a braced arch which is fixed to pins at A, B, C and D. If the members ED, FD and FC are removed, the structure becomes a cantilever and is statically determinate.

If the forces in the members be represented by F_1, F_2 and F_3 respectively, then considering S, u_1, u_2 and u_3 as representing the forces in the various members of the determinate cantilever truss due to external loading and unit forces on members ED, FD and FC respectively, method of consistent deformation with application of unit load gives as under :

$$\left.\begin{aligned}
&\sum \frac{Su_1L}{AE} + F_1 \sum \frac{u_1^2L}{AE} + F_2 \sum \frac{u_1u_2L}{AE} + F_3 \sum \frac{u_1u_3L}{AE} = 0\\
&\sum \frac{Su_2L}{AE} + F_1 \sum \frac{u_1u_2L}{AE} + F_2 \sum \frac{u_2^2L}{AE} + F_3 \sum \frac{u_2u_3L}{AE} = 0\\
\text{and}\quad &\sum \frac{Su_3L}{AE} + F_1 \sum \frac{u_1u_3L}{AE} + F_2 \sum \frac{u_2u_3L}{AE} + F_3 \sum \frac{u_3^2L}{AE} = 0
\end{aligned}\right] \tag{7.26}$$

Solving Eq. (7.26) the redundant forces F_1, F_2 and F_3 are obtained.

It may be more convenient, however, if any three members such as GL, JL and JK are considered as redundant. The arch then is converted into two cantilevers. Similar equations as Eq. (7.26) can also be formed for solution of the redundant forces.

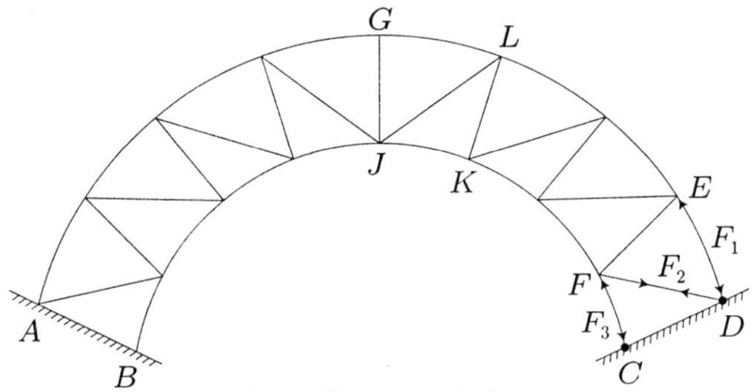

Fig. 7.32: Braced Arch

7.13 Fixed Arch

A fixed arch having both ends fixed is statically indeterminate to the third degree as shown in Fig. 7.33(c). Considering the forces H_A and V_A and moment M_A acting at A as redundants and removing them, the arch can be reduced to a determinate cantilever arch free at A as shown in Fig. 7.33(b). The structural behaviour of the fixed arch is identical with that of

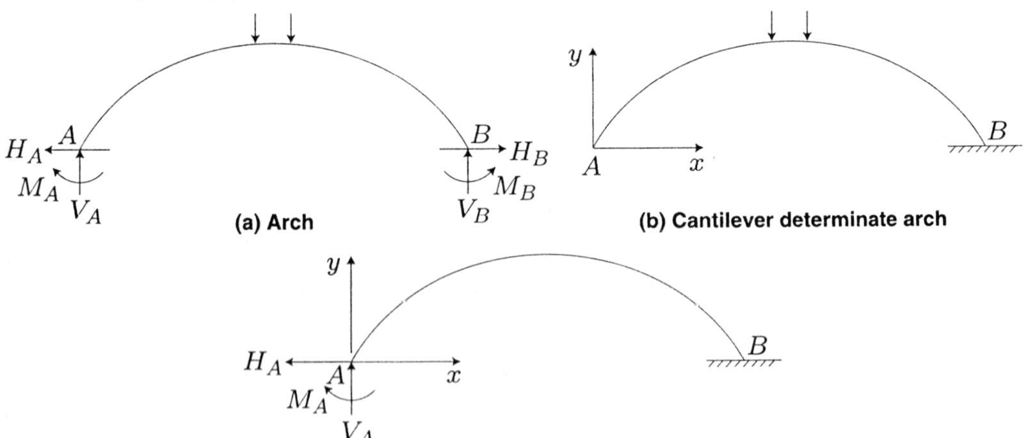

(a) Arch **(b) Cantilever determinate arch** **(c) Redundants in Arch**

Fig. 7.33: Fixed Arch

the determinate cantilever arch, provided the net deformations at A of the cantilever arch corresponding to H_A, V_A and M_A due to the external loads and H_A, V_A and M_A all acting simultaneously are zero as in the case of the fixed arch. Thus in this case, the governing conditions of geometry are

$$\Delta_V = \text{vertical displacement at } A = 0$$

$$\Delta_H = \text{horizontal displacement at } A = 0$$

and $\theta = \text{rotation at } A = 0.$

Total deformation in any direction at A of the cantilever arch is the algebraic sum of the deformations in that direction due to applied loads, V_A, H_A and M_A, shown in Figs. 7.33(b) and 7.33(c). The deformations at A due to these four loading cases can be determined separately and then added to obtain the net deformation. For each loading cases, the deformation corresponding to any redundant can be obtained using unit load method or any other suitable method. In the determinate cantilever arch, if

M_s = bending moment at any point due to applied loads, m_v = bending moment at the point due to unit vertical force at A acting in the direction of V_A as shown in Fig. 7.34(a), m_H = bending moment at the point due to unit horizontal force at A acting in the direction of H_A as shown in Fig. 7.34(b), and m_θ = bending moment at the point due to unit moment at A acting in the direction of M_A as shown in Fig. 7.34(c), then considering A as origin, x positive to the right, y positive upwards and bending moment positive if it creates tension inside,

$$m_v = +x, m_H = +y, \text{ and } m_\theta = +1.$$

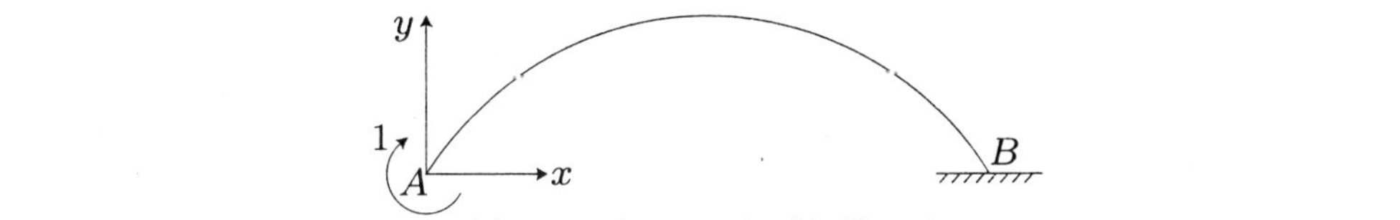

(a) Determinate arch with $V_A = 1$

(b) Determinate arch with $H_A = 1$

(c) Determinate arch with $M_A = 1$

Fig. 7.34: Unit forces in determinate arch

Using unit load method, deformations at A of the determinate arch due to applied loads in the direction of V_A, H_A and M_A are respectively,

$$\Delta_{VO} = \sum \frac{M_s m_v ds}{EI} = \sum \frac{M_s x ds}{EI}$$

$$\Delta_{HO} = \sum \frac{M_s m_H ds}{EI} = \sum \frac{M_s y ds}{EI}$$

$$\theta_o = \sum \frac{M_s m_\theta ds}{EI} = \sum \frac{M_s ds}{EI}.$$

Similarly, deformations at A due to V_A acting alone are

$$\Delta_{v_1} = \sum \frac{(m_v V_A)(m_v)ds}{EI} = V_A \sum \frac{x^2 ds}{EI}$$

$$\Delta_{H_1} = \sum \frac{(m_v V_A)(m_H)ds}{EI} = V_A \sum \frac{xyds}{EI}$$

and $$\theta_1 = \sum \frac{(m_v V_A)(m_\theta)ds}{EI} = V_A \sum \frac{xds}{EI}.$$

Deformations at A due to H_A acting alone are

$$\Delta_{v_2} = \sum \frac{(m_H H_A)(m_v)ds}{EI} = H_A \sum \frac{xyds}{EI}$$

$$\Delta_{H_2} = \sum \frac{(m_H H_A)(m_H)ds}{EI} = H_A \sum \frac{y^2 ds}{EI}$$

$$\theta_2 = \frac{(m_H H_A)(m_\theta)ds}{EI} = H_A \sum \frac{yds}{EI},$$

those due to M_A acting alone being

$$\Delta_{v_3} = \sum \frac{(m_\theta M_A)(m_v)ds}{EI} = M_A \frac{xds}{EI}$$

$$\Delta_{H_3} = \sum \frac{(m_\theta M_A)(m_H)ds}{EI} = M_A \frac{yds}{EI}$$

$$\theta_3 = \sum \frac{(m_\theta M_A)(m_\theta)ds}{EI} = M_A \sum \frac{ds}{EI}.$$

Deformations at A of the cantilever arch due to all the above four loading cases corresponding to V_A, H_A and M_A are obtained by summation and they must be equal to zero as per the governing conditions of geometry.

Hence,
$$\left.\begin{aligned} \Delta_V &= \sum \frac{M_s xds}{EI} + V_A \sum \frac{x^2 ds}{EI} + H_A \sum \frac{xyds}{EI} + M_A \sum \frac{xds}{EI} = 0 \\ \Delta_H &= \sum \frac{M_s yds}{EI} + V_A \sum \frac{xyds}{EI} + H_A \sum \frac{y^2 ds}{EI} + M_A \sum \frac{yds}{EI} = 0 \\ \theta &= \sum \frac{M_s ds}{EI} + V_A \sum \frac{xds}{EI} + H_A \sum \frac{yds}{EI} + M_A \sum \frac{ds}{EI} = 0 \end{aligned}\right] \quad (7.27)$$

Solving Eq. (7.27), values of V_A, H_A and M_A can be determined.

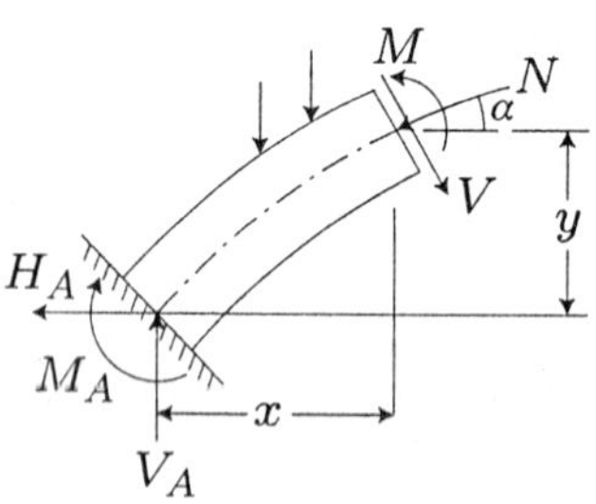

Fig. 7.35: Fixed arch

The bending moment, radial shear and normal thrust at any point having co-ordinates x, y as shown in Fig. 7.35 with positive directions of M_A, V_A and H_A as shown are respectively,

$$\left.\begin{aligned} M &= M_s + M_A + V_A \cdot x + H_A y \\ V &= V_s \cos\alpha + H_A \sin\alpha \\ N &= V_s \sin\alpha - H_A \cos\alpha, \end{aligned}\right] \quad (7.28)$$

where V_s is transverse shear in a simple beam of the same span as the arch due to applied loads and considered positive when acting upwards on the left of the section, α being the slope of the tangent to the arch at the point.

7.14 Elastic Centre Method for Solution of Fixed Arch

In 1886, Carl Culmann (1821–1882), a German, published his book on graphic statics. In his work he demonstrated that the analysis of hingeless arches, i.e., fixed arches can be considerably simplified if the redundent reaction components are considered to act at the centroid of the elastic areas. The point is known as the elastic centre or sometimes as the 'neutral point'.

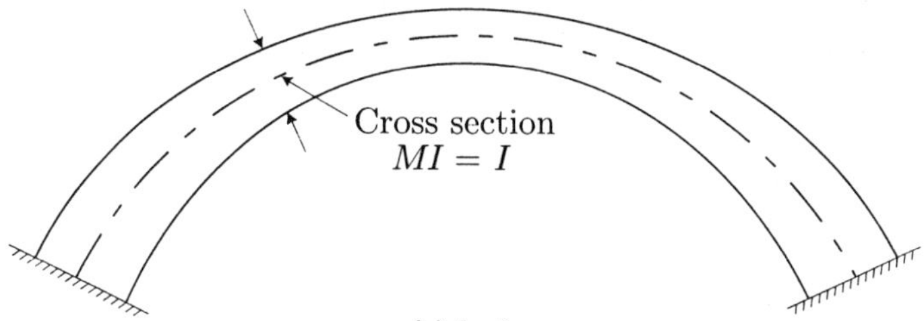

(a) Arch

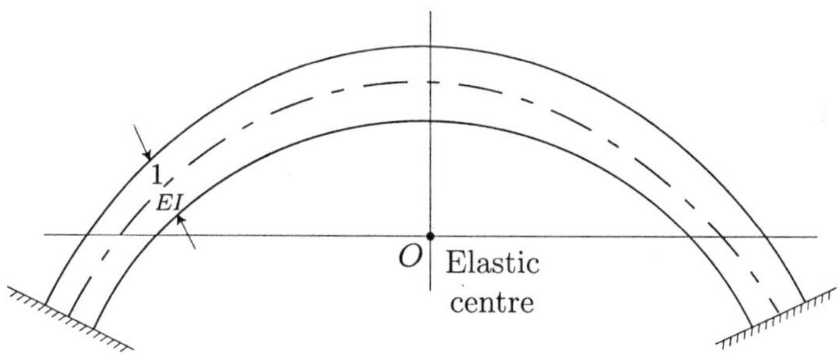

(b) Elastic centre

Fig. 7.36: Elastic centre of a fixed arch

The elastic centre then may be defined as the centroid of elastic quantities $\frac{ds}{EI}$ of any structure, where d_s is a very small length of the axis of the structure. In case of a fixed arch, the elastic centre is the centroid of an elastic area having its axis identical with that of the arch, width at any point being numerically equal to $\frac{1}{EI}$ value of the arch section at that point as shown in Fig. 7.36(b).

If the co-ordinate axis passes through the elastic centre, i.e., origin is taken at the elastic centre, then

$$\sum \frac{xds}{EI} = 0, \text{ and } \sum \frac{yds}{EI} = 0.$$

If a rigid arm AO be connected between the points A of the arch and its elastic centre O and the arm is fixed at O removing the fixed support at A as shown in Fig. 7.37, the arch will also behave as fixed arch with A and B fixed. Theb total deformations at O can be obtained from Eq. (7.27), taking O as origin and replacing M_A, V_A and H_A by M_O, V_O and

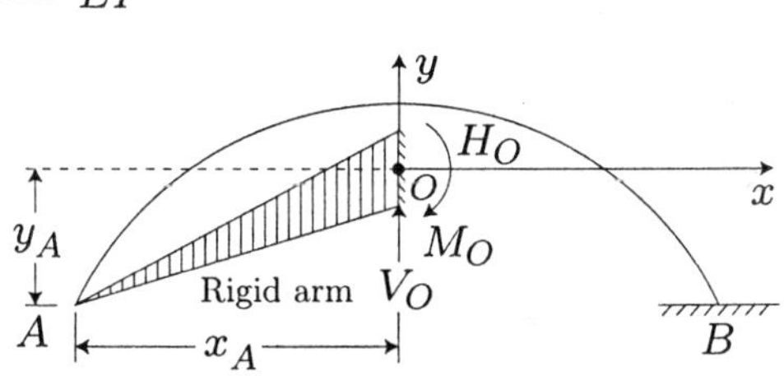

Direction of Reactions at *O* indicates positive directions with cantilever B.M. as negative

Fig. 7.37: Method of elastic centre for fixed arch

H_O respectively which act at the point O as shown in Fig. 7.37. Each of these deformations is zero as O is fixed. Hence, with O as origin, deformations at O are given by

$$\left.\begin{aligned} \Delta_V &= \sum \frac{M_s x ds}{EI} + V_o \sum \frac{x^2 ds}{EI} + H_o \sum \frac{xyds}{EI} = 0 \\ \Delta_H &= \sum \frac{M_s y ds}{EI} + V_o \sum \frac{xyds}{EI} + H_o \sum \frac{y^2 ds}{EI} = 0 \\ \text{and} \quad \theta &= \sum \frac{M_s ds}{EI} + M_o \sum \frac{ds}{EI} = 0, \end{aligned}\right] \tag{7.29}$$

since with elastic centre O as the origin,

$$\sum \frac{xds}{EI} = 0 \text{ and } \sum \frac{yds}{EI} = 0,$$

After evaluation of the integrals, Eq. (7.29) can be solved for M_o, V_o, and H_o. The bending moment at any point E having co-ordinates x, y is then given by

$$M_E = M_s + M_o + V_o \cdot x + H_o \cdot y, \tag{7.30}$$

with appropriate sign for x and y.

The bending moment and vertical and horizontal components of reaction at A of the fixed arch are

$$\left.\begin{aligned} M_A &= M_o + V_o \cdot x_A + H_o \cdot y_A \\ V_A &= V_o \\ \text{and} \quad H_A &= H_o, \end{aligned}\right] \tag{7.31}$$

where x_A and y_A are co-ordinates of the point A with reference to elastic centre O as origin with appropriate sign.

The radial shear and normal thrust at any point can be obtained from Eq. (7.28). The positive direction of the moment and reactions will be as shown in Fig. 7.37 with cantilever B.M. as negative.

7.15 Symmetrical Fixed Arch

For a symmetrical fixed arch having y-axis as axis of symmetry shown in Fig. 7.38, $\sum \frac{xyds}{EI} = 0$, with elastic centre as origin and Eq. (7.29) reduces to

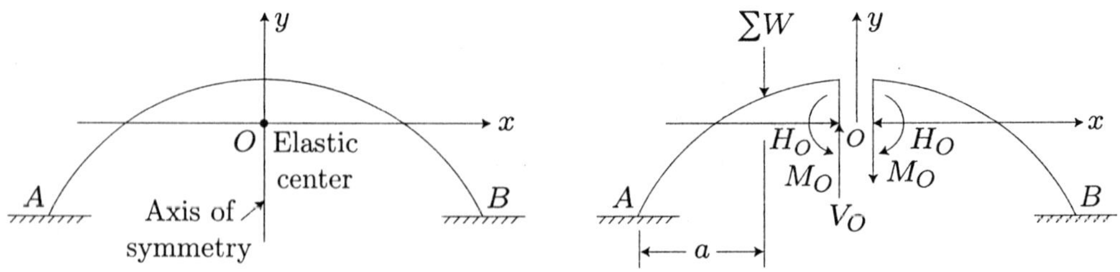

(a) Elastic centre **(b) Elastic centre method with cut at centre**

Fig. 7.38: Symmetrical fixed arch

$$\left.\begin{aligned}\sum\frac{M_s x ds}{EI}+V_o\sum\frac{x^2 ds}{EI}&=0\\ \sum\frac{M_s y ds}{EI}+H_o\sum\frac{y^2 ds}{EI}&=0\\ \sum\frac{M_s ds}{EI}+M_o\sum\frac{ds}{EI}&=0,\end{aligned}\right] \tag{7.32}$$

$$\left.\begin{aligned}\text{whence}\quad M_o&=-\frac{\sum\frac{M_s ds}{EI}}{\sum\frac{ds}{EI}}\\ V_o&=-\frac{\sum\frac{M_s x ds}{EI}}{\sum\frac{x^2 ds}{EI}}\\ \text{and}\quad H_o&=-\frac{\sum\frac{M_s y ds}{EI}}{\sum\frac{y^2 ds}{EI}}\end{aligned}\right] \tag{7.33}$$

In case of the symmetrical arch, however, it is better to cut the arch at the centre assuming that a rigid arm is connected to the part of the frame on each side of the cut. Both the rigid arms extend down to the elastic centre as shown in Fig. 7.38(b) with positive direction of V_o, H_o and M_o as indicated. In case of unsymmetrical arch, however, one rigid area connected with the end of the arch as in Fig. 7.37 is more convenient.

In case of Fig. 7.38(b), with a cut at center and rigid arms as shown with directions of M_o, V_o and H_o as indicated,

$$\left.\begin{aligned}M_A&=M_o-V_o x_A+H_o y_A-\sum W(a),\\ V_A \text{ from } \sum V&=0 \text{ in } CA, \text{ and } H_A=-H_o\end{aligned}\right] \tag{7.33a}$$

7.16 Unsymmetrical Fixed Arch

For an unsymmetrical fixed arch as shown in Fig. 7.39, $\sum\frac{xyds}{EI}\neq 0$. Eq. (7.19) cannot be reduced as in case of a symmetrical arch. In such case, solving Eq. (7.29) gives

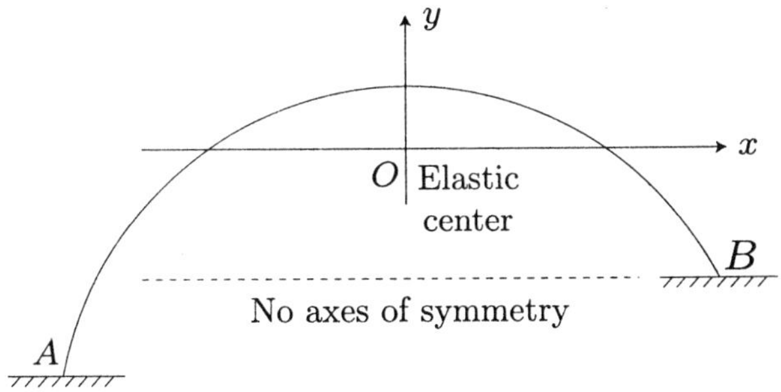

Fig. 7.39: Unsymmetrical fixed arch

$$\left.\begin{aligned}
M_o &= -\frac{\sum \frac{M_s ds}{EI}}{\sum \frac{ds}{EI}} \\
V_o &= -\frac{\sum \frac{M_s x ds}{EI} - \left(\sum \frac{M_s y ds}{EI}\right)\left(\frac{\sum \frac{xyds}{EI}}{\sum \frac{y^2 ds}{EI}}\right)}{\sum \frac{x^2 ds}{EI} - \frac{\left(\sum \frac{xyds}{EI}\right)^2}{\sum \frac{y^2 ds}{EI}}} \\
H_o &= -\frac{\sum \frac{M_s y ds}{EI} - \left(\sum \frac{M_s x ds}{EI}\right)\left(\frac{\sum \frac{xyds}{EI}}{\sum \frac{x^2 ds}{EI}}\right)}{\sum \frac{x^2 ds}{EI} - \frac{\left(\sum \frac{xyds}{EI}\right)^2}{\sum \frac{x^2 ds}{EI}}}
\end{aligned}\right] \tag{7.34}$$

Taking

$$\left.\begin{aligned}
\sum \frac{x^2 ds}{EI} &= I_y \\
\sum \frac{y^2 ds}{EI} &= I_x \\
\sum \frac{xyds}{EI} &= I_{xy} \\
\sum \frac{M_s ds}{EI} &= P \\
\sum \frac{ds}{EI} &= A \\
\sum \frac{M_s x ds}{EI} &= M_y \\
\text{and} \quad \sum \frac{M_s y ds}{EI} &= M_x,
\end{aligned}\right] \tag{7.35}$$

Eq. (7.34) can be written in a simple form as

$$\left.\begin{aligned}
M_o &= -\frac{P}{A} \\
V_o &= -\frac{M_y - M_x \frac{I_{xy}}{I_x}}{I_y - \frac{I_{xy}^2}{I_x}} \\
H_o &= -\frac{M_x - M_y \cdot \frac{I_{xy}}{I_y}}{I_x - \frac{I_{xy}^2}{I_y}}
\end{aligned}\right] \tag{7.36}$$

From Eq. (7.36), it is evident that the terms A, I_x, I_y and I_{xy} actually denote the area, moments of inertia and product of inertia respectively of the plane figure having axis identical with the axis of the fixed arch and width at any point equal to $\frac{1}{EI}$ value of the arch at that point as shown in Fig. 7.36(b), values of I_x, I_y and I_{xy} being calculated with the elastic centre as origin. Considering M_s at any point as the load intensity on the plane figure as mentioned above, terms P, M_x and M_y denote the total load and moments of the load about the x- and y- axes respectively.

7.17 Determination of the Position of Elastic Centre

The elastic centre can be obtained considering that with respect to this point,

$$\sum \frac{xds}{EI} = 0 \quad \text{and} \quad \sum \frac{yds}{EI} = 0,$$

which implies that it may be considered as the centroid of all elemental quantities $\frac{ds}{EI}$ of the entire arch. The co-ordinates (x_o, y_o) of the elastic centre with respect to any chosen axes can be obtained from

$$x_o = \frac{\sum \frac{x'ds}{EI}}{\sum \frac{ds}{EI}} \quad \text{and} \quad y_o = \frac{\sum \frac{y'ds}{EI}}{\sum \frac{ds}{EI}}, \tag{a}$$

where (x', y') are co-ordinates of an element ds of the arch rib with respect to the chosen axis. The above method is elaborated below for symmetrical and unsymmetrical arches.

(i) Elastic Centre for a Symmetrical Arch

For a symmetrical arch as shown in Fig. 7.40, the elastic centre lies on the axis of symmetry and vertically below the crown.

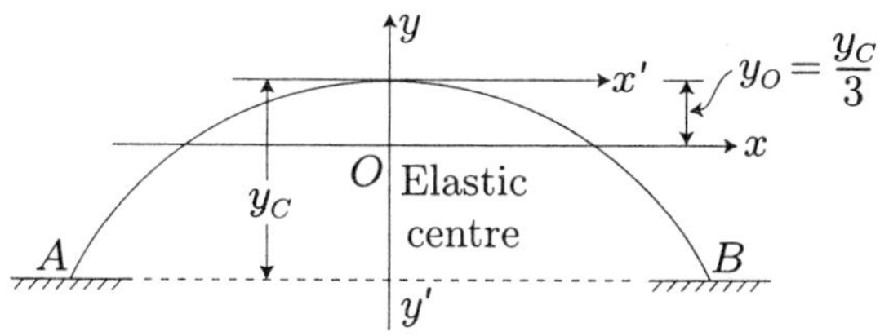

Fig. 7.40: Elastic centre for symmetrical parabolic arch

If y_o be the distance of the elastic centre below the crown, with the crown as origin and co-ordinates x' and y' positive to the right and downwards respectively,

$$y_o = \frac{\sum \frac{y'ds}{EI}}{\sum \frac{ds}{EI}}.$$

For a parabolic arch with moment of inertia $I = I_c \sec \alpha$, the above relation reduces to

$$y_o = \frac{\sum y'dx'}{\sum dx'}.$$

Putting $y' = \frac{4y_c}{L^2}(x')^2$,

$$y_o = \frac{2\int_0^{L/2} \frac{4y_c}{L^2}(x')^2 dx'}{2\int_0^{L/2} dx'} = \frac{y_c}{3}. \tag{7.37}$$

With elastic centre O as origin and co-ordinates x and y positive to the right and upwards respectively,

$$y = y_o - \frac{4y_c}{L^2}(x^2) = \frac{1}{3}y_c - \frac{4y_c}{L^2}(x^2). \tag{7.38}$$

(ii) Elastic Centre for Unsymmetrical Arch

The location of elastic centre of an unsymmetrical arch as shown in Fig. 7.41 also can be obtained using Eq. (a). For an unsymmetrical parabolic arch with $I = I_c \sec \alpha$ and horizontal distances of the crown a and b from the left and right springings respectively such that $a > b$, the co-ordinates x_o and y_o of the elastic centre with the crown C as origin and co-ordinates x' and y' positive to the left and downwards respectively are

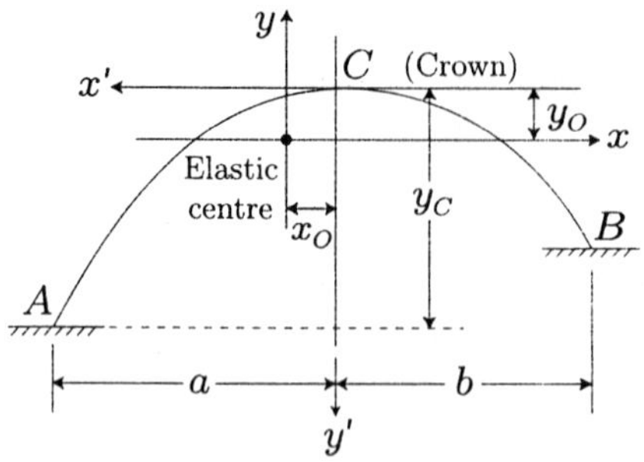

Fig. 7.41: Elastic centre for unsymmetrical parabolic

$$x_o = \frac{\sum x'dx'}{\sum dx'} = \frac{\int_{-b}^{a} x'dx'}{\int_{-b}^{a} dx'} = \frac{a-b}{2},$$

$$y_o = \frac{\sum y'dx'}{\sum dx'} = \frac{\int_{-b}^{a} \frac{4y_c}{(2a)^2}(x')^2dx'}{\int_{-b}^{a} dx'}$$

$$= \frac{y_c(a^3+b^3)}{3a^2(a+b)}. \tag{7.39}$$

With the elastic centre O as origin and co-ordinates x and y positive to the right and upwards respectively, the co-ordinates (x, y) of any element ds are related as

$$y = y_o - \frac{4y_c}{(2a)^2}(x - x_o)^2$$

i.e., $$y = \frac{y_c(a^3+b^3)}{3a^2(a+b)} - \frac{y_c}{a^2}\left(x - \frac{a-b}{2}\right)^2. \tag{7.40}$$

Example 7.11. *A parabolic arch with $I = I_c \sec\alpha$ is symmetrical and fixed at ends having span* 48 m *and central rise of* 6 m. *The arch is subjected to a concentrated load of* 20 kN *at* 12 m *horizontally from left springing, as shown in Fig. 7.42(a). Obtain the horizontal thrust and bending moment at supports.*

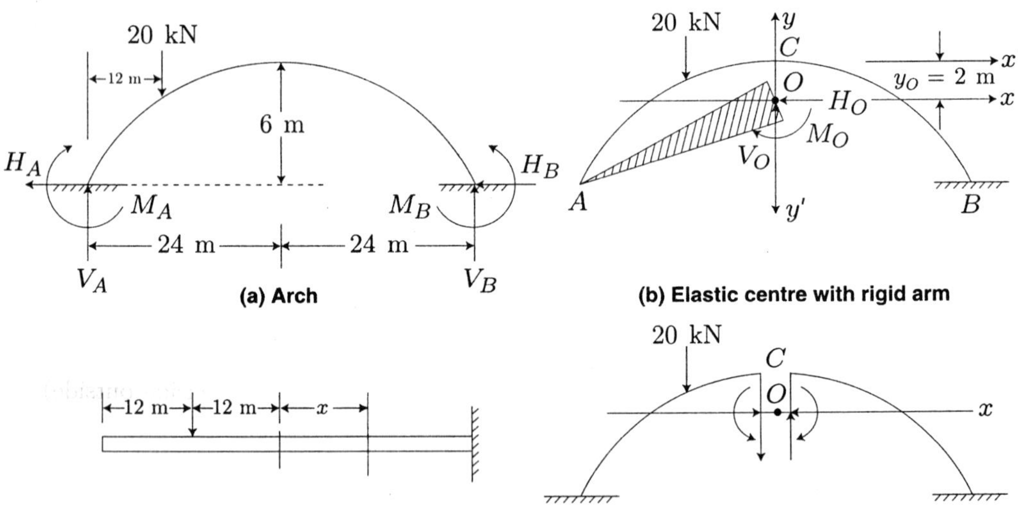

Fig. 7.42: For Example 7.11

Solution : (a) With one rigid arm as in Fig. 7.42(b),

The distance y_o of the elastic centre O as shown in Fig. 7.42(b) below the crown is

$$y_o = \frac{y_c}{3} = \frac{6}{3} = 2 \text{ m}.$$

With the elastic centre O as origin, co-ordinates x and y of an element ds of arch are given by

$$y = y_o - \frac{4y_c}{L^2}x^2 = 2 - \frac{(4)(6)}{48^2}x^2 = 2 - \frac{x^2}{96}$$

$$\sum x^2 dx = \int_{-24}^{+24} x^2 dx = 9216$$

$$\sum y^2 dx = \int_{24}^{+24} \left(2 - \frac{x^2}{96}\right)^2 dx = 153.6$$

$$\sum dx = \int_{-24}^{+24} dx = 48.$$

The bending moment M_s due to determinate condition is obtained with reference to Fig. 7.42(c), and is given by

$$M_s = -20(x + 12),$$

which holds good from $x = -12$ m to $x = 24$ m

$$\sum M_s dx = \int_{-12}^{+24} -20(x+12)dx = -12960$$

$$\sum M_s x dx = \int_{-12}^{+24} [-20(x+12)(x)(dx) = -155520$$

$$\sum M_s y dx = \int_{-12}^{+24} [-20(x+12)] \left(2 - \frac{x^2}{96}\right) dx = 3240$$

$$V_o = -\frac{\sum M_s x dx}{\sum x^2 dx} = -\frac{-15520}{9216} = 16.88 \text{ kN (upward)} \uparrow$$

$$H_o = -\frac{\sum M_s y dx}{\sum y^2 dx} = -\frac{3240}{153.6} = -21.1 \text{ kN (towards right)} \longrightarrow$$

$$M_o = -\frac{\sum M_s dx}{\sum dx} = -\frac{-12,960}{48} = +270.0 \text{ kNm}.$$

With the elastic centre as origin, x positive to the right and y positive upwards, the bending moment at any point (x, y) of the arch is given by

$$M = M_s + M_o + V_o x + H_o y = M_s + 270 + 16.88x - 21.1y.$$

For the left springing A,

$$x = -24, y = -4 \text{ and } M_s = 0$$

$$M_A = 270 + (16.88)(-24) - (21.1)(-4) = -50.72 \text{ kNm (causing tension outside)}$$

$$H_A = H_o = -21.1 \text{ kN (towards right)} \longrightarrow$$

$$V_A = V_o = 16.88 \text{ kN (upward)} \uparrow$$

For the right springing B,

$$x = +24, y = -4$$

and $M_s = -(20)(36) = -720$ kNm

$$M_B = -720 + 270 + (16.88)(24) - (21.1)(-4)$$

$$= +39.5 \text{ kN-m (causing tension inside)}$$

$$H_B = -H_A = +21.1 \text{ kN (towards left)} \longleftarrow$$

$$V_B = 20 - V_A = 20 - 16.88 = 3.12 \text{ kN (upwards)} \uparrow$$

(b) With two rigid arms, with cut at centre C as in Fig. 7.42(d). As before characteristic properties will be the same.

With reference to Fig. 7.42(d),

for part CA, $\sum M_s x dx = \int_{-24}^{-12} [20(x+12)](x)dx = 28800$

and for Part CB, $\sum M_s x dx = 0$, since $M_s = 0$

$$V_o = -\frac{28800}{9216} = -3.125 \text{ kN, i.e., upwards} \uparrow.$$

For part CA, $\sum M_s y dx = \int_{-24}^{-12} 20(x+12)\left(2 - \frac{x^2}{96}\right) dx = 3240.$

Hence, $H_o = -\dfrac{3240}{153.6} = -21.1$ kN (towards left) $\longleftarrow$

Also $\sum M_s dx = \int_{-24}^{-12} 20(x+12)dx = -1440$, whence $M_o = -\dfrac{-1440}{4.8} = 30.$

Values of reactions at A then are

$$M_A = 240 - 21.1(6-2) - 3.125(24) - 30$$

$$= 50.6, \text{ i.e., causing tension on outside.}$$

$$H_A = -H_o = 21.1 \text{ kN (towards right)} \longrightarrow$$

$$V_A = 20 - 3.125 = 16.875 \text{ (upwards)} \uparrow.$$

Example 7.12. *An unsymmetric parabolic arch with* $I = I_c \sec\alpha$ *has left and right springings* 6 m *and* 1.5 m *below the crown respectively, the span being* 36 m. *The arch is subjected to a concentrated load of* 20 kN *at the crown as shown in Fig. 7.43(a). Obtain the horizontal thrust and support reactions.*

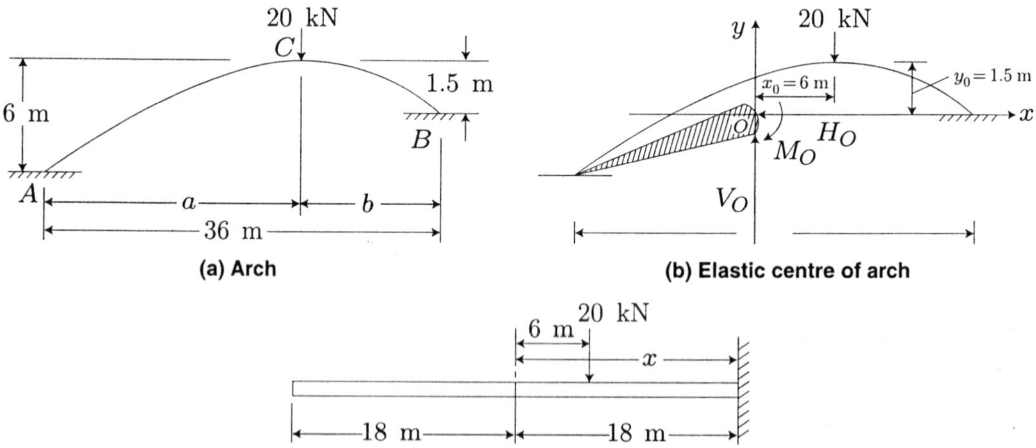

(a) Arch

(b) Elastic centre of arch

(c) Determinate condition

Fig. 7.43: For Example 7.12

Solution : If a and b be the horizontal distances of the left and right springings from the crown, then

$$b = 36 - a$$

and from the properties of parabola,

$$\frac{6}{1.5} = \left(\frac{a}{b}\right)^2 = \left(\frac{a}{36-a}\right)^2,$$

whence $a = 24$ m and $b = 12$ m.

If x_o and y_o by the horizontal and vertical distances of the elastic centre from the crown, then

$$x_o = \frac{a-b}{2} = \frac{24-12}{2} = 6 \text{ m}$$

$$y_o = \frac{y_c(a^3+b^3)}{3a^2(a+b)} = \frac{(6)(24^3+12^3)}{(3)(24)^2(24+12)} = 1.5 \text{ m}.$$

With respect to elastic centre as origin, the co-ordinates (x, y) of an element of arch are given by

$$y = y_o - \frac{4y_c}{(2a)^2}(x - x_o)^2, \quad \text{whence} \quad y = 1.5 - \frac{(4)(6)}{(48)^2}(x-6)^2 = 1.5 - \frac{(x-6)^2}{96}.$$

$$\sum dx = \int_{-18}^{+18} dx = 36 \text{ m}$$

$$\sum x^2 dx = \int_{-18}^{+18} x^2 dx = \frac{(18)^3 + (18)^3}{3} = 3888$$

$$\sum y^2 dx = \int_{-18}^{+18} \left[1.5 - \frac{1}{96}(x-6)^2\right]^2 dx = 97.2$$

$$\sum xy dx = \int_{-18}^{+18} \left[1.5 - \frac{1}{96}(x-6)^2\right] x dx = 486.$$

The bending moment M_s for determinate condition can be obtained from Fig. 7.43(c). From $x = -18$ to $x = 6$ m, $M_s = 0$ and from $x = 6$ m to $x = 18$ m,

$$M_s = -20(x-6)$$

$$\sum M_s dx = \int_6^{18} -(20)(x-6) dx = -1440$$

$$\sum M_s x dx = \int_6^{18} [-20(x-6)] x dx = -20160$$

$$\sum M_s y dx = \int_6^{18} [-20(x-6)] \left[1.5 - \frac{(x-6)^2}{96}\right] dx = -1080.$$

$$M_o = -\frac{\sum M_s dx}{\sum dx} = -\frac{-1440}{36} - +40 \text{ kNm}$$

$$V_o = -\frac{\sum M_s x ds - \sum M_s y dx \left(\frac{\sum xy dx}{\sum y^2 dx}\right)}{\sum x^2 dx - \frac{(\sum xy dx)^2}{\sum y^2 dx}}$$

$$= -\frac{-20160 - (-1080)\left(\frac{486}{97.2}\right)}{3884 - \frac{(486)^2}{972}} = +10.123 \text{ kN}$$

$$H_o = -\frac{\sum M_s y dx - \sum M_s x dx \left(\frac{\sum xydx}{\sum x^2 dx}\right)}{\sum y^2 dx - \frac{(\sum xydx)^2}{\sum x^2 dx}}$$

$$= -\frac{-1080 - (-20160)\left(\frac{486}{3888}\right)}{97.2 - \frac{(486)^2}{3888}} = -39.506 \text{ kN}.$$

With respect to the elastic centre as origin, $M = M_s + M_o + V_o x + H_o y$.

For the left springing A,

$$x = -18, y = -4.5 \text{ and } M_s = 0$$

$$M_A = 40 + (10.123)(-18) + (-39.506)(-4.5)$$

$$= +35.563 \text{ kNm (causing tension inside)}$$

$$H_A = H_o = -39.506 \text{ kN (towards right)} \longrightarrow$$

$$V_A = V_o = 10.123 \text{ kN (upwards)} \uparrow.$$

For the right springings,

$$x = +18, y = 0 \text{ and } M_s = -240 \text{ kNm}$$

$$M_B = -240 + 40 + (10.123)(18) + (-39.506)(0)$$

$$= -17.786 \text{ kN-m (causing tension outside)}$$

$$H_B = -H_A = 39.506 \text{ kN (towards left)} \longleftarrow$$

$$V_B = 20 - 10.123 = 9.877 \text{ kN(upward)} \uparrow.$$

7.18 Effect of Yielding of Supports of Fixed Arch

Yielding of supports induces additional moments and forces in a fixed arch. Yielding may consist of linear displacement or rotational yielding or both. For application of elastic centre method, the displacements of the elastic centre corresponding to those at supports are required to be evaluated. This is, however, a case of simple geometry as the arch support and the elastic centre are presumed to be connected by a rigid arm. For the support A of the fixed arch AB in Fig. 7.44(a) yielding by

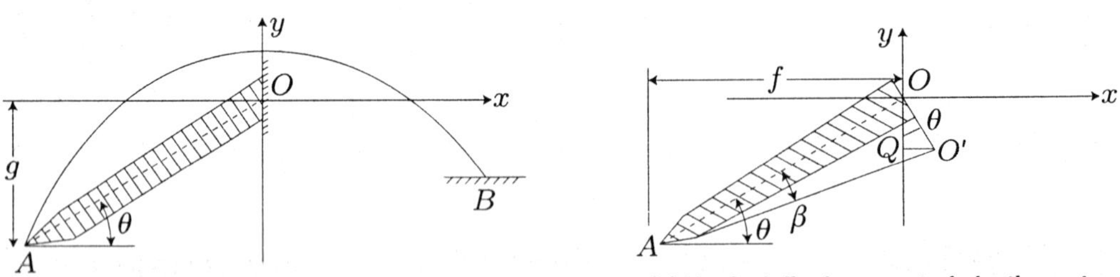

(a) Arch **(b) Vertical displacement of plastic centre**

Fig. 7.44: Yielding of supports of fixed arch

(i) rotation of β radian in the clockwise direction,

(ii) horizontal displacement h to the left, and

(iii) vertical displacement v upwards,

consider the horizontal and vertical distances of the support A from the elastic centre as 'f' to the left and 'g' downwards respectively. Due to rotation β at A, there will be rotation and horizontal and vertical displacements of the elastic centre O as shown in Fig. 7.44(b). AO' is the position of the arm OA after rotation. Since AO is a rigid arm, the rotation at O will be the same as that at A, i.e., β in the clockwise direction. The horizontal displacement of O due to rotation β only at A is $O'Q = OO' \sin\theta = (OA)(\beta)\sin\theta = g\beta$, towards right. The vertical displacement of O due to rotation β only at $A = OQ = OQ' \cos\theta = (OA)(\beta)\cos\theta = f\beta$, downward.

Due to horizontal displacement 'h' at A to the left there will be horizontal displacement 'h' only at the elastic centre O to the left. Due to the upward vertical displacement 'v' at A, there will be upward vertical displacement 'v' at the elastic centre O. Thus the total displacements at the elastic centre are

(i) clockwise rotation β

(ii) horizontal displacement $(h - g\beta)$ to the left and

(iii) vertical displacement $(v - f\beta)$ upwards.

(a) Unsymmetrical Arch. Since the displacements of the elastic centre are not zero, the conditions of geometry given by Eq. (7.29) for an unsymmetrical arch are modified as follows :

$$\Delta_v = \sum \frac{M_s x ds}{EI} + V_o \sum \frac{x^2 ds}{EI} + H_o \sum \frac{xy ds}{EI} = v - f\beta$$

$$\Delta_H = \sum \frac{M_s y ds}{EI} + V_o \sum \frac{xy ds}{EI} + H_o \sum \frac{y^2 ds}{EI} = h - g\beta$$

$$\theta = \sum \frac{M_s ds}{EI} + M_o \sum \frac{ds}{EI} = \beta.$$

Using notations as given by Eq. (7.35), the above equations can be written as under :

$$M_y + V_o I_y + H_o I_{xy} = v - f\beta$$

$$M_x + V_o I_{xy} + H_o I_x = h - g\beta$$

$$P + M_o A = \beta.$$

Solving

$$\left.\begin{aligned} M_o &= -\frac{P}{A} + \frac{\beta}{A} \\ V_o &= -\frac{M_y - M_x \cdot \frac{I_{xy}}{I_x}}{I_y - \frac{(I_{xy})^2}{I_x}} + \frac{(v - f\beta) - (h - g\beta)\frac{I_{xy}}{I_x}}{I_y - \frac{(I_{xy})^2}{I_x}} \\ H_o &= -\frac{M_x - M_y \frac{I_{xy}}{I_y}}{I_x - \frac{(I_{xy})^2}{I_y}} + \frac{(h - g\beta) - (v - f\beta)\frac{I_{xy}}{I_y}}{I_x - \frac{(I_{xy})^2}{I_y}} \end{aligned}\right] \quad (7.41)$$

where the terms A, I_x, I_y, I_{xy}, P, M_x and M_y are as given by Eq. (7.35).

Due to yielding only without any external load, Eq. (7.41) reduces to

$$\left.\begin{aligned} M_o &= \frac{\beta}{A} \\ V_o &= \frac{(v - f\beta) - (h - g\beta)\frac{I_{xy}}{I_x}}{I_y - \frac{(I_{xy})^2}{I_x}} \\ H_o &= \frac{(h - g\beta) - (v - f\beta)\frac{I_{xy}}{I_y}}{I_x - \frac{(I_{xy})^2}{I_y}} \end{aligned}\right] \tag{7.42}$$

The effect for any single displacement can be obtained from Eq. (7.42) taking zero for the other displacements.

(b) For Symmetrical Arch. For a symmetrical arch $\sum \dfrac{xyds}{EI} = I_{xy} = 0$ and using the condition in Eq. (7.41), the values M_o, V_o and H_o acting at the elastic centre are obtained.

Thus,

$$\left.\begin{aligned} M_o &= -\frac{P}{A} + \frac{\beta}{A} \\ V_o &= -\frac{M_y}{I_y} + \frac{v - f\beta}{I_y} \\ H_o &= -\frac{M_x}{I_x} + \frac{h - g\beta}{I_x} \end{aligned}\right] \tag{7.43}$$

where the terms A, I_x, I_y, P, M_x and M_y are as given by Eq. (7.35).

For yielding of supports only and no external load Eq. (7.43) reduces to

$$\left.\begin{aligned} M_o &= \frac{\beta}{A} \\ V_o &= \frac{v - f\beta}{I_y} \\ H_o &= \frac{h - g\beta}{I_x} \end{aligned}\right] \tag{7.44}$$

Eq. (7.44) includes the effect of clockwise rotation β, leftward horizontal displacement h and upward vertical displacement v of the left support A. For the effect of any single displacement, the non-existent ones are taken to be zero in Eq. (7.44).

Equations developed can also be used with two rigid arms with cut at centre as in Fig. 7.38(b).

Example 7.13. *If the support A of the symmetrical parabolic fixed arch of Example 7.11 suffers a rotational yielding of* 4° *in the clockwise direction and horizontal displacements of* 20 mm *to the left and vertical displacement of* 30 mm *downwards while the external load acts, what will be the values of the reactive moments and forces at the supports? Take flexural rigidity at crown as* $EI_c = 2(10^3)$ *kN-m*^{2}.

Solution : (a) Considering one rigid arm connected to end :

Since $\quad I = I_c \sec\alpha$

$$A = \sum \frac{ds}{EI} = \sum \frac{dx \sec\alpha}{EI_c \sec\alpha} = \sum \frac{dx}{EI_c} = \frac{1}{EI_c} \sum dx.$$

$$\text{Similarly, } I_y = \sum \frac{x^2 ds}{EI} = \frac{1}{EI_c} \sum x^2 dx$$

$$I_x = \sum \frac{y^2 ds}{EI} = \frac{1}{EI_c} \sum y^2 dx$$

$$P = \sum \frac{M_s ds}{EI} = \frac{1}{EI_c} \sum M_s dx$$

$$M_x = \sum \frac{M_s y ds}{EI} = \frac{1}{EI_c} \sum M_s y dx$$

$$M_y = \sum \frac{M_s x ds}{EI} = \frac{1}{EI_c} \sum M_s x dx.$$

Using the values of integrals already calculated in Example 7.11.

$$A = \frac{1}{(2)(10)^3}(48) = (24)(10)^{-3}$$

$$I_x = \frac{1}{(2)(10)^3}(153.6) = 76.8(10)^{-3}$$

$$I_y = \frac{1}{(2)(10)^3}(9216) = 4608(10)^{-3}$$

$$P = \frac{1}{(2)(10)^3}(-12960) = -6480(10)^{-3}$$

$$M_x = \frac{1}{(2)(10)^3}(3240) = 1620(10)^{-3}$$

$$M_y = \frac{1}{(2)(10)^3}(-155520) = -77760(10)^{-3}.$$

In this case, $\beta = \frac{4\pi}{180}$, $v = -\frac{30}{1000}$ m, $h = \frac{20}{1000}$ m

$f = 24$ m and $g = 6$ m.

From Eq. (7.43),

$$M_o = -\frac{P}{A} + \frac{\beta}{A} = -\frac{-6480(10)^{-3}}{24(10)^{-3}} + \frac{\frac{4}{180}(\pi)}{(24)(10)^{-3}} = +272.91 \text{ kNm}$$

$$V_o = -\frac{M_y}{I_y} + \frac{v - f\beta}{I_y} = -\frac{-77760(10)^{-3}}{4608(10)^{-3}} + \frac{-0.03 - (6)\left(\frac{4}{180}\right)(\pi)}{(4608)(10)^{-3}}$$

$$= 16.50 \text{ kN (upward)} \uparrow$$

$$H_o = -\frac{M_x}{I_x} + \frac{h - g\beta}{I_x} = -\frac{1620(10)^{-3}}{76.8(10)^{-3}} + \frac{0.02 - (6)\left(\frac{4}{180}\right)(\pi)}{(76.8)(10)^{-3}}$$

$$= -26.29 \text{ kN (towards right)} \longrightarrow.$$

At the left springing A,

$$M_A = M_s + M_o + V_o x + H_o y = 0 + 272.91 + (16.50)(-24) + (-26.29)(-4)$$

$$= -17.93 \text{ kNm (causing tension outside)}$$

$$H_A = H_o = -26.29 \text{ kN (towards right)} \longrightarrow$$

$$V_A = V_o = 16.5 \text{ kN (upward)} \uparrow$$

At the right springing B,

$$M_B = -720 + 272.91 + (16.5)(24) + (-26.29)(-4)$$

$$= +54.07 \text{ kNm (causing tension inside)}$$

$$H_B = H_A = 26.29 \text{ kN (towards left)} \longleftarrow$$

$$V_B = 20 - V_A = 20 - 16.5 = 3.5 \text{ kN (upward)} \uparrow.$$

(b) Considering two rigid arms with a cut at centre.

Characteristics of section will be as in (a).

Also, $$\frac{\sum M_s x dx}{EI_c} = \frac{28800}{2(10)^3} = 14.4$$

$$\frac{\sum M_s y dx}{EI_c} = \frac{3240}{2(10)^3} = 1.62$$

$$\frac{\sum M_s dx}{EI_c} = -\frac{1440}{2(10)^3} = -0.722.$$

With values of β, v, h, f and g as in case (a),

$$M_o = -\frac{(-0.722)}{24(10)^{-3}} + \frac{\frac{4}{180}(\pi)}{(24)(10)^3} = +32.99$$

$$V_o = -\frac{14.40}{4608(10)^{-3}} + \frac{0.03 - (24)\left(\frac{4}{180}\right)(\pi)}{4608(10^{-3})} = -3.50 \text{ kN (upward)} \uparrow$$

and $$H_o = -\frac{1.62}{768(10)^{-3}} + \frac{0.02 - (6)\left(\frac{4}{180}\right)(\pi)}{76.8(10)^{-3}} = -26.29 \text{ kN (towards left)} \longleftarrow$$

It follows that $V_A = 20 - 3.50 = 16.50$ kN (upward) $\uparrow$

$$H_A = H_o = 26.29 \text{ (towards right)} \longrightarrow$$

$$M_A = 240 - [32.99 + 3.50(24) + 26.29(6 - 2)]$$

$$= 17.85 \text{ kNm (causing tension outside)}.$$

Example 7.14. *If the left support A of the fixed arch of Example 7.12 yields by a clockwise rotation of* 4°, *horizontal displacement of* 20 mm *to the left and vertical displacement of* 30 mm *downwards under the applied load of* 20 kN *at the crown, what will be the additional reactive forces and moments at the supports? Given,* $EI_c = (2)(10)^3$ *kN-m*^{2}.

Solution : Due to yielding only, the moment and forces at the elastic centre are

$$M_o = \frac{\beta}{A}$$

$$V_o = \frac{(v - f\beta) - (h - g\beta)\frac{I_{xy}}{I_x}}{I_y - \frac{(I_{xy})^2}{I_x}}$$

$$H_o = \frac{(h - g\beta) - (v - f\beta)\frac{I_{xy}}{I_y}}{I_x - \frac{(I_{xy})^2}{I_y}}$$

where $$A = \sum \frac{ds}{EI},\ I_x = \sum \frac{y^2 ds}{EI}, I_y = \sum \frac{x^2 ds}{EI} \text{ and } I_{xy} = \sum \frac{xyds}{EI}.$$

Since $\quad I = I_c \sec\alpha,\ A = \sum \dfrac{ds}{EI} = \sum \dfrac{dx \sec\alpha}{EI_c \sec\alpha} = \dfrac{1}{EI_c} \sum dx.$

Similarly, $I_x = \dfrac{1}{EI_c} \sum y^2 dx,\ I_y = \dfrac{1}{EI_c} \sum x^2 dx \quad \text{and} \quad I_{xy} = \dfrac{1}{EI_c} \sum xy dx.$

Using the values of integrals as calculated in Example 7.12,

$$A = \frac{1}{2(10)^3}(36) = (18)(10)^{-3}$$

$$I_x = \frac{97.2}{(2)(10)^3} = (48.6)(10)^{-3}$$

$$I_y = \frac{3888}{2(10)^3} = (1944)(10)^{-3}$$

$$I_{xy} = \frac{486}{(2)(10)^3} = (243)(10)^{-3}.$$

In the case, $\beta = \left(\dfrac{4}{180}\right)(\pi)$ radian

$$v = -\frac{30}{1000}\text{ m} = -0.03\text{ m}$$

$$h = \frac{20}{1000}\text{ m} = 0.02\text{ m}$$

$$f = 18\text{ m}, g = 4.5\text{ m}$$

$$M_o = \frac{\frac{4}{180}(\pi)}{(18)(10)^{-3}} = +3.8785\text{ kNm (causing tension inside)}$$

$$V_o = \frac{\left[-0.03 - (18)\left(\frac{4}{180}\right)(\pi)\right] - \left[0.02 - (4.5)\left(\frac{4}{180}\right)(\pi)\right]\frac{(243)(10)^{-3}}{(48.6)(10)^{-3}}}{(1944)(10)^{-3} - \frac{(243)^2(10)^{-6}}{(48.6)(10)^{-3}}}$$

$$= 0.2526\text{ kN (upward)} \uparrow$$

$$H_o = \frac{\left[0.02 - (4.5)\left(\frac{4}{180}\right)(\pi)\right] - \left[-0.03 - 18\left(\frac{4}{180}\right)(\pi)\right]\frac{(243)(10)^{-3}}{(1944)(10)^{-3}}}{(48.6)(10)^{-3} - \frac{(243)^2(10)^{-6}}{(1944)(10)^{-3}}}$$

$$= -7.3158\text{ kN (to the right)} \longrightarrow.$$

Additional reactive moment and forces at support A are

$$M_A = M_o + (V_o)x + (H_o y) = +3.8785 + (0.2526)(-18) + (-7.3158)(-4.5)$$

$$= +32.2528\text{ kNm (causing tension inside)}$$

$$V_A = V_o = +0.2526\text{ kN (upward)} \uparrow$$

$$H_A = H_o = -7.3158\text{ kN (towards right)} \longrightarrow.$$

Additional reactive moment and forces at support B are

$$M_B = 3.8785 + (0.2526)(18) + (-7.3158)(0)$$

$$= +8.4253\text{ kN-m (causing tension inside)}$$

$$V_B = -V_A = -0.2526\text{ kN (downward)} \downarrow$$

$$H_B = -H_A = 7.3158\text{ kN (towards left)} \longleftarrow.$$

7.19 Effect of Temperature Change in Fixed Arch

(a) Symmetrical Arch. If the left support A of the fixed arch AB of Fig. 7.45(a) be made free as shown in Fig. 7.45(b), the point A will be displaced horizontally by $L\alpha t$ to the left due to a rise of temperature of $t°$, where L is horizontal span of arch and α is the thermal coefficient of expansion. To bring it to the original position, therefore, a horizontal displacement of $L\alpha t$ has to be applied towards the right. Hence, this case may be treated as horizontal yielding of magnitude $L\alpha t$ of left support towards the right. Then the values of M_o, V_o and H_o at the elastic centre due to effect of temperature change alone can be obtained from Eq. (7.44), taking h as $-L\alpha t$; $v = 0$, and $\beta = 0$.

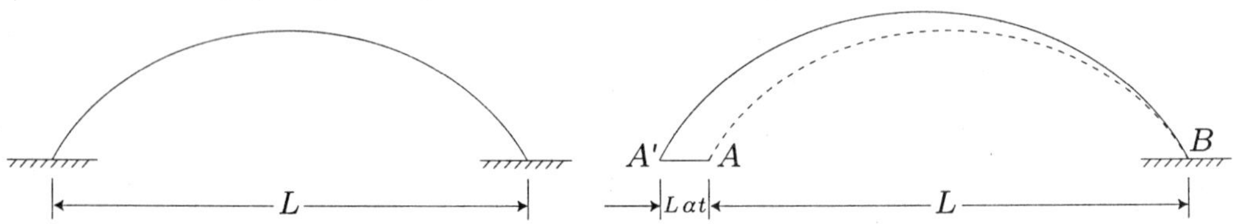

(a) Arch (b) Temperature effect

Displacement of arch if end *A* be free

Fig. 7.45: Temperature change in symmetrical fixed arch

$$\text{Thus, } M_o = 0;\ V_o = 0 \ \text{ and } \ H_o = -\frac{L\alpha t}{I_x} = -\frac{L\alpha t}{\sum \frac{y^2 ds}{EI}}. \tag{7.45}$$

For a temperature drop of $t°$, the value of h will be taken as $+L\alpha t$.

(b) Unsymmetrical Arch. For a temperature rise of $t°$ in the unsymmetrical arch AB of Fig. 7.46(a), the displacements of the left support A which is made free as shown in Fig. 7.46(b) will be $L\alpha t$ horizontally towards the left and $L'\alpha t$ vertically downwards where L and L' are the horizontal and vertical distances between the supports, the right support being at higher elevation than the left support and α being the coefficient of thermal expansion.

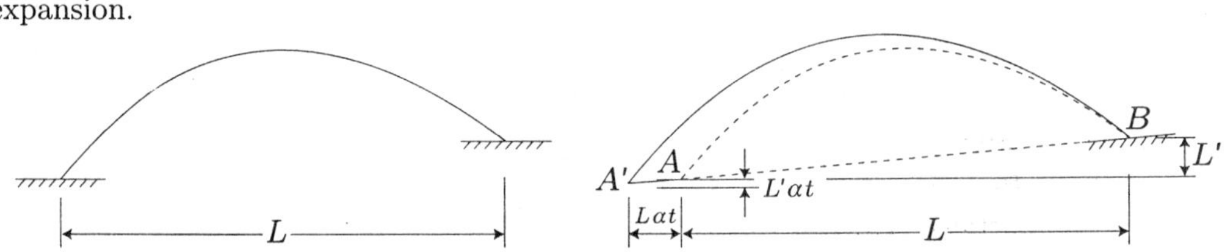

(a) Arch (b) Temperature effect

Displacement of arch of end *A* be free

Fig. 7.46: Temperature change in symmetrical fixed arch

To bring the point A to its original position, a rightward horizontal displacement of $L\alpha t$ and an upward vertical displacement of $L'\alpha t$ are required at this point. The effect due to these displacements can be determined treating the case as that of yielding of support A by $L\alpha t$ horizontally towards the right and $L'\alpha t$ vertically upwards. Thus the values of M_o, V_o and H_o acting at the elastic centre are obtained from Eq. (7.42), with h as $-L\alpha t$, v as $+L'\alpha t$ and β as 0. Thus it follows that,

$$\left.\begin{aligned} M_o &= 0, \\ V_o &= \frac{L'\alpha t + (L\alpha t)\frac{I_{xy}}{I_x}}{I_y - \frac{(I_{xy})^2}{I_x}} \\ H_o &= \frac{-L\alpha t - (L'\alpha t)\frac{I_{xy}}{I_y}}{I_x - \frac{(I_{xy})^2}{I_y}} \end{aligned}\right] \tag{7.46}$$

For a temperature drop of $t°$, the sign of V_o and H_o will be reversed.

7.20 Effect of Shrinkage in Fixed Arch

Shrinkage is a phenomenon usually associated with concrete structures resulting from the decrease in volume of concrete with drying. In indeterminate structures, shrinkage is prevented due to support restraints resulting in secondary moments and forces. In case of a reinforced concrete fixed arch, the shortening of span due to free shrinkage is prevented and its effect is thus similar to a temperature drop. The values of M_o, V_o and H_o acting at the elastic centre due to shrinkage alone can be obtained from Eqns. (7.45) and (7.46), as the case may be, putting a suitable negative value for t, so that shortening due to temperature drop is equivalent to that due to shrinkage. Usually the effect of shrinkage in concrete is considered equivalent to a temperature drop of 15 °F(≈ -9.5 °C).

7.21 Effect of Rib-Shortening in Fixed Arch

The arch rib is subjected to an axial compression in addition to bending moment and shear force. The axial compression causes elastic shortening of the rib. If σ be the uniform compressive stress, the free shortening of span is $\sigma L/E$, where L is the span length and E the modulus of elasticity. This shortening is possible if any of the supports has freedom for movement along the span and in that case no additional stresses will be induced. In case of a fixed arch, however, the elastic shortening due to axial compression is prevented and hence additional moments and forces are induced. The effect of rib-shortening is thus similar to a temperature drop and can be obtained from Eqs. (7.45) and (7.46) as the case may be replacing (αt) by $\left(-\frac{\sigma}{E}\right)$.

In practical case, where the section of the arch is non-uniform and the axial compression also varies from section to section, the compressive stress is not uniform throughout the centre length of the arch. It is, however, customary to assume a uniform stress of suitable magnitude for the purpose of analysis.

Example 7.15. *Determine the reactive forces and moment at the supports of the unsymmetrical fixed arch of Example 7.12 due to (i) a temperature rise of* 40 °F, *(ii) shrinkage equivalent to a temperature drop of* 15 °F *and (iii) rib-shortening caused by an average compressive stress of* 2 N/mm^2, *if* $EI_c = (4)(10)^5$ kN-m^2, $E = (10)^4$ kN/m^2. *Assume coefficient of thermal expansion* α *as* $(6)(10)^{-6}$/°F.

Solution : Horizontal span of the arch, $L = 36$ m

Vertical distance between supports, $L' = 4.5$ m

Since $I = I_c \sec\alpha$,

$$I_x = \sum \frac{y^2 ds}{EI} = \frac{1}{EI_c}\sum y^2 dx$$

$$I_y = \sum \frac{x^2 ds}{EI} = \frac{1}{EI_c}\sum x^2 dx$$

$$I_{xy} = \sum \frac{xyds}{EI} = \frac{1}{EI_c}\sum xydx.$$

Putting the values of the integrals as calculated in Example 7.12,

$$I_x = \frac{97.2}{(4)(10)^5} = (24.3)(10)^{-5}$$

$$I_y = \frac{3888}{(4)(10)^5} = (972)(10)^{-5}$$

$$I_{xy} = \frac{486}{(4)(10)^5} = (121.5)(10)^{-5}.$$

(i) Due to temperature rise

From Eq. (7.46),

$$M_o = 0$$

$$V_o = \frac{L'\alpha t + (L\alpha t)\left(\frac{I_{xy}}{I_x}\right)}{I_y - \frac{(I_{xy})^2}{I_x}} = \frac{(4.5)(6)(10)^{-6}(40) + (36)(6)(10)^{-6}(40)\left(\frac{121.5}{24.3}\right)}{(972)(10)^{-5} - \frac{(121.5)^2(10)^{-10}}{(24.3)(10)^{-5}}} = +12.05 \text{ kN}$$

$$H_o = \frac{-(36)(6)(10)^{-6}(40) - (4.5)(6)(10)^{-6}(40)\left(\frac{121.5}{972}\right)}{24.3(10)^{-5} - \frac{(121.5)^2(10)^{-10}}{(972)(10)^{-5}}} = -96.30 \text{ kN}.$$

At support A,

$$M_A = M_o + V_o \cdot x + H_o \cdot y$$

$$= 0 + (12.05)(-18) + (-96.3)(-4.5) = +216.45 \text{ kNm} \quad \text{(causing tension inside)}$$

$$V_A = V_o = 12.05 \text{ kN (upward)} \uparrow$$

$$H_A = H_o = -96.30 \text{ kN (towards right)} \longrightarrow.$$

At support B,

$$M_B = 0 + (12.05)(18) + (-96.3)(0) = (216.9) \text{ kNm} \quad \text{(causing tension inside)}$$

$$H_B = -H_A = 96.30 \text{ kN (towards left)} \longleftarrow$$

$$V_B = -V_A = -12.05 \text{ kN (downwards)} \downarrow.$$

(ii) Due to shrinkage

The reaction components due to shrinkage which is equivalent to a temperature drop of 15 °F can be obtained by multiplying the results of (i) by $-\frac{15}{40}$ i.e. $-\frac{3}{8}$.

(iii) Due to rib-shortening

The reaction components due to rib-shortening caused by an average compressive of 2 N/mm^2 can be obtained by multiplying the reslts of (i) by

$$-\frac{\sigma/E}{\alpha t}, \text{ i.e., } -\frac{2/(10)^4}{(6)(10)^{-6}(40)}, \text{ i.e., } -\frac{5}{6}.$$

7.22 Influence Lines for Symmetrical Parabolic Fixed Arch with $I = I_c \sec\alpha$

Fig. 7.47(a) shows a symmetrical parabolic arch subjected to a concentrated unit load at $x = kL$, where L is the span of the arch, and k, a numerical factor varying from zero to 0.5, x being measured horizontally from the elastic centre.

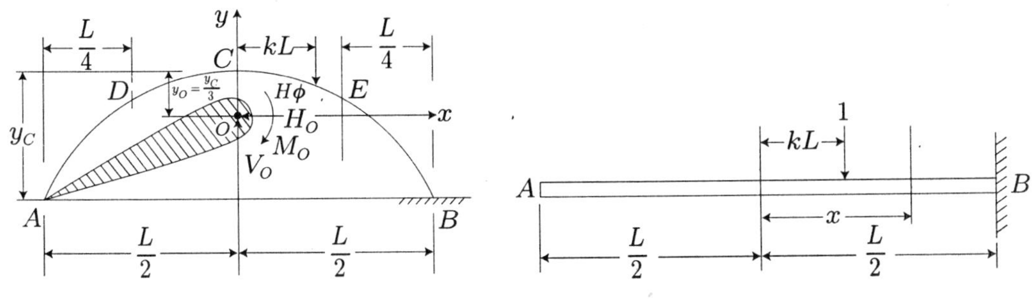

(a) Arch (b) Determinate conditions

Fig. 7.47: I.L. for symmetrical parabolic fixed end arch

Values of M_o, V_o and H_o acting at the elastic centre due to the loading condition can be obtained.

$$\left.\begin{aligned} \text{Since}\quad I &= I_c \sec\alpha, \\ M_o &= -\frac{\sum M_s dx}{\sum dx} \\ V_o &= -\frac{\sum M_s x ds}{\sum x^2 dx} \\ H_o &= -\frac{\sum M_s y dx}{\sum y^2 dx} \end{aligned}\right] \tag{a}$$

In Section 7.17, it has been shown that for such a case the elastic centre is at $\frac{y_c}{3}$ below the crown vertically where y_c is the height of crown from the springings and co-ordinates (x, y) of a point on the arch are related as

$$y = \frac{y_c}{3} - \frac{4y_c}{L^2}x^2. \tag{b}$$

Taking the elastic centre as the origin, for the case,

$$\left.\begin{aligned} \sum dx &= \int_{-L/2}^{+L/2} dx = L \\ \sum x^2 dx &= \int_{-L/2}^{+L/2} x^2 dx = \frac{1}{12}L^3 \\ \sum y^2 dx &= \int_{-L/2}^{+L/2} \left[\frac{1}{3}y_c - \frac{4y_c}{L^2}\cdot x^2\right]^2 dx = \frac{4}{45}y_c^2 L \end{aligned}\right\} \tag{7.47}$$

The bending moment M_s is obtained from the determinate condition shown in Fig. 7.47(b) which gives

$$M_s = 0 \qquad \text{for } x = -\frac{L}{2} \text{ to } +kL$$

and
$$= -(x - kL) \text{ for } x = +kL \text{ to } +\frac{L}{2}.$$

Hence
$$\sum M_s ds = \int_{kL}^{L/2} -(x - kL)dx = -\frac{L^2}{8}(1 - 4k + 4k^2)$$

$$\sum M_s x dx = \int_{kL}^{L/2} -(x - kL)x dx = -\frac{L^3}{24}[1 - 3k + 4k^3]$$

$$\sum M \cdot y dx = \int_{kL}^{L/2} [-(x - kL)] \left[\frac{y_c}{3} - \frac{4y_c}{L^2} \times x^2\right] dx = \frac{y_c L^2}{48}[1 - 8k^2 + 16k^4].$$

Using the values of these integrals in Eq. (a),

$$M_o = \frac{L}{8}(1 - 2k)^2$$

$$V_o = \frac{1}{2}(1 + k)(1 - 2k)^2$$

and
$$H_o = -\frac{15}{64} \cdot \frac{L}{y_c}(1 - 4k^2)^2. \tag{7.48}$$

The bending moment at any point on the arch having co-ordinates (x, y) with respect to the elastic centre as origin is given by

$$M = M_s + M_o + V_o \cdot x + H_o \cdot y,$$

positive value, which indicates moment causing tension inside. On the basis of the above equation, bending moments M_A, M_B, M_C, M_D and M_E at the supports A and B, crown C and quarter points D and E are as given in Table 7.5.

Table 7.5. Values of bending moments for unit load at kL from crown

Point	x	y	M_s	Bending Moment
A	$-\frac{L}{2}$	$-\frac{2}{3}y_c$	0	$M_A = M_o - \frac{L}{2}(V_o) - \frac{2}{3}y_c(H_o)$
B	$+\frac{L}{2}$	$-\frac{2}{3}y_c$	$-\left(\frac{L}{2} - kL\right)$	$M_B = -\left(\frac{L}{2} - kL\right) + M_o + \frac{L}{4}(V_o) - \frac{2}{3}y_c(H_o)$
C	0	$+\frac{1}{3}y_c$	0	$M_c = M_o + \frac{1}{3}y_c(H_o)$
D	$-\frac{L}{4}$	$+\frac{1}{12}y_c$	0	$M_D = M_o - \frac{L}{4}(V_o) + \frac{1}{12}y_c(H_o)$
E	$+\frac{L}{4}$	$+\frac{1}{12}y_c$	$-\left(\frac{L}{4} - kL\right)$ for $k \le \frac{1}{4}$ and zero for $k \ge \frac{1}{4}$.	$M_E = -\left(\frac{L}{4} - kL\right) + M_o + \frac{1}{4}(V_o) + \frac{1}{12}y_c(H_o)$ for $k \le \frac{1}{4}$ $= M_o + \frac{L}{4}(V_o) + \frac{1}{12}y_c(H_o)$, for $k \ge \frac{1}{4}$

Values of M_o, V_o and H_o and then those of M_A, M_B, M_C, M_D and M_E can be obtained for various positions of the unit load between B and C by putting appropriate values of k

in the respective expressions. The values of these moments are tabulated in Table 7.6 for some particular values of k.

Table 7.6. Values of Bending Moments for Various Positive Values of k

Value of k	M_o	V_o	H_o	M_A	M_B	M_C	M_D	M_E
0.0	0.125 L	0.500	−0.234 L/y_c	0.031L	+0.031 L	+0.047 L	−0.020 L	−0.020 L
0.1	0.08 L	0.352	−0.216 L/y_c	0.048 L	0.0	+0.008 L	−0.026 L	0.0
0.2	0.045 L	0.216	−0.165 L/y_c	0.047 L	−0.037 L	−0.010 L	−0.023 L	+0.035 L
0.25	0.031 L	0.156	−0.132 L/y_c	0.041 L	−0.053 L	−0.013 L	−0.019 L	+0.059 L
0.3	0.02 L	0.104	−0.096 L/y_c	0.032 L	−0.064 L	−0.0121 L	−0.014 L	+0.038 L
0.4	0.005 L	0.028	−0.030 L/y_c	0.0113 L	−0.0607 L	−0.0051 L	−0.0045 L	+0.0095 L
0.5	0.0	0.0	0.0	0.0	0.0	0.0	0.0	0.0

Knowing that the values of M_A and M_D for the unit load at kL from the crown on one side are the same as those of M_B and M_E respectively for the unit load at kL from the crown on the other side and *vice versa* and that the values of M_C are the same for the unit load at kL and $-kL$ because of symmetry, the values of M_A, M_B, M_C, M_D and M_E for unit loads acting between A and C can be obtained from the calculated values of the moments for the unit load acting between B and C. On the basis of what is stated above, the values of the moment for $k = 0$ to -0.5 are found and given in Table 7.7.

Table 7.7. Values of Bending Moments for Various Negative Values of k

Values of k	M_A	M_B	M_C	M_D	M_E
0.0	+0.031 L	+0.031 L	+0.047 L	−0.020 L	−0.020 L
−0.1	0.0	+0.048 L	+0.008 L	0.0	0.026 L
−0.2	−0.037 L	+0.047 L	−0.010 L	+0.035 L	−0.023 L
−0.25	−0.053 L	+0.041 L	−0.013 L	+0.059 L	−0.019 L
−0.3	−0.064 L	+0.032 L	−0.012 L	+0.038 L	−0.014 L
−0.4	−0.0607 L	+0.0113 L	−0.0051 L	+0.0095 L	−0.0045 L
−0.5	0.0	0.0	0.0	0.0	0.0

Plotting the values M_A, M_B, M_C, M_D and M_E as calculated above the respective influence lines are obtained as shown in Fig. 7.48(a) to 7.48(c).

Knowing that $H_A = H_o$, $H_B = -H_A$ and the value of H_A for unit load at $-kL$ from crown is identical to that of H_B for unit load at kL from crown, the values of H_A and H_B are obtained from values of H_o as given in Table 7.6 for various values of k and given in Table 7.8.

Table 7.8. Values of H_A (or H_B) for Various Values of k

Value of k	Value of $H_A\left(\frac{y_c}{L}\right)$ or $H_B\left(\frac{y_c}{L}\right)$	Value of k	Value of $H_A\left(\frac{y_c}{L}\right)$ or $H_B\left(\frac{y_c}{L}\right)$
−0.50	0.0	+0.10	0.216
−0.40	0.0304	+0.20	0.165
−0.30	0.096	+0.25	0.132
−0.25	0.132	+0.30	0.096
−0.20	0.165	+0.40	0.0304
−0.100	0.216	+0.50	0.0
−0.06	0.234		

Note : When positive, H_A is directed to the right and H_B is directed to the left.

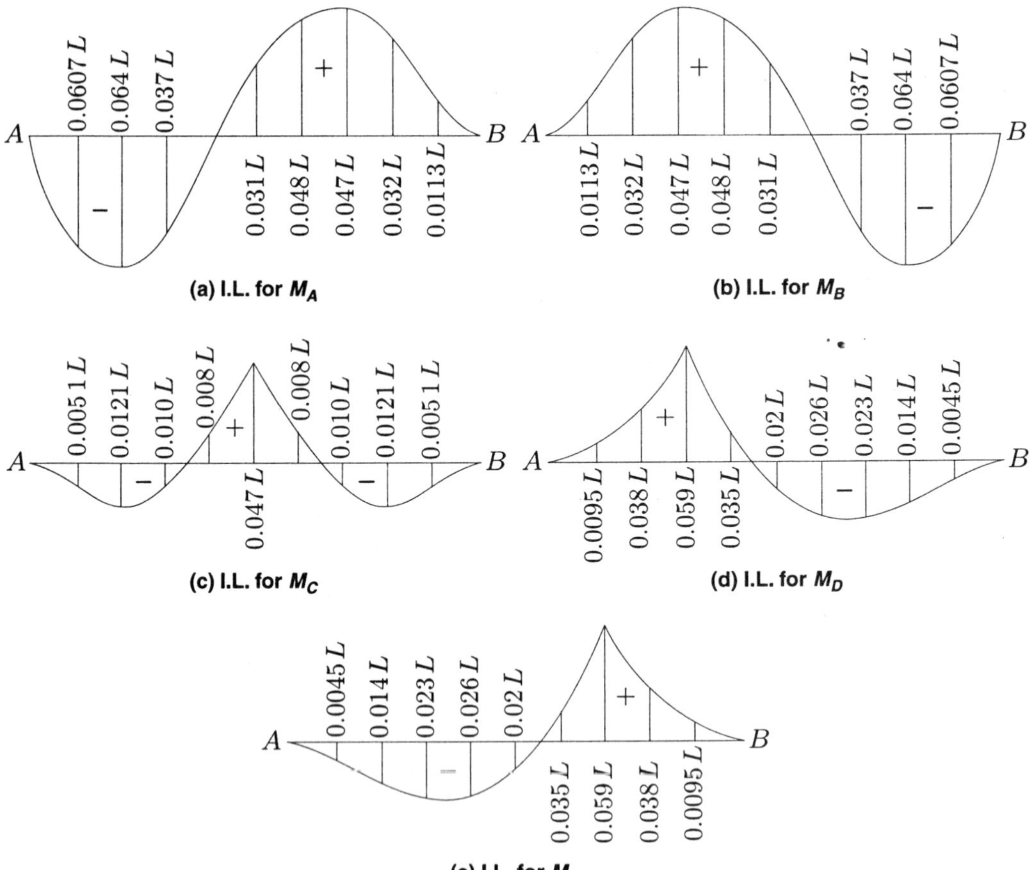

Fig. 7.48: I.L. for B.M. in fixed arch

Plotting the values of Table 7.8, the influence line for H_A or H_B can be obtained as shown in Fig. 7.49(a).

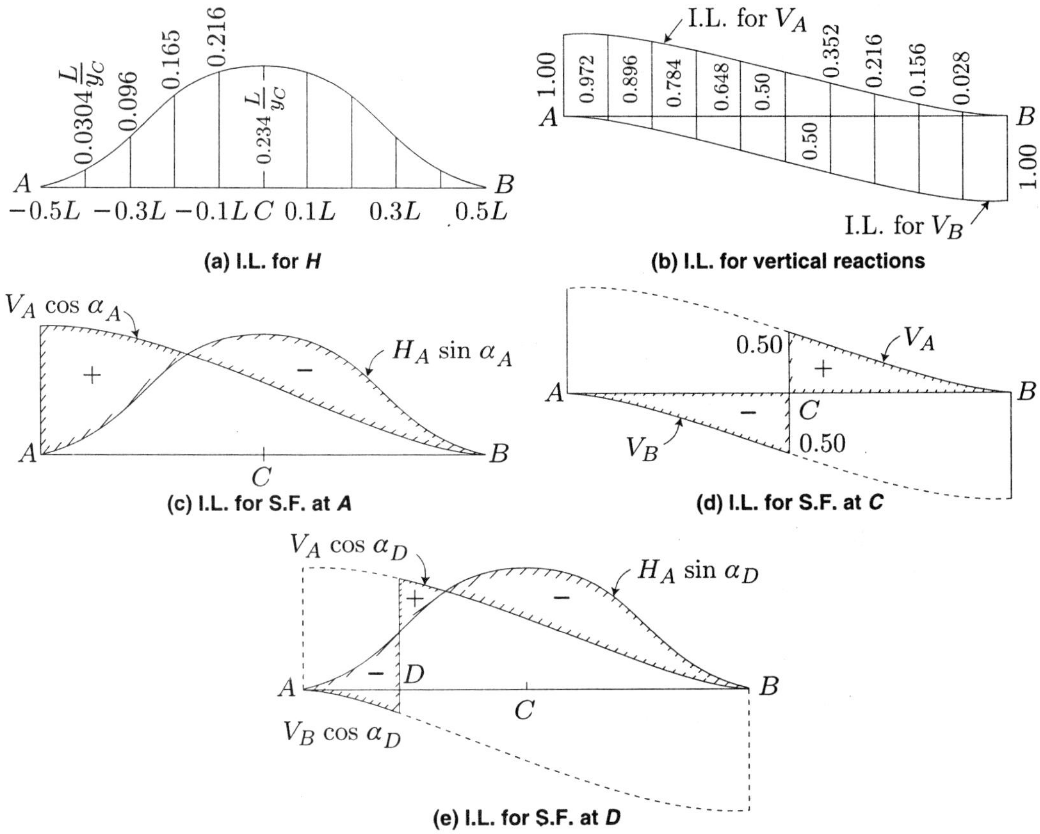

Fig. 7.49: I.L. for *H*, reactions and S.F. for fixed arches

As $V_A = V_o$, $V_B = 1 - V_A$ and the value of V_A for the unit load at $-kL$ is identical to that of V_B for the unit load at kL, values of V_A and V_B can be obtained from those of V_o as given in Table 7.6 for various values of k as in Table 7.9.

Table 7.9. Values of V_A and V_B for Various Values of k

Values of k	V_A	V_B	Value of k	V_A	V_B
−0.5	1.000	0.0	+0.1	0.352	0.648
−0.4	0.972	0.028	+0.2	0.216	0.784
−0.3	0.896	0.104	+0.25	0.156	0.844
−0.25	0.844	0.156	+0.3	0.104	0.896
−0.2	0.784	0.216	+0.4	0.028	0.972
−0.1	0.648	0.352	+0.5	0.0	1.000
0.0	0.500	0.500			

The influence lines for V_A and V_B are obtained by plotting values from Table 7.9 and are shown in Fig. 7.49(b).

Considering H_A and H_B positive when directed inward, V_A and V_B positive when directed upward and angle α made by the tangent to the arch at the point concerned with the horizontal positive when anticlockwise, as shown in Fig. 7.50. The radial shear force, which is taken positive when directed upward on the left of the section at different sections of the arch is given by

S.F. at $A = V_A \cos\alpha_A - H_A \sin\alpha_A$, for unit load anywhere in the arch

S.F. at $C = +V_A$ for unit load acting between B and C and

$= -V_B$ for unit load acting between A and C

S.F. at $D = V_A \cos\alpha_D - H_A \sin\alpha_D$ for unit load acting between B and D and

$= -V_B \cos\alpha_D - H_B \sin\alpha_D$ for unit load acting between A and B.

Influence lines for radial shear force at any section of the arch can be obtained by using the influence lines for V_A, V_B, H_A and H_B multiplying the IL ordinates by sine or cosine of the slope as the case may be and superimposing in appropriate manner. The influence lines for radial shear force at A, C and D are shown in Figs. 7.49(d) to 7.49(e).

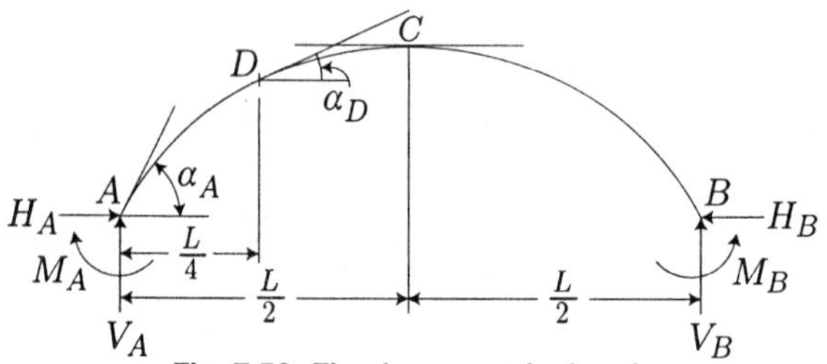

Fig. 7.50: Fixed symmetrical arch

With the positive direction of forces and slopes as indicated in Fig. 7.50, the normal thrust N at different sections of the arch is given below :

At A, $N = V_A \sin\alpha_A + H_A \cos\alpha_A$, for unit load anywhere on the arch

At C, $N = H_A$

and at D, $N = V_A \sin\alpha_D + H_A \cos\alpha_D$, for unit load acting between B and D and

$= -V_B \sin\alpha_D + H_B \cos\alpha_D$, for unit load acting between A and D.

Hence the influence lines for normal thrust at any section of the arch can be obtained using the influence lines for V_A, V_B, H_A and H_B as shown in Figs. 7.51(a) to 7.51(c) for the sections at A, C and D.

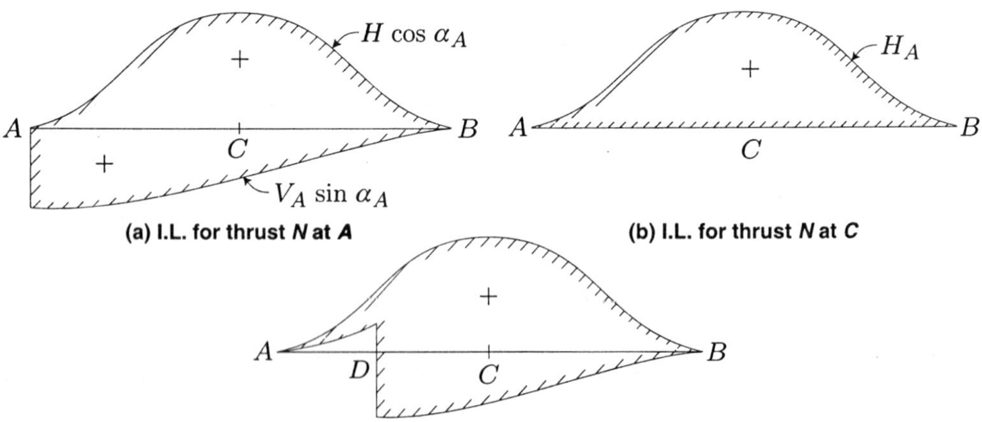

Fig. 7.51: I.L. for normal thrust in fixed arch

7.23 Influence lines for Unsymmetrical Parabolic Fixed Arch with $I = I_c \sec\alpha$.

The procedure laid down in the previous Section is also applicable to unsymmetrical arches with the following modifications :

(i) Distances x_o and y_o of the elastic centre from crown are to be obtained from Eq. (7.39).

(ii) As $\sum xydx$ is not zero for such an arch, the value of the integral and that of $\sum M_s xydx$ have to be evaluated.

(iii) The values of M_o, V_o and H_o have to be calculated from Eq. (7.34).

(iv) Two sets of values of M_o, V_o, and H_o have to be determined for two positions of the unit load, viz., (a) at kL from the crown and (b) at $-kL$ from the crown.

The values of M_o, V_o and H_o and those of M_A, M_o, M_B, etc. are then obtained for various values of k.

7.24 Numerical Method of Solution for Fixed Arch

In the previous Sections, the variation of the arch cross section has been taken to follow a particular pattern so that moment of inertia I for a parabolic arch was considered as varying as $I_c \sec\alpha$. For such cases, direct evaluation of integrals given by Eq. (7.35) has been possible. But practical arches very often do no follow exactly parabolic form and the assumed relation $I = I_c \sec\alpha$ and for such practical arches direct evaluation of the associated integrals is practically impossible. For such cases of fixed arches, where the variation of section cannot be represented by suitable relationship, numerical method proves to be expedient, though the values obtained by this method are approximate. However, increasing the number of divisions, into which the arch is to be divided, the results may be brought to any desired accuracy.

In this method, the curved length of the arch is divided into a number of equal segments. Alternately, the segments may be of unequal length such that their projected lengths Δx on the horizontal are equal. The later method is simple and easy to work. Each segment of arch is, however, considered to be straight and the centroid of each segment may be taken at the middle of its projected length. The method is discussed in detail below.

(a) Unsymmetrical Arch. The arch is divided into several segments such that their horizontal projections Δx are equal. It is convenient to make some full number of divisions on each side of the crown as shown in Fig. 7.52 where AB, BC, CD, DE, EF and FG are the segments in which the arch is divided. The centroid of each segment is approximately taken at the middle of its horizontal projection. In Fig. 7.52, the centriods are marked as 1, 2,. . ., 6.

$$ds = \sqrt{(\Delta x)^2 + (\Delta y)^2}, \tag{7.49}$$

where Δx and Δy are the projected lengths of the segment in horizontal and vertical planes respectively. Table 7.10 shows systematic calculation of segment lengths. x' and y' are the co-ordinates of the division points A, B, . . ., etc. with respect to the crown as the origin and obtained from the given shape of the arch. The differences in values of x' and y' between the two ends of a segment give the values of Δx and Δy respectively for that segment.

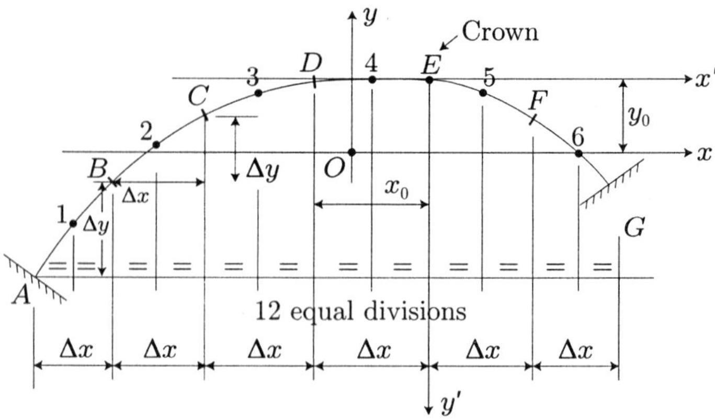

Fig. 7.52: Unsymmetrical fixed arch

Table 7.10. Determination of Segment Length

Points	x'	y'	Segment	Δx	Δy	ds
(1)	(2)	(3)	(4)	(5)	(6)	(7)
A			–			
B			AB			
C			BC			
D			CD			
E			DE			
F			EF			
G			FG			

In the next step, the position of the elastic centre is to be determined, that is, its horizontal and vertical distances x_o and y_o respectively with respect to the crown are to be determined. The numerical calculations for this can be done in a tabular form as shown in Table 7.11.

Table 7.11. Determination of Elastic Centre

Centroid	x'	y'	t	$r = \frac{I}{I_c}$	ds	$\frac{ds}{r}$	$\frac{x'ds}{r}$	$\frac{y'ds}{r}$
(1)	(2)	(3)	(4)	(5)	(6)	(7)	(8)	(9)
1								
2								
3								
4								
5								
6								
Total						$\sum \frac{ds}{r}$	$\sum \frac{x'ds}{r}$	$\sum \frac{y'ds}{r}$

$$\sum \frac{ds}{EI} = \frac{1}{EI_c}\sum \frac{ds}{r};\ \sum \frac{x'ds}{EI} = \frac{1}{EI_c}\sum \frac{x'ds}{r};\ \sum \frac{y'ds}{EI} = \frac{1}{EI_c}\sum \frac{y'ds}{r}.$$

$$x_o = \frac{\sum \frac{x'ds}{EI}}{\sum \frac{ds}{EI}} = \frac{\sum \frac{x'ds}{r}}{\sum \frac{ds}{r}};\ y_o = \frac{\sum \frac{y'ds}{EI}}{\sum \frac{ds}{EI}} = \frac{\sum \frac{y'ds}{r}}{\sum \frac{ds}{r}}.$$

The thickness t and other dimensions of the arch section at the centroid of any segment can be obtained from the given condition of variation of the arch section. From the section dimensions, the moments of inertia I of the arch section at the centroids of the segments are calculated. The ratio r of the moment of inertia I of the arch section at the centroid of a segment to the moment of inertia I_c at the crown is then calculated and entered in column 5 of Table 7.11. The summations of the values in columns 7, 8 and 9 give the required integrals for evaluation of x_o and y_o as shown above.

Calculations of integrals involved in the determination of M_o, V_o and H_o as given in Eqn. (7.34) are shown in Table 7.12. In this Table, x and y of columns (2) and (3) respectively indicate co-ordinates of the segment centroids with respect to the elastic centre as origin.

Table 7.12. Evaluation of Integrals for Determination of M_o, V_o and H_o

Centroid	x	y	$\frac{ds}{r}$	M_s	$\frac{x^2ds}{r}$	$\frac{y^2ds}{r}$	$\frac{xyds}{r}$	$\frac{M_sds}{r}$	$\frac{M_sxds}{r}$	$\frac{M_syds}{r}$
(1)	(2)	(3)	(4)	(5)	(6)	(7)	(8)	(9)	(10)	(11)
1										
2										
3										
4										
5										
6										
Total			$\sum \frac{ds}{r}$		$\sum \frac{x^2ds}{r}$	$\sum \frac{y^2ds}{r}$	$\sum \frac{xyds}{r}$	$\sum \frac{M_sds}{r}$	$\sum \frac{M_sxds}{r}$	$\sum \frac{M_syds}{r}$

Having obtained M_o, V_o and H_o from Eq. (7.34), the bending moment, radial shear and normal thrust at any section of the arch can be determined from Eqs. (7.28) and (7.30).

$$\sum \frac{ds}{EI} = \frac{1}{EI_c}\sum \frac{ds}{r};\ \sum \frac{x^2ds}{EI} = \frac{1}{EI_c}\sum \frac{x^2ds}{r};\ \sum \frac{y^2ds}{EI} = \frac{1}{EI_c}\sum \frac{y^2ds}{r}$$

$$\sum \frac{xyds}{EI} = \frac{1}{EI_c}\sum \frac{xyds}{r};\ \sum \frac{M_sds}{EI} = \frac{1}{EI_c}\sum \frac{M_sds}{r};$$

$$\sum \frac{M_sxds}{EI} = \frac{1}{EI_c}\sum \frac{M_sxds}{r};\ \sum \frac{M_syds}{EI} = \frac{1}{EI_c}\sum \frac{M_syds}{r}.$$

(b) Symmetrical Arch. For the symmetrical arch the procedure is the same as that for unsymmetrical arch with the only difference that in this case calculation of $xyds/r$ in column 8 of Table 7.12 is not necessary and that because of symmetry the summation calculations can be done for the half length of the arch only. The values of M_o, V_o and H_o are calculated using Eq. (7.33).

Example 7.16. *Fig. 7.53(a) shows an unsymmetrical parabolic fixed arch of rectangular section having uniform width b, but variable thickness which is given by*

$$t = \left(0.6 + \frac{|x|}{7}\right) m,$$

where x is horizontal distance of the section from the crown. Determine the reactive forces and moments at the supports.

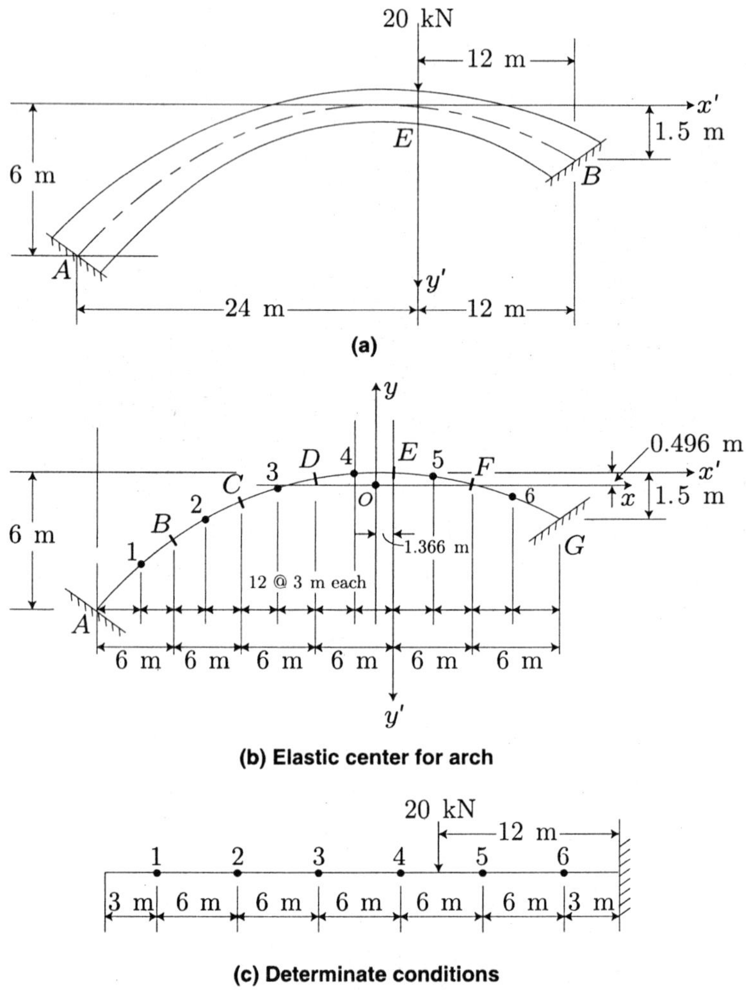

Fig. 7.53: For Example 7.16

Solution : The arch is divided into six segments AB, BC, CD, DE, EF and FG as shown in Fig. 7.53(b), the centroids of the segments being marked as 1, 2, 3, 4, 5 and 6. The co-ordinates of the points 1, 2, 3, ..., 6 with respect to the crown as the origin are obtained from

$$y' = \frac{4y_c}{L^2}(x')^2$$

i.e., $$y' = \frac{(4)(6)}{(48)^2}(x')^2 = \frac{(x')^2}{96}.$$

Calculation of segment length ds is shown in Table 7.13.

Table 7.13. Calculation of Segment Length ds for Example 7.16

Points	$x'(m)$	$y'(m)$	Segment	$\Delta x(m)$	$\Delta y(m)$	$ds(m)$
A	−24	6.0	–	–	–	–
B	−18	3.375	AB	6	2.625	6.549
C	−12	1.500	BC	6	1.875	6.286
D	−6	0.375	CD	6	1.125	6.105
E	0.0	0.0	DE	6	0.375	6.012
F	+6	0.375	EF	6	0.375	6.012
G	+12	1.500	FG	6	1.125	6.105

Tables 7.14 and 7.15 show determination of elastic centre and the evaluation of various integrals involved in determination of M_o, V_o and H_o, the values of M_s being obtained from the determinate condition shown in Fig. 7.53(c).

Table 7.14. Determination of Elastic Centre

Centroid of segments	x'(m)	y'(m)	t'(m)	r	ds(m)	$\frac{ds}{r}$	$\frac{x'ds}{r}$ (m^2)	$\frac{y'ds}{r}$ (m^2)
1	−21	4.596	3.600	216.0	6.549	0.0303	−0.6363	0.1393
2	−15	2.344	2.743	95.5	6.286	0.0658	−0.9870	0.1542
3	−9	0.844	1.886	31.1	6.105	0.1963	−1.7667	0.1657
4	−3	0.094	1.029	5.0	6.012	1.2024	−3.6072	0.1130
5	+3	0.094	1.029	5.0	6.012	1.2024	+9.6072	0.1130
6	+9	0.844	1.886	31.1	6.105	0.1963	+1.7667	0.1657
Total						2.8935	−1.6233	0.8509

Note : $r = \dfrac{I}{I_o} = \left(\dfrac{d}{d_o}\right)^8$; $d_c = 0.600$ m

$$x_o = \frac{\sum \frac{x'ds}{r}}{\sum \frac{ds}{r}} = \frac{-1.6233}{2.8935} = -0.561 \text{ m}$$

$$y_o = \frac{\sum \frac{y'ds}{r}}{\sum \frac{ds}{r}} = \frac{0.8509}{0.8935} = 0.294 \text{ m.}$$

Table 7.15. Evaluation of Integrals

Centroid of segments	x	y	$\frac{ds}{r}$	M_s (kN-m)	$\frac{x^2ds}{r}$	$\frac{y^2ds}{r}$	$\frac{xyds}{r}$	$\frac{M_sds}{r}$	$\frac{M_sxds}{r}$	$\frac{M_syds}{r}$
1	−20.439	−4.302	0.0303	0	12.658	0.561	+2.664	0	0	0
2	−14.439	−2.05	0.0658	0	13.718	0.277	+1.948	0	0	0
3	−8.439	−0.55	0.1963	0	13.980	0.059	+0.911	0	0	0
4	−2.439	+0.20	1.2024	0	7.153	0.048	−0.587	0	0	0
5	+3.561	+0.20	1.2024	−60.0	15.247	0.018	+0.856	−72.144	−256.9	−14.43
6	+9.561	−0.55	0.1963	−180.0	17.944	0.059	−1.032	−35.334	−337.8	+19.43
Total			2.8935		80.70	1.052	4.76	−107.48	−594.7	+5.0

From Eq. (7.34),

$$M_o = -\frac{-107.48}{2.8935} = +37.15 \text{ kNm}$$

$$V_o = -\frac{-594.7 - (5)\left(\frac{4.76}{1.052}\right)}{80.7 - \frac{(4.76)^2}{1.052}} = +10.43 \text{ kN}$$

$$H_o = -\frac{5.0 - (-594.7)\left(\frac{4.76}{80.7}\right)}{1.052 - \frac{(4.76)^2}{80.7}} = -51.97 \text{ kN.}$$

For the left springing $A, x = -22.634$ m, $y = -5.504$ m and $M_s = 0$

$$M_A = M_s + M_o + V_o.x + H_o.y$$

$$= 0 + 37.15 + (10.43)(-22.634) + (-51.97)(-5.504)$$

$$= 87.12 \text{ kNm (causing tension inside)}$$

$$H_A = H_o = -51.97 \text{ kN (towards right)} \longrightarrow$$

$$V_A = V_o = 10.43 \text{ kN (upwards)} \uparrow.$$

For the right springing G,

$$x = +13.366 \text{ m}, y = -1.004 \text{ m}$$

and $M_s = -240$ kNm

$$M_G = -240 + 37.15 + (10.43)(13.366) + (-51.97)(-1.00)$$

$$= -11.26 \text{ kNm (causing tension outside)}$$

$$H_G = -H_A = 51.97 \text{ kN (towards left)} \longleftarrow$$

$$V_G = 20 - V_A = 20 - 10.43 = 9.57 \text{ kN (upwards)} \uparrow.$$

7.25 Fixed Portal Frame

The methods described for fixed arch including the elastic centre method are applicable to fixed portal frame also. The elastic centre may be located with respect to any support as the origin as shown in Fig. 7.54(a) from Eq. (a) of Section 7.17.

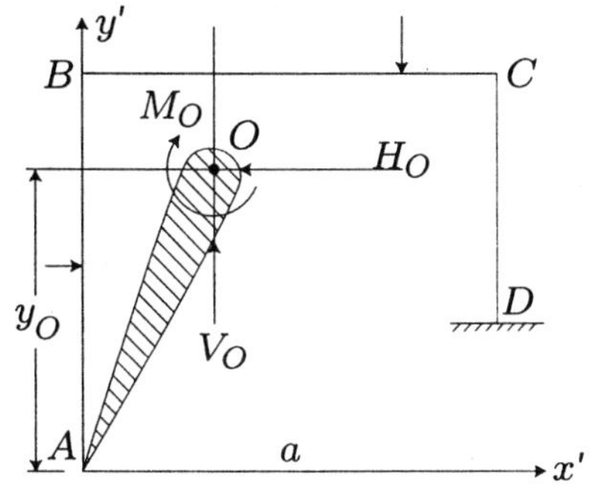

(a) Elastic centre in fixed portal frame **(b) Determinate cantilever portal frame**

Fig. 7.54: Fixed Portal frame

The general equations for M_o, V_o and H_o acting at elastic centre are the same as Eq. (7.34). Determinate bending moment M_s will be obtained from the cantilever portal frame shown in Fig. 7.54(b) for the applied loads. In case of portal frames having prismatic vertical and horizontal members, the evaluation of integrals is comparatively easier.

As mentioned in case of hingeless arch, if the frame is symmetrical, it is convenient to solve it by cutting the horizontal member at its mid-point and by assuming that a rigid arm is attached to the part of the frame on each side of the cut. For unsymmetrical frame, however, rigid arm attached to one end is more convenient.

Example 7.17. *Analyse the frame shown in Fig. 7.55(a), using elastic centre approach.*

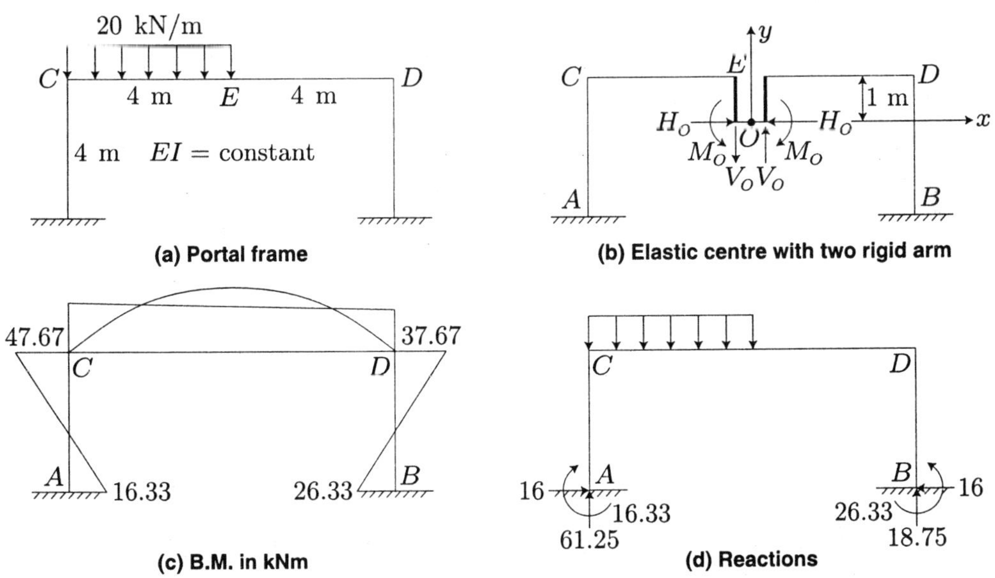

Fig. 7.55: For Example 7.17

Solution : Since the frame is symmetrical structurally, the frame is cut at mid-point E of the member BC and two adjacent edges of these cut are connected to the elastic centre by means of two rigid arms. The position of the elastic centre is as shown in Fig. 7.55(b) with its distance at

$$\overline{x} = \frac{4(2)(2)}{2(4)+1(8)} = 1 \text{ m from } BC.$$

With elastic centre O as origin,

$$\sum x^2 dx = 2[4(4)^2] + \frac{(8)^2}{12} = 170.67$$

$$\sum y^2 dx = 2\left[\frac{(4)^3}{12} + 4(2-1)^2\right] + 8(1)^2 = 26.67$$

and $\sum dx = 2(4) + 8 = 16.$

B.M. M_s due to determinate condition is obtained from Fig. 7.55(b).

$$\sum M_s ds = \int_{-4}^{0} -20\left(\frac{(x)^2}{2}\right) dx + (-160)(4) = -853.33$$

$$\sum M_s x dx = \int_{-4}^{0} -20\left(\frac{x^2}{2}\right) x dx + (-160)(4)(-4) = +3200$$

$$\sum M_s y dx = \int_{-4}^{0} -20\left(\frac{x^2}{2}\right)(1)(dx) + (-160)(4)(-1) = +426.67.$$

Solving $M_o = -\dfrac{-853.33}{16} = +53.33$ kNm

$$V_o = -\frac{+3200}{170.67} = -18.75 \text{ kN (upward)} \uparrow$$

$$H_o = -\frac{+426.67}{26.67} = -16 \text{ kN (towards left)} \longleftarrow.$$

It follows that

$$M_A = 55.33 + 16(4-1) + 18.75(4) - \frac{20(4)^2}{2}$$

$$= 18.33 \text{ kN (Tension inside)}$$

$$V_A = (20)(4) - 18.75 = 61.25 \text{ kN} \uparrow$$

$$H_A = -H_o = 16 \text{ kN (towards right)} \longrightarrow.$$

Also $M_o = 53.33 - 16(1) + 18.75(4) - \dfrac{20(4)^2}{2}$

$$= -47.67 \text{ kNm (tension outside)}$$

$$M_B = 53.33 + 16(3) - 18.75(4) = 26.33 \text{ (tension inside)}$$

$$M_D = 33.33 - 16(1) - 18.75(4) = -37.67 \text{ (tension outside)}$$

$$V_B = 18.75 \text{ kN} \uparrow$$

$$H_B = 16 \text{ kN (towards left)} \longleftarrow.$$

B.M. and Reaction diagrams are shown in Fig. 7.55.

Problems

7.1 A two-hinged circular arch has a span of 40 m and rise of 8 m with both supports at the same level. It is loaded with a concentrated load of 100 kN at the centre. Calculate the horizontal thrust and draw the B.M. diagram. If the temperature rises by 40 °C, calculate the change in horizontal thrust and change in bending stress. Coefficient of expansion $\alpha = 11(10)^{-6}$ per °C and $E = 200$ kN/mm^2. The section is 2000 mm deep throughout the length.

7.2 Derive an expression for the horizontal thrust of a symmetrical two-hinged parabolic arch of span L and rise d, if a concentrated load P be placed at a distance kL from one end. Assume $I = I_c \sec\alpha$.

Use the expression to obtain the influence lines for horizontal thrust of a similar arch having span =10 (rise). Indicate the values at various points.

7.3 (a) What are the effects of temperature rise and rib-shortening on the magnitude of horizontal thrust in case of a two-hinged arch.

(b) A symmetrical two-hinged parabolic arch having secant variation of moment of inertial has a central rise y_c and a span L. Show that H for a rise of temperature of t is given by

$$\frac{15EI_c\alpha t}{8y_o^2}.$$

Thus show that due to a rise of temperature of 20 °C the horizontal thrust is the same as that due to a central point load equal to $\frac{192EI_c\alpha}{y_c(L)}$, where α is the coefficient of expansion.

7.4 A triangular two-hinged arch of uniform EI has a span L with the apex point at a horizontal distance 'a' from left-hand support and 'b' from the right-hand support, the span of the arch being $(a+b) = L$. The rise of the arch is 'h'. The arch is subjected to a concentrated load P at the apex. Calculate values of horizontal thrust and indicate the B.M. diagram. Also obtain the radial shear and normal thrust diagram.

If the arch be subjected to u.d.l. of magnitude w/unit length, obtain the B.M. diagram.

7.5 A two-hinged symmetrical parabolic arch of span 30 m and rise of 3 m is subjected to u.d.l. on the left-half of the span of magnitude 30 kN/m along with a concentrated load of 100 kN at the crown. Assume secant variation of the moment of inertia. Determine the horizontal thrust and obtain B.M. diagram indicating salient values.

7.6 A parabolic arch having secant variation of M.I. with span 30 m, rise 3 m is subjected to a concentrated vertical load of magnitude 300 kN at 10 m from left-hand support. One of the supports is a hinge, the other being a roller. A tie of area 3000 mm^2 connects the arch at the two supports. Determine the force in the tie-rod and obtain the B.M. diagrams, considering rib-shortening effect. Assume area of rib as 20000 mm^2, $I = 5(10)^8 \sec\alpha$ mm^4, $E_{arch} = 16$ kN/m^2 and $E_{tie} = 200$ kN/mm^8.

7.7 A symmetrical parabolic arch AB of span 20 m and rise 4 m is pinned at both ends A and B. The arch is subjected to a horizontal u.d.l. of magnitude 20 kN/m from left-hand A to the crown. Calculate the horizontal thrust, if rib-shortening effects are not considered. Draw the B.M. diagram.

7.8 If right-hand support of the arch in Problem 7.6 moves out by 5 mm, calculate the horizontal thrust and draw the B.M. diagram.

7.9 Draw influence lines for (*a*) horizontal thrust (*b*) B.M. at crown (*c*) Radial S.F. at $\frac{1}{4}$ span and (*d*) Normal thrust at $\frac{1}{4}$ span, for a two-hinged parabolic arch of span 30 m and rise of 5 m. Assume $I = I_c \sec\alpha$. Also evaluate the magnitude of the functions when a u.d.l. of magnitude 10 kN/m is placed on the entire span of the arch.

7.10 The reaction locus for a two-hinged spandrel arch is given by

$$y = (y_c + d)[1.3 - 0.2\sin(k360° - 90°)].$$

Making use of this formula, construct influence lines for the arch shown in Fig. P-7.1 for (*a*) Horizontal Thrust (*b*) Forces in members marked (1), (2), (3) and (4). Also determine the maximum forces in members due to a D.L. of 20 kN/m and L.L. of 40 kN/m of length larger than span together with a knife edge load of magnitude 500 kN.

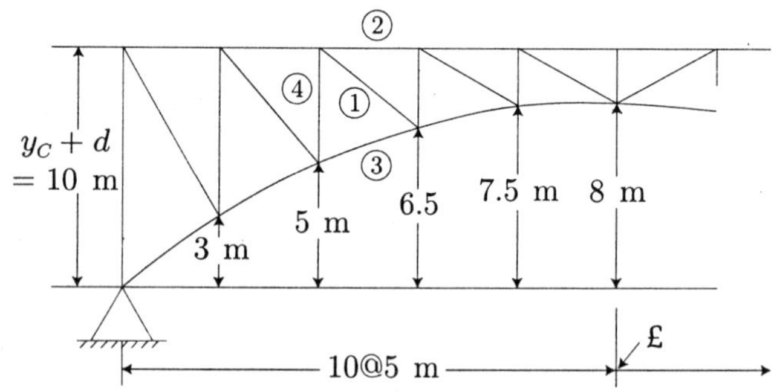

Fig. P-7.1

7.11 A semicircular arch of radias R with the two ends hinged is subjected to a concentrated load W at the centre. Obtain the horizontal thrust, if the arch has a constant cross section. Also draw the B.M. diagram.

7.12 Derive expressions for the moment, shear and thrust at the crown of a symmetrical parabolic fixed arch having $I = I_c \sec\alpha$. Neglect effect of rib-shortening. Consider the arch as loaded with vertical u.d.l. on the left half of the span.

If the span of the arch be 30 m and rise 6 m and u.d.l. on the left half of the span be 50 kN/m, calculate the values and draw B.M. diagram indicating salient values.

7.13 (a) Explain the procedure for analysis of a fixed arch by the elastic centre method.

(b) Considering a parabolic arch having $I = I_c \sec\alpha$, obtain expressions for the position of the elastic centre. Also deduce the equations for analysis of the arch in the case.

7.14 Obtain expressions for a symmetrical parabolic fixed arch of span L, having $I = I_c \sec \alpha$ for the following cases :

(a) Temperature effect

(b) Shrinkage effect

(c) Rib-shortening effect

(d) Horizontal yield of a support

and (e) Vertical yield of a support.

7.15 A symmetrical fixed ended parabolic arch of span 48 m and rise 6 m is subjected to a temperature increase of of 40 °C. Assume $\alpha = 12(10)^{-8}$ per °C, $I_c = 2(10)^{10}$ mm^4. Calculate the value of the horizontal thrust due to the temperature change. Also draw the resulting B.M. diagram.

What will be the effect due to shrinkage if it is considered as equivalent to a temperature change of 10 °C.

Considering stress in the ribs due to normal thrust as 2 N/mm^2, calculate the horizontal thrust due to rib-shortening.

7.16 An unsymmetrical parabolic arch of span 36 m has its left-hand support below the right-hand one, the crown being at a distance of 24 m from the left-hand support. In the case, $I = I_c \sec \alpha$. If the arch be subjected to a concentrated load of magnitude 150 kN at 12 m from left-hand support, obtain the magnitude of the horizontal thrust, the vertical reactions and moments at supports. Also draw the B.M. diagram. Consider two cases of supports, viz.,

(a) Both ends hinged (b) Both ends fixed.

7.17 Explain qualitatively the effect of :

(a) Rib-shortening (b) Fall of temperature and (c) Movement of supports.

In case of fixed ended arch, what will be the typical B.M. diagram for the above causes in case of a symmetrical parabolic arch.

Derive expressions to obtain moment, thrust and shear at crown of the fixed arch due to above causes.

7.18 Determine all reaction components induced on the symmetrical circular fixed arch shown in Fig. P-7.2 due to each of the following causes. The arch is also subjected to u.d.l. of magnitude 20 kN/m on the full span.

(a) Temperature rise of 30 °C.

(b) Temperature drop of 50 °C.

(c) Shrinkage equivalent to a temperature drop of 10 °C.

(d) Rib-shortening equivalent to an axial stress of 2 N/mm^2 throughout the rib.

(e) Rotational slip of 0.001 radians counterclockwise at right-hand support.

(f) Horizontal displacement of 5 mm to the right of support B.

(g) Vertical settlement of 6 mm downward at support B.

Assume $\alpha = 10(10)^{-6}$ per °C and $E = 15$ kN/m^2.

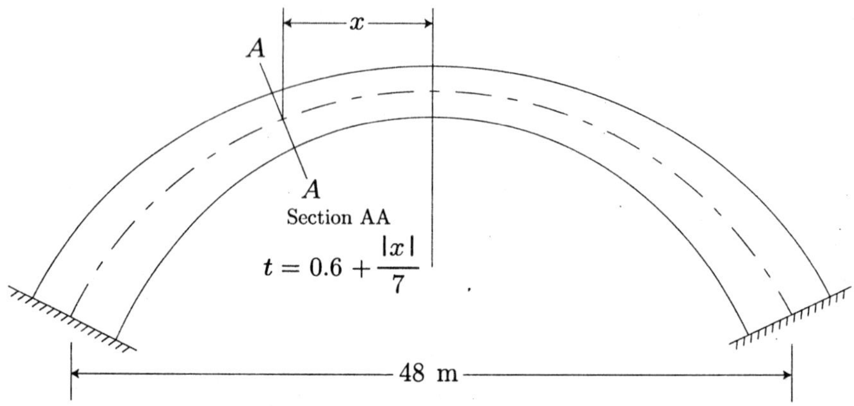

Fig. P-7.2

7.19 Determine the effects in Problem 7.18 for an unsymmetrical fixed-end arch obtained from Fig. P-7.2 with span of 48 m having left-hand support A below the right-hand support B, if the crown be at a distance 18 m from B, the vertical distances from crown A and B being 30 m and 6 m respectively.

7.20 Analyse the rigid frames shown in Fig. P-7.3.

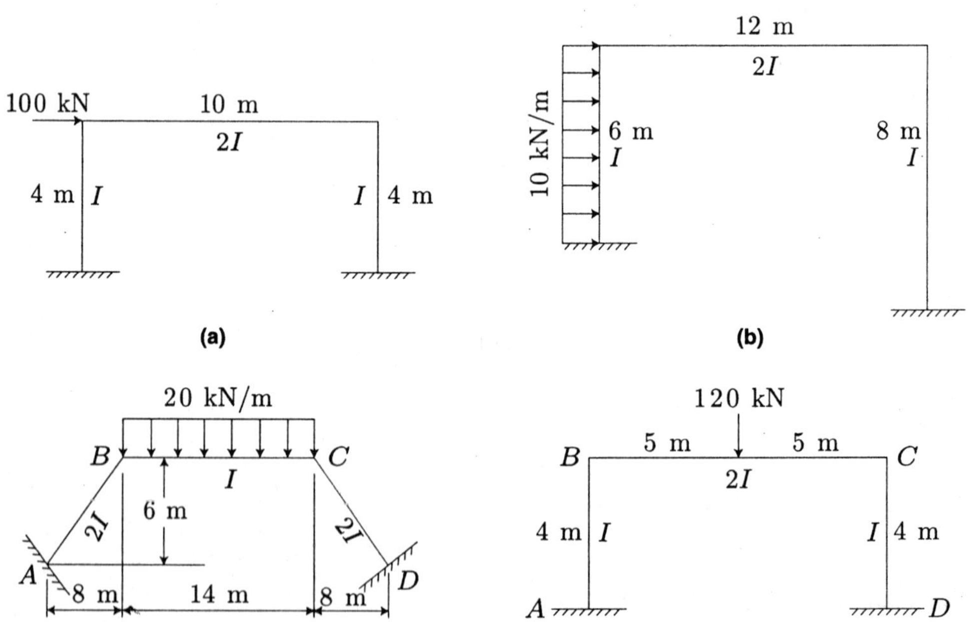

Fig. P-7.3 (Contd.)

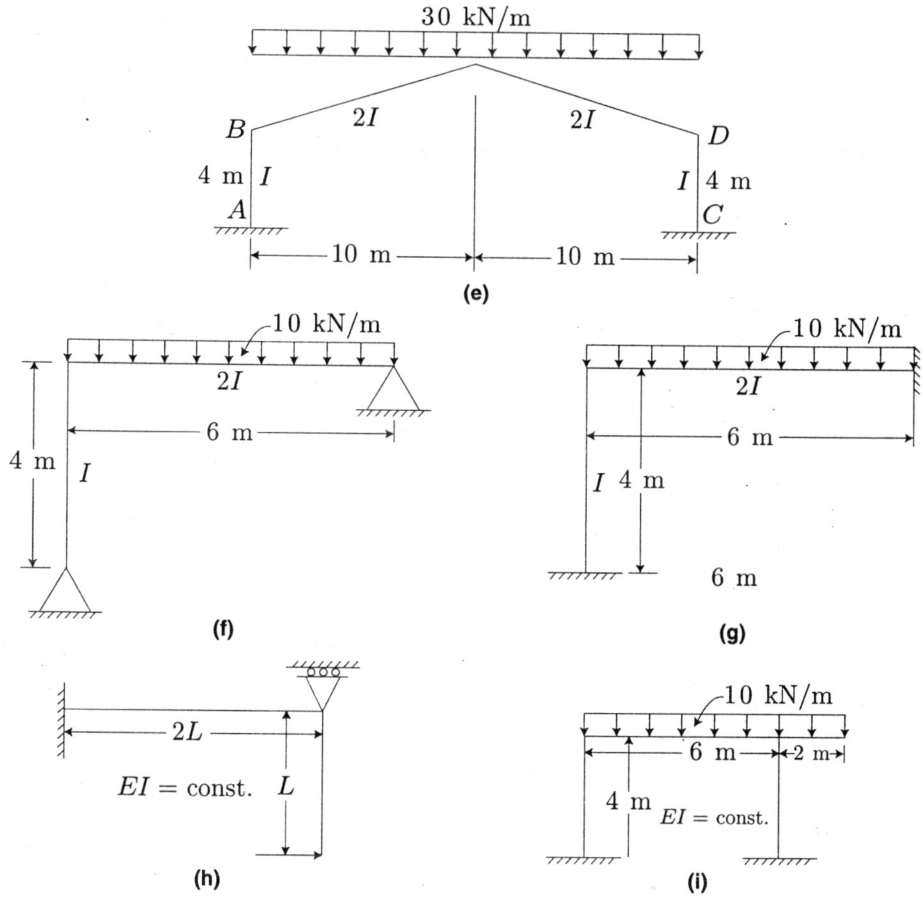

Fig. P-7.3

7.21 A parabolic arch of span 100 m and a central rise of 25 m is hinged at ends. The arch carries uniformly distributed load 10 kN/m over the left half of the span. Find the values of the horizontal thrust and the maximum B.M. over the arch. Assume $I = I_c \sec \alpha$.

How will the magnitude of the horizontal thrust be modified if one of supports be a roller, the arch being connected by a tie at the support level. The tie is of the same material as the arch. Assume area of cross section of the tie as 1000 mm^2.

7.22 A parabolic arch of span L and having a central rise y_c is hinged at its ends and is subjected to uniformly distributed load of magnitude w per unit length over the whole span. Assuming $I = I_c \sec \alpha$, calculate the magnitude of the horizontal thrust and show that B.M. at any section of the arch is zero.

7.23 A rectangular rigid frame $ABCD$ has sides $AB = CD = 1.20$ m and $AD = BC = 1.80$ m long. Loads P act at the centre of AB and CD and at right angles to them. Similarly, loads Q act at the centre of AD and BC at right angles to them. If EI from the frame be constant, calculate the ratio of P and Q for which the distance between centres of AD and BC will remain the same.

7.24 Discuss about use of Numerical method for solution of fixed-end arches.

7.25 A symmetrical arch rib of parabolic form spans 70 m with a rise of 8 m at the crown. Cross section varies such that $I = I_c \sec\alpha$. The rib is subjected to a temperature rise of 60 °F. Determine the B.M. diagram in the arch with end conditions as :

(a) encastre (b) hinged.

Assume $\alpha = 6(10)^{-6}$ per °F and $EI_c = 72(10)^4$ kN-m^2.

7.26 A parabolic symmetrical arch of span L has a rise of h at crown. The arch is 'encastre at A', the left-hand support. If the arch cross section is such that $I = I_c \sec\alpha$, obtain deflections at the other end B, if this is free when the arch is subjected to a concentrated vertical load at crown.

Analyse the arch, if the support at end B be :

(a) a roller (b) a hinge (c) fixed.

Draw B.M. diagrams for the cases.

Chapter 8

Method of Slope Deflection

8.1 Introduction

The Slope Deflection method has been used extensively for many years for analysis of all types of Statically Indeterminate Beams and Rigid Frames. The method was introduced by Prof. G. A. Maney in 1915 in a University of Minnesota Engineering Publication, as a general method for use in analysis of structures. This method extended the use of equations originally proposed by Prof. Mandrella in 1878 and Prof. Möhr in 1892 for calculation of secondary stresses in trusses.

The Slope Deflection method became a standard technique long before the generalised classification of the methods of analysis into force and displacement methods.

The method involves the solution of a set of simultaneous equations. Thus it is often long and tedious and sometimes impractical, specially for highly indeterminate structures, unless computational aids are used. However, the solution of a large number of linear equations is cumbersome and time consuming unless electronic computers are available. The method also forms the basis of other analytical methods like Moment Distribution, Kani's method and also Kloueck's Distribution of Deformation method. A proper understanding of the method is, therefore, necessary and desirable.

Basic Slope Deflection equation expresses the moments at the ends of a member as the superposition of the end moments due to external loads on the member with the ends assumed restrained and of end moments caused by the actual end displacements. The solution of these equations gives the unknown joint displacements, which are then substituted in the original slope deflection equations to give the end moments. The method takes into consideration flexural deformations of members only but neglects shear and axial deformations. Advantages connected with the method are manifold as

(i) the method is convenient for structures with small numbers of indeterminacies.
(ii) the method is helpful in understanding the moment distribution method.
(iii) the method is a special case of the displacement method widely used nowadays.
(iv) the method is useful in understanding the behaviour of the structure as it is possible to obtain the elastic curve after the slopes and deflection at joints are determined by the method.

8.2 Slope Deflection Method

In the slope deflection method, all joints are considered rigid, i.e., the angles between members at the joints are considered not to change with application of loads. Thus, the joints at the interior supports of statically indeterminate beams can be considered as 180° rigid joints and the joints in the right-angled rigid frames as 90° rigid joints. When beams or rigid frames are deformed, the rigid joints are considered to rotate as a whole, which means that the angles between tangents to the various parts of the elastic curve meeting at a joint remain the same as those in the original undeformed structure. Deflections due to shear and axial stresses are not considered as the effects are small in case of beams and rigid frames.

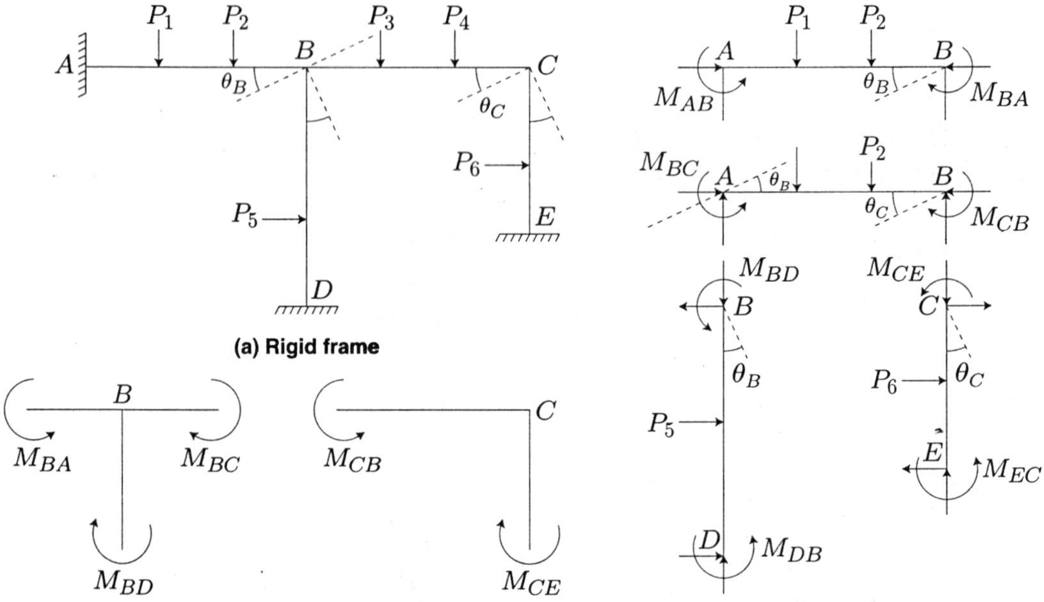

(a) Rigid frame

(c) Free body diagrams of joints

(b) Free body diagrams of members

Fig. 8.1: Rigid frame

Fig. 8.1(a) shows a rigid frame subjected to external loading, the frame being statically indeterminate to the sixth degree. The free body diagrams of the frame are shown in Fig. 8.1(b). At any one end of each member there are three reaction components, direct pull or thrust, and shear and end moment. End moments acting at ends A and B of a member are denoted by M_{AB} and M_{BA}. Counterclockwise end moments on members or clockwise joint moments at ends are considered as positive. Using the slope deflection equations, the end moments of each member can be expressed in terms of the end rotations and the loading acting on the member. Considering that all the joints of the frame remain in their original locations, the eight end moments in this case can be expressed in terms of the two unknown joint rotations. Free body diagrams of all the joints are shown in Fig. 8.1(c). The action of the member on the joint consists of a force in the direction of the axis of the member, a force perpendicular to this axis and a moment, each being the opposite of the action of the joint on the member. In Fig. 8.1(c), only the moments are indicated. For equilibrium, summation of all moments acting on each joint must be zero, whence for joint B,

$$M_{BA} + M_{BC} + M_{BD} = 0 \tag{8.1a}$$

and for joint C,

$$M_{CB} + M_{CE} = 0. \tag{8.1b}$$

Eqs. 8.1(a) and 8.1(b) are necessary and sufficient for determining the values of end rotations θ_B and θ_C. All end moments can then be found by substituting the known joint rotations into the slope deflection equations. Using statics, the direct force, end shear and moment diagrams can be obtained.

Analysis of statically indeterminate structures must satisfy both statics and geometry. In the slope deflection method, the conditions of geometry are satisfied by considering the joint rotations as the unknowns. Conditions of statics requiring that sum of the moments acting on each joint be zero are used to solve for the joint rotations.

8.3 Derivation of Slope Deflection Equations

Fig. 8.2(a) shows a prismatic beam of length L and having moment of inertia I, subjected to a system of loads. For span AB, shown in Fig. 8.2(a) it is required to express end moments M_{AB} and M_{BA} in terms of the end rotations θ_A and θ_B, the relative lateral displacement Δ_{AB} and the applied loads. The end rotations are considered positive and shown in the counterclockwise directions with the end moments counterclockwise as positive.

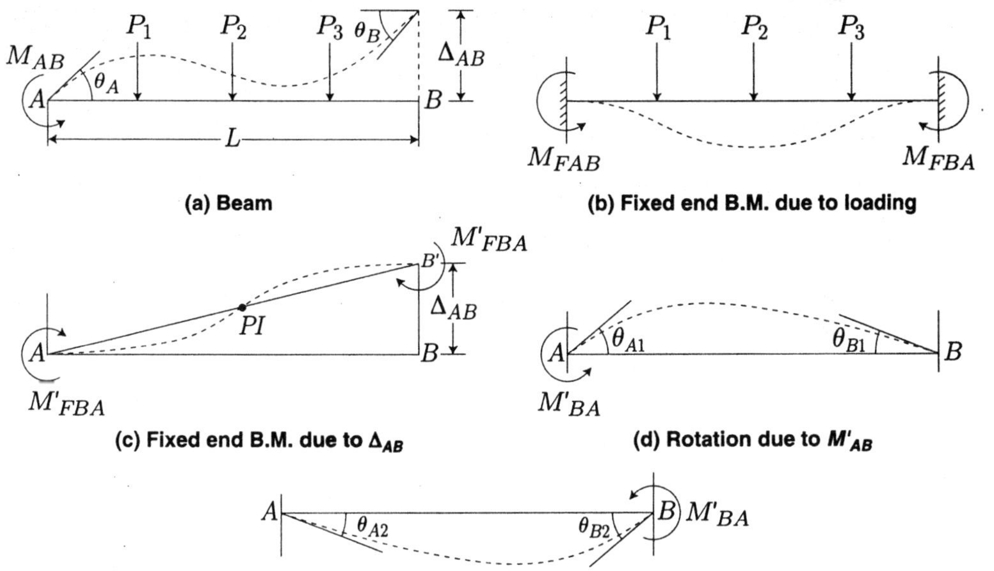

(a) Beam

(b) Fixed end B.M. due to loading

(c) Fixed end B.M. due to Δ_{AB}

(d) Rotation due to M'_{AB}

(e) Rotation due to M'_{BA}

Fig. 8.2: Derivation of slope deflection equations

With the applied loading of the member, the fixed end moments M_{FAB} and M_{FBA} together with the additional fixed end moments M'_{FAB} and M'_{FBA} for unequal movements of supports in a direction perpendicular to the axis of the member are induced to act on the members for keeping the tangents at the ends fixed as shown in Figs. 8.2(b) and 8.2(c). Additional end moments M'_{AB} and M'_{BA} should be such as to cause rotations θ_A and θ_B respectively at ends A and B. If θ_{A_1} and θ_{B_1} be the end rotations caused by M'_{AB}, and θ_{A_2} and θ_{B_2} due to M'_{BA} as shown in Figs. 8.2(d) and 8.2(e), the conditions required of geometry are

$$\left.\begin{aligned} \theta_A &= \theta_{A_1} - \theta_{A_2} \\ \text{and } \theta_B &= \theta_{B_1} - \theta_{B_2}. \end{aligned}\right] \tag{8.2}$$

By superposition,

$$\left.\begin{aligned} M_{AB} &= M_{FAB} + M'_{FAB} + M'_{AB} \\ \text{and } M_{BA} &= M_{FBA} + M'_{FBA} + M'_{BA}. \end{aligned}\right] \tag{8.3}$$

From conjugate beam method,

$$\left.\begin{aligned} \theta_{A_1} &= \frac{M'_{AB}L}{3EI}; \qquad \theta_{B_1} = -\frac{M'_{AB}L}{6EI} \\ \theta_{A_2} &= -\frac{M'_{BA}L}{6EI}; \quad \theta_{B_2} = \frac{M'_{BA}L}{3EI}. \end{aligned}\right] \tag{8.4}$$

and $\triangle_{AB}$, i.e., relative displacements of ends A and B is zero,

$$\left.\begin{aligned} \theta_A &= \frac{M'_{AB}L}{3EI} - \frac{M'_{BA}L}{6EI} \\ \theta_B &= \frac{M'_{BA}L}{3EI} - \frac{M'_{AB}L}{6EI}. \end{aligned}\right] \tag{8.5}$$

Solving Eqns. (8.5) for M'_{AB} and M'_{BA},

$$\left.\begin{aligned} M'_{AB} &= \frac{2EI}{L}[2\theta_A + \theta_B] \\ \text{and} \quad M'_{BA} &= \frac{2EI}{L}[\theta_A + 2\theta_B]. \end{aligned}\right\} \tag{8.6}$$

Substituting Eq. (8.6) into Eq. (8.3),

$$\left.\begin{aligned} M_{AB} &= M_{FAB} + \frac{2EL}{L}[2\theta_A + \theta_B] \\ M_{BA} &= M_{FBA} + \frac{2EI}{L}[\theta_A + 2\theta_B]. \end{aligned}\right\} \tag{8.7}$$

If ψ_{AB} be the angle measured from the original direction of member AB to the line joining the displaced joints, and considered positive when counterclockwise, from Fig. 2(b),

$$\psi_{AB} = -\frac{\triangle_{AB}}{L}.$$

Using moment area theorem, as change of slope between tangents A and B given by the area of $\frac{M}{EI}$ diagram between A and B is zero,

$$M'_{FAB} = M'_{FBA}.$$

From second moment area theorem, deflection of B' beyond tangent at A, i.e., $\triangle_{AB}$ equals the moment of $\frac{M}{EI}$ diagram between A and B about B, whence

$$M'_{FAB} = M'_{FBA} = -\frac{6EI\triangle_{BA}}{L^2} = -\frac{6EI\psi_{AB}}{L}. \tag{8.8}$$

Substituting Eqs. (8.6) and (8.8) into Eq. (8.7),

$$\left.\begin{aligned} M_{AB} &= M_{FAB} + \frac{2EI}{L}\left[2\theta_A + \theta_B - \frac{3\triangle_{AB}}{L}\right] \\ \text{and} \quad M_{BA} &= M_{FBA} + \frac{2EI}{L}\left[\theta_A + 2\theta_B - \frac{3\triangle_{AB}}{L}\right]. \end{aligned}\right\} \tag{8.9}$$

It may be noted here that the counterclockwise moments acting at ends of the members $(M_{FAB}, M_{FBA}, M'_{FAB}, M'_{FBA}, M_{AB}, M_{BA})$ are positive with counterclockwise rotations θ_A and θ_B being positive.

In the slope deflection equations, if the coefficient $\frac{2EI}{L}$ values for all spans are made n times smaller, the effect will be only to make θ values n times larger, while the products of the expressions $\frac{2EI}{L}$ and $(2\theta_{\text{near}} + \theta_{\text{far}})$ or the values of the end moments remain unchanged. If the absolute magnitudes of the θ values are not of direct interest, relative values of $\frac{2EI}{L}$ can be used. Thus, if the relative values of $\frac{I}{L}$ are considered as the relative stiffness K, the slope deflection equations for cases with no relative displacements of ends become

$$\left.\begin{aligned} M_{AB} &= M_{FAB} + K_{AB}[2\theta_A + \theta_B] \\ \text{and} \quad M_{BA} &= M_{FBA} + K_{BA}[\theta_A + 2\theta_B]. \end{aligned}\right\} \tag{8.10}$$

8.4 Application of Slope Deflection Method to Continuous Beam

The procedure for analysis of continuous beams by slope deflection method is as follows :

(i) Determine the fixed end moments in each span.

(ii) Express all end moments in each span in terms of the fixed end moments and the end rotations using the slope deflection equations.

(iii) Write down equilibrium equations with the rotations at the supports as unknown, by applying the conditions that the sum of the end moments acting on the ends of the two members meeting at the support should be zero.

(iv) Determine rotations by simultaneously solving the above set of the equilibrium equations.

(v) Substitute the rotations back into the slope deflection equations and compute the end moments.

(vi) Determine all reactions, obtain the shear force and bending moment diagrams.

The fixed end moments for various cases of loading in case of prismatic members are given in Section 9.4 of Chapter 9.

Example 8.1. *Analyse the continuous beam shown in Fig. 8.3(a), using slope deflection method.*

Solution : Relative stiffness : $K_{BA} : K_{BC} = 1 : 1$.

Fixed-end moments :

$$M_{FAB} = +\frac{wL^2}{12} = \frac{20(6)^2}{12} = 60 \text{ kNm}$$

$$M_{FBA} = -60 \text{ kNm}$$

$$M_{FBC} = +\frac{PL}{8} = +\frac{120(6)}{8} = 90 \text{ kNm}$$

$$M_{FCB} = 90 \text{ kNm}.$$

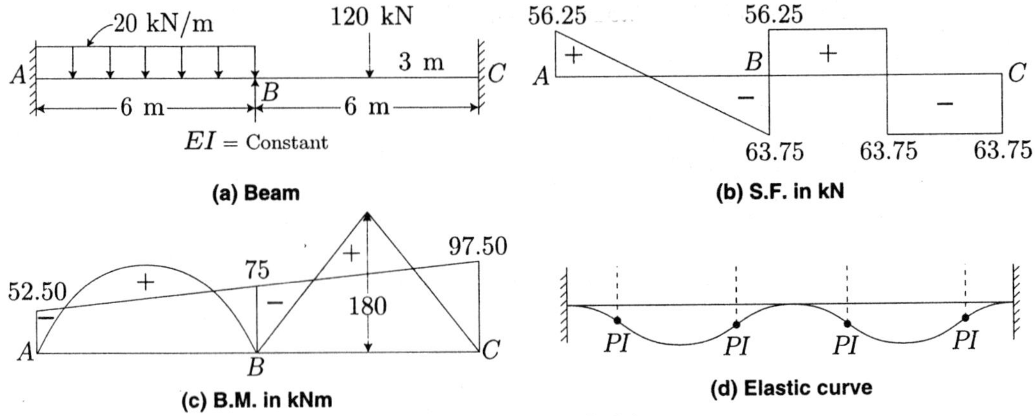

Fig. 8.3: For Example 8.1

Slope deflection equations :

Using modified slope deflection equation as in Eq. (8.10), the following expressions for the end moments are

$$M_{AB} = 60 + 1(2\theta_A + \theta_B)$$

$$M_{BA} = -60 + 1(\theta_A + 2\theta_B)$$

$$M_{BC} = 90 + 1(2\theta_B + \theta_C)$$

$$M_{CB} = -90 + 1(\theta_B + 2\theta_C)$$

Joint conditions :

Joint A, $\theta_A = 0$,

Joint B, $M_{BA} + M_{BC} = 0$,

Joint C, $\theta_C = 0$,

For Joint B, it follows that,

$$30 + 1(4\theta_B) = 0, \quad \text{whence } \theta_B = -\frac{30}{4}.$$

Computation of end moments

$$M_{AB} = +60 + 1\left(-\frac{30}{4}\right) = 52.50 \text{ kNm}$$

$$M_{BA} = -60 + 1\left(-\frac{60}{4}\right) = -75 \text{ kNm} = -M_{BC}$$

$$M_{CB} = -90 + 1\left(-\frac{30}{4}\right) - 97.5 \text{ kNm}$$

Reactions at supports :

At support A, $R_A = \dfrac{20(6)}{2} - \dfrac{75 - 52.50}{6} = 56.25$ kN

At support C, $R_C = \dfrac{120}{2} + \dfrac{97.50 - 75}{6} = 63.75$ kN.

Also $\quad R_B = (20)(6) + 120 - (56.25 + 63.75) = 120$ kN.

S.F. and B.M. diagrams and the elastic curves are shown in Fig. 8.3.

Example 8.2. *Analyse the beam shown in Fig. 8.4(a) using slope deflection method.*

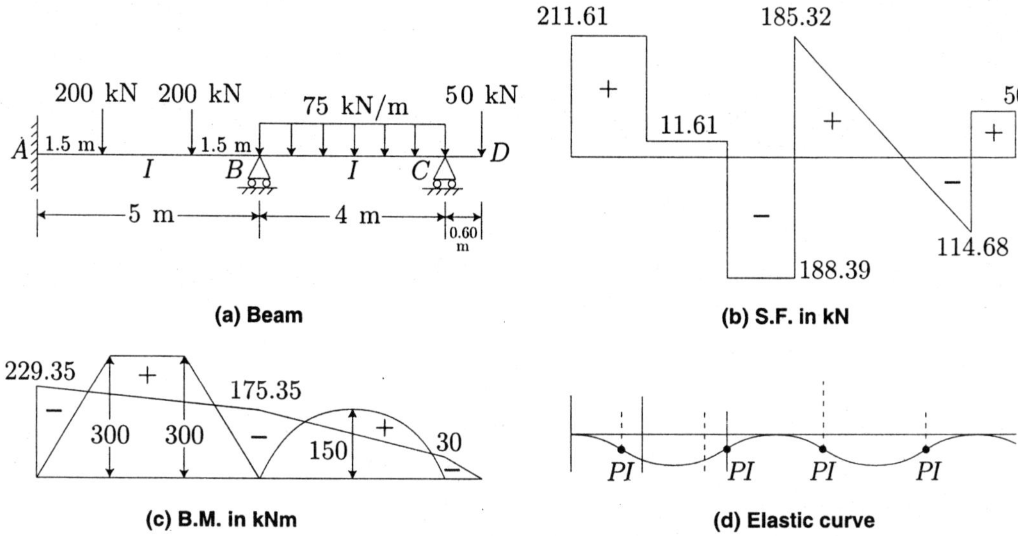

Fig. 8.4: For Example 8.2

Solution : *Relative stiffness :*

for $AB = \dfrac{I}{5}$

for $BC = \dfrac{I}{4}$.

Hence $K_{AB} : K_{BC} = 4 : 5$.

Fixed-end moments :

$$M_{FAB} = \frac{200(1.5)(3.5)^2}{(5)^2} + \frac{200(1.5)^2(3.5)}{(5)^2} = +210 \text{ kNm}$$

$$M_{FBA} = -210 \text{ kNm}$$

$$M_{FBC} = \frac{75(4)^2}{12} = 100 = -M_{FCB}$$

$$M_{FCD} = 50(0.60) = 30 \text{ kNm}.$$

Slope deflection equations :

From Eq. (8.10),

$$M_{AB} = 210 + 4(2\theta_A + \theta_B)$$

$$M_{BA} = -210 + 4(\theta_A + 2\theta_B)$$

$$M_{BC} = 100 + 5(2\theta_B + \theta_C)$$

$$M_{CB} = -100 + 5(\theta_B + 2\theta_C)$$

$$M_{CD} = 30.$$

Joint conditions :

Joint A, $\theta_A = 0$,

Joint B, $M_{BA} + M_{BC} = 0$, whence $-110 + 18\theta_B + 5\theta_C = 0.$ (i)

Joint C, $M_{CD} + M_{CB} = 0$, whence $-70 + 5\theta_B + 10\theta_C = 0.$ (ii)

Solving Eqs. (i) and (ii),

$$\theta_B = \frac{150}{31} \text{ and } \theta_C = \frac{142}{31}.$$

Computation of end moments :

$$M_{AB} = 210 + 4\left(\frac{150}{31}\right) = 229.35 \text{ kNm}$$

$$M_{BA} = -210 + 8\left(\frac{150}{31}\right) = -171.29 \text{ kNm}$$

$$M_{BC} = 100 + 5\left[\frac{2(150)}{31} + \frac{142}{31}\right] = 171.29 \text{ kNm}$$

$$M_{CB} = -100 + 5\left[\frac{150}{31} + \frac{284}{31}\right] = -30 = -M_{CD}.$$

Reactions at supports :

$$R_A = 200 + \frac{229.35 - 171.29}{5} = 211.61 \text{ kN}$$

$$R_C = 50 + \frac{75(4)}{2} - \frac{171.29 - 30}{4} = 164.68 \text{ kN}$$

$$R_B = 400 + 300 + 50 - (211.61 + 164.68) = 373.71 \text{ kN}$$

S.F. and B.M. diagrams and the elastic curves are shown in Fig. 8.4.

Example 8.3. *Analyse the beam shown in Fig. 8.5(a) using Slope Deflection Method.*

Solution : *Relative stiffness for members :*

$$K_{AB} = \frac{I}{9}$$

$$K_{BC} = \frac{I}{12} \quad \text{and} \quad K_{CD} = \frac{I}{6}.$$

Hence, $K_{AB} : K_{BC} : K_{CD} = 4 : 3 : 6.$

Fixed-end moments :

$$M_{FAB} = \frac{250(6)^2(3)}{(9)^2} = 333.33 \text{ kNm}$$

$$M_{FBA} = -\frac{250(3)^2(6)}{(9)^2} = -166.67 \text{ kNm}$$

$$M_{FBC} = \frac{25(12)^2}{12} + \frac{100(9)^2(3)}{(12)^2} + \frac{100(3)^2(9)}{(12)^2} = 525 \text{ kNm} = -M_{FCB}$$

$$M_{FCD} = \frac{100(6)^2}{12} = 300 \text{ kNm} = -M_{FDC}.$$

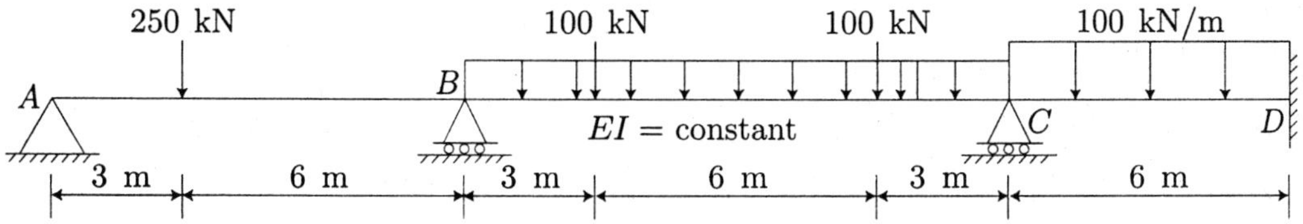

(a) Beam

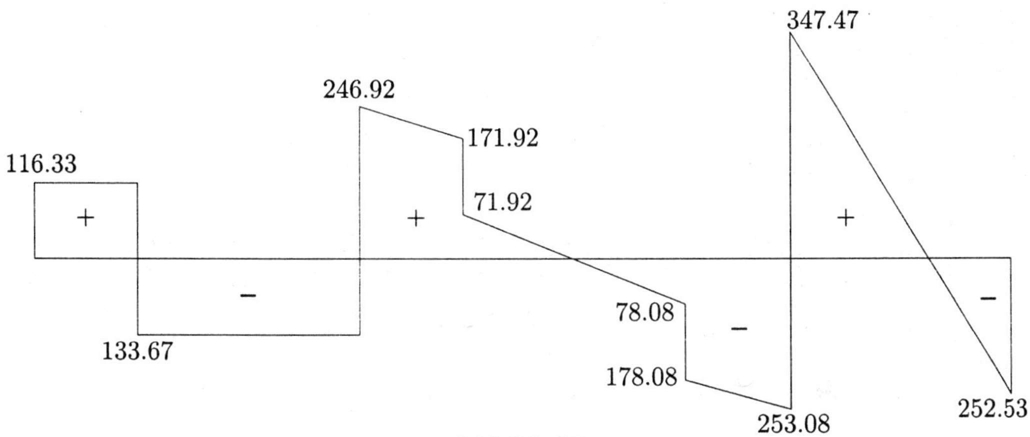

(b) S.F. in kN

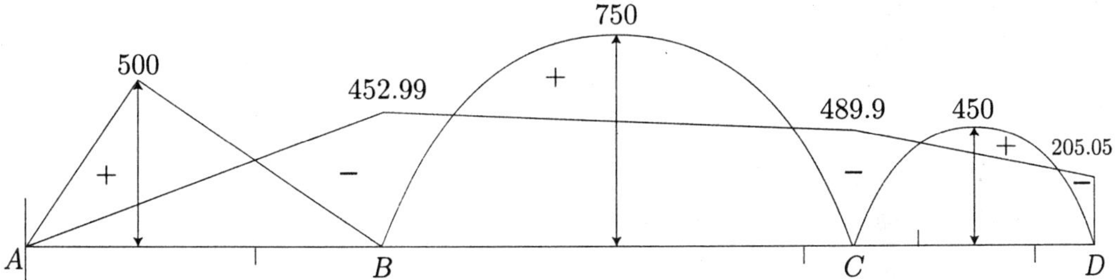

(c) B.M. in kNm

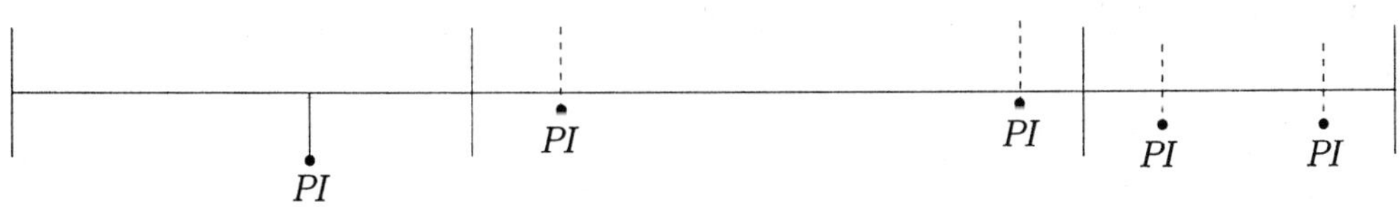

(d) Elastic curve

Fig. 8.5: For Example 8.3

Slope deflection equations :

$$M_{AB} = 333.33 + 4(2\theta_A + \theta_B)$$

$$M_{BA} = -166.67 + 4(\theta_A + 2\theta_B)$$

$$M_{BC} = 525 + 3(2\theta_B + \theta_C)$$

$$M_{CB} = -525 + 3(\theta_B + 2\theta_C)$$

$$M_{CD} = 300 + 6(2\theta_C + \theta_D)$$

$$M_{DC} = -300 + 6(\theta_C + 2\theta_D).$$

Joint conditions :

Joint A, $M_{AB} = 0$, whence $333.33 + 8\theta_A + 4\theta_B = 0$ (i)

Joint B, $M_{BA} + M_{BC} = 0$, whence $358.33 + 4\theta_A + 14\theta_B + 3\theta_C = 0$ (ii)

Joint C, $M_{CB} + M_{CD} = 0$, whence $-225 + 3\theta_B + 18\theta_C + 6\theta_D = 0$ (iii)

Joint D, $\theta_D = 0$.

From Eqs. (i), (ii) and (iii),

$$\left.\begin{aligned} 8\theta_A + 4\theta_B &= -333.33 \\ 4\theta_A + 14\theta_B + 3\theta_C &= -358.33 \\ 3\theta_B + 18\theta_C + 6\theta_D &= 225. \end{aligned}\right] \quad \text{(iv)}$$

It may be noted that if a diagonal is drawn downward to the right on the left side of Eq. (iv) as indicated, the coefficients on the diagonal are predominant in their own equations, but other coefficients are symmetrical with respect to this diagonal. The predominant coefficients are also twice the sum of the coefficients of the other two rotations in the particular equation.

Solving Eq. (iv) simultaneously, using Cramer's rule or any other convenient method,

$$\theta_A = -31.695$$

$$\theta_B = -19.942$$

and $\theta_C = 15.825$.

Computation of end moments :

$$M_{AB} = 333.33 + 4[-2(31.695) - 19.942] \approx 0$$

$$M_{BA} = -166.67 + 4(-31.625 - 39.884) = -452.99 \text{ kNm}$$

$$M_{CB} = -525 + 3(-19.942 + 31.65) = -489.88 \text{ kNm}$$

$$M_{CD} = 300 + 6(31.65 + 0) = 489.9 \text{ kNm}$$

$$M_{DC} = -300 + 6(15.825) = -205.05 \text{ kNm}.$$

Reactions at supports :

$$R_A = \frac{250(6)}{9} - \frac{452.99}{9} = 116.33 \text{ kN}$$

$$R_B = \frac{250(3)}{9} + \frac{452.99}{9} + \frac{25(12)}{2} + 100 - \frac{489.88 - 452.99}{12} = 380.59 \text{ kN}$$

$$R_C = \frac{25(12)}{2} + 100 + \frac{489.88 - 452.99}{12} + \frac{100(6)}{2} + \frac{489.9 - 205.05}{6} = 600.55 \text{ kN}$$

$$R_D = \frac{100(6)}{2} - \frac{489.9 - 205.05}{6} = 252.53 \text{ kN}.$$

S.F. and B.M. diagrams and elastic curve are shown in Fig. 8.5.

Example 8.4. *Analyse the continuous beam in Fig. 8.6(a), with the support at B setting by 5 mm, if for the beam,* $E = 2(10)^5$ N/mm^2 *and* $I = 36(10)^6$ mm^4.

Solution : In this case, absolute values of E and I have to be used.

For the case, $EI = 72(10)^{11}$ Nmm^2 = $72(10)^5$ Nm^2 = $72(10)^2$ kNm^2.

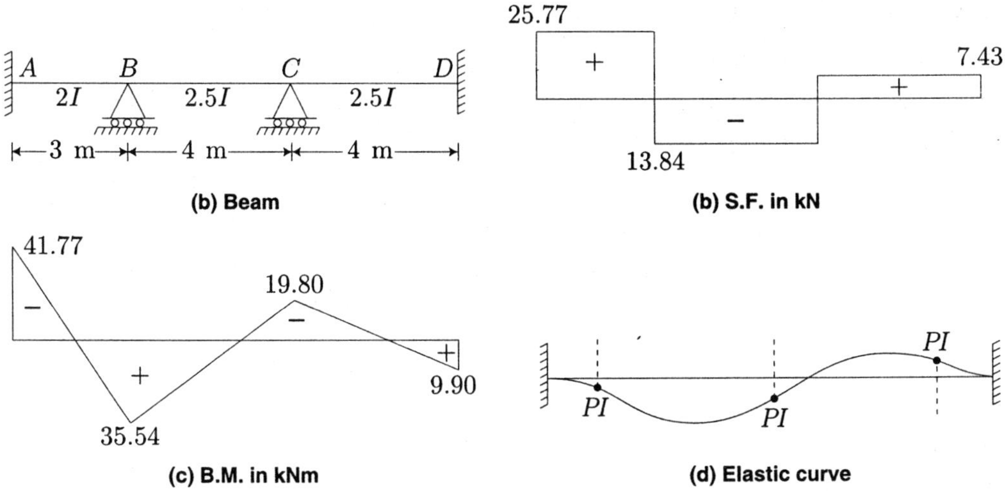

Fig. 8.6: For Example 8.4

Slope deflection equations :

$$M_{AB} = \frac{2(2EI)}{3}\left[2\theta_A + \theta_B + \frac{3(5)}{3000}\right] = \frac{2(144)(10)^2}{3}\left[\theta_B + \frac{5}{1000}\right], \text{as } \theta_A = 0$$

$$= 9600\left[\theta_B + \frac{5}{1000}\right]$$

$$M_{BA} = \frac{2(144)(10)^2}{3}\left[2\theta_B + \frac{5}{1000}\right] = 9600\left[2\theta_B + \frac{5}{1000}\right]$$

$$M_{BC} = \frac{2(2.5)(72)(10)^2}{4}\left[2\theta_B + \theta_C - \frac{3(5)}{4000}\right] = 9000\left[2\theta_B + \theta_C - \frac{15}{4000}\right]$$

$$M_{CB} = 9000\left[\theta_B + 2\theta_C - \frac{15}{4000}\right]$$

$$M_{CD} = 9000[2\theta_C], \text{as } \theta_D = 0$$

$$M_{DC} = 9000[\theta_C].$$

Joint conditions :

Joint B, $M_{BA} + M_{BC} = 0$, whence

$$372000\theta_B + 9000\theta_C + 14.25 = 0 \quad \text{(i)}$$

Joint C, $M_{CE} + M_{CD} = 0$, whence

$$9000\theta_B + 36000\theta_C - 33.75 = 0. \quad \text{(ii)}$$

Solving Eqs. (i) and (ii), $\theta_B = -\dfrac{90.75}{139800} = -64.91(10)^{-5}$

and $\quad \theta_C = +109.97(10)^{-5}$.

Computation of end moments :

$$M_{AB} = 9600\left(\theta_B + \frac{5}{1000}\right) = 41.77 \text{ kNm}$$

$$M_{BA} = 9600\left(2\theta_B + \frac{5}{1000}\right) = +35.54 \text{ kNm}$$

$$M_{BC} = 9000\left(2\theta_B + \theta_C - \frac{15}{4000}\right) = -35.54 \text{ kNm}$$

$$M_{CB} = 9000\left(\theta_B + 2\theta_C - \frac{15}{4000}\right) = -19.80 \text{ kNm}$$

$$M_{CD} = 9000(2\theta_C) = +19.80 \text{ kNm}$$

$$M_{DC} = 9000(\theta_C) = +9.90 \text{ kNm.}$$

Reaction at supports :

$$R_A = \frac{41.77 + 35.54}{4} = 25.77 \text{ kN}$$

$$R_B = 25.77 + \frac{35.54 + 19.80}{4} = 39.61 \text{ kN}$$

$$R_C = 13.84 + \frac{19.80 + 5.90}{4} = 21.27 \text{ kN}$$

$$R_D = 7.43 \text{ kN.}$$

S.F., B.M. diagrams and elastic curve are shown in Fig. 8.6.

Example 8.5. *Analyse the continuous beam in Fig. 8.7(a), if the support at A rotates through 0.25 radians in anticlockwise direction. Assume that EI has a constant value of 32000 k-Nm.*

Solution : In this case $EI = 32000$ kNm.

Slope deflection equations are as under :

$$M_{AB} = \frac{2(32000)}{8}(2\theta_A + \theta_B) = 8000(0.50 + \theta_B)$$

$$M_{BA} = 8000(0.25 + 2\theta_B)$$

$$M_{BC} = 8000(2\theta_B + \theta_C)$$

$$M_{CB} = 8000(\theta_B + 2\theta_C)$$

$$M_{CD} = 8000(2\theta_C + \theta_D)$$

$$M_{DC} = 8000(\theta_C + 2\theta_D).$$

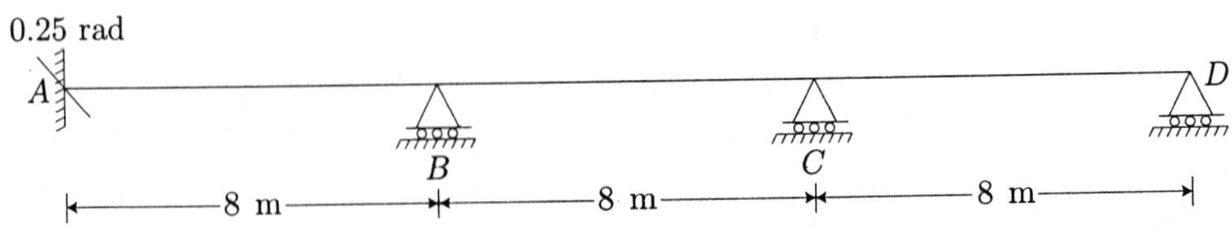

(a) Beam

Fig. 8.7: For Example 8.5 (Contd.)

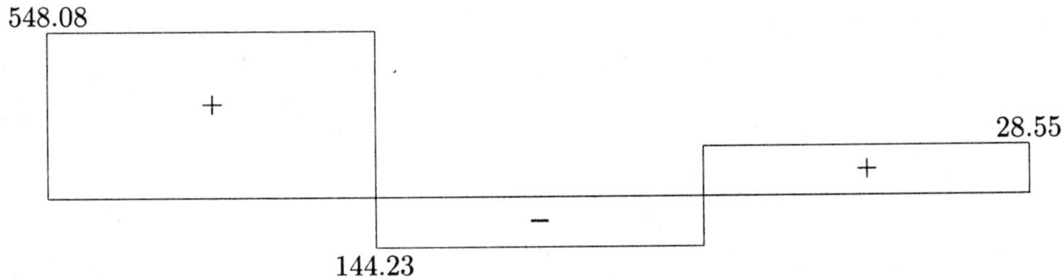

(b) S.F. in kN

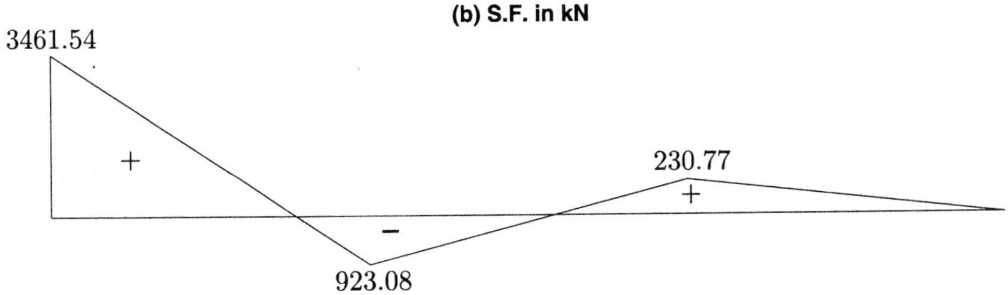

(c) B.M. in kNm

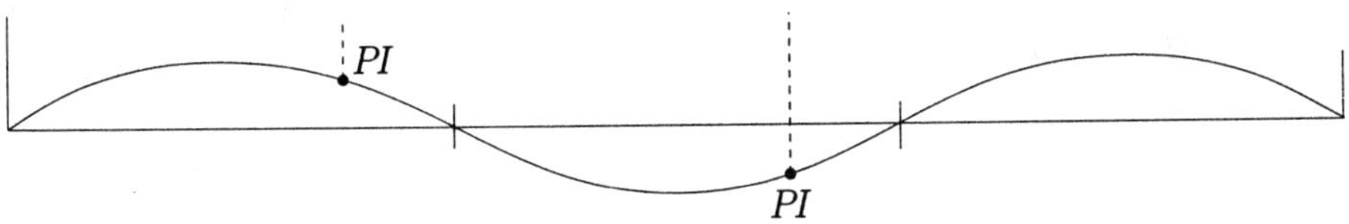

(d) Elastic curve

Fig. 8.7: For Example 8.5

Joint conditions :

Joint B, $M_{BA} + M_{BC} = 0$, whence $320000\theta_B + 80000\theta_C + 2000 = 0$. (i)

Joint C, $M_{CB} + M_{CD} = 0$, whence $80000\theta_B + 3200\theta_C + 80000\theta_D = 0$. (ii)

Joint D, $M_{DC} = 0$, whence $80000\theta_C + 160000\theta_D = 0$. (iii)

Solving Eqs. (i), (ii) and (iii),

$$\theta_B = -673.076(10)^{4}$$

$$\theta_C = 192.307(10)^{-4}$$

$$\theta_D = -96.153(10)^{-4}.$$

Calculation of end moments :

$$M_{AB} = 8000(0.50 - 0.0673076) = 3461.54 \text{ kNm}$$

$$M_{BA} = 8000(0.25 - 0.1346152) = 923.08 \text{ kNm}$$

$$M_{BC} = 8000[2(-673.076) + 192.307]10^{-4} = -923.05 \text{ kNm}$$

$$M_{CB} = 8000[-673.076 + 2(192.307)10^{-4} = -230.77 \text{ kNm}$$

$$M_{CD} = 8000[2(192.307) - (96.153)]10^{-4} = 230.77 \text{ kNm}$$

$$M_{DC} = 0.$$

Reactions at supports :

$$R_A = \frac{3461.54 + 923.08}{8} = 548.08 \text{ kN}$$

$$R_B = 548.08 + \frac{923.08 + 230.77}{8} = 692.31 \text{ kN}$$

$$R_C = 144.23 + \frac{230.77}{8} = 173.08 \text{ kN}$$

$$R_D = 28.85 \text{ kN}.$$

S.F., B.M. diagrams and elastic curve are shown in Fig. 8.7.

Example 8.6. *Analyse the continuous beam in Fig. 8.8(a), if the support C sinks by 10 mm. Assume EI as* 48000 kN-m^2.

Solution : In this case, $EI = 48000$ kNm^2.

Fixed-end moments :

$$M_{FAB} = +\frac{20(10)^2}{12} + \frac{40(4)^2(6)}{(10)^2} = 205.07 \text{ kNm}$$

$$M_{FBA} = -\frac{20(10)^2}{12} - \frac{40(6)^2(4)}{(10)^2} = -224.27 \text{ kNm}$$

$$M_{FBC} = +\frac{30(4)^2}{12} = 40 \text{ kN-m} = -M_{FCB}$$

$$M_{FCD} = M_{FDC} = 0$$

$$M_{FDE} = 40 \text{ kN-m}.$$

Slope deflection equations :

$$M_{AB} = 205.07 + \frac{2(10)(48000)}{10}(\theta_B), \text{as } \theta_A = 0 = 205.07 + 960000\theta_B$$

$$M_{BA} = -224.27 + 1920000\theta_B$$

$$M_{BC} = 40 + \frac{2(3)(48000)}{4}\left[2\theta_B + \theta_C + \frac{3(10)}{4000}\right] = 580 + 1440000\theta_B + 720000\theta_C$$

$$M_{CB} = 500 + 720000\theta_B + 1940000\theta_C$$

$$M_{CD} = -160 + 640000\theta_C + 320000\theta_D$$

$$M_{DC} = -160 + 320000\theta_C + 640000\theta_C$$

$$M_{DE} = 40 \text{ kNm}.$$

Joint conditions :

Joint B, $M_{BA} + M_{BC} = 0$, whence

$$3360000\theta_B + 720000\theta_C + 355.73 = 0. \qquad \text{(i)}$$

Joint C, $M_{CB} + M_{CD} = 0$, whence

$$720000\theta_B + 2080000\theta_C + 320000\theta_D + 340 = 0 \qquad \text{(ii)}$$

Joint D, $M_{DC} + M_{DE} = 0$, whence

$$0(\theta_B) + 320000\theta_C + 640000\theta_D - 120 = 0. \qquad \text{(iii)}$$

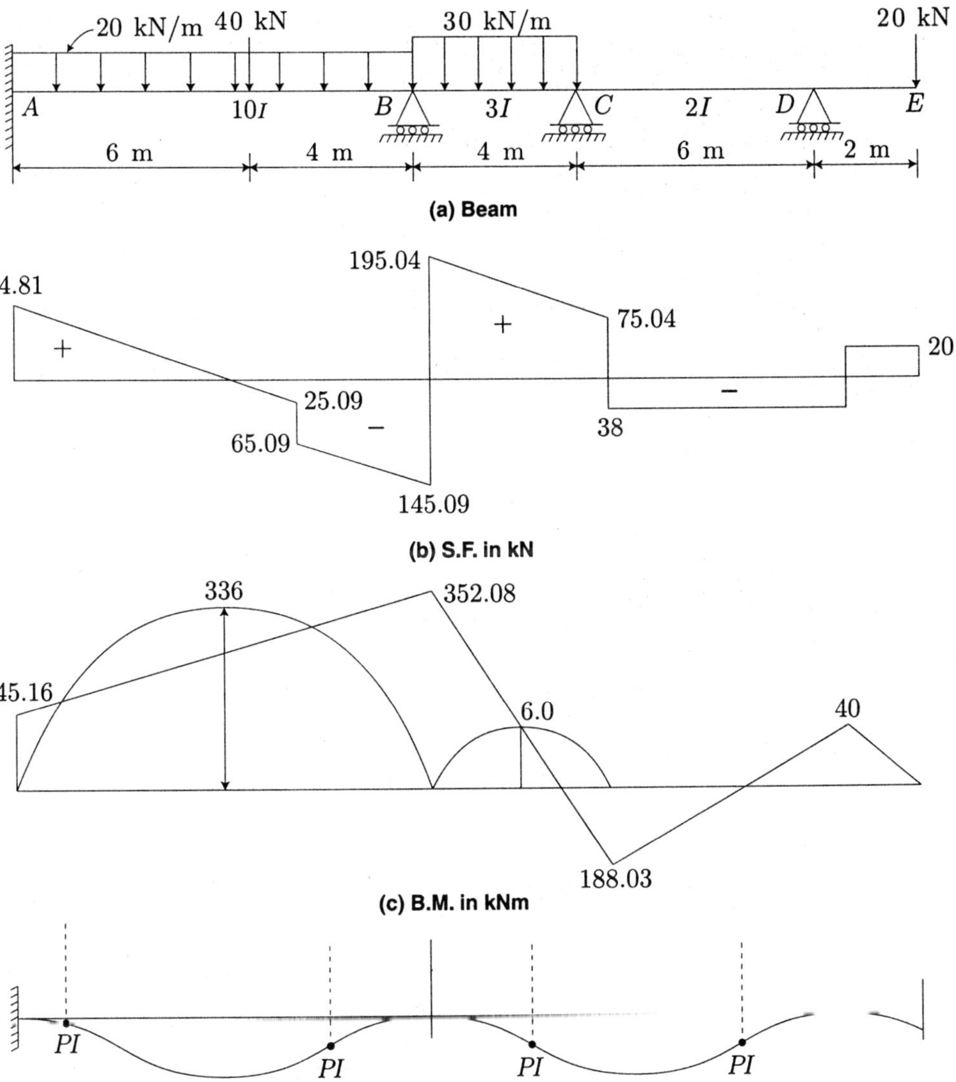

Fig. 8.8: For Example 8.6

Solving Eqs. (i),(ii) and (iii), simultaneously,

$$\theta_B = -6.657(10)^{-4}$$

$$\theta_C = -18.336(10)^{-4}$$

$$\theta_D = 27.918(10)^{-4}.$$

Calculation of end moments :

$$M_{AB} = 205.07 - 63.91 = 141.16 \text{ kNm}$$

$$M_{BA} = -224.27 - 127.81 = -352.08 \text{ kNm}$$

$$M_{BC} = 580 - 95.86 - 132.02 = 352.12 \text{ kNm}$$

$$M_{CB} = 500 - 47.93 - 264.04 = 188.08 \text{ kNm}$$

$$M_{CD} = -160 - 117.35 + 89.34 = -188.02 \text{ kNm}$$

$$M_{DC} = -160 - 58.67 + 178.68 \approx -40 \text{ kNm}.$$

Reactions at support :

$$R_A = \frac{20(10)}{2} + \frac{40(4)}{10} - \frac{352.08 - 141.16}{10} = 94.91 \text{ kN}$$

$$R_B = \frac{20(10)}{2} + \frac{40(6)}{10} + \frac{30(4)}{2} + \frac{352.08 - 141.16}{10} + \frac{352.12 + 188.03}{4} = 340.13 \text{ kN}$$

$$R_C = \frac{30(4)}{2} - \frac{352.12 + 188.03}{4} - \frac{188.03 - 40}{6} = -113.04 \text{ kN}$$

$$R_D = 20 + \frac{188.03 + 40}{6} = 58.00 \text{ kN}.$$

S.F., B.M. diagrams and elastic curve are shown in Fig. 8.8.

8.5 Application of Slope Deflection Method to Frames having No Joint Movement

The modified Slope Deflection equations as given in Eq. (8.10) can be used to analyse statically indeterminate frames, wherein all joints remain fixed in location during deformation. Axial deformation in the members due to direct stress is neglected in the analysis, but direct stresses together with shear and movements have to be considered in design of the members. Steps to be followed are the sams as in the analysis of continuous beams. Applications of the method for the case are illustrated by Examples.

Example 8.7. *Analyse the rigid frame shown in Fig. 8.9 using slope deflection method.*

Solution : Joints A and C are fixed. Joint B must be at 6 m to the left of C and at 6 m above A and, therefore, it must remain stationary. Thus all three joints cannot move and relative displacements for all members are zero.

Relative stiffness :

$$K_{AB} = \frac{I}{6}$$

$$K_{BC} = \frac{2I}{6}$$

Hence, $K_{AB} : K_{BC} = 1 : 2$.

Fixed-end moments :

$$M_{FAB} = \frac{200(6)}{8} = 150 \text{ kNm} = -M_{FBA}$$

$$M_{FBC} = \frac{60(6)^2}{12} = 180 \text{ kNm} = -M_{FCB}$$

$$M_{FBD} = -80 \text{ kNm}.$$

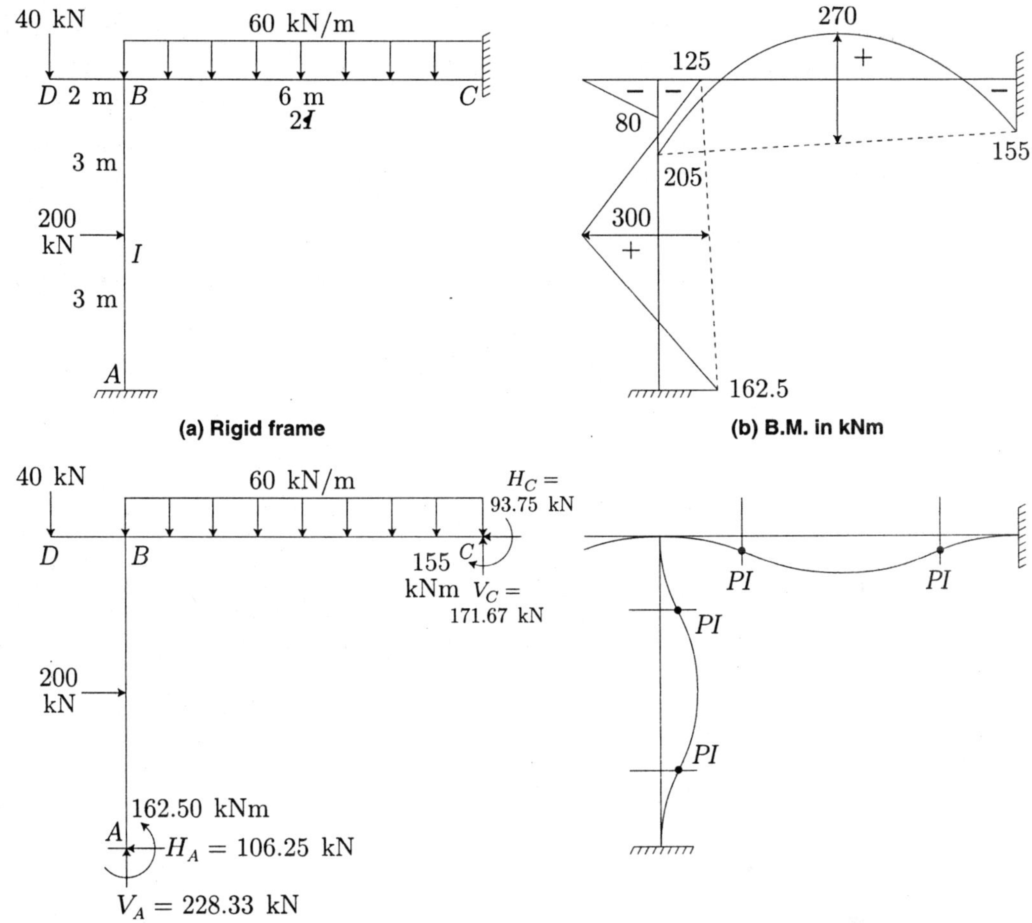

(a) Rigid frame **(b) B.M. in kNm**

(c) Reactions **(d) Elastic curve**

Fig. 8.9: For Example 8.7

Slope deflection equations are as under :

$$M_{AB} = 150 + 1(\theta_B), \text{ as } \theta_A = 0$$

$$M_{BA} = -150 + 2\theta_B,$$

$$M_{BC} = 180 + 2\theta_B, \quad \text{since } \theta_C = 0$$

$$M_{CB} = -180 + \theta_B$$

and $$M_{BD} = -80 \text{ kN-m}.$$

Joint conditions :

Joint B, $M_{BA} + M_{BC} + M_{BD} = 0$, whence $-50 + 4\theta_B = 0$, whence $\theta_B = +12.5$.

Calculation of end moments :

$$M_{AB} = 150 + 12.5 = 162.5 \text{ kNm}$$

$$M_{BA} = -150 + 25 = -125 \text{ kNm}$$

$$M_{BC} = 180 + 25 = 205 \text{ kNm}$$

$$M_{CB} = -180 + 25 = -155 \text{ kNm}$$

$$M_{BD} = -80 \text{ kNm}.$$

Reactions at ends

At A, $\quad H_A = 100 + \dfrac{162.5 - 125}{6} = 106.25 \text{ kN} \leftarrow \text{(towards left)}$

At C, $\quad H_C = 200 - 106.25 = 93.75 \text{ kN}$

$$V_C = \frac{60(6)}{2} - \frac{205 - 155}{6} = 171.67 \text{ kN} \uparrow \text{(upwards)}$$

At A, $\quad V_A = 400 - 171.67 = 228.33 \text{ kN} \uparrow \text{(upwards)}.$

B.M. diagram, Reactions and elastic curve are shown in Fig. 8.9.

Example 8.8. *Analyse the Rigid Frame shown in Fig. 8.10(a), using Slope Deflection Method.*

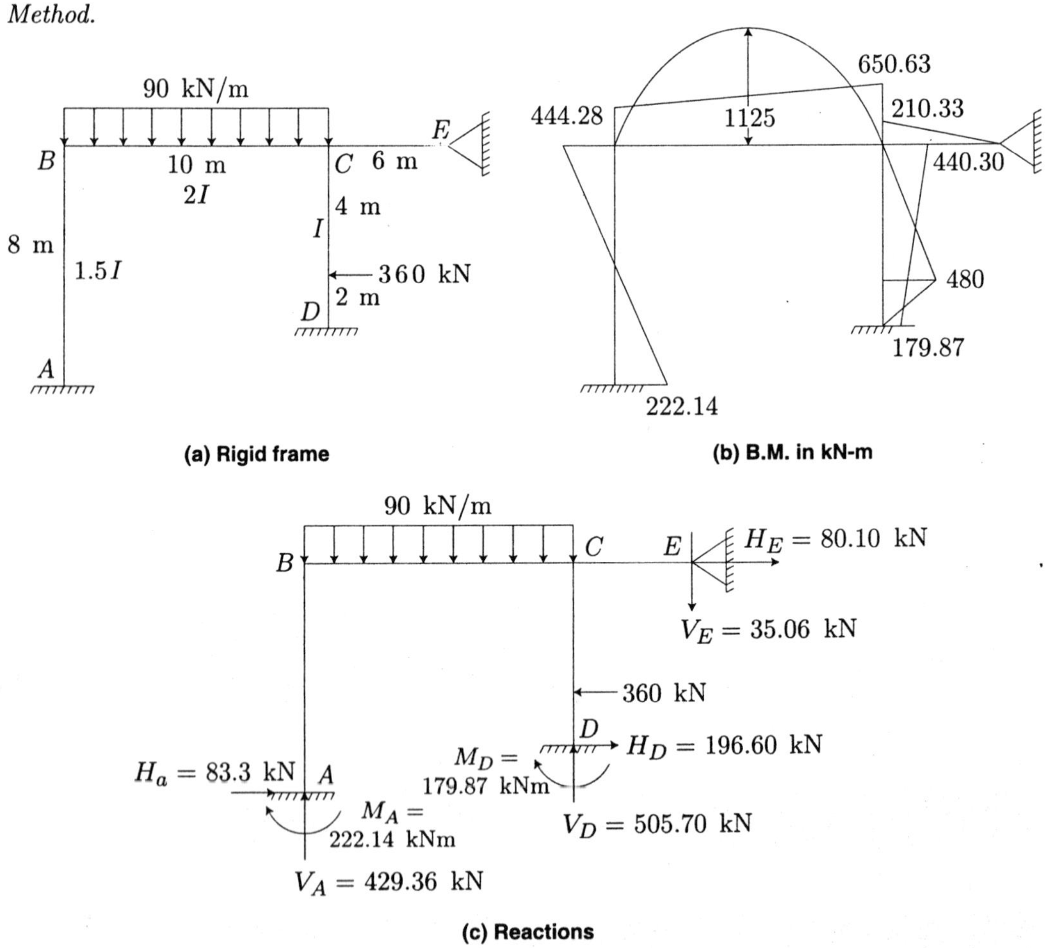

(a) Rigid frame **(b) B.M. in kN-m**

(c) Reactions

Fig. 8.10: For Example 8.8

Solution : Relative stiffnesses for members are as under :

$$K_{AB} = \frac{1.5I}{8}, \text{ i.e., } \frac{1.5}{8} \text{ with } I = 1$$

$$K_{BC} = \frac{2I}{10}, \text{ i.e., } \frac{2}{10}$$

$$K_{CD} = \frac{I}{6}, \text{ i.e., } \frac{1}{6}$$

$$K_{CE} = \frac{I}{6}, \text{ i.e., } \frac{1}{6}$$

Fixed-end moments :

$$M_{FBC} = \frac{90(10)^2}{12} = 750 \text{ kN-m} = -M_{FCB}$$

$$M_{FCD} = \frac{360(2)^2(4)}{6}^2 = 160 \text{ kNm}$$

$$M_{FDC} = -\frac{360(2)(4)^2}{(6)^2} = -320 \text{ kN-m.}$$

Slope deflection equation with $EI = 1$,

$$M_{AB} = \frac{2(1.5)}{8}(\theta_B) = \frac{3}{8}\theta_B, \text{ as } \theta_A = 0$$

$$M_{BA} = \frac{3}{4}\theta_B \quad .$$

$$M_{BC} = 750 + \frac{2(2)}{10}(2\theta_B + \theta_C) = 750 + \frac{4}{5}\theta_B + \frac{2}{5}\theta_C$$

$$M_{CB} = -750 + \frac{2}{5}\theta_B + \frac{4}{5}\theta_C$$

$$M_{CE} = \frac{2(1)}{6}(2\theta_C + \theta_E) = \frac{2}{3}\theta_C + \frac{1}{3}\theta_E$$

$$M_{EC} = \frac{1}{3}\theta_C + \frac{2}{3}\theta_E$$

$$M_{CD} = 160 + \frac{2}{3}\theta, \text{ as } \theta_D = 0$$

$$M_{DC} = -320 + \frac{1}{3}\theta_C.$$

Joint conditions :

Joint B, $M_{BA} + M_{BC} = 0$, whence

$$1.55\theta_B + 0.40\theta_C + 0(\theta_E) + 750 = 0 \quad \text{(i)}$$

Joint C, $M_{CB} + M_{CE} + M_{CD} = 0$, whence

$$0.40\theta_B + 2.133\theta_C + 0.333\theta_E - 590 = 0 \quad \text{(ii)}$$

Joint E, as $M_{EC} = 0$,

$$0(\theta_B) + 0.333\theta_C + 0.667\theta_E = 0. \quad \text{(iii)}$$

Solving Eqs. (i), (ii) and (iii) simultaneously,

$$\theta_B = -592.36$$

$$\theta_C = 420.395$$

$$\theta_E = 210.198.$$

Calculation of end moments :

$$M_{AB} = -222.14 \text{ kNm}$$

$$M_{BA} = -444.28 \text{ kNm}$$

$$M_{BC} = -444.28 \text{ kNm}$$

$$M_{CB} = -650.63 \text{ kNm}$$

$$M_{CD} = 444.30 \text{ kNm}$$

$$M_{DC} = -179.87 \text{ kNm}$$

$$M_{CE} = 210.33 \text{ kNm}$$

$$M_{EC} = 0 \text{ kNm}.$$

B.M. diagrams and Reactions are shown in Fig. 8.10.

Example 8.9. *Analyse the frame in Fig. 8.11(a) using Slope Deflection method.*

Solution : The frame and loadings are structurally symmetrical about its vertical axis. The frame, therefore, will not have any sidesway. Joint rotations at B and C will be identical in magnitude, but opposite in sign, whence $\theta_B = -\theta_C$. Taking advantage of this the number of unknowns may be reduced.

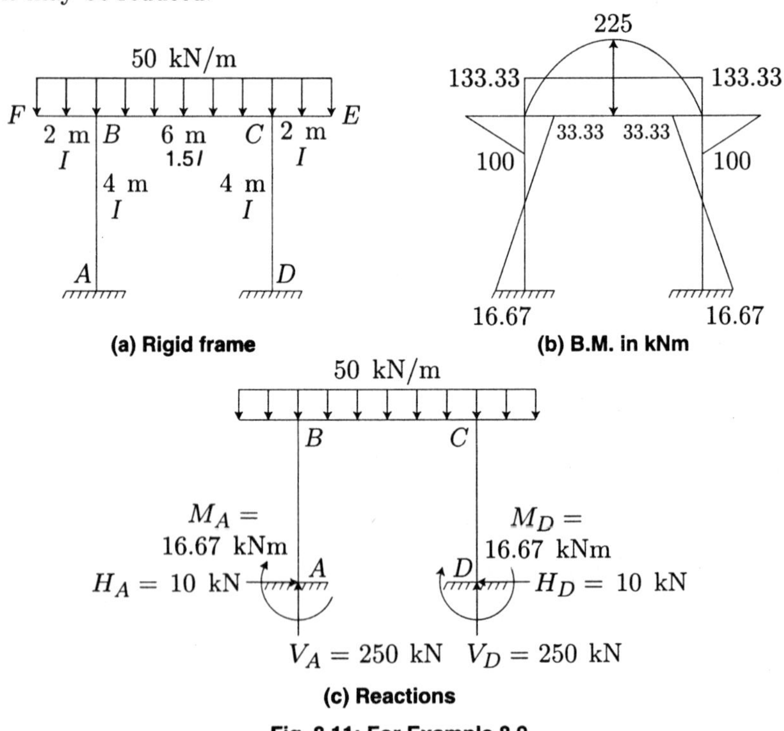

Fig. 8.11: For Example 8.9

Relative stiffness of member are as under :

$$K_{AB} = \frac{I}{4}, \text{ i.e., 1 taking } I = 4$$

$$K_{BC} = \frac{1.5I}{6}, \text{ i.e., 1}$$

$$K_{CD} = \frac{I}{4}, \text{ i.e., 1.}$$

Fixed-end moments :

$$M_{FBC} = \frac{50(6)^2}{12} = 150 \text{ kNm} = -M_{FCE}$$

$$M_{FCE} = \frac{50(2)^2}{2} = 100 \text{ kNm}$$

$$M_{FBF} = -100 \text{ kNm.}$$

Slope deflection equations are as under :

$$M_{AB} = \theta_B, \text{as } \theta_A = 0$$

$$M_{BA} = 2\theta_B$$

$$M_{BC} = 150 + 2\theta_B + \theta_C = 150 + \theta_B, \text{as } \theta_B = -\theta_C$$

$$M_{CB} = -150 + \theta_B + 2\theta_C = -150 - \theta_B$$

$$M_{CE} = 100 \text{ kN-m}$$

$$M_{BF} = -100 \text{ kNm.}$$

Joint conditions :

Joint B, $M_{BA} + M_{BC} + M_{BF} = 0$, whence

$$3\theta_B + 50 = 0, \text{ whence } \theta_B = -\frac{50}{3} = -\theta_C.$$

Calculation of end moments :

$$M_{AB} = -\frac{50}{3} = -16.67 \text{ kNm}$$

$$M_{BA} = -\frac{100}{3} = -33.33 \text{ kNm}$$

$$M_{BC} = 150 - \frac{50}{3} = -133.33 \text{ kNm}$$

$$M_{CB} = -150 + \frac{50}{3} = -133.33 \text{ kNm.}$$

B.M. diagram and Reactions are shown in Fig. 8.11.

8.6 Application of Slope Deflection Method to Frames having Joint Movement (Sidesway)

When loads are applied to statically indeterminate frames, there are cases in which some joints move unknown distances although usually in known directions. Consider a rigid frame as in Fig. 8.12(a). The joints A, B and C are fixed but the joints D, E and F may move equal distances in the horizontal direction. This horizontal movement is generally called sidesway. If the amount of sidesway to the right be as indicated by dotted line in Fig. 4.12(a), then

$$\psi_{AD} = \frac{\Delta}{H_1}, \quad \psi_{BE} = \frac{\Delta}{H_2} \quad \text{and} \quad \psi_{CF} = \frac{\Delta}{H_3}.$$

Slope deflection Eq. (8.9) will then have to be used for members AD, BE and CF. As there are one more unknown denoted by Δ, over and above the three rotations θ_B, θ_E and θ_F, it is necessary to seek another condition to cope with the unknown sidesway Δ. Applying the equations of statics to the free bodies of members AD, BE and CF, as in Fig. 8.12(b).

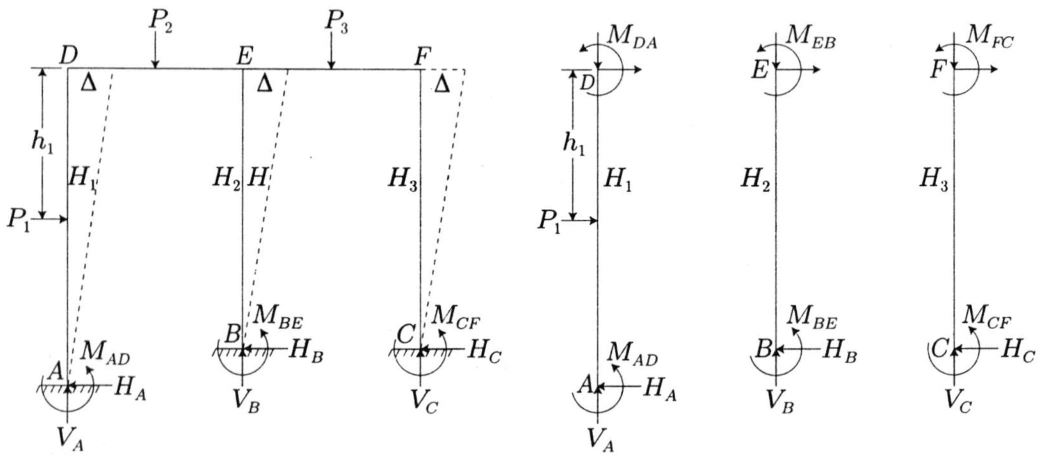

(a) Rigid frame **(b) Free body diagram of *AD, BE* and *CF***

Fig. 8.12: Rigid frame with sidesway

$$H_A = \frac{P_1 h_1}{H_1} + \frac{M_{AD} + M_{DA}}{H_1}$$

$$H_B = \frac{M_{BE} + M_{EB}}{H_2}$$

and $$H_C = \frac{M_{CF} + M_{FC}}{H_3}.$$

Applying statics, from $\sum H = 0$, for the whole frame,

$$P_1 - H_A - H_B - H_C = 0. \tag{8.11}$$

Eq. (8.11) is usually called the shear equation, which furnishes the extra condition corresponding to the unknown Δ.

If a frame of two storeys as in Fig. 8.13(a) be considered, Joints A, B and C are fixed. Joints D, E and F may move an amount Δ_1 horizontally to the right and the joint G, H and I an amount Δ_2. As $\Delta_2 > \Delta_1$, it follows that,

$$\psi_{GD} = \psi_{HE} = \psi_{IF} = \frac{\Delta_2 - \Delta_1}{H_2}$$

and $$\psi_{DA} = \frac{\Delta_1}{H_1}, \quad \psi_{EB} = \frac{\Delta_1}{H_3} \text{ and } \psi_{FC} = \frac{\Delta_1}{H_4}.$$

Slope deflection equations as in Eq. (8.9) are to be used for all vertical members . It is also necessary to have two more conditions corresponding to the unknowns $\triangle_1$ and $\triangle_2$. From free body diagrams of DG, EH and FI as in Fig. 8.13(b),

$$H_D = \frac{M_{DG} + M_{GD}}{H_2};\ H_E = \frac{M_{EH} + M_{HE}}{H_2}$$

and $$H_F = \frac{M_{FI} + M_{IF}}{H_2}.$$

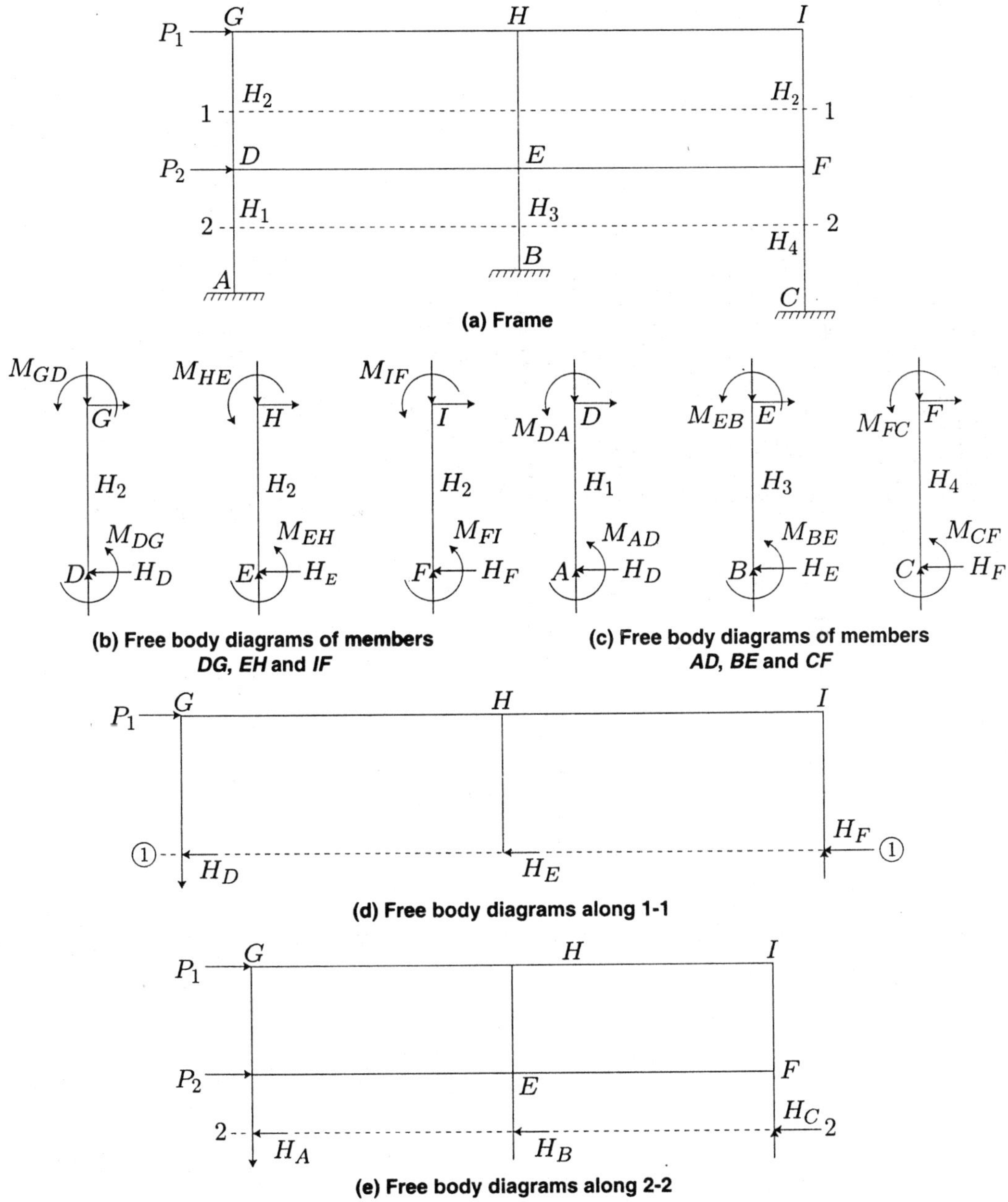

(a) Frame

(b) Free body diagrams of members *DG, EH* and *IF*

(c) Free body diagrams of members *AD, BE* and *CF*

(d) Free body diagrams along 1-1

(e) Free body diagrams along 2-2

Fig. 8.13: Two-storey frame with sidesway

The shear conditions for the upper storey referring to the free body shown in Fig. 8.13(d), gives

$$P_1 - H_D - H_E - H_F = 0. \tag{8.12}$$

Similarly, from free body diagram of AD, BE and CF as in Fig. 8.13(c),

$$H_A = \frac{M_{AD} + M_{DA}}{H_1}; H_E = \frac{M_{BE} + MEB}{H_3}$$

and $$H_C = \frac{M_{CF} + M_{FC}}{H_4}.$$

The shear condition for the lower storey with reference to the free body shown in Fig. 8.13(e),

$$(P_1 + P_2) - H_A - H_B - H_C = 0. \tag{8.13}$$

Eqs. (8.12) and (8.13) furnish the two extra conditions, corresponding to the two extra unknowns $\triangle_1$ and $\triangle_2$.

Applications of the slope deflection method to the analysis of statically indeterminate frames, wherein some joints are displaced during deformation will be illustrated by Examples.

Example 8.10. *Analyse the rigid frame in Fig. 8.14(a) using slope deflection method.*

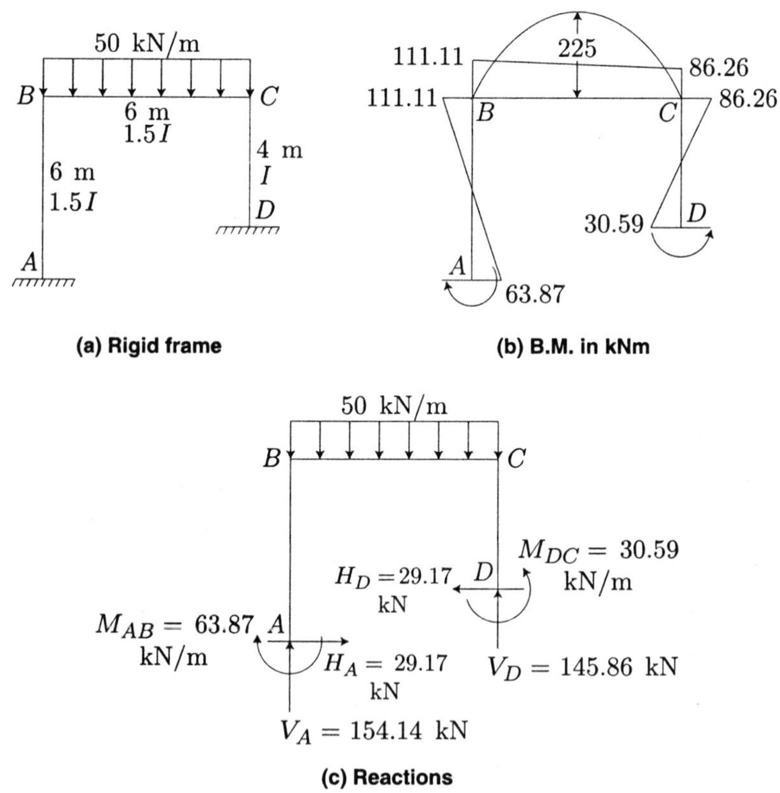

(a) Rigid frame **(b) B.M. in kNm** **(c) Reactions**

Fig. 8.14: For Example 8.10

Solution : In this case, relative stiffness can be used.

Relative stiffness :

$$K_{AB} = \frac{1.5I}{6}, \text{ say, } 1$$

$$K_{BC} = \frac{1.5I}{6}, \text{ i.e., } 1$$

$$K_{AD} = \frac{I}{4}, \text{ i.e., } 1.$$

Fixed end moments :

$$M_{FBC} = \frac{50(6)^2}{12} = 150 \text{ kNm} = -M_{FCB}.$$

Slope deflection equations :

$$M_{AB} = \theta_B - \frac{3\triangle}{6} = \theta_B - 0.5\triangle, \text{ as } \theta_A = 0$$

$$M_{BA} = 2\theta_B - 0.5\triangle$$

$$M_{BC} = 150 + 2\theta_B + \theta_C$$

$$M_{CB} = -150 + \theta_B + 2\theta_C$$

$$M_{CD} = 2\theta_C - 0.75\triangle, \text{ as } \theta_D = 0$$

$$M_{DC} = \theta_C - 0.75\triangle.$$

Joint condition :

Joint B, $M_{BA} + M_{BC} = 0$, whence

$$4\theta_B + \theta_C - 0.5\triangle + 150 = 0 \qquad \text{(i)}$$

Joint C, $M_{CB} + M_{CD} = 0$, whence

$$\theta_B + 4\theta_C - 0.75\triangle \quad 150 = 0. \qquad \text{(ii)}$$

Shear equation :

For horizontal equilibrium of the frame as a whole $\sum H = 0$, whence

$$\frac{M_{BA} + M_{AB}}{6} + \frac{M_{CD} + M_{DC}}{4} = 0, \text{whence}$$

$$6\theta_B + 9\theta_C - 6.5\triangle = 0. \qquad \text{(iii)}$$

Solving Eqs. (i), (ii) and (iii) simultaneously,

$$\theta_B = -47.26$$

$$\theta_C = +55.50$$

$$\triangle = +33.22.$$

Positive value of $\triangle$ indicates that frame sways on the left so that angle ψ is anticlockwise(positive).

Calculation of end moments :

$$M_{AB} = -47.26 - 0.5(33.22) = -63.87 \text{ kNm}$$

$$M_{BA} = 2(-47.26) - 0.5(33.22) = -111.11 \text{ kNm}$$

$$M_{BC} = 150 + 2(-47.26) + 55.50 = 110.98 \text{ kNm}$$

$$M_{CB} = -150 + (-47.26) + 2(55.50) = -86.26 \text{ kNm}$$

$$M_{CD} = 2(55.50) - 0.75(33.22) = 86.09 \text{ kNm}$$

$$M_{DC} = 55.50 - 0.75(33.22) = 30.59 \text{ kNm}.$$

B.M. diagram and Reactions are shown in Fig. 8.14.

Example 8.11. *Analyse the Rigid Frame shown in Fig. 8.15(a) using Slope Deflection Method.*

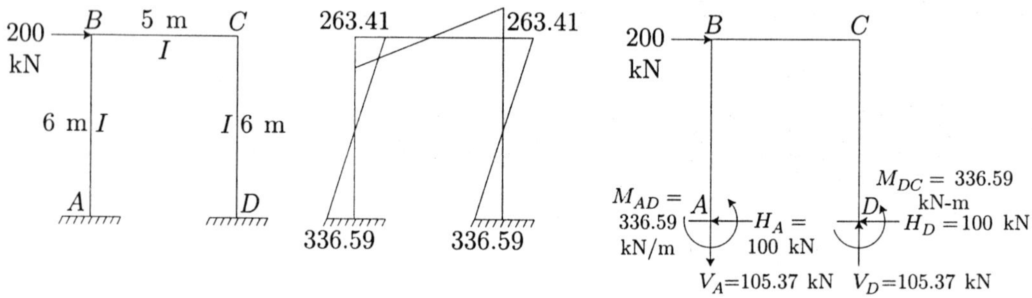

(a) Frame **(b) B.M. in kNm** **(c) Reactions**

Fig. 8.15: For Example 8.11

Solution : Since there are no intermediate loads on any of the frame members all fixed end moments are zero. Also from anti-symmetry of the structure, rotations at B and C are of the same magnitude and sign, i.e., $\theta_B = \theta_C$. This reduces the number of indeterminate joint rotation to one in place of two.

Slope deflection equations :

Considering EI as 30, and $\triangle$ as the sidesway,

$$M_{AB} = 10\theta_B - 5\triangle$$

$$M_{BA} = 20\theta_B - 5\triangle$$

$$M_{BC} = 36\theta_B, \text{ as } \theta_B = \theta_C.$$

Joint conditions :

Joint B, $M_{BA} + M_{BC} = 0$, whence

$$56\theta_B - 5\triangle = 0. \qquad \text{(i)}$$

Shear equation :

From $$\frac{M_{AB} + M_{BA}}{6} + \frac{M_{CD} + M_{DC}}{6} - 200 = 0,$$

$$3\theta_B - \triangle = 60 \qquad \text{(ii)}$$

Solving Eqs. (i) and (ii), $\theta_B = -\dfrac{300}{14}$

and $\qquad \triangle = -\dfrac{3360}{41}$

i.e., sway is towards right, so that ψ, i.e., angle due to sidesway is clockwise.

Calculation of end moments :

$$M_{AB} = -\frac{3000}{41} + \frac{16800}{41} = 336.59 \text{ kNm}$$

$$M_{BA} = -\frac{6000}{41} + \frac{16800}{41} = 263.41 \text{ kNm.}$$

B.M. diagrams and Reactions are shown in Fig. 8.13.

Example 8.12. *Analyse the Rigid Frame shown in Fig. 8.16(a), using Slope Deflection method.*

Solution : The frame will be subjected to sideway due to unsymmetric nature of the frame and unsymmetric loading.

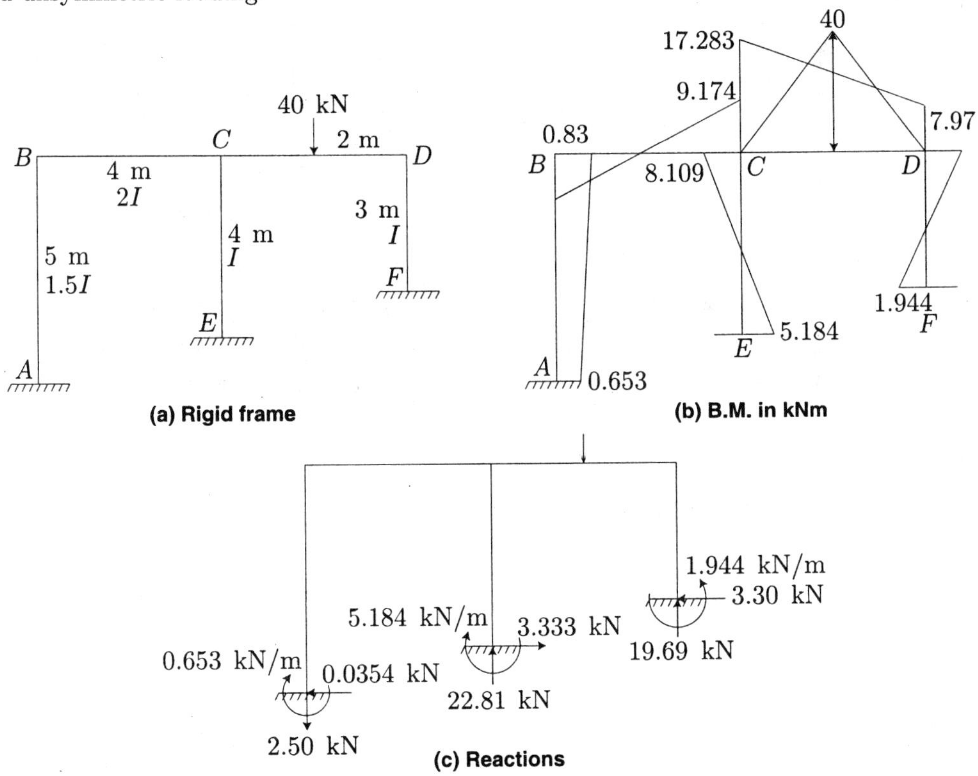

(a) Rigid frame **(b) B.M. in kNm** **(c) Reactions**

Fig. 8.16: For Example 8.12

Relative stiffness :

$$K_{AB} = \frac{1.5I}{5}, \text{ i.e., 18 taking } I = 60$$

$$K_{BC} = \frac{2I}{4}, \text{ i.e., 30}$$

$$K_{CD} = \frac{2I}{4}, \text{ i.e., } 30$$

$$K_{CE} = \frac{I}{4}, \text{ i.e., } 15$$

$$K_{DF} = \frac{I}{3}, \text{ i.e., } 20.$$

Fixed-end moments :

$$M_{FCD} = \frac{40(4)}{8} = 20 \text{ kN-m} = -M_{FDC}.$$

Slope deflection equations are

$$M_{AB} = 36\left(\theta_B - \frac{3\triangle}{5}\right)$$

$$M_{BA} = 36\left(2\theta_B - \frac{3\triangle}{5}\right)$$

$$M_{BC} = 60(2\theta_B + \theta_C)$$

$$M_{CB} = 60(\theta_B + 2\theta_C)$$

$$M_{CD} = 20 + 60(2\theta_C + \theta_D)$$

$$M_{DC} = -20 + 60(\theta_C + 2\theta_D)$$

$$M_{CE} = 30\left(2\theta_C - \frac{3\triangle}{4}\right)$$

$$M_{EC} = 30\left(\theta_C - \frac{3\triangle}{4}\right)$$

$$M_{DF} = 40\left(2\theta_D - \frac{3\triangle}{3}\right)$$

$$M_{FD} = 40\left(\theta_D - \frac{3\triangle}{3}\right).$$

Joint conditions :

For Joint B, $M_{BA} + M_{BC} = 0$, whence

$$192\theta_B + 60\theta_C - 21.6\triangle = 0 \tag{i}$$

For Joint C, $M_{CB} + M_{CD} + M_{CE} = 0$, whence

$$60\theta_B + 200\theta_C + 60\theta_D - \frac{90\triangle}{4} + 20 = 0 \tag{ii}$$

For Joint D, $M_{DC} + M_{DF} = 0$, whence

$$60\theta_C + 200\theta_D - 40\triangle - 20 = 0. \tag{iii}$$

From shear equation, $\sum H = 0$, i.e.,

$$\frac{M_{AB} + M_{BA}}{5} + \frac{M_{CE} + M_{EC}}{4} + \frac{M_{DF} + M_{FD}}{3} = 0, \text{ whence}$$

$$21.6\theta_B + 22.5\theta_C + 40\theta_D - 46.57\triangle = 0. \tag{iv}$$

Solving Eqs. (i), (ii), (iii) and (iv),

$$\theta_B = 0.0421;\ \theta_C = -0.0975;\ \theta_D = 0.149$$

$$\triangle = 0.1004.$$

Calculation of end moments :

$$M_{AB} = -0.653 \text{ kNm } M_{BA} = 0.863 \text{ kNm}$$

$$M_{BC} = -0.798 \text{ kNm } M_{CB} = -9.174 \text{ kNm}$$

$$M_{CD} = 17.28 \text{ kNm } M_{DC} = -7.97 \text{ kNm}$$

$$M_{CE} = -8.109 \text{ kNm } M_{EC} = -5.184 \text{ kNm}$$

$$M_{DF} = 7.904 \text{ kNm } M_{FD} = 1.944 \text{ kNm}$$

B.M. diagrams and Reactions are shown in Fig. 8.16.

Example 8.13. *Analyse the rigid frame shown in Fig. 8.17(a), using Slope Deflection method.*

Solution : *Relative stiffness :*

For AB and $CD = \dfrac{3I}{3}$, i.e., 2 taking $I = 2$

For AC and $BD = \dfrac{2I}{4}$, i.e., 1

For $CE = \dfrac{4I}{4}$, i.e., 2

For $DF = \dfrac{2I}{2}$, i.e., 2.

As no load is applied on any member between end joints, there is no fixed end moments for any member.

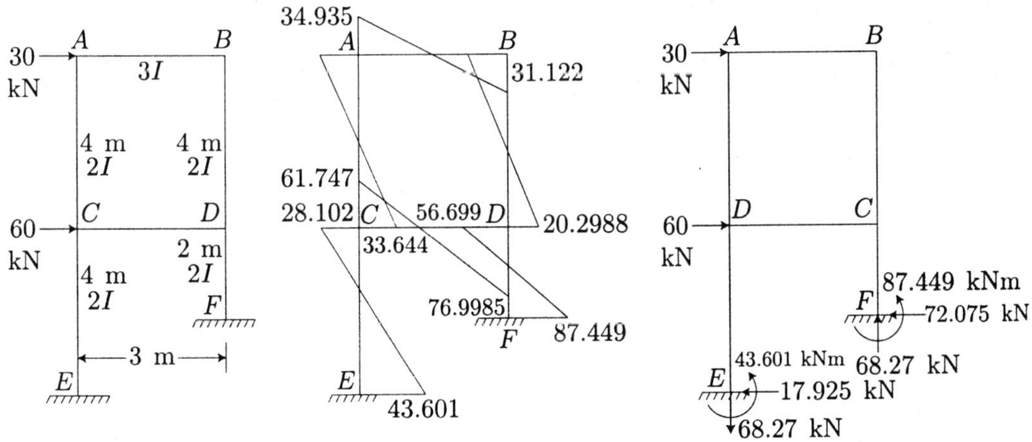

(a) Rigid frame **(b) B.M. in kNm** **(c) Reactions**

Fig. 8.17: For Example 8.13

Relative value of $\psi_{AC} = \dfrac{\triangle_2 - \triangle_1}{4} = \psi_2$, say

Relative value of $\psi_{BD} = \dfrac{\triangle_2 - \triangle_1}{4} = \psi_2$,

Relative value of $\psi_{CE} = \dfrac{\triangle_1}{4} = \psi_1$, say

Relative value of $\psi_{DF} = \dfrac{\triangle_1}{2} = 2\psi_1$.

Slope deflection equations are as under :

$$M_{AB} = 0 + 2(2\theta_A + \theta_B) = 4\theta_A + 2\theta_B$$

$$M_{BA} = 2\theta_A + 4\theta_B$$

$$M_{CD} = 4\theta_C + 2\theta_D$$

$$M_{DC} = 2\theta_C + 4\theta_D$$

$$M_{AC} = 2\theta_A + \theta_C - \psi_2$$

$$M_{CA} = \theta_A + 2\theta_C - \psi_2$$

$$M_{BD} = 2\theta_B + \theta_D - \psi_2$$

$$M_{DB} = \theta_B + 2\theta_D - \psi_2$$

$$M_{CE} = 4\theta_C - 2\psi_1$$

$$M_{EC} = 2\theta_C - 2\psi_1$$

$$M_{DF} = 4\theta_D - 4\psi_1$$

$$M_{FD} = 2\theta_D - 4\psi_1.$$

Joint conditions :

Joint A, $M_{AB} + M_{AC} = 0$, whence

$$6\theta_A + 2\theta_B + \theta_C - \psi_2 = 0 \quad \text{(i)}$$

Joint B, $M_{BA} + M_{BD} = 0$, whence

$$2\theta_A + 6\theta_B - \theta_D - \psi_2 = 0 \quad \text{(ii)}$$

Joint C, $M_{CA} + M_{CD} + M_{CE} = 0$, whence

$$\theta_A + 10\theta_C + 2\theta_D - \psi_2 - 2\psi_1 = 0 \quad \text{(iii)}$$

Joint D, $M_{DB} + M_{DC} + M_{DF} = 0$, whence

$$\theta_B + 2\theta_C + 10\theta_D - \psi_2 - 4\psi_1 = 0. \quad \text{(iv)}$$

From shear condition,

$$\frac{M_{AC} + M_{CA}}{4} + \frac{M_{BD} + M_{DB}}{4} - 30 = 0,$$

$$\theta_A + \theta_B + \theta_C + \theta_D - \frac{4}{3}\psi_2 - 40 = 0. \quad \text{(v)}$$

From $$\frac{M_{CE}+M_{EC}}{4}+\frac{M_{DF}+M_{FD}}{2}-90=0,$$

$$2\theta_C+4\theta_D-\frac{20}{3}\psi_1-120=0. \qquad \text{(vi)}$$

Solving Eqs. (i), (ii), (iii), (iv), (v) and (vi),

$$\theta_A=-6.458;\ \theta_B=-4.5515;\ \theta_C=-7.7493;$$

$$\theta_D=-15.375;\ \psi_1=-29.55 \text{ and } \psi_2=-55.10.$$

Calculation of end moments (kNm) :

$$M_{AB}=-34.935 \text{ kNm}$$

$$M_{BA}=-31.122 \text{ kNm}$$

$$M_{CD}=-61.747 \text{ kNm}$$

$$M_{DC}=-76.9985 \text{ kNm}$$

$$M_{AC}=34.935 \text{ kNm}$$

$$M_{CA}=33.6438 \text{ kNm}$$

$$M_{BD}=31.122 \text{ kNm}$$

$$M_{DB}=20.2988 \text{ kNm}$$

$$M_{CE}=28.1025 \text{ kNm}$$

$$M_{EC}=43.601 \text{ kNm}$$

$$M_{DF}=56.699 \text{ kNm}$$

$$M_{FD}=87.449 \text{ kNm}.$$

B.M. diagram and Reactions are shown in Fig. 8.17.

8.7 Analysis of Statically Indeterminate Frames Due to Yielding of Supports

The induced reactions and the shear and moments in the members of a statically indeterminate frame due to yielding of supports can be found by the slope deflection method. Considering the frames in Fig. 8.18, it is required to determine all end moments due to rotational slip of α radians of joint D and a horizontal movement Δ_H and a vertical settlement Δ_V at joint E. Slope deflection equations as in Eq. (8.9) have to be used in such a case. In this case, $\theta_D=+\alpha$ radians, $\theta_E=0$ and $\theta_F=0$.

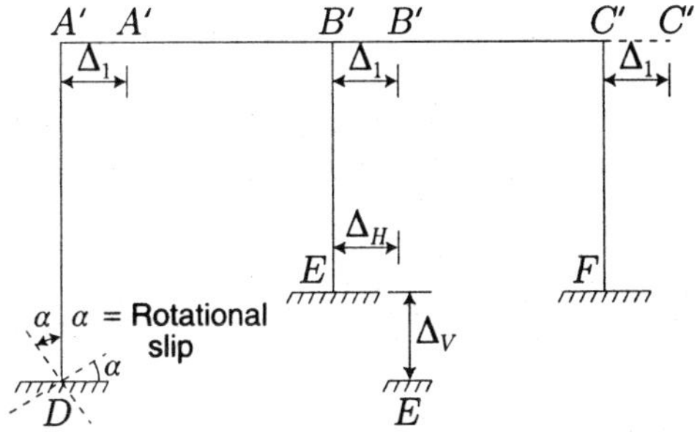

Fig. 8.18: Yielding of supports of rigid frame

Rotations $\theta_A, \theta_B, \theta_C$ and $\triangle_1$ are the unknowns in this case. Values of ψ are as under :

$$\psi_{AB} = -\frac{\triangle_V}{L_{AB}}, \ \psi_{BC} = +\frac{\triangle_V}{L_{BC}}$$

$$\psi_{AD} = -\frac{\triangle_1}{L_{AD}}, \ \psi_{BE} = -\frac{\triangle_1 - \triangle_H}{L_{BE}}$$

and $\psi_{CF} = -\frac{\triangle_1}{L_{CF}}$.

The required conditions are furnished by the three joint conditions and the shear condition. Of course, generally the effect of one yielding component only is investigated at one time.

Example 8.14. *Using slope deflection method analyse the rigid frame shown in Fig. 8.19(a) if joint D has a rotational slip of 0.002 radians clockwise and a vertical settlement of 12 mm at joint D.*

Solution : Values of $\frac{2EI}{L}$:

For AB, $\frac{2EI}{L} = \frac{2(200)(3)(10)^5}{(7.5)(1000)(1000)} = 16$ kNm

For BC, $\frac{2EI}{L} = \frac{2(200)(6)(10)^5}{(6)(1000)(1000)} = 40$ kNm

For CD, $\frac{2EI}{L} = \frac{2(200)(3)(10)^5}{(5)(1000)(1000)} = 24$ kNm.

Values of ψ in radians considering sidesway as $\triangle$,

$$\psi_{AB} = -\frac{\triangle}{7.5(1000)} = -\frac{\triangle}{7500}$$

$$\psi_{BC} = -\frac{12}{6(1000)} = -\frac{1}{500}$$

$$\psi_{CD} = -\frac{\triangle}{5000}.$$

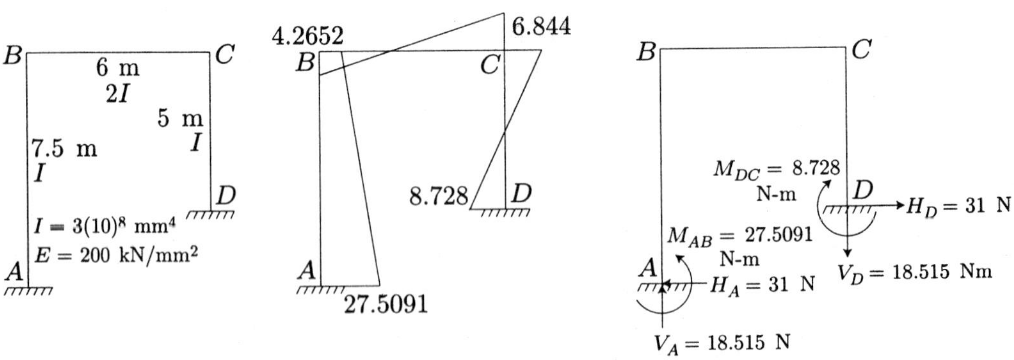

(a) Frame (b) B.M. in Nm (c) Reactions in N

Fig. 8.19: For Example 8.14

Slope deflection conditions

$$M_{AB} = 16\theta_B + 0.0064\triangle$$

$$M_{BA} = 32\theta_B + 0.0064\triangle$$

$$M_{BC} = 80\theta_B + 40\theta_C + 0.24$$

$$M_{CB} = 40\theta_B + 80\theta_C + 0.24$$

$$M_{CD} = 48\theta_C + 0.0144\triangle - 0.048$$

$$M_{DC} = 24\theta_C + 0.0144\triangle - 0.096.$$

Two joint conditions and one shear condition :

Joint B, $M_{BA} + M_{BC} = 0$, whence

$$112\theta_B + 40\theta_C + 0.0064\triangle + 0.24 = 0. \quad \text{(i)}$$

Joint C, $M_{CB} + M_{CD} = 0$,

$$40\theta_B + 128\theta_C + 0.0144\triangle + 0.192 = 0. \quad \text{(ii)}$$

Shear, $\dfrac{M_{AB} + M_{BA}}{7.5} + \dfrac{M_{CD} + M_{DC}}{5} = 0$, whence

$$6.4\theta_B + 14.4\theta_C + 0.0074666\triangle - 0.0288 = 0. \quad \text{(iii)}$$

Solving Eqs. (i), (ii) and (iii),

$$\theta_B = -0.0019859$$

$$\theta_C = -0.0019215$$

and $\triangle = 9.2630556.$

Computation of end moments :

$$M_{AB} = 0.0275091 \text{ kNm} = 27.5091 \text{ Nm}$$

$$M_{BA} = -0.0042652 \text{ kNm} = -4.2652 \text{ Nm}$$

$$M_{BC} = 0.004268 \text{ kNm} = -M_{BA} = 4.268 \text{ Nm}$$

$$M_{CB} = 0.006844 \text{ kNm} = 6.844 \text{ Nm}$$

$$M_{CD} = -0.006844 \text{ kNm} = 6.844 \text{ Nm}$$

$$M_{DC} = -0.008728 \text{ kNm} = -8.728 \text{ Nm}.$$

B.M. diagram and reactions are shown in Fig. 4.19.

Problems

8.1 Analyse the continuous beams shown in Fig. P-8.1 by Slope Deflection method. Obtain the B.M. and S.F. diagrams and also draw the elastic curve.

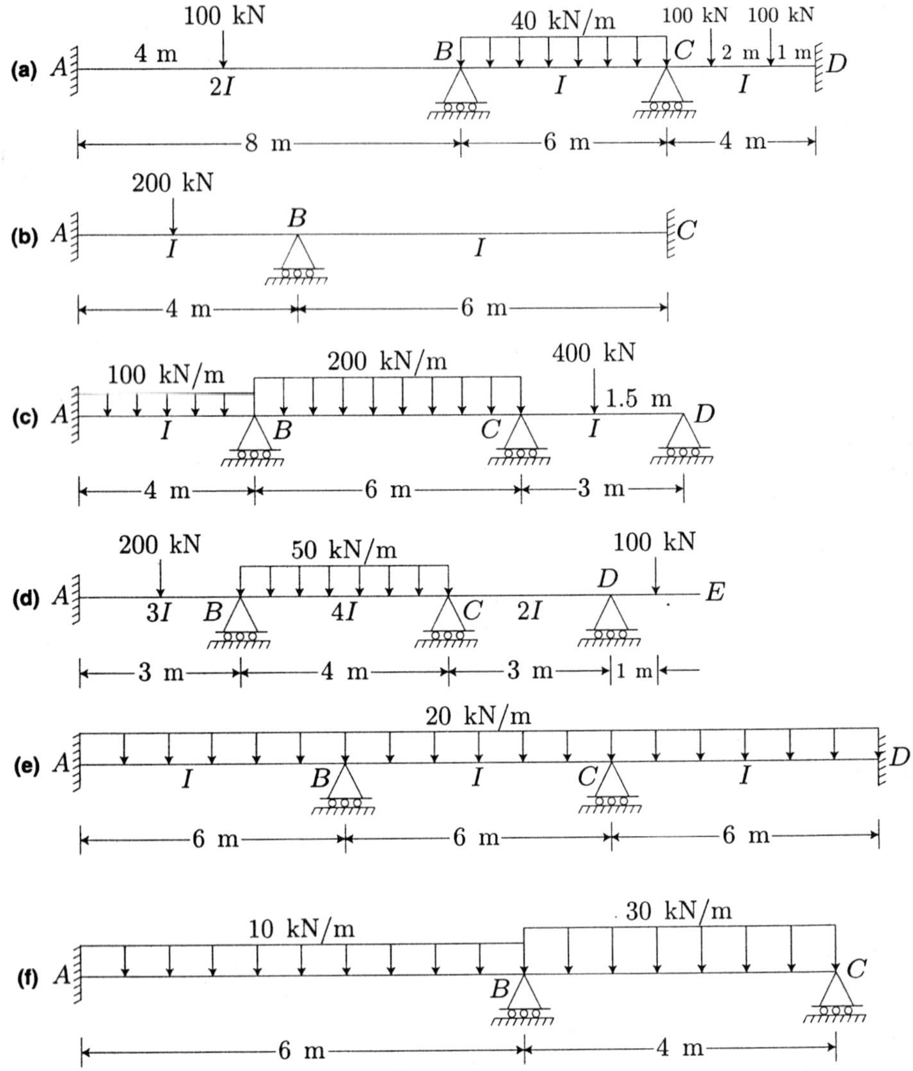

EI = constant

Fig. P-8.1

8.2 Analyse the continuous beams shown in Fig. P-8.2 by Slope Deflection method, if the joints are subjected to rotations and/or supports settle as indicated.

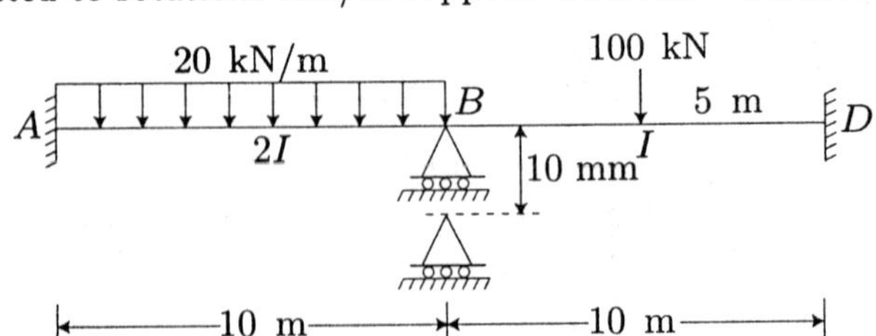

$EI = 8000$ kN/m^2

(a)

Fig. P-8.2 (Contd.)

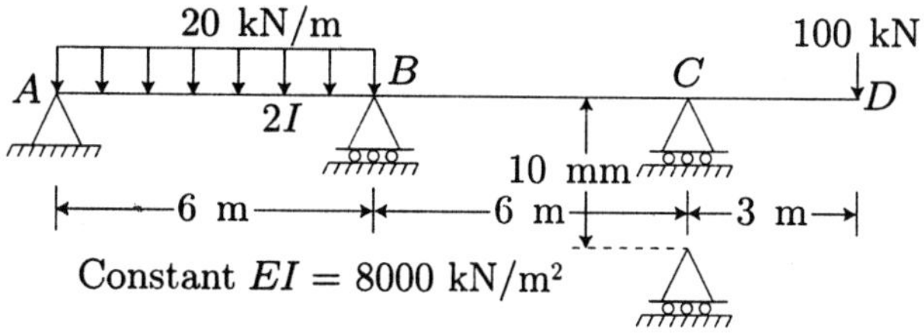

(b)

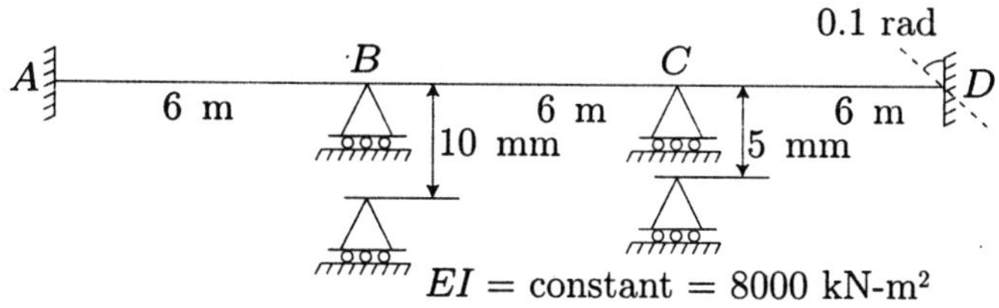

(c)

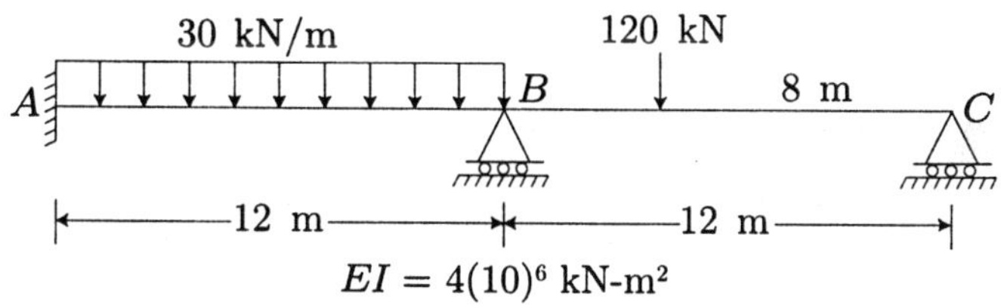

(d)

Fig. P-8.2

8.3 Analyse the rigid frames shown in Fig. P-8.3, using Slope Deflection method. Obtain B.M. diagram and Reactions and draw the elastic curve.

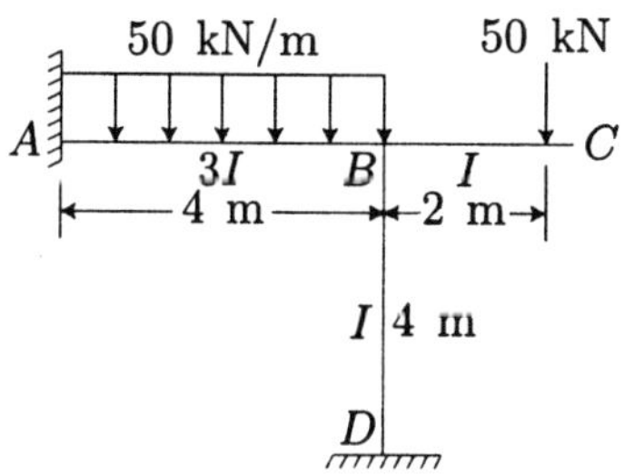

(a)

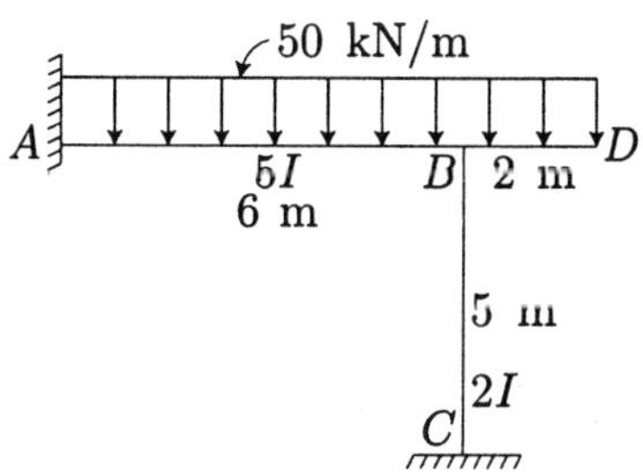

(b)

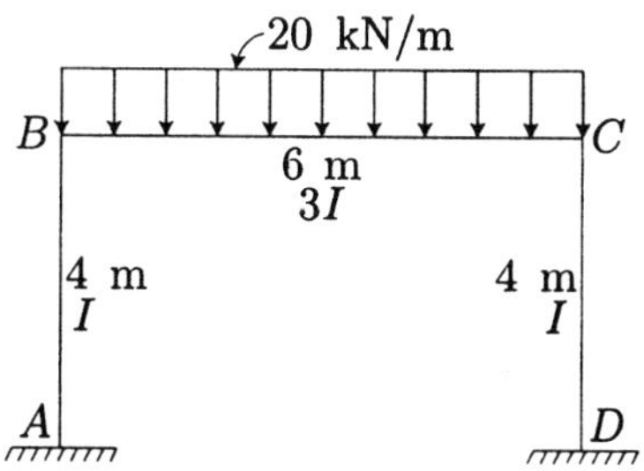

(c)

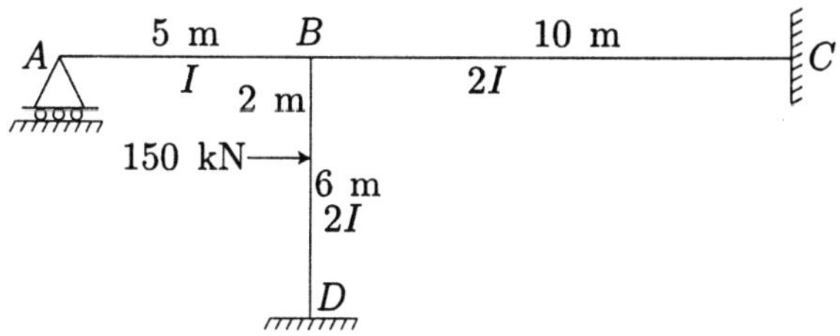

(d)

Fig. P-8.3 (Contd.)

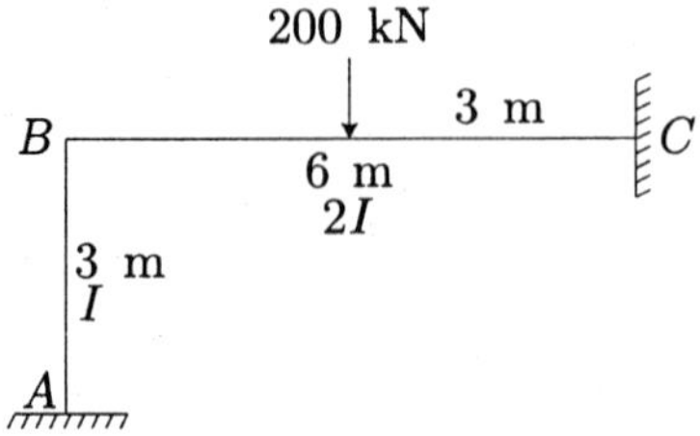

(e)

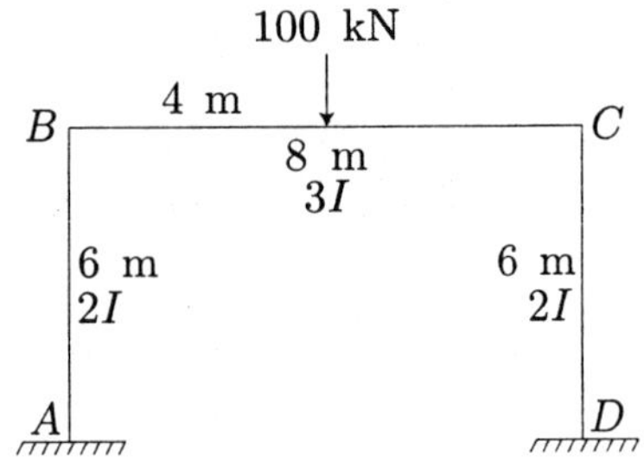

(f)

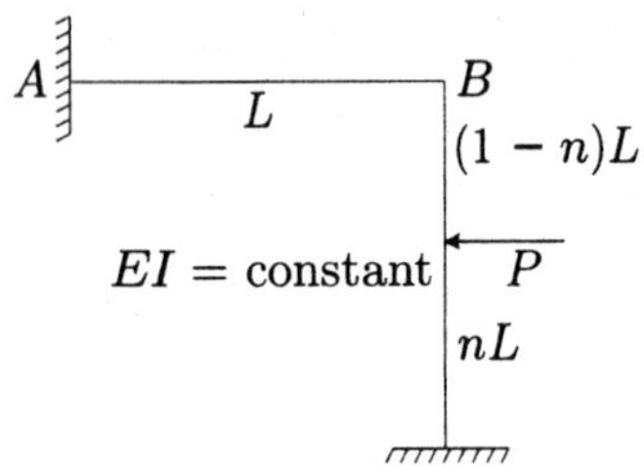

(g)

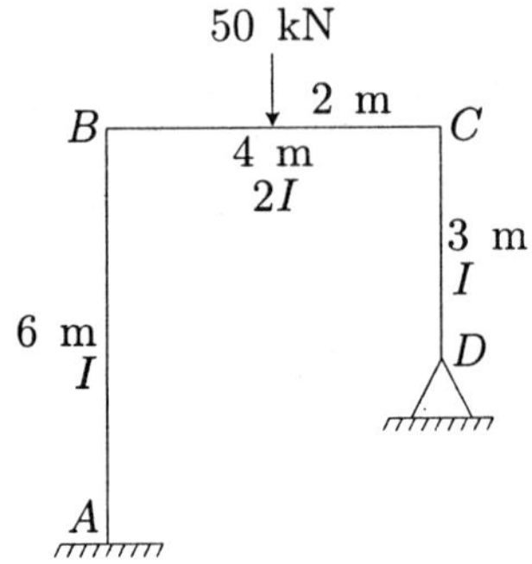

(h)

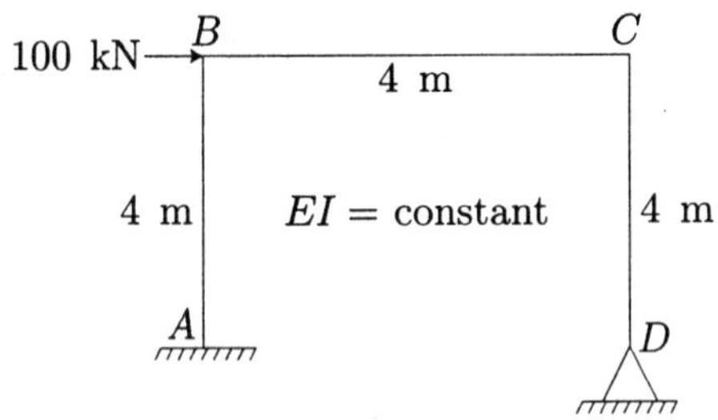

(i)

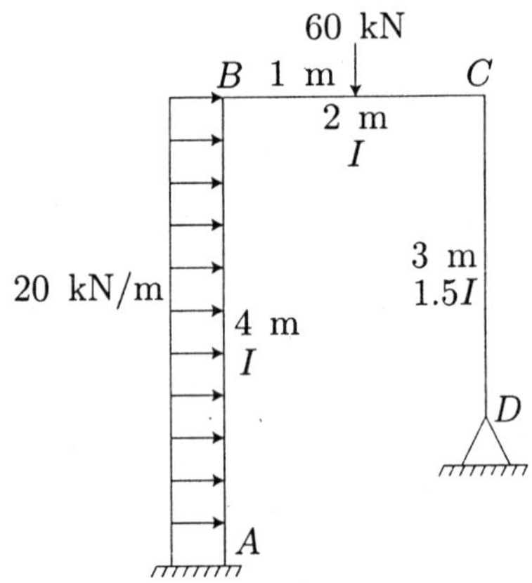

(j)

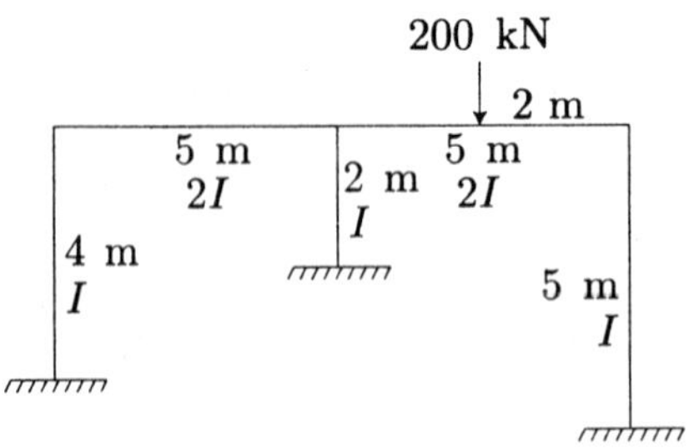

(k)

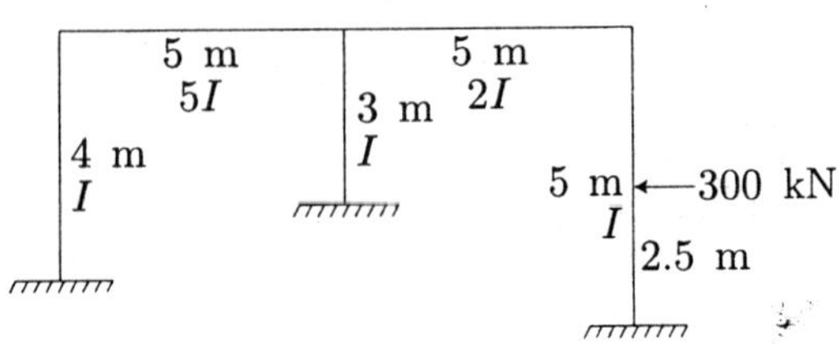

(l)

Fig. P-8.3 (Contd.)

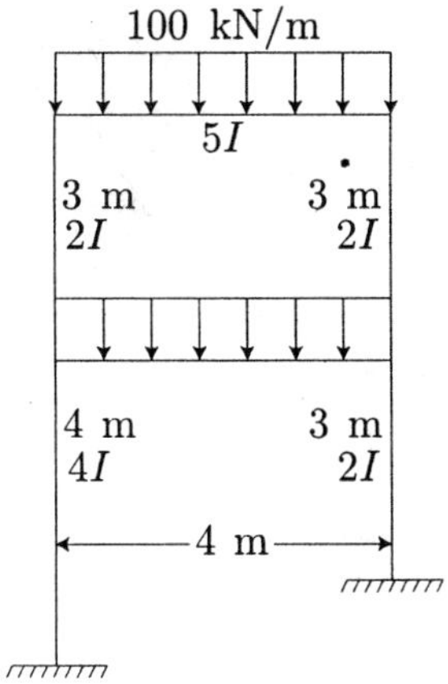

(m)

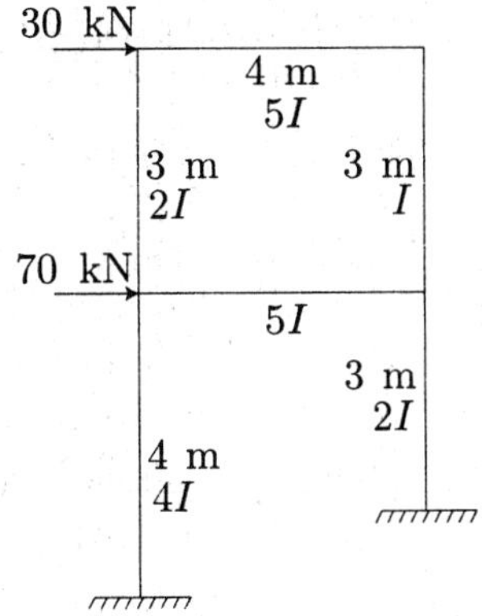

(n)

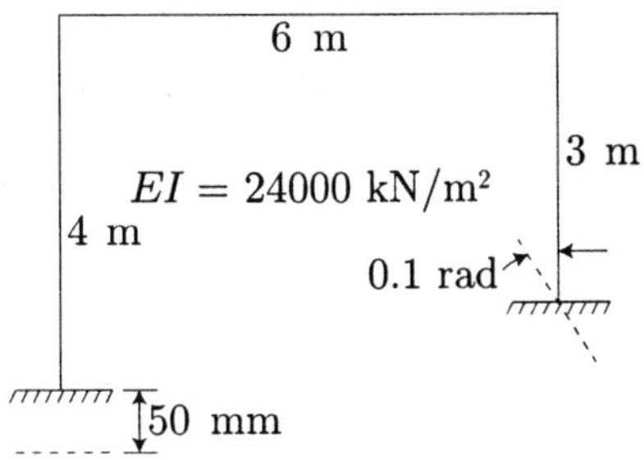

(o)

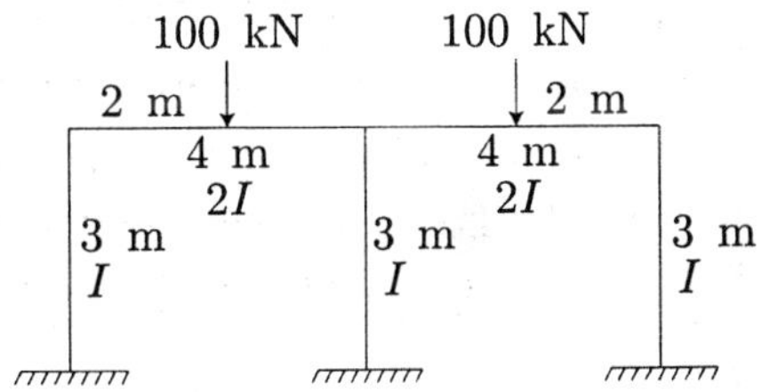

(p)

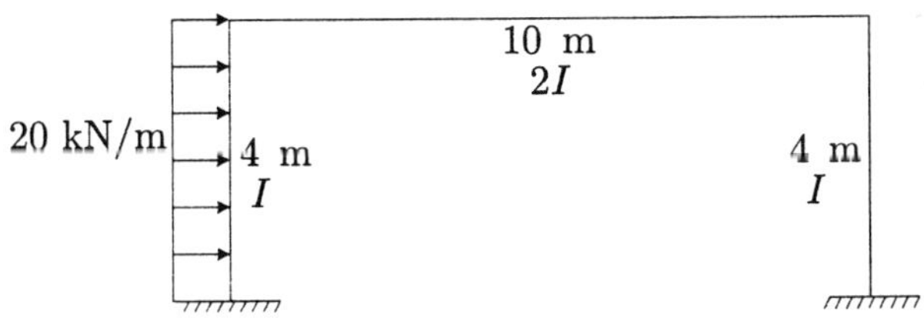

(q)

Fig. P-8.3

Chapter 9

Method of Moment Distribution

9.1 Introduction

The moment distribution method also known as Hardy Cross method is an ingeneous and convenient method of analysis of rigid jointed structure. This method was presented by Prof. Hardy Cross in an article in Transactions, ASCE (Vol. 96, paper 1793) in the year 1932 and in several prior publications. Actually Prof. Hardy Cross had been teaching the method to his students in the Yale University for several years before publishing the method. The method can be considered as the most important contribution to structural analysis in the twentieth century.

The method of moment distribution does not involve simultaneous equations. It essentially consists in solving the simultaneous equations of the Slope Deflection method by successive approximations. It has the advantage of consisting of a series of cycles each converging on the precise final result, as a result of which the series can be terminated whenever the degree of precision required by the problem is reached.

Considering the fundamental slope deflection as in Eqs. (8.7) and (8.9) in Section 8.3, it is observed that the moment acting on the ends of a member is the algebraic sum o four separate effects :

(i) The moment due to applied loads on the member, if it was a fixed-end beam, i.e., the fixed-end moment denoted by F.E.M.

(ii) The moment caused by the rotation of the tangent of the elastic curve at the near end.

(iii) The moment caused by the rotation of the tangent to the elastic curve at the far end.

(iv) The moment caused by the rotation of the chord of the elastic curve joining the two ends of the members.

Since the total end moment is the superposition of these four effects, it is suggested that it might be possible to allow them to take place separately and thus arrive at the total resulting from all four acting simultaneously.

9.2 General Description of Moment Distribution Method

The fundamental principle of moment distribution can be illustrated by referring to the structure shown in Fig. 9.1, which is the type of structure Hardy Cross used in his paper. The supports at A, C, D and E are fixed, and are unyielding and have no joint rotation. The members are rigidly joined together at B, so that angles between the members do not change when the joints rotate under load.

The first step in the moment distribution procedure consists of applying an imaginary clamp at joint B, which locks the joint against rotation. When all joints are thus locked against rotation, each member is in the condition of a fixed-end beam. As loads are applied to the structure, fixed-end moments are developed at the ends of the members.

The next step is to release the locked joint, in this case joint B, when the external constraint must exert a moment on the structure called the unbalanced moment, in order to prevent joint rotation. This unbalanced moment is equal to the sum of the fixed end moments at the joint, i.e., equal to the sum of the fixed end moments in members BA, BC, BD and BE.

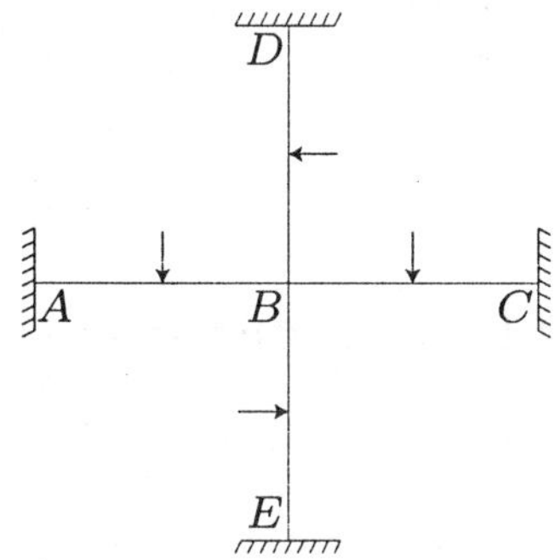

Fig. 9.1: Rigid structure

The next step in the moment distribution procedure is the removal of the external constraints. Since the external moment applied at a joint is zero when there is no constraint at the joint and no applied couple acting as a load, the removal of a constraint can be accomplished by acting to the joint, a moment equal and opposite to the unbalanced moment. Due to the balancing moments, moments will also be developed at the far ends of each of the members meeting at the joint. For example the balancing moment, which is applied to the structure at joint B, will result in moment in members, BA, BC, BD and BE. In effect, the balancing moment is divided into components, called distributed moments, each acting on one of the members meeting at the joint. Since all of the members, meeting at the joint, when the balancing moment is applied, the proportion of the balancing moment, which is distributed to any particular member will depend on the resistance of that member to rotation. Each member meeting at a joint acts as rotational spring in resisting rotation of the joint, and since the amount of rotation is the same for all members, the moment distributed to each member will be in proportion to the rotational stiffness of that member. If the rotational stiffnesses are known, the fraction of the total balancing moment which is distributed to any particular member can be found by dividing the stiffness of that member by the sum of all the stiffnesses. For example, if member BA, BC, BD and BE are identical and thus have the same stiffnesses, one-fourth of the balancing moment will be distributed to each. On the other hand, if the stiffness of BA is twice that of the others, the moment distributed to BA will be 0.40 of the balancing moment while 0.20 of the balancing moment will act on the other members. The factor by which the balancing moment is multiplied in order to obtain the distributed moment is called the distribution factor (D.F.) for the member.

The addition of a distributed moment to one end of a member usually involves a moment at the other end, if the latter is restrained against rotation. Thus moments will be induced at end A, C, D and E during the process of balancing moment at B. The end of the member, where the joint is being balanced is usually considered as the near end of the member, the other end being called the far end. When moment is distributed to the near end of a member, 'carry over moment' (C.O.M.) is induced at the far end of the member, the carry over moment for a far end which is pinned being zero. The factor by which the distributed moment is multiplied in order to obtain the carry over moment is called the 'carry over factor' (C.O.F.).

After the moments at a joint have been distributed, the joint is balanced, i.e., it remains in equilibrium and no external constraint is required. However, there may be other joints in the structure which are still locked by external means. The next step, therefore, is to relock the joint, which has just been balanced and then proceed to unlock the next joint of the structure. The process of moment distribution is repeated at that joint, resulting in carry over moments to other joints. Thus, the process of moment distribution is cyclic in nature and continues until the balanced moments at the joints are so small as to be negligible.

The process of locking and unlocking each joint of the structure once only is referred to as one cycle of moment distribution. If the moment distribution calculations are continued indefinitely, the results will be theoretically exact.

Moment distribution is thus not an approximate method, but is a method of successive approximations by which the exact results can be approached with any desired degree of accuracy. In practice, the cycles of calculations are continued only until desired degree of accuracy in the results is obtained. Usually, three or more cycles are enough for fairly accurate results.

Process of moment distribution for a structure in which joint translation is prevented involves the following steps :

(a) Calculate stiffness factors, distribution factors (D.F.) and carry over factors (C.O.F.) for all members.

(b) Lock all the joints against rotation.

(c) Calculate fixed-end moments (F.E.M.).

(d) Unlock a selected joint first and calculate unbalanced moments at the joint.

(e) Calculate distributed moments for the near ends of the members meeting at the joint.

(f) Calculate carry over moments at the far ends of the members.

(g) Relock joint and select next joint to be unlocked. Repeat steps, (d), (e) and (f).

(h) Continue the procedure till all the joints have been balanced once.

(i) Repeat steps (d) to (h) in a cyclic fashion until balanced moments are negligible.

(j) Add the distributed and carry over moments of all cycles to the F.E.M. at any member and obtain final moments in the member-end.

In order to carry out the steps listed above, it is necessary to adopt a sign convention for moments and to obtain values of the various quantities mentioned in the general description of the method.

The complete distribution can be carried out by one of the three techniques :

(i) *Unlocking joint at a time.* The joint with the largest unbalanced moment is usually selected first and the moment distributed and carried over. By successive selection of the largest unbalanced moment distribution and carried over, the unbalance converges to zero.

(ii) *Releasing several joints simultaneously.* Instead of releasing one joint at a time, several joints can be balanced simultaneously. The joints selected for a particular balance need to be chosen judiciously, so that carry over moments do not go to a joint which has been released simultaneously. The cycle is repeated till unbalanced moments become negligible.

(iii) *Releasing all joints.* All joints can be balanced simultaneously and the carry over moments transferred to their appropriate joints as unbalanced moments. These in turn are balanced and carry over moments transferred. The procedure is repeated till the unbalanced moment is everywhere zero.

9.3 Sign Convention for Moments

In structural analysis, there are two sign conventions for moments, one from the deformation point of view as in B.M. sign convention, the other being from statics sign convention, which considers whether the moments applied at a joint are clockwise or anticlockwise.

Bending moments in continuous beams are usually considered positive when the top of the beam is in compression. Bending moment signs on rigid frames are obtained from this convention applicable in case of the horizontal member.

In carrying out moment distribution calculations, the direction of the end moments are of immediate interest rather than the way in which moment deforms the members. It is, therefore, customary to adopt the statics sign convention, based on direction of joint moments. A moment acting at a joint is considered positive when in the clockwise direction and negative when in the anticlockwise direction. Fig. 9.2 indicates the direction of the sign conventions.

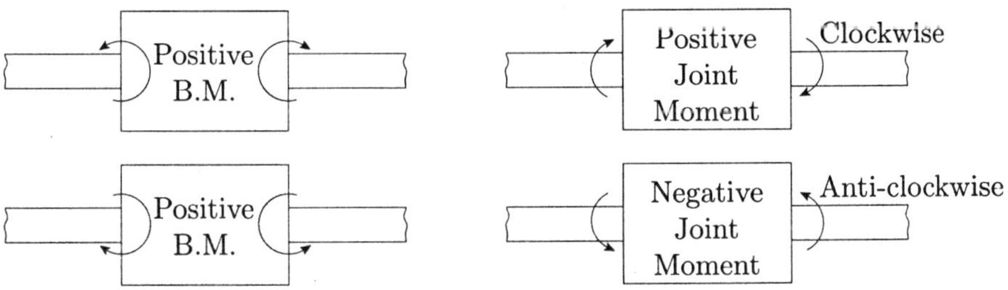

(a) B.M. sign convention **(b) Joint sign convention**

Fig. 9.2: Sign convention for moments

9.4 Fixed End Moments (F.E.M.)

The first step in the moment distribution procedure is to lock the joints against rotation and determine the fixed-end moment resulting from the loads. Such moments can be obtained by well known method of analysis such as moment area method. The fixed-end moments (F.E.M.) for prismatic members for some usual cases of loading can be obtained from Fig. 9.3.

The formulae are based on sign convention for moment distribution procedure adopted in Section 9.3.

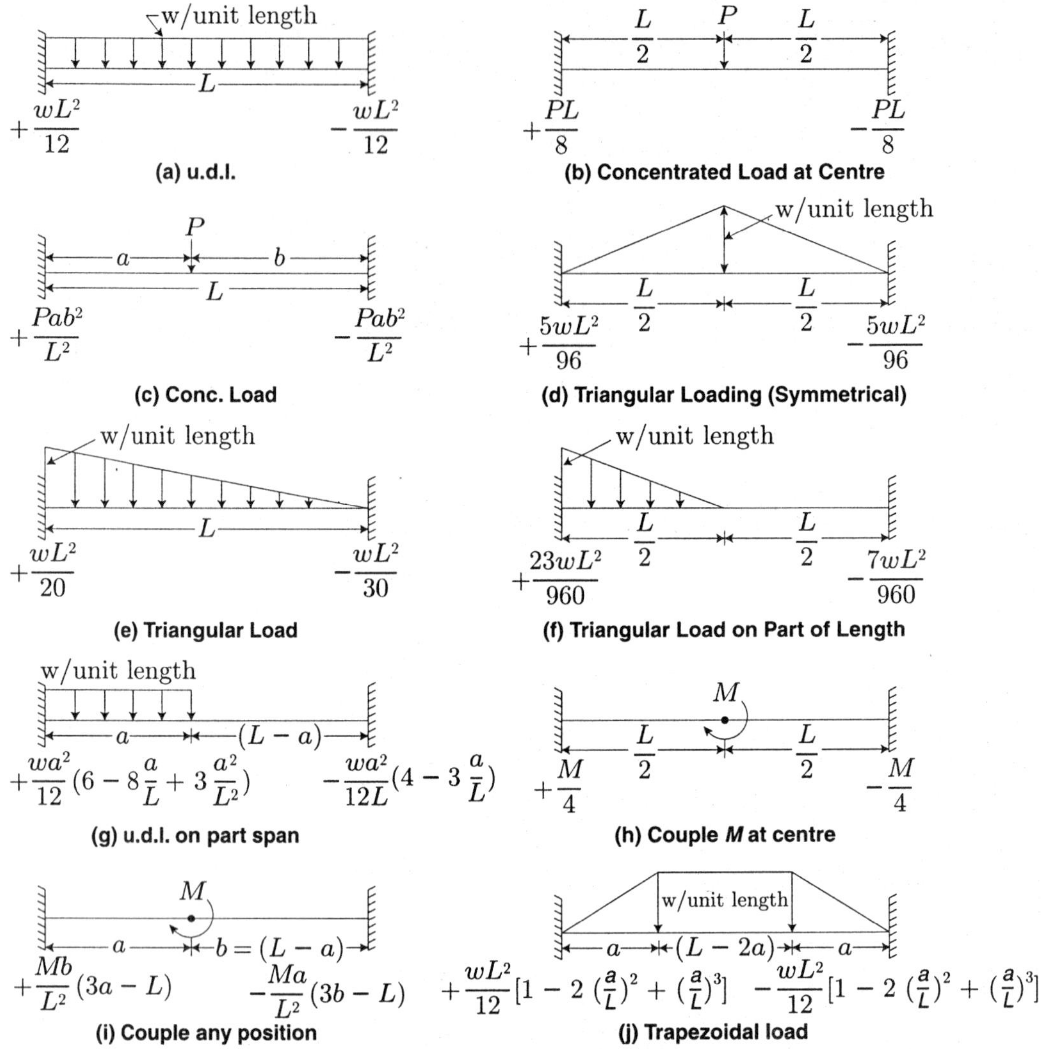

Fig. 9.3: Standard cases of Fixed end Moments

9.5 Rotational Stiffness, Distribution Factor (D.F.) and Carry Over Factors (C.O.F.)

The balancing moment is distributed to the members meeting at a joint of the structure in proportion to their rotational stiffness, the rotational stiffness being a measure of the resistance of the beam to rotation at one end. The absolute rotational stiffness K is defined as the value of the moment required to rotate the near end of the beam through a unit angle. Thus, the absolute stiffness of beam AB in Fig. 9.4(a) is given by

$$K_{AB} = \frac{M_A}{\theta_A}, \tag{9.1a}$$

where θ_A is the angle of rotation caused by the moment M_A. For θ_A = unity, $K_{AB} = M_A$. The actual value of the stiffness, which has units of force times length, depends on the dimensions of the beam and the conditions of support at the far end B.

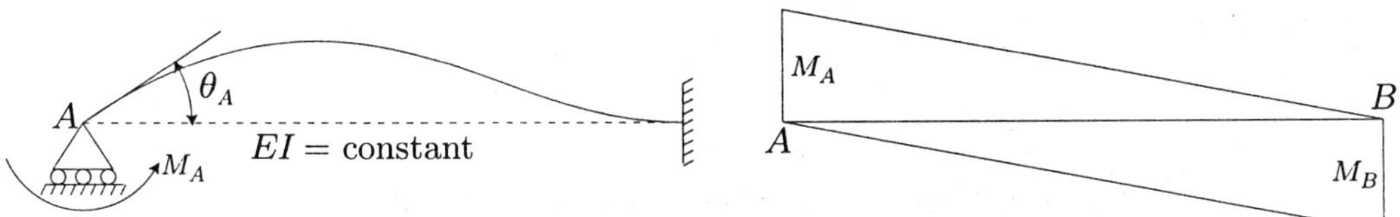

(a) Beam **(b) B.M. diagram**

Fig. 9.4: Development of stiffness and C.O.F.

In order to derive expressions for rotational stiffnesses for various end conditions, beam AB with A as the near end and B as the far end which is fixed, is considered. Moment M_A is applied at the near end A and as adjacent end A rotates through an angle θ_A, end moment M_B is induced at the far end B. Considering the B.M. pattern in Fig. 9.4(b), using second moment area theorem,

$$\frac{1}{EI}\left[\frac{1}{2}M_A L\left(\frac{L}{3}\right)+\frac{1}{2}M_B(L)\left(\frac{2L}{3}\right)\right]=0,$$

whence $M_B = -\dfrac{M_A}{2}$.

Considering joint sign convention as used in moment distribution method,

$$\frac{M_B}{M_A}=\frac{1}{2}, \tag{9.1b}$$

which is called the carry over factor denoted by C.O.F.

Using first moment area theorem,

$$\frac{1}{EI}\left[\frac{M_A L}{2}-\frac{M_A}{2}\frac{L}{2}\right]=\frac{M_A L}{4EI}=\theta_A,$$

whence $$K_{AB}=\frac{M_A}{\theta_A}=\frac{4EI}{L}, \tag{9.2}$$

which is the stiffness factor for far end fixed.

The stiffness of a member with far end hinged can be considered as the 'modified stiffness'. Considering beam AB in Fig. 9.5(a) with A as the near end, where moment M_A is applied, the moment at the far end which is hinged being zero, the carry over factor for this case is zero.

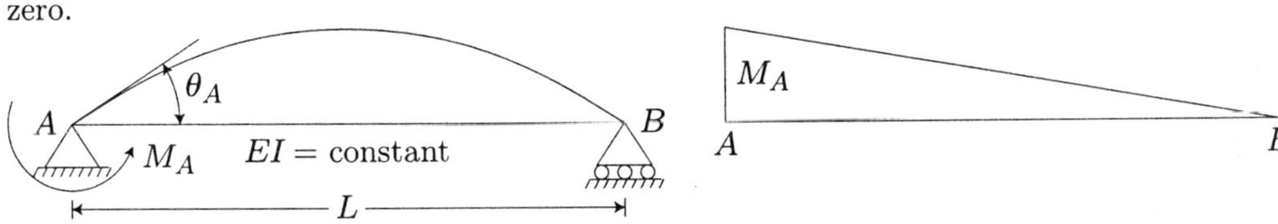

(a) Beam **(b) B.M. diagram**

Fig. 9.5: Stiffness for member with far end hinged

Using second moment area theorem

$$\theta_A = \frac{1}{EIL}\left[\frac{M_A L}{2}\frac{2}{3}L\right] = \frac{M_A L}{3EI},$$

whence modified stiffness, i.e., stiffness with far end hinged is given by

$$\frac{M_A}{\theta_A} = \frac{3EI}{L}. \tag{9.3}$$

The 'distribution factor' is the fraction of the total balancing moment which is distributed to a particular member. Since each member meeting at a joint rotates through the same angle, it follows that the distribution factor is the ratio of the rotational stiffness K of the member under consideration to the sum of the stiffnesses of all members meeting at the joint. This can be expressed by a general formula,

$$\text{Distribution factor, D.F.} = \frac{K}{\sum K},$$

where K is the stiffness of the particular member and $\sum K$ is the sum of all member stiffnesses. It is evident that either the absolute or relative stiffness factors may be used in this expression and it is apparent also that the sum of all distribution factors for the members meeting at a joint must be a unity.

As an example, the distribution factor for member BA at joint B of the rigid frame in Fig. 9.1 is calculated from

$$\text{D.F.}_{BA} = \frac{K_{BA}}{K_{BA} + K_{BC} + K_{BD} + K_{BE}}.$$

9.6 Check on Moment Distribution

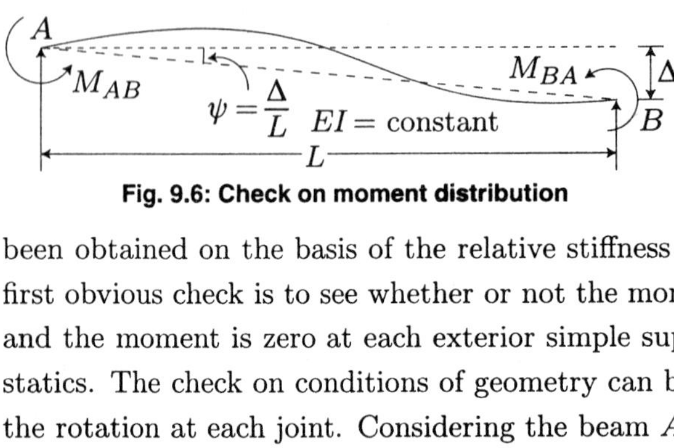

Fig. 9.6: Check on moment distribution

The moment distribution table begins with relative stiffness and fixed-end moments and ends with final end moments. A check can be made to ensure that the correct end moments have been obtained on the basis of the relative stiffness and fixed end moments. Of course, the first obvious check is to see whether or not the moments are balanced at each interior joint and the moment is zero at each exterior simple support. This check is on the condition of statics. The check on conditions of geometry can be made by finding the relative values of the rotation at each joint. Considering the beam AB shown in Fig. 9.6, the rotation θ_B at B is given by

$$\theta_B = \frac{M_{AB}L}{6EI} - \frac{M_{BA}L}{3EI} - \frac{M_{FAB}}{6EI}L + \frac{M_{FBA}}{3EI}L + \Psi$$
$$= \frac{L}{6EI}[M_{AB} - M_{FAB}] - \frac{2L}{6EI}[M_{BA} - M_{FBA}] + \Psi.$$

Considering $(M_{AB} - M_{FAB})$ as ΔM_{AB} and $(M_{BA} - M_{FBA})$ as ΔM_{BA},

$$\left.\begin{aligned} \theta_B &= \frac{L}{6EI}[\Delta M_{AB} - 2\Delta M_{Ba}] + \Psi \\ \text{Similarly, } \theta_A &= \frac{L}{6EI}[\Delta M_{BA} - 2\Delta M_{AB}] + \Psi. \end{aligned}\right] \tag{9.4}$$

Example 9.1. *Analyse the continuous beam shown in Fig. 9.7(a) by the moment distribution method. Draw the B.M. and S.F. diagrams.*

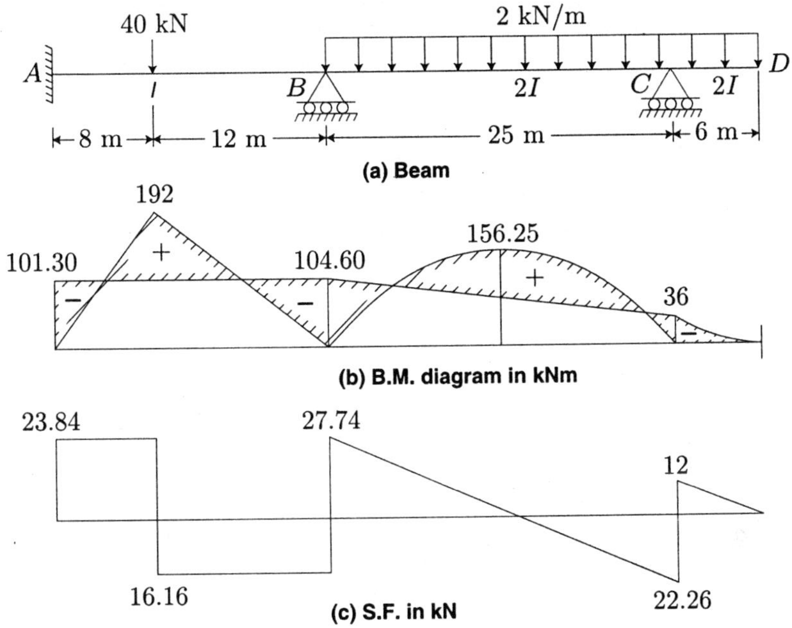

Fig. 9.7: For Example 9.1

Solution : *Rotational stiffness*

For fixed support $= \infty$

For member AB = For member $BA = \dfrac{4EI}{20}$.

For member BC = For member $CB = \dfrac{8EI}{25}$.

For cantilever $CD = 0$.

Distribution factors

Joint A, $DF_{AB} = \dfrac{\frac{4EI}{20}}{\frac{4EI}{20} + \infty} = 0$

Joint B For BA, $DF_{BA} = 0.384$, For BC, $DF_{BC} = 0.616$; $\sum = 1.00$

Joint C For CB, $DF_{CB} = 1.00$, For CD, $DF_{CD} = 0$; $\sum = 1.00$

C.O.F.

$$\text{C.O.F.}_{BA} = \text{C.O.F}_{BC} = \text{C.O.F.}_{CB} = \frac{1}{2}.$$

Fixed end moments

$$M_{FAB} = +\frac{40(12)^2(8)}{(20)^2} = +115.20 \text{ kNm}$$

$$M_{FBA} = -\frac{40(8)^2(12)}{(20)^2} = -76.80 \text{ kNm}$$

$$M_{FBC} = -M_{FCB}\frac{2(25)^2}{12} = 104.20 \text{ kNm}$$

$$M_{FCD} = +\frac{2(6)^2}{2} = +36 \text{ kNm.}$$

The moment distribution is carried on in a Tabular form in Table 9.1 which follows a schematic pattern dependent on the structure.

Joints A, B and C are first locked against rotation and the fixed end moments are recorded on the first line of calculations. In the next step, all joints are simultaneously unlocked and the resulting unbalanced moment calculated. The unbalanced moment at A is +115.20 kNm so that the balancing moment is −115.20, which is distributed to member AB in terms of $D.F_{AB}$ as in Table 9.1. The unbalanced moment at joint B and C are

Table 9.1. For Example 9.1

Joint	A	B		C		
Member	AB	BA	BC	CB	CD	
K	$\frac{4EI}{20}$	$\frac{4EI}{20}$	$\frac{8EI}{25}$	$\frac{8EI}{25}$	0	
D.F.	0	0.384	0.616	1.0	0	
C.O.F.	←	0.50	← 0.50	→		
F.E.M.	+115.20	−76.80	+104.20	−104.20	+36	1st
Bal.	0	−10.50	−16.90	+68.20	0	cycle
C.O.M.	−5.25	0	+34.10	−8.45	0	2nd
Bal.	0	−13.10	−28.00	+8.45	0	cycle
C.O.M.	−6.55	0	+4.22	−10.50	0	3rd
Bal.	0	−1.60	−2.62	+10.50	0	cycle
C.O.M.	−0.80	0	+5.25	−1.30	0	4th
Bal.	0	−2.05	−3.20	+1.30	0	cycle
C.O.M.	−1.03	0	+0.65	−1.60	0	5th
Bal.	0	−0.25	−0.40	+1.60	0	cycle
C.O.M.	−0.12	0	+0.80	−0.20	0	6th
Bal.	0	−0.30	−0.50	+0.20	0	cycle
C.O.M.	−0.15	0	+0.10	−0.25	0	7th
Bal.	0	0	−0.10	+0.25	0	cycle
C.O.M.	0					
Total moments Jt. sign Convention	+101.30	−104.60	+104.60	−36	+36	
B.M. sign convention	−101.30	−104.60	−104.60	−36	−36	

(−76.80 + 104.20), i.e. +27.40 kNm, and (−104.20 + 36), i.e., −68.20 kNm respectively. Balancing moments of −27.40 kNm in joint B and −68.20 kNm in joint C are distributed as 0.384(−27.40), i.e., −10.50, 0.616(−27.40), i.e., −16.90, and +1(68.20), i.e., 68.20 in members BA, BC and CB respectively. It is to be noted that the summation of such distributed moments at a joint shall be equal to the balancing moments at the joint. As after the balancing, the moments at a joint are exactly balanced, i.e., summation of the moments on the members at the joint is zero, it is customary to draw a line after each balancing operation.

The carry over moments shown in the first line of the second cycle are $+\frac{1}{2}$ times the balancing moments placed at the far ends of the respective members in the preceding line. These carry over moments are the new unbalances, which are balanced in the same manner as before. The process is repeated all through the various cycles.

It is observed here that the total end moments at all places are not materially affected after the seventh cycle of moment distribution, which suggests that no further moment distribution is needed. By adding the moments in the respective columns, the total end moments are obtained.

It is to be noted that the last step in the fixed end at A is the carry over moments.

The bending moments and shear force diagrams are worked out and drawn in Figs. 9.7(b) and 9.7(c) respectively.

The checking of the moments obtained by the moment distribution procedure can be done as shown in Table 9.2 considering Ψ as zero as there is no relative settlement of the ends of members.

Table 9.2. For Example 9.1

Joint	A	B		C		D
Final B.M., kN-m		+101.30	−104.60	+104.60	−36	+36
Length		20 m		25 m		
M.I.		I		$2I$		
F.E.M. in kN-m		+115.20	−76.80	+104.20	−104.20	+36
ΔM in kN-m		−13.82	−27.80	+0.40	+68.20	0
$2\Delta M$ in kN-m		−27.64	−55.60	+0.80	+136.40	0
$\Delta M_{far} - 2\Delta M_{near}$ in kN-m		−0.16	+41.78	+67.40	−136.0	0
$\theta = \frac{L}{6EI}(\Delta M_{far} - 2\Delta M_{near})$		$\frac{-0.53}{EI}$	$\frac{+139.27}{EI}$	$\frac{+140.42}{EI}$	$\frac{-283.3}{EI}$	
			← Check →			

9.7 Modification for Pin Ended Ends with Moments Statically Determinate

Modification can be used in such cases to shorten the process of calculations of end moments by moment distribution. Such modification consists in keeping such joints unlocked throughout the process and modify the stiffness factor, fixed end moments and carry over moments.

Thus, for the case of a beam AB with end A hinged and B fixed, the modified fixed-end moment taking actual end conditions into consideration is illustrated in Fig. 9.8.

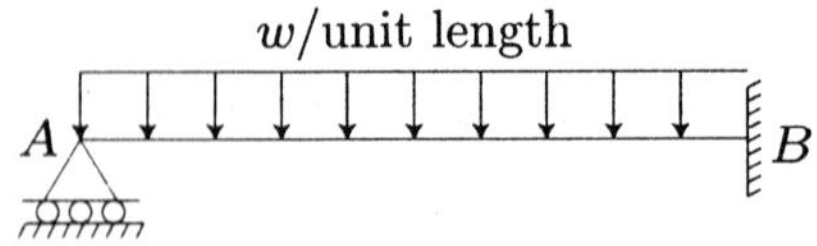

DF	1	0
COF	←	$\frac{1}{2}$
FEM	$+\frac{wL^2}{12}$	$-\frac{wL^2}{12}$
Bal	$-\frac{wL^2}{12}$	0
COM	0	$-\frac{wL^2}{24}$
Bal	0	0
Σ	0	$-\frac{wL^2}{8}$

Fig. 9.8: Modified FEM

Use of moment distribution procedure with the concept will be illustrated taking Example 9.1.

Example 9.2. *Solve Example 9.1 using modification for end C, where moment is statically determinate.*

Solution : The fixed-end moments at end A and B of beam will be the same as in Example 9.1. Modified F.E.M. for beam BCD at ends B and C have to be obtained. Calculations of the modified F.E.M. for the case are shown in Fig. 9.9.

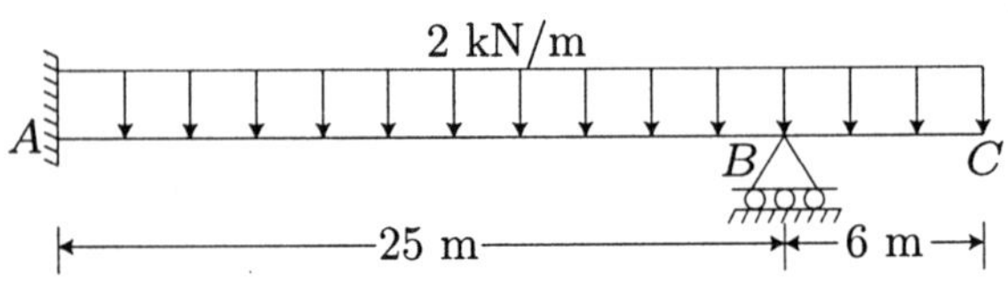

DF	0	1	0
COF	← $\frac{1}{2}$		
FEM	+104.20	−104.20	+36
Bal	0	+68.20	
COM	+34.20	0	
Bal	0	0	
Modified FEM	+138.30	−36	+36

Fig. 9.9: For Example 9.2

Modified stiffness for beam BC at B is

$$K_{BC} = \frac{3E(2I)}{25} = \frac{6EI}{25}, \quad \text{while} \quad K_{CB} = \frac{4E(2I)}{25} = \frac{8EI}{25}.$$

C.O.F. from B to $C = 0$.

Moment distribution is illustrated in Table. 9.3.

Table 9.3. For Example 9.1

Joint	*A*		*B*		*C*	
Member	*AB*	*BA*	*BC*	*CB*	*CD*	
K	$\frac{4EI}{20}$	$\frac{4EI}{20}$	$\frac{6EI}{25}$	$\frac{8EI}{25}$	0	
D.F.	0	0.455	0.545	1.00	0	
C.O.F.	←	0.50	←	0.50		
F.E.M.	+115.20	−76.80	+138.30	−36	+36	1st cycle
Bal.	0	−27.90	−33.60	0	0	
C.O.M.	−13.95	0	0	0	0	2nd cycle
Bal.	0	0	0	0	0	
Resulting Moments (Jt. sign)	+101.25	−104.70	+104.70	−36	+36	

The example illustrates the advantage of using the modified concept. The values obtained in this case are exact and not approximate since no further cycles are possible.

Example 9.3. *Analyse the continuous beam shown in Fig. 9.10 using three schemes as follows :*

(a) *Locking B and C and unlocking B and C consecutively always relocking after each step.*

(b) *Locking B and C, unlock B and then unlock C without relocking B.*

(c) *Locking C only and unlocking C.*

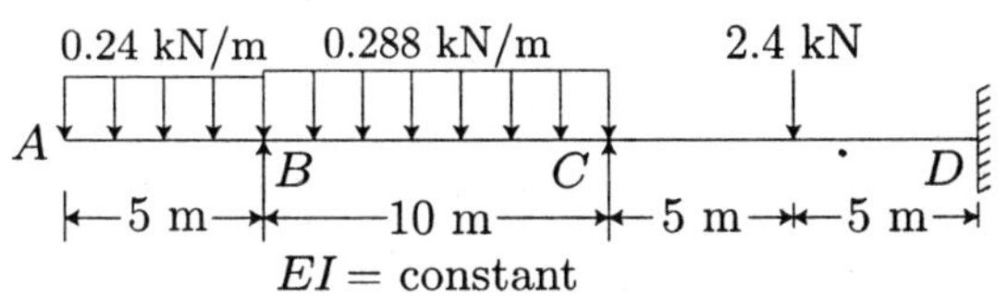

Fig. 9.10: For Example 9.3

Solution : Tables 9.4(a), 9.4(b) and 9.4(c) illustrate the various schemes.

Table 9.4(a) First Scheme for Example 9.3

Joint	*A*	*B*		*C*		*D*
Member	*AB*	*BA*	*BC*	*CB*	*CD*	*DC*
Rel. K		0	$\frac{1}{10}$	$\frac{1}{10}$	$\frac{1}{10}$	$\frac{1}{10}$
D.F.		0	1	0.50	0.50	1
C.O.F.			← 0.50	→	0.50	→
FEM		−3	+2.40	−2.40	+3	−3
Bal.		0	+0.60	−0.30	−0.30	0
COM		0	−0.15	+0.30	0	−0.15
Bal.		0	+0.15	−0.15	−0.15	0
COM		0	−0.07	+0.07	0	−0.07
Bal.		0	+0.07	−0.03	−0.04	0
COM		0				−0.02
$\sum$ Final Moment		−3	+3	−2.51	+2.51	−3.24

Table 9.4(b) Second Scheme for Example 9.3

Joint	A	B		C		D	
Member	AB	BA	BC	CB	CD	DC	
Rel. K		0	$\frac{1}{10}$	$\frac{3}{10}$	$\frac{4}{10}$	$\frac{1}{10}$	
D.F.		0		$\frac{3}{7}$	$\frac{4}{7}$	1	
C.O.F.			0.50 ⟶		—0.50 ⟶		
FEM		−3	+2.40	−2.40	+3	−3	Lock C and D
Bal.		0	+0.60	0	0	0	Unlock B
COM				+0.30	0	0	
Bal.				−0.39	−0.51		
COM						−0.26	Unlock C
Final Moment		−3	+3	−2.49	+2.49	−3.26	

Table 9.4(c) Third Scheme for Example 9.3

Joint	A	B		C		D	
Member	AB	BA	BC	CB	CD	DC	
Rel. K		0	$\frac{1}{10}$	$\frac{3}{10}$	$\frac{4}{10}$	$\frac{1}{10}$	
D.F.		0	1	$\frac{3}{7}$	$\frac{4}{7}$	1	
C.O.F.			—0.50 ⟶		—0.50 ⟶		
FEM		−3	+3	$(-2.40 + 0.30)$ $= -2.10$	+3	−3	Lock C
Bal.			0	−0.39	−0.51	0	Unlock C
COM			0	0	0	−0.26	
Final Moment		−3	+3	−2.49	+2.49	−3.26	

Checking of B.M. is shown in Table 9.5. For the beam EI = constant.

Table 9.5. For Example 9.3

Joint	B		C		D
Final BM	−3	+3	−2.49	+2.49	−3.26
Length		10 m		10 m	
F.E.M.	−3	+2.40	−2.40	+3	−3
ΔM	0	+0.60	−0.09	−0.51	−0.26
$2\Delta M$		+1.20	−0.18	−1.02	−0.52
$\Delta M_{far} - 2\Delta M_{near}$		−1.29	+0.78	+0.76	+0.01
$\theta = \frac{L}{6EI}(\Delta M_{far} - 2\Delta M_{near})$		−2.15	+1.3	+1.267	+0.017
			Check		

9.8 Effect of Yielding and Rotation of Supports

The moment distribution method can be used for analysing statically indeterminate beams due to yielding and rotation of supports. The physical concept involved is that the joints are first locked against rotation and then displaced to conform with the amount of yielding. The locking moments acting at the ends of each member will be the fixed end moments. The other steps involved are exactly similar as those due to applied loadings.

Fig. 9.11(a) shows a member AB with fixed ends, the end B having deflected by an amount Δ without rotation of joints. The resulting fixed end moment can be obtained using moment area method.

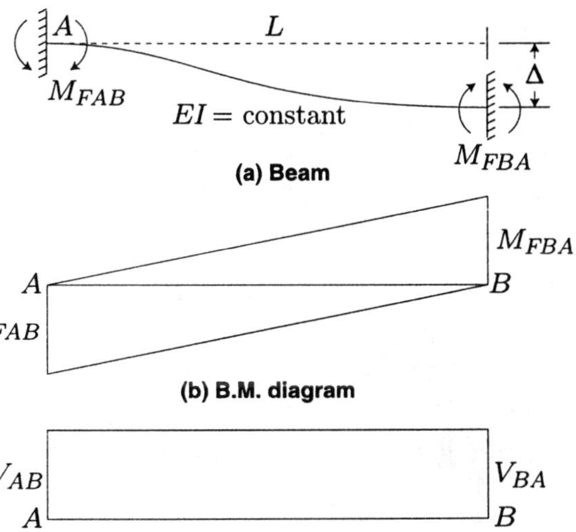

Fig. 9.11: Relative settlements of support

From first moment area theorem, referring to Fig. 9.11(b), as there is no relation of joints, $M_{FAB} = -M_{FBA}$.

From second moment area theorem,

$$\frac{1}{EI}\left[-\frac{1}{2}M_{FAB}(L)\left(\frac{L}{3}\right)+\frac{1}{2}M_{FAB}(L)\frac{2L}{3}\right] = \Delta,$$

whence $M_{FAB} = \dfrac{6EI\Delta}{L^2} = 6EI\dfrac{\Psi}{L}$, (9.5a)

where $\Psi = \dfrac{\Delta}{L}$.

Also $R_{AB} = \dfrac{12EI\Delta}{L^3} = -R_{BA}$. (9.5b)

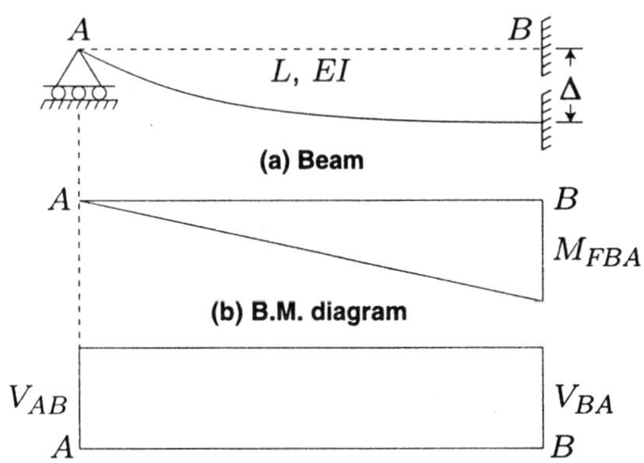

Fig. 9.12: Effect of relative settlement

If the end A be hinged, the fixed end moment at B may be obtained using similar procedure. Thus referring to Fig. 9.12, and using second moment area theorem,

$$\frac{1}{2}M_{FBA}(L)\left(\frac{2L}{3}\right) = \Delta,$$

whence $M_{FBA} = \dfrac{3EI\Delta}{L^2} = 3EI\dfrac{\Psi}{L}$, (9.6a)

where $\Psi = \dfrac{\Delta}{L}$.

As before, $R_{AB} = -R_{BA} = \dfrac{3EI\Delta}{L^3}$ (9.6b)

In case of rotation of joints for far end fixed, the magnitude of the fixed end moment is given by $\frac{4EI}{L}\theta$. If the far end is hinged, the fixed end moment is $\frac{3EI}{L}\theta$. Figs. 9.13 and 9.14 illustrate the cases.

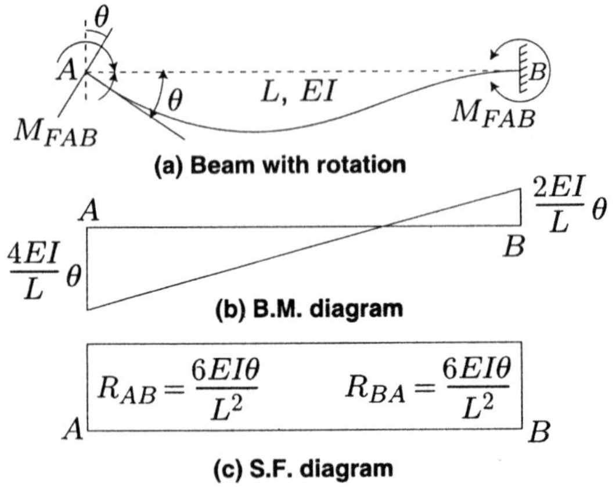

Fig. 9.13: Effect of rotation at ends

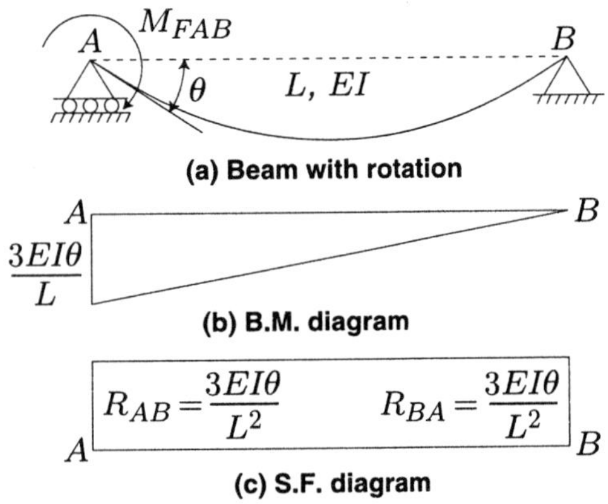

Fig. 9.14: Effect of rotation at ends

Example 9.4. *Using moment distribution method, analyse the continuous beam shown in Fig. 9.15 if support B goes down by* 10 mm *and joint A rotates by 0.006 radians clockwise. Assume* $I = I(10^8)$ mm^4 *and* $E = 2(10)^5$ N/mm^2.

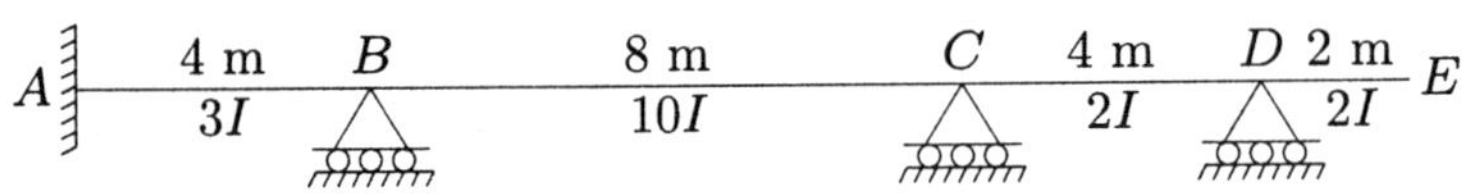

Fig. 9.15: For Example 9.4

Solution : Fixed end moments in this case are due to rotation of joint A and settlement of B.

Due to rotation of 0.006 clockwise of joint A,

$$M_{FAB} = -\frac{4E(3I)}{L}(\theta_A) = -\frac{4(2)(10)^5(3)(10^8)(0.006)}{4000}$$

$$= -36(10)^7 \text{ Nmm} = -360 \text{ kNm.}$$

$$MF_{3A} = -180 \text{ kNm.}$$

Due to relative settlement of B by 10 mm,

$$M_{FAB} = +\frac{6E(3I)\Delta}{L^2} = \frac{6(2)(10)^5(3)(10)^8(10)}{(4000)^2}$$

$$= 2.25(10)^8 \text{ Nmm} = 225 \text{ kNm} = M_{FBA}.$$

Also $$M_{FBC} = -\frac{6(2)(10)^5(10)(10)^8(10)}{(8000)^2} = -1.875(10)^8 \text{ Nmm}$$

$$= -187.5 \text{ kNm} = M_{FCB}.$$

Resultant $M_{FAB} = -360 + 225 = -135$ kNm

$$M_{FBA} = -180 + 225 = +45 \text{ kNm}$$

$$M_{FBC} = M_{FCB} = -187.5 \text{ kNm}$$

$$M_{FCD} = M_{FDC} = 0.$$

The moment distribution procedure using modified stiffness for members CD at end C is shown in the tabular form in Table 9.6.

Table 9.6. For Example 9.4

Joint	A		B		C		D
Member	AB	BA	BC	CB	CD	DC	DE
K	$\frac{12EI}{4}$	$\frac{12EI}{4}$	$\frac{40EI}{8}$	$\frac{40EI}{8}$	$\frac{6EI}{4}$	$\frac{8EI}{4}$	0
DF	0	0.375	0.625	0.77	0.23	1.0	
C.O.F.	⟵ 0.50		0.50⟶		⟵ 0.50—		
F.E.M.	−135.00	+45.00	−187.50	−187.50	0	0	
Bal.		+53.44	+89.06	+144.38	+43.12	0	
C.O.M.	+26.72	0	+72.19	+44.53	0	0	
Bal.	0	−27.07	−45.12	−34.29	−10.25	0	
C.O.M.	−13.54	0	−17.15	−22.56	0	0	
Bal.	0	+6.43	+10.72	+17.37	+5.19	0	
C.O.M.	+3.22	0	+8.69	+5.36	0	0	
Bal.	0	−3.26	−5.43	−4.13	−1.23	0	
C.O.M.	−1.63	0	−2.07	−2.71	0	0	
Bal.	0	+0.78	+1.29	+2.09	+0.62	0	
C.O.M.	+0.39	0	+1.10	+0.65	0	0	
Bal.	0	−0.41	−0.69	−0.50	−0.15	0	
C.O.M.	−0.2						
Final moments (Joint sign convention)	−120.94	+74.91	−74.91	−37.31	+37.3	0	

Example 9.5. *Analyse the continuous beam shown in Fig. 9.16 using moment distribution procedure. Consider support B sinks by* 5 mm *and support C by* 12.5 mm, $I = 6(10^8)$ mm^4 *and* $E = 2(10)^5$ N/mm^2.

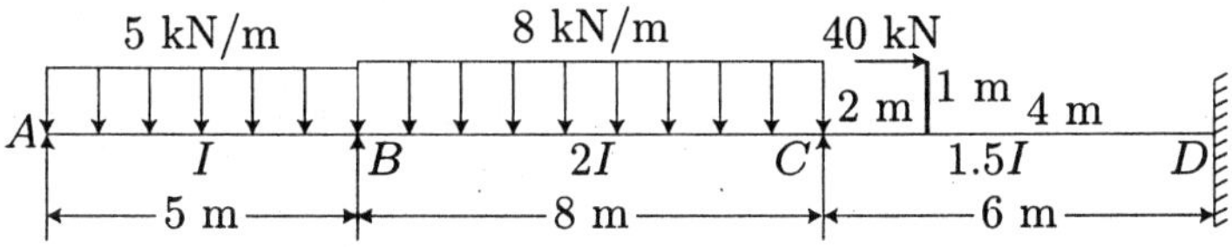

Fig. 9.16: For Example 9.5

Solution : *Fixed end moments*

For AB

$$M_{FAB} = 0$$

$$M_{FBA} = -\frac{wL^2}{8} + \frac{3EI\Delta}{L^2} = -\frac{(5)(5)^2}{8} + \frac{(3)(2)(10)^5(6)(10)^8(5)}{(5)^2(1000)^2(1000)^2} \text{ kNm}$$

$$= +5640 \text{ kNm.}$$

For BC $M_{FBC} = +\frac{wL^2}{12} + \frac{6EI\Delta}{L^2} = +\frac{(8)(8)^2}{12} + \frac{(6)(2)(10)^5(2)(6)(10)^8(12.5-5)}{(8)^2(1000)^2(1000)^2}$ kN-m

$= +42.7 + 168.7 = +211.4$ kNm

$M_{FCB} = -42.7 + 168.7 = +126.0$ kNm.

For $CD, M_{FCD} = +\frac{Mb}{L^2}(3a - L) - \frac{6EI\Delta}{L^2} = \frac{(40)(4)(6-6)}{(6)^2} - \frac{(6)(2)(10)^5(1.5)(6)(10)^8(12.5)}{(6)^2(1000)^2(1000)^2}$

$= -375.0$ kNm.

$M_{FCD} = -\frac{(40)(2)(12-6)}{(6)^2} - 375.0 \text{ kNm} = -388.3$ kNm.

The moment distribution procedure using modified stiffness for member BA at end B is shown in tabular form in Table 9.7.

Table 9.7. For Example 9.5

Joint	A		B		C		D
Member	AB	BA	BC	CB	CD	DC	
K	$\frac{4EI}{5}$	$\frac{3EI}{5}$	$\frac{8EI}{8}$	$\frac{8EI}{8}$	$\frac{6EI}{6}$	$\frac{6EI}{6}$	
D.F.	1	0.375	0.625	0.5	0.5	0	
C.O.F.	← 0.0 →		← 0.50 →		← 0.50 →		
F.E.M.	0	+56.4	+211.4	+126.0	−375.0	−388.3	
Bal.	0	−100.4	−167.4	+124.5	+124.5	0	
C.O.M.	0	0	+62.3	−83.7	0	+62.3	
Bal.		−23.4	−38.9	+41.9	+41.9	0	
C.O.M.	0	0	+21.0	−19.5	0	+21.0	
Bal.	0	−7.9	−13.1	+9.8	+9.8	0	
C.O.M.	0	0	+4.9	−6.6	0	+4.9	
Bal.	0	−1.8	−3.1	+3.3	+3.3	0	
C.O.M.	0	0	+1.7	−1.6	0	+1.7	
Bal.	0	−0.6	−1.1	+0.8	+0.8	0	
Final Moment	0	−77.7	+77.7	+194.9	194.9	−298.0	

9.9 Structures with Hinges

It is not unusual to find structures which contain hinges located at intermediate points along the members, i.e., at locations, other than the joints of the structure.

Fig. 9.17(a) shows a beam AB with a hinge at a distance αL from the left end. The bending moment pattern with end B fixed as moment M_A is applied at end A is shown in Fig. 9.17(b).

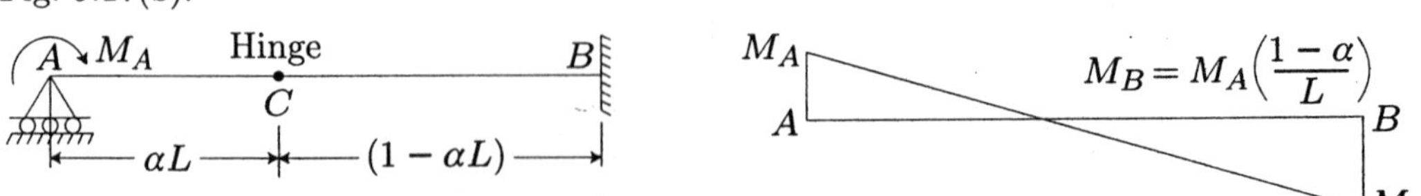

(a) Beam (b) B.M. diagram

Fig. 9.17: Beam with hinges

For stiffness, the rotation at A, i.e., θ_A has to be a unity.

From CB, using moment area method,

$$\Delta_C = \frac{M_B}{3EI}(l-\alpha)^2 L^2 = \frac{M_A}{3EI}\frac{(l-\alpha)^3}{\alpha}L^2$$

From AB, using moment area method,

$$\theta_A = \frac{1}{\alpha L}\left[\frac{M_A}{3EI}\frac{(1-a)^3}{\alpha}L^2 + \frac{M_A}{3EI}\alpha^2 L^2\right] = \frac{M_A L}{3EI}\left[\frac{3\alpha^2 - 3\alpha + 1}{\alpha^2}\right],$$

whence stiffness $K_{AB} = \dfrac{3EI}{L}\dfrac{\alpha^2}{3\alpha^2 - 3\alpha + 1}$. (9.7)

Also C.O.F. $= \dfrac{M_B}{M_A} = \dfrac{1-\alpha}{\alpha}$. (9.8)

The carry over factor is zero when $\alpha = 1$, and equals $\frac{1}{2}$ when $\alpha = \frac{2}{3}$. As α approaches zero, the carry over factor approaches infinity.

Fixed end moment for uniformly distributed load is shown in Fig. 9.18.

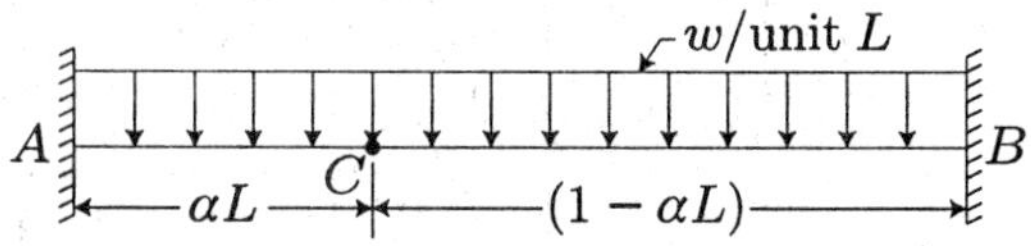

(a) Beam

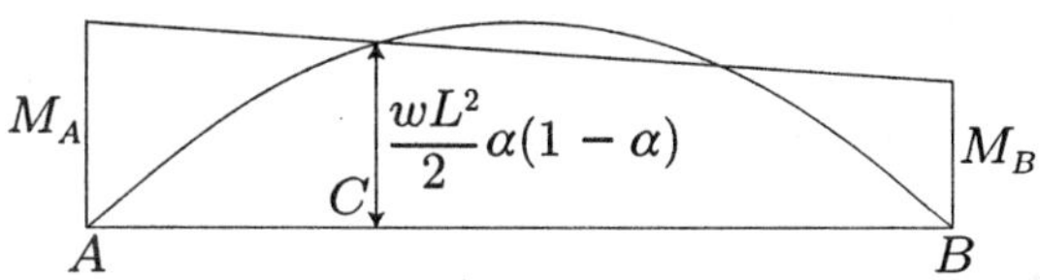

(b) B.M. diagram

Fig. 9.18: F.E.M. for Beam with hinges

Equating deflection at point C from CB and CA and considering linear relationship between M_A and M_B, i.e., from

$$M_B = M_A - \frac{1}{\alpha}\left[M_A - \frac{wL^2}{2}\alpha(1-\alpha)\right],$$

values of M_A and M_B can be obtained.

$$\left.\begin{aligned} \text{Thus}\quad M_A &= \frac{wL^2}{8}\frac{\alpha(6\alpha^2 - 8\alpha + 3)}{3\alpha^2 - 3\alpha + 1} \\ \text{and}\quad M_B &= \frac{wL^2}{8}\frac{(1-\alpha)(6\alpha^2 - 4\alpha + 1)}{3\alpha^2 - 3\alpha + 1}. \end{aligned}\right] \quad (9.9)$$

For the case, when hinge is at end A, i.e., $\alpha = 0$, $M_A = 0$ and $M_B = \dfrac{wL^2}{8}$.

For hinge at B, i.e., $\alpha = 1$,

$$M_A = \frac{wL^2}{8} \text{ and } M_B = 0.$$

For hinges at centre, i.e., $\alpha = \dfrac{1}{2}$, $M_A = M_B = \dfrac{wL^2}{8}$.

For concentrated load placed on such hinged beams, the fixed end moments can be similarly obtained. Thus referring to Fig. 9.19, for $\beta \leq \alpha$.

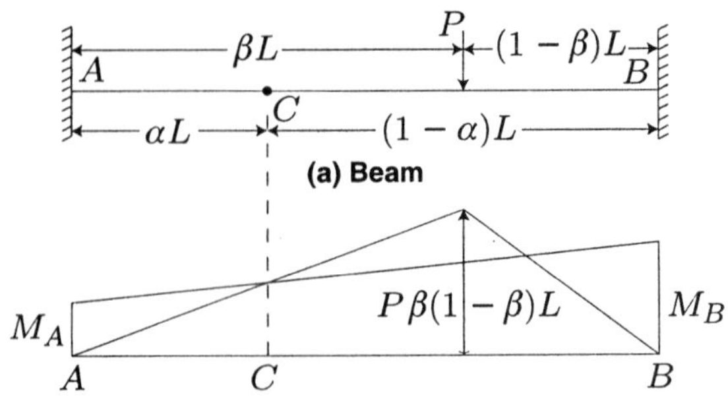

Fig. 9.19: F.E.M. for Beam with hinges

$$\left.\begin{aligned} M_A &= PL\left[\beta - \frac{\beta^2\alpha(3\alpha-\beta)}{2(3\alpha^2-3\alpha+1)}\right] \\ \text{and}\quad M_B &= PL\left[\frac{\beta^2(1-\alpha)(3\alpha-\beta)}{2(3\alpha^2-3\alpha+1)}\right]. \end{aligned}\right\} \qquad (9.10a)$$

For $\beta \geq \alpha$, β is replaced by $(1-\beta)$ and α by $(1-\alpha)$ in the above expressions and

$$\left.\begin{aligned} M_A &= PL\left[\frac{\alpha(1-\beta)^2(\beta+2-3\alpha)}{2(3\alpha^2-3\alpha+1)}\right] \\ \text{and}\quad M_B &= PL\left[(1-\beta) - \frac{(1-\beta)^2(1-\alpha)(\beta+2-3\alpha)}{2(3\alpha^2-3\alpha+1)}\right]. \end{aligned}\right\} \qquad (9.10b)$$

Fixed end moments caused due to joint displacement Δ is given by

$$\left.\begin{aligned} M_A &= \frac{3EI\Delta}{L^2}\left[\frac{\alpha}{3\alpha^2-3\alpha+1}\right] \\ \text{and}\quad M_B &= \frac{3EI\Delta}{L^2}\left[\frac{(1-\alpha)}{(3\alpha^2-3\alpha+1)}\right]. \end{aligned}\right\} \qquad (9.11)$$

9.10 Application of Moment Distribution Method to the Analysis of Statically Indeterminate Frame without Any Sidesway or Joint Movement

The application of the moment distribution method wherein no joint movement or sidesway is involved does not present any significant problem. The procedure is very similar to that of beams discussed in the previous sections, except that in case of frame there may frequently be more than two members meeting at a joint. The unbalanced moment at any such joint is distributed to the ends of the members meeting at that joint in the ratio of their relative stiffness.

There are two ways in which the work for moment distribution may be arranged. One of these is the tabular form in which all members meeting at the same joint are grouped together. The other is the figure form in which the moment distribution is done on the figure. The method of the tabular form is suggested as the more convenient form.

Example 9.6. *Analyse the rigid frames shown in Fig. 9.20, using moment distribution method.*

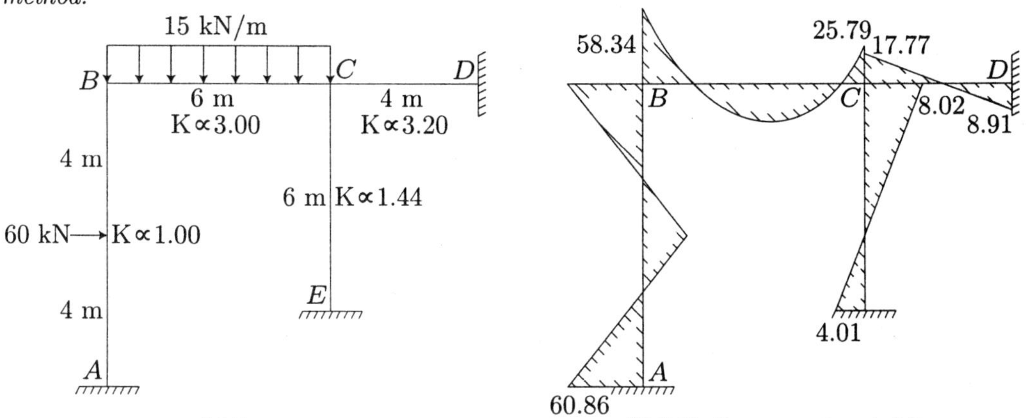

(a) Frame (b) B.M. diagram, values in kN-m

Fig. 9.20: For Example 9.6

Solution : *Relative stiffness :*

$$K_{AB} : K_{BA} : K_{BC} : K_{CB} : K_{CD} : K_{DC} : K_{CE} : K_{EC}$$
$$= 1 : 1 : 3 : 3 : 3.2 : 3.2 : 1.44 : 1.44$$

Distribution factors :

At point A, $DF_{AB} = 0.$

At point B, $DF_{BA} = \frac{1}{1+3} = 0.25$

$$DF_{BC} = \frac{3}{1+3} = 0.75.$$

At joint C, $DF_{CB} = \frac{3}{3+3.20+1.44} = \frac{3}{7.64} = 0.39$

$$DF_{CD} = \frac{3.20}{7.64} = 042.$$

$$DF_{CE} = \frac{1.44}{7.64} = 0.19.$$

At joint D, $DF_{DC} = 0.$

At joint E, $DF_{EC} = 0.$

Fixed end moments

$$M_{FAB} = +\frac{(60)(8)}{8} = +60 \text{ kNm}$$

$$M_{FBA} = -60 \text{ kNm}$$

$$M_{FBC} = +\frac{(15)(6)^2}{12} = +45 \text{ kNm}$$

$$M_{FCB} = -45 \text{ kNm}$$

$$M_{FCD} = M_{FDC} = M_{FCE} = M_{FEC} = 0.$$

The moment distribution procedure is shown in Table 9.8. Fig. 9.20(b) shows B.M. diagram drawn on the tension side.

Table 9.8. For Example 9.6

Joint	A	B		C			D	E
Member	AB	BA	BC	CB	CE	CD	DC	EC
Rel. Stiffness	1	1	3	3	1.44	3.2	3.2	1.44
Dist. Factor	0	0.25	0.75	0.39	0.19	0.42	0	0
C.O.F.	← 0.50		← 0.50 →		0.50 →	0.50 →		
F.E.M.	+60.0	−60.0	+45.0	−45.0	0.0	0.0	0.0	0.0
Bal.	0.0	+3.75	+11.25	+17.55	+8.55	+18.90	0.0	0.0
C.O.M.	+1.90	0.0	+8.80	+5.60	0.0	0.0	+9.5	+4.30
Bal.	0.0	−2.20	−6.60	−2.18	−1.07	−2.35	0.0	0.0
C.O.M.	−1.10	0.0	−1.09	−3.30	0.0	0.0	−1.20	−0.50
Bal.	0.0	+0.27	+0.82	+1.29	+0.62	+1.39	0.0	0.0
C.O.M.	+0.14	0.0	+0.65	+0.41	0.0	0.0	+0.70	+0.31
Bal.	0.0	−0.16	−0.49	−0.16	−0.08	−0.17	0.0	0.0
C.O.M.	−0.08							−0.04
Final end moments	+60.86	−58.34	+58.34	−25.79	+8.02	+17.77	+8.91	+4.07
Check $\frac{\text{Check}}{\Delta M}$	+0.86	+1.66	+13.34	+19.21	+8.02	+17.77	+8.91	+4.07
$(\Delta M)_{far}$	+1.66	+0.86	+19.21	+13.34	+4.07	+8.91	+17.77	+8.02
$-2(\Delta M)_{near}$	−1.72	−3.32	−26.68	−38.42	−16.04	−35.54	−17.82	−8.14
Sum	−0.06	−2.46	−7.47	−25.08	−11.97	−26.63	−0.05	−0.12
$\theta_{rel} = \frac{\text{Sum}}{K_{rel}}$	−0.06	−2.46	−2.49	−8.36	−8.31	−8.32	−0.02	−0.08
	≈ 0						≈ 0	≈ 0
	Check	← Check →		← Check →			Check	Check

Example 9.7. *Analyse the symmetrical rigid frame shown in Fig. 9.21, by the moment distribution method.*

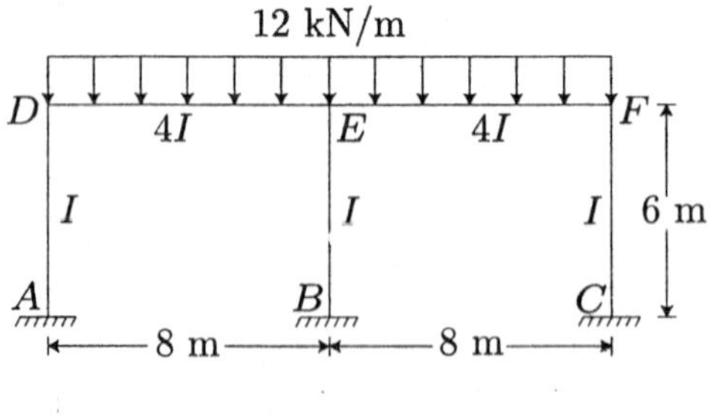

(a) Frame

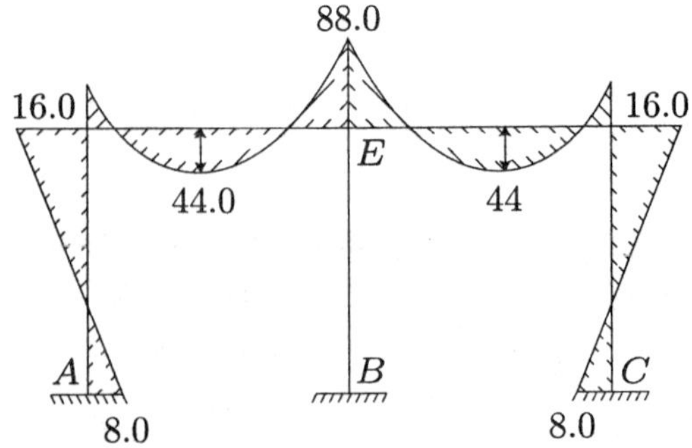

(b) B.M. in kNm

Fig. 9.21: For Example 9.7

Solution : *Relative stiffness*

$$K_{AD} : K_{DA} : K_{DE} : K_{ED} : K_{EB} : K_{BE} : K_{EF} : K_{FE} : K_{FC} : K_{CF}$$

$$= \frac{4EI}{6} : \frac{4EI}{6} : \frac{16EI}{8} : \frac{16EI}{8} : \frac{4EI}{6} : \frac{4EI}{6} : \frac{16EI}{8} : \frac{16EI}{8} : \frac{4EI}{6} : \frac{4EI}{6}$$

$$= 1 : 1 : 3 : 3 : 1 : 1 : 3 : 3 : 1 : 1.$$

Distribution factors : $\text{D.F.} = \frac{K}{\sum K}$.

Joint	A	D		E			B	F		C
Member	AD	DA	DE	ED	EB	EF	BE	FE	FC	CF
D.F.	$\frac{1}{\infty}$ $= 0$	$\frac{1}{1+3}$ $= 0.25$	$\frac{3}{1+3}$ $= 0.75$	$\frac{3}{3+3+1}$ $= 0.426$	$\frac{1}{3+3+1}$ $= 0.148$	$\frac{1}{3+3+1}$ $= 0.426$	$\frac{1}{\infty}$ $= 0$	$\frac{3}{3+1}$ $= 0.75$	$\frac{1}{3+1}$ $= 0.25$	$\frac{1}{\infty}$ $= 0$

Fixed end moments

$$M_{FDE} = -M_{FED} = +\frac{wL^2}{12} = +\frac{(12)(8)^2}{12} = +64.0 \text{ kNm}$$

$$M_{FEF} = -M_{FFE} = +\frac{wL^2}{12} = +64.0 \text{ kNm}.$$

Moment distribution is shown in Table 9.9.

Table 9.9. For Example 9.7

Joint	A	D		E			B	F		C
Member	AD	DA	DE	ED	EB	EF	BE	FE	FC	CF
Rel. K.	1	1	3	3	1	3	1	3	1	1
D.F.	0	0.25	0.75	0.426	0.148	0.426	0	0.75	0.25	0
F.E.M.	0	0	+64.00	−64.00	0	+64.00	0	−64.00	0	0
Bal	0	−16.00	48.00	0	0	0	0	+48.00	+16.00	0
C.O.M.	−8.00	0	0	−24.00	0	+24.00	0	0	0	+8.00
Bal.	0	0	0	0	0	0	0	0	0	0
Final end moments	−8.00	−16.00	+16.00	−88.00	0	+88.00	0	−16.00	+16.00	+8.00
Check :										
ΔM	−8.00	−16.00	−48.00	−240.0	0.0	+24.00	0.0	+48.00	+16.00	+8.00
$(\Delta M)_{far}$	−16.00	−8.00	−24.00	−48.00	0.0	+48.00	0.0	+24.00	+8.00	+16.00
$-2(\Delta M)_{near}$	+16.00	+32.00	+96.00	+48.00	0.0	−48.00	0.0	−96.00	−32.00	−16.00
Sum	00	+24.00	+72.00	0.0	0.0	0.0	0.0	−72.00	−24.00	0.0
$\theta_{rel} = \frac{\text{Sum}}{K_{rel}}$	00	+24.00	+24.00	0.0	0.0	0.0	0.0	−24.00	−24.00	0.0
	Check	⟵ Check ⟶		⟵ Check ⟶			Check	⟵ Check ⟶		Check

The bending moment diagram is drawn on tension side in Fig. 9.21(b)

Example 9.8. *Analyse the frame shown in Fig. 9.22, by the moment distribution method.*

Solution : The frame shows a rigid frame supporting a uniform load. There is a pin connection at the centre of member DE. The carry over factor for member DE is 1 and the stiffness factors is $\frac{3EI}{L}$, i.e., $\frac{3EI}{5}$. The fixed end moments in member DE caused by the uniform load are

$$M_{FDE} = -M_{FED} = \frac{wL^2}{8} = 37.5 \text{ kNm}.$$

Also for beam EC, $M_{FEC} = \dfrac{12(3.75)^2}{8} = 21.09$ kNm and $M_{FCE} = 0$.

The moment distribution factors for the remaining members of the structure are obtained in the usual way, after which the distribution of moments can be carried out as shown in Table 9.10; Fig. 9.22(b) shows the BM diagram.

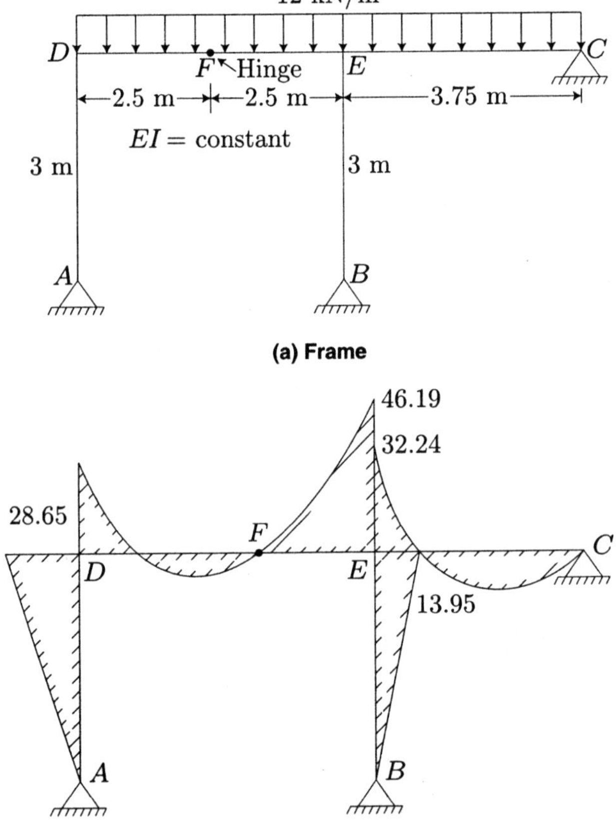

(a) Frame

(b) B.M. in kNm

Fig. 9.22: For Example 9.8

9.11 Application of Moment Distribution Method to the Analysis of Statically Indeterminate Frames with Sidesway or Joint Movement

This type includes such rigid frames, which undergo not only flexure of members when subjected to loads, but also an inplane lateral movement of certain joints with the result that the frame as a whole sways sidewise. Such side-sways may be due to

(i) Unsymmetric bent

(ii) Unsymmetric loading and

(iii) Lateral loading.

Table 9.10. For Example 9.8

Joint	*D*		*E*		
Member	*DA*	*DE*	*ED*	*EB*	*EC*
K (Stiffness)	1	$\frac{3}{5}$	$\frac{3}{5}$	1.0	$\frac{3}{3.75}$
D.F.	0.625	0.375	0.25	0.417	0.333
C.O.F.		←	1 →		
F.E.M.	0	+37.50	−37.50	0	+21.09
Bal.	−23.44	−14.06	+4.10	+6.84	+5.47
C.O.M.	0	+4.10	−14.06	0	0
Bal.	−2.56	−1.54	+3.52	+5.86	+4.68
C.O.M.	0	+3.52	−1.54	0	0
Bal.	−2.20	−1.32	+0.39	+0.64	+0.51
C.O.M.	0	+0.39	−1.32	0	0
Bal.	−0.24	−0.15	+0.33	+0.55	+0.44
C.O.M.	0	+0.33	−0.15	0	0
Bal.	−0.21	−0.12	+0.04	+0.06	+0.05
Final moment (Jt. sign. convention)	−28.65	+28.65	−46.19	+13.95	+32.24

The application of the moment distribution method to the analysis of statically indeterminate frames, in which sidesways or joint movements are involved consists of the following steps :

Step I. The joints are held against side-sway. The fixed end moments as caused by the applied loading are distributed and a first set of balanced end moments are obtained.

Step II. The unloaded frame is assumed to have a certain amount of side-sway, which will cause a set of fixed end moments. These fixed end moments are then distributed, and a second set of balanced end moments are obtained.

Step III. The final set of end moments can be obtained by adding the first set to the product of a ratio and the second set, the ratio being determined by use of shear condition.

Fig. 9.23(a) shows a rigid frame subjected to sidesway due to loading as well as unsymmetric characteristics of the bent. The given frame is equivalent to the sum of the frame shown in Fig. 9.23(b), and subsequently to those in Fig. 9.23(c).

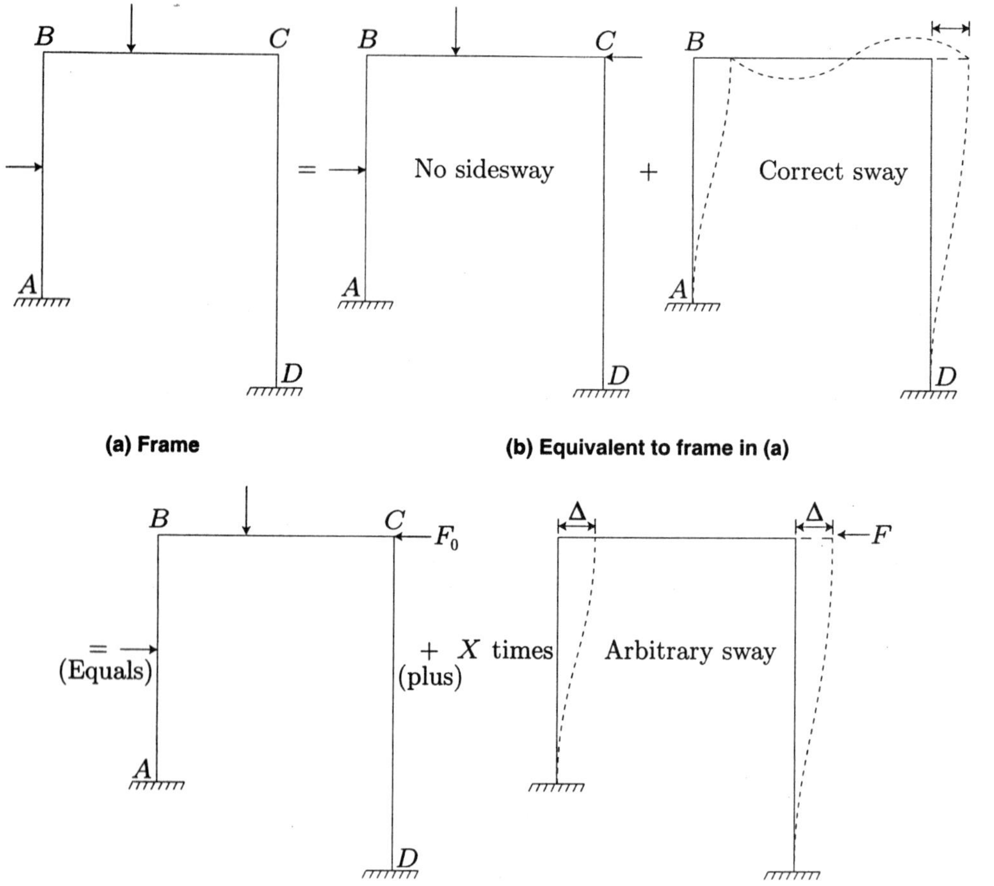

(c) Equivalent to (b)

Fig. 9.23: Frames with sidesway

The number of unknown factors depend on the number of degrees of freedom, which in case of frame in Fig. 9.23 is one. Two stages of moment distribution are necessary in such cases. For the first stage of moment distribution, i.e., for no sway, the magnitude of the lateral force F_o necessary to prevent sidesway is obtained from statics.

The second moment distribution is obtained on the basis of arbitrary sway as shown in Fig. 9.23(c). The magnitude of the unbalanced force F_1 is also obtained from statics. As no force exists at C in the actual structure, since there is no support at C,

$$F_o + X(F_1) = 0, \text{ whence } X = -\frac{F_o}{F_1}. \tag{9.12}$$

The resultant B.M. is obtained by addition of the moments. Thus,

Final $BM = BM$ for no sway $+X$ (B.M. for arbitrary sway).

For the sway moment distribution, the fixed end moments are obtained as illustrated in Fig. 9.24.

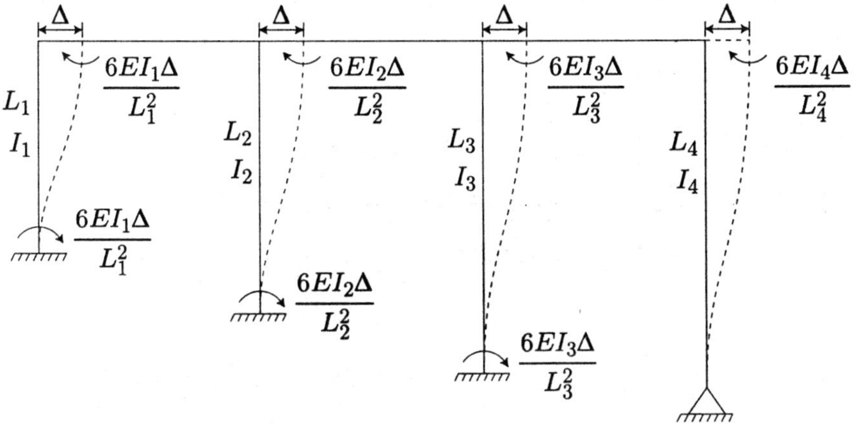

Fig. 9.24: Fixed end moments for sway moment distribution

Example 9.9. *Analyse the rigid frame shown in Fig. 9.25(a) by moment distribution method. Obtain the B.M. diagram.*

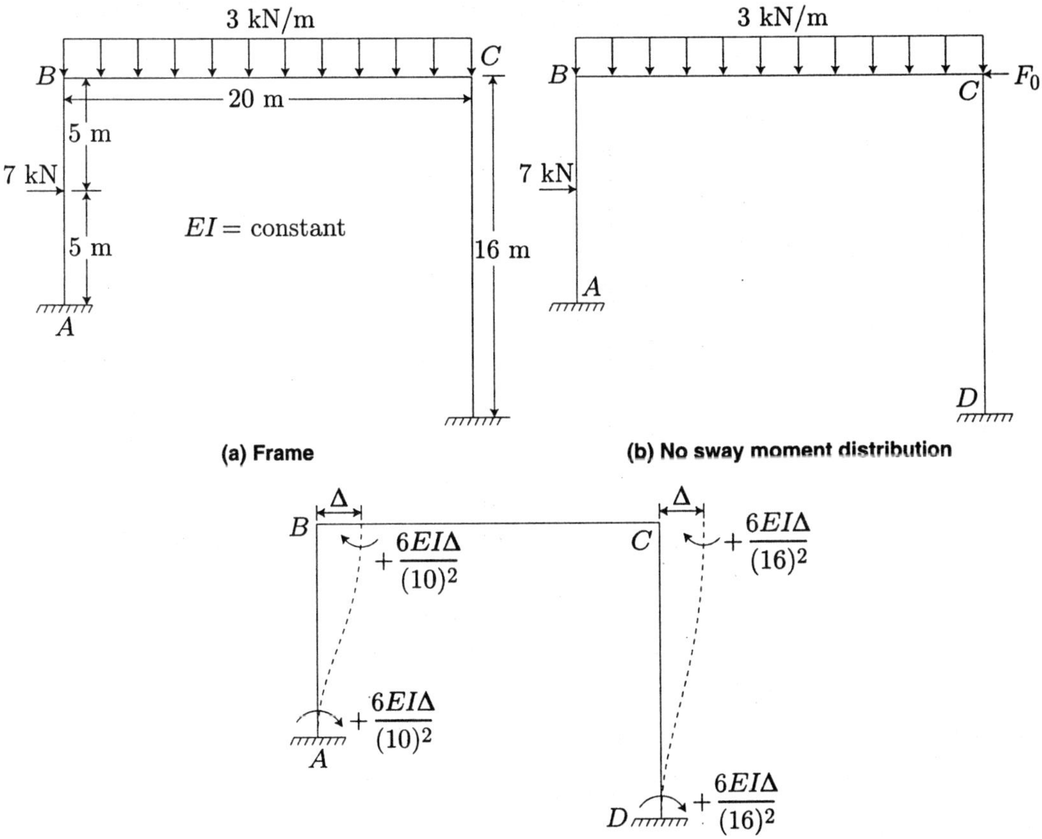

Fig. 9.25: For Example 9.9

Solution : *Stiffness* :

$$K_{BA} = \frac{4EI}{10}$$

$$K_{BC} = K_{CB} = \frac{4EI}{20}$$

$$K_{CD} = \frac{4EI}{16}.$$

Distribution factors $\left.\begin{aligned} DF_{BA} &= 0.667 \\ DF_{BC} &= 0.333 \end{aligned}\right] \sum = 1.00$

$$\left.\begin{aligned} DF_{CB} &= 0.444 \\ DF_{CD} &= 0.556 \end{aligned}\right] \sum = 1.00$$

Moment distribution for no sway, assuming that a restraint is applied at C as in Fig. 9.25(b) is shown in Table 9.11.

Table 9.11. Non-Sway Moment Distribution for Example 9.9

Joint	A	B		C		D
Member	AB	BA	BC	CB	CD	DC
Rel K	$\frac{1}{10}$	$\frac{1}{10}$	$\frac{1}{20}$	$\frac{1}{20}$	$\frac{1}{16}$	$\frac{1}{16}$
D.F.	0	0.667	0.333	0.444	0.556	0
C.O.F.	← 0.50		←	0.50 →	0.50	→
F.E.M.	+8.75	−8.75	+100	−100	0	0
Bal.	0	−60.80	−30.45	+44.40	+55.60	0
C.O.M.	−30.40	0	+22.20	−15.22	0	+27.80
Bal.	0	−14.80	−7.40	+6.76	+8.46	0
C.O.M.	−7.40	0	+3.38	−3.70	0	+4.23
Bal.	0	−2.24	−1.14	+1.64	+2.06	0
C.O.M.	−1.12	0	+0.82	−0.57	0	+1.03
Bal.	0	−0.54	−0.28	+0.25	+0.32	0
C.O.M.	−0.27	0	+0.12	−0.14	0	+0.10
Bal.	0	−0.08	−0.04	+0.06	+0.08	0
C.O.M.	−0.04	–	–	–	–	+0.04
Resulting B.M. in Joint Sign. Convention.	−30.48	−87.21	+87.21	−66.52	+66.52	+33.20
B.M. Sign. Convention.	+30.48	−87.21	−87.21	−66.52	−66.52	+33.20

Fixed end moments

$$M_{FAB} = +\frac{7(10)}{8} = +8.75 \text{ kNm} = -M_{FBA}$$

$$M_{FBC} = +\frac{3(20)^2}{12} = +100 \text{ kNm} = -M_{FCB}.$$

For moment distribution on account of side-sway, the loads and the restraint at C are removed. The joints are relocked and the frame is allowed to sway to the right through an arbitrary amount Δ as shown in Fig. 9.25(a).

Fixed-end moments then are

$$M_{FAB} = M_{FBA} = +\frac{6EI\Delta}{(10)^2} = +1000 \text{ kNm (say)}.$$

Then $$M_{FCD} = M_{FDC} + \frac{6EI\Delta}{(16)^2} = +391 \text{ kNm}.$$

The moment distribution due to sidesway is shown in Table 9.12.

Table 9.12. Sway Moment Distribution for Example 9.9

Joint	A	B		C		D
Member	AB	BA	BC	CB	CD	DC
D.F.	0	0.667	0.333	0.444	0.556	0
C.O.F.	←	0.50	← 0.50	→	0.50	→
F.E.M.	+1000	+1000	0	0	+391	+391
Bal.	0	−667	−333	−174	−217	0
C.O.M.	−334	0	−87	−167	0	−109
Bal.	0	+58	+29	+74	+93	0
C.O.M.	+29	0	+37	+15	0	+47
Bal.	0	−25	−12	−7	−8	0
C.O.M.	−13	0	−3.50	−6	0	−4
Bal.	0	+2.30	+1.20	+2.70	+3.30	0
C.O.M.	+1.15	0	+1.35	+0.60	0	+1.65
Bal.	0	−0.90	−0.45	−0.27	0.33	0
C.O.M.	−0.45					−0.17
Final moments Jt. Sign. Convention.	+682.70	+367.40	−367.40	−261.97	+261.97	+327.98
Final moment B.M. Sign Convention.	−682.70	+367.40	+367.40	−261.97	−291.97	+327.48

The magnitude of the restraint force F_o and F_1 in Figs. 9.25(b) and 9.25(c) can be obtained from considerations of statics of the frame shown in Figs. 9.26(a) and 9.26(b) respectively.

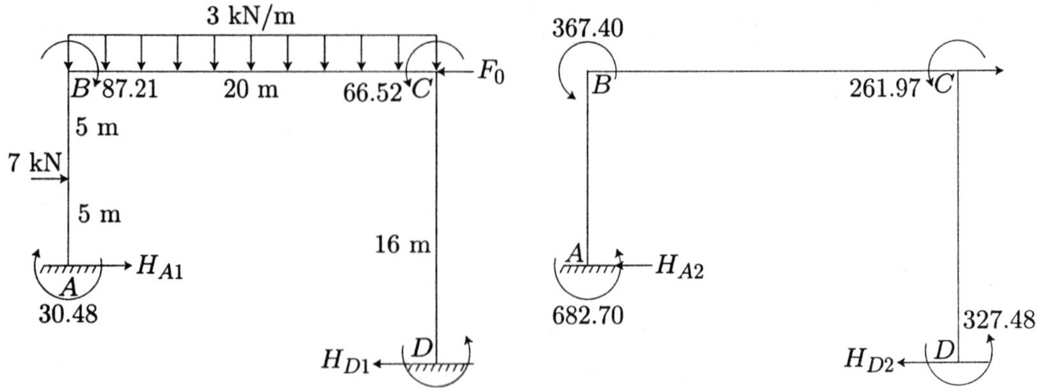

(a) End moments for non-sway moment distribution **(b) End moments for sway moment distribution**

Fig. 9.26: For Example 9.9

From Fig. 9.26(a) considering statics of column AB,

$$H_{A_1}(10) - 30.48 + 7(5) = 87.21, \quad \text{whence} \quad H_{A_1} = 8.27 \text{ kN} \longrightarrow$$

From statics of column CD,

$$H_{D_1}(16) - 33.26 = 66.52, \quad \text{whence} \quad H_{D_1} = 6.24 \text{ kN} \longleftarrow$$

$$\text{Unbalanced Shear} = 8.27 + 7 - 6.24 = 9.03 \text{ kN} \longrightarrow$$

$$= -F_0,$$

whence $F_0 = 9.03 \text{ kN} \longleftarrow$.

From Fig. 9.26(b), similarly,

$$H_{A_2} = 105.01 \text{ kN}, \longleftarrow$$

and $H_{D_2} = 36.84 \text{ kN} \longleftarrow$

Hence $F_1 = 105.01 + 36.84 = 141.85 \text{ kN} \longrightarrow$.

Then $9.03 + X(-141.85) = 0, \quad \text{whence} \quad X = 0.06366.$

Resulting B.M. diagram and deflected shape are shown in Figs. 9.27(a) and 9.27(b) respectively.

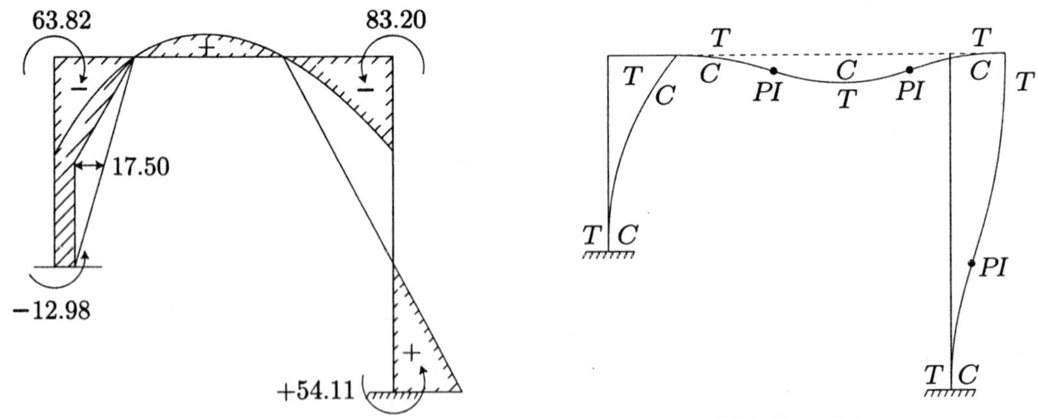

(a) B.M. in kNm **(b) Deflected shape**

Fig. 9.27: For Example 9.9

Example 9.10. *Analyse the rigid frame shown in Fig. 9.28(a) using moment distribution method.*

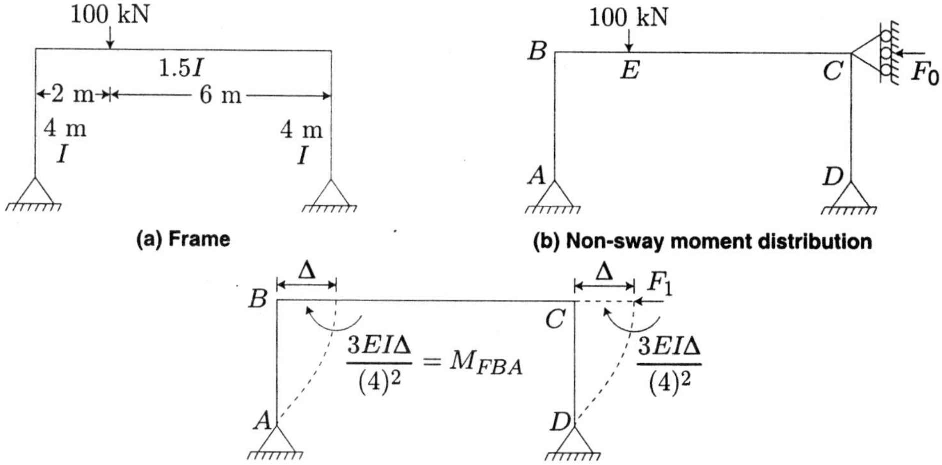

(a) Frame **(b) Non-sway moment distribution**

(c) Sway moment distribution

Fig. 9.28: For Example 9.10

Solution : Moment distribution will be accomplished in two stages :

(i) Non-sway moment distribution for applied loads with a lateral support at C as shown in Fig. 9.28(b) and (ii) Sway moment distribution for an arbitrary sway as shown in Fig. 9.28(c).

Stiffness

$$K_{AB} = \frac{4EI}{4} = EI; \quad K_{BA} = \frac{3EI}{4}; \quad K_{BC} = \frac{4E(1.5I)}{8} = \frac{3}{4}EI; \quad K_{CB} = \frac{3}{4}EI$$

$$K_{CD} = \frac{3EI}{4}; \quad K_{DC} = \frac{4EI}{4} = EI.$$

$$K_{AB} : K_{BA} : K_{BC} : K_{CB} : K_{CD} : K_{DC} = 1.0 : 0.75 : 0.75 : 0.75 : 0.75 : 1.$$

Distribution factors :

Joint A : $DF_{AB} = 1.0$

Joint B: $DF_{BA} = \dfrac{\frac{3}{4}}{\frac{3}{4} + \frac{3}{4}} = 0.5$

$DF_{BC} = 0.5$

Joint C: $DF_{CB} = \dfrac{\frac{3}{4}}{\frac{3}{4} + \frac{3}{4}} = 0.5$

$DF_{CD} = 0.5$

Joint D : $DF_{DC} = 1.0.$

Fixed-end moments for non-sway moment distribution

Assuming a lateral support at C, so as to prevent sway as shown in Fig. 9.28(b)

$$M_{FBC} = +\frac{(100)(6)^2(2)}{(8)^2} = +112.5 \text{ kNm}$$

$$M_{FCB} = -\frac{(100)(2)^2(6)}{(8)^2} = -37.5 \text{ kNm}.$$

The distribution of moments for non-sway condition is shown in Table 9.13.

Table 9.13. Non-sway Moment Distribution for Example 9.10

Joint	A	B		C		D
Member	AB	BA	BC	CB	CD	DC
Stiffness Distribution Factor C.O.F.	EI 1.0	$\frac{3}{4}EI$ 0.5	$\frac{3}{4}EI$ 0.5 ← 0.50	$\frac{3}{4}EI$ 0.5 →	$\frac{3}{4}EI$ 0.5	EI 1.0
F.E.M.	0	0	+112.50	−37.50	0	0
Bal.	0	−56.25	−56.25	+18.75	+18.75	0
C.O.	0	0	+9.38	−28.13	0	0
Bal.	0	−4.69	−4.69	+14.07	+14.06	0
C.O.	0	0	+7.04	−2.34	0	0
Bal.	0	−3.52	−3.52	+1.17	+1.17	0
C.O.	0	0	+0.59	−1.76	0	0
Bal.	0	−0.29	−0.29	+0.88	+0.88	0
C.O.	0	0	+0.44	−0.14	0	0
Bal.	0	−0.22	−0.22	+0.07	+0.07	
Final End Moments (Joint sign convention)	0	−64.97	+64.97	−34.93	+34.93	0
Final End Moments (Tension inside positive)	0	−64.97	−64.97	−34.93	−34.93	0

Fixed-end moments for sway moment distribution

With joints B and C clamped, an arbitrary sway is produced by a force F_1 at C as shown in Fig. 9.28(c). For this case,

$$M_{FBA} = +\frac{3EI\Delta}{(4)^2} = +\frac{3EI\Delta}{16}$$

$$M_{FCD} = +\frac{3EI\Delta}{16}.$$

Putting $EI\Delta = \dfrac{1600}{3}$ kNm^3,

$$M_{FBA} = M_{FCD} = +100 \text{ kNm}.$$

Distribution of this fixed end moments for an arbitrary sway is shown in Table 9.14.

Table 9.14. Distribution of Sway Moments for Example 9.10

Joint	A	B		C		D
Member	AB	BA	BC	CB	CD	DC
Stiffness	EI	$\frac{3}{4}EI$	$\frac{3}{4}EI$	$\frac{3}{4}EI$	$\frac{3}{4}EI$	EI
D.F.	1.0	0.5	0.5	0.5	0.5	1.0
C.O.F.			← 0.50	→		
F.E.M.	0	+100.00	0	0	+100.00	0
Bal.	0	−50.00	−50.00	−50.00	−50.00	0
C.O.	0	0	−25.00	−25.00	0	0
Bal.	0	+12.50	+12.50	+12.50	+12.50	0
C.O.	0	0	+6.25	+6.25	0	0
Bal.	0	−3.12	−3.13	−3.13	−3.12	0
C.O.	0	0	−1.56	−1.56	0	0
Bal.	0	+0.78	+0.78	+0.78	+0.78	0
C.O.	0	0	+0.38	+0.38	0	0
Bal.	0	−0.19	−0.19	−0.19	−0.19	0
Final End Moments (Joint Sign Convention)	0	+59.97	59.97	−59.97	+59.97	0
Final End Moments (Tension inside positive)		+59.97	+59.97	−59.97	−59.97	

From Fig. 9.29(a) using equilibrium condition $\sum H = 0,\ F_0 = H_{A_1} + H_{D_1}$

and considering equilibrium of columns AB and CD,

$$H_{A_1} = +\frac{64.97}{4} \text{ kN} = +16.24 \text{ kN} \quad \text{and} \quad H_{D_1} = -\frac{34.93}{4} \text{ kN} = -8.73 \text{ kN}.$$

Hence $F_0 = 16.24 - 8.73 = 7.51$ kN.

Similarly, from Fig. 9.29(b) using equilibrium condition $\sum H = 0,\ F_1 = H_{A_2} + H_{D_2}$,

and considering equilibrium of columns AB and CD,

$$H_{A_2} = -\frac{59.97}{4} = -15.0 \text{ kN} \quad \text{and} \quad H_{D_2} = -\frac{59.97}{4} = -15.0 \text{ kN}.$$

Hence $F_1 = -15 - 15 = -30$ kN.

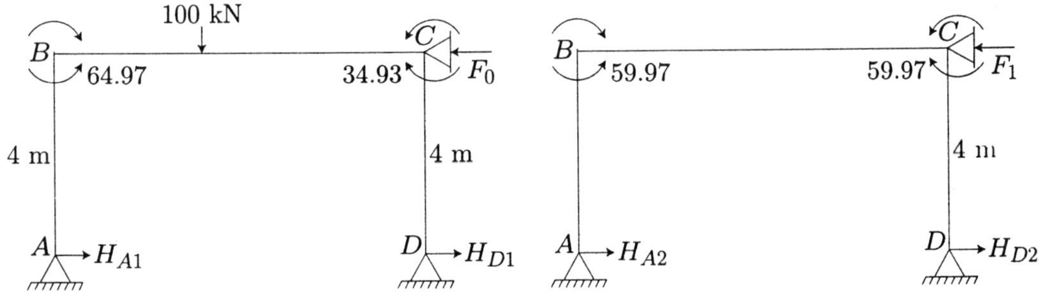

(a) End moments for non-sway moment distribution **(b) End moments for sway moment distribution**

Fig. 9.29: For Example 9.10

Assuming that the actual frame shown in Fig. 9.28(a) is equivalent in behaviour to sum of the case shown in Fig. 9.28(b) and X times the case shown in Fig. 9.28(c) and also considering that there is no external force acting at C in the actual frame,

$$F_0 + XF_1 = 0$$

i.e., $$7.51 + X(-30) = 0,$$

whence $$X = \frac{7.51}{30} = 0.25.$$

The sway moment at a member end will then be X times that obtained in Table 9.14 and the resulting end moments are as given in Table 9.15.

Table 9.15. Final End Moments for Example 9.10

Member	AB	BA	BC	CB	CD	DC
Non-sway B.M.	0	−64.97	−64.97	−34.93	−34.93	0
Sway B.M.	0	+14.99	+14.99	−14.99	−14.99	0
Final End Moments (Tension inside positive)	0	−49.98	−49.98	−49.92	−49.92	0

The B.M. diagram is drawn on tension side in Fig. 9.30(a) and the deflected shape is qualitatively drawn in Fig. 9.30(b)

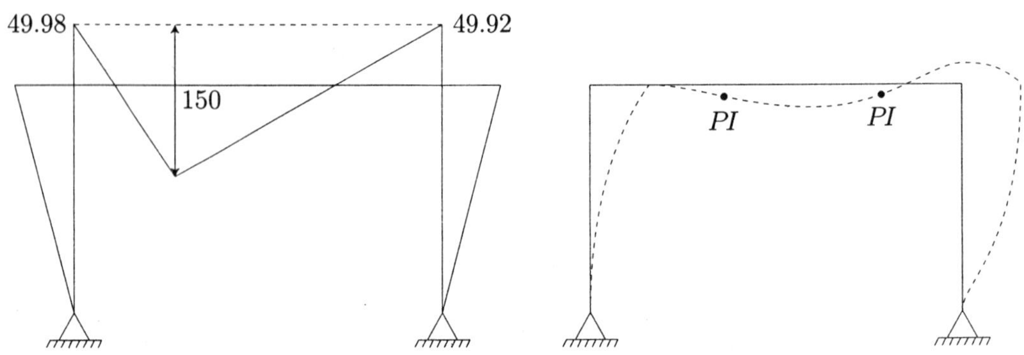

(a) B.M. in kNm **(b) Deflected shape**

Fig. 9.30: For Example 9.10

Example 9.11. *Analyse the frame shown in Fig. 9.31(a) using moment distribution procedure.*

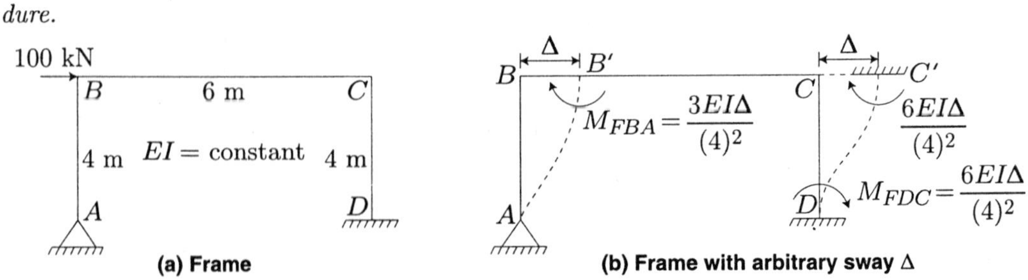

(a) Frame **(b) Frame with arbitrary sway Δ**

Fig. 9.31: For Example 9.11

Solution : In this case, since the fixed-end moments due to applied loads are nonexistent, there is no non-sway moment distribution. Sway moment distribution only will be performed for arbitrary sway Δ as shown in Fig. 9.31(b), produced by a force F_1 applied at C.

Stiffness :

$$K_{AB} = \frac{4EI}{4} = EI; \quad K_{BA} = \frac{3EI}{4} = 0.75EI$$

$$K_{BC} = \frac{4EI}{6} = \frac{2}{3}EI; \quad K_{CB} = \frac{4EI}{6} = \frac{2}{3}EI;$$

$$K_{CD} = \frac{4EI}{4} = EI \quad \text{and} \quad K_{DC} = \frac{4EI}{4} = EI$$

Distribution factors;

Joint A, $DF_{AB} = \dfrac{1}{1} = 1.0$

Joint B, $DF_{BA} = \dfrac{0.75}{2/3 + 0.75} = 0.53$

$$DF_{BC} = \frac{2/3}{2/3 + 0.75} = 0.47$$

Joint C, $DF_{CB} = \dfrac{2/3}{2/3 + 1} = 0.4$

$$DF_{CD} = \frac{1}{2/3 + 1} = 0.6$$

Joint D, $DF_{DC} = \dfrac{1}{\infty} = 0.$

Fixed-end moments for sway moment distribution

$$M_{FBA} = +\frac{3EI\Delta}{(4)^2} = +\frac{3}{16}EI\Delta$$

$$M_{FCD} = +\frac{6EI\Delta}{(4)^2} = +\frac{3}{8}EI\Delta$$

$$M_{FDC} = +\frac{6EI\Delta}{(4)^2} = +\frac{3}{8}EI\Delta.$$

Taking $EI\Delta = \dfrac{1600}{3} \cdot \text{kNm}^3$,

$$M_{FBA} = +100 \text{ kNm}$$

$$M_{FCD} = +200 \text{ kNm}$$

$$M_{FDC} = +200 \text{ kNm}.$$

The distribution of sway moments is shown in Table 9.16.

Table 9.16. Sway Moment Distribution for Example 9.11

Joint	A	B		C		D
Member	AB	BA	BC	CB	CD	DC
Stiffness	EI	$0.75EI$	$\frac{2}{3}EI$	$\frac{2}{3}EI$	EI	EI
D.F.	1.0	0.53	0.47	0.4	0.6	0
C.O.F.			0.50 ⟶		⟵	0.50 ⟶
F.E.M.		+100.00			+200.00	+200.00
Bal.	0	−53.00	−47.00	−80.00	−120.00	0
C.O.	0	0	−40.00	−23.50	0	−60.00
Bal.	0	+21.20	+18.80	+9.40	+14.10	0
C.O.	0	0	+4.70	+9.40	0	−7.05
Bal.	0	−2.49	−2.21	−3.76	−5.64	0
C.O.	0	0	−1.88	−1.10	0	−2.82
Bal.	0	+1.00	+0.88	+0.44	+0.66	0
C.O.						−0.33
Final end moment (Joint moment sign)	0	+66.71	−66.71	−89.12	+89.12	+129.80
Final end moment taking tension inside positive	0	+66.71	+66.71	−89.12	−89.12	+129.80

Determination of force at joint C :

From Fig. 9.32, the horizontal force F_1 acting at C causing the arbitrary sway is obtained from

$$F_1 = H_A + H_D.$$

Forces H_A and H_D acting at supports A and D respectively are obtained from consideration of equilibrium of columns AB and CD respectively.

Thus, $$H_A = +\frac{66.71}{4} = +16.68 \text{ kN} \longleftarrow.$$

and $$H_D = +\frac{89.12 + 129.8}{4} = \underleftarrow{+54.73 \text{ kN}},$$

whence $F_1 = 16.68 + 54.73 = 71.41$ kN $\longrightarrow$.

Determination of ratio :

Considering that the actual case is equivalent to X times the arbitrary case shown in Fig. 9.32,

$(X)(F_1) = 100$, which is the magnitude of the applied load,

and $$X = \frac{100}{F_1} = \frac{100}{71.41} = 1.40.$$

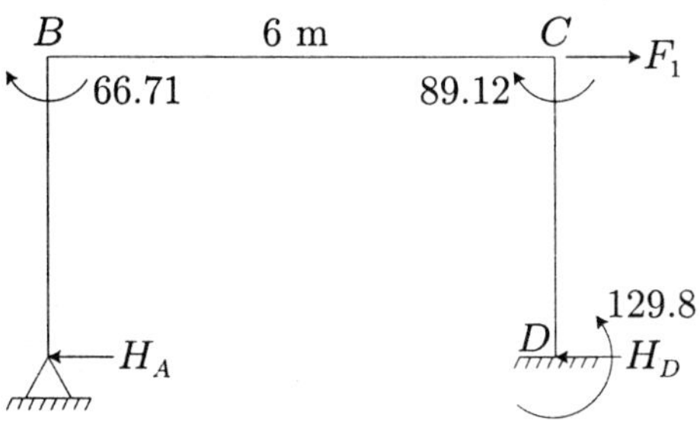

(a) Frame with end moments

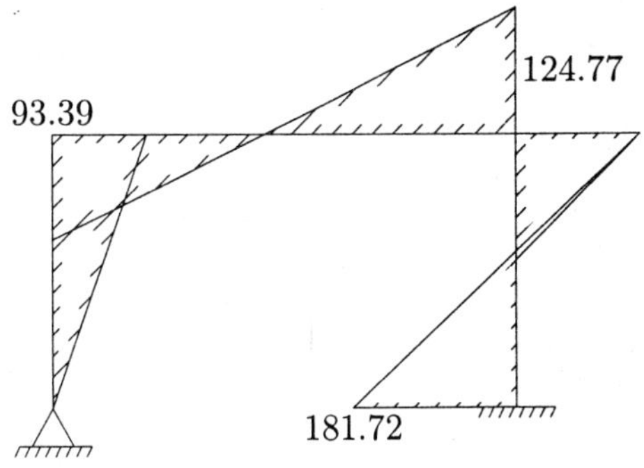

(b) B.M. in kNm

Fig. 9.32: For Example 9.11

Hence the actual end moments are 1.40 times that obtained for the arbitrary sway condition shown in Fig. 9.32. Table 9.17 shows the final moments.

Table 9.17. Final Moments for Example 9.11

Member	AB	BA	BC	CB	CD	DC
Final end moments taking tension inside positive, kNm	0	+93.39	+9.339	−124.77	−124.77	+181.72

Fig. 9.32(b) shows B.M. diagram drawn on tension side and Fig. 9.33 shows the deflected shape of the frame.

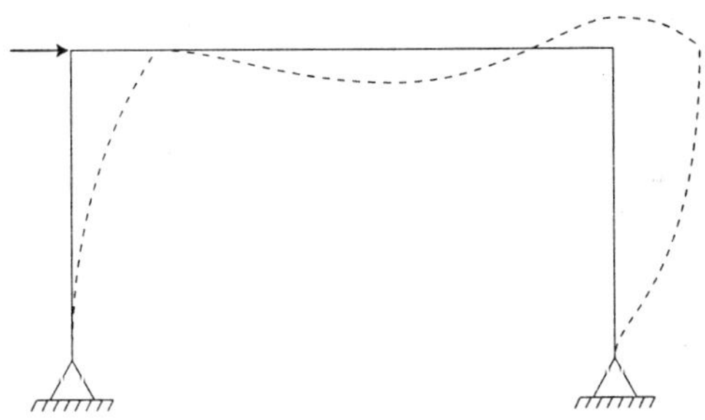

Fig. 9.33: Deflected shape for Example 9.11

Example 9.12. *Analyse the two-bay frame in Fig. 9.34(a) using moment distribution procedure.*

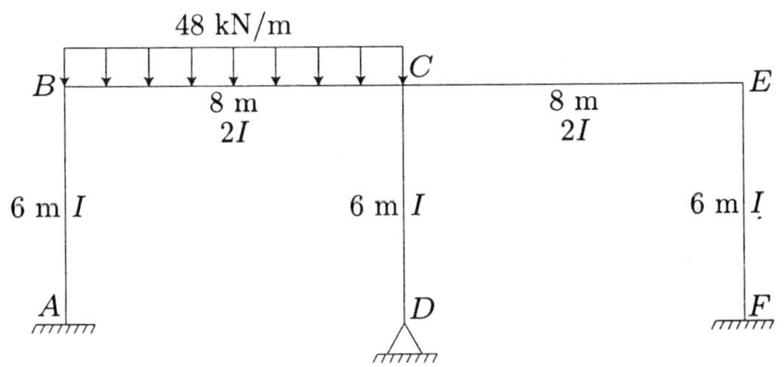

(a) Frame with loading

Fig. 9.34: For Example 9.12 (Contd.)

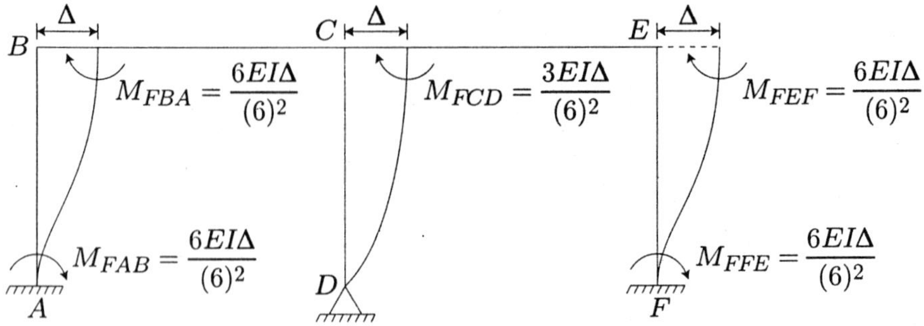

(b) Frame with arbitrary sway

Fig. 9.34: For Example 9.12

Solution : *Stiffness and distribution factors*

Member	Actual stiffness	Relative stiffness	Distribution Factor
AB	$\frac{4EI}{6} = \frac{2}{3}EI$	4	$\frac{4}{4+\infty} = 0$
BA	$\frac{4EI}{6} = \frac{2}{3}EI$	4	$\frac{4}{6+4} = 0.4$
BC	$\frac{4E(2I)}{8} = EI$	6	$\frac{6}{6+4} = 0.6$
CB	$\frac{4E(2I)}{8} = EI$	6	$\frac{6}{6+3+6} = 0.4$
CD	$\frac{3EI}{6} = \frac{1}{2}EI$	3	$\frac{3}{6+3+6} = 0.2$
CE	$\frac{4E(2I)}{8} = EI$	6	$\frac{6}{6+3+6} = 0.4$
DC	$\frac{4EI}{6} = \frac{2}{3}EI$	4	$\frac{4}{4} = 1$
EC	$\frac{4E(2I)}{8} = EI$	6	$\frac{6}{6+4} = 0.6$
EF	$\frac{4EI}{6} = \frac{2}{3}EI$	4	$\frac{4}{6+4} = 0.4$
FE	$\frac{4EI}{6} = \frac{2}{3}EI$	4	$\frac{4}{4+\infty} = 0$

Fixed-end moments due to applied loads

$$M_{FBC} = +\frac{(48)(8)^2}{12} = +256 \text{ kNm} = -M_{FCB}.$$

Fixed-end moments due to sway :

For an arbitrary sway, Δ as shown in Fig. 9.34(b)

$$M_{FBA} = M_{FAB} = +\frac{6EI\Delta}{(6)^2} = +\frac{EI\Delta}{6}$$

$$M_{FCD} = +\frac{3EI\Delta}{(6)^2} = +\frac{EI\Delta}{12}$$

$$M_{FEF} = M_{FFE} = +\frac{6EI\Delta}{(6)^2} = +\frac{EI\Delta}{6}.$$

Putting $EI\Delta = 1200 \text{ kNm}^3$,

$$M_{FBA} = M_{FAB} = M_{FEF} = M_{FFE} = +200 \text{ kNm}$$

$$M_{FCD} = +100 \text{ kNm}.$$

The distributions of non-sway and sway moments are shown in Tables 9.18 and 9.19 respectively.

Table 9.18. Distribution of Non-sway Moments for Example 9.12

Joint	*A*	*B*		*C*			*D*	*E*		*F*
Member	*AB*	*BA*	*BC*	*CB*	*CD*	*CE*	*DC*	*EC*	*EF*	*FE*
Rel. Stiffness	4	4	6	6	3	6	4	6	4	4
D.F.	0	0.4	0.6	0.4	0.2	0.4	1	0.6	0.4	0
C.O.F.	← 0.50 →		← 0.50 →			←	0.50	→	0.50 →	
F.E.M.			+256.0	−256.0						
Bal.		−102.4	−153.6	+102.4	+51.2	+102.4				
C.O.	−51.2	0.0	+51.2	−76.8	0.0	0.0		+51.2	0.0	
Bal.	0	−20.50	−30.7	+30.7	+15.4	+30.7		−30.7	−20.5	
C.O.	−10.3		+15.4	−15.4		−15.4		+15.4		−10.3
Bal.	0.0	−6.2	−9.2	+12.3	+6.2	+12.3		−9.2	−6.2	
C.O.	−3.1		+6.2	−4.6		−4.6		+6.2		−3.1
Bal.	0.0	−2.5	−3.7	+3.7	+1.8	+3.7		−3.7	−2.5	
C.O.	−1.3		+1.9	−1.9		−1.9		+1.9		−1.3
Bal.		−0.8	−1.1	+1.5	+0.8	+1.5		−1.1	−0.8	
C.O.	−0.4									−0.4
End Moment with joint moment sign	−66.3	−132.4	+132.4	−204.1	+75.4	+128.7	0	+30.0	−30.0	−15.1

Table 9.19. Distribution of Sway Moments for Example 9.12

Joint	*A*	*B*		*C*			*D*	*E*		*F*
Member	*AB*	*BA*	*BC*	*CB*	*CD*	*CE*	*DC*	*EC*	*EF*	*FE*
D.F.	0	0.4	0.6	0.4	0.2	0.4	1.0	0.6	0.4	0
C.O.F.	← 0.50 →		← 0.50 →			←	0.50	→	0.50 →	
F.E.M.	+200.0	+200.0			+100.0				+200.00	+200.0
Bal.	0	−80.0	−120.0	−40.0	−20.0	−40.0		−120.0	−80.0	0
C.O.	−40.0		−20.0	−60.0		−60.0		−20.0		−40.0
Bal.	0	+8.0	+12.0	+48.0	+24.0	+48.0		+12.0	+8.0	
C.O.	+4.0		+24.0	+6.0		+6.0		+24.0		+4.0
Bal.	0	−9.6	−14.4	−4.8	−2.4	−4.8		−14.4	−9.6	
C.O.	−4.8		−2.4	−7.2		−7.2		−2.4		−4.8
Bal.	0	+1.0	+1.4	+5.8	+2.8	+5.8		+1.4	+1.0	
C.O.	+0.5		+2.9	+0.7		+0.7		+2.9		+0.5
Bal.	0	−1.2	−1.7	−0.6	−0.2	−0.6		−1.7	−1.2	
C.O.	−0.6									−0.6
End Moment with joint moment sign	+159.1	+118.2	−118.2	−52.1	+104.2	−52.1	0	−118.2	+118.2	+159.1

Determination of forces at joint E :

The horizontal force F_0, required at E for non-sway condition as shown in Fig. 9.35(a) is obtained from $F_0 = H_A + H_D + H_F$.

H_A, H_D and H_F are obtained from consideration of equilibrium of columns AB, CD and EF respectively as

$$H_A = \frac{132.4 + 66.3}{6} = 33.12 \text{ kN} \longrightarrow$$

$$H_D = -\frac{75.4}{6} = -12.6 \text{ kN} \longleftarrow$$

$$H_F = \frac{30.0 + 15.1}{6} = 7.5 \text{ kN}, \longrightarrow$$

whence $F_0 = 33.1 - 12.6 + 7.5 = 28.0 \text{ kN} \longleftarrow$.

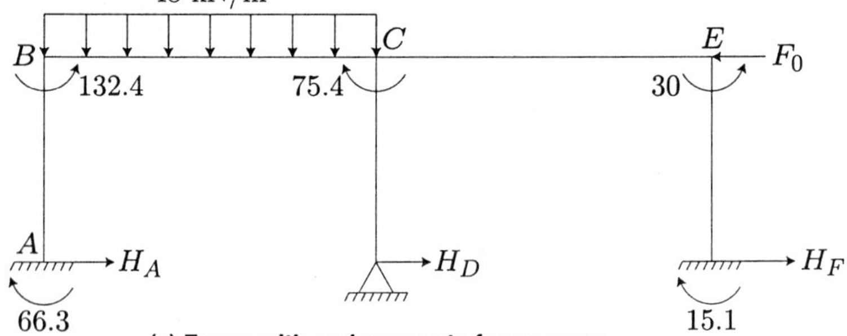

(a) Frame with end moments for no sway

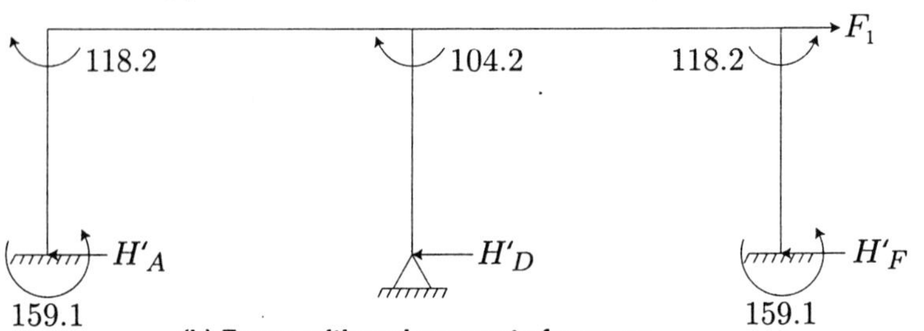

(b) Frame with end moments for sway

Fig. 9.35: For Example 9.12

The horizontal force F_1 required to be applied at E for the arbitrary sway as shown in Fig. 9.35(b) is obtained from

$$F_1 = H'_A + H'_D + H'_F.$$

H'_A, H'_D and H'_F are obtained from equilibrium of columns AB, CD and EF respectively as below :

$$H'_A = \frac{118.2 + 159.1}{6} = 46.2 \text{ kN} \longleftarrow$$

$$H'_D = \frac{104.2}{6} = 17.4 \text{ kN} \longleftarrow$$

$$H'_F = \frac{118.2 + 159.1}{6} = 46.2 \text{ kN} \longleftarrow.$$

Hence $F_1 = 46.2 + 17.4 + 46.2 = 109.8 \longrightarrow$.

Determination of ratio :

Considering that the actual sway is X times the arbitrary sway as assumed in Fig. 9.35(b) and that the net force at E in the actual frame is zero,

$$XF_1 = F_0$$

i.e.,
$$X = \frac{F_0}{F_1} = \frac{28}{109.8} = 0.255.$$

The actual end moments are obtained by adding the values of Table 9.18 to X times those of Table 9.19 and given in Table 9.20. Figs. 9.36(a) and 9.36(b) show respectively B.M. diagram and deflected shape.

Table 9.20. Actual End Moments for Example 9.12

Member	AB	BA	BC	CB	CD	CE	DC	EC	EF	FE
Non-sway B.M., kN-m	−66.3	−132.4	+132.4	−204.1	+75.4	+128.7	0	+30.0	−30.0	−15.1
Sway B.M., kN-m	+40.6	+30.1	−30.1	−13.3	+26.6	−13.3	0	−30.1	+30.1	+40.6
Final B.M., kN-m with joint moment sign	−25.7	−102.3	+102.3	−217.4	+102.0	+115.4	0	−0.1	+0.1	+25.5

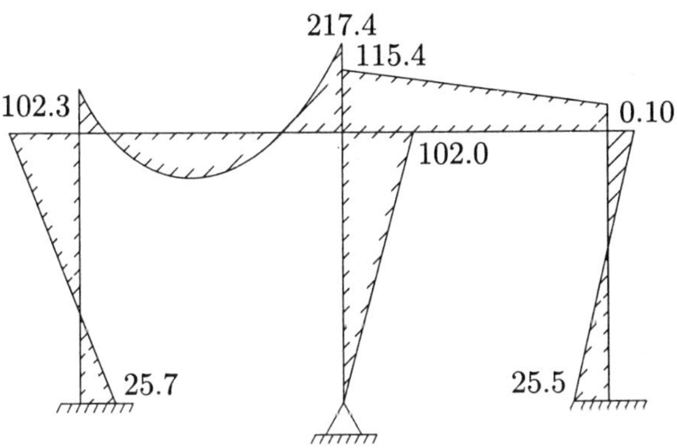

(a) B.M. in kNm

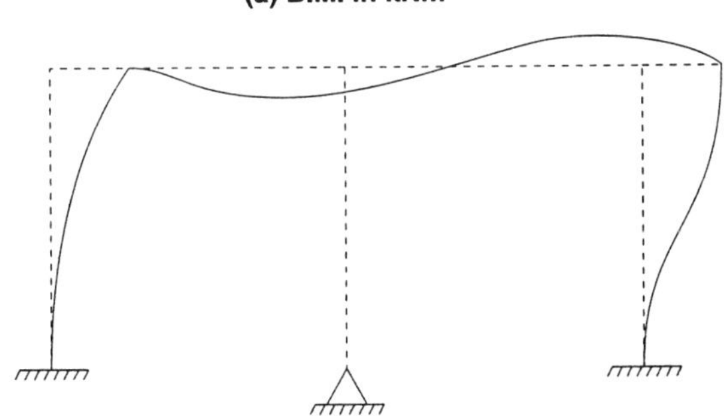

(b) Deflected shape

Fig. 9.36: For Example 9.12

9.12 Frames with Inclined Vertical Members

The method of superposition described in the preceding Section for rectangular frames having one degree of freedom of joint translation can be used in the analysis of frames having other arrangement of the members. For example, if the frames has inclined members in place of vertical members as shown in Fig. 9.37(a), it is first of all supported against translation as in Fig. 9.37(b). After the moment distribution calculations are carried out, the reaction F_0 at the support is calculated.

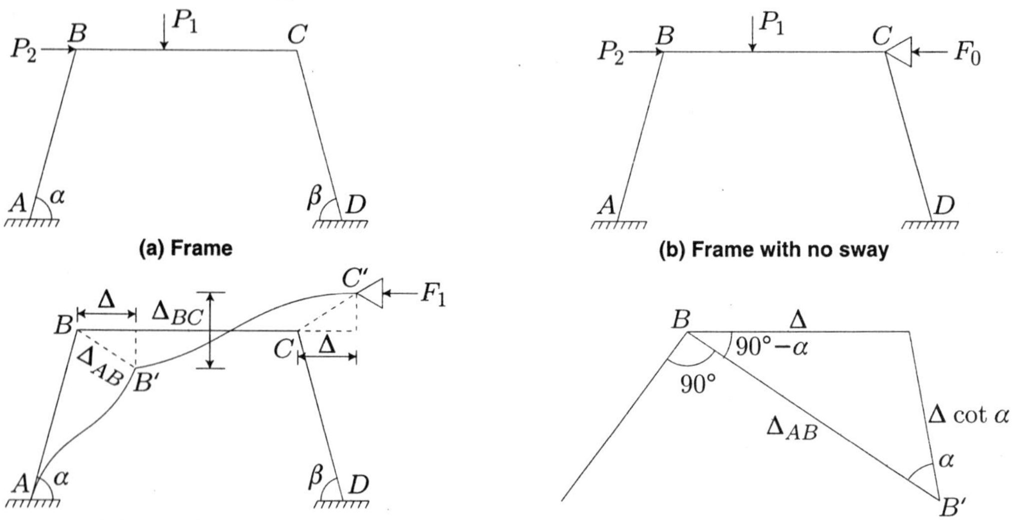

Fig. 9.37: Frame with inclined member

The second analysis gives the moments due to an arbitrary side-sway Δ as in Fig. 9.37(c). If F_1 be the force corresponding to this arbitrary sway and the actual sway be X times the arbitrary sway, then,

$$F_0 + X(F_1) = 0, \text{ whence } X = -\frac{F_0}{F_1}. \tag{9.12}$$

A new feature of this analysis is the determination of the fixed end moments due to assumed joint displacements. When the frame is displaced horizontally to the right by the amount Δ, as shown in Fig. 9.37(c), fixed end moments will be developed in all the members, when the joints are held against rotation. The magnitude of these moments depends on the relative displacements of the ends of the members. For example, the fixed end moments in member AB can be found if the displacement BB' of joint B is obtained either by Williot diagram or analytically. The use of Williot diagram obtaining the displacements is shown in Fig. 9.37(d). The relative displacement of the ends of members AB, CD and BC are given by

$$\left.\begin{aligned} \Delta_{AB} &= \Delta \csc\alpha \\ \Delta_{CD} &= \Delta \csc\beta \\ \text{and} \quad \Delta_{BC} &= \Delta\cot\beta + \Delta\cot\alpha \\ &= \Delta(\cot\alpha + \cot\beta). \end{aligned}\right] \tag{9.13}$$

Fixed-end moments due to such relative displacements are given by

$$\left.\begin{aligned} M_{FAB} &= \frac{6EI_{AB}\Delta AB}{L_{AB}^2} = M_{FBA} \\ M_{FCD} &= \frac{6EI_{CD}\Delta_{CD}}{L_{CD}^2} = M_{FDC} \\ \text{and} \quad M_{FBC} &= -\frac{6EI_{BC}\Delta_{BC}}{L_{BC}^2} = M_{FCB} \end{aligned}\right] \tag{9.14}$$

The signs of moments can be determined by inspection from the deflected shape of the structure with the joints locked against rotation as in Fig. 9.37(c).

If a numerical value for the horizontal joint displacement Δ is assumed, then corresponding values of the fixed end moments can be used in the calculations. On the other hand, for the purpose of finding bending moments in the structure, it is sufficient to assume relative values of the fixed end moments.

Example 9.13. *Analyse the frame in Fig. 9.38(a) using moment distribution method.*

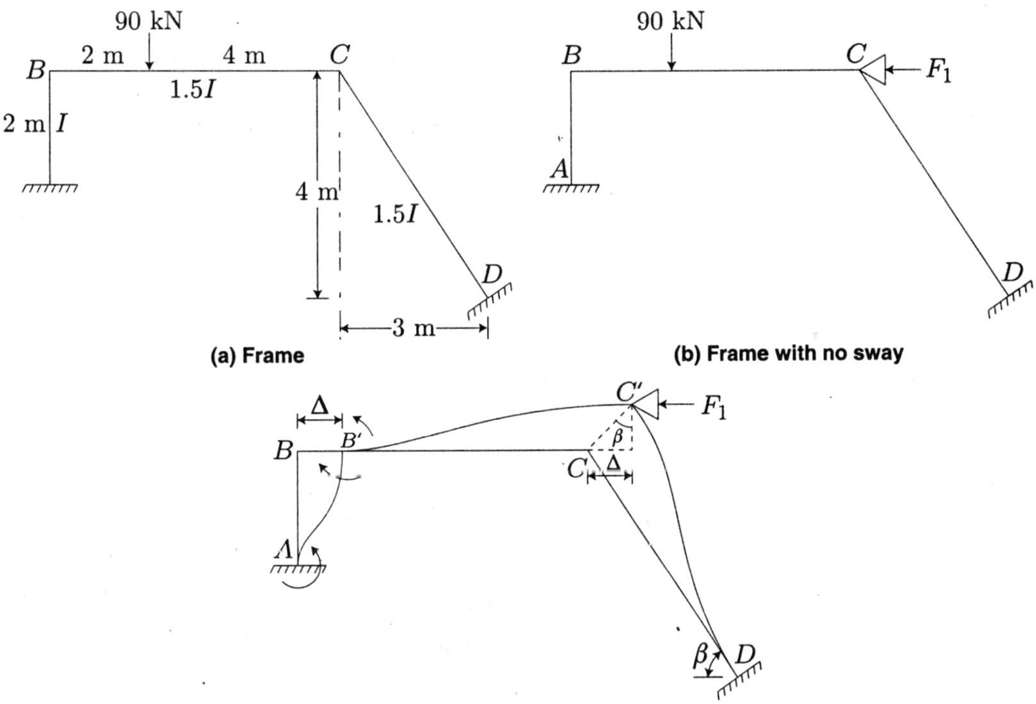

Fig. 9.38: For Example 9.13

Solution : *Stiffness values :*

$$K_{AB} = K_{BA} = \frac{4EI}{2} = 2EI$$

$$K_{BC} = K_{CB} = \frac{4E(1.5I)}{6} = EI$$

$$K_{CD} = K_{DC} = \frac{4E(1.5I)}{5} = 1.2EI.$$

Distribution factor :

$$\left.\begin{aligned} DF_{BA} &= \frac{2}{1+2} = 0.667 \\ DF_{BC} &= \frac{1}{1+2} = 0.333 \end{aligned}\right] \sum = 1.00$$

$$\left.\begin{aligned} DF_{CB} &= \frac{1}{1+1.2} = 0.455 \\ DF_{CD} &= \frac{1.2}{1+1.2} = 0.545 \end{aligned}\right] \sum = 1.00$$

Moment distribution for no sway is carried on considering the frame in Fig. 9.38(b) provided with a restraint at C developing a reaction F_0 as shown. For sway, moment distribution is carried on for an arbitrary sway Δ caused by a force F_1 at C as shown in Fig. 9.38(c).

Fixed end moment due to applied loads for no-sway moment distribution :

$$M_{FBC} = +\frac{(90)(2)(4)^2}{(6)^2} = +80.0 \text{ kNm}$$

$$M_{FCB} = -\frac{(90)(2)^2(4)}{(6)^2} = -40.0 \text{ kNm}.$$

Fixed-end moment due to arbitrary sway :

Due to an arbitrary sway Δ towards right as shown in Fig. 9.38(c), the relative displacements between the ends of the members are

$$\Delta_{AB} = BB' = \Delta$$

$$\Delta_{CD} = CC' = \Delta \csc\beta = \frac{5}{4}\Delta$$

$$\Delta_{BC} = C'C'' = \Delta \cot\beta = \frac{3}{4}\Delta$$

$$M_{FAB} = M_{FBA} = +\frac{6EI_{AB}\Delta_{AB}}{(L_{AB})^2} = \frac{6EI\Delta}{(2)^2} = \frac{3}{2}EI\Delta$$

$$M_{FCD} = M_{FDC} = +\frac{6EI_{CD}\Delta_{CD}}{(L_{CD})^2} = \frac{6E(1.5I)\left(\frac{5}{4}\Delta\right)}{(5^2)} = +\frac{9}{20}EI\Delta$$

$$M_{FBC} = M_{FCB} = -\frac{6EI_{BC}\Delta_{BC}}{(L_{BC})^2} = -\frac{6E(1.5I)(\frac{3}{4}\Delta)}{(6)^2} = \frac{3}{16}EI\Delta.$$

Putting $EI\Delta = 160$ kN-m^3,

$$M_{FAB} = M_{FBA} = +240 \text{ kNm}$$

$$M_{FCD} = M_{FDC} = +72 \text{ kNm}$$

$$M_{FBC} = M_{FCB} = -30 \text{ kNm}.$$

No-sway and sway moment distributions are shown in Tables 9.21 and 9.22 respectively.

Table 9.21. No-sway Moment Distribution for Example 9.13

Joint	A	B		C		D
Member	AB	BA	BC	CB	CD	DC
Stiffness	$2EI$	$2EI$	EI	EI	$1.2EI$	$1.2EI$
D.F.	0	0.667	0.333	0.455	0.545	0
C.O.F.	←	0.50	← 0.50	→	0.50	→
F.E.M.			+80.0	−40.0		
Bal.		−53.4	−26.6	+18.2	21.8	
C.O.	−26.7		+9.1	−13.2		+10.9
Bal.		−6.1	−3.0	+6.0	+7.2	
C.O.	−3.1		+3.0	−1.5		+3.6
Bal.		−2.0	−1.0	+0.7	+0.8	
C.O.	−1.0					+0.4
Final end moment (Joint sign Convention)	−30.8	−61.5	+61.5	−29.8	+29.8	+14.9

Table 9.22. Sway Moment Distribution for Example 9.13

Joint	A	B		C		D
Member	AB	BA	BC	CB	CD	DC
Stiffness	$2EI$	$2EI$	EI	EI	$1.2EI$	$1.2EI$
D.F.	0	0.667	0.333	0.455	0.545	0
C.O.F.	← 0.50		←	0.50 →		0.50 →
F.E.M.	+240.0	+240.0	−30.0	−30.0	+72.0	+72.0
Bal.	0	−140.0	−70.0	−19.0	23.0	0
C.O.	−70.0	0	−9.5	−35.0	0	−11.5
Bal.	0	+6.3	+3.2	+15.9	+19.1	0
C.O.	+3.2	0	+8.0	+1.6	0	+9.6
Bal.	0	−5.3	−2.7	−0.7	−0.9	0
C.O.	−2.					−0.5
Final end moment (Joint sign Convention)	+170.5	+101.0	−101.0	−67.2	+67.2	+69.6

Determination of forces at joint C.

The reaction F_0 at C required for no sway condition may be obtained from consideration of equilibrium of the frame as a whole and taking moment of all forces about the point 0, at which the members AB and DC meet when produced as shown in Fig. 9.39(a). The force F_0 is obtained from

$$F_0(OB) - H_A(OA) + H_D(OD) + 30.8 - 14.9 + (90)(2) = 0. \quad \text{(i)}$$

H_A and H_D are obtained from equilibrium of columns AB and CD respectively from

$$H_A = \frac{61.5 + 30.8}{2} = 46.15 \text{ kN} \quad \text{and} \quad H_D = \frac{29.8 + 14.9}{5} = 8.94 \text{ kN}.$$

From geometry, $OB = BC \tan\beta = (6)\left(\frac{4}{3}\right) = 8$ m.

$$OC = \sqrt{8^2 + 6^2} = 10 \text{ m}$$

$$OD = OC + CD = 10 + 5 = 15 \text{ m}$$

$$OA = OB + BA = 8 + 2 = 10 \text{ m.}$$

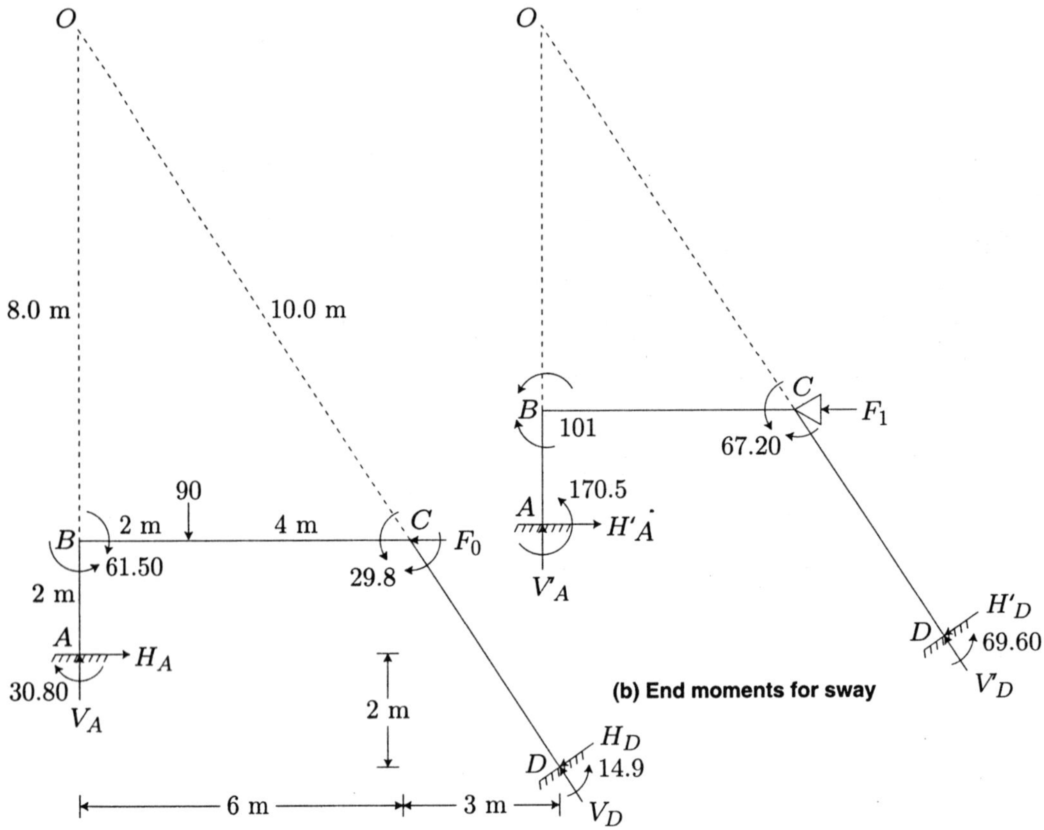

(a) End moments for no sway

Fig. 9.39: For Example 9.13

Putting appropriate values in Eq. (i), and solving, $F_0 = 16.44$ kN.

The force F_1 corresponding to the arbitrary sway $\Delta = \frac{160}{EI}$, is obtained from equilibrium of the frame as shown in Fig. 9.39(b). By taking moment of all forces about O,

$$F_1(OB) - H'_A(OA) + H'_D(OD) - 170.5 - 69.6 = 0. \quad \text{(ii)}$$

Considering equilibrium of columns AB and CD,

$$H'_A = \frac{170.5 + 101.0}{2} = -135.75 \text{ kN}$$

and $$H'_D = \frac{69.6 + 67.2}{5} = +27.36 \text{ kN.}$$

Putting appropriate values in Eq. (ii) and solving, $F_1 = -190.98$ kN.

Determination of ratio :

Assuming that the actual sway is X times the arbitrary sway and considering that there is no external force at C in the actual frame,

$$F_0 + XF_1 = 0$$

i.e., $16.44 + X(-190.98) = 0$, whence $X = \dfrac{16.44}{190.98} = 0.08607$.

The net end moments are obtained by adding X times the end moments of Table 9.22 to those of Table 9.21 as shown in Table 9.23.

Table 9.23. Net End Moments for Example 9.13

Member	AB	BA	BC	CB	CD	DC
No-sway moment, kNm	−30.8	−61.5	+61.5	−29.8	+29.8	+14.9
Sway moment, kNm	+14.7	+8.7	−8.7	−5.8	+5.8	+6.0
Net moment, kNm with joint moment sign convention	−16.1	−52.8	+52.8	−35.6	+35.6	+20.9

Fig. 9.40 shows B.M. diagram drawn on tension side.

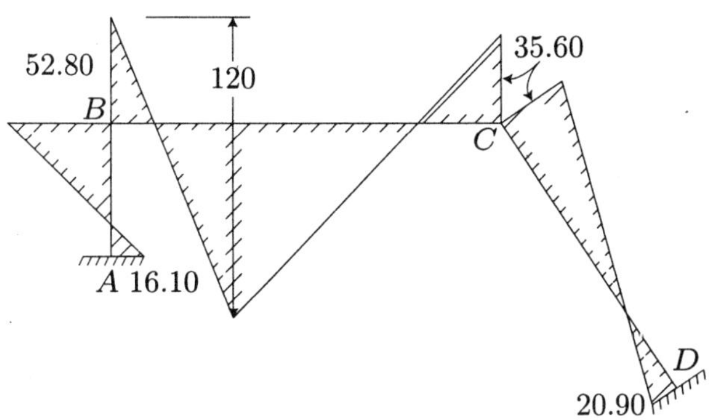

Fig. 9.40: B.M. in kNm

9.13 Frames with Several Degrees of Freedom of Joint Translation

The number of degrees of freedom for joint translation in a rigid frame structure is equal to the number of independent joint translation, which can be given to the structure. For example, in case of a two-storeyed rectangular frame as in Fig. 9.41(a) joints F and D can be given horizontal displacements which are independent of one another, those at E and C being similar as at F and D.

The rigid frame with several degrees of joint translation can be analysed by moment distribution by a simple extension of the superposition method used in case of frames with one degree of freedom. The procedure is illustrated in Figs. 9.41(b), 9.41(c) and 9.41(d). The frame is first supported against joint translation by artificial supports with loads acting as in Fig. 9.41(b). The moments in the frame under these conditions can be found without difficulty by moment distribution, and the reactions F_{o_1} and F_{o_2} can be calculated. The frame is then displaced at joint F by an arbitrary amount Δ_1 as in Fig. 9.41(c). The fixed end moments in this frame are determined, balancing process is carried out and the reactions F_1' and F_2' are calculated.

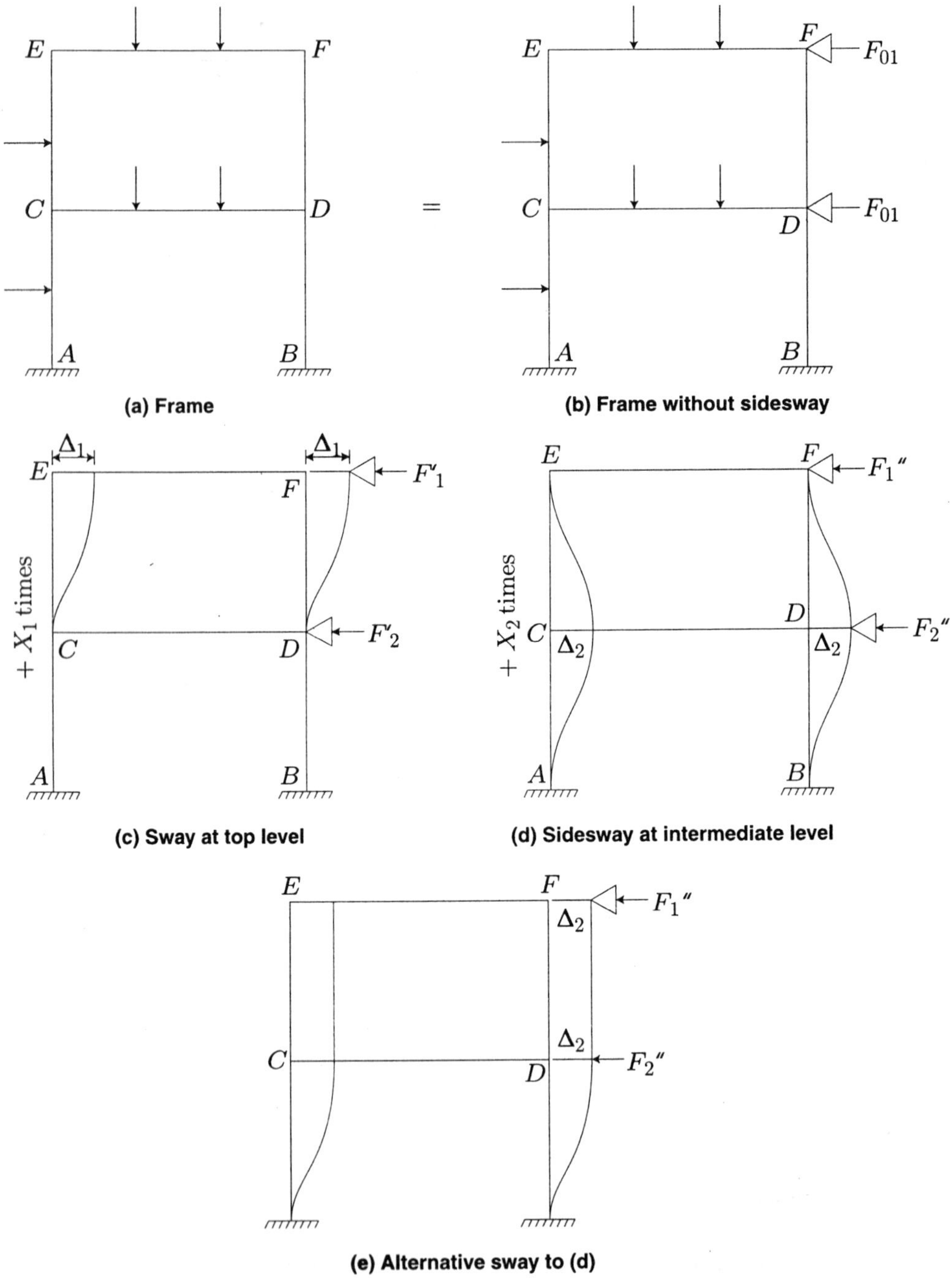

Fig. 9.41: Two-storeyed rectangular frame

A similar procedure is followed for an arbitrary displacement Δ_2 at joint D. The essential feature concerning the joint displacements Δ_1 and Δ_2 is that when the joint displacements of the frames in Figs. 9.41(c) and 9.41(d) are superimposed, there must be a total displacement of joint F which is independent of the total displacement of joint D.

The forces and moments in the original frame in Fig. 9.41(a) can now be obtained by superposition of the results of the analysis of the frames in Figs. 9.41(b), 9.41(c) and 9.41(d). Thus,

$$\left.\begin{aligned} &F_{o_1} + X_1F_1' + X_2F_1'' = 0 \\ \text{and} \quad &F_{o_2} + X_1F_2' + X_2F_2'' = 0. \end{aligned}\right] \tag{9.15}$$

After the constants X_1 and X_2 are obtained, the resulting moments can be calculated. This method of analysis can be used equally well in case of irregular frames and frames with higher degrees of freedom as shown in Figs. 9.42 and 9.43 respectively.

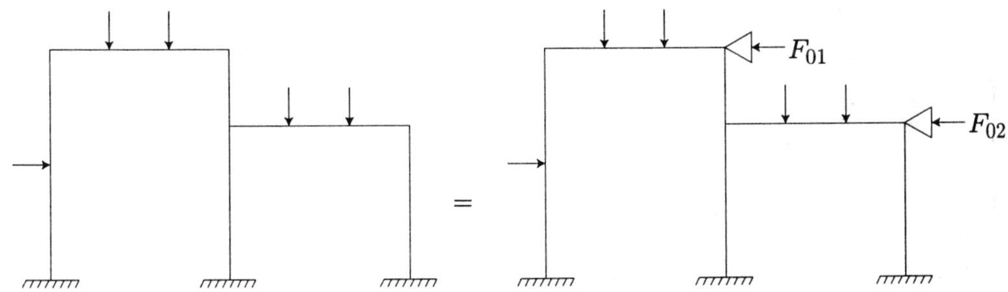

(a) Frame (b) Frame without sidesway

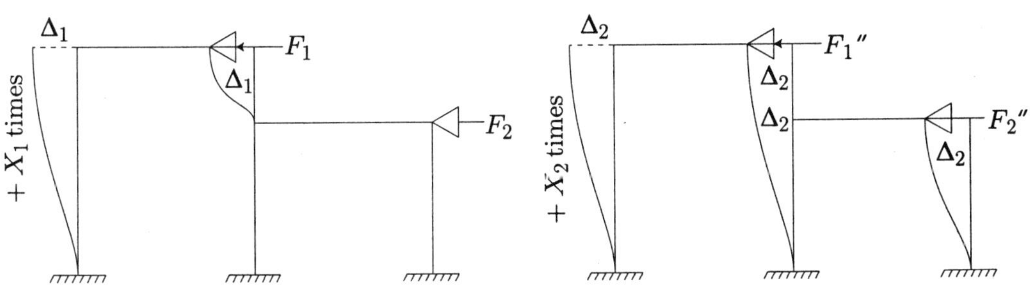

(c) Frame with sway at top level (d) Frame with sway at intermediate level

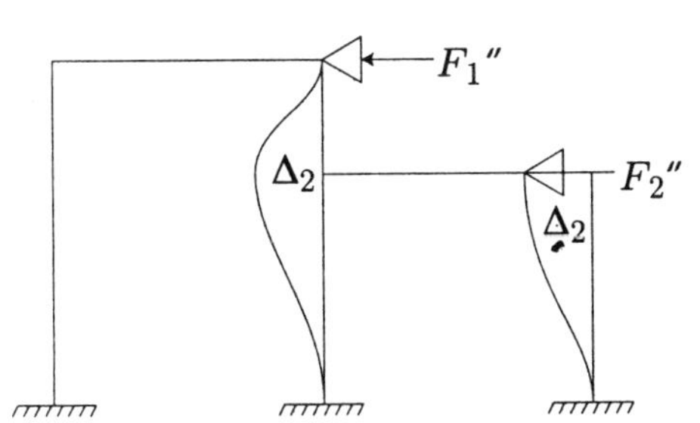

(e) Alternative sway pattern of (d)

Fig. 9.42: Irregular frames

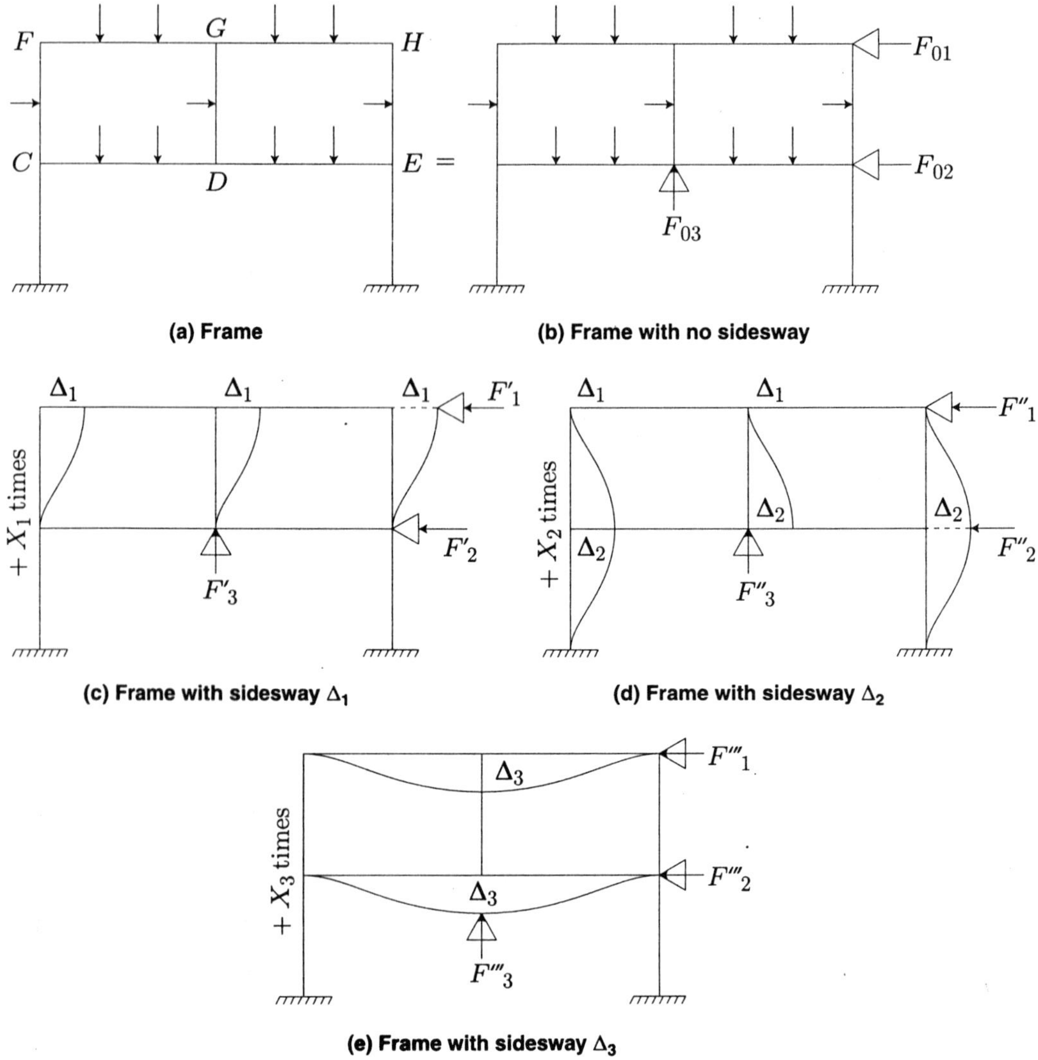

Fig. 9.43: Frames with multiple sway

In case of the frame in Fig. 9.42, there will be two simultaneous equations for the unknown constants X_1 and X_2. For the frame in Fig. 9.43, the relationship developed will involve three unknown constants. Equations for the case are as below :

$$\left.\begin{aligned} &F_{o_1} + X_1 F_1' + X_2 F_1'' + X_3 F_1''' = 0 \\ &F_{o_2} + X_1 F_2' + X_2 F_2'' + X_3 F_2''' = 0 \\ \text{and} \quad &F_{o_3} + X_1 F_3' + X_2 F_3'' + X_3 F_3''' = 0 \end{aligned}\right] \tag{9.16}$$

Other examples of rigid frame structures having more than one degree of freedom are shown in Fig. 9.44. The position of the reactive forces to prevent translation and keep the frame in position are indicated in the figures by dashed line.

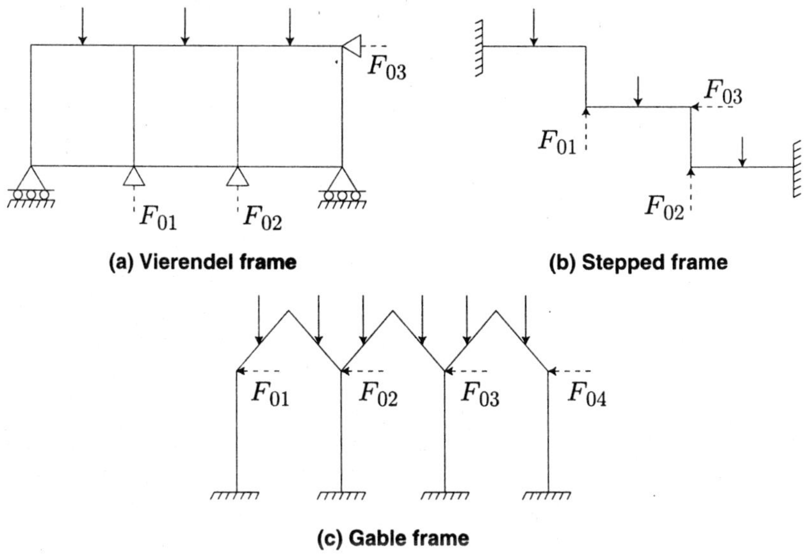

Fig. 9.44: Structures with several degrees of freedom for joint translation

9.14 Gable Frames

The steps required for the analysis of a single span gable frame by moment distribution are shown in Fig. 9.45. Since the frame has two degrees of freedom for joint translation two artificial supports are necessary to prevent the joints from moving. The first of the three moment distribution analysis required for the frame is the one with no joint translation. The fixed-end moments are those resulting from the applied loads and any other applied effects. For the inclined members, the component of the load normal to the axis of the member will produce fixed-end moments.

Fig. 9.45(c) shows the gable frames with an arbitrary horizontal displacement Δ_1 at joint C. This displacement results in fixed-end moments in members AC, CE and ED. The magnitude of these moments depends on the relative displacements of the ends of the members, which can be obtained from Williot diagram in Fig. 9.45(d)

The relative displacement Δ_{1EC} of E with respect to C and the displacement Δ_{1ED} of E with respect to D are calculated as under :

$$\left.\begin{aligned} \Delta_{1EC} &= \Delta_1 \frac{\sin(90° - \beta)}{\sin(\alpha + \beta)} \\ \text{and} \quad \Delta_{1ED} &= \Delta_1 \frac{\sin(90° - \alpha)}{\sin(\alpha + \beta)}. \end{aligned}\right] \tag{9.17}$$

The relative displacements $\Delta_{1CA} = \Delta_1$. With all relative displacements expressed in terms of Δ_1, the fixed end moments can be calculated readily for any assumed value of Δ_1.

For effect of displacement Δ_2 as in Fig. 9.45(e), the displacements Δ_{2EC} and Δ_{2ED} can be obtained from Fig. 9.45(f).

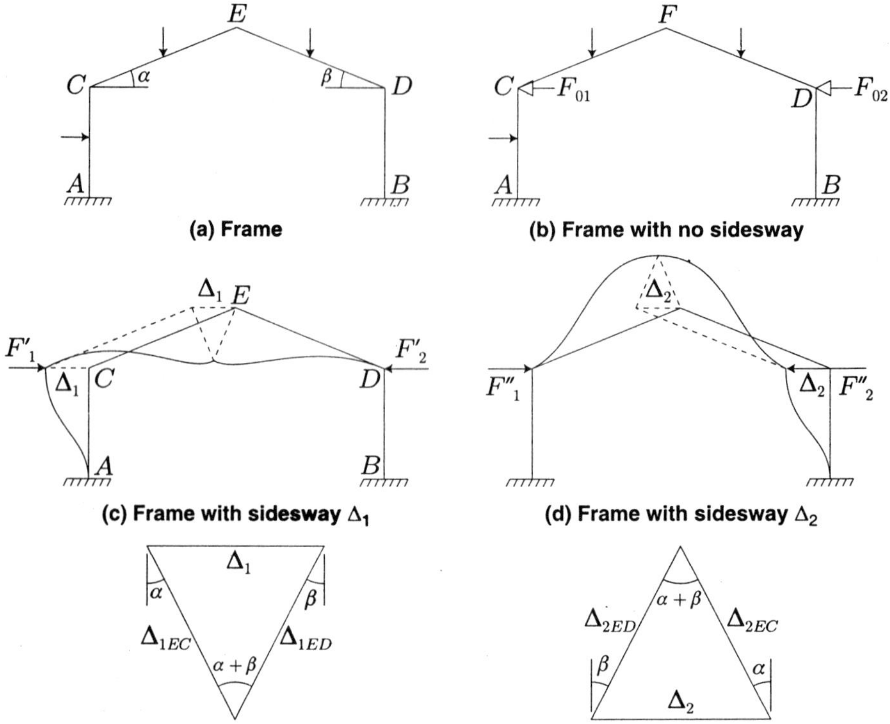

(a) Frame (b) Frame with no sidesway

(c) Frame with sidesway Δ_1 (d) Frame with sidesway Δ_2

(e) Displacement diagram for (c) (f) Displacement diagram for (e)

Fig. 9.45: Analysis of gabled frame

$$\left.\begin{aligned} \text{Thus,}\quad \Delta_{2EC} &= \Delta_2 \frac{\sin(90^\circ - \beta)}{\sin(\alpha + \beta)} \\ \text{and}\quad \Delta_{2ED} &= \Delta_2 \frac{\sin(90^\circ - \alpha)}{\sin(\alpha + \beta)}. \end{aligned}\right] \tag{9.18}$$

From these values of relative displacements, fixed end moments are obtained and moment distribution analysis is carried out. For values of F_{01}, F_{02}, F_1', F_2', F_1'' and F_2'', simultaneous equations are formed for obtaining values of unknown constants.

Instead of introducing independent displacements at joints C and D, it is also possible to use displacements as in Fig. 9.46.

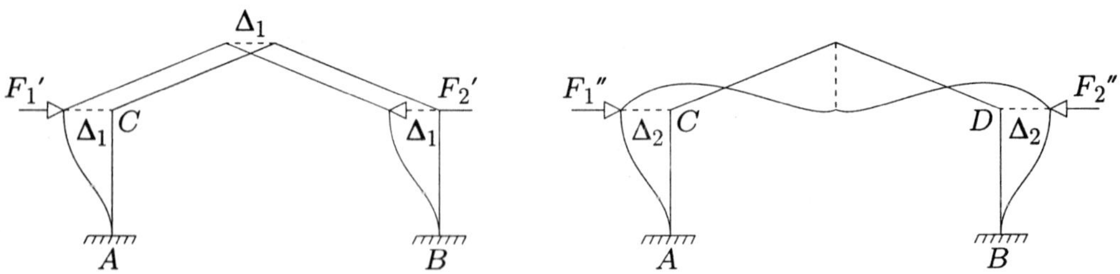

(a) Sidesway Δ_1 in frame (b) Sidesway Δ_2 in frame

Fig. 9.46: Alternative sidesway pattern in gabled frame

In case of a symmetrical cable frame, one set of displacements at C and D will be enough as from symmetry, the displacements at C and D will be the same.

In general, lateral forces developed at the joints which are restrained against sway, such as F_1 and F_2 at joints B and D in Fig. 9.47(a) where P_1, P_2, H_1 and H_2 are the applied loads can be obtained by considering equilibrium of portions AB, ABC and CDE and DE, as shown, in Figs. 9.47(b), 9.47(c), 9.47(d) and 9.47(e).

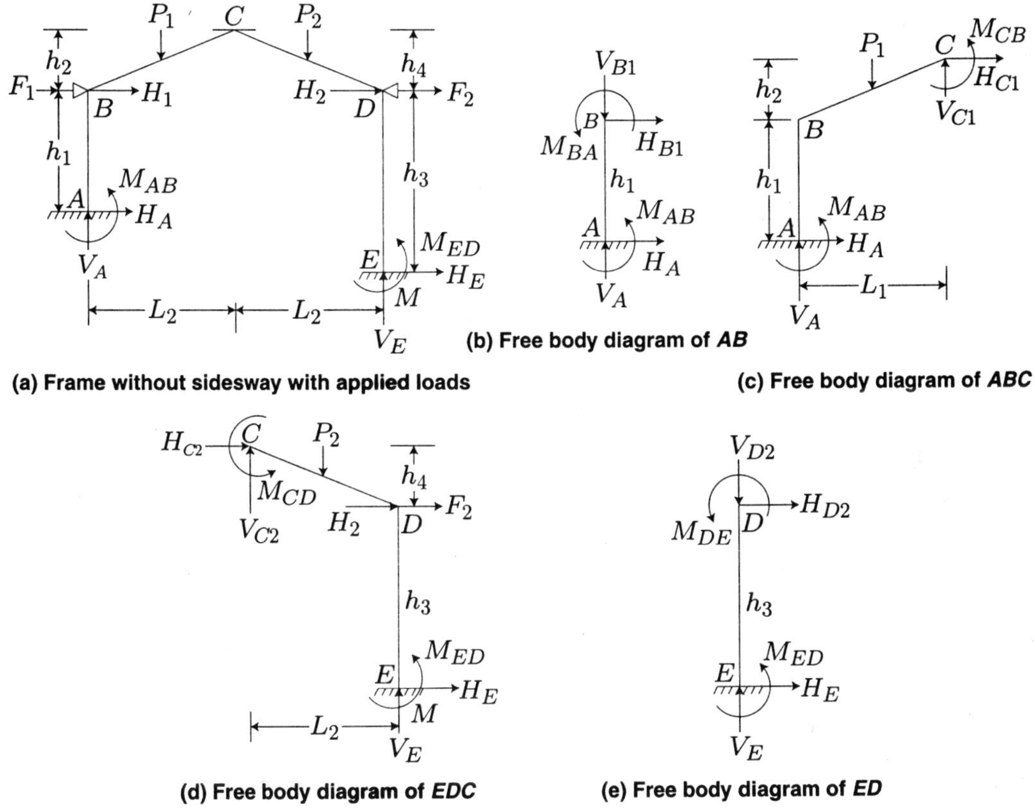

Fig. 9.47: Analysis of gabled frame

Considering equilibrium of portion AB and DE as in Figs. 9.47(b), 9.47(c).

$$H_A = -\frac{M_{AB} + M_{BA}}{h_1} \tag{a}$$

$$H_E = -\frac{M_{DE} + M_{ED}}{h_3}. \tag{b}$$

Considering equilibrium of the entire frame as shown in Fig. 9.47(a) and using conditions $\sum H = 0$ and $\sum V = 0$,

$$F_1 + F_2 + H_1 + H_2 + H_A + H_E = 0 \tag{c}$$

$$V_A + V_E = P_1 + P_2. \tag{d}$$

Considering equilibrium of portions ABC and CDE as shown in Figs. 9.47(c) and 9.47(d) and using condition $\sum M_C = 0$,

$$(F_1 + H_1)(h_2) + H_A(h_1 + h_2) - V_A(L_1) + M_{AB} + M_{CB} + M_{ext.1=0} \tag{e}$$

and $$(F_2 + H_2)(h_4) + H_E(h_3 + h_4) + V_E(L_2) + M_{ED} + M_{CD} + M_{ext.2=0}, \tag{f}$$

where $M_{ext.1}$ and $M_{ext.2}$ are respectively the moments of applied vertical loads P_1 and P_2 about C in Figs. 9.47(c) and 9.47(d). Solving Eqs. (a) to (f), the values of the forces F_1 and F_2 can be obtained.

Example 9.14. *Analyse the two-storeyed frame shown in Fig. 9.47(a), using moment distribution.*

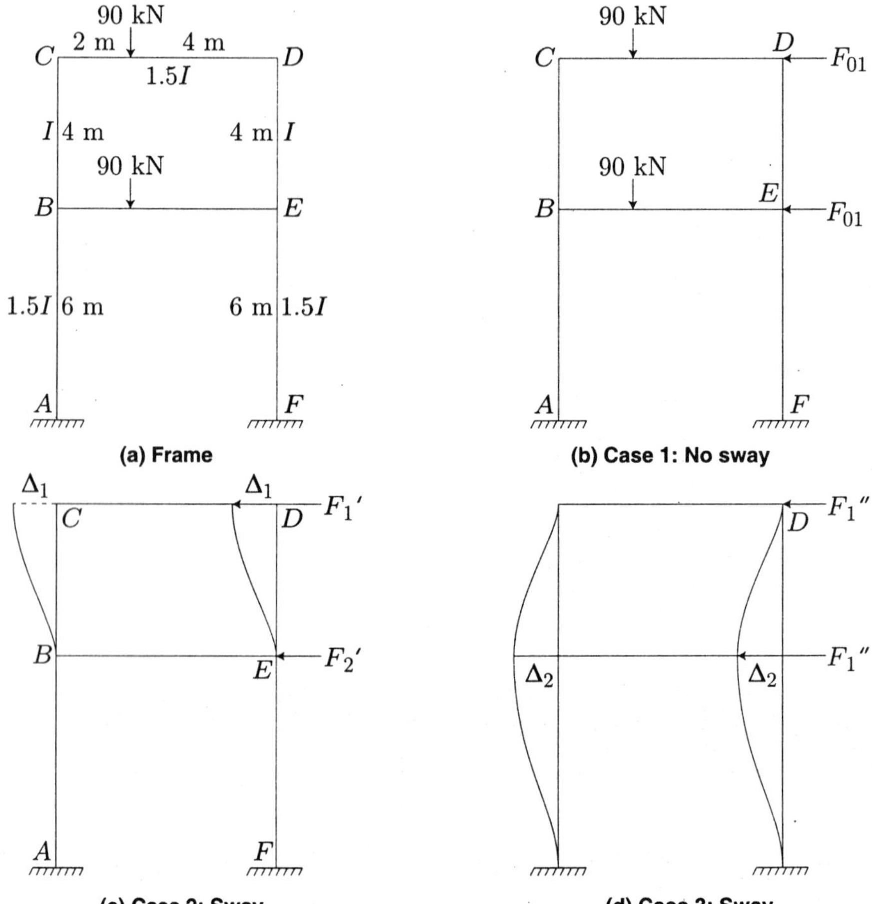

Fig. 9.48: For Example 9.14

Solution : The three cases of loading for which moment distribution operations will be carried out are :

Case 1. Applied loads with lateral supports provided at D and E or no sway as shown in Fig. 9.48(b).

Case 2. Arbitrary sway Δ_1 at C and D with support at E for no sway at E as shown in Fig. 9.48(c).

Case 3. Arbitrary sway Δ_2 at B at E with support at D for no sway at D as shown in Fig. 9.48(d).

Stiffness :

$$K_{AB} = K_{BA} = K_{EF} = K_{FE} = \frac{4E(1.5I)}{6} = EI$$

$$K_{BC} = K_{CB} = K_{DE} = K_{ED} = \frac{4EI}{4} = EI$$

$$K_{BE} = K_{EB} = K_{CD} = K_{DC} = \frac{4E(1.5I)}{6} = EI.$$

Distribution factors :

Joint A : $DF_{AB} = \dfrac{1}{1+\infty} = 0$

Joint B : $DF_{BA} = \dfrac{1}{1+1+1} = \dfrac{1}{3}$

$$DF_{BC} = \frac{1}{3}$$

$$DF_{BE} = \frac{1}{3}.$$

Joint C : $DF_{CB} = \dfrac{1}{1+1} = \dfrac{1}{2}$

$$DF_{CD} = \frac{1}{2}.$$

Joint D : $DF_{DC} = \dfrac{1}{2}$

$$DF_{DE} = \frac{1}{2}.$$

Joint E : $DF_{ED} = \dfrac{1}{3}$

$$DF_{EB} = \frac{1}{3}$$

$$DF_{EF} = \frac{1}{3}.$$

Joint F : $DF_{FE} = 0.$

Fixed-end moments :

(i) For Case 1 : With reference to Fig. 9.48(b),

$$M_{FCD} = +\frac{(90)(2)(4)^2}{(6)^2} = +80.0 \text{ kNm}$$

$$M_{FDC} = -\frac{(90)(2)^2(4)}{(6)^2} = -40.0 \text{ kNm}$$

$$M_{FBE} = +80.0 \text{ kNm}$$

$$M_{FEB} = -40.0 \text{ kNm}.$$

(ii) For Case 2 : With reference to Fig. 9.48(c)

$$M_{FBC} = M_{FCB} = -\frac{6EI\Delta_1}{(4)^2} = -\frac{3}{8}EI\Delta_1$$

$$M_{FDE} = M_{FED} = -\frac{6EI\Delta_1}{(4)^2} = -\frac{3}{8}EI\Delta_1.$$

Putting $EI\Delta_1 = 160 \text{ kNm}^3$

$$M_{FBC} = M_{FCB} = M_{FDE} = M_{FED} = -60.0 \text{ kNm}.$$

(iii) For Case 3 : With reference to Fig. 9.48(d),

$$M_{FAB} = M_{FBA} = -\frac{6E(1.5I)\Delta_2}{(6)^2} = -\frac{1}{4}EI\Delta_2$$

$$M_{FBC} = M_{FCB} = +\frac{6EI\Delta_2}{(4)^2} = +\frac{3}{8}EI\Delta_2$$

$$M_{FDE} = M_{FED} = +\frac{6EI\Delta_2}{(4)^2} = +\frac{3}{8}EI\Delta_2$$

$$M_{FEF} = M_{FFE} = -\frac{6E(1.5I)\Delta_2}{(6)^2} = -\frac{1}{4}EI\Delta_2.$$

Putting $EI\Delta_2 = 160$ kNm^3

$$M_{FAB} = M_{FBA} = M_{FEF} = M_{FFE} = -40.0 \text{ kNm}$$

$$M_{FBC} = M_{FCB} = M_{FDE} = M_{FED} = +60.0 \text{ kNm.}$$

Moment distribution operations for the above three cases are shown in Tables 9.24, 9.25 and 9.26 respectively.

Determination of lateral forces at D and E :

Table 9.24. Moment Distribution for Case 1 for Example 9.14

Joint	A	B			C		D		E			F
Member	AB	BA	BE	BC	CB	CD	DC	DE	ED	EB	EF	FE
Stiffness	EI	EI	EI	EI	EI	EI	EI	EI	EI	EI	EI	EI
D.F.	0	$\frac{1}{3}$	$\frac{1}{3}$	$\frac{1}{3}$	$\frac{1}{2}$	$\frac{1}{2}$	$\frac{1}{2}$	$\frac{1}{2}$	$\frac{1}{3}$	$\frac{1}{3}$	$\frac{1}{3}$	0
C.O.F.	← 0.50 (AB–BA)		← 0.50 → (BE–EB)	← 0.50 → (BC–CB)		← 0.50 → (CD–DC)		← 0.50 → (DE–ED)			0.50 → (EF–FE)	
F.E.M.			+80.0			+80.0	−40.0			−40.0		
Bal.	0	−26.7	−26.6	−26.7	−40.0	−40.0	+20.0	+20.0	+13.4	+13.3	+13.3	0
C.O.	−13.4		+6.7	−20.0	−13.4	+10.0	−20.0	+6.7	+10.0	−13.3		+6.7
Bal.	0	+4.4	+4.5	+4.4	+1.7	+1.7	+6.6	+6.7	+1.1	+1.1	+1.1	0
C.O.	+2.2		+0.6	+0.9	+2.2	+3.3	+0.9	+0.6	+3.4	+2.2		+0.5
Bal.	0	−0.5	−0.5	−0.5	−2.8	−2.7	−0.8	−0.7	−1.9	−1.9	−1.8	0
C.O.	−0.3		−1.0	−1.4	−0.3	−0.4	−1.4	−1.0	−0.4	−0.3		−0.9
Bal.		+0.8	+0.8	+0.8	+0.4	+0.3	+0.2	+1.2	+0.2	+0.3	+0.2	
C.O.	+0.4											+0.1
Final End Moment	−11.1	−22.2	+64.4	−42.5	−52.2	+52.2	−33.5	+33.5	+25.8	−38.6	+12.8	+6.4

Table 9.25. Moment Distribution for Case 2 for Example 9.14

Joint	A	B			C		D		E			F
Member	AB	BA	BE	BC	CB	CD	DC	DE	ED	EB	EF	FE
Stiffness	EI	EI	EI	EI	EI	EI	EI	EI	EI	EI	EI	EI
D.F.	0	$\frac{1}{3}$	$\frac{1}{3}$	$\frac{1}{3}$	$\frac{1}{2}$	$\frac{1}{2}$	$\frac{1}{2}$	$\frac{1}{2}$	$\frac{1}{3}$	$\frac{1}{3}$	$\frac{1}{3}$	0
C.O.F.	←——	0.50	←	← 0.	50 →	← 0.50 (BE–EB)	0.50→	← 0.	50 →	→	0.50	→
F.E.M.				−60.0	−60.0			−60.0	−60.0			
Bal.		+20.0	+20.0	+20.0	+30.0	+30.0	+30.0	+30.0	+20.0	+20.0	+20.0	
C.O.	+10.0	0	+10.0	+15.0	+10.0	+15.0	+15.0	+10.0	+15.0	+10.0	0	+10.0
Bal.	0	−8.4	−8.3	−8.3	−12.5	−12.5	−12.5	−12.5	−8.3	−8.3	−8.4	0
C.O.	−4.2	0	−4.2	−6.3	−4.2	−6.3	−6.3	−4.2	−6.3	−4.2	0	−4.2
Bal.	0	+3.5	+3.5	+3.5	+5.3	+5.2	+5.2	+5.3	+3.5	+3.5	+3.5	0
C.O.	+1.8	0	+1.8	+2.7	+1.8	+2.6	+2.6	+1.8	+2.7	+1.8	0	+1.8
Bal.	0	−1.5	−1.5	−1.5	−2.2	−2.2	−2.2	−2.2	−1.5	−1.5	−1.5	0
C.O.	−0.8	0	−0.8	−1.1	−0.8	−1.1	−1.1	−0.8	−1.1	−0.8	0	−0.8
Bal.	0	+0.7	+0.6	+0.6	+1.0	+0.9	+0.9	+1.0	+0.6	+0.6	+0.7	
C.O.	+0.4											+0.4
Final End Moment	+7.2	+14.3	+21.1	−35.4	−31.6	+31.6	+31.6	−31.6	−35.4	+21.1	+14.3	+7.2

Table 9.26. Moment Distribution for Case 3 for Example 9.14

Joint	A	B			C		D		E			F
Member	AB	BA	BE	BC	CB	CD	DC	DE	ED	EB	EF	FE
Stiffness	EI	EI	EI	EI	EI	EI	EI	EI	EI	EI	EI	EI
D.F.	0	$\frac{1}{3}$	$\frac{1}{3}$	$\frac{1}{3}$	$\frac{1}{2}$	$\frac{1}{2}$	$\frac{1}{2}$	$\frac{1}{2}$	$\frac{1}{3}$	$\frac{1}{3}$	$\frac{1}{3}$	0
C.O.F.	← 0.50		←	← 0.50	→	← 0.50 (0.50 BE–EB)	→	← 0.50	→	→	0.50	→
F.E.M.	−40.0	−40.0		+60.0	+60.0			+60.0	+60.0		−40.0	−40.0
Bal.	0	−6.7	−6.6	−6.7	−30.0	−30.0	−30.0	−30.0	−6.7	−6.6	−6.7	0
C.O.	−3.4	0	−3.3	−15.0	−3.4	−15.0	−15.0	−3.4	−15.0	−3.3	0	−3.4
Bal.	0	+6.1	+6.1	+6.1	+9.2	+9.2	+9.2	+9.2	+6.1	+6.1	+6.1	0
C.O.	+3.1	0	+3.1	+4.6	+3.1	+4.6	+4.6	+3.1	+4.6	+3.1	0	+3.1
Bal.	0	−2.6	−2.5	−2.6	−3.9	−3.8	−3.9	−3.8	−2.6	−2.5	−2.6	0
C.O.	−1.3	0	−1.3	−2.0	−1.3	−2.0	−1.9	−1.3	−1.9	−1.3	0	−1.3
Bal.	0	+1.1	+1.1	+1.1	+1.7	+1.6	+1.6	+1.6	+1.1	+1.1	+1.0	0
C.O.	+0.6	0	+0.6	+0.9	+0.6	+0.8	+0.8	+0.6	+0.8	+0.6	0	+0.5
Bal.	0	−0.5	−0.5	−0.5	−0.7	−0.7	−0.7	−0.7	−0.5	−0.5	−0.4	0
C.O.	−0.3											−0.2
Final End Moment	−41.3	−42.6	−3.3	+45.9	+35.3	−35.3	−35.3	+35.3	+45.9	−3.3	−42.6	−41.3

The lateral forces at joint D for the three cases are determined considering equilibrium of the upper storey as shown in Fig. 9.49(a) and those at joint E considering equilibrium of the lower storey as shown in Fig. 9.49(b). Thus,

$$F_{01} = H_B + H_E = -\frac{M_{BC} + M_{CB}}{L_{BC}} - \frac{M_{DE} + M_{ED}}{L_{DE}}$$

$$= -\frac{-42.5 - 52.2}{4} - \frac{33.5 + 25.8}{4} = +8.9 \text{ kN} \longleftarrow.$$

Similarly, $F_1' = -\frac{-35.4 - 31.6}{4} - \frac{-31.6 - 35.4}{4} = +33.5 \text{ kN} \longleftarrow$

$$F_1'' = -\frac{45.9 + 35.3}{2} - \frac{35.3 + 45.9}{4} = -40.6 \text{ kN} \longrightarrow.$$

Again, $F_{02} = H_A + H_F - F_{01} = -\frac{M_{AB} + M_{BA}}{L_{AB}} - \frac{M_{EF} + M_{FE}}{L_{EF}} - F_{01}$

$$= -\frac{-11.1 - 22.0}{6} - \frac{12.8 + 6.4}{6} - 8.9 = -6.6 \text{ kN} \longrightarrow.$$

Similarly, $F_2' = H_A + H_F - F_1' = -\frac{7.2 + 14.3}{6} - \frac{14.3 + 7.2}{6} - 33.5 = -40.7 \text{ kN} \longrightarrow$

$$F_1'' = H_A + H_F - F_1'' = \frac{-41.3 - 42.6}{6} - \frac{-42.6 - 41.3}{6} - (-40.6) = +68.6 \text{ kN} \longrightarrow.$$

Determination of ratio :

Assuming that the actual sway at the level of CD is X_1 times the arbitrary sway considered in Case 2 and that at the level of BE is X_2 times the arbitrary sway considered in Case 3, and considering that in actual frame there are no lateral forces at joint D and E,

$$F_{01} + X_1(F_1') + X_2(F_1'') = 0 \quad \text{and} \quad F_{02} + X_1(F_2') + X_2(F_2'') = 0.$$

Putting the calculated values,

$$8.9 + X_1(33.5) + X_2(-40.6) = 0 \quad \text{and} \quad -6.6 + X_1(-40.8) + X_2(68.6) = 0.$$

Solving, $X_1 = -0.52965$ and $X_2 = +0.21912$.

Net moments are obtained by summing together with values of Case 1, X_1 times the values of Case 2 and X_2 times the values of Case 3, as shown in Table 9.27.

Table 9.27. Net End Moments for Example 9.14

Member	AB	BA	BE	BC	CB	CD	DC	DE	ED	EB	EF	FE
Case 1, kN-m	−11.1	−22.0	+64.4	−42.5	−52.2	+52.2	−33.5	+33.5	+25.8	−38.6	+12.8	+6.4
X_1 times, Case 2 kN-m	−3.8	−7.6	−11.2	+18.7	+16.8	−16.8	−16.8	+16.8	+18.7	−11.2	−7.6	−3.8
X_2 times, Case 3 kN-m	+9.0	+9.3	+0.8	−10.1	−7.7	+7.7	+7.7	−7.7	−10.1	+0.8	+9.3	+9.0
Net End Moment kN-m	−5.9	−20.3	+54.0	−33.9	−43.1	+43.1	−42.6	+42.6	+34.4	−49.0	+14.5	+11.6

Fig. 9.49(c) shows *B.M.* diagram drawn on tension side.

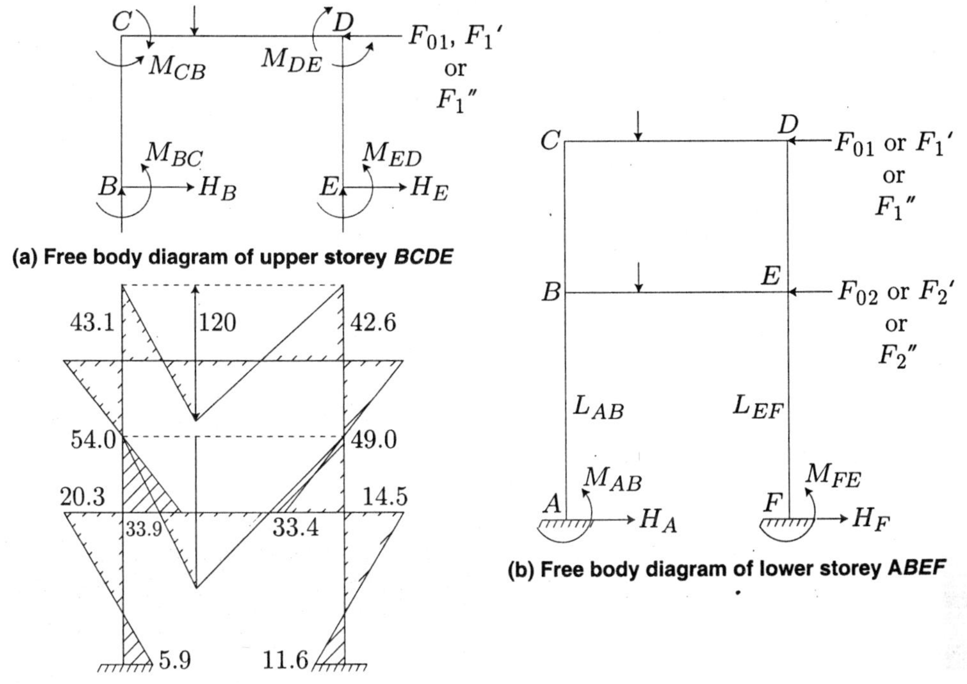

(a) Free body diagram of upper storey *BCDE*

(b) Free body diagram of lower storey A*BEF*

(c) B.M. in kNm

Fig. 9.49: For Example 9.14

Example 9.15. *Analyse the single bay gable frame shown in Fig. 9.50(a) using moment distribution.*

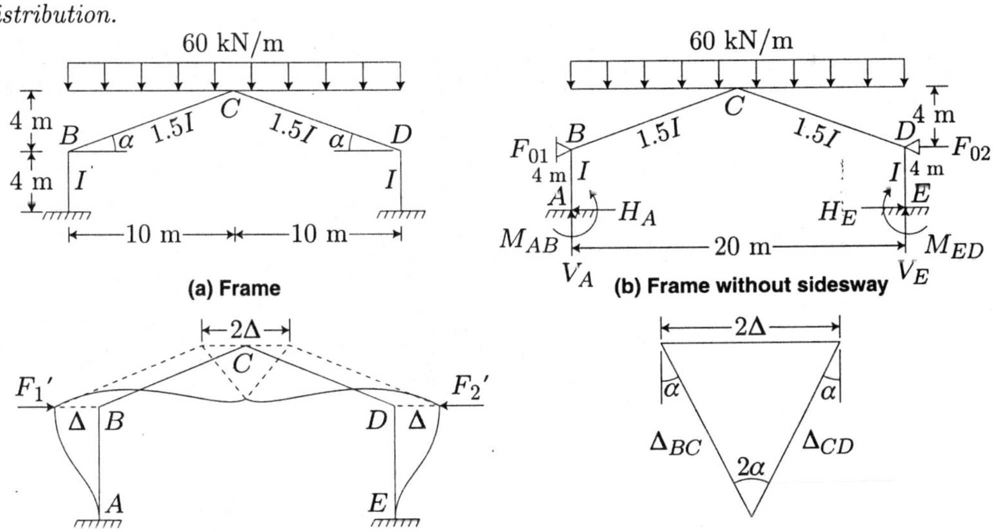

(a) Frame

(b) Frame without sidesway

(c) Frame with sidesway

(d) Displacements of joint *C*

Fig. 9.50: For Example 9.15

Solution : As the gable frame as well as the applied loading is symmetrical, the displacements at Joints *B* and *D* are the same, but in opposite directions. For this case, moment distribution operations will be carried on for two cases.

Case 1. The gable frame subjected to applied loads and provided with lateral supports at B and D to prevent sway as shown in Fig. 9.50(b).

Case 2. The gable frame subjected to equal and opposite arbitrary sway at B and D as shown in Fig. 9.50(c).

Stiffness :

$$K_{AB} = K_{BA} = \frac{4EI}{4} = EI$$

$$K_{BC} = K_{CB} = \frac{4E(1.5I)}{10.77} = 0.557EI$$

$$K_{CD} = K_{DC} = 0.557EI$$

$$K_{DE} = K_{ED} = \frac{4EI}{4} = EI.$$

Relative stiffness :

$$K_{AB} = K_{BA} = K_{DE} = K_{ED} = 1.0$$

$$K_{BC} = K_{CB} = K_{CD} = K_{DC} = 0.557.$$

Distribution factors :

Joint A: $DF_{AB} = 0$

Joint B: $DF_{BA} = \dfrac{1}{1 + 0.557} = 0.642$

$$DF_{BC} = \frac{0.557}{1 + 0.557} = 0.358$$

Joint C: $DF_{CB} = \dfrac{0.557}{0.557 + 0.557} = 0.5$

$$DF_{CD} = 0.5$$

Joint D: $DF_{DC} = \dfrac{0.557}{1 + 0.557} = 0.358$

$$DF_{DE} = \frac{1}{1 + 0.557} = 0.642$$

Joint D: $DF_{ED} = 0.$

Fixed-end moments :

(i) For Case 1:

$$M_{FBC} = +\frac{wL^2}{12} = +\frac{(60)(10)^2}{12} = +500 \text{ kNm}$$

$$M_{FCB} = -500 \text{ kNm}$$

$$M_{FCD} = +500 \text{ kNm}$$

$$M_{FDC} = -500 \text{ kNm}.$$

(ii) For Case 2:

For an imaginary sway at joints B and D as shown in Fig. 9.50(c) the relative lateral displacements of members BC and CD are obtained from Fig. 9.50(d) as

$$\Delta_{BC} = \Delta_{CD} = \frac{2\Delta}{\sin 2\alpha} \cdot \sin(90^\circ - \alpha),$$

where $\alpha = \tan^{-1}\frac{4}{10} = 21.8^\circ.$

Hence $\Delta_{BC} = \Delta_{CD} = \dfrac{2\Delta}{\sin 43.6^\circ} \cdot \sin 68.2^\circ = 2.693\Delta$

$$M_{FAB} = M_{FBA} = -\frac{6EI\Delta}{(4)^2} = -\frac{3}{8}EI\Delta$$

$$M_{FBC} = M_{FCB} = +\frac{6E(1.5I)\Delta_{BC}}{(L_{BC})^2}$$

$$= +\frac{6E(1.5I)(2.693\Delta)}{(10.77)^2} = +0.209EI\Delta$$

$$M_{FCD} = M_{FDC} = -\frac{6E(1.5I)(2.693\Delta)}{(10.77)^2} = -0.209EI\Delta$$

$$M_{FDE} = M_{FED} = +\frac{6EI\Delta}{(4)^2} = +\frac{3}{8}EI\Delta.$$

Putting $EI\Delta = 1000$ kNm^3,

$$M_{FAB} = M_{FBA} = -375 \text{ kNm}$$

$$M_{FBC} = M_{FCB} = +209 \text{ kNm}$$

$$M_{FCD} = M_{FDC} = -209 \text{ kNm}$$

$$M_{FDE} = M_{FED} = +375 \text{ kNm}.$$

Moment distribution operation for Case 1 and Case 2 are shown in Tables 9.28 and 9.29 respectively.

Table 9.28. Moment Distribution for Case 1 for Example 9.15

Joint	*A*	*B*		*C*		*D*		*E*
Member	*AB*	*BA*	*BC*	*CB*	*CD*	*DC*	*DE*	*ED*
Rel. Stiffness	1.0	1.0	0.557	0.557	0.557	0.557	1.0	1.0
D.F.	0	0.642	0.358	0.5	0.5	0.358	0.642	0
C.O.F.	←+0.50—	—	←+0.50—	→	←+0.50—	→	←+0.50—	→
F.E.M.			+500.0	−500.0	+500.0	+500.0		
Bal.		−321.0	−179.0	0	0	+179.0	+321.0	
C.O.	−160.5		0	−89.5	+89.5	0		+160.5
Bal.	0	0	0	0	0	0	0	0
Final End Moment	−160.5	−321.0	+321.0	−589.5	+589.5	−321.0	+321.0	+160.5

Table 9.29. Moment Distribution for Case 2 for Example 9.15

Joint	A	B		C		D		E
Member	AB	BA	BC	CB	CD	DC	DE	ED
Rel. Stiffness	1.0	1.0	0.557	0.557	0.557	0.557	1.0	1.0
D.F.	0	0.642	0.358	0.50	0.50	0.358	0.642	0
C.O.F.	←+0.50→		←+0.50→		←+0.50→		0.50→	
F.E.M.	−375.0	−375.0	+209.0	+209.0	−209.0	−209.0	+375.0	+375.0
Bal.		+106.6	+59.4	0	0	−59.4	−106.6	
C.O.	+53.3		0	+29.7	−29.7	0		−53.3
Bal.		0	0	0	0	0	0	
Final End Moment	−321.7	−268.4	+268.4	+238.7	−238.7	−268.4	+268.4	+321.7

Determination of lateral forces at B and D :

From symmetry, $F_{01} = F_{02}$ in case 1

and $F_1' = F_2'$ in case 2.

The force F_{01} in case 1 can be obtained from consideration of equilibrium of portions AB and ABC as shown in Figs. 9.51(a) and 9.51(b).

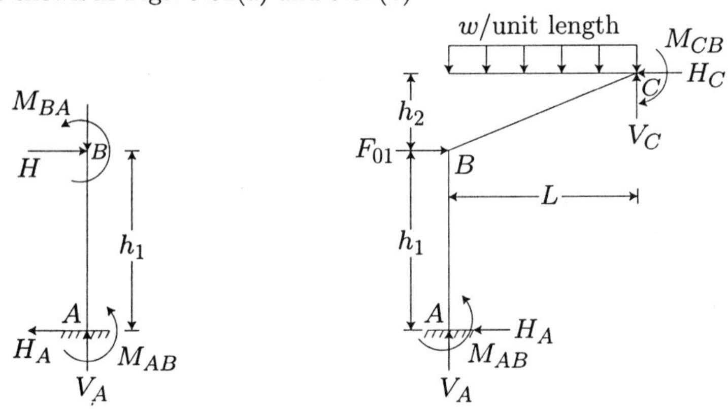

(a) Free body diagram of *AB* **(b) Free body diagram of *ABC***

Fig. 9.51: For Example 9.15

From Fig. 9.51(a), $H_A = \dfrac{M_{AB} + M_{BA}}{h_1}$ (i)

From Fig. 9.51(b), using condition $\sum M_C = 0$,

$$F_{01} = \frac{1}{h_2}\left[H_A(h_1 + h_2) + V_A(L) - M_{AB} - M_{CB} - \frac{wL^2}{2}\right]. \quad \text{(ii)}$$

In case I, Fig. 9.50(b), from symmetry, $V_A = \dfrac{(60)(20)}{2} = 600$ kN.

From Eqn. (i), $H_A = \dfrac{-160.5 - 321.0}{4} = -120.375.$

Hence, From Eq. (ii),

$$F_{01} = \frac{1}{4}\left[(-120.375)(4+4) + (600)(10) - (-160.5)(+589.5) - \frac{(60)(10)^2}{2}\right]$$

$$= 696.75 \text{ kN}.$$

The force F_1' in case 2 also can be obtained from Eqns. (i) and (ii), using values of moments from Table 9.29 and taking applied loads as non-existent. In case 2 from symmetry,

$$V_1 = 0$$

$$H_A = \frac{-321.7 - 268.4}{4} = -147.525 \text{ kN},$$

whence $F_1' = \frac{1}{4}[(-147.525)(4+4) + 321.7 - 238.7] = -274.3$ kN.

Determination of Ratio :

Assuming that the actual sway is X_1 time the arbitrary away considered in Case 2 and considering that net force at B is zero,

$$F_{01} + XF_1' = 0, \quad \text{whence} \quad X = -\frac{F_{01}}{F_1'}.$$

In this case, $X = -\dfrac{696.75}{-274.3} = +2.54$.

Net End Moments :

Net end moments are obtained by adding X_1 times the values of Case 2 to those of Case 1 as shown in Table 9.30.

Table 9.30. Net End Moments of Example 9.15

Member	AB	BA	BC	CB	CD	DC	DE	ED
Moment from Case 1, kN-m	−160.5	−321.0	+321.0	−589.5	+589.5	−321.0	+321.0	+160.5
X_1 times moment from Case 2, kN-m	−817.2	−681.8	+681.8	+606.3	−606.3	−681.8	+681.8	+817.2
Net End Moments kN-m	−977.7	−1002.8	+1002.8	+16.8	−16.8	−1002.8	+1002.8	−977.7

The bending moment diagram is drawn on tension side in Fig. 9.52.

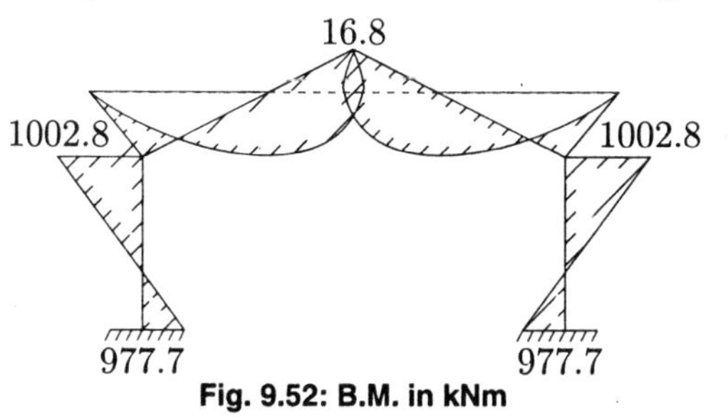

Fig. 9.52: B.M. in kNm

Example 9.16. *Analyse the Gable frame shown in Fig. 9.58(a) using moment distribution.*

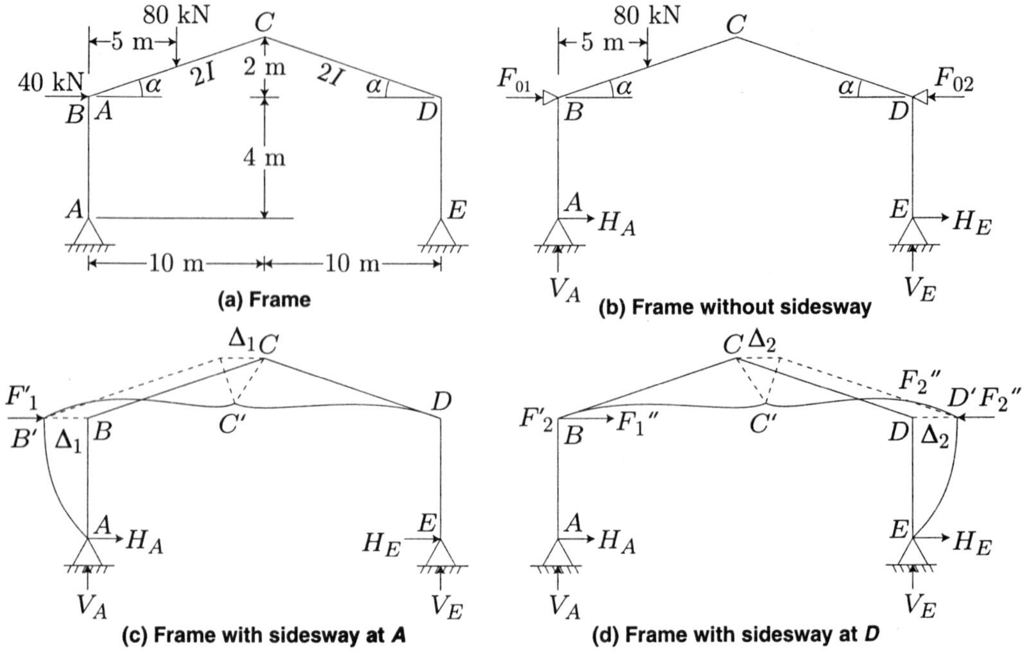

Fig. 9.53: For Example 9.16

Solution : The three cases of moment distribution operations for this problem are :

Case 1. Given gable frame with applied loads and provided with lateral supports at B and D to prevent sway as shown in Fig. 9.53(b).

Case 2. Given gable frame with arbitrary sway Δ_1 at B and provided with lateral support at D to prevent sway at D as shown in Fig. 9.53(c), and

Case 3. Given gable frame with arbitrary sway Δ_2 at D and provided with lateral support at B to prevent sway at B as shown in Fig. 9.53(d).

Assuming that the values of actual sway at B and D are sum of X_1 times that in Case 2 and X_2 times that in Case 3,

$$F_{01} + X_1 F_1' + X_2 F_1'' = 40 \qquad \text{(a)}$$

and $$F_{02} + X_1 F_2' + X_2 F_2'' = 0. \qquad \text{(b)}$$

Stiffness :

$$K_{AB} = \frac{4EI}{4} = EI$$

$$K_{BA} = \frac{3EI}{4} = 0.75EI$$

$$K_{BC} = K_{CB} = K_{CD} = K_{DC} = \frac{4E(2I)}{\sqrt{10^2 + 2^2}} = 0.7845EI$$

$$K_{DE} = \frac{3EI}{4} = 0.75EI$$

$$K_{ED} = \frac{4EI}{4} = EI.$$

Distribution Factors :

Joint A, $DF_{AB} = 1.0$

Joint B, $DF_{BA} = \dfrac{0.75}{0.75 + 0.7845} = 0.489$

$$DF_{BC} = \frac{0.7845}{0.75 + 0.7845} = 0.511$$

Joint C, $DF_{CB} = \dfrac{0.7845}{(0.7845)(2)} = 0.5$

$$DF_{CD} = 0.5,$$

Joint D, $DF_{DC} = \dfrac{0.7845}{0.75 + 0.7845} = 0.511$

$$DF_{DE} = \frac{0.75}{0.75 + 0.7845} = 0.489$$

Joint E, $DF_{ED} = 1.0.$

Fixed-end Moments in Case 1,

$$M_{FBC} = +\frac{PL}{8} = \frac{(80)(10)}{8} = +100 \text{ kNm}$$

$$M_{FCB} = -100 \text{ kNm}.$$

Fixed-end Moments for Case 2,

$$M_{FBA} = -\frac{3EI\Delta_1}{(4)^2} = -\frac{3}{16}EI\Delta_1$$

$$M_{FBC} = M_{FBC} = +\frac{6E(2I)\sin(90^\circ - \alpha)\Delta_1}{(10^2 + 2^2)\sin(\alpha + \alpha)}$$

$$= +0.294EI\Delta_1, \text{ with } \alpha = \tan^{-1}\frac{2}{10}$$

$$M_{FCD} = M_{FDC} = -\frac{6E(2I)\sin(90^\circ - \alpha)\Delta_1}{(10^2 + 2^2)\sin(\alpha + \alpha)} = -0.294E\Delta_1.$$

Putting $EI\Delta_1 = 1000 \text{ kNm}^3$,

$$M_{FBA} = -187.5 \text{ kNm}$$

$$M_{FBC} = M_{FCB} = +294 \text{ kNm}$$

$$M_{FCD} = M_{FDC} = -294 \text{ kNm}.$$

Fixed-end Moments for Case 3,

Putting $EI\Delta_2 = 1000 \text{ kN-m}^3$,

$$M_{FBC} = M_{FCB} = +294 \text{ kNm}$$

$$M_{FCD} = M_{FDC} = -294 \text{ kNm}$$

$$M_{FDE} = +187.5 \text{ kNm}.$$

Moment distribution operations for three cases are shown in Tables 9.31, 9.32 and 9.33 respectively.

Table 9.31. Moment Distribution for Case 1 of Example 9.16

Joint	A	B		C		D		E
Member	AB	BA	BC	CB	CD	DC	DE	ED
Rel. Stiffness	1.0	0.75	0.7845	0.7845	0.7845	0.7848	0.75	1.0
D.F.	1.0	0.489	0.511	0.5	0.5	0.511	0.489	1.0
C.O.F.			← +0.50	→	← +0.50	→		
FEM			+100.0	−100.0				
Bal.		−49.0	−51.0	+50.0	+50.0			
C.O.			+25.0	−25.5	0	+25.0		
Bal.		−12.2	−12.8	+12.8	+12.7	−12.8	−12.2	
C.O.			+6.4	−6.4	−6.4	+6.4		
Bal.		−3.1	−3.3	+6.4	+6.4	−3.3	−3.1	
C.O.			+3.2	−1.6	−1.6	+3.2		
Bal.		−1.5	−1.7	+1.6	+1.6	−1.7	−1.5	
C.O.			+0.8	−0.8	−0.8	+0.8		
Bal.		−0.4	−0.4	+0.8	+0.8	−0.4	−0.4	
Final end moment	0	−66.2	+66.2	−62.7	+62.7	+17.2	−17.2	0

Table 9.32. Moment Distribution for Case 2 of Example 9.16

Joint	A	B		C		D		E
Member	AB	BA	BC	CB	CD	DC	DE	ED
Rel. Stiffness	1.0	0.75	0.7845	0.7845	0.7845	0.7845	0.75	1.0
D.F.	1.0	0.489	0.511	0.5	0.5	0.511	0.489	1.0
C.O.F.			← +0.5	→	←	+0.5 →		
F.E.M.		−187.5	+294.0	+294.0	−294.0	−294.0		
Bal.		−52.1	−54.4	0	0	+150.2	+143.8	
C.O.			0	−27.2	+75.1	0		
Bal.		0	0	−24.0	−23.9	0	0	
C.O.			−12.0	0	0	−12.0		
Bal.		+5.9	+6.1	0	0	+6.1	+5.9	
C.O.			0	+3.1	+3.1	0		
Bal.		0	0	−3.1	−3.1	0	0	
C.O.			−1.6	0	0	−1.6		
Bal.		+0.8	+0.8	0	0	+0.8	+0.8	
Final end moments	0	−232.9	+232.9	+242.8	−242.8	−150.5	+150.5	0

Table 9.33. Moment Distribution for Case 3 of Example 9.16

Joint	A	B		C		D		E
Member	AB	BA	BC	CB	CD	DC	DE	ED
Rel. Stiffness	1.0	0.75	0.7845	0.7845	0.7845	0.7845	0.75	1.0
D.F.	1.0	0.489	0.511	0.5	0.5	0.511	0.489	1.0
C.O.F.			← +0.5	→	← +0.5	→		
F.E.M.			+294.0	+294.0	−294.0	−294.0	+187.5	
Bal.		−143.8	−150.2	0	0	+54.4	+52.1	
C.O.			0	−75.1	+27.2	0		
Bal.		0	0	+23.9	+24.0			
C.O.			+12.0	0	0	+12.0		
Bal.		−5.9	−6.1	0	0	−6.1	−5.9	
C.O.			0	−3.1	−3.1	0		
Bal.		0	0	+3.1	+3.1	0	0	
C.O.			+1.6	0	0	+1.6		
Bal.		−0.8	−0.8	0	0	−0.8	−0.8	
Final end moments	0	−150.5	+150.5	+242.8	−242.8	−232.9	+232.9	0

Determination of Lateral Forces at B and D : The lateral forces F_{01} and F_{02} at B and D respectively for Case 1 (no sway) can be obtained from consideration of equilibrium of portions AB, ABC, CDE and DE as shown in Figs. 9.54(a), 9.54(b), 9.54(c) and 9.54(d) in addition to that of the entire frame. From Fig. 9.54(a),

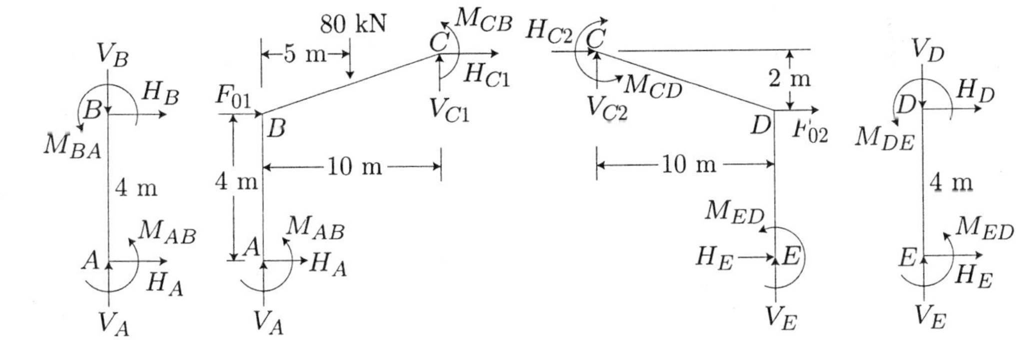

(a) Free body diagram for *AB* **(b) Free body diagram for *ABC*** **(c) Free body diagram for *EDC*** **(d) Free body diagram for *ED***

Fig. 9.54: For Example 9.16

$$H_A = -\frac{M_{AB} + M_{BA}}{4} = -\frac{0 - 66.2}{4} = 16.55 \text{ kN.}$$

From Fig. 9.54(d), $H_E = -\dfrac{M_{DE} + M_{ED}}{4} = -\dfrac{-17.2 + 0}{4} = +4.3$ kN.

From Fig. 9.54(b) using condition, $\sum M_C = 0$,

$$F_{01}(2) + H_A(6) - V_A(10) + (80)(5) + M_{AB} + M_{CB} = 0,$$

whence $F_{01}(2) + (16.55)(6) - V_A(10) + (80)(5) + 0 + (-62.7) = 0$

or, $F_{01} - 5V_A + 218.3 = 0.$ (i)

From Fig. 9.54(c), using condition $\sum M_C = 0$,

$$F_{02}(2) + H_E(6) + V_E(10) + M_{ED} + M_{CD} = 0,$$

whence $2F_{02} + (4.3)(6) + 10V_E + 0 + (62.7) = 0$

or, $F_{02} + 5V_E + 44.25 = 0.$ (ii)

Considering equilibrium of entire gable frame for Case 1 and using conditions,

$$\sum V = 0 \text{ and } \sum H = 0,$$

$$V_A + V_E = 80 \quad \text{(iii)}$$

and $F_{01} + F_{02} + H_A + H_E = 0.$ (iv)

From Eq. (iv), $F_{01} + F_{02} + 16.55 + 4.30 = 0$,

whence $F_{01} + F_{02} + 20.85 = 0.$ (v)

Subtracting Eq. (ii) from Eq. (i) and eliminating V_A and V_E from Eq. (iii),

$$F_{01} - F_{02} = 225.95. \quad \text{(vi)}$$

Solving Eqs. (v) and (vi),

$$F_{01} = 102.55 \text{ kN} \longrightarrow$$

$$F_{02} = -123.40 \text{ kN} \longleftarrow.$$

In the same procedure as above and using values of moments from Tables 9.32 and 9.33 respectively, the lateral forces F_1' and F_2' for case 2 and F_1'' and F_2'' for Case 3 are obtained.

For Case 2 :

$$H_A = -\frac{0 - 232.9}{4} = +58.225 \text{ kN}$$

$$H_E = -\frac{150.5 + 0}{4} = -37.625 \text{ kN}.$$

From $\sum H = 0$ for the entire frame,

$$F_1' + F_2' + H_A + H_E = 0$$

i.e., $F_1' + F_2' + 58.225 - 37.625 = 0$,

whence $F_1' + F_2' + 20.6 = 0.$ (vii)

From $\sum V = 0$ for the entire frame,

$$V_A + V_E = 0. \quad \text{(viii)}$$

Using $\sum M_c = 0$ on the portion ABC, taken as free body,

$$F_1'(2) + H_A(6) - V_A(10) + M_{AB} + M_{CB} = 0$$

i.e., $2F_1' + (58.225)(6) - 10V_A + 0 + 242.8 = 0$,

whence $F_1' - 5V_A + 296.075 = 0.$ (ix)

Using $\sum M_C = 0$ on the portion CDE taken as free body,

$$F_2'(2) + H_E(6) + V_E(10) + M_{ED} + M_{CD} = 0$$

i.e., $2F_2' + (-37.625)(6) + (-V_A)(10) + 0 + (-242.8) = 0,$

whence $F_2' - 5V_A - 234.275 = 0.$ (x)

Solving Eqs. (vii), (ix) and (x),

$$\left.\begin{aligned} F_1' &= -275.475 \text{ kN} \\ F_2' &= +254.875 \text{ kN.} \end{aligned}\right] \qquad \text{(d)}$$

For Case 3 :

$$H_A = -\frac{0 - 150.5}{4} = +37.625 \text{ kN}$$

$$H_E = -\frac{232.9 + 0}{4} = -58.225 \text{ kN.}$$

From $\sum H = 0$ for the entire gable frame,

$$F_1'' + F_2'' + H_A + H_E = 0$$

i.e., $F_1'' + F_2'' + 37.625 - 58.225 = 0,$

whence $F_1'' + F_2'' - 20.6 = 0.$ (xi)

From $\sum V = 0$ for the entire gable frame,

$$V_A + V_E = 0. \qquad \text{(xii)}$$

Considering equilibrium of portion ABC and using $\sum M_C = 0,$

$$F_1''(2) + H_A(6) - V_A(10) + M_{AB} + M_{CB} = 0$$

i.e., $F_1''(2) + (37.625)(6) - V_A(10) + 0 + 242.8 = 0,$

whence $F_1'' - 5V_A + 234.275 = 0.$ (xiii)

Considering equilibrium of portion CDE and using condition $\sum M_C = 0,$

$$F_2''(2) + H_E(6) + V_E(10) + M_{ED} + M_{CD} = 0$$

i.e., $F_2''(2) + (-58.225)(6) + (-V_A)(10) + 0 + (-242.8) = 0,$

whence $F_2'' - 5V_A - 296.075 = 0.$ (xiv)

Solving Eqs. (xi), (xiii) and (xiv),

$$\left.\begin{aligned} F_1'' &= -254.875 \text{ kN} \\ F_2'' &= +275.475 \text{ kN.} \end{aligned}\right] \qquad \text{(e)}$$

Determination of Ratios :

Using the values from Eqs. (c), (d) and (e) in Eqs. (a) and (b),

$$102.55 + X_1(-275.475) + X_2(-254.875) = 0$$

and $$-123.40 + X_1(254.875) + X_2(275.475) = 0.$$

Solving, $X_1 = -0.293$ and $X_2 = +0.7191.$

Net End Moments : The net end moments are obtained by adding X_1 times the moments of Case 2 and X_2 times the moments of Case 3 to those of Case 1 as shown in Table 9.34.

Table 9.34. Net End Moments for Example 9.16

Member	AB	BA	BC	CB	CD	DC	DE	ED
Moment from Case 1, kN-m	0	−66.2	+66.2	−62.7	+62.7	+17.2	−17.2	0
X_1 times moment from Case 2, kN-m	0	+68.2	−68.2	−71.1	+71.1	+44.1	−44.1	0
X_2 times moment from Case 3, kN-m	0	−108.2	+108.2	+174.6	−174.6	−167.5	+167.5	0
Net End Moments kN-m	0	−106.2	+106.2	+40.8	−40.8	−106.2	+106.2	0

Fig. 9.55 shows B.M. diagram drawn on tension side for the gable frame.

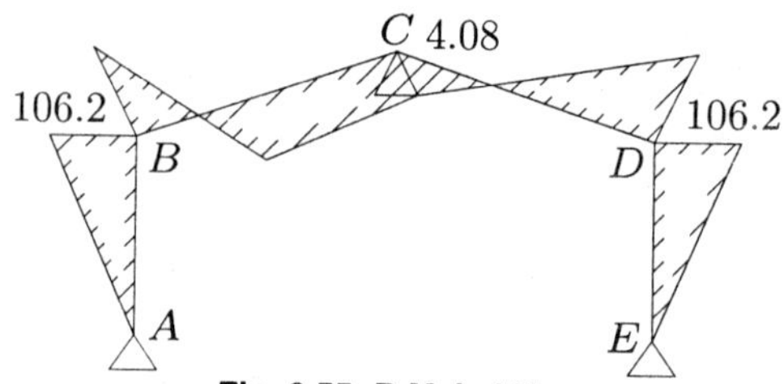

Fig. 9.55: B.M. in kNm

9.15 Symmetrical and Antisymmetrical Cases

The moment distribution analysis of a symmetrical structure can be simplified if the loading in the structure is either symmetrical or antisymmetrical. In such cases, it is convenient in many instances to distribute moments for only one-half of the structure. This is accomplished by making use of the stiffness factors for beams with symmetrical or anti-symmetrical end moments.

A structure may be symmetrical both in its geometry and construction about a support, as in case of a continuous beam or multi-storeyed buildings with an even number of span or bays. In such cases with symmetrical loading, these will be no joint rotation one way or the other in the beams passing over or through the central supports. In such cases, it is necessary to consider only one half of the structure with the ends of members at the centre line being taken as fixed in position and direction.

Another type of symmetry occurs when a symmetrical structure has its central line equally between supports, as in case of a continuous beam or a multi-storeyed frame with an odd number of spans or bays. In such cases with symmetrical loading, the adjacent joint on either side of the centre line and at the same level undergoes equal and opposite angular displacement. It is then necessary to analyse one half of the structure, if a modified stiffness, considering the symmetry and nature of angular displacements can be deduced for the central members.

Fig. 9.56 shows a symmetrical beam in which moments at ends A and B, i.e., M_{AB} and M_{BA} are equal in magnitude, the rotation at each ends being the same in magnitude but of different direction.

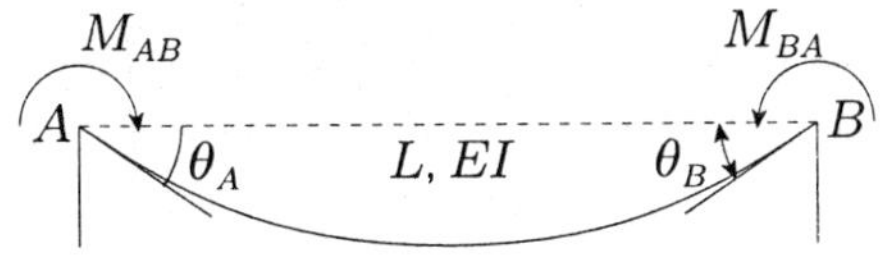

Fig. 9.56: Symmetrical case

For the case, the rotation,

$$\theta_A = \frac{M_{AB}L}{3EI} + \frac{M_{BA}L}{6EI} = \frac{M_{AB}L}{2EI},$$

$$\text{whence stiffness} = \frac{M_{AB}}{\theta_A} = \frac{2EI}{L}. \tag{9.19}$$

The carry over factor for the case, i.e., C.O.F. is zero.

Antisymmetrical behaviour occurs with a structure symmetrical in geometry and construction, but with pattern of loading which produces equal joint rotation of the same sign and magnitude in corresponding parts of the structure. Symmetrical loading on a symmetrical structure produces symmetrical behaviour and reversal of the sense of loading on one half of the structure results in an antisymmetrical behaviour.

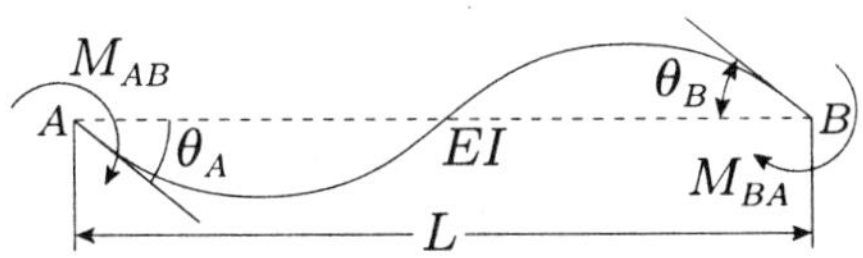

Fig. 9.57: Antisymmetrical case

Fig. 9.57 shows a symmetrical beam in which moments at ends A and B are equal in magnitude but different in sign from BM point of view. The rotations θ_A and θ_B at ends are the same in magnitude and sign. In this case,

$$\theta_A = \frac{M_{AB}L}{3EI} - \frac{M_{BA}L}{6EI} = \frac{M_{AB}L}{6EI},$$

$$\text{whence stiffness} = \frac{M_{AB}}{\theta_A} = \frac{6EI}{L}. \tag{9.20}$$

The carry over factor for the case, i.e., C.O.F. is zero.

A symmetrical structure may be solved in separate stages for component loadings, as the loading can be replaced by symmetrical and antisymmetrical loadings. Thus it is possible to solve such structure in two stages.

Example 9.17. *Analyse the symmetrical beam in Fig. 9.58 by moment distribution procedure using advantage of symmetry.*

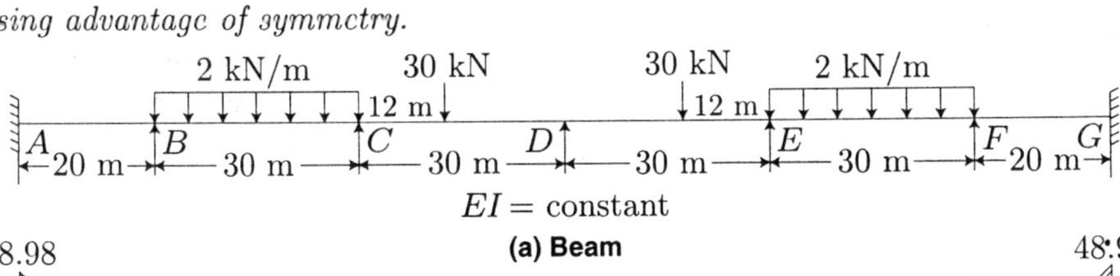

(a) Beam

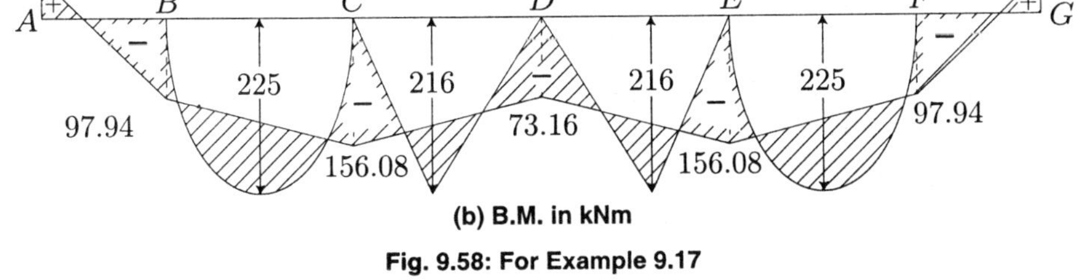

(b) B.M. in kNm

Fig. 9.58: For Example 9.17

Solution : This is a beam symmetrical in both geometry and loadings. The symmetry is about the support D. Hence only one half of the beam is to be analysed considering support D as fixed.

The moment distribution is shown in Table 9.35.

Table 9.35. Moment Distribution for Example 9.17

Joint	A	B		C		D
Member	AB	BA	BC	CB	CD	DC
Rel. K	$\frac{4}{20}$	$\frac{4}{20}$	$\frac{4}{30}$	$\frac{4}{30}$	$\frac{4}{30}$	$\frac{4}{30}$
D.F.	0	0.60	0.40	0.50	0.50	0
C.O.F.	←	0.50 —	← 0.50	→	— 0.50	→
F.E.M.	0	0	+1.50	−1.50	+129.6	−86.4
Bal.	0	−90	−60	+10.20	+10.20	0
C.O.M.	−45	0	+5.10	−30	0	+5.10
Bal.	0	−3.06	−2.04	+15	+15	0
C.O.M.	−1.53	0	+7.50	−1.02	0	+7.50
Bal.	0	−4.50	−3.00	+0.51	+0.51	0
C.O.M.	−2.25	0	+0.25	−1.50	0	+0.25
Bal.	0	−0.15	−0.10	+0.75	+0.75	0
C.O.M.	−0.08	0	+0.38	−0.05	0	+0.38
Bal.	0	−0.23	−0.15	+0.03	+0.32	0
C.O.M.	−0.12	0	0	0	0	+0.01
Joint moments (Jt. sign convention)	−48.98	−97.94	+97.94	−156.08	+156.08	−73.16

B.M. Diagram for the beam is shown in Fig. 9.58(b) obtained by superimposing the simple beam bending moment diagram over the support moment diagram.

Example 9.18. *Analyse the symmetrical frame shown in Fig. 9.59(a), using moment distribution technique.*

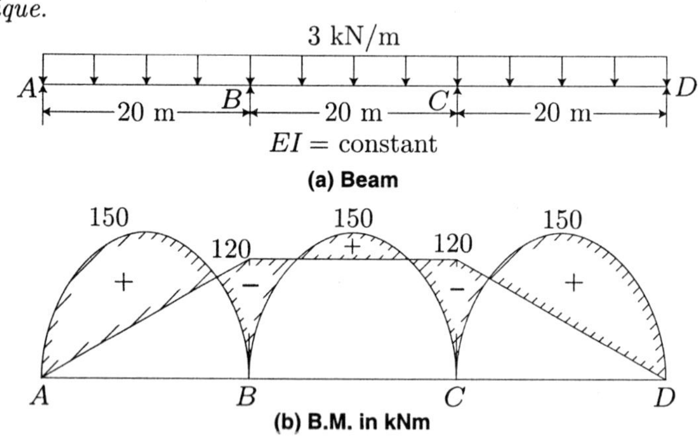

Fig. 9.59: For Example 9.18

Solution : The continuous beam is symmetrical about the centre of beam BC. Taking advantage of symmetry, only one half has to be analysed. Stiffness of member BC is $\frac{2EI}{20}$ and C.O.F. = 0. The moment distribution is illustrated in Table 9.36. B.M. diagram is shown in Fig. 9.59(b).

Table 9.36. Moment Distribution for Example 9.18

Joint	A	B		C
Member	AB	BA	BC	CB
Rel. K		$\frac{3}{20}$	$\frac{2}{20}$	$\frac{2}{20}$
D.F.		$\frac{3}{5}$	$\frac{2}{5}$	
C.O.F.	← 0.50 →		← 0 →	
F.E.M.	0	−150	+100	−100
Bal.	0	+30	+20	−20
Total moment (Jt. sign convention)	0	−120	+120	−12

Moment distribution with the total beam would require four cycles.

Example 9.19. *Analyse the symmetrical frame shown in Fig. 9.60 by moment distribution procedure.*

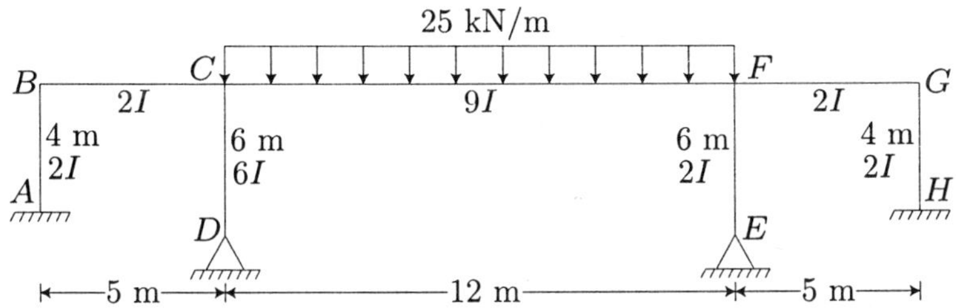

Fig. 9.60: For Example 9.19

Solution : The frame is symmetrical about centre of horizontal member CF. Hence, if it is analysed as a symmetrical case, stiffness of member $CF = \frac{2E(I)}{12} = \frac{3EI}{2}$ and carry over factor for CF is zero.

The moment distribution is shown in table 9.37.

Table 9.37. Moment Distribution for Example 9.19

Joint	A	B		C			D
Member	AB	BA	BC	CB	CF	CD	DC
Rel. K		$\frac{8EI}{4}$	$\frac{8EI}{5}$	$\frac{8EI}{5}$	$\frac{3EI}{2}$	$\frac{18EI}{6}$	
D.F.		0.56	0.44	0.26	0.25	0.49	
C.O.F.	← 0.50		← 0.50 →			← 0 →	
F.E.M.	0	0	0	0	+300	0	0
Bal.	0	0	0	−78	−75	−147	0
C.O.M.	0	0	−39	0	0	0	0
Bal.	0	+21.84	+17.16	0	0	0	0
C.O.M.	+10.92	0	0	+8.58	0	0	0
Bal.	0	0	0	−2.23	−2.14	−4.11	0
C.O.M.	0	0	−1.12	0	0	0	0
Bal.	0	+0.63	+0.49	0	0	0	0
C.O.M.	+0.32	0	0	+0.25	0	0	0
Bal.	0	0	0	−0.07	−0.06	−0.12	0
Total moment (Jt. sign convention)	+11.24	+22.47	−22.47	−71.47	+222.8	−151.23	0

Example 9.20. *Analyse the silo structure shown in Fig. 9.61(a) using moment distribution.*

Solution : The structure has two axes of symmetry from point of view of geometry and loading. It is, therefore, possible to analyse only one quarter of the frame, making use of both axes of symmetry. All structural members cut by the axis of symmetry have a stiffness of $\frac{2EI}{L}$ and other members have a stiffness of $\frac{4EI}{L}$. The moment distribution is illustrated in Table 9.38.

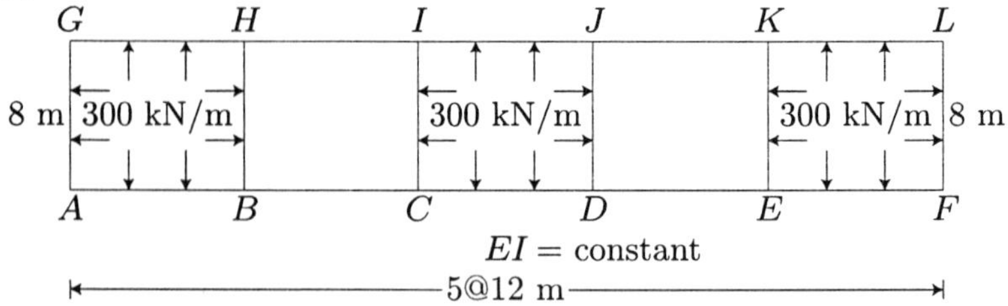

(a) Beam

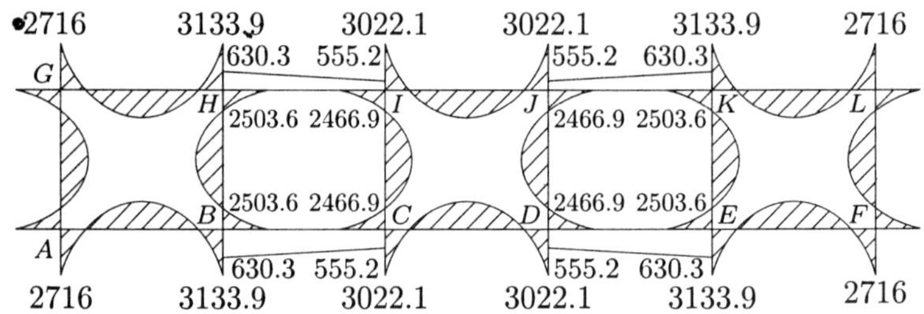

(b) B.M. in kNm

Fig. 9.61: For Example 9.20

Fig. 9.61(b) shows B.M. diagram for the frame drawn on compression side.

Example 9.21. *Analyse the symmetrical beam in Fig. 9.62, loaded antisymmetrically using moment distribution method.*

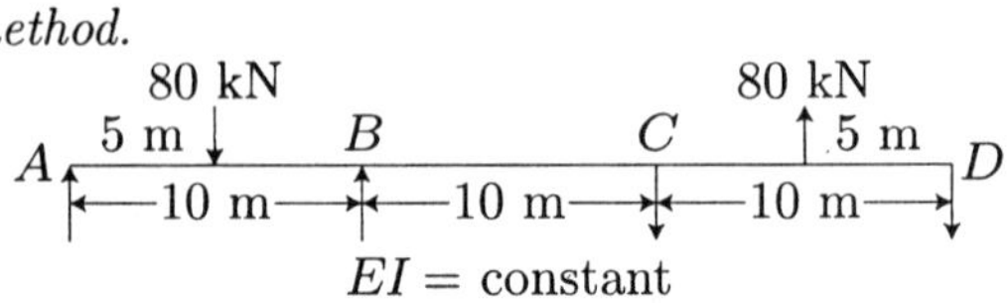

(a) Beam

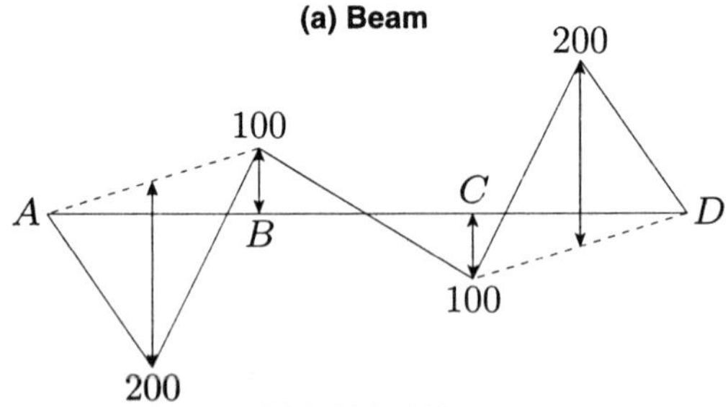

(b) B.M. in kNm

Fig. 9.62: For Example 9.20

Solution : In this case, the structure is antisymmetrical and stiffness of $BC = \frac{6EI}{10}$, stiffness of BA and CB being $\frac{3EI}{10}$. The moment distribution is illustrated in Table 9.39. Fig. 9.62(b) shown B.M. diagram drawn on tension side.

Table 9.38. Moment Distribution for Example 9.20

Joint	G		H			I		
Member	GA	GH	HG	HB	HI	IH	IC	IJ
K	$\frac{2EI}{8}$	$\frac{4EI}{12}$	$\frac{4EI}{12}$	$\frac{2EI}{8}$	$\frac{4EI}{12}$	$\frac{4EI}{12}$	$\frac{2EI}{8}$	$\frac{2EI}{12}$
D.F.	0.429	0.571	0.364	0.272	0.364	0.445	0.333	0.222
C.O.F.		← 0.5	⟶	0	←0.5	⟶	0	0
F.E.M.	+1600	−3600	+3600	−1600	0	0	+1600	−3600
Bal.	+858	+1142	−728	−544	−728	+890	+666	+444
C.O.M.	0	−364	+571	0	+445	−364	0	0
Bal.	+156	+208	−370	−276	−370	+162	+121	+81
C.O.M.	0	−185	+104	0	+81	−185	0	0
Bal.	+79	+106	−67	−51	−67	+82	+62	+41
C.O.M.	0	−33.5	+53	0	+41	−33.5	0	0
Bal.	+14.4	+19.1	−34.2	−25.6	−34.2	+14.9	+11.2	+7.4
C.O.M.	0	−17.1	+9.6	0	+7.5	−17.1	0	0
Bal.	+7.3	+9.8	−6.2	−4.7	−6.2	+7.6	+5.7	+3.8
C.O.M.	0	−3.1	+4.9	0	+3.8	−3.1	0	0
Bal.	+1.3	+1.8	−3.2	−2.3	−3.2	+1.4	+1.0	+0.7
Final End Moments, kN-m	+2716.0	−2716.0	+3133.9	−2503.6	−630.6	+555.2	+2466.9	−3022.1

Table 9.39. Moment Distribution for Example 9.21

Joint	A	B	
Member	AB	BA	BC
Rel. K		$\frac{3EI}{10}$	$\frac{6EI}{10}$
D.F.		$\frac{1}{3}$	$\frac{2}{3}$
C.O.F.	0	0	0
F.E.M., kNm	0	−150	0
Bal.		+50	+100
C.O.M.	0	0	0
Total moments in kNm (Joint sign convention)	0	−100	+100

Example 9.22. *Analyse the symmetrical frame in Fig. 9.63, for antisymmetrical pattern due to side sway.*

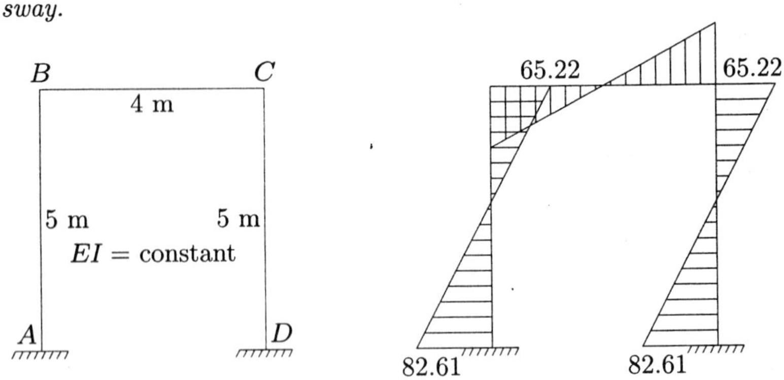

(a) Frame (b) B.M in kNm

Fig. 9.63: For Example 9.22

Solution : Due to antisymmetry stiffness for $BC = \frac{6EI}{4}$

Stiffness for BA and $CD = \frac{4EI}{5}$.

Half of the frame up to centre of BC is to be analysed by moment distributions. The procedure is shown in Table 9.40. B.M. diagram is shown in Fig. 9.63(b).

Table 9.40. Moment Distribution for Example 9.22

Member	AB	BA	BC
Rel. K		$\frac{5}{4}$	$\frac{6}{4}$
D.F.		$\frac{8}{23}$	$\frac{15}{23}$
C.O.F.	← 0.50 —		0
F.E.M.	+100	+100	0
Bal.	0	−34.78	−65.22
COM	−17.39	0	0
Total End Moments	+82.61	+65.22	−65.22

9.16 Superposition of Loading

In case a structure is symmetrical from consideration of geometry, it is advantageous to make use of the axis of symmetry. This can be done by dividing the loading into two parts— a symmetrical load and an antisymmetrical load, the algebraic sum of both the loadings being equal to the actual loading. Example of such division of loading are given in Fig. 9.64.

In the symmetrical loaded case, these is no sway as the frame is symmetrical in geometry. Analysis of half the frame can be done by taking advantage of the use of symmetry. In the antisymmetrically loaded case, the axis of symmetry of the frames can be used to reduce computation. Only one half of the frame has to be analysed, the functions in the other half being of the same magnitude and sign (considering joint sign convention).

Computations for antisymmetrical cases can be further reduced with use of Cantilever Moment Distribution method as discussed later in Section 9.17.

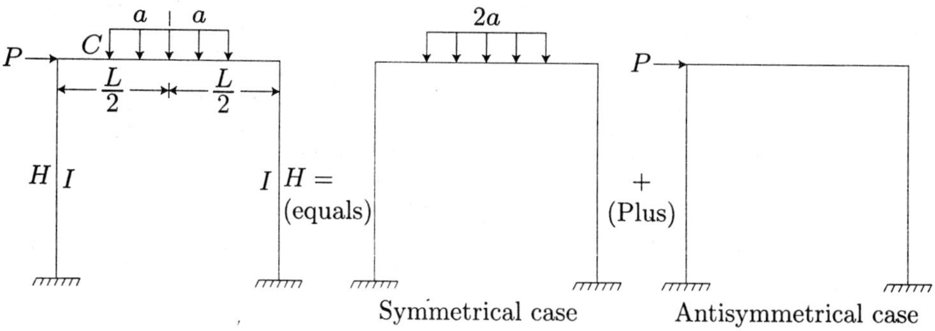

(a) Vertical and horizontal loading (unsymmetrical)

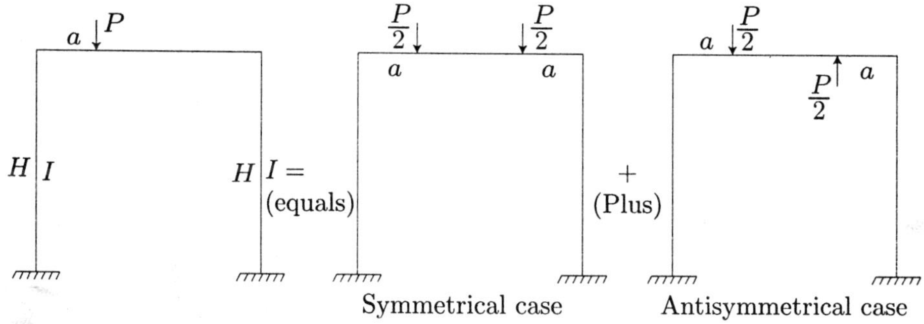

(b) Vertical load (unsymmetrical + antisymmetrical)

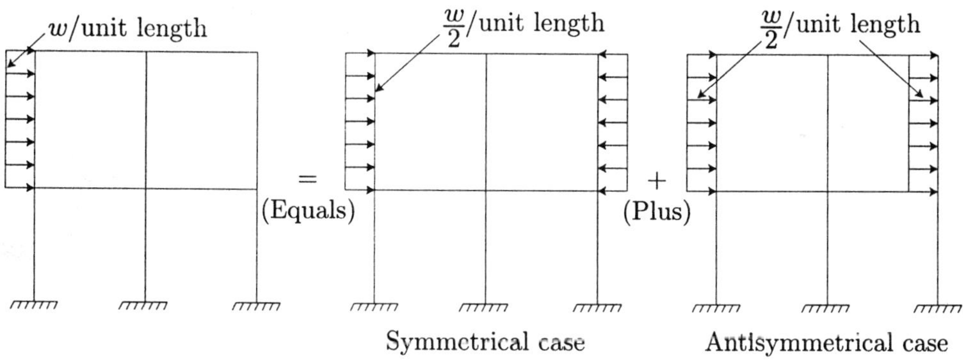

(c) Horizontal load (symmetrical + antisymmetrical)

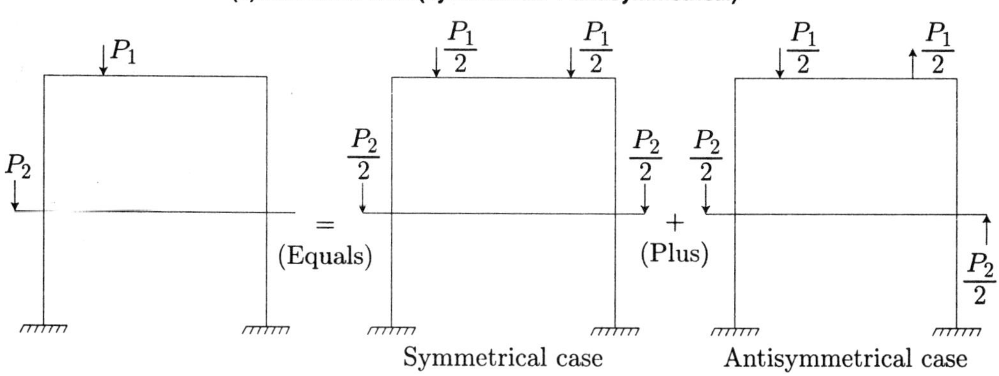

(d) Vertical load (symmetrical + antisymmetrical)

Fig. 9.64: Superposition of loading

9.17 Cantilever Moment Distribution

The earlier demonstration of the simpler method of Cantilever Moment Distribution was included in a Text book by Leuthers in the year 1946. The method was probably developed in the Continent between 1940 to 1942 and was introduced in U.K. in the year 1950 by Naylor. This method is a special case of Cross Moment Distribution and its use is limited to symmetrical structures supporting antisymmetric loads. The method is also known as No Shear Method or Naylor's method.

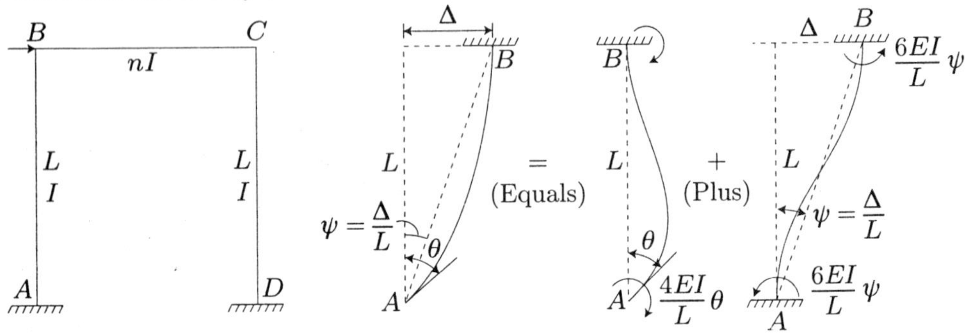

(a) Beam **(b) Stiffness of *AB*** **(c) Equivalence of (b)**

Fig. 9.65: Stiffness for No Shear

Fig. 9.65(a) shows a symmetrical structure antisymmetrically loaded. The member may have the deflected shape shown in Fig. 9.65(b), if end B is allowed to sway by Δ without rotation of joint, when a rotation θ occurs at end A due to application of moment there. The moment at the two ends A and B for the member can be obtained by summing up the moments in the member as in Fig. 9.65(c).

For no shear in AB, $M_A = M_B$ and from Fig. 9.65(c),

$$\frac{6EI}{L}\psi - \frac{4EI}{L}\theta = \frac{2EI}{L}\theta - \frac{6EI}{L}\psi, \quad \text{and} \quad \psi = \frac{\theta}{2}.$$

$$\text{Moment at } A \text{ and } B \text{ then} = \frac{\frac{4EI}{L} - \frac{2EI}{L}}{2}\theta = \frac{EI}{L}\theta.$$

$$\text{Stiffness for the member } AB = \frac{EI}{L} \text{ and C.O.F.} = -1.$$

The value of stiffness and C.O.F. can also be obtained from consideration of moment distribution. Since the frame is symmetrical, the load P will produce fixed end moments at BA and CD which are the same and hence equal moments will be distributed to members BA and CD if these joints are balanced simultaneously. This means that joints B and C will rotate through the same angles and member BA and CD will deform identically. Under these conditions neither member offers any support to the other and each one behaves as a cantilever with the upper end free. Hence the rotational stiffness of members BA and CD is that of a beam having no elastic support whatsoever, thus giving a rotational stiffness of $\frac{EI}{L}$. The carryover factor for such a member is similarly obtained as C.O.F. $= -1$.

If the far end is pinned both the stiffness and carry over factors are zero.

Another simplification arises due to symmetry of the structure and the antisymmetry of the loading. The stiffness of the member perpendicular to the axis of symmetry such as BC can be taken as $\frac{6EI}{L}$, where L is the length of the member. There is no carry over factor for such members and it is only necessary to balance moments for one half of the structure.

The procedure of solution for this type of structure involves the following steps :

(i) Stiffness of a member subject to translation, i.e., parallel to axis of symmetry is $\frac{EI}{L}$ and the carry over factor is -1, when the far end is fixed. For a member with far end hinged, these factors are both zero.

(ii) Stiffness of each member without translation, i.e., for members perpendicular to the symmetry axis are $\frac{6EI}{L}$ and the carry over is zero.

(iii) Fixed moments induced by sway are the fixed-end moments assuming the joints translate, but do not rotate and are determined in the same manner.

(iv) Moments are balanced for one half of the structure only.

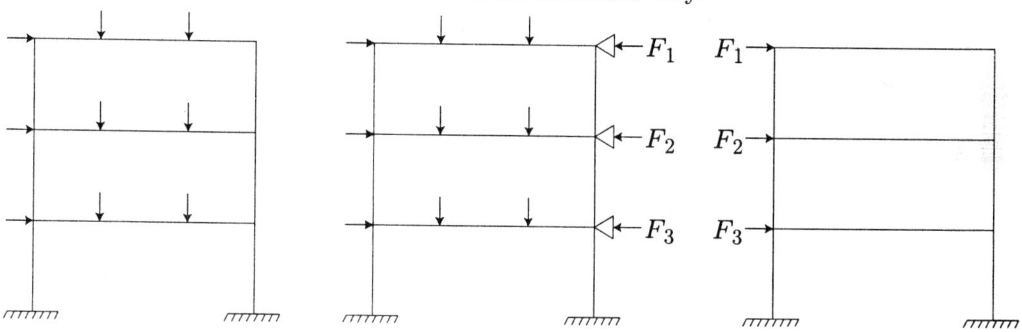

(a) Beam **(b) Frame with no sway** **(c) Frame with lateral load**

Fig. 9.66: Analysis for symmetrical loading

It is possible to take advantage of the cantilever moment distribution during one or more stages in the analysis of a structure with unsymmetrical loads. For example, in analysing the symmetrical three storeyed frame of Fig. 9.66(a), calculations can be performed in two parts. First the joints are prevented from translating and a standard moment distribution analysis is made [Fig. 9.66(b)] and the reactions F_1, F_2 and F_3 are determined. These forces are then reversed in direction and treated as loads acting in the same frame as shown in Fig. 9.66(c). The analysis of the frame in Fig. 9.66(c) can be done using cantilever moment distribution and the resulting moments are combined with the previous ones to give the moments in the structure of Fig. 9.66(a).

This technique requires two sets of moment distribution calculations as compared to the four sets required in analysis of structure with three degrees of freedom. The use of cantilever moment distribution also eliminates the need of solving simultaneous equations.

Similar advantages can be obtained by resolving the original loading into two sets of loads, one of which is symmetrical and the other antisymmetrical. The moments due to symmetrical as well as antisymmetrical loading can be found by moment distribution of one half of the structures. The moments in structure can be obtained by combining the moments obtained for the two cases.

Example 9.23. *Analyse the rectangular frame shown in Fig. 9.67(a) using cantilever moment distribution.*

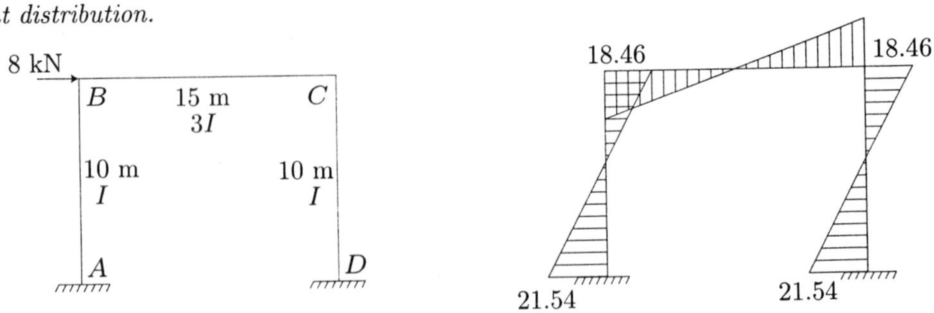

(a) Beam (b) B.M in kNm

Fig. 9.67: For Example 9.23

Solution : Stiffness of member $BA = \dfrac{EI}{10}$

Stiffness of member $BC = \dfrac{18EI}{15}$

C.O.F. for $BA = -1.0$.

The moment distribution in one half of the frame is shown in Table 9.41.

Table 9.41. Moment Distribution for Example 9.23

Member	AB	BA	BC
Rel. K		$\frac{EI}{10}$	$\frac{18EI}{15}$
D.F.		$\frac{1}{13}$	$\frac{12}{13}$
C.O.F.	⟵ −1 ——		0
F.E.M.	+20	+20	0
Bal.	0	−1.54	−18.46
C.O.M.	+1.54	0	0
Total moment kN-m	+21.54	+18.46	−18.46

Fig. 9.67(b) shows the B.M. diagram drawn on tension side for the frame.

Example 9.24. *Analyse the rigid Frame shown in Fig. 9.68, using cantilever moment selfdistribution, Values of $\frac{EI}{L}$ for members are as shown.*

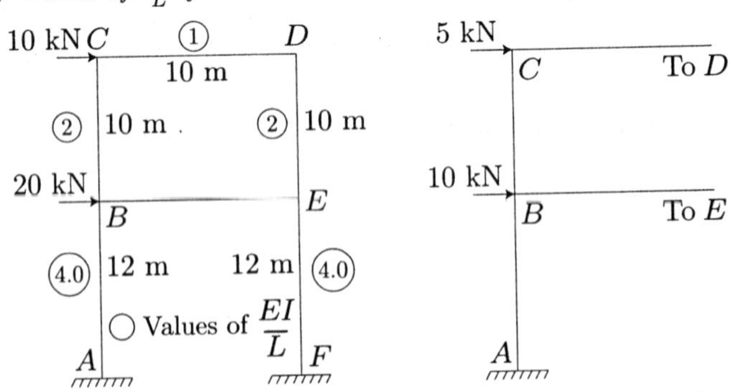

(a) Frame (b) Half frame with loading

Fig. 9.68: For Example 9.24 (Contd.)

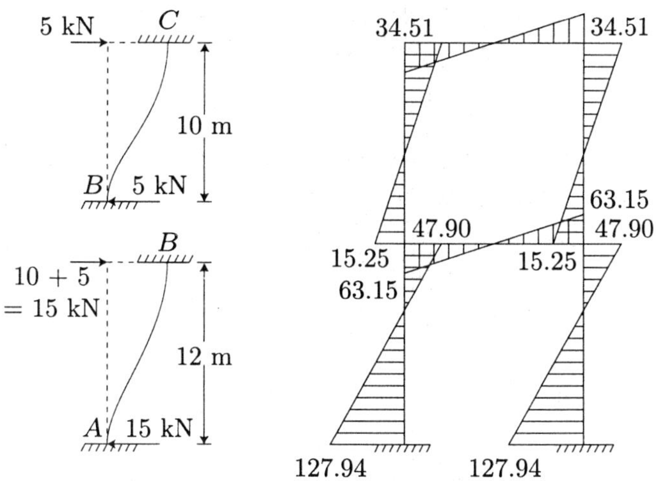

(c) Fixed end moments **(d) B.M. in kN**

Fig. 9.68: For Example 9.24

Solution : Using symmetry, one half of the structure as shown in Fig. 9.68(b) will be analysed. Hence in this case,

$$K_{CD} = K_{BE} = \frac{6EI}{L} \approx (6)(1) \approx 6$$

$$K_{BA} = \frac{EI}{L} \approx 4$$

and $$K_{CB} = \frac{EI}{L} \approx 2.$$

The carry over factors are :

$$COF_{BA} = COF_{BC} = COF_{CB} = -1.0$$

$$COF_{CD} = COF_{BE} = 0.$$

The fixed end moments are obtained considering loading cases in Fig. 9.68(c). The nodal loads are equally divided between the two vertical members AC and DF as shown. The horizontal force acting at ends of BC is 5 kN, while that at ends of BA is 15 kN which is sum of all forces above B.

$$M_{FAB} = M_{FBA} = +(15)\left(\frac{12}{2}\right) = +90 \text{ kNm}$$

$$M_{FBC} = M_{FCB} = +(5)\left(\frac{10}{2}\right) = +25 \text{ kNm}.$$

The moment distribution procedure is shown in Table 9.42.

B.M. diagram is shown in Fig. 9.68(d).

Table 9.42. Moment Distribution for Example 9.24

Joint	A	B			C	
Member	AB	BA	BC	BE	CB	CD
Rel. K		4	2	6	2	6
D.F.		$\frac{1}{3}$	$\frac{1}{6}$	$\frac{1}{2}$	$\frac{1}{4}$	$\frac{3}{4}$
C.O.F.	← −1 →		← −1 →			
F.E.M.	+90.00	+90.00	+25.00	0	+25,00	0
Bal.	0	−38.33	−19.17	−57.50	−6.25	−18.75
C.O.M.	+38.33	0	+6.25	0	+19.17	0
Bal.	0	−2.08	−1.04	−3.13	−4.79	−14.38
C.O.M.	−2.08	0	+4.79	0	+1.04	0
Bal.	0	−1.60	−0.80	−2.39	−0.26	−0.78
C.O.M.	+1.60	0	+0.26	0	+0.80	0
Bal.	0	−0.09	−0.04	−0.13	−0.20	−0.60
C.O.M.	+0.09	−	−	−	−	−
Final Moments kN-m	+127.94	+47.90	+15.25	+63.15	+34.51	−34.51

Example 9.25. *Analyse the Vierendel Truss shown in Fig. 9.69(a) using cantilever moment distribution. Values within small circles indicate $\frac{EI}{L}$ values of the members.*

Solution : The structure has an axis of symmetry about horizontal axis passing through the centres of vertical members and using this symmetry, stiffness of members are

$$K_{AB} = K_{BA} = K_{DE} = K_{ED} = \frac{EI}{L} \propto 15$$

$$K_{AD} = K_{DA} = K_{BE} = K_{EB} = \frac{6EI}{L} \propto 60$$

$$K_{CF} = K_{FC} = \frac{6EI}{L} \propto 75$$

$$K_{BC} = K_{CB} = K_{EF} = K_{FE} = \frac{EI}{L} \propto 25.$$

The carryover factors are :

$$COF_{AB} = COF_{BC} = -1.0$$

$$COF_{AD} = COF_{BE} = COF_{CF} = 0.$$

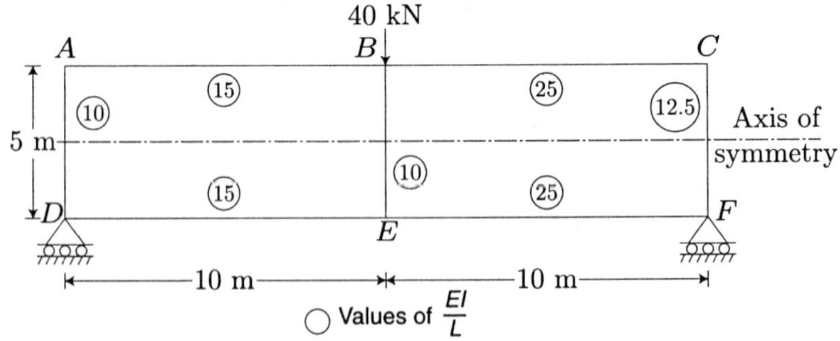

(a) Frame

Fig. 9.69: For Example 9.25 (Contd.)

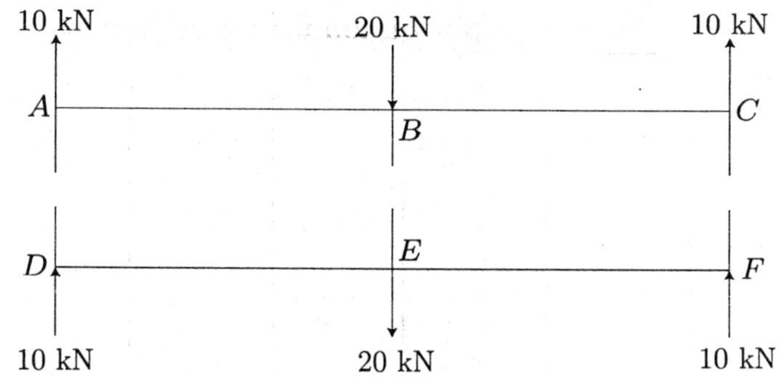

(b) Two halves of frame

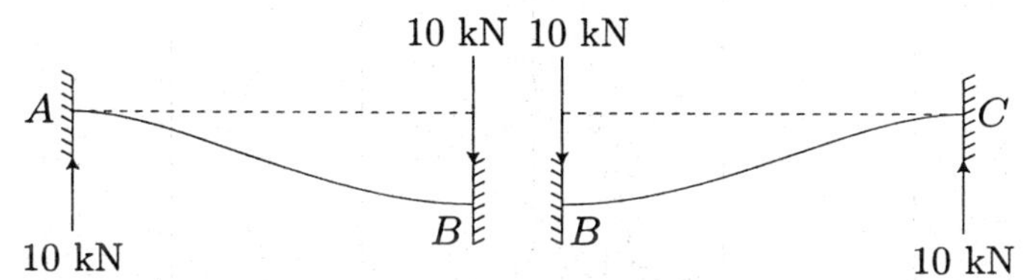

(c) Fixed end B.M.

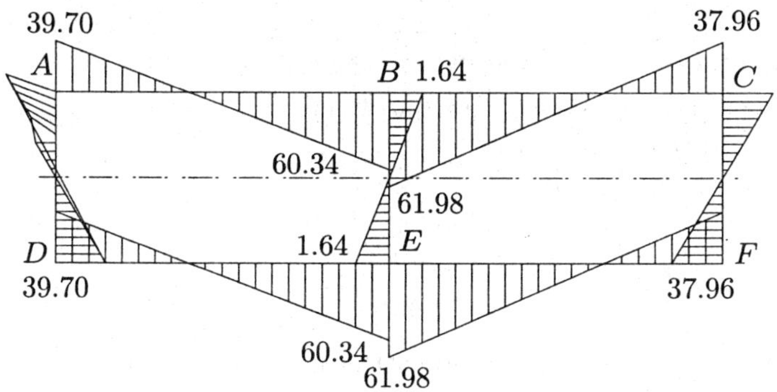

(d) B.M. in kNm

Fig. 9.69: For Example 9.25

For the fixed end moments, applied load of 40 kN is equally distributed between two halves as shown in Fig. 9.69(b). Then the load of 20 kN acting at B is equally divided between members AB and BC as shown in Fig. 9.69(c) in conformity with the proportion of support reactions. Then

$$M_{FAB} = M_{FBA} = +(10)\left(\frac{10}{2}\right) = +50 \text{ kNm}.$$

$$M_{FBC} = M_{FCB} = -(10)\left(\frac{10}{2}\right) = -50 \text{ kNm}.$$

The moment distribution procedure for the upper half is shown in Table 9.43. The B.M. diagram is shown in Fig. 9.69(d) drawn on tension side.

Table 9.43. Moment Distribution for Example 9.25

Joint	A		B			C	
Member	AD	AB	BA	BE	BC	CB	CF
Rel. K	60	15	15	60	25	25	75
D.F.	0.80	0.20	0.15	0.60	0.25	0.25	0.75
C.O.F.	0	$\leftarrow -1$	$\rightarrow$	0	$\leftarrow -1$	$\rightarrow$	0
F.E.M.	0	+50	+50	0	−50	−50	0
Bal.	−40	−10	0	0	0	+12.50	+37.50
C.O.M.	0	0	+10	0	−12.50	0	0
Bal.	0	0	+0.38	+1.5	+0.62	0	0
C.O.M.	0	−0.38	0	0	0	−0.62	0
Bal.	+0.30	+0.08	0	0	0	+0.16	+0.46
C.O.M.	0	0	−0.08	0	−0.16	0	0
Bal.	0	0	+0.04	+0.14	+0.06		
Final Moments, kN-m	−39.7	+39.7	+60.34	+1.64	−61.98	−37.96	+37.96

Example 9.26. *Analyse the Rigid frame shown in Fig. 9.70(a).*

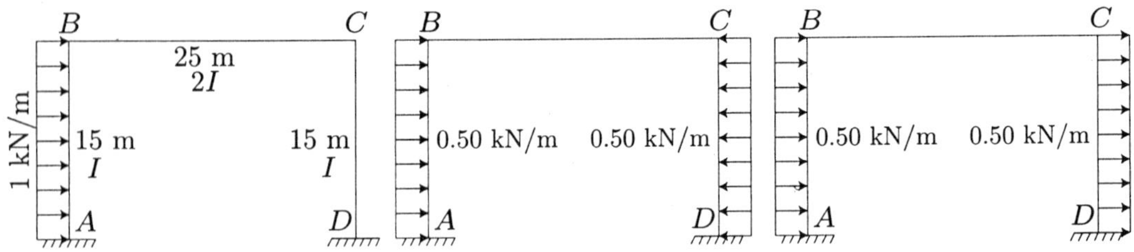

(a) Frame **(b) Symmetrical case** **(c) Antisymmetrical case**

Fig. 9.70: For Example 9.26

Solution : The loading in the structure may be resolved into the symmetrical and antisymmetrical loadings shown in Figs. 9.70(b) and 9.70(c). For the symmetrical loading there is no side-sway and it is only necessary to distribute moments for one half of the structure. In this case, the rotational stiffness for BA is $\frac{4EI}{15}$, that for BC being $\frac{2E(2I)}{25}$, i.e., $\frac{4EI}{25}$. COF from B to A is $\frac{1}{2}$, that for BC being zero. The fixed-end moments are obtained as in normal cases for no joint translation and moment distribution calculations are done in tabular form, in Table 9.44.

In the antisymmetrical case as in Fig. 9.70(c), the analysis is done by cantilever moment distribution. Stiffness for member BA is $\frac{EI}{15}$, that for BC being $\frac{6E(2I)}{25}$, i.e., $\frac{12EI}{25}$. The fixed-end moments for vertical members are

$$M_{FAB} = \frac{wL^2}{12} + \frac{wL}{2}\left(\frac{L}{2}\right) = \frac{wL^2}{3}$$

and $$M_{FBA} = -\frac{wL^2}{12} + \frac{wL}{2}\left(\frac{L}{2}\right) = \frac{wL^2}{6}.$$

After distributing moments for the symmetrical and antisymmetrical loadings, the moments in the original frame are found by superimposing results. Table 9.44 shows the moment distribution in tabular form.

Table 9.44. For Example 9.26

(a) For symmetrical loading

Member	AB	BA	BC
Rel K		$\frac{4EI}{15}$	$\frac{4EI}{25}$
D.F.	← 0.50	$\frac{5}{8}$	$\frac{3}{8}$
C.O.F.			0
F.E.M.	+9.38	−9.38	0
Bal.	0	+5.87	+3.51
C.O.M.	+2.94	0	0
Bal.	0	0	0
Total moments	+12.32	−3.51	+3.51

(b) For antisymmetrical loading

AB	BA	BC
	$\frac{EI}{15}$	$\frac{12EI}{25}$
←	$\frac{5}{41}$	$\frac{36}{41}$
−1		0
+37.50	+18.75	0
0	−2.29	−16.46
+2.29	0	0
0	0	0
+39.79	+16.46	−16.46

Resulting BM

$M_{AB} = +52.11$ kNm $\quad M_{BA} = +12.95$ kNm

$M_{BC} = -12.95$ kNm $\quad M_{CB} = -19.97$ kNm

$M_{CD} = +19.97$ kNm $\quad M_{DC} = +27.47$ kNm.

Example 9.27. *Analyse the Vierendel Truss shown in Fig. 9.71 using cantilever moment distribution.*

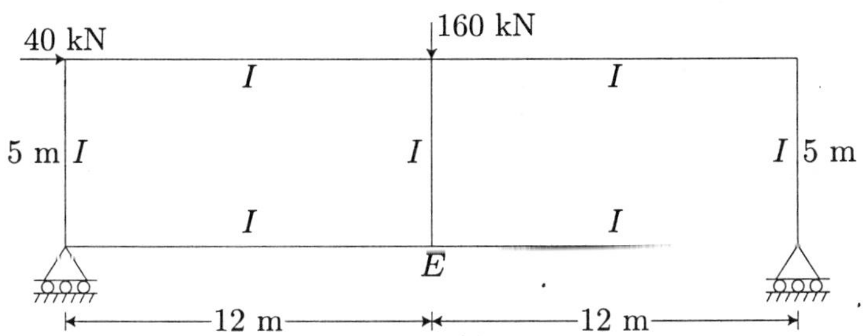

Fig. 9.71: For Example 9.27

Solution : The structure is symmetrical about a horizontal axis through the centre of the vertical member.

$$\text{Stiffness of } AB = \frac{EI}{L} \approx \frac{1}{12}$$

$$\text{Stiffness of } AD = \frac{6EI}{L} \approx \frac{6}{5}.$$

For F.E.M.,

$$\text{Reaction at } D = \frac{430}{6} \text{ kN}$$

$$\text{Reaction at } F = \frac{530}{6} \text{ kN.}$$

$$M_{FAB} = +\frac{1}{2}\left(\frac{430}{6}\right)\left(\frac{12}{2}\right) = +215 = M_{FBA}$$

$$M_{FAD} = +\frac{1}{3}(40)\left(\frac{5}{2}\right) = +33.3 = M_{FDA}$$

$$M_{FBC} = -\frac{1}{2}\left(\frac{530}{6}\right)\left(\frac{12}{2}\right) = -265 = M_{FCB}.$$

The moment distribution is shown in Table 9.45.

Table 9.45. Moment Distribution for Example 9.27

Joint	A		B			C	
Member	*AD*	*AB*	*BA*	*BE*	*BC*	*CB*	*CF*
Rel. K	$\frac{6}{5}$	$\frac{1}{12}$	$\frac{1}{12}$	$\frac{6}{5}$	$\frac{1}{12}$	$\frac{1}{12}$	$\frac{6}{5}$
D.F.	$\frac{72}{77}$	$\frac{5}{77}$	$\frac{5}{82}$	$\frac{72}{82}$	$\frac{5}{82}$	$\frac{5}{77}$	$\frac{72}{77}$
C.O.F.		← −1 →			← −1 →		
F.E.M.	+33.3	+215.0	+215.0	+33.3	−265.0	−265.0	+33.3
Bal.	−232.2	−16.1	+1.0	+14.7	+1.0	+15.0	+216.7
C.O.M.	0	−1.0	+16.1	0	−150	−1.0	0
Bal.	+0.90	+0.10	−0.10	−0.90	−0.10	+0.10	+0.90
Total Moments	−198.0	+198.0	+232.0	+47.1	−279.1	−250.9	+250.9

9.18 Single Bay Symmetrical Frame with Inclined Columns

A method similar to the cantilever method for multi-storeyed rectangular single bay portal frame can be developed for symmetrical single bay multistoreyed frames with sloping columns.

Method of determining the stiffness of a sloping column is illustrated with the frame shown in Fig. 9.72(a). Two stiffness K_{CD} and K_{DC} for the member CD are determined to illustrate the method of determining the stiffness of a sloping member.

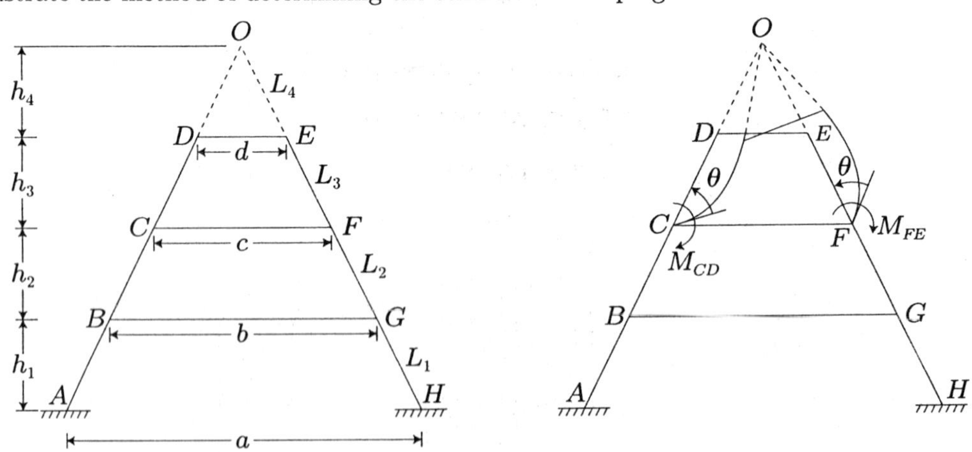

(a) Frame **(b) Deflected shape**

Fig. 9.72: Single bay symmetrical frame with inclined members

To determine K_{CD}, the ends C and F of members CD and EF are pinned at joints C and F respectively. Triangular part ODE is considered to be infinitely stiff. If joints B, C, F, and G are restrained against translation and equal moments M_{CD} and M_{FE} are applied, the deflected form of the structure is as shown in Fig. 9.72(b). Joints C and F have a rotation θ each and deflection is in the form of a rotation about pole O. Stiffness K_{CD} is the value of M_{CD} when θ is a unity.

From the conjugate beam for CDO shown in Fig. 9.73(a), the slope at C can be obtained from the conjugate beam in Fig. 9.73(c). Thus,

$$R_C = \theta_C = \frac{1}{L_3+L_4}\frac{M_{CD}}{EI}\left[\frac{L_4}{L_3+L_4}L_3\left(L_4+\frac{L_3}{2}\right) + \left(1-\frac{L_4}{L_3+L_4}\right)\frac{L_3}{2}\left(L_4+\frac{2}{3}L_3\right)\right]$$

$$= \frac{M_{CD}}{EI}\frac{L_3}{(L_2+L_4)^2}\left[\frac{L_3^2}{3}+L_3L_4+L_4^2\right].$$

With $L_4 = \dfrac{d}{c-d}(L_3)$, $\theta_C = \dfrac{M_{CD}}{3EI}L_3\left[1+\dfrac{d}{c}+\left(\dfrac{d}{c}\right)^2\right]$.

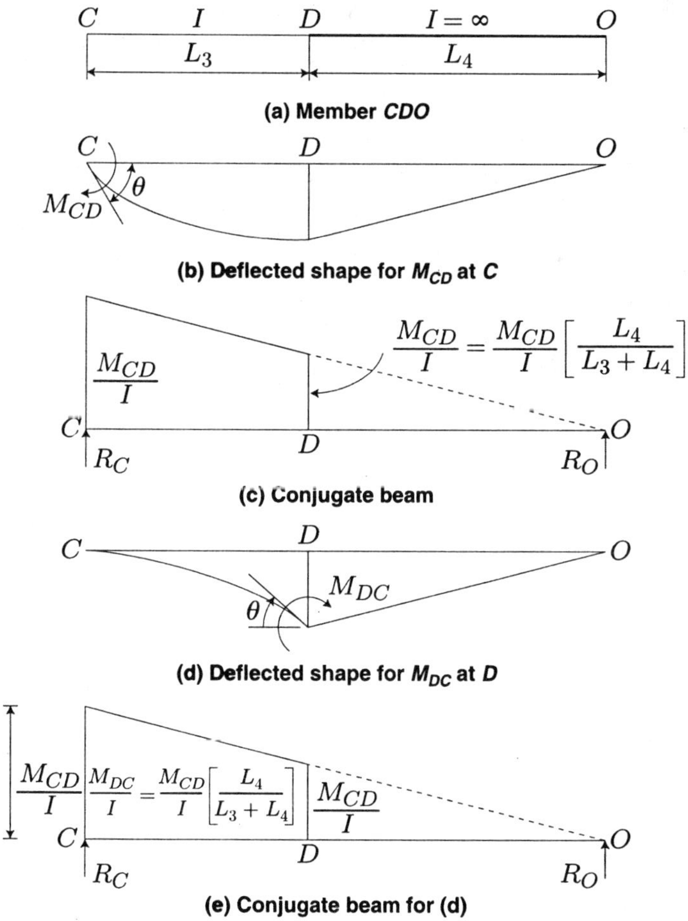

(a) Member *CDO*

(b) Deflected shape for M_{CD} at *C*

(c) Conjugate beam

(d) Deflected shape for M_{DC} at *D*

(e) Conjugate beam for (d)

Fig. 9.73: Stiffness and carry over factors

Stiffness K_{CD} is given for $\theta_C = 1$, whence

$$K_{CD} = \frac{3EI}{L_3}\left[\frac{1}{1+\left(\frac{d}{c}\right)+\left(\frac{d}{c}\right)^2}\right]. \tag{9.21}$$

Carry over factor from C to D is given by

$$C_{CD} = -\frac{M_{DC}}{M_{CD}} = -\frac{L_4}{L_3+L_4} = \frac{d}{c}. \tag{9.22}$$

Stiffness K_{DC} can be obtained in a similar manner, with ends D and E of the members CD and EF hinged at joints D and E, and moments M_{DC} and M_{EF} applied, with the same restriction on translation as before. From Figs. 9.87(d) to 9.87(f), it follows that,

$$K_{DC} = \frac{3EI}{L_3}\left[\frac{1}{1+\left(\frac{c}{d}\right)+\left(\frac{c}{d}\right)^2}\right] \tag{9.23}$$

and $C_{DC} = -\dfrac{c}{d}$. (9.24)

Fixed end moments due to sidesway of the frame considered can be obtained as shown in Fig. 9.74. In this case, imaginary pins are placed at the ends B and G of members BC and FG. With all joints restrained against rotation and joints B and G restrained against translation also, external forces are applied. The frame deflects in the shape as in Fig. 9.74(a). Fixed-end moments M_{FBC} for the applied loading is the value of moment applied at B to member BC, which makes θ equal to zero.

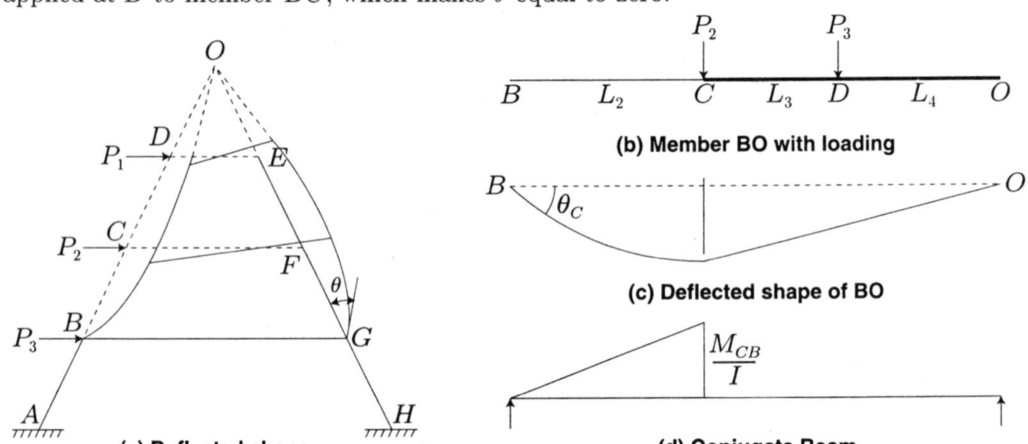

Fig. 9.74: Fixed end moments due to sidesway

Fig. 9.74(b) shows the pin-jointed member BO subjected to external loading. From the conjugate beam in Fig. 9.74(d), M_{CB} is the statically determinate bending moment at C. Length CO is considered infinitely stiff ($I = \infty$), and magnitude of θc is given by

$$\theta_c = \frac{1}{L_2+L_3+L_4}\left[\frac{M_{CB}}{EI}\frac{L_2}{2}\left(\frac{L_2}{3}+L_3+L_4\right)\right]$$

Fixed-end moment M_{FBC} is then given by

$$M_{FBC} = \theta_c(K_{BC}) \tag{9.25}$$

and $M_{FCB} = -M_{CB} + M_{FBC}(C_{BC})$ (9.26)

Example 9.28. *Analyse the frame shown in Fig. 9.75.*

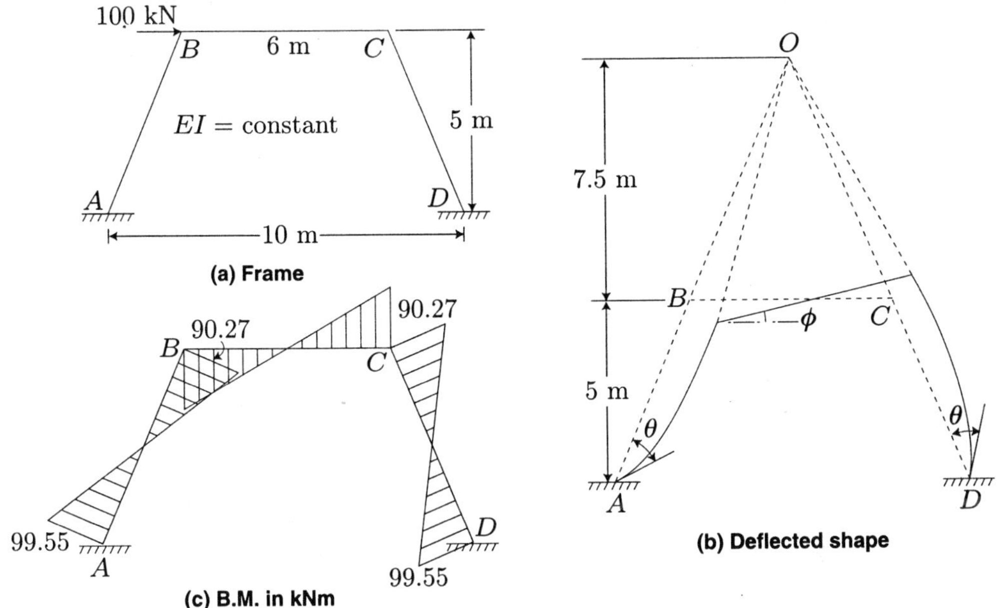

Fig. 9.75: For Example 9.28

Solution: Length $L_{AB} = \sqrt{5^2 + 2^2} = \sqrt{29} = 5.385$ m

$$L_{BO} = \sqrt{(7.5)^2 + (3)^2} = \sqrt{65.25} = 8.078 \text{ m.}$$

Stiffness :
$$K_{AB} = \frac{3}{\sqrt{29}}\left[\frac{1}{1 + \frac{6}{10} + \left(\frac{6}{10}\right)^2}\right] = 0.284$$

$$K_{BA} = \frac{3}{\sqrt{29}}\left[\frac{1}{1 + \frac{10}{6} + \left(\frac{10}{6}\right)^2}\right] = 0.102$$

$$K_{BC} = \frac{6EI}{L} = \frac{6}{6} = 1.00.$$

D.F.
$$DF_{BA} : DF_{BC} = 0.102 : 1$$

$$DF_{BA} = 0.093 \quad \text{and} \quad DF_{BC} = 0.907$$

C.O.F.
$$C_{BA} = -\frac{10}{6} = -1.667.$$

Fixed-end moments

$$M_{BA} = \frac{100}{2}\left(\frac{7.5}{12.5}\right)(5) = 150 \text{ kNm}$$

$$\theta_c = \frac{150\sqrt{29}}{2}\left[\frac{\frac{\sqrt{29}}{3} + \sqrt{65.25}}{\sqrt{29} + \sqrt{65.25}}\right] = 296.18$$

$$M_{FAB} = 296.18(K_{AB}) = 84.12 \text{ kNm}$$

$$M_{FBA} = 150 - 84.12\left(\frac{6}{10}\right) = 99.53 \text{ kNm.}$$

Moment distribution is shown in Table 9.46.

Table 9.46. Moment Distribution for Example 9.28

Joint	A	B	
Member	AB	BA	BC
D.F.		0.093	0.907
C.O.F.	← −1.667		
F.E.M.	+84.12	+99.53	0
Bal.		−9.26	−90.27
C.O.M.	+15.43	0	0
Bal.	0	0	0
Σ	+99.55	+90.27	−90.27

Example 9.29. *Analyse the frame shown in Fig. 9.76 of height 18 m in each storey.*

Solution : The analysis is carried on in two steps :

(a) Symmetrical loading as shown in Fig. 9.76(b). The distribution of moment for this is shown in Table 9.47.

(b) Antisymmetrical loading as shown in Fig. 9.76(c). The stiffness and carry over factors are calculated as in Section 9.16. The distribution of moment for this is shown in Table 9.48.

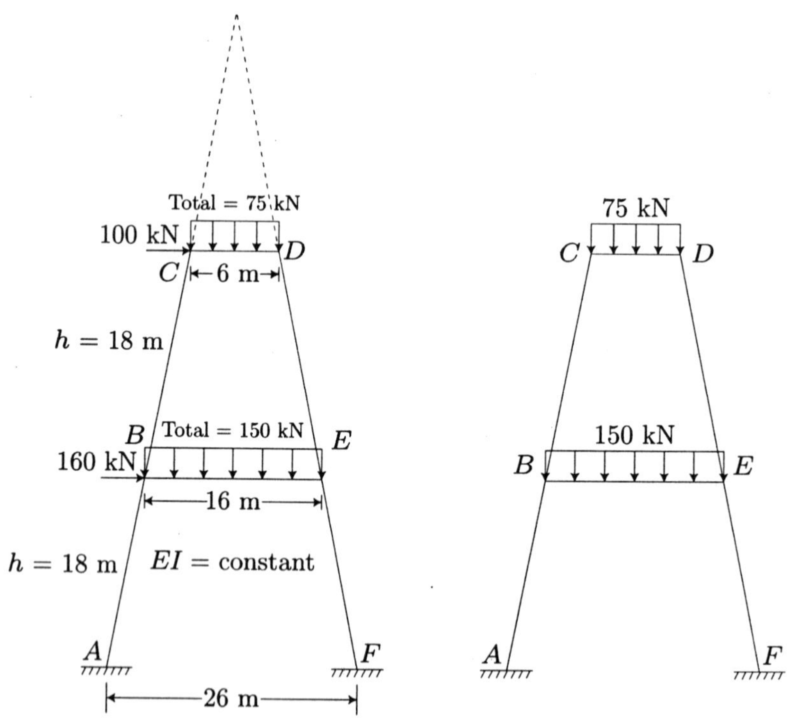

(a) Frame (b) Symmetrical case

Fig. 9.76: For Example 9.29 (Contd.)

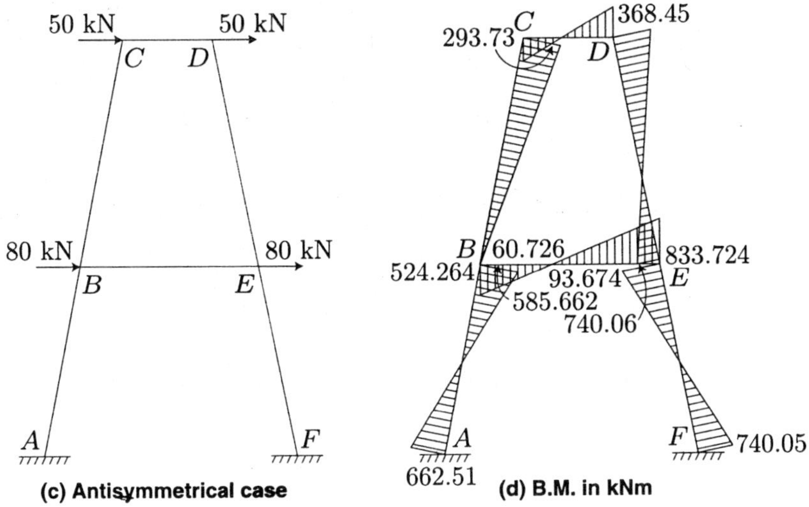

Fig. 9.76: For Example 9.29

Final moments are obtained in Table 9.49 and shown in Fig. 9.76(d).

Symmetrical case :

Stiffness :

$$K_{BA} = K_{BC} = \frac{4}{\sqrt{349}} = 0.214$$

$$K_{CD} = \frac{2}{6} = 0.333$$

$$K_{BE} = \frac{2}{16} = 0.125.$$

D.F. :

$$DF_{BA} = DF_{BC} = 0.387$$

$$DF_{BE} = 0.226$$

$$DF_{CB} = 0.391$$

$$DF_{CD} = 0.609.$$

C.O.F. : 0.50 everywhere.

F.E.M. :

$$M_{FCD} = +37.50 = -M_{FDC}$$

$$M_{FBE} = +200 = -M_{FEB}.$$

Antisymmetrical case :

Stiffness :

$$K_{CD} \approx \frac{6}{6} = 1; \quad K_{BE} \approx \frac{6}{16} = 0.375$$

$$K_{BC} \approx \frac{3}{\sqrt{349}} \left[\frac{1}{1 + \frac{6}{16} + \left(\frac{6}{16}\right)^2}\right] = 0.106$$

$$K_{CB} \approx \frac{3}{\sqrt{349}} \left[\frac{1}{1 + \frac{16}{6} + \left(\frac{16}{6}\right)^2}\right] = 0.0149$$

$$K_{AB} \approx \frac{3}{\sqrt{349}}\left[\frac{1}{1+\frac{16}{26}+\left(\frac{16}{26}\right)^2}\right] = 0.0805$$

$$K_{BA} \approx \frac{3}{\sqrt{349}}\left[\frac{1}{1+\frac{26}{16}+\left(\frac{26}{16}\right)^2}\right] = 0.0305$$

$$DF_{BA} = 0.0596; \quad DF_{BE} = 0.7332; \quad DF_{BC} = 0.2072$$

$$DF_{CB} = 0.015; \quad DF_{CD} = 0.985$$

C.O.F. : $C_{BA} = -\dfrac{26}{16} = -1.625$

$$C_{AB} = -\frac{16}{26} = -0.615$$

$$C_{BC} = -\frac{6}{16} = -0.375$$

$$C_{CB} = -\frac{16}{6} = -2.667.$$

Fixed-end moments :

$$M_{CB} = \frac{50(10.80)}{28.80}(18) = 337.50 \text{ as } \frac{3}{5}(18) = 10.80$$

$$\theta_{CB} = \frac{337.50(\sqrt{349})}{2}\left(\frac{\frac{\sqrt{349}}{3}+\sqrt{125.64}}{\sqrt{349}+\sqrt{125.64}}\right) = 1838.99.$$

Table 9.47. Symmetrical Case for Example 9.29

Joint	A	B			C	
Member	AB	BA	BE	BC	CB	CD
D.F.		0.387	0.226	0.387	0.391	0.609
C.O.F.	← 0.50 ——			← 0.50 ——	→	
F.E.M.			+200			+37.50
Bal.		−77.40	−45.20	−77.40	−14.66	−22.84
C.O.M.	−38.70	0	0	−7.33	−38.70	0
Bal.	0	+2.84	+1.66	+2.83	+15.13	+23.57
C.O.M.	+1.42	0	0	+7.67	+1.42	0
Bal.	0	−2.97	−1.73	−2.97	−0.55	−0.87
C.O.M.	−1.49	–	–	–	–	–
Σ	−38.77	−77.53	+154.73	−77.20	−37.36	+37.36

$$M_{FBC} = 1838.99(0.106) = 194.93$$

$$M_{FCB} = 337.50 - 194.93(0.375) = 264.40$$

$$M_{BA} = \left[\frac{50(10.80)}{46.80}+\frac{80(28.80)}{46.80}\right](18) = 1093.85$$

$$\theta_{BA} = \frac{1093.85(\sqrt{349})}{2}\left(\frac{\frac{\sqrt{349}}{3} + \sqrt{349} + \sqrt{125.64}}{2\sqrt{349} + \sqrt{125.64}}\right) = 7597.56$$

$$M_{FAB} = 7597.56(0.0805) = 611.60$$

$$M_{FBA} = 1093.85 - 0.615(611.60) = 717.72.$$

Table 9.48. Antisymmetry Case for Example 9.29

Joint	*A*	*B*			*C*	
Member	*AB*	*BA*	*BE*	*BC*	*CB*	*CD*
D.F.		0.0596	0.7332	0.2072	0.015	0.985
				— 0.375 →		
C.O.F.	←	−1.625 —		←	−2.667 —	
F.E.M.	+611.60	+717.72	–	+194.93	+264.40	
Bal.	0	−54.39	−669.15	−189.10	−3.97	−260.43
C.O.M.	+88.38	0	0	+10.588	+70.91	0
Bal.	0	−0.631	−7.763	−2.194	−1.064	−69.846
C.O.M.	+1.025	0	0	+2.838	+0.822	0
Bal.	0	−0.169	−2.081	−0.588	−0.013	−0.810
C.O.M.	+0.275	–	–	–	–	–
$\sum$	+701.28	+662.53	−678.994	+16.474	+331.09	−331.09

Table 9.49. Final B.M. for Example 9.29

Member	*AB*	*BA*	*BE*	*BC*	*CB*	*CD*
From Table 9.47	−38.77	−77.53	+154.73	−77.20	−37.36	+37.36
From Table 9.48	+701.28	+662.53	−678.994	+16.474	+331.09	−331.09
Final B.M.	+662.51	+585.00	−524.264	−60.726	+293.73	−293.73
Member	*FE*	*EF*	*EB*	*ED*	*DE*	*DC*
From Table 9.47	+38.77	+77.53	−154.73	+77.20	+37.36	−37.36
From Table 9.48	+701.280	+662.53	−678.994	+16.474	+331.09	−331.09
Final Moments	+740.05	+740.060	−833.724	+93.674	+368.45	−368.45

Problems

9.1 (a) Define rotational stiffness and carry over factor and obtain their values for a prismatic member. Show that the modified stiffness when far end is hinged is $\frac{3}{4}$th of the rotational stiffness with far end fixed.

(b) A beam *ABCD* is supported on rollers at *B* and *C*, fixed at *D*, and free at *A*. The portion *BC* carries a.u.d.l. of 20 kN/m, portion *CD* carries a concentrated load of 100 kN at 3 m from *D*, and another concentrated load of 50 kN acts at the free end *A*. The moment of inertia is *I* for *AB* and 2*I* for *BCD*. Lengths of *AB*, *BC* and *CD* are 3 m, 8 m and 6 m respectively. Obtain the B.M. and S.F. diagrams using moment distribution method, and sketch the deflected shape.

9.2 The beam of constant EI in Fig. P-9.1 is subjected to a dead load and live load of intensities w and $2w$ per unit length respectively. Obtain the maximum moment diagram considering the following four cases of loading :

(i) D.L. on AD and L.L. on AB

(ii) D.L. on AD and L.L. on AC

(iii) D.L. on AD and L.L. on BC

(iv) D.L. on AD and L.L. on CD

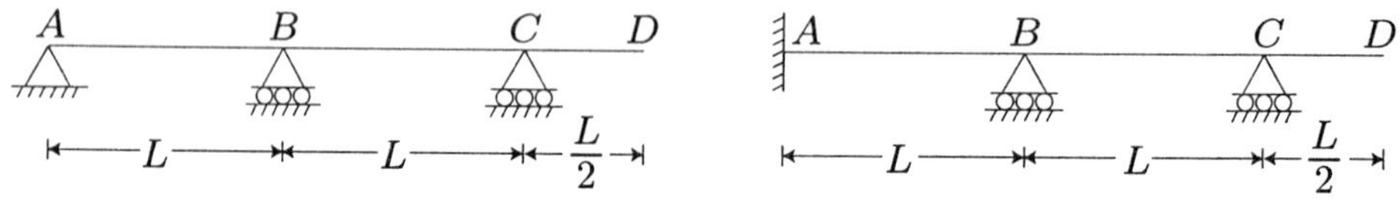

Fig. P-9.1

9.3 Obtain the B.M. and S.F. diagram for the beam $ABCD$ shown in Fig. P-9.2 considering EI as constant. Also sketch the deflected shape.

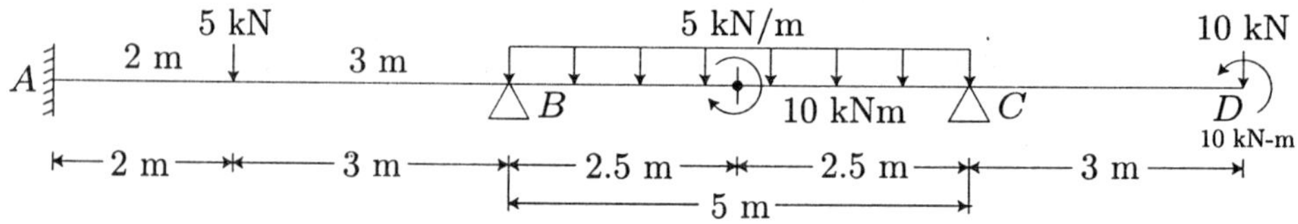

Fig. P-9.2

9.4 Obtain the B.M. and S.F. diagram for the beam shown in Fig. P-9.3. Sketch the elastic curve.

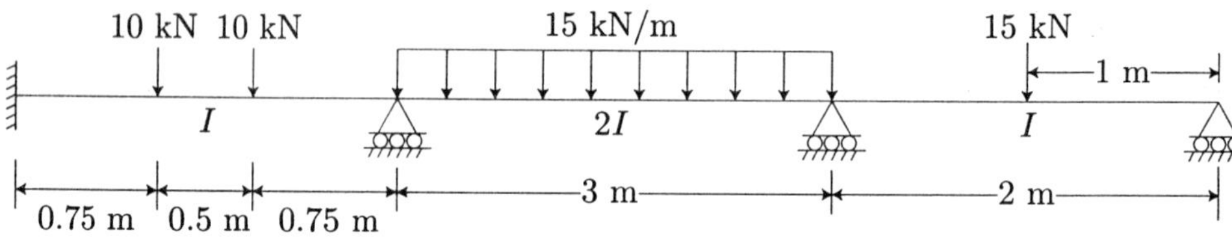

Fig. P-9.3

9.5 Solve Problems 8.1 to 8.3 in Chapter 8 by using Moment distribution method.

9.6 Use moment distribution procedure to analyse the frame in Fig. P-9.4, if FEM due to the loading shown be as follows with appropriate sign.

$$M_{FAB} = M_{FBA} = 150; \quad M_{FBC} = 200$$

$$M_{FCB} = 285; \; M_{FCD} = M_{FDC} = 0$$

$$M_{FCE} = 200; \quad M_{FEC} = 200.$$

Consider values given in O as the $\frac{I}{L}$ values for the members.

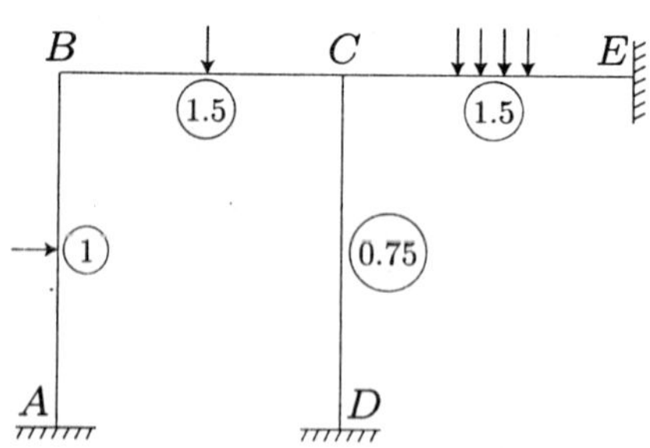

Fig. P-9.4

9.7 Find the B.M. in the symmetrical frame shown in Fig. P-9.5. using moment distribution method replacing the loading by equivalent symmetrical and antisymmetrical loadings. Express moment in terms of $\frac{PL}{100}$.

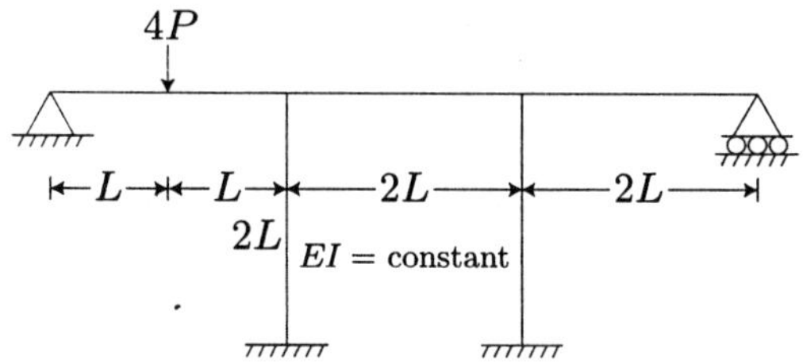

Fig. P-9.5

9.8 Analyse the frames shown in Fig. P-9.6 using moment distribution method.

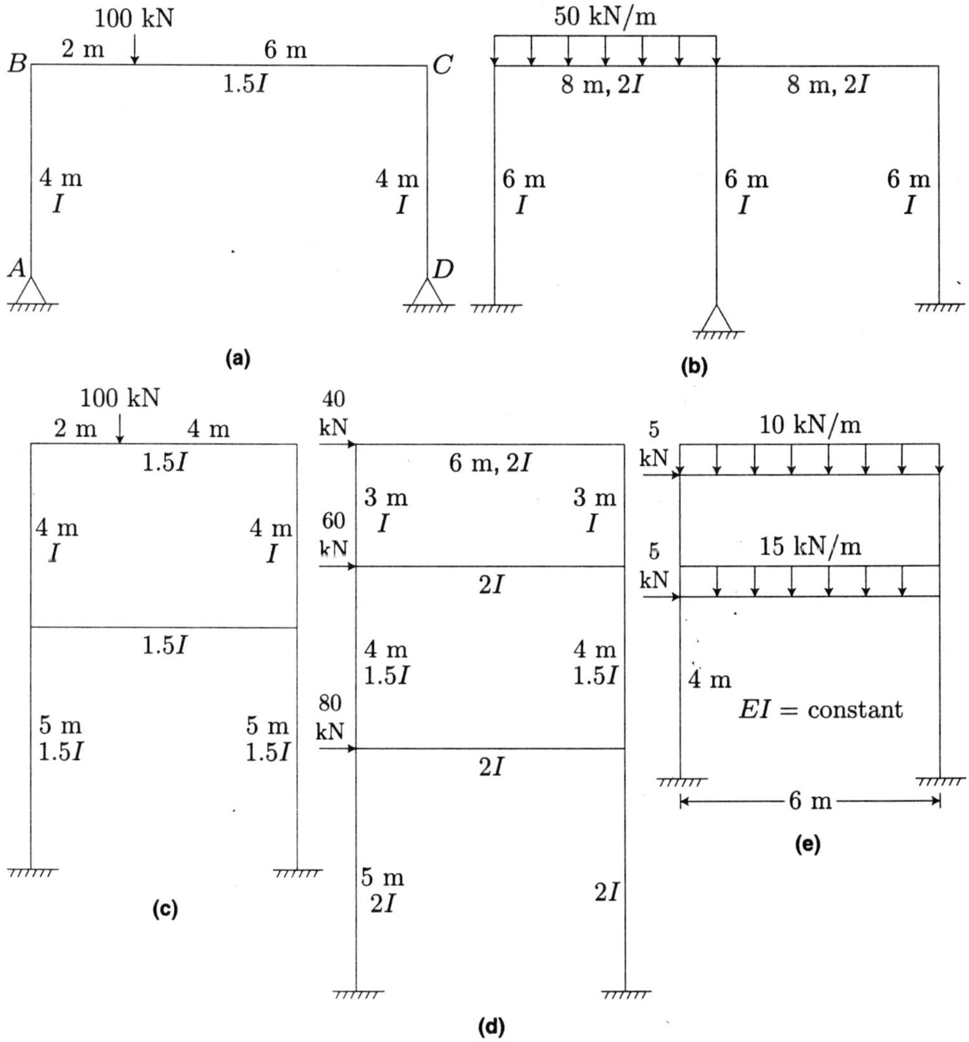

Fig. P-9.6 (Contd.)

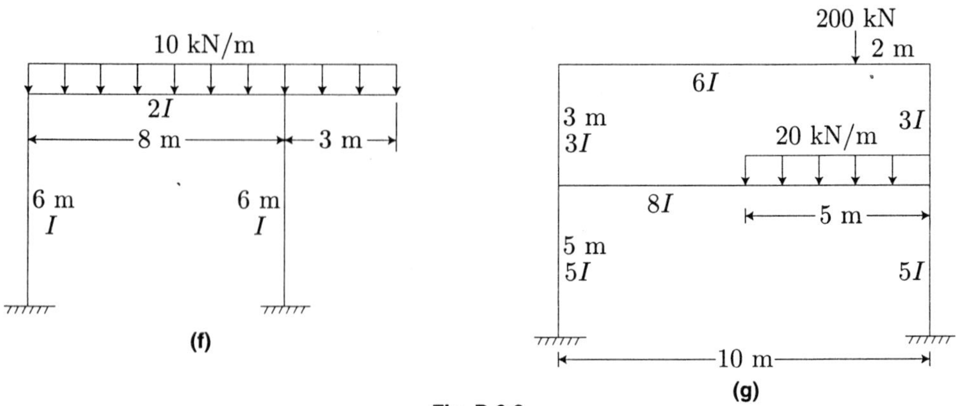

Fig. P-9.6

9.9 Analyse frame in Fig. P-9.6(d), using cantilever moment distribution method.

9.10 Analyse frame in Figs. P-9.6(a), P-9.6(b), P-9.6(c) and P-9.6(e) using equivalent symmetrical and antisymmetrical loadings.

9.11 Analyse the frames shown in Fig. P-9.7 using moment distribution techniques.

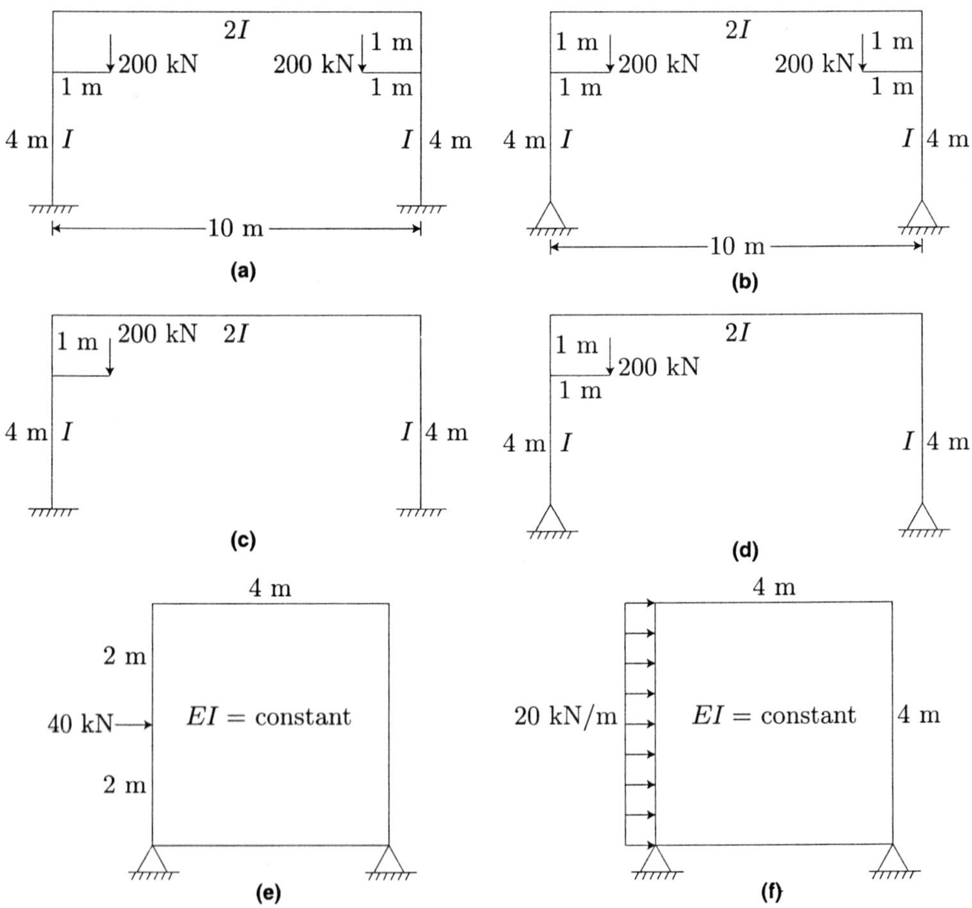

Fig. P-9.7

9.12 Analyse the frames shown in Fig. P-9.8 using moment distribution method.

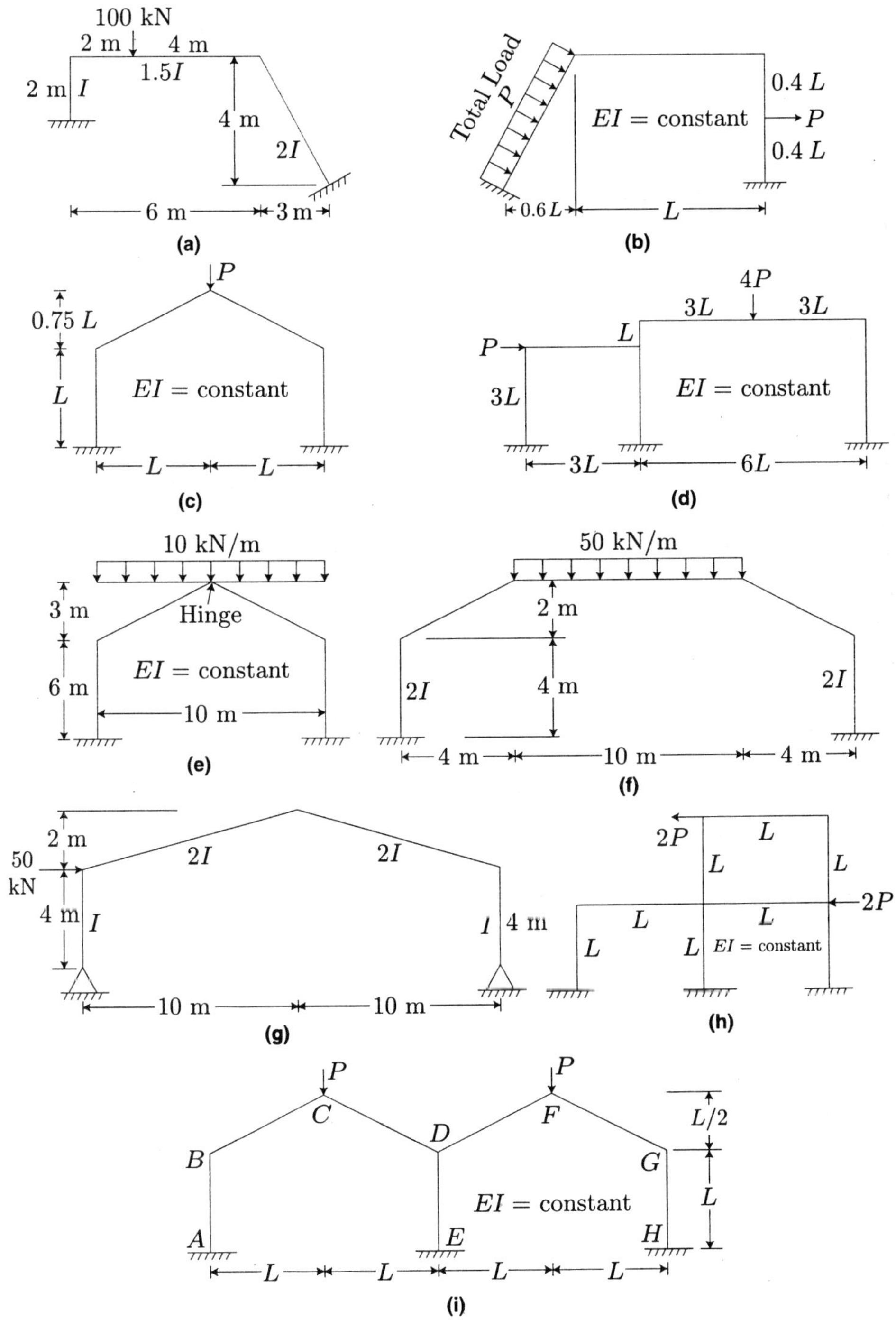

Fig. P-9.8

9.13 Fig. P-9.9 shows concrete frames supporting an elevated water tank. Wind pressure causes the loads shown in the frame. Assuming that the tank elements are of infinite rigidity compared to the rigidity of horizontal bracing beams, find the B.M. in the frame.

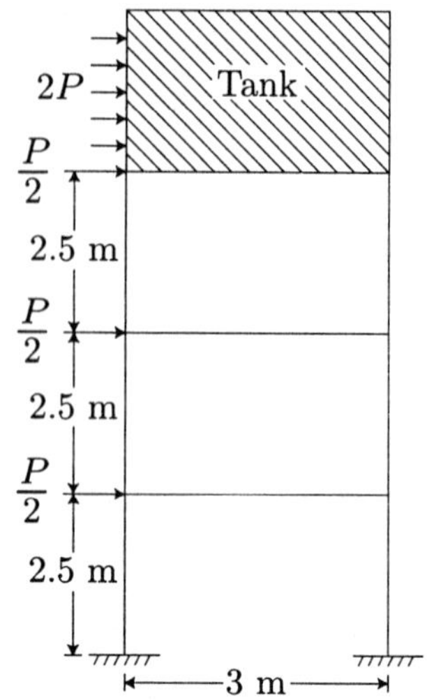

Fig. P-9.9

9.14 Find the bending moment diagram for the Vierendel girder shown in Fig. P-9.10. Assume EI as constant.

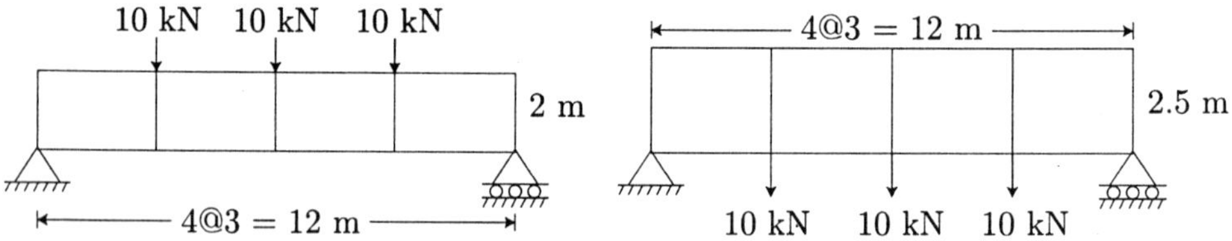

Fig. P-9.10

9.15 Find the B.M. diagram in the building frame shown in Fig. P-9.11. Values in circle indicate $\frac{I}{L}$ values for the members.

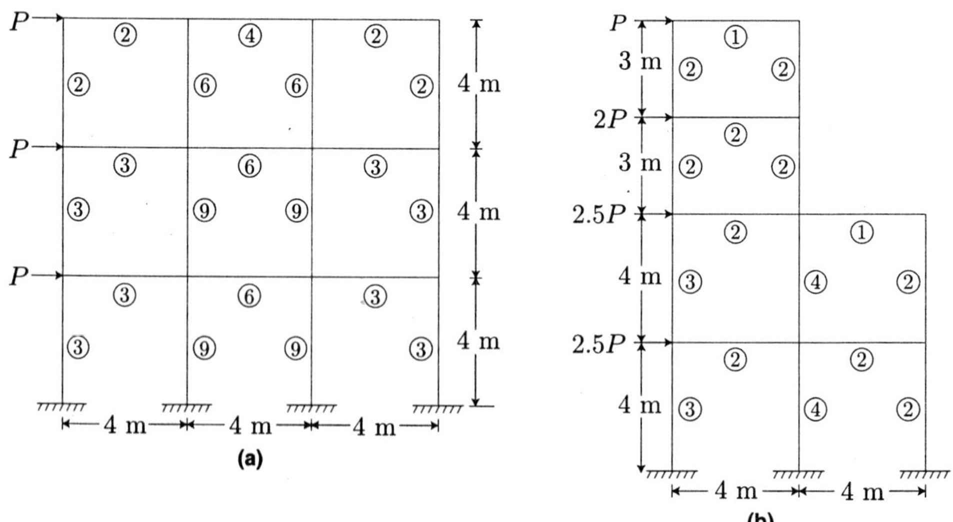

Fig. P-9.11

9.16 Fig. P-9.12 shows a bridge frame subjected to a horizontal force P at the deck level. Assume that the deck has an infinite rigidity.

Find the B.M. and S.F. at the bottom of the piers.

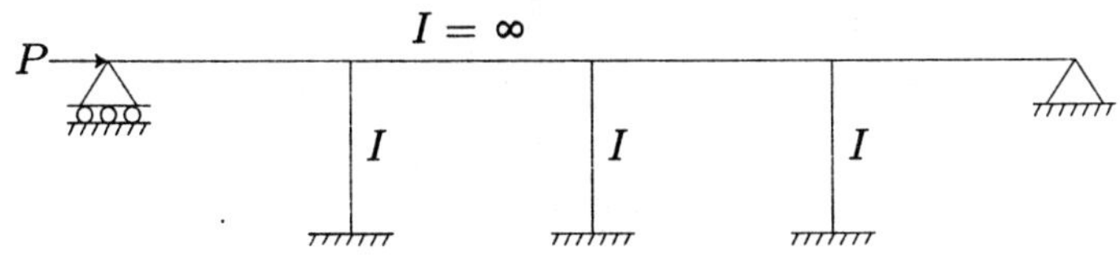

(a)

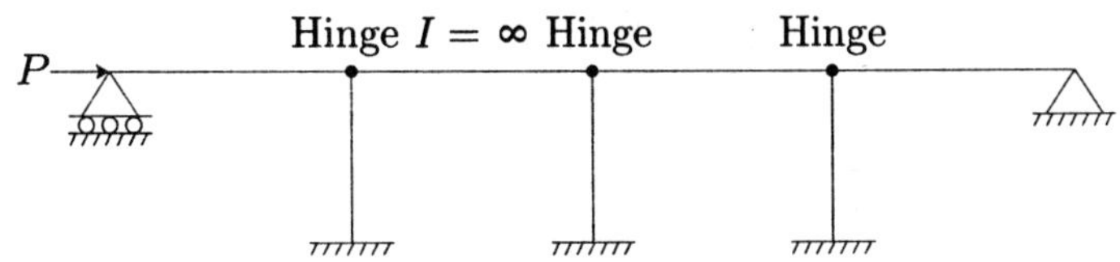

(b)

Fig. P-9.12

Chapter 10

Method of Column Analogy

10.1 Introduction

The method of Column Analogy was introduced by Prof. Hardy Cross of the University of Illinois, U.S.A. in the year 1932 for analysis of statically indeterminate structures having indeterminacies not exceeding three. The column analogy is a mathematical identity between the moments produced by continuity in a beam, bent or arch and the fibre stresses in a short column eccentrically loaded. The advantage of the analogy is the same as that of the analogy of moment areas or, stress areas or angle weights. It reduces certain procedures in the determination of bending moments in continuous structures to a routine work with which structural engineers are familiar, as it deals with the stress analysis of eccentrically loaded column sections.

The column analogy is a tool and a mechanical device and as such is not specially illuminating and helpful in understanding behaviour of the structure. It is necessary that results from column analogy are interpreted in the more familiar terms of pressure lines and deformed structures. The column analogy simply does with precision what these methods do somewhat crudely.

The analysis of structures which are statically indeterminate to the second and third degree involves solution of two or three simultaneous equations, if methods of consistent deformation be used. With the method of column analogy, however, the solution of such structures can be obtained directly.

Rigid jointed portal frames, single bay gable bents, closed rigid frames or box structures and curved structures like fixed arches, indeterminate to the third degree, can be easily solved by this method. For frames with sway, no special solutions are required. Significant advantage may be realised in analysing members with variable moment of inertia or structures with asymmetrically arranged members. Probably the most extensive use of the method is in the computation of fixed-end moments, stiffness and carry over factors for members with variable moment inertia.

10.2 The Column Analogy

Fig. 10.1 shows cross section of a column with two reference axes in the x and y directions. As the column cross section is subjected to a compressive load W with eccentricity x_w and y_w from the axes, the stress σ developed in the column at any point can be expressed as,

$$\sigma = C + mx + ny \tag{10.1}$$

where C, m and n are constants, x and y being the distances of the element of area dA, for which the stress is calculated, from the y- and x-axes respectively.

The magnitudes of the compressive load W, the moments M_x and M_y about the x- and y-axes respectively are given by

$$\left.\begin{aligned} W &= \int \sigma dA, \\ M_x &= \int \sigma y dA \\ \text{and} \quad M_y &= \int \sigma x dA. \end{aligned}\right] \tag{10.2}$$

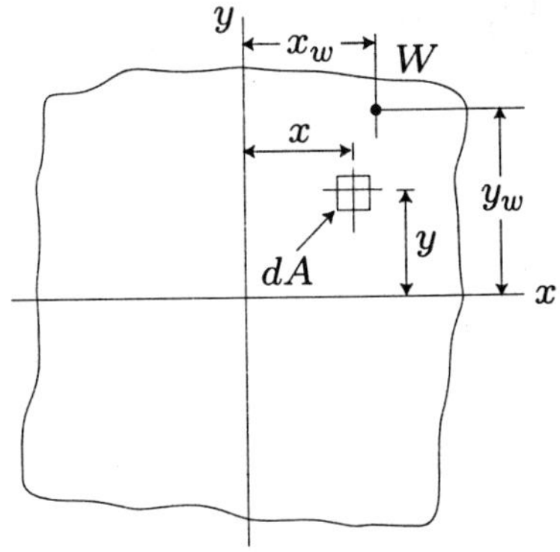

Fig. 10.1: Column cross section

Values of W, M_x and M_y can be rewritten with Eqn. (10.1). Thus,

$$\left.\begin{aligned} W &= C\int dA + m\int xdA + n\int ydA, \\ M_x &= C\int ydA + m\int xydA + n\int y^2dA, \\ M_y &= C\int xdA + m\int x^2dA + n\int xydA. \end{aligned}\right] \tag{10.3}$$

If reference axes pass through the centroid of section, then $\int xdA = 0, \int ydA = 0$

and $\quad W = CA.$ (10.4a)

$$M_x = m\int xydA + n\int y^2dA = mI_{xy} + nI_{xx} \tag{10.4b}$$

$$M_y = m\int x^2dA + n\int xydA = mI_{yy} + nI_{xy}, \tag{10.4c}$$

where I_{xx} and I_{yy} are the moment of intertia about the xx- and yy- axes, I_{xy} being the product of intertia.

Solving Eqns. 10.4(a) to 10.4(c),

$$C = \frac{W}{A}$$

$$m = \frac{M_y I_{xx} - M_x I_{xy}}{I_{xx}I_{yy} - (I_{xy})^2} \quad \text{and} \quad n = \frac{M_x I_{yy} - M_y I_{xy}}{I_{xx}I_{yy} - (I_{xy})^2},$$

$$\text{whence} \quad \sigma = \frac{W}{A} + \frac{M_y I_{xx} -. M_x I_{xy}}{I_{xx}I_{yy} - (I_{xy})^2}x + \frac{M_x I_{yy} - M_y Ixy}{I_{xx}I_{yy} - (I_{xy})^2}y. \tag{10.5}$$

With one or both axes of symmetry $I_{xy} = 0$ and Eq. (10.5) reduces to

$$\sigma = \frac{W}{A} + \frac{M_y}{I_{yy}}x + \frac{M_x}{I_{xx}}y. \tag{10.6}$$

Fig. 10.2 shows a frame $ABCD$ fixed at ends A and D subjected to external loads. The frame is indeterminate to the third degree. Considering V_D, H_D and M_D as the redundant reactions produced at D by the given system of loads, the deformation of the basic determinate structure after removal of the redundant reactions is shown in Fig. 10.2(b). The rotation, horizontal deflection and vertical deflection at the free end D are denoted by $\theta_D, \triangle H_D$, and $\triangle V_D$ respectively. If M_s denotes the statically determinate bending moment diagram due to external loads as in Fig. 10.2(b), then from unit load method,

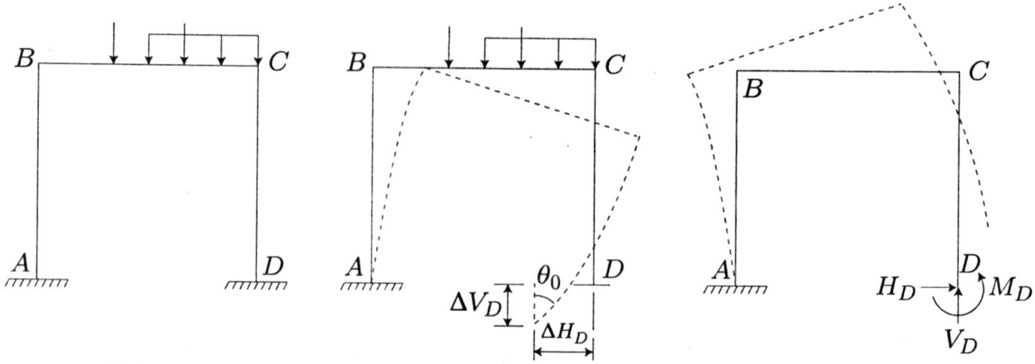

(a) Frame (b) M_S diagram (c) M_i diagram

Fig. 10.2: Frame fixed at ends

$$\left.\begin{aligned}\theta_D &= \int_A^D \frac{M_s ds}{EI} \\ \triangle H_D &= \int_A^D \frac{M_s y ds}{EI} \\ \text{and} \quad \triangle V_D &= \int_A^D \frac{M_s x ds}{EI}.\end{aligned}\right] \tag{10.7}$$

From compatibility of deformations at D of the original structure in Fig. 10.2(a), the deformations produced at D, by subjecting the basic structure to these redundants above [as in Fig. 10.2(c)] must be equal in magnitude but opposite in direction to those given by Eq. (10.7). If M_i represents the bending moment in the basic structure due to the redundants M_D, H_D and V_D, then

$$\left.\begin{aligned}\int_A^D \frac{M_s ds}{EI} &= \int_A^D \frac{M_i ds}{EI} \\ \int_A^D \frac{M_s y ds}{EI} &= \int_A^D \frac{M_i y ds}{EI} \\ \text{and} \quad \int_A^D \frac{M_s x ds}{EI} &= \int_A^D \frac{M_i x ds}{EI}.\end{aligned}\right] \tag{10.8}$$

Eqs. (10.2) and (10.8) can be considered as analogous such that,

$$W = \int_A^D \frac{M_i ds}{EI} \tag{10.9a}$$

$$M_x = \int_A^D \frac{M_i y ds}{EI} \tag{10.9b}$$

$$\text{and} \quad M_y = \int_A^D \frac{M_i x ds}{EI}. \tag{10.9c}$$

From the analogy it also follows that,

$$\sigma = M_i \quad \text{and} \quad \frac{ds}{EI} = dA$$

It is evident that,

(i) The moment M_i caused in a frame due to redundant forces is analogous to the stress σ in case of a column,

(ii) $\frac{ds}{EI}$ is analogous to dA, and width of the analogous column anywhere is equal to quantity $\frac{1}{EI}$ of the corresponding point of the actual structure.

(iii) The load, in the column is given by the M_s diagram acting along the width of the analogous column. Such loading is referred to as the analogous load.

From the above analogy, the name Column Analogy has come. If a frame be considered as an analogous column (short column) section of width $\frac{I}{EI}$, which is subjected to loads corresponding to the bending moment caused by the externally applied loads on the determinate form, then the axial stress at any point in the analogous column corresponds to the moment caused in the frame at that point, by the redundant forces. Eq. (10.5) can by rewritten in terms of the analogous column parameter as

$$M_i = \frac{W}{A} + \left[\frac{M_y I_{xx} - M_x I_{xy}}{I_{xx} I_{yy} - (I_{xy})^2}\right] x + \left[\frac{M_x I_{yy} - M_y I_{xy}}{I_{xx} I_{yy} - (I_{xy})^2}\right] y \tag{10.10a}$$

$$= \frac{W}{A} + \frac{M'_y}{I'_{yy}} x + \frac{M'_x}{I'_{yy}} y, \tag{10.10b}$$

where $M_x, M_y, I_{xx}, I_{yy}, I_{xy}$ are with reference to the centroidal reference axes of the analogous column and $M'_x = M_x - M_y\left(\frac{I_{xy}}{I_{yy}}\right)$; $M'_y = M_y - M_x\frac{I_{xy}}{I_{xx}}$; $I'_{xx} = I_{xx}\left[1 - \frac{(I_{xy})^2}{I_{xx}I_{yy}}\right]$ and$I'_{yy} = I_{yy}\left[1 - \frac{(I_{xy})^2}{I_{xx}I_{yy}}\right]$.

If these axes are the principal axes of the column, I_{xy} is zero and Eqns. (10.10a) and (10.10b) reduce to

$$M_i = \frac{W}{A} + \frac{M_y}{I_{yy}} x + \frac{M_x}{I_{xx}} y. \tag{10.10c}$$

The net bending moment at any section in the structure is given by

$$M = M_s - M_i. \tag{10.11}$$

This is obvious since the net moment anywhere in an indeterminate structure is obtained by superimposing the end moment diagram to the free moment diagram.

Values of the reaction V and H may be obtained similarly as for values of bending moment. Thus,

$$V = V_s - V_i \tag{10.12}$$

and $$H = H_s - H_i, \tag{10.13}$$

where V_s and V_i refer to the vertical reactions for the determinate case and the case of the redundant reactions, while H_s and H_i refer to the horizontal reactions for the corresponding cases.

Values of V_i and H_i are also obtained from

$$\left.\begin{aligned} V_i &= \frac{\partial M_i}{\partial x} = \frac{M_y I_{xx} - M_x I_{xy}}{I_{xx} I_{yy} - (I_{xy})^2} = \frac{M'_y}{I'_{yy}} \\ \text{and} \quad H_i &= \frac{\partial M_i}{\partial y} = \frac{M_x I_{yy} - M_y I_{xy}}{I_{xx} I_{yy} - (I_{xy})^2} = \frac{M'_x}{I'_{xx}}. \end{aligned}\right] \tag{10.14}$$

It is necessary to establish a sign convention for computation works. The sign convention includes :

(i) Loading on top of the column is downward for positive M_s which causes compression on the outside.

(ii) Upward pressure on the bottom of the column is positive.

(iii) Resultant moment at any point is positive, if it causes compression on the outside.

10.3 Method of Column Analogy Applied to Straight Members

The method of column analogy can be used to determine the fixed-end moments on a straight member when it is subjected to loads in the transverse direction. Fig. 10.3 shows a straight beam fixed at both ends and subjected to external loading. With the method of consistent deformation and choosing a simple beam AB as the basic determinate beam, the conditions from which M_A and M_B can be determined, are

(i) Change of slope between A and $B = 0$, i.e., area of Fig. 10.3(b)= area of Fig. 10.3(c).

(ii) Deflection of B from tangent at $A = 0$, i.e., moment of area of Fig. 10.3(b) about $B =$ moment of area of Fig. 10.3(c) about B.

If an imaginary short column with a cross section as shown in Fig. 10.3(e) be subjected to the loading on top and the pressure on bottom, as shown in Figs. 10.3(e) and 10.3(f) respectively, it is obvious that this column is in equilibrium due to the two conditions stated above.

As $\sum\left(\frac{M_i}{EI}\right) dx$ in Figs. 10.3(e) and 10.3(f) is the same as $\sum M_i(dx)\frac{1}{EI}$ in Figs. 10.3(h) and 10.3(i), the upward reactions over any division dx at the base of the respective columns are also the same.

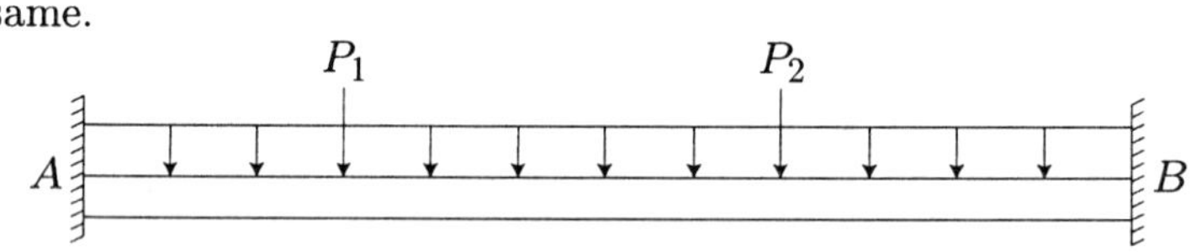

(a) Beam

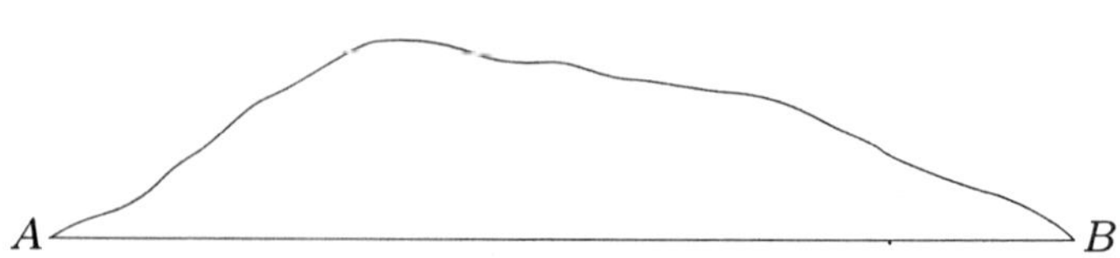

(b) diagram

Fig. 10.3: Method of column analogy (Contd.)

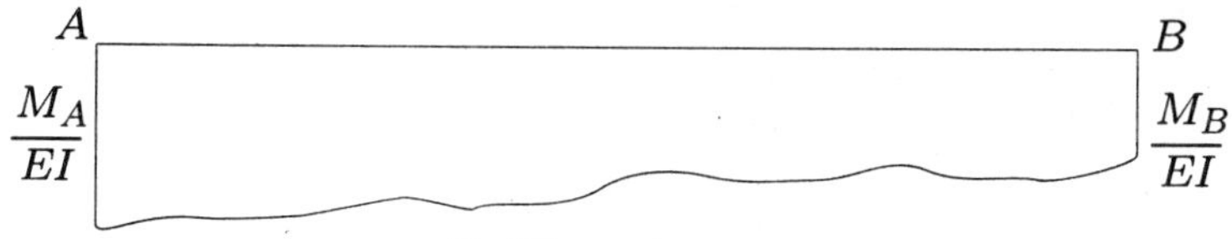

(c) $\frac{\text{End Moments}}{EI}$ diagram

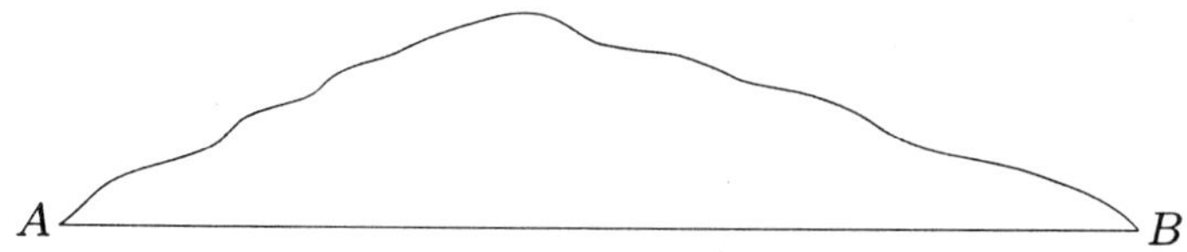

(d) Loading on analogous

Column (e) $\frac{M_S}{EI}$ diagram as in (b)

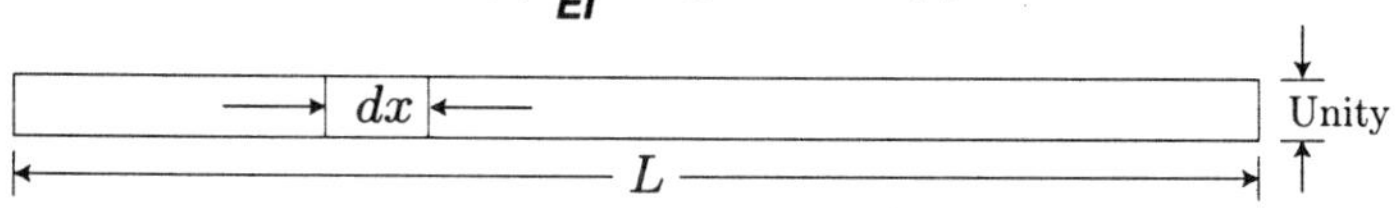

(e) Analogous column of height unity

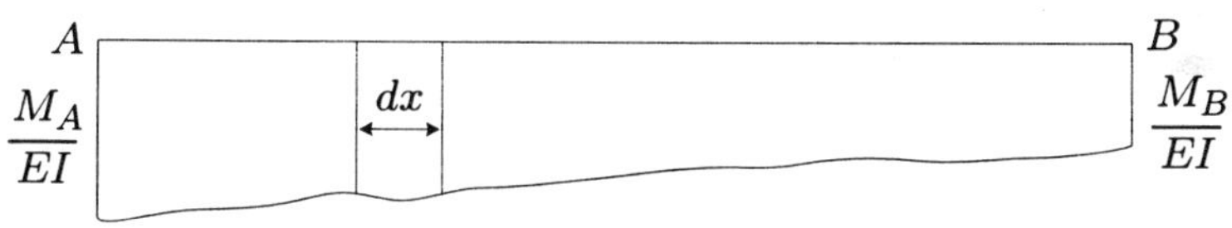

(f) $\frac{M_i}{EI}$ diagram: Pressure on bottom of analogous column

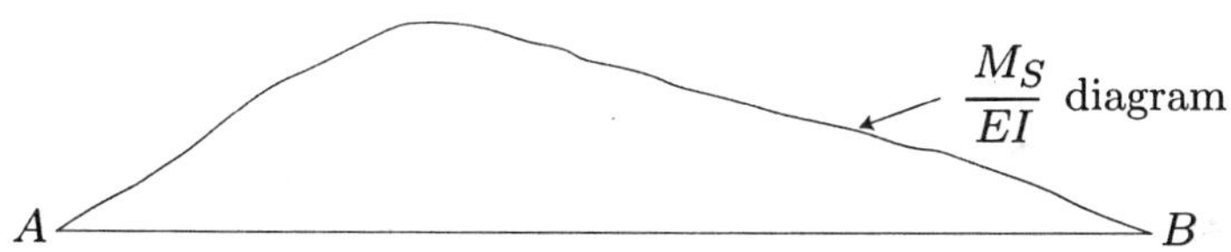

(g) Loading on analogous column (h)

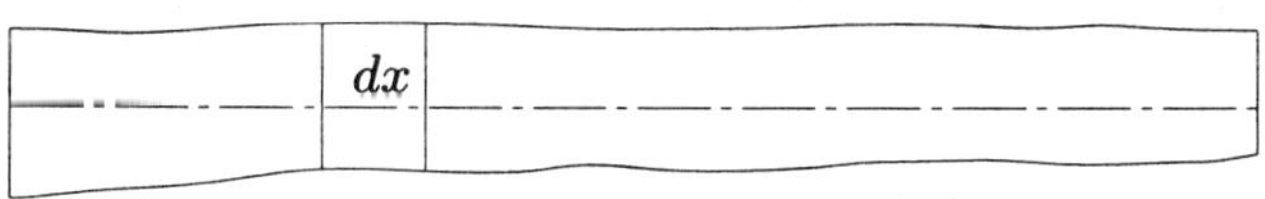

(h) Analogous column width = $\frac{1}{EI}$

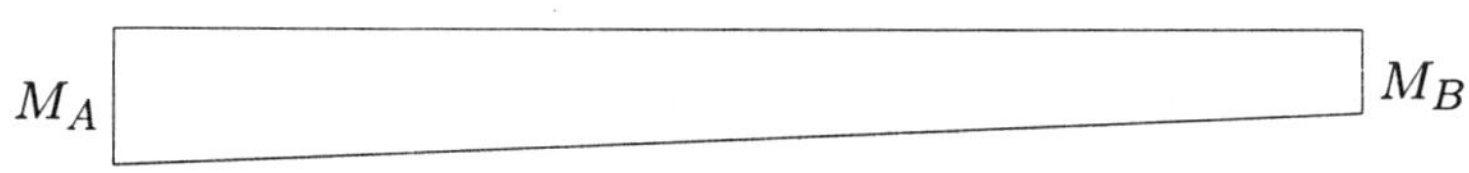

(i) M_i diagram: Pressure on bottom of Analogous, column

Fig. 10.3: Method of column analogy

10.4 The Column Analogy Method

Steps involved in analysing indeterminate structure by the Column Analogy method are as under:

(i) Remove suitable redundants to reduce the given structure to a basic determinate and stable structure and draw the $\frac{M_i}{EI}$ diagram.

(ii) Obtain the analogous column such that its central line is identical in shape and length to that of the actual structure and its width anywhere along its centre line equals $\frac{1}{EI}$ at the corresponding point in the actual structure.

(iii) Calculate the geometrical characteristics, viz., area, centroid, I_{xx}, I_{yy} and I_{xy} of the analogous column.

(iv) Load the analogous column with the $\frac{M_s}{EI}$ diagram of the basic structure. The load acts in line with the column centre line and is perpendicular to its cross section.

(v) Calculate the stresses in the analogous column thus determining indeterminate moments M_i in the actual structure.

(vi) Calculate the net or actual bending moment anywhere in the actual structure by superimposing the two moments M_s and M_i.

The various steps of the method provide a definite routine for flexural computation. The computations of the properties of the section may be tabulated and by proper arrangement much work may be saved.

In using the column analogy the arrangements of tabulation shown in Tables 10.1 and 10.2 are recommended. If the ordinates x, y, x_1, y_1 are given the proper signs and if the proper sign, according to the convention stated, be given to the static moment M_s, all signs in these Tables follow automatically.

Steps in the procedure are numbered in the Tables. In the first and third columns the properties of the sections and of the loads are filled up. The computation of the derived properties are written in the second and fourth columns. The totals of characteristics of the properties of sections and loads are given in step marked (5) in Table 10.1. Centroid is located as in step marked (6) and correction to centroid and characteristics are done in steps marked (7) and (8) in Table 10.1. For unsymmetrical structures correction for unsymmetry has to be done as shown in step marked (9) in Table 10.2. Corrections to characteristics taking effect of dissymmetry in step marked (9) is done as shown in step marked (10) in Table 10.2.

Values of x, y and x_1, y_1 are taken from any convenient pair of axes in the beginning. Co-ordinates are to be taken positive upward and to the right.

It is to be noted that the analysis of unsymmetrical sections is not specially more difficult than that for the symmetrical sections. Only two additional steps marked (9) and (10) in Table 10.2 are required.

10.5 Choice of Typical Determinate Moment Curves

Any moment pattern statically consistent with the loads may be used in applying the column analogy. Some of moment curves, however, are more convenient for analysis than the actual choice depending not only on the structure but also on the pattern of loading. Some of the typical cases are illustrated in Fig. 10.4.

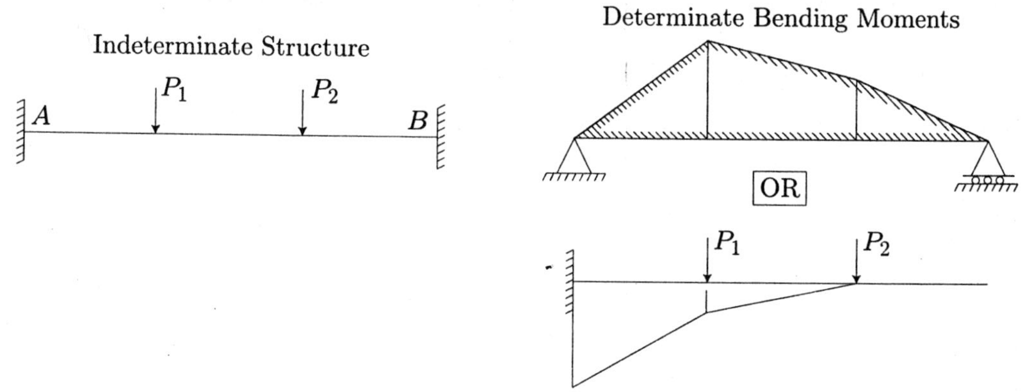

(a) Fixed ended beam

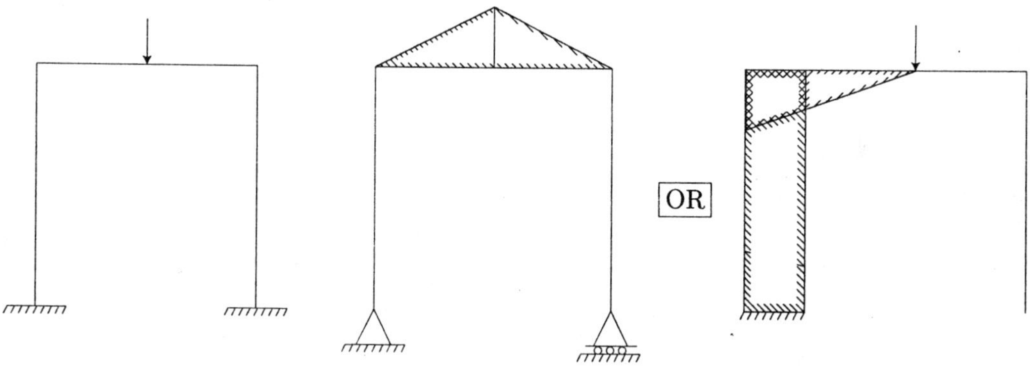

(b) Rigid frame with ends fixed

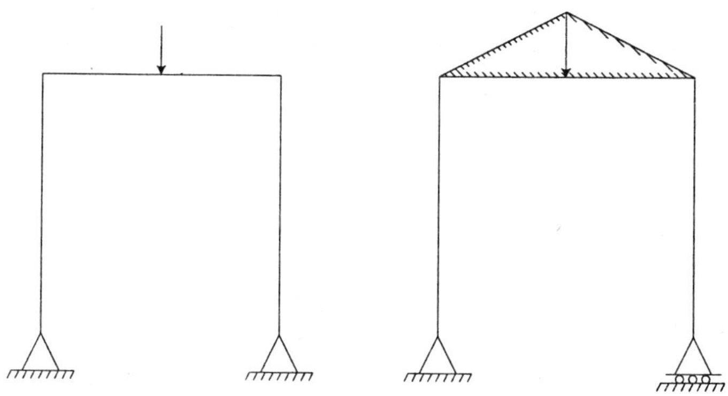

(c) Rigid frame with hinged ends

Fig 10.4: Typical Determinate Moment curves (Contd.)

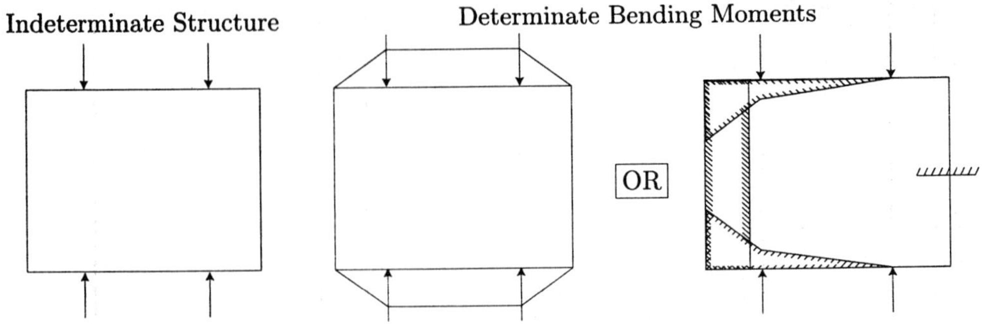

(d) Box frame

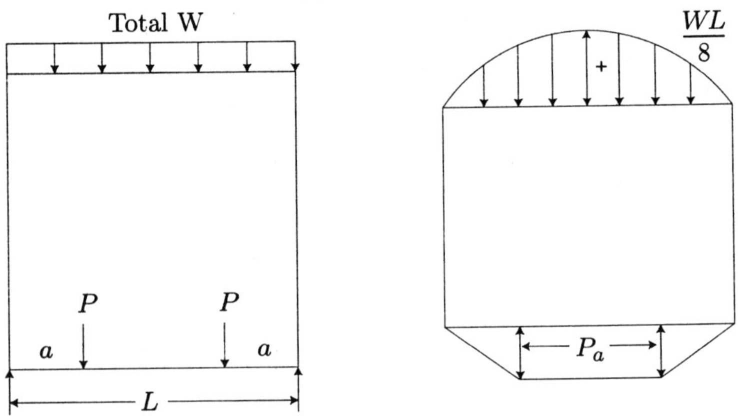

(e) Box frame

Fig 10.4: Typical Determinate Moment curves

Table 10.1. Column Analogy Tabular Form for Symmetrical Structures

<table>
<tr><td colspan="10">Properties of Section</td><td colspan="5">Loads of Section</td></tr>
<tr><td colspan="6" rowspan="2">(1) Given Properties</td><td colspan="4">(2) Computed Properties</td><td colspan="3" rowspan="2">(3) Given</td><td colspan="2" rowspan="2">(4) Computed</td></tr>
<tr><td colspan="2">Statical Moments</td><td colspan="2">Products of Inertia</td></tr>
<tr><td>Member</td><td>Length
L</td><td>Width
$\frac{1}{EI}$</td><td>Area
$a = \frac{L}{EI}$</td><td>x</td><td>y</td><td>ax</td><td>ay</td><td>$ax^2 + i_{yy}$</td><td>$ay^2 + i_{xx}$</td><td>Load</td><td colspan="2">Coordinates of Load</td><td colspan="2">Moments</td></tr>
<tr><td></td><td></td><td></td><td></td><td></td><td></td><td></td><td></td><td></td><td></td><td>$W_s = M_s(a)$</td><td>x_1</td><td>y_1</td><td>W_0x_1</td><td>W_0y_1</td></tr>
<tr><td colspan="3">(5) Totals</td><td>A</td><td></td><td></td><td>$A\overline{x}$</td><td>$A\overline{y}$</td><td></td><td></td><td>W</td><td></td><td></td><td>√</td><td>√</td></tr>
<tr><td colspan="3">(6) Locate centroid</td><td>$\overline{x} = \frac{A\overline{x}}{A}$
$\overline{y} = \frac{A\overline{y}}{A}$</td><td>√</td><td>√</td><td></td><td></td><td></td><td></td><td colspan="3"></td><td></td><td></td></tr>
<tr><td colspan="8">(7) Corrections to centroid</td><td>$A\overline{x}^2$</td><td>$A\overline{y}^2$</td><td colspan="3"></td><td>$W\overline{x}$</td><td>$W\overline{y}$</td></tr>
<tr><td colspan="8">(8) Subtract corrections</td><td>I_{yy}</td><td>I_{xx}</td><td colspan="3"></td><td>M_y</td><td>M_x</td></tr>
<tr><td colspan="3">(9) Compute moments or Forces by formulas</td><td colspan="12">$M_i = \frac{W}{A} + \frac{M_y}{I_{yy}}x + \frac{M_x}{I_{xx}}y$ $M = M_s - M_i$ Intercepts of Indeterminate Force
$V_i = \frac{M_y}{I_{yy}}$ $V = V_s - V_i$ $x_2 = -\frac{\frac{W}{A}}{\frac{M_y}{I_{yy}}}$ and $y_1 = \frac{\frac{W}{A}}{\frac{M_x}{I_{xx}}}$
$H_i = \frac{M_x}{I_{xx}}$ $H = H_s - H_i$</td></tr>
</table>

Table 10.2. Column Analogy Tabular Form for Unsymmetrical Sections

Properties of Section											Loads of Section				
(1) Given Properties						(2) Computed Properties					(3) Given			(4) Computed	
						Statical Moments		Products of Inertia							
Member	Length	Width	Area	x	y	ax	ay	ax^2+i_{yy}	ay^2+i_{xx}	$axy+i_{xy}$	Load	Co-ordinates of Load		Moments	
	L	$\frac{1}{EI}$	$a=\frac{L}{EI}$												
											$W_s = M_s(a)$	x_1	y_1	W_0x_1	W_0y_1
(5) Totals			A			$A\overline{x}$	$A\overline{y}$	√	√	√	W			√	√
(6) Locate centroid			$\overline{x}=\frac{A\overline{x}}{A}$	√											
			$\overline{y}=\frac{A\overline{y}}{A}$		√										
(7) Corrections to centroid								$A\overline{x}^2$	$A\overline{y}^2$	$A\overline{x}\,\overline{y}$				$W\overline{x}$	$W\overline{y}$
(8) Subtract corrections								I_{yy}	I_{xx}	I_{xy}				M_y	M_x
(9) Corrections for unsymmetry								$\frac{I_{xy}^2}{I_{xx}}$	$\frac{I_{xy}^2}{I_{yy}}$					$\frac{M_x}{I_{xx}} - I_{xy}$	$\frac{M_y}{I_{yx}} - I_{xy}$
(10) Subtract corrections								I'_{yy}	I'_{xx}					M'_y	M'_x

(11) Compute moments or forces by

$$M_i = \frac{W}{A} + \frac{M'_y}{I'_{yy}}x + \frac{M'_x}{I'_{xx}}y$$

$$V_i = \frac{M'_y}{I'_{yy}}, \quad H_i = \frac{M'_x}{I'_{xx}}$$

$$M = M_s - M_i$$
$$V = V_s - V_i$$
$$H = H_s - H_i$$

Intercepts of Intermediate Forces

$$x_i = \frac{\frac{W}{A}}{\frac{M'_y}{I'_{yy}}}; \quad y_i = \frac{\frac{W}{A}}{\frac{M'_x}{I'_{xx}}}$$

10.6 Analogous Columns

The analogous column is a short column with a cross section in plan where centre line is identical in shape and length to that of the real frame under investigation. The width of the column section anywhere is $\frac{1}{EI}$ of the corresponding point of the actual structure. At a hinge point the flexural rigidity is zero and hence $\frac{1}{EI}$ is infinite. For fixed support again flexural rigidity is infinity and $\frac{1}{EI}$ is zero. Such characteristics of the support conditions are

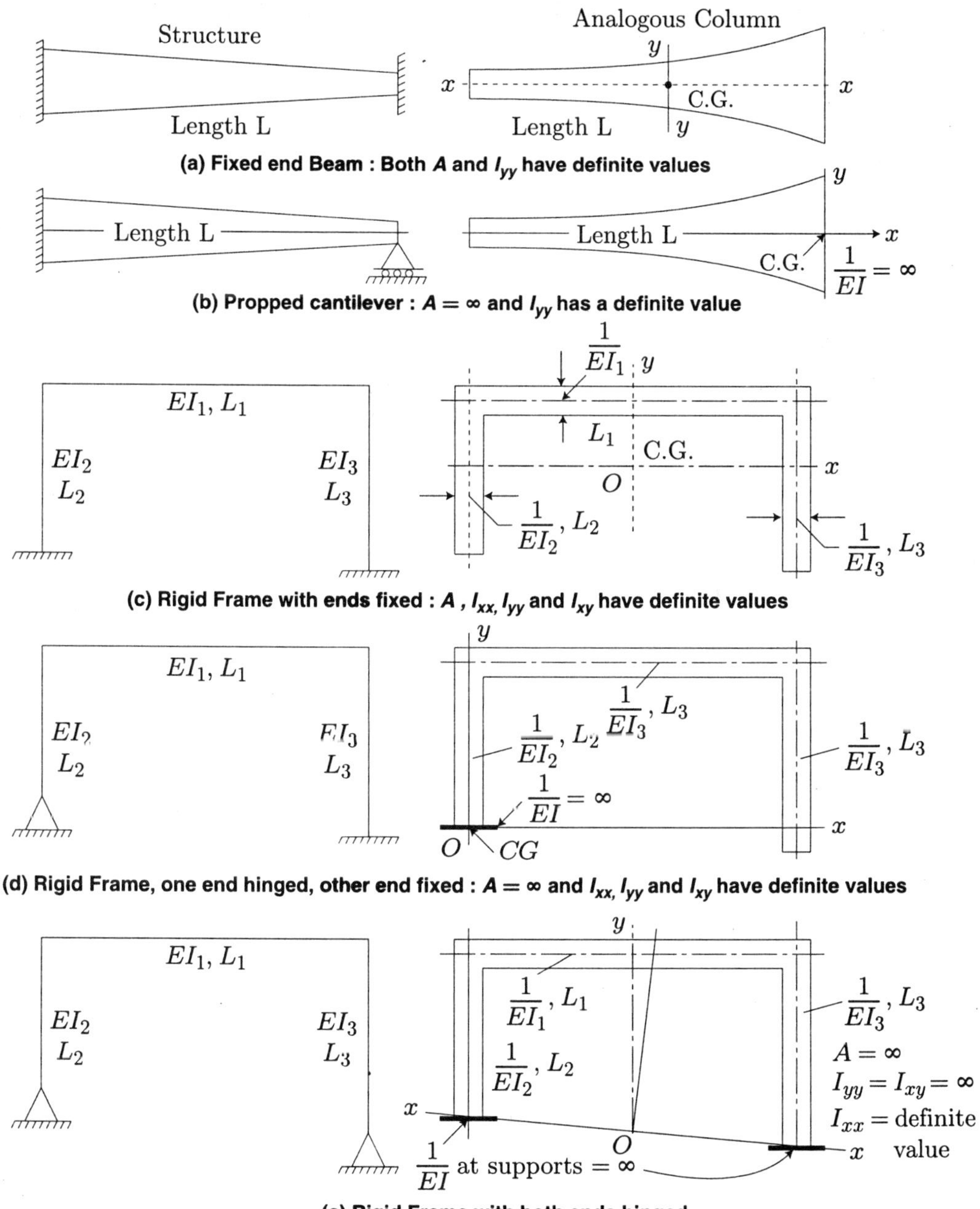

Fig. 10.5: Characteristics of analogous columns

to be considered in arriving at the properties of the analogous column. Typical analogous columns and centroidal axes for various cases are shown in Fig. 10.5.

All properties of beams of irregular section may be found by dividing the analogous column section into lengths so short that their moment of inertia about their own centroids may be neglected. Since in most cases, a large part of the beam is of uniform or nearly uniform section, this part can be treated as one section.

In calculating properties of analogous column, characteristics of geometrical data as given in Fig. 10.6 are sometimes useful, particularly when the frame is a gabled one and for cases of members of non-prismatic section.

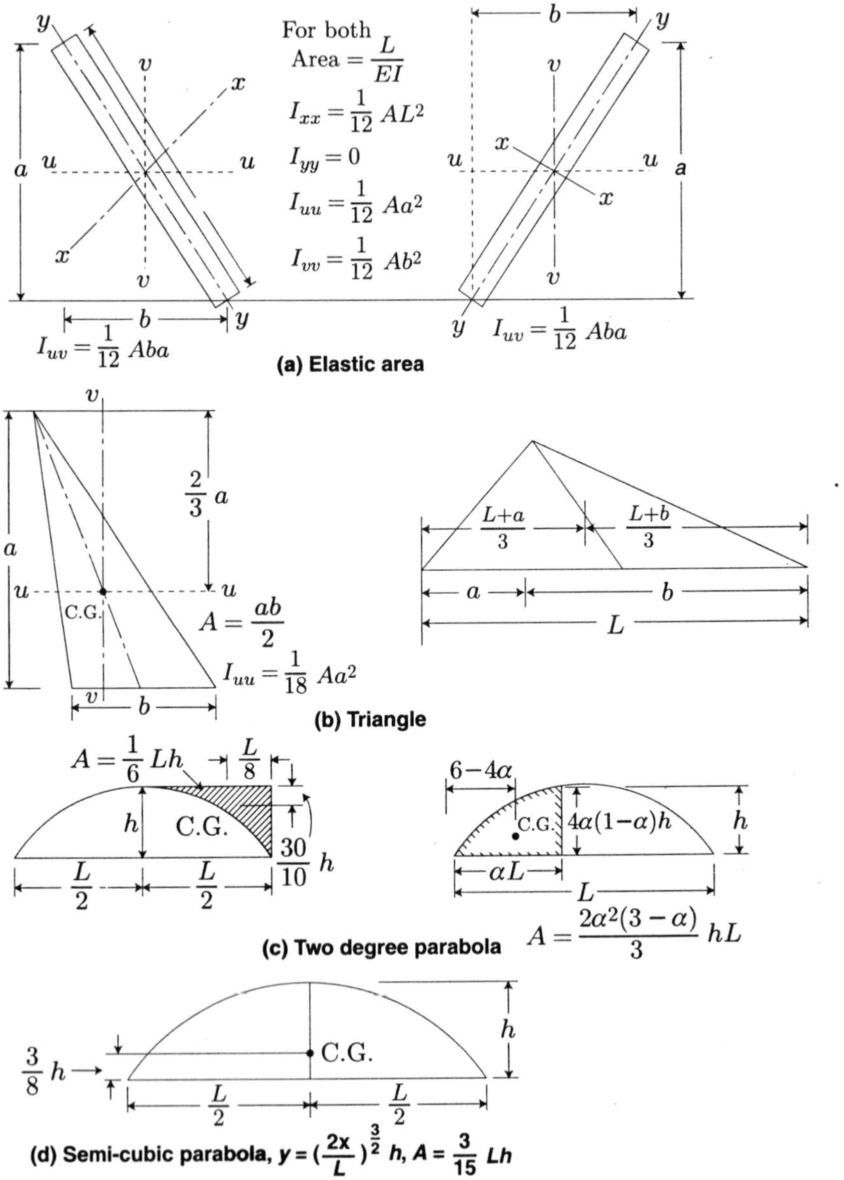

Fig. 10.6: Characteristics of Geometrical Data

Example 10.1. *Analyse the prismatic section beam shown in Fig. 10.7(a) by Column Analogy method.*

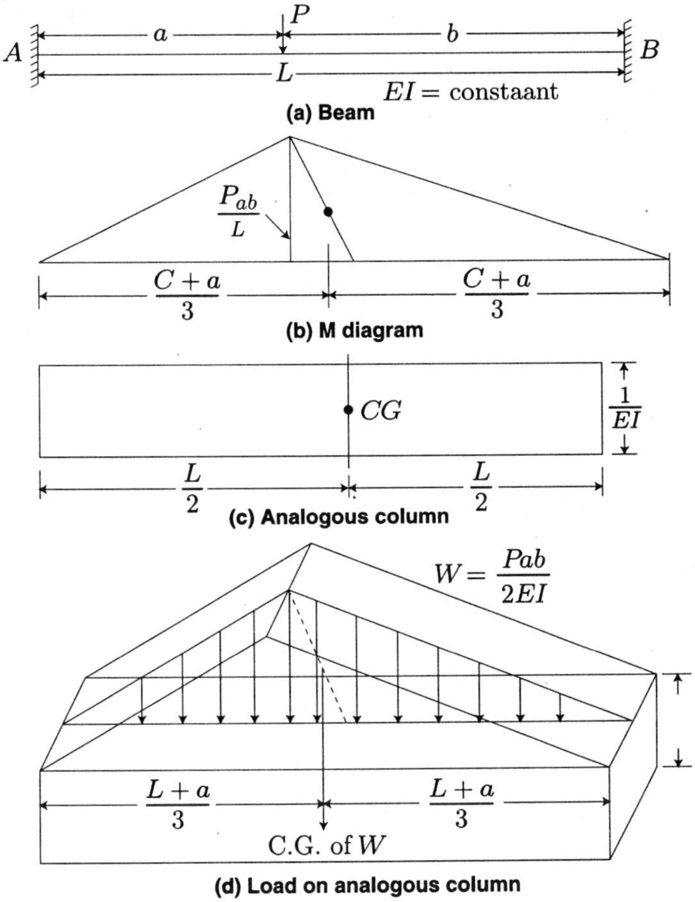

Fig. 10.7: For Example 10.1

Solution : Properties of Analogous column, Fig. 10.7(c).

$$A = \frac{L}{EI};\ I_{yy} = \frac{L^3}{12EI}.$$

From Fig. 10.7(d), Load $W = \frac{1}{2}\frac{Pab}{L}\left(\frac{L}{EI}\right) = \frac{Pab}{2EI}.$

According to sign convention, sign of W is positive and can be considered as acting downwards.

Eccentricity of Load W, i.e., $e = \frac{L}{2} - \frac{L+a}{3} = \frac{L-2a}{6},\ \frac{W}{A} = \frac{Pab}{2EI}\left(\frac{1}{\frac{L}{EI}}\right) = \frac{Pab}{2L}.$

For the case $a < b$, both $e = \frac{L-2a}{6}$ and for point $A, y = \frac{L}{2}$ are $+Ve$

and $$\frac{My}{I} = \frac{\frac{Pab}{2EI}\left(\frac{L-2a}{6}\right)}{\frac{L^3}{12EI}}\frac{L}{2} = \frac{Pab}{2L^2}(L-2a)$$

$$M_i = \frac{W}{A} + MyI \quad \text{and} \quad M_A = M_s - M_i = 0 - \left[\frac{Pab}{2L} + \frac{Pab}{2L^2}(L-2a)\right] = -\frac{Pa^2b}{L^2}.$$

For point B, $y = -\dfrac{L}{2}$ and , $\dfrac{My}{I} = -\dfrac{Pab}{2L^2}(L-2a)$,

whence $M_B = M_s - M_i = 0 - \left[\dfrac{Pab}{2L} - \dfrac{Pab}{2L^2}(L-2a)\right] = -\dfrac{Pa^2b}{L^2}$.

For the case $a > b$, $e = \dfrac{L-2a}{6}$ is $-ve$ and $y = \pm\dfrac{L}{2}$,

and $M_A = -\dfrac{Pab^2}{L^2}$ and $M_B = -\dfrac{Pa^2b}{L^2}$.

Example 10.2. *Determine the bending moments in the beam shown in Fig. 10.8(a) by Column Analogy method.*

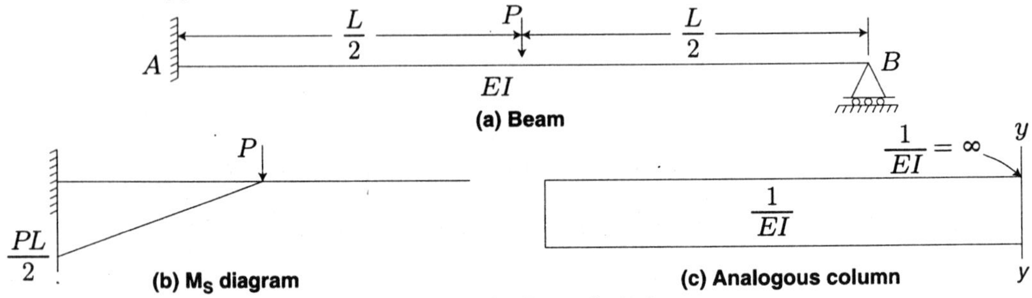

Fig. 10.8: For Example 10.2

Solution : In this case, area of analogous column = infinity.

C.G. passes through the support at B and $I_{yy} = \dfrac{L^3}{3EI}$.

$$\text{Load } W = -\frac{1}{2}\left(\frac{PL}{2}\right)\left(\frac{L}{2EI}\right) = -\frac{PL^2}{EI}$$

and eccentricity of load $= \left(L - \dfrac{1}{3}\dfrac{L}{2}\right) = \dfrac{5L}{6}$.

$$M_A = M_s - M_i = -\frac{PL}{2} - \left[-\frac{PL^2}{8EI}\left(\frac{1}{\infty}\right) - \frac{PL^2}{8EI}\left(\frac{5L}{6}\right)\frac{1}{\frac{L^3}{3EI}}\right](L)$$

$$= -\frac{PL}{2} + \frac{5PL}{16} = -\frac{3PL}{16}.$$

Example 10.3. *Determine the fixed end moments for the beam shown in Fig. 10.9(a).*

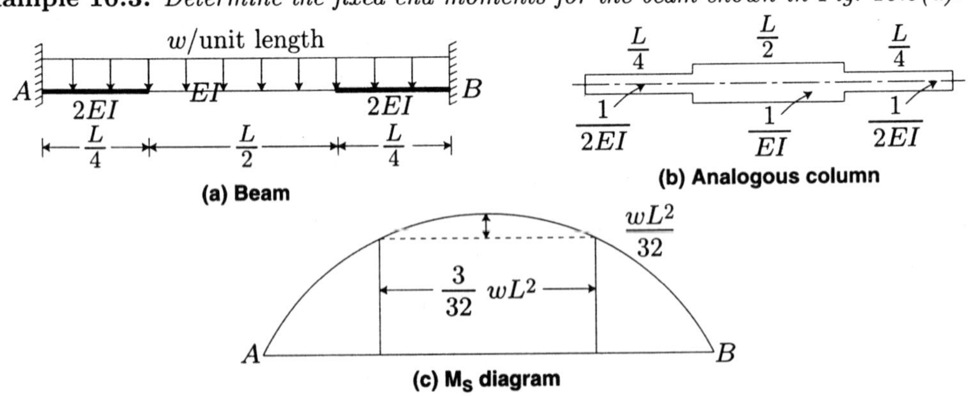

Fig. 10.9: For Example 10.3

Solution : Fig. 10.9(b) shows the analogous column.

$$\text{Area of analogous column} = \frac{L}{8EI} + \frac{L}{2EI} + \frac{L}{8EI} = \frac{3}{4}\frac{L}{EI}$$

$$I_{yy} = 2\left[\frac{1}{2EI}\frac{\left(\frac{L}{4}\right)^3}{12} + \frac{L}{8EI}\left(\frac{3L}{8}\right)^2\right] + \frac{1}{EI}\frac{\left(\frac{L}{2}\right)^2}{12} = \frac{3L^3}{64EI}$$

$$W = \frac{2}{3}\left(\frac{wL^2}{32}\right)\left(\frac{L}{2EI}\right) + \frac{3wL^2}{32}\left(\frac{L}{2EI}\right) + \frac{2}{3}\frac{wL^2}{8}\left(\frac{L}{2EI}\right) - \left[\frac{2}{3}\left(\frac{wL^2}{32}\right)\left(\frac{L}{4EI}\right) + \frac{3wL^2}{32}\left(\frac{L}{4EI}\right)\right]$$

$$= \frac{9}{128}\left(\frac{wL^3}{EI}\right).$$

$$M_A = M_B = 0 - \frac{9}{128}\left(\frac{wL^3}{\frac{3}{4}L}\right) = -\frac{3wL^2}{32}.$$

Example 10.4. *Analyse the beam shown in Fig. 10.10(a) by Column Analogy method.*

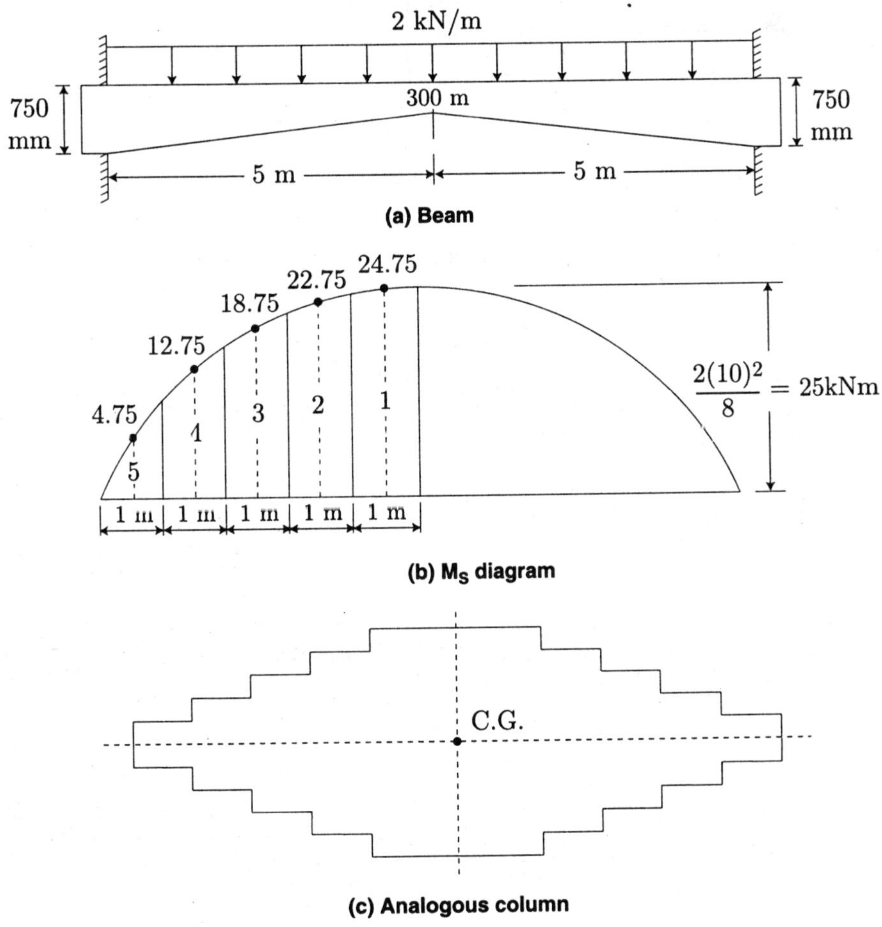

Fig. 10.10: For Example 10.4

Solution : Table 10.3 gives the solution of the beam in tabular form.

Table 10.3. For Example 10.4

Section	Length m	d in m	x (m)	a (m^2)	$(ax^2 + i_x)$	M_s	$W = M_{sa}$	
1	1	0.345	±0.50	24.35	$6.09 + 2.03$	+24.75	+602.66	$a = \frac{L}{d^3}$
2	1	0.435	±1.50	12.15	$27.34 + 1.01$	+22.75	+276.41	taking E as
3	1	0.525	±2.50	6.91	$43.19 + 0.58$	+18.75	+129.56	unity and as
4	1	0.615	±3.50	4.30	$52.68 + 0.36$	+12.75	+54.83	$I = \frac{bd^3}{12} = d^3$
5	1	0.705	±4.50	2.85	$57.71 + 0.24$	+4.75	+13.54	with $b = 12$
			Σ	50.56	191.22	Σ	1077	

Total Area, $A = 101.12$; M.I., $I = 382.49$ and $W = 2154$

$$\text{At } A \text{ and } B,\ M_i = +\frac{W}{A} = \frac{2(1077)}{101.12} = 21.30$$

$$M_A = M_s - M_i = -21.30 \text{ kN-m} = M_B.$$

Example 10.5. *Analyse the beam in Fig. 7.11(a), by Column Analogy Method for end moments due to unit rotation at one end.*

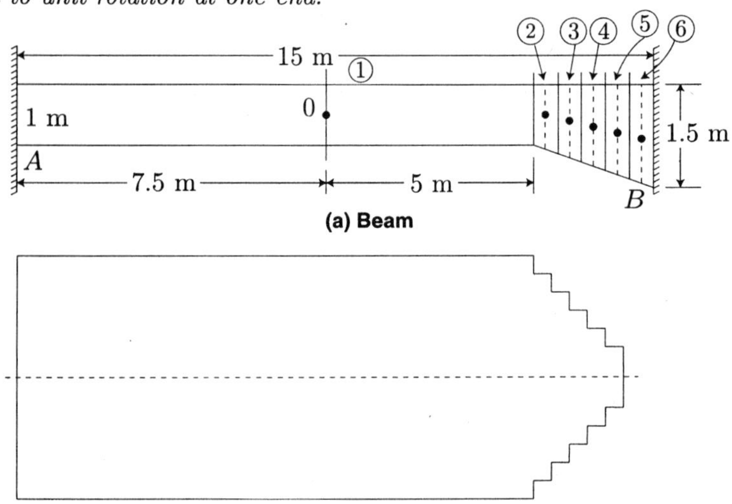

(a) Beam

(b) Analogous column

Fig. 10.11: For Example 10.5

Solution : Table 10.4 showing solution of the beam is made with respect to a point 'O' as shown in Fig. 10.11.

For unit rotation at A, applying unit load on analogous column at$A$,

$$M_{iA} = \frac{1}{13.884} + \frac{(-7.5 + 0.526)(-7.5 + 0.526)}{229.11} = +0.284$$

$$M_{iB} = \frac{1}{13.884} + \frac{(-7.5 + 0.526)(15 - 7.5 + 0.526)}{229.11} = -0.172$$

and $\quad M_A = -M_{iA} = -0.284$ and $M_b = -M_{iB} = +0.172.$

For unit rotation at B, applying unit load on the analogous column at B,

$$M_{iB} = \frac{1}{13.884} + \frac{(7.5+0.526)(7.5+0.526)}{229.13} = +0.353$$

and $$M_{iA} = \frac{1}{13.884} + \frac{(7.5+0.526)(-7.5+0.526)}{229.13} = -0.172.$$

Table 10.4. For Example 10.5

Section	Length (m)	d (m)	x (m)	a (m^2)	ax	$ax^2 + i_x$	Remarks
1	12.5	1	−1.25	12.50	−15.625	19.55	
						+162.76	
2	0.50	1.05	+5.25	0.432	+2.268	11.91	
						+0.00	$i_x \approx 0.00$
3	0.50	1.15	+5.75	0.329	+1.892	10.88	
						+0.00	$i_x \approx 0.00$
4	0.50	1.25	+6.25	0.256	+1.600	10.00	$a = \frac{L}{d^3}$ taking
						+0.00	E as unity and b as 12
5	0.50	1.35	+6.75	0.203	+1.370	9.25	
						+0.00	$i_x \approx 0.00$
6	0.50	1.45	+7.25	0.164	+1.189	8.62	
						+0.00	$i_x \approx 0.00$
				13.884	−7.30	232.97	
			−0.526			−3.84	$\overline{x} = -\frac{7.30}{13.884}$
			$\sum$	13.884		229.13	

Example 10.6. *Analyse the portal frame shown in Fig. 10.12(a) by the method of Column Analogy.*

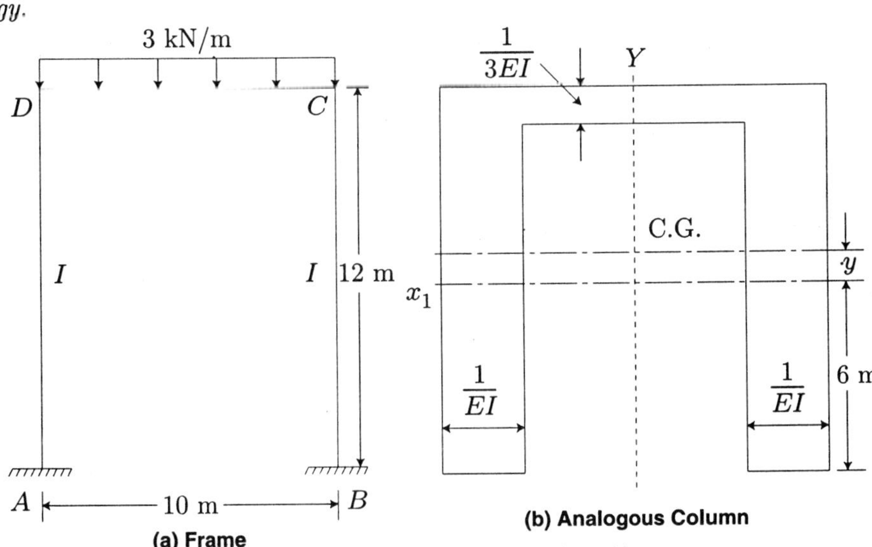

Fig. 10.12: For Example 10.6 (Contd.)

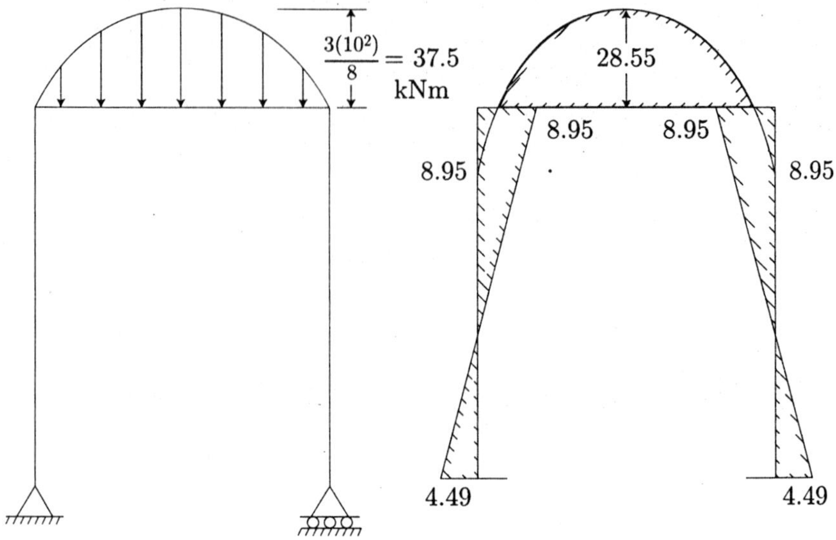

(c) M_S Diagram (d) BM Diagram: in kNm

Fig. 10.12: For Example 10.6

Solution : For the analogous column shown in Fig. 10.12(b),

Area $A = \dfrac{2(12)}{EI} + \dfrac{10}{3EI} = \dfrac{27.33}{EI}$.

To obtain C.G.,

$$y = \frac{\frac{10}{3EI}(6)}{\frac{27.33}{EI}} = 0.73 \text{ m}$$

$$I_{xx} = I_{x_1x_1} - Ay^2 = 2\left(\frac{1}{EI}\right)\frac{(12)^3}{12} + \frac{10}{3EI}(6)^2 - \frac{27.33}{EI}(0.73)^2 = \frac{393.50}{EI}.$$

M_s diagram is shown in Fig. 10.12(c).

Total load on analogous column

i.e., $W = \dfrac{2}{3}(37.50)\dfrac{10}{3EI} = \dfrac{250}{3EI}$.

Also $M_x = \dfrac{250}{3EI}[6 - 0.73] = \dfrac{1317.5}{3EI}; M_y = 0$

Then $M_i = \dfrac{W}{A} + \dfrac{M_x}{I_{xx}}y = \dfrac{250(EI)}{3EI(27.33)} + \dfrac{1317.5}{3EI}\left(\dfrac{EI}{393.50}\right)y = 3.05 + 1.12y$.

At A and $D, M_{iA} = M_{iD} = -4.49$

At B and $C, M_{iB} = M_{iC} = +8.95$

B.M. at A and $D, M_A = M_D = M_s - M_i = +4.49$ kNm.

B.M. at B and $C, M_B = M_C = M_s - M_i = -8.95$ kNm.

Resulting B.M. diagram is shown in Fig. 10.12(d).

Example 10.7. *Analyse the frame shown in Fig. 10.13(a) by the method of Column Analogy.*

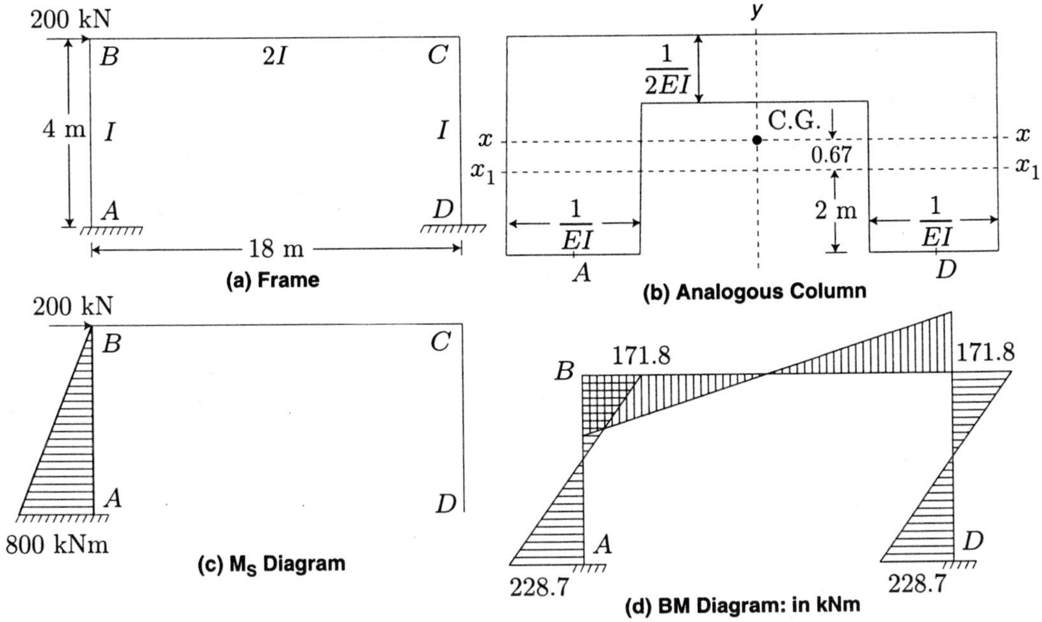

Fig. 10.13: For Example 10.7

Solution : For the analogous column shown in Fig. 10.13(b),

Area $A = \frac{8}{2EI} + \frac{2(4)}{EI} = \frac{12}{EI}$.

For C.G., $y = \frac{\frac{8}{2EI}(2)}{\frac{12}{EI}} = 0.67$ m.

Also $I_{xx} = I_{x_1x_1} - Ay^2 = 2\left(\frac{1}{EI}\right)\frac{4^3}{12} + \frac{8}{2EI}(2)^2 - \frac{12}{EI}(0.67)^2 = \frac{21.28}{EI}$

$$I_{yy} = \frac{1}{2EI}\frac{(8)^3}{12} + 2\frac{4}{EI}(4)^2 = \frac{149.33}{EI}$$

Total load on analogous column,

$$W = \frac{1}{2}\frac{(4)}{EI}(800) = -\frac{1600}{EI}$$

$$M_x = -\frac{1600}{EI}\left(-2.67 + \frac{3}{4}\right) = +\frac{2133}{EI}$$

and $M_y = -\frac{1600}{EI}(-4) = +\frac{6400}{EI}$.

$$M_i = -\frac{1600}{12} + \frac{2133}{21.28}y + \frac{6400}{149.33}x = -133.3 + 100.2y + 42.9x$$

At A, $M_A = -800 - [-133.3 + 100.2(-2.67) + 42.9(-4)] = -228.7$ kNm

$M_B = 0 - [-133.3 + 100.2(1.33) + 42.9(-4)] = -171.8$ kNm$= -M_C$.

B.M. diagram is shown in Fig. 10.13(d).

Example 10.8. *Analyse the two-hinged symmetric frame shown in Fig. 10.14(a) by the method of Column Analogy.*

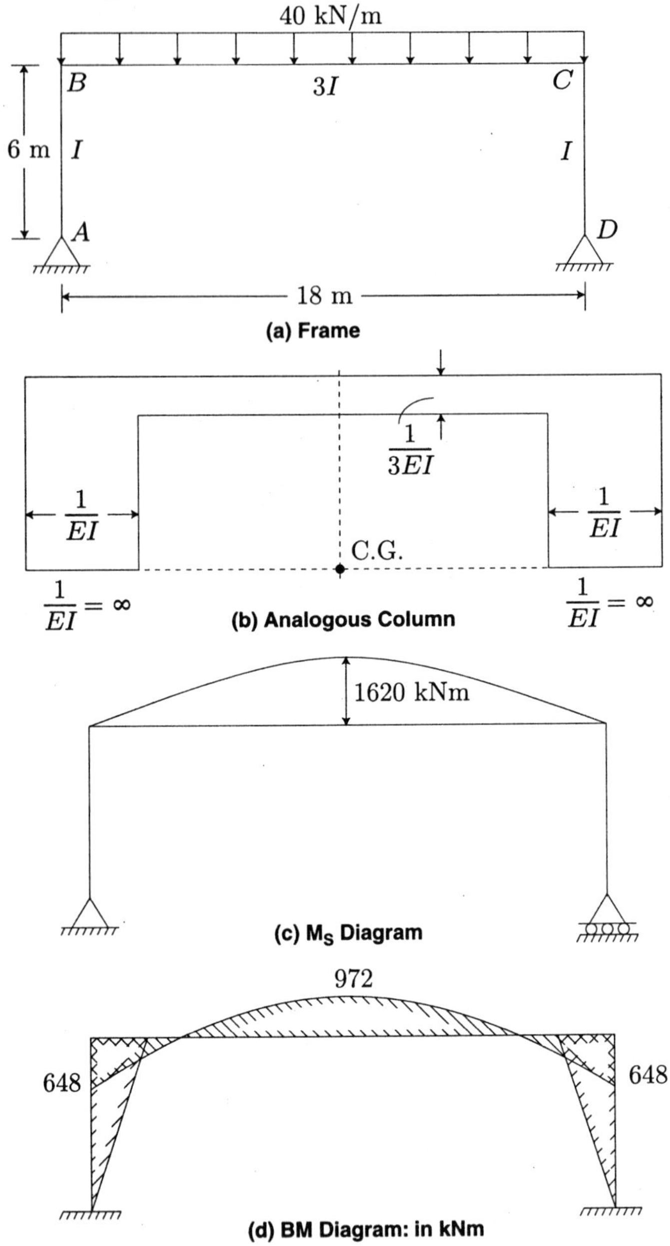

Fig. 10.14: For Example 10.8

Solution : Fig. 10.14(b) shows the analogous column.

Area of analogous column = infinity

$$I_{xx} = \frac{18}{3EI}(6)^2 + 2\frac{(6)^3}{12EI} + \frac{2(6)}{EI}(3)^2 = \frac{360}{EI} \text{ and } I_{yy} = \text{ infinity.}$$

Load on analogous column $= W = \dfrac{2}{3}(1620)(18)\dfrac{1}{3EI} = \dfrac{6480}{EI}$.

$$M_x = \frac{6480}{EI}(6) = \frac{38880}{EI}.$$

Then $M_B = M_s - M_i = 0 - \left[\dfrac{38880}{360}(6)\right] = -648$ kNm.

B.M. diagram is shown in Fig. 10.14

Example 10.9. *Analyse frame shown in Fig. 10.15(a) by the method of Column Analogy.*

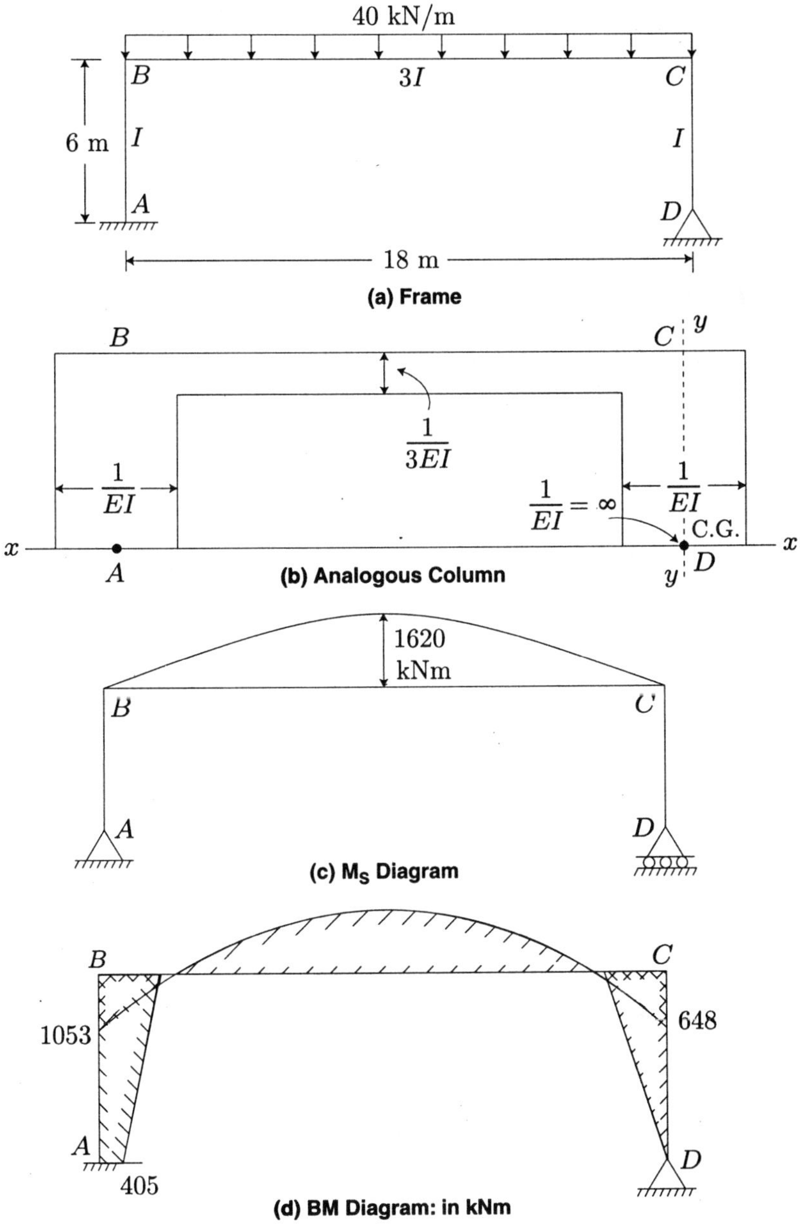

Fig. 10.15: For Example 10.9

Solution : The analogous column is shown in Fig. 10.15(b).

Area of analogous column= infinity

$$I_{xx} = \frac{360}{EI} \text{ and } I_{yy} = \frac{1}{3EI}\frac{(18)^3}{3} + \frac{6}{EI}(18)^2 = \frac{2592}{EI}.$$

Load on analogous column $W = \dfrac{6480}{EI}$

$$M_x = \frac{38880}{EI} \quad \text{and} \quad M_y = \frac{6480}{EI}(-9) = -\frac{58320}{EI}$$

Also $M_i = \dfrac{38880}{EI}\left(\dfrac{EI}{360}\right)y + \left(-\dfrac{58320}{EI}\right)\left(\dfrac{EI}{2592}\right)x = 108.0y - 22.5x$

$$M_A = 0 - [108.0(0) - 22.5(-18)] = -405.0 \text{ kNm}$$

$$M_B = 0 - [108.0(6) - 22.5(-18)] = -1053.0 \text{ kNm}$$

$$M_C = 0 - [108.(0) - 22.5(0)] = -648.0 \text{ kNm}$$

$$M_D = 0 - [108(0) - 22.5(0)] = 0.$$

B.M. diagram is shown in Fig. 10.15(d).

Example 10.10. *Analyse the closed frame in Fig. 10.16(a) by the method of Column Analogy.*

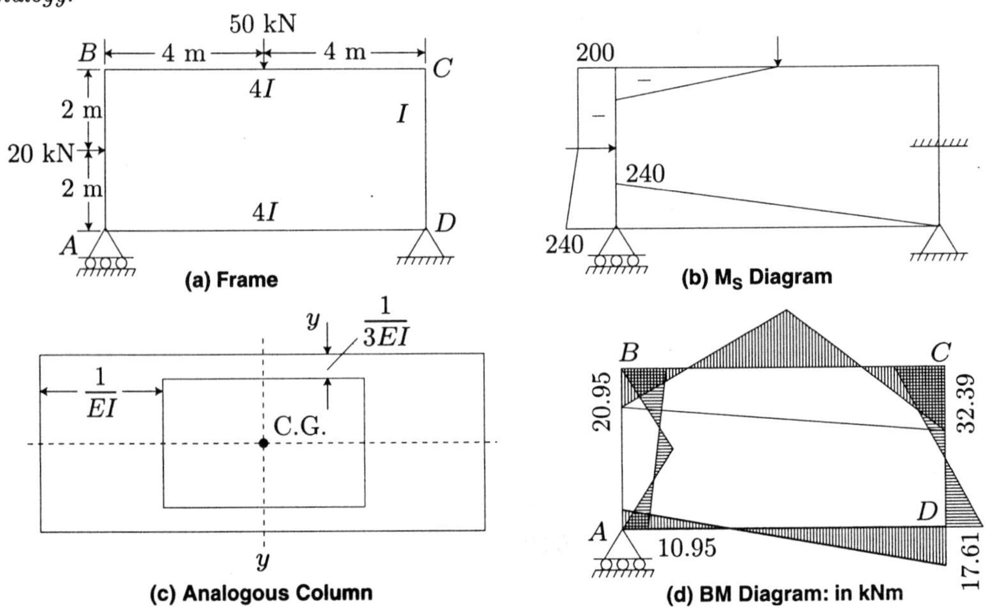

Fig. 10.16: For Example 10.10

Solution : The analogous column is shown in Fig. 10.16(c).

$$\text{Area of Analogous Column} = 2\left[\frac{8}{4EI} + \frac{4}{EI}\right] = \frac{12}{EI}$$

$$I_{xx} = 2\frac{1}{EI}\frac{(4)^3}{12} + 2\left(\frac{8}{4EI}\right)(2)^2 = \frac{26.67}{EI}$$

$$I_{yy} = 2\left(\frac{4}{EI}\right)(4)^2 + 2\left(\frac{1}{4EI}\right)\frac{(8)^3}{12} = \frac{149.33}{EI}.$$

Load on analogous column

$$W = -\left[\frac{1}{2}(200)(4)\frac{1}{4EI} + (200)(2)\left(\frac{1}{EI}\right) + \frac{200+240}{2}(2)\frac{1}{EI} + \frac{1}{2}(240)(8)\frac{1}{4EI}\right] = -\frac{1180}{EI}$$

$$M_y = +\left[\frac{1}{2}(200)(4)\frac{1}{4EI}\left(4-\frac{4}{3}\right) + (200)\frac{(4)}{EI}(4) + \frac{1}{2}(40)\left(\frac{2}{EI}\right)(4) + \frac{1}{2}(240)\frac{8}{4EI}\left(4-\frac{8}{3}\right)\right]$$

$$= +\left[\frac{800}{3EI} + \frac{3200}{EI} + \frac{160}{EI} + \frac{320}{EI}\right] = \frac{3946.67}{EI}$$

$$M_x = -\left[\frac{1}{2}(200)(4)\frac{1}{4EI}(2) + \frac{1}{2}(40)\left(\frac{2}{EI}\right)\left(-\frac{4}{3}\right) + \frac{1}{2}(240)\frac{8}{4EI}(-2)\right] = +\frac{333.33}{EI}.$$

Then $M_i = -\dfrac{1180}{EI}\left(\dfrac{EI}{12}\right) + \dfrac{333.33}{EI}\left(\dfrac{EI}{26.67}\right)y + \dfrac{3946.67}{EI}\left(\dfrac{EI}{149.33}\right)x$

$$= -98.33 + 12.5y + 26.43x$$

$$M_A = -240 - [-98.33 + 12.5(-2) + 26.43(-4)] = -10.95$$

$$M_B = -20.95;\ M_C = -32.39 \text{ and } M_D = +17.61.$$

B.M. diagram is shown in Fig. 10.16(d).

Example 10.11. *Analyse the gabled frame in Fig. 10.17(a) by the method of Column Analogy.*

Solution : Fig. 10.17(b) shows the analogous column.

Area of analogous column $A = 2\left(\dfrac{4}{EI}\right) + 2\left(\dfrac{6.32}{2EI}\right) = \dfrac{14.32}{EI}.$

Distance of C.C. from base $y_0 = \dfrac{1}{14.32}\left[2(4)(2) + \dfrac{2(6.32)}{2}\left(4+\dfrac{2}{2}\right)\right] = 3.32$ m.

$$I_{xx} = \frac{2}{EI}\left[\frac{4^3}{12} + 4(3.32-2)^2 + \frac{(6.32)}{2}\frac{(2)^2}{12} + \frac{6.32}{2}(5-3.32)^2\right] = \frac{44.55}{EI}$$

$$I_{yy} = \frac{2}{EI}\left[4(6)^2 + \frac{6.32}{2}\frac{(6)^2}{12} + \frac{6.32(6-3)^2}{2}\right] = \frac{363.84}{EI}.$$

Load on the analogous column is shown in Fig. 10.17(c),

Total load $W = 2(2/3)\dfrac{(360)(6.32)}{2EI} = \dfrac{1516.80}{EI}.$

$$M_x = \frac{1516.80}{EI}\left[4 - 3.32 + \frac{5}{8}(2)\right] = \frac{2927.42}{EI}$$

$$M_y = 0$$

$$M_A = 0 - \left[\frac{1516.80}{14.32} + \frac{2927.42}{EI}\left(\frac{EI}{44.55}\right)(-3.32)\right] = +112.24$$

$$M_B = 0 - \left[\frac{1516.80}{14.32} + \frac{2927.42}{44.55}(4 - 3.32)\right] = -150.61$$

$$M_C = 360 - \left[\frac{1516.80}{14.32} + \frac{2927.42}{44.55}(2 + 4 - 3.32)\right] = 77.97.$$

B.M. diagram is shown in Fig. 10.17(d).

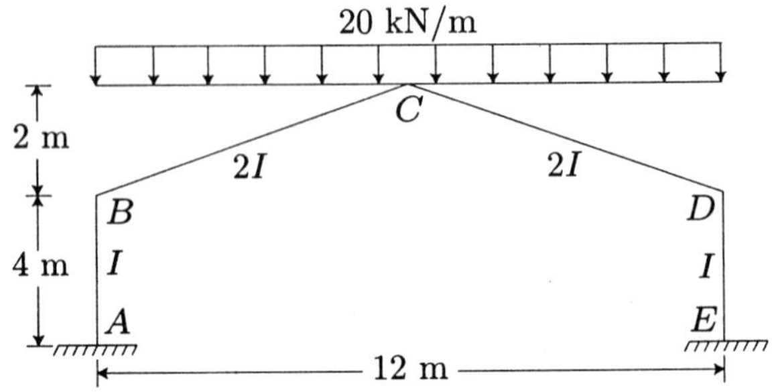

(a) Frame

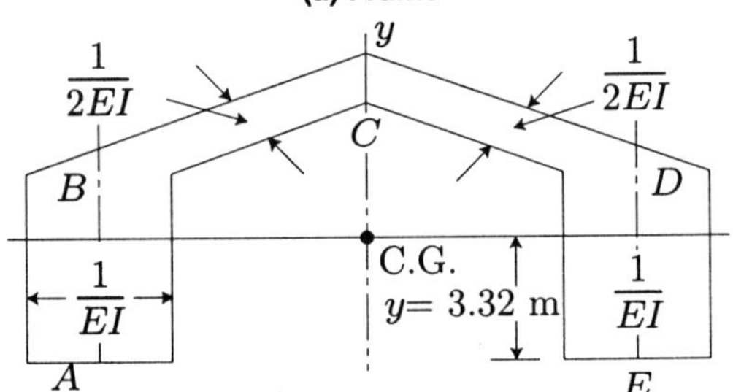

(b) Analogous Column

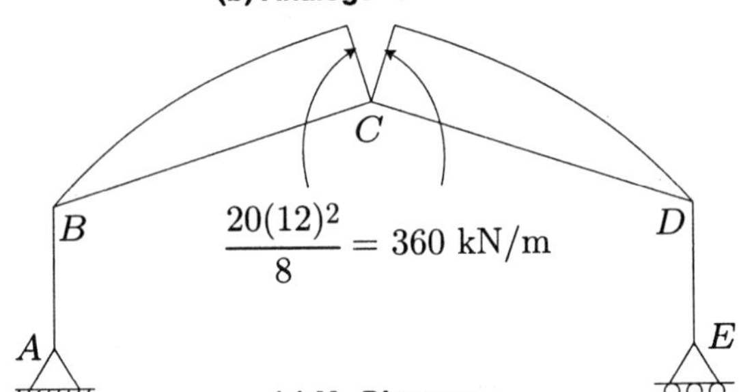

(c) M_S Diagram

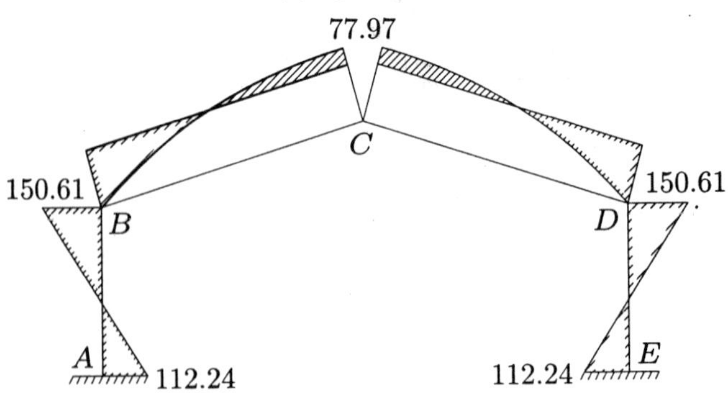

(d) BM Diagram: in kNm

Fig. 10.17: For Example 10.11

Example 10.12. *Analyse frame in Fig. 10.18(a) by the method of Column Analogy.*

Solution : Fig. 10.18(b) shows the analogous column.

$$\text{Area of analogous column} = \frac{8}{4EI} + \frac{4}{2EI} + \frac{4}{3EI} = \frac{16}{3EI}.$$

C.G. of Column as shown in Fig. 10.18(b).

The characteristics of the analogous column are given in Table 10.5 following patterns given in Table 10.2. Distance x and y in Table are measured from axes taken, i.e., along AD and AB.

$M.I.$ of members about its own axes are :

For member AB, $i_{xx} = \frac{1}{2EI}\left(\frac{4^3}{12}\right) = \frac{8}{3EI}$

$$i_{yy} = 0; i_{xy} = 0.$$

For member BC, $i_{xx} = 0; i_{yy} = \frac{1}{4EI}\left(\frac{8^3}{12}\right) = \frac{32}{3EI}$

$$i_{xy} = 0.$$

For member CD, $i_{xx} = \frac{1}{3EI}\left(\frac{4^3}{12}\right) = \frac{16}{9EI}$

$$i_{yy} = i_{xy} = 0.$$

Referring to Table 10.2 and characteristics of analogous columns from Table 10.6 in the problem;

$$M_i = -\frac{400}{16}(3) + \frac{1263.71}{62.25}x + \frac{98.73}{9.386}y = -75 + 20.30x + 10.52y$$

At A, $M_A = M_s - M_i = -160 - [-75 + 20.30(-3.50) + 10.52(-2.75)] = 14.98\ kNm$

At B, $M_B = -160 - [-75 + 20.30(-3.58) + 10.52(1.25)] = -27.10$ kNm

At C, $M_C = 0 - [-75 + 20.30(4.50) + 10.52(1.25)] = -29.50$ kNm

At D, $M_D = 0 - [-75 + 20.30(4.50) + 10.52(-2.75)] = 12.58$ kNm.

B.M. diagram is shown in Fig. 10.18(d).

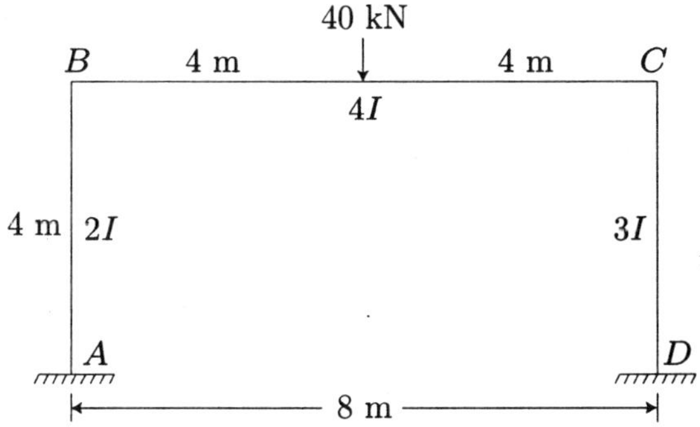

(a) Frame

Fig. 10.18: For Example 10.12 (Contd.)

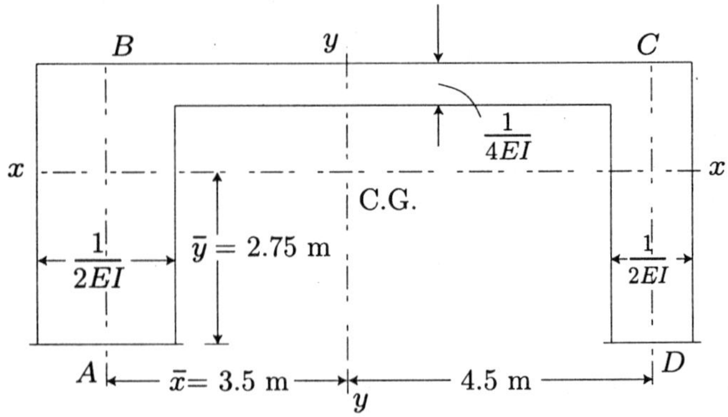

(b) Analogous Column

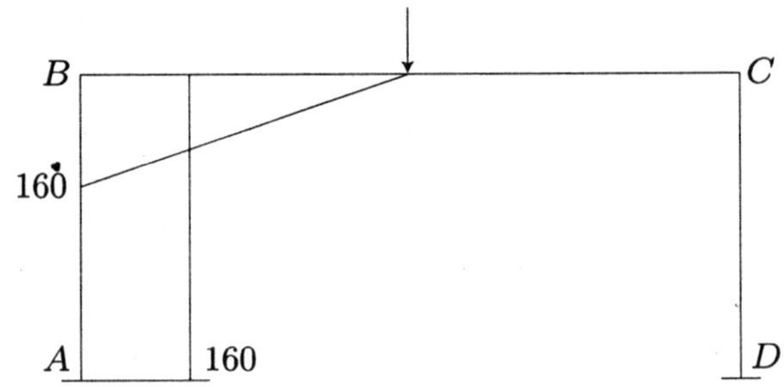

(c) M_S Diagram

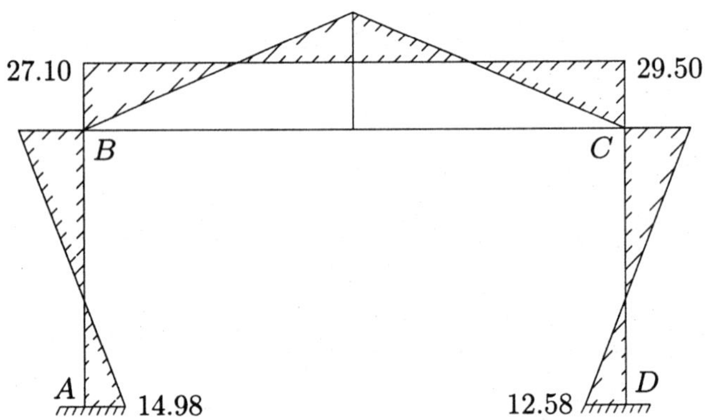

(d) BM Diagram: in kNm

Fig. 10.18: For Example 10.12

Table 10.5. For Example 10.12

Properties of Section											Loads of Section				
Given Properties						Computed Properties					Given			Computed	
						Statical Moments		Products of Inertia							
Member	Length L(m)	Width $\frac{1}{EI}$	Area $a = \frac{L}{EI}$	x	y	ax	ay	$ax^2 + i_{yy}$	$ay^2 + i_{xx}$	$axy + i_{xy}$	Load	Coordinates of Load		Moments	
											$W_0 = M_s a$	x_1	y_1	$W_0 x_1$	$W_0 y_1$
AB	4	$\frac{1}{2EI}$	$\frac{4}{2EI}$	0	2	0	$\frac{4}{EI}$	0	$\frac{32}{3EI}$	0	$-\frac{640}{2EI}$	0	2	0	$-\frac{640}{EI}$
BC	8	$\frac{1}{4EI}$	$\frac{8}{4EI}$	4	4	$\frac{8}{EI}$	$\frac{8}{EI}$	$\frac{128}{3EI}$	$\frac{32}{EI}$	$\frac{32}{EI}$	$-\frac{320}{4EI}$	+1.33	4	$-\frac{106.67}{EI}$	$\frac{320}{EI}$
CD	4	$\frac{1}{3EI}$	$\frac{4}{3EI}$	8	2	$\frac{32}{3EI}$	$\frac{8}{3EI}$	$\frac{256}{3EI}$	$\frac{64}{9EI}$	$\frac{64}{3EI}$	0	–	–	–	–
Totals			$\frac{16}{3EI}$			$\frac{56}{3EI}$	$\frac{44}{3EI}$	$\frac{384}{3EI}$	$\frac{49.78}{EI}$	$\frac{53.33}{EI}$	$-\frac{400}{EI}$			$-\frac{106.67}{EI}$	$-\frac{960}{EI}$
Locate centroid $\bar{x} = \frac{A\bar{x}}{A}$, $\bar{y} = \frac{A\bar{y}}{A}$				$\frac{56}{16}$	$\frac{44}{16}$										
Corrections to centroid								$\frac{65.33}{EI}$	$\frac{40.33}{EI}$	$\frac{51.33}{EI}$				$-\frac{1400}{EI}$	$-\frac{1100}{EI}$
Subtract corrections								$\frac{62.67}{EI}$	$\frac{9.45}{EI}$	$\frac{2.00}{EI}$				$\frac{1293.33}{EI}$	$\frac{140}{EI}$
Corrections for unsymmetry								$\frac{0.42}{EI}$	$\frac{0.064}{EI}$					$\frac{29.62}{EI}$	$\frac{41.27}{EI}$
Subtract corrections								$\frac{62.25}{EI}$	$\frac{9.386}{EI}$					$\frac{1263.71}{EI}$	$\frac{98.73}{EI}$
								I'_{yy}	I'_{xx}	I_{xy}				M'_y	M'_x

Example 10.13. *Analyse frame in Fig. 10.19(a) using the method of Column Analogy.*

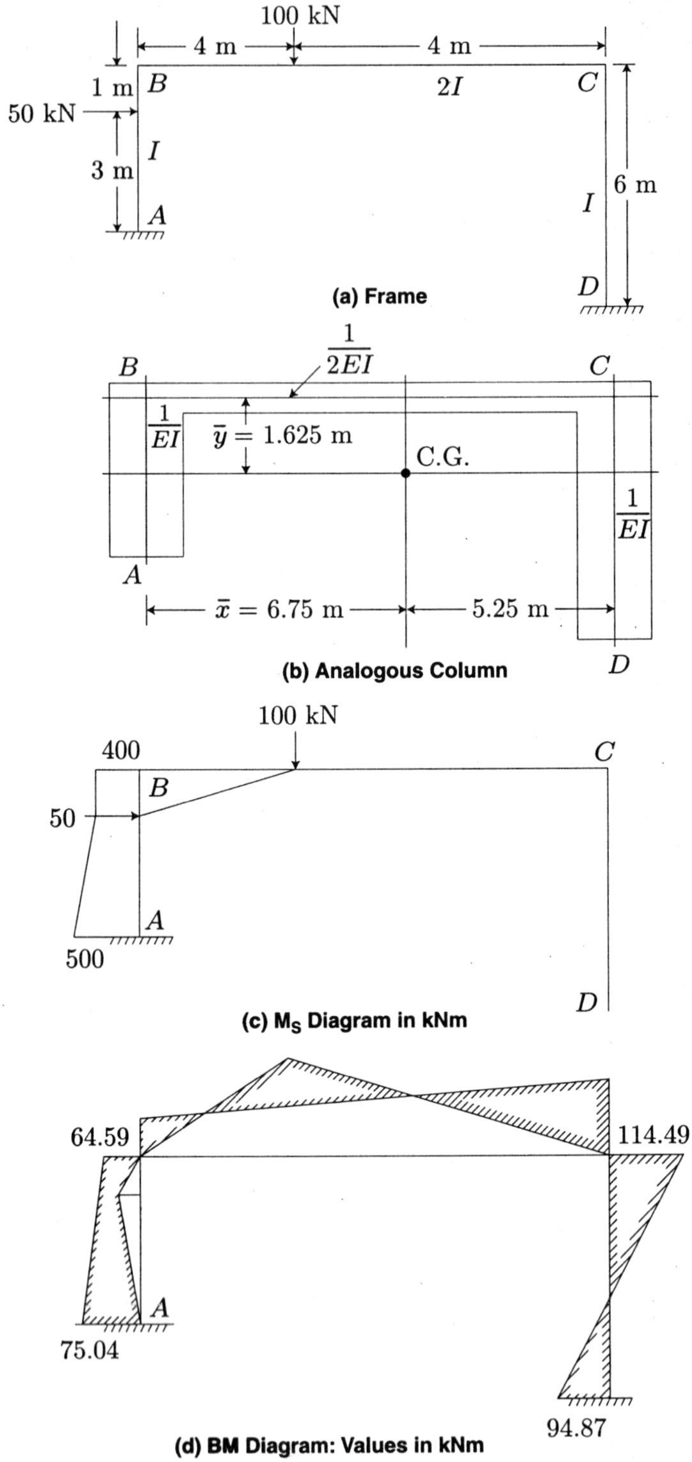

Fig. 10.19: For Example 10.13

Table 10.6. For Example 10.13

Properties of Section											Loads of Section				
Given Properties						Computed Properties					Given			Computed	
						Statical Moments		Products of Inertia			Load	Coordinate of Load		Moments	
Member	Length L(m)	Width $\frac{1}{EI}$	Area $a = \frac{L}{EI}$	x	y	ax	ay	$ax^2 + i_{yy}$	$ay^2 + i_{xx}$	$axy + i_{xy}$	$W_0 = M_2 a$	x_1	y_1	$W_0 x_1$	$W_0 y_1$
AB	4	$\frac{1}{EI}$	$\frac{4}{EI}$	0	−2	0	$\frac{8}{EI}$	0	$\frac{64}{3EI}$	0	$-\frac{1825}{EI}$	0	−2.123	0	$+\frac{3875}{EI}$
BC	12	$\frac{1}{2EI}$	$\frac{6}{EI}$	6	0	$\frac{36}{EI}$	0	$\frac{228}{EI}$	0	0	$-\frac{400}{EI}$	+1.33	0	$-\frac{533.33}{EI}$	0
CD	6	$\frac{1}{EI}$	$\frac{6}{EI}$	12	−3	$\frac{72}{EI}$	$-\frac{18}{EI}$	$\frac{864}{EI}$	$\frac{72}{EI}$	$-\frac{216}{EI}$	0	–	–	–	–
Total			$\frac{16}{EI}$			$\frac{108}{EI}$	$-\frac{26}{EI}$	$\frac{1152}{EI}$	$\frac{280}{3EI}$	$-\frac{216}{EI}$	$\frac{2225}{EI}$			$-\frac{533.33}{EI}$	$+\frac{3875}{EI}$
Locate centroid $\bar{x} = \frac{A\bar{x}}{A}$, $\bar{y} = \frac{A\bar{y}}{A}$				6.75	−1.625										
Corrections to centroid $A\bar{x}^2$, $A\bar{y}^2$ and $A\bar{x}\,\bar{y}$								$\frac{729}{EI}$	$\frac{42.25}{EI}$	$-\frac{175.5}{EI}$				$-\frac{15018.35}{EI}$	$+\frac{3615.625}{EI}$
Subtract corrections								$\frac{423}{EI}$	$\frac{51.08}{EI}$	$-\frac{40.5}{EI}$				$+\frac{14485.42}{EI}$	$+\frac{259.375}{EI}$
Corrections for unsymmetry $\frac{I_{xy}^2}{I_{xx}}$ and $\frac{I_{xy}^2}{I_{yy}}$								$\frac{32.11}{EI}$	$\frac{3.89}{EI}$	–	$\frac{M_x}{I_{xx}} I_{xy}$ and $\frac{M_y}{I_{yy}} I_{xy}$			$-\frac{205.65}{EI}$	$-\frac{1386.902}{EI}$
Subtract corrections								$\frac{391.89}{EI}$	$\frac{47.19}{EI}$	–				$\frac{14691.07}{EI}$	$\frac{1646.277}{EI}$
								I'_{yy}	I'_{xx}	I_{xy}				M'_y	M'_x

Solution : Fig. 10.19(b) shows the analogous column.

Area of analogous column $= \frac{4}{EI} + \frac{12}{2EI} + \frac{6}{EI} = \frac{16}{EI}$.

C.G. of column is shown in Fig. 10.19(b).

For member $AB, i_{xx} = \frac{(4)^3}{12EI} = \frac{16}{3EI}$; $i_{yy} = 0, i_{xy} = 0$.

For member $BC, i_{xx} = 0$; $i_{yy} = \frac{1}{2EI}\frac{(12)^3}{12} = \frac{72}{EI}$; $i_{xy} = 0$.

For member $CD, i_{xx} = \frac{(6)^3}{12EI} = \frac{18}{EI}$; $i_{yy} = 0; i_{xy} = 0$.

The characteristics of the analogous column are shown in Table 10.6 following patterns given in Table 10.2. Distances x and y in Table are measured with respect to axes taken along BC and AB respectively.

Referring to Table 10.2 and from Table 10.7 in the problem,

$$M_i = -\frac{2225}{16} + \frac{14691.07}{391.89}x + \frac{1646.277}{47.19}y = -139.06 + 37.488x + 34.886y$$

At A, $M_A = M_s - M_i = -550 - [-139.06 + 37.488(-6.750) + 34.886(-2.375)]$

$= -75.04$ kNm

At B, $M_B = -400 - [-139.06 + 37.488(-6.75) + 34.886(1.625)] = -64.59$ kNm

At C, $M_C = 0 - [-139.06 + 37.488(5.25) + 34 > 886(1.625)] = -114.49$ kNm

At D, $M_D = 0 - [-139.06 + 37.488(5.25) + 34.886(-4.375)] = +94.87$ kNm.

B.M. diagram is shown in Fig. 7.19(d).

Example 10.14. *Analyse the portal frame in Fig. 10.20(a) by Column Analogy method.*

Solution : Fig. 10.20(b) shows the analogous column.

Area of analogous column=infinity.

Also $I_{yy} = \infty; I_{xy} = 0; \tan\theta = \frac{1}{5}$ and $\cos\theta = 0.98$ and $\sin\theta = 0.196$

$$I_{xx} = \frac{1}{EI}\frac{(4)^3}{3}(\cos^2\theta) + \frac{1}{2EI}\frac{(10)^3}{12}\sin^2\theta + \frac{10}{2EI}(5\cos\theta)^2 + \frac{1}{EI}\frac{(6)^3}{3}\cos^2\theta = \frac{211.29}{EI}.$$

Fig. 10.20(c) shows the M_s diagram.

On AB, $W_1 = \frac{1}{2}\left(\frac{1}{EI}\right)(400)(4) = \frac{800}{EI}$, acting at $\frac{8}{3}$ m from A.

On BC, $W_2 = \frac{1}{2EI}[(400)(4) + \frac{1}{2}(480 - 400)(4) + \frac{1}{2}(480)(6)] = \frac{1600}{EI}$.

C.G. of load on BC at $\overline{x}$ distance for A, where

$$\overline{x} = \frac{800(2) + 180\left(\frac{8}{3}\right) + 720(4+2)}{1600} = 3.833 \text{ m}.$$

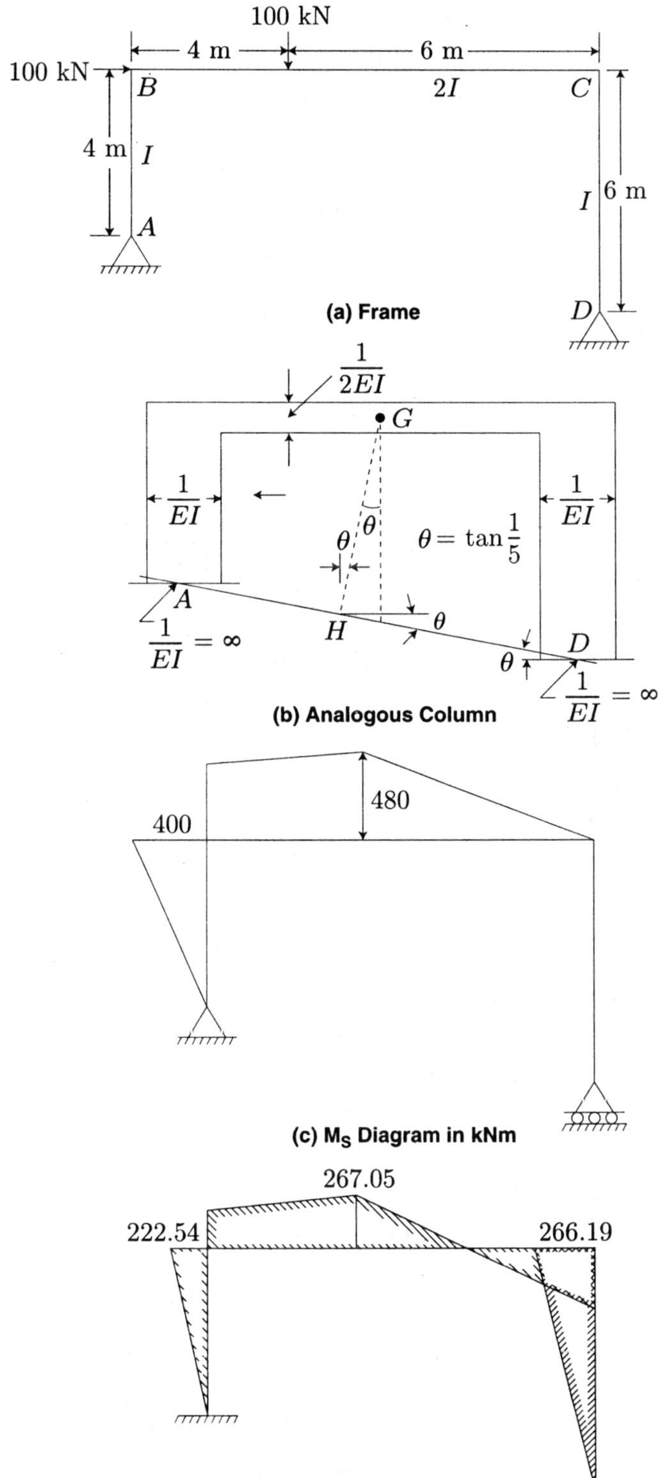

(a) Frame

(b) Analogous Column

(c) M_S Diagram in kNm

(d) BM Diagram: in kNm

Fig. 10.20: For Example 10.14

Vertical distance at the C.G. from xx,

$$= 4 + \frac{(6-4)}{10}(3.833) = 4.767 \text{ m}.$$

Perpendicular distance$= 4.767 \cos\theta = 4.671$ m

$$W = W_1 + W_2 = \frac{2400}{EI}$$

$$M_x = \frac{800}{EI}\left(\frac{8}{3}\right)\cos\theta + \frac{1600}{EI}(4.671) = \frac{9564.27}{EI}$$

$$M_i = \frac{9564.27}{211.29}y = 45.27y.$$

B.M. at B, $M_B = 400 - [45.27(4)(\cos\theta)] = 222.54$ kNm

B.M. at C, $M_C = 0 - [45.27(6)(\cos\theta)] = -266.19$ kNm.

B.M. diagram is shown in Fig. 10.20(d).

Example 10.15. *Analyse the frame in Fig. 10.21(a) by the method of Column Analogy.*

Solution : Fig. 10.21(b) shows the analogous column with the location of C.G. on the support at D, since $\frac{1}{EI}$ at support D is infinity.

Area of analogous column $A = \infty$

$$I_{xx} = \frac{1}{2EL}\frac{(6)^3}{12} + \frac{4.5}{1.5EI}(3)^2 + \frac{1}{EI}\frac{(3)^3}{3} = \frac{45}{EI}$$

$$I_{yy} = \frac{6}{2EI}(4.5)^2 + \frac{1}{1.5EI}\frac{(4.5)^3}{12} + \frac{4.5}{1.5EI}\frac{(4.5)^2}{(2)^2} = \frac{81}{EI}$$

$$I_{xy} = \frac{6}{2EI}(0)(-4.5) + \frac{4.5}{1.5EI}(-2.25)(3) + \frac{3}{EI}(0)(1.5) = -\frac{20.25}{EI}$$

$$M_x = \frac{1}{2}(300)\frac{4.5}{1.5EI}(+3) = +\frac{1350}{EI};\ M_y = \frac{1}{2}(300)\frac{4.5}{1.5EI}\left(-\frac{4.5+3}{3}\right) = -\frac{1125}{EI}$$

$$I'_{xx} = \frac{45}{EI} - \frac{I_{xy}^2}{I_{yy}} = \frac{39.94}{EI} \text{ and } I_{yy} - \frac{I_{xy}^2}{I_{xx}} = \frac{71.89}{EI} = I'_{yy}$$

$$M'_x = M_x - M_y\frac{I_{xy}}{I_{yy}} = \frac{1068.75}{EI};\ M'_y = M_y - M_x\frac{I_{xy}}{I_{xx}} = -\frac{517.5}{EI}$$

$$M_i = 0 + \frac{M'_x}{I'_{xx}}y + = \frac{M'_y}{I'_{yy}}x = 26.76y - 7.198x.$$

At A, $M_A = 0 - [26.76(-3) - 7.198(-4.5)] = 47.89$ kNm.

At B, $M_B = 0 - [26.76(3) - 7.198(-4.50)] = -112.67$ kNm

At C, $M_C = 0 - [26.76(3) - 7.198(0)] = -80.28$ kNm

At D, $M_D = 0$.

B.M. diagram is shown in Fig. 10.21(d).

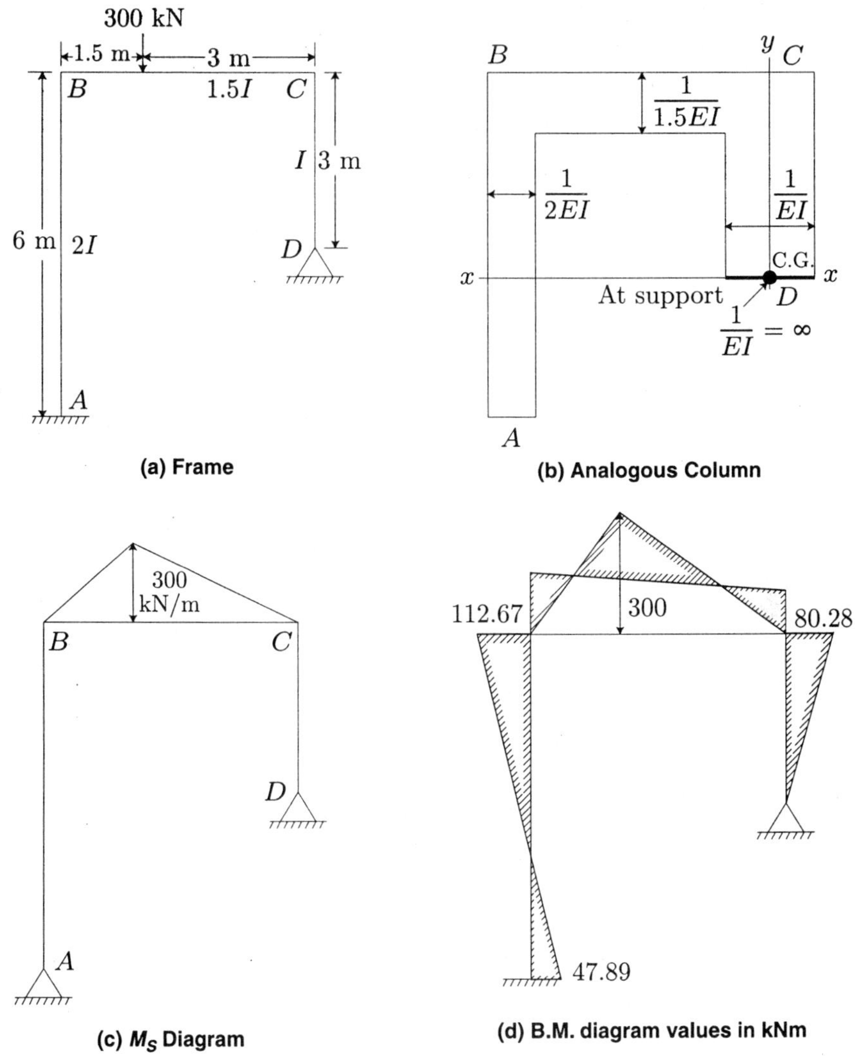

Fig. 10.21: For Example 10.15

Example 10.16. *Analyse frame of constant EI in Fig. 10.22(a) by Column Analogy method.*

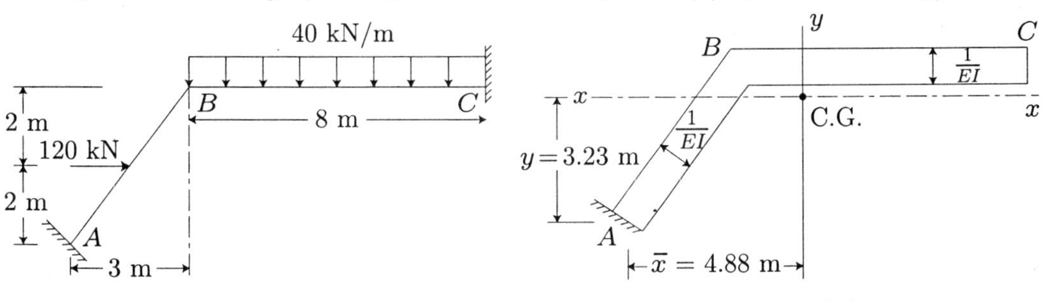

Fig. 10.22: For Example 10.16 (Contd.)

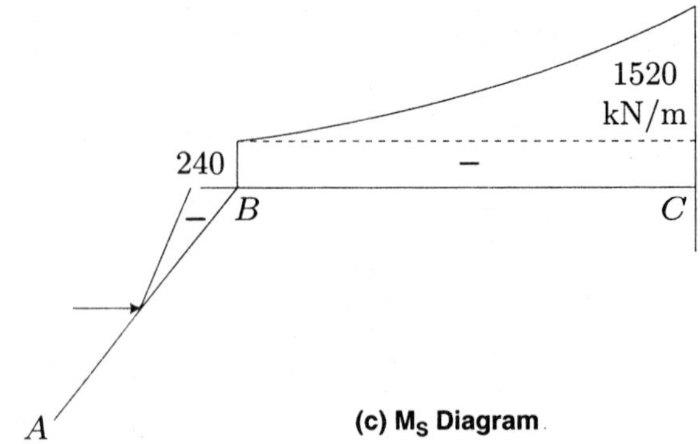

(c) M_S Diagram

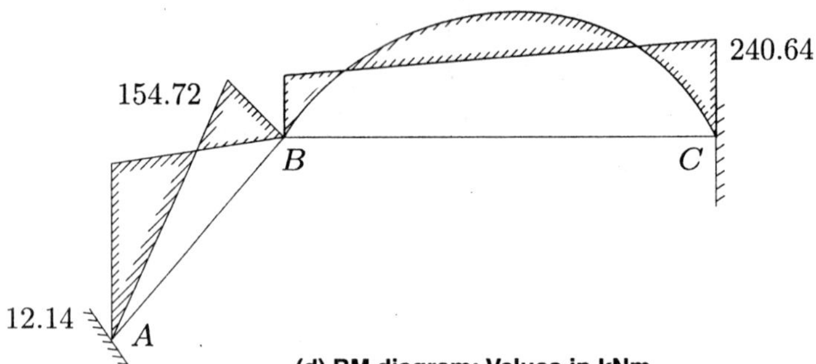

(d) BM diagram: Values in kNm

Fig. 10.22: For Example 10.16

Solution : Fig. 10.22(b) shows the analogous column.

Area of analogous column $A = \dfrac{13}{IE}$.

C.G. of analogous column is shown in Fig. 10.22(b).

$$\bar{y} = \frac{5(2) + 8(4)}{13} = 3.23 \text{ m.}$$

$$\bar{x} = \frac{5(1.5) + 8(7)}{13} = 4.88 \text{ m.}$$

$$I_{xx} = \frac{5}{EI}\frac{(4)^2}{12} + \frac{5}{EI}(2 - 0.77)^2 + \frac{8}{EI}(0.77)^2 = \frac{18.97}{EI}$$

$$I_{yy} = \frac{5}{EI}\frac{(3)^2}{12} + \frac{5}{EI}(4.88 - 1.5)^2 + \frac{1}{EI}\frac{(8)^3}{12} + \frac{8}{EI}(7 - 4.88)^2 = \frac{139.49}{EI}$$

$$I_{xy} = \frac{5}{EI}\frac{(4)(3)}{12} + \frac{5}{EI}(-1.23)(-3.38) + \frac{8}{EI}(0.77)(2.12) = \frac{38.85}{EI}$$

$$W = \frac{1}{2}\frac{2.50}{EI}(-240) + \frac{8}{EI}(-240) + \frac{8}{EI}\left(\frac{11}{3}\right)(-1280) = -\frac{5633.33}{EI}$$

$$M_y = \frac{1}{2}\frac{2.50}{EI}(-240)(-4.88 + 1.5 + 1) + \frac{8}{EI}(-240)(6.12 - 4)$$
$$+\frac{8}{EI}\left(\frac{1}{3}\right)(-1280)(6.12 - 2) = -\frac{17419.33}{EI}$$

$$M_x = \frac{1}{2}\frac{2.50}{EI}(-240)\left(0.77 - \frac{2}{3}\right) + \frac{8}{EI}\frac{8}{EI}(-240)(0.77)$$

$$+\frac{8}{EI}\left(\frac{1}{3}\right)(-1280)(0.77) = -\frac{4137.67}{EI}$$

$$I'_{xx} = \frac{18.97}{EI} - \frac{I_{xy}^2}{I_{yy}} = \frac{8.15}{EI}$$

$$I'_{yy} = \frac{139.49}{EI} - \frac{I_{xy}^2}{I_{xx}} = \frac{59.93}{EI}$$

$$M'_x = -\frac{4137.67}{EI} - M_y\frac{I_{xy}}{I_{yy}} = \frac{713.87}{EI}$$

$$M'_y = -\frac{17419.33}{EI} - M_x\frac{I_{xy}}{I_{xx}} = -\frac{8945.50}{EI}$$

$$M_i = -\frac{5633.33}{13} + \frac{713.87}{8.15}y - \frac{8945.50}{59.93}x = -433.33 + 87.59y - 149.26x.$$

At A, $M_A = 0 - [-433.33 + 87.59(-4 + 0.77) - 149.26(-4.88)] = -12.14$ kNm.

At B, $M_B = -240 - [-433.33 + 87.59(0.77) - 149.26(-4.88 + 3)] = -154.72$ kNm.

At C, $M_C = -1520 - [-433.33 + 87.59(0.77) - 149.26(6.12)] = 240.64$ kNm.

B.M. Diagram is shown in Fig. 10.22(d).

Example 10.17. *Analyse the unsymmetrical bent in Fig. 10.23(a) by use of Column Analogy method.*

Solution : Fig. 10.23(b) shows the analogous column.

$$\text{Area of analogous column} = \frac{5}{EI} + \frac{5}{EI} + \frac{4}{1.5EI} + \frac{10}{1.5EI} + \frac{5}{EI} = \frac{24.33}{EI}.$$

C.G. of the analogous column is shown in Fig. 10.23(b).

$$I_{xx} = \frac{1}{EI}\frac{(5)^3}{12} + \frac{5}{EI}(-1.493)^2 + \frac{5}{EI}\frac{(3)^2}{12} + \frac{5}{EI}(4 - 1.493)^2$$

$$+\frac{4}{1.5EI}(5.50 - 1.493)^2 + \frac{10}{1.5EI}\frac{(6)^2}{12} + \frac{10}{1.5EI}(5.50 - 1.493 - 3)^2$$

$$+\frac{1}{EI}\frac{(5)^2}{12} + \frac{5}{EI}(-5.50 + 2.50 - 1.493)^2 = \frac{237.66}{EI}$$

$$I_{yy} = \frac{5}{EI}(-8 + 0.356)^2 + \frac{5}{EI}\frac{(4)^2}{12} + \frac{5}{EI}(-6 + 0.356)^2 + \frac{4}{1.5EI}(-2 + 0.356)^2$$

$$+\frac{10}{1.5EI}\frac{(8)^2}{12} + \frac{10}{1.5EI}(4.356)^2 + \frac{5}{EI}(+8.356)^2 = \frac{976.47}{EI}$$

$$I_{xy} = \frac{5}{EI}(-1.439)(-7.644) + \frac{5}{EI}\frac{(3)(4)}{12} + \frac{5}{EI}(2.507)(-5.644)$$

$$+\frac{4}{1.5EI}(4.007)(-1.644) - \frac{10}{1.5EI}\frac{(8)(6)}{12}$$

$$+\frac{10}{1.5EI}(1.007)(4.356) + \frac{5}{EI}(-4.493)(8.356) = -\frac{211.39}{EI}$$

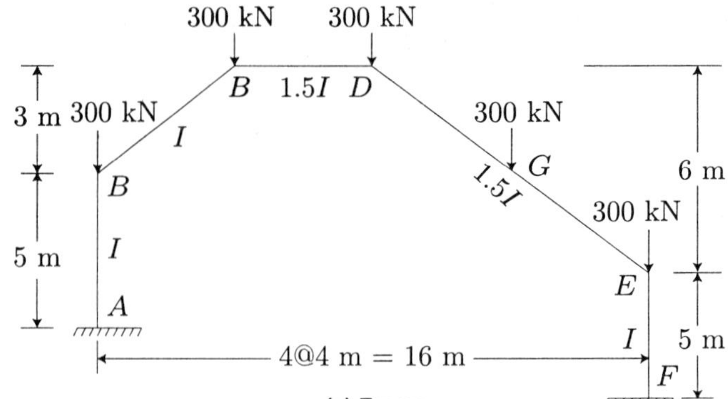

(a) Frame

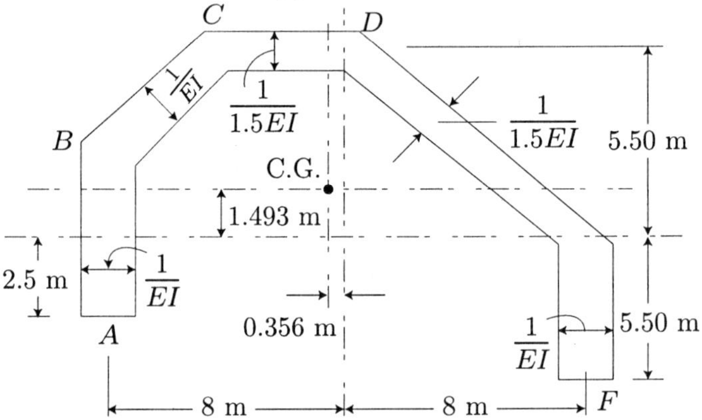

(b) Analogous Column

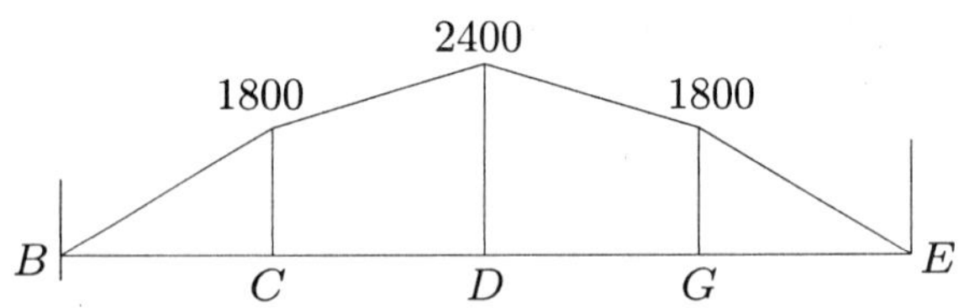

(c) M_S Diagram in kNm

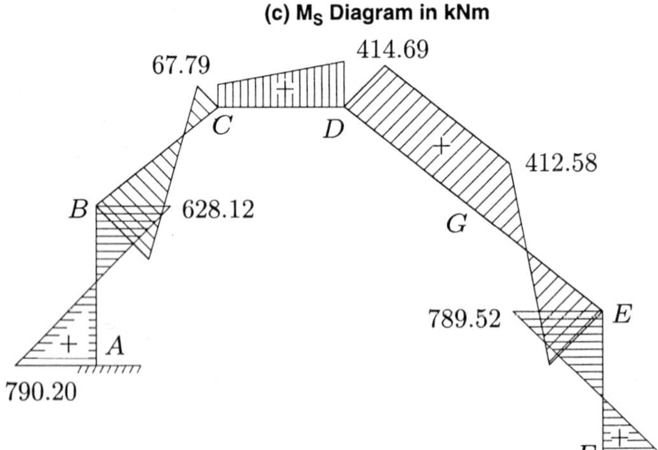

(d) B.M. diagram: Values in kNm

Fig. 10.23: For Example 10.17

$$W = \frac{1}{2}(1800)\frac{5}{EI} + \frac{1800+2400}{2}\frac{4}{1.5EI} + \left(\frac{2400+1800}{2} + \frac{1800}{2}\right)\frac{5}{1.5EI} = \frac{20.100}{EI}$$

$$M_x = \frac{4500}{EI}(4.50-1.493) + \frac{5600}{EI}(5.50-1.493) + \frac{5}{1.5EI}\left[\frac{1}{2}(2400)(5.50-1.493-1)\right.$$
$$\left. + \frac{1}{2}(1800)(5.50-1.493-2) + \frac{1}{2}(1800)(5.50-1.493-4)\right] = +\frac{54040.70}{EI}$$

$$M_y = \frac{4500}{EI}\left[-8+0.356+\frac{8}{3}\right] + \frac{4}{1.5EI}\left[\frac{1}{2}(1800)\left(-\frac{8}{3}+0.356\right)\right.$$
$$\left. + \frac{1}{2}(2400)\left(-\frac{4}{3}+0.356\right)\right]$$
$$+ \frac{5}{1.5EI}\left[\frac{1}{2}(2400)\left(\frac{4}{3}+0.356\right) + \frac{1}{2}(1800)\left(\frac{8}{3}+0.356\right)\right.$$
$$\left. + \frac{1}{2}(1800)\left(\frac{16}{3}+0.356\right)\right] = \frac{1822.27}{EI}$$

$$I'_{xx} = \frac{237.66}{EI} - \frac{I_{xy}^2}{I_{yy}} = \frac{191.90}{EI}$$

$$I'_{yy} = \frac{976.47}{EI} - \frac{I_{xy}^2}{I_{xx}} = \frac{788.45}{EI}$$

$$M'_x = \frac{54040.90}{EI} - M_y\frac{I_{xy}}{I_{yy}} = \frac{54435.19}{EI}$$

$$M'_y = \frac{1822.70}{EI} - M_x\frac{I_{xy}}{I_{xx}} = \frac{49889.95}{EI}$$

$$M_i = \frac{20.100}{24.33} + \frac{54435.19}{191.90}y + \frac{49889.95}{788.45}x = 826.14 + 283.664y + 63.275x.$$

B.M. at $A = 0 - [826.14 + 283.664(-2.50 - 1.493) + 63.275(-8 + 0.356)] = +790.20$ kNm

B.M. at $B = 0 - [826.14 + 283.664(2.50 - 1.493) + 63.276(-8 + 0.356)] = 628.12$ kNm

B.M. at $C = 1800 - [826.14 + 283.664(5.50 - 1.493) + 63.276(-4 + 0.356)] = 67.79$ kNm

B.M. at $D = 2400 - [826.14 + 283.664(5.50 - 1.490) + 63.276(0.356)] = 414.69$ kNm

B.M. at $E = 0 - [826.14 + 283.664(-0.50 - 1.493) + 63.276(8.356)] = -789.52$ kNm

B.M. at $F = 0 - [826.14 + 283.664(-5.50 - 1.493) + 63.276(8.356)] = 628.79$ kNm

B.M. at $G = 1800 - [826.14 + 283.664(2.50 - 1.493) + 63.276(4.356)] = 412.58$ kNm.

B.M. diagram is shown in Fig. 7.23(d).

10.7 Comparison of Column Analogy Method with Method of Elastic Centre

The basic equations used in the development of the Column Analogy method may be interpreted in a slightly different way, as in the case of Elastic Centre method of solution.

The elastic centre O shown in Fig. 10.24 is referred as the centre of the $\frac{1}{EI}$ area (i.e., the analogous column.)

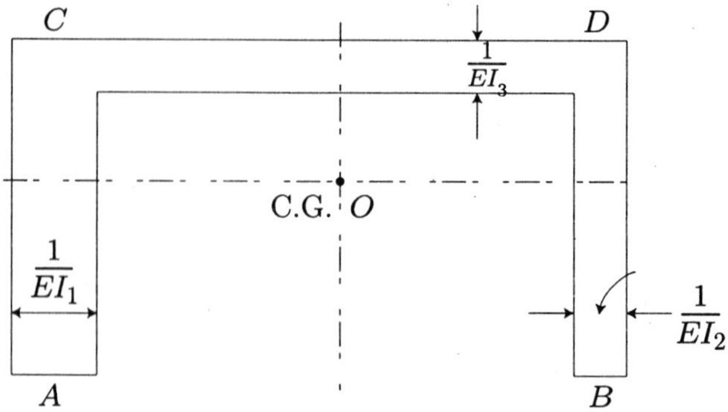

Fig. 10.24: Elastic centre

Considering the case of a frame with fixed ends shown in Fig. 10.25(a), from the method of consistent deformation, taking M_B, H_B and V_B as the redundants and considering that the support at B does neither yield nor rotate,

$$\int \frac{M_s ds}{EI} + M_B \int \frac{ds}{EI} + H_B \int \frac{y' ds}{EI} - V_B \int \frac{x' ds}{EI} = 0 \tag{10.15a}$$

$$\int \frac{M_s x' ds}{EI} + M_B \int \frac{x' ds}{EI} + H_B \int \frac{x' y' ds}{EI} - V_B \int \frac{x'^2 ds}{EI} = 0 \tag{10.15b}$$

and $$\int \frac{M_s y' ds}{EI} + M_B \int \frac{y' ds}{EI} + H_B \int \frac{y'^2 ds}{EI} - V_B \int \frac{x' y' ds}{EI} = 0, \tag{10.15c}$$

where x' and y' refer to values, with point B as the origin of co-ordinates.

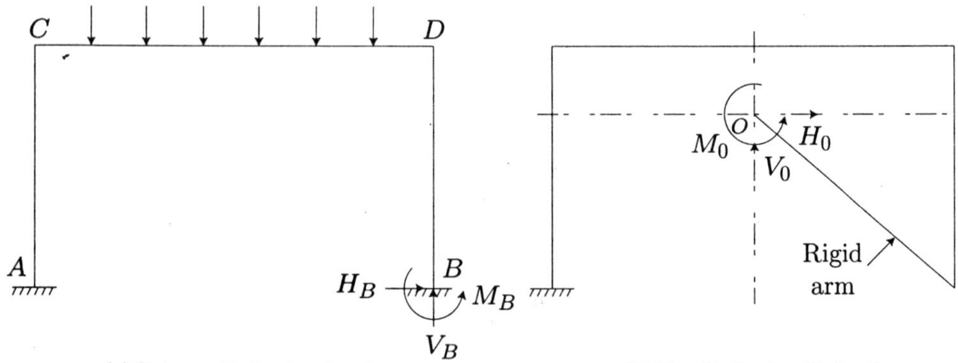

(a) Frame with Redundancies **(b) Elastic Centre Method**

Fig. 10.25: Frame with Fixed ends

If the origin of co-ordinates be transferred to the elastic centre 0, then Eqs. (10.15a) to (10.15c) are simplified since $\int \frac{xds}{EI}$ and $\int \frac{xds}{EI}$ being the statical moments of area become zero. If the axes through 0 are the principal axes, then $\int \frac{xyds}{EI}$ also becomes zero.

The transfer of forces from B to 0 is made a rigid arm incapable of any deformation connecting 0 and B as shown in Fig. 10.25(b). The horizontal and vertical deflections and the rotation at 0 are also zero. Eqs. (10.15a) to (10.15c) can then be written as

$$\int \frac{M_s ds}{EI} + M_0 \int \frac{ds}{EI} = 0 \tag{10.16a}$$

$$\int \frac{M_s x ds}{EI} + H_0 \int \frac{xyds}{EI} - V_0 \int \frac{x^2 ds}{EI} = 0 \tag{10.16b}$$

and $$\int \frac{M_s y ds}{EI} + H_0 \int \frac{y^2 ds}{EI} - V_0 \int \frac{xyds}{EI} = 0. \tag{10.16c}$$

Using notations as in case of Column Analogy method in Section 10.2,

$$W = \int \frac{M_s ds}{EI} = \text{area of } \frac{M_s}{EI} \text{ diagram}$$

$$M_x = \int \frac{M_s y ds}{EI} = \text{moment of } \frac{M_s}{EI} \text{ area about x-axis through 0}$$

and $$M_y = \int \frac{M_s x ds}{EI} = \text{moment of } \frac{M_s}{EI} \text{ area about y-axis through 0.}$$

Eqs. 10.16(a) to 10.16(c) then can be rewritten as

$$W + M_0(A) = 0$$

$$M_y + H_0 l_{xy} - V_0 I_{yy} = 0 \quad \text{and} \quad M_x + H_0 I_{xx} - V_0 Ixy = 0.$$

Solving, $$M_0 = -\frac{W}{A}, \tag{10.17a}$$

$$H_0 = -\left[\frac{M_x I_{yy} - M_y I_{xy}}{I_{xx} I_{yy} - (I_{xy})^2}\right] \tag{10.17b}$$

and $$V_0 = +\left[\frac{M_y I_{xx} - M_x I_{xy}}{I_{xx} I_{yy} - (I_{xy})^2}\right]. \tag{10.17c}$$

Considering the frame in Fig. 10.25(a), bending moment at any point can be written as

$$M = M_s + M_B + H_B y' - V_B x'$$

$$= M_s + M_0 + H_0 y - V_0 x.$$

Substituting from Eqs. 10.17(a) to 10.17(c),

$$M = M_s - \left[\frac{W}{A} + \frac{M_x I_{yy} - M_y I_{xy}}{I_{xx} I_{yy} - (I_{xy})^2} y + \frac{M_y I_{xx} - M_x Ixy}{I_{xx} I_{yy} - (I_{xy})^2} x\right]$$

$$= M_s - M_i \text{ as in case of Column Analogy .}$$

It is thus seen that the mathematical basis of the elastic centre method and the column analogy method is identical.

10.8 Stiffness and Carry over Factors by the Column Analogy Method

One of the most important use of column analogy method is to obtain the stiffness and carry over factors for members. Such determination of stiffness and carry over factors for prismatic or non-prismatic beams is quite easily achieved by the method.

The stiffness factor of a span AB is defined as the moment required to be applied at A in order to rotate the tangent at A by one radian, when the end B is fixed. The carry over factor (C.O.F.) from A to B is the ratio of the fixed-end moment at B to the moment applied at A under the above conditions.

Fig. 10.26(a) shows a beam AB, with moment M_A applied at A, the loading and reaction diagram on the corresponding conjugate beam being shown in Figs. 10.26(b) and 10.26(c) respectively. The reaction to the conjugate beam is only θ_A at A, the end B being fixed, whence $\theta_B = 0$.

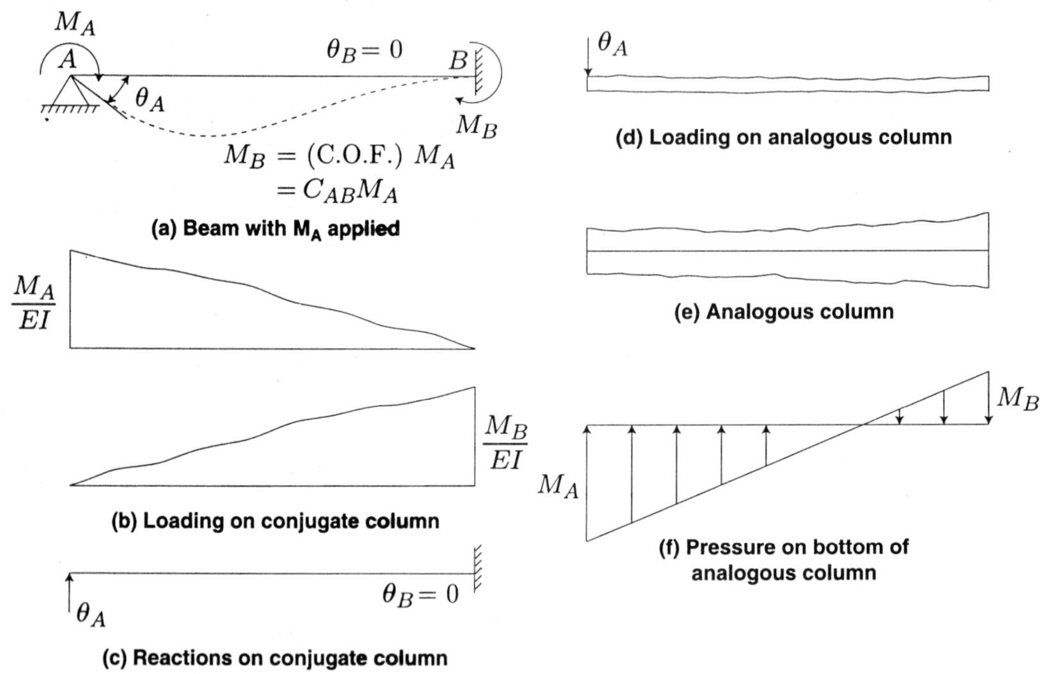

Fig. 10.26: Developments of stiffness and C.O.F

Considering M_A and M_B as the pressure due to loading on the analogous column for the beam, stiffness and carry over factors can be obtained with θ_A as a concentrated load at end A of the analogous column as shown in Fig. 10.26(d). The stiffness at A of the beam will be the value of M_A when rotation at A of the beam is unity which means that $\theta_A =$ unity . It is, therefore, evident that in order to find the stiffness at A and carry over factor, i.e., M_B/M_A, it is only necessary to apply $\theta_A =$ unity at A on the analogous column for fixed-ended beam AB. For value of stiffness at B of the beam AB, it is necessary to apply $\theta_B =$ unity at B on the analogous column.

Considering a member AB of length L, the stiffness at an end A, i.e., K_{AB} is given by $\frac{1}{\text{Area}} + \frac{1(e)^2}{\text{Area }(r^2)} \doteq \frac{1}{\text{Area}}\left[1 + \frac{e^2}{r^2}\right]$, where area of analogous column is considered, e being distance of load θ_A at A from the C.G. of the analogous column and r the radius of gyration of the analogous column cross section. The moment at the other end is given by $\frac{1}{\text{Area}}\left[1 - \frac{e(L-e)}{r^2}\right]$ and equals $K_{AB}C_{AB}$, where C_{AB} is the carry over factor from A to B.

Similarly, K_B at end B is given by $\frac{1}{\text{Area}}\left[1+\frac{(L-e)^2}{r^2}\right]$, the moment at the end A being given by $\frac{1}{\text{Area}}\left[1-\frac{e(L-e)^2}{r^2}\right]$, which is equal to $K_{BA}C_{BA}$, where C_{BA} is the carry over factor from B to A.

This implies that $K_{AB}C_{AB} = K_{BA}C_{BA}$.

In order to find values of modified stiffness, a beam with the far end hinged has to be considered with $\theta =$ unity, applied at the end where stiffness is required. Values of modified stiffness at an end and moments generated due to relative settlement of ends without rotation may be calculated using the scheme shown in Figs. 10.27(a) and 10.27(b).

	A	$\longleftarrow C_{BA}$ B
F. E. M. due to $\theta_A =$ unity	$+K_{AB}$	$+C_{AB}K_{AB}$
Balance	0	$-C_{AB}K_{AB}$
C.O.M.	$-C_{AB}C_{BA}K_{AB}$	0
Balance	0	0
Σ	$K_{AB}[1-C_{AB}C_{BA}]$	0

(a) Modified stiffness

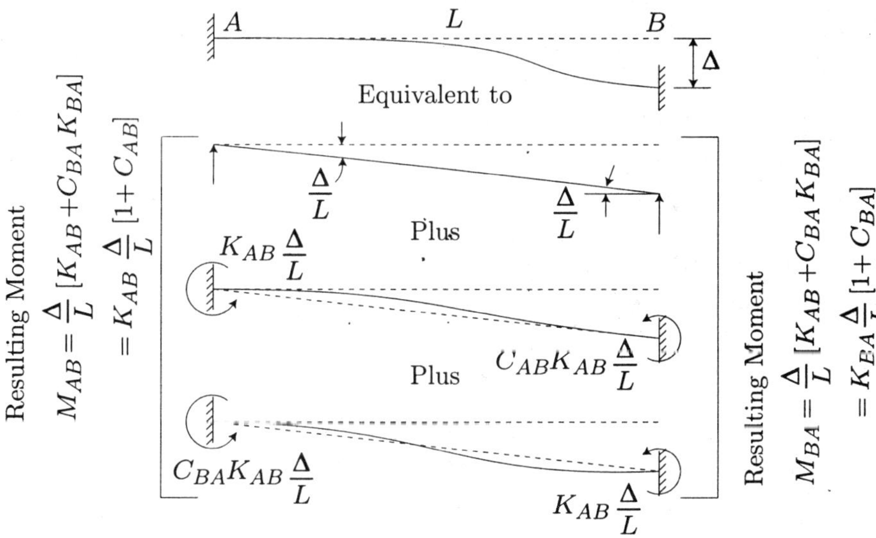

(b) Moments due to relative settlement

Fig. 10.27: Scheme for modified stiffness and moments due to relative settlement

In the schemes of Fig. 10.27, K_{AB}, K_{BA} denote the stiffness at ends A and B respectively, C_{AB} and C_{BA} being the carry over factors, from A to B and B to A respectively.

In obtaining stiffness and carry over factors for beams with haunched ends a scheme as developed by Hardy Cross can also be used.

Fig. 10.28 shows a beam with loading at ends, for which the analogous column is shown in Fig. 10.28(b). From the characteristics of the analogous column shown in Fig. 10.28(b), $I = I_{cg} = Ar^2$, where $A =$ area of analogous column and $r =$ radius of gyration.

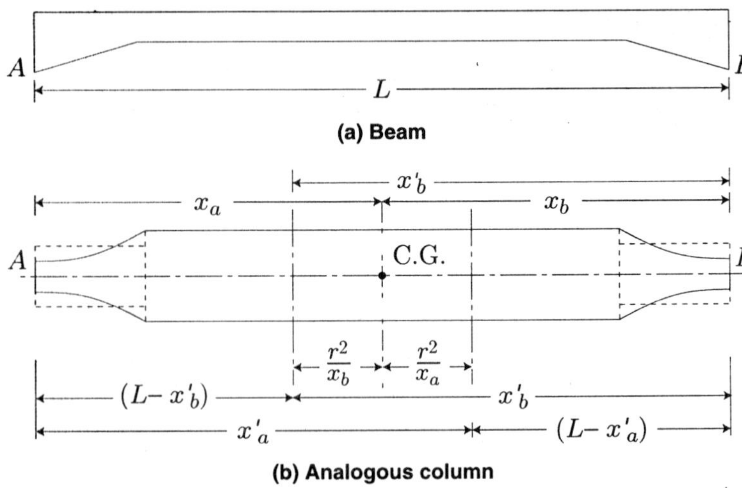

Fig. 10.28: Beam with haunches

Stiffness at A, i.e., $K_{AB} = \dfrac{1}{A} + \dfrac{1(x_a)}{I}(x_a) = \dfrac{1}{I}[r^2 + x_a^2] = \dfrac{1}{I}\left[\dfrac{r^2}{x_a} + x_a\right] x_a$

$$= \frac{x_a x'_a}{I}.$$

In this relation x'_a is the distance of one of the kern points from A as shown in Fig. 10.28(b).

C.O.F., $C_{AB} = \dfrac{-\frac{1}{I}[r^2 - x_a x_b]}{\frac{1}{I}[r^2 + x_a^2]} = \dfrac{(L - x'_a)x_a}{x'_a x_a} = \dfrac{L - x'_a}{x'_a}.$

Similarly, $K_{BA} = \dfrac{x'_b x_b}{I}$ and $C_{BA} = \dfrac{L - x'_b}{x'b}.$

These relationship can be used for calculations of stiffness and carry over factors once characteristics of the analogous column are obtained.

From the values of K_{AB}, K_{BA}, C_{AB} and C_{BA} it is observed that the relationship $K_{AB}C_{AB} = K_{BA}C_{BA}$ holds good, since

$$x_a(L - x'_a) = x_a\left(x_b - \frac{r^2}{x_a}\right) = x_a x_b - r^2$$

and $x_b(L - x'_b) = x_b\left(x_a - \frac{r^2}{x_b}\right) = x_a x_b - r^2.$

Hardy Cross recommends that for beam with haunches near supports as shown in Fig. 10.29(a), an equivalent curve for $\frac{1}{EI}$ may be considered as shown in Fig. 10.29(c). Calculations of stiffness and carry over factors and fixed end moments may be done on the basis of the equivalent curve of $\frac{1}{EI}$ of the beam and considering the end portion as having very large values of EI for which $\frac{1}{EI}$ becomes zero. Two alternative values of L' are indicated on Fig. 10.29(c).

Stiffness at A is given by

$$K_{AB} = \frac{EI}{L'}\left[1 + 12\left(\frac{x_a}{L'}\right)^2\right].$$

C.O.F., $C_{AB} = \dfrac{1 - 12\frac{x_a x_b}{L'^2}}{1 + 12\left(\frac{x_a}{L'}\right)^2}$.

Similarly, $K_{BA} = \dfrac{EI}{L'}\left[1 + 12\left(\dfrac{x_b}{L'}\right)^2\right]$ and $C_{BA} = \dfrac{1 - 12\frac{x_a x_b}{L'^2}}{1 + 12\left(\frac{x_b}{L'}\right)^2}$.

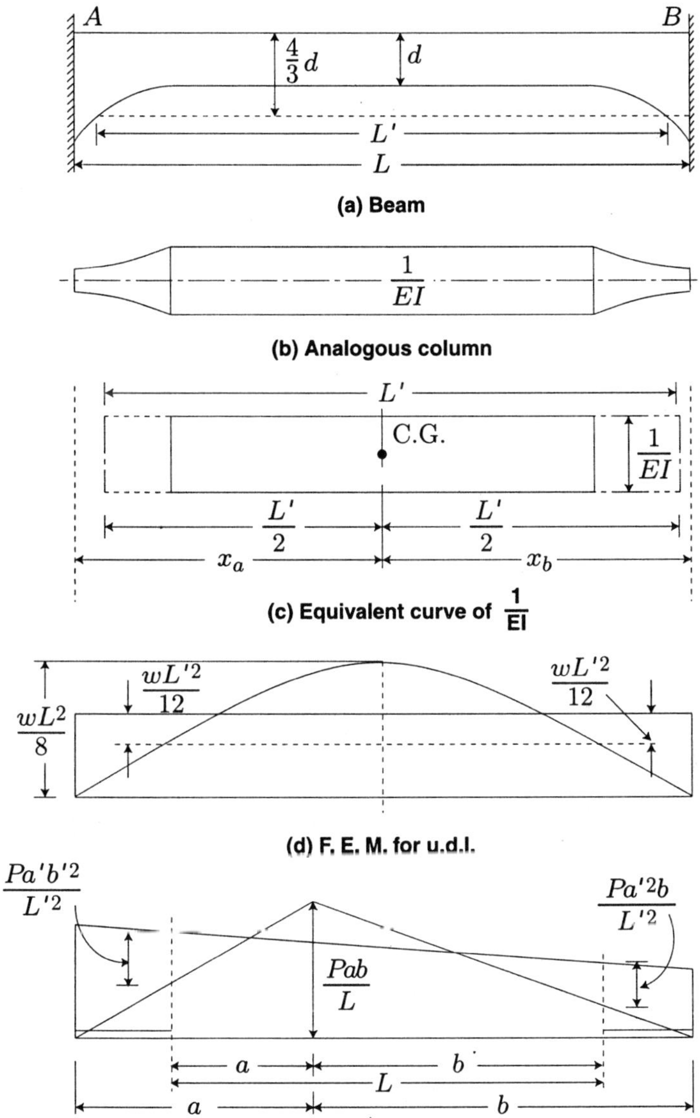

(a) Beam

(b) Analogous column

(c) Equivalent curve of $\frac{1}{EI}$

(d) F. E. M. for u.d.l.

(e) F. E. M. for concentrated load

Fig. 10.29: Approximate analysis of haunched beams

Example 10.18. *Using the method of column analogy, determine the stiffness and carry over factors for a beam of prismatic section.*

Solution : Characteristics of the analogous column are shown in Fig. 10.30(b) and the loading in Fig. 10.30(c).

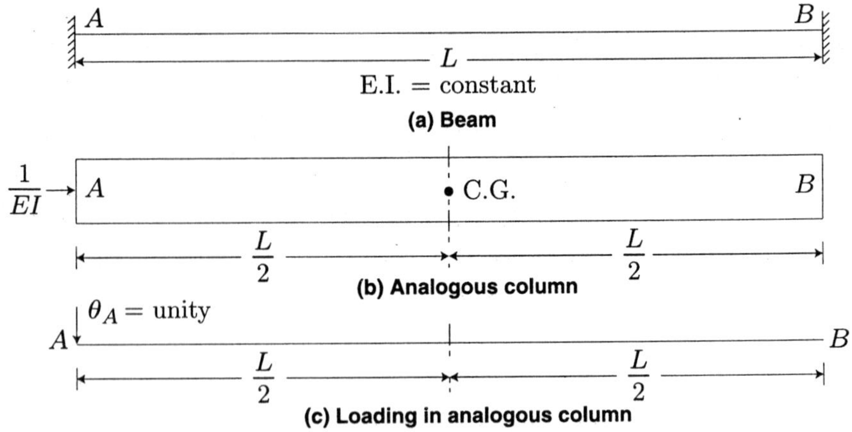

Fig. 10.30: For Example 10.18

Area $A = \dfrac{L}{EI}; I = \dfrac{L^3}{12EI}$

$$M_A = \frac{1}{\left(\frac{L}{EI}\right)} + \frac{1\left(\frac{L}{2}\right)}{\left(\frac{L^3}{12EI}\right)}\left(\frac{L}{2}\right) = \frac{4EI}{L}$$

$$M_B = \frac{1}{\left(\frac{L}{EI}\right)} - \frac{1\left(\frac{L}{2}\right)}{\left(\frac{L^3}{12EI}\right)}\left(\frac{L}{2}\right) = -\frac{2EI}{L}.$$

C.O.F. by moment distribution sign convention gives $C_{AB} = \dfrac{\frac{2EI}{L}}{\frac{4EI}{L}} = \dfrac{1}{2}$.

Example 10.19. *Determine the stiffness factor at A and B and carry over factors from A to B and from B to A for the beam shown in Fig. 10.31. From these values calculate the modified stiffness and the fixed end moment due to relative displacement of ends by* $\triangle$.

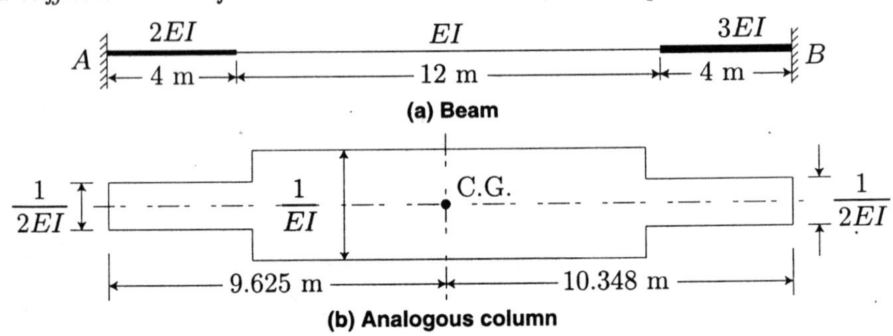

Fig. 10.31: For Example 10.19

Solution : Fig. 10.31(b) shows the analogous column.

Area of analogous column $= \dfrac{15.33}{EI}$.

C.G. of analogous column is shown in Fig. 10.31(b)

$$I_{yy} = \frac{1}{2EI}\left(\frac{4^3}{12}\right) + \frac{4}{2EI}(9.652-2)^2 + \frac{1}{3EI}\left(\frac{4^3}{12}\right)$$

$$+\frac{4}{3EI}(20-9.652-2)^2 + \frac{1}{EI}\frac{(12)^3}{12} + \frac{12}{EI}(10-9.652)^2 = \frac{359.93}{EI}.$$

With $\theta_A = 1$ at A, $M_A = \left[\dfrac{1(EI)}{15,33} + \dfrac{1(-9.652)^2}{359.33}EI\right]\dfrac{L}{L}$.

With $L = 20$, $M_A = 6.49\dfrac{EI}{L} = K_{AB}$.

Also $M_B = \left[\dfrac{1(EI)}{15.33} + \dfrac{(-9.652)(10.348)EI}{359.33}\right]\dfrac{20}{L} = -4.254\dfrac{EI}{L}$

$$C_{AB} = \frac{4.254}{6.49} = 0.655.$$

For stiffness at B considering $\theta_B = 1$ at B,

$$M_B = \left[\frac{1(EI)}{15.33} + \frac{(10.348)^2}{359.33}EI\right]\frac{20}{L} = 7.265\frac{EI}{L} = K_{BA}$$

$$M_A = \left[\frac{1(EI)}{15.33} + \frac{(10.348)(-9.652)}{359.33}EI\right]\frac{20}{L} = -4.254\frac{EI}{L}$$

$$C_{BA} = 0.586.$$

Modified stiffness at A, $= K_{AB}[1 - C_{AB}C_{BA}] = 3.999\dfrac{EI}{L}$

Modified stiffness at B, $= K_{BA}[1 - C_{AB}C_{BA}] = 4.476\dfrac{EI}{L}$.

Fixed-end moments at ends due to settlement of end B by Δ, without rotation of supports are given by

$$M_{FAB} = K_{AB}[1 + C_{AB}]\frac{\Delta}{L} = 10.741EI\frac{\Delta}{L^2}$$

and $M_{FBA} = K_{BA}[1 + C_{BA}]\dfrac{\Delta}{L} = 11.522EI\dfrac{\Delta}{L^2}$.

Example 10.20. *Determine the stiffness and carry over factors of a beam for which the moment of inertia varies linearly from 2I at A to I at B.*

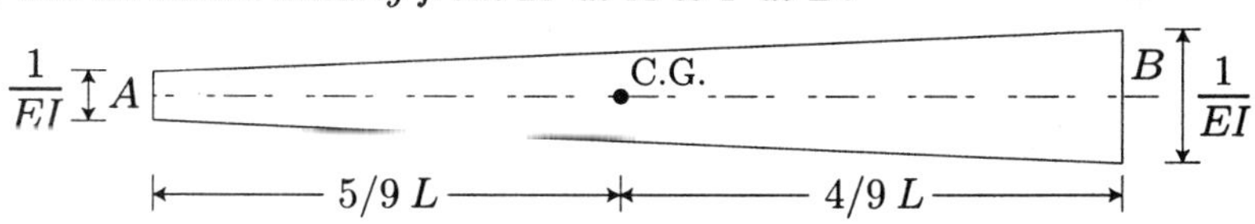

Fig. 10.32: For Example 10.20

Solution : Fig. 10.32 shows the analogous column.

Area of analogous column $= \dfrac{1}{2}\left[\dfrac{1}{2EI} + \dfrac{1}{EI}\right]L = \dfrac{3}{4}\dfrac{L}{EI}$

C.G. of analogous column is shown in Fig. 10.32.

$$I_{yy} = \frac{1}{2EI}\frac{L^3}{12} + \frac{L}{2EI}\left(\frac{5}{9}L - \frac{L}{2}\right)^2 + \frac{1}{18}\left(\frac{1}{2EI}\right)\left(\frac{L}{2}\right)L^2 + \frac{L}{4EI}\left(\frac{2}{3}L - \frac{5}{9}L\right)^2 = \frac{39L^3}{648EI}.$$

With θ_A = unity at A,

$$M_A = \left[\frac{1(EI)}{\frac{3}{4}(L)} + \frac{1\left(-\frac{5}{9}L\right)^2(EI)}{39L^3}(648)\right] = 6.46\frac{EI}{L}$$

and $M_B = \left[\dfrac{1(EI)}{\frac{3}{4}(L)} + \dfrac{1\left(-\frac{5}{9}L\right)\left(\frac{4}{9}L\right)(EI)}{39L^3}(648)\right] = -2.769\dfrac{EI}{L}$.

Carry over factors, $C_{AB} = \dfrac{2.769}{6.46} = 0.429.$

For stiffness at B, with $\theta_B =$ unity,

$$M_B = \left[\frac{1(EI)}{\frac{3}{4}L} + \frac{1\left(\frac{4}{9}L\right)^2 (EI)}{39L^3} - (648)\right] = 4.615\frac{EI}{L}$$

$$M_A = \left[\frac{4}{3}\frac{EI}{L} + \frac{1\left(\frac{4}{9}L\right)\left(-\frac{5}{9}L\right)(EI)}{39L^3}(648)\right] = -2.769\frac{EI}{L}$$

$C_{AB} = 0.60.$

Example 10.21. *Determine the stiffness and carry over factors for the beam in Fig. 10.10(a). Use the Table in the solution of Example 10.4.*

Solution : In this case, it is necessary to find the moment necessary to give unit rotation by applying a unit load at A on the analogous column.

Then Moment at $A = \dfrac{1}{101.12} + \dfrac{(-5)(-5)}{2(191.22)} = +0.0753.$

Also Moment at $B = \dfrac{1}{101.12} + \dfrac{(-5)(+5)}{2(191.22)} = -0.0554.$

Then stiffness at $A = +0.0753$ and $COF_{AB} = \dfrac{0.0554}{0.0753} = 0.7368.$

Example 10.22. *Determine the stiffness and fixed-end moment due to u.d.l. in the case of the haunched beam shown in Fig. 10.33(a) using approximate method suggested by Hardy Cross.*

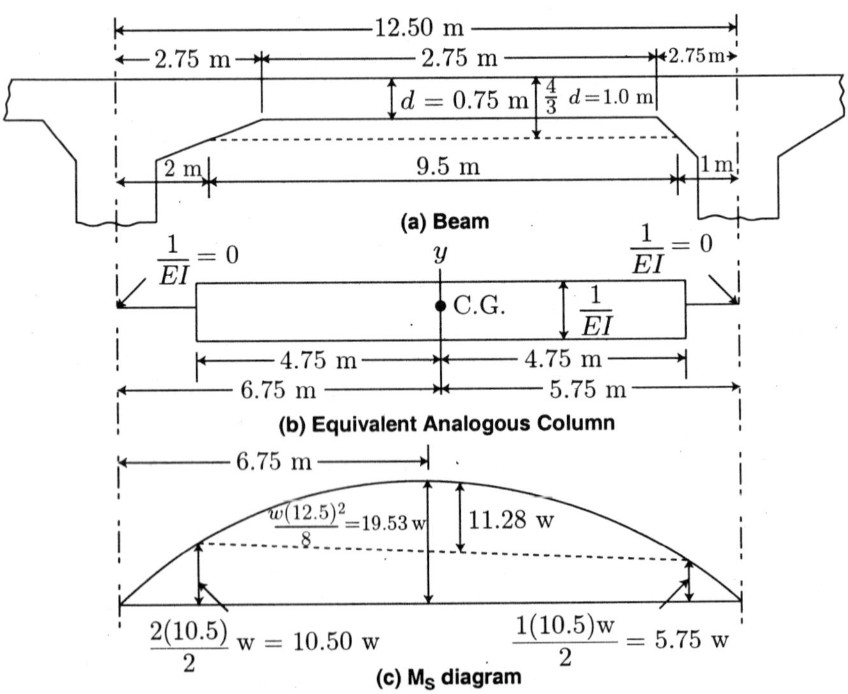

Fig. 10.33: For Example 10.22

Solution : Fig. 10.33(b) shows the analogous column on the basis of the approximate procedure suggested by Hardy Cross.

Area of analogous column $= \dfrac{9.5}{EI}$

$$I_{yy} = \frac{1}{EI}\frac{(9.5)^3}{12} = \frac{71.448}{EI}.$$

$$M_A = \left[\frac{1(EI)}{9.5} + \frac{1(-6.75)^2}{71.448}EI\right]\frac{12.50}{L} = 12.605\frac{EI}{L}$$

$$M_B = \left[\frac{1(EI)}{9.5} + \frac{1(-6.75)(5.75)}{71.448}EI\right]\frac{12.50}{L} = -5.475\frac{EI}{L}$$

and $C_{AB} = 0.434$.

For stiffness at B, with $\theta_B =$ unity,

$$M_B = \left[\frac{1(EI)}{9.5} + \frac{(5.75)^2}{71.448}EI\right]\frac{12.50}{L} = 7.10\frac{EI}{L}$$

$$M_A = -5.475\frac{EI}{L} \quad \text{and} \quad C_{BA} = 0.771.$$

Fig. 10.33(c) shows the M_s diagram.

$$W = \frac{2}{3}\left(\frac{11.28w}{EI}\right)(9.50) + \frac{10.50 + 5.75w}{2}\left(\frac{9.50}{EI}\right) = \frac{1}{EI}[71.45w + 77.19w]$$

$$= \frac{148.64w}{EI}$$

$$M = \frac{1}{2}\frac{(10.50w)(9.50)}{EI}\left[-4.75 + \frac{9.50}{3}\right] + \frac{1}{2}\frac{(5.75w)(9.50)}{EI}\left[4.75 - \frac{9.50}{3}\right] = -\frac{35.724w}{EI}$$

$$M_A = 0 - \left[\frac{148.64w}{9.50} - \frac{35.724w}{71.448}(-6.75)\right] \doteq 19.02w$$

$$M_B = 0 - \left[\frac{148.64w}{9.50} - \frac{35.724w(5.75)}{71.448}\right] = -12.77w.$$

Problems

10.1 What is Column Analogy Method? How does it differ from moment distribution method?

10.2 Explain the Column Analogy Method and state the type of structures to which it is applicable. Suggest sign convention regarding moments.

10.3 What is the difference between column analogy method and elastic centre method. Illustrate by suitable sketches.

10.4 Using Column Analogy Method, obtain the B.M. diagram of a propped cantilever of uniform cross section and of length 12 m fixed at A and propped at B when loaded with two concentrated loads of magnitudes 100 kN and 80 kN placed at 4 m and 8 m respectively from A.

10.5 Determine the B.M. diagram for the beams of constant EI shown in Fig. P-10.1 using Column Analogy Method.

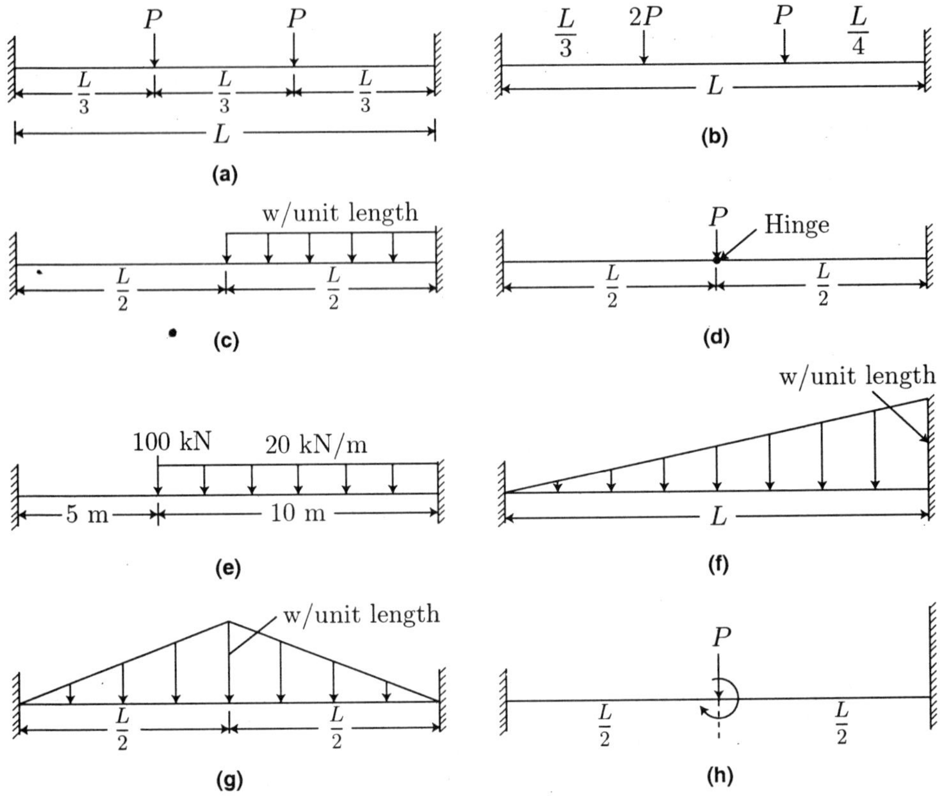

Fig. P-10.1

10.6 Obtain the F.E.M., stiffness factors and carry over factors, at ends A and B of the members shown in Fig. P-10.2.

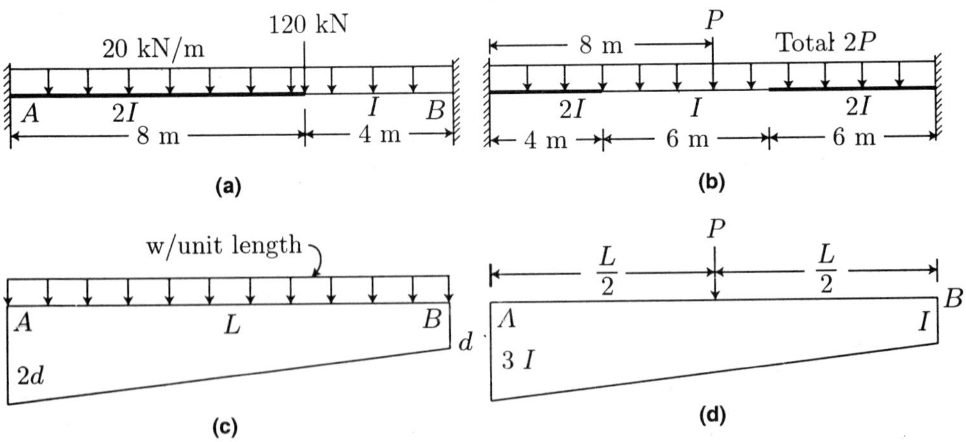

Fig. P-10.2

10.7 Analyse the portal frames shown in Fig. P-10.3 using Column Analogy Method.

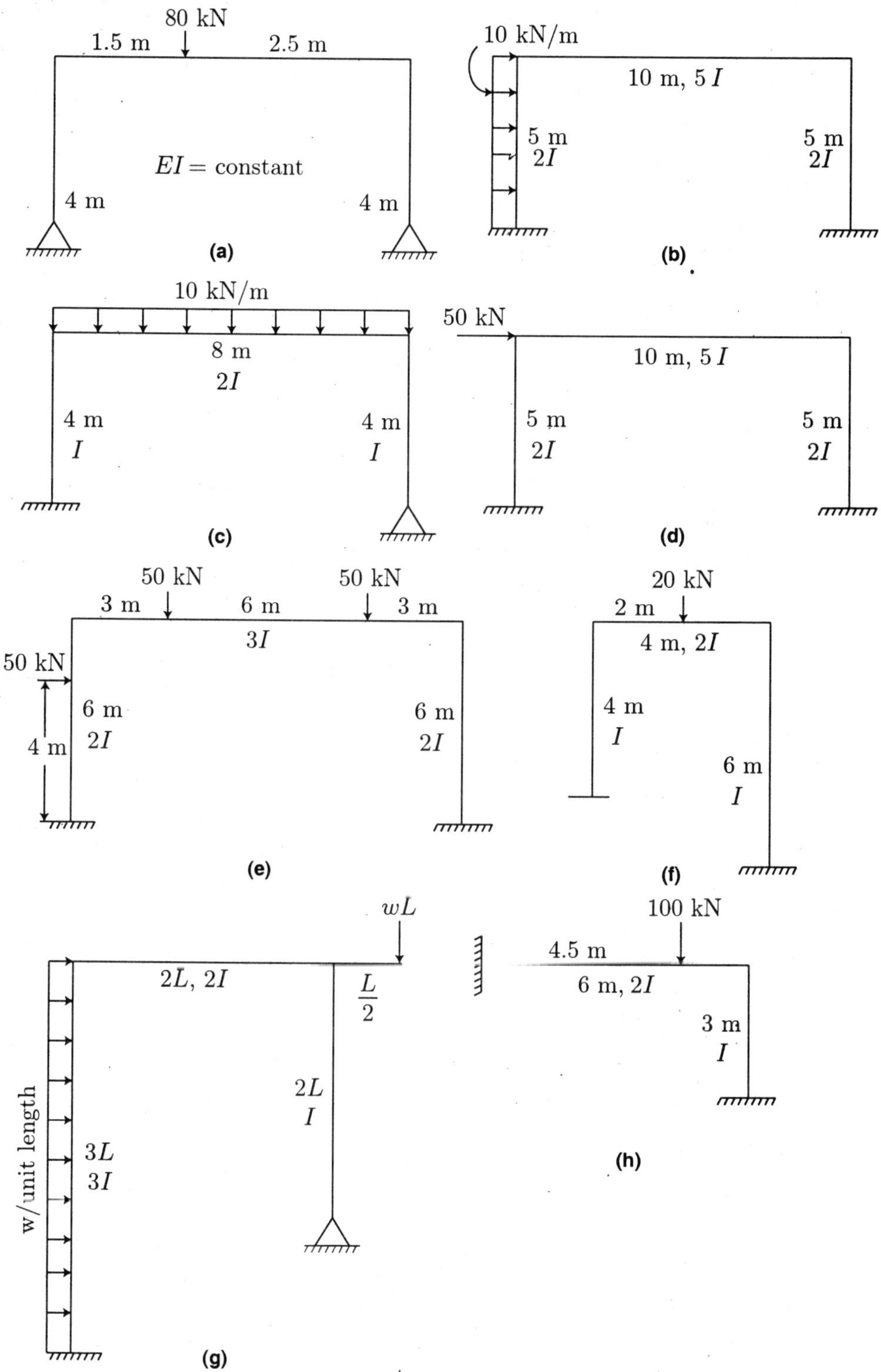

Fig. P-10.3 (Contd.)

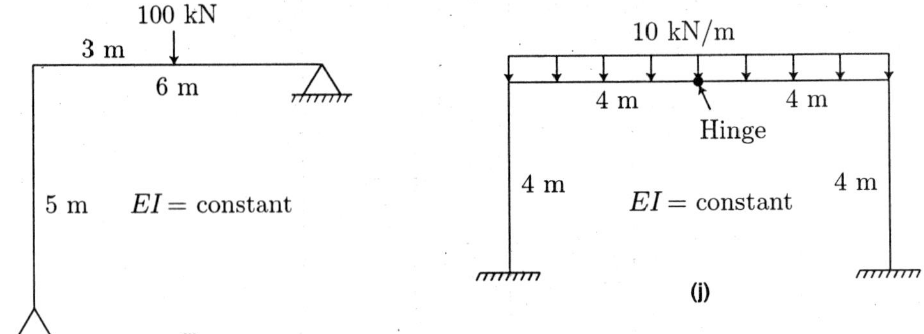

Fig. P-10.3

10.8 Analyse the frames shown in Fig. P-10.4 using Column Analogy Method.

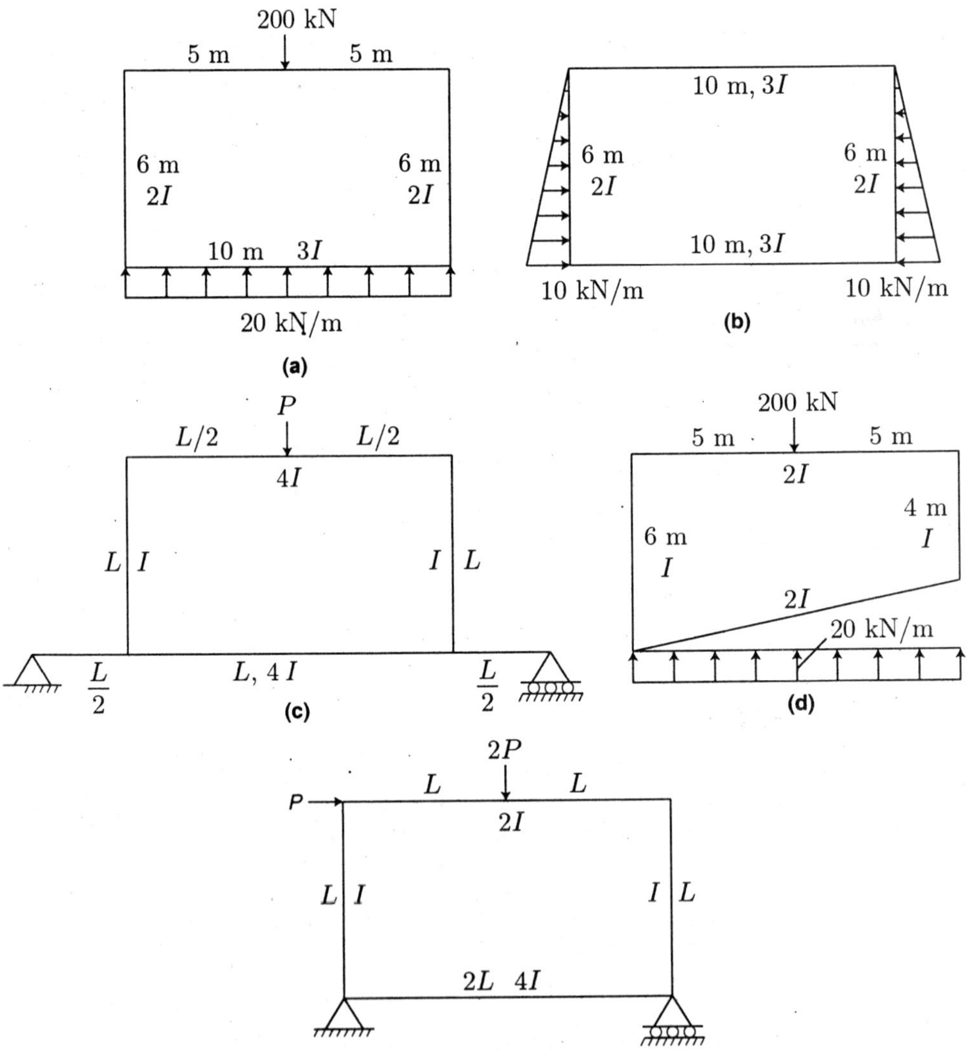

Fig. P-10.4

10.9 Analyse the Gable frames shown in Fig. P-10.5 using the method of Column Analogy.

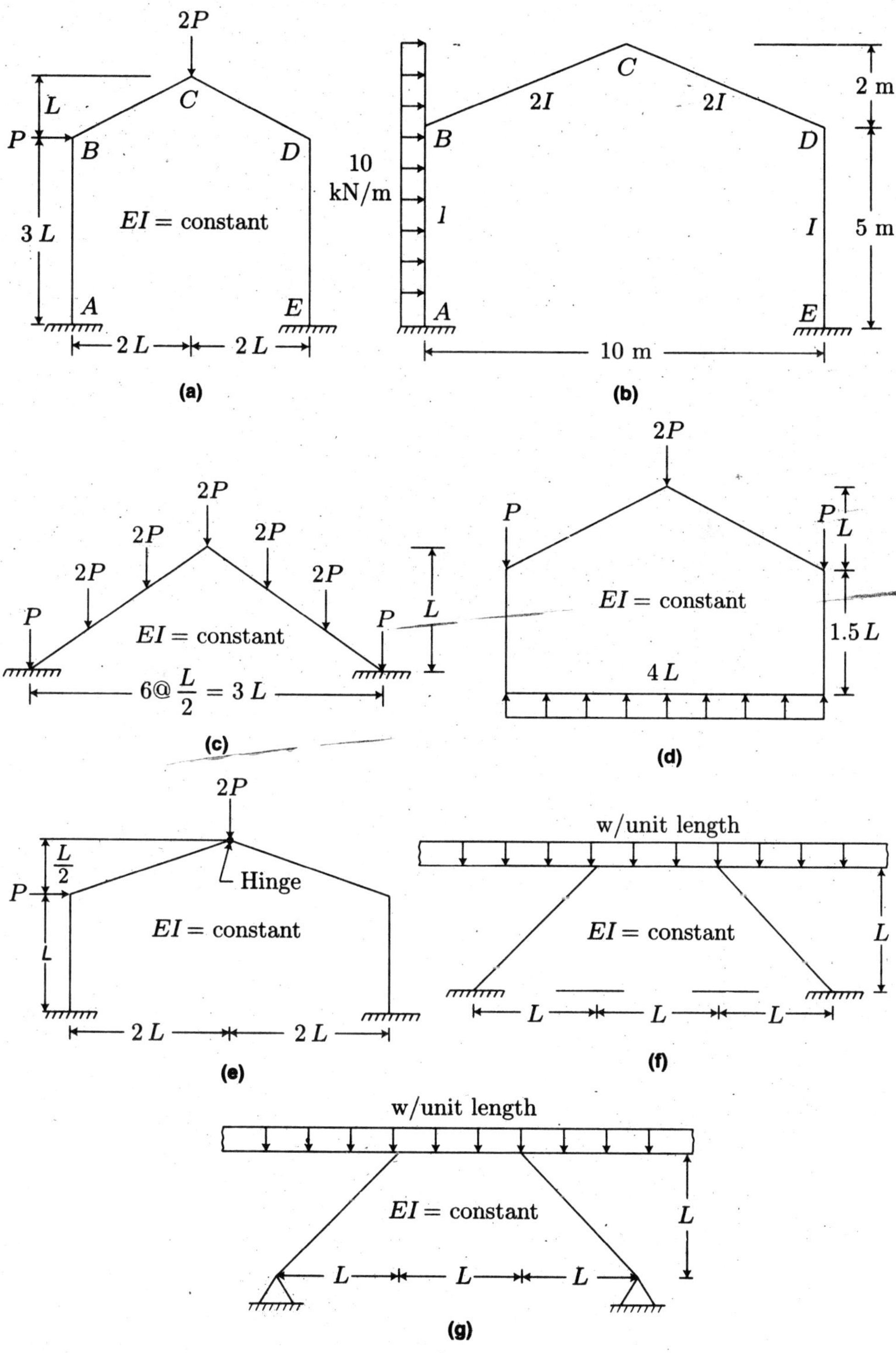

Fig. P-10.5

10.10 Calculate the carry over factor for the span BC of the propped cantilever ABC shown in Fig. P-10.6. Hence, draw the complete B.M. diagram for the beam ABC showing the numerical values of the principal ordinates. Obtain also the reaction at B.

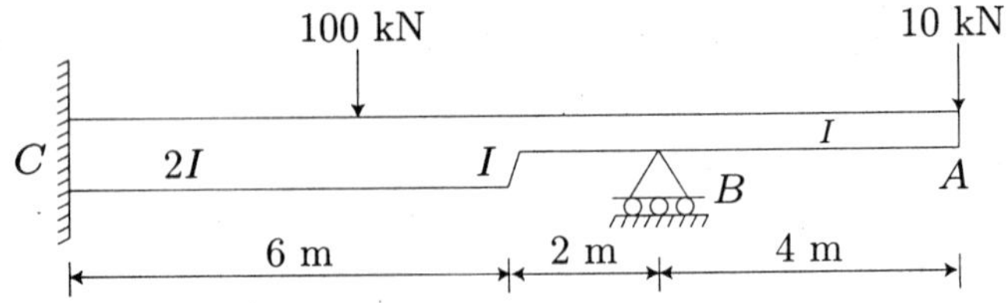

Fig. P-10.6

10.11 A beam $ABCD$ of total length 12 m has the following characteristics.

Length AB=3 m $\quad M.I. = 1.5I$

Length BC=6 m $\quad M.I. = 1I$

Length CD=3 m $\quad M.I. = 2I$

(a) Using Column Analogy Method, calculate the stiffness and carry over factors for the beam AD, at ends, when far end is fixed.

(b) Using values in (a), obtain the modified stiffness for the member at ends, when far end is hinged.

(c) Using values in (a) and (b), obtain values of moments at ends, if there is a relative settlement of an amount $\triangle$ of the supports, without any rotation of joints.

10.12 Obtain the rotational stiffness and the carry over factors for the beam shown in Fig. 10.7.

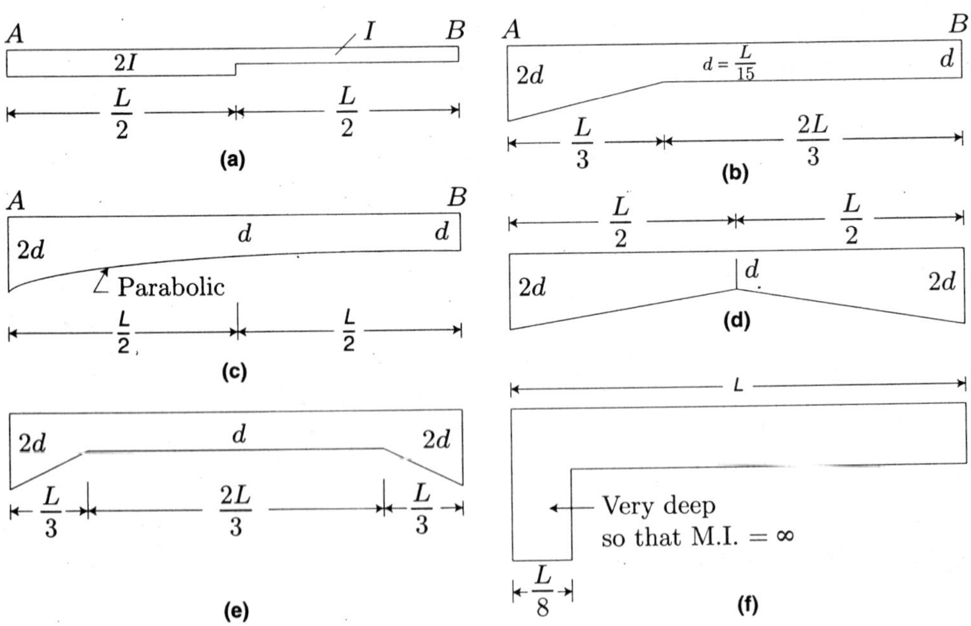

Fig. P-10.7

Chapter 11

Kani's Method

11.1 Introduction

The rotation contribution method of structural analysis was presented by Gasper Kani of Germany in the year 1957. The method offers a well organised method of iteration based on Gauss-Siedel iteration procedure for the solution of the slope deflection equations, without explicitly writing them down. With this properly organised procedure, the slope deflection equations can be solved fairly rapidly.

This method essentially consists of a single, simple numerical operation performed repeatedly at the joints of a structure in a chosen sequence. Results of any desired accuracy may be obtained by performing this operation a sufficient number of times using the required number of significant digits. Kani's method is specially useful for the analysis of multi-storeyed frames. It has the advantage of simplicity, speed, economy of time, labour and space and accuracy.

Kani's method of structural analysis is similar to Hardy Cross method of Moment Distribution, as both of them use similar iterative procedure. However, whereas the moment distribution method obtains the unknown moments of structural members by iterating their increments, Kani's method iterates these unknown themselves. Compared to the moment distribution method this method has the following distinct advantages:

(a) The method is error eliminating as computational errors, even if occurring, are progressively corrected by the repeated application of certain simple equations.

(b) The method is very rapid. The errors in the values of rotation contribution obtained at any joint in successive rounds of calculation converage very rapidly. Moreover, sway and nonsway calculations are carried out in a single continuous calculation operation, which results in considerable saving of time and efforts.

(c) It is possible to introduce changes in load or member lengths or flexural rigidities of members required in design office. Values of contributions obtained as correct values for the initial case may be taken as the starting trial values for the altered cases, and calculation work continued. This requires much less time than that required for a completely new analysis, which would be necessary in case of other methods.

(d) Checking is less time consuming as only the final values of contributions have to be checked.

11.2 Application to Continuous Beams

(a) Member without relative displacement at ends. The end moments of a structural member consist of three components, viz., fixed-end moments due to applied loading, a component involving rotation of joint and a third one for relative displacement of the ends of the member. The last two components called Rotation and Displacement contributions respectively are calculated by Kani's method by successive approximations with repeated application of certain derived equations.

This method of breaking up the end moments is illustrated diagrammatically in Fig. 11.1 for the case of a continuous beam of constant EI and considering no joint translation. For the beam ij with end restraint and no joint translation due to loadings shown in Fig. 11.1(a) superimposing the cases of the beam in Figs. 11.1(b), 11.1(c) and 11.1(d),

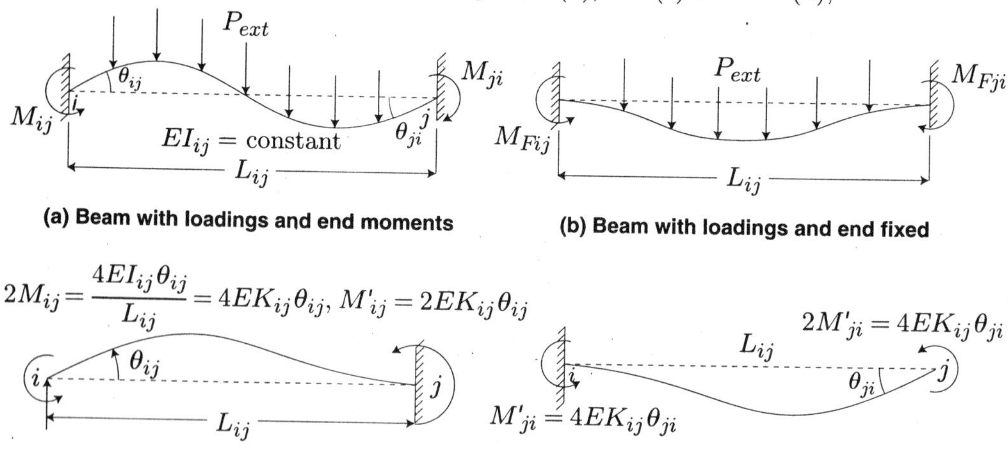

(a) Beam with loadings and end moments **(b) Beam with loadings and end fixed**

(c) Beam with rotation θ_{ij} at end i and end j fixed **(d) Beam with rotation θ_{ji} at end j and end i fixed**

Fig. 11.1: Continuous beam without settlement

$$\left.\begin{aligned} M_{ij} &= M_{F_{ij}} + 2M'_{ij} + M'_{ji} \\ \text{and} \quad M_{ji} &= M_{F_{ji}} + M'_{ij} + 2M'_{ji}. \end{aligned}\right] \tag{11.1}$$

where M_{ij} and M_{ji} are the end moments,

$M_{F_{ij}}$ and $M_{F_{ji}}$ are the fixed end moments

and M'_{ij} and M'_{ji} are the rotation contributions at joints i and j respectively.

In the above, an end moment is considered positive, it if acts in an anticlockwise direction, i.e., if it tends to rotate a member in anticlockwise direction.

For a moment equilibrium at any joint i, the algebraic sum of the end moments of all the members meeting there must be zero, so that $\sum M_{ij} = 0$,

$$\text{whence } \sum M_{F_{ij}} + 2\sum M'_{ij} + \sum M'_{ji} = 0. \tag{11.2}$$

$$\text{From Eq. (11.2), } \sum M'_{ij} = -\frac{1}{2}\left[\sum M_{F_{ij}} + \sum M'_{ji}\right]. \tag{11.3}$$

Eq. (11.3) determines the algebraic sum of the rotation contributions of all members meeting at joint i. Considering members $i-1$, $i-2$, $\cdots$ meeting at the joint i, for members of primatic sections,

$$\begin{aligned}\sum M_{ij} &= 2EK_{i-1}\theta_i + 2EK_{i-2}\theta_i + \ldots \\ &= 2E\theta_i \sum K_{ij},\end{aligned} \tag{11.4}$$

where θ_i is the rotation of the members at the joint i,

$\sum K_{ij}$ is the sum of the values of K, i.e., $\dfrac{I}{L}$ of the member meeting at the joint.

Since $M'_{ij} = 2EK_{ij}\theta_i$, from Eq. (10.4), it follows that,

$$\begin{aligned}M'_{ij} &= \frac{K_{ij}}{\sum K_{ij}} \sum M'_{ij} \\ &= -\frac{1}{2}\frac{K_{ij}}{\sum K_{ij}} \left[\sum M_{F_{ij}} + \sum M'_{ji}\right].\end{aligned} \tag{11.5}$$

The factor $-\dfrac{1}{2}\dfrac{K_{ij}}{\sum K_{ij}}$ is called the rotation factor (R.F.) for member ij.

Eqs. (11.1) and (11.5) form the basis of Kani's solution of frames not subjected to lateral displacements or sway. Eq. (11.5) determines the contribution of the rotation θ_{ij} at the end i of a member $i-j$, in terms of the restraint moment at i and the contribution of the rotation θ_{ij} of the far end j. Eq. (11.1) gives the value of the end moment M_{ij} in terms of the rotation contributions of the two ends and the fixed-end moment at i. Magnitudes of rotation contribution M'_{ij} and M'_{ji} must be determined before M_{ij} can be evaluated. This is accomplished by Gauss-Seidell iteration procedure.

The fixed-end moments for the members calculated directly from the loadings $M_{F_{ij}}$ at joint i is a known value. If trial values are assumed for the rotation contributions at far ends j, the individual rotation contributions at the near end can be calculated from Eq. (11.5), which is repeatedly applied to the various joints of the structure, one joint at a time being considered as the near joint i. The rotation contributions calculated at a joint are taken as the trial values when the joint becomes a far one to some other near joint.

The simple procedure is repeatedly carried out at the various joints. As the calculation work proceeds, increasingly more accurate values of rotation contributions become available at the various joints. The final test of accuracy is the occurrence of nearly the same values of rotation contributions for every member in two successive cycles of calculations, as this indicates that Eq. (11.5) is satisfied by the values of rotation contributions at every joint. With these accurate values of rotation contributions, the end moments are then calculated from Eq. (11.1), applied at every joint.

It is customary to take the trial value of the rotation contribution in the first cycle of operation as zero.

(b) Members with relative lateral displacements of ends. In case in addition to end effects in Fig. 11.1 relative joint translation of the ends is Δ_{ij} as shown in Fig. 11.2, Eq. (11.1) can be rewritten as,

$$M_{ij} = M_{F_{ij}} + 2M'_{ij} + M'_{ji} + M''_{ij}, \tag{11.1a}$$

where M''_{ij} is the displacement contribution and given by $\dfrac{6EI_{ij}\Delta_{ij}}{L^2_{ij}}$.

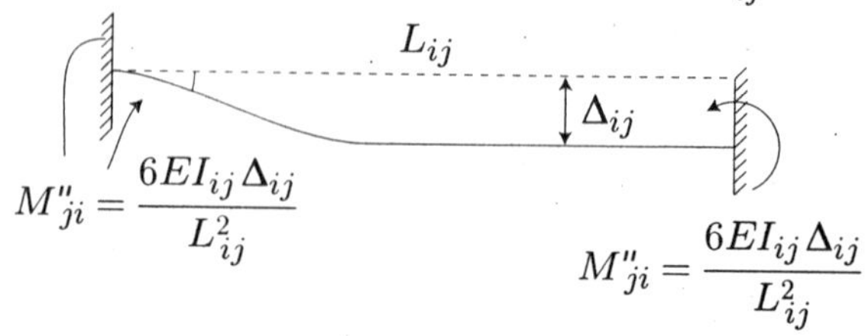

Fig. 11.2: Member with relative settlement at ends

In such a case, $M'_{ij} = -\dfrac{1}{2}\dfrac{K_{ij}}{\sum K_{ij}}\left[\sum M_{F_{ij}} + \sum M'_{ji} + \sum M''_{ij}\right]$ (11.5a)

If Δ_{ij} is known M''_{ij} is known.

If not known and Δ_{ij} is related to reactions, solution has to be done by successive approximations. The procedure of the applications of the rotation contribution method is similar as in the previous case.

(c) Effect of end conditions. When an end is fixed, the rotation contribution is zero, there being no rotation at the end.

If far end of a member is pinned or hinged to a support for analysis, it is convenient to replace the end by a fixed end with value of modified K, i.e., K' for the member, this being taken as $\frac{3}{4}\frac{I}{L}$ of the actual member. Alternatively, the value of K' can be calculated considering a beam with far end fixed, the span length being taken as $\frac{4}{3}L$, whence $K' = \frac{I}{\frac{4}{3}L} = \frac{3}{4}\frac{I}{L} = \frac{3}{4}K$. This follows from the consideration that if a moment M be applied at a joint, the far end being hinged, the slope at the end for a prismatic section is given by $\frac{M}{3EK}$ in place of $\frac{M}{4EK}$ for the far end being fixed.

In such a case, it is necessary to calculate the fixed end moments for actual end conditions. The member with end hinged is then replaced by a member with end fixed with value of K being replaced by $K' = \frac{3}{4}K$. The fixed end offers the convenience of zero rotation contribution, the rotation contributions being calculated for the substitute structure as usual. Moments at all joints except the hinged end are calculated as usual. At the hinged end no calculation is done for the moment, this being taken as zero.

Cantilever portion of a beam can be conveniently dealt with by regarding this as a member with $K = 0$, this being from the consideration that for such part of the beam length L is infinity.

If at a support, moments are statically known the fixed end moments in the adjacent span can be calculated on the basis of the known moments at the joint and in proceeding with the application of Kani's method, the value of K for the adjacent span is replaced by $\frac{3}{4}K$ as in case of the member with end hinged.

(d) Symmetry in case of Beam. Symmetry of beam can be about an axis passing over supports in case of even number of bays or about an axis passing through centres of beams as in case of odd number of bays.

It is obvious that in the first case, the joints at the axis of symmetry do not rotate and as a result the joints on the axis of symmetry behave as fixed joints. One half of the beam can, therefore, be analysed considering the joint on the axis of symmetry as fixed.

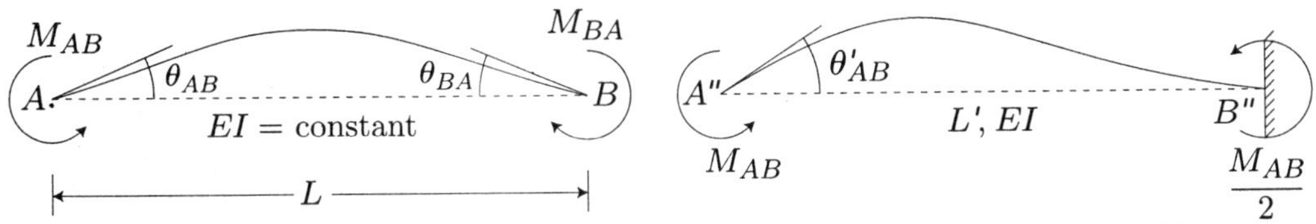

(a) Member with $\theta_{AB} = \theta_{BA}$ **(b) Replacement of beam in (a)**

Fig. 11.3: Beam with symmetry

Considering the case of a beam AB shown in Fig. 11.3(a) through the centre of which the symmetry axis passed, the slope θ_{AB} at the end A in terms of the end moments M_{AB} and M_{BA} is given by

$$\theta_{AB} = \frac{M_{AB}L}{3EI} + \frac{M_{BA}.L}{6EI},$$

as both these rotations are anticlockwise.

$$\theta_{AB} = \frac{M_{AB}L}{2EI}, \text{ anticlockwise,}$$

as from symmetry $M_{AB} = M_{BA}$ in magnitude.

Fig. 11.3(b) represents a substitute beam AB with B fixed which is supposed to replace the beam AB. Such a substitution is justified only if AB and AB' are structurally similar at A.

For AB and AB' to be structurally similar at A, a given moment applied at A in each of the two cases should cause the same amount of rotation in both the cases.

If a moment M_{AB} is applied at A of beam AB', a moment $\frac{M_{AB}}{2}$ is induced at B'. Rotation θ'_{AB} in beam AB' is given by

$$\theta'_{AB} = \frac{M_{AB}L'}{3EI'} - \frac{M_{AB}L'}{12EI'} = \frac{M_{AB}L'}{4EI'}.$$

For structural similarity between AB and AB' at A, $\theta_{AB} = \theta'_{AB}$,

whence $\frac{I'}{L'} = \frac{I}{2L}$, i.e., $K' = \frac{K}{2}$.

This means that half of the structure can be replaced by fixed ends for beam cut by the symmetry axis, the K' values of the substitute beam being taken as half the K values of the actual beam.

11.3 Application to Frames without Lateral Sway

Symmetrical right-angled frames with only vertical and horizontal members, subjected to symmetrical vertical loading do not have any lateral sway. Their analysis by Kani's method does not differ materially from that of the continuous beams, in the sense that calculations are done by Eqs. (11.1) and (11.5). Most of the cases of frames in actual practice in which wind effect is negligible come under this category. Symmetry, which is essential so that lateral sway does not occur can be made use of, to limit the calculation work to one half of the frame. As in case of continuous beams, such symmetry may be two-fold, viz., (i) Frames with even member of bays in which symmetry axis passes through the columns and (ii) Frames with odd member of bays in which symmetry axis passes through centres of beams. Treatment in either of the cases is similar as in case of continuous beam discussed in previous Section.

11.4 Steps in Kani's Method

The modus operandi for successive approximation is illustrated in Fig. 11.4(b) with reference to frame in Fig. 11.4(a). The following steps can be used in applying the method to frames having no sidesway.

Step 1. A line diagram of the frame in Fig. 11.4(a) is shown in Fig. 11.4(b) with double square or circular boxes at each joint as shown. Each joint is considered fixed and the relevant fixed end moments M_F are calculated and written at the respective ends above the corresponding members. Thus M_{FAB} is written about member AB on the right side of double square boxes at joint A as shown in Fig. 11.4(b).

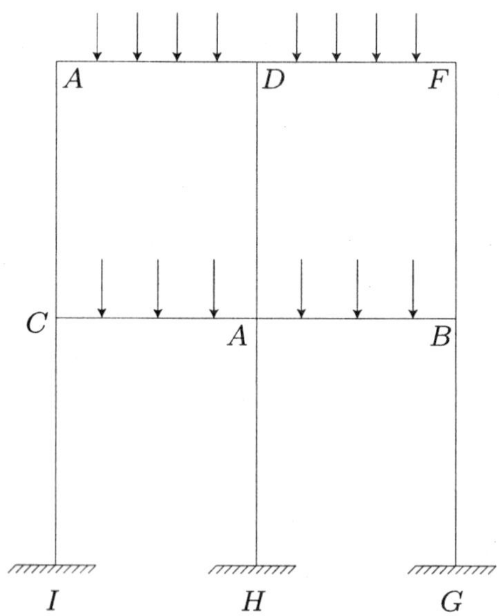

(a) Frame

Fig. 11.4: Symmetrical frames without sidesway (Contd.)

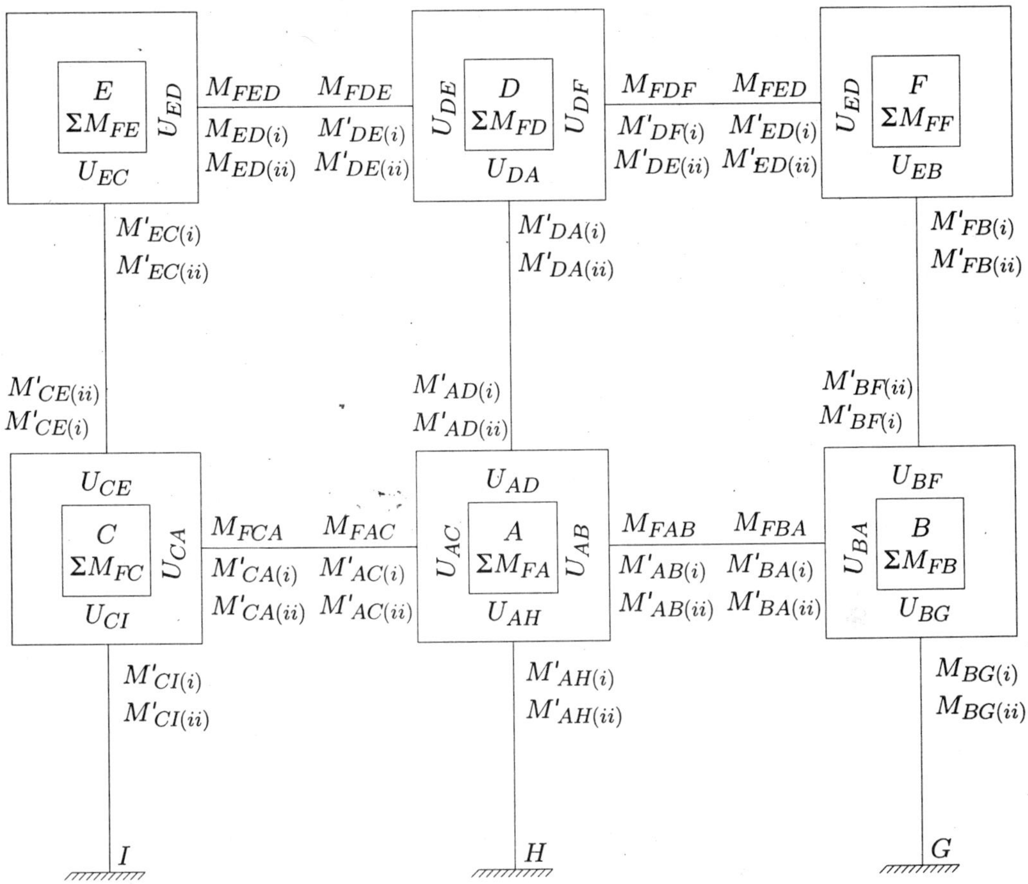

(b) Schemes for Kani's Method

Fig. 11.4: Symmetrical frames without sidesway

Step 2. The algebraic sum of the fixed-end moments at each joint which is the restraint moment is then calculated and entered inside the inner square of each joint. Thus the value entered at joint A is

$$\sum M_{FA} = M_{FAB} + M_{FAC}.$$

Step 3. The value of the rotation factor for each member is then calculated from the formula $U_{AB} = -\frac{1}{2}\frac{K_{AB}}{\sum K_{AB}}$. These values of R.F.C., i.e., U are entered in the respective components in the double box as shown in Fig. 11.4(b). It should be noted that the summation of rotation factor at a joint is $-\frac{1}{2}$.

It should be noted that since no rotation can occur at joints G, H and I, no rotation factors are entered against these joints.

Step 4. the rotation contributions are then calculated at joint by successive approximations using Eq. (11.5). The calculation of the rotation components may be started from any joint. Thus starting from joint A, and member AB, relation used is $M'_{AB_{(i)}} = U_{AB} \sum M_{FA}$, taking initial values of M'_{BA}, M'_{CA} and M'_{DA} as zeros.

This value of $M'_{AB_{(i)}}$ is taken into account, when evaluating the first approximate value of $M'_{BA_{(i)}}$. Thus,

$$M'_{BA_{(i)}} = U_{BA}\left[\sum M_{FB} + \sum M'_{AB_{(i)}}\right].$$

In this manner, the first approximate values of all rotation components are evaluated. As soon as the value of a rotation component is found, it is immediately entered at the appropriate member end as shown in Fig. 11.4(b).

After the first approximate values of all the rotation components have been found and entered in their proper places, the second approximate values are computed, taking into account the first approximate values of rotation components of all the far ends connected to each joint. Thus the second approximate value of $M'_{AB_{(ii)}}$ is evaluated from

$$\begin{aligned} M'_{AB_{(ii)}} &= U_{AB}\left[\sum M_{FA} + \sum M'_{BA_{(i)}}\right] \\ &= U_{AB}\left[\sum M_{FA} + M'_{BA_{(i)}} + M'_{CA_{(i)}} + M'_{DA_{(i)}}\right]. \end{aligned}$$

In this way, the second approximate values of all the rotation components are found and thereafter the third approximate values and so on, until steady values are obtained, i.e., values which do not differ much from the immediately preceding approximations. Such values which are usually found after two or three cycles are then accepted as final values of the rotation components.

The results of these series of computations are shown schematically in Fig. 11.4(b). The final values may be shown within brackets to indicate that these are the accepted values.

Step 5. The fifth and last step is the computation of the final end moments at each end of every member. For example, the final end moment M_{AB} is given by

$$M_{AB} = M_{FAB} + 2M'_{AB_{(n)}} + M'_{BA_{(n)}}.$$

The application of the method of iteration can best be illustrated by working out numerical Examples.

Example 11.1. *Analyse the continuous beam shown in Fig. 11.5 by Kani's method.*

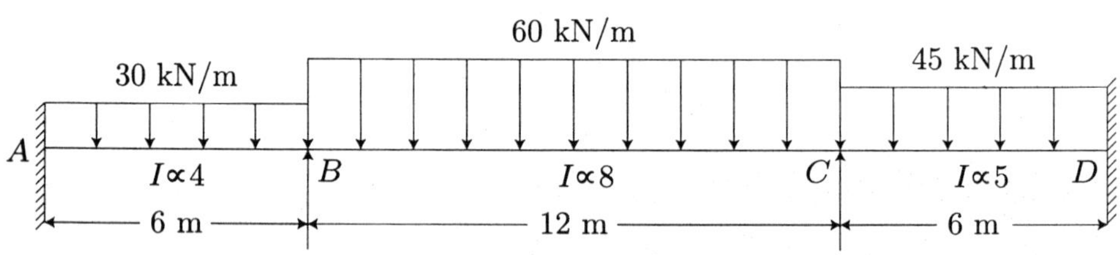

Fig. 11.5: For Example 11.1

Solution : Step 1. The fixed end moments are calculated considering clockwise moments on joints as positive.

$$M_{FAB} = \frac{30(6)^2}{12} = 90 \text{ kNm} = -M_{FBA}$$

$$M_{FBC} = \frac{60(12)^2}{12} = 720 \text{ kNm} = -M_{FCB}$$

$$M_{FCD} = \frac{45(6)^2}{12} = 135 \text{ kNm} = -M_{FDC}.$$

Step 2. Calculate restraint moment at each joint.

Thus, $\sum M_{FB} = -90 + 720 = 630$ kNm

$$\sum M_{FC} = -720 + 135.0 = -585.00 \text{ kNm.}$$

Step 3. Calculate rotation factors at joints B and C.

For member BA, $K_{BA} \approx \frac{4}{6} = \frac{2}{3}$

$$\left.\begin{matrix} BC,\ K_{BC} \\ CB,\ K_{CB} \end{matrix}\right\} \approx \frac{8}{12} = \frac{2}{3}$$

$$CD,\ K_{CD} \approx \frac{5}{6} = \frac{5}{6}.$$

Rotation factors at joint B are as under:

$$\left.\begin{matrix} \text{For } BA,\ U_{BA} = -\frac{1}{2}\frac{\frac{2}{3}}{\frac{2}{3}+\frac{2}{3}} = -0.25 \\ \text{For } BC,\ U_{BC} = -0.25. \end{matrix}\right] \sum = -0.50$$

Rotation factors at joints C are as under:

$$\left.\begin{matrix} \text{For } CB,\ U_{CB} = -\frac{1}{2}\frac{\frac{2}{3}}{\frac{2}{3}+\frac{5}{6}} = -0.220 \\ \text{For } CD,\ U_{CD} = -0.280. \end{matrix}\right] \sum = -0.50$$

Step 4. Calculate rotation components by using Eq. (11.1). The various cycles of calculations involved in evaluating the various approximate values of the rotation contributions are shown below starting from joint B.

Joint B	1st cycle	2nd cycle	3rd cycle	4th cycle
$\sum M_{FB}$	+630.0	+630.0	+630.0	+630.0
M'_{AB}	0.00	0.00	0.00	0.00
M'_{CB}	0.00	+163.5	+172.34	+172.83
Total:	+630.0	+793.5	+802.34	+802.83
Value of M'_{BA}	−157.5	−198.38	−200.58	−200.70
Value of M'_{BC}	−157.5	−198.38	−200.58	−200.70
Joint C	**1st cycle ↓**	**2nd cycle ↓**	**3rd cycle ↓**	**↓ 4th cycle**
$\sum M_{FC}$	−585.0	−585.0	−585.0	−585.0
M'_{BC}	−157.5	−198.38	−200.58	−200.70
M'_{DC}	0.00	0.00	0.00	0.00
Total:	−742.5	−783.38	−785.58	−785.70
M'_{CB}	+163.50	+172.34	+172.83	+172.85
M'_{CD}	+207.75	+219.35	+219.96	+219.99

The completed scheme of calculations is shown in Fig. 11.6(a), from which it can be seen that the evaluated values at the fourth cycle are reasonably correct, as they are very close to accurate values.

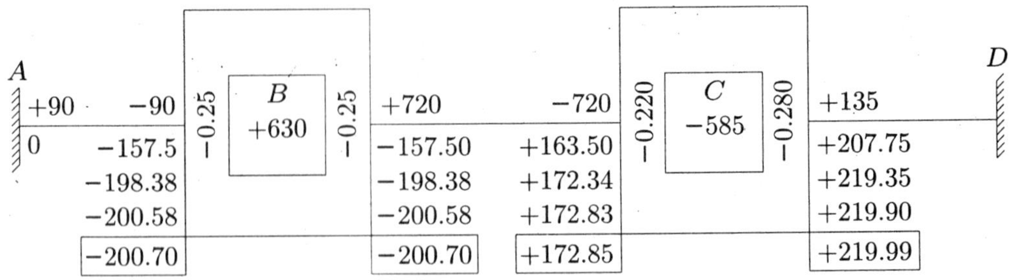

(a) Schematic Diagram. Moments in kN-m. Final rotation contributions are given inside

	A	B	B	C	C	D
M_F	+90	−90	+720	−720	+135	−135
$2M'_{ij}$	0.00	−401.40	−401.14	+345.70	+439.98	0.00
M'_{ji}	−200.70	0.00	+172.85	−200.70	0.00	+219.99
M_{ij}	−110.70 $=M_{AB}$	−491.4 $=M_{BA}$	+491.49 $=M_{BC}$	−575.0 $=M_{CB}$	+574.98 $=M_{CD}$	+84.99 $=M_{DC}$

(b) Final end moments in kN-m

Fig. 11.6: For Example 11.1

Example 11.2. *Determine the moments at the supports of the beam shown in Fig. 11.7, if the support C yields by 15 mm.*

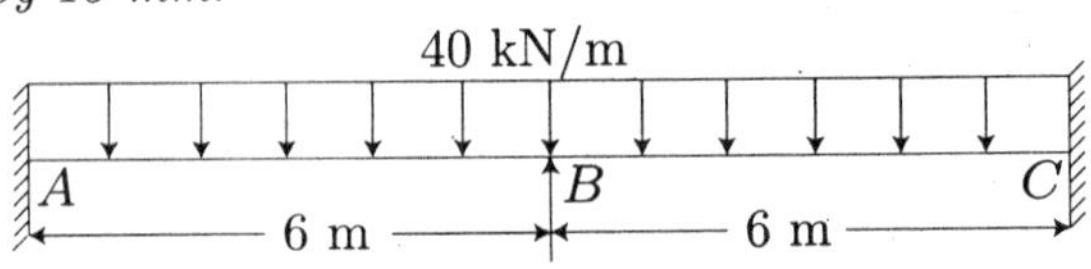

Fig. 11.7: For Example 11.2

Solution : Step 1. F.E.M. due to load

$$M_{FAB} = \frac{40(6)^2}{12} = 120 \text{ kNm} = -M_{FBA} = M_{FBC} = -M_{FCB}.$$

Due to settlement at C by 15 mm,

$$M'_{FBC} = M'_{FCB} = +\frac{6FI\Delta}{L^2} = \frac{6(200)(4)(10)^7(15)}{(6)^2(1000)^2(1000)} = +20.0 \text{ kNm}.$$

Total F.E.M.s in beam BC are :

$$M_{FBC} + M'_{FBC} = 140 \text{ kNm and } M_{FCB} + M'_{FCB} = -100 \text{ kNm}.$$

Step 2. Restraint moment at joint $B = -120 + 140 = 20$ kNm.

Step 3. Since K values for both beams are the same, rotation factors

$$U_{BA} = U_{BC} = -0.25.$$

Step 4. Calculated values of rotation contributions are shown in Fig. 11.8(b). Since values of M'_{AB} and M'_{CB} are both zeros, only one cycle is sufficient.

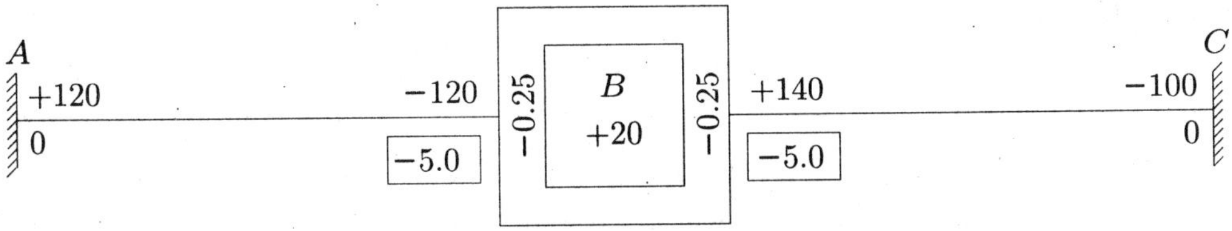

(a) Schematic Diagram: Moments in kN-m

A	B		C
+120	−120	+140	−100
0.0	−10	−10	0
−5.0	0.00	0.00	−5.0
+115	−130	+130	−105

(b) Final end moments in kN-m

Fig. 11.8: For Example 11.2

Step 5. Final bending moments are shown in Fig. 11.8(b).

Example 11.3. *Determine the moments of the beam in Fig. 11.9, if the support at B settles by 1 mm per kN of reaction. Assume, $E = 200$ kN/mm^2 and $I = 4(10)^8$ mm^4.*

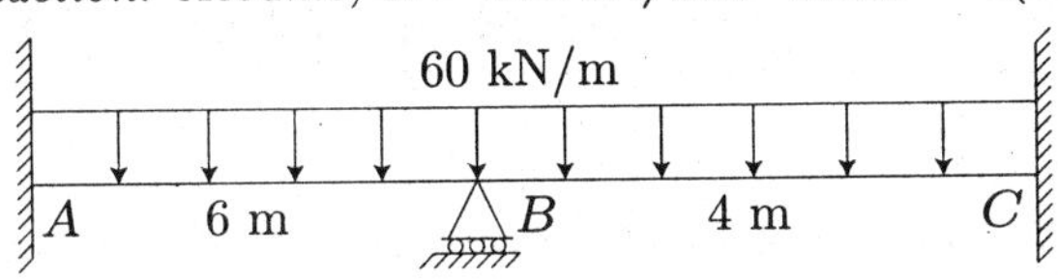

Fig. 11.9: For Example 11.3

Solution : In this case it is convenient to separate the non-settlement calculation from the settlement calculation.

Considering no settlement at B,

$$M_{FAB} = \frac{60(6)^2}{12} = 180 \text{ kNm} - M_{FBA}$$

and $$M_{FBC} = \frac{(60)(4)^2}{12} = 80 \text{ kNm} = -M_{FCB}.$$

Considering settlement of Δ at B,

$$M'_{FAB} = M'_{FBA} = \frac{6EI\Delta}{L^2} = \frac{EI\Delta}{6} = \text{say, } 4\alpha\Delta,$$

and $$M'_{FBC} = M'_{FCB} = -\frac{6EI\Delta}{L^2} = -\frac{3EI\Delta}{8} - 9\alpha\Delta.$$

Resulting F.E.M.s are

at A, $M_{FAB} + M'_{FAB} = 180 + 4\alpha\Delta$

at B, $M_{FBA} + M'_{FBA} = -180 + 4\alpha\Delta$

$M_{FBC} + M'_{FBC} = 80 - 9\alpha\Delta$

at C, $M_{FCB} + M'_{FCB} = -80 - 9\alpha\Delta.$

Rotation factors at joint B are,

$$U_{BA} = -0.20 \text{ and } U_{BC} = -0.30.$$

Calculations of rotation contributions and resulting end moments by Kani's method are shown in Fig. 11.10.

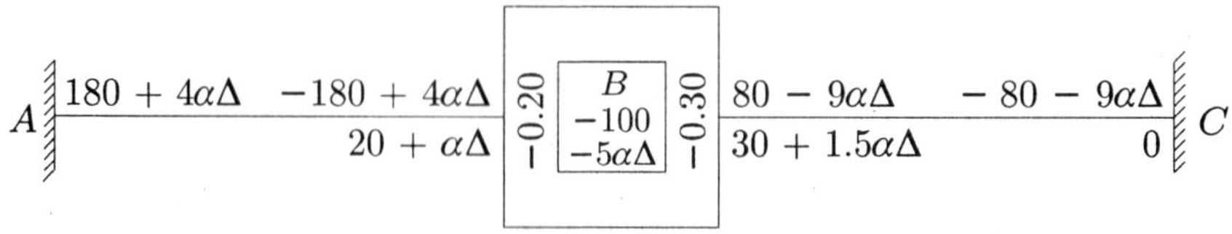

(a) Schematic diagram

A	B		C
$180 + 4\alpha\Delta$	$-180 + 4\alpha\Delta$	$80 - 9\alpha\Delta$	$-80 - 9\alpha\Delta$
0	$40 + \alpha\Delta$	$60 + 3\alpha\Delta$	0
$20 + \alpha\Delta$	0	0	$30 + 1.5\alpha\Delta$
$200 + 5\alpha\Delta$	$-140 + 6\alpha\Delta$	$140 - 6\alpha\Delta$	$-50 - 7.5\alpha\Delta$

(b) Final end moments

Fig. 11.10: For Example 11.3

$$\text{Reaction } R_B = R_{BA} + R_{BC} = \frac{-200 - 5\alpha\Delta + 140 - 6\alpha\Delta + (60)(6)(3)}{6} + \frac{-50 - 7.5\alpha\Delta + 140 - 6\alpha\Delta + (60)(4)(2)}{4}$$

$$= \frac{1020 - 11\alpha\Delta}{6} + \frac{570 - 13.5\alpha\Delta}{4}$$

$$= \frac{3750 - 62.5\alpha\Delta}{12} \uparrow.$$

As support B settles by 1 mm per kN,

$$\Delta = \frac{3750 - 62 \cdot 5\alpha\Delta}{12}(0.001) = (0.3125 - 0.00521\alpha\Delta).$$

As $4\alpha = \dfrac{EI}{6}$,

$$\alpha = \frac{EI}{24} = \frac{200(4)(10)^8}{24(1000)^2} = \frac{(10)^4}{3},$$

whence $\Delta = 0.01701 \approx 0.0170$ m.

Resulting B.M. then are as below:

$$M_{AB} = 200 + 5\alpha\Delta = 200 + 5\frac{(10)^4}{3}(0.0170) = 483.33 \text{ kNm}.$$

$$M_{BA} = -140 + 6\alpha\Delta = -140 + 6\frac{(10)^4}{3}(0.0170) = 200 \text{ kNm}.$$

$$M_{BC} = 140 - 6\alpha\Delta = 140 - 6\frac{(10)^4}{3}(0.0170) = -200 \text{ kNm}.$$

$$M_{CB} = -50 - 7.5\alpha\Delta = -50 - 7.5\frac{(10)^4}{3}(0.0170) = -475 \text{ kNm}.$$

Example 11.4. *Analyse the beam shown in Fig. 11.11 by Kani's method.*

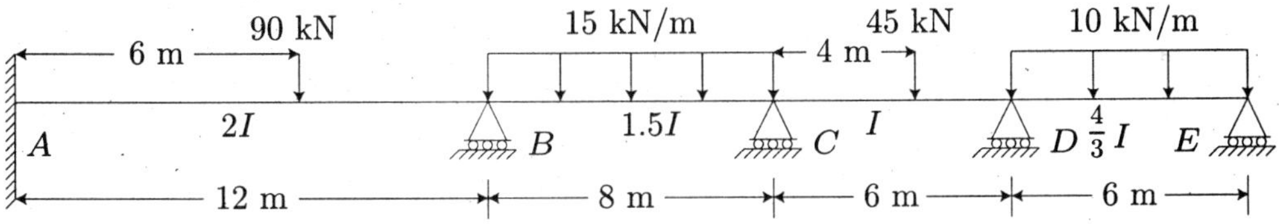

Fig. 11.11: For Example 11.4

Solution : Advantage of hinged end at E can be taken to reduce the number of cycles. Fixed end moments in kNm are as below:

$$M_{FAB} = +80; \quad M_{FBA} = -160,$$

$$M_{FBC} = 80 = -M_{FCB},$$

$$M_{FCD} = +20; \quad M_{FDC} = -40,$$

$$M_{FDE} = +\frac{10(6)^2}{8} = +45$$

Rotation factors are as under:

$$\text{At } B, \quad \left.\begin{aligned} U_{BA} &= -\tfrac{1}{2}\frac{1/6}{1/6+1.5/8} = -0.235 \\ U_{BC} &= -0.265 \end{aligned}\right] \sum = -0.50$$

$$\text{At } C, \quad \left.\begin{aligned} U_{CB} &= -\tfrac{1}{2}\frac{1.5/8}{1.5/8+1/6} = -0.265 \\ U_{CD} &= -0.235 \end{aligned}\right] \sum = -0.50$$

$$\text{At } D, \quad \left.\begin{aligned} U_{DC} &= -\tfrac{1}{2}\frac{1/6}{1/6+1/6} = -0.25 \\ U_{DE} &= -0.25 \end{aligned}\right] \sum = -0.50, \text{ as } \frac{\frac{4}{3}I}{6}\left(\frac{3}{4}\right) = \frac{I}{6}.$$

Other steps starting from joint D are shown in Fig. 11.12(a). Final end moments are given in Fig. 11.12(b).

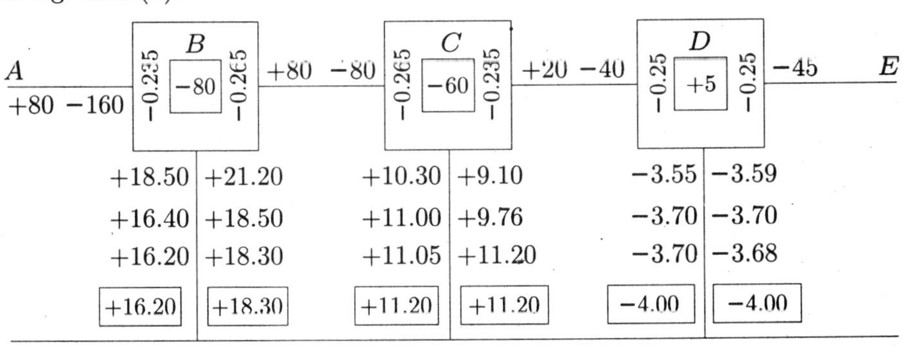

(a) Schematic diagram

Final Rotation Contribution Marked □

Fig. 11.12: For Example 11.4 (Contd.)

	A	B	B	C	C	D	D	E
M_F	+80	−160	+80	−80	+20	−40	+45	0
$2M'_{ji}$	0.00	+32.40	+36.60	+22.40	+22.40	−8.00	−8.40	
M'_{ij}	+16.20	0.00	+11.20	+18.30	−4.00	+11.20	0.00	
M_{ij}	+96.20	−127.60	+127.80	−39.30	+38.40	−36.80	+36.60	0

(b) Final B.M. in kNm

Fig. 11.12: For Example 11.4

Example 11.5. *Analyse the continuous beam shown in Fig. 11.13.*

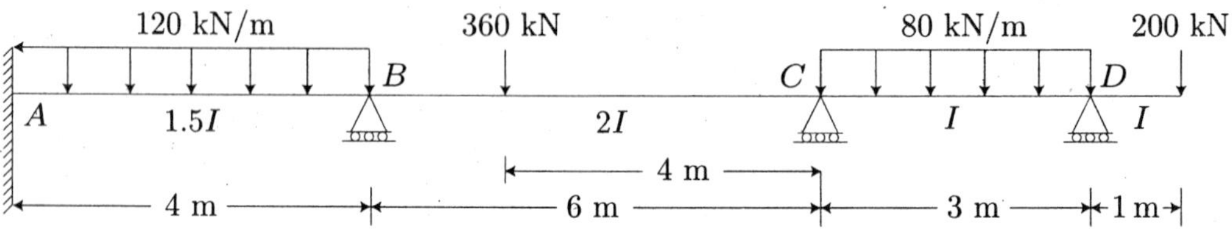

Fig. 11.13: For Example 11.5

Solution : Advantage of known B.M. at D can be taken to reduce the number of cycles. Fixed-end moments in kNm are :

$$M_{FAB} = -M_{FBA} = 160; \quad M_{FBC} = 320; \quad M_{FCB} = -160$$

$$M_{FCD} = 60 - \frac{1}{2}(200 - 60) = -10; \quad M_{FDC} = -200 = -M_{FDE}.$$

Rotation factors are as below :

At B, $$U_{BA} = -\frac{1}{2}\frac{1.5/4}{(1.5/4) + 2/6} = -0.265$$

$$U_{BC} = -\frac{1}{2}\frac{2/6}{(1.5/4) + 2/6} = -0.235$$

At C, $$U_{CB} = -\frac{1}{2}\frac{2/6}{(2/6) + (\frac{1}{4})} = -0.286$$

$$U_{CD} = -\frac{1}{2}\frac{3/4(1/3)}{(2/6) + \frac{1}{4}} = -0.214.$$

Other steps starting from joint B are shown in Fig. 11.14(a). Final moments are given in Fig. 11.14(b).

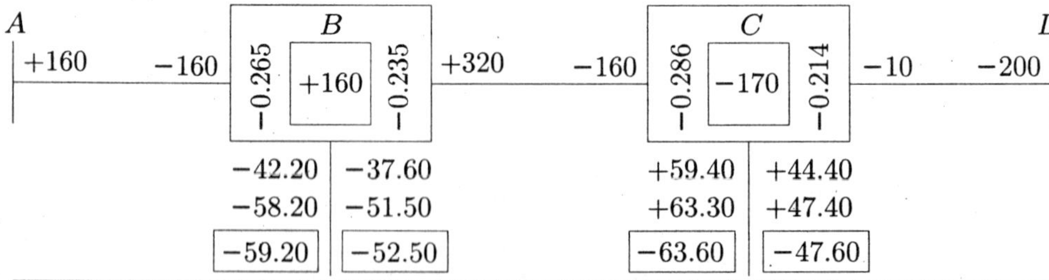

(a) Schematic diagram
Final Rotation Contribution Marked □

Fig. 11.14: For Example 11.5 (Contd.)

	A		B		C		D
M_F	+160	−160	+320	−160	−10	−200	+200
$2M'_{ji}$	0.00	−118.4	−105	+127.20	+95.20		
M'_{ij}	−59.20	0.00	+63.60	−52.50	0.00		
M_{ij}	+100.80	−278.40	+278.60	−85.30	+85.20	−200	+200

(b) Find end moments in kNm

Fig. 11.14: For Example 11.5

Example 11.6. *Analyse the frame shown in Fig. 11.15 by Kani's method.*

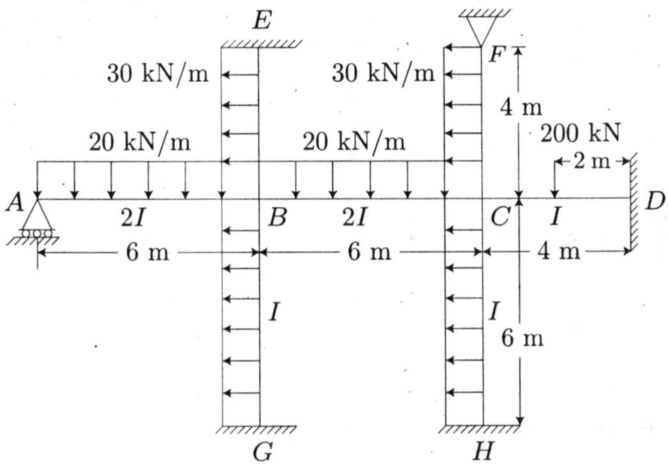

Fig. 11.15: For Example 11.6

Solution: Fixed-end bending moments in kNm are :

$$M_{FBA} = -90; \quad M_{FBC} = +60 = -M_{FCB}$$

$$M_{FCD} = 100 = -M_{FDC}; \quad M_{FBE} = -40 = -M_{FEB}$$

$$M_{FCF} = -60; \quad M_{FBG} = +90 = -M_{FGB}$$

$$M_{FCH} = +90 = -M_{FBC}.$$

Rotation contribution are as below :

At B, $$U_{BA} = -\frac{1}{2}\frac{(3/4)(2/6)}{(3/4)(2/6) + 1/4 + 1/6 + 2/6} = -0.125$$

$$U_{BC} = -\frac{1}{2}\frac{1/3}{(3/4)(1/3) + 1/4 + 1/6 + 2/6} = -0.167$$

$$U_{BE} = -\frac{1}{2}\frac{1/4}{(3/4)(1/3) + 1/4 + 1/6 + 2/6} = -0.125$$

$$U_{BG} = -\frac{1}{2}\frac{1/6}{(3/4)(1/3) + 1/4 + 1/6 + 2/6} = -0.083$$

At C, $U_{CD} = -0.133$; $U_{CB} = -0.177$; $U_{CF} = -0.10$; $U_{CH} = -0.09$.

Fig. 11.16 shows the steps continued to three cycles starting from joint C.

Final moments are obtained by equations

$$M_{ij} = M_{F_{ij}} + 2M'_{ij} + M'_{ji}.$$

Thus, $M_{AB} = 0$

$M_{BA} = -90 - 1.96 + 0 = -91.96$ kNm

$M_{BC} = +60 - 2.62 - 12.16 = 45.22$ kNm

$M_{CB} = -60 - 24.32 - 1.31 = -85.63$ kNm

$M_{CD} = +100 - 18.28 - 0 = +81.72$ kNm

$M_{BE} = -40 - 1.96 = -41.96$ kNm

$M_{BG} = +90 - 1.30 = +88.70$ kNm

$M_{CF} = -60 - 13.74 - 0 = -73.94$ kNm

$M_{CH} = +90 - 12.36 - 0 = +77.64$ kNm

$M_{DC} = -100 + 0 - 9.14 = -109.14$ kNm

$M_{EB} = +40 - 0.98 = 39.02$ kNm

$M_{GB} = -90 - 0.65 = -90.65$ kNm

$M_{FC} = 0$

$M_{HC} = -90 - 6.18 = -96.18.$ kNm.

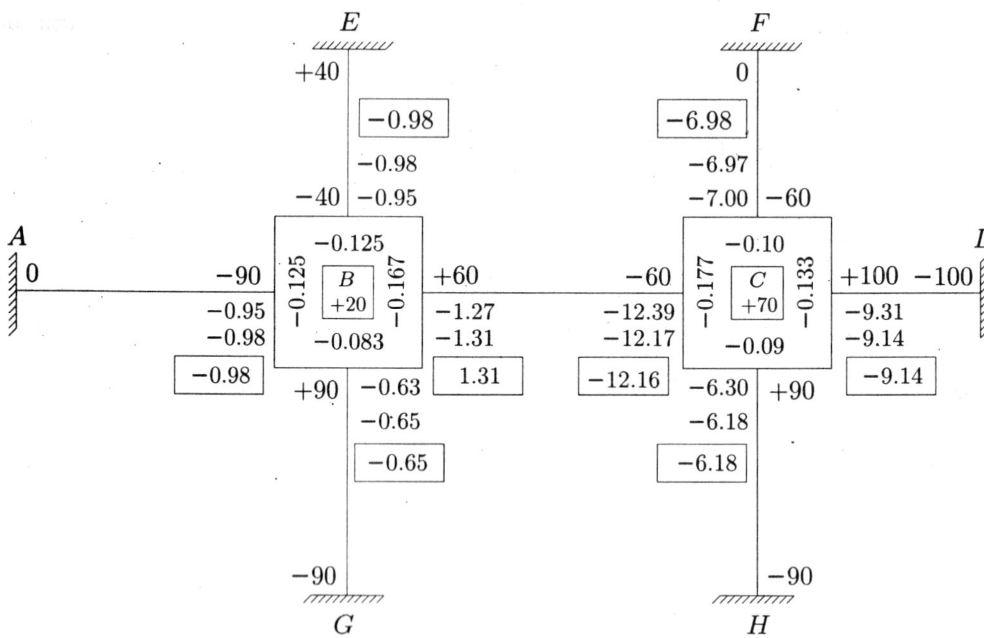

Fig. 11.16: Schematic diagram starting from joint C for Example 11.6

Example 11.7. *Analyse the frame in Fig. 11.7(a) by Kani's method, considering symmetry.*

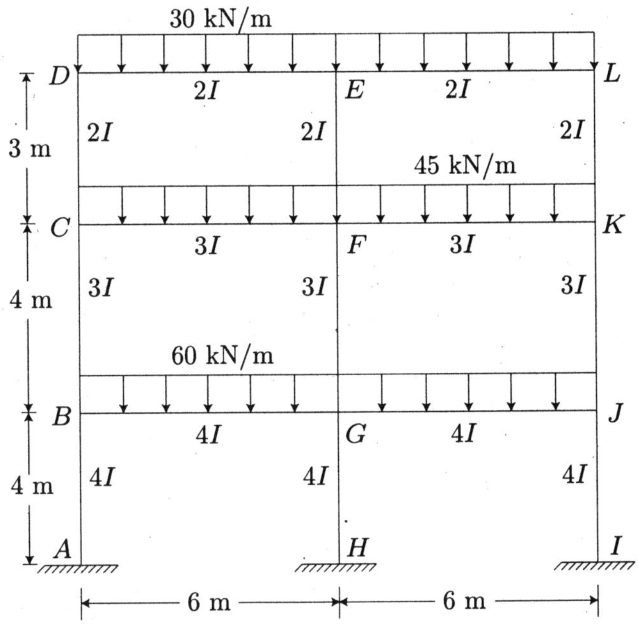

(a) Frame

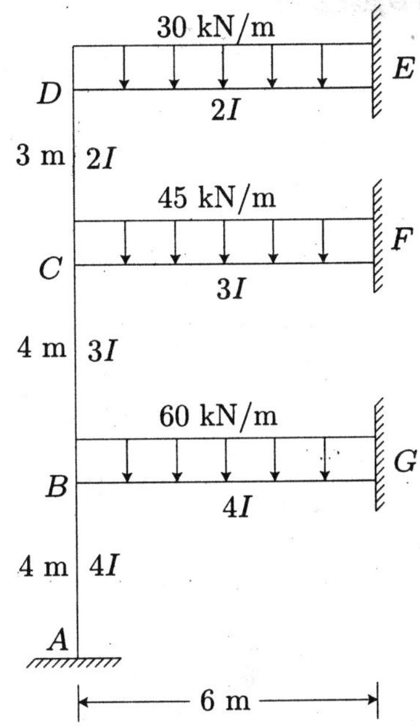

(b) Reduced half of frame for analysis

Fig. 11.17: For Example 11.7

Solution : The fráme is symmetrical about the central line of columns. The joints on the axis of symmetry behave as clamped joints. The frame can be solved by considering only one half of the frame as shown in Fig. 11.17(b).

In Fig. 11.18(a), the rotation contributions are calculated step by step, starting from point D, considering half of the frame shown in Fig. 11.17(b). The end moments in the frame are the calculated as shown in Fig. 11.18(b).

From these moments for half the frame, those for the whole frame can be decided on the basis of symmetry as shown in Fig. 11.18(c).

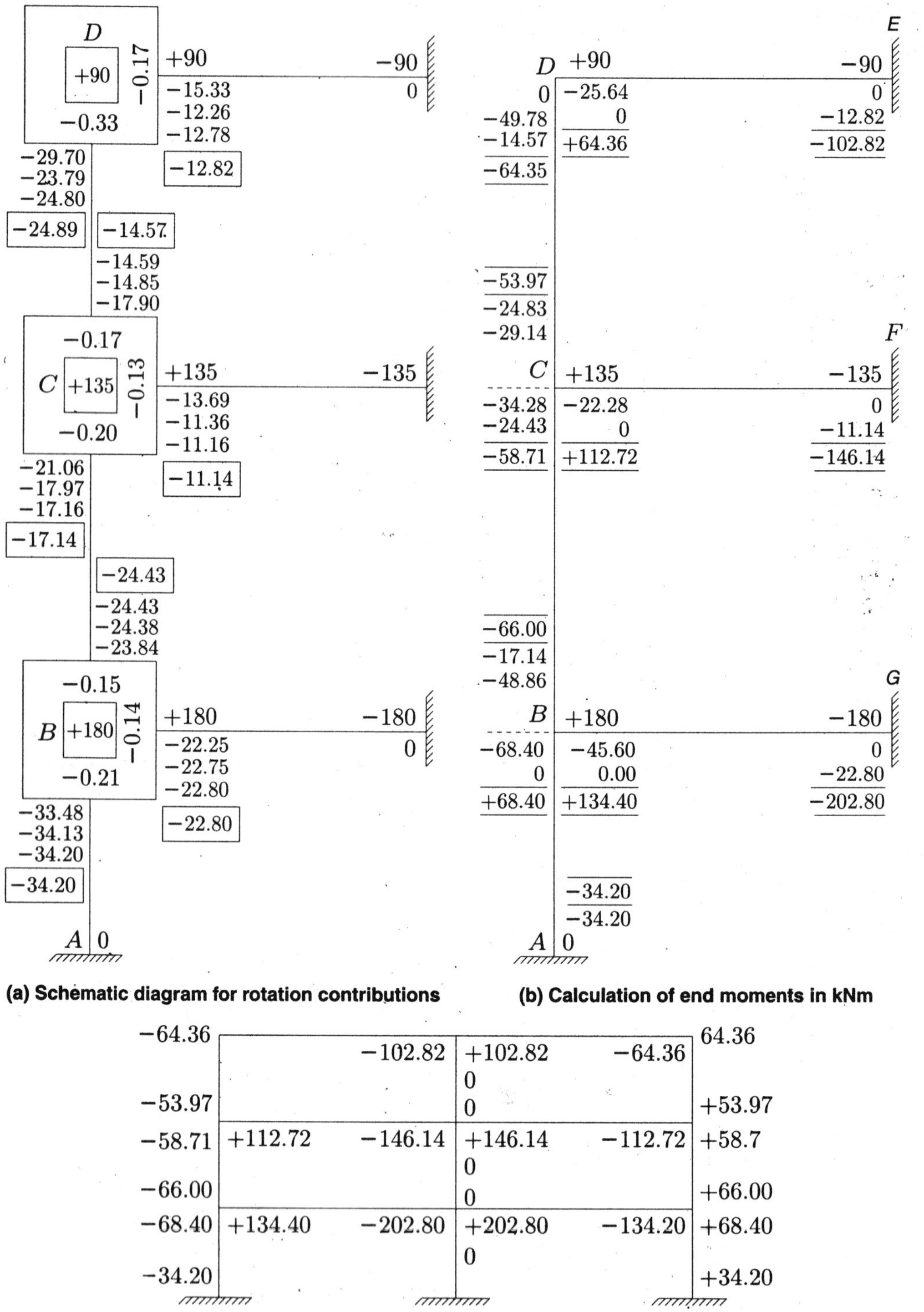

(a) Schematic diagram for rotation contributions

(b) Calculation of end moments in kNm

(c) Resulting end moments in kNm

Fig. 11.18: For Example 11.7

Example 11.8. *Analyse the frame shown in Fig. 11.19(a) by Kani's method taking advantage of symmetry.*

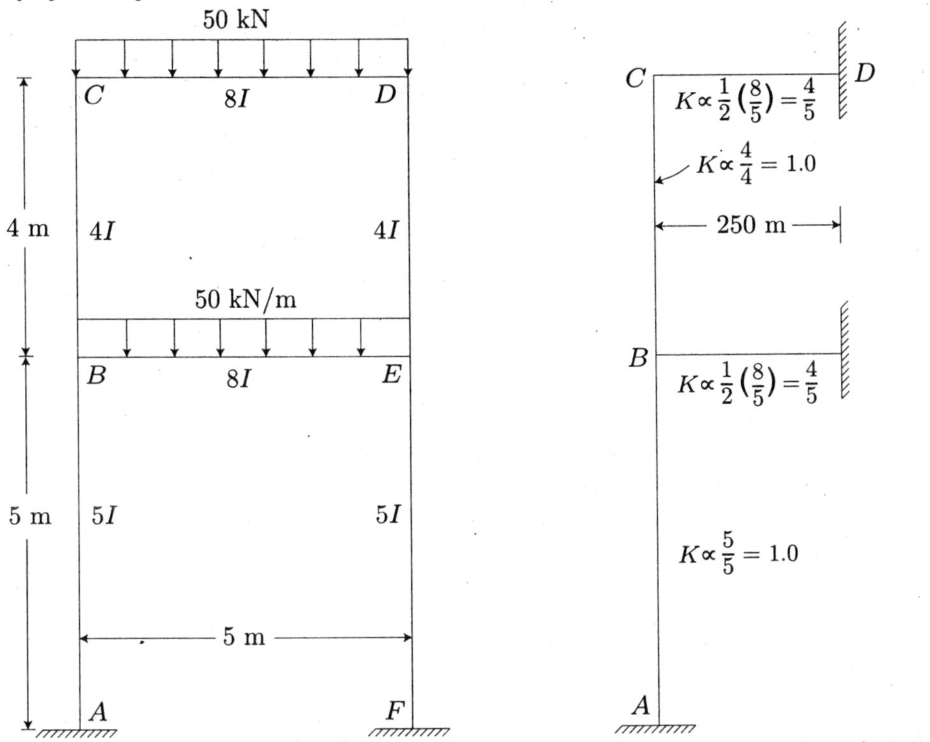

(a) Frame (b) Reduced half of frame for analysis

Fig. 11.19: For Example 11.8

Solution : The frame shown in Fig. 11.19(a) can be solved by considering only half the frame as shown in Fig. 11.19(b).

Calculations of contributions for various cycles are shown in Fig. 11.20, starting from joint C, then to joint B.

Resulting moments in kNm are as under :

$$M_{CD} = 104.17 - 30.58 - 0 = 64.59 = -M_{DC}$$

$$M_{CB} = 0 - 50.38 - 14.22 = -64.60 = -M_{DE}$$

$$M_{BC} = 0 - 28.44 - 25.19 = -53.63 = -M_{ED}$$

$$M_{BE} = 104.17 - 22.12 - 0 = +82.05 = -M_{EB}$$

$$M_{BA} = 0 - 28.44 - 0 = -28.44 = -M_{EF}$$

$$M_{AB} = 0 - 0 - 14.22 = -14.22 = -M_{FE}.$$

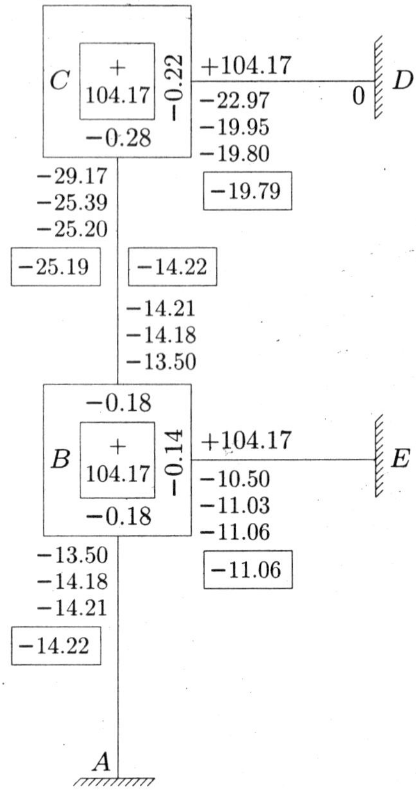

Fig. 11.20: Schematic diagram for calculation of rotation contribution

11.5 Frames with Sidesway

Frames which are either structurally unsymmetric or symmetric with asymmetric loads undergo both joint rotations and joint displacements (side sway). Such frames with fixed bases will be discussed for four separate cases, viz., for vertical and horizontal loading at nodal points only, for general case of loading and for cases of columns with unequal heights. A separate case will be considered for columns with hinged bases. The discussions are valid for rectangular frames with horizontal and vertical members and not for frames with inclined members.

Case I. *Frames of equal height with Lateral Sway.*

Fig. 11.21(a) shows a frame with columns having equal height in each floor. The loads cause the frame to undergo flexure as well as side-sway. The dotted lines indicate the deflected shape of the frame due to loading. The analysis may be carried on in two steps as is done in case of moment distribution.

For analysis a no-sway solution can be made considering artificial restraints at nodal points along the storey heights as shown in Fig. 11.21(b). The loads then cause only flexure in the frame. The analysis of this case can be done by Eqs. (11.1) and (11.5) and follows the same steps as presented in the previous section. The resultant restraining forces as shown in Fig. 11.21(b) may be calculated from the end moments of the columns.

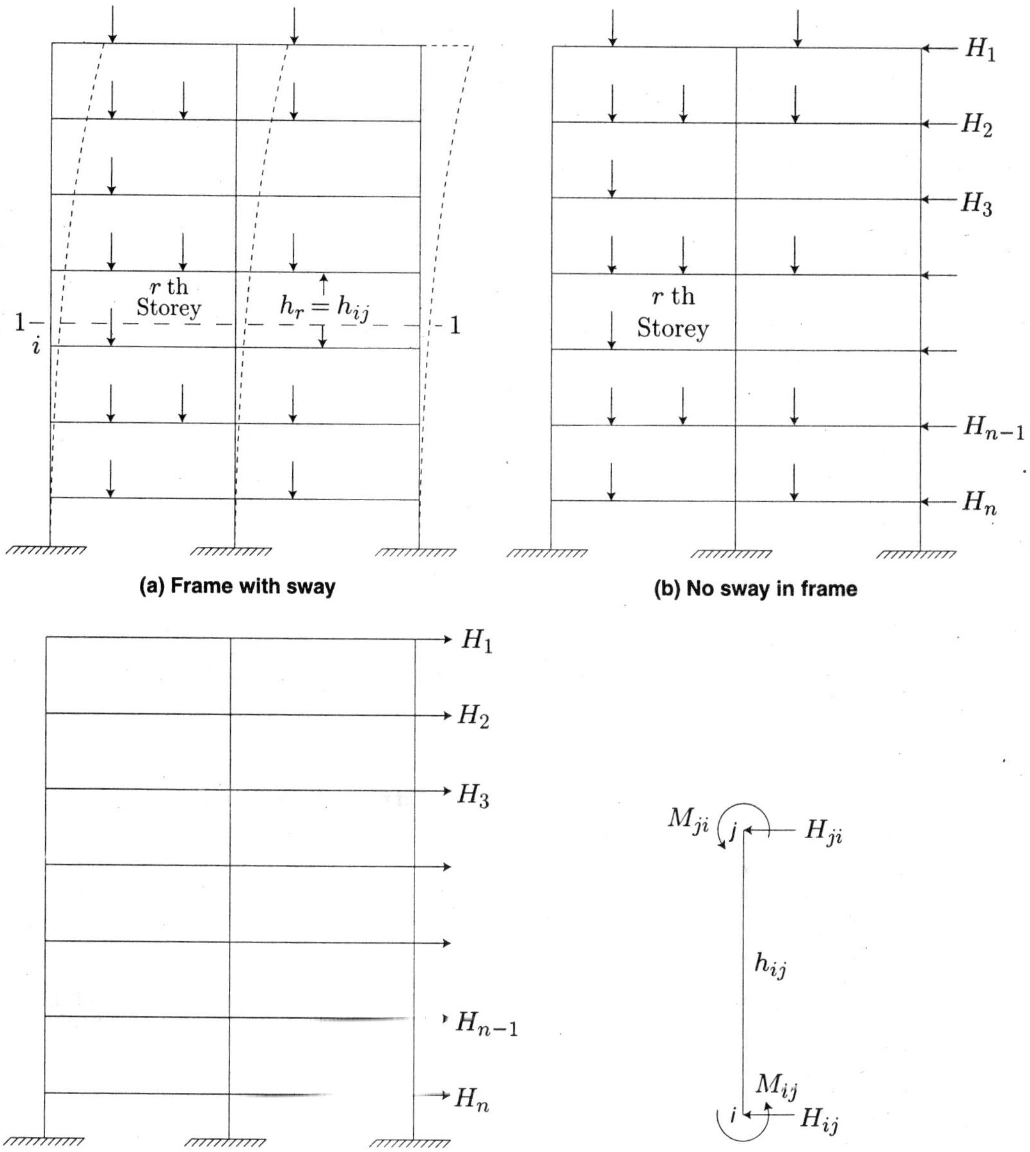

(a) Frame with sway **(b) No sway in frame**

(c) Auxiliary sway frame **(d) Positive signs for moments and shears**

Fig. 11.21: Frames with lateral sway

Since the restraining forces in Fig. 11.21(b) do not actually exist in the given frame, their effect may be eliminated by applying an auxiliary force system as in Fig. 11.21(c), the forces being cqual in magnitude but opposite in directions to their counterparts in Fig. 11.21(b). The total solution of the given frame may be obtained as the algebraic sum of the solutions of the frame in Figs. 11.21(b) and 11.21(c).

For the horizontal equilibrium of the part of the frame above a section $1-1$ in the rth storey, the algebraic sum of the column shears must be zero, i.e., $\sum_r H_{ij} = 0$, where $\sum_r H_{ij}$ denotes the sum of shears for all the column in the rth storey.

Considering H_{ij} as the shear in the column $i - j$ of the rth storey of height h_r, with positive direction as shown in Fig. 11.21(d).

$$H_{ij} = \frac{M_{ij} + M_{ji}}{h_r}. \tag{11.6}$$

For column ij,

$$M_{ij} = M_{F_{ij}} + 2M'_{ij} + M'_{ji} + M''_{ij} \quad \text{and} \quad M_{ji} = M_{F_{ji}} + M'_{ij} + 2M'_{ji} + M''_{ji},$$

where M''_{ij} and M''_{ji} represent the displacement contributions.

For cases where loads are applied at the nodal points only, fixed-end moments

$$M_{F_{ij}} = M_{F_{ji}} = 0, \text{ and as } M''_{ji} = M''_{ji},$$

$$\left.\begin{aligned} M_{ij} &= 2M'_{ij} + M'_{ji} + M''_{ij} \\ \text{and} \quad M_{ji} &= M'_{ij} + 2M'_{ji} + M''_{ij}. \end{aligned}\right] \tag{11.7}$$

From equilibrium of the column ij of height h_r subjected to horizontal nodal force H_{ij} at the nodal points,

$$M_{ij} + M_{ji} + H_{ij}h_r = 0. \tag{11.8}$$

For all column of same height in the same rth storey,

$$\sum_r M_{ij} + \sum_r M_{ji} + \sum_r H_{ij}h_r = 0.$$

From Eqs. (6.7) and (6.8),

$$3\sum_r M'_{ij} + 3\sum_r M'_{ji} + 2\sum_r M''_{ij} + \sum_r H_{ij}h_r = 0. \tag{11.9}$$

Since no nodal force exists in case of vertical loading only, $\sum_r H_{ij} = 0$ and from Eq. (11.9) it follows that,

$$\sum_r M''_{ij} = -3/2\left[\sum_r M'_{ij} + \sum_r M'_{ji}\right]. \tag{11.10}$$

This indicates that the sum of the displacement contributions of all the columns in the rth storey is 1.50 times that of the rotation contributions of the two ends of the same column.

Since $\quad M''_{ij} = \dfrac{6EK_{ij}\Delta_{ij}}{h_r},$

$$\sum M''_{ij} = \frac{6E\Delta_{ij}}{h_r}\sum K_{ij},$$

whence $\quad \dfrac{6E\Delta_{ij}}{h_r} = \dfrac{\sum M''_{ij}}{\sum K_{ij}}.$

Hence $\quad M''_{ij} = \dfrac{\sum M''_{ij}}{\sum K_{ij}} \cdot K_{ij},$

which, with substitution from Eq. (11.10), reduces to

$$M''_{ij} = -\frac{3}{2}\frac{K_{ij}}{\sum K_{ij}}\left[\sum_r M'_{ij} + \sum_r M'_{ji}\right], \tag{11.11}$$

where $\sum K_{ij}$ denotes the sum of the K-values of all the columns in the rth storey.

With the help of Eq. (11.11), the individual displacement contribution in a given storey can be calculated. The factor $-\frac{3}{2}\frac{K_{ij}}{\sum_r K_{ij}}$ can be referred to as the 'displacement factor' (D.F.), denoted by D_{ij}. The sum of the displacement factors for any storey is obviously $-\frac{3}{2}$.

Eqs. (11.1a), (11.3a) and (11.3) represent the method of solution of the problems for frames with lateral sway and zero nodal forces. The correct values of all the rotation contributions and displacement contributions are those satisfying Eq. (11.3a) at all the joints and Eq. (11.11) at all the storeys. They are determined by Gauss-Seidel iteration method as before. Usually the initial start is made from calculation of rotation contributions for a joint. Calculations for the other joints moving in a clockwise or anticlockwise direction are then completed. The calculation of displacement contributions for columns of various storeys are then carried on.

Case II. *Frames with Lateral Sway–Horizontal Loading at Nodal points and Vertical Loading.*

In this case, horizontal loads acting as concentrated loads at beam column junctions only will be considered.

As in Case I, the solution may be done in two stages, viz., solution for no sway and solution for sway. Restraining forces may be applied at storey heights to stop side-sway of frames. Since such restraining forces do not actually exist, they may be eliminated by applying the auxiliary force system as in Case I.

Eq. (11.9) in Case I can be used in this case also which gives

$$3\sum_r M'_{ij} + 3\sum_r M'_{ji} + 2\sum_r M''_{ij} + \left(\sum_r H_{ij}\right) r = 0. \tag{11.9a}$$

In this case, $\sum_r H_{ij}$ represent the storey shear at the r-th storey as shown in Figs. 11.21(c) and 11.21(d) and given by

$$\sum_r H_{ij} = H_1 + H_2 + \ldots + H_r.$$

As nodal forces exist, $\sum_r H_{ij} \neq 0$ and from Eq. (11.9), it follows that,

$$M''_{ij} = -\frac{3}{2}\frac{K_{ij}}{\sum_r K_{ij}}\left[\frac{\sum_r H_{ij} h_r}{3} + \sum_r M'_{ij} + \sum_r M'_{ji}\right]. \tag{11.12}$$

If there is no horizontal loading on the frame, then $\sum H_{ij} = 0$ and Eq. (11.11) follows.

The calculation operation follows Eqs. (11.5) and (11.12). The correct values of all the rotation contributions and the displacement contributions are determined by iteration. End moments are then calculated from Eq. (11.5a).

Since anticlockwise moments are considered positive, horizontal forces should be taken as positive if the algebraic sum of such forces above the storey is towards left.

Case III. *Frames with Lateral Sway–General case of Loading.*

In this case any possible nature of horizontal and vertical loading will be discussed. As lateral sway is involved, Eqs. (11.1a) and (11.5a) are applicable at the joints of the frame. An equation based on the shear condition at each storey is to be derived for this case.

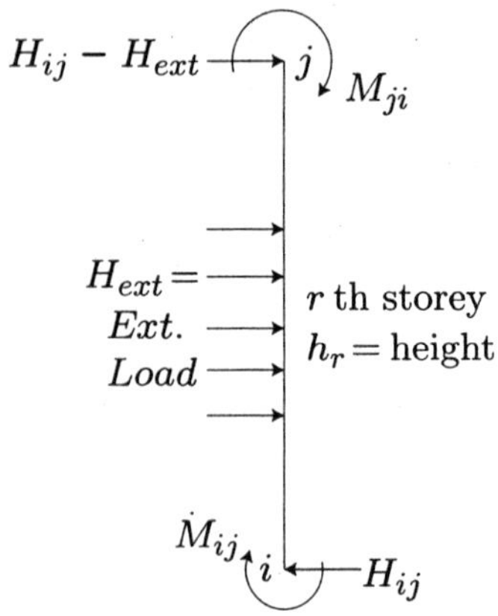

Fig. 11.22: Column in *r* th storey

Considering the case of a column of rth storey of a frame, as in Fig. 11.22, taking moments about i,

$$M_{ij} + M_{ji} + M_{ext} + H_{ij}h_r = 0. \tag{11.13}$$

Substituting from Eq. (11.7),

$$M_{F_{ij}} + M_{F_{ji}} + 3M'_{ij} + 3M'_{ji} + 2M''_{ij} + M_{ext} + H_{ij}h_r = 0. \tag{11.13a}$$

For all the columns in the rth storey,

$$\sum_r M_{F_{ji}} + \sum_r M_{F_{ij}} + 3\sum_r M'_{ij} + 3\sum_r M'_{ji} + 2\sum M''_{ij} + \sum_r M_{ext} + \sum_r H_{ij}h_r = 0, \tag{11.14}$$

where $\sum_r H_{ij}$ is the storey shear at the base of rth storey.

If ends i and j of the column be fixed, taking moments about i,

$$M_{F_{ij}} + M_{F_{ji}} + M_{ext} + R_{F_{ij}}(h_r) = 0,$$

where $R_{F_{ij}}$ is the fixed end reaction at i, i.e., at bottom of the column.

For all the columns in the rth storey,

$$\sum_r M_{F_{ij}} + \sum_r M_{F_{ji}} + \sum_r M_{ext} + h_r \sum_r R_{F_{ij}} = 0. \tag{11.15}$$

Eq. (11.15) expresses relation between the sum of the fixed-end moments, the sum of the fixed-end reactions at the lower ends of the columns and the sum of the moments due to external horizontal load on column, i.e., M_{ext}. Substracting Eq. (11.15) from Eq. (11.14)

$$3\sum_r M'_{ij} + 3\sum_r M'_{ji} + 2\sum_r M''_{ij} + \sum_r H_{ij}h_r - h_r \sum_r R_{F_{ij}} = 0. \tag{11.16}$$

Analysis of such frames loaded with general case of vertical loading has to be done in two stages. Figs. 11.23(b) and 11.23(c) show the two stages of loading for the general case shown in Fig. 11.23(a).

In stage 1, only the non-nodal horizontal loadings and the vertical loadings are taken into consideration. In addition, certain imaginary horizontal point loads will be applied at the various nodes of the frame on one face so as to make the horizontal force at the r-th storey for the first stage denoted by $\sum_r H_{ij1}$.

The imaginary point load applied at each nodal point should be equal to and in the same direction as the sum of the fixed-end reactions at the level of that node for all the columns immediately above and below the particular node. Thus, referring to Fig. 11.23(b) at the base of the rth storey the nodal point load is $\sum_r R_{F_{ij}} + \sum_{r-1} R_{F_{ig}}$.

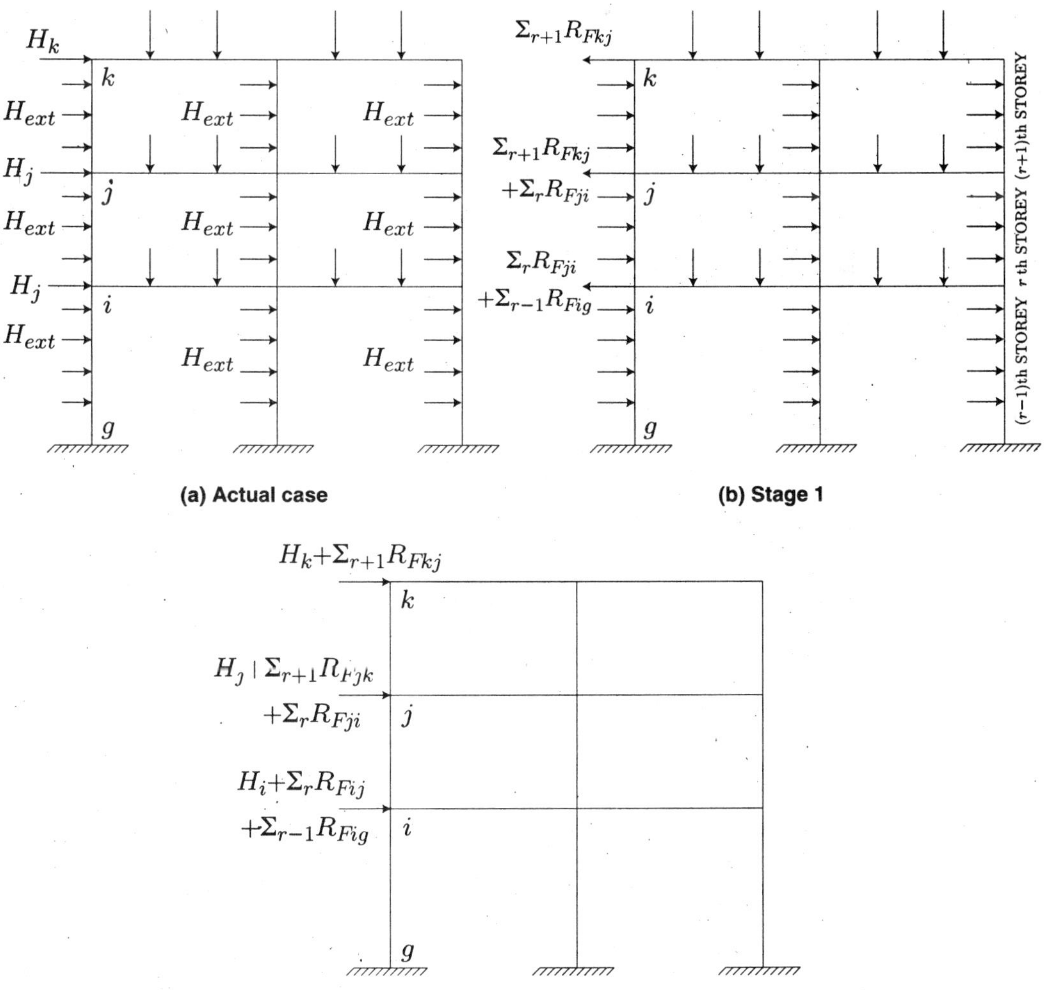

(a) Actual case **(b) Stage 1**

(c) Stage 2

Fig. 11.23: General case of loading of frames

In the second stage, the actual horizontal nodal loading if any on the frame is taken into account and in addition loads equal and opposite to the imaginary loads introduced in stage 1 are applied. Combination of stage 1 and stage 2 corresponds to the actual case.

Since Eq. (11.16) is the general equation or any case of loading it can be applied to stage 1. But since at stage 1, $\sum_r H_{ij} = \sum_r R_{F_{ij}}$, the last two terms of the left-hand side of Eq. (11.16) vanish when the equation is applied to stage 1. Eq. (11.16) then reduces to

$$3\sum_r M'_{ij1} + 3\sum_r M'_{ji1} + 2\sum_r M''_{ij1} = 0, \tag{11.17}$$

where suffix 1 indicates that the values correspond to stage 1.

For stage 2, the loading is only at nodal points, so that there are no fixed end moments or fixed end reactions at the column ends. Consequently the last term of Eq. (11.16) vanishes and Eq. (11.16) is modified as

$$3\sum_r M'_{ij2} + 3\sum_r M'_{ji2} + 2\sum_r M''_{ij2} + \sum_r H_{ij2}h_r = 0, \tag{11.18}$$

where suffix 2 indicates that the values correspond to stage 2.

From Eqs. (11.17) and (11.18), then,

$$3\sum_r M'_{ij} + 3\sum_r M'_{ji} + 2\sum_r M''_{ij} + \sum_r H_{ij2}h_r = 0,$$

whence $$\sum_r M''_{ij} = -\frac{3}{2}\left[\frac{\sum_r H_{ij2}h_r}{3} + \sum_r M'_{ij} + \sum_r M'_{ji}\right]. \tag{11.19}$$

Proceeding as in case of Eqs. (11.11) and (11.12), Eq. (11.19) can be written as

$$M''_{ij} = D_{ij}\left[\frac{\sum_r H_{ij2}h_r}{3} + \sum_r M'_{ij} + \sum_r M'_{ji}\right], \tag{11.20}$$

where $$D_{ij} = -\frac{3}{2}\frac{K_{ij}}{\sum K_{ij}}.$$

Eq. (11.20) defines the calculation operation at every storey in the general case of loading. It is obviously similar to Eq. (11.12) for the case of horizontal loads at nodal points only.

Whereas the term $\sum_r H_{ij}$ in Eq. (11.12) represents the actual storey shear at the rth storey, the term $\sum_r H_{ij2}$ in Eq. (11.20), represents the storey shear at the base of the storey as calculated for stage 2 only and not the actual storey shear.

Case IV. *Frames with unequal lengths with fixed ends.*

The calculation of rotation contributions in all stages of frames of this class does not differ from the previous cases in any respect. The governing equations remain unchanged for the calculation of the displacement contributions of the columns in storey with constant heights as well. However, the computation of the displacement contributions of the columns in a storey with varying heights needs special consideration. Taking h_{ij} as the height of a column in r-th storey, Eq. (11.19) can be rewritten as

$$\sum_r \frac{M''_{ij}}{h_{ij}} = -\frac{3}{2}\left[\frac{\sum_r H_{ij2}}{3} + \sum_r\left(\frac{M'_{ij}}{h_{ij}} + \frac{M'_{ji}}{h_{ji}}\right)\right]. \tag{11.21}$$

This equation is applicable to the displacement contributions of the columns of the rth storey, the column heights for the storey being unequal. Considering h_r as the height of an arbitrary chosen column for the rth storey, and h_{ij} as the height of another column in the same storey with fixed ends,

$$\sum_r M''_{ij}\frac{hr}{h_{ij}} = -\frac{3}{2}\left[\frac{\sum_r H_{ij2}h_r}{3} + \sum_r h_r\left(\frac{M'_{ij}}{h_{ij}} + \frac{M'_{ji}}{h_{ij}}\right)\right] \tag{11.21a}$$

Taking $\frac{h_r}{h_{ij}}$ as α_{ij}, a factor for the column $i-j$ on rth storey, Eq. (11-21) can be rewritten as

$$\sum_r \alpha_{ij}M''_{ij} = -\frac{3}{2}\left[\frac{\sum_r H_{ij2}h_r}{3} + \sum_r \alpha_{ij}(M'_{ij} + M'_{ji})\right]. \tag{11.21b}$$

Since $M''_{ij} = \dfrac{6EK_{ij}\Delta}{h_{ij}} \propto \dfrac{K_{ij}}{h_{ij}} \propto \alpha_{ij}K_{ij}$,

and $\dfrac{M''_{ij}}{\sum_r \alpha_{ij}M''_{ij}} = \dfrac{\alpha_{ij}K_{ij}}{\sum_r \alpha_{ij}^2 K_{ij}}$, it follows that,

$$M''_{ij} = -\frac{3}{2}\frac{\alpha_{ij}K_{ij}}{\sum_r \alpha_{ij}^2 K_{ij}}\left[\frac{\sum_r H_{ij2}h_r}{3} + \sum_r \alpha_{ij}(M'_{ij} + M'_{ji})\right], \tag{11.22}$$

where $-\dfrac{3}{2}\dfrac{\alpha_{ij}K_{ij}}{\sum_r \alpha_{ij}^2 K_{ij}}$ is the displacement factor.

If all columns are of the same height, α_{ij} = unity and displacement factor becomes $-\frac{3}{2}\frac{K_{ij}}{\sum_r K_{ij}}$ as in Eq. (11.12).

Eq. (11.12) along with Eqs. (11.1a) and (11.5a) form the basis of the analysis of frames with storeys having columns of unequal heights.

Case V. *Columns with hinged ends.* Such columns usually pertain to first storey columns. A column with a hinged base may be replaced for analysis by a substitute column with fixed end, provided the real column and the substitute column are structurally similar at the top end of the column. For structural similarity, the following two conditions must be fulfilled.

(i) A moment M applied at the tops of the real and the substitute columns cause equal rotations at the top in both cases.

(ii) A lateral sway Δ occurring at the top of the columns must cause equal fixed-end moments at top.

If a moment M be applied at the top of the real (hinged) column as shown in Fig. (11.24) the rotation at the top $= \dfrac{Mh_{ij}}{3EI_{ij}}$.

If a moment M be applied at the top of the substitute column shown in Fig. 11.24(b), the rotation at top $= \dfrac{Mh'_{ij}}{3EI'_{ij}} - \dfrac{M}{2}\dfrac{h'_{ij}}{6EI'_{ij}} = \dfrac{Mh'_{ij}}{4EI'_{ij}}$.

From the first condition, $\dfrac{Mh_{ij}}{3EI_{ij}} = \dfrac{Mh'_{ij}}{4EI'_{ij}}$,

whence $\dfrac{I'_{ij}}{h^r_{ij}} = \dfrac{3}{4}\dfrac{I_{ij}}{h_{ij}}$, i.e., $K'_{ij} = \dfrac{3}{4}K_{ij}$.

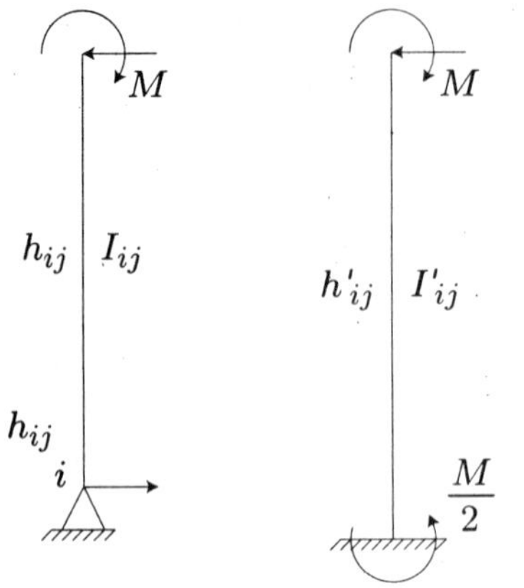

(a) Hinged column (b) Fixed column

Fig. 11.24: Columns

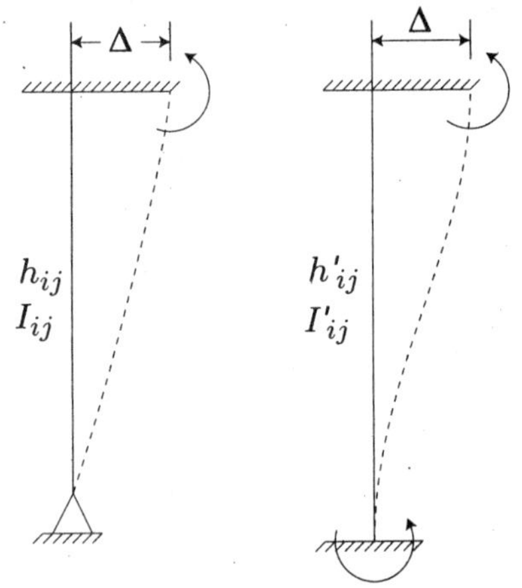

(a) Hinged column (b) Fixed column

Fig. 11.25: Sway in columns

If there is a lateral sway Δ at top of the real column as in Fig. 11.25(a), fixed-end moment at the top $= \dfrac{3EI_{ij}\Delta}{h_{ij}^2}$.

For a lateral sway Δ at top of the substitute column as in Fig. 6.25(b), fixed-end moment at top $= \dfrac{6EI'_{ij}\Delta}{h'_{ij^2}}$.

From the second condition,

$$\frac{3EI_{ij}\Delta}{h_{ij}^2} = \frac{6EI'_{ij}\Delta}{h'_{ij^2}}, \quad \text{whence} \quad \frac{K_{ij}}{h_{ij}} = \frac{2K'_{ij}}{h'_{ij}}.$$

From first condition, $K'_{ij} = \dfrac{3}{4}K_{ij}$, so that $h'_{ij} = \dfrac{3}{2}h_{ij}$.

Thus, if the substitute column is provided with the K-value equal to $\frac{3}{4}K_{ij}$ of actual column and height equal to $\frac{3}{2}h_{ij}$, the replacement does not affect the rest of the structure.

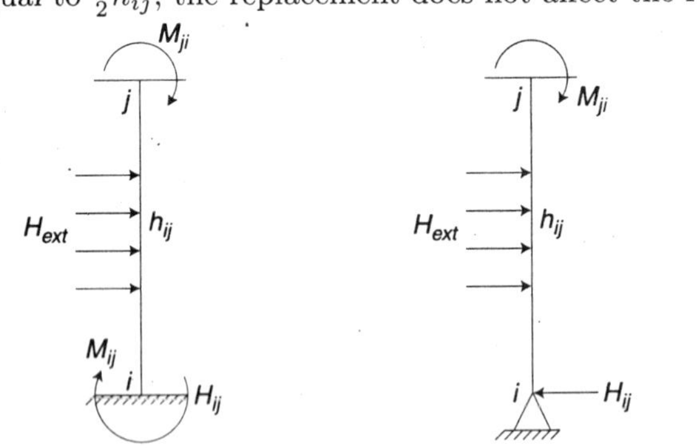

(a) Fixed ended column (b) Column with hinged base

Fig. 11.26: Columns

Considering fixed end column subjected to external horizontal load as in Fig. (11.26a), as for fixed base $M'_{ij} = 0$, from Eq. 6.13(a), $M_{F_{ij}} + M_{F_{ji}} + 3M'_{ji} + 2M''_{ij} + M_{ext} + H_{ij}h_{ij} = 0.$

It follows that, $$\frac{M_{F_{ij}}}{h_{ij}} + \frac{M_{F_{ji}}}{h_{ij}} + \frac{3M'_{ji}}{h_{ij}} + \frac{2M''_{ij}}{h_{ij}} + \frac{M_{ext}}{h_{ij}} + H_{ij} = 0. \tag{11.23}$$

For the column with hinged base as in Fig. 11.26(b), similarly considering it replaced by substitute column,

$$M_{F_{ji}} + 2M'_{ji} + M''_{ji} + M_{ext} + H_{ij}(h_{ij}) = 0,$$

whence $$\frac{M_{F_{ij}}}{h_{ij}} + \frac{2M'_{ji}}{h_{ij}} + \frac{M''_{ji}}{h_{ij}} + \frac{M_{ext}}{h_{ij}} + H_{ij} = 0.$$

If h'_{ij} be the height of the substitute column, with $h'_{ij} = \frac{3}{2}h_{ij}$,

$$\frac{3}{2}\frac{M_{F_{ji}}}{h'_{ij}} + 3\frac{M'_{ji}}{h'_{ij}} + \frac{3}{2}\frac{M''_{ji}}{h'_{ij}} + \frac{3}{2}\frac{M_{ext}}{h'_{ij}} + H_{ij} = 0. \tag{11.24}$$

Eq. (11.23) corresponds to a column which is really fixed at the bases. Eq. (11.24) corresponds to a column which is in fact not fixed at bases, but has been considered as fixed at bases as a result of the introduction of substitute column.

Considering Eqs. (11.23) and (11.24), the general equation for the equilibrium of the column can be written as

$$\frac{pM_{F_{ij}}}{h_{ij}} + \frac{nM_{F_{ji}}}{h_{ij}} + \frac{3M'_{ji}}{h_{ij}} + \frac{2mM''_{ji}}{h_{ij}} + \frac{nM_{ext}}{h_{ij}} + H_{ij} = 0, \tag{11.25}$$

where values of m, n, and p depend on the end conditions.

For fixed base column $m = n = p = 1$, h_{ij} being the actual height of the column and for substitute columns for hinged bases, $m = \frac{3}{4}$, $n = \frac{3}{2}$, $p = 0$, h_{ij} being the height of the substitute column.

Considering all columns in the 1st storey where base conditions may vary,

$$\sum_1 \frac{pM_{F_{ij}}}{h_{ij}} + \sum_1 \frac{nM_{F_{ji}}}{h_{ij}} + 3\sum_1 \frac{M'_{ji}}{h_{ij}} + 2\sum_1 \frac{mM''_{ji}}{h_{ij}} + \sum_1 \frac{nM_{ext}}{h_{ij}} + \sum_1 H_{ij} = 0. \tag{11.26}$$

Similarly, the general equilibrium equation for columns fixed at top and subjected to horizontal external loads can be written as

$$\frac{pM_{F_{ij}}}{h_{ij}} + \frac{nM_{F_{ji}}}{h_{ij}} + \frac{nM_{ext}}{h_{ij}} + R_{F_{ij}} = 0, \tag{11.27}$$

where n, p and h_{ij} have similar values as in case of Eq. (11.25), depending on the support condition at base and $R_{F_{ji}}$ refers to the reactions at bottom considering top as fixed.

Adding up the terms of such equations for all the columns of the first storey,

$$\sum_1 \frac{pM_{F_{ij}}}{h_{ij}} + \sum_1 \frac{nM_{F_{ji}}}{h_{ij}} + \sum_1 \frac{nM_{ext}}{h_{ij}} + \sum_1 R_{F_{ij}} = 0. \tag{11.28}$$

Subtracting Eq. (11.28) from Eq. (11.26),

$$3\sum_1 \frac{M'_{ji}}{h_{ij}} + 2\sum_1 \frac{mM''_{ji}}{h_{ij}} + \sum_1 H_{ij} - \sum R_{F_{ij}} = 0.$$

Visualising the frame problem in two stages as discussed in case III, for stage 1, $\sum H_{ij1} = \sum_1 R_{F_{ij1}}$, and for stage 2, $\sum_1 R_{F_{ij2}} = 0$, so that,

$$3\sum_1 \frac{M'_{ji}}{h_{ij}} + 2\sum \frac{mM''_{ji}}{h_{ij}} + \sum_1 H_{ij2} = 0$$

$$\text{i.e.,} \quad \sum_1 \frac{mM''_{ji}}{h_{ij}} = -\frac{3}{2}\left[\frac{\sum H_{ij2}}{3} + \sum_1 \frac{M'_{ji}}{h_{ij}}\right].$$

Multiplying all the terms of this equation by the reference height h_1, with $\frac{h_1}{h_{ij}} = \alpha_{ij}$,

$$\sum_1 m(\alpha_{ij})(M''_{ji}) = -\frac{3}{2}\left[\frac{\sum H_{ij}h_1}{3} + \sum_1 \alpha_{ij}M'_{ji}\right]. \tag{11.29}$$

The individual displacement contribution is $M''_{ji} = \frac{6EI_{ij}\Delta}{h_{ij}^2} = \frac{6EK_{ij}\Delta}{h_1}\alpha_{ij}$.

Combining, $\dfrac{M''_{ji}}{\sum_1 m\alpha_{ij}M''_{ji}} = \dfrac{K_{ij}\alpha_{ij}}{\sum_1 mK_{ij}\alpha_{ij}^2}$

Hence
$$M''_{ji} = -\frac{3}{2}\frac{\alpha_{ij}K_{ij}}{\sum_1 mK_{ij}\alpha_{ij}^2}\left[\frac{\sum H_{ij2}h_1}{3} + \sum_1 \alpha_{ij}M'_{ji}\right], \tag{11.30}$$

where Displacement Factor $= -\dfrac{3}{2}\dfrac{\alpha_{ij}K_{ij}}{\sum_1 mK_{ij}\alpha_{ij}^2}$.

Eq. (11.30) is the general displacement contribution equation for the first storey, applicable to both types of columns. In Eq. (11.30), the terms used have the following definitions.

h_1 is the arbitrary chosen reference height for the first storey.

α_{ij} is the reduction factor $\frac{h_1}{h_{ij}}$, where h_{ij} is the actual height for columns with fixed base and substitute height for columns with hinged base.

$\sum_1 H_{12}$ is the storey shear for the first storey for stage 2 of general loading. The fixed end reactions included in $\sum_1 H_{12}$ correspond to the actual end conditions of the columns of the first storey.

m is the end condition factor. Its value is 1.00 for columns with fixed base and 0.75 for substitute columns.

K_{ij} is $\frac{I_{ij}}{h_{ij}}$ for column with fixed base and $\frac{3}{4}\frac{I_{ij}}{h_{ij}}$ for columns with hinged base.

Example 11.9. *Analyse the frame in Fig. 11.27 by Kani's method.*

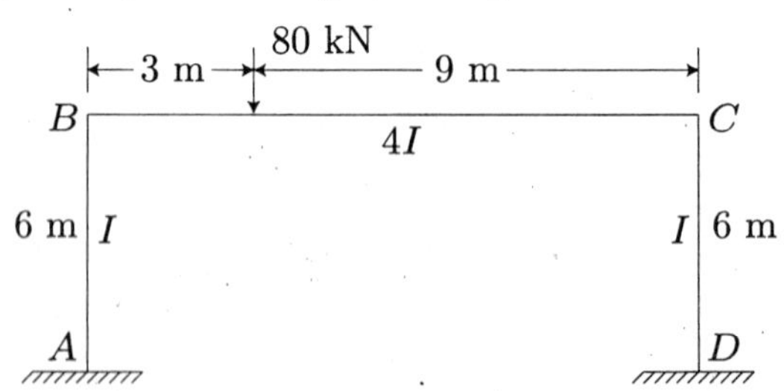

Fig. 11.27: For Example 11.9

Solution : In this case there is no horizontal load and the base is fixed. It corresponds to case I of frames with lateral sway.

Fixed-end moments :

$$M_{FBC} = +35 \text{ kNm}; \quad M_{FCB} = -45 \text{ kNm}.$$

Rotation factors:

Joints B and C

For BA, CD, $U_{BA} = U_{CD} = -\frac{1}{2}\frac{\frac{1}{6}}{\frac{1}{3}+\frac{1}{6}} = -\frac{1}{6}$

For BC, CB, $U_{BC} = U_{CB} = -\frac{1}{2}\frac{\frac{1}{3}}{\frac{1}{3}+\frac{1}{6}} = -\frac{1}{3}$

$$\sum = -\frac{1}{2}.$$

Displacement factors:

For columns AB and DC,

$$D_{AB} = D_{DC} = -\frac{3}{2}\frac{\frac{1}{6}}{\frac{1}{6}+\frac{1}{6}} = -\frac{3}{4}$$

$$D_{AB} + D_{DC} = -\frac{3}{2}.$$

Calculation of rotational and displacement contributions are shown below starting from joint B.

Joint B	1st cycle	2nd cycle	3rd cycle	4th cycle
$\sum M_{FB}$	+135.0	+135.0	+135.0	+135.0
M''_{AB}	0.00	+5.60	+9.20	+10.40
M'_{AB}	0.00	0.00	0.00	0.00
M'_{CB}	0.00	+30.00	+33.20	+31.80
Total	+135.00	+170.60	+177.40	+177.20
M'_{BA}	−22.50	−28.40	−29.60	−29.60
M'_{BC}	−45.0	−56.80	−59.20	−59.20
Joint C	↓	↓	↓	↓
$\sum M_{FC}$	−45.0	−45.00	−45.00	−45.00
M''_{DC}	0.00	+5.40	+9.20	+10.40
M'_{DC}	0.00	0.00	0.00	0.00
M'_{BC}	−45.0	−56.80	−59.20	−59.20
Total	−90	−96.40	−95.00	−93.80
M'_{CD}	+15.0	+16.10	+15.90	+15.60
M'_{CB}	+30.0	+32.20	+31.80	+31.30
Storey calculation	↓	↓	↓	↓
M'_{AB}	0.00	0.00	0:00	0.00
M'_{BA}	−22.50	−28.40	−29.60	−29.60
M'_{CD}	+15.0	+16.10	+15.90	+15.60
M'_{DC}	0.00	0.00	0.00	0.00
Total	−7.50	−12.30	−13.70	−14.00
M''_{BA}, M''_{CD}	+5.60	+9.20	+10.30	+10.50

Figs. 11.28(a) and 11.28(b) indicate schematically the calculation of rotation and displacement contributions and the calculation of final end moments.

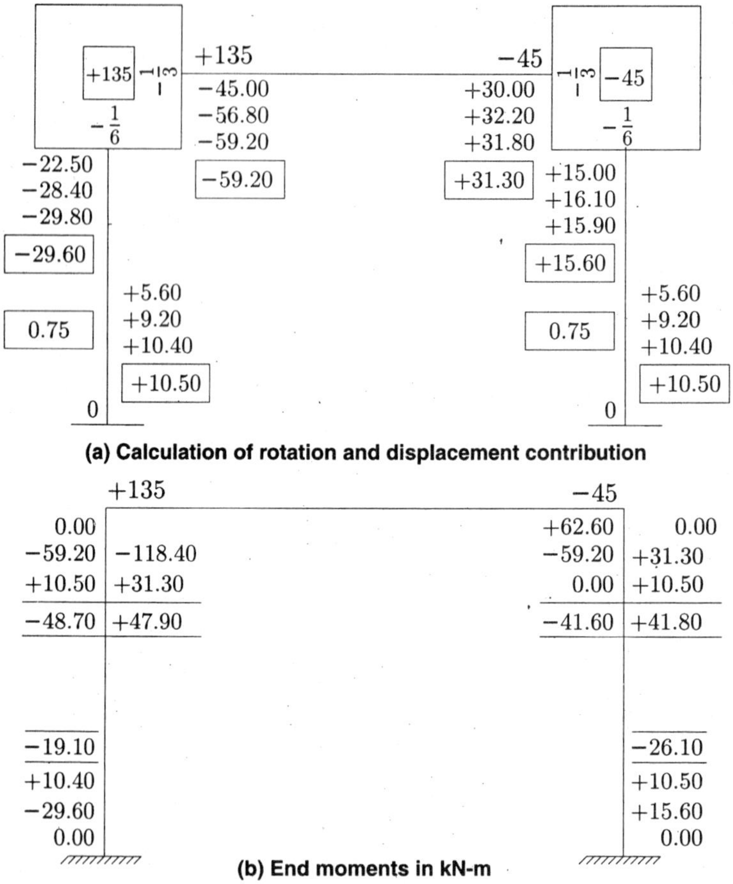

Fig. 11.28: For Example 11.9

Example 11.10. *Analyse the frame shown in Fig. 6.29 using Kani's method.*

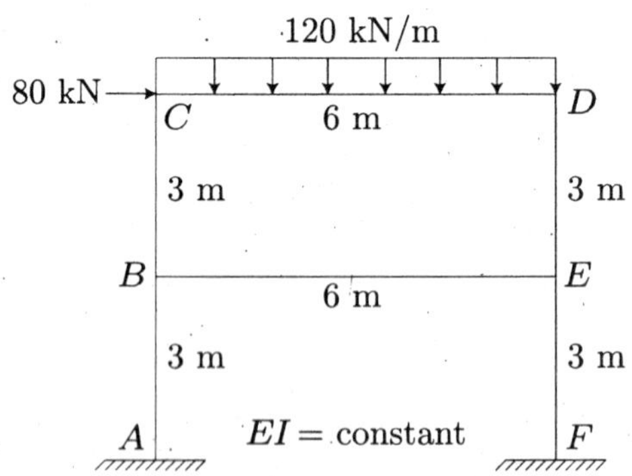

Fig. 11.29: For Example 11.10

Solution : *Fixed-end moments*:

$$M_{FCD} = \frac{120(6)^2}{12} = 360 \text{ kNm} = -M_{FDC}.$$

Rotation factors are shown in Fig. 11.30(a).

Displacement factors (D.F.) :

For columns in first storey, For BA and $EF = -\frac{3}{2}\frac{\frac{1}{2}}{\frac{1}{3}+\frac{1}{3}} = -0.75.$

For column in second storey, For BC and $ED = -0.75.$

Storey shear : At 2nd storey, $H_2 = \sum H_{ij} = -80$ kN

At 1st storey, $H_1 = \sum H_{ij} = -80$ kN.

Calculations of rotational and displacement contributions are shown below dealing with joints in the following order:

Joints C, D, E, B, second storey, first storey.

This completes one cycle.

Fig. 11.30(a) shows schematically the calculations of rotation and displacement contributions and Fig. 11.30(b) shows the calculation of end moments.

Joint C	1st cycle	2nd cycle	3rd cycle	4th cycle
$\sum M_{FC}$	+360	+360	+360	+360
M'_{DC}	0	+70	+70.98	+66.54
M'_{BC}	0	+26.80	+17.40	+5.28
M''_{BC}	0	+45.90	+105.48	+144.71
Total	+360	+502.70	+533.86	+576.53
Joint D	↓	↓	↓	↓
$\sum M_{FD}$	−360	−360	−360	−360
M'_{CD}	−60	−83.78	−92.31	−96.10
M'_{ED}	0	−28.00	−52.43	−66.71
M''_{DE}	0	+45.90	+105.48	+144.71
Total	−120	−425.88	−399.26	−378.10
Joint E	↓	↓	↓	↓
$\sum M_{FE}$	0	0	0	0
M'_{DE}	+140	+141.96	+133.10	+126.00
M'_{BE}	0	+13.40	+8.70	+2.64
M'_{FE}	0	0	0	0
M''_{DE}	0	+45.90	+105.48	+144.71
M''_{FE}	0	+60.90	+86.27	+106.07
Total	+140	+262.16	+333.55	+379.42

Joint B	↓	↓	↓	↓
$\sum M_{FB}$	0	0	0	0
M'_{CB}	−120	−167.57	−184.62	−192.18
M'_{EB}	−14	−26.22	−33.56	−37.94
M'_{AB}	0	0	0	0
M''_{BC}	0	+45.90	+105.48	+144.71
M''_{AB}	0	+60.90	+86.27	+106.07
Total	−134	−86.99	−26.43	−20.66
Second storey :	↓	↓	↓	↓
$\frac{H_r h_r}{3}$	−80	−80	−80	−80
M'_{CB}	−120	−167.57	−184.62	−192.18
M'_{DE}	+140	+141.96	+133.10	+126.00
M'_{BC}	+26.80	+17.40	+5.28	−4.14
M'_{ED}	−28.00	−52.43	−66.71	−75.88
Total	−61.20	−140.64	−192.95	−226.20
1st storey	↓	↓	↓	↓
$\frac{H_r h_r}{3}$	−80	−80	−80	−80
M'_{BA}	+26.80	+17.40	+5.28	−4.14
M'_{EF}	−28.00	−52.43	−66.71	−75.88
M'_{AB}	0	0	0	0
M'_{FE}	0	0	0	0
Total	−81.20	−115.03	−141.43	−160.02
	↓	↓	↓	↓
Joint C	5th cycle	6th cycle	7th cycle	8th cycle
$\sum M_{FC}$	+360	+360	+360	+360
M'_{DC}	+63.00	+60.72	+59.30	+58.40
M'_{BC}	−4.14	−10.52	−14.66	−17.28
M''_{BC}	+169.65	+185.29	+195.12	+201.27
Total	588.51	+595.49	+599.76	+602.39
Joint D	↓	↓	↓	↓
$\sum M_{FD}$	−360	−360	−360	−360
M'_{CD}	−98.08	−99.25	−99.96	−100.40
M'_{ED}	−75.88	−81.81	−85.58	−87.96
M''_{DE}	+169.55	+185.29	+195.12	+201.27
Total	−364.31	−355.77	−350.42	−347.09

Joint E	↓	↓	↓	↓
$\sum M_{FE}$	0	0	0	0
M'_{DE}	+121.44	+118.60	+116.80	+115.70
M'_{BE}	−2.07	−5.26	−7.33	−8.64
M'_{FE}	0	0	0	0
M''_{DE}	+169.65	+185.29	+195.12	+201.27
M''_{FE}	+120.02	+129.25	+135.18	+138.93
Total	+409.04	+427.88	+439.77	+447.26
Joint B	↓	↓	↓	↓
$\sum M_{FB}$	0	0	0	0
M'_{CB}	−196.16	−198.50	−199.92	−200.80
M'_{EB}	−40.90	−42.79	−43.98	−44.73
M'_{AB}	0	0	0	0
M''_{BC}	+169.65	+185.29	+195.12	+201.27
M''_{AB}	+120.02	+129.25	+135.18	+138.93
Total	+52.61	+73.25	+86.40	+94.67
2nd storey	↓	↓	↓	↓
$\frac{H_r h_r}{3}$	−80.00	−80.00	−80.00	−80.00
M'_{CB}	−196.16	−198.50	−199.92	−200.80
M'_{DE}	+121.44	+118.60	+116.82	+115.70
M'_{BC}	−10.52	−14.66	−17.28	−18.93
M'_{ED}	−81.81	−85.58	−87.96	−89.45
Total	−247.05	−260.14	−268.36	−273.48
1st storey	↓	↓	↓	↓
$\frac{H_r h_r}{3}$	−80.00	−80.00	−80.00	−80.00
M'_{BA}	−10.52	−14.66	−17.28	−18.93
M'_{EF}	−81.81	−85.58	−87.96	−89.45
M'_{AB}	0	0	0	0
M'_{FE}	0	0	0	0
Total	−172.33	−180.24	−185.24	−188.38

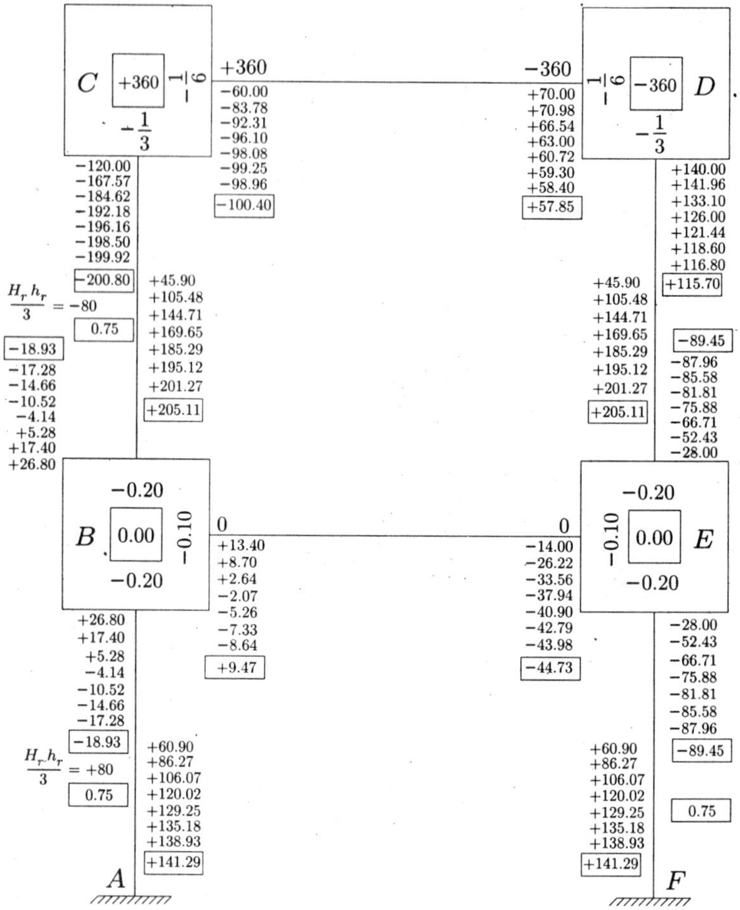

(a) Schematic diagram for contributions

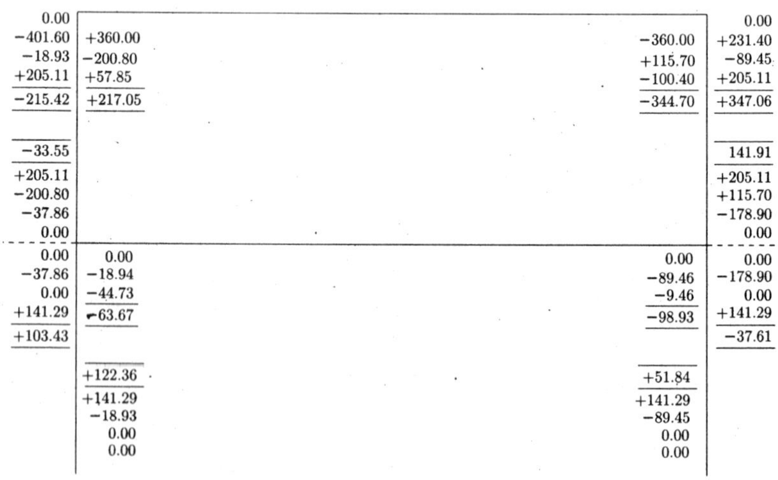

(b) End moments in kN-m

Fig. 11.30: For Example 11.10

Example 11.11. *Analyse the frame shown in Fig. 11.31 using Kani's method.*

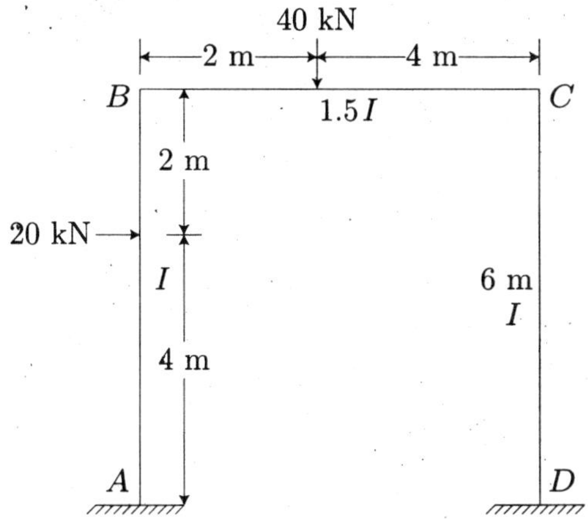

Fig. 11.31: For Example 11.11

Solution : Calculations of rotation factors and fixed-end moments are to be made as usual. The storey shear at the base of the storey is made using column AB as fixed at ends.

Thus $\quad M_{FAB} = +8.89$ kN-m and $M_{FBA} = -17.78$ kNm.

R_{FBA} is obtained from, $17.78 - R_{FBA}(6) + 20(4) = 8.89$,

whence $R_{FBA} = 14.815 \underset{\longleftarrow}{\text{kN}}$ towards left and $H_r = 14.815 \underset{\longleftarrow}{\text{kN}}$ towards right $= \sum H_{ij}$

Also $\quad M_{FBC} = +28.80$ kNm and $M_{FCB} = -19.20$ kNm.

Calculations of rotation and displacement contributions are shown in tabular forms and illustrated in Fig. 11.32(a) and the order followed in carrying in the calculation are as follows :

Member BA and CD (storey); then Joint B and then Joint C.

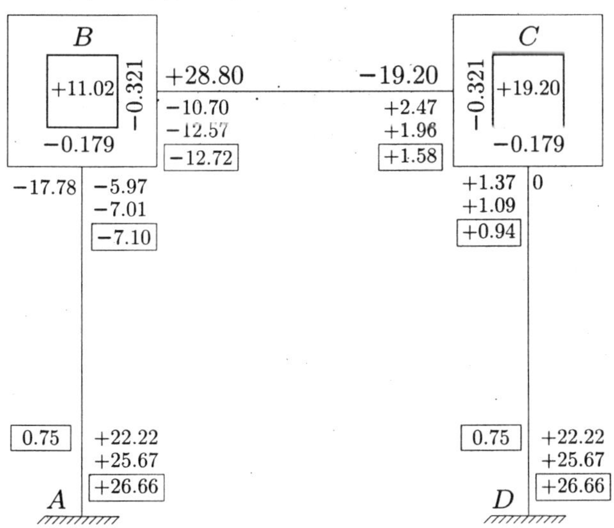

(a) Schematic diagrams for contributions

Fig. 11.32: For Example 11.11 (Contd.)

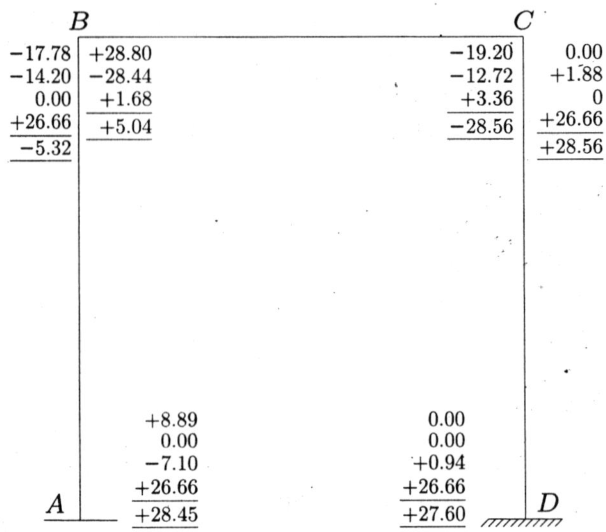

(b) End moments in kNm

Fig. 11.32: For Example 11.11

		1st cycle	2nd cycle	3rd cycle
Sway	$\frac{H_r h_r}{3}$	−29.63	−29.63	−29.63
BA	M'_{BA}	0	−5.97	−7.01
CD	M'_{CD}	0	+1.37	+1.09
	M'_{AB}	0	0	0
	M'_{DC}	0	0	0
Total		−29.63	−34.23	−35.55
		↓	↓	↓
Joint B	$\sum M_{FB}$	+11.02	+11.02	+11.02
	M'_{CB}	0.0	+2.47	+1.96
	M'_{AB}	0.0	0.00	0.00
	M''_{AB}	+22.22	+25.67	+26.66
Total		+33.34	+39.16	+39.64
		↓	↓	↓
Joint C	$\sum M_{FC}$	−19.20	−19.20	−19.20
	M'_{BC}	−10.70	−12.57	−12.72
	M'_{DC}	0.00	0.00	0.00
	M''_{DC}	+22.22	+25.67	+26.66
Total		−7.68	−6.10	−5.26

Example 11.12. *Analyse the frame in Fig. 11.33 using Kani's method.*

Solution : For member AB with B as fixed and A hinged,

$$M_{BBA} = \left[\frac{5(3)^2(1.5)}{(4.5)^2} + \frac{1}{2}\frac{5(1.5)^2(3)}{(4.5)^2}\right] = -4.167 \text{ kN-m}$$

$$R_{FBA} = \frac{5(3)}{4.5} + \frac{4.167}{4.5} = 4.259 \text{ kN (towards left)} \longleftarrow.$$

and $H_r = H_{r2} = 4.259$ kN (towards right) $\longrightarrow$.

Also $M_{FBC} = +5.625 = -M_{FCB}$.

Equivalent height

For BA, $h'_{AB} = 1.5(4.5) = 6.75$ m

$$h'_{CD} = 1.5(3.5) = 5.25 \text{ m}.$$

With $h_r = 6.75$ m, $\alpha_{AB} = 1$ and $\alpha_{CD} = 1.286$.

R.F. Joint B

For $BC = -\frac{1}{2}\frac{\frac{1}{2}}{\frac{1}{2} + \frac{3}{4}\left(\frac{1}{4.5}\right)} = -0.375.$

For $BA = -0.125$.

Joint C, For $CB = -0.35$

For $CD = -0.15$.

D.F. For both AB and $CD, m = \frac{3}{4}$.

$$\text{For } BA = -\frac{3}{2}\frac{\alpha_{ij}K_{ij}}{\sum mK_{ij}\alpha_{ij}^2} = -\frac{3}{2}\frac{1\left(\frac{3}{4}\right)\left(\frac{1}{4.5}\right)}{\left(\frac{3}{4}\right)^2\left(\frac{1}{4.5}\right)(1)^2 + \left(\frac{3}{4}\right)^2\frac{1}{3.5}(1.286)^2} = -0.64$$

$$\text{For } CD = -\frac{3}{2}\frac{1.286\left(\frac{3}{4}\right)\left(\frac{1}{3.5}\right)}{\left(\frac{3}{4}\right)^2\left(\frac{1}{4.5}\right)(1)^2 + \left(\frac{3}{4}\right)^2\left(\frac{1}{3.5}\right)(1.286)^2} = -1.06,$$

$$\frac{H_{r2}h_r}{3} = -4.259\left(\frac{6.75}{3}\right) = -9.583.$$

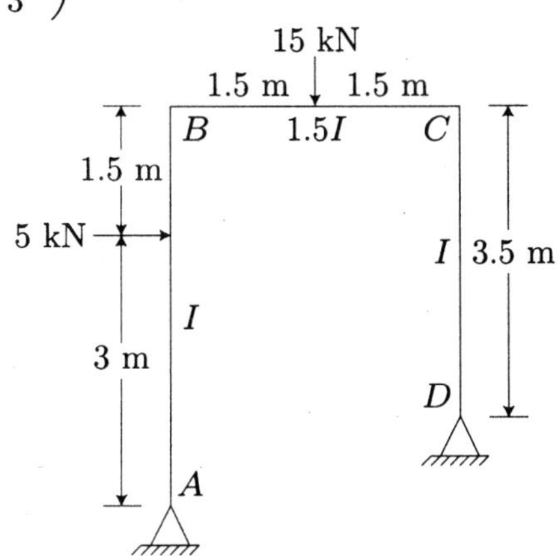

Fig. 11.33: For Example 11.12

Calculations of rotation and displacement contributions are shown in Fig. 11.34(a) and the order followed in carrying on the calculations are in the following sequence :

Member BA and CD (Sway)

Joint B

Joint C

Calculations of contributions are shown below :

		1st cycle	2nd cycle	3rd cycle	4th cycle
Sway	$\frac{H_{r2}h_r}{3}$	−9.583	−9.583	−9.583	−9.583
	$\alpha_{AB}M'_{BA}$	0	−0.937	−0.963	−0.926
	$\alpha_{CD}M'_{CD}$	0	−0.331	−0.576	−0.653
Total		−9.583	−10.851	−11.122	−11.162
	Joint B	↓	↓	↓	↓
	$\sum M_{FB}$	+1.366	+1.366	+1.366	+1.366
	M'_{CB}	0	−0.602	−1.045	−1.185
	M'_{AB}	0	0	0	0
	M''_{AB}	+6.133	+6.945	+7.118	+7.144
Total		+7.499	+7.709	+7.409	+7.325
	Joint C	↓	↓	↓	↓
	$\sum M_{FC}$	−5.625	−5.625	−5.625	−5.625
	M'_{BC}	−2.812	−2.891	−2.778	−2.747
	M'_{DC}	0	0	0	0
	M''_{CD}	+10.158	+11.502	+11.789	+11.832
Total		+1.721	+2.986	+3.386	+3.460

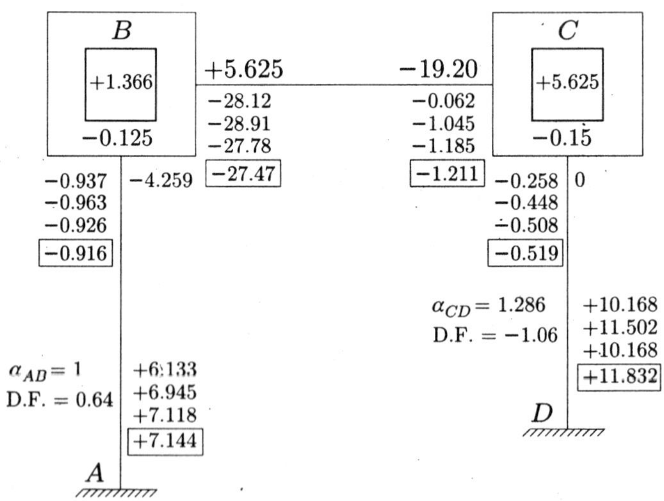

(a) Schematic diagrams for contributions

Fig. 11.34: For Example 11.12 (Contd.)

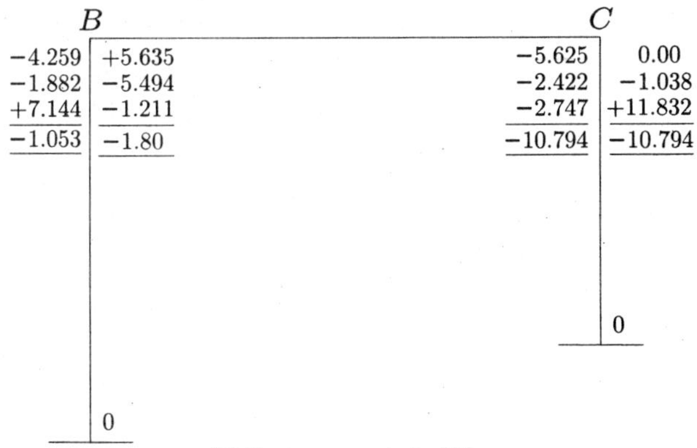

(b) End moments in kNm

Fig. 11.34: For Example 11.12

Example 11.13. *Analyse the frame in Fig. 11.35 using Kani's method.*

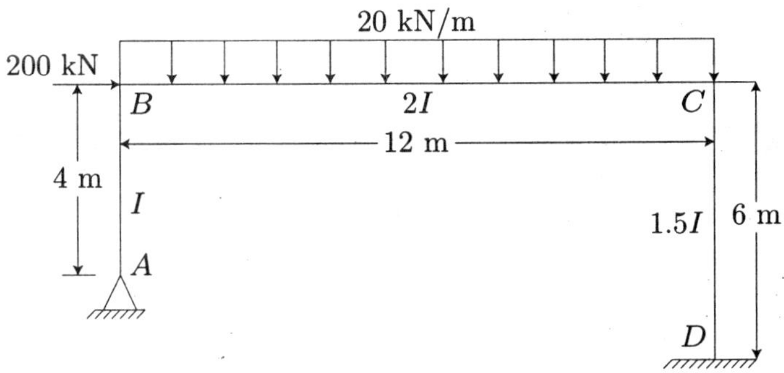

Fig. 11.35: For Example 11.13

Solution : For member BC, $M_{FBC} = 240$ kNm $= -M_{FCB}$.

For member BA equivalent height $= \frac{3}{2}(4) = 6$.

With $h_r = 6$ m, $\alpha_{AB} - \alpha_{CD} = 1$, $H_r \frac{h_r}{3} = -400$.

R.F.

Joint B

$$\text{For } BC = -\frac{1}{2}\frac{\frac{1}{6}}{\frac{1}{6}+\frac{3}{4}\left(\frac{1}{4}\right)} = -0.2353$$

For $BA = -0.2647$.

Joint C

For $CB = -0.20$

For $CD - -0.30$.

D.F.

$$\text{For } BA = -\frac{3}{2}\frac{1\left(\frac{1}{4}\right)\left(\frac{3}{4}\right)}{\left(\frac{3}{4}\right)^2(1)^2\left(\frac{1}{4}\right)+1(1)^2\left(\frac{1}{4}\right)} = -0.72.$$

$$\text{For } CD = -\frac{3}{2}\frac{1\left(\frac{1}{4}\right)}{\left(\frac{3}{4}\right)^2(1)^2\left(\frac{1}{4}\right)+1(1)^2\left(\frac{1}{4}\right)} = -0.96.$$

Calculations of displacement and rotation contributions are shown in tabular form:

	1st cycle	2nd cycle	3rd cycle	4th cycle	5th cycle
Sway $H_r \frac{h_r}{3}$	−400	−400	−400	−400	−400
$BA\ \alpha_{AB} M'_{BA}$	0	−139.76	−166.48	−172.06	−173.36
$CD\ \alpha_{CD} M'_{CD}$	0	−5.928	−40.761	−57.00	−62.94
Total	−400	−545.688	−607.241	−629.06	−636.30
	↓	↓	↓	↓	↓
Joint B	1st cycle	2nd cycle	3rd cycle	4th cycle	5th cycle
$\sum M_{FB}$	+240	+240	+240	+240	+240
M'_{CB}	0	−3.952	−27.174	−38.00	−41.96
M'_{AB}	0	0	0	0	0
M''_{AB}	+288	+392.90	+437.21	+452.92	+458.14
Total	+528	+628.948	+650.036	+654.92	+656.18
Joint C	↓	↓	↓	↓	↓
$\sum M_{FC}$	−240	−240	−240	−240	−240
M'_{BC}	−124.24	−147.99	−152.95	−154.10	−154.40
M'_{DC}	0	0	0	0	0
M''_{DC}	+384.00	+523.86	+582.95	+603.90	+610.85
Total	+19.76	+135.87	+190.00	+209.80	+216.45

Figs. 11.36(a) and 11.36(b) indicate the values of the rotation and distribution contributions for several cycles and the end moments respectively.

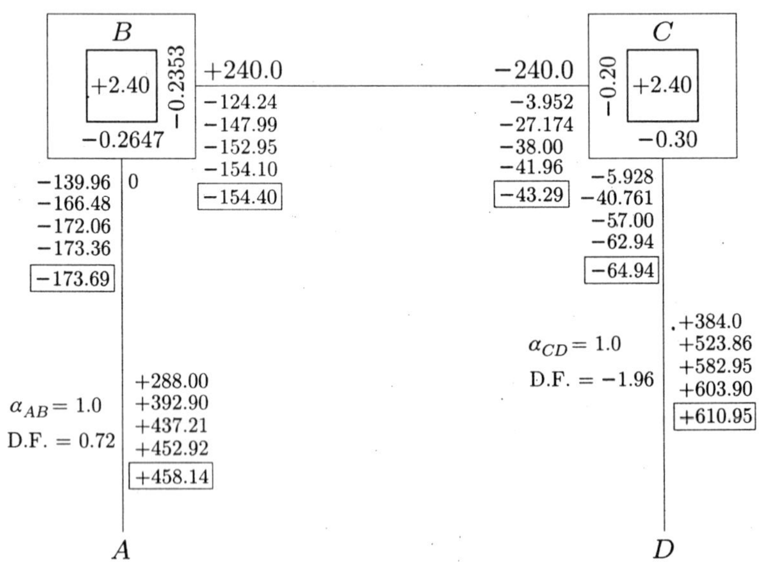

(a) Schematic diagrams for contributions

Fig. 11.36: For Example 11.13 (Contd.)

(b) End moments in kN-m

Fig. 11.36: For Example 11.13

Example 11.14. *Analyse the frame in Fig. 11.37 using Kani's method.*

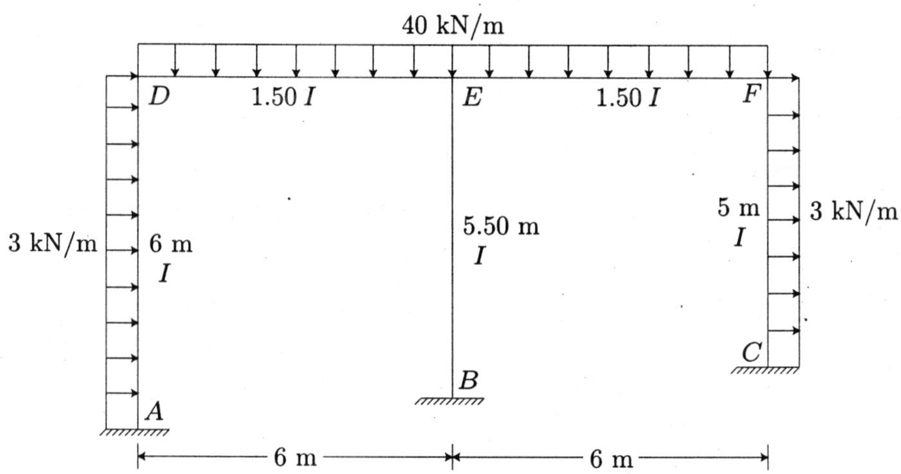

Fig. 11.37: For Example 11.14

Solution : For members DE and EF,

$$M_{FDE} = +120 \text{ kNm} = -M_{FED} = M_{FEF} = -M_{FFE}.$$

For member AD,

$$M_{FAD} = 9 \text{ kNm} = -M_{FDA}$$

$$R_{FDA} = \frac{3(6)}{2} = 9 \text{ kN (towards left)} \longleftarrow$$

For member CF,

$$M_{FCF} = 6.25 \text{ kNm} = -M_{FFC}$$

$$R_{FFC} = \frac{3(5)}{2} = 7.50 \text{ kN (towards left)} \longleftarrow.$$

With $h_r = 6$ m,

$$\alpha_{AD} = \frac{6}{6} = 1.00, \ \alpha_{BE} = \frac{6}{5.5} = 1.091 \ \text{ and } \ \alpha_{CF} = \frac{6}{5} = 1.20$$

$$\frac{H_r h_r}{3} = -33 \text{ kNm}.$$

R.F. Joint D Joint E Joint F

Joint D	Joint E	Joint F
For $DA = -0.20$	For $ED = -0.183$	For $FE = -0.28$
For $DE = -0.30$	For $EF = -0.183$	For $ED = -0.22$
	For $EB = -0.134$	

D.F.

$$\text{For } AD = -\frac{3}{2}\frac{1\left(\frac{1}{6}\right)}{(1)^2\left(\frac{1}{6}\right)+\frac{1}{5.5}(1.091)^2+\frac{1}{5}(1.20)^2} = -0.373$$

$$\text{For } BE = -\frac{3}{2}\frac{1.091\left(\frac{1}{5.5}\right)}{(1)^2\left(\frac{1}{6}\right)+(1.091)^2\frac{1}{5.5}+(1.20)^2\left(\frac{1}{5}\right)} = -0.443$$

$$\text{For } CF = -\frac{3}{2}\frac{1.20\left(\frac{1}{5}\right)}{(1)^2\frac{1}{6}+(1.091)^2\left(\frac{1}{5.5}\right)+(1.20)^2\left(\frac{1}{5}\right)} = -0.536.$$

Calculations for displacement and rotation contributions are shown in tabular forms:

	1st cycle	2nd cycle	3rd cycle	4th cycle
Sway $\frac{H_r h_r}{3}$	−33	−33	−33.0	−33
DA $\alpha_{AD}M'_{DA}$	0	−24.66	−25.02	−24.23
EB $\alpha_{BE}M'_{EB}$	0	+3.27	−0.52	−1.17
FC $\alpha_{CF}M'_{FC}$	0	+27.58	+29.71	+29.64
Total	−33	−26.81	−28.83	−28.76
Joint D	↓	↓	↓	↓
$\sum M_{FD}$	+111.00	+111.00	+111.00	+111.00
M'_{ED}	0	+4.09	−0.66	−1.45
M'_{AD}	0	0	0	0
M''_{AD}	+12.31	+10.00	+10.75	+10.73
Total	+123.31	+125.09	+121.09	+120.28
Joint E		↓	↓	↓
$\sum M_{FE}$	0	0	0	0
M'_{DE}	−36.99	−37.53	−36.33	−36.08
M'_{FE}	0.0	+29.25	+31.51	−31.43
M'_{BE}	0.0	0.0	0.0	0.0
M''_{BE}	+14.62	+11.88	+12.77	+12.74
Total	−22.37	+3.60	+7.95	+8.09
Joint F		↓	↓	↓
$\sum M_{FE}$	−126.25	−126.25	−126.25	−126.25
M'_{EF}	+4.09	−0.66	−1.45	−1.48
M'_{CF}	0	0	0	0
M''_{EF}	+17.63	+14.37	+15.45	+15.42
Total	−104.47	−112.54	−112.25	−112.31

Fig. 11.38 shows the rotation and displacement contributions and the values of the end moments.

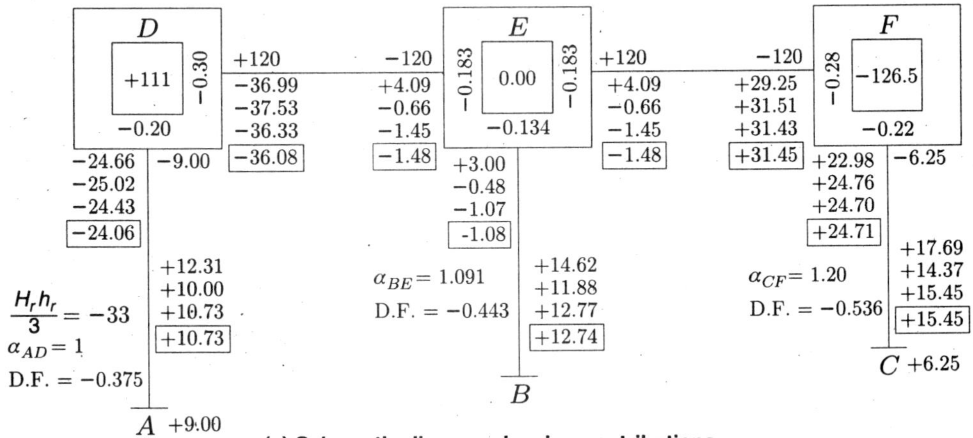

(a) Schematic diagram showing contributions

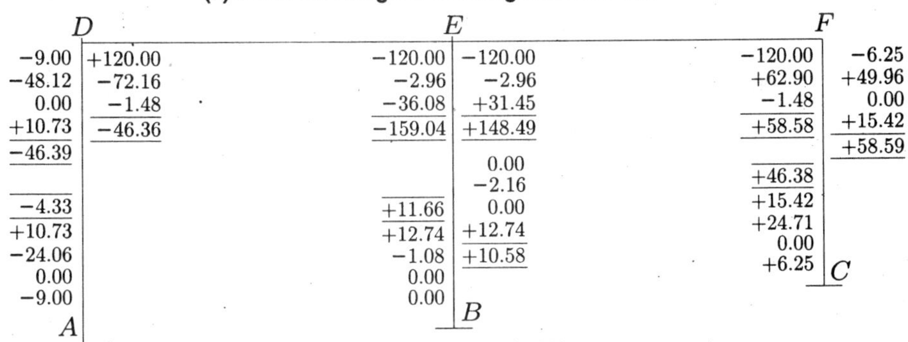

(b) End moments in kNm

Fig. 11.38: For Example 11.14

Problems

11.1 Analyse the structures shown in Fig. P-11.1 by Kani's method.

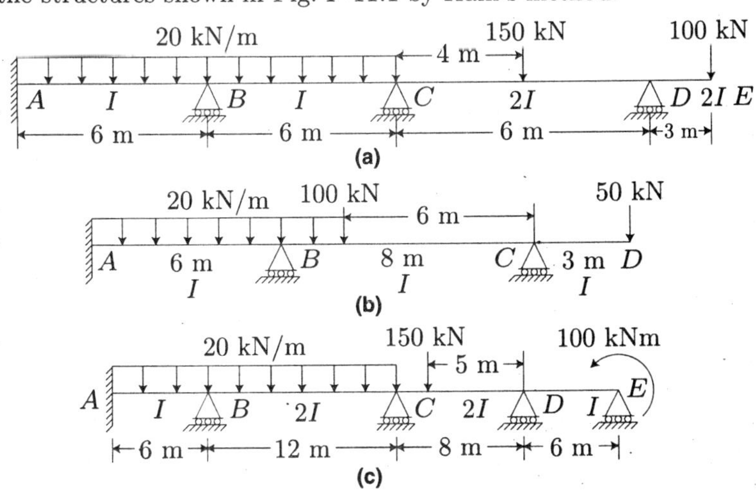

Fig. P-11.1

11.2 Analyse the structures in Fig. P-11.1 if support at A rotates by 0.001 radians and support C settles by 10 mm. Take $E = 2(10)^5$ N/mm^2 and $I = 5(10)^7$ mm^4.

11.3 Analyse the structures in Fig. P-11.1 if support at C settles by 1 mm per 1 kN of reaction. Assume E and I as in Problem 11.2.

11.4 Analyse the frames in Fig. P-11.2 using Kani's method.

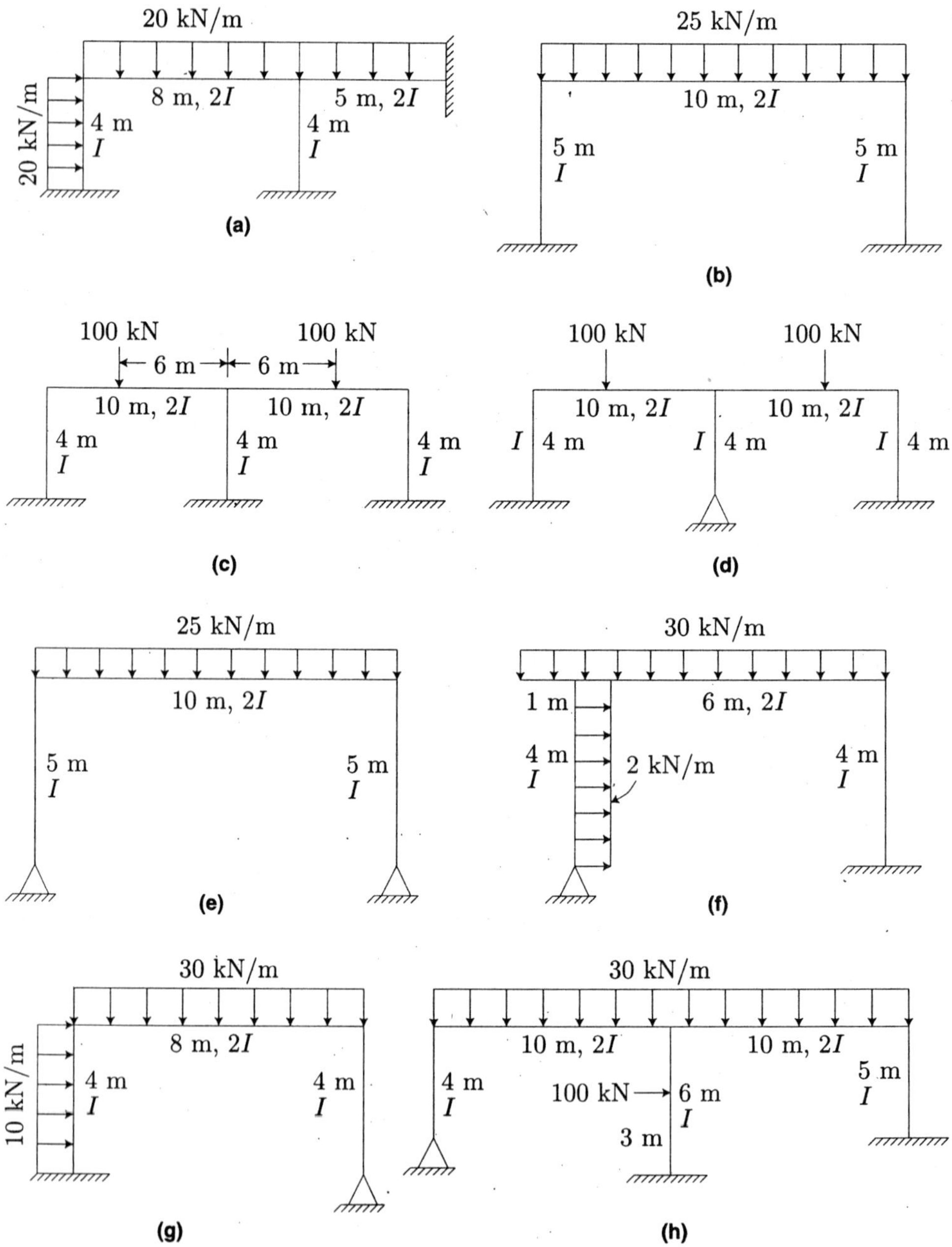

Fig. P-11.2 (Contd.)

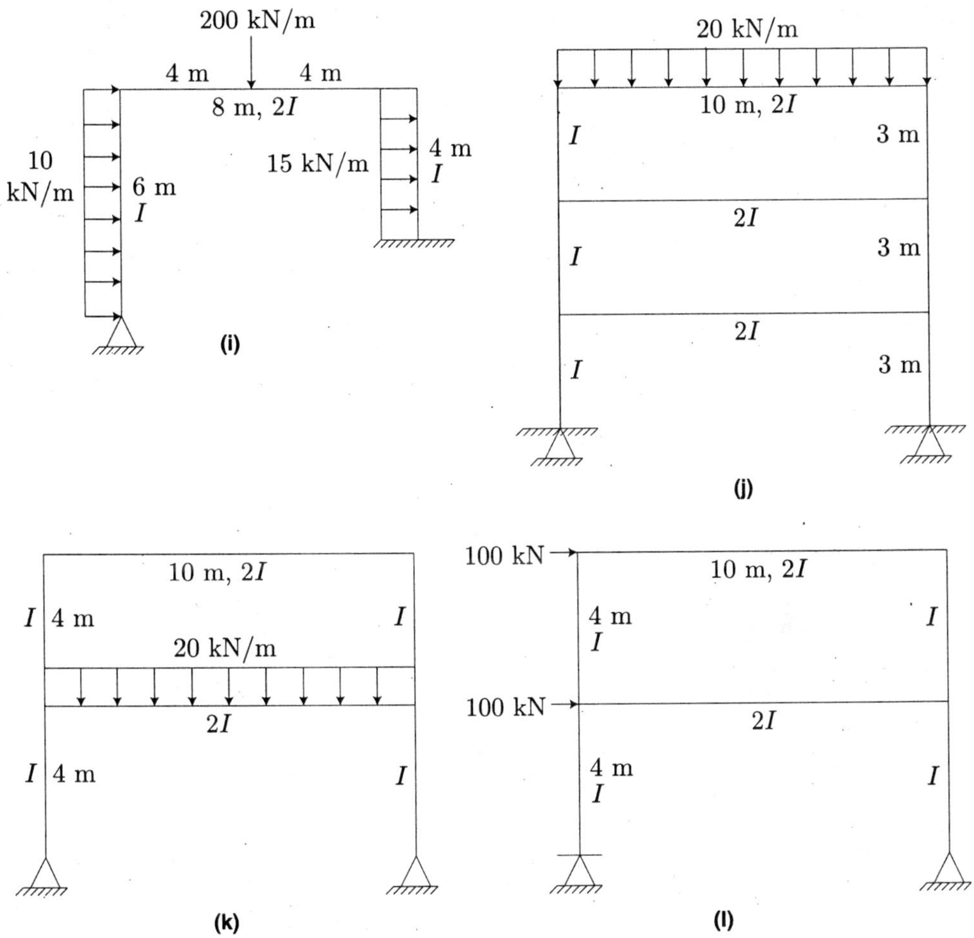
200 kN/m
4 m
4 m
8 m, 2I
10
kN/m
6 m
I
15 kN/m
4 m
I
(i)
20 kN/m
10 m, 2I
I
3 m
2I
I
3 m
2I
I
3 m
(j)
10 m, 2I
I
4 m
I
20 kN/m
2I
I
4 m
I
(k)
100 kN
10 m, 2I
4 m
I
I
100 kN
2I
4 m
I
I
(l)

Fig. P-11.2

Chapter 12

Unsymmetrical Bending and Curved Beams, Rings and Bow Girder

12.1 Introduction to Unsymmetrical Bending

The flexure formula $M = \sigma\frac{I}{y}$ is based on the assumption that the neutral axis for any cross section of the beam is perpendicular to the plane of the loads. The condition requires that the plane of the loads shall contain one of the principal axis of symmetry is always a principal axis, bending caused by loads that do not lie in a plane containing a principal axis of each cross section usually is call unsymmetrical bending.

Some beams such as roof purlins are subjected to loads whose planes make large angles with principal planes as indicated in Fig. 12.1 whereas many beams and structures and machines are subjected to loads whose planes are inclined only slightly to principal plane. But even small inclinations of loading may be important for the section of certain shapes. For example, the shape of most of the steel rolled sections, such as I sections and Channel sections used for beams is such as to give maximum resistance to bending that the stresses developed are in accordance with the flexural formula.

A section that is well designed to resist elastic bending due to symmetrical bending may be poorly designed to resist bending due to unsymmetrical bending.

12.2 Stresses Due to Unsymmetrical Bending

For determining the effect of unsymmetrical bending, the following assumptions are made :

(a) Loads pass through the shear centre so that no torsion is produced.

(b) Loads are perpendicular to the longitudinal axis so that there is no axial component of the loads producing direct stress on a section.

(c) Supports of the beam allow rotation of the beam in any longitudinal plane.

(d) The beam is prismatic, i.e., it has the same cross section throughout its entire length.

(e) The plane section before bending remains plane after bending.

Fig. 12.1 shows the cross section of a prismatic beam, subjected to a bending moment M in any arbitrary longitudinal plane. If NA be the neutral axis for the given bending moment, then from the assumption of plane section before bending remaining plane after bending, the strain in any longitudinal fibre is proportional to its distance from the neutral axis and consequently the bending stress is also proportional to its distance from the neutral axis. If v be the distance of a point from the neutral axis, then stress at the point may be given by,

$$\sigma = Kv,$$

where K is a constant. The force on an elementary area dA surrounding the point is σdA and then the total force on the entire area is $\int_A \sigma dA$ which must be equal to zero from the assumption of no axial longitudinal force acting on the beam.

With $\sigma = Kv$ the above expression becomes

$$\int_A KvdA = 0, \int_A vdA = 0 \text{ as } K \neq 0.$$

Hence $A\overline{v} = 0$,

where $\overline{v}$ is the distance of the centroid of the entire area A from the neutral axis. Since the area A is not zero, $\overline{v} = 0$.

This implies that the neutral axis passes through the centroid of the section, though its inclination with the plane of bending may vary.

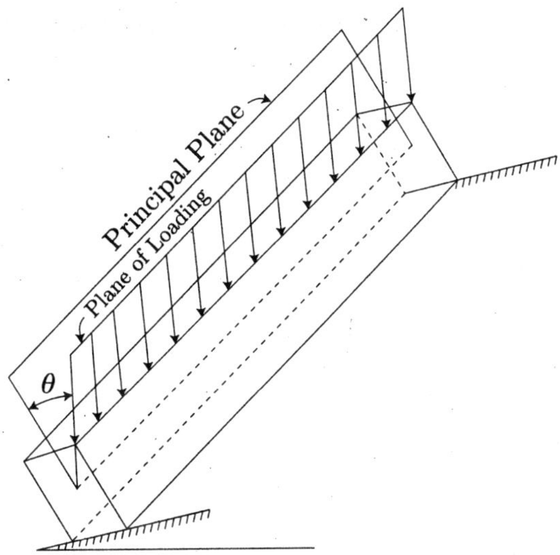

Fig. 12.1: Unsymmetrically loaded beam

Since the cross sections of the beam rotate about the neutral axis, bending of the longitudinal axis of the beam takes place in a longitudinal plane perpendicular to the neutral axis.

In Fig. 12.2(b) x-x and y-y are any two orthogonal (mutually perpendicular) axes passing through the centroid O of the cross section. A load P acts on the beam at a clockwise angle θ with the y-y axis producing bending moment M at the section in the plane of the load. The bending moment M may be represented by a vector at right angles to the line of action of P, i.e., the moment vector will be inclined at an angle θ with x-x axis as shown in the figure. The vector-components M_x and M_y of M in the directions of x-x and y-y respectively are :

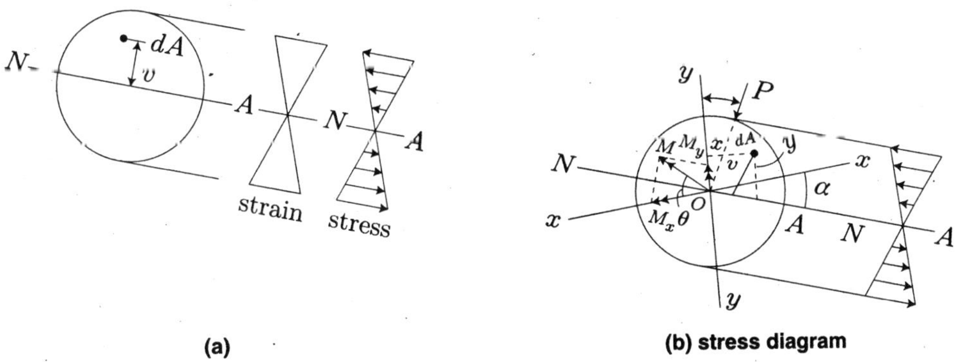

Fig. 12.2: Stresses in section

$$M_x = M\cos\theta \quad \text{(a)}$$

and $M_y = M\sin\theta$, (b)

which give magnitudes of the moments about x-x and y-y axes respectively. M_x and M_y may, however, be obtained as bending moments created by the components $P\cos\theta$ and $P\sin\theta$ respectively of the load P along y-y and x-x axes respectively.

For the assumed loading condition, NA is taken as the neutral axis making a clockwise angle α with the x-x axis and passing through the centroid O. If x, y be the co-ordinates of an elemental area dA with respect to x-x and y-y axes, the perpendicular distance v of the element from the neutral axis NA is obtained as

$$v = x \sin\alpha + y\cos\alpha. \tag{c}$$

Since the bending stress at any point is directly proportional to its distance from the neutral axis, the stress acting on dA may be given as

$$\sigma = Kv = K(x\sin\alpha + y\cos\alpha), \tag{d}$$

where K is a constant.

The normal force on dA

$$\sigma \cdot dA = K(x\sin\alpha + y\cos\alpha)dA. \tag{e}$$

For equilibrium of the section, summation of moments of the forces on all such elements over the entire section about x-x axis equals to M_x,

$$\text{i.e., } M_x = \int_A \sigma \cdot y dA \tag{f}$$

and summation of moments of the forces on all such elements over the entire section about y-y axis equals to M_y,

$$\text{i.e., } M_y = \int_A \sigma \cdot x dA. \tag{g}$$

From Eq. (d),

$$\begin{aligned} M_x &= \int_A K(x\sin\alpha + y\cos\alpha)ydA = K\left[\sin\alpha \int_A xydA + \cos\alpha\int_A y^2 dA\right] \\ &= K[I_{xy}\sin\alpha + I_x\cos\alpha], \end{aligned} \tag{h}$$

where I_x and I_{xy} are moment of inertia of the section about x-x axis and product of inertia of the section about x-x and y-y axis.

From, Eqs. (d) and (g),

$$M_y = K[I_y \sin\alpha + I_{xy}\cos\alpha], \tag{i}$$

where I_y is the moment of inertia of the cross section about y-y axis.

With $K = \dfrac{\sigma}{x\sin\alpha + y\cos\alpha}$, Eqs. (g) and (h) become respectively

$$\begin{aligned} M_x &= \sigma \cdot \frac{I_{xy}\sin\alpha + I_x\cos\alpha}{x\sin\alpha + y\cos\alpha} \\ &= \sigma \cdot \frac{I_{xy}\tan\alpha + I_x}{x\tan\alpha + y} \end{aligned} \tag{j}$$

and

$$\begin{aligned} M_y \sigma &\cdot \frac{I_y\sin\alpha + I_{xy}\cos\alpha}{x\sin\alpha + y\cos\alpha} \\ &= \sigma \cdot \frac{I_y\tan\alpha + I_{xy}}{x\tan\alpha + y}. \end{aligned} \tag{k}$$

From Eqns. (a) and (b),

$$\tan\theta = \frac{M_y}{M_x} = \frac{I_{xy} + I_y\tan\alpha}{I_x + Ixy\tan\alpha}, \tag{12.1}$$

whence $\tan\alpha = \dfrac{I_{xy} - I_x \tan\theta}{I_{xy}\tan\theta - I_y}$. (12.2)

With this value of $\tan\alpha$, Eq. (j) becomes

$$M_x = \sigma \frac{I_{xy}\left(\frac{I_{xy}-I_x\tan\theta}{I_{xy}\tan\theta - I_y}\right) + I_x}{x\left(\frac{I_{xy}-I_x\tan\theta}{I_{xy}\tan\theta - I_y}\right) + y} = \sigma \cdot \frac{(I_{xy})^2 - I_x \cdot I_y}{x(I_{xy} - I_x\tan\theta) + y(I_{xy}\tan\theta - I_y)}.$$

$$\therefore \qquad \sigma = \frac{M_x[x(I_{xy} - I_x\tan\theta) + y(I_{xy}\tan\theta - I_y)]}{(I_{xy})^2 - I_x \cdot I_y}.$$

Substituting $\tan\theta = \dfrac{M_y}{M_x}$ and simplifying, the above equation reduces to

$$\sigma = \frac{M_x\left(I_y - I_{xy}\cdot\frac{M_y}{M_x}\right)}{I_xI_y - (I_{xy})^2}\cdot y + \frac{M_x\left(I_x\cdot\frac{M_y}{M_x} - I_{xy}\right)}{I_xI_y - (I_{xy})^2}\cdot x$$

$$= \frac{M_x - M_y\left(\frac{I_{xy}}{I_y}\right)}{I_x - I_{xy}\left(\frac{I_{xy}}{I_y}\right)}\cdot y + \frac{M_y - M_x\left(\frac{I_{xy}}{I_x}\right)}{I_y - I_{xy}\left(\frac{I_{xy}}{I_x}\right)}\cdot x. \qquad (12.3)$$

This can be taken as the general expression for bending stress.

Sign conventions in Eq. (12.3) are as follows :

(i) M_x considered positive if it produces tension at points having positive y-co-ordinates.

(ii) M_y considered positive if it produces tension at points having positive x-co-ordinates.

(iii) σ is tenstle if positive and compressive, if negative, and

(iv) θ is the angle made by the plane of loads with y-y axis and may be clockwise or anticlockwise, but the sign of $\tan\theta$ is obtained from relation $\tan\theta = \frac{M_y}{M_x}$, where M_y and M_x have signs as described above. The angle α is the angle between the neutral axis and the x-x axis and it is clockwise, if the sign of $\tan\alpha$ obtained from Eq. (12.2) is positive and anticlockwise, if $\tan\alpha$ is negative.

It may be noted that in the above derivation the inclination of the load θ is measured from y-y axis, while the inclination of neutral axis α is measured from the other axis, i.e., x-x axis. Usually the angle α is in the same sense as θ, that is, if θ is clockwise from y-y axis, α also is clockwise from x-x axis.

It is again emphasised that Eq. (12.3) gives bending stress at any point (x, y) with reference to any pair of centroidal orthogonal axes x-x and y-y.

12.3 Bending Stress with Reference to the Principal Axes

With respect to the principal axes the product of inertia of a closed area is zero. If u-u and v-v be the principal axes of the area shown in Fig. 12.3(a), then the product of inertia, $I_{uv} = 0$.

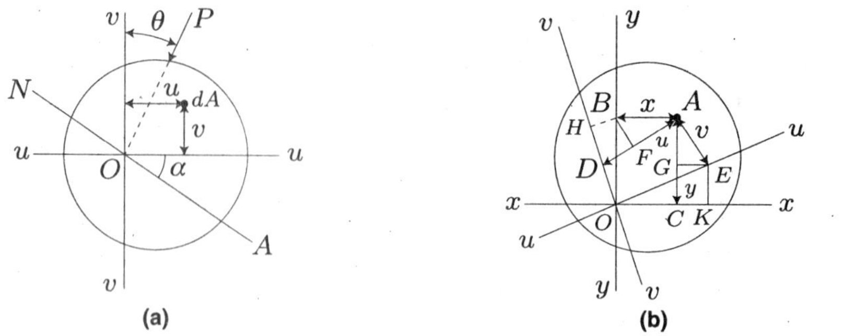

Fig. 12.3: Stresses in section

If the applied load P is inclined to v-v axis at an angle θ and the corresponding neutral axis n-n is inclined to u-u axis at an angle α, then using Eqs. (12.1) and (12.2) and as $I_{uv} = 0$, the angles of inclination can be given as

$$\tan\theta = \frac{I_v}{I_u}\tan\alpha \tag{12.4}$$

and $$\tan\alpha = \frac{I_u}{I_v}\tan\theta. \tag{12.5}$$

Substituting, u, v, M_u, M_v, I_u, I_v and I_{uv} for x, y, M_x, M_y, I_x, I_y and I_{xy} in Eq. (12.3) and as $I_{uv} = 0$, the stress σ with reference to principal axes is given by

$$\sigma = \frac{M_u}{I_u}\cdot v + \frac{M_v}{I_v}\cdot u, \tag{12.6}$$

in which M_u and M_v are the moments about the principal axes u-u and v-v, I_u and I_y are the moments of inertia about u-u and v-v axes and u, v are the co-ordinates of the point with reference to the same axes.

In order to use Eq. (12.6), the co-ordinates (u, v) of the point with respect to the principal axes should be determined. With reference to Fig. 12.3(b) the co-ordinates (u, v) of the point A with respect to the principal axes u-u and v-v can be obtained from the known co-ordinates (x, y) of the same point with respect to any arbitrary pair of orthogonal axes x-x and y-y as

$$u = AD = AF + FD = AF + BH = x\cdot\cos\theta + y\cdot\sin\theta$$

and $$v = AE = OD = OH - DH = OH - BF = y\cdot\cos\theta - x\cdot\sin\theta,$$

where θ is the angle made by u-u axis with x-x in the anticlockwise direction.

From Eqs. (12.4) and (12.5), it is found that, if θ is zero, then α also is zero, i.e., if the load P acts along v-v axis, then u-u axis becomes the neutral axis. From this it is concluded that if a bending moment acts in a plane containing one principal axis, the other principal axis becomes the neutral axis and deflection occurs in the plane of loading. For such a case, the bending stress σ equals $\frac{M_u}{I_u}v\cdot$ or $\frac{M_v}{I_v}\cdot u$ as the case may be, i.e., the theory of simple bending becomes valid. Hence, for applicability of simple beam theory, the load must act in a plane containing a principal axis.

Since the axis of symmetry is always a principal axis, loads acting in a plane of symmetry produces simple bending, the neutral axis being perpendicular to the axis of symmetry.

12.4 Flexure Modulus

With reference to Fig. 12.3, if the bending moment due to P be M in the plane of load, i.e., in a plane inclined at an angle θ with the v-v axis, then,

$$M_u = M\cos\theta \quad \text{and} \quad M_y = M\sin\theta.$$

Using these values in Eq. (12.6), the bending stress at any point (u, v) is given by

$$\sigma = \frac{M\cos\theta}{I_u}\cdot v + \frac{M\sin\theta}{I_y}\cdot u = M\left[\frac{I_v\cos\theta\cdot v + I_u\sin\theta\cdot u}{I_u\cdot I_v}\right]$$

$$= \frac{M}{\left[\frac{I_u\cdot I_v}{I_v\cos\theta\cdot v + I_u\sin\theta\cdot u}\right]} = \frac{M}{Z_f},$$

where $$Z_f = \frac{I_u\cdot I_v}{(I_v\cos\theta\cdot v + I_u\sin\theta\cdot u)},$$

which is called 'flexure modulus' and is the section modulus for biaxial bending.

Example 12.1. *An angle section* 80 mm × 60 mm × 10 mm *is used as a simple beam on a span of 4 m with its longer leg vertical as shown in Fig. 12.4(a). The beam is subjected to a uniformly distributed vertical load of* 400 N/m *including self weight, assumed to pass through the shear centre. Determine the position of neutral axis and calculate the bending stress at points A, C and E.*

Solution :

Sectional Properties

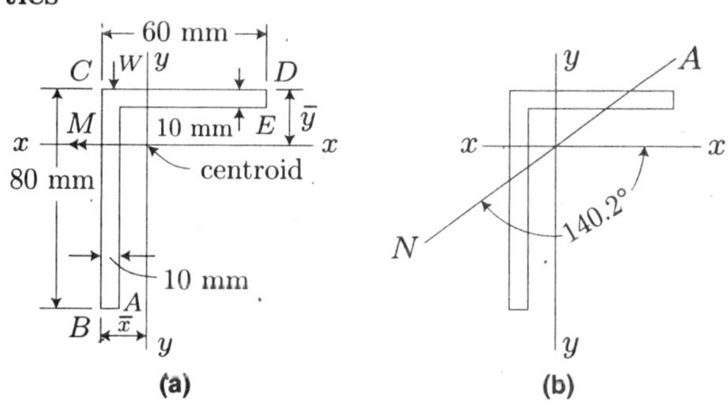

Fig. 12.4: For Example 12.1

Area $A = (80 + 50)10 = 1300 \text{ mm}^2$

$$\overline{x} = \frac{60(10)(30) + 70(10)(5)}{1300} = 16.5 \text{ mm}$$

$$\overline{y} = \frac{80(10)(40) + 50(10)(5)}{1300} = 26.5 \text{ mm}$$

$$I_x = \left[\frac{1}{12}(60)(80^3) + 60(80)(40 - 26.5)^2\right] - \left[\frac{1}{12}(50)(70^3) + 50(70)(35 - 16.5)^2\right]$$

$$= 807758 \text{ mm}^4$$

$$I_y = \frac{1}{12}(80)(10^3)(80)(10)(16.5 - 5)^2 + \frac{1}{12}(10)(50^3) + (10)(50)(35 - 16.5)^2$$

$$= 387758 \text{ mm}^4.$$

The co-ordinates of the centroids of two component rectangles 80 mm $\times$ 10 mm and 50 mm$\times$80 mm with reference to the centroidal axes x-x and y-y with their positive direction as indicated in the figure are $(-11.5, -13.5)$, and (18.5,21.5) respectively.

Hence, $I_{xy} = 80(10)(-11.5)(-13.5) + 50(10)(18.5)(21.5) = 323075 \text{ mm}^4$.

Neutral Axis

In this case the plane of load is parallel to y-y axis and so the angle θ between the plane of load and y-y axis is zero. If α be the angle made by the neutral axis with x-x axis in clockwise direction, then knowing $\tan\theta$ to be zero,

$$\tan\alpha = \frac{I_{xy} - I_x\tan\theta}{I_{xy}\tan\theta - I_y} = \frac{32.31}{-38.78} = -0.83316$$

and $\quad -\tan\alpha = 0.83316 = \tan 39.8°,$

whence $\tan(180° - \alpha) = \tan 39.8° \quad$ and $\quad \alpha = 180° - 39.8 = 140.2°.$

The position of the neutral axis is shown in Fig. 12.4(b).

Bending stress :

Bending stress at any point (x, y) is given by

$$\sigma = \frac{M_x - M_y\left(\frac{I_{xy}}{I_y}\right)}{I_x - \frac{I_{xy}^2}{I_y}} \cdot y + \frac{M_y - M_x\left(\frac{I_{xy}}{I_x}\right)}{I_y - \frac{I_{xy}}{I_x}} \cdot x,$$

where M_x and M_y are the bending moments about x-x and y-y axes respectively, caused by the components of the applied loads parallel to y-y and x-x axes respectively. In this problem the applied uniformly distributed load is parallel to the y-y axis and it has no component parallel to x-x axis. Hence for the middle section of the beam,

$$M_y = 0 \quad \text{and} \quad M_x = -\frac{400(4^2)}{8} = -800 \text{ Nm}.$$

Negative sign indicates that M_x creates compression at points having positive y-co-ordinates. M_x and M_y also can be obtained using the concept of moment vector. In this case the maximum bending moment $M = \frac{wl^2}{8} = 800$ N-m acts in the plane of load w, i.e., in a plane parallel to y-y axis. Hence the moment vector M is directed perpendicular to y-y axis, i.e., along x-x axis. Since angle θ between the moment vector and the x-x axis zero,

$$M_x = M\cos\theta = M = 800 \text{ Nm (negative)}$$

$$M_y = M\sin\theta = 0.$$

Co-ordinates in mm are as below :

For A, $\quad -6.5$ and $\quad -53.5$

For C, $\quad -16.6$ and $\quad -26.5$

For E, $\quad +43.5$ and $\quad -16.5$.

$$I_x - \frac{I_{xy}^2}{I_y} = 807758 - \frac{(323075)^2}{387758} = 538576 \text{ mm}^4$$

$$I_y - \frac{I_{xy}^2}{I_x} = 387758 - \frac{(323075)^2}{807758} = 258539 \text{ mm}^4$$

$$\frac{I_{xy}}{I_x} = \frac{323075}{807758} = 0.4,$$

and $$\sigma = \frac{M_x - M_y\left(\frac{I_{xy}}{I_y}\right)}{I_x - \sigma\frac{I_{xy}^2}{I_y}} \cdot y + \frac{M_y - M_x\left(\frac{I_{xy}}{I_x}\right)}{I_y - \frac{I_{xy}^2}{I_x}.} \cdot x$$

$$= \frac{-800(10^3)}{538576} \cdot y + \frac{-(-800)(10^3)(0.4)}{258539} \cdot x$$

$$= -1.4854y + 1.2377x.$$

Then $\sigma_A = -1.4854(-53.5) + 1.2377(-6.5) = 71.4 \text{ N/mm}^2$ (Tension)

$\sigma_C = -1.4854(26.5) + 1.2377(-16.5) = -59.8 \text{ N/mm}^2$ (Compression)

$\sigma_E = -1.4854(16.5) + 1.2377(43.5) = 29.3 \text{ N/mm}^2$ (Tension).

Example 12.2. *Solve the problem in Example 12.1 using Principal axes.*

Solution :

Principal Axes

If u-u and v-v are the principal axes, u-u axis making an angle θ with the x-x axis as shown in Fig. 12.5. Then using the values of I_x, I_y and I_{xy} as obtained in Example 12.1,

$$\tan 2\theta = -\frac{2I_{xy}}{I_x - I_y} = -\frac{2(323075)}{807758 - 387758} = -1.584524 - \tan 2\theta$$

$$= 1.584524 = \tan 56.98°,$$

whence $180° - 2\theta - 56.98°$ and $\theta = 61.51°$

Hence the angle which v-v axis makes with the x-x axis

$$= \theta + 90° = 61.51° + 90° = 151.51°.$$

$$\therefore \quad I_u = \frac{I_x + I_y}{2} - \frac{I_x - I_y}{2}\cos 2\theta - I_{xy}\sin 2\theta$$

$$= \frac{807758 + 387758}{2} + \frac{807758 - 387758}{2}\cos 123.02°$$
$$-323075 \sin 123.02°$$

$$= 597758 - 114436 - 270892 = 212430 \text{ mm}^4.$$

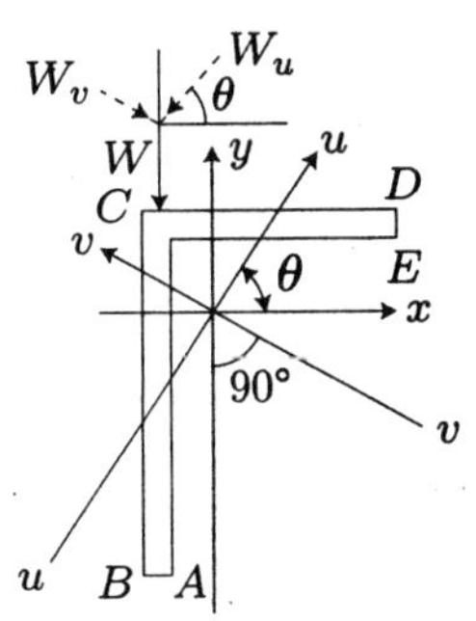

Fig. 12.5: For Example 12.2

$$I_y = \frac{I_x + I_y}{2} + \frac{I_x - I_y}{2}\cos 2(90° + \theta) - I_{xy}\sin 2(90° + \theta)$$

$$= \frac{I_x + I_y}{2} - \frac{I_x - I_y}{2}\cos 2\theta + I_{xy}\sin 2\theta$$

$$= 597758 + 114436 + 270892 = 983086 \text{ mm}^4.$$

Bending Moments

The components of uniformly distributed vertical load $w = 400$ N/m along v-v and u-u axes are respectively

$$w_v = w \cdot \cos\theta = (400\cos 61.51°) \text{ N/m} = 190.80 \text{ N/m}$$

and $w_u = w \cdot \sin\theta = (400\sin 61.51°) \text{ N/m} = 351.56 \text{ N/m}.$

Bending moments M_u and M_v about u-u and v-v axes respectively are due to the loads w_v and w_u. From the directions of the components w_v and w_u (Fig. 12.5), it is evident that M_u creates compression at points having positive v- co-ordinates and M_v also creates compression at points having positive u- co-ordinates. Hence, according to the sign convention used in this chapter, M_u and M_v are negative.

$$M_u = -\frac{w_v l^2}{8} = -\frac{190.8(4^2)}{8} \text{ Nm} = -381.6 \text{ Nm}$$
$$= -381600 \text{ Nmm}.$$
$$M_v = -\frac{w_u l^2}{8} = -\frac{351.56(4^2)}{8} \text{ N-m} = -703.12 \text{ Nm}$$
$$= -703120 \text{ Nmm}.$$

Co-ordinates of Points

The angle θ between x-x and u-u axes, $= 61.51°$

$$\sin\theta = \sin 61.51° = 0.8789$$
$$\cos\theta = \cos 61.51° = 0.4770.$$

Coordinates are	u	v
For A	−50.12	−19.81 mm
For C	+15.42	+27.14 mm
For E	+35.25	−30.36 mm.

Stresses

$$\sigma = \frac{M_u}{I_u} \cdot v + \frac{M_v}{I_v} \cdot u = \frac{-381600}{212430} \cdot v + \frac{-703120}{983086} \cdot u$$
$$= -1.79636v - 0.71522u$$

$$\therefore \quad \sigma_A = -1.79636(-19.81) - 0.71522(-50.12) = +71.4 \text{ N/mm}^2$$
$$\sigma_C = -1.79636(+27.14) - 0.71522(+15.42) = -59.8 \text{ N/mm}^2$$
$$\sigma_E = -1.79636(-30.36) - 0.71522(+35.25) = +29.3 \text{ N/mm}^2.$$

Example 12.3. *The channel section of Fig. 12.6(a) is the cross section of a cantilever beam of span 2 m and subjected to a concentrated load P at the free end at an angle* 30° *with the y-y axis passing through the shear centre of the section. Determine the value of P if the allowable stresses in tension and compression are* 150 N/mm^2 *and* 100 N/mm^2 *respectively.*

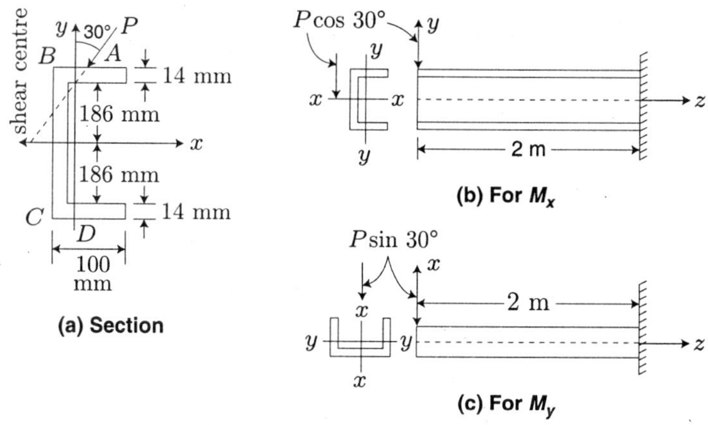

Fig. 12.6: For Example 12.3

Solution :

Sectional Properties

Since x-x is the axis of symmetry, it is a principal axis and the perpendicular axis y-y is the other principal axis.

$$\overline{x} = \frac{400(8)(4) + 2(92)(14)(54)}{400(8) + 2(92)(14)} = 26.3 \text{ mm}$$

$$I_x = \frac{1}{12}(100)(400^3) - \frac{1}{12}(92)(372^3) = 1.38662(10^8) \text{ mm}^4$$

$$I_y = \frac{1}{12}(400)(8^3) + 400(8)(22.3^2) + 2\left[\frac{1}{12}(14)(92^3) + 14(92)(27.7^2)\right]$$

$$= 5401872 \text{ mm}^4.$$

Bending Moments

With reference to Figs. 12.6(b) and 12.(c),

$$M_x = P \cdot \cos 30°(2000) = 1732P \text{ in Nmm}$$

and $M_y = P \cdot \sin 30°(2000) = 1000P$ in Nmm

In this case both M_x and M_y are positive as they create tension at points having positive y- co-ordinates and positive x- co-ordinates respectively.

The points where stresses are greatest can be easily determined by inspection. The farthest points from the origin are the four corner points A, B, C and D. Of these four distant points, at A tension is created by both M_x and M_y and at C compression is created by both M_x and M_y, but at points B or D stresses of opposite nature are created by both M_x and M_y. Hence the point A is critical in tension and point C is critical in compression. The co-ordinates of A and C are as below :

For A, +73.7 +200 mm

For C, −26.3 −200 mm.

Bending Stresses

$$\sigma = \frac{M_x}{l_x} \cdot y + \frac{M_y}{I_y} \cdot x = \frac{1732P}{1.38662(10^8)} \cdot y + \frac{1000P}{5401872} \cdot x$$

$$= 1.249(10^{-5})P \cdot y + 1.85(10^{-4})P \cdot x$$

$$\sigma_A = 1.249(10^{-5})(200)P + 1.85(10^{-4})(73.7)P = 0.0161P \text{ (tension)}$$

$$\sigma_C = 1.249(10^{-5})P(-200) + 1.85(10^{-4})P(-26.3) = -7.3635(10^{-3})P \text{ (compression)}.$$

Safe value of *P*

Equating σ_A to permissible tension, $0.0161P = 150$.

whence $P = 9,317 \text{ N} = 9.317 \text{ kN}$.

Equating σ_C to permissible compression, $7.3635(10-3)P = 100$

whence $P = 13,580 \text{ N} = 13.580 \text{ kN}$.

Hence the safe value of P will be smaller of the above two values, i.e.,

$$P_{safe} = 9.317 \text{ kN}.$$

Example 12.4. *Determine the angle θ made by the plane of load with the y-y axis so that the neutral axis of the z-section shown in Fig. 12.7 coincides with the x-x axis. Assume that the load passes through the shear centre.*

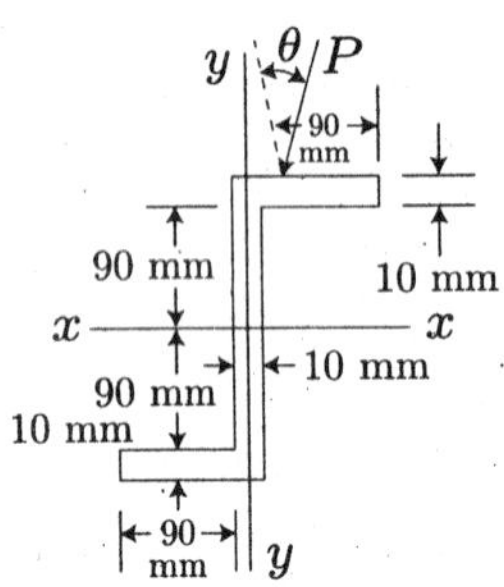

Fig. 12.7: For Example 12.4

Solution :

$$I_x = \left[\frac{1}{12}(90)(10^3) + 90(10)(95^2) + \frac{1}{3}(10)(100^3)\right](2)$$

$$= 2,292,667 \text{ mm}^4$$

$$I_y = \frac{1}{12}(200)(10^3) + 2\left[\frac{1}{12}(10)(90^3) + 10(90)(50^2)\right]$$

$$= 5,731,667 \text{ mm}^4$$

$$I_{xy} = 2(10)(90)(50)(95) = 855(10^4) \text{ mm}^4.$$

If α be the angle made by the neutral axis with x-x axis, then

$$\tan\alpha = \frac{I_{xy} - I_x \tan\theta}{I_{xy}\tan\theta - I_y}.$$

If neutral axis coincides with x-x axis, then $\alpha = 0$ and $\tan\alpha$ also is zero. Hence,

from $$0 = \frac{855(10^4) - 2292667\tan\theta}{855(10^4)\tan\theta - 5731667}.$$

$$\tan\theta = 3.7292812 \quad \text{and} \quad \theta = 74.99°.$$

12.5 Deflection of Beams Due to Unsymmetrical Bending

For a load acting parallel to a principal axis, the other principal axis becomes the neutral axis. Since deflection occurs perpendicular to the neutral axis, deflection due to a load acting parallel to a principal axis takes place in the plane of loading.

In case of an unsymmetrical bending, the applied load is resolved into two parts parallel to the principal axes and deflections in the directions of the principal axes due to the corresponding components of the load are calculated as in the case of simple beam. The final deflection is obtained as the vectorial sum of the above deflections as shown in Fig. 12.8. The direction of the final deflection must be normal to the neutral axis.

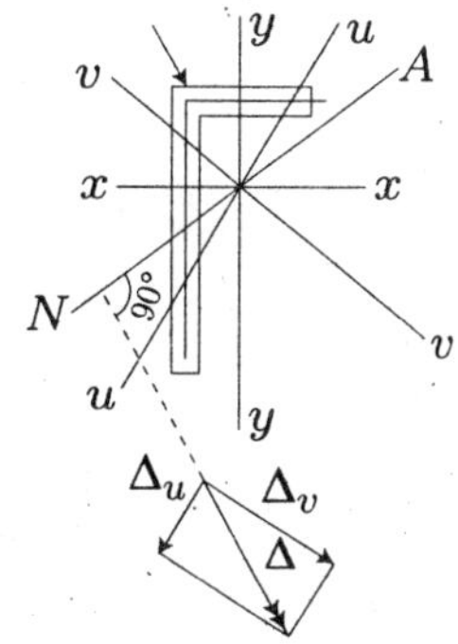

Fig. 12.8: Deflection

Example 12.5. *Determine the central deflection of the Beam of Example 12.1. Assume* $E = 2(10^5)$ N/mm^2.

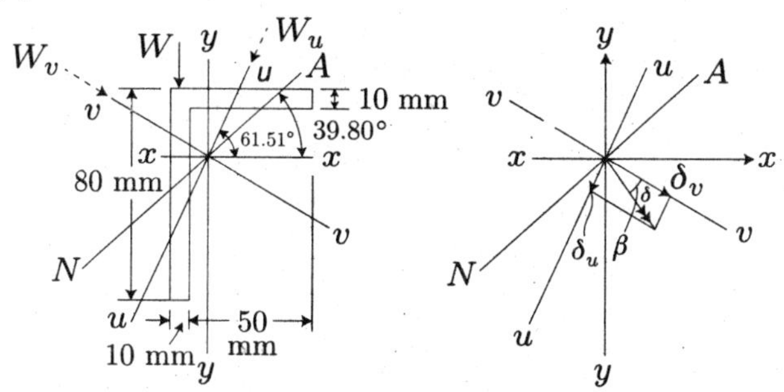

(a) Section **(b) Deflection**

Fig. 12.9: For Example 12.5

Solution : The directions of the principal axes and the moments of inertia about these axes are as obtained in Example 12.2 and shown in Fig. 12.5.

Principal axes u-u and v-v are inclined to x-x axis at 61.51° and 28.49° respectively. As in Example 12.1, the neutral axis makes with x-x axis an angle of 140.2° in the clockwise direction or (180° − 140.20°) = 39.8° in the anticlockwise direction.

Also $I_u = 212430$ mm^4

$I_v = 983086$ mm^4

$w_u = 351.56$ N/m $= 0.35156$ N/mm

$w_v = 190.8$ N/m $= 0.1908$ N/mm.

The deflection along u-u axis is

$$\delta_u = \frac{5}{384}\left(\frac{w_u L^4}{EI_v}\right) = \frac{5}{384}\left(\frac{0.35156 \times 4000^4}{2 \times 10^5 \times 983086}\right) \text{ mm} = 5.96 \text{ mm}.$$

The deflection along $v - v$ axis is

$$\delta_v = \frac{5}{384}\left(\frac{w_v L^4}{EI_u}\right) = \frac{5}{384}\left(\frac{0.1908 \times 4000^4}{2(10^5)(212430)}\right) \text{ mm} = 14.97 \text{ mm}.$$

The resultant deflection δ is the vector sum of δ_u and δ_v

and $\qquad \delta = \sqrt{\delta_u^2 + \delta_v^2} = \sqrt{(5.96)^2 + (14.79)^2} = 16.11$ mm.

The inclination β of the resultant defection δ is given by

$$\tan\beta = \frac{\delta_u}{\delta_v} = \frac{5.96}{14.97} = 0.39813, \quad \text{whence} \quad \beta = 21.71°.$$

The angle between the neutral axis (which makes 39.8° with x-x axis in the anticlockwise direction) and the direction of the resultant deflection

$$= 50.2° + 39.8° = 90°.$$

Example 12.6. *A beam of trapezoidal section as shown in Fig. 12.10(a) is simply supported on an effective span of 4 m and is subjected to a central vertical load (parallel to y-y axis) of* 10 kN. *Calculate the stresses at corners and resultant deflection at the middle of the beam. Assume* $E = 1(10^4)\text{N/mm}^2$ *and also assume that the load passes through the shear centre.*

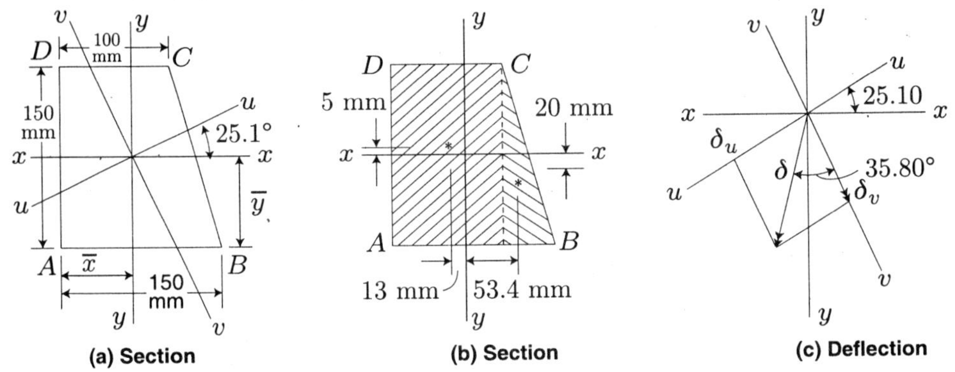

Fig. 12.10: For Example 12.6

Solution :

Sectional Properties

Sectional area,

$$A = \frac{100+150}{2}(150)\ \text{mm}^2 = 18750\ \text{mm}^2$$

$$\overline{X} = \frac{100(150)(50) + \frac{1}{2}(150)(50)\left[\frac{1}{3}(50)+100\right]}{18750} = 63.3\ \text{mm}$$

$$\overline{Y} = \frac{100(150)(75) + \frac{1}{2}(150)(50)(\frac{1}{3})(150)}{18750} = 70\ \text{mm}$$

$$I_x = \frac{1}{12}(100)(150^3) + 100(150)(5^2) + \frac{1}{36}(50)(150^3) + \frac{1}{2}(50)(150)\left(70 - \frac{150}{3}\right)^2$$

$$= 34,687,500\ \text{mm}^4$$

$$I_y = \frac{1}{12}(150)(100^3) + 150(100)(63.3-50)^2 + \frac{1}{36}(150)(50^3) + \frac{1}{2}(50)(150)\left[100 + \frac{1}{3}(50) - 63.3\right]^2$$

$$= 26,354,167\ \text{mm}^4.$$

With reference to Fig. 12.10(b),

$$I_{xy} = 100(150)(-13.3)(+5) + \frac{1}{2}(50)(150)(53.4)(-20) = -5,002,500 \text{ mm}^4.$$

If θ be the angle made by principal axis u-u with x-x axis in anticlockwise direction, then

$$\tan 2\theta = -\frac{2Ixy}{I_x - I_y} = +\frac{2(5002500)}{34,687,500 - 26,354,167} = 1.2006.$$

Then $2\theta = \tan^{-1} 1.2006 = 50.2°, \quad \therefore \quad \theta = 25.1°.$

$$\cos\theta = 0.9055688 \quad \text{and} \quad \sin\theta = 0.4241994.$$

$$I_u = \frac{I_x + I_y}{2} + \frac{I_x - I_y}{2}\cos 2\theta - I_{xy}\sin 2\theta$$

$$= \frac{34687500 + 26354167}{2} + \frac{34687500 - 26354167}{2}\cos 50.2° - (-5002,500)\sin 50.2°$$

$$= 30,520,834 + 2,667,124 + 3,843,338 = 37,031,296 \text{ mm}^4.$$

$$I_y = \frac{I_x + I_Y}{2} + \frac{I_x - I_y}{2}\cos(180° + 50.2°) - I_{xy}\sin(180° + 50.2°).$$

$$= 30,520,834 - 2,667,124 - 3,843,338 = 24,010,372 \text{ mm}^4$$

Bending moments

The plane of load makes angle 25.1° with v-v axis. Hence the components of the load along u-u and v-v axes are :

$$P_u = P.\sin 25.1° = (10)\sin 25.1° = 4.242 \text{ kN}$$

$$P_v = P.\cos 25.1° = (10)\cos 25.1° = 9.056 \text{ kN}.$$

$$\therefore \quad M_u = -\frac{P_v \cdot L}{4} = -\frac{9.056(4)}{4} \text{ kN-m} = -9.056 \text{ kN-m}$$

$$M_v = -\frac{P_u \cdot L}{4} = -\frac{4.242(4)}{4} \text{ kN-m} = -4.242 \text{ kN-m}$$

The negative sign of the bending moments M_u and M_v indicate that they create compression at points having positive v- and u- co-ordinates respectively.

Co-ordinates of the corner points are as under :

	(u)	(v)
For A,	−87.01	−36.54 mm
For B,	+48.82	−100.17 mm
For C,	+67.17	+54.88 mm
For D,	−23.38	+97.3 mm

Bending Stresses

$$\sigma = \frac{M_u}{I_u} \cdot v + \frac{M_v}{I_v} \cdot u = \frac{-9.056(10^6)}{37,031,296} \cdot v + \frac{-4.242(10^6)}{24,010,372} \cdot u$$

$$= -0.24455v - 0.17667u$$

$$\therefore \quad \sigma_A = 0.24455(-36.54) - 0.17667(-87.01) = +24.3 \text{ N/mm}^2 \text{(tension)}$$

$$\sigma_B = 0.24455(-100.17) - 0.17667(+48.82) = +15.90 \text{ N/mm}^2 \text{(tension)}$$

$$\sigma_C = -0.24455(+54.88) - 0.17667(+67.17) = -25.3 \text{ N/mm}^2 \text{(compression)}$$

$$\sigma_D = -0.24455(+97.30) - 0.17667(-23.38) = -19.7 \text{ N/mm}^2 \text{(compression)}.$$

Deflection

$$\delta_u = \frac{P_u L^3}{48EI_v} = \frac{4.242(10^3)(4000^3)}{48(1)(10^4)(24010372)} = 23.56 \text{ mm}$$

$$\delta_v = \frac{P_v L^3}{48EI_u} = \frac{9.056(10^3)(4000^3)}{48(1)(10^4)(37,031,296)} = 32.61 \text{ mm}.$$

Resultant deflection is

$$\delta = \sqrt{\delta_u^2 + \delta_v^2} = \sqrt{(23.56)^2 + (32.61)^2} = 40.23 \text{ mm}.$$

The angle β between the direction of the resultant deflection and v-v axis is given by

$$\tan\beta = \frac{\delta_y}{\delta_u} = \frac{23.56}{32.61} = 0.722478,$$

whence $\beta = 35.8°$.

$\therefore$ the angle between the directions of the resultant deflection and y-y axis $= 35.8° - 25.1° = 10.7°$.

12.6 Bending of Curved Beams by Couples in the Plane of Curvature

The theory of simple bending developed for straight beams is not applicable to beams having curved axis. To develop pure bending theory for curved beams, following assumptions are made.

(i) The beam has a longitudinal plane of symmetry and uniform cross section.

(ii) The applied pure couples act in the longitudinal plane of symmetry.

(iii) Plane sections before bending remain plane after bending.

(iv) The material of the beam obeys Hooke's Law so that relation between stress and strain is linear.

Fig. 12.11(a) shows a curved beam subjected to bending couples M in the plane of curvature. Beams of large and small curvatures are commonly distinguished on the basis of the ratio of the depth of the section d in the plane of curvature to the radius R_m of the centroidal axis. If $d/R_m \leq 0.2$, the beam is taken to be of small curvature, otherwise it is taken to be large. This distinction is, however, purely conventional. Straight beam formula may be used for beams of small curvature to a good approximation.

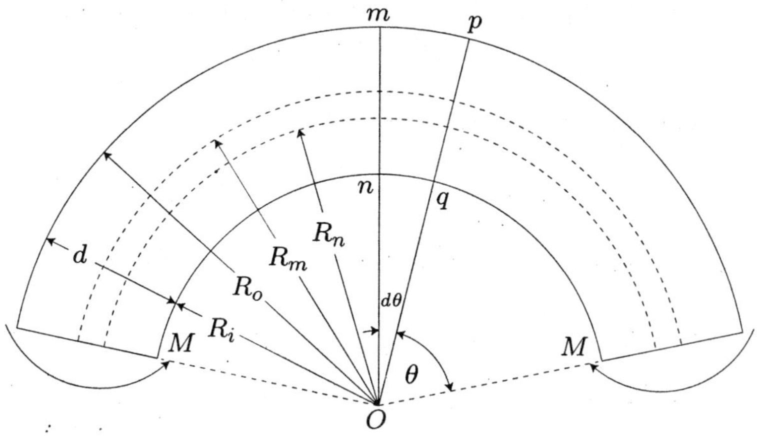

(a) Curved beam

Since bending is due to end couples, the bending moment at any section is the same and is not a function of θ. All elementary length segments of the beam must, therefore, deform in exactly similar fashion. Isolating an elementary portion of the beam between two adjacent sections mn and pq, an enlarged view of this portion is shown in Fig. 12.11(b). Considering the deformation of the element $mpqn$ due to bending, the top layer, mp will be in tension and the bottom layer, nq in compression. Because of continuous deformation, a layer cd between the two extreme layers must be neutral. All the layers above the neutral surface are in tension and below it in compression. To help simplify the geometry of deformation,

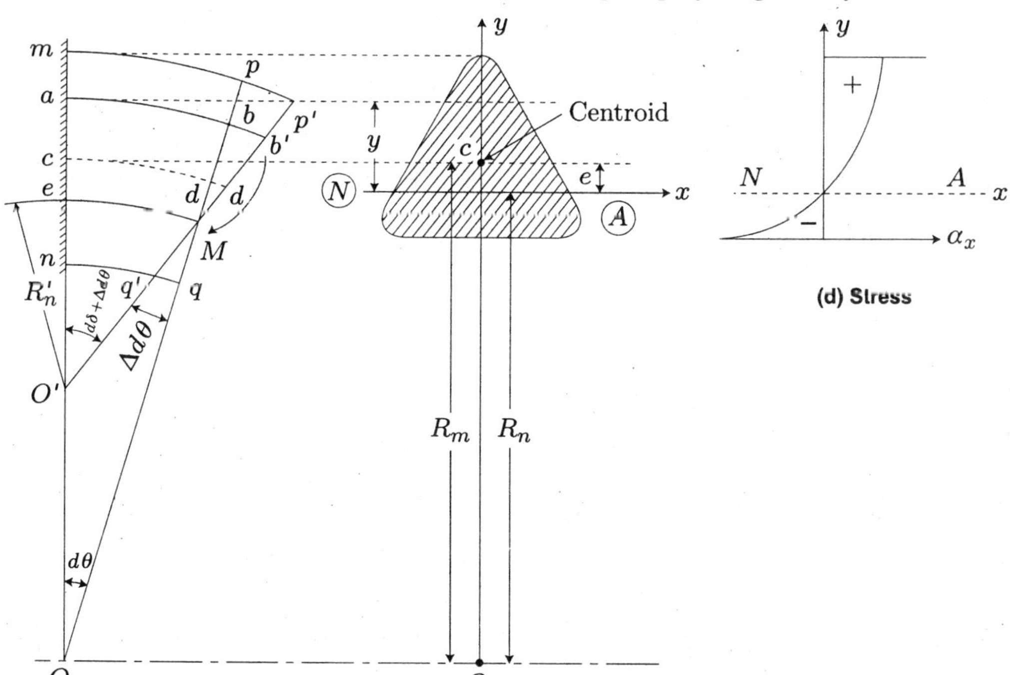

(d) Stress

(b) Strain **(c) Section**

Fig. 12.11: Curved beam subject to pure bending

from the familiar assumption that plane section before bending remains plane after bending, the section pq remaining plane, rotates with respect to mn through a small angle $d\theta$ and occupies the position $p'q'$. The cross section as shown in Fig. 12.11(c) rotates about a straight line N.A. which is the trace of the neutral surface cd on the cross-sectional plane. From an argument of symmetry, N.A. is perpendicular to the axis of symmetry of the cross section. The neutral axis N.A. does not necessarily pass through the centroid and is shifted relative to it towards the centre of curvature.

The basic assumption about the deformation is the same as that for a straight beam. However, there is a marked difference between the two cases. Unlike a straight beam, the initial undeformed length of beam fibres depends upon the distance from the centre of curvature and varies from a small value $nq = R_i d\theta$ at the bottom layer to a large value $mp = R_0 d\theta$ at the top. As a result of this, although total deformations (elongation or contraction) of the fibres have been assumed to vary linearly with the distance from neutral surface, the normal strains do not vary so. Consequently the normal stresses also, for an elastic beam, are not linear. This is in contrast to the normal stress distribution in case of a straight beam, which formulate the so-called Winkler-Bach theory for the curved beams. For this purpose, a set of axes X, Y, Z at the section pq [Fig. 12.11(c)] is taken along the neutral axis and the Y-axis along the axis of symmetry of the cross section. Obviously then, Z-axis is normal to the section. The bending moment at the section is about X-axis and is positive. Considering an arbitrary layer ab at a distance of y above the neutral surface, the deformed position of the layer ab is ab' as shown in Fig. 12.11(b). Then the normal strain in a fibre on the layer ab',

$$\epsilon_z = \frac{ab' - ab}{ab} = bb'/ab = \frac{y(\Delta d\theta)}{(R_n + y)d\theta}. \tag{12.7}$$

Here, $\Delta d\theta$ is the angle of rotation of the section, $d\theta$ being the angle subtended at the centre of curvature by the elementary portion $mpqn$ of the beam and R_n is the radius of curvature of the neutral surface in the undeformed beam. In Eq. (12.7), it is assumed that $db = db' = y$, i.e., the distance y remains unaltered during the process of bending. But, strictly speaking, this is not so. The fibres undergo radial displacement also, as there exists interaction between the adjacent fibres in the form of the radial forces. The distance y, therefore, does not remain the same. For an approximate solution, this change is ignored. The principal difference between the present approximate solution and the exact solution in Theory of Elasticity lies in the presence of radial stress which has been assumed to be absent here. The solution so obtained though

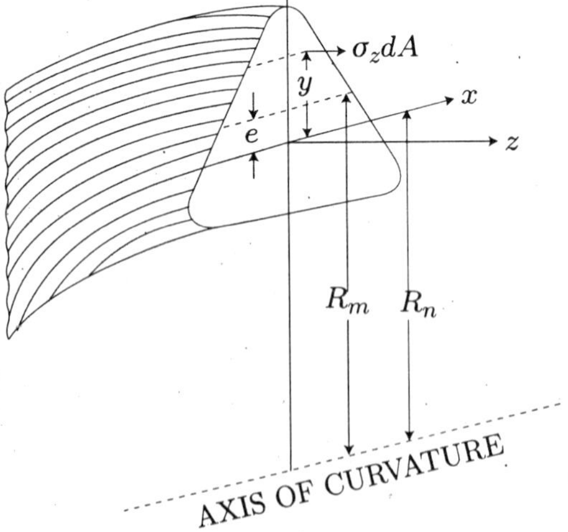

Fig. 12.12: Streses in curved beam

violates radial equilibrium as well as the compatibility condition, is adequate for practical purposes. Considering an elemental area dA on the cross section at a distance y from the N.A. as shown in Fig. 12.12, denoting the normal stress on this elemental area by σ_z and assuming all other stress components to be zero, Hooke's law gives

$$\sigma_z = E\epsilon_z. \tag{12.8}$$

Substituting for ϵ_x from Eq. (12.7),

$$\sigma_z = \frac{E(\Delta d\theta)}{d\theta} \frac{y}{(R_n + y)}. \tag{12.9}$$

The neutral surface cd has a radius of curvature, R'_n after deformation [Fig. 12.11(b)]

and $\quad cd = R_n d\theta = R'_n(d\theta + \Delta d\theta)$

and $$\frac{\Delta d\theta}{d\theta} = R_n\left(\frac{1}{R'_n} - \frac{1}{R_n}\right). \tag{12.10}$$

From Eq. (12.10), Eq. (12.9)

gives $$\sigma_z = ER_n\left(\frac{1}{R'_n} - \frac{1}{R_n}\right)\frac{y}{(R_n + y)}. \tag{12.11}$$

Since E, R_n and R'_n are constants for a given beam subjected to a constant bending moment, Eq. (12.11) gives

$$\sigma_z = \frac{Cy}{(R_n + y)}, \tag{12.12}$$

where C is a constant.

Since the normal stress distribution is due to a bending moment alone, the resultant of the elemental normal forces, $dF = \sigma_z dA$ on the cross section must be zero, i.e. $\int_A \sigma_z dA = 0$. Using Eq. (12.12),

$$C\int_A \frac{y}{R_n + y} dA = 0$$

or, $$\int_A \frac{y}{R_n + y} dA = 0. \tag{12.13}$$

With $R_n + y = r$, Eq. (12.13) gives $\int_A \frac{r - R_n}{r} dA = 0$,

whence $$R_n = \frac{A}{\int_A \frac{dA}{r}}. \tag{12.13a}$$

That neutral axis does not pass through the centroid of the section as obvious from Eq. (12.13). Also, the resultant moment of the elemental normal forces, $dF(= \sigma_z dA)$, about the X-axis must give the B.M. on the section. Thus,

$$\int_A (\sigma_z dA)y = M$$

i.e., $$C\int_A \frac{y^2}{R_n + y} dA = C\int_A \frac{y^2 + R_n y - R_n y}{R_n + y} dA = M$$

and $C\int_A y dA = M$, as $\int \frac{y dA}{R_n + y} = 0$ from Eq. (12.13).

As $CAe = M$,

$$C = \frac{M}{Ae}, \tag{12.14}$$

where A is the area of the cross section and e is the distance of the N.A. from the centroid.

Substituting the value of the constant C in Eq. (12.12),

$$\sigma_z = \frac{M_y}{Ae(R_n + y)}. \tag{12.15}$$

Note that positive B.M. causes positive normal stress (tensile) on the layers of positive y- co-ordinate and so, in the expression $\int_A y dA = Ae$, used in Eqn. (12.14), e must always be positive. This clearly means that the shift of the N.A. with respect to the centroid is always towards the centre of curvature.

The normal stress distribution across the depth of the section is nonlinear as shown in Fig. 12.13. The stress diagram is a hyperbola with one of its asymptotes coinciding with the axis of curvature. The stress is always maximum (numerically) at the extreme layer of concave side or intrados of the beam.

To obtain stress values from Eq. (12.13) it is necessary to know the radius of curvature of the neutral axis, i.e. R_n for a given cross section.

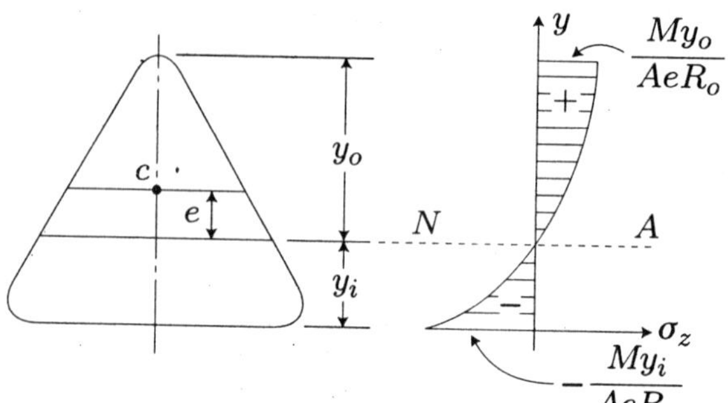

(a) Section (b) Stresses

Fig. 12.13: Stresses in curved beam

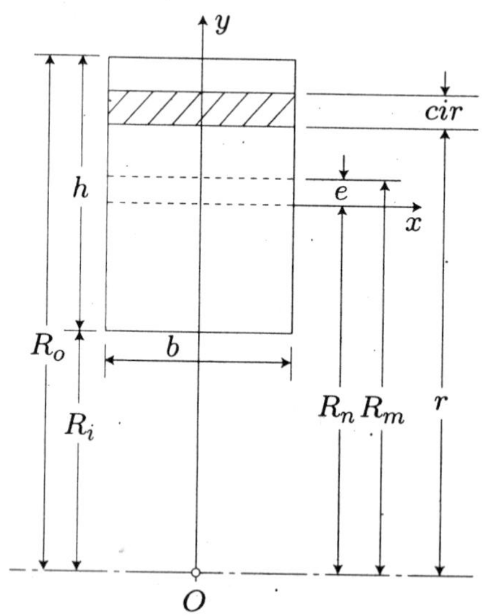

Fig. 12.14: Rectangular section

Fig. 12.14 shows a rectangular cross section of width, b and depth, h, with R_m as a radius of curvature of the centroidal axis. The neutral axis is shifted towards the centre of curvature and its radius of curvature is R_n. From Eqn. (12.13),

$$\int_A \frac{R_n + y - R_n}{R_n + y} dA = 0 \quad \text{and} \quad A - R_n \int_A \frac{dA}{R_n + y} = 0$$

or,
$$R_n = \frac{A}{\int_A \frac{dA}{R_n+y}}, \tag{12.16}$$

where A is the area of cross section. Introducing a new variable $r = R_n + y$, we have :

$$\int_A \frac{dA}{R_n + y} = \int_{R_m - h/2}^{R_m + h/2} \frac{bdr}{r} = b \ln \frac{R_m + h/2}{R_m - h/2}. \tag{12.17}$$

Substituting Eq. (12.17), Eq. (12.16) gives

$$R_n = \frac{h}{\ln \frac{R_m + h/2}{R_m - h/2}}. \tag{12.18}$$

Similar equations for various cross sections have been listed in Table 12.1.

The quantity $e = R_m - R_n$, which gives the shift of N.A., is generally very small. It must be determined very accurately because even a small error in its value will considerably affect the value of stress determined from Eq. (12.15).

Table 12.1. Values of R_n for Various Sections

Section	R_n
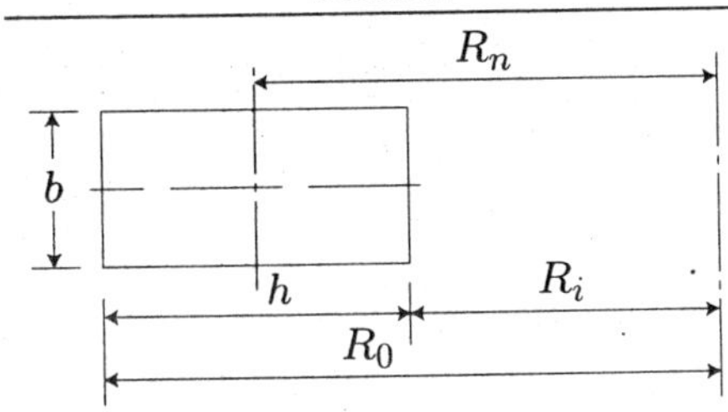 I *Rectangular section*	$R_n = \frac{A}{\ln\left(\frac{R_o}{R_i}\right)^b} = \frac{h}{\ln \frac{R_o}{R_i}}$, where $A = bh$.
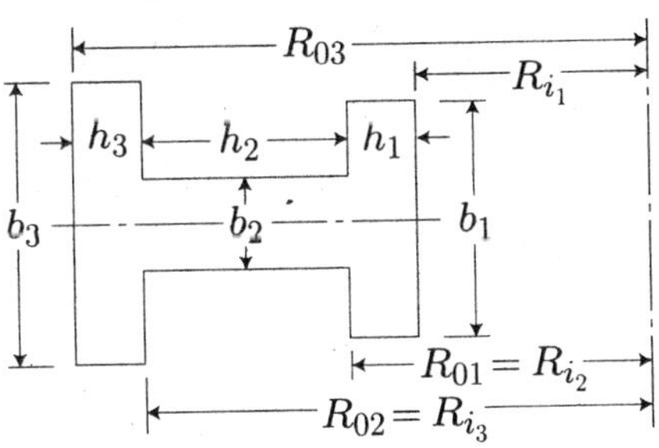 II *Section consisting of several rectangles in series*	$R_n = \frac{A}{\ln\left[\left(\frac{R_{o1}}{R_{i1}}\right)^{b_1}\left(\frac{R_{o2}}{R_{i2}}\right)^{b_2}\left(\frac{R_{o3}}{R_{i3}}\right)^{b_3}\right]}$, where $A = b_1h_1 + b_2h_2 + b_3h_3$.
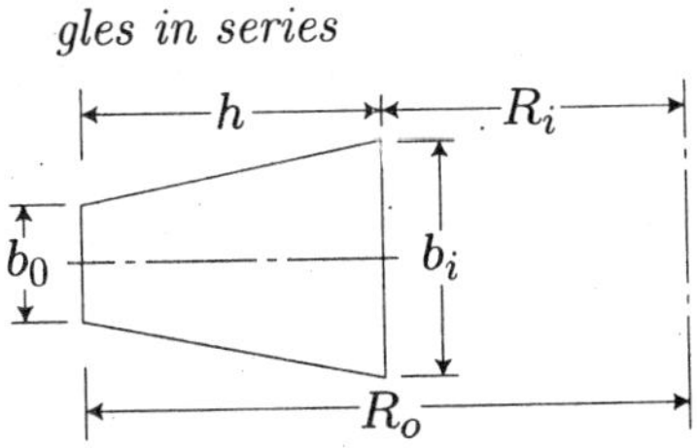 III *Trapezoidal section*	$R_n = \frac{A}{\left(b_i + \frac{b_i - b_o}{h} R_i\right) \ln \frac{R_o}{R_i} - (b_i - b_o)}$, where $A = \frac{1}{2}(b_i + b_o)h$.

Contd.

Table 12.1 (Contd.)

Section	R_n
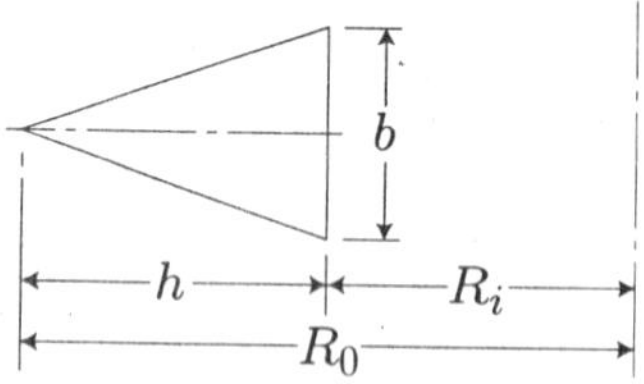 IV *Triangular section*	$R_n = \dfrac{\frac{1}{2}h^2}{R_o \ln\left(\frac{R_o}{R_i}\right) - h}.$
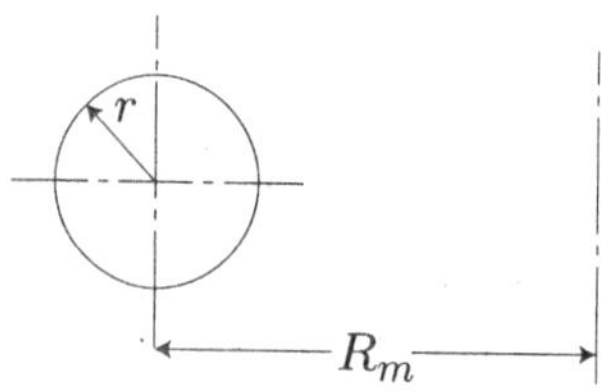 V *Circular section*	$R_n = \dfrac{1}{2}[R_n + \sqrt{R_m^2 - r^2}]$ $= \dfrac{r^2}{2[R_m - \sqrt{R_m^2 - r^2}]}.$
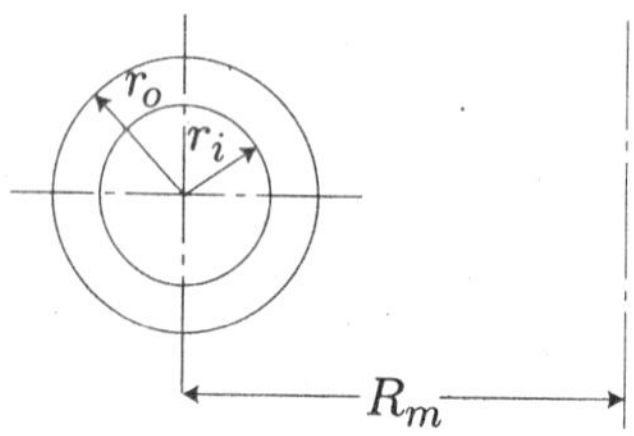 VI *Hollow circular section*	$R_n = \dfrac{1}{2}[\sqrt{R_m^2 - r_o^2}]$
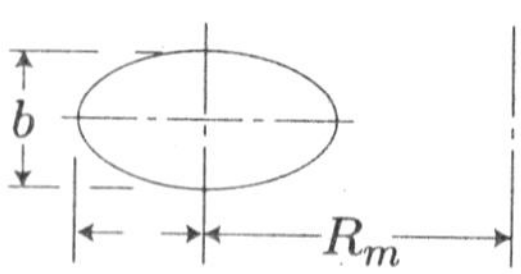 VII *Elliptical section*	$R_n = [R_m + \sqrt{R_m^2 - a^2}].$
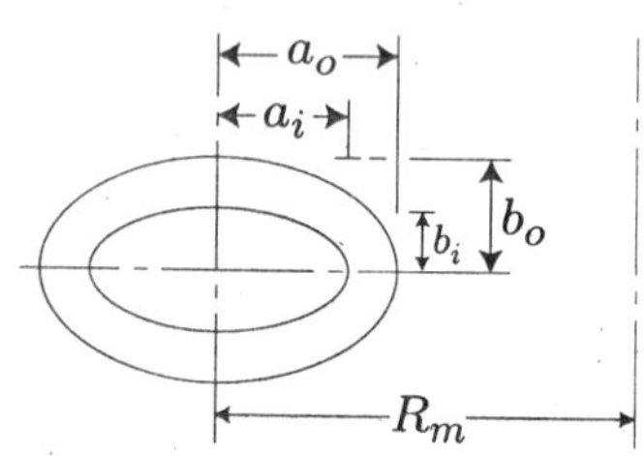 VIII *Hollow elliptical section*	$R_n = \dfrac{\frac{1}{2}(a_o b_o - a_i b_i)}{\frac{b_o}{a_o}(R_m - \sqrt{R_m^2 - a_o^2}) - \frac{b_i}{a_i}\left(R_m - \sqrt{R_m^2 - a_i^2}\right)}.$

Example 12.7. *Fig. 12.15 shows a punch press. Determine the limiting value of the punching force, P, if the normal stress is not to exceed* 40 MN/m^2 *anywhere. The section at AB has been shown in Fig. 12.15(b) (All the dimensions are in mm.)*

Solution : The method of sections are used to show the internal force factors at the critical section AB as shown in Fig. 12.15(c). A normal force, P (tensile in nature and a bending moment, $M_b = 2PR_m$ are acting on the section (R_m is the radius of curvature of the centroidal axis). The normal force P will cause a uniform tensile stress, P/A, over the section whereas the B.M., M_b, will cause a hyperbolic stress distribution, the maximum intensity being tensile at the intardos, B. These stress distributions have been shown across the depth of the section in Fig. 12.15(c). The normal stress on the section obviously occurs at the intrados and from the principle of superposition :

$$\sigma_{\max} = \frac{P}{A} + \frac{2PR_m(R_n - R_i)}{AR_i(R_m - R_n)}, \tag{a}$$

where A is the area of cross section, R_n is the radius of curvature of the neutral axis, and R_i is the radius of curvature of the intrados.

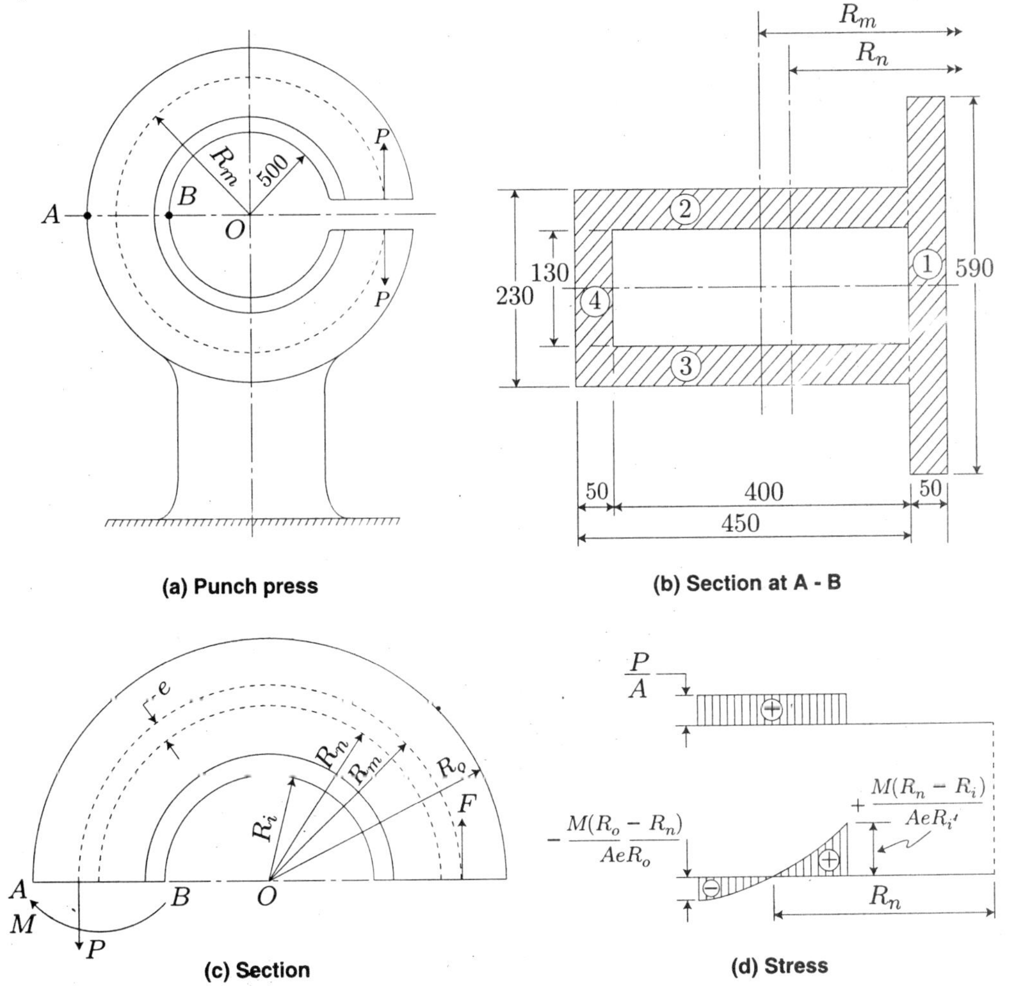

Fig. 12.15: For Example 12.7

The quantities A, R_m, R_n and R_i are the properties of the cross section and can be computed. $\sigma_{\max}$ is also given to be 40 MN/m^2. Thus the only one unknown, the punching force P in Eqn. (a) can be easily determined.

To compute R_m and R_n the given section is divided into four rectangular parts, namely (1), (2), (3) and (4) as shown in Fig. 12.15(b). For the sake of convenience, the useful data for these parts; are tabulated in Table 12.2.

Table 12.2. Useful data for computation of $R_m + R_n$

Sl. No.	Item	Rectangle (1)	Rectangle (2)	Rectangle (3)	Rectangle (4)
1.	A, area in mm^2	$590(50) = 29,500$	$400(50) = 20,000$	$400(50). = 20,000$	$230(50) = 11,500$
2.	r, radius of curvature of the centroid in mm	$500 + \frac{50}{2} = 525$	$500 + 50 + \frac{400}{2} = 750$	$500 + 50 + \frac{400}{2} = 750$	$500 + 450 + \frac{50}{2} = 975$
3.	b, width in mm	590	50	50	230
4.	r_i, radius of curvature of innermost fibre in mm	500	$500 + 50 = 550$	$500 + 50 = 550$	$500 + 450 = 950$
5.	r_o, radius of curvature of outermost fibre in mm	$500 + 50 = 550$	$550 + 400 = 950$	$550 + 400 = 950$	$950 + 50 = 1000$
6.	Integral $\int_A \frac{dA}{r} = b \ln \frac{r_o}{r_i}$	$590 \ln \frac{550}{500} = 56.2330$	$50 \ln \frac{950}{550} = 27.3272$	$50 \ln \frac{950}{550} = 27.3272$	$230 \ln \frac{1000}{950} = 11.7974$

Now, from the definition of the centroid,

$$R_m = \frac{(A_1\bar{r}_1 + A_2\bar{r}_2 + A_3\bar{r}_3 + A_4\bar{r}_4)}{(A_1 + A_2 + A_3 + A_4)}$$

Using the relevant value from the table above,

$$R_m = \frac{29,500(525) + 20,000(750) + 20,000(750) + 11,500(975)}{29,500 + 20,000 + 20,000 + 11,500} = 700 \text{ mm.}$$

Recalling the expression for R_n from section 12.6,

$$R_n = \frac{A}{\int_A \frac{dA}{r}} = \frac{A_1 + A_2 + A_3 + A_4}{\int_{A1} \frac{dA}{r} + \int_{A2} \frac{dA}{r} + \int_{A3} \frac{dA}{r} + \int_{A4} \frac{dA}{r}}$$

$$= \frac{A_1 + A_2 + A_3 + A_4}{b_1 \ln \frac{r_{o1}}{R_{i_1}} + b_2 \ln \frac{r_{o2}}{R_{i_2}} + b_3 \ln \frac{r_{o3}}{R_{i_3}} + b_4 \ln \frac{r_{o4}}{R_{i_4}}}.$$

Substituting the proper values from the Table,

$$R_n = \frac{29,500 + 20,000 + 20,000 + 11,500}{56.2330 + 27.3272 + 27.3272 + 11.7974} = 81,000/122.6848 = 660.228 \text{ mm}$$

$$\sigma_{\max} = 40(10^6) \text{ N/m}^2$$

$$A = 81,000 \text{ mm}^2 = 81000(10^{-6}) \text{ m}^2$$

$$R_m = 700 \text{ mm} = 0.7 \text{ m}$$

$$R_n = 660.228 \text{ mm} = 0.6602 \text{ m}$$

$$R_i = 500 \text{ mm} = 0.5 \text{ m}$$

$$(R_n - R_i) = (0.6602 - 0.5) = 0.1602 \text{ m}$$
$$(R_m - R_n) = (0.7 - 0.6602) = 0.0398 \text{ m}.$$

Substituting these values in Eq. (a),

$$40(10^6) = P\left[\frac{1}{81000(10^{-6})} + \frac{2 \times 0.7 \times 0.1602}{81000(10^{-6})(0.5)(0.0398)}\right]$$

and $P = 264.051(10^{-3})$ N $= 264.051$ kN.

12.7 Curved Beams under Transverse Loads in the Plane of Curvature

A section of a curved beam under transverse loads acting in the plane of curvature is in general subjected to a bending moment, an axial force and a shearing force. Fig. 12.16(a) shows the simply supported curved beam subjected to transverse loads P_1, P_2, P_3, etc. Considering the portion of the beam on the left of the section 11 as a free body as shown in Fig. 12.16(b).

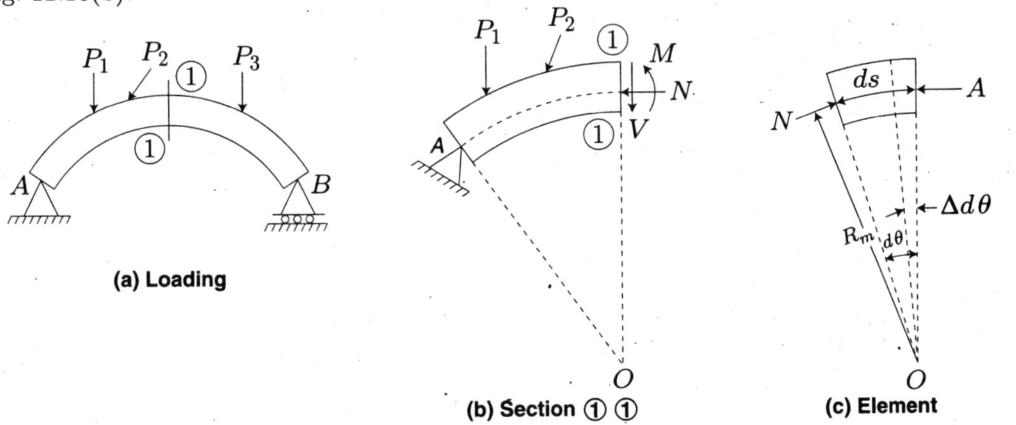

Fig. 12.16: Curved beam

The bending moment M, shearing force V and axial force N acting at section 1-1 can be determined from consideration of equilibrium of the free body. The stresses due to M can be determined as previously. The normal force N produces uniform stress over the cross section. Hence the total stress due to combined effect of bending moment and normal force is given by

$$\sigma = \frac{M \cdot y}{Ae(R_n + y)} + \frac{N}{A}. \tag{12.19}$$

The normal stress due to axial force N is uniform over the section but since length of longitudinal fibres varies and is proportional to their distance from the centre of curvature, the change in length of longitudinal fibres is also proportional to their distance from the centre of curvature, so that the section remains plane after deformation. The change in length of an element ds along the neutral surface in Fig. 12.16(c),

$$\Delta ds = \frac{N ds}{AE} \tag{a}$$

and this change in length due to axial force which is assumed to be compressive decreases the initial angle $d\theta$ and is at the centre of curvature by $\Delta d\theta$ and is given by

$$\Delta d\theta = \frac{Nds}{AER_n}, \tag{b}$$

which does not, however, change the curvature.

The shear stress distribution across the section produced by radial shear V is considered to be similar to that in a straight member,

i.e.,
$$\tau = \frac{VQ}{Ib}, \tag{12.20}$$

where for the purpose of calculating Q and I, the centroidal axis (which is same as the neutral axis for a straight beam) is considered.

12.8 Stresses in Crane Hooks

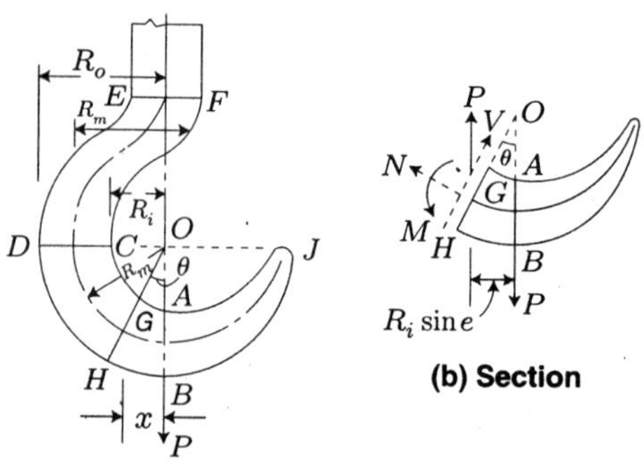

(b) Section

(a) Hook

Fig. 12.17: Crane hook

Crane hooks behave as curved beams having large curvature of the axis. The applied load usually passes through the centre of curvature as in Fig. 12.17(a), the section AB being in the same plane of the load and subjected to radial shear only whereas section EF which is normal to the load is subjected to direct tension only. The load is normal to the section CD, but eccentric and so this section is subjected to bending moment in addition to direct tension and no shearing force. Any other section between AB and EF is subjected to bending moment, shear force and direct tension. Section HG whose plane makes an angle θ with the plane of load is under the action of a bending moment $M = PR_m \sin\theta$, a shear force $V = P\cos\theta$ and normal force $N = P\sin\theta$ as in Fig. 12.17(b), where R_m is the radius of curvature of the axis of the bar. The values of M and N becomes maximum when θ is $90°$, i.e., at section CD, for which $M = PR_m$ and $N = P$, but shear force V is zero. The portion AJ is free from any stress. Thus, so far normal stress is concerned section CD is critical and the maximum tensile stress occurs at the inner point C where the stresses due to direct force P and bending moment M are additive being of the same nature, maximum compressive stress occurring at the other extreme point D.

Maximum tensile stress

$$= \frac{MC_1}{Ae(R_n - C_1)} + \frac{P}{A} \tag{12.21a}$$

and maximum compressive stress

$$= \frac{MC_2}{Ae(R_m + C_2)} - \frac{P}{A}, \tag{12.21b}$$

where C_1 and C_2 are distances from the neutral axis.

Example 12.8. *A crane hook of circular cross section of diameter 80 mm has axis curved in the form of a circular arc of radius 110 mm. Determine the maximum tensile and com-*

pressive stresses, if a load $P = 20$ kN is suspended from the hook with its line of action passing through the centre of curvature.

Solution : In this case $R_m = 110$ mm,

$$R_i = 110 - \frac{80}{2} = 70 \text{ mm},$$

$$R_o = 110 + \frac{80}{2} = 150 \text{ mm}, \quad \text{as } a = 40 \text{ mm}$$

area $A = \pi a^2 = 5026.5 \text{ mm}^2$.

The distance of the neutral axis from the centre of curvature of the hook is

$$R_n = \frac{a^2}{2[R_m - \sqrt{R_m^2 - a^2}]} = \frac{40^2}{2[110 - \sqrt{110^2 - 40^2}]} = 106.23 \text{ mm}$$

$$e = R_m - R_n = 110 - 106.23 = 3.77 \text{ mm}.$$

The distances of the extreme fibres on the inner and outer sides respectively from the centroid of the cross section are

$$h_1 = \frac{80}{2} = 40 \text{ mm} \quad \text{and} \quad h_2 = 40 \text{ mm}$$

and their distances from the neutral axis are

$$C_1 = h_1 - e = 40 - 3.77 = 36.23 \text{ mm} = y_t$$

and $C_2 = h_2 + e = 40 + 3.77 = 43.77 \text{ mm} = y_b$.

Maximum bending moment $M = PR_m = 20(10^3)(110) = 22(10^5)$ N-mm

$$\text{Maximum tensile stress} = \frac{MC_1}{A_e(R_n - C_1)} + \frac{P}{A} = \frac{22(10^5)(36.23)}{5026.5(3.77)(106.23 - 36.23)} + \frac{20(10^5)}{5026.5}$$

$$= 60.09 + 3.98 = 64.07 \text{ N/mm}^2$$

$$\text{Maximum compressive stress} = \frac{MC_2}{Ae(R_n + C_2)} - \frac{P}{A}$$

$$= \frac{22(10^5)(43.77)}{5026.5(3.77)(106.23 + 43.77)} - \frac{20(10^3)}{5026.5}$$

$$= 33.88 - 3.98 = 29.9 \text{ N/mm}^2.$$

12.9 Beams Curved in Plan

Beams usually consist of single straight line in plan. Whilst this is generally true of most bemas in practice, yet occassionally a beam occurs which is non-linear or curved in plan. Such curved beams are used to support circular reservoirs, curved balconies and curved ramps. In general, such cases will occur where a non-linear or curved structure is loaded at right angles to the plane of the structure. It is necessary to determinate the resultant actions at any section of beam which is curved in plan. In order to ensure stability such beams must be built in at one end.

The simplest case is that of a horizontal cantilever beam curved in plan as shown in Fig. 12.18 fixed at end A and carrying a vertical point load P at the free end. The tangent of the slope of the beam at A is considered to be normal to the wall to which it is fixed. If the line BE be drawn from B perpendicular to the wall as in Fig. 12.18, the resultant actions at A will consist of :

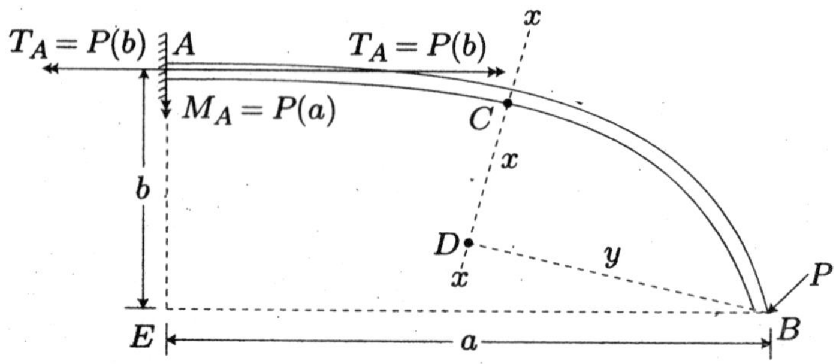

Fig. 12.18: Representation of bending moment and Twisting moment

(a) A shear force P tending to move the beam downwards at the wall section.

(b) A bending moment M_A of magnitude $P(a)$ tending to bend the beam convex up-wards at the wall section.

(c) A twisting moment or torque T_A of magnitude $P(b)$ tending to twist the beam anti-clockwise at the wall section.

The presence of the twisting moment differentiates the problem from that of the straight beam.

The bending moments and the twisting moments (or torque) on the beam at A (or at A of member) may be represented vectorially by the right-hand (or left-hand) rule as shown in Fig. 12.8. The right-hand rule states that if the thumb of the right hand points in the direction of the section, the bent fingers indicates the direction of rotation.

At any other section such as at C, the resultant actions are similar. Through C a normal XX is drawn. From B, a perpendicular BD is drawn to XX. It the distances BD and CD be y and x, the bending moments and twisting moments at C are $P(y)$ and $P(x)$ respectively. The section at C must be designed to resist both the bending and twisting moments.

In the case of the cantilever, these values are statically determinate. If the curve AB be the quadrant of a circle of radius as shown in Fig. 12.19, calculation can be done taking the centre of the circle O as the origin and angle BOC as θ as shown in Fig. 12.19.

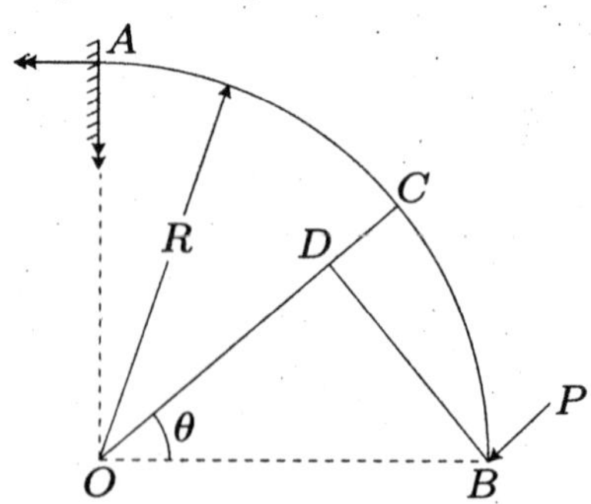

Fig. 12.19: Circular curved beam with concentrated load

Then $BM, M_c = P(BD) = PR\sin\theta$.

Twisting moment $T_c = P(DC) = PR(1 - \cos\theta)$

and S.F. at C.$V_c = P$.

In this case, at A, magnitude of both B.M. and Twisting moments are PR.

If instead of a concentrated load P, the beam carries a uniformly distributed load over its length as shown in Fig. 12.20, the load acting on a small element at E which is α from OC is $wRd\alpha$.

The bending moment at C due to this is $wRd\alpha(DE)$, i.e., $wR^2\sin\alpha d\alpha$, the twisting moments is $wRd\alpha(DC)$, i.e., $wR^2(1-\cos\alpha)d\alpha$ and the shear force is $wRd\alpha$.

The total resultant actions at C are as below:

$$M_c = wR^2\int_0^\theta \sin\alpha d\alpha = wR^2(1-\cos\theta)$$

$$T_c = wR^2\int_0^\theta (1-\cos\alpha)d\alpha = wR^2(\theta - \sin\theta)$$

and shear force

$$V_c = wR\int_0^\theta d\alpha = wR\theta.$$

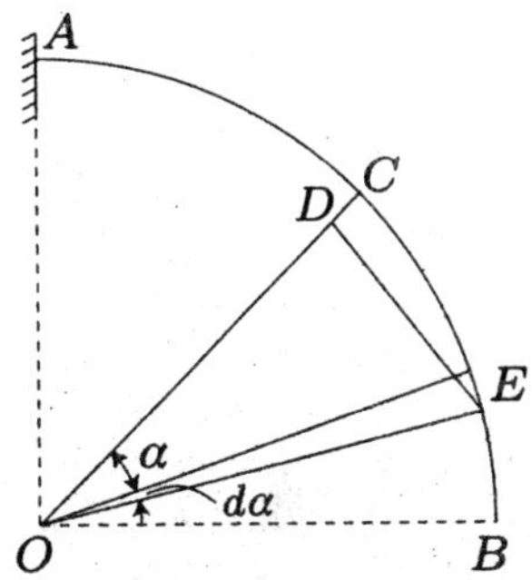

Fig. 12.20: Circular curved beam with uniformly distributed load

Some other types of linearly bent cantilever beams are illustrated in Fig. 12.21.

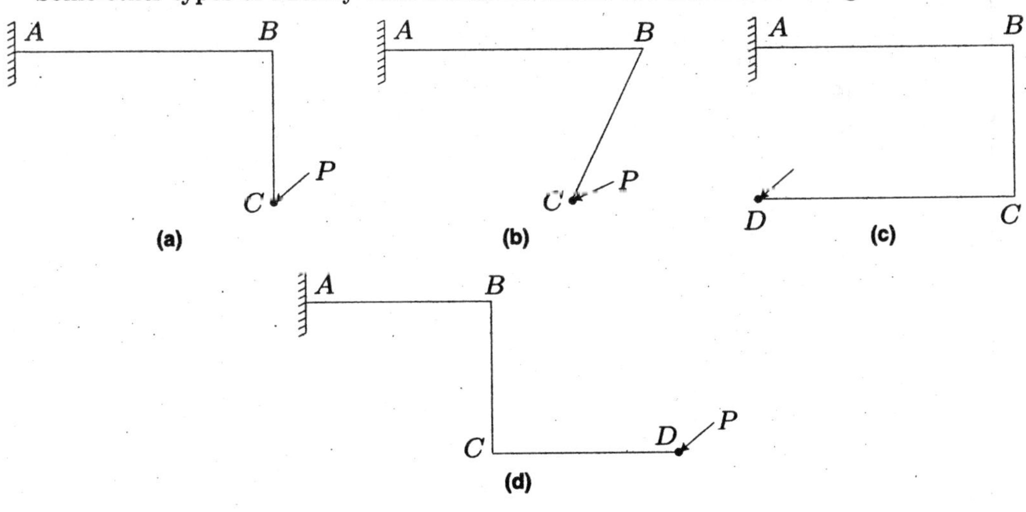

Fig. 12.21: Linearly bent cantilever beams

12.10 Statically Indeterminate Cases of Curved Beams in Plan

Fig. 12.18, shows a statically determinate case, in which the beam is fixed in all directions at A and free at B. Any type of support at B makes the beam statically indeterminate.

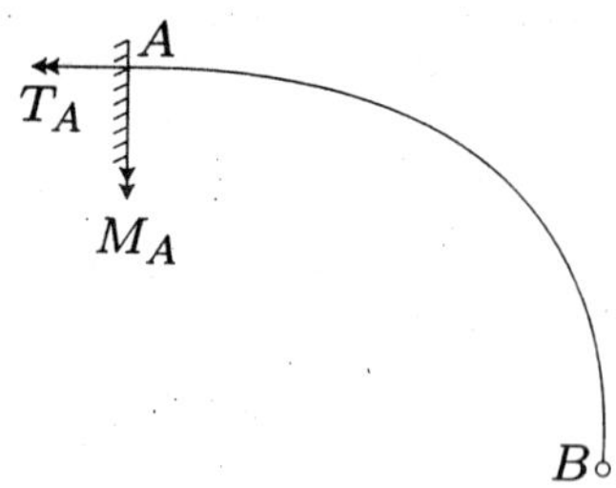

Fig. 12.22: Curved beam with spherical seating

The simples support at B is a spherical seating which will permit rotation in any direction but prevent any vertical deflection at B. This gives the member shown in Fig. 12.22, in which the reactions are a shear force, a bending moment and a twisting moment at A and a shear force at B. There are four reactions in all and as static equilibrium equations give only three of these, the fourth must be found by strain energy, slope deflection on any other method, used in redundant frame analysis.

The more usual types of support is shown in Fig. 12.23 where the end B is completely built in, thus providing not only restraint against deflection but also against change of slope in any direction. There are, therefore, three reactions at B—a shear force, a bending moment and a twisting moment. These combined with similar reactions at A give six reactions in all, thus making the structure statically indeterminate to the third degree.

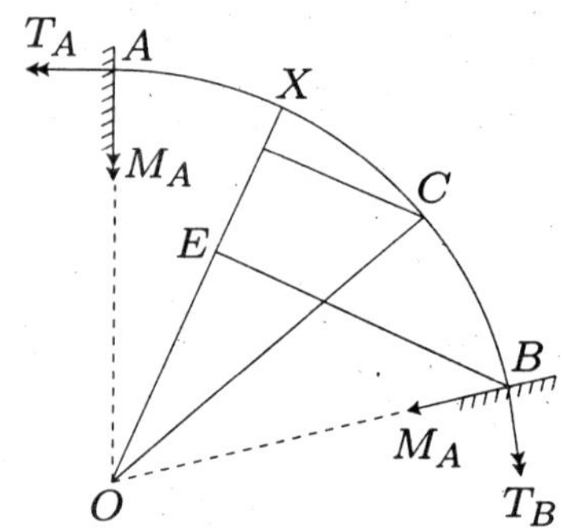

Fig. 12.23: Curved beam with both ends built in

A slope deflection method of solution to this redundant problem was published by A.H. Gibson and E.C. Ritchie in the year 1914, for the particular case of the bow girder, circular in plan. It can be extended to the general case of a girder of any plan form, but the integrations must then be dealt with graphically.

A solution based on strain energy method was given by A.J.S. Pippard and F.L.Barrow in the year 1926. This method will be discussed here as it is considered as the neater solution.

As with all redundant structures, the properties of the section must be known before the analysis can be done. For analysis both the torsional and flexural rigidities are required.

For the bow girder shown in Fig. 12.23, if M_B, T_B and V_B are three indeterminate reactions at B, from strain energy theorem considering perfect fixity at B, it follows that,

$$\frac{\partial U}{\partial M_B} = \frac{\partial U}{\partial T_B} = \frac{\partial U}{\partial V_B} = 0, \tag{12.22}$$

where $U =$ total strain energy in the girder.

The total strain energy U consists of the energy due to flexure, torsion and shear. The last one can generally be neglected, but as the energies due to flexure and torsion are relatively of the same order, effect of one cannot be neglected relative to the other. If U_f and U_t are the strain energies due to flexure and torsion respectively, then

$$\left.\begin{aligned} &\frac{\partial U}{\partial M_B} = 0, \quad \text{i.e.,} \quad \frac{\partial U_f}{\partial M_B} + \frac{\partial U_t}{\delta M_B} = 0 \\ &\frac{\partial U}{\partial T_B} = 0, \quad \text{i.e.,} \quad \frac{\partial U_f}{\partial T_B} + \frac{\partial U_t}{\partial T_B} = 0 \\ \text{and} \quad &\frac{\partial U}{\partial V_B} = 0, \quad \text{i.e.,} \quad \frac{\partial U_f}{\partial V_B} + \frac{\partial U_t}{\partial V_B} = 0. \end{aligned}\right] \tag{12.23}$$

If M_x and T_x are the bending moment and twisting moment at any section, taking EI and GJ as the flexural and torsional rigidities, relations in Eq. (12.23) can be expressed as

$$\int \frac{M_x}{EI}\frac{\partial M_x}{\partial M_B}ds + \int \frac{T_x}{GJ}\frac{\partial T_x}{\partial M_B}ds = 0 \tag{12.24}$$

$$\int \frac{M_x}{EI}\frac{\partial M_x}{\partial T_B}ds + \int \frac{T_x}{GJ}\frac{\partial T_x}{\partial T_B}ds = 0 \tag{12.25}$$

and $$\int \frac{M_x}{EI}\frac{\partial M_x}{\partial V_B}ds + \int \frac{T_x}{GJ}\frac{\partial T_x}{\partial V_B}ds = 0. \tag{12.26}$$

Solving Eqs. (12.24), (12.25) and (12.26) simultaneously, the indeterminacies M_B, T_B and V_B can be obtained.

In obtaining expressions for M_x, T_x and V_x, a convention of sign must be adopted. M_x will be considered positive for a sagging bending moment. A positive torque at X is the one that tends to turn the section clockwise and positive shear force at X is one that tends to raise the viewed section.

Case I. Circular Bow Girder with Concentrated Load

Fig. 12.24 represents a circular bow girder of uniform section firmly built in at A and B and carrying a load P as shown. The angle subtended at the centre of the arc is ϕ and load acts at an angular distance α from OB. Considering M_B, T_B and V_B as the three indeterminate reactions at B at an angular distance θ from OB, i.e. at X,

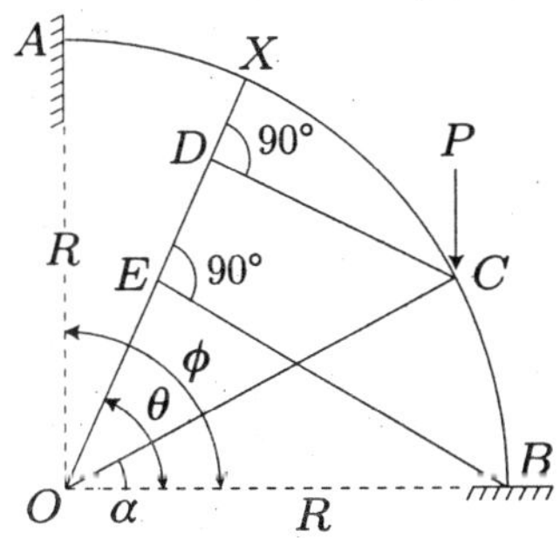

Fig. 12.24: Circular Bow Girder with concentrated load

$$\left.\begin{aligned} M_x &= M_B\cos\theta - T_B\sin\theta - V_B(BE) + P(CD) \\ T_x &= M_B\sin\theta + T_B\cos\theta - V_B(EX) + P(DX). \end{aligned}\right] \tag{12.27}$$

For $\theta \le \alpha$ terms with P are to be omitted.

From Eq. (12.27),

$$\frac{\partial M_x}{\partial M_B} = \cos\theta; \frac{\partial T_x}{\partial M_B} = \sin\theta$$

$$\frac{\partial M_x}{\partial T_B} = -\sin\theta; \frac{\partial T_x}{\partial T_B} = \cos\theta$$

$$\frac{\partial M_x}{\partial V_B} = -BE = -R\sin\theta$$

$$\frac{\partial T_x}{\partial V_B} = -EX = -R[1-\cos\theta].$$

Also, $CD = R\sin(\theta - \alpha)$ and $DX = R[1 - \cos(\theta - \alpha)]$.

As EI and GJ are constants Eqs. (12.24) (12.25) and (12.26) then become

$$\frac{1}{EI}\left[\int_0^\alpha \{M_B\cos\theta - T_B\sin\theta - V_B(R\sin\theta)\}(\cos\theta)(Rd\theta) + \int_\alpha^\phi \{M_B\cos\theta - T_B\sin\theta - V_B(R\sin\theta) + PR\sin(\theta-\alpha)\}(\cos\theta)(Rd\theta)\right] + \frac{1}{GJ}\left[\int_0^\alpha \{M_B\sin\theta + T_B\cos\theta - V_B(R)(1-\cos\theta)\}(\sin\theta)(Rd\theta) + \int_\alpha^\phi \{M_B\sin\theta + T_B\cos\theta - V_B(R)(1-\cos\theta) + P(R)[1-\cos(\theta-\alpha)\}(\sin\theta)(Rd\theta)]\right] = 0$$

i.e.,

$$\frac{1}{EI}\left[\int_0^\phi \{M_B\cos\theta - T_B\sin\theta - V_B(R\sin\theta)\}(\cos\theta)(Rd\theta) + \int_\alpha^\phi \{PR\sin(\theta-\alpha)(\cos\theta)(Rd\theta)\}\right] + \frac{1}{GJ}\left[\int_0^\phi \{M_B\sin\theta + T_B\sin\theta - V_B(R)(1-\cos\theta)\}(\sin\theta)(Rd\theta) + \int_\alpha^\phi \{PR[1-\cos(\theta-\alpha)](\sin\theta)(Rd\theta)\}\right] = 0. \quad (12.28)$$

$$\frac{1}{EI}\left[\int_0^\phi \{M_B\cos\theta - T_B\sin\theta - V_B(R)(\sin\theta)\}(-\sin\theta)(Rd\theta) + \int_\alpha^\phi PR\sin(\theta-\alpha)(-\sin\theta)(Rd\theta)\right] + \frac{1}{GJ}\left[\int_0^\phi \{M_B\sin\theta + T_B\cos\theta - V_BR(1-\cos\theta)\}(\cos\theta)(Rd\theta) + \int_\alpha^\phi PR\{1-\cos(\theta-\alpha)\}(\cos\theta(Rd\theta)\right] = 0 \quad (12.29)$$

and

$$\frac{1}{EI}\left[\int_0^\phi \{M_B\cos\theta - T_B\sin\theta - V_B(R\sin\theta)\}(-R\sin\theta)(Rd\theta) + \int_\alpha^\phi PR\sin(\theta-\alpha)(-R\sin\theta)(Rd\theta)\right] + \frac{1}{GJ}\left[\int_0^\phi \{M_B\sin\theta + T_B\cos\theta - V_BR(1-\cos\theta)\}\{-R(1-\cos\theta)\}(Rd\theta) + \int_\alpha^\phi PR\{1-\cos(\theta-\alpha)\}\{-R(1-\cos\theta)\}(Rd\theta)\right] = 0. \quad (12.30)$$

Integrating Eqs. (12.28), (12.29) and (12.30),

$$M_B\left[\frac{1}{EI}(2\phi - \sin 2\phi) + \frac{1}{GJ}(2\phi - \sin 2\phi)\right] + T_B\left[\left(\frac{1}{EI} - \frac{1}{GJ}\right)(\cos 2\phi - 1)\right] + V_BR\left[\frac{1}{EI}(\cos 2\phi - 1) + \frac{1}{GJ}(4\cos 2\phi - \cos 2\phi - 3)\right]$$

$$- PR\left[\frac{1}{EI}\{\cos(2\phi-\alpha)+2(\phi-\alpha)\sin\alpha-\cos\alpha\}\right.$$

$$\left.+\frac{1}{GJ}\{4\cos\phi-3\cos\alpha+2(\phi-\alpha)\sin\alpha-\cos(2\phi-\alpha)\right]=0 \qquad (12.31)$$

$$M_B\left(\frac{1}{EI}-\frac{1}{GJ}\right)(\cos 2\phi-1)+T_B\left[\frac{1}{EI}(2\phi-\sin 2\phi)+\frac{1}{GJ}(2\phi+\sin 2\phi)\right]$$

$$+V_BR\left[\frac{1}{EI}(2\phi-\sin 2\phi)-\frac{1}{GJ}(4\sin\phi-\sin 2\phi-2\phi)\right]$$

$$-PR\left[\frac{1}{EI}\{2(\phi-\alpha)\cos\alpha+2\cos\phi\sin(\alpha-\phi)\}\right.$$

$$\left.-\frac{1}{GJ}\{4\cos\phi-3\sin\alpha-2(\phi-\alpha)\cos\alpha-\sin(2\phi-\alpha)\}\right]=0 \qquad (12.32)$$

and $$M_B\left[\frac{1}{EI}(\cos 2\phi-1)+\frac{1}{GJ}(4\cos\phi-\cos 2\phi-3)\right]$$

$$+T_B\left[\frac{1}{EI}(2\phi-\sin 2\phi)-\frac{1}{GJ}(4\sin\phi-\sin 2\phi-2\phi)\right]$$

$$+\frac{V_BR}{4}\left[\frac{1}{EI}(2\phi-\sin 2\phi)+\frac{1}{GJ}(6\phi-8\sin\phi+\sin 2\phi)\right]$$

$$-PR\left[\frac{1}{EI}2(\phi-\alpha)\cos\alpha+2\cos\phi\sin(\alpha-\phi)\right.$$

$$+\frac{1}{GJ}\{4(\phi-\alpha)-4\sin\phi+3\sin\alpha+2(\phi-\alpha)\cos\alpha$$

$$\left.-4\sin(\phi-\alpha)+\sin(2\phi-\alpha)\}\right]=0. \qquad (12.33)$$

Combining Eqs. (12.32) and (12.33),

$$4M_B(1-\cos\phi)+4T_B\sin\phi+4V_BR(\phi-\sin\phi)$$

$$-4PR\{(\phi-\alpha)-\sin(\phi-\alpha)\}=0. \qquad (12.33a)$$

These are the general equations for the circular arc bow girder with a concentrated load at any position.

These equations cannot be readily solved in algebraic terms but, if definite values of EI, GJ, ϕ and α are given, the values of M_B, T_B and R_B can be evaluated.

As a specific case, considering a semicircular bow girder carrying a concentrated load at centre, values of ϕ and α are π and $\pi/2$ respectively.

Eqs. (12.31), (12.32) and (12.33a) then reduce to

$$2\pi M_B\left[\frac{1}{EI}+\frac{1}{GJ}\right]-\frac{8V_BR}{GJ}-PR\left[\frac{\pi}{EI}-\frac{4-\pi}{GJ}\right]=0.$$

$$2\pi T_B+2\pi V_BR-2PR=0 \quad \text{and} \quad -2M_B+\pi V_BR-PR\left(\frac{\pi}{2}-1\right)=0.$$

Solving, $V_B=\dfrac{P}{2}$: $M_B=\dfrac{PR}{2}$ and $T_B=PR\left(\dfrac{1}{\pi}-\dfrac{1}{2}\right)$.

The bending moment and twisting moment diagrams for the specific case are shown in Fig. 12.25.

For the specific case, however, it is not necessary to consider V_B as unknown since from symmetry $V_B = \frac{P}{2}$.

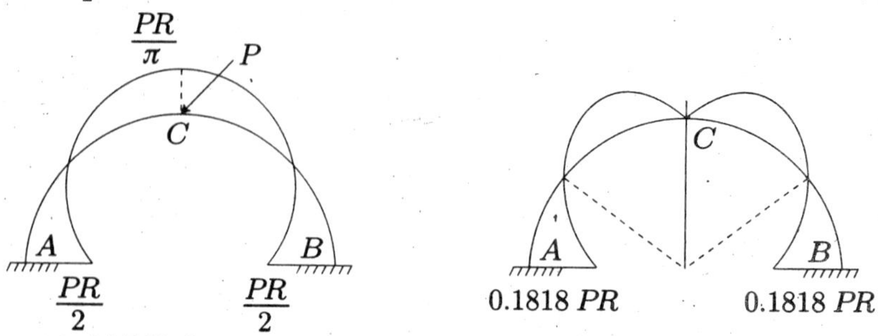

(a) B.M. diagram **(b) Torque diagram**

Fig. 12.25: Semicircular Bow Girder with concentrated load at centre

Case II. Circular Arc Bow Girder with Distribution Load

If the girder carries a uniformly distributed load of intensity w per unit length, the from of equations naturally differs from those of a point load. The coefficients of the terms containing M_B, T_B and R_B in Eqns. (12.27), (12.28) and (12.29) remain unchanged but the terms in P are replaced by terms in w.

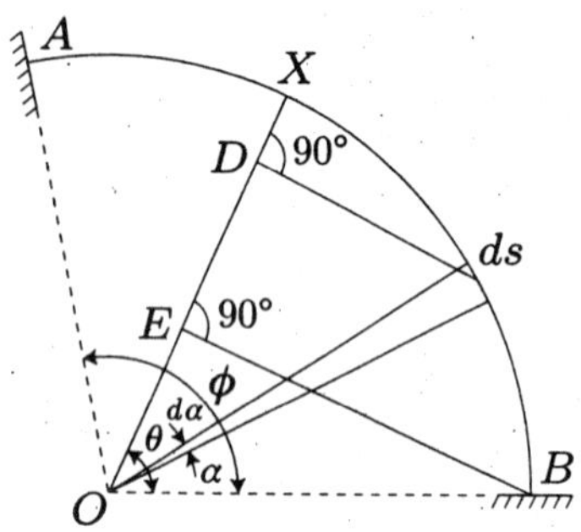

Fig. 12.26: Circular Bow Girder with uniformly distributed load

Considering the girder shown in plan in Fig. 12.26, carrying a uniformly distributed load of w per unit length, if ΔM_x and ΔT_x be the bending moment and twisting moment at any point X due to the element of load wds at C, the angular distance of C from OB being α, then

$$\Delta M_x = wds(CD) = wR^2 \sin(\theta - \alpha)d\alpha \text{ and}$$

$$\Delta T_x = wds(DX) = wR^2[1 - \cos(\theta - \alpha)]d\alpha.$$

The bending moment and twisting moment at X due to distributed load between B and X are then

$$M_x = wR^2 \int_0^\theta \sin(\theta - \alpha)d\alpha = wR^2(1 - \cos\theta)$$

and $$T_x = wR^2 \int_0^\theta [1 - \cos(\theta - \alpha)]d\alpha = wR^2(\theta - \sin\theta).$$

The terms with P in Eqs. (12.28), (12.29) and (12.30) are to be replaced by the following. In Eqs. (12.28),

$$wR^2 \left[\frac{1}{EI} \int_0^\phi (1 - \cos\theta)\cos\theta(d\theta) + \frac{1}{GJ} \int_0^\phi (\theta - \sin\theta)\sin\theta(d\theta) \right]$$

$$= \frac{wR^2}{4} \left[\frac{1}{EI}(4\sin\phi - 2\phi - \sin 2\phi) + \frac{1}{GJ}(4\sin\phi - 2\phi + \sin 2\phi - 4\phi\cos\phi) \right].$$

In Eqn. (12.29),

$$wR^2\left[\frac{1}{EI}\int_0^{\phi}(1-\cos\theta)(\theta-\sin\theta)d+\frac{1}{GJ}\int_0^{\phi}(\theta-\sin\theta)\cos\theta\cdot d\theta\right]$$

$$=\frac{wR^2}{4}\left[\frac{1}{EI}(4\cos\phi-\cos 2\phi-3)+\frac{1}{GJ}(\cos 2\phi+4\cos\phi-5+4\phi\sin\phi)\right].$$

In Eqn. (12.30),

$$wR^2\left[\frac{1}{EI}\int_0^{\phi}(1-\cos\theta)(-\sin\theta)d\theta+\frac{1}{GJ}\int_0^{\phi}(\theta-\sin\theta)(\cos\theta-1)d\theta\right]$$

$$=\frac{wR^2}{4}\left[\frac{1}{EI}(4\cos\phi-\cos 2\phi-3)+\frac{1}{GJ}(\cos 2\phi-4+4\phi\sin\phi-2\phi^2)\right].$$

In this case, of course, there is no need of the third expression with w, since Eqn. (12.30) is not necessary, as V_B is known from symmetry, it being $\frac{wR\phi}{2}$.

The equations do not give a simple solution in general term even though R_B is known from symmetry. However, values of bending moment and twisting moment are given by

$$M_x = M_B\cos\theta - T_B\sin\theta - \frac{wR^2\phi}{2}\sin\theta + wR^2(1-\cos\theta)$$

and $$T_x = M_B\sin\theta + T_B\cos\theta - \frac{wR^2\phi}{\theta}(1-\cos\theta) + wR^2(\theta-\sin\theta).$$

The maximum twisting moment is obtained by differentiating T_x with respect to θ and equating to zero giving,

$$M_B\cos\theta - T_B\sin\theta - \frac{wR^2\phi}{2}\sin\theta + wR^2(1-\cos\theta) = 0,$$

for the value of θ for maximum value of T_x.

The expression on the left-hand side of the above equation is the same as the expression for M_x. Hence, for a maximum value of T_x, the condition is that $M_x = 0$. Thus, the maximum twisting moment occurs at the point of contraflexure.

For the special case when $\phi = \pi$, i.e., for a semicircular girder, the equations reduce to

$$2M_B\left[\frac{1}{EI}+\frac{1}{GJ}\right]-\frac{8V_BR}{GJ}-2\pi wR^2\left[\frac{1}{EI}-\frac{1}{GJ}\right]=0$$

$$2\pi T_B + 2\pi V_BR - 8wR^2 = 0 \quad \text{and} \quad -8M_B + 4\pi V_BR + 2wR^2(4-\pi^2) = 0.$$

Solving, $V_B = \dfrac{\pi wR}{2}$; $M_B = wR^2$

and $$T_B = wR^2\left[\frac{4}{\pi}-\frac{\pi}{2}\right].$$

Case III. Bow Girder having Symmetry about an Axis

In case, a bow girder is symmetrical and also symmetrically loaded about one axis, the analysis is considerably simplified. Fig. 12.27 shows a bow girder which is symmetrical about CD, the loading also being symmetrical about this line as shown in the figure.

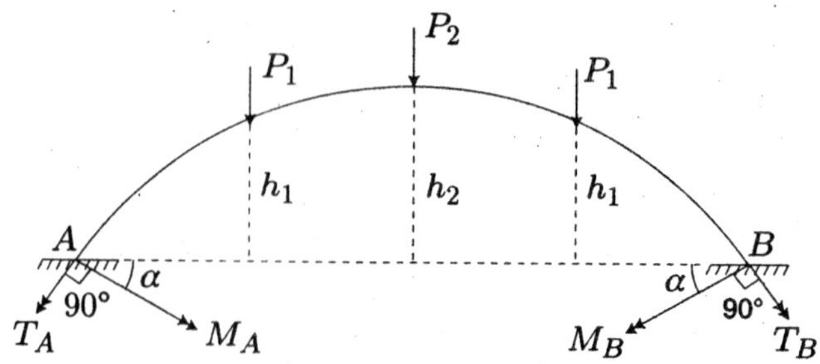

Fig. 12.27: Symmetrical Bow Girder

Reactions R_A and R_B, bending moments M_A and M_B as well as torsion moments T_A and T_B are equal. Equating vertical forces on the girder,

$$V_A + V_B = (2P_1 + P_2), \quad \text{whence} \quad V_A = V_B = \frac{2P_1 + P_2}{2}.$$

Taking moments about the line AB, $(M_A + M_B)\cos\alpha - (T_A + T_B)\sin\alpha = 2P_1h_1 + P_2h_2$,

$$\text{whence} \quad M_A = M_B = \frac{2P_1h_1 + P_2h_2}{2}\sec\alpha + T_B\tan\alpha,$$

α being the angle between AB and the normal to the girder at A and B.

The only indeterminate term for the case is only the twisting moment T_B. Equations derived from conditions $\frac{\partial U}{\partial M_B} = 0$ and $\frac{\partial U}{\partial V_B} = 0$ are not required and for obtaining the solution, it is only necessary to substitute the above values of M_B and R_B in the equation $\frac{\partial U}{\partial T_B} = 0$.

Applying these results in case of a circular bow girder of uniform section, subtending an angle ϕ at the centre and subjected to a single concentrated load P at centre,

$$\alpha = \frac{\pi - \phi}{2} \quad \text{and} \quad V_A = V_B = \frac{P}{2}$$

$$M_B = \frac{PR}{2}\left(1 - \cos\frac{\phi}{2}\right)\operatorname{cosec}\frac{\phi}{2} + T_B\cot\frac{\phi}{2}.$$

On substituting for V_B and M_B and α, Eqs. (12.32) becomes

$$2T_B\left[\frac{1}{EI}(\phi + \sin\phi) + \frac{1}{GJ}(\phi - \sin\phi)\right] + PR\left(\frac{1}{EI} + \frac{1}{GJ}\right)\left(1 - \cos\frac{\phi}{2}\right)\left(2\sin\frac{\phi}{2} - \phi\right) = 0.$$

For the special case of the semicircular girder, $\phi = \pi$, and

$$V_B = \frac{P}{2}$$

$$T_B = PR\left(\frac{1}{\pi} - \frac{1}{2}\right) \quad \text{and} \quad M_B = \frac{PR}{2}.$$

Another way of taking advantage of symmetry is to consider the bending moment M_C, twisting moment T_C and shear force V_C at C, the section of symmetry as the indeterminate reactions. From symmetry, it follows that, $T_C = 0$ and $V_C = 0$, so that M_C is the only unknown, which can be obtained from $\frac{\partial U}{\partial M_C} = 0$.

Case IV. Bow Girder of Any Shape

In case of bow girders of shapes which do not follow definite patterns, as in case of circular arc, the necessary integrals cannot be solved by direct mathematical methods and it is necessary to obtain their values graphically.

Fig. 12.28(a) shows a Bow girder AB with a concentrated load at C. Eqns. (12.28),(12.29) and (12.30) can be expressed as

$$\frac{1}{EI}\left[\int_B^A \{M_B \cos^2\theta ds - \frac{T_B}{2}\sin 2\theta ds - V_B(BE)\cos\theta ds\} + P\int_C^A (CD)\cos\theta ds\right]$$
$$+\frac{1}{GJ}\left[\int_B^A \left\{M_B \sin^2\theta ds + \frac{T_B}{2}\sin 2\theta ds - V_B(EX)\sin\theta ds\right\}\right.$$
$$\left.+P\int_C^A (DX)\sin\theta ds\right] = 0 \qquad (12.34)$$

$$\frac{1}{EI}\left[\int_B^A \left\{-\frac{M_B}{2}\sin 2\theta ds + T_B\sin^2\theta ds + T_B \sin^2\theta ds + V_B(BE)\sin\theta ds\right\}\right.$$
$$\left.-P\int_C^A (CD)\sin\theta ds\right] + \frac{1}{GJ}\left[\int_B^A \left\{\frac{M_B}{2}\sin 2\theta ds + T_B\cos^2\theta ds - V_B(EX)\cos\theta ds\right\}\right.$$
$$\left.+P\int_C^A (DX)\cos\theta ds\right] = 0 \qquad (12.35)$$

and $$\frac{1}{EI}\left[\int_B^A \{-M_B(BE)\cos\theta ds + T_B(BE)\sin\theta ds + V_B(BE)^2 ds\} - P\int_C^A (CD)(BE)ds\right]$$
$$+\frac{1}{GJ}\left[\int_B^A \{-M_B(EX)\sin\theta ds - T_B(EX)\cos\theta ds + V_B(EX)^2 ds\}\right.$$
$$\left.-P\int_C^A (DX)(EX)ds\right] = 0. \qquad (12.36)$$

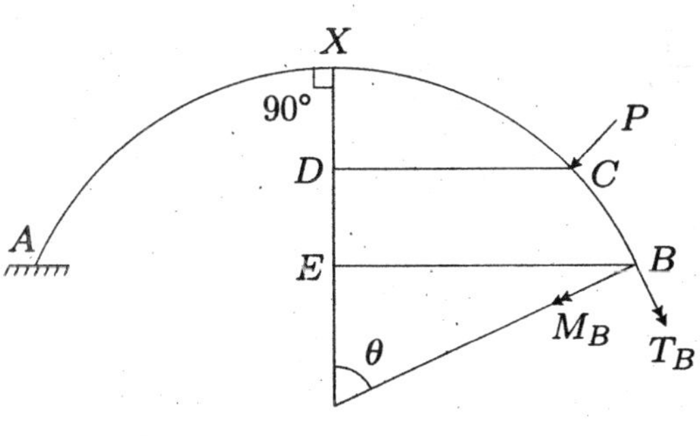

(a) Bow Girder with a concentrated load

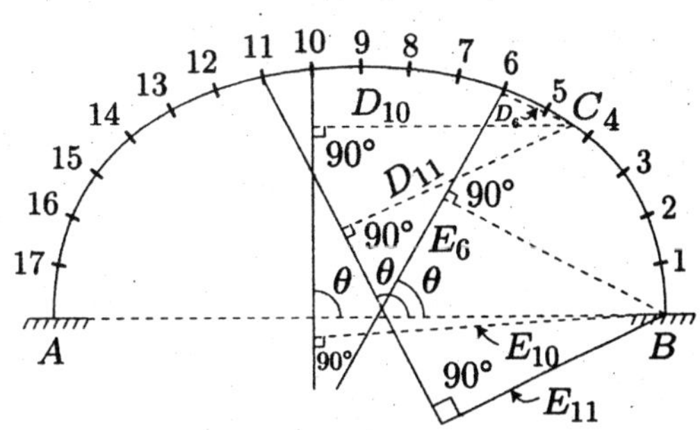

(b) Graphical method

Fig. 12.28: Bow Girder with any shape

The terms inside the integral are dependent on the shape of the girders and are the same whatever is the position of the load. If more than one load acts on the girder, each will appear in the several equations. For example, if loads denoted by $P_1, P_2, \ldots, P_n$ act at points denoted by $1, 2, \ldots, n$, then the term giving the coefficient of load P, i.e., Eq. (12.34) has to be replaced by

$$P_1\left[\frac{1}{EI}\int_1^A C_1D_1\cos\theta ds+\frac{1}{GJ}\int_1^A D_1X\sin\theta ds\right]$$

$$+P_2\left[\frac{1}{EI}\int_2^A C_2D_2\cos\theta ds+\frac{1}{GJ}\int_2^A D_2X\sin\theta ds\right]+\cdots$$

$$+P_n\left[\frac{1}{EI}\int_n^A C_nD_n\cos\theta ds+\int_n^A D_nX\sin\theta ds\right],$$

where $C_1D_1, \cdots, C_nD_n$ and $D_1X \cdots D_nX$ represent the lengths corresponding to CD, DX drawn for $1, 2, \ldots, n$. Similar terms replace the coefficients of P in the other two equations.

In using the graphical method, the parameter is divided into a number of equal parts, which would be as small as practicable, since the greater the member, the more accurate will be the results obtained. Fig. 12.28(b) shows 18 equal divisions of the bow girders, the parts being marked 1 to 17. At each of these points on the diagram, normals to the curve are drawn and from B perpendiculars to each of these normals are drawn. In Fig. 12.28(b) several such normals and perpendiculars for points 6,10 and 11 are shown. Perpendicular from C, the point of application of the load are also drawn on individual normals as shown in Fig. 12.28(a). Values of θ for the points 6,10 and 11 are also shown in the figure. Since the terms in E appear between C and A the perpendicular from C are only required for points on the curve living between C and A.

Eqs. (12.34), (12.35) and (12.36) show that the integrals to be evaluated involve a number of trigonometrical functions of θ and the lengths BX, EX, CD and DX.

The various terms of the integrals involved are as follows :

(i) $\int_B^A \cos^2\theta ds$; (ii) $\int_B^A \sin^2\theta ds$; (iii) $\int_B^A \sin 2\theta ds$;

(iv) $\int_B^A BE\cos\theta ds$; (v) $\int_B^A (EX)\sin\theta ds$; (vi) $\int_B^A BE\sin\theta ds$;

(vii) $\int_B^A EX\cos\theta ds$; (viii) $\int_B^A (BE)^2 ds$; (ix) $\int_B^A (EX)^2 ds$;

(x) $\int_C^A (CD)\cos\theta ds$; (xi) $\int_C^A (DX)\sin\theta ds$; (xii) $\int_C^A (CD)\sin\theta ds$;

(xiii) $\int_C^A (DX)\cos\theta ds$; (xiv) $\int_C^A (CD)BE ds$; (xv) $\int_C^A (DX)(EX) ds$;

Simpson's rule for even number of division gives the area enclosed by a boundary as

$$A_S = \frac{h}{3}[y_0 + 4y_1 + 2y_2 + 4y_3 + 2y_4 + \cdots + 2y_{n-2} + 4y_{n-1} + y_n],$$

where $y_0, y_1, y_2, \cdots, y_n$ are the lengths of a series of equally spaced parallel chords and h their distance apart.

In this case, $h = ds$. The values of trigonometrical functions may be obtained from trigonometrical Tables. Calculation of summations may be done in Tabular form as shown in Table 12.3, taking $\int_B^A BE \sin\theta ds$.

Table 12.3. Table For Evaluation of Integral

Point	BE	θ	$\sin\theta$	$BE\sin\theta$	Area by Simpson's Rule	
					Multiplication factor	$BE\sin\theta$ (Col.6)
1	2	3	4	5	6	7
B					1	
1					4	
2					2	
17					4	
B					1	
$\int_B^A BE \sin\theta ds = \frac{ds}{3}\left(\sum \text{ in column } 7\right)$						$\sum$

In case of terms with limits between C and A, as for $\int_C^A (CD)\cos\theta ds$, only points between C and A are to be considered and calculations are done as in case of Table 12.3. If there is odd number of divisions between C and A, Simpsons $\frac{3}{8}$th rule which gives areas as

$$A = \frac{3}{8}h[y_0 + 3y_1 + 3y_2 + \cdots + 3y_{n-1} + y_n] \text{ can be used.}$$

Substituting the values of the integrals in the relevant equations and solving simultaneously, the indeterminate reactions can be obtained.

If the bow girder has symmetry, then advantages of symmetry help in reducing the number of indeterminate reactions.

12.11 Moment Distribution Method of Solution

Beams subjected to vertical load and framing into girders generally exert twisting moment on the girder and the system is usually subjected to combination of torsion and bending. Moment distribution method can be extended for solution of such cases.

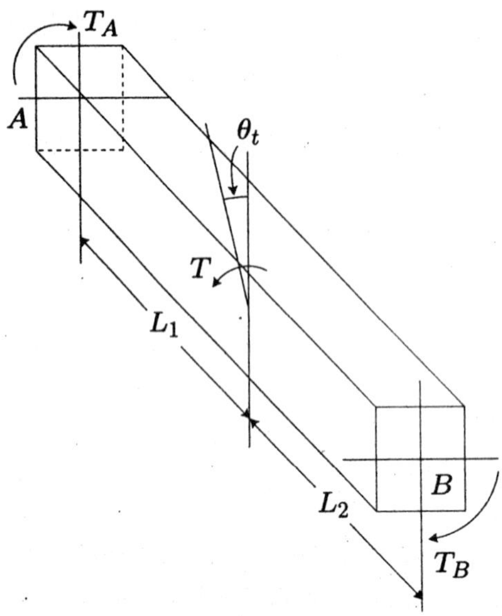

Fig. 12.29: Bar under twisting moment

Fig. 12.29 shows a bar of solid cross section with an external twisting moment. The bar is supported at the ends in such a way that there can be no rotation and no displacement of the end cross section. The reactions T_A and T_B are obtained from the consideration that the rotation of sections an infinitesimal distance to either side of T must be the same.

Thus, $$\theta_t = \frac{T_A L_1}{GJ} = \frac{T_B L_2}{GJ}. \tag{12.37a}$$

From equilibrium $T = T_A + T_B$. (12.37b)

Solving for T_A and T_B,

$$T_A = \frac{TL_2}{L_1 + L_2}$$

and $$T_B = \frac{TL_1}{L_1 + L_2}. \tag{12.37c}$$

In moment distribution T_A and T_B represent the fixed end twisting moments due to applied twisting moment T acting somewhere along the length of the bar. The distribution factors can be different in terms of the particular value of T which make the rotation θ_t unity. From Eq. (12.37c), this leads to,

$$\theta_t = \frac{T_A L_1}{GJ} = \frac{TL_1 L_2}{GJ(L_1 + L_2)} = 1 \text{ radian},$$

whence $T = GJ\dfrac{L_1 + L_2}{L_1 L_2}$.

The part of T, which is distributed to each side of the bar is given by

$$T_A = \frac{TL_2}{L_1 + L_2} = \frac{GJ}{L_1} \tag{12.37d}$$

and $$T_B = \frac{TL_1}{L_1 + L_2} = \frac{GJ}{L_2}. \tag{12.37e}$$

The torque stiffness factors by definition are

$$K_{TCA} = \frac{GJ}{L_1} \tag{12.38a}$$

and $$K_{TCB} = \frac{GJ}{L_2}. \tag{12.38b}$$

The torque carry over factors to ends A and B are both unity provided these ends are restrained against rotation.

For the case of a beam CE framing into a girder AB and CO as shown in Fig. 12.30, if joints C and E are initially considered as constrained against rotation, i.e., $\theta_t = 0$, fixed-end moments will be developed in beam CE. If joint C be released in accordance with moment distribution procedure, joint C will rotate until the fixed moment at C is balanced by the change in bending moment on CE and twisting moments T_{BA} and T_{BC} developed in the girder AB. The bending moment stiffness factor for CE is given by $\left(\frac{4EI}{L}\right)_{CE} = K_{CE}$.

Thus $$M_{CE} = M_{FCE}\,\frac{K_{CE}}{K_{TCA}+K_{TCB}+K_{CE}} \quad (12.39a)$$

$$T_{CA} = M_{FCE}\,\frac{K_{TCA}}{K_{TCA}+K_{TCB}+K_{CE}} \quad (12.39b)$$

and $$T_{CB} = M_{FCE}\,\frac{K_{TCB}}{K_{TCA}+K_{TCB}+K_{CE}}. \quad (12.39c)$$

Values of the distribution factors are given as

$$\text{D.F. for } M_{CE},\ \text{i.e., } D.F._{CE} = \frac{K_{CE}}{K_{TCA}+K_{TCB}+K_{CE}}$$

$$\text{D.F. for } T_{CA},\ \text{i.e., } D.F._{TCA} = \frac{K_{TCA}}{K_{TCA}+K_{TCB}+K_{CE}}$$

and $$\text{D.F. for } T_{CB},\ \text{i.e., } D.F._{TCB} = \frac{K_{TCB}}{K_{TCA}+K_{TCB}+K_{CE}}.$$

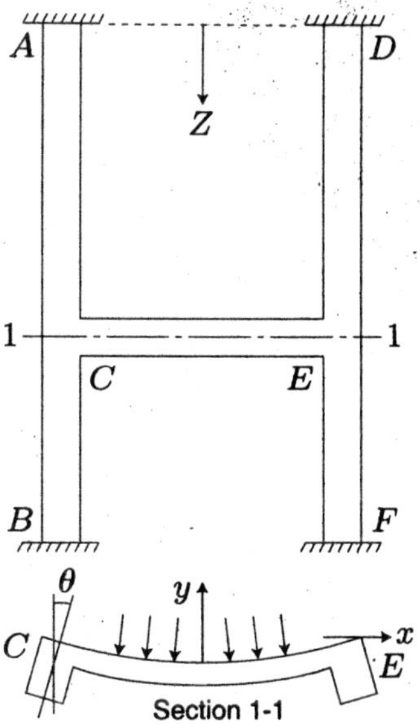

Fig. 12.30: Beam supported on girder

For vertical loads on span CE as in Fig. 12.30, the moment distribution procedure applied in accordance with usual practice will finally result in a balance between bending moments in CE and twisting moments in AB and DF. Moments induced by rotations in plane parallel to the X-Y plane in Fig. 12.30 will be balanced. It will then be to consider rotations in planes parallel to the Y-Z plane. These will result in bending moments in girder AB and DF. If the loading is not symmetric, twisting moments will also be induced in beam CE. These may or may not be of the same order of magnitude as the twisting moments are considered in the original moment distribution. To obtain their values and effects, the moment distribution procedure must be applied in planes parallel to the X-Y and Y-Z planes and the final results are obtained by superposition of effects. In addition to this, there may be some effects due to differential settlement of supports.

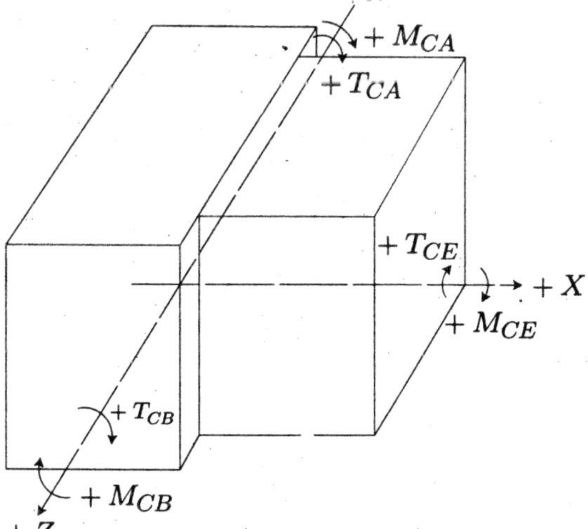

Fig. 12.31: Sign convention for moments

As sign convention, clockwise joint moments will be considered positive and consequently twisting moments will be considered positive if they act clockwise on a positive face of a free body of the joint and if they act anticlockwise on a negative face as shown in Fig. 12.31.

Use of moment distribution procedure for finding bending moments and torsion in frames will be illustrated by means of Examples.

Example 12.9. *Fig. 12.32(a) shows a bent in plan with support at A and C fixed. The bent is subjected to a vertical load of 100 kN at D, centre of AB. Analyse the bent and draw B.M., S.F. and Torsional moment diagram. Assume $GJ = 0.8EI$.*

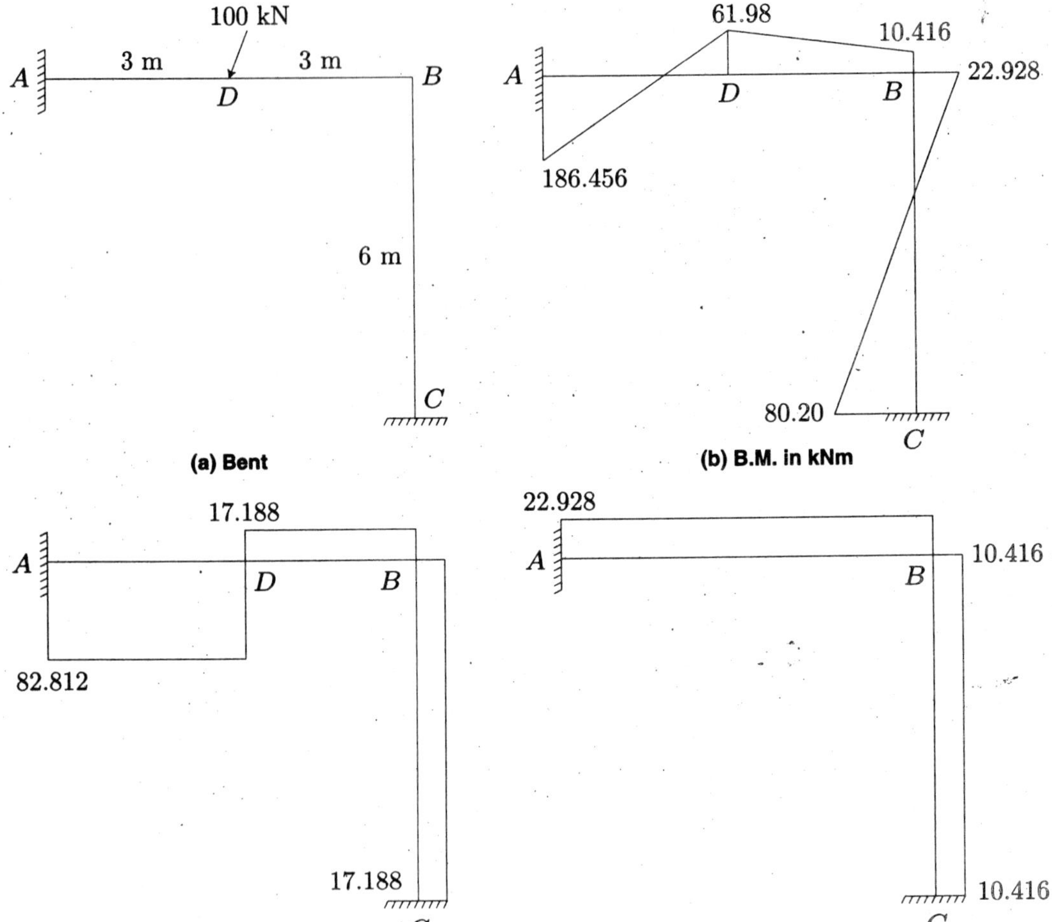

(a) Bent (b) B.M. in kNm

(c) S.F. in kN (d) Twisting moment in kNm

Fig. 12.32: For Example 12.9

Solution : Considering the reactions act C, i.e., M_C, T_C and R_C as the redundants, expressions for the B.M. and Torsional moments can be obtained. Thus :

For CB, $M = M_C - V_C(x)$, whence $\dfrac{\partial M}{\partial M_C} = 1; \dfrac{\partial M}{\partial T_C} = 0$

and $\dfrac{\partial M}{\partial V_C} = (-x)$ and $T = T_C$, for CB

whence $\dfrac{\partial T}{\partial M_C} = 0; \dfrac{\partial T}{\partial T_C} = 1$ and $\dfrac{\partial T}{\partial V_C} = 0.$

For, BD, $M = T_C + V_C(x)$, whence $\dfrac{\partial M}{\partial M_C} = 0, \dfrac{\partial M}{\partial T_C} = 1$ and $\dfrac{\partial M}{\partial V_C} = (x)$

and $T = T_C - 6V_C$, whence $\dfrac{\partial T}{\partial M_C} = 0, \dfrac{\partial T}{\partial T_C} = 1$ and $\dfrac{\partial T}{\partial V_C} = -6$

For $DA, M = T_C + V_C(x) - 100(x-3)$, whence

$$\frac{\partial M}{\partial M_C} = 0; \frac{\partial M}{\partial T_C} = 1 \quad \text{and} \quad \frac{\partial M}{\partial V_C} = (x)$$

and $T = M_C - 6V_C$, whence $\dfrac{\partial T}{\partial M_C} = 1$; $\dfrac{\partial T}{\partial T_C} = 1$ and $\dfrac{\partial T}{\partial V_C} = (-6)$.

From $\dfrac{\partial U}{\partial M_C} = 0$, it follows that,

$$\int_0^6 \frac{[M_c - V_c x]}{EI}(dx) + \int_0^6 \frac{M_C - 6V_C}{GJ} dx = 0,$$

whence $10.8M_C - 50.4V_C = 0$. (i)

From $\dfrac{\partial U}{\partial T_C} = 0$, it follows that,

$$\int_0^3 \frac{T_C + V_C x}{EI}(1)dx + \int_3^6 \frac{T_C + V_C x - 100(x-3)}{EI}(1)dx + \int_0^6 \frac{T_C}{GJ}(1)dx = 0,$$

whence $10.80T_C + 14.40V_C - 360 = 0$. (ii)

From $\dfrac{\partial U}{\partial V_C} = 0$, it follows that,

$$\int_0^6 \frac{(M_C - V_C x)(-x)dx}{EI} + \int_0^3 \frac{[T_C + V_C x](x)dx}{EI}$$

$$+ \int_3^6 \frac{[T_C + V_c x - 100(x-3)](+x)(dx)}{EI} + \int_0^6 \frac{[M_C - 6V_C](-6)dx}{GJ} = 0,$$

whence $-50.40M_C + 14.40T_C + 231.2V_C - 180 = 0$. (iii)

Solving, $V_c = 17.188$ kN, $M_C = 80.20$ kN-m and $T_C = 10.416$ kN-m.

Resulting B.M., S.F. and Torsion diagrams are shown in Figs. 12.32(b), 12.32(c) and 12.32(d).

Example 12.10. *The rigid frame AB in plan shown in Fig. 12.33(a), is supported on a spherical bearing capable of providing a vertical reaction at B. Obtain the BM, SF and Twisting Moment diagrams if it is subjected to vertical u.d.l. of w/unit length. Calculate the angle of twist at B.*

Solution : The vertical reaction at B, i.e., V_B is the onlv indeterminate in this case, since $M_B = 0$ and $T_B = 0$.

For $BC,\ M = -\dfrac{wx^2}{2} - V_B(x)$ and $\dfrac{dM}{dV_B} = -(x)$

$T = 0$.

For $CA, M = wL(x) + \dfrac{wx^2}{2} - V_B(x)$ and $\dfrac{dM}{dV_B} = -(x)$

$T = \dfrac{wL^2}{2} - V_B(L)$ and $\dfrac{dT}{dV_B} = (-L)$.

From $\dfrac{dU}{dV_B} = 0,$

$$\int_0^L \frac{\left[\frac{wx^2}{2} - V_B(x)\right]}{EI}(-x)dx + \int_0^L \frac{\left[wLx + \frac{wx^2}{2} - V_B(x)\right]}{EI}(-x)dx + \int_0^L \frac{\left(\frac{wL^2}{2} - V_B L\right)(-L)dx}{GJ} = 0,$$

whence with $\dfrac{GJ}{EI} = 0.80,\ \dfrac{4.6V_B}{3} - \dfrac{11.6wL}{12} = 0$

and $V_B = 0.63wL$

$V_A = 2wL - 0.63wL = 1.37wL$

$M_A = -0.87wL^2$

and $T_A = -0.13wL^2.$

Resulting B.M. S.F. and Twisting Moment diagrams are shown in Figs. 12.33(b) to 12.33(d).

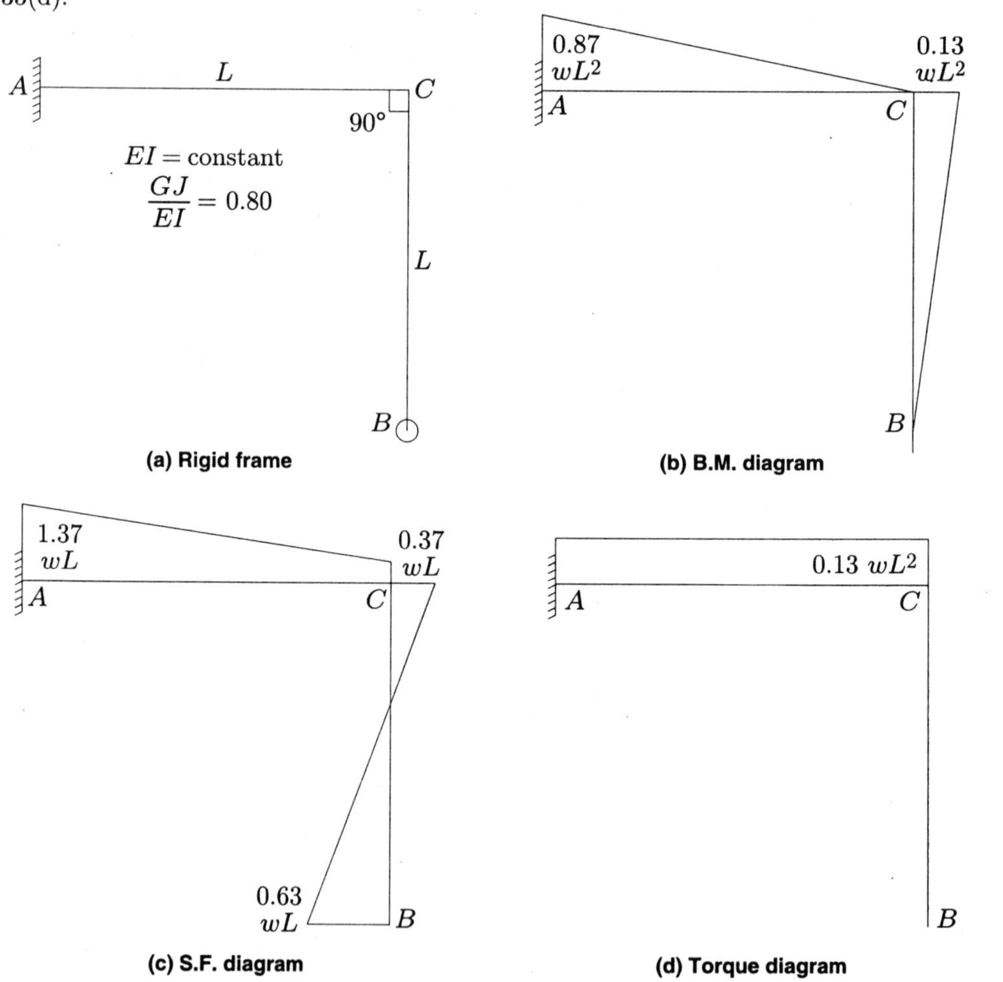

Fig. 12.33: For Example 12.10

For angle of twist at B, considering a unit twisting couple at B.

In BC, $t = 1$ and $m = 0$

In CA, $t = 0$ and $m = 1$.

$$\text{Angle of twist} = \sum \frac{Mmdx}{EI} + \sum \frac{Ttds}{GJ} = \frac{1}{EI} \int_0^L \frac{\left[wLx + \frac{wx^2}{2} - V_B(x)\right](1)dx}{EI}$$

$$= 0.352 \frac{wL^3}{EI}.$$

Example 12.11. *The horizontal rigid bent Girder ABCD in plan consists of circular cross section fixed at ends A and D and is shown in Fig. 12.34. A vertical concentrated load P is placed at E, on the centre of BC. If $G = 0.40E$, analyse the bent and calculate the vertical deflection at E of the bent.*

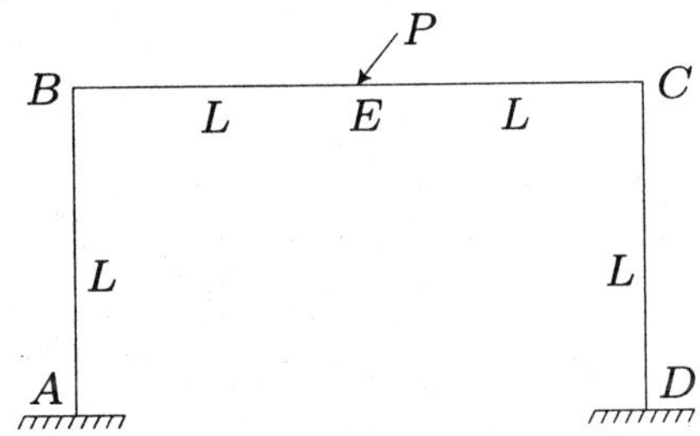

Fig. 12.34: For Example 12.11

Solution : As E is a point of symmetry, shear and torsion there will be zero. If M_E be the moment at E,

For $\quad BE,\ M_x = M_E - \dfrac{P}{2}(x)$ and $T_x = 0$.

For $\quad BA,\ M_y = -\dfrac{Py}{2}$ and $T_y = M_E - \dfrac{PL}{2}$.

From $\quad \dfrac{\partial V}{\partial M_E} = 0$,

$$2\left[\int_0^L \frac{\left[M_E - \frac{P}{2}(x)\right]dx}{EI} + \int_0^L \frac{\left[M_E - \frac{PL}{2}\right]dy}{GJ}\right],$$

whence $\quad \dfrac{M_E(L) - \frac{PL^2}{4}}{EI} + \dfrac{M_E(L) - \frac{PL^2}{2}}{GJ} = 0.$

With $EI = E\left(\frac{\pi}{4}\right)(r)^4$ and $GJ = 0.4E\left(\frac{\pi}{2}\right)(r)^4$, it follows that,

$$M_E(L) - 3.50PL^2 = 0, \quad \text{whence} \quad M_E = \frac{7}{18}PL.$$

For $\quad BC,\ M_x = \dfrac{7}{18}PL - \dfrac{P}{2}(x)$

$T_x = 0$.

For $\quad BA,\ M_y = -\dfrac{Py}{2}$ and $T_y = M_E - \dfrac{PL}{L} = -\dfrac{PL}{9}$.

$$\Delta_E \text{ from, } \frac{\partial U}{\partial P} = 2\left[\int_0^L \frac{P\left(\frac{7}{18}L - \frac{x}{2}\right)^2 dx}{EI} + \int_0^L \frac{\left(-\frac{Py}{2}\right)\left(-\frac{y}{2}\right)dy}{EI} + \int_0^L \frac{\left(-\frac{PL}{9}\right)\left(-\frac{L}{9}\right)dy}{GJ}\right]$$

$$= \frac{90}{324}\frac{PL^3}{EI}, \text{ with } J = 2I$$

and $\quad G = 0.40E$.

Example 12.12. *Fig. 12.35 shows a horizontal rigid bent ACDB in plan subjected to a vertical concentrated load of W at the centre of CD. Obtain an expression for the Twisting moment at one of the supports, if the bent be fixed at the supports.*

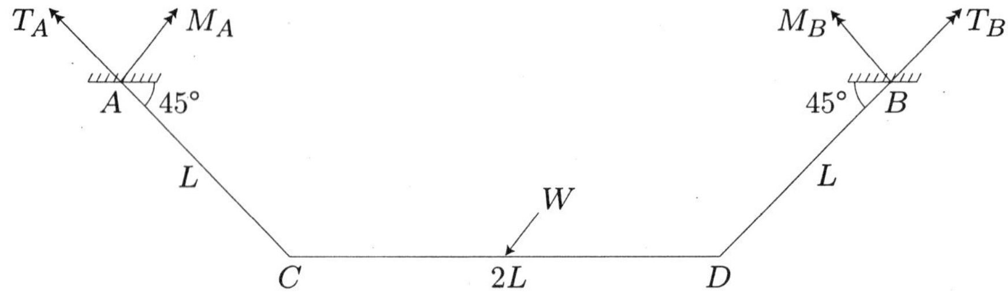

Fig. 12.35: For Example 12.12

Solution : From symmetry, $V_A = V_B = \dfrac{W}{2}$ and $M_A = M_B$ and $T_A = T_B$.

Taking moments about AB,

$$(M_A + M_B)\cos 45^\circ = W\frac{L}{\sqrt{2}} + (T_A + T_B)\sin 45^\circ,$$

whence $\quad M_A = M_B = \dfrac{WL}{2} + T_A.$

In members AC and BD, $M = M_A - \dfrac{W}{2}(x) = \dfrac{WL}{2} + T_A - \dfrac{W}{2}(x)$

and $\quad T = T_A.$

$$\frac{\partial M}{\partial T_A} = 1.$$

In member CE, $M = \left[M_A - \dfrac{W}{2}L\right]\cos 45^\circ + T_A \sin 45^\circ - \dfrac{W}{2}x$

$$= \left[\frac{WL}{2} + T_A - \frac{WL}{2}\right]\cos 45^\circ + T_A\sin 45^\circ - \frac{W}{2}x$$

$$= T_A\sqrt{2} - \frac{W}{2}x, \quad \text{whence} \quad \frac{\partial M}{\partial T_A} = \sqrt{2}.$$

Also $\quad T = \left[\dfrac{WL}{2} + T_A - \dfrac{W}{2}L\right]\sin 45^\circ - T_A \sin 45^\circ = 0 \quad \text{and} \quad \dfrac{\partial T}{\partial T_A} = 0.$

From $\quad \sum \dfrac{M}{EI}\dfrac{\partial M}{\partial T_A}dx + \sum \dfrac{T}{GJ}\dfrac{\partial T}{\partial T_A}dx = 0,$

$$\frac{2}{EI}\left[\int_0^L \left(\frac{WL}{2} + T_A - \frac{W}{2}x\right)(1)dx\right.$$

$$\left. + \int_0^L \left(T_A\sqrt{2} - \frac{W}{2}x\right)\sqrt{2}dx\right] + \frac{2}{GJ}\left[\int_0^L T_A(1)dx\right] = 0.$$

Simplifying, $T_A L\left[\dfrac{3}{EI} + \dfrac{1}{GJ}\right] = \dfrac{WL^2}{4EI}[\sqrt{2} - 1].$

Taking $\dfrac{EI}{GI}$ as α, $T_A = \dfrac{WL}{4}\dfrac{\sqrt{3} - 1}{(3 + \alpha)}.$

Example 12.13. *Fig. 12.36 shows a horizontal rigid bent in plan, fixed at supports A and B. The bent is subjected to a vertical load of 10,000 N at C. If the bent be made up of circular rods and $G = 0.4E$, obtain magnitude of M_A, T_A and V_A.*

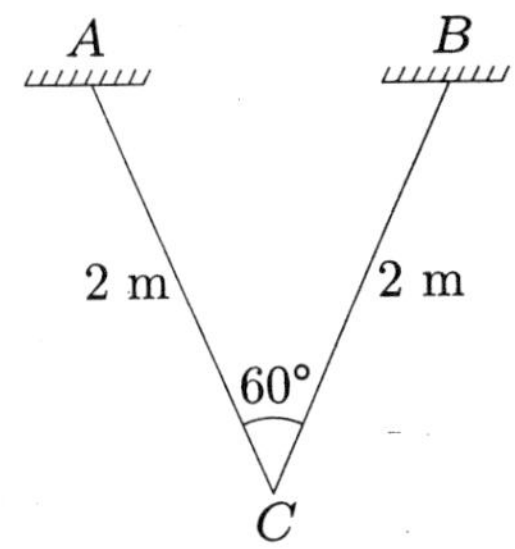

Fig. 12.36: For Example 12.13

Solution : From symmetry $V_A = V_B = \dfrac{10,000}{2} = 5000$ N.

As $M_A = M_B$ and $T_A = T_B$, from symmetry, taking moments about AB gives

$$(M_A + M_B)\cos 30° = 10000(2\cos 30°) + (T_A + T_B)\sin 30°,$$

whence $$M_A = M_B = 10000 + \frac{T_A}{\sqrt{3}}.$$

In member AC and BC,

$$M = M_A - 5000(x) = 10,000 + \frac{T_A}{\sqrt{3}} - 5000(x) \quad \text{and} \quad T = T_A.$$

From $$\sum \frac{M}{EI}\frac{\partial M}{\partial T_A}dx + \sum \frac{T}{GJ}\frac{\partial T}{\partial T_A}dx = 0,$$

$$\frac{2}{EI}\int_0^2 [M_A - 5000x]\frac{1}{\sqrt{3}}dx + \frac{2}{GJ}\int_0^2 \frac{T_A}{GJ}(1)dx = 0.$$

Simplifying, $$\frac{1}{EI}\left[11547 + \frac{2}{3}T_A - 5773.5\right] + \frac{2T_A}{GJ} = 0.$$

For circular bars, $J = 2I$ and as $G = 0.40E$. It follows that

$$T_A = -1823.2 \text{ Nm}$$

$$T_C = T_A \quad \text{and} \quad M_A = 10000 + \frac{T_A}{\sqrt{3}} = 8947.4 \text{ N-m}$$

and $$M_C = 8947.4 - 5000(2) = -1052.6 \text{ N-m}.$$

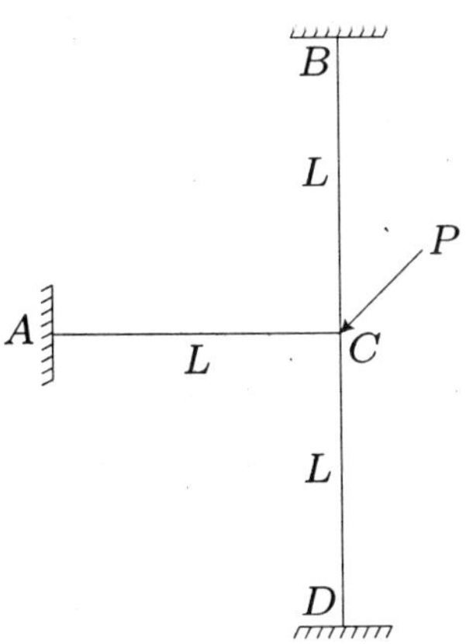

Fig. 12.37: For Example 12.14

Example 12.14. *Fig 12.37 shows a horizontal rigid bent in plan rigidly fixed at supports A, B and D. The bent is subjected to a vertical concentrated load of magnitude W as shown. If $\frac{EI}{GJ} = \alpha$, obtain expressions for twisting moments, BM and load carried by members AC.*

Solution : Assuming twisting moment in CB and CD as $\frac{T_C}{2}$ each. BM at C on beam BCD as M_C and load carried by AC as P,

Moment in AC at $C = T_C$

For $\quad CA,\ M = T_C - P(x) \quad$ and $\quad T = 0,$

so that $$\frac{\partial M}{\partial M_C} = 0$$

and $$\frac{\partial M}{\partial T_c} = 1; \quad \frac{\partial M}{\partial P} = (-x) \quad \text{and} \quad \frac{\partial T}{\partial M_C} = \frac{\partial T}{\partial T_C} = \frac{\partial T}{\partial P} = 0.$$

For $\quad CB,\ M = M_C - \dfrac{W-P}{2}(x) \quad \text{and} \quad T = \dfrac{T_c}{2},$

so that, $\dfrac{\partial M}{\partial M_C} = 1,\ \dfrac{\partial M}{\partial T_C} = 0, \quad \text{and} \quad \dfrac{\partial M}{\partial P} = \left(\dfrac{x}{2}\right)$

and $\quad \dfrac{\delta T}{\partial M_C} = 0,\ \dfrac{\partial T}{\partial T_C} = \dfrac{1}{2} \quad \text{and} \quad \dfrac{\partial T}{\partial P} = 0.$

From $\quad \sum \dfrac{M}{EI}\dfrac{\partial M}{\partial M_C}dx + \sum \dfrac{T}{GJ}\dfrac{\partial T}{\partial M_C}dx = 0,$

$$\frac{1}{EI}\left[\int_0^L (T_C - Px)(O)dx + 2\int_0^L \left(M_C - \frac{W-P}{2}x\right)(1)dx = 0,\right.$$

whence $2M_C + \dfrac{PL}{2} = \dfrac{WL}{2}.$ (i)

From $\quad \sum \dfrac{M}{EI}\dfrac{\partial M}{\partial T_C}dx + \sum \dfrac{T}{GJ}\dfrac{\partial M}{\partial T_C}dx = 0,$

$$\frac{1}{EI}\int_0^L (T_C - Px)(1)dx + \frac{2}{GJ}\int_0^L \frac{T_C}{2}\left(\frac{1}{2}\right)dx = 0,$$

whence $T_C\left(1 - \dfrac{\alpha}{2}\right) - \dfrac{PL}{2} = 0.$ (ii)

From $\quad \sum \dfrac{M}{EI}\dfrac{\partial M}{\partial P}dx + \sum \dfrac{T}{GJ}\dfrac{\partial T}{\partial P}dx = 0,$

$$\frac{1}{EI}\left[\int_0^L (T_C - Px)(-x)dx + 2\int_0^L \left(M_C - \frac{W-P}{2}x\right)\left(\frac{x}{2}\right)dx = 0,\right.$$

whence $M_C - T_C + PL = \dfrac{WL}{3}.$ (iii)

From (i) $2M_c + \dfrac{PL}{2} = \dfrac{WL}{2}.$

From (iii) $2M_C - 2T_C + 2PL = \dfrac{2}{3}WL,$

whence $2T_C - \dfrac{3}{2}PL = -\dfrac{WL}{6}.$ (iv)

From (ii), $3T_C\left(1 + \dfrac{\alpha}{2}\right) - \dfrac{3}{2}PL = 0.$ (v)

From (iv) and (v), $T_C = \dfrac{WL}{3(2+3\alpha)}$ = B.M. at C in CA.

Torsion in CB and $CD = \dfrac{T_C}{2} = \dfrac{WL}{6(2+3\alpha)}.$

From $\quad 2T_C - \dfrac{3}{2}PL = -\dfrac{WL}{6},\ P = \dfrac{W}{9}\left[\dfrac{7+2\alpha}{3+2\alpha}\right].$

Also $\quad M_C = \dfrac{1}{4}(W-P)L = \dfrac{WL}{9}\left(\dfrac{5+4\alpha}{3+2\alpha}\right).$

B.M. at $B = M_B = \dfrac{WL}{9}\left(\dfrac{5+4\alpha}{3+2\alpha}\right) - \dfrac{W-P}{2}L = -\dfrac{WL}{9}\left[\dfrac{5+4\alpha}{3+2\alpha}\right].$

Example 12.15. *The three members shown in Fig. 12.38 are mutually perpendicular and of tubular section. Taking J as 2I and G as 0.40E, analyse the frame by moment distribution.*

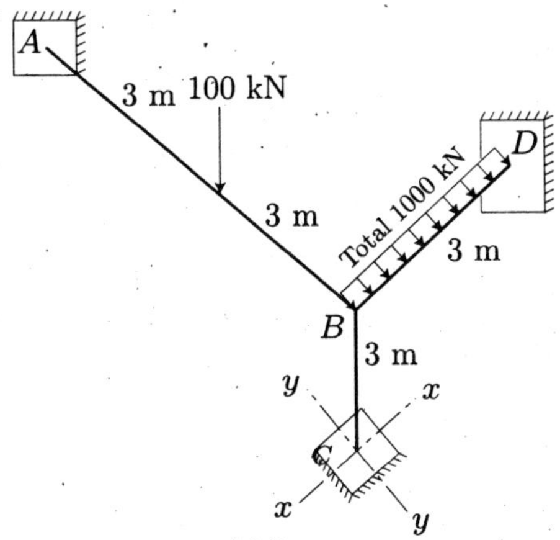

(a) Frame

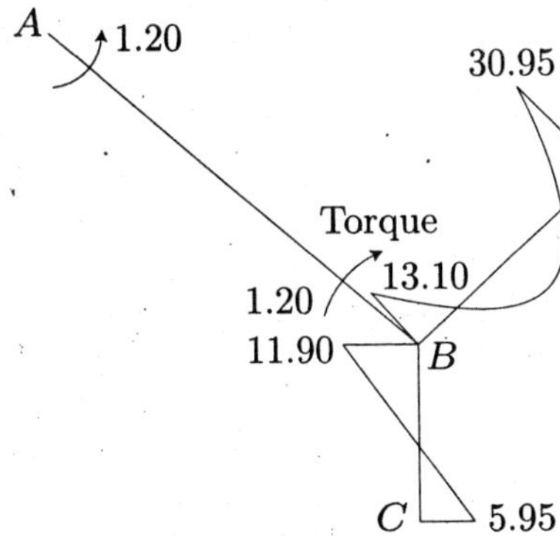

(b) For *CBD*-values in kNm

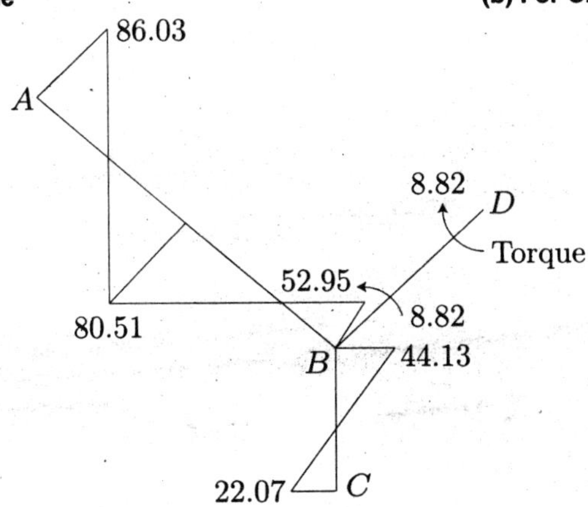

(c) For *ABC*-values in kNm

Fig. 12.38: For Example 12.15

Solution : For frame ABC with bending of BC about one axis xx,

$$K_{BA} = \frac{4EI}{6}; K_{BC} = \frac{4EI}{3}; K_{TBD} = \frac{GJ}{3} = \frac{0.8EI}{3}.$$

Thus, $K_{BA} : K_{BC} : K_{TBD} = 2 : 4 : 0.80$,

whence $DF_{BA} = \dfrac{5}{17}; DF_{BC} = \dfrac{10}{17}$ and $DF_{TBD} = \dfrac{2}{17}$.

$$M_{FBA} = -75 \text{ kNm} = -M_{FAB}.$$

For frame CBD, with bending of BC about axis yy,

$$K_{BC} = \frac{4EI}{3}; K_{BD} = \frac{4EI}{3} \text{ and } K_{TBA} = \frac{0.8EI}{6},$$

whence $DF_{EC} = \dfrac{10}{21}; DF_{BD} = \dfrac{10}{21}$ and $DF_{TBA} = \dfrac{1}{21}$.

Also $M_{FBD} = +25$ kNm.

The distribution procedure is shown in Table 12.4.

Table 12.4. For Example 12.15

	Frame ABC						Frame CBD					
Member	AB (M)	BA	BC (M_x)	BD (T)	DB	CB	AB (T)	BA (T)	BC (M_y)	BD	DB	CB
D.F.	—	$\frac{5}{17}$	$\frac{10}{17}$	$\frac{2}{17}$	—	—	—	$\frac{1}{21}$	$\frac{10}{21}$	$\frac{10}{21}$	—	—
C.O.F.				— 1 →						— 0.50 →		
	←	0.50 —	— 0.50			→	←	1 —	— 0.50			→
F.E.M.	+75	−75	0	0	0	0	0	0	0	+25	−25	0
Bal.	0	+22.05	+44.13	+8.82	0	0	0	−1.20	−11.90	−11.90	0	0
C.O.M.	+11.03	0	0	0	+8.82	+22.07	−1.20	0	0	0	−5.95	−5.95
Bal.	0	0	0	0	0	0	0	0	0	0	0	0
Moments in kNm	+86.03	−52.95	+44.13	+8.82	+8.82	+22.07	−1.20	−1.20	−11.90	+13.10	−30.95	−5.95

In the table, (T) means torque, (M_x) and (M_y) being bending moments about x-x and y-y axes.

Resulting B.M. and twisting moment diagrams are shown in Figs. 12.38(b) and 12.38(c).

Example 12.16. *A single vertical concentrated load of* 240 kN *acts at the centre of beam CE of Fig. 12.39(a). Using moment distribution, find the resulting bending moments and twisting moments. Assume G as* 0.40*E*.

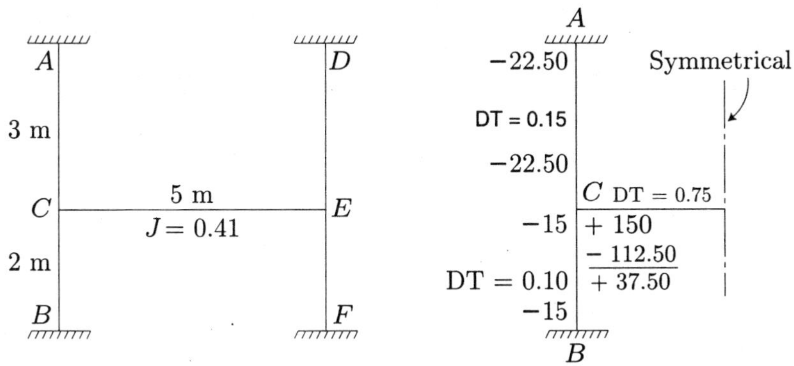

(a) Frame (b) Moment distribution

Fig. 12.39: For Example 12.16

Solution : Taking advantage of symmetry,

$$K_{CE} = \left(\frac{2EI}{L}\right) = \frac{2EI}{5} = \frac{6EI}{15}$$

$$K_{TCB} = \left(\frac{GJ}{L}\right) = \frac{0.40E}{2}(0.41) = \frac{1.20EI}{15}$$

$$K_{TCA} = \frac{0.40E}{3}(0.40I) = \frac{0.80EI}{15}$$

F.E.M. $M_{FCE} = \dfrac{(240)(5)}{8} = 150 \text{ kNm} = -M_{FEC}.$

The moment distribution procedure for the case is shown in Fig. 11.22(b) considering no settlements of C.

Values of moments and torques are

$$M_{CE} = +37.50 \text{ kNm}$$

$$T_{CA} = -22.50 \text{ kNm}$$

and $T_{CB} = -15$ kNm.

The C.O.F. for torque is 1, whence

$$T_{AC} = -22.50 \text{ kNm} \quad \text{and} \quad T_{BC} = -15 \text{ kNm}.$$

Due to the given load, the frame will deflect in the Y-Z plane.

From symmetry, $\Delta_C = \Delta_B; \theta_{CE} = \theta_{EC}$ and $\theta_{CB} = \theta_{EF}$.

It follows that, (i) There is not twist in beam CE.

(ii) The effect of the symmetric load is distributed equally to both girders.

Each girder can then be analysed as a beam subjected to 120 kN load at C and E.

If A (or D) and B (or F) are fixed, the B.M. at A and B are

$$\frac{120(3)^2(2)}{(3)^2}, \text{ i.e., } 86.4 \text{ kNm and}$$

$$\frac{120(2)^2(3)}{(5)^2}, \text{ i.e.,} 57.6 \text{ kNm respectively,}$$

the B.M. at C and E being $\dfrac{120(2)(3)}{(5)} - \left[57.6 + 28.8\left(\dfrac{3}{5}\right)\right] = 69.12$ kNm.

Example 12.17. *A vertical load of* 240 kN *acts on the ends of span* CE *in Fig. 12.40(a). Using moment distribution, find the resulting B.M. and torque at* A, B, C, E, F *and* D. *Assume* $G \doteqdot 0.40E$.

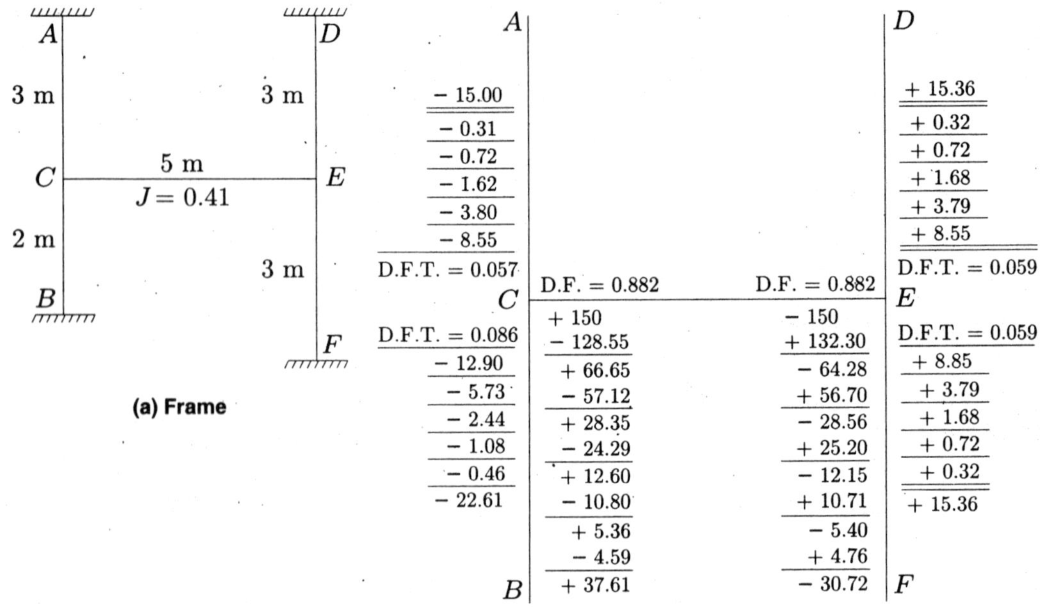

(a) Frame **(b) Moment distribution**

Fig. 12.40: For Example 12.17

Solution : Stiffness factors

$$K_{CE} = \frac{4EI}{L} = \frac{4EI}{5} = \frac{12EI}{15}$$

$$K_{TCA} = K_{TED} = \frac{0.4I(0.4E)}{3} = \frac{0.8EI}{15}$$

$$K_{TCB} = \frac{0.40I(0.4E)}{2} = \frac{1.2EI}{15}$$

D.F. Joint C Joint E

$$DF_{CE} = 0.857 \quad DF_{EC} = 0.882$$

$$DF_{TCA} = 0.057 \quad DF_{TEF} = DF_{TED} = 0.059$$

$$DF_{TCB} = 0.086$$

$$\sum = 1.000$$

F.E.M. $M_{FCE} = +\dfrac{240(5)}{8} = +150 \text{ kNm} = -M_{FEC}$.

The primary moment distribution is shown in Fig. 12.40(b). The moments and reactions in beam CD is shown in Fig. 12.41(a).

The reaction between CE and the girder will now be computed for the equilibrium requirements on CE and the effects on the girders will be considered. If the beam CE were perfectly flexible in torsion, its two ends will freely rotate through angles θ_{CE} and θ_{EC} respectively, so as to be compatible with the slopes of the elastic curves of the girders at their points of juncture with the beam. The beam CE, however, offers resistance to torsion as given by

$$T = \left(\frac{GJ}{L}\right)_{CE} (\theta_{CE} - \theta_{EC}).$$

It is necessary to solve for θ_{CE} in terms of the loads applied to girder AB and to solve for θ_{EC} in terms of the load applied to girder DF. Using unit load method, if a and b locate the point of application of T, at which θ is measured,

$$EI\theta = \frac{1}{3}\frac{Pa^3b + Ta^3}{(a+b)^2} - \frac{1}{3}\frac{Pab^3 - Tb^3}{(a+b)^2}.$$

Considering Figs. 12.41(a) to 12.41(c),

$$EI_{AB}\theta_{CE} = \frac{1}{3}\frac{121.378(3)^3(2) - T(3)^3}{(5)^2} - \frac{1}{3}\frac{121.378(3)(2)^3 + T(2)^3}{(5)^2},$$

whence $\theta_{CE} = \dfrac{48.55 - 0.4679T}{EI_{AB}}$.

Also $EI_{DF}\theta_{EC} = \dfrac{1}{3}\dfrac{118.62(3)^4 + T(3)^3}{(6)^2} - \dfrac{1}{3}\dfrac{118.62(3)^4 - T(3)^3}{(6)^2},$

whence, $\theta_{EC} = \dfrac{0.5T}{EI_{DF}}$.

In the case, $EI_{AD} = EI_{DF} = EI$, and $\theta_{CE} - \theta_{EC} = \dfrac{48.55 - 0.967T}{EI}$,

$$T = \frac{GJ}{L}(\theta_{CE} - \theta_{EC}) = \frac{0.4E(0.4I)}{5}\left(\frac{48.55 - 0.967T}{EI}\right) = 155.36 - 0.0309T,$$

whence $T = 1.507$ kNm.

The effect of the torque on the final moments and twists in the case can be considered as negligible. The solution is obtained by superposition of moments and twists shown in Fig. 12.41(a) to 12.41(c).

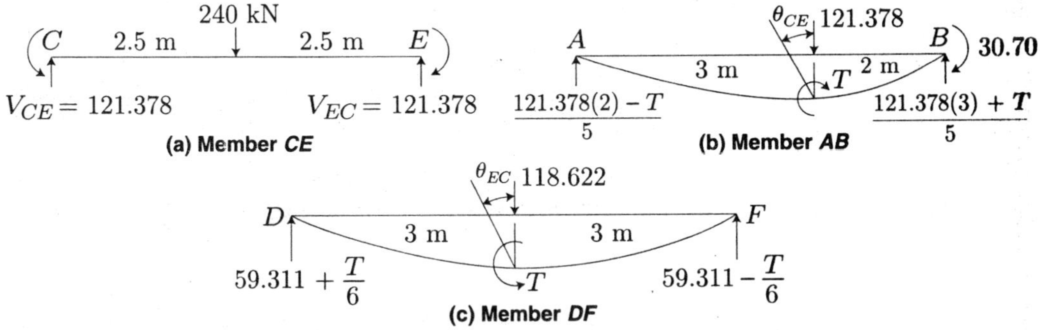

Fig. 12.41: For Example 12.18

12.12 Bow Girder with Intermediate Supports

In some cases, a bow Girder is supported at points along its span by columns or by cantilevers. Such supports may provide restraint of two types.

Case I. The columns may be connected to the girders so as to effectively restraint them against vertical moment and change of slope, as occurs in case of monolithic construction, if columns are of such dimensions so that their shortenings and changes in slope under loads on the beams are negligible. The girder can then be considered as consisting of a number of seperate units which can be analysed as done in the preceding Sections.

Case II. In this type, the connection between the columns and beam may be such that little or no restraint is imposed on the slope of the latter and the vertical movement only is affected. This may be entirely or partially prevented depending on the elastic properties of the supports. In such case, the magnitudes of the vertical reactions $V_1, V_2, \ldots, V_n$ from the supports may be determined from the conditions,

$$\frac{\partial U}{\partial V_1} = \Delta_1$$

$$\frac{\partial U}{\partial V_2} = \Delta_2$$

$$\cdots\cdots\cdots\cdots\cdots$$

$$\cdots\cdots\cdots\cdots\cdots$$

$$\frac{\partial U}{\partial V_n} = \Delta_n,$$

where U = total strain energy and $\Delta_1, \Delta_2, \ldots, \Delta_n$ are the movement of the support columns,

For very rigid supports $\Delta_1, \Delta_2, \ldots, \Delta_n$ may be considered as zero.

Such equations together with the usual ones for determination of the reactive forces at the end of the girder, viz.

$$\frac{\partial U}{\partial M_B} = \frac{\partial U}{\partial T_B} = \frac{\partial U}{\partial V_B} = 0$$

enable a complete analysis of the girder.

12.13 Circular Beams Supported Symmetrically

Fig. 12.42(a) shows a complete circular beam supported symmetrically on columns, the beam being thus continuous and forming a closed loop. The angle subtended by two consecutive columns is 2θ. When the beam is subjected to uniformly distributed loads w per unit length, due to symmetry, the vertical reaction at each columns will be the same, the magnitude being $2wR\theta$. Also the shear force and twisting moment at the centre of each span will be zero. Besides, the twisting moments at the supports will be zero.

Shear force at each support= $wR\theta$.

The bending moment M_0 at each end will be hogging in nature and it can be vectorially represented as shown in Fig. 12.42(b). Moment about AB gives

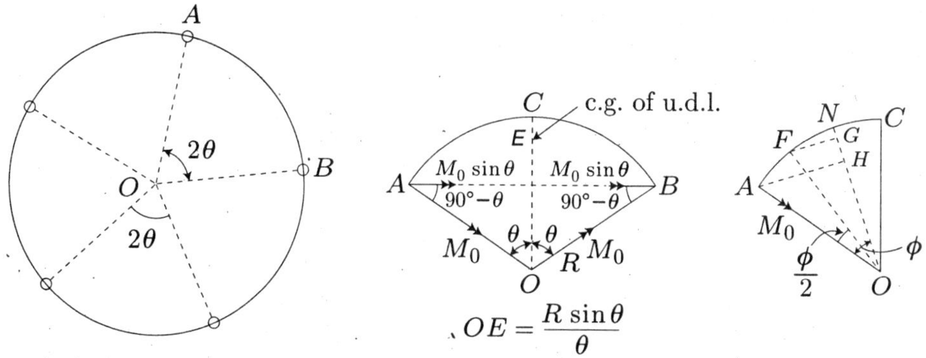

(a) Circular beam **(b) Moments in *AB*** **(c) Part of *AB***

Fig. 12.42: Circular beams symmetrically supported

$$2M_0 \sin\theta = wR(2\theta)[OE - OD] = wR(2\theta)\left[\frac{R\sin\theta}{\theta} - R\cos\theta\right],$$

whence $M_0 = wR^2[1 - \theta\cot\theta]$. (12.39)

This is negative.

At any other point N as in Fig. 12.42(c),

$$\text{S.F. } V_\phi = \text{S.F. at } A - \text{load on AN} = wR(\theta - \phi). \quad (12.40)$$

Bending moment

$$M_\phi = M_0\cos\phi + wR\cdot\theta(AH) - wR\phi(FG)$$

$$= wR^2[\theta\sin\phi + \theta\cot\theta\cdot\cos\phi - 1]. \quad (12.41)$$

Twisting moment,

$$T_\phi \;\; M_0\sin\phi + wR\theta(NH) - wR(NG)$$

$$= wR^2[\theta\cos\phi - \theta\cot\theta\sin\phi - (\theta - \phi)]. \quad (12.42)$$

Location of maximum Twisting Moment is obtained

from $\dfrac{dT\phi}{d\phi} = 0$, whence $[\theta\sin\phi + \theta\cot\theta\cos\phi - 1] = 0$. (12.43)

This also gives the location of the point of contraflexures which indicates that maximum twisting moment occurs at the point of contraflexure.

For design purposes, the values of support moment M_0, midspan moment M_{centre} and the maximum twisting moment T_{max} are important. They can be represented by the expressions,

$$M_0 = -C_1(wR^2)(2\theta) \text{ (hogging)}$$

$$M_{centre} = +C_2(wR^2)(2\theta) \text{ (sagging)}$$

and $T_{max} = C_3wR^2(2\theta)$.

Values of coefficients C_1, C_2 and C_3 depend on the values of 2θ and can be taken from Table 12.5, which also gives the value of ϕ where maximum twisting moment occurs.

Table 12.5. Coefficients for Bending Moments and Twisting Moments in Circular Beams

Nos. of supports	2θ	C_1	C_2	C_3	Value of ϕ for T_{max}
4	90°	0.137	0.070	0.021	$19\frac{1}{4}^\circ$
5	72°	0.108	0.054	0.014	$15\frac{1}{4}^\circ$
6	60°	0.089	0.045	0.009	$12\frac{3}{4}^\circ$
7	$51\frac{3}{7}^\circ$	0.077	0.037	0.007	$10\frac{3}{4}^\circ$
8	45°	0.066	0.030	0.005	$9\frac{1}{2}^\circ$
9	40°	0.060	0.027	0.004	$8\frac{1}{2}^\circ$
10	36°	0.054	0.023	0.003	$7\frac{1}{2}^\circ$
12	30°	0.045	0.017	0.002	$6\frac{1}{4}^\circ$

The distribution of shear force, bending moment between supports are shown in Fig. 12.43.

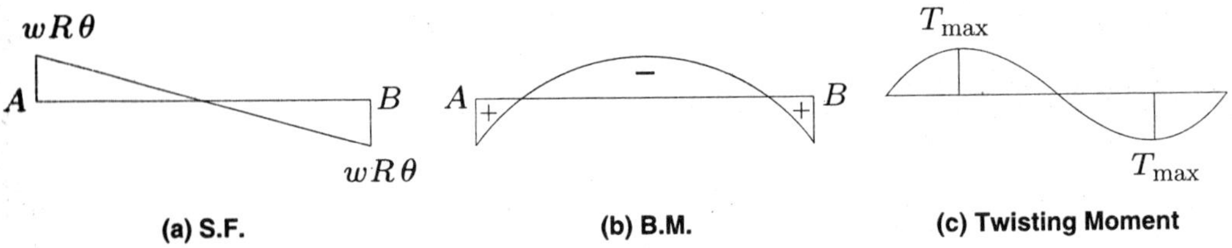

Fig. 12.43: Distribution of S.F., B.M. and Twisting Moments between support

The values of B.M. and Twisting Moment can also be expressed as

$$M = wR^2\left[\frac{\theta}{\sin\theta}\cos\phi - 1\right]$$

and $$T = wR^2\left[\frac{\theta}{\sin\theta}\sin\phi - \phi\right]$$

with co-ordinates as shown in Fig. 12.44.

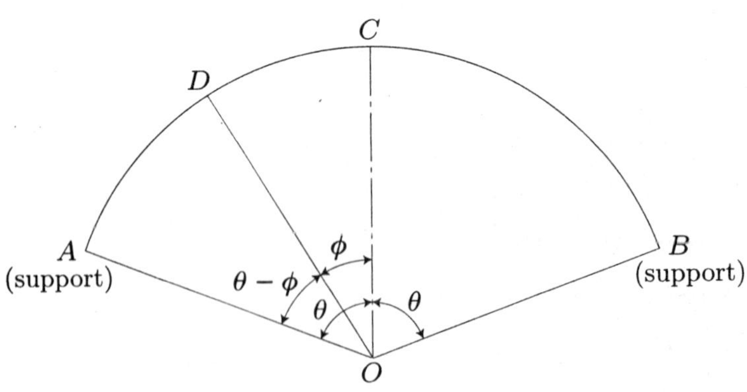

Fig. 12.44: Circular beam

12.14 Semicircular Beam Simply Supported on Three Equally Spaced Supports

Fig. 12.45 shows the case of a semicircular beam ABC simply supported over end columns A and B and continuous over a central column at C. End moments at A and B are evidently zero, i.e., $M_A = M_B = 0$.

From symmetry reaction at A, i.e., V_A and reaction at B, i.e., V_B will be the same, i.e., $V_A = V_B$.

Taking moment about the tangent at C,

$$V_A(R) + V_B(R) = wR\pi\left(R - \frac{2R}{\pi}\right),$$

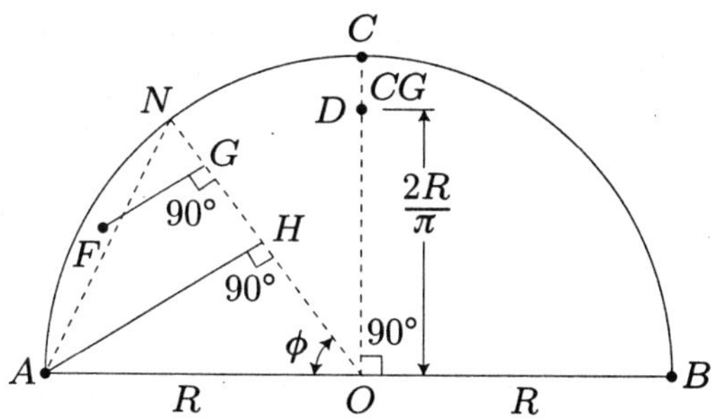

Fig. 12.45: Semicircular beam simply supported on equally spaced supports

whence $V_A = \dfrac{wR}{2}(\pi - 2)$.

Then reaction at C, i.e., $V_C = wR\pi - wR(\pi - 2) = 2wR$.

Shear force at any point N is given by

$$V_\phi = \frac{wR}{2}(\pi - 2) - wR\phi = wR\left[\frac{\pi}{2} - \phi - 1\right].$$

This is valid for $0 \leq \phi \leq \dfrac{\pi}{2}$.

Bending moment $M\phi = V_A(AH) - wR\phi(FG)$

$$= \frac{wR}{2}(\pi - 2)(R\sin\phi) - wR\phi\frac{R\sin^2\frac{\phi}{2}}{\frac{\phi}{2}}$$

$$= wR^2\left[\frac{\pi - 2}{2}\sin\phi - 2\sin^2\frac{\phi}{2}\right].$$

At $\quad \theta = \dfrac{\pi}{2}$, B.M. $M_C = wR^2\left[\dfrac{\pi - 2}{2} - 1\right] = -0.429wR^2$ (hogging).

Maximum positive bending moment occurs at value of ϕ given by

$$\frac{dM_\phi}{d\phi} = 0,$$

whence $\dfrac{\pi - 2}{2}\cos\phi = 2\sin\dfrac{\phi}{2}\cos\dfrac{\phi}{2}$.

Value of $\phi = 29°44' = 0.52$ radians

$$M_{max} = 0.1514wR^2 \text{ (sagging)}.$$

Twisting moment T_ϕ is given by

$$T_\phi = V_A(NH) - wR(NG) = wR^2\left[\frac{\pi - 2}{2} - \frac{\pi - 2}{2}\cos\phi - \phi + \sin\phi\right].$$

At $\quad \phi = \dfrac{\pi}{2}, T_C = 0$.

For maximum T_ϕ, $\frac{dT_\phi}{d\phi} = 0$, whence $\frac{\pi - 2}{2} \sin\phi - 1 \cos\phi = 0$.

Solving, $\phi = 59.44° = 1.037$ radians.

$$T_{max} = 0.1045wR^2.$$

For $\phi = 59.44°, M_\phi = 0$, which means that T_{max} occurs at the point of contraflexure.

The distribution of shear force, bending moment and twisting moment are shown in Fig. 12.46.

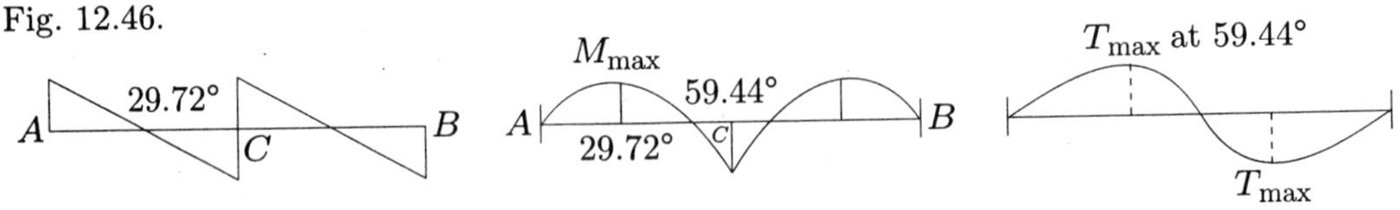

(a) S.F. **(b) B.M.** **(c) Twisting moment**

Fig. 12.46: Distribution of S.F., B.M. and Twisting Moment

Example 12.18. *Obtain distribution of bending moment, shear force and torque in case of a circular bow gird r of radius 4 m continuous over six supports uniformly spaced and subjected to uniformly distributed load of* 50 kN/m.

Solution : In this case, $2\theta = \frac{360°}{6} = 60°$, so that $\theta = 30°$,

From Table 12.5,

$$C_1 = 0.089$$

$$C_2 = 0.045$$

and $C_3 = 0.009$

$$M_0 = -C_1 wR^2(2\theta) - 0.089(50)(4)^2(2)(0.5236) = 74.60 \text{ kNm}.$$

At centre of span,

$$M_C = +0.045(50)(4)(2)(0.5236) = 37.70 \text{ kN-m}.$$

Distribution of M_ϕ is obtained from

$$M_\phi = wR^2[\sin\phi + \theta\cot\theta\cos\phi - 1]$$

and is shown in Fig. 12.47(a).

Shear force $V_\phi = wR(\theta - \phi)$ and $wR\theta = (50)(4)(0.5237) = 104.74$ kN

Distribution of S.F. is linear and is shown in Fig. 12.47(b).

Maximum Torque $= 0.009wR^2(2\theta) = 7.544$kNm at $\left(12\frac{3}{4}^\circ\right)$ from supports.

Torque at centre = 0.

Distribution of Torque is given by

$$T_\phi = wR^2[\theta\cos\phi - \theta\cot\theta\sin\phi - (\theta - \phi)]$$

and is shown in Fig. 12.47(c).

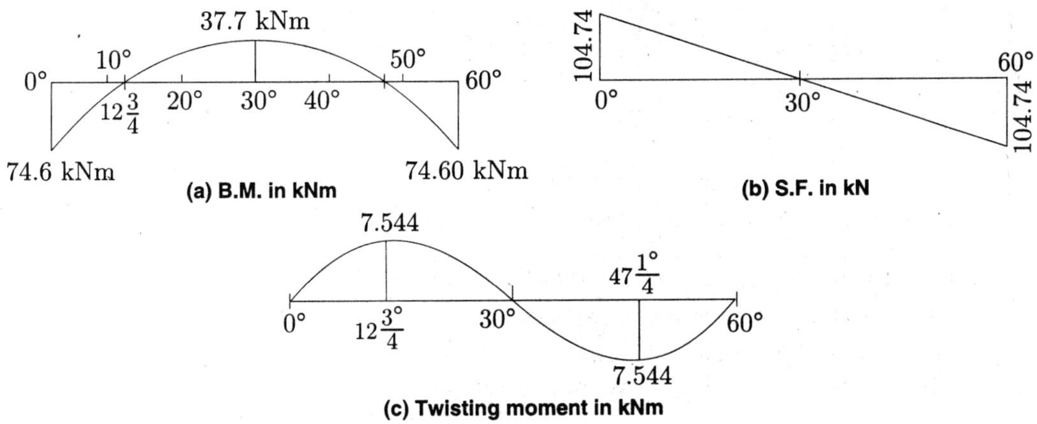

Fig. 12.47: For Example 12.18

12.15 Curved Beam Simply Supported at Ends and Continuous Over Two Equally Spaced Intermediate Supports

Fig. 12.48, shows a circular beam continuous over four supports which are equally spaced, each span subtending an angle of 90° at centre.

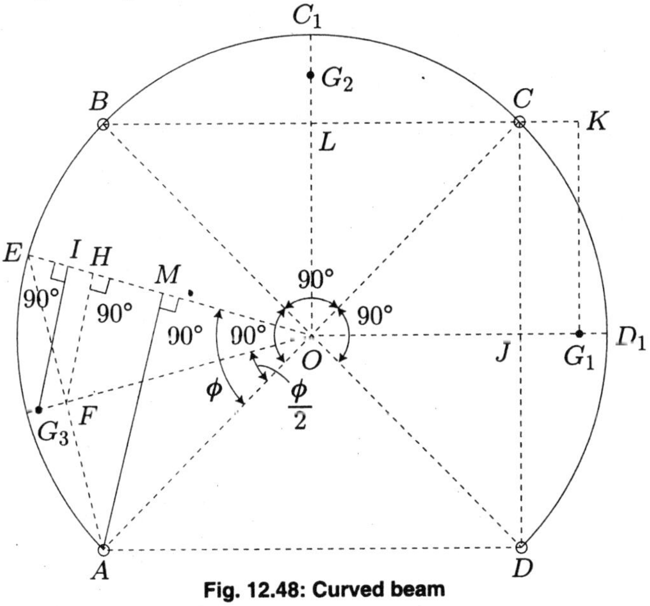

Fig. 12.48: Curved beam

Considering the part CDD_1, load on it is $wR\frac{\pi}{2}$, with its CG at G_1, the perpendicular G_1K on line BC being equal to CJ.

From geometry $CD = \dfrac{AC}{\sqrt{2}} = \sqrt{2}R$,

where R is the radius of the circle,

Also $\quad CJ = \dfrac{CD}{2} = \dfrac{R}{\sqrt{2}}.$

Load on $BCC_1 = wR\frac{\pi}{2}$ with its CG at G_2,

where $$OG_2 = \frac{R\sin\frac{\pi}{4}}{\frac{\pi}{4}}.$$

Also $$OL = CJ = \frac{R}{\sqrt{2}} \quad \text{and} \quad LG_2 = OG_2 - OL = \frac{R\sin\frac{\pi}{4}}{\frac{\pi}{4}} - \frac{R}{\sqrt{2}}.$$

From Symmetry,

Reaction at column A, i.e., V_A = reaction at column D, i.e., $V_D = V_1$ and reaction at column B, i.e. V_B = reaction at column C, i.e., $V_C = V_2$.

Moments of all forces about line BC gives

$$2V_1(CD) = 2\left(wR\frac{\pi}{2}\right)CJ - \left(wR\frac{\pi}{2}\right)(G_2L),$$

whence $$V_1 = wR\left(\frac{3\pi}{8} - \frac{1}{2}\right) = 0.678wR.$$

From equilibrium,

$$2(V_1 + V_2) = wR\left(\frac{3}{2}\pi\right), \quad \text{whence } V_2 = wR\left(\frac{3}{8}\pi + \frac{1}{2}\right) = 1.678wR.$$

Considering point E at angle ϕ from OA as shown in Fig. 12.48,

Load on $AE = wR\phi$, with its C.G. at point G_3, such that,

$$OG_3 = \frac{R\sin\frac{\phi}{2}}{\frac{\phi}{2}}.$$

Also $$G_3I = OG_3\sin\frac{\phi}{2} = \frac{R\sin^2\frac{\phi}{2}}{\frac{\phi}{2}} \quad \text{and} \quad OI = OG_3\cos\frac{\phi}{2} = \frac{R\sin\phi}{\phi}.$$

Also $$FH = OF\sin\frac{\phi}{2} = R\cos\frac{\phi}{2}\sin\frac{\phi}{2} \quad \text{and} \quad EI = OE - OI = R\left(1 - \frac{\sin\phi}{\phi}\right).$$

Distance $AM = R\sin\phi, OM = R\cos\phi$ and $EM = R(1-\cos\phi)$.

B.M. at E is the moment about E of all forces on the left of it, whence for $\phi < 90°$, as $W_\phi = wR\phi$,

$$M_\phi = V_1(AM) - W_\phi(G_3I) = wR^2\left[0.678\sin\phi - 2\sin^2\frac{\phi}{2}\right]$$
$$= wR^2[0.678\sin\phi - 1 + \cos\phi].$$

For $\phi > 90°$,

$$M_\phi = wR^2[0.678\sin\phi - 1 + \cos\phi] + V_2R\sin\left(\phi - \frac{\pi}{2}\right)$$
$$= wR^2[0.678\sin\phi - 1 + \cos\phi] + 1.678wR^2\sin\left(\phi - \frac{\pi}{2}\right).$$

Twisting moment at point E is the moment about tangent at E of all forces on one side of this. Thus for $\phi < 90°$,

$$M_t\phi = V_1(EM) + W_\phi(EI) = wR^2[\phi + 0.678\cos\phi - \sin\phi - 0.678].$$

For $\phi > 90°$,

$$M_t\phi = wR^2[\phi + 0.678\cos\phi - \sin\phi - 0.678] - 0.678wR^2\left[R - R\cos\left(\phi - \frac{\pi}{2}\right)\right]$$
$$= wR^2[\phi + 0.678\cos\phi + 0.678\sin\phi - 2.356].$$

SF at a point is given by

$$V_\phi = V_1 - W_\phi = (0.678 - \phi)wR, \quad \text{for } \phi < 90°$$
$$= wR(2.356 - \phi), \quad \text{for } \phi > 90°.$$

12.16 Semicircular Beam with Slab

Fig. 12.49 shows a semicircular beam ACB, fixed at ends A and B and supporting a semi-circular slab. Since the length of the slab is much greater than the width, it may be considered to behave as a one-way slab.

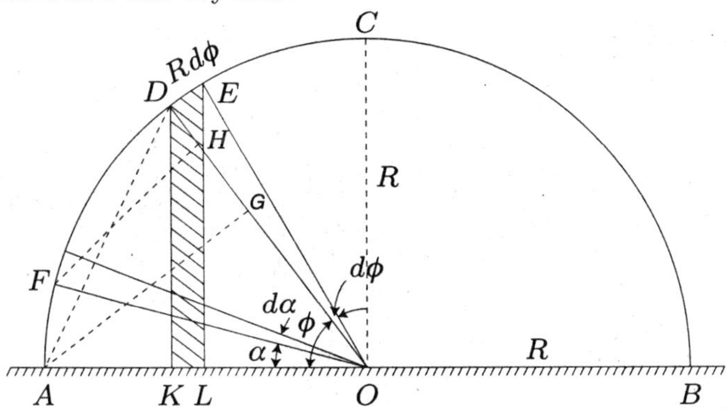

Fig. 12.49: Semicircular beam with slab

From symmetry, reaction components at A and B will be the same.

From Fig. 12.49, the load transferred to element DE of the beam will be equal to half the load on the shaded area $DELK$.

From geometry, $DK = R\sin\phi$ and $OK = R\cos\phi$.

Also $\quad KL = DE\sin\phi = R\sin\phi d\phi.$

Load on shaded area $\approx (DK)(KL) = wR^2\sin^2\phi d\phi.$

Uniformly distributed load $w\phi$ on the element $DE = \dfrac{\frac{1}{2}wR^2\sin^2\phi d\phi}{Rd\phi} = \dfrac{1}{2}wR\sin^2\phi.$

Equating vertical forces and the balancing moment about AB, vertical force F_0 and moment M_0 can be determined.

Thus, $F_0 = \text{load on portion } AC = \displaystyle\int_0^{\pi/2} w_\phi R d_\phi = \int_0^{\pi/2} \frac{wR^2\sin^2\phi}{2} Rd\phi = \frac{\pi}{8}wR^2$

and $\quad M_0 = \displaystyle\int_0^{\pi/2}\left(\frac{1}{2}wR\sin^2\phi\right)Rd\phi(R\sin\phi) = \frac{wR^2}{3}.$

Twisting moment M_{0t} cannot be determined statically and the problem is statically indeterminate to the first degree. The twisting moment M_{0i} is considered acting clockwise at A.

Load on element at H is given by

$$\delta W_\alpha = w\alpha R d\phi = \frac{1}{2} w R \sin^2\alpha (R d\alpha).$$

Also $\quad FH = R\sin(\phi - \alpha);\ OH = R\cos(\phi - \alpha)$

$$AG = R\sin\phi \quad \text{and} \quad OG = R\cos\phi$$

$$M_\phi = F_0(AG) = M_0\cos\phi + M_{0t}\sin\phi - \int_0^\phi (\delta w_\alpha) FH$$

$$= \frac{\pi}{8} w R^2 (R\sin\phi) - \frac{wR^3}{3}\cos\phi + M_{0t}\sin\phi - \int_0^\phi \frac{1}{2} w R \sin^2\alpha R d\alpha R \sin(\phi - \alpha)$$

$$= \frac{wR^2}{3}\left[\frac{3\pi}{8}\sin\phi - \frac{1}{2}(1 + \cos^2\phi)\right] + M_{0t}\sin\phi.$$

Similarly, twisting moment at D equals the moment of all forces to the left of it about tangent at D.

$$M_{\phi t} = -\frac{\pi}{8} w R^2 R(1 - \cos\phi) + \frac{wR^3}{3}\sin\phi + M_{0t}\cos\phi$$
$$+ \frac{1}{2}\int_0^\phi \frac{1}{2} w R \sin^2\alpha R d\alpha R\{1 - \cos(\phi - \alpha)\},$$

whence $\quad M_{\phi t} = wR^3\left[\frac{\pi}{8}(\cos\phi - 1) + \frac{\phi}{4} + \frac{1}{24}\sin 2\phi\right] + M_{0t}\cos\phi.$

From $\quad \dfrac{\partial U}{\partial M_{0t}} = 0,\ \displaystyle\int_0^{\pi/2} -\frac{M_\phi}{EI}\frac{\partial M_\phi}{\partial M_{0t}} R d\phi + \int_0^{\pi/2} \frac{M_{\phi t}}{GJ}\frac{\partial M_{\phi t}}{\partial M_{0t}} R d\phi = 0.$

Also $\quad \dfrac{\partial M_\phi}{\partial M_{0t}} = \sin\phi \quad \text{and} \quad \dfrac{\partial M_{\phi t}}{\partial M_{0t}} = \cos\phi.$

It follows that,

$$\frac{R}{EI}\int_0^{\pi/2} \sin\phi\left[\frac{wR^3}{3}\left\{\frac{3\pi}{8}\sin\phi - \frac{1}{2}(1 + \cos^2\phi)\right\} + (M_{0t}\sin\phi) d\phi\right.$$
$$\left. + \frac{R}{GJ}\int_0^{\pi/2} \cos\phi\left[wR^3\left\{\frac{\pi}{8}(\cos\phi - 1) + \frac{\phi}{4} + \frac{1}{24}\sin 2\phi\right\} + M_{0t}\cos\phi\right] d\phi = 0.$$

Solving, $\quad M_{0t} = wR^3\left[\frac{\pi}{8} - \frac{8}{9\pi}\right]$

Minus sign indicates that the twisting moment at A is in anticlockwise direction. Knowing values of M_{0t}, magnitudes of M_ϕ and $M_{\phi t}$ can be easily evaluated. Thus at $\phi = \frac{\pi}{2}$,

$$M_C = \frac{wR^3}{3}\left[\frac{3\pi}{8}\sin\frac{\pi}{2} - \frac{1}{3}\left(1 + \cos^2\frac{\pi}{2}\right) - wR^3\left[\frac{\pi}{8} - \frac{8}{9\pi}\right]\sin\frac{\pi}{2}\right.$$

$$= wR^2\left[\frac{8}{9\pi} - \frac{1}{6}\right]$$

and $\quad M_{C_t} = wR^3\left[\frac{\pi}{8}\left(\cos\frac{\pi}{2} - 1\right) + \frac{\pi}{8} + \frac{1}{24}\sin\pi\right] - wR^2\left[\frac{\pi}{8} - \frac{8}{9\pi}\right]\cos\frac{\pi}{2}$

$$= 0, \text{ as expected .}$$

12.17 Torsion Factors

For statically indeterminate beams, curved in plan, torsion factors J is to be used for the determination of torsional moment at any section. Torsion factor is also sometimes known as the rotational constant. Its value for some common shares are given below :

(a) For Circular Section, $J = \dfrac{\pi d^4}{32}$, where $d =$ diameter of the circle.

(b) For Elliptical Section, $J = \dfrac{\pi a^2 b^2}{a^2 + b^2}$, where a and b are the semi-axis of the ellipse.

(c) For Equilateral Triangular Section, $J = \dfrac{a^4\sqrt{3}}{80}$, where $a =$ side of the triangle.

(d) For Hollow Circular Section,

$$J = \frac{4A^2 t}{p}, \quad \text{where} \quad A = \text{area enclosed}$$

$t =$ thickness of tube

and $p =$ perimeter of the tube

(e) For Rectangular Section of width b and depth d, various values are suggested.

(i) *Approximate expressions* :

For $\dfrac{d}{b} < 16, J = \dfrac{b^3 d^3}{3.58(b^2 + d^2)}$.

For $\dfrac{d}{b} > 16, J = \dfrac{b^3 d}{3}\left(1 - 0.63\dfrac{b}{d}\right)$.

(ii) *St. Venant's expression* :

$$J = \frac{\alpha d b^2}{16}, \quad \text{where} \quad \alpha \approx \frac{16}{4} - 3.36\left(\frac{b}{d}\right)\left[1 - \frac{b^4}{12d^4}\right].$$

(iii) *Theoretical solution by Timoshenko and Goodier* : Values of J as given are compared with J values for (i) and (ii) in Table 12.6.

Table 12.6. Values of J

$\dfrac{d}{b}$	From Approximate expression	From St. Venant's expression	From Timoshenko and Goodier
1.0	$0.1397d^4$	$0.1406d^4$	$0.1406d^4$
1.2	$0.0954d^4$	$0.0961d^4$	$0.0956d^4$
1.5	$0.0573d^4$	$0.0580d^4$	$0.0580d^4$
2.0	$0.0285d^4$	$0.0287d^4$	$0.0285d^4$
2.5	$0.0159d^4$	$0.0160d^4$	$0.0158d^4$

Problems

12.1 A bar circular in cross section of radius r is bent in the form of a rectangle $ABCD$ held in the horizontal plane and provided with fixed supports at A and D as shown in Fig. P-12.1. A vertical uniformly distributed load w/unit length acts on the part BC. Obtain the B.M., torsional moment and S.F. diagram, if $G = 0.40E$. Calculate also the maximum deflection.

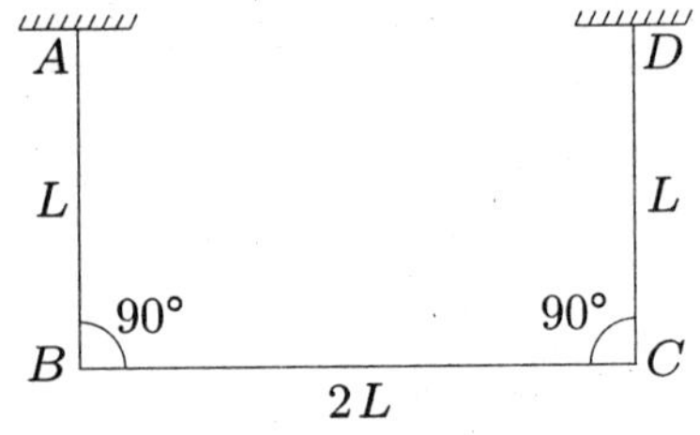

Fig. P-12.1

12.2 A horizontal frame fixed at ends A and D is shown in Fig. P-12.2. The parts AB and CD are circular in cross section of diameter d, the members BC being rigid. A vertical load P acts at a distance nL from B, as shown in figure. Obtain the reactions at A and draw B.M., S.F. and Torsional moment diagram.

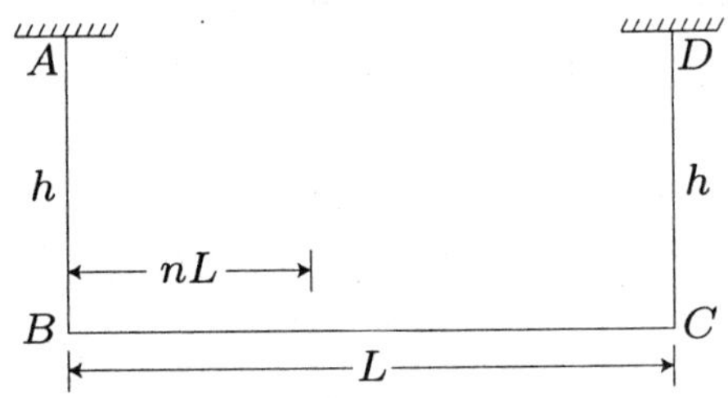

Fig. P-12.2

12.3 A circular bow girder of radius R is horizontal and is fixed at ends A and B, the angle subtended by the ends at centre being 120°. A concentrated vertical load P acts at centre. Obtain the reactions at ends, if $G = 0.4E$ and $J = 2I$.

12.4 A circular bow girder of radius R is horizontal and is fixed at ends, the angle subtended by the ends at centre being 90°. If a vertical uniformly distributed load acts on the girder, obtain the reactions at ends. The girder is circular in cross section and $G = 0.40E$.

12.5 The bent girder shown in Fig. P-12.3 is fixed at ends and carries a uniformly distributed load of w per unit length on the straight part BC. If the flexural rigidity be α times the torsional rigidity, determine the reactions at supports. Also draw the B.M. and Torsional Moment diagrams.

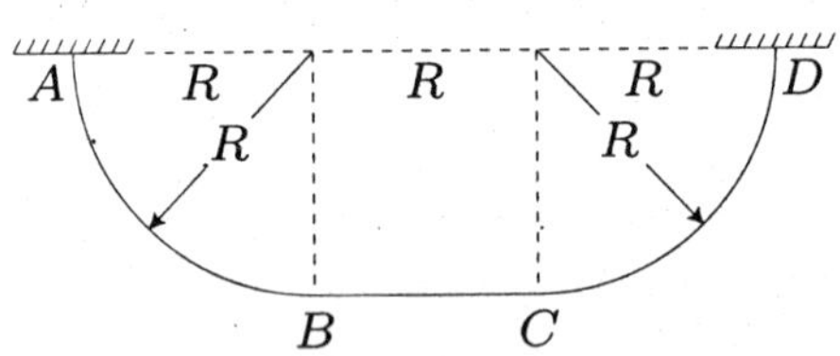

Fig. P-12.3

12.6 Solve the bow girder in Fig. P-12.3 if a concentrated vertical load P acts at the centre of BC.

12.7 A horizontal circular bow girder of radius 6 m is continuous over eight supports. It caries a vertical uniformly distributed load of magnitudes 100 kN/m. Obtain the B.M., Torsional Moment and S.F. diagrams for one span indicating the critical values.

12.8 A semicircular beam is simply supported on three equally spaced columns. Show that the maximum bending moment and the maximum twisting moments are equal to $0.429wR^2$ and $0.1045wR^2$ respectively, when w is the uniformly distributed load on the beam.

12.9 A vertical beam is in the form of a full continuous circle in plan with a radius of 3 m and is supported continuously on five supports. The beam carries a uniformly distributed load of 50 kN/m inclusive of its own weight. Determine the bending moment, twisting moment and shear force at salient locations. Also plot the B.M., T.M. and S.F. diagrams.

12.10 A beam is curved in plan in the form of an arc of a circle with radius of 5 m and central angle of 90°. The beam carries a superimposed load of 20 kN/m and is fixed at both ends. If the section of the beam is rectangular having depth of 750 mm and width of 350 mm, obtain the bending moment and twisting moment diagrams for the beam. Assume $G = 0.4E$.

12.11 If the channel section in Example 12.3 be used as a simply supported beam of span 4 m, determine the magnitude of the concentrated. P kN, at centre of the span if allowable spaces in tension and compression be 160 N/mm^2 and 120 N/mm^2 respectively. Assume, load P is applied along the shear centre.

12.12 A curved bar of cross section as shown in Fig. P-12.4 is subjected to pure bending. Compute the width b_0 so that the extreme fibres be stressed equally (in magnitude). All dimensions are in mm.

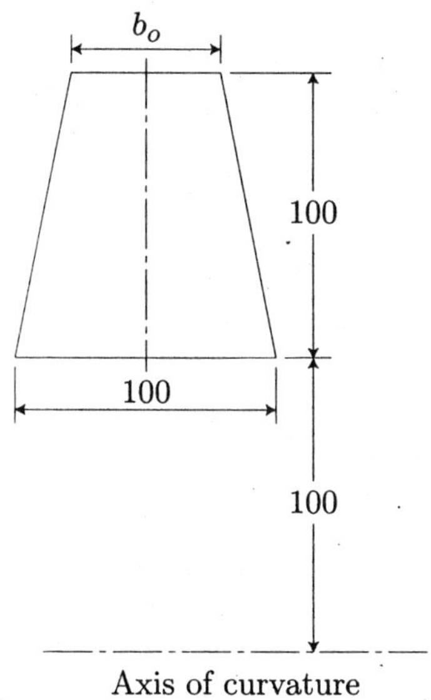

Fig. P-12.4

12.13 The principal section of a crane hook is trapezoidal and 70 mm deep in the plane of curvature. Its width at the intrados is 40 mm and at the extrados, 10 mm. The centre of curvature of intrados and extrados lies in the plane of the cross section and is 30 mm from the intrados. It carries a load of 20 kN. The load line passes through the centre of curvature. Compute the maximum intensity of tensile stress developed in the hook.

12.14 A curved bar of rectangular cross section, 20 mm deep and 40 mm wide is loaded as shown in Fig. P-12.5. Determine the maximum tensile and compressive stresses.

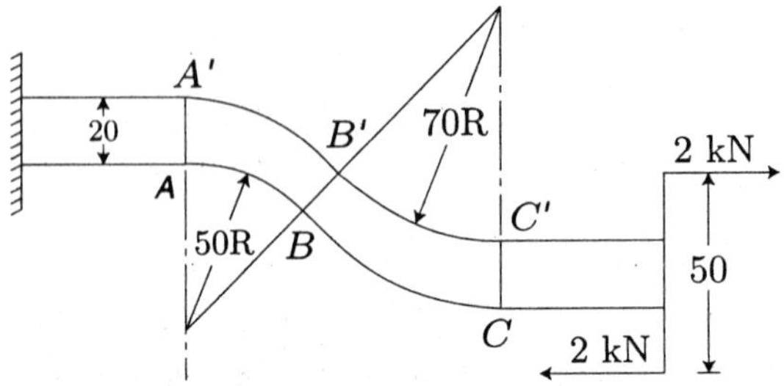

Fig. P-12.5

12.15 A semicircular curved bar is loaded as shown in Fig. P-12.6(a) and has the trapezoidal section given in Fig. P-12.6(b). Calculate the stress at the point A, if $a = b = h = 40$ mm and $F = 5$ kN.

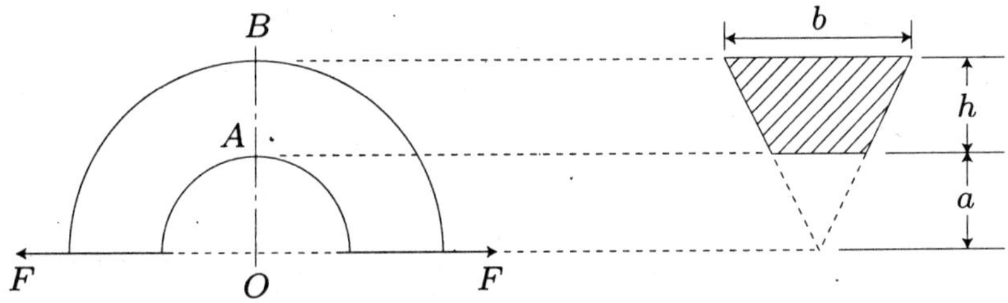

(a) Curved bar **(b) Section at *A-B***

Fig. P-12.6

Chapter 13

Influence Lines for Indeterminate Structures

13.1 Introduction

The concept of influence line was introduced by E. Winkler in the year 1867. In the year 1886, the important principle by means of which influence lines for both determinate and indeterminate structures are determined quite easily was enunciated by Professor Muller Breslau. An improved version of the method of Maxwell and Mohr was published. In developing the method, Maxwell's theorem of reciprocal displacements was utilised. This principle is the basis for determining influence lines for indeterminate structures.

13.2 Müller-Breslau Principle

Müller-Breslau's principle provides a suitable method for both qualitative and quantitative determination of influence line diagrams for various functions in a structure. It is possible to sketch the shape of the influence line diagrams and determine accurately enough for design purposes the loading pattern of the structures in order to produce the maximum effect. The principle also enables computation of ordinates of influence lines and likewise is the basis for certain indirect methods of model analysis.

Müller-Breslau's principle states that ordinates of the influence line for any function such as reaction, shear, bending moment, torsion, fibre stress, etc. in an element of a structure are proportional to those of the deflection curve drawn for the structure, obtained by removing the restraints corresponding to that element from the structure and introducing in its place a corresponding deformation into the primary structure which remains. This principle is applicable to any type of structures like beams, frames, and to both determinate and indeterminate structures. For indeterminate structures, this principle is linked to structures, the material of which is elastic and follows Hooke's law.

For the case of a two-span continuous beam ABC shown in Fig. 13.1(a), the influence line for vertical reaction at B can be plotted after the reaction has been evaluated for unit vertical load applied successively at various locations along the structure. The deflected shape and the reaction R_B of the continuous beam ABC due to unit load placed at D is shown in Fig. 13.1(b), the vertical deflection at B being zero.

The support at B of the beam ABC is then removed and the beam is acted upon by a vertical force of any magnitude, say P at B. The beam will then deflect as shown in Fig. 13.1(c). From Maxwell-Betti Theorem, referring to Figs. 13.1(b) and 13.1(c), whence

$$R_B\,(\Delta B_1) - 1\,(\Delta D_1) = P(0) \quad \text{and} \quad R_B = 1\frac{\Delta D_1}{\Delta B_1}. \tag{13.1}$$

From Eq. (13.1), it is apparent that the reaction R_B when the unit vertical load is at D is proportional to the deflection ΔD_1 at that point. The shape of the influence line of R_B is, therefore, the same, as the shape of the elastic curve of the structure acted upon by a force P at B. The magnitude of the influence line ordinate at any point D can be obtained by dividing the deflection at that point on the elastic curve by the deflection at point B.

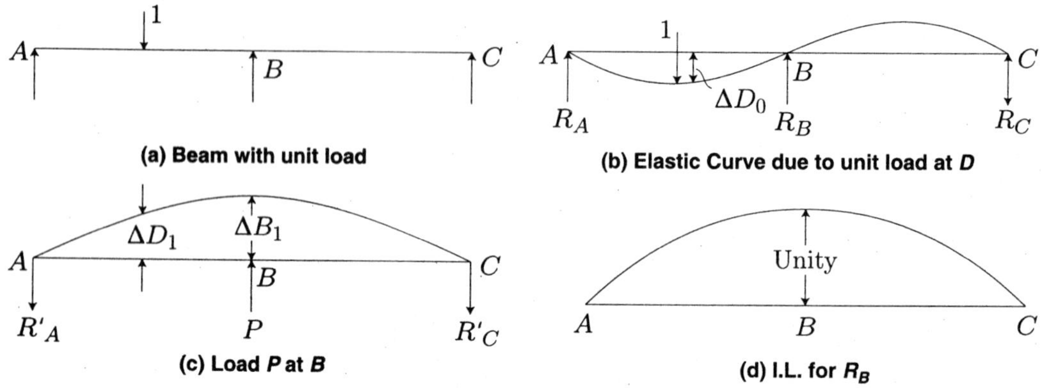

Fig. 13.1: I.L. for two span continuous beam

The validity of the principle can be demonstrated similarly for any other functions. Thus, for $I.L.$ of M_B, from Figs. 13.2(a) and 13.2(b),

$$M_B = -1\frac{\Delta D_1}{\theta B_1}. \tag{13.2}$$

For S.F. from Figs. 13.2(a) and 13.2(c),

$$\text{S.F. at } E = -1\frac{\Delta D_1}{\Delta E_1}. \tag{13.3}$$

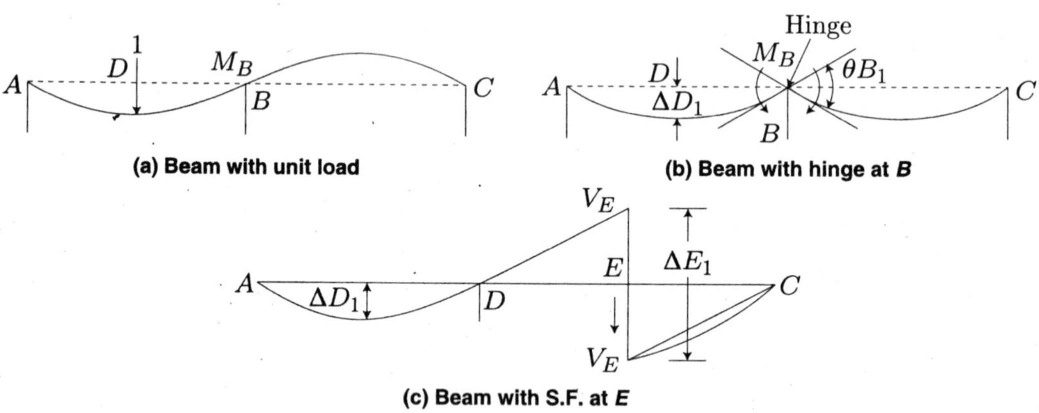

Fig. 13.2: I.L. for continuous beam

The Müller-Breslau principle suggests two different methods by which influence lines for redundant stresses and reactions can be obtained. The first of these is the experimental method requiring the use of models. The second method is of course by computation.

For obtaining influence line it is necessary to remove the redundant and to introduce a displacement in the structure at the point of action and in the direction of the redundant. When the influence line is being determined experimentally by use of a model, the magnitude of the impressed displacement is arbitrarily selected for convenience, the magnitude of the action causing the displacement being of no importance. When, however, the influence line ordinates are to be computed, it is convenient to apply a unit action.

13.3 Qualitative Influence Lines by Müller-Breslau's Principle

Müller-Breslau's principle is suitable for determining the qualitative influence line diagram, which are indispensable in obtaining the correct loading pattern for maximum values of various functions in various types of structures.

If it is necessary to obtain the qualitative influence line for positive moment M_{AB} at the centre of span AB of the continuous beam in Fig. 13.3(a), it is necessary to introduce a hinge at that point and apply couples, on each side of the hinge to produce a unit angular displacement as indicated in Fig. 13.3(b). The shape of the deformed structure shown in Fig. 13.3(b) is the qualitative influence line for bending moment at the centre of span AB. The influence line indicates that for maximum positive moments at the section, the spans AB, CD and EF have to be loaded.

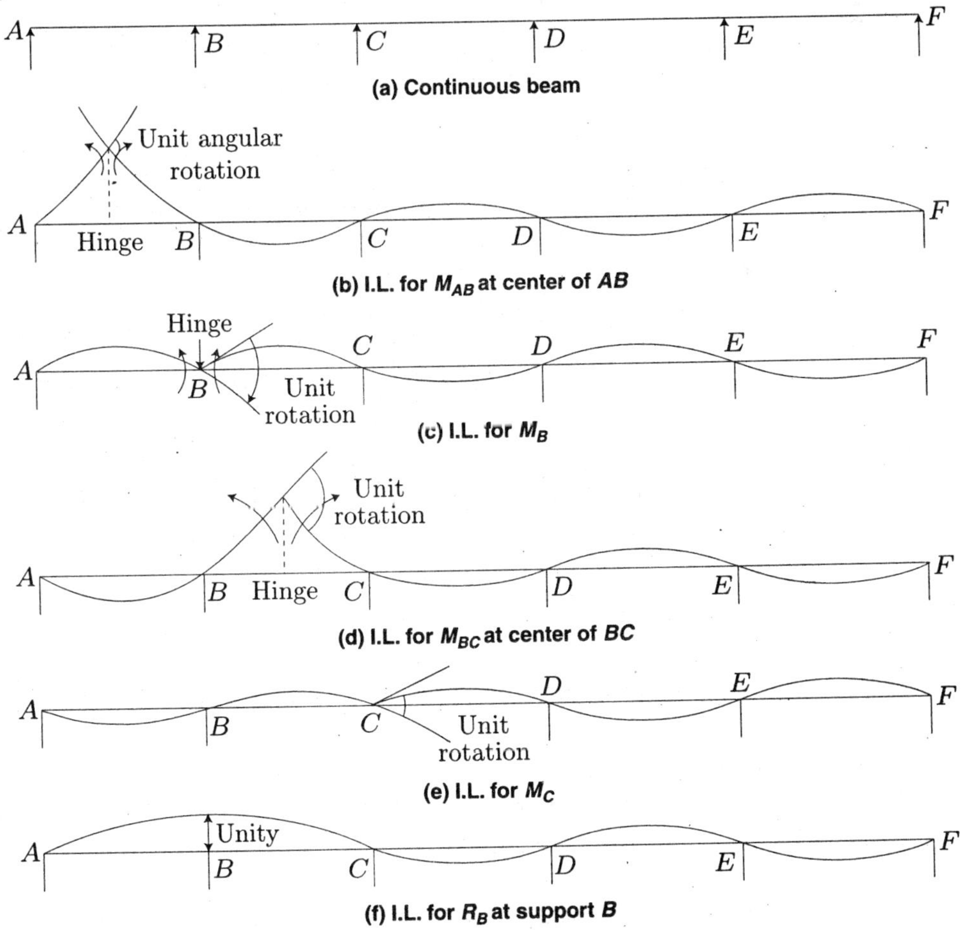

Fig. 13.3: Qualitative influence diagrams (Contd.)

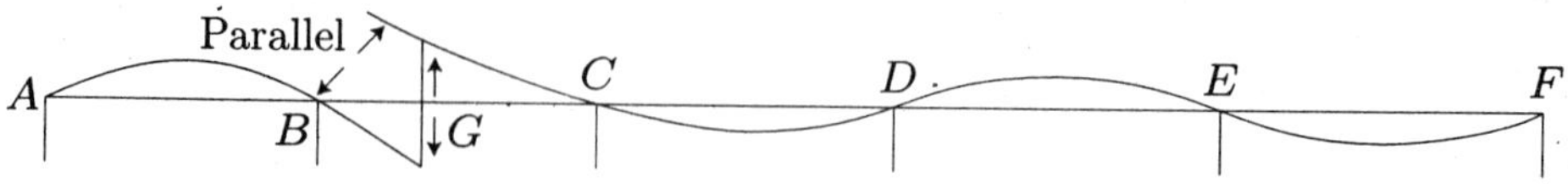

(g) I.L. for S.F. at *G* in span *BC*

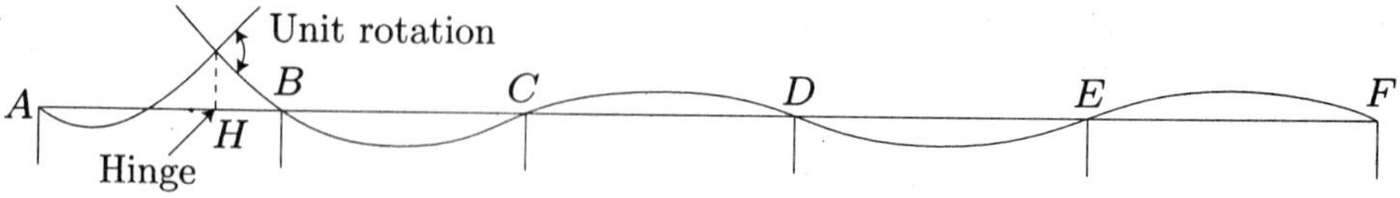

(h) I.L. for M_M at *H* in *AB*

Fig. 13.3: Qualitative influence diagrams

For influence line diagram for negative moments M_B at support B, it is necessary to introduce a hinge at point B and apply couples on each side of the hinge, so as to produce unit rotation. The deflected shape of the beam in Fig. 13.3(c) is the qualitative influence line diagram for B.M. at B. It is evident that for maximum negative moment at C, spans AB, BC and DE have to be loaded.

Similarly, the influence lines for B.M. M_{BC} at centre of span BC and B.M. M_C at support C can be obtained as shown in Figs. 13.3(d) and 13.3(e).

The same method enables sketching influence lines for reactions and shear forces as illustrated in Figs. 13.3(f) and 13.3(g). In applying a shearing deformation the displaced ends must remain parallel, the shear sections may rotate, but the two surfaces must rotate the same.

The same procedure should not be applied at sections near the support. Between a certain point and the nearer support, the influence line takes the shape as shown in Fig. 13.3(h). Such sections sometimes called 'fixed points' are not far from the quarter points. In cases of influence lines for such sections, the qualitative diagram cannot be sketched.

The qualitative influence line is particularly valuable in the analysis of building frames. Fig. 13.4(a) shows the influence line for positive moment at the centre of span $C4$-$D4$. The influence line for negative B.M. at left end of span $C4$-$D4$ is also shown in Fig. 13.4(b). From Fig. 13.4(b), it is evident that for maximum negative moment some small part of the right end of the beams along 3 and 5 is not to be loaded. In addition, levels 2 and 6, a short loaded length is theoretically required at the right end. For practical analysis, however, levels 3 and 5 in bay CD are entirely loaded, while levels 2 and 6 for the same bay are unloaded.

The influence line for moment causing tension on the right side at the top of the column $C3$-$C4$ is shown in Fig. 13.4(c).

Such influence line diagrams indicate the theoretical bending patterns for maximum values of the functions. For practical analysis, however, the effect of the load in a given bay on a stress component in a member two or three bays apart is negligible. The usual practice, therefore, is to analyse large building frames by dividing them into sections. In analysing girders at a particular load, it is permissible to consider that all columns both above and below the girder level are fixed at their far end. Sometimes, the ends of girders two bays away from the span of the section being investigated may also be considered to be fixed. This simplifies the analysis even though the accuracy is slightly affected.

13.4 Quantitative Influence Lines

For obtaining the quantitative influence line diagrams, it is necessary to compute the influence line ordinates. Various techniques based on use of Reciprocal theorem including Maxwell-Betty theorem. Conjugate beam method, Moment distribution methods and Distribution of slope discontinuity at supports can be used for computation of the influence line ordinates. Suitability of a particular technique depends on the type of structure and the number of indeterminacy of the structure being dealt with.

(A) With Reciprocal Theorem for Computation of Influence Lines. The reciprocal theorem is specially useful in the construction of influence line for redundant reactions.

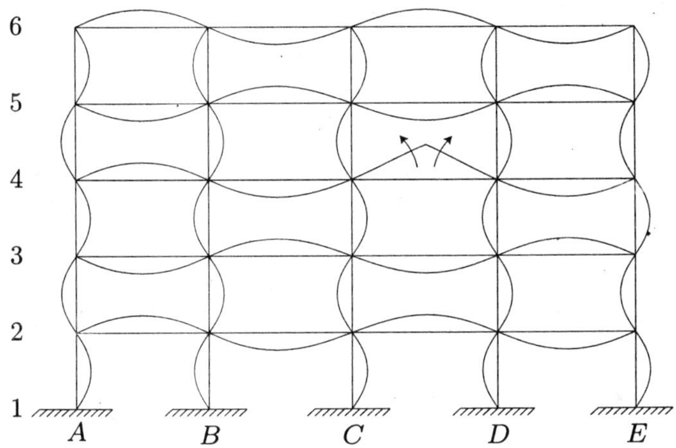

(a) I.L. for B.M. at centre of C_4-D_4

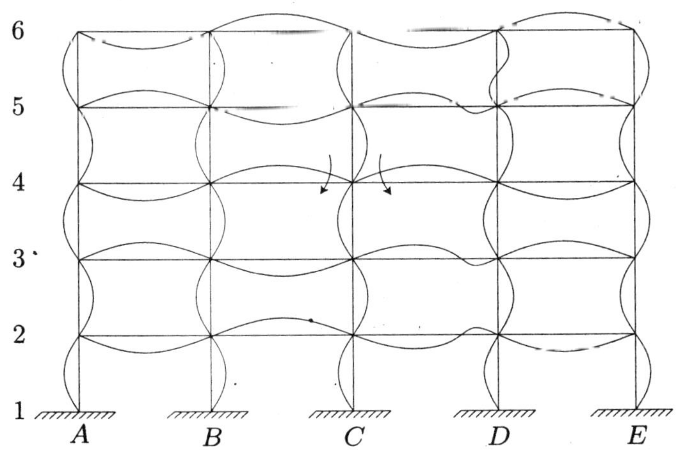

(b) I.L. for B.M. at left of C_4D_4

Fig. 13.4: Qualitative I.L. for Building Frame (Contd.)

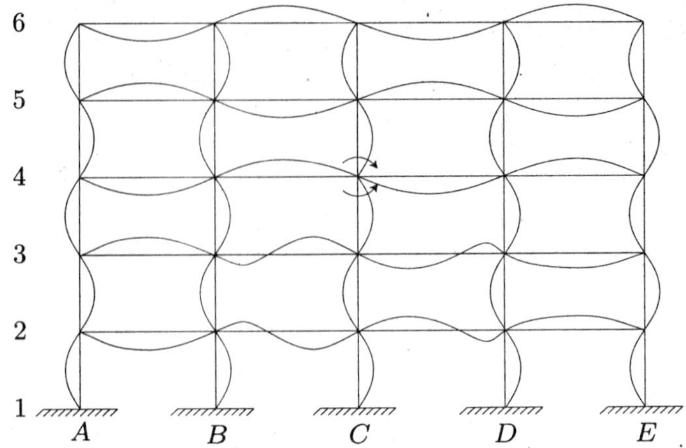

(c) I.L. for B.M. on top of column C_3-C_4

Fig. 13.4: Qualitative I.L. for Building Frame

Considering the continuous beam of two spans as shown in Fig. 13.5(a) in applying the reciprocal theorem, two conditions of loading of the given structure have to be considered. For the first one, the actual loading by unit load is considered as shown in Fig. 13.5(a). For influence line of reaction at B, i.e., R_B, the second loading pattern is shown in Fig. 13.5(b) in which the unit load is removed and in place of support at B, an external unit force acts at B. The second case in Fig. 13.5(b) is statically determinate and the elastic curve for the part AB is given by the relationship,

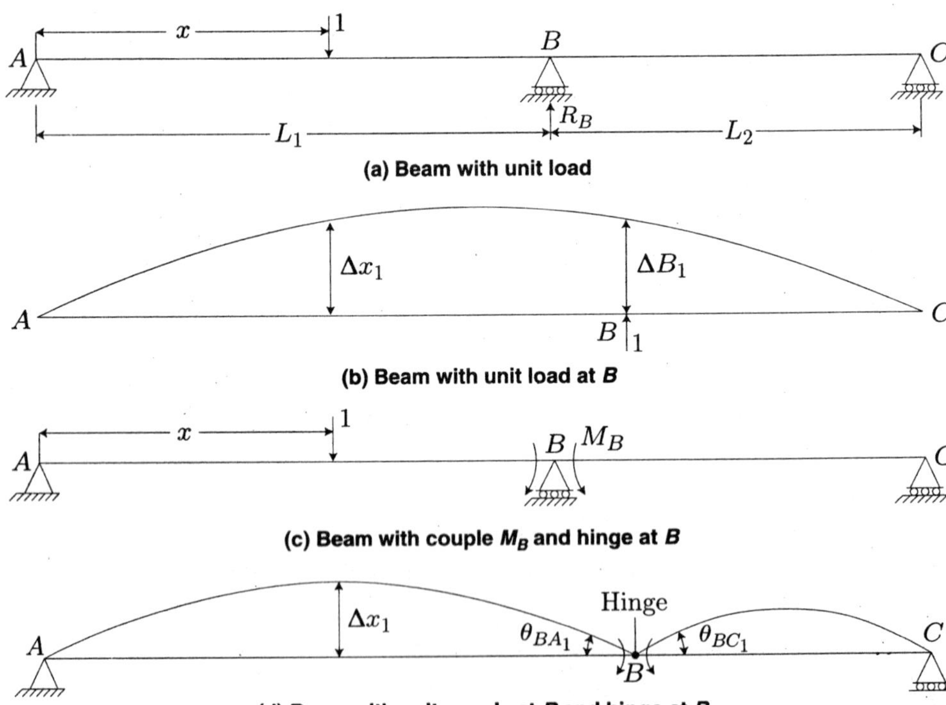

Fig. 13.5: Use of Reciprocal Theorem for I.L.

$$\Delta_x = 1\frac{L_2 x}{6(L_1+L_2)EI}[(L_1+L_2)^2 - L_2^2 - x^2].$$

A similar expression can be established for the right part CB. The deflection under the load R_B is obtained by substituting $x = L_1$ in Eqn. (13.9) and is given by

$$\Delta_{B_1} = 1\frac{L_1^2 L_2^2}{3(L_1+L_2)EI}.$$

Applying the reciprocal theorem,

$$R_B(\Delta B_1) - 1(\Delta_x) = 0,$$

whence $$R_B = \frac{\Delta_x}{\Delta_{B_1}} = \frac{3(L_1+L_2)EI}{L_1^2 L_2^2}\Delta_x.$$

It is seen that as the unit load is moving along the beam AC, the reaction at B varies in the same proportion as the ordinates Δ_x of the deflection curve ABC calculated for a unit load at B, the deflection curve being the influence line for R_B. From the influence line diagram, the reaction R_B for any system of vertical loads, P_1, P_2, $P_3, \cdots$ can be calculated as $R_B = \frac{1}{\Delta_B}[P_1\Delta_{x1} + P_2\Delta_{x2} + P_3\Delta_{x3} + \ldots]$, whence x_1, x_2, $x_1, \ldots$ are the ordinates of the influence line ABC corresponding to the forces P_1, P_2, $P_3, \ldots$

If it is desired to obtain influence line for bending moment M_B at the intermediate support B of the beam in Fig. 13.5(b), the beam is first provided with a hinge at the section B and two equal and opposite couples M_B as shown in Fig. 13.5(d). From reciprocal theorem,

$$M_B[\theta_{BA_1} + \theta_{BC_1}] - 1(\Delta_{x1}) = 0,$$

whence $$M_B = 1\frac{\Delta_{x1}}{\theta_{BA_1} + \theta_{BC_1}}.$$

From expressions of Δ_{x1}, θ_{BA_1} and θ_{BC_1} the influence coefficient for unit load on AB is given by

$$M_B = \frac{x[L_1^2 - x^2]}{2L_1(L_1+L_2)}.$$

The same expression can be used for the right span CB, if L_1 and L_2 are interchanged and x is measured from C.

Maxwell-Betti theorem (sometimes called the generalised Maxwell's law) which states that in any elastic structure following Hooke's law, the external work done by a system of forces P_m during deformation caused by a system of forces P_n is equal to the work done by the P_n system during the deformation caused by the P_m system, can also be used for obtaining influence line. The use of the theorem in finding out influence line are illustrated by suitable Examples.

Example 13.1. *Obtain an influence line diagram for the vertical reaction at the end B of a propped cantilever beam AB shown in Fig. 13.6(a).*

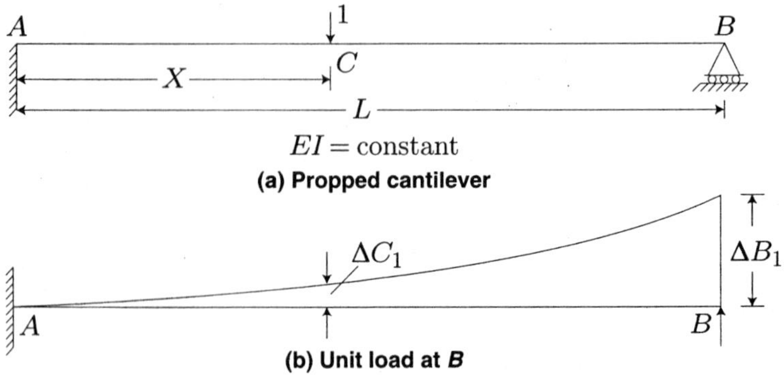

Fig. 13.6: For Example 13.1

Solution : In this case equation for elastic curve for load in Fig. 13.6(b) is given by

$$\frac{(3L-x)x^2}{6EI}, \quad \text{whence} \quad \Delta_{C_1} = \frac{(3L-x)x^2}{6EI}.$$

Also $\Delta_{B_1} = \dfrac{L^3}{3EI}$.

From Reciprocal theorem, $R_B = \dfrac{(3L-x)x^2}{2L^3}$.

Example 13.2. *Using reciprocal theorem, derive an expression for influence coefficient for end moment M_{AB} in case of a beam AB of constant EI with built-in ends. Where should a load P be placed on the beam to produce the maximum value of this end moment? Also obtain an expression for the moment at centre, i.e., M_C.*

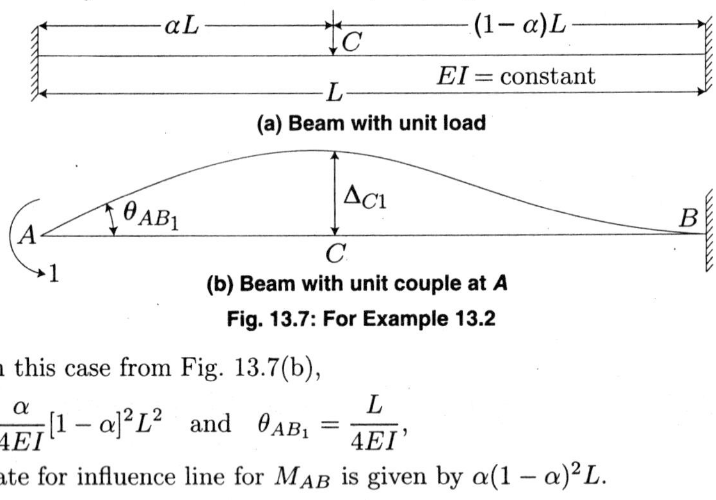

Fig. 13.7: For Example 13.2

Solution : In this case from Fig. 13.7(b),

$$\Delta_{C_1} = \frac{\alpha}{4EI}[1-\alpha]^2L^2 \quad \text{and} \quad \theta_{AB_1} = \frac{L}{4EI},$$

whence ordinate for influence line for M_{AB} is given by $\alpha(1-\alpha)^2L$.

For maximum value of $M_{AB}, \alpha = \dfrac{1}{3}$.

Maximum ordinate at $\alpha L = \dfrac{L}{3}$ is $\dfrac{1}{3}\left[1-\dfrac{1}{3}\right]^2 L = \dfrac{4}{27}L$.

Load at αL for $\alpha < \dfrac{1}{2}$,

$$M_C = \frac{1}{2(1-\alpha)}(1-\alpha)\alpha L - \frac{1}{2}[\alpha(1-\alpha)^2 + \alpha^2(1-\alpha)]L = \frac{L}{2}[\alpha]^2.$$

Example 13.3. *Obtain influence line for the reaction at B for the propped cantlever shown in Figs. 13.8(a) and 13.8(b).*

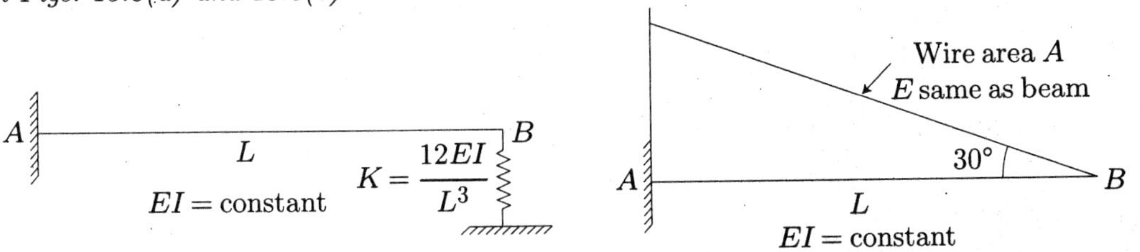

(a) Propped cantilever with spring support **(b) Beam with wire support**

Fig. 13.8: For Example 13.3

Solution : (i) Considering a unit load applied at a distance x from A, the deflection of a cantilever of length L and of constant EI is given by $\frac{x^2}{2EI}\left(L-\frac{x}{3}\right)$. When the unit load is placed at end B of the cantilever, deflection at end $= \frac{L^3}{3EI}$.

Using Maxwell-Betti's theorem,

$$R_B\frac{L^3}{3EI} = 1\frac{x^2}{2EI}\left(L-\frac{x}{3}\right) - 1\frac{R_BL^3}{1^2EI},$$

whence $R_B = \dfrac{6}{5}\dfrac{x^2}{L^3}\left[L-\dfrac{x}{3}\right]$.

(ii) With similar procedure as in (i), the deflection at end B of a cantilever as a unit load is placed at a distance x from the free end is given by $\frac{(L-x)^2}{2EI}\left(\frac{2L+\alpha}{3}\right)$. Considering T as the tension in the wire with Maxwell-Betti's Theorem,

$$T\sin 30^\circ\frac{L^3}{3EI} = 1\frac{(L-x)^2}{2EI}\left(\frac{2L+x}{3}\right) - 1\frac{4TL}{\sqrt{3}AE},$$

whence $T = \dfrac{A(L-x)^2(2L+x)}{8\sqrt{3}IL + AL^3}$.

Example 13.4. *Obtain the influence line for the reaction R_B and R_C of the beam shown in Fig. 13.9.*

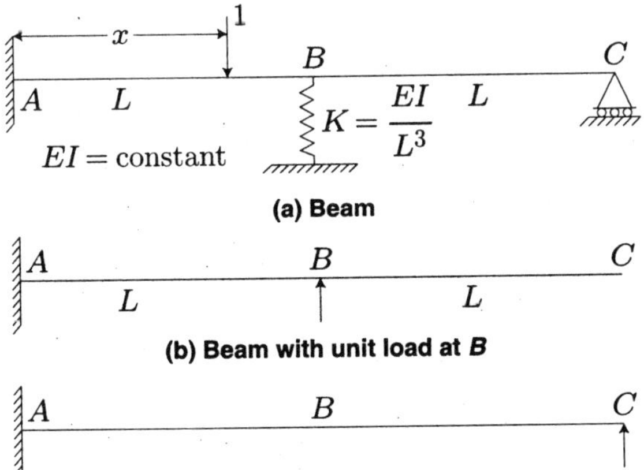

(a) Beam

(b) Beam with unit load at *B*

(c) Beam with unit load at *C*

Fig. 13.9: For Example 13.4

Solution : For unit load placed at a distance x from A of a cantilever beam of length $2L$, the deflection at a distance L and $2L$ for the cantilever beam are to be obtained from two curves.

For $x < L$,

Deflection at a distance L from fixed end $A = \dfrac{x^2}{2EI}\left(L - \dfrac{x}{3}\right)$.

Deflection at a distance $2L$ from fixed end $A = \dfrac{x^2}{2EI}\left(2L - \dfrac{x}{3}\right)$.

For $x > L$,

Deflection at a distance L from fixed end $= \dfrac{L^2(3x - L)}{6EI}$.

Deflection at a distance $2L$ from fixed end $= \dfrac{x^2}{2EI}\left(2L - \dfrac{x}{3}\right)$.

For unit load at a distance L from fixed end A,

Deflection at the point of application of load $= \dfrac{L^3}{3EI}$.

Deflection at free end at a distance $2L$ from fixed end $= \dfrac{5}{6}\dfrac{L^3}{EI}$. (i)

For unit load at a distance $2L$ from fixed end A,

Deflection at a distance L from fixed end $= \dfrac{5L^3}{6EI}$.

Deflection at free end $= \dfrac{8L^3}{3EI}$. (ii)

Using Maxwell-Betti Theorem, combining loading in Figs. 13.9(a) and 13.9(b),

$$R_B\frac{L^3}{3EI} + R_C\frac{5L^3}{6EI} = 1\frac{x^2}{2EI}\left(L - \frac{x}{3}\right) - 1\frac{R_B L^3}{EI}.$$

Similarly, combining Eqns. (i) and (ii),

$$R_B\frac{5L^3}{6EI} + R_C\frac{8L^3}{3EI} = 1\frac{x^2}{2EI}\left(2L - \frac{x}{3}\right).$$

Solving the above equation simultaneously,

$$R_C = \frac{3x^2}{103L^3}[11L - x] \quad \text{and} \quad R_B = \frac{x^2}{103L^3}[18L - 11x].$$

(B) With Conjugate Beam Method :

This probably provides the earliest method for computing influence line ordinates for continuous beams.

Example 13.5. *Compute the ordinates at intervals of $\frac{L}{4}$ of the influence line for R_A for the two span continuous beam of Fig. 13.10(a) for which EI is constant.*

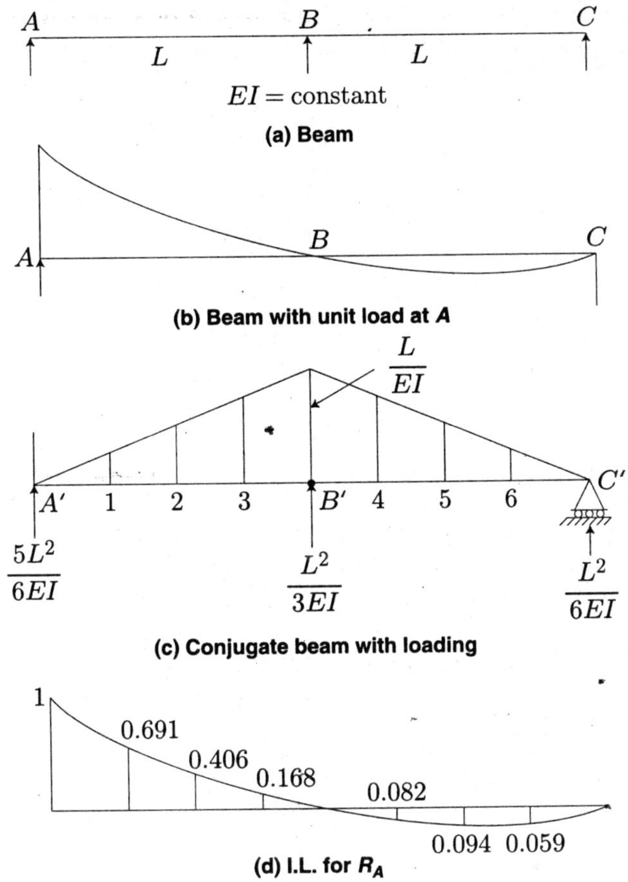

Fig. 13.10: For Example 13.5

Solution : The redundant reaction at A is removed and a unit vertical force is applied as in Fig. 13.10(b). The conjugate beam for computing the deflection at various sections of the beam is shown in Fig. 13.10(c). The moments of the conjugate beam at various portions are as under :

$$M_6 = \frac{15}{384}\frac{L^3}{EI}; \qquad M_5 = \frac{L^3}{16EI}$$

$$M_4 = \frac{7L^3}{128EI}; \qquad M'_B = 0$$

$$M_3 = \frac{43}{384}\frac{L^3}{EI}; \qquad M_2 = \frac{13}{48}\frac{L^3}{EI}$$

$$M_1 = \frac{177}{384}\frac{L^3}{EI}; \qquad M'_A = \frac{2}{3}\frac{L^3}{EI}.$$

The moments computed above are proportional to the deflection at the corresponding sections of the beam. Since for influence line ordinates of R_A, the deflection at A is to be unity, the influence line ordinates for other sections along the beam are obtained by dividing the conjugate beam moments by M'_A, i.e., $\frac{2}{3}\frac{L^3}{EI}$. The resulting influence line is shown in Fig. 13.10(d).

Example 13.6. *Compute the ordinates at intervals of $\frac{L}{4}$ of the influence line for moments at the midspan of BC for the beam shown in Fig. 13.11.*

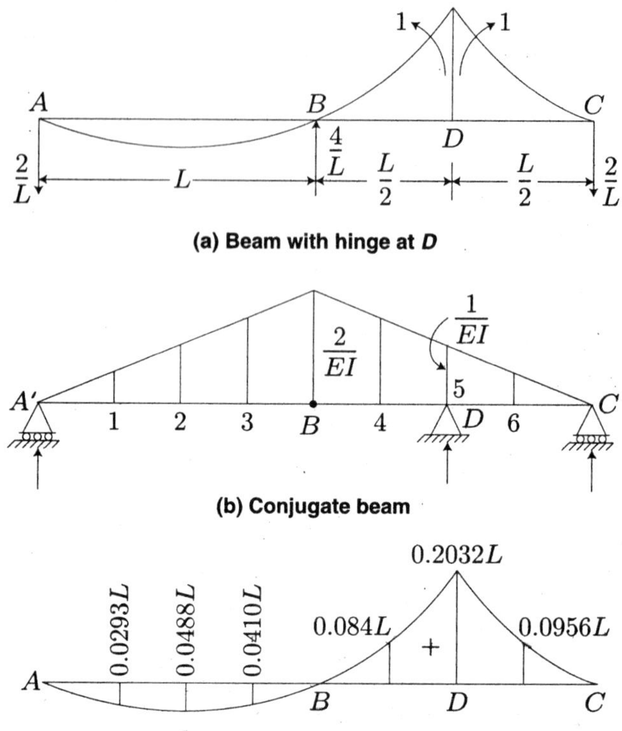

(c) I.L. for M_D

Fig. 13.11: For Example 13.6

Solution : The influence line diagram for M_D is obtained by removing restraint at D and providing a pin there. Unit couples are applied on the beam on each side of the pin. The modified beam is shown in Fig. 13.11(a). The conjugate beam is shown in Fig. 13.11(b). The moments of the conjugate beam at various points are as under:

$$M_1 = +\frac{0.0781L^2}{EI};\ M_2 = +0.1248\ \frac{L^2}{EI}$$

$$M_3 = +0.1094\frac{L^2}{EI};\ M'_B = 0;\ M_4 = -0.2242\ \frac{L^2}{EI}$$

$$M_5 = -0.5418\frac{L^2}{EI};\ M_6 = -0.2552\frac{L^2}{EI}\quad \text{and}\quad M'_C = 0.$$

Also $R_5 = 2.667\dfrac{L}{EI}$.

The value of influence line ordinate at each of the sections is computed by dividing each moment by $\theta = 2.667\frac{L}{EI}$. The resulting influence line is shown in Fig. 13.11(c).

Example 13.7. *Compute the ordinates of the influence line for shear at midpoint of BC for the beam shown in Fig. 13.12(a). Length AB = BC = L and EI = constant.*

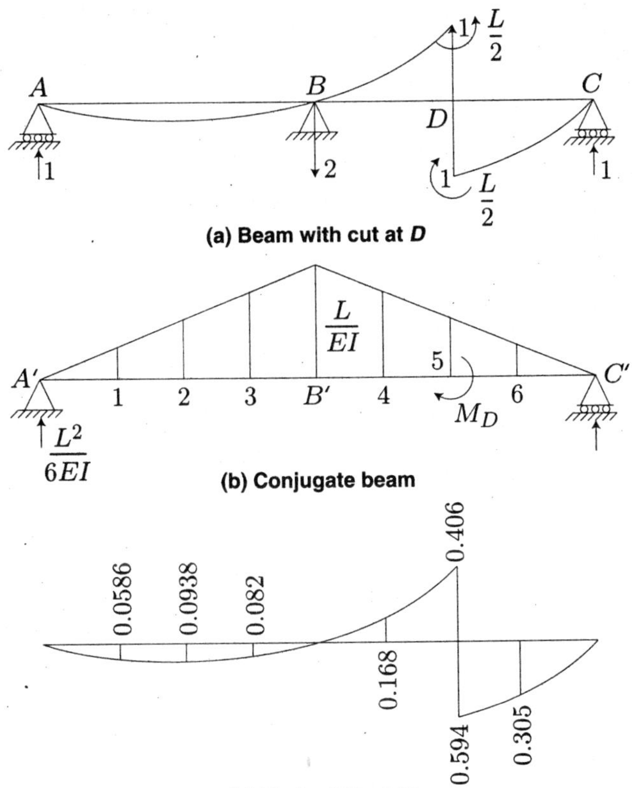

Fig. 13.12: For Example 13.7

Solution : To find the influence line for S.F. at D, the shearing restraint at D has to be removed without impairing the moment resistance. This is accomplished by cutting the beam at D and inserting a sliding device permitting a relative vertical displacement between the two beam-ends adjacent to the cut, but which will require that these two ends be tangent to a common slope in the distorted beam. The deflected shape and the induced reactions and couples are as shown in Fig. 13.12(a). The conjugate beam with the loading and reactions is shown in Fig. 13.12(b). Moment M_D is required in the conjugate beam as at D the two parts of the beam deflect relative to each other. Since the tangent to the deflected parts of the beam at D must be parallel, the shear force in the conjugate beam just at the right of the section must be the same as at the left of the section. The correct sense of M_D is determined by inspection and indicated in Fig. 13.12(b). Values of M_D is given by

$$M_D = \frac{L^2}{6EI}(1.5L) - \frac{1}{2}\left(\frac{L}{EI}\right)(L)\left(\frac{L}{2}+\frac{L}{3}\right) - \frac{L}{2EI}\left(\frac{L}{2}\right)\left(\frac{L}{4}\right)$$
$$-\frac{1}{2}\frac{L}{2EI}\left(\frac{L}{2}\right)\left(\frac{L}{3}\right) - \frac{5L^2}{6EI}\left(\frac{L}{2}\right) + \frac{1}{2}\left(\frac{L}{2EI}\right)\left(\frac{L}{2}\right)\left(\frac{L}{6}\right)$$
$$= -\frac{2}{3}\frac{L^3}{EI}.$$

Moments in the conjugate beam at indeterminate sections are as below:

$$M_1 = +\frac{15}{384}\frac{L^3}{EI}, \quad M_2 = +\frac{1}{16}\frac{L^2}{EI}$$

$$M_3 = +\frac{7}{128}\frac{L^3}{EI}, \quad M'_B = 0$$

$$M_4 = -\frac{43}{384}\frac{L^3}{EI}, \quad M_{5C} = -\frac{13}{48}\frac{L^3}{EI}$$

$$M_{5R} = +\frac{19}{48}\frac{L^3}{EI}, \quad M_6 = +\frac{39}{192}\frac{L^3}{EI}.$$

The above moments are to be divided by relative deflections of the two ends at D as represented by M_D, and ordinates thus obtained give the influence line for shear force at D, as shown in Fig. 13.12(c).

The influence lines for shear force can be obtained more easily by statics from the influence lines for reactions. This is also the case of influence lines for moments.

Example 13.8. *Compute the ordinates of the influence line for reaction at B of the three-span continuous beam of Fig. 13.13(a).*

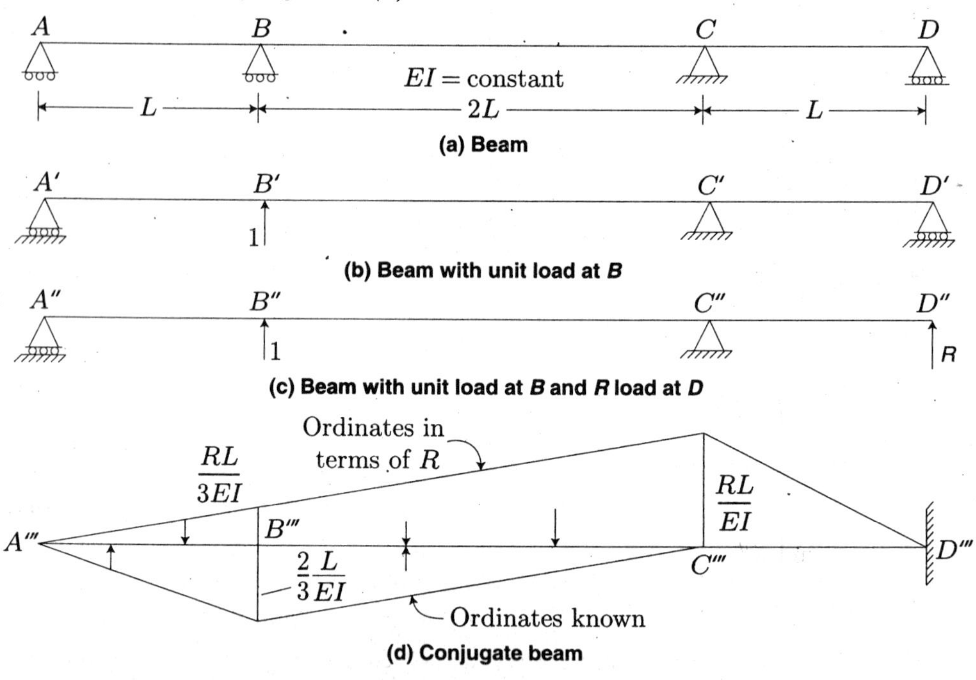

(a) Beam

(b) Beam with unit load at B

(c) Beam with unit load at B and R load at D

(d) Conjugate beam

Fig. 13.13: For Example 13.8

Solution : To obtain the influence line ordinates for R_B, the support at B is first removed and unit vertical force applied there, the resulting structure being shown in Fig. 13.13(b). The support at D' of the structure in beam in Fig. 13.13(b) is then removed and an upward force R replacing the reaction is taken as shown in Fig. 13.13(c). The loaded conjugate beam is shown in Fig. 13.13(d). Since the beam at D' of Fig. 13.13(b) cannot deflect, the moment at D'' of the conjugate beam must be zero. From this condition, magnitude of R can be evaluated. The value of R thus obtained is used to obtain the magnitude of the moment ordinates on the conjugate beam which were initially expressed in terms of R. Finally, the moments in the conjugate beam at the centres of the various segments are computed.

From Fig. 13.13(d),

$$R_A''' = \frac{1}{2}\frac{RL}{EI}(3L)\left(\frac{1}{3}\right) - \frac{1}{6L}\left(\frac{2}{3}\frac{L}{EI}\right)(3L)\left(\frac{5L}{3}\right)$$

$$= \uparrow \frac{RL^2}{2EI} - \frac{5L^2}{9EI}\downarrow$$

$$R_C''' = \uparrow \frac{RL^2}{EI} - \frac{4L^2}{9EI}\downarrow.$$

From $M_D''' = \left(\frac{RL^2}{EI} - \frac{4L^2}{9EI}\right)L + \frac{1}{2}\frac{RL}{EI}(L)\left(\frac{2}{3}L\right) = 0,\ R = \frac{1}{3}.$

and $R_A''' = \frac{7L^2}{18EI}$ and $R_C''' = -\frac{2L^2}{9EI}.$

For deflections at various points at interval of $\frac{L}{2}$, the bending moments at the conjugate beam are calculated.

Thus at $x = 0$, i.e., at A''', $\Delta_0 =$ B.M. in conjugate beam $= 0$.

At $x = \frac{L}{2}$ from end A''',

$$\Delta_1 = \frac{7L^2}{18EI}\left(\frac{L}{2}\right) - \frac{1}{2}\left[\frac{L}{3EI} - \frac{L}{18EI}\right]\left(\frac{L}{2}\right)\left(\frac{L}{6}\right) = \frac{79L^3}{432EI}.$$

At $x = L$ from end A''',

$$\Delta_2 = \frac{7L^2}{18EI}(L) - \frac{1}{2}\left(\frac{5L}{9}\right)(L)\left(\frac{L}{3}\right) = \frac{16L^3}{54EI}.$$

At $x = \frac{3L}{2}$ from end A''',

$$\Delta_3 = \frac{7L^2}{18EI}\left(\frac{3L}{2}\right) + \frac{1}{2}\left(\frac{L}{6EI}\right)\left(\frac{3L}{2}\right)\left(\frac{L}{2}\right) - \frac{1}{2}\left(\frac{2}{3}\frac{L}{EI}\right)(L)\left(\frac{5L}{6}\right)$$
$$- \left(\frac{L}{2EI}\right)\left(\frac{L}{2}\right)\left(\frac{L}{4}\right) - \frac{1}{2}\left(\frac{L}{6EI}\right)\left(\frac{L}{2}\right)\left(\frac{L}{3}\right) = \frac{42L^2}{144EI}.$$

At $x = 2L$ from end A''',

$$\Delta_4 = \frac{7L^2}{18EI}(2L) + \frac{1}{2}\left(\frac{2L}{9EI}\right)(2L)\left(\frac{2}{3}L\right) - \frac{1}{2}\left(\frac{2}{3}\frac{L}{EI}\right)(L)\left(\frac{4L}{3}\right)$$
$$- \left(\frac{L}{3EI}\right)(L)\left(\frac{L}{2}\right) - \frac{1}{2}\left(\frac{L}{3EI}\right)(L)\left(\frac{2L}{3}\right) = \frac{11L^3}{54EI}.$$

At $x = \frac{5L}{2}$ from end A''',

$$\Delta_5 = \frac{7L^2}{18EI}\left(\frac{5L}{2}\right) + \frac{1}{2}\left(\frac{5L}{18EI}\right)\left(\frac{5L}{2}\right)\left(\frac{5L}{6}\right) - \frac{1}{2}\left(\frac{2}{3}\frac{L}{EI}\right)\left(\frac{11L}{6}\right)$$
$$- \left(\frac{L}{6EI}\right)\left(\frac{3L}{2}\right)\left(\frac{3L}{4}\right) - \frac{1}{2}\left(\frac{L}{2EI}\right)\left(\frac{3L}{2}\right)(L) = \frac{56L^3}{452EI}.$$

At $x = 3L$ from A''', $\Delta_6 = 0$.

At $x = \frac{7L}{2}$ from A''',

$$\Delta_7 = -\frac{L^2}{9EI}\left(\frac{L}{2}\right) + \frac{L}{6EI}\left(\frac{L}{2}\right)\left(\frac{L}{4}\right) + \frac{1}{2}\left(\frac{L}{6EI}\right)\left(\frac{L}{2}\right)\left(\frac{L}{3}\right) = -\frac{L^3}{48EI}.$$

At $x = 4L$ from A''', $\Delta_8 = 0$.

Hence, ordinates for IL at various points are as below:

At $\quad A = 0$

At $\quad \frac{L}{2}$ from $A = \frac{79}{432}\left(\frac{54}{16}\right) = 0.617$

At $\quad L$ from $A = \frac{16}{54}\left(\frac{54}{16}\right) = 1$

At $\quad \frac{3L}{2}$ from $A = \frac{42}{144}\left(\frac{54}{16}\right) = 0.984$

At $\quad 2L$ from $A = \frac{11}{54}\left(\frac{54}{16}\right) = 0.688$

At $\quad \frac{5L}{2}$ from $A = \frac{56}{432}\left(\frac{54}{16}\right) = 0.4375$

At $\quad 3L$ and $4L$ from $A = 0$.

At $\quad \frac{7L}{2}$ from $A = \frac{1}{48}\left(\frac{54}{16}\right) = 0.00703.$

(C) Influence Line by Moment Distribution :

Influence lines can be obtained most suitably by use of moment distribution. Various techniques of computing influence lines involve methods in which (i) bending moment diagrams for obtaining influence lines ordinate are obtained by moment distribution procedure, (ii) moments at supports are calculated separately for various positions of unit load on the span by moment distribution and (iii) superposition of the effects of fixed end moments, by moment distribution procedure is made.

(i) *Bending moment diagrams obtained by Moment Distribution procedure.* The method will be illustrated by means of an Example which follows.

Example 13.9. *Obtain an influence line at 1 m intervals for the moment at support B of the beam shown in Fig. 13.14.*

Solution : The moment distribution steps carried on by introducing a hinge at B, and applying unit moment these are illustrated in Table 13.1. The stiffness of member CB and DE is taken as the modified stiffness.

The $\frac{M}{EI}$ diagram is drawn for the resulting B.M.s thus calculated and is shown in Fig. 13.14(b).

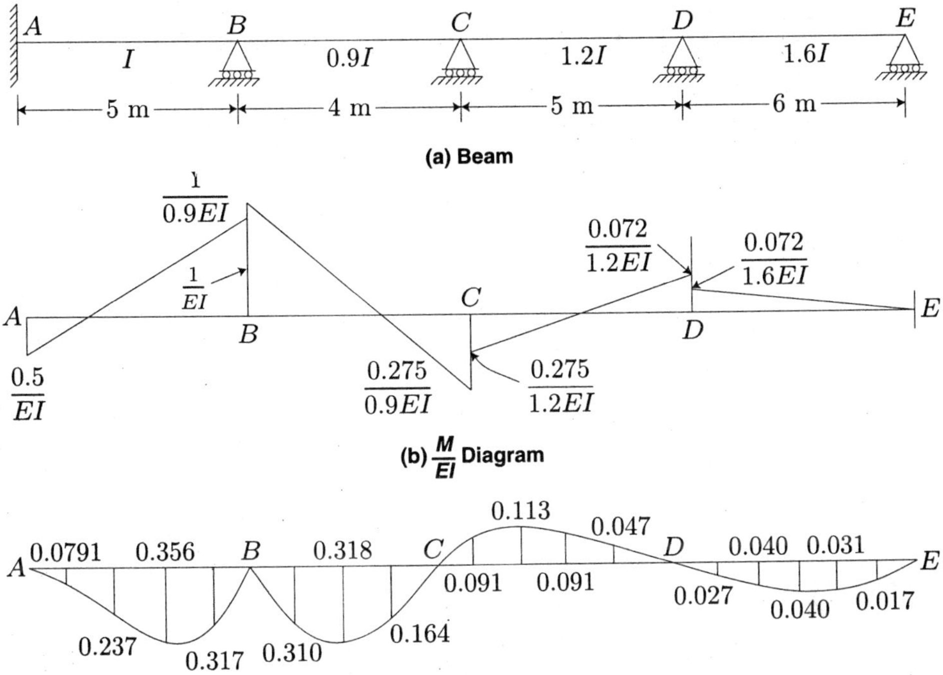

Fig. 13.14: For Example 13.9

Table 13.1. For Example 13.9

A	B		C			D	E	
	AB	BA	BC	CB	CD	DC	DE	ED
Rel. K				0.675	0.96	0.96	0.80	
D.F				0.413	0.987	0.595	0.455	
C.O.F.	← 0.50 —		— 0.50 →		← 0.50 →		← 0.50 —	
F.E.M.	−0.50	−1	+1	+0.50	0	0	0	0
Bal.	0	0	0	−0.207	−0.293	0	0	0
C.O.M.	0	0	0	0	0	−0.147	0	0
Bal.	0	0	0	0	0	+0.080	+0.067	0
C.O.M.	0	0	0	0	+0.040	0	0	0
Bal.	0	0	0	−0.017	−0.023	0	0	0
C.O.M.	0	0	0	0	0	−0.012	0	0
Bal.	0	0	0	0	0	+0.007	+0.005	0
C.O.M.	0	0	0	0	+0.0030	0	0	0
Bal.	0	0	0	−0.0010	−0.0010	0	0	0
$\sum$	−0.50	−1	+1	+0.275	−0.275	−0.072	+0.072	0

Values of deflection at 1 m intervals from A, and the magnitude of θ_{BA} and θ_{BC} are calculated. The total change in slope at point B is given by $(\theta_{BA}+\theta_{BC})=\alpha$.

Since the influence line for M_B will correspond to unit value of α, the ordinates of deflection at every segment have to be divided by α to get the ordinates for influence line diagram. The I.L. diagram thus obtained is shown in Fig. 13.14(c).

From Fig. 13.14(b),

$$EI(\theta_{BA}+\theta_{BC}) = \left[\frac{1}{2}(1)(5)-\frac{1}{2}(0.50)(5)+\left\{\frac{1}{2}\,\frac{1}{0.90}(4)\frac{2}{3}(4) - \frac{1}{2}\,\frac{0.275}{0.90}(4)\left(\frac{1}{3}\right)(40)\right\}\frac{1}{4}\right]$$

$$= 1.25 + 1.28 = 2.53.$$

For span AB,

$$EI\Delta = \frac{1}{2}\left(\frac{x}{5}\right)\left(\frac{x^2}{3}\right)-\frac{1}{2}\left(\frac{x}{2}\right)\left(\frac{2x}{3}\right)-\left(\frac{1}{2}-\frac{x}{10}\right)\frac{x^2}{6}=\frac{x^3}{20}-\frac{x^2}{4}.$$

For $EI(\theta_{BA}+\theta_{BC})=$ unity,

$$\text{Ordinate} = \frac{1}{2.53}\left[\frac{x^3}{20}-\frac{x^2}{4}\right].$$

For span BC,

$$EI\Delta = -1.28x+1.1\frac{x^3}{3}+\left[1.1-\frac{1.1}{4}x\right]\left(\frac{x^2}{6}\right)-0.305\left(\frac{x}{4}\right)\left(\frac{x^2}{6}\right)$$

$$= -1.28x+0.556x^2-0.0585x^3.$$

With $EI(\theta_{BA}+\theta_{BC})=$ unity,

$$\text{Ordinate} = \frac{1}{2.53}\left[-1.28x+0.556x^2-0.0585x^3\right].$$

Similar calculations are made for other spans and resulting ordinates are plotted in 13.14(c) to give the I.L. diagram.

Example 13.10. *Draw the influence line for B.M. at E of the beam ABCD shown in 13.15(a).*

Table 13.2. For Example 13.10

	AB	BA	BC	CB	CD	DC
D.F.				0.60	0.40	
C.O.F.			← 0.50 →			
	0	+4.00	0	0	0	0
Bal.	0	0	−4.00	0	0	0
C.O.M.	0	0	0	−2.00	0	0
Bal	0	0	0	+1.20	+0.80	0
Σ		+4.00	−4.00	−0.80	+0.80	0

Solution : A hinge is inserted at E and equal and opposite moments M_E = unity applied there as shown in Fig. 13.15(b). The resulting moment diagram is obtained by moment distribution, as shown in Table 13.2. The conjugate beam $A'B'C'D'$ with loading is shown in Fig. 13.15(b).

In the conjugate beam,

$$EIR'_D = \frac{1}{3}\left[\frac{0.80(9)}{2}\right] = 1.20 \downarrow.$$

Also $EIR'_A = 61.61 \downarrow$ and $EIR'_E = 91.31 \uparrow$.

Moments are calculated at sections 1.50 m apart. From end A such EI (moments) are given by

$M_1 = 61.725;\; M_E = 124.22;\; M_2 = 95.39$

$M_3 = 68.81;\; M_4 = 45.24;\; M_5 = 25.41$

$M_6 = 10.08;\; M'_B = 0;\; M_7 = -4.50;\; M_8 = -4.80$

$M_9 = -2.70;\; M'_C = 0;\; M_{10} = 1.88;\; M_{11} = 2.70$

$M_{12} = 2.70;\; M_{13} = 2.13;\; M_{14} = 1.17.$

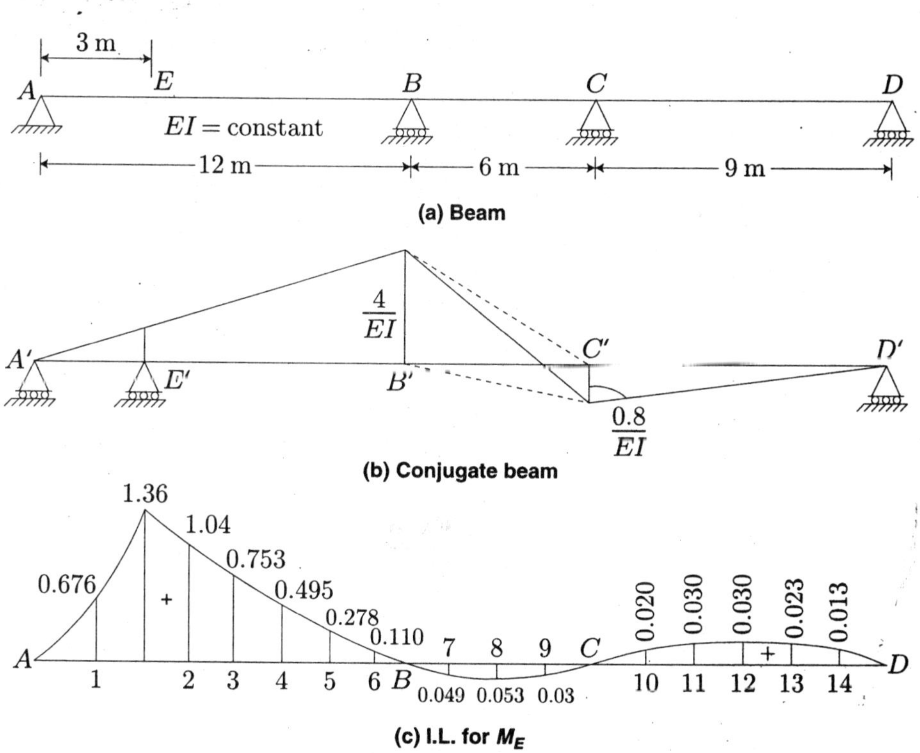

Fig. 13.15: For Example 13.10

The ordinates of the influence line diagram at various sections are obtained by dividing the value of moments in conjugate beam by R'_E. The influence line is indicated in Fig. 13.15(c).

(ii) *Use of Moment Distribution considering various positions of unit load.* The most direct procedure is to place a unit load at a series of points. Calculate the fixed end moments or moments due to the load in the various positions, determine the moments at the supports from the data of the figure and finally calculate by statics, the shear, moment or reaction for which the influence line is being drawn.

Example 13.11. *Obtain influence lines for bending moments at support C and centre of span BC of the beam shown in Fig. 13.16.*

Solution : Value of M_C for identical position of load between B, C and D, C will be same. For positions of load between A, B and E, D will vary linearly. Maximum value of M_C for loads between A, B and E, D will occur when load is at A and E.

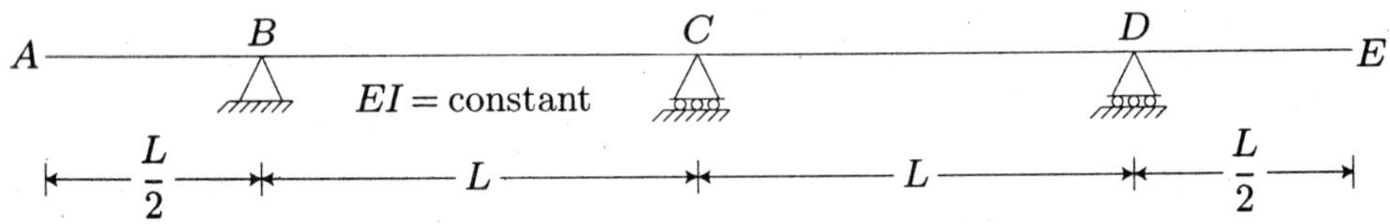

(a) Beam

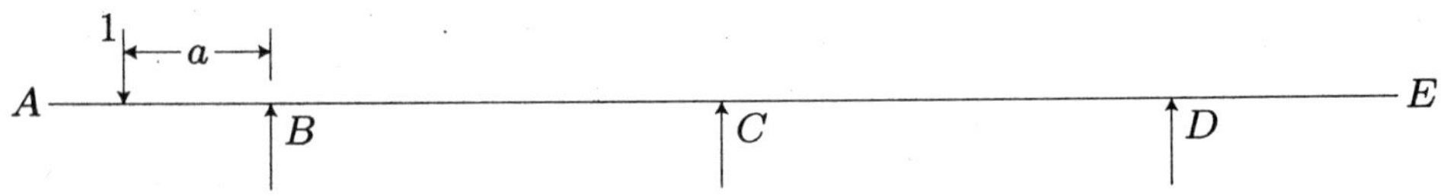

(b) Loading for *MD* in Table 15.3a

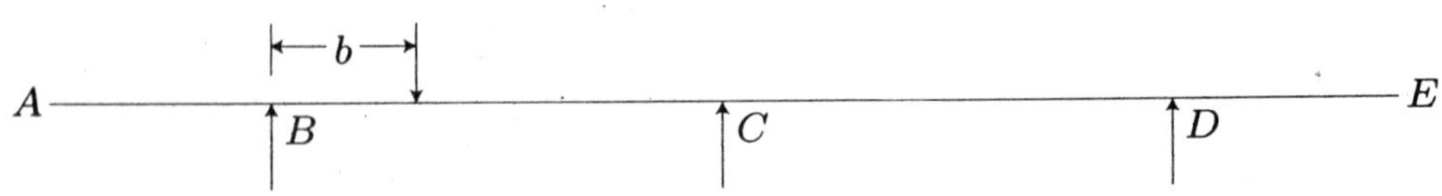

(c) Loading for *MD* in Table 15.3b

Fig. 13.16: For Example 13.11

For position of load between A and B, B.M. at centre of BC, i.e.,

$$M_{BC} = \frac{M_B + M_C}{2}.$$

For positions of load between B, C,

$$M_{BC} = \text{Simply supported B.M. at centre of } BC + \frac{M_C}{2}.$$

For position of load between C and D,

$$M_{BC} = \frac{M_C}{2}.$$

Table 13.3 shows application of moment distribution for load positions between AB and BC, respectively.

Table 13.3(a) shows the moment distribution for unit load at distance a from B as in Fig. 13.16(a) and Table 13.3(b) gives the moment distribution for unit load at distance b from B on span BC.

Table 13.3. For Example 13.11

Table 13.3(a) For unit load as in Fig. 13.16(a)

		B		C		D
D.F.	0	1	$\frac{1}{2}$	$\frac{1}{2}$	1	0
C.O.F.		← 0.50 →		← 0.50		
F.E.M.	$-a$	0	0	0	0	0
Bal.	0	$+a$	0	0	0	0
C.O.M.	0	0	$+\frac{a}{2}$	0	0	0
Bal.	0	0	$-\frac{a}{4}$	$-\frac{a}{4}$	0	0
	$-a$	$+a$	$+\frac{a}{4}$	$-\frac{a}{4}$	0	0

Table 13.3(b) For unit load at distance *b* from *B* on span *BC*, as shown in Fig. 13.16(c)

		B		C		D
D.F.	0	1	$\frac{1}{2}$	$\frac{1}{2}$	1	0
C.O.F.		0.50 →		← 0.50		
F.E.M.	0	$+\frac{bc^2}{L^2}$	$-\frac{cb^2}{L^2}$	0	0	
Bal.	0	$-\frac{bc^2}{L^2}$	$+\frac{cb^2}{2L^2}$	$+\frac{cb^2}{2L^2}$	0	
C.O.M.	0	0	$-\frac{bc^2}{2L^2}$	0	0	
Bal.	0	0	$+\frac{bc^2}{4L^2}$	$+\frac{bc^2}{4L^2}$	0	
	0	0	$-\frac{bc}{L^2}\left[\frac{b}{2}+\frac{c}{4}\right]$	$+\frac{bc}{L^2}\left[\frac{b}{2}+\frac{c}{4}\right]$	0	0

Values of M_C and M_{BC} for various values of a and b can be obtained and are plotted. With values at $\frac{L}{4}$ intervals, the pattern of influence line diagrams for M_C and M_{BC} are shown in Figs. 13.17(a) and 13.17(b).

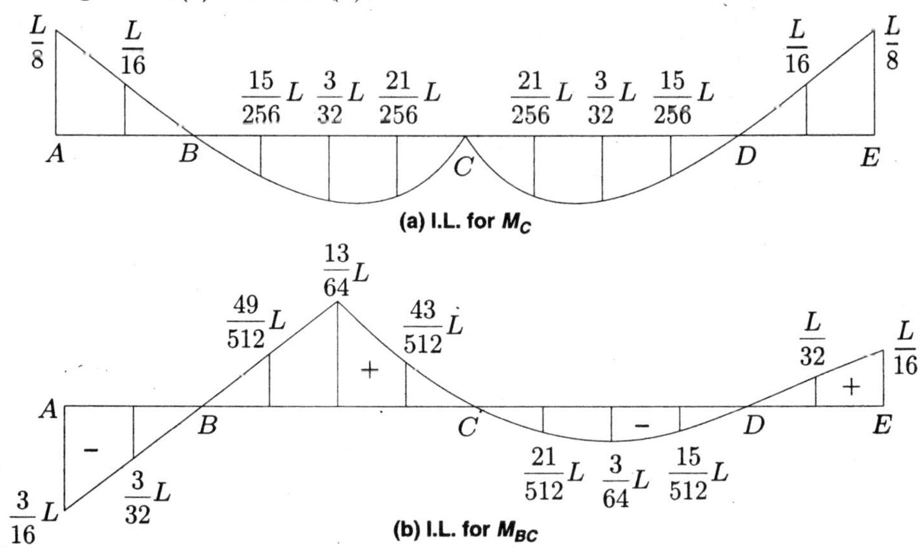

(a) I.L. for M_C

(b) I.L. for M_{BC}

Fig. 13.17: For Example 13.11

In case of continuous beam of large numbers of span, the calculation of bending moment for various positions of unit load have to be done separately in terms of specific values of b and c relative to L, i.e., $b = \frac{L}{4}$, $c = \frac{3L}{4}$, $b = \frac{L}{2}$, $c = \frac{L}{2}$; $b = \frac{3L}{4}$ $c = \frac{L}{4}$ and so on.

Example 13.12. *Compute influence line ordinates for various functions in case of a continuous beam shown in Fig. 13.18(a) at intervals shown in the figure.*

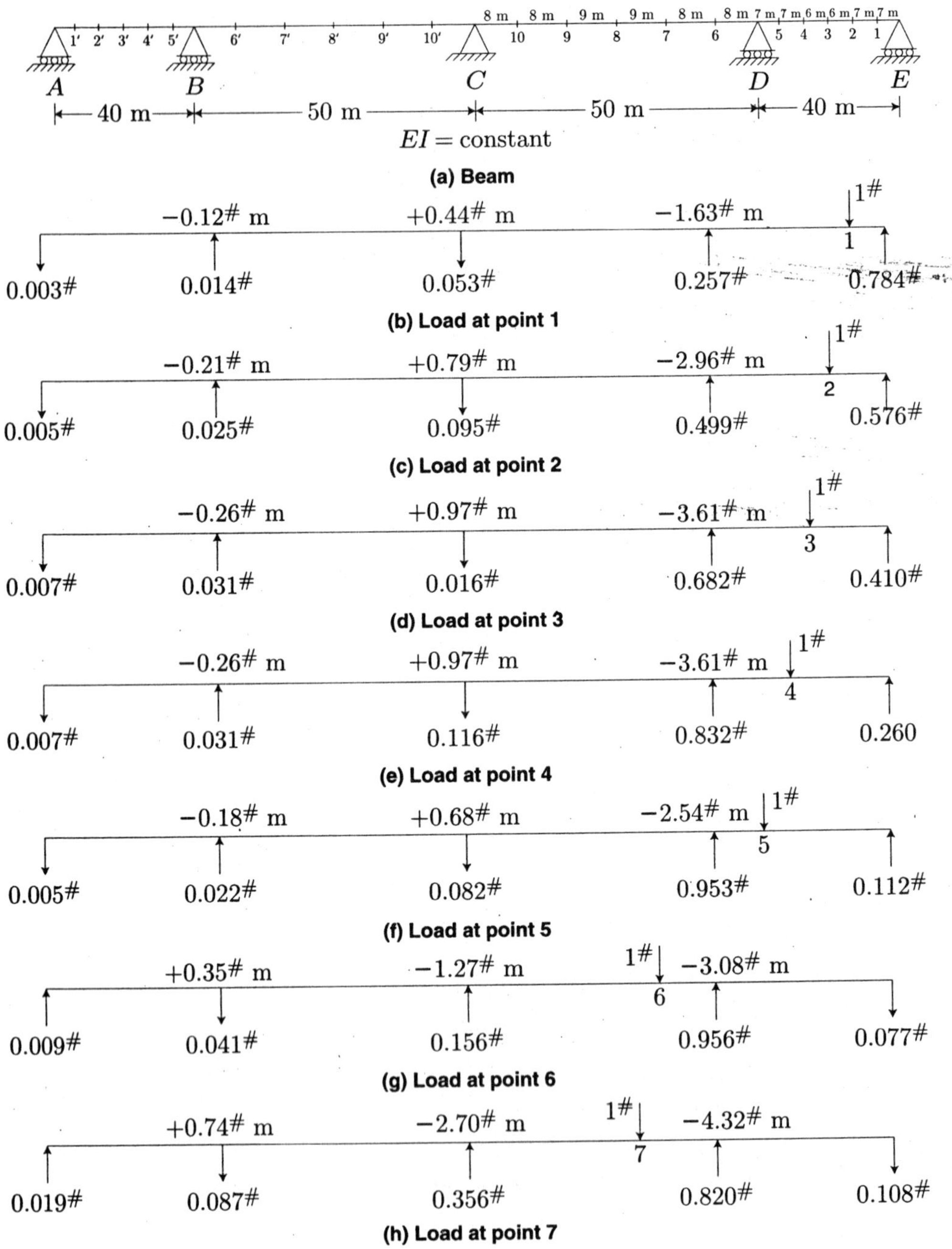

Fig. 13.18: For Example 13.12 (Contd.)

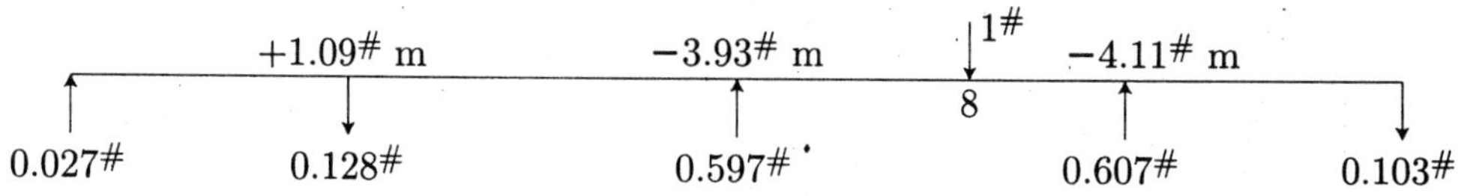

(i) Load at point 8

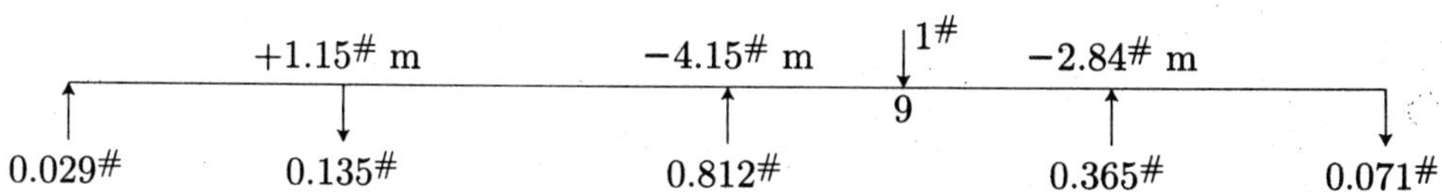

(j) Load at point 9

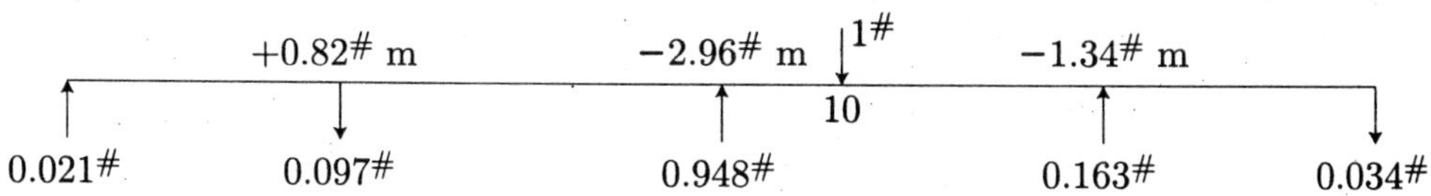

(k) Load at point 10

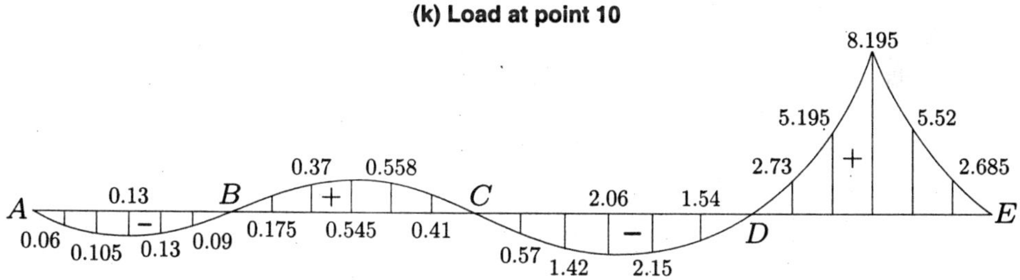

(l) I.L. for M_{DE}

Fig. 13.18: For Example 13.12

Solution : Figs. 13.18(a) to 13.18(k) show the values of support moments and reactions at supports, obtained in terms of moment distribution procedure for various positions of load from position 1 to position 10. The data in the figures can be used to calculate influence line for various functions, one of which is shown in Fig. 13.18(f). In arriving at the values of influence line ordinates advantage of symmetry has been utilised and calculations by moment distribution for load positions spans AB and BC are not necessary.

Typical calculations for influence line ordinates for B.M. at centre of DE, M_{DE} is illustrated below.

For loads in identical positions as points 1 to 10 in span AB and BC, magnitude of B.M., $M_{DE} = \frac{1}{2} M_D$.

Thus ordinates at position identical to 1 to 10 in spans AB and BC are

Load at $1'$ ordinate $= -0.06$	Load at $6' = +0.175$
Load at $2'$ ordinate $= -0.105$	Load at $7' = +0.370$
Load at $3' = -0.13$	Load at $8' = +0.545$
Load at $4' = -0.13$	Load at $9' = +0.558$
Load at $5' = -0.09$	Load at $10' = +0.410$.

For loads in span DC, B.M. $M_{DE} = \frac{1}{2} M_D$.

Thus ordinates for

Load at 6 = −1.54 Load at 9 = −1.42

Load at 7 = −2.16 Load at 10 = −0.67.

Load at 8 = −2.06

For loads in span ED

B.M. M_{DE} = simply supported B.M. at centre of $DE + \frac{1}{2} M_D$.

Thus ordinates for

Load at 1 = 3.500 − 0.815 = 2.685 Load at 4 = 7.000 − 1.805 = 5.195

Load at 2 = 7.000 − 1.48 = 5.520 Load at 5 = 3.500 − 1.27 = 2.230.

Load at 3 = 10 − 1.805 = 8.195

(iii) *Influence Line obtained by Superposition of the Effects of Fixed end moments.* This method is very useful for determining influence lines for the end moments of indeterminate beams or frames. It involves using moment distribution to determine the separate effects of each of the fixed end moments of the various loaded members. These separate effects can be superimposed to give the total end moments. When properly organised, it is a very effective method of obtaining influence lines for the end moments. This method involves applying a unit fixed-end moment in turn at each end of every member and computing the resulting end moments by moment distribution procedure. The fixed-end moments developed by unit load placed in turn at each of the various load points are then calculated. The data from the two sides of support moments calculated above are then combined to find the end moments throughout caused by a unit load at each of the various load points of the structure.

From the influence lines for the end moments, other influence lines can be computed by statics.

Example 13.13. *Using method of moment distribution compute the ordinates of the influence lines for B.M. at support B, at 1 m intervals for the beam in Fig. 13.19.*

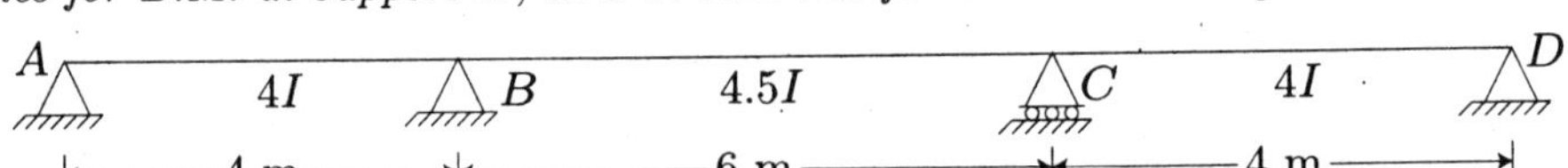

Fig. 13.19: For Example 13.13

Solution : The resulting moments at supports have been calculated and shown in Table 13.4, when M_{FBA} of +100 units is considered at member BA. Values of resulting moments then are

$$M_{BA} = +46.69 = -M_{BC};\ M_{CB} = 13.28 = -M_{FCD}.$$

Similarly, for M_{FBC} of +100,

$$M_{BA} = -53.2 = -M_{BC};\ M_{CB} = -13.3 = -M_{FCD}.$$

For M_{FCB} of +100,

$$M_{BA} = +13.3 = -M_{BC};\ M_{CB} = +53.2 = M_{CD}.$$

For M_{FCD} of +100,

$$M_{BA} = +13.3 = -M_{BC};\ \text{and } M_{CB} = -46.8 = -M_{FD}.$$

In terms of F.E.M., B.M. M_{BA} can be written as

$$M_{BA} = 0.4669 M_{FBA} - 0.532 M_{FBC} + 0.133 M_{FCB} + 0.133 M_{FCD}.$$

Table 13.4. For Example 13.13

	A	B		C		D
	AB	*BA*	*BC*	*CB*	*CD*	*DC*
DF		0.50	0.50	0.50	0.50	
C.O.F.	—— 0.50 ⟶		⟵ 0.50 ⟶		⟵ 0.50 ——	
F.E.M.	0	+100	0	0	0	0
Bal.	0	−50	−50	0	0	0
C.O.M.	0	0	0	−25	0	0
Bal.	0	0	0	+12.5	+12.50	0
C.O.M.	0	0	+6.25	0	0	0
Bal.	0	−3.12	−3.13	0	0	0
C.O.M.	0	0	0	−1.56	0	0
Bal.	0	0	0	+0.78	+0.78	0
C.O.M.	0	0	+0.39	0	0	0
Bal.	0	−0.19	−0.20	0	0	0
$\sum$	0	+46.69	−46.69	−13.28	+13.28	0

The fixed-end moments on supports are then obtained for various positions of unit load. For span *AB* and *DC*,

	M_{FAB}, M_{FDC}	M_{FBA}, M_{FCD}	Modified M_{FBA} / Modified M_{FCD}
$x = 1$ m from *A* or *D*,	$+\frac{9}{16}$	$-\frac{3}{16}$	$-\frac{3}{16} - \frac{1}{2}\left(\frac{9}{16}\right) = -\frac{15}{32}$
2 m from *A* or *D*,	$+\frac{1}{2}$	$-\frac{1}{2}$	$-\frac{3}{4}$
3 m from *A* or *D*,	$+\frac{3}{16}$	$-\frac{9}{16}$	$-\frac{21}{32}$
4 m from *A* or *D*,	0	0	0

For span *BC*	M_{FBC}	M_{FCB}
1 m for *B*	$+\frac{25}{36}$	$-\frac{5}{36}$
2 m	$+\frac{32}{36}$	$-\frac{16}{36}$
3 m	$+\frac{3}{4}$	$-\frac{3}{4}$
4 m	$+\frac{16}{36}$	$-\frac{32}{36}$
5 m	$+\frac{5}{36}$	$-\frac{25}{36}$

Table 13.5 shows the magnitude of the influence line diagrams for moment at B.

Table 13.5. For Example 13.13

For AB		For BC		For CD	
Location	M_{FBA} (Mod.)	M_{FBC}	M_{FCB}	M_{FDC} (Mod.)	$M_{BA} = 0.4669\ M_{FBA} - 0.523\ M_{FBC} + 0.133\ M_{FCB} + 0.133\ M_{FCD}$
A	0	0	0	0	0
1	$-\frac{15}{32}$	0	0	0	−0.219
2	$-\frac{3}{4}$	0	0	0	−0.350
3	$-\frac{21}{32}$	0	0	0	−0.307
B	0	0	0	0	0
B	0	0	0	0	0
5	0	$+\frac{25}{36}$	$-\frac{5}{36}$	0	−0.382
6	0	$+\frac{32}{36}$	$-\frac{16}{36}$	0	−0.524
7	0	$+\frac{3}{4}$	$-\frac{3}{4}$	0	−0.492
8	0	$+\frac{16}{36}$	$-\frac{32}{36}$	0	−0.351
9	0	$+\frac{5}{36}$	$-\frac{25}{26}$	0	−0.165
10C	0	0	0	0	0
C	0	0	0	0	0
11	0	0	0	$+\frac{21}{32}$	+0.087
12	0	0	0	$+\frac{3}{4}$	+0.100
13	0	0	0	$+\frac{15}{32}$	+0.062
14D	0	0	0	0	0

Example 13.14. *Superimposing effects of fixed moments prepare an influence table for the end moments of the frame in Fig. 13.20(a) with ordinate at every 1 m interval, for external vertical load only.*

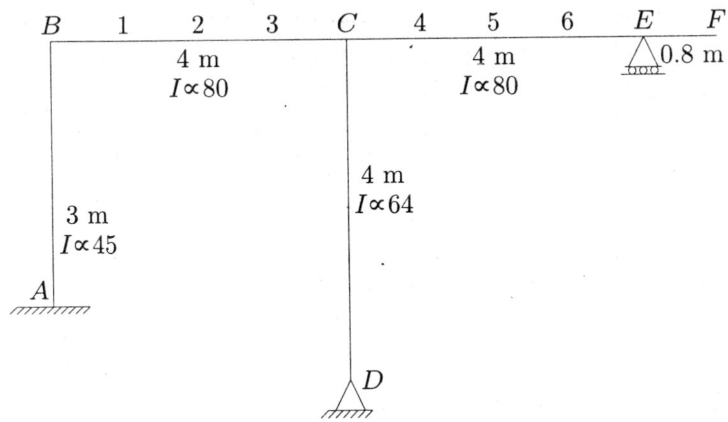

(a) Frame

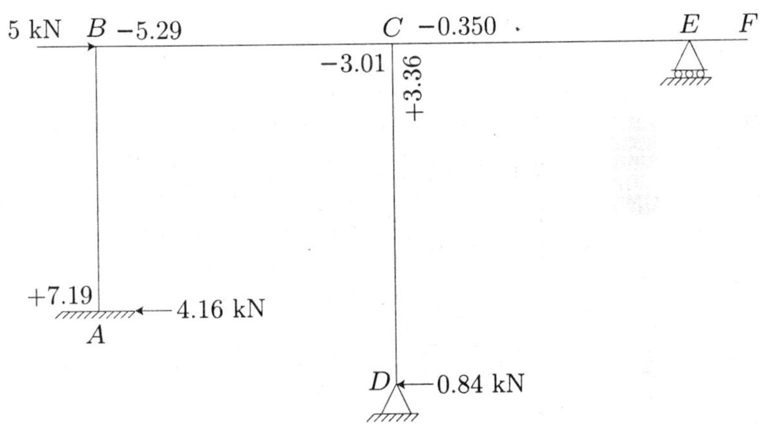

(b) Sidesway Moments

Fig. 13.20: For Example 13.14

Solution : The solution involves consideration of separately fixed-end moments of magnitude 1 at each end of the horizontal member and resulting moments are obtained by moment distribution without considering side-sway. The resulting unbalanced shear for the case is then calculated and side-sway effects from Fig. 13.21(b) is then included to get the final resulting moments. The procedure for fixed end moments at end B of member BC and at end C of member CB is shown in Figs. 13.21(a) and 13.21(b).

It may be mentioned that actually one moment distribution solution for each joint of the frame is necessary. The effect of F.E.M. of unity on any other members meeting at that point can be determined by inspection as in Figs. 13.21(c), 13.21(d) and 13.21(e).

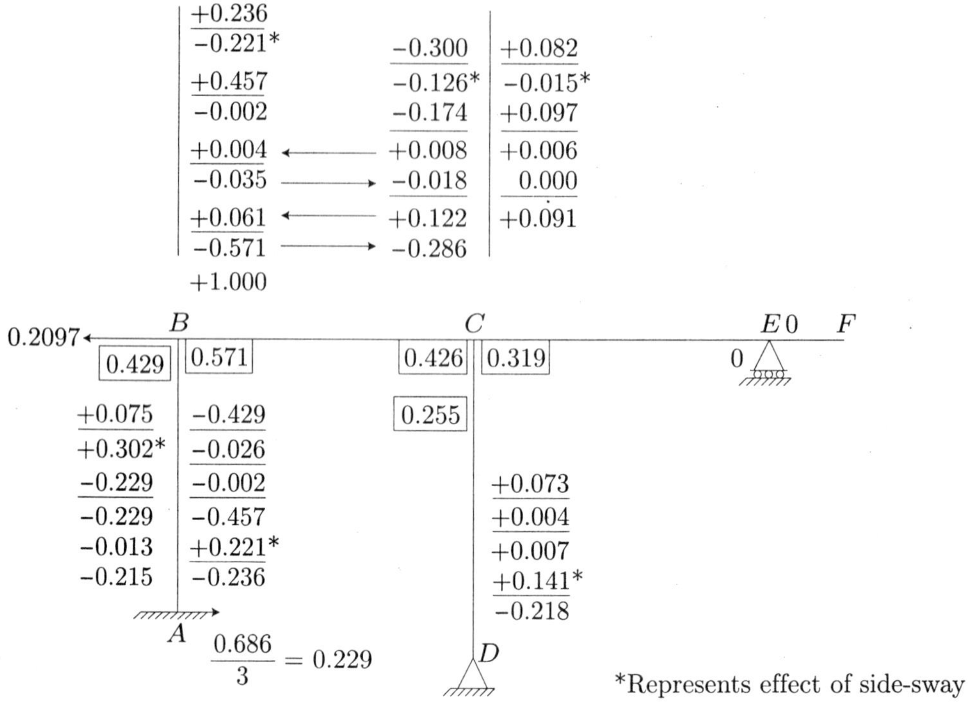

(a) For F.E.M. +1.00 at *B* of *BC*

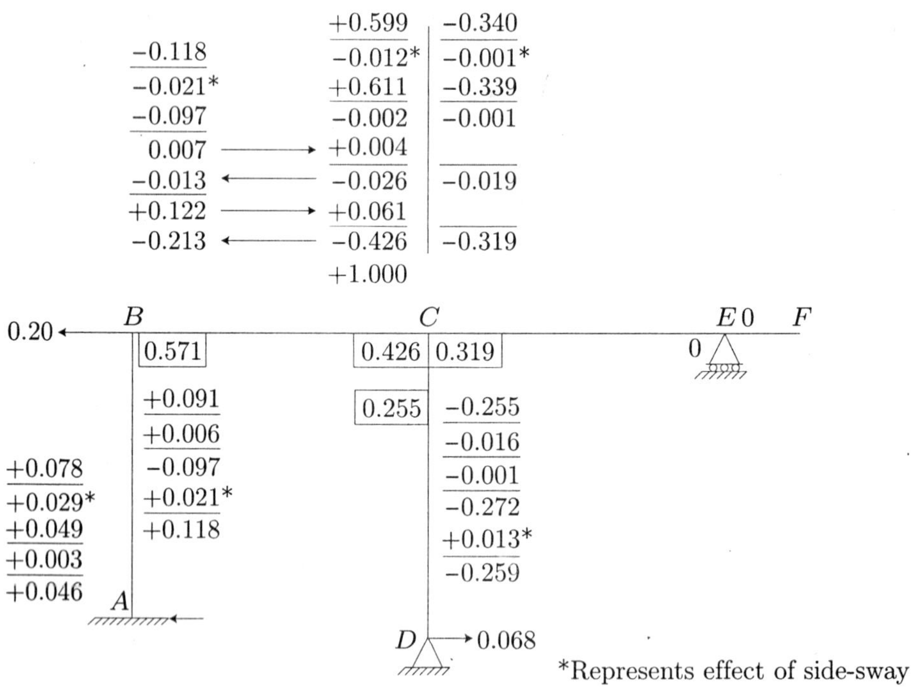

(b) F.E.M. +1.00 at *C* of *CB*

Fig. 13.21: Moment distribution for F.E.M. (Contd.)

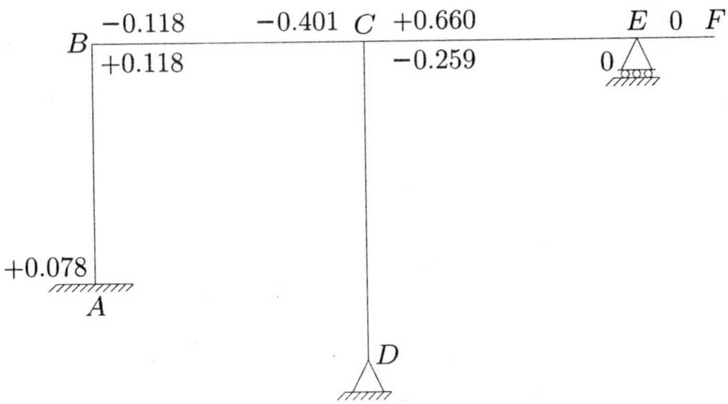

(c) F.E.M. +1.00 at *C* of *CE*

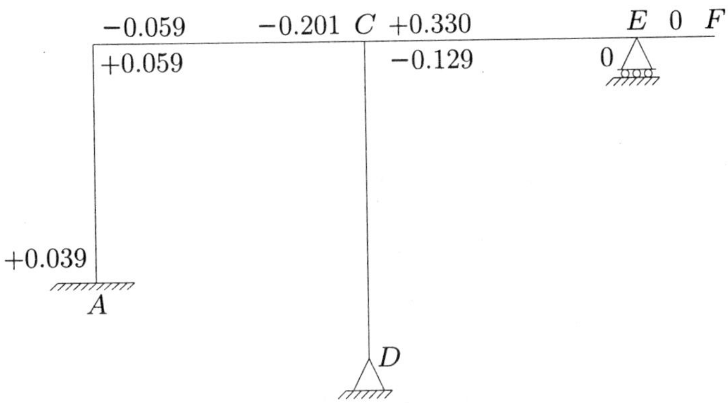

(d) F.E.M. +1.00 at *E* of *EC*

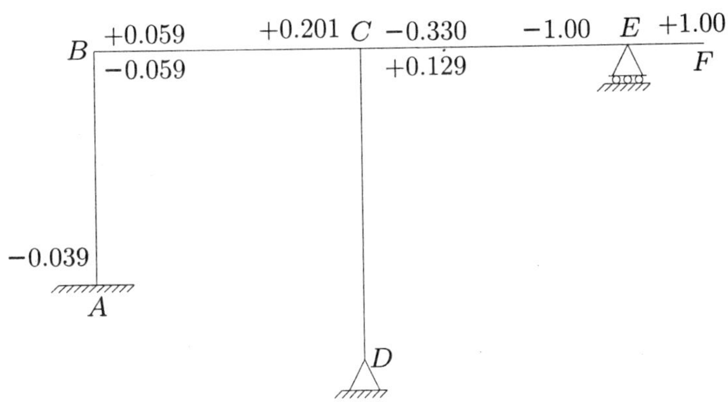

(e) F.E.M. +1.00 at *E* of *EF*

Fig. 13.21: Moment distribution for F.E.M.

Table 13.6 gives the summary of the end moments caused by F.E.M. of +1.00 at various ends and magnitudes for F.E.M. for load position at points 1 to F.

Table 13.6. For Example 13.14

Table 13.6(a) Summary of End Moments

F.E.M. +100 at	M_{AB}	M_{BC}	M_{CB}	M_{CE}	M_{CD}	M_{EC}
BC	+0.073	+0.236	−0.300	+0.082	+0.218	0
CB	+0.078	−0.118	+0.599	−0.340	−0.259	0
CE	+0.078	−0.118	−0.401	+0.660	−0.259	0
EC	+0.039	−0.059	−0.201	+0.330	−0.129	0
EF	−0.039	+0.059	+0.201	−0.330	+0.129	−1.00

Table 13.6(b) F.E.M. due to unit load at various point

Load at	F.E.M. at				
	BC	CB	CE	EC	EF
1	+0.5625	−0.1875			
2	+0.5000	−0.5000			
3	+0.1875	−0.5625			
4	–	–	+0.5625	−0.1875	
5	–	–	+0.5000	−0.5000	
6	–	–	+0.1875	−0.5625	
F	0	0	0	0	+0.80

Influence line coefficients for M_{BC} in terms of the initial fixed-end moments are given by

$$M_{BC} = +0.236M_{FBC} - 0.119M_{FCB} - 0.119M_{FCE} + 0.059M_{FEC} + 0.059M_{FEF}.$$

The values of the fixed-end moments are substituted in the above relationship and the influence line ordinates are computed in Table 13.7.

Influence line ordinates for other end moments may be obtained similarly. Once the influence lines for end moments are obtained, ordinates of influence lines for reactions, shears or moments at intermediate sections may be easily obtained by statics.

If effect of horizontal forces on AB and DC are to be considered similar effects of M_{FAB}, M_{FBA}, and M_{FDC} have to be included.

(iv) *Superposition of Deflection patterns for simply supported beam.* The influence line diagrams for fixed ended beams can also be obtained by superposition of the deflection pattern of simply supported beam subjected to end moments.

Consider the case of a simply supported beam AB of span L and of constant EI. The values of deflection due to end moments $M_{AB} = 1$, and $M_{BA} = 0$ is given by

$$y = \frac{L^2}{6EI}\left[2\frac{x}{L} - 3\frac{x^2}{L^2} + \frac{x^3}{L^3}\right], \tag{13.4}$$

where x is measured from end A and L is the length of member.

Table 13.7. Influence Coordinates for Example

Point	M_{FBC}	M_{FCB}	M_{FCE}	M_{FEC}	M_{FEF}	$+0.236$ M_{FBC}	-0.118 M_{FCB}	-0.118 M_{FCE}	$+0.059$ M_{FEC}	$+0.059$ M_{FEF}	M_{BC}
B	–	–	–	–	–						
1	+0.5625	−0.1875	–	–	–	+0.1328	+0.0221	–	–	–	+0.1549
2	+0.5000	−0.5000	–	–	–	+0.1180	+0.0590	–	–	–	+0.1770
3	+0.1875	−0.5625	–	–	–	+0.0443	+0.0664	–	–	–	+0.1112
C	–	–	–	–	–	–	–	–	–	–	0
4	–	–	+0.5625	−0.1875	–	–	–	−0.0664	−0.0110	–	−0.0774
5	–	–	+0.5000	−0.5000	–	–	–	−0.0590	−0.0295	–	−0.0885
6	–	–	+0.1875	−0.5625	–	–	–	−0.0221	−0.0332	–	−0.0553
E	–	–	–	–	–	–	–	–	–	–	0
F	–	–	–	–	+0.80	–	–	–	–	+0.042	+0.0472

Values of deflections at different values of $\frac{x}{L}$ are given in Table 13.8.

Table 13.8. Deflection due to unit clockwise moment at left-hand end A, i.e., $M_{AB} = 1$

$\frac{x}{L}$	0	0.10	0.20	0.30	0.40	0.50	0.60	0.70	0.80	0.90	1.0
$10^4 \frac{yL^2}{EI}$	0	285	480	595	640	625	560	455	320	165	0

Deflection y due to $M_{BA} = 1$ and $M_{AB} = 0$ is given by,

$$y = -\frac{L^2}{6EI}\left[x - \left(\frac{x}{L}\right)^3\right] \tag{13.5}$$

Value of the deflection at various values of $\frac{x}{L}$ are given in Table 13.9.

Table 13.9. Deflection due to unit clockwise moment at right-hand end B, i.e., $M_{BA} = 1$

$\frac{x}{L}$	0	0.10	0.20	0.30	0.40	0.50	0.60	0.70	0.80	0.90	1.0
$10^4 \frac{yL^2}{EI}$	0	−165	−320	−455	−560	−625	−640	−595	−480	−285	0

To find the influence line for the end moment M_{AB} of the fixed ended beam in Fig. 13.22(a), a hinge is introduced at A and an anticlockwise moment is applied to produce a unit rotation at end A [Fig. 13.22(b)]. The moment must be equal in magnitude to the end rotational stiffness K_{AB}, the corresponding end moment at B being $C_{AB}\ K_{AB}$, where C_{AB} is the carry over factor from A to B. The deflection corresponding to the bending moment diagram in Fig. 13.22(c) is the required influence line for M_{AB}.

When the beam has a constant flexural rigidity EI and length L, the end moments at A and B are $\frac{4EI}{L}$ and $\frac{2EI}{L}$ respectively. The values of the deflections for the case considering moments M_{AB} and M_{BA} as anticlockwise are given by

$$y = \frac{L^2}{6EI}\left[M_{AB}\left\{2\frac{x}{L} - 3\left(\frac{x}{L}\right)^2 + \left(\frac{x}{L}\right)^3\right\} - M_{BA}\left\{\frac{x}{L} - \left(\frac{x}{L}\right)^3\right\}\right], \tag{13.6}$$

where x is measured from left-hand A.

The resulting equations for influence line for the two end moments are given by $-\frac{x}{L^2}(L-x)^2$ and $\frac{x^2(L-x)}{L^2}$ for M_{AB} and M_{BA} respectively.

Superposition of the deflection caused by an end moment of $-\frac{4EI}{L}$ at A (with zero moment at B) and of the deflections caused by an end moment of $-\frac{2EI}{L}$ at B (with zero moment at A) gives the required influence line. This is shown in Table 13.10.

As the beam is symmetrical, the influence ordinates of end moments M_{BA} can be obtained from those of M_{AB} by reversing the sign and the order as shown in Table 13.11.

The influence lines for the two end moments are plotted in Fig. 13.22(d).

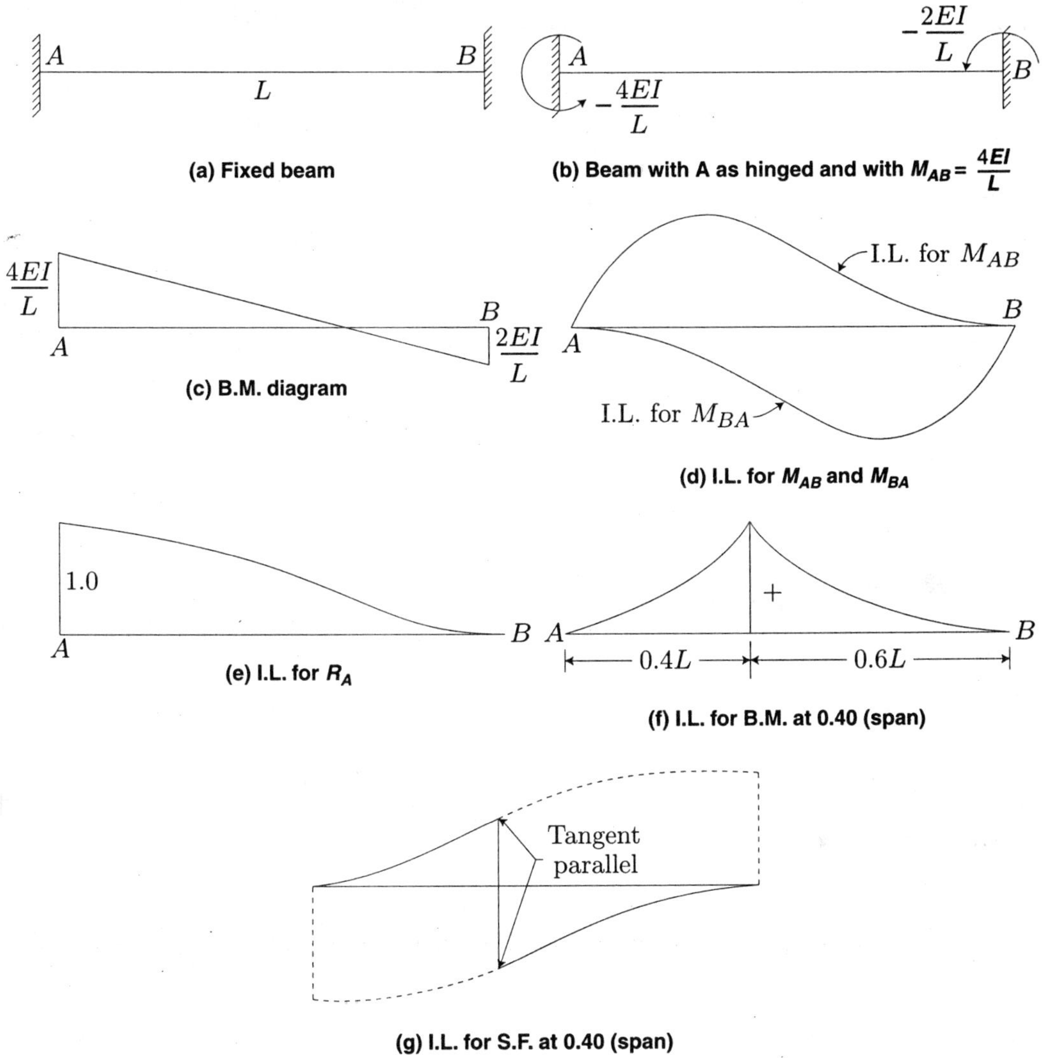

(a) Fixed beam

(b) Beam with A as hinged and with $M_{AB} = \frac{4EI}{L}$

(c) B.M. diagram

(d) I.L. for M_{AB} and M_{BA}

(e) I.L. for R_A

(f) I.L. for B.M. at 0.40 (span)

(g) I.L. for S.F. at 0.40 (span)

Fig. 13.22: I.L. for fixed ended beam

Table 13.10. Calculation of ordinate of the influence line for the end moment M_{AB}

Distance for left hand end	0.1 L	0.2 L	0.3 L	0.4 L	0.5 L	0.6 L	0.7 L	0.8 L	0.9 L	L
Deflection due to end moment of $\frac{4EI}{L}$ at A	−0.114 L	−0.192 L	−0.238 L	−0.256 L	−0.250 L	−0.224 L	−0.182 L	−0.128 L	−0.066 L	0
Deflection due to end moment of $\frac{2EI}{L}$ at B	0.033 L	0.064 L	0.091 L	0.112 L	0.125 L	0.128 L	0.119 L	0.096 L	0.057 L	0
Influence ordinates for M_{AB}	−0.081 L	−0.128 L	−0.147 L	−0.144 L	−0.125 L	−0.096 L	−0.063 L	−0.032 L	−0.009 L	0

Table 13.11. Ordinate for influence line for end moment M_{BA}

Distance for Left hand end	0.1 L	0.2 L	0.3 L	0.4 L	0.5 L	0.6 L	0.7 L	0.8 L	0.9 L	L
I.L. ordinates for M_{BA}	0.009 L	0.032 L	0.063 L	0.096 L	0.125 L	0.144 L	0.147 L	0.128 L	0.081 L	0

Table 13.12. Ordinates for I.L. of R_A

Distance from left hand end	0.0	0.1 L	0.2 L	0.3 L	0.4 L	0.5 L	0.6 L	0.7 L	0.8 L	0.9 L	L
αR_{AS}	1.00	0.900	0.800	0.700	0.600	0.500	0.400	0.300	0.200	0.100	0
$\frac{-\alpha M_{AB}}{L}$	0	0.081	0.128	0.147	0.144	0.125	0.096	0.063	0.032	0.009	0
$\frac{-\alpha M_{BA}}{L}$	0	−0.009	−0.032	−0.063	−0.096	−0.125	−0.144	−0.147	−0.128	−0.081	0
I.L. for R_A	1.00	0.972	0.896	0.784	0.648	0.500	0.352	0.216	0.104	0.028	0

Table 13.13. Ordinates for I.L. of B.M. at $x = 0.4$L

Distance for Left hand end	0.1 L	0.2 L	0.3 L	0.4 L	0.5 L	0.6 L	0.7 L	0.8 L	0.9 L	L
αM_S	0.060 L	0.120 L	0.180 L	0.240 L	0.200 L	0.160 L	0.120 L	0.080 L	0.040 L	0
$0.6\alpha M_{AB}$	−0.049 L	−0.077 L	−0.088 L	−0.086 L	−0.075 L	−0.058 L	−0.038 L	−0.019 L	−0.005 L	0
$-0.040\alpha M_{BA}$	−0.004 L	−0.013 L	−0.025 L	−0.038 L	−0.050 L	−0.058 L	−0.059 L	−0.051 L	−0.032 L	0
I.L. for M at $x = 0.4\ L$	0.007 L	0.030 L	0.067 L	0.116 L	0.075 L	0.044 L	0.023 L	0.010 L	0.003 L	0

The reaction R_A may be expressed as

$$R_A = R_{AB} - \frac{M_{AB} + M_{BA}}{L}, \tag{13.7}$$

where R_{AS} is the statically determinate reaction of the beam AB, if simply supported.

The equation can be written as

$$\alpha R_A = \alpha R_{AS} - \frac{1}{L}[\alpha M_{AB} + \alpha M_{BA}], \tag{13.8}$$

where α is the influence ordinate of action indicated by the subscript.

The influence line of R_{AS} is a straight line with ordinate 1 at A and zero at B. The calculation of the influence line ordinates for the reaction R_A is shown in Table 13.12 and the influence line is plotted in Fig. 13.22(e).

Similarly, influence ordinate for the bending moment at any section of distance x from left-hand end is given by

$$\alpha M = \alpha M_S + \frac{(L-x)}{L}\alpha M_{AB} - \frac{x}{L}\alpha M_{BA}, \tag{13.9}$$

where αM, and αM_S denote the influence ordinate for the bending moment at the section for a beam with fixed ends and simply supported respectively. The influence line for M_S at distance x is given by $\frac{x(L-x)}{L}$.

Ordinates αM for a section $x = 0.4\ L$ are calculated in Table 13.12 and plotted in Fig. 13.22(f).

The influence line ordinates αV for shear force at any section is similarly calculated from

$$\alpha V = \alpha V_S - \frac{1}{L}[\alpha M_{AB} + \alpha M_{BA}], \tag{13.10}$$

where αV_S is the influence ordinate for the shear at the same section in a simply supported beam. The influence line for shear of a section $x = 0.4\ L$ can be obtained with the help of IL diagrams for R_A and R_B and is shown in Fig. 13.22(g)

(v) *Influence line by distribution of slope discontinuity at support.* Considering continuity of a structure over supports, the slope must be the same on all the members meeting at the point. For the beam in Fig. 13.23, the discontinuity at support B is denoted by

$$\theta_{BA} + \theta_{BC} = \theta.$$

For continuity at B, the slope has to be the same as indicated. This requires additional rotations of α_{BA} and α_{BC}, such that

$$\alpha_{BA} + \alpha_{BC} = \theta. \tag{13.11}$$

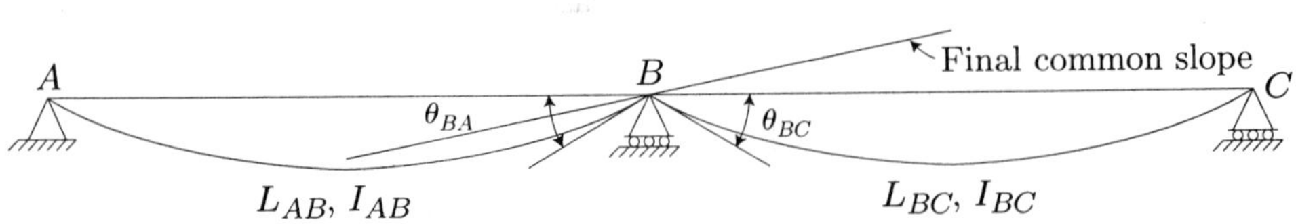

Fig. 13.23: Continuity at supports

Besides continuity equilibrium has to be maintained, i.e.,

$$M_{BA} = M_{BC} = M. \tag{13.12}$$

Considerations of rotations due to moments of a member with far end hinged as in Fig. 13.23, give

$$\left.\begin{aligned} \alpha_{BA} &= \frac{L_{AB}}{3EI_{AB}}\ M_{BA} \\ \text{and} \quad \alpha_{BC} &= \frac{L_{BC}}{3EI_{BC}}\ M_{BC}. \end{aligned}\right] \tag{13.13}$$

The coefficients $\frac{L}{3EI}$ are the flexibilities of the members, i.e, it is the rotation produced at an end of a member by a unit moment ($M = 1$), applied there when the far end is hinged.

Taking f_{AB} as $\dfrac{L_{AB}}{3EI_{AB}}$ and f_{BC} as $\dfrac{L_{BC}}{3EI_{BC}}$,

Eq. (13.11) can be rewritten as

$$\left.\begin{aligned} \alpha_{AB} &= f_{AB}\ M \\ \text{and} \quad \alpha_{BC} &= f_{BC}\ M. \end{aligned}\right] \tag{13.14}$$

Adding $\alpha_{AB} + \alpha_{BC} = (f_{AB} + f_{BC})M,$ (13.15)

whence $\theta = [f_{AB} + f_{BC}]M$ from Eq (13.11),

Substituting this value in Eq. (13.12) it follows that,

$$\left.\begin{aligned} \alpha_{AB} &= \frac{f_{AB}}{f_{AB} + f_{BC}}\ \theta \\ \text{and} \quad \alpha_{BC} &= \frac{f_{BC}}{f_{AB} + f_{BC}}\ \theta. \end{aligned}\right] \tag{13.16}$$

Factors $\frac{f_{AB}}{f_{AB}+f_{BC}}$ and $\frac{f_{BC}}{f_{AB}+f_{BC}}$, by which the total discontinuity at B are to be multiplied to get the additional angle changes θ_{AB} and α_{BC} for achieving continuity are the distribution factors denoted by r_{AB} and r_{BC}.

The method is analogous to moment distribution method. As one end of a member is rotated through α the far end which is hinged rotates through $\frac{\alpha}{2}$, so that carry over factor (C.O.F.) is $\frac{1}{2}$.

When far end is fixed, flexibility of the member is $\frac{L}{4EI}$ i.e., $\frac{3}{4}$ the flexibility of the member with far end hinged and the C.O.F. $= 0$.

In calculating the ordinates of influence line for B.M. on S.F. for any given section of a continuous beam, the member is split into its symmetrical and anti-symmetrical parts as shown in Fig. 13.24, the end slopes for the two cases being $\frac{\theta_A+\theta_B}{2} = \theta$, and $\frac{\theta_A-\theta_B}{2} = \theta_{AS}$, where θ_A and θ_B denote the rotation of the member AB and supports A and B respectively.

Deflection for the case in Fig. 9.24(a) for the symmetrical case is given by

$$y_s = \theta_s Lk(1-k) = C_s.\theta_s.L,$$

where $C_s = k(1-k)$.

For the case in Fig. 9.24(b), deflection is given by

$$y_{AS} = \theta_{AS}.L\left(\frac{n}{2} - 2n^3\right) = C_{AS}\theta_{AS}L, \tag{13.17}$$

where $C_{AS} = \left[\frac{n}{2} - 2n^3\right]$.

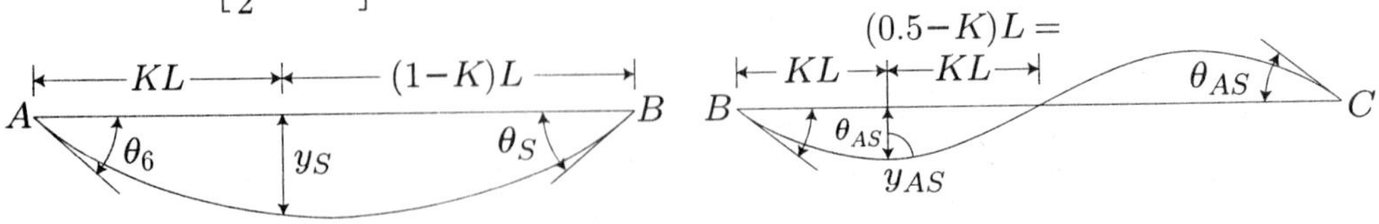

(a) Symmetrical case

(b) Anti-symmetrical case

Fig. 13.24: Deflection and slope for symmetrical and anti-symmetrical case

Values of C_S and C_{AS} are calculated and given in Tables 9.14 and 9.15.

Table 13.14. Value of $C_S = k(1 - k)$

k	0.10	0.20	0.30	0.40	0.50	0.60	0.70	0.80	0.90
C_S	0.01	0.16	0.21	0.24	0.25	0.24	0.21	0.16	0.01

k	$\frac{1}{8}$	$\frac{1}{4}$	$\frac{3}{8}$	$\frac{1}{2}$	$\frac{5}{8}$	$\frac{3}{4}$	$\frac{7}{8}$
C_s	$\frac{7}{64}$	$\frac{3}{16}$	$\frac{15}{64}$	$\frac{1}{4}$	$\frac{15}{64}$	$\frac{3}{16}$	$\frac{7}{64}$

Table 13.15. Value of $C_{AS} = \left[\frac{n}{2} - 2n^3\right]$

n	0.10	0.20	0.30	0.40	0.50	−0.10	−0.20	−0.30	−0.40
C_{AS}	0.048	0.084	0.096	0.072	0	−0.72	−0.096	−0.084	−0.048

$\left(\frac{1}{2} - k\right)$	$\frac{1}{8}$	$\frac{1}{4}$	$\frac{3}{8}$	$\frac{1}{2}$	$-\frac{1}{8}$	$-\frac{1}{4}$	$-\frac{3}{8}$.
C_{AS}	0.0586	0.0934	0.0820	0	−0.0820	−0.0934	−0.0586

with respect to k

k	0.10	0.20	0.30	0.40	0.50	0.60	0.70	0.80	0.90
C_{AS}	0.072	0.096	0.089	0.048	0	−0.048	−0.084	−0.096	−0.072

k	$\frac{1}{8}$	$\frac{1}{4}$	$\frac{3}{8}$	$\frac{1}{2}$	$\frac{5}{8}$	$\frac{3}{4}$	$\frac{7}{8}$
C_{AS}	0.0820	0.0934	0.0586	0	−0.0586	−0.0934	−0.0820

In order to calculate the influence line ordinates for the case of a continuous beam the following steps are followed :

(i) Span AB of length L, which contains the section P for which influence line is required, is considered to be simply supported at its ends. A unit angle change is introduced at the section P. This results in a triangular deflection of the beam with end slopes which can be considered as $\frac{x}{L}$ and $\frac{x}{L-x}$, where x denotes the distance from one end. This deflection is called the determinate deflection.

(ii) Since AB is not simply supported, but continuous with adjoining spans, the continuity is to be restored. This is done by removing the slope discontinuity or incompatibilities at A and B by successive approximation.

(iii) Starting with additional slopes at each support resulting from restoration of continuity (i.e., excluding the initial determinate end slopes of 0.25 and 0.75 as in Example 13.15), the resulting deflections are calculated. These are indeterminate deflections. For span AB, the determinate deflections are to be added to these to get the total deflection.

The method is illustrated by an Example which follows.

Example 13.15. *Obtain influence lines for bending moment at P of the continuous beam shown in Fig. 13.25 using distribution of slope discontinuity.*

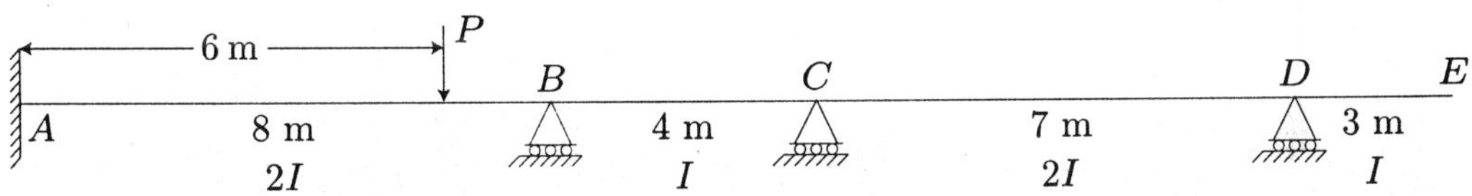

Fig. 13.25: For Example 13.15

Solution : The method of solution is illustrated in the Tables 13.16 and 13.17, the first one showing calculation of support slopes and the second one calculation of influence line ordinates. Slope and deflections are considered positive in downward direction.

For member BA, flexibility with far end fixed $= \frac{3}{4}$ (flexibility with far end hinged).

Table 13.16. Calculation of Support Slopes

	A	*B*		*C*		*D*		*E*
		AB	*BA*	*BC*	*CB*	*CD*	*DC*	*DE*
Relative flexibilities $\propto \frac{I}{L}$	0	$\frac{3(8)}{4(2)} = 3$		4	4	3.5	3.50	∞
D.F.	0	1	$\frac{3}{7}$	$\frac{4}{7}$	$\frac{8}{15}$	$\frac{7}{15}$	0	1
C.O.F.		0.50		0.50				
Unit angle change given at *P*, assuming no continuity with adjoining span.		+0.2500	+0.7500	0	0	0	0	0
Bal.		−0.2500	−0.3214	−0.4286	0	0	0	0
C.O.			−0.1250	0	−0.2143	0	0	0
Bal.			+0.0536	+0.0714	+0.1142	+0.1001	0	0
C.O.			0	+0.0571	+0.0357	0	+0.0500	
Bal.			−0.0245	−0.0326	−0.0190	−0.0167	0	−0.0500
C.O.			0	−0.0095	−0.0163		−0.0084	–
Bal.			+0.0041	+0.0054	+0.0087	+0.0076	0	+0.0084
Total change in slope due to restoration of continuity		−0.2500	−0.4132	−0.3368	−0.0910	+0.0910	+0.0416	−0.0416

Table 13.17. Calculation of Influence Line Ordinates

	Deflection			Slopes			
	Indeterminate Deflection		Determinate Deflection $y_0 \longrightarrow$	Antisymmetrical parts of slope $\theta_{AS} \longrightarrow$	Symmetrical parts of slope $\theta_S \longrightarrow$	Slope θ from Table	
	Antisym-metrical $y_{AS} = C_{AS}\theta_{AS}L$	Symmetrical $y_S = C_S\theta_S L$					
				+0.0816	−0.3316	−0.2500	*A*
+0.0636	+0.0610	−0.4974	+0.500				2 m
+0.3368	0	−0.6632	+1.00				4 m
+0.9416	−0.0610	−0.4974	+1.500				6 m
+0.4063	−0.0535	−0.2902	+0.750				7 m
0	0	0	0	−0.0816	−0.3316	−0.4132	*B*
	0	0	0	−0.1229	−0.2139	−0.3368	8 m
−0.2063	−0.0459	−0.1604	0				1 m
−0.2139	0	−0.2139	0				2 m
			0				
−0.1145	+0.0459	−0.1604	0				3 m
			0				
0	0	0	0	+0.1229	−0.2139	−0.0910	*C*
	0	0	0	+0.0247	+0.0663	+0.0910	4 m
+0.1031	+0.0161	+0.0870	0				1.75 m
			0				
+0.1160	0	+0.1160	0				3.50 m
+0.0709	−0.0161	+0.0870					5.25 m
0	0	0	0	0.0247	+0.0663	+0.0416	*D*
					–	−0.0416	7 m
							E
							3 m

13.5 Influence Lines for Plane Frames

The influence lines for shear or bending moment at any section of a member can be determined from the influence lines for end moments by simple equations of statics.

The influence line for the end moment M_{BC} in the frame of Fig. 13.26(a) can be obtained by Müller-Breslau's principle, and is given by the deflected shape of the frame corresponding to a unit angular discontinuity at end BC. Unit angular rotation is introduced at end BC without other displacements at the joints as shown in Fig. 13.26(b). The end moments corresponding to the configuration are K_{BC} and $C_{BC}\,K_{BC}$, where K_{BC} is the end rotational stiffness and C_{BC}, the carry over factor from B to C.

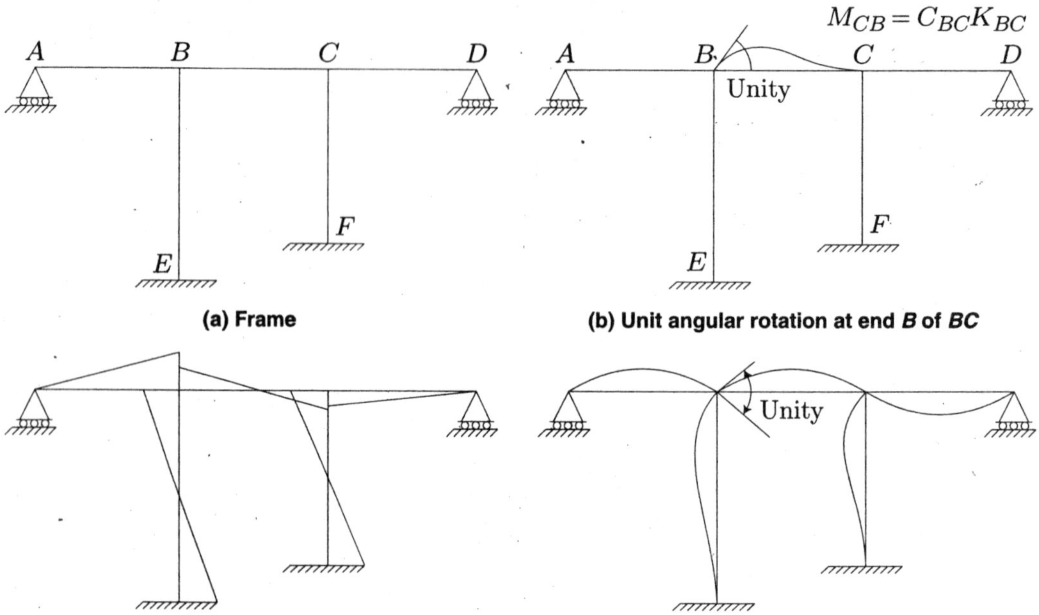

(a) Frame

(b) Unit angular rotation at end *B* of *BC*

(c) B.M. Diagram corresponding to elastic line in (d)

(d) I.L. for M_{BC}

Fig. 13.26: I.L. for plane frames

Joint rotations and joint translation, if any, are allowed to take place and the corresponding moments at ends of the members are obtained by moment distribution. The corresponding bending moment diagrams will be linear as shown in Fig. 13.26(c). The deflections which are the influence line ordinates are obtained by superposition of the deflections due to end moments.

For prismatic members values given in Tables 13.8 and 13.9 of Section 13.4 may be used. For members of variables I, influence line ordinates of the moment at a fixed end of a member with the other end hinged may be used or calculated. To obtain the deflection due to a unit couple applied at one end these ordinates should be divided by the modified stiffness at the fixed end, while the other end is hinged.

The shape of the influence line for the end moment M_{BC} for the frame considered is shown in Fig. 13.26(d). The ordinates plotted on the vertical members BE and CF can be used to find magnitudes of M_{BC}, if horizontal load be applied to either of the columns.

Example 13.16. *Obtain the influence line for end moment M_{BA} for the bridge frame shown in Fig. 13.27(a), subjected to vertical loading. Find the influence linc ordinates for B.M., M_G at centre of AB and of shear force V_N at a point just left of B. The relative values of I are shown in the figure.*

Solution : A unit rotation in an anticlockwise direction is introduced at end B of BA as shown in Fig. 13.27(b). The corresponding end moments are $M_{BA} = \frac{3EI_{AB}}{L_{AB}} = \frac{1.86EI}{L}$ and zero for all the other ends. These are the initial fixed-end moments for which a moment distribution is carried out in Table 13.18. The deflections of members AB, BC and CD due to the final end moments are calculated in Table 13.19 at 0.30, 0.50 and 0.70 of each span with the help of the deflection ordinates in Tables 13.8 and 13.9. These deflections which are the influence ordinates of the end moments M_{BA} are plotted in Fig. 13.27(c).

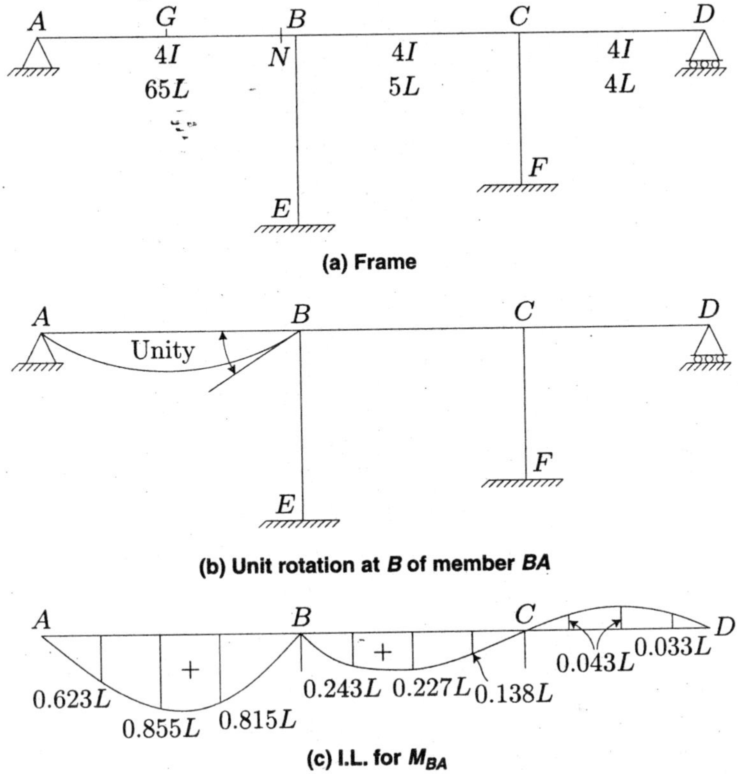

(a) Frame

(b) Unit rotation at *B* of member *BA*

(c) I.L. for M_{BA}

Fig. 13.27: For Example 13.16

Table 13.18. For Example 13.16

	B			C		
	BA	BE	BC	CB	CF	CD
D.F.	0.29	0.21	0.50	0.39	0.24	0.37
C.O.F.			⟵ 0.50	⟶		
F.E.M. Bal.	−1.85 +0.53	0 +0.39	0 +0.93	0 0	0 0	0 0
C.O.M. Bal.			0	+0.47 −0.18	0 −0.12	0 −0.17
C.O.M. Bal.	0 +0.02	0 +0.02	−0.09 +0.05	0	0	0
C.O.M. Bal.	0	0	0	+0.03 −0.01	0 −0.01	0 −0.01
Σ	−1.30	+0.41	+0.89	+0.31	−0.13	−0.18

Table 13.19. Ordinates of influence Line for end moment M_{BA}.
(From Tables 13.8 and 13.9)

Deflections due to end moment at	Member AB			Member BC			Member CD		
	$0.3(6.5L)$	$0.5(6.5L)$	$0.7(6.5L)$	$0.3(5L)$	$0.5(5L)$	$0.7(5L)$	$0.3(4L)$	$0.5(4L)$	$0.7(4L)$
Left-hand end	0	0	0	$0.331L$	$0.348L$	$0.253L$	$-0.043L$	$-0.045L$	$-0.033L$
Right-hand end	$+0.623L$	$0.855L$	$0.815L$	$-0.088L$	$-0.121L$	$-0.115L$	0	0	0
I.L. ordinate	$0.623L$	$0.855L$	$0.815L$	$0.243L$	$0.227L$	$0.138L$	$-0.043L$	$-0.045L$	$-0.033L$

The ordinates of the influence lines for M_G and V_N are determined by superposition of Eqs. (13.9) and (13.10) as calculated in Tables 13.20 and 13.21. The influence lines are plotted in Figs. 13.28(a) and 13.28(b).

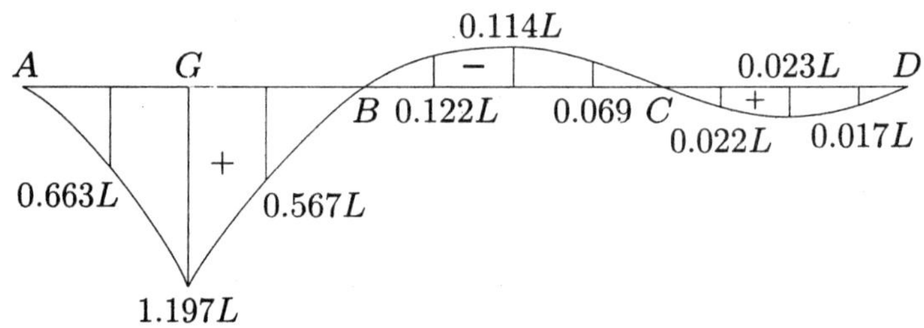

(a) I.L. for M_G

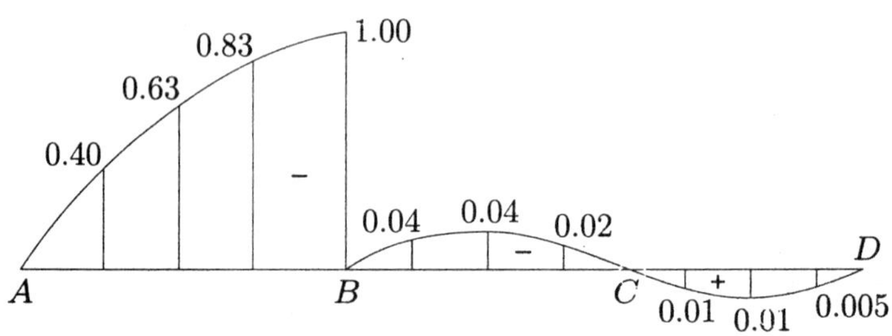

(b) I.L. for V_N

Fig. 13.28: For Example 13.16

Table 13.20. Ordinates for Influence Lines for B.M. M_G at G

Influence	Member AB			Member BC			Member CD		
Coefficient	$0.3(6.5L)$	$0.5(6.5L)$	$0.7(6.5L)$	$0.3(5L)$	$0.5(5L)$	$0.7(5L)$	$0.3(4L)$	$0.5(4L)$	$0.7(4L)$
αM_S	$0.975L$	$1.625L$	$0.975L$	0	0	0	0	0	0
$-\frac{1}{2}\alpha M_{BA}$	$-0.312L$	$-0.428L$	$-0.408L$	$-0.122L$	$-0.114L$	$-0.069L$	$0.022L$	$0.023L$	$0.017L$
I.L. Ordinate	$0.663L$	$1.197L$	$0.567L$	$-0.122L$	$-0.114L$	$-0.069L$	$0.022L$	$0.023L$	$0.017L$

Table 13.21. Ordinates for Influence Line V_N

Influence	Member AB				Member BC				Member CD			
Coefficient	$0.3(6.5L)$	$0.5(6.5L)$	$0.7(6.5L)$	$6.5L$	$0.3(5L)$	$0.5(5L)$	$0.7(5L)$	$5L$	$0.3(4L)$	$0.5(4L)$	$0.7(4L)$	$4L$
αV_S	−0.30	−0.50	−0.70	−1.00	0	0	0	0	0	0	0	0
$-\frac{1}{6.5L}(\alpha M_{BA})$	−0.10	−0.13	−0.13	0	−0.04	−0.04	−0.02	0	0.01	0.01	0.005	0
I.L. Ordinate	−0.40	−0.63	−0.83	−1.00	−0.04	−0.04	−0.02	0	0.01	0.01	0.005	0

Problems

13.1 Obtain the Influence line diagrams for R_A, R_B, R_C and M_D for a continuous beam, ABC of two span having length $AB = 2L$, and $BC = 3L$, if EI is constant.

13.2 A propped cantilever AB of length L and of uniform EI is fixed at end A and is supported on a spring having stiffness $\frac{L^3}{12EI}$ at the other end B. Obtain influence line for R_B and plot indicating values at $\frac{L}{4}$ intervals. Also obtain from the influence line of R_B, the influence line for B.M. and S.F. at mid-span of the beam.

13.3 A beam AB of length L and of uniform cross section is fixed at end A and is supported at the other end by a tie inclined at 30° to the horizontal. Obtain an expression for the tension in the tie as a unit load moving over the span AB.

13.4 A two-span continuous beam ABC is hinged at A, continuous over the support B, and rests on a roller at C. Length $AB = 6$ m and $BC = 4$ m, M.I. of $AB = I$ and of $BC = 1.5I$. With the help of Müller-Breslau's principle, obtain the influence lines for B.M., M_B and vertical reaction at B, indicating ordinates at 1 m intervals.

13.5 A beam AB of length L is of uniform cross section and is fixed at the two ends A and B. Obtain an expression for influence line for the moment M_A. Plot the influence line diagrams showing the maximum ordinates. From the influence line for M_A, obtain an expression for the reaction R_A and plot the diagram showing values at $\frac{L}{4}$ apart.

13.6 A beam ABC of uniform cross section is encastre at A and rests on a unyielding support at B, located at a distance L from A. BC is overhung. If the distance of a unit rolling load from B be αL, show that the IL for the reaction at B is give by $\left(I - \frac{3\alpha}{2} + \frac{\alpha^3}{2}\right)$, for load between B and A and $\left(1 + \frac{3\alpha}{2}\right)$, for load between B and C. Plot the influence line diagram for the reaction at B.

Also obtain from IL of R_B the IL diagrams for M_A and M_B.

13.7 Using Müller-Breslau's principle obtain influence lines for the end moments and the moment at C of the beam shown in Fig. P-13.1.

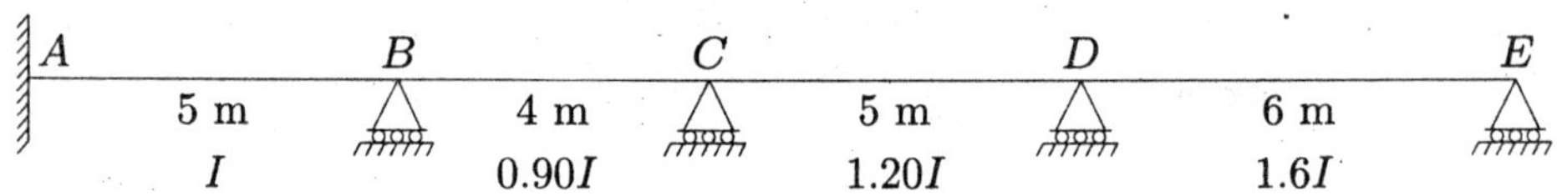

Fig. P-13.1

13.8 Obtain influence line for M_A and M_B of the continuous beam shown in Fig. P-13.2.

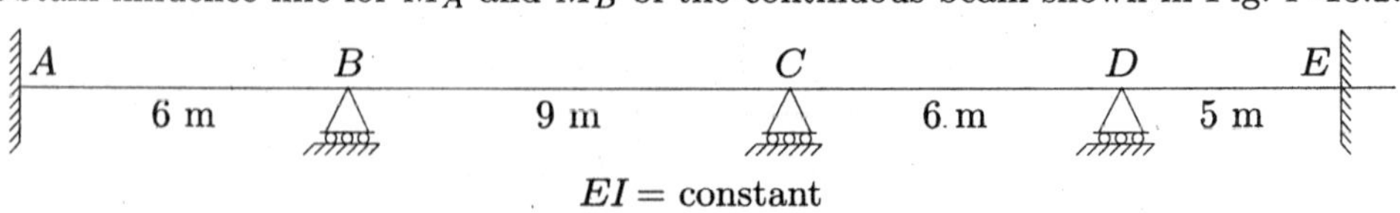

Fig. P-13.2

13.9 Using Moment Distribution techniques, obtain influence line diagrams for M_B and M_A of the beams shown in Fig. P-13.3 and indicate values at $\frac{L}{5}$ intervals.

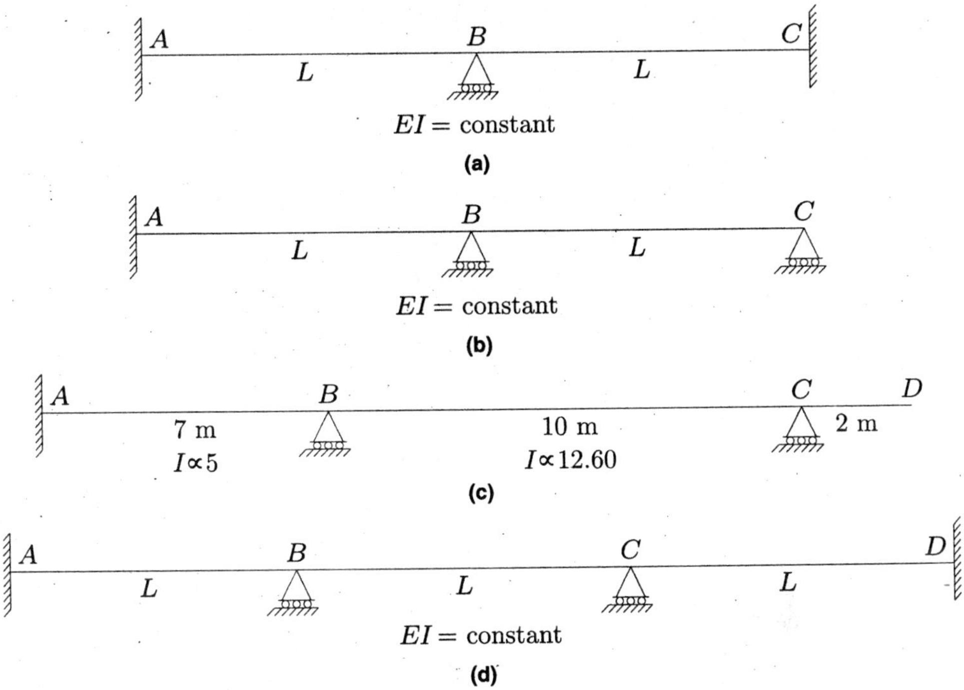

Fig. P-13.3

13.10 A two span continuous prismatic beam ABC is totally fixed at A and supported at B and C on elastic supports of stiffness $\frac{EI}{L^3}$, where EI is the flexural rigidity of the beam and L is the length of each span. Obtain influence lines for R_B, R_C and the end moment M_A. Indicate supports at $L/2$ intervals.

13.11 A three-span continuous beam $ABCD$ of prismatic section is simply supported at ends A and D and continuous over elastic supports having stiffness of 200 kN/m at B and C. Length AB = 6 m, BC = 10 m and CD = 4 m. If EI for the beam is 15000 kN-m^2, obtain influence lines for the reactions.

13.12 Obtain influence line for the horizontal reactions as vertical loads move over the horizontal member of the frame shown in Fig. P-13.4.

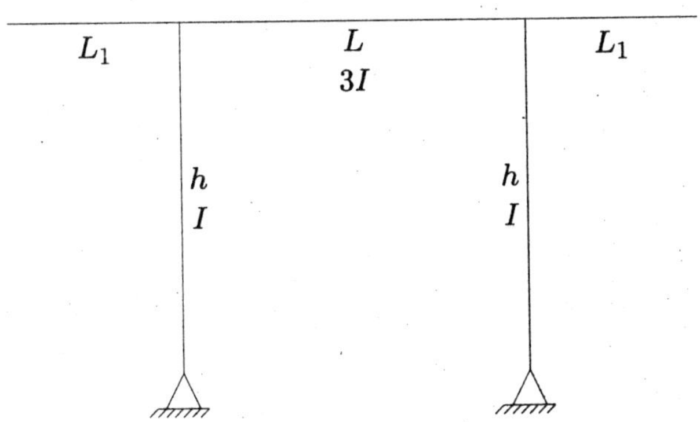

Fig. P-13.4

13.13 For the bridge frame shown in Fig. P-13.5, calculate the influence ordinates at $0.3L$, $0.5L$ and $0.7L$ of each span and draw the influence line for the following actions.

(a) B.M. at B on member AB.

(b) B.M. at centre of AB.

(c) S.F. at B on member AB.

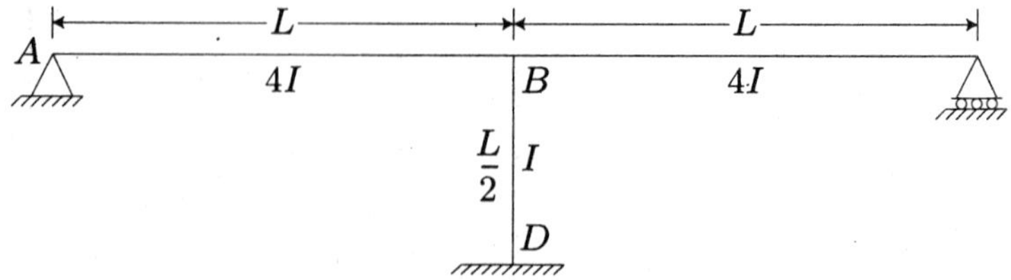

Fig. P-13.5

13.14 If the hinge support at A be replaced by a roller support obtain IL for end moment at D.

Chapter 14

Suspension Cables and Suspension Bridges

14.1 Introduction

A method of reducing bending moments in case of long span structures consists in providing partial support at points along the span by means of a system of cables, as in case of suspension bridges. Such bridges are preferable when its span is more than 200 metres for a roadway or 300 metres for light traffic way. Depending on the nature and magnitude of the traffic suspension bridges may be of two types:

(a) *Unstiffened suspension bridges* : For light traffic the roadway is carried directly by hangers suspended from the cables, which being flexible throughout cannot resist any moment and can adopt any shape under the loads. The hangers being large in number, the load transmitted can be considered as uniformly distributed through the hangers. The central sag or dip of the cable usually varies from $\frac{1}{10}$th to $\frac{1}{15}$th of the span.

(b) *Stiffened suspension bridges* : For heavier traffic, the roadway is supported on stiffening girder or truss on cables by closely spaced hangers. They are usually erected in a manner such that all the dead load is carried by the cable. When live load is applied, tension in the hangers transfers a large portion of the live load to the cable. The stiffening girder or truss is not subjected to dead load moments and the live load moments are substantially reduced. For long span structures, this is of particular importance, since most of the load is carried by the cable in tension, which is an highly efficient manner of carrying loads.

Besides suspension bridges, cables are also used for cable car system and for permanent guys on structures such as Derricks and Radio Towers. They are also used for temporary guys during erection.

The cable supporting a load uniform per unit length of the cable such as its own weight takes the shape of a catenary, but unless the sag of the cable is large in proportion to its length the shape taken may often be assumed to be parabolic and the analysis becomes largely simplified.

14.2 Analysis of General Cable Shape

Fig. 14.1(a) shows the general case of a cable supported at two points A and B which are not at the same elevation and acted on by a system of vertical loads $W_1, W_2, \ldots, W_n$. The cable is assumed to be perfectly flexible, so that the bending moment at any point on the cable must be zero and the cable will assume the shape of a polygonal curve depending on the magnitude and position of the loads. Only axial tension present at any section and tensions T_A and T_B at supports A and B may be resolved into vertical and horizontal components.

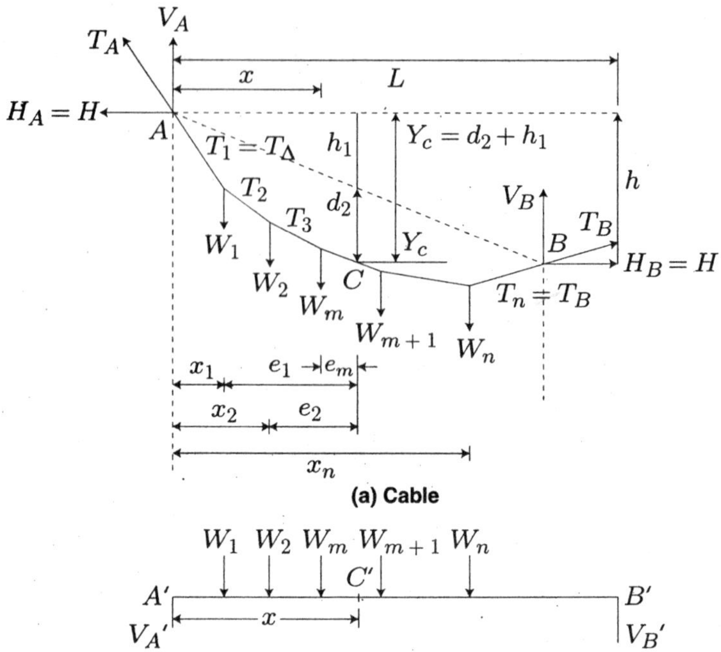

(a) Cable

(b) Beam

Fig. 14.1: Cable—general case

From $\sum F_X = 0,\ H_A = H_B = H$

From $\sum F_Y = 0,\ V_A + V_B = \sum_1^n W_n$

From $\sum M_B = 0$ gives $V_A(L) - \sum_1^n W_n(L - x_B) - Hh = 0,$

where x_n = horizontal distance of load W_n from support A.

From $\sum M_C = 0,\ V_A(x) - \sum_1^m W_m e_m - H(y_c) = 0$

From $\sum M_B = 0$ and $\sum M_C = 0,$

$$\sum_1^n \frac{W_n(L - x_n)}{L} + \frac{Hh}{L} = \sum_1^m \frac{W_m e_m}{x} + \frac{HY_c}{x},$$

whence $H\left[Y_c - \dfrac{hx}{L}\right] \dfrac{x}{L}\sum_1^n W_n(L - x_n) - \sum_1^m W_m e_m.$

Taking $h_1 = \dfrac{hx}{L}$ and $d_2 = (Y_c - h_1)$, as in Fig. 14.1(a), and denoting

$$M' = \frac{x}{L}\sum_1^n W_n(L - x_n) - \sum_1^m W_m e_m. \tag{14.1a}$$

From $\quad H(d_2) = M' \quad$ and $\quad H = \dfrac{M'}{d_2}$

as $$Y_c - \frac{hx}{L} = d_2. \tag{14.1b}$$

The distance d_2 is the sag of the cable. If the sag d_2 of a cable is known, the horizontal component of reaction, i.e., H, the horizontal tension may be computed from Eq. (14.1b). The vertical components V_A and V_B of the reaction can be obtained from

$$V_A = \frac{1}{L}\sum_1^n W_n(L - x_n) + \frac{Hh}{L} \tag{14.1c}$$

and $$V_B = \sum_1^n W_n - \frac{1}{L}\sum_1^n W_n(L - x_n) - \frac{Hh}{L}. \tag{14.1d}$$

For a simple beam $A'B'$ of horizontal span L subjected to the same loading as the cable as shown in Fig. 14.1(b), reactions V'_A and V'_B are equal to

$$V'_A = \frac{1}{L}\sum_1^n W_n(L - x_n) \tag{14.2a}$$

and $$V'_B = \sum_1^n W_n - \frac{1}{L}\sum_1^n W_n(L - x_n). \tag{14.2b}$$

At point C' of the beam moment is equal to M' as indicated in Eq. (14.1a).

Also $$V_A = V'_A + \frac{Hh}{L} \tag{14.2c}$$

and $$V_B = V'_B - \frac{Hh}{L}. \tag{14.2d}$$

When supports A and B of the cable are at the same elevation, i.e., $h = 0$, then $V_A = V'_A$ and $V_B = V'_B$.

Tension in the cable at end A is given by $T_A = \sqrt{V_A^2 + H^2}$ and that at end B is given by $T_B - \sqrt{V_B^2 + H^2}$. Value of tension T_1 in the cable between support A and first load W_1 will equal to T_A as shown in Fig. 14.1(a). On either side of the points of the application of load W_1, the cable is acted on by tension T_1 and T_2 and these two together with load W_1 are in equilibrium. Then,

$$T_2 = \sqrt{(V_A - W_1)^2 + H^2}.$$

Values of tension in the other portions of the cable may be obtained similarly.

The shape of the loaded cable will be the same as the force polygon for the given load system drawn form one support to the other with a polar distance equal to the horizontal component H of the pull in the cable.

14.3 Uniformly Loaded Cable

For uniform loading on the horizontal length applied over the entire span of the cable is important as it is substantially the type of cable loading occurring in suspension bridges and

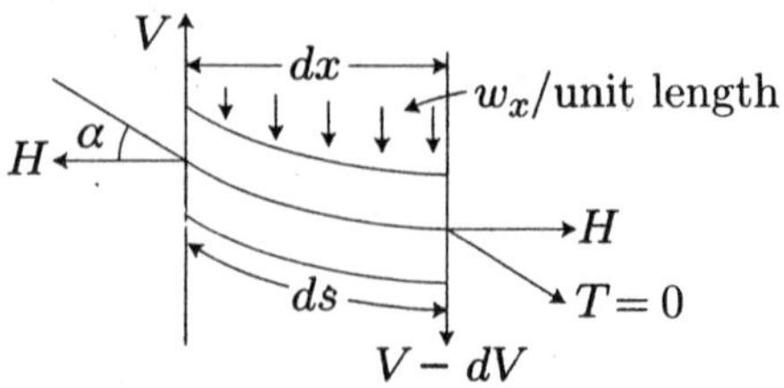

Fig. 14.2: Element of a cable under u.d.l.

as a cable carrying only its self weight can also be treated approximately, on the assumption that the dead weight is uniform on horizontal length.

For a small length ds of the cable subjected to uniformly distributed load w_x per unit horizontal length and kept in equilibrium under the actions of the forces from

$$\sum F_y = 0,\ V - w_x(dx) = V - dV, \quad \text{whence} \quad \frac{dV}{dx} = w_x.$$

As $V = H\tan\alpha = H\dfrac{dy}{dx},\ H\dfrac{d^2y}{dx^2} = \dfrac{w_x}{H}.$

This is the general equations of cable curve.

Case I. *Cable supported at same level (Symmetrical Cable).* For the case of a cable supported at the same level at A and B as in Fig. 14.3 and loaded with uniform distributed load of magnitude w per unit length along the horizontal length, equation of the cable will be represented by $\frac{d^2y}{dx^2} = \frac{x}{H}$, where H is the horizontal component of the tension in the cable. With origin at A, and x- and y- axes as shown,

$$-\frac{d^2y}{dx^2} = \frac{w}{H} \quad \text{and} \quad \frac{dy}{dx} = -\frac{w}{H}x + C_1,$$

whence $y = \dfrac{w}{H}\dfrac{x^2}{2} + C_1x + C_2.$

From boundary conditions, $\dfrac{dy}{dx} = 0$ at $x = \dfrac{L}{2},\ C_1 = +\dfrac{wL}{2H}.$

At $x = 0,\ y = 0,\ C_2 = 0,$

complete solution is $y = -\dfrac{wx^2}{2H} + \dfrac{wLx}{2H}$

For $\quad x = \dfrac{L}{2},\ y = d_c = d, \quad \text{whence} \quad H = +\dfrac{wL^2}{8d} = \dfrac{\text{B.M. at centre}}{\text{Cenral dip}}.$

Shape of the cable is then given by

$$y = -\frac{4dx^2}{L^2} + \frac{4dLx}{L^2} = \frac{4dx}{L^2}(L - x). \tag{14.3}$$

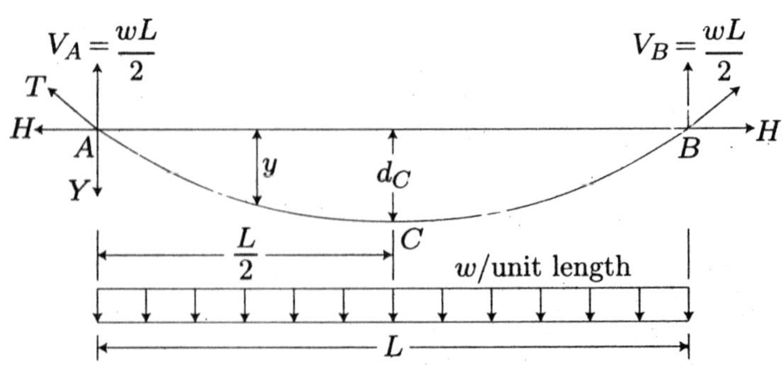

Fig. 14.3: Cable supported at same level

This represents a parabola which denotes that the cable when subjected to uniformly distributed load on the horizontal length assumes the shape of a parabola.

With horizontal tension H known, the tension T at any point in the cable at a horizontal distance x from left support at A is given by

$$T = H\sec\theta = H\frac{ds}{dx}.$$

From Eq (14.3), $\dfrac{dy}{dx} = \dfrac{4d}{L^2}[L - 2x]$

and $$\frac{ds}{dx} = \sqrt{1+\left(\frac{dy}{dx}\right)^2} = \left[1+\left(\frac{4d}{L}-\frac{8dx}{L^2}\right)^2\right]^{\frac{1}{2}} = \left[1+\frac{16d^2}{L^2}+\frac{64d^2x^2}{L^4}-\frac{64d^2x}{L^3}\right]^{\frac{1}{2}}.$$

Also $$T = H\left[1+16\frac{d^2}{L^2}+64\frac{d^2x^2}{L^4}-64\frac{d^2x}{L^3}\right]^{\frac{1}{2}}. \tag{14.4a}$$

Maximum value of tension,

$$T_{max} = H\left[1+16\frac{d^2}{L^2}\right]^{\frac{1}{2}} \tag{14.4b}$$

at $x = 0$, i.e., at origin.

Value of T_{max} may also be obtained from

$$T = \sqrt{V^2+H^2} = \sqrt{\left(\frac{wL}{2}\right)^2+\left(\frac{wL^2}{8d}\right)^2}$$

$$= \frac{wL}{2}\sqrt{1+\frac{L^2}{16d^2}}. \tag{14.4c}$$

Length S of the cable may be obtained from

$$S = \int_0^L \frac{ds}{dx}dx$$

$$= \int_0^L \left[1+16\frac{d^2}{L^2}+64\frac{d^2x^2}{L^4}-64\frac{d^2x}{L^3}\right]^{\frac{1}{2}} dx. \tag{14.5a}$$

Expanding by binomial theorem and considering only a few terms,

$$S = \int_0^L \left[(1)^{-\frac{1}{2}}+\frac{1}{2}(1)^{-\frac{1}{2}}\left(16\frac{d^2}{L^2}\right)+\frac{1}{2}(1)^{-\frac{1}{2}}\left(64\frac{d^2x^2}{L^4}\right)-\frac{1}{2}(1)^{-\frac{1}{2}}\left(64\frac{d^2x}{L^3}\right)+\cdots\right]dx$$

$$\approx L\left[1+8\frac{d^2}{L^2}+\frac{32}{3}\frac{d^2}{L^2}-16\frac{d^2}{L^2}\right]$$

$$= L\left[1+\frac{8}{3}\frac{d^2}{L^2}\right]. \tag{14.5b}$$

Due to change of temperature, the cable changes length. Such change in length is given by

$$\delta S = S\alpha t = \frac{16}{3}\frac{d}{L}\delta d, \text{ from Eq. (14.5b)},$$

whence $$\delta d = \frac{3}{16}\frac{S\alpha t}{n}, \tag{14.6a}$$

where $n = \dfrac{d}{L}$.

Change in value of H is given by

$$\delta H = -\frac{wL^2}{8d^2}\delta d = -\frac{H}{d}\delta d$$

$$= -\frac{H}{d}\left(\frac{3}{16}\right)\frac{S\alpha t}{n}. \tag{14.6b}$$

This denotes that due to increase in temperature, the value of H reduces due to increase of the central dip.

Case II. *Cables supported at different levels (unsymmetrical cables).*

Considering a cable ACB supported at different levels at A and B and carrying a uniformly distributed load, if C be the lowest point of the cable as in Fig. 14.4, the horizontal thrust at A will be equal to the horizontal thrust at B, since the cable is supporting only vertical loads.

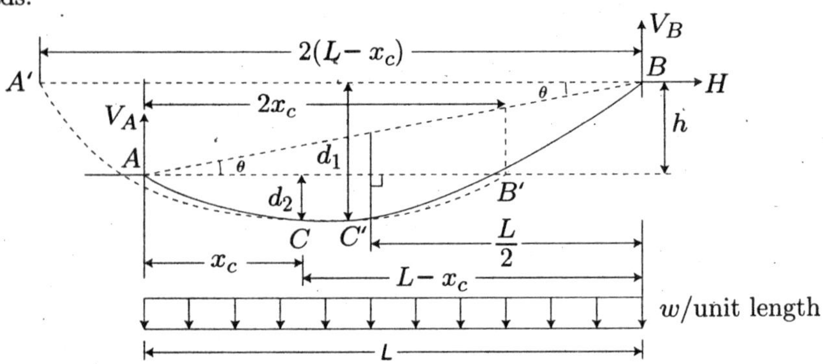

Fig. 14.4: Cable supported at different levels

Referring to Fig. 14.4,

$$H = \frac{w(2x_c)^2}{8d_2}, \text{ for cable } ACB,$$

$$= \frac{w[2(L - x_c)]^2}{8d_1} \text{ for cable } A_1CB,$$

whence
$$\frac{x_c}{L - x_c} = \sqrt{\frac{d_2}{d_1}}. \tag{14.7}$$

Eq. (14.7) gives the location of the lowest point of the cable, after which value of H can be calculated.

Vertical reaction at A and B may be obtained from $\sum M_B = 0$ and $\sum M_A = 0$ respectively.

Thus $V_A = \frac{wL}{2} - H\frac{h}{L}$ and $V_B = \frac{wL}{2} + H\frac{h}{L}$.

Resulting tensions T_A and T_B at A and B are given by

$$T_A = \sqrt{V_A^2 + H^2} \tag{14.8a}$$

and
$$T_B = \sqrt{V_B^2 + H^2}. \tag{14.8b}$$

Approximate values of T in terms of H can be calculated from

$$T_A = H\left[1 + 16\frac{d^2}{L^2} + \tan^2\theta - 8\frac{d}{L}\tan\theta\right]^{\frac{1}{2}} \tag{14.8c}$$

and $$T_B = H\left[1 + 16\frac{d^2}{L^2} + \tan^2\theta + 8\frac{d}{L}\tan\theta\right]^{\frac{1}{2}}. \tag{14.8d}$$

A and B are supports at the lower and upper levels respectively.

As in this case V_B is larger than V_A, maximum tension in the cable will be at the higher support at B.

Length of the cable ACB may be obtained by adding length of the cable AC and BC.

Thus Length of cable $= \frac{1}{2}$ [Length of cable ACB_1+ Length of cable A_1CB]

$$= L + \frac{8}{3}\left[\frac{d_2^2}{4x_c} + \frac{d_1^2}{4(L - x_c)}\right]$$

14.4 Unstiffened Suspension Bridges

In this case the roadway is suspended directly from the cables by means of closely spaced hangers. The simplified theory for such suspension bridges is obtained assuming that

(a) Cable does not change shape under load,

(b) Weight of the cable is negligible,

(c) Suspenders (hangers) produce uniformly distributed load on cables, and

(d) Cable is perfectly flexible and cannot carry any bending moment.

The analysis of unstiffened suspension bridges can be done on the basis of previous discussions.

14.5 Anchor Cables

The cables carrying the hangers are supported on two towers on both of its sides. The cables after passing over the supporting tower are anchored into huge mass of concrete. Two types of arrangement usually adopted for passing the cable over the supporting towers are

(a) Continuous Cable (Pulley Type)

(b) Non-continuous Cable (Saddle Type on Rollers)

(a) Continuous Cable (Pulley Type)

In this case the suspension cable is passed over a frictionless guide pulley fixed on the top of supporting tower and then anchored down as shown in Fig. 14.5(a). In such a case, the tension in suspension cable is equal to that in the anchor cable.

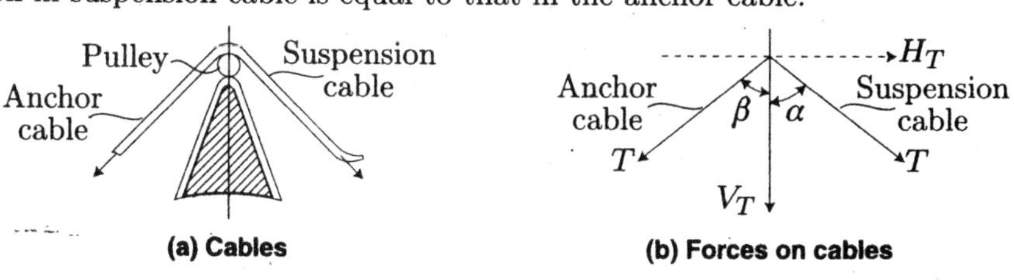

Fig. 14.5: Continuous cable

In Fig. 14.5(b), if α and β be the angles which the anchor cable and the suspension cable make with the vertical, the magnitude of the vertical force on the tower will be given by

$$V_T = T(\cos\beta + \cos\alpha).$$

Horizontal force acting on the tower will be

$$H_T = T(\sin\beta - \sin\alpha).$$

Moment at the base of tower of height h will be

$$M = H_r(h).$$

(b) Non-continuous Cable (Saddle Type on Roller)

In this case, the suspension cable and the anchor cable are attached to a saddle mounted on rollers on the top of the tower as in Fig. 14.6. The tension in the cables will not be the same, but there will be no horizontal force on the tower.

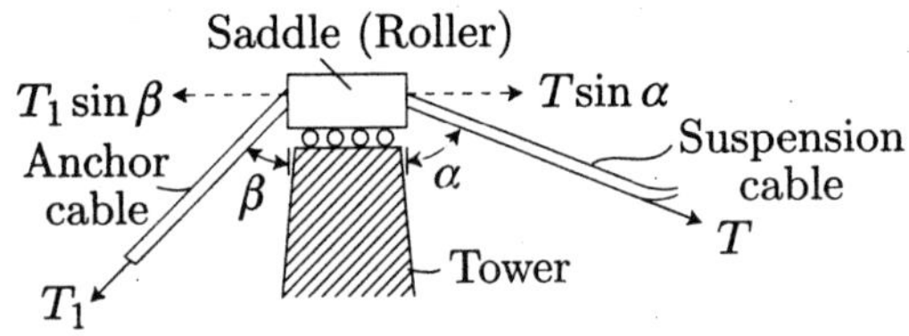

Fig. 14.6: Non-continuous cable (Roller Type)

Referring to Fig. 14.6,

$$T_1 \sin\beta = T\sin\alpha.$$

Tension in anchor cable, i.e., $T_1 = T\,\dfrac{\sin\alpha}{\sin\beta}$.

Vertical force on Tower, i.e., $V_T = T\cos\alpha + T_1\cos\beta$.

Example 14.1. *Determine the component of reactions and the shape for the cable shown in Fig. 14.7 for which dip at F is known.*

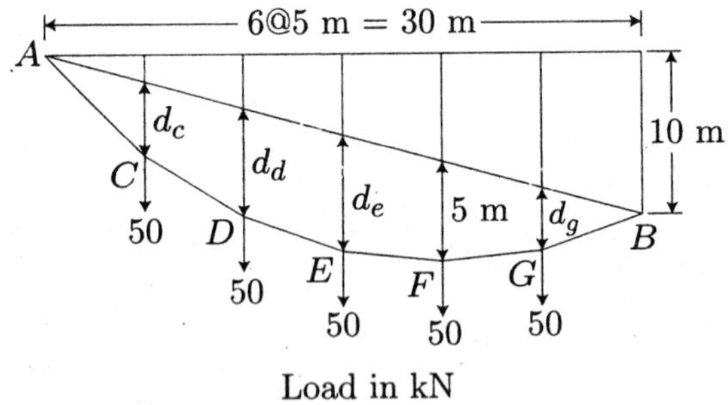

Fig. 14.7: For Example 14.1

Solution : From an imaginary beam corresponding to horizontal projection of the cable, from Eqs. (14.1a) and (14.2a),

$$V'_A = \frac{1}{30}[50(5)(1+2+3+4+5)] = 125 \text{ kN}$$

$$M' = 125(20) - 50(5)(1+2+3) = 1000 \text{ kNm}.$$

From Eqs. (14.1b), (14.2c) and (14.2d),

$$H = \frac{1000}{5} = 200 \text{ kN}$$

$$V_A = 125 + \frac{200(10)}{30} = 191.67 \text{ kN}$$

$$V_B = 250 - 125 - \frac{200(10)}{30} = 58.33 \text{ kN}.$$

The shape of the cable may be obtained from the considerations that bending moments at all points of the cable are zero. Thus,

$$d_c = \frac{M'}{H} = \frac{1}{200}(125)(5) = 3.125 \text{ m}$$

$$d_d = \frac{1}{200}[(125)(10) - 50(5)] = 5 \text{ m}$$

$$d_e = \frac{1}{200}[(125)(15) - (50)(5)(1+2)] = 5.625 \text{ m}$$

$$d_g = \frac{1}{200}[(125)(25) - (50)(5)(1+2+3+4)] = 3.125 \text{ m}.$$

Example 14.2. *A foot-bridge of 30 m span and of width 3 m is suspended between piers at the same level. The platform is supported by twenty equidistant hangers attached to each of the two cables, the central dip of the cable being 3 m. Load on the footway including self weight, weight of cables, etc. may be taken as 10 kN/m². Find the cross-sectional area of the hanger and the cable, if allowable stress be 120 N/m². Also obtain the length of the cable.*

Solution : Total load = 3(30)(10) = 900 kN.

Load on each hanger $= \dfrac{900}{20(2)} = 22.50$ kN.

Cross-sectional area of hanger $= \dfrac{22.50(1000)}{120} = 187.50 \text{ mm}^2$.

Horizontal thrust in cable, i.e.,

$$H = \frac{wL^2}{8d} = \frac{10\left(\frac{3}{2}\right)(30)^2}{(8)(3)} = 562.50 \text{ kN}$$

Vertical reaction at support, i.e., $V = \dfrac{wL}{2} = 225$ kN

$$T_{max} = \sqrt{V^2 + H^2} = 605.83 \text{ kN}.$$

Cross-sectional area of cables, $= \dfrac{T_{max}}{120} = 5048.583 \text{ mm}^2$.

Length of cable $= L + \dfrac{8}{3}\dfrac{d^2}{L} = 30.80$ m.

Example 14.3. *A foot-bridge of width 3 m is carried over a river of span 90 m by means of two cables. The supports are 3 m and 12 m higher than the lowest point of the cable. If the roadway carries a uniformly distributed load of* 20 kn/m² *including self-weight, calculate* *the tension at the two supports of the cable. If the anchor cables make an angle of 30 degree with the horizontal, calculate forces on the tower, when*

(a) Cables run over a pulley device.

(b) Cables are attached to saddles resting on rollers at top of the tower.

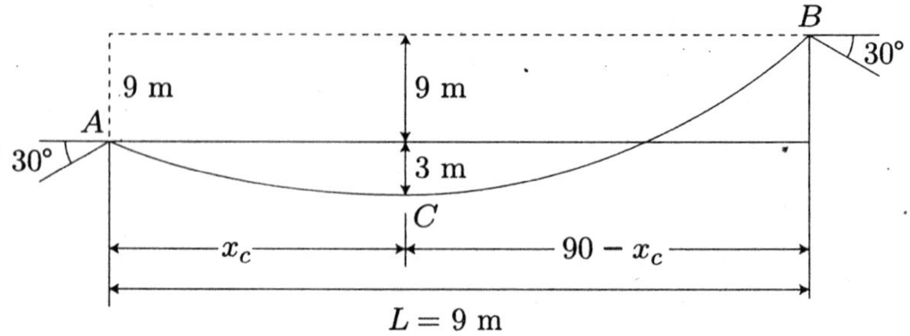

Fig. 14.8: For Example 14.3

Solution : From Fig. 14.8,

$$\frac{x_c}{90 - x_c} = \sqrt{\frac{3}{12}} = \frac{1}{2},$$

whence $x_c = 30$ m and $90 - x_c = 60$ m.

$$\text{Load on each cable} = \frac{20(3)}{2} = 30 \text{ kN/m}.$$

$$\text{Value of } H = \frac{(30)(30)^2}{2(3)} = 4500 \text{ kN}.$$

Also $$V_A = \frac{(30)(90)}{2} - \frac{(4500)(9)}{(90)} = 900 \text{ kN}$$

$$V_B = (30)(90) - 900 = 1800 \text{ kN}$$

$$T_A = \sqrt{V_A^2 + H^2} = 4589.12 \text{ kN}$$

$$T_B = \sqrt{V_B^2 + H^2} = 4846.65 \text{ kN}.$$

Forces on Towers

(a) Guide pulley device.

Tension on Anchor cable at $A = 4589.12$ kN

Tension on Anchor cable at $B = 4846.65$ kN.

Horizontal force on Tower

$$\text{at } A = 4500 - (4589.12)(\cos 30°) = 525.705 \text{ kN}$$

$$\text{at } B = 4500 - (4846.65)(\cos 30°) = 302.67 \text{ kN}.$$

Vertical force on Tower

$$\text{at } A = (900) + (4589.12)(\sin 30°) = 3194.58 \text{ kN}$$

$$\text{at } B = (1800) + (4846.65)(\sin 30°) = 4223.325 \text{ kN}.$$

(b) Saddle Type on Roller

Horizontal force on Tower = zero.

Tension on anchor cable at A, i.e., $T_{A1} = \dfrac{4500}{\cos 30°} = 5196.152 = T_{B1}$.

Vertical force on Tower

$$\text{at } A = (900) + (5196.152)\sin 30° = 3498.076 \text{ kN}$$

$$\text{at } B = 1800 + (5196.152)\sin 30° = 4398.076 \text{ kN}.$$

14.6 Guyed Structure

Cables are sometimes used as guys. Fig. 14.9(a) shows a cable used as a guy in holding the mast vertically, when it is subjected to a vertical load of 1000 kN. From $\sum M_B = 0$, $30(H) = (1000)(3)$, whence $H = 100$ kN.

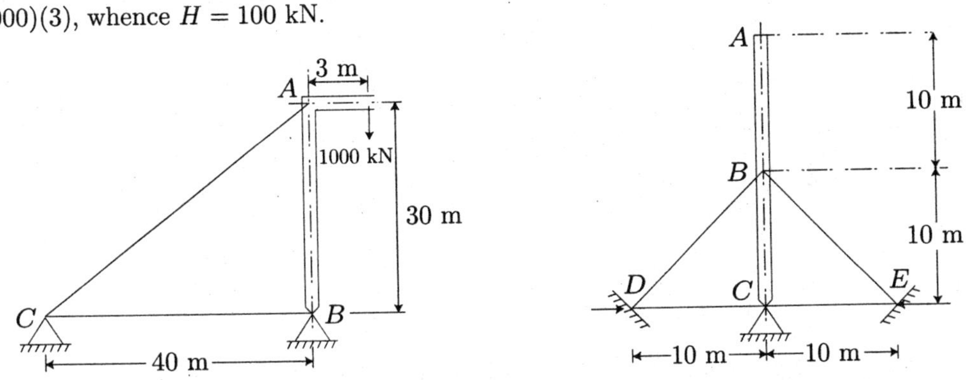

(a) Cable guy **(b) Guyed most**

Fig. 14.9: Cable guys

This value of H must be developed for holding the mast vertically. Assuming the cable shape to be parabolic and acted upon by a uniform load of 100 N/m due to self-weight of the cable, the cable sag 'd' will be given by

$$(100)(1000) = \frac{wL^?}{8d} = \frac{(100)(40)^2}{8d}, \quad \text{whence} \quad d = \frac{1600}{8000} = 0.20 \text{ m}.$$

The maximum tension in the guy will occur at A'' and using Eq. (14.8c),

$$T_{max} = 100\left[1 + 16\left(\frac{0.20}{40}\right)^2 + \left(\frac{3}{4}\right)^2\right]^{\frac{1}{2}} = 125 \text{ kN}.$$

The value of T_{max} may also be obtained approximately by considering the cable to be a straight line along AC. The tension in the cable will then be given by

$$100\left(\frac{5}{4}\right) = 125 \text{ kN}.$$

When the sag of cable is small, the value of cable tension calculated by the approximate method will not involve much error.

In most cases, however, it is necessary to have more than one guy on a mast or other guyed structure, in order that the mast may be held against overturning in more than one direction. Under these conditions, the guys are adjusted so that they have a certain amount of initial tension. Considering Fig. 14.9(b), if the initial tension on each cable be adjusted to 85 kN in order to prevent any sag of the cable when there is no load on the mast, horizontal component of this tension is $\frac{85000}{1.414} \approx 60104$ N $= 60.104$ kN.

When the load of 50 kN, is applied at the top of the mast, a horizontal force of 100 kN, acting to the left must be applied at B by the two guys. It may be assumed as an approximation that this force of 100 kN is provided by increasing the horizontal component of tension in guy BD by 50 kN and decreasing the horizontal component of tension in guy BE by the same amount so that,

$$H_{BD} = +110 \text{ kN} \quad \text{and} \quad H_{BE} = +\ 10 \text{ kN}.$$

The maximum guy tension then equals, $+110(1.414) = +155.54$ kN.

Actually when the load of 50 kN is applied at the top of mast, point B moves to the right. Guy BC permits this movement since it undergoes additional elastic stretch and its sag decreases. The sag in guy BD increases and the length of the guy shortens elastically. An exact analysis taking into account elastic change of length and sag changes can be made, but is relatively complicated. The approximate method discussed is more practical for usual design problems.

14.7 Suspension Bridges with Stiffeners

For a suspension bridge usually constructed in a manner the dead loads are carried entirely by the cables. A large portion of the dead load comes from the roadways and is considered uniform horizontal length. Such cables are parabolic under dead load only. If live load is applied, partially to give maximum forces in members, the cables tend to change their shape. In order to prevent local changes of slope in the roadway, due to live load from being too large, the floor beams of the floor system are usually framed into stiffening girders or trusses which in turn are supported by hangers connected to the cables. Stiffening girders or trusses can have two hinges at the support or can have three hinges, two at the supports and one in between the two supports. A bridge having stiffened girder or truss with three hinges is statically determinate, whereas the one with stiffened girder or truss with two hinges is statically indeterminate.

A simple solution of the stress distribution in a stiffened suspension bridge can only be found by adopting the following assumptions.

(i) Hangers are so adjusted that under the dead weight of the bridge, the cable assumes a parabolic shape and the whole of the dead weight is carried direct to the cables through the hangers. The stiffening girder or truss thus is not subjected to any stress due to dead load.

(ii) The cable retains the parabolic shape under all conditions of live load, so that the load on the cables due to any pattern of live load is uniformly distributed, i.e., all the hangers in a given span have equal tensions.

14.8 Suspension Bridge with Three-Hinged Stiffening Truss

Fig. 14.10(a) shows a suspension bridge having three hinged stiffening trusses, with a hinge at some intermediate point. The truss members will not be subjected to any force due to the dead weight of the truss and roadway is considered uniformly distributed. The members of the truss will be subjected to stress, when the live load applied is of any nature except uniformly distributed load covering the full span.

The method of calculation of forces in various truss members considered will be illustrated by assuming a concentrated live load of 200 kN acting as in Fig. 14.10(a).

From free body diagram of the suspended bridge in Fig. 14.10(b), from $\sum M_E = 0$,

$$(V_A + V_D)(100) - (200)(75) = 0, \text{ whence } V_A + V_D = 150 \text{ kN} \uparrow.$$

From Fig. 14.10(c), for moment about F,

$$(V_A + V_D)(50) + H(10) - H(20) - (200)(25) = 0, \quad \text{whence,} \quad H = 250 \text{ kN}.$$

The maximum cable tension occurs at A and B and is given by Eq. (14.4b), whence

$$T_{max} = 250\left(1 + \frac{16}{100}\right)^{\frac{1}{2}} = 269.258 \text{ kN}.$$

If load in each hanger be F, equivalent uniform load will be $\frac{F}{5}$

From $H = \frac{wL^2}{8d}$, $250 = \frac{F(100)^2}{5(8)(10)}$, whence $F = 10$ kN.

Once the hanger tension is obtained the forces in the members of the truss may be analysed from Fig. 14.10(d).

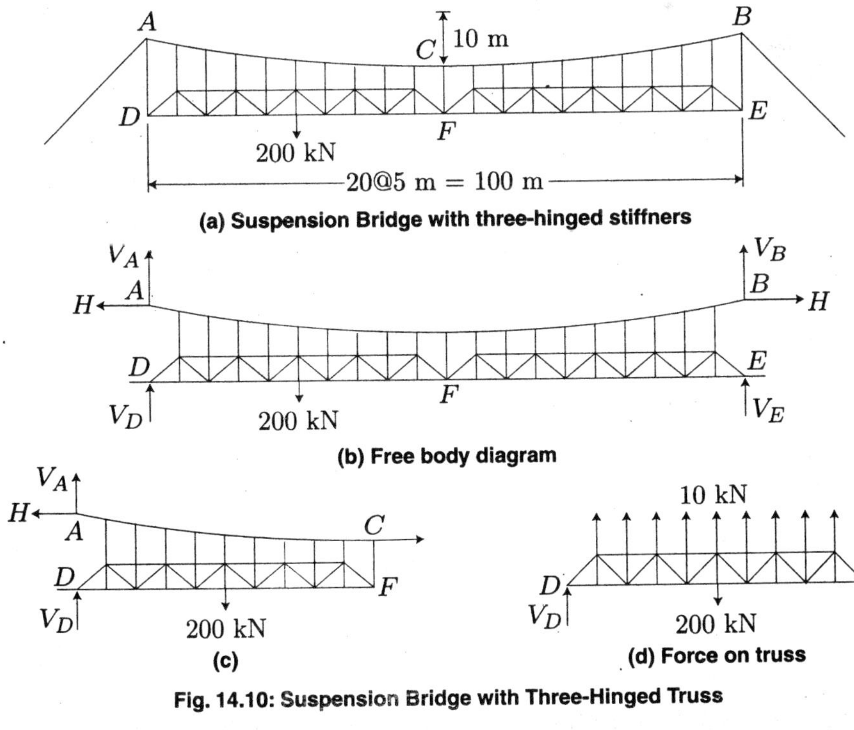

Fig. 14.10: Suspension Bridge with Three-Hinged Truss

14.9 Suspension Bridge with Three-Hinged Stiffening Girder

The stiffening girder will not be carrying any load due to the dead weight of the structure which can be considered as uniformly distributed. The effect of moving live load has to be considered.

Fig. 14.11(a) shows a suspension bridge with a three-hinged stiffening girder AB with the third hinge at C. The reactions from the cable loads will consist of horizontal and vertical components. The total dead weight of the bridge is equivalent to a uniformly distributed load w_0 on the horizontal length.

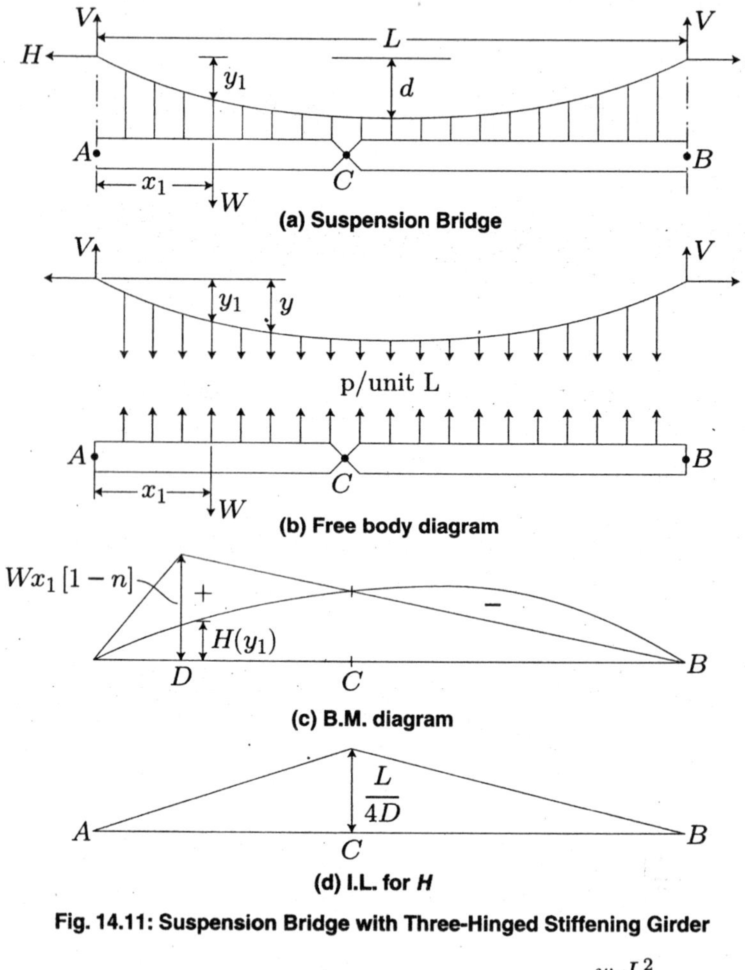

Fig. 14.11: Suspension Bridge with Three-Hinged Stiffening Girder

Horizontal, component of tension in the cable will be, $H_0 = \dfrac{w_0 L^2}{8d}$.

The tension in each hanger will be $\dfrac{w_0 L}{N}$,
where N is the number of hangers,

Considering the cable form as parabolic, the equation of the cable is given by

$$y = \frac{4d}{L^2}x(L-x),$$

where x is measured from the support.

If a load W is placed on the bridge at a distance x_1 from A, it causes equal tension in each hanger and conditions of loading of the cable and girder are as in Fig. 14.11(b).

The bending moment in the girder at a section distance nL from A, for $nL > x_1$, is given by

$$\begin{aligned} M &= \frac{W(L-x_1)}{L}nL - W(nL - x_1) + \frac{n^2L^2}{2} - \frac{pL}{2}(nL) \\ &= W\left[n(L-x_1) - (nL - x_1)\right] + \frac{pnL^2}{2}(n-1) \\ &= Wx_1[1-n] + \frac{pnL^2}{2}(n-1). \end{aligned}$$

As moment at any point in the cable is zero,

$$V(nL) - Hy - p\frac{n^2L^2}{2} = 0, \text{ whence with } V = \frac{pL}{2},$$

$$H = p\frac{nL^2}{2y}(1-n).$$

The bending moment in girder may then be expressed as

$$M = Wx_1(1-n) - Hy \tag{14.9a}$$

$$= \text{B.M. in beam} = Hy. \tag{14.9b}$$

The resulting bending moment diagram is illustrated in Fig. 14.11(c), the magnitude of the bending moment at point of third hinge at C being zero.

Case I. Third hinge at centre, i.e., $\frac{L}{2}$ from A.

As $M_C = 0$, from Eq. (14.9a),

$$Wx_1\left[1 - \frac{1}{2}\right] - Hd = 0,$$

whence $$H = \frac{Wx_1}{2d} = \frac{pL^2}{8d} \tag{14.10a}$$

and $$p = \frac{4Wx_1}{L^2}. \tag{14.10b}$$

Case II. Third hinge at $\frac{L}{3}$ from A.

For this case,

$$Wx_1\left[1 - \frac{1}{3}\right] - Hy = 0,$$

whence $$H = \frac{2Wx_1}{3y}.$$

For cable curve given by $y = \frac{4d}{L^2}x(L-x)$,

$$y = \frac{8}{9}d, \text{ at } x = \frac{L}{3}$$

and $$H = \frac{3Wx_1}{4d} = \frac{pL^2}{8d}, \tag{14.10c}$$

whence $$p = \frac{6Wx_1}{L^2}. \tag{14.10d}$$

Case III. Third hinge at $\frac{L}{4}$ from A.

For the case, $H = \dfrac{Wx_1}{d}$ (14.10e)

and $p = \dfrac{8W_{x_1}}{L^2}$. (14.10f)

The influence line diagram for H may be obtained from values of H. The influence line diagram for H will be in the form of a triangle with the maximum ordinate $\frac{L}{4d}$ at the location of third hinge as in Fig. 14.11(d).

Having calculated 'p', the shearing force diagrams for the girder can be obtained.

For $nL \leq x_1$,

S.F. at $nL = \frac{W(L-x_1)}{L} - \frac{pL}{2} + pnL =$ Beam S.F. $-\frac{pL}{2} + pnL$, the actual magnitude depending on the location of the third hinge. Thus, for third hinge at centre,

$$\text{S.F.} = \frac{W}{L}(L - x_1) - \frac{2W_{x_1}L}{L^2} + \frac{4W_{x_1}nL}{L^2} = \frac{W}{L}[L - 3x_1 + 4nx_1]. \tag{14.11a}$$

For third hinge at $\dfrac{L}{3}$ and $\dfrac{L}{4}$ from A,

$$\text{S.F.} = \frac{w}{L}[L - 4x_1 + 6nx_1]$$

and $= \dfrac{W}{L}[L - 5x_1 + 8nx_1]$ respectively.

For $nL > x_1$,

$$\text{S.F. at } nL = \frac{W(L - x_1)}{L} - W - \frac{pL}{2} + pnL.$$

For third hinge at centre,

$$\text{S.F.} = \frac{W}{L}[4nx_1 - 3x_1]. \tag{14.11b}$$

Also S.F. $= \dfrac{W}{L}[6nx_1 - 4x_1]$ for third hinge at $\dfrac{L}{3}$ from A.

$= \dfrac{W}{L}[8nx_1 - 5x_1]$, for third hinge at $\dfrac{L}{4}$ from A.

For the case, when the third hinge is at centre of span, the shear force diagram as represented by Eqs. (14.11a) and (14.11b) will vary according as $x_1 < \frac{L}{3}$ or $x_1 > \frac{L}{3}$.

The two cases are shown in Figs. 14.12(a) and 14.12(b).

The bending moment at any section of the stiffening girder is shown in Fig. 14.11(c). It is evident that the maximum positive bending moment at a section will occur when the concentrated load is at that section, the magnitude being given by

$$\frac{Wx_1}{L}(L - x_1) - Hy_1$$

For the third hinge at centre,

$$+\text{ Ve B.M.} = \frac{Wx_1}{L}(L - x_1) - \frac{Wx_1}{2d}\frac{4d}{L^2}x_1(L - x_1)$$

$$= \frac{Wx_1}{L}(L - x_1) - \frac{2W}{L^2}x_1^2(L - x_1).$$

For maximum positive B.M.,

$$\frac{dM}{dx} = 0, \quad \text{whence} \quad 1 - \frac{2x_1}{L} - \frac{4x_1}{L} + \frac{6x_1^2}{L^2} = 0,$$

which gives $x_1 = +0.211L$, the location of the maximum positive B.M. in the girder under effect of a moving concentrated load, the value of the maximum B.M. being $+0.096WL$.

The maximum negative B.M. will occur midway between C and B.

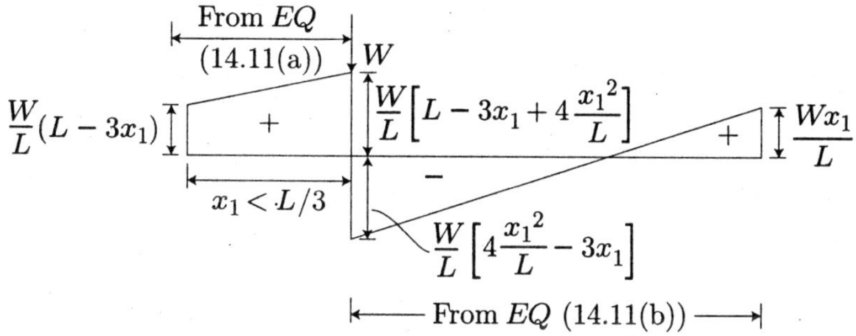

(a) S.F. for load $x_1 < L/3$

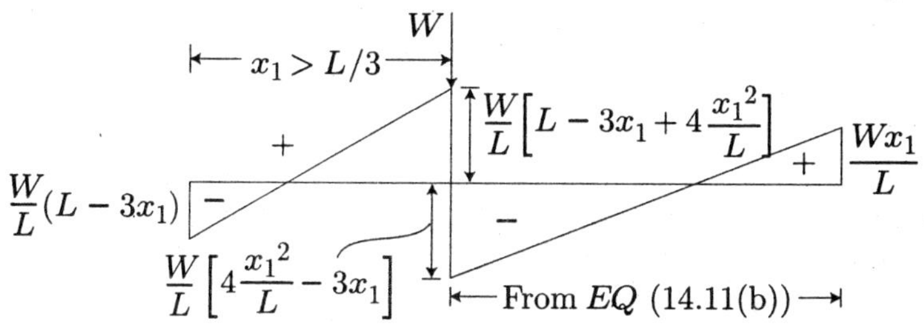

(b) S.F. for load $x_1 > L/3$

Fig. 14.12: S.F. in Three-Hinged Stiffening Girder

For concentrated load as in Fig. 14.11, with the third hinge is at centre, the negative B.M. midway between C and B becomes

$$\frac{Wx_1}{L}(L - x_1)\frac{L}{4(L - x_1)} - \frac{Wx_1}{2d}\frac{3}{4}d = -\frac{Wx_1}{8}.$$

This indicates that the maximum value of negative B.M. occurs when the concentrated load is at the centre.

Magnitude of maximum negative B.M. $= -\dfrac{WL}{16}$.

Fig. 14.12 indicates that maximum positive shear force in a section occurs just at the left of the section, when the concentrated load is placed there. Magnitude of this S.F. is given by

$$\frac{W}{L}\left[L - 3x_1 + \frac{4x_1^2}{L}\right].$$

For maximum value of positive S.F.,

$$\frac{d(\text{S.F.})}{dx_1} = 0, \text{ which gives } -3 + \frac{8x_1}{L} = 0, \text{ whence } x_1 = \frac{3}{8}L.$$

Magnitude of the maximum positive S.F. is given by $+\dfrac{7}{16}W$.

The maximum negative S.F. at a section occurs just on the right of the section, when the load is placed there. Magnitude of this maximum negative S.F. is $\frac{W}{L}\left[\frac{4x_1^2}{L} - 3x_1\right]$.

For maximum S.F., $x_1 = \dfrac{3}{8}L$

and magnitude of the maximum negative S.F. is $-\dfrac{9}{16}W$.

14.10 Influence Lines for Suspension Bridge with Three-Hinged Stiffening Girder

The effect of a distributed load on the suspension bridge is best studied by the use of influence line diagram. At any section of the bridge, influence lines required are

(a) Influence lines for horizontal thrust H on cable

(b) Influence lines for shear force on girder and

(c) Influence lines for bending moment on girder.

(a) *I.L. for H.*

The influence line for the horizontal thrust is illustrated in Fig. 14.11(d).

(b) *I.L. for S.F. on girder.*

At any section of the girder, distance nL from left hand support, S.F. = Beam S.F. $-p\left(\frac{L}{2} - nL\right)$,

Now $\quad p\left[\dfrac{L}{2} - nL\right] = \dfrac{pl^2}{8d}\dfrac{8d}{L^2}\left[\dfrac{L}{2} - nL\right] = \dfrac{pL^2}{8d}\dfrac{4d}{L}[1-2n] = H\tan\theta,$

where θ = slope of cable at distance nL from left-hand support A.

Resultant S.F. then = Beam S.F. $-H\tan\theta$

$$= \text{Beam S.F. } -H\frac{4d}{L}(1-2n).$$

The influence line diagrams obtained from above for the case when the third hinge is at centre can be represented in Figs. 14.13(a) to 14.13(c) and as shown, the nature depending on the distance of the section from the left-hand support A.

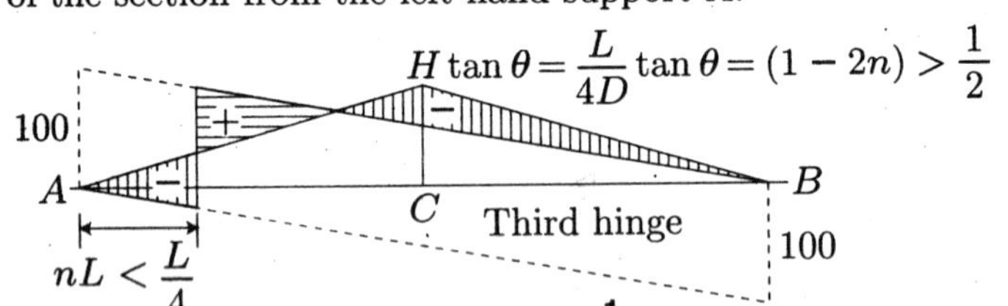

(a) I.L. for S.F. for $n < \frac{1}{4}$

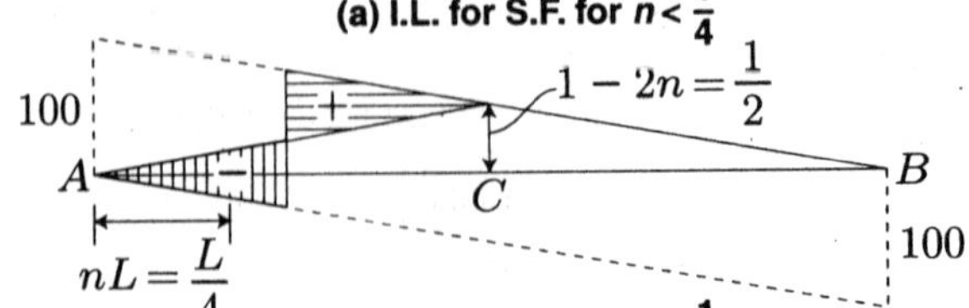

(b) I.L. for S.F. for $n = \frac{1}{4}$

Fig. 14.13: S.F. in Three-Hinged Stiffening Girder (Contd.)

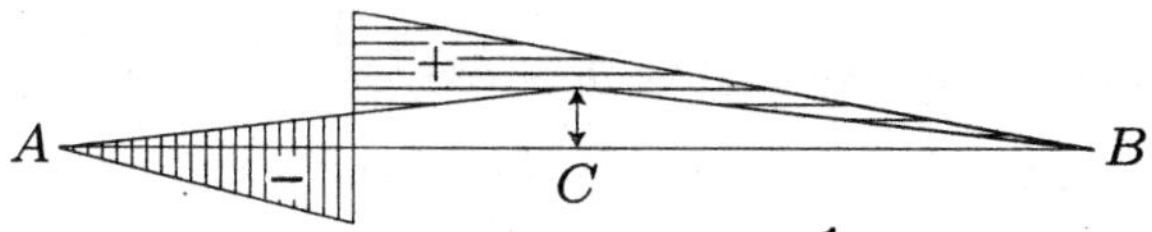

(c) I.L. for S.F. for $n > \frac{1}{4}$

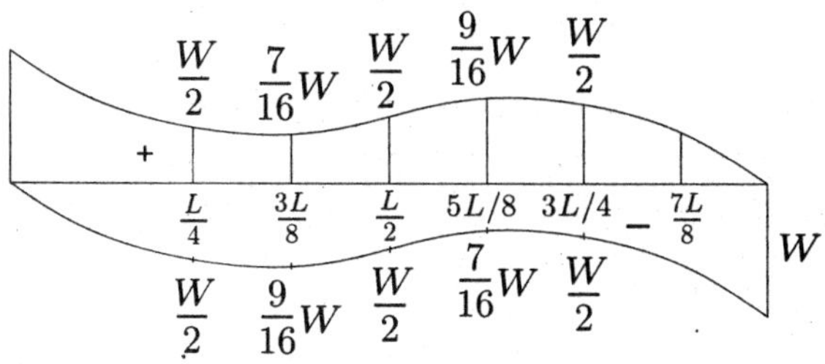

(d) Max S.F. diagram

Fig. 14.13: S.F. in Three-Hinged Stiffening Girder

The maximum S.F. diagram along the stiffening girder as a concentrated load rolls over the girder is shown in Fig. 14.13(d).

(c) *I.L. for B.M. on girder*

At any section in the stiffening girder,

B.M. = B.M. in beam $-Hy$.

The influence line for B.M. in beam will be a triangle having maximum ordinate of $\frac{x(L-x)}{L}$ at the section, whereas the influence line for Hy will be another triangle having maximum ordinate of $\frac{L}{4d}y$, where y is the dip of the cable at the section.

From shape of cable given by $y = \frac{4d}{L^2}x(L-x)$, the maximum ordinate of Hy at location of the third hinge will also be $\frac{x}{L}(L-x)$.

The influence diagram for bending moment on stiffening girder thus obtained is shown in Fig. 11.14(a) for the case where the third hinge is at centre. As the areas of the two triangles are same, the resulting bending moment at any section of the girder for uniformly distributed load on the whole span will be zero.

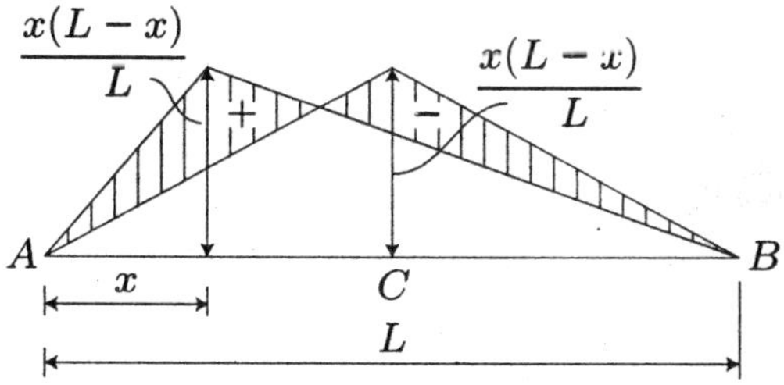

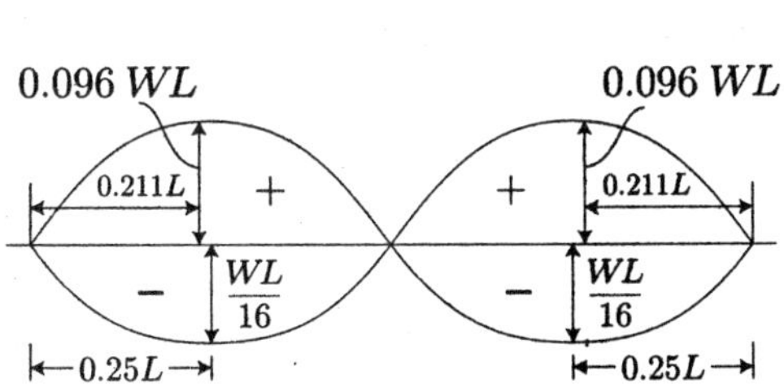

(a) I.L. for B.M.

(b) Max B.M. diagram

Fig. 14.14: B.M. in Three-Hinged Stiffening Girder

Fig. 14.14(b) shows the maximum bending moment diagram at various sections of the girder as a concentrated load W rolls over the span.

Example 14.4. *Obtain the S.F. and B.M.diagrams for the suspension bridge shown in Fig. 14.15(a).*

Solution : The cable may be considered as uniformly loaded with $p = \dfrac{F}{5}$.

Value of $p = \dfrac{4Wx_1}{L^2} = \dfrac{4(500)(20)}{(50)(50)} = 16$ kN/m,

whence $F = 50$ kN and $H = \dfrac{Wx_1}{2d} = \dfrac{(500)(20)}{2(10)} = 500$ kN.

Vertical force on cable, $V_A = 5F = 400 = V_B$

$$T_{max} = \sqrt{V_A^2 + H^2} = \sqrt{(400)^2 + (500)^2} = 640.312 \text{ kN.}$$

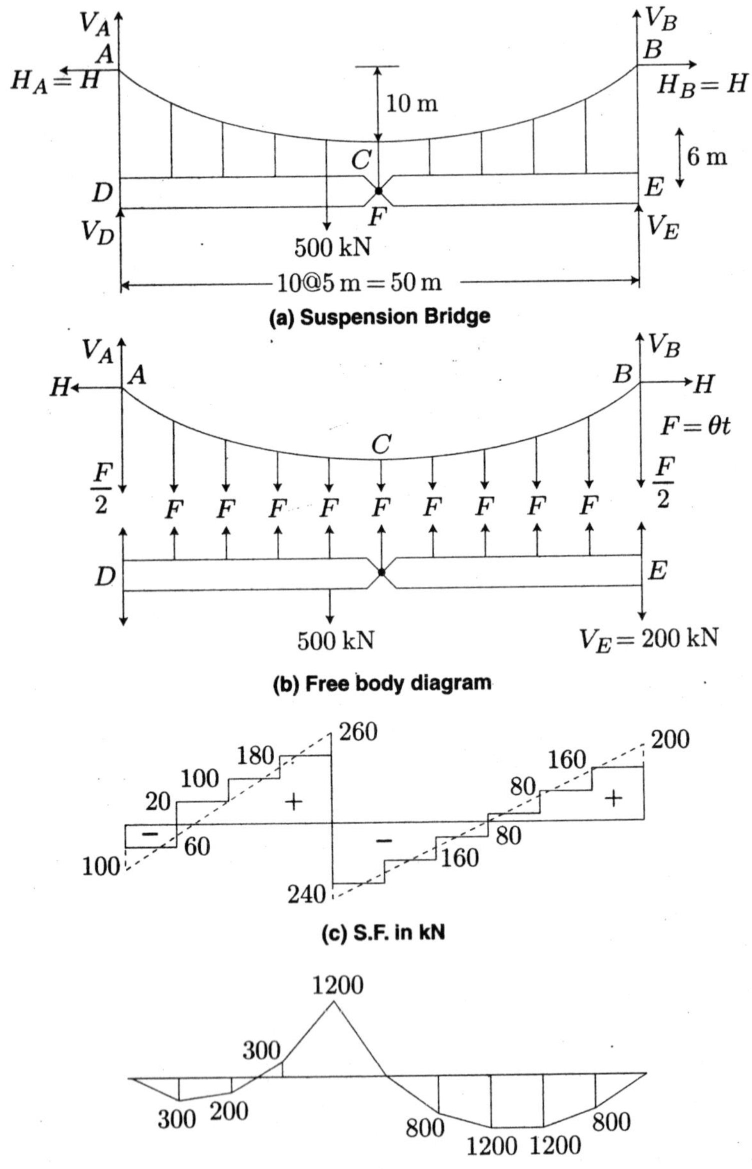

(a) Suspension Bridge

(b) Free body diagram

(c) S.F. in kN

(d) B.M. in kN-m

Fig. 14.15: For Example 14.4

The girder will be subjected to forces as shown in Fig. 14.15(b) and S.F. and B.M. diagrams on the basis of actual loads are illustrated in Fig. 14.15(c) and 14.15(d) respectively. In the figures, the S.F. and B.M. diagrams on girders as obtained on the basis of uniformly distributed load on girder are shown by dashed lines.

Example 14.5. *A symmetrical suspension cable of 160 m span and 20 m dip at centre is stiffened by a three hinged girder with the third hinge at centre. The D.L. is 5 kN/m. Determine the maximum values of H, maximum value of T, and the maximum B.M. in the girder, when a concentrated load of 200 kN rolls over the span.*

Solution : From influence line diagram for H,

$$\text{Maximum } H = \frac{1}{2}\frac{L}{4d}(160)(5) + (200)\frac{L}{4d} = 1200 \text{ kN}.$$

Magnitude of u.d.l. due to concentrated load

$$= \frac{4Wx_1}{L^2} = \frac{4(200)(80)}{(160)^2} = 2.50 \text{ kN/m}.$$

$$\text{Vertical Reaction on cable} = (5 + 2.50)\frac{160}{2} = 600 = V.$$

$$T_{max} = \sqrt{H^2 + V^2} = 1341.64 \text{ kN}.$$

$$\text{By approximation, } T_{max} = H\sqrt{1 + 16\frac{d^2}{L^3}} = 1341.64 \text{ kN}.$$

No B.M. on girder due to uniformly distributed deadload.

Maximum positive B.M. = 0.096WL = 3072 kNm, when concentrated load is at 0.211L, i.e., 33.76 m from supports.

Maximum negative B.M. $= -\frac{WL}{16} = -2000$ kNm, at 40 m from ends when concentrated load is at centre.

Example 14.6. *A suspension cable of span L, in the shape of a symmetrical parabola is strengthened by a three hinged stiffening girder, with the third hinge at centre. Calculate the maximum S.F. in the girder at the supports and the centre, and the maximum B.M. at quarter span, when uniformly distributed load of magnitude w/unit length and larger than the span moves over the span.*

Solution : From characteristics of the I.L. diagram in Fig. 14.16(a),

$$\text{Maximum positive S.F.} = \frac{1}{2}(1)\left(\frac{L}{3}\right)w = +\frac{wL}{6}.$$

$$\text{Maximum negative S.F.} = -\left[\frac{1}{2}(L)(1) - \frac{1}{2}(L)\left(\frac{2}{3}\right)\right]w = -\frac{wL}{6}.$$

From Fig. 14.16(b),

$$\text{Maximum positive S.F.} = \frac{1}{2}\left(\frac{L}{2}\right)\left(\frac{1}{2}\right)w = +\frac{wL}{8}.$$

$$\text{Maximum negative S.F.} = -\frac{wL}{8}.$$

From Fig. 14.16(c),

$$\text{Maximum positive B.M.} = \frac{1}{2}\left[(L)\left(\frac{3L}{16}\right) - L\left(\frac{3L}{20}\right)\right]w = +\frac{3}{160}wL^2.$$

$$\text{Maximum negative B.M.} = -\frac{1}{2}\left[(L)\left(\frac{3L}{16}\right) - L\left(\frac{3L}{20}\right)\right]w = -\frac{3}{160}wL^2.$$

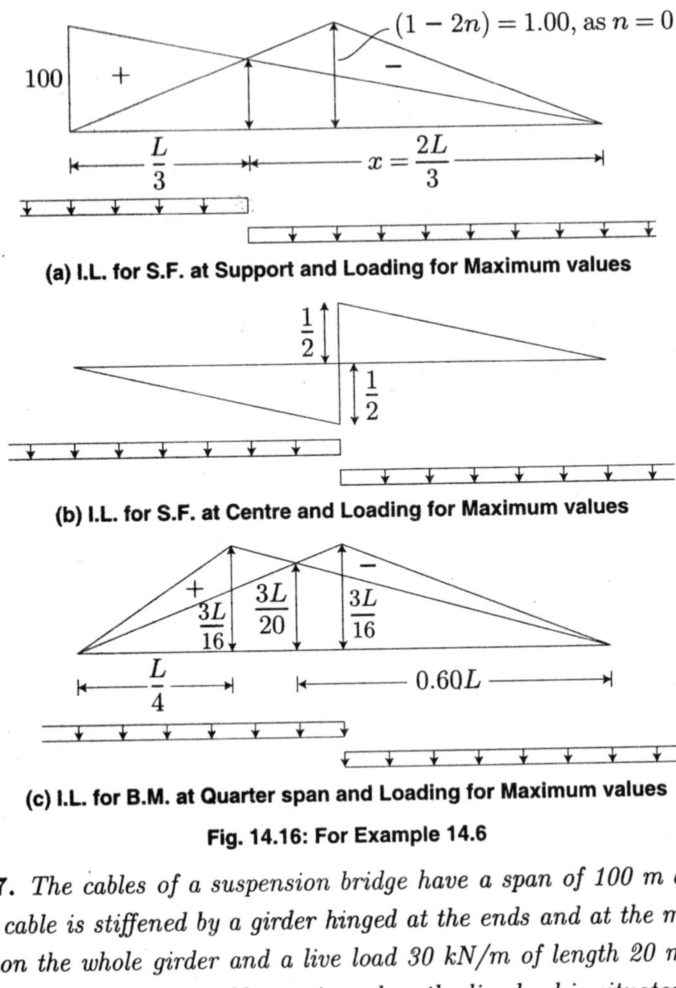

(a) I.L. for S.F. at Support and Loading for Maximum values

(b) I.L. for S.F. at Centre and Loading for Maximum values

(c) I.L. for B.M. at Quarter span and Loading for Maximum values

Fig. 14.16: For Example 14.6

Example 14.7. *The cables of a suspension bridge have a span of 100 m and a central dip of 10 m. Each cable is stiffened by a girder hinged at the ends and at the middle. A.D.L. of 20 kN/m acts on the whole girder and a live load 30 kN/m of length 20 m moves over the girder. Determine the maximum cable tension when the live load is situated on the left-hand half of the stiffening girder such that the right-hand end of the live load is just over the central hinge. Also obtain the S.F. and B.M. diagrams of the girders.*

Solution : In this case $w_d = 20$ kN/m and $w_L = 30$ kN/m, 20 m long.

$$M_F \text{ in girder} = \text{B.M. in simply supported beam } -H(d) = V_E(50) - w_d\frac{(50)^2}{2} - H(10)$$

$$= \left[\frac{(600)(40)}{100} + \frac{(20)(100)}{2}\right](50) - 20\frac{(50)^2}{2} - H(10) = 0,$$

whence $H = 3700$ kN.

If $p = u.d.l.$ on cable due to D.L. and L.L., from $H = \dfrac{pL^2}{8d}$,

$$p = \frac{(3700)(8)(10)}{(100)(100)} = 29.60 \text{ kN/m}.$$

$$\text{Vertical reaction in cable} = V = \frac{29.60(100)}{2} = 1480 \text{ kN}.$$

$$T_{max} = \sqrt{V^2 + H^2} = \sqrt{(1480)^2 + (3700)^2} = 3985.02 \text{ kN}.$$

Also from $T_{max} = H\sqrt{1 + 16\dfrac{d^2}{L^2}}$

$$T_{max} = 3700\sqrt{1 + \frac{16}{100}} = 3985.02 \text{ kN}.$$

S.F. and B.M. in the girder will occur only due to L.L.

Horizontal thrust due to L.L. = 1200 kN.

Load pL/unit length on the cable due to L.L. = 9.60 kN/m.

S.F. on girder = S.F. on simple beam − S.F. due to pL on girder.

B.M. on girder = B.M. on simple beam − B.M. due to pL on girder or from B.M. on beam $-H(y)$.

Resulting S.F. and B.M. diagrams are shown in Figs. 14.17(b) and 14.17(c) respectively.

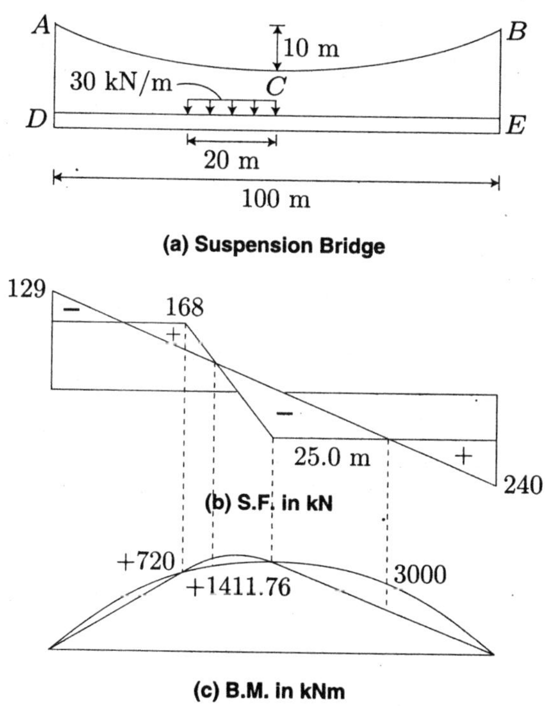

(a) Suspension Bridge

(b) S.F. in kN

(c) B.M. in kNm

Fig. 14.17: For Example 14.7

Problems

14.1 Find the shape of the cable subjected to concentrated loads as shown in Figs. P-14.1(a) to 14.1(c). Calculate also the maximum tension in each cable and the length of the cable.

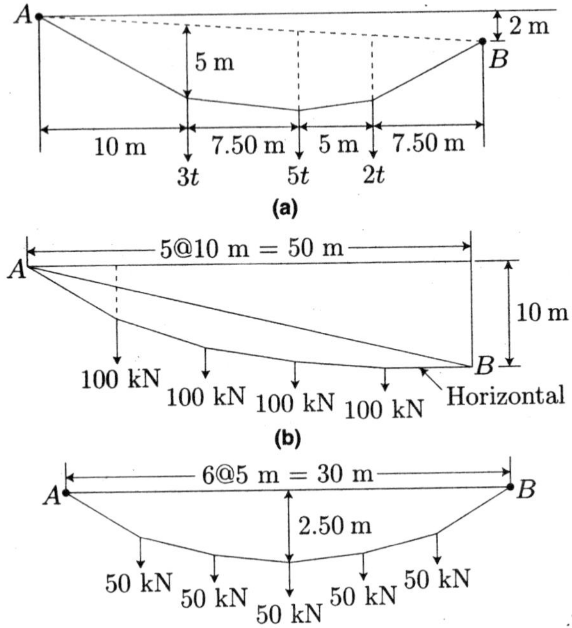

Fig. P-14.1

14.2 A suspension cable of span 30 m and central dip 3 m is carrying a uniformly distributed load of 20 kN/m. Find :

(a) The sectional area of the cable if allowable stress in cable be 1200 kg/cm^2.

(b) the length of the cable and

(c) maximum slope of the cable.

14.3 A suspension cable of span 100 m and central dip 10 m carries a uniformly distributed load of 10 kN/m. Find the maximum and minimum tensions in the cable and the maximum slope of the cable.

14.4 Calculate the sectional area of the cables in Problem 14.1, if allowable stresses be 120 N/mm^2.

14.5 A pipe line of length 60 m weighing 1200 kN is carried over a river by means of a suspension cable, the level of support at one end being 9 m below the other, the level of the lowest point of the cable being 3 m below the lower support. If the pipe line be considered to have no flexural rigidity, calculate the position of the lowest point, the horizontal component of tension in the cable, the maximum tension in the cable and the maximum slope of the cable.

14.6 A cable of span 100 m and a central dip of 10 m is carrying a load of 20 kN/m of horizontal length. Calculate the change in central dip and the increase in the cable tension when there is a reduction of temperature of 20 °C. Assume $\alpha = 1(10)^{-5}$ per °C.

14.7 Find the maximum horizontal span of a cable so that under its own weight, stress does not exceed 120 N/mm^2, if the central dip be $\frac{1}{10}$th of the span. Assume density of steel as 78 kN/m^3.

14.8 A suspension bridge is carried by two cables which are supported by vertical towers 60 m apart. The tops of these towers are at the same level and the cables have a central sag of 6 m. The cable passes over the towers on saddles which are free to move horizontally on rollers. The anchor-cables are inclined at 45° with the vertical. If the dock hangers and live loads are equivalent to u.d.l. of 15 kN/m run of the span calculate the maximum tensile in each cable and pressure on each pier.

14.9 A pipe line CD weighing 1500 kN is carried over a river by means of a suspension cable as shown in Fig. P-14.2. The lowest point of the cable is 3 m below A and 12 m below B. If the pipe line may be considered to have no flexural rigidity, calculate the position of the lowest point and the horizontal component of the tension in the cable.

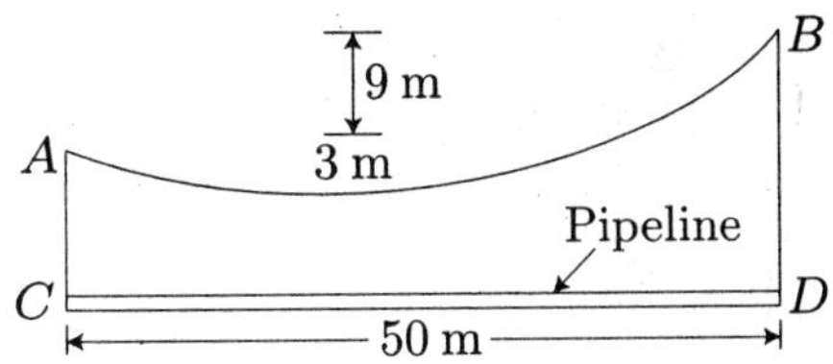

Fig. P-14.2

14.10 An unstiffened suspension bridge spans a horizontal distance of 25 m between piers A and B, A being 1 m below the level of B and the lowest point of the cables 4 m below A. The cables and anchor chains are attached to saddles and are thus free to move horizontally on the piers and the anchor cables are inclined at 45° and may be assumed straight. Each cable carries a load of 15 kN/m horizontally, calculate :

(a) the tension at the lowest point of the cables,

(b) the tension at the higher pier B, and

(c) thrust on the pier B.

14.11 One section of a suspension bridge has a span of 30 m with towers of 5 m and 8 m effective height above the road level. Determine the slope of the cables at each tower and the force in each cable at these points, if there is a uniform load of 15 kN as per metre run covering the bridge.

14.12 A bridge cable slung between two piers 30 m apart carries a load of 22.5 kN/m of the span. The tops of the piers are at the same level and the cable at its lowest point sags 3 m below this level. Calculate the maximum and minimum values of the cable tension.

At each end, the back stays (assumed straight) are inclined to the vertical at 60°. Compare the tensions in the back stay under the following conditions : (a) with the cable supported on frictionless rollers on the tops of the piers, and (b) with ends of the cable (and back stays) attached to saddles carried on rollers on the tops of the piers.

14.13 A suspension cable is suspended from two piers 200 m apart one support being 5 m above the other. The cable carries a.u.d. load of 10 kN/m and has its lowest point 10 m below the lower support. The ends of the cable are attached to saddles on rollers at the top of the pier and the back stays which may be assumed straight are inclined at 60° to the vertical. Determine (a) the maximum tension in the cable, (b) the tension in the back stays and (c) the thrust on each pier.

14.14 A three-hinged segmental arch of span 30 m and rise 6 m carries a *u.d.d.l.* of 30 kN/m of the span and *u.d.l.l.* of 30 kN/m of the span over left half of the span. Draw B.M. diagram and find :

(i) the thrust at the hinges.

(ii) the maximum B.M.

(iii) the thrust and radial shear at the section where the B.M. is maximum.

14.15 A three-hinged arch consists of two quadrants of circle as shown in Fig. P-14.3. Determine the maximum bending moment on the arch and the vertical and horizontal forces on the pin at the crown, when it carries a concentrated load as shown.

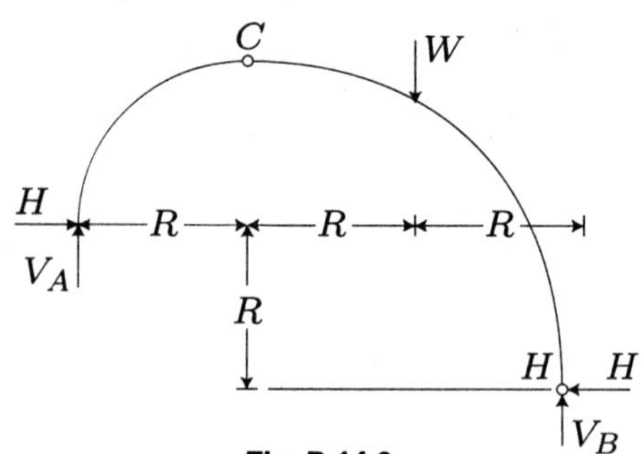

Fig. P-14.3

14.16 A three-hinged arch is formed as shown in Fig. P-14.4 from two segments. Determine the maximum bending moment.

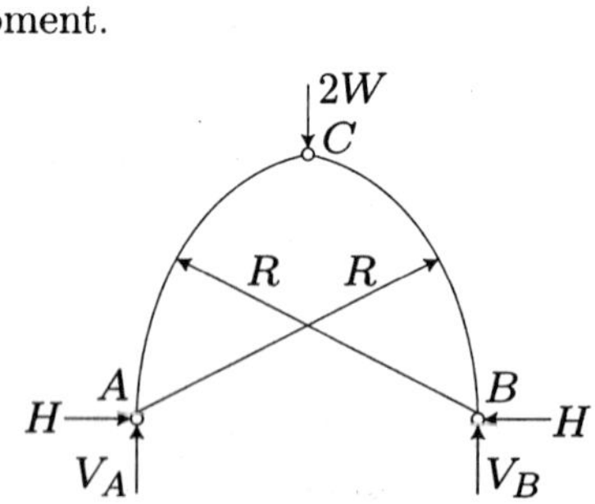

Fig. P-14.4

14.17 A parabolic three-hinged arch is hinged at each abutment and at the crown. The rise to the crown is $\frac{1}{6}$ of the span and left-hand half of the arch is covered by a load of w per metre run.

Find the value of H, the bending moment at each quarter point and the resultant thrust on the top hinge and its direction.

14.18 A three-pinned segmental arch has a span of 40 m and a rise of 10 m. The arch is loaded with 60 kN/m run over the whole span and with an additional load of 30 kN/m run over half the span only from one abutment. Find the B.M. at quarter points and the horizontal thrust.

14.19 A symmetrical three-hinged parabolic arch has span of 24 m and a rise of 6 m. The arch is loaded with a point load of 16 tonnes at a horizontal distance of 6 m from the left-hand support and u.d.l. of 15 kN/m run covering the right half of the arch. Determine the bending moments in the arch at two-span points.

14.20 In a three-hinged circular arch the span is 40 m and rise is 5 m. It carries a uniform load of 15 kN/m of span on the whole length and point loads of 100, 150 and 200 kN at 12 m, 16 m and 28 m from left-hand hinge. Calculate the horizontal thrust and reactions at the hinge. Also draw the B.M.D.

14.21 A 3-hinged parabolic arch has a span of 12 m rise of 2.4 m. It carries a load of 15 kN/m run over the left half of the span, find the bending moment, normal thrust and radial shear at the hinges and the quarter span points.

14.22 The deck of a three-hinged segmental arch rib is carried on spandrel columns. The span of the arch is 48 m and the rise at the crown is 10 m and the spandrel columns are spaced at 8 m centres. Determine the bending moment, radial shear and normal thrust at the left hand quarter span section when a concentrated load of 160 kN is placed over the section.

14.23 A pair of roof rafters span 12 m, rise 3 m. as in Fig. P-14.5 butt against each other the ridge. Their feet butt into rebates at the same level in two walls. There is no horizontal tie. Assuming, the rafters to be pin-jointed at the apex and the feet, calculate the bending moments on the rafters at the points of application of each of the 30 kN loads.

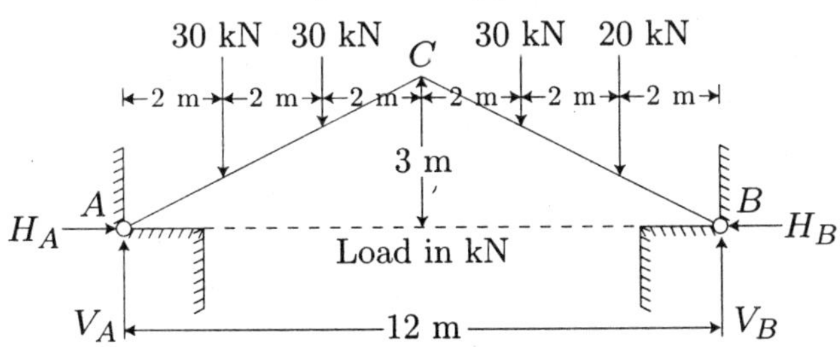

Fig. P-14.5

14.24 A three-hinged arch rib in the form of a circular arc has a span of 20 m and a rise of 5 m. The rib carries a concentrated load of 320 kN at 6 m from the left-hand end. Determine (a) the end reaction, (b) the bending moment at the load point.

14.25 Fig. P-14.6 shows a spandrel braced arch subjected to a concentrated load of 100 kN at G. Find the force in the member FG.

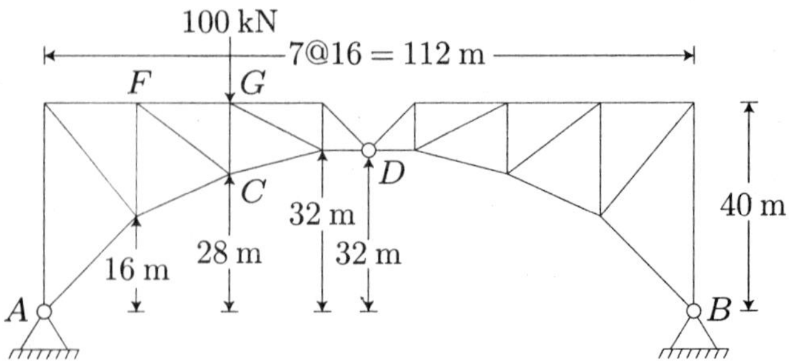

Fig. P-14.6

14.26 Fig. P-14.7 shows three-hinged arch truss. Determine the reactions at A and B and the forces in the members DE, DH and GH.

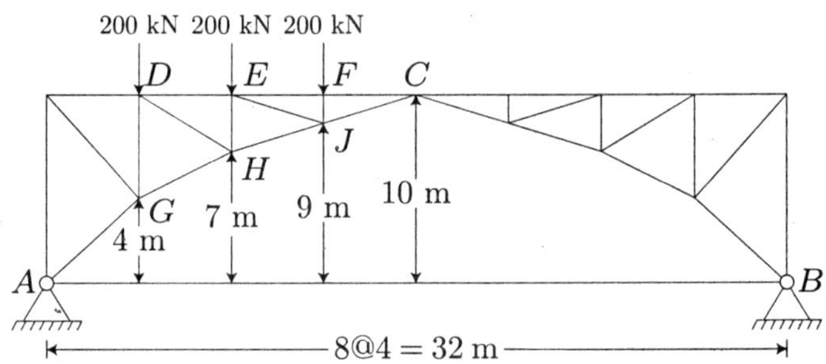

Fig. P-14.7

14.27 A suspension bridge of 300 m has a central dip of 30 m. It carries by means of two cables a total load of 3000 kN uniformly distributed over the whole span. Determine the maximum tension in the cable and obtain the cross section of the cables if allowable stress be 120 N/mm^2. Also calculate the forces in the tower, when the anchor chain makes in angle of 45° with the vertical, if

(a) cables run over a pulley drive.

(b) cables are attached to saddles resting on rollers on top of the towers.

14.28 A suspension cable is suspended from two piers 200 m apart, one support being 5 m above the other. The cable carries a u.d.l. of 15 kN/m and has its lowest point 10 m below the lower support. Ends of the cable are attached to saddles on rollers at top of piers and the anchors are inclined at 60° to the vertical. Calculate :

(a) Maximum tension in cable

(b) Tension in the anchor cable

(c) Forces in the piers.

14.29 A thin uniform flexible cable of weight w per unit length connects two points which are at the same level. If the central dip 'd' be small with respect to the span L, show that the difference between the greatest and the least tension is wd.

14.30 Obtain the S.F. and B.M. diagrams for the suspension bridge with three hinged stiffening girders shown in Figs. P-14.8(a) and P-14.8(b).

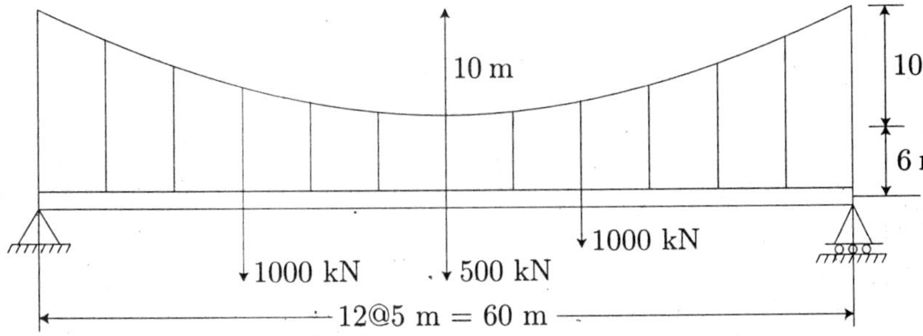

(a)

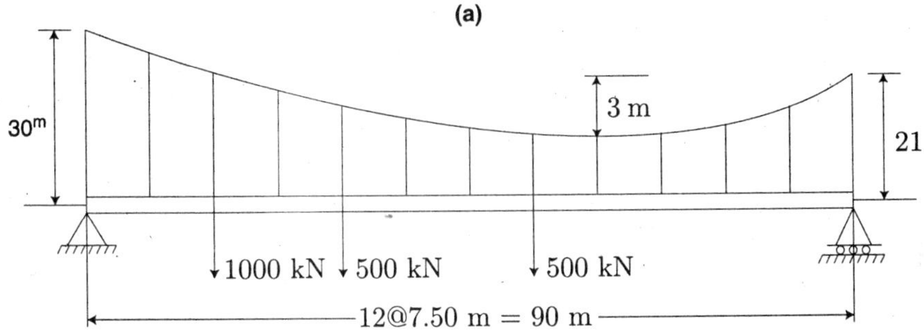

(b)

Fig. P-14.8

14.31 A suspension bridge three-hinged stiffening girder with the third hinge at centre has to carry a L.L. of w/unit length of infinite length which can occupy any part of the span. Show that

(a) Maximum positive bending moment of magnitude 0.01883 wL^2 occurs at a distance 0.234 L from the support of the girder, when the loaded length from the same support end is 0.395 L.

(b) Maximum negative bending moment of magnitude 0.01883 wL^2 occurs at a distance of 0.234 L from the support of the girder when the girder is loaded for a distance of 0.605 L from the other support.

(c) Maximum positive and negative shear force of magnitude $\pm\frac{wL}{16}$ occurs at the support when the loaded length is $\frac{2}{3}L$ and $\frac{L}{3}$ respectively.

14.32 Two symmetrical suspension cables having a parabolic form have span of 250 m and support two equal three-hinged girders, hinged with the third hinge at centre. The cable has a central dip of 25 m. The width of the roadway is 8 m and it carries a D.L. of 5 kN/m over the whole length. The L.L. on the girder from the roadway is 10 kN/m and of length 125 m. Find the maximum tension on the cable and the maximum B.M. and S.F. at 50 m and 150 m from the left-hand support.

14.33 A suspension bridge with two three-hinged stiffening girders has a span of 100 m and a central dip of 10 m and weighs 2000 kN. The bridge is supported by two cables and the girder is to be designed to carry a rolling load of 20 kN/m and of indefinite length. The rolling load may cover all or any part of the bridge. Using a working stress of 120 N/mm^2. Determine

(a) the cross-sectional area of each cable

(b) the necessary sectional modules of the stiffening girders

(c) suitable cross section of hangers placed 5 mc/c.

14.34 A suspension cable has a span L and a central dip d and is stiffened by a girder which is hinged to the abutments and at a point $\frac{L}{4}$ from the left hand abutment. Show that if $\frac{d}{L}$ be small the maximum tension in the cable when a unit load traverses the girder is approximately $\frac{L}{4d}$.

14.35 A cable is suspended from two points separated horizontally by a distance of 200 m and a vertical distances of 15 m. The maximum dip of the cable below the lower support is 5 m. It is stiffened by a girder which is hinged at both ends and also at a point vertically below the lowest point of the cable. Calculate the maximum tension in the cable, when the girder carries a u.d.l. of 30 kN/m over the whole span. Also obtain the B.M. and S.F. diagrams in the girder, when concentrated load of 100 kN is placed on the hinge points between supports.

14.36 Solve Problem 14.33, when the live load is a concentrated load of magnitude 200 kN moving over the girder. If the girders are supported from the cables by hangers 5 mc/c, obtain influence line diagrams for the force on the hangers.

14.37 A suspension bridge is similar to that shown in Fig. 14.10(a).

(a) If a live load of 20 kN/m acts over the entire bottom chord, compute the maximum cable tension in the suspended span and the maximum bottom chord force in the stiffening truss.

(b) Construct influence lines for (*i*) horizontal component of cable tension, (*ii*) hanger tension (*iii*) force in the stiffening truss top chord member in the fourth panel from left lower and (*iv*) force in the stiffening truss diagonal of the third panel from left tower.

(c) The D.L. is 50 kN/m. The L.L. consists of u.d.l. of 2t/m and a concentrated load of 200 kN. Obtain cross-sectional area of cable and hanger, if allowable stress is 50 N/mm$^-$.

14.38 Obtain influence line diagrams for S.F. and B.M. in case of a three-hinged stiffening girder with the third hinge at 0.40 L from one of the supports. Assume a symmetrical parabolic shape of the cable profile having a dip d at centre.

14.39 A suspension of span L hanging in the shape of a symmetrical parabola is strengthened by a stiffening girder pinned at the abutments and at the centre. If a uniformly distributed load longer than the span moves over the girder, find the maximum positive and negative S.F. at the centre of the span.

14.40 Fig. P-14.9 shows a suspension bridge stiffened with three-hinged stiffening truss. Find maximum force in member a due to a concentrated load W moving over the bridge.

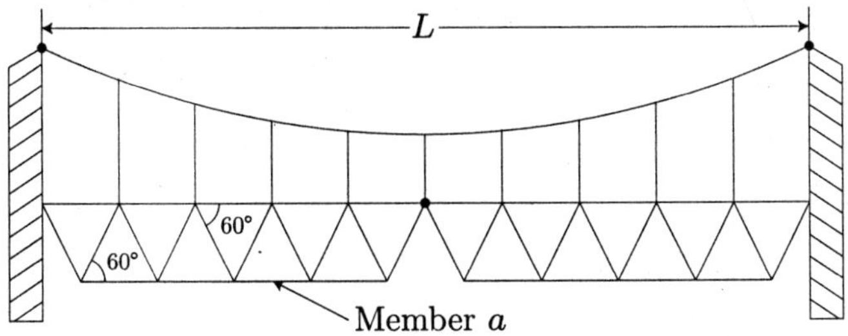

Fig. P-14.9

Chapter 15

Approximate Analysis of Statically Indeterminate Structures

15.1 Introduction

The analysis of every structures is approximate as it is necessary to make certain assumption in order to carry out the analysis. It may, therefore, be said that there is no such thing as exact analysis.

However, if proper judgement is considered in making the analysis of a given structure, the resultant error will be small. The structural analysis based on the usual assumptions which underline structural theory is often called exact, although strictly speaking the term is not used correctly. It is however, a convenient term to use as it is desirable to distinguish between analysis based on the usual assumption which are relatively exact and analysis based on further assumption which introduces some errors and which, therefore, is approximate.

Approximate analysis for a given structural problem does not refer to any particular set of assumptions and resultant approximations. Particular approximate method to be used under any giver circumstances will depend on the time available for the analysis and degree of accuracy necessary.

For a good structural analysis it is necessary to be familiar with the actions of statically indeterminate structures so as to be able to set up assumption not usually covered in carrying structural analysis.

In this Chapter a number of approximate solutions for some common types of statically indeterminate structures are given. A knowledge of such methods is of importance and procedure outline will serve as a basis for making intelligence assumption permitting simplified approximate analysis of other types of statically indeterminate structures.

15.2 Importance of Approximate Method in Analysis of Statically Indeterminate Structures

In statically indeterminate structures the stress analysis depends of the elastic properties of members. However, in actual practice the configuration and complexity of the structure may be such that exact methods of analysis are either not available or impractical to apply. Also the time required to carry out analysis may be so great that an exact solution must be abandoned. In such cases and also for meeting time schedule, use of approximate method of analysis may be proper to you. Some of the usual types of structures are discussed in the following sections.

15.3 Industrial Building Bents—Portal Truss

Fig. 15.1(a) shows an ordinary industrial bent, one storey in height and one aisle in width. The columns for such bents are generally identical and the bases of such bents may be fixed or hinged. Analysis of such bents when subjected to vertical dead loads which are usually symmetrically applied can be done on the basis of vertical reactions obtained as in case of simply supported ends, considering that horizontal reactions are zero for vertical loads.

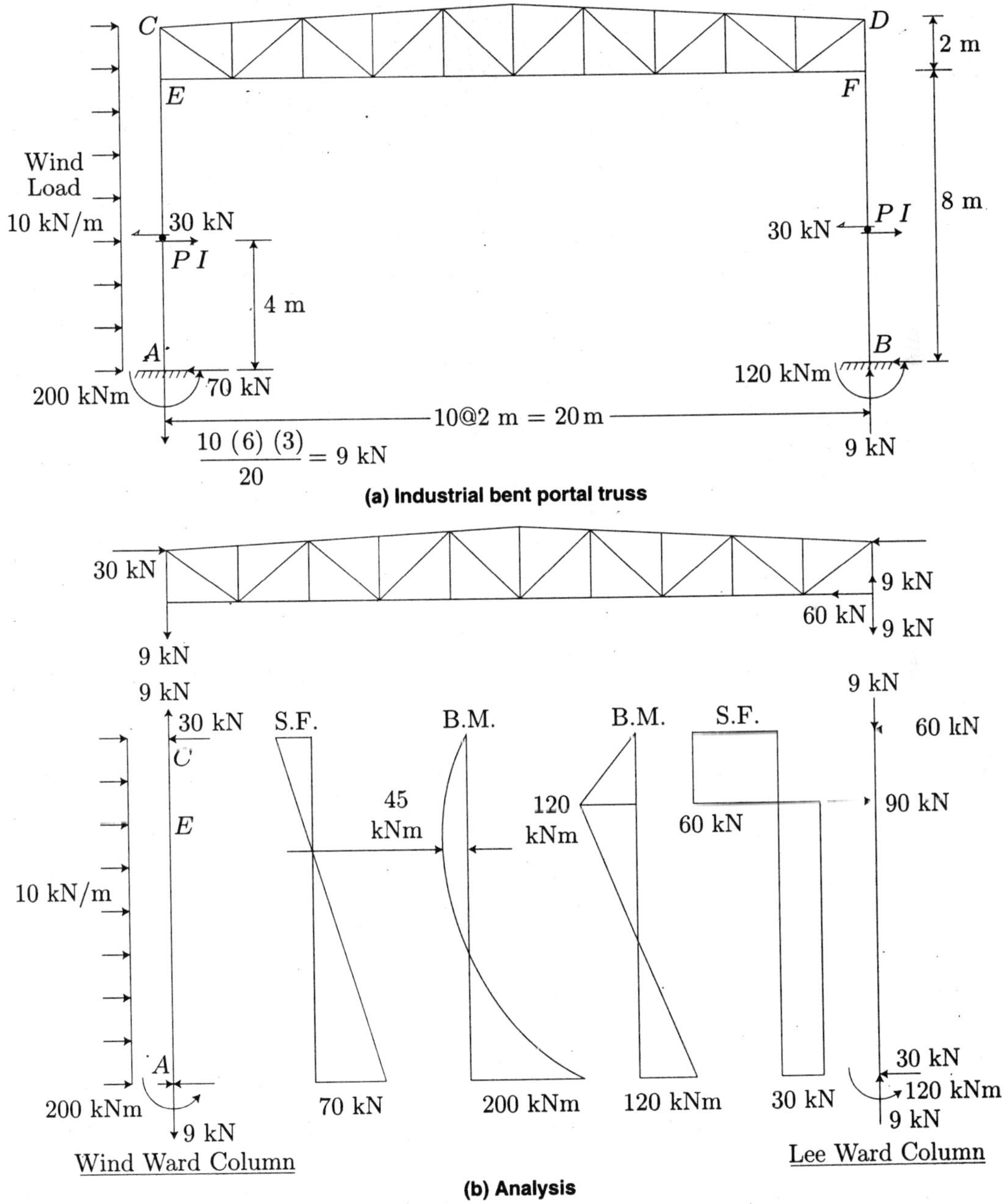

Fig. 15.1: Analysis of Industrial bents

When such bents are subjected to lateral loads (as in case of wind pressure) an approximate analysis may have to be carried out on the basis of assumptions identical with the ones made for the portal. The usual assumptions made are dependent on the conditions of column bases and may be stated as follows :

(a) Points of inflections (points of zero bending moment) shall be assumed halfway between the bottom of the column and the underside of the truss, in case column bases are fixed. For partially fixed column, such points of inflection may be taken at $\frac{1}{3}$rd the column height from bottom. For column base hinged, the base gives points of zero bending moment.

(b) Horizontal reaction, i.e., shear force on columns at the points of inflection are zero for vertical loads and are equal for the horizontal component of the loads.

Figs. 15.1(a) and 15.1(c) show the analysis of the bent with column base fixed when subjected to lateral loads due to wind. The forces on the various members of the truss may be obtained on the basis of the free body diagram in Fig. 15.1(b).

Analysis of such bents with column bars hinged may be done on the basis of equal horizontal reactions at the bases of the columns.

When the columns are alike the assumption of equal distribution of shear force among them may be considered fairly correct. If the columns differ much in section, this assumption may be considerably in error and requires modification. In such cases a more accurate analysis may be made by distributing the shear force in accordance with the stiffness of the columns.

15.4 Mill Building Bents

A roof truss attached to its supporting columns is commonly called a *bent.* The vertical faces of mill buildings are generally covered with side sheets supported on *girts*, which are attached to the columns. The action of girt is similar to purlins. The wind pressure on the vertical face of the building is carried by the girts to the columns, just as the purlins bring their loads to the roof truss. Therefore, horizontal forces shall be acting on the windward column, at the points where girts are connected to the columns.

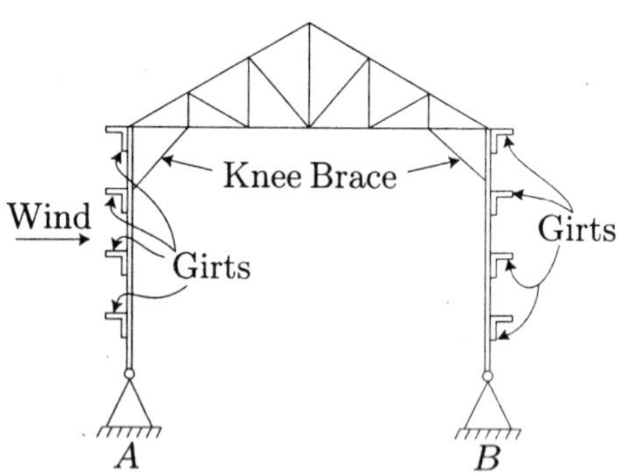

Fig. 15.2: Mill bent hinged at bases

Since it is assumed that the roof truss is pin-connected to columns, knee braces are most essential for lateral stability of a bent subjected to wind loads.

Columns of a bent may be hinged, partially fixed or fixed at the base.

(a) Hinged Column Bases as in Fig. 15.2

In such a bent, there will be four unknown reactions (horizontal and vertical) under any loading. When whole of the bent is taken as free body there are three equations of statics.

Horizontal reactions at the hinges are zero under vertical loads only. Therefore, analysis of bent under vertical loads is similar as in an ordinary truss.

When the bent is subjected to horizontal loads, e.g., wind loads, etc., the horizontal component of the reactions at the support can be considered equal when two columns are of same section and height.

With the reactions determined, the rest of analysis can be done.

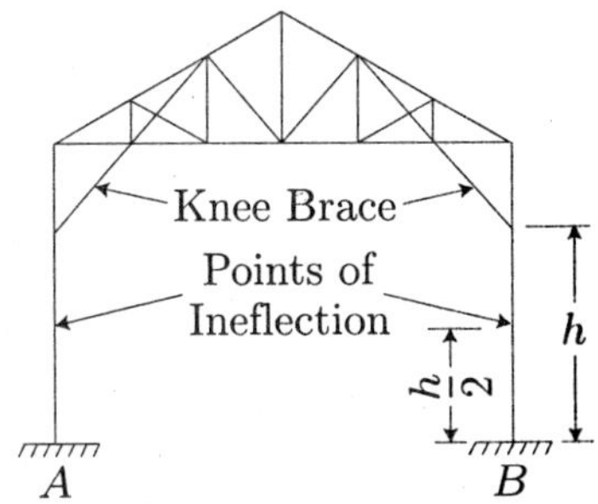

Fig. 15.3: Mill bent fixed at bases

(b) Fixed Columns Bases as in Fig. 15.3

When the lower ends of the bent are fixed or partially fixed, there will be three reaction components (a horizontal force, a vertical force, and a restraining moment) at each support or total of six unknowns. Therefore, three more conditions have to be assumed or provided in addition to the three equations of statics.

Assumptions

(i) Horizontal reactions are zero for vertical loads and are equal for inclined and horizontal loads.

(ii) Assume points of contraflexure in the two columns. The points of contraflexure are at $h/2$ from the base for rigidly fixed columns of equal height and section (here 'h' is the distance from the root of the knee brace to the base of the column).

For partially fixed columns, points of contraflexure may be assumed at $\frac{h}{3}$ from the base.

Since points of contraflexure are equivalent to hinges, the method of analysis of the portion of the bent above the points of inflection will be same as that of a bent with hinged supports.

Example 15.1. *Determine the forces due to dead load and wind load in the columns of the Knee braced bent hinged at base shown in Fig. 15.4.*

Centre to centre distance of bents = 6 m.

Dead load consists of

(i) Weight of roofing, rafters and purlins – 730 N/m^2 of roof surface

(ii) Weight of bracing system = 50N/m^2 of roof surface.

(iii) Weight of truss = $20 + 6.6L$ per square metre of horizontal surface.

Wind pressure $p = 1400$ N/m^2 of vertical surface.

Solution : Dead Load

$$L_0U_1 = \frac{1}{4}\sqrt{10^2 + 5^2} = 2.795 \approx 2.80 \text{ m}$$

Weight of truss $= 20 + 6.6L = 152$ N/m^2

Panel Load

(i) Due to roofing rafters and purlins $(730)(2.8)(6) = 12,264$ N

Taken as 12.28 kN

(ii) Due to bracing system $= 50(2.8)(6) = 840$ N $= 0.84$ kN.

(iii) Due to weight of truss $= 152(6)\left(\frac{20}{8}\right) = 2280 \text{ N} = 2.28 \text{ kN}.$

Panel load $= 12.28 + 0.84 + 2.28 = 15.40$ kN.

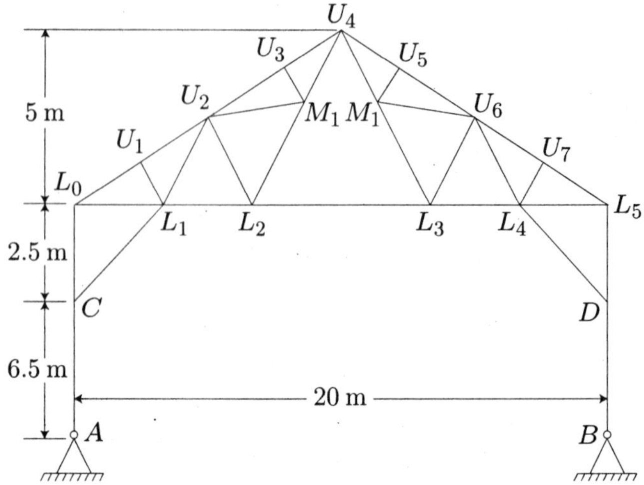

Fig. 15.4: For Example 15.1

When only vertical loads are acting on the bent, the horizontal reactions at the case is assumed to be zero. Therefore, columns are under the axial stresses and dead load stresses in knee braces are zero. The stresses in other members of the bent due to vertical loads are, therefore, identical with those of the truss supported on walls. Therefore, the dead load forces can be determined and are shown in Fig. 15.5.

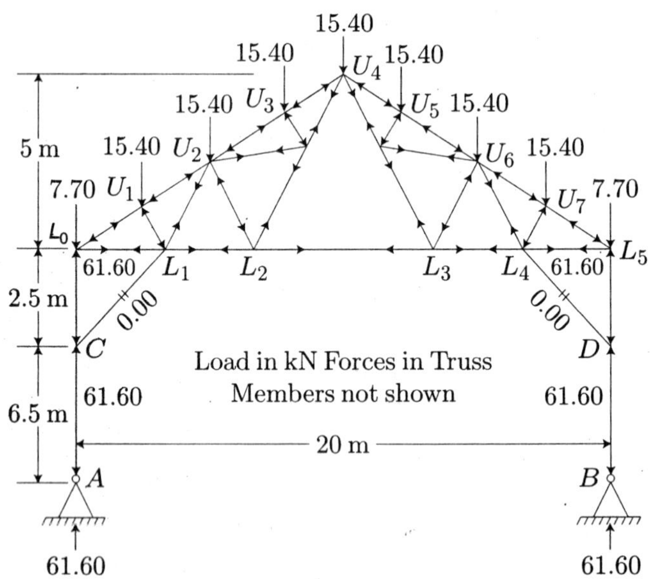

Fig. 15.5: D.L. forces in column in kN

Wind Load. It is assumed that the wind is blowing from left to right.

The windward column ACL_0 is under uniformly distributed loads. There is no wind on leeward column BDL_5. Wind loads on truss are assumed to act at panel points.

Wind load on column

$$ACL_0 = 1400(6) = 8400 \text{ N} = 8.40 \text{ kN}.$$

Normal pressure on roof

$$p_n = p \cdot \frac{2\sin\theta}{1+\sin^2\theta}$$

$$\sin\theta = \frac{1}{\sqrt{5}}$$

$$p_n = 140\frac{2\left(\frac{1}{\sqrt{5}}\right)}{\left(1+\frac{1}{5}\right)} = 1043.50 \text{ N} \approx 1.045 \text{ kN/m}^2.$$

Wind load per panel $= 1.045(2.80)(6) = 17.556$ kN

$$\approx 17.60 \text{ kN}.$$

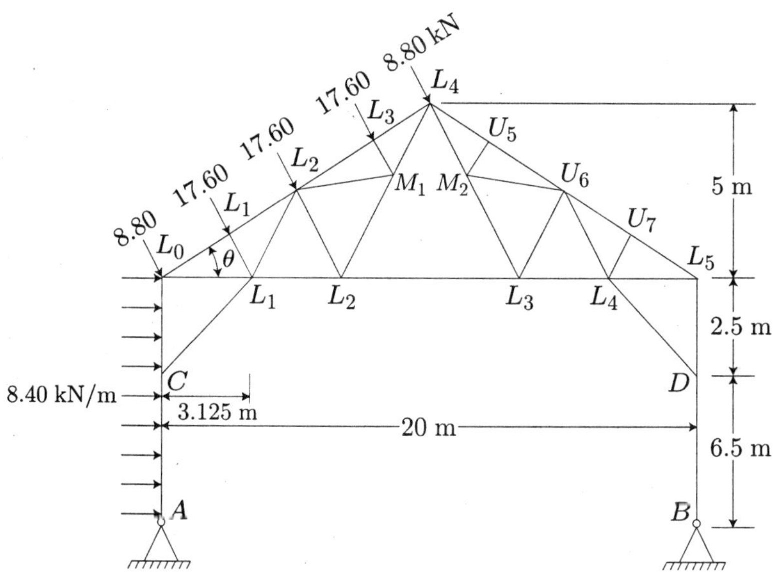

Fig. 15.6: W.L. in truss in kN

The loads are shown in Fig. 15.7. Resultant of wind loads on the roof acts at panel point U_2.

Resultant wind load $W = 4(17.60) = 70.40$ kN

Vertical component of wind load

$$W_v = W\cos\theta = 70.40\left(\frac{2}{\sqrt{5}}\right) = 62.96 \text{ kN}.$$

Horizontal component of wind load

$$W_h = W\sin\theta = 70.40\left(\frac{1}{\sqrt{5}}\right) = 31.484 \text{ kN}.$$

Since columns are hinged at the bases, horizontal reactions at A and B are assumed to be equal.

Total horizontal loads on the bent $= 8.40(9) + 31.484 = 107.084$ kN.

Therefore, $H_A = H_B = \dfrac{107.084}{2} = 53.542$ kN.

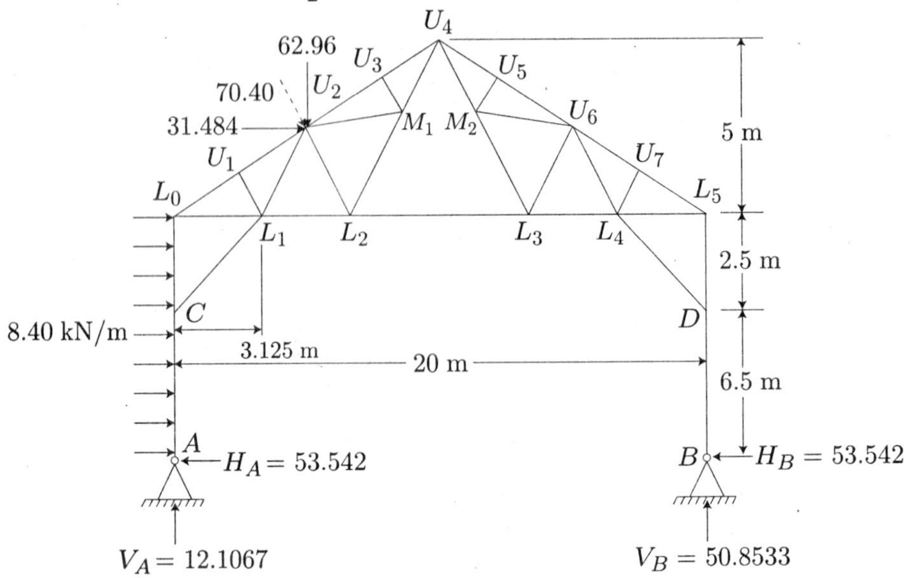

Fig. 15.7: Reactions in kN

In order to determine vertical reactions at A and B, take the moments of all forces about $A. \sum M_A = 0$.

$$20V_B = 314.80 + 362.066 + 340.20$$

$$V_B = 50.8533 \text{ kN}$$

$$V_A = 12.1067 \text{ kN}$$

$$L_0L_1 = 1.325 \text{ m.}$$

Right-Hand Column. The free body diagram of the right-hand column is drawn as shown in Fig. 15.8.

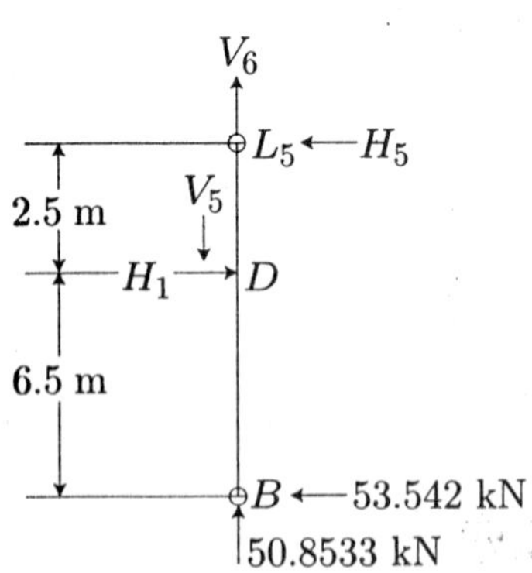

Fig. 15.8: Right-hand column

Taking moments about L_s, horizontal component of the force in knee brace can be determined.

$$H_1(2.50) - 53.542(9) = 0$$

$$H_1 = 192.7512 \text{ kN}$$

Vertical component of the force in knee brace

$$V_1 = 192.7512\frac{2.5}{3.125} = 154.20096 \text{ kN.}$$

Force in knee brace L_4D

$$F_{L_4D} = 246.842 \text{ kN.}$$

Horizontal force at $L_5 = H_5 = 139.2092$ kN.

Vertical force at $L_5 = V_5 = 103.34766$ kN.

The thrust, shear force and bending moment diagrams for the right-hand columns are shown in Fig. 15.9.

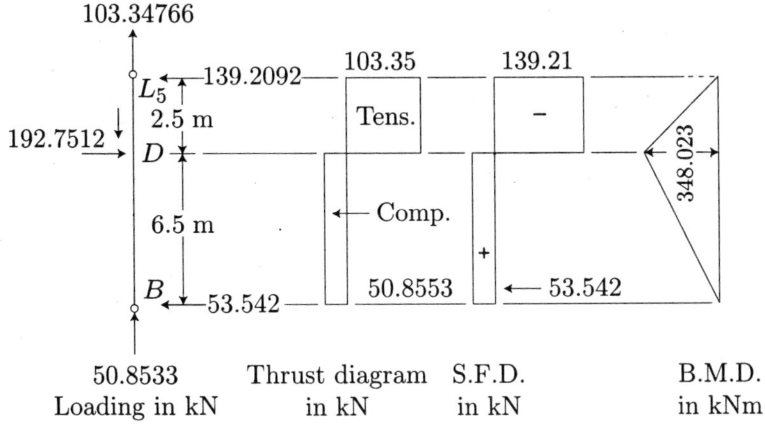

Fig. 15.9: Right-hand column

Left-Hand Column. The free body diagram of left hand column is drawn as shown in Fig. 15.10.

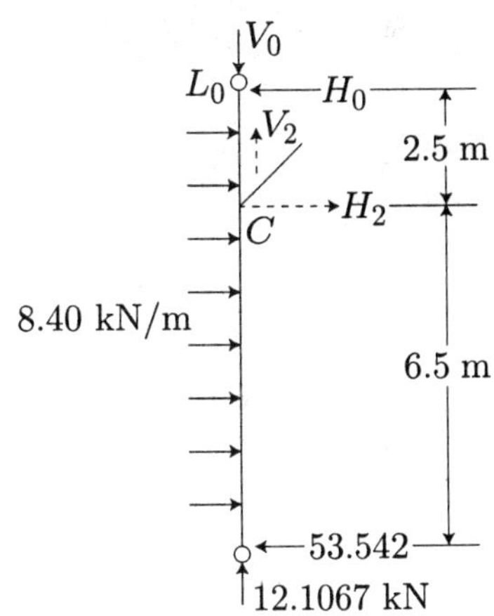

Fig. 15.10: Left-hand column

Taking moments about L_0

$$H_2 = 56.6712 \text{ kN} \longrightarrow$$

$$V_2 = 45.33696 \text{ kN} \uparrow$$

$$H_0 = 78.729 \text{ kN} \longleftarrow$$

$$V_0 = 57.44366 \text{ kN} \downarrow.$$

The thrust, shear force and bending moment diagrams for the left-hand column is shown in Fig. 15.11.

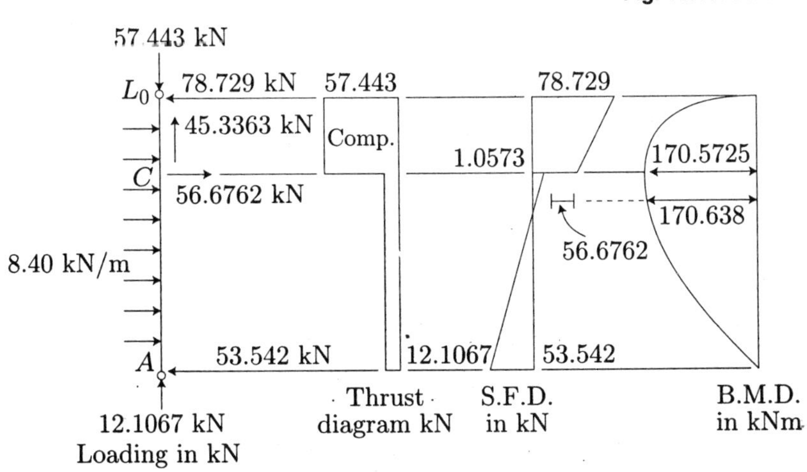

Fig. 15.11: Left-hand column

Example 15.2. *Determine the forces due to dead load and wind load in the columns of knee braced bent fixed at the base as shown in Fig. 15.12. Assume points of inflection at midway from base of column to knee brace. Assume loads as in Example 15.1.*

Centre to centre distance of bents = 6 m.

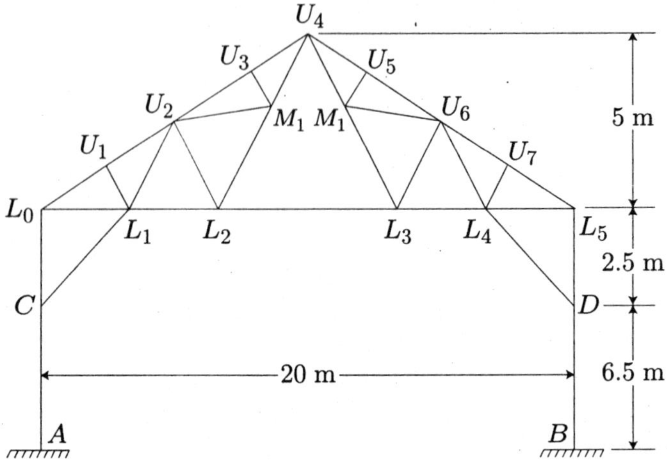

Fig. 15.12: For Example 15.2

Solution : Dead Load Forces. The dead load is same as in Example 15.1. The dead load forces in various members are shown in Fig. 15.13.

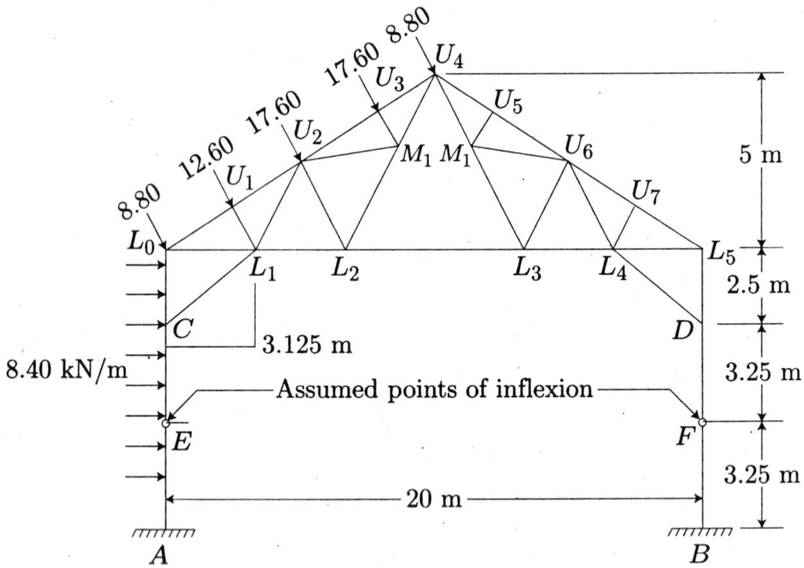

Fig. 15.13: W.L. in truss in kN

Wind Load. It is assumed that wind is blowing from left to right only.

$$L_0U_1 = \frac{1}{4}\sqrt{10^2 + 5^2} = 2.8 \text{ m}.$$

Wind load on windward column $ACL_0 = 8.40$ kN/m.

Normal wind pressure on roof = 1.045 kN/m^2.

Wind load per panel = 17.60 kN.

Loads are shown in Fig. 15.13.

Horizontal and Vertical Reactions : Total wind load on truss = 70.40 kN acting at mid-panel point U_2.

Vertical component of wind load = 62.96 kN

Horizontal component of wind load = 31.484 kN

Total horizontal load on the bent = 107.084 kN

Therefore, $H_A = H_B = 53.542$ kN.

Free body diagram above points of inflexion : The structure above the points of inflexion is considered for determining forces in the various members, as shown in Fig. 15.14.

Taking moments about left inflexion point E,

$V_2 = 35.67$ kN ↑

and $V_1 = 27.29$ kN ↑.

Horizontal reaction at E = 26.242 kN

Horizontal reaction at F = 53.542 kN.

Moments at fixed bases

$M_B = 174.0115$ kNm

$M_A = 129.649$ kNm.

All the reactions are shown in Fig. 15.14.

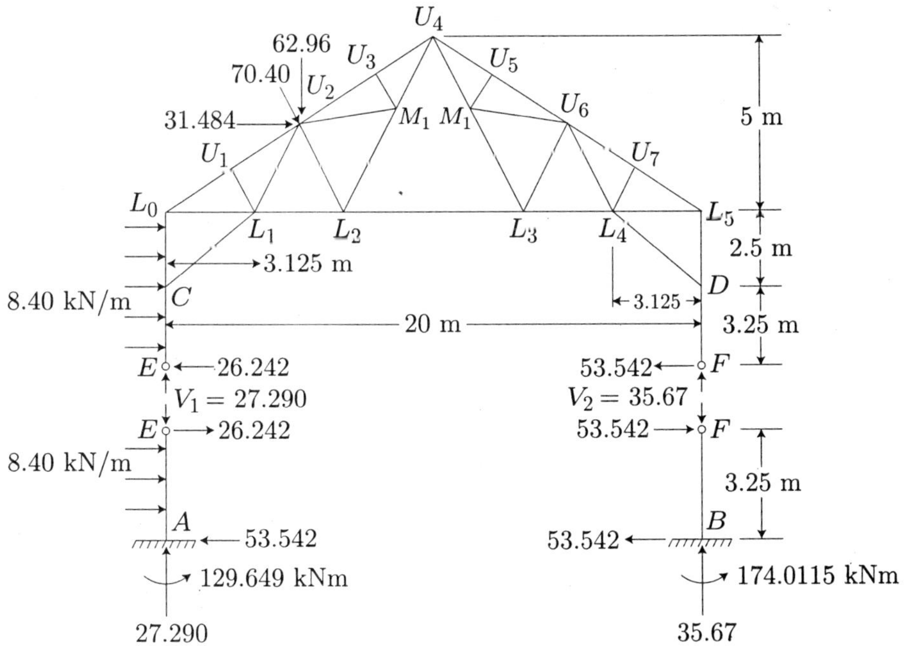

Fig. 15.14: Reactions in kN

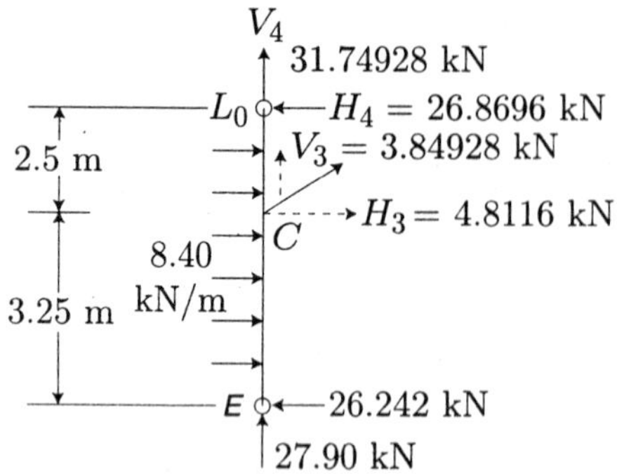

Fig. 15.15: Left-hand column

Left-hand column : The free body diagram of the left-hand column above the point of inflexion E is drawn as shown in Fig. 15.15.

$$L_0L_1 = 3.125 \text{ m}$$

Taking moments about L_0, top of column force in knee can be determined.

$$H_3 = 4.8116 \text{ kN} \quad \text{and} \quad V_3 = 3.84928 \text{ kN}.$$

Force in knee brace CL_1

$$F_{CL_1} = 6.1618545 \text{ kN}$$

$$V_4 = 31, 74928 \text{ kN} \quad \text{and} \quad H_4 = 26.8696 \text{ kN}.$$

Thrust, S.F., and B.M. diagrams for the left-hand column are drawn as shown in Fig. 15.16.

Point of zero S.F. occurs at 0.12595 m below C, which is 6.374 m from base of the column.

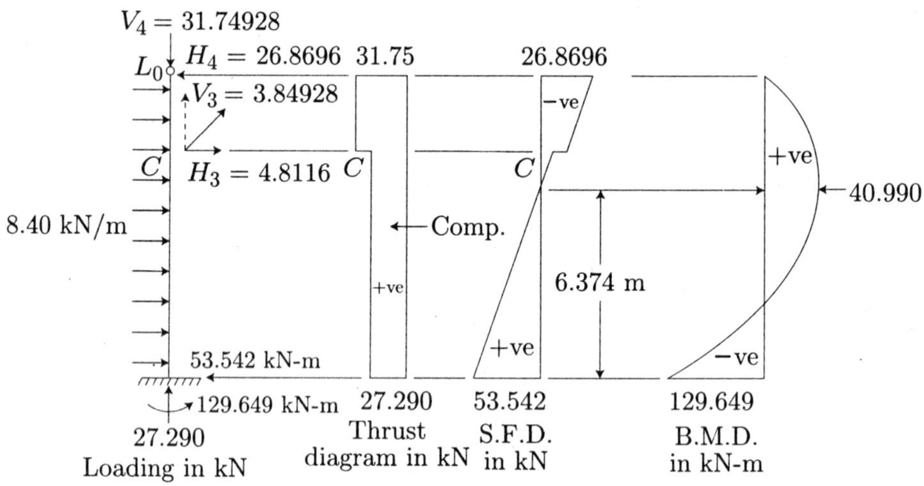

Fig. 15.16: Left-hand column

Right-Hand Column : The free body diagram of the right-hand column above the point of inflexion F is shown in Fig. 15.17. The force in knee brace is determined by taking moments about L_5.

Taking moments about L_5,

$$H_5 = 123.1466 \text{ kN} \qquad V_5 = 98.51728 \text{ kN} \downarrow.$$

For in knee brace L_4D,

$$F_{L_4D} = 157.7046 \text{ kN}$$

$$V_6 = 62.81728 \text{ kN} \uparrow \quad H_6 = 69.6046 \text{ kN} \longleftarrow.$$

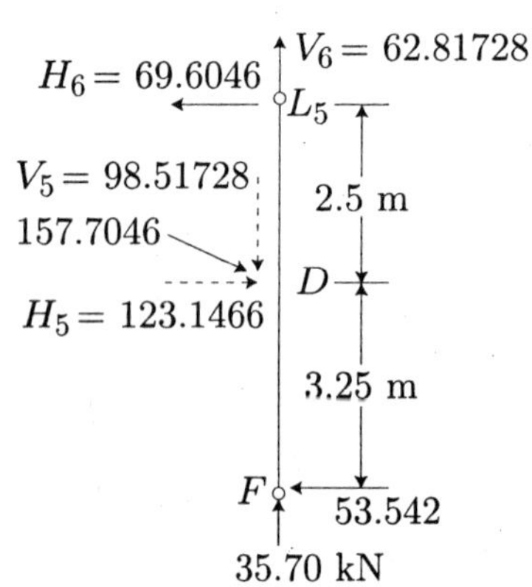

Fig. 15.17: Right-hand column

The thrust, shear force and bending moment diagrams for right-hand column BDL_5 are shown in Fig. 15.18.

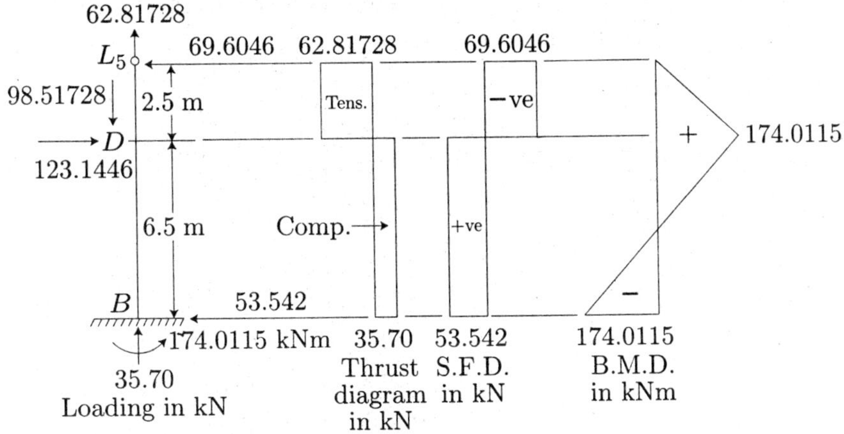

Fig. 15.18: Right-hand column

15.5 Multiple Storeyed Bent

Fig. 15.19 shows a multistoreyed single axle bent for an industrial building subjected to uniformly distributed wind pressure. The effect of the lateral forces in the case may be estimated in the same manner as for a single storeyed building, except that it is necessary to assume position of points of inflection in columns of each storey as indicated in the figure. The shear force on each place of inflection is assumed as divided equally between the columns cut by that place. The calculations of shear force and bending moment in the columns then follows as in case of a single storeyed bent.

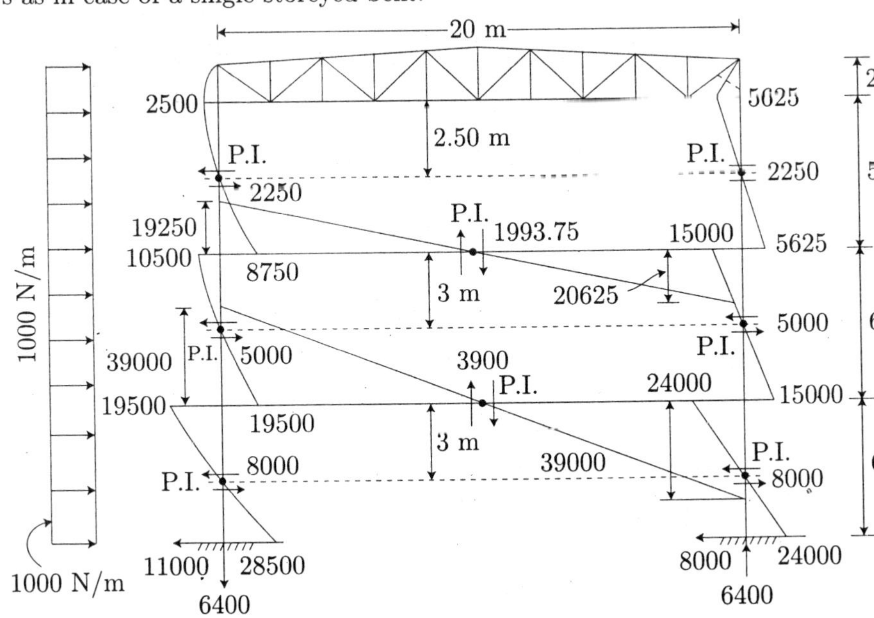

Fig. 15.19: Multiple storeyed bent. Load in N,S,F in N and B.M. in Nm

In obtaining effects of lateral forces in multistoreyed bents, it is usual to disregard the depth of the floors and treat the girders as a line having no depth and consider the distance from the top of one floor to the top of the floor above as the storey height. It is also sometimes customary to assume that the wind pressure is delivered to the bents as a series of concentrated loads applied at the floor lines, instead of a uniformly distributed load applied along the full height of the columns. This is equivalent to treating the wall construction as a series of vertical beams having reactions at the floor lines and the floors as horizontal beams, which transmit these load to the columns. The wall construction for most multistoreyed building is such that this practice is probably a fairly accurate representation of the action of wind pressure on such buildings. The points of inflection on the girders then occur at their centres.

15.6 Application to Building Frames

The general method of approach used in the analysis of the three storey bent of Fig. 15.19 can be widely used in an extended form in the analysis of building frames under the effect of wind pressure. When a building bent is acted upon by the lateral loads, there will be points of inflection near the centres of each girder and each column. The assumption that points of inflection occur at the midpoints of all members is, therefore, reasonable.

The building bents usually are composed of several narrow aisles instead of a single wide one and may be considered to be made of a series of aisles, the total lateral load being divided among the aisles. The resulting shear force and bending moments on the members which are common to two aisles are then combined together. As an exterior column corresponds to the vertical leg of a single aisle and an interior column to vertical legs of two aisle, it is reasonable to assume that interior columns carry twice the shear force of the exterior columns.

The Portal method for analysis of building frames subject to lateral loads, which are considered concentrated at the floor level has been developed based on the above considerations. According to this method, the frame is analysed as a whole and the following simplifying assumptions are made :

(a) The point of inflection of each column is at the midheight of the storey.

(b) The point of inflection of each girder is at its centre.

(c) The total horizontal shear force on each storey is divided among the columns of that storey in a manner such that each interior column carries twice as much shear force as each exterior column. Alternatively, the shear force may be divided among the columns in proportion of the shear spans of columns (column panel). The method, with this alternate assumption regarding distribution of shear among columns is sometimes also known as "Modified Portal Method".

Fig. 15.20 illustrates the application of the portal method in the case of a building frame subjected to lateral loads due to wind pressure of 6 kN/m. The following discussion refers to Fig. 15.20, where the results of the portal method of analysis are given on the basis of division of shear force among columns considering that interior columns carry twice as much shear force as each exterior column.

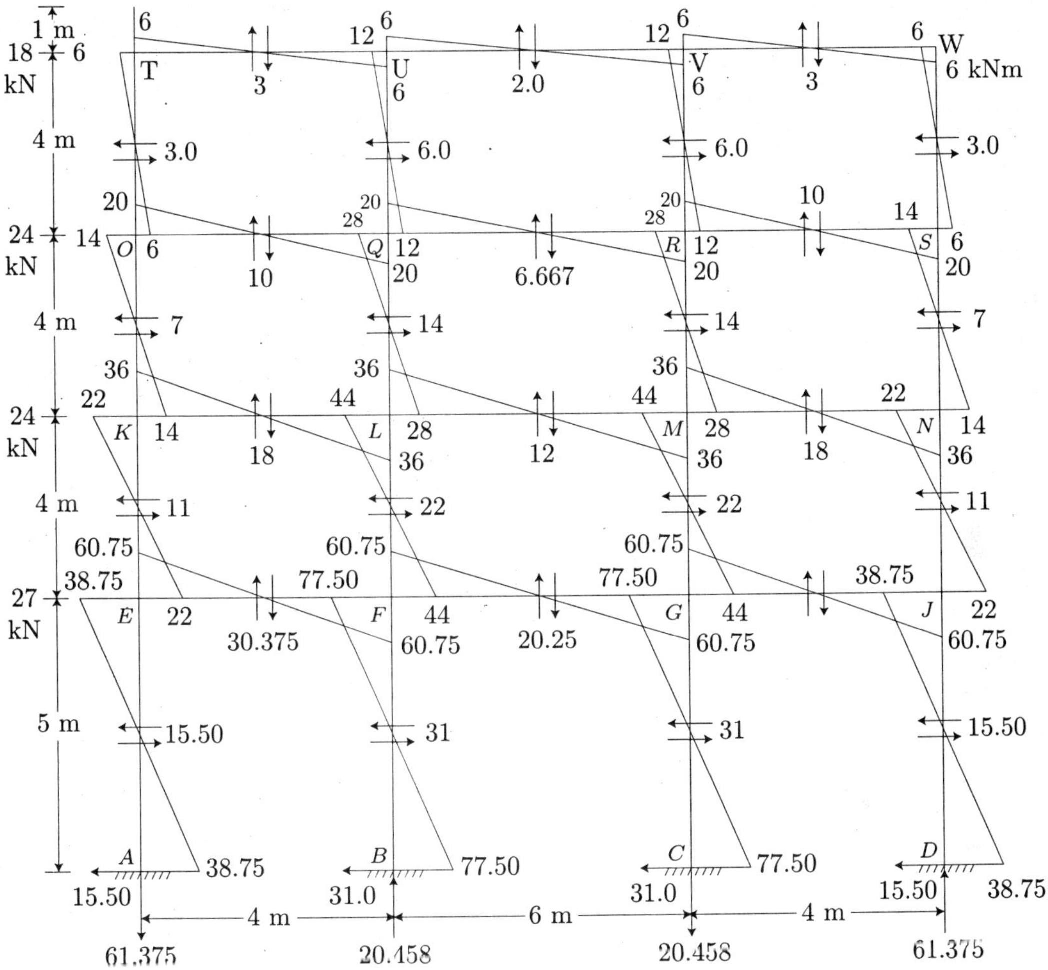

Loading and S.F. in kN; B.M. kN/m

Fig. 15.20: Portal method in building frame

Column shear forces. In accordance with assumption (c), if shear force in exterior column of a given storey be H, then shear force in the interior column of the same storey will be $2H$. For the topmost storey,

$$H + 2H + 2H + H = 18 \text{ kN},$$

whence, $H = 3$ kN and $2H = 6$ kN.

With alternative assumptions on the basis of column spans, proportion of shear force on columns are, $2 : (2+3) : (3+2) : 2$, i.e., $1 : 2.5 : 2.5 : 1$ and S.F. in the columns are $\frac{18}{7}$ on exterior column and $\frac{45}{7}$ on interior ones.

For the lower floor,

$6H = 93$ kN and $H = 15.50$ kN and $2H = 31$ kN.

With alternative assumptions S.F. on exterior column is $\frac{93}{7}$ kN and S.F. on interior column is $\frac{232.50}{7}$ kN.

Column moments : In accordance with assumption (a), the bending moment at the centre of each of the 7 columns is zero. Hence each end moment for a given column equals the shear force on the column multiplied by half the length of that column. For example, bending moment at the top and bottom of the topmost exterior column equals $3(2) = 6$ kNm and that at the interior column at the same level is $6(2) = 12$ kNm.

Girder moments : From the consideration of equilibrium for any joint, the sum of the column end moments equals the sum of the girder end moments. This relation may be used to determine the girder end moments since the column end moments have already been evaluated. For example, at joint E sum of column moments is 60.75 kN. The bending moment in the girder EF at E equals 60.73 kNm. Since from assumption (a), there is a point of inflection at the centre of girder the end moment on girder EF at F also equals 60.75 kNm. Equating girder moments to column moments at joint F shows that the end moment at F of girder FG also equals 60.75 kNm.

Continuing similar procedure, the girder moments at all levels may be obtained.

Girder Shear Force. Taking moments about one end of the girder the shear force in the girder is given by $\frac{2M}{L}$, where M is the bending moment at the end of girder and L is the length of the girder.

For example, shear force in girder EF is given by $\frac{2(60.75)}{2}$, i.e., 30.375 kN.

Column Axial Forces : The axial forces in the columns may be obtained by summing up from the top of the columns, the shear force of the girders applied as axial force to the columns.

Thus, axial force on exterior column at A is given by, $(3.0 + 10.0 + 18.0 + 30.375)$, i.e., 61.375 kN.

Girder axial forces, while not usually important in design, may be obtained in a similar manner by summing up from one end of the girder the shear force of the columns applied as axial force on the girders. The effects of the lateral concentrated loads shall be included to find the resulting axial forces on girders.

If the alternative assumption of distribution of shear over columns be considered, the shear force taken by the four columns (exterior : interior : interior : exterior) will be in the proportion of $\frac{4}{2} : \left(\frac{4}{2} + \frac{6}{2}\right) : \left(\frac{4}{2} + \frac{6}{2}\right) : \frac{4}{2}$, i.e., $1 : 2.5 : 2.5 : 1$. The shear force on column at the topmost storey will then be $\left(\frac{18}{7}\right)$ kN on the exterior columns and $\left[\frac{2.5(18)}{7}\right]$, i.e., $\frac{45}{7}$ kN on the interior columns. Other calculations can be made as in case of Fig. 15.21 obtained on the basis of the other assumptions for distribution of shear.

The Cantilever method is also used for lateral load analysis of frames. It is an approximate method. In this method the structure is rendered statically determinate by means of two assumptions.

(a) Hinges occur in the middle of all columns and beam of the structure.

(b) Along an horizontal plane passing through the column hinges, the vertical column loads are those that would be present if the structure were acting as a vertical cantilever beam. If areas of columns are known, then they will be considered. Otherwise column area is taken as unity. Then the formula

$$f = \frac{My}{I}$$

is used to determine the vertical load. Subsequently the equations of equilibrium $\sum M = 0$, $\sum H = 0$ and $\sum V = 0$, are applied to the different parts of the structure to determine the stress resultants in the various members.

Advantages of the Method

(i) If column areas are known, they may be used in a realistic manner.

(ii) It is said that for buildings with 25 to 35 storeys with moderate height/width ratios, this method provides satisfactory results.

Disadvantages of the Method

(i) It is very tedious. If mistake is committed in the middle, the whole process will have to be repeated.

(ii) It can be used for analysing only for frames with rectangular configuration with rectilinear elements.

Example 15.3. *Analyse the single bay frame shown in Fig. 15.21 by the Cantilever method.*

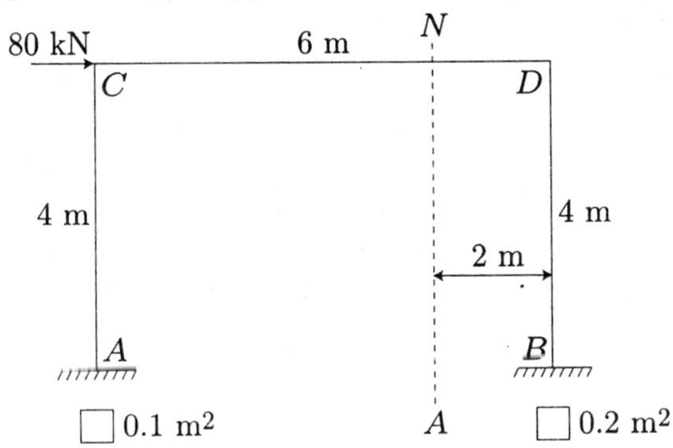

Fig. 15.21: For Example 15.3

Solution : Taking moment in Fig. 15.21 about B,

$$\overline{x} \text{ from B} = \frac{[0.1(6) + 0.2(0)]}{0.30} = 2 \text{ m}.$$

Consider lines at the middle of all members as in Fig. 15.22.

$$\therefore \; IN_A = Ad^2 = 0.1[4]^2 + 0.2[2]^2 = 2.4 \text{ m}^4.$$

Moment of the force $= (80)2 = 160$ kNm.

$$\text{Axial force in the left column} = \left[\frac{M}{I}y\right] \text{(area)}$$

$$= \frac{160}{2.40}(4)(0.1) = 26.667 \text{ kN}.$$

Axial force in the right column $= \dfrac{160}{2.40}(2)(0.2) = 26.667$ kN.

From Fig. 15.22(c) using $\sum M = 0$, $H = 40$ kN,

Moment at $C = 80$ kNm (tension inside) $=$ Moment at D.

Moment at $A = 80$ kNm (tension outside).

Moment at $B = 80$ kNm (tension inside).

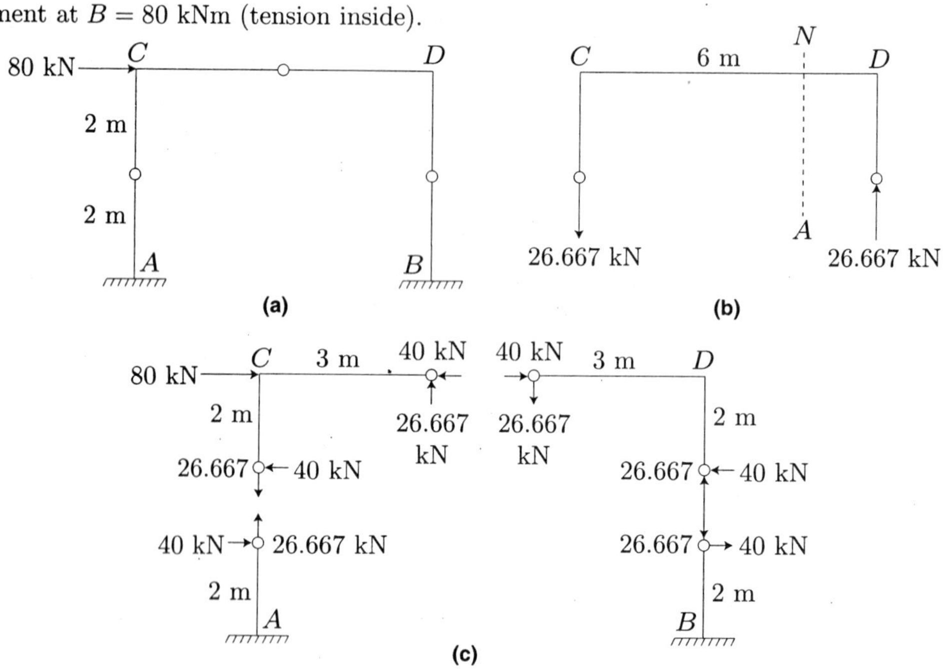

Fig. 15.22: For Example 15.3

Example 15.4. *Analyse the two storeyed frame shown in Fig. 15.23 by the Cantilever method.*

Solution : Taking moment about the right end $\overline{x} = 9$ m

$$I_{NA} = 1(7)^2 + 1(3)^2 + 1(1)^2 + 1(9)^2 = 140 \text{ m}^4.$$

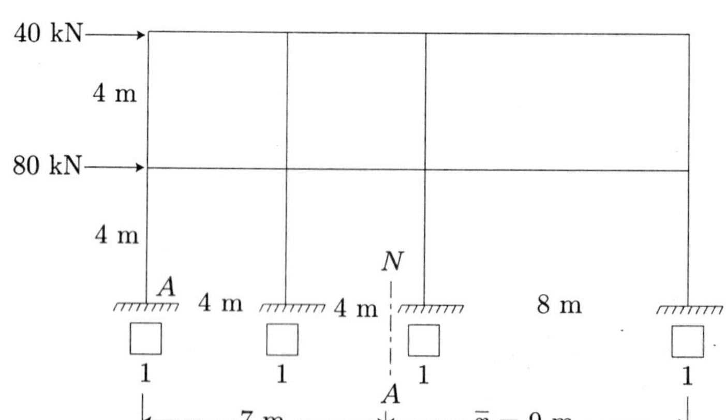

Fig. 15.23: For Example 15.4

In Fig. 15.24, the external moment about the horizontal plans passing through the hinges.

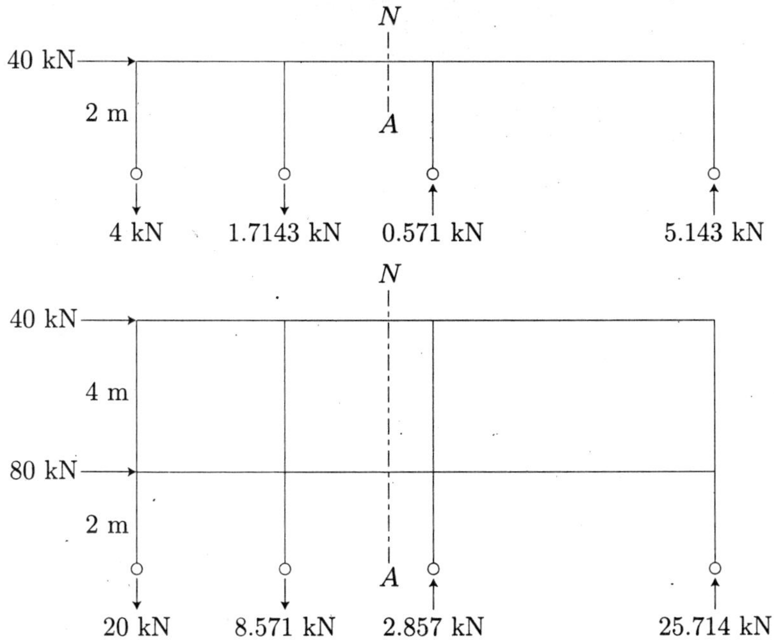

Fig. 15.24: Computation of axial forces

Moment = 40(2) = 80 kNm.

Moment about the second plane = 40(6) + 80(2) = 400 kNm.

First column axial force $= \frac{My}{I}(A) = \frac{80(7)}{140}(1) = 4.0$ kN.

Second column axial force $= \frac{80(3)}{140}(1) = 1.7143$ kN.

Third column axial force $= \frac{80(1)}{140} = 0.571$ kN.

Fourth column axial force $= \frac{80(9)}{140} = 5.143$ kN.

These are shown in Fig. 15.24.

Bottom floor :

First column axial force $= \frac{400(7)}{140} = 20$ kN.

Second column axial force $= \frac{400(3)}{140} = 8.571$ kN.

Third column axial force $= \frac{400(1)}{140} = 2.857$ kN.

Fourth column axial force $= \frac{400(9)}{140} = 25.714$ kN.

These are shown in Fig. 15.25.

Consider the left top most sketch in Fig. 15.25.

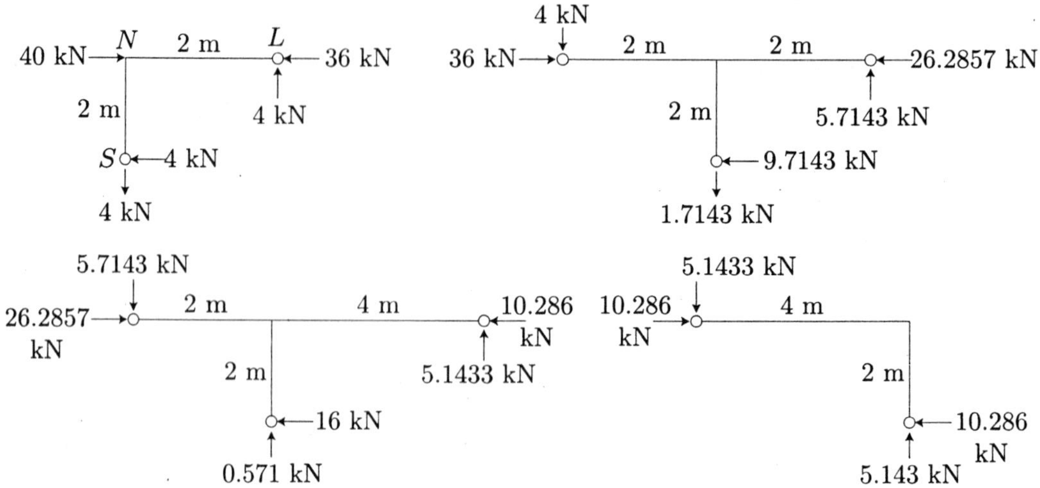

Fig. 15.25: For Example 15.4

Consider $\sum H = 0, \sum V = 0$ and $\sum M = 0$

From $\sum V = 0$, vertical force at $L = 4.0$ kN.

From $\sum M = 0$ about 5, $H = 36$ kN.

B.M. at $N = 8$ kNm (tension being inside).

From the right top member in Fig. 15.25, the process is repeated for all the segments from left to right in the topmost storey.

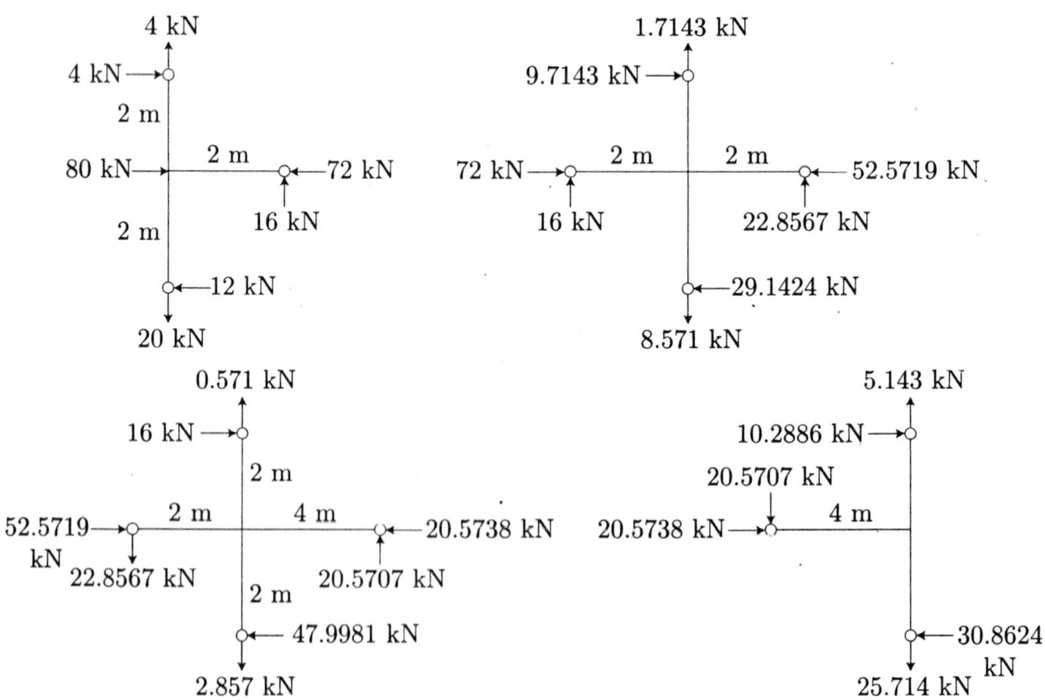

Fig. 15.26: Reactions in kN

In any segment moment at the joint is equal to the transverse force multiplied by lever arm.

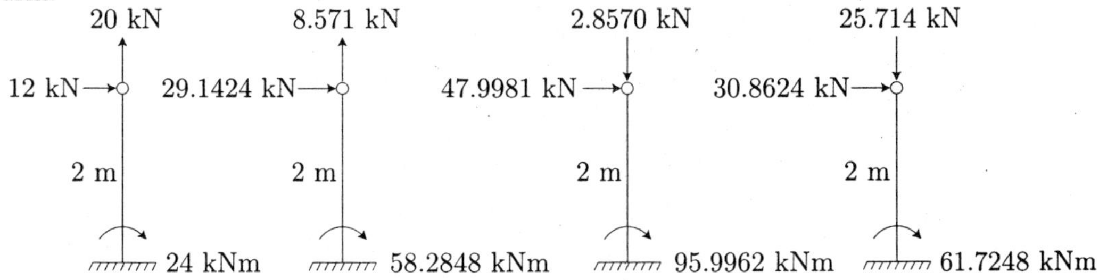

Fig. 15.27: Forces and moments at bottom

Example 15.5. *Analyse the two storeyed frame shown in Fig. 15.28 by the Cantilever method.*

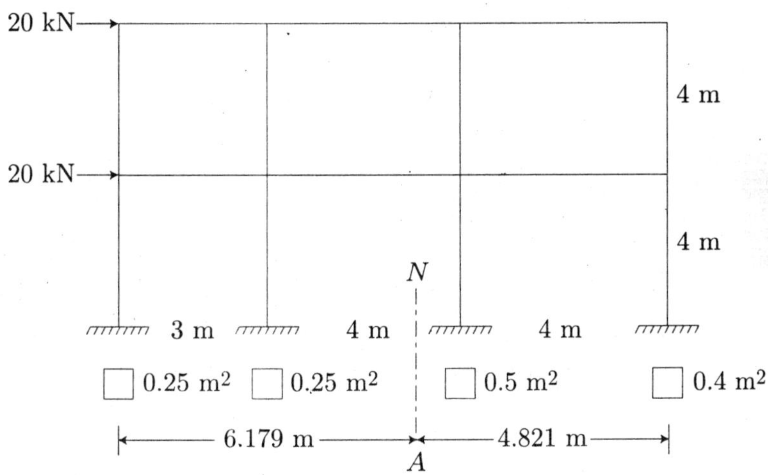

Fig. 15.28: For Example 15.5

Solution : The axial forces are shown in Fig. 15.29.

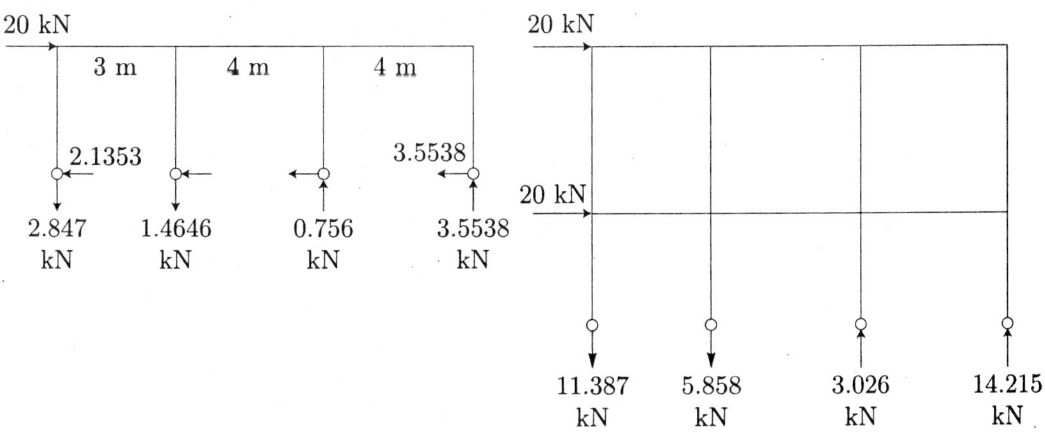

Fig. 15.29: Axial forces in kN

The pattern of moments are shown in Fig. 15.30.

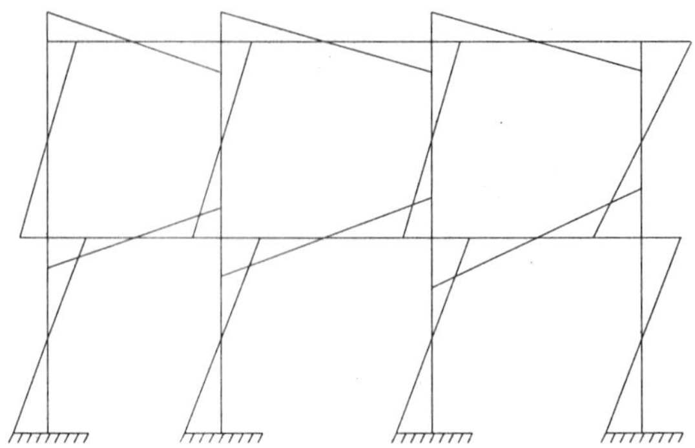

Fig. 15.30: Pattern of column moment

In any column, top and bottom moments will be same. In any joint sum of column moments is equally divided between the beams meeting at the joint.

15.7 Building Frames under Vertical Loading

A building frame consists primarily of girders, which carry vertical loads coming from slabs and in turn transfer the loads to the columns. If such a frame be constructed with girders simply supported at their supports on columns, the structure will be statically determinate, but would have little resistance against horizontal forces, such as winds loads which it may have to resist. A building frame, therefore, is made rigid by making joints between girder and columns rigid. Such a building frame is statically indeterminate to a degree equal to three times the numbers of girders.

The girders of the building frames are normally subjected to uniformly distributed load. If the ends of girders are perfectly fixed the points of inflection on the girder would be located at a distance 0.21 L from each end. If the ends are hinged, the points of zero bending moments will be at the ends. For the actual conditions of partial fixity of support which is between the two extreme conditions, the points of inflection may be assumed to lie somewhere between the two extreme values of 0.21 L and 0.00 L from the ends of the beam, depending on the elastic properties of the members. Solution of building bents also indicates that the axial force on girders subjected to vertical loading is also small.

On the basis of the above considerations analysis of building bent having small variation in the girder spans of adjacent bays and subjected to uniformly distributed vertical loading may be done by means of statics if the following simplifying assumptions be made.

(a) The axial force on the girder is negligible.

(b) Points of inflection on the girder near end supports and intermediate supports may be taken as occurring from one tenth to one twelfth and one sixth to one eighth of the span respectively as shown in Fig. 15.31(a).

Bending moment diagrams on a girder of the building frame having a span of 6 m, on the basis of position of P.I. as shown, is illustrated in Fig. 15.31(b). Since the shear force acting on the girders are equal to the vertical forces applied to the columns by girders the axial forces in the column are easily found by summing up the girder shear forces from the top of the column down to the column section under consideration.

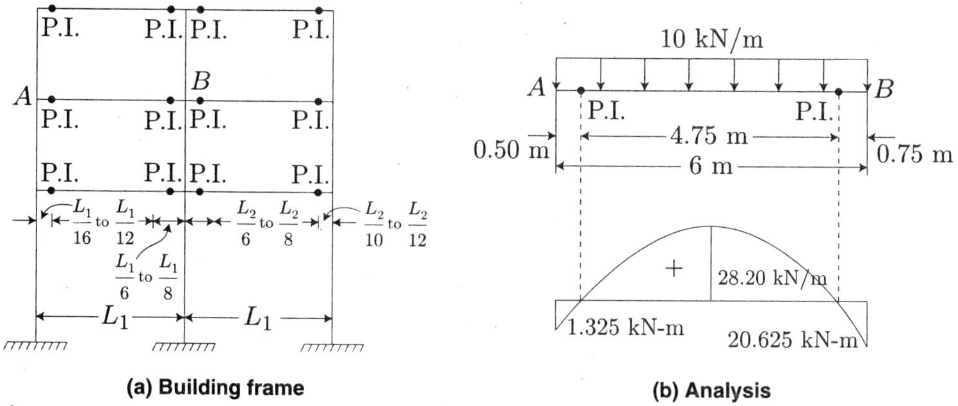

Fig. 15.31: Approximate analysis of building frame

For maximum axial force in a column, bays on both sides of the column must be loaded. For interior cases, therefore, the moment on columns are small and often neglected in design. For exterior columns, however, the resulting moments are larger as girders exist only on one side of the column. In such cases, the resulting girder end moments is distributed between the columns at the joint, in proportion to their stiffness.

15.8 Multiple System Trusses

Bar force in the members of a statically indeterminate truss can in certain cases be obtained, by subdividing the indeterminate truss to two or more statically determinate trusses. The statically determinate trusses are separately analysed, the bar forces in the original truss being obtained by adding algebraically the forces in the component truss. This method, which is an approximation is called the method of superposition or method of subdivision and its accuracy depends on the type of truss and the loading.

Fig. 15.32(a) shows an example of such a truss, the subdivided statically determinate trusses being shown in Fig. 15.32(b), Due to live loads, such diagonal members may be subjected to compressive forces, which may exceed the tensile forces. This would tend to produce compression in these diagonal members, a condition that must be avoided particularly in cases where the member when designed for resulting maximum tensile force is incapable of carrying the compressive force due to large slenderness ratio. In such a case instead of changing the diagonal member, it is sometimes desirable to provide counterbracing in the form of a second diagonal member, having a slope opposite to that of the main tension diagonal. The counterbracing member will have a large slenderness ratio and will not carry any compressive load and hence no load due to the dead loads only, since it will buckle slightly under its action. When the resulting compressive load in the main tension diagonal causes it to buckle, it cannot carry any load and the counterbracing member comes into action, carrying the resulting load as a tension member.

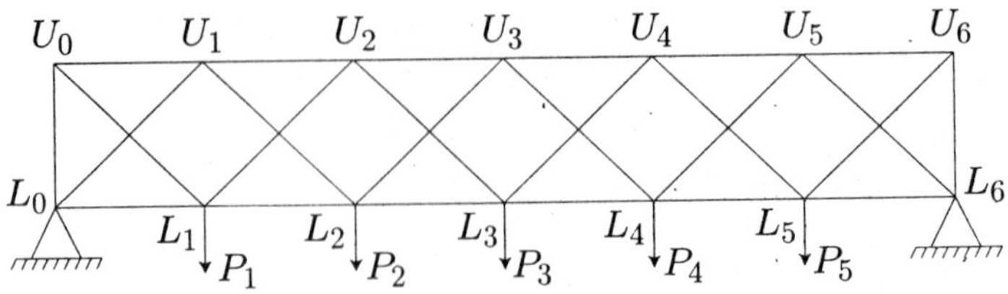

(a) Building frame

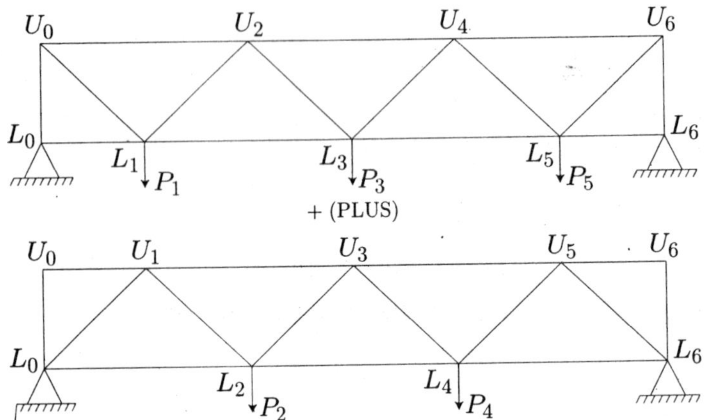

(b) Analysis

Fig. 15.32: Method of subdivision in truss

In designing a truss with tension diagonals, it should first be determined as to which panels require counterbracing. They are most likely to be necessary in the panels nearest the centre of the truss. This will be evident from analysis of the truss in Art 10.5, wherein it is seen that the panels on either side of the centre requires counterbracing as the tension diagonal members in those panels are also subjected to compressive load, when live load effects are considered.

When a truss has counterbracing, the determination of maximum bar forces in members influenced by the action of counters must be approached by trial. Each position of the live load that may reasonably lead to the maximum bar force desired must be investigated. If diagonals, capable of carrying tension only have been assumed to be in action, the results of any investigation are invalidated unless the computed bar forces in such members indicate that they actually carry tension.

15.9 Lateral Forces in Bridge Trusses

Fig. 15.33 shows diagramatically an elevation of the bridge truss, a plan of its top lateral bracing truss and a plan of its bottom bracing truss. The function of lateral bracing trusses, is to resist forces transverse to the axis of the structure. Such forces are due to the action of wind on the structure, and on the vehicle or train, and also lateral vibration caused by the swaying of the locomotive and train.

The pressure of the wind on the structure itself is delivered to the lateral trusses at the panel points by the vertical truss member is acting as beams. For example, the diagonal member U_2L_3 or the vertical member U_3L_3 of the main truss when pressed against by the wind must have reactions at the top and bottom at the panel points and presumably each reaction is equal to one half the total wind force against the member. Top and bottom chord members must also contribute horizontal load at the panel points, the wind force on the top chord resisted by the top lateral truss, the wind force on the bottom chord being resisted by the bottom lateral truss. Wind pressure against the floor system of the truss evidently must be resisted by the lateral truss to which the floor connects the bottom lateral truss in this case and in other through bridges. Wind pressure against the vehicle or train is supported by the floor system and thus by the lateral truss on the side of the floor system.

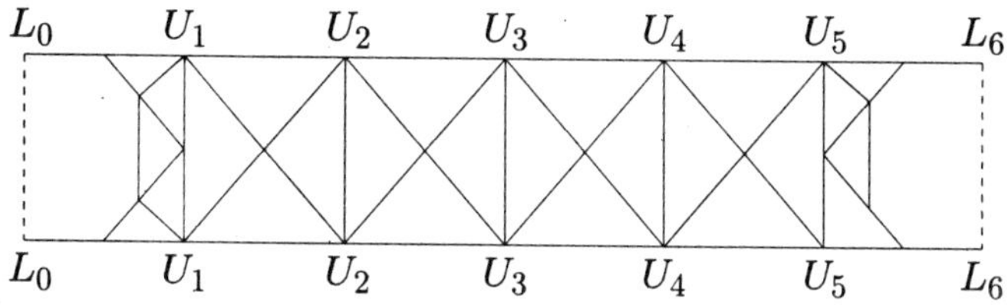

(a) Plan of lateral bracing truss

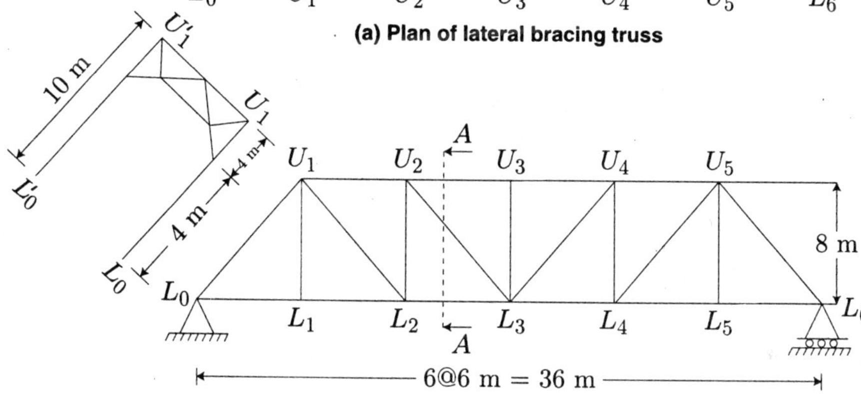

(b) Elevation of main truss

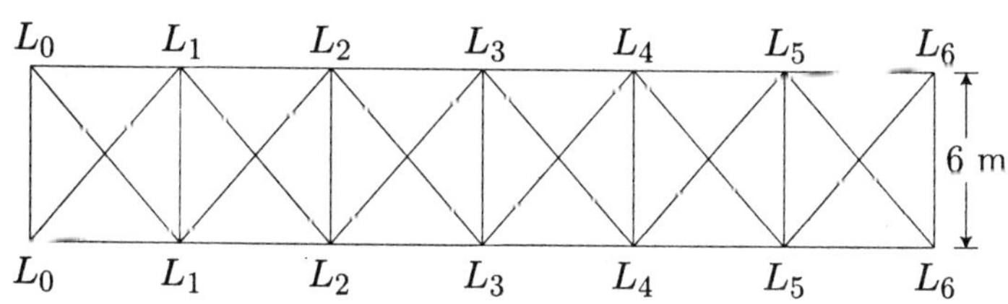

(c) Plan of bottom lateral truss

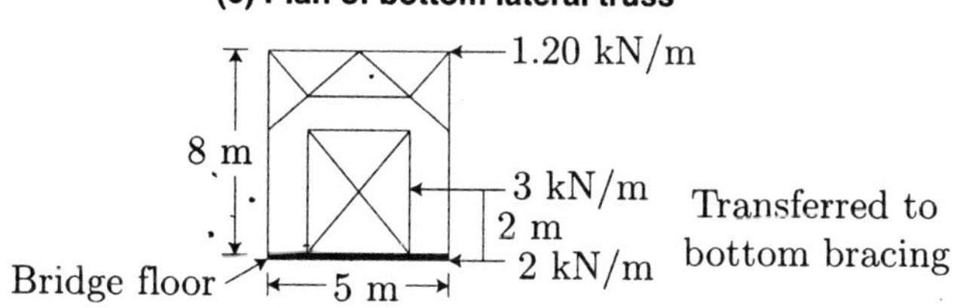

(d) Section *AA*

Fig. 15.33: Bridge truss

The determination of the magnitude of the loads to which lateral trusses are finally subjected from these various sources requires a knowledge of the exposed area of the truss members, floor system and vehicle or train as well as the amount of pressure to be expected from the wind and a reasonable estimate of the effect of lateral vibration. The exposed area of the structure can be determined accurately only after its design and the wind pressure and then lateral effects on the vehicle or train are generally somewhat arbitrarily fixed in accordance with prevailing practice.

15.10 General Analysis of Lateral Bracing Truss

To illustrate the procedure of analysis of the lateral bracing truss, the system illustrated in Fig. 15.33 will be considered and analysed. For the purpose of analysis, the lateral forces will be assumed as follows :-

Lateral wind force from main truss transferred to top lateral truss = 1.20 kN/m.

Lateral wind force on bottom lateral truss (wind force from main truss plus effect of wind on floor = 2 kN/m.

Wind on vehicle and effect of lateral vibration transferred to bottom lateral truss = 3 kN/m applied at 1.50 m above floor level in case of highway bridges.

These forces will be treated as moving loads.

The top lateral truss shown in Fig. 15.33 is a complete truss only between the top chord panel points U_1 and U_5. Beyond these points, the diagonals must be omitted to allow the clearance necessary for the passage of the vehicle. The top lateral truss may be treated as extending from U_1 to U_5 having at these points its reactions which are provided by rigid frames consisting of the end posts of the vertical trusses and portal frames between them as shown in Fig. 15.33. The reactions of the lateral truss at the top of the portal frames are transferred to the masonry support by the ability of the end posts to resist transverse forces in bending. The top chords of the main trusses serve also as chords in the lateral truss, the only additional members necessary to form the lateral truss being the diagonals $U_1\ U_2'$, $U_1'\ U_2$, etc. and the struts U_1U_1', U_2U_2', etc.

The bottom lateral truss is similarly formed as the top lateral truss except that it extends from L_0 to L_6, i.e., all through the span of the bridge. As in the case of the top lateral truss, the bottom lateral truss also only needs the diagonals and the struts for its formation, the bottom chords of the main trusses serving as the chords of the lateral truss.

Both the top and bottom lateral trusses are statically indeterminate as there are more members than are possible for calculations of forces by the laws of statics. It is, however, the usual practice to make approximate analysis of such statically indeterminate lateral trusses. Such approximate methods of analysis are based on certain simplifying assumptions, which make it possible to calculate the forces with the help of statics alone. The following two simplifying assumptions represent the two methods of analysis usually used :

(a) Diagonal in the truss may be considered as having large slenderness ratio so as to make them incapable of resisting compression. On the basis of this it may be assumed that the diagonals which would be in compression for a given loading are not acting.

(b) Diagonal in the truss may be considered as capable of taking both compression and tension. The usual assumption made is that the shear in any panel is resisted equally by the two diagonal one acting in tension and the other in compression and this gives the necessary additional information to enable the calculation of forces.

In case of analysis of the top lateral truss, the first procedure is adopted, as the diagonals in such a truss, however, are likely to be so slender as to be incapable of resisting compression. For the lateral loads applied as in Fig. 15.34(a) the top lateral truss consists of the members shown in full line, the diagonals in dotted lines being out of action. For loads in the opposite direction, the dotted diagonals come into action in tension, and the others are considered as out of action. Analysis of the truss on the basis of wind loads on all the panels, is shown in Fig. 15.35(b).

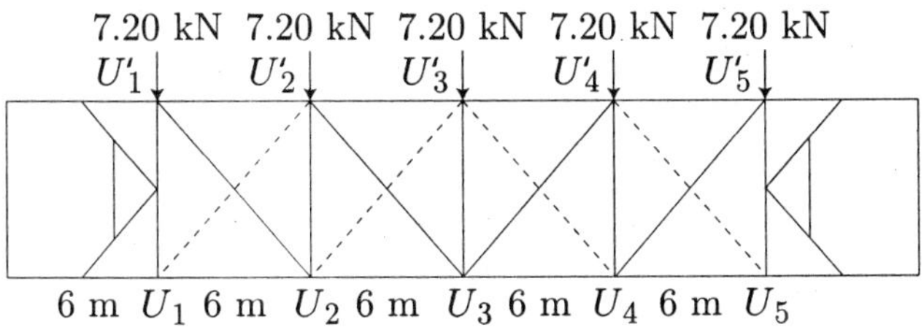

Truss with loading

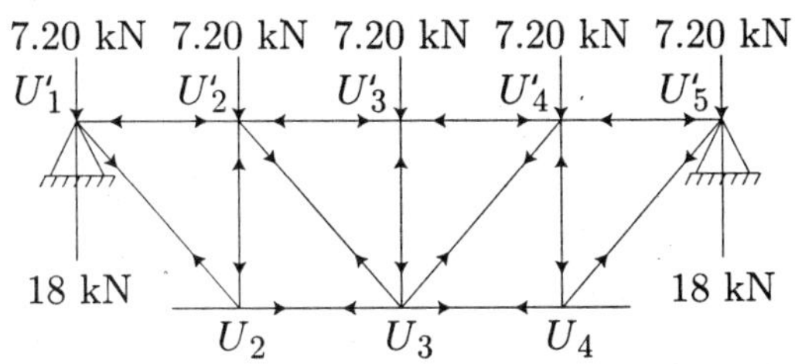

(a) Lateral truss analysis

Bar Forces

Design Forces With Wind Load From Any Direction :

$U_1'U_2$, $U_5'U_4$ $+ (18 - 7.20)\sqrt{2} = +15.274$ +15.274 kN

$U_1'U_2'$, $U_5'U_4'$ $- 15.274\left(\frac{1}{\sqrt{2}}\right) = -10.800$ −10.800 kN

$U_2'U_2$, $U_4'U_4$ $- 15.274\left(\frac{1}{\sqrt{2}}\right) = -10.800$ −10.800 kN

U_2U_3, U_4U_3 $15.274\left(\frac{1}{\sqrt{2}}\right) = +10.800$ +10.800 kN (−14.400) kN

$U_2'U_3$, $U_4'U_3$ $+ (10.80 - 7.20)\sqrt{2} = +5.091$ +5.091 kN

$U_2'U_3'$, $U_4'U_3'$ $- \left(10.80 + \frac{5.091}{\sqrt{2}}\right) = -14.400$ −14.400 kN (+10.800) kN

$U_3'U_3 - 07.20$ −7.20 kN

(b) Approximate analysis

Fig. 15.34: Analysis of lateral truss

The bottom lateral truss may be treated in the same manner as the top lateral truss, if its diagonals are too slender to resist compression. However, the diagonals of the bottom bracing truss are usually stiffer and capable of resisting compression due to the larger magnitude of the lateral forces and the support provided by the stringers. The second procedure is, therefore, usually adopted in the analysis of bottom lateral truss. Analysis of bottom lateral truss on the basis of wind forces on all panel points is illustrated in Fig. 15.35. Forces in chord members in a panel, will be equal and opposite and may be calculated by taking a section through the panel in which the chord in question occurs and moments are taken about the intersection of the diagonals. As indicated in Fig. 15.35, design bar forces in diagonal and chord members may be tension or compression depending on the direction of wind loads but the magnitudes will be the same as calculated.

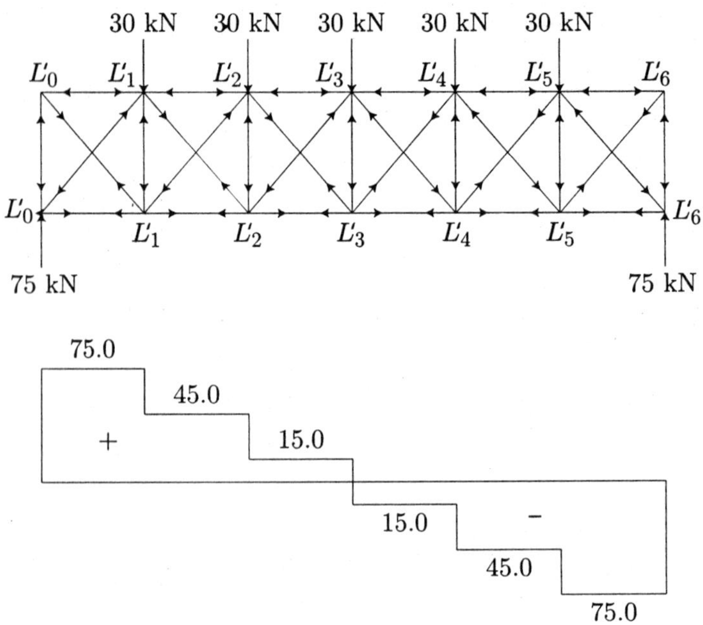

Fig. 15.35: Bottom lateral truss

Forces in Members :

$$L_0L_1'; L_6L_5' \quad -75\left(\frac{1}{\sqrt{2}}\right) = -53.03$$

$$L_0'L_1; L_6'L_5 \quad +75\left(\frac{1}{\sqrt{2}}\right) = +53.03$$

$$L_2'L_2'; L_5L_4' \quad -45\left(\frac{1}{\sqrt{2}}\right) = -31.82$$

$$L_1'L_2; L_5'L_4 \quad 45\left(\frac{1}{\sqrt{2}}\right) = +31.82$$

$$L_2L_3'; L_4L_3' \quad -15\left(\frac{1}{\sqrt{2}}\right) = -10.60$$

$L_2'L_3;\ L_4'L_3 \quad +15\left(\frac{1}{\sqrt{2}}\right) = +10.60$

$L_0L_0';\ L_6L_6' \quad \left(-75+\frac{53.03}{\sqrt{2}}\right) = -37.50$

$L_0L_1;\ L_6L_5\ 53.03\left(\frac{1}{\sqrt{2}}\right) = +37.50$

$L_0'L_1';\ L_6'L_5' \quad -53.03\left(\frac{1}{\sqrt{2}}\right) = -37.50$

$L_1'L_1;\ L_5'L_5 \quad -53.03\left(\frac{1}{\sqrt{2}}\right) + 3182\left(\frac{1}{\sqrt{2}}\right) = 15.00$

$L_1L_2;\ L_5L_4\ 37.50 + 53.03\left(\frac{1}{\sqrt{2}}\right) + 31.82\left(\frac{1}{\sqrt{2}}\right) = 97.50$

$L_1'L_2';\ L_5'L_4' \quad -97.50$

$L_2'L_2;\ L_4'L_4 \quad -31.82\left(\frac{1}{\sqrt{2}}\right) + 10.60\left(\frac{1}{\sqrt{2}}\right) = -15.00$

$L_2L_3;\ L_4L_3\ 92.50 + (31.82 + 10.60)\frac{1}{\sqrt{2}} = +127.495$

$L_2'L_3';\ L_4'L_3' \quad -127.495$

$L_3'L_3 \quad -15.00$

Design forces are obtained by combinations of bar forces for similar members. Design forces for diagonal and chord members may be compression or tension, but magnitude will be the same as calculated.

15.11 Analysis of End Portal Frame

The top lateral bracing truss is considered supported against the portal frame at the top, and hence carries the reactions at end of the top lateral truss. This reaction must be transferred to the bottom by the ability of the frame to resist transverse forces. In the portal frame shown in Fig. 10.5, the end posts of the main trusses forming the columns of the portal frame are subjected to bending moment and shear force as well as axial force in resisting the lateral truss reaction, over and above the axial forces as members of the main trusses. The other members that comprise the truss at top of the portal frame are considered as pin-connected and carrying axial force only. Such a structure is statically indeterminate to the third degree and for analysis, the following three assumptions are usually made.

(a) Point of inflection are assumed (i) midway between the bottom of the post and the bottom of the portal knee brace, i.e., $h/2$ from base, for columns fixed at base.

(ii) $h/3$ from base, for partially fixed columns, and

(iii) at base for columns hinged at base.

(b) Shear force on the plane of inflection is divided equally between the two columns.

Analysis of the portal frame for bridge truss of Fig. 15.36 on the basis of these simplifying assumptions is illustrated in Fig. 15.36. For end portal frame for a wider bridge, the method of analysis is similar as illustrated in Fig. 15.36.

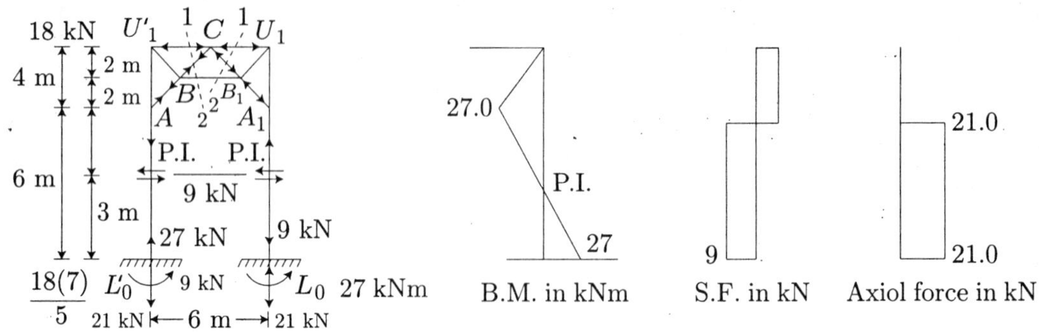

$$AB,\ A_1B_1,\ BU'_1,\ B_2U_1,\ BC,\ B_1C = \sqrt{2^2 + 1.5^2} = 2.50 \text{ m}$$

Fig. 15.36: Portal frame truss

Bar forces : (Refer Fig. 15.36)

AB, BC $\quad +\dfrac{9.0(7)}{4}\dfrac{2.50}{1.50} = +26.25 \quad$ From $\sum M_{U'_1} = 0$

A_1B_1, B_1C $\quad = -26.25 \quad$ From $\sum M_{U_1} = 0$

$BB_1 \quad \dfrac{(9.0)(7) - 21(3)}{2} = 0 \quad$ From $\sum M_C = 0$, Section 1-1

$CU'_1 \quad -\dfrac{(9)(3)}{4} - 18.0 = -24.75 \quad$ From $\sum M_A = 0$, Section 2-2

$CU_1 \quad +\dfrac{9(3)}{4} = +6.75 \quad$ From $\sum M_{A1} = 0$, Section 1-1

$BU'_1 \quad = 0 \quad$ From Equilibrium at joint B and B_1.

$BU_1 \quad = 0$

15.12 Bar Forces in Chord Members and on End Members of Main Truss Due to Lateral Load

In analysis of end portal frame in Fig. 15.36, a direct force of 21 kN was found in the columns $U'_1L'_0$ and U_1L_0. The axial force is tensile in the windward column and compressive in the leeward column. The horizontal component of this inclined force is furnished by the bottom chord of main truss, since one end of the structure is supported on rollers. The magnitude of this axial force is $21\left(\frac{6}{10}\right) = 12.60$ kN and will be constant throughout the bottom chord of the main truss and is evidently tension in the leeward chord and compression in the windward one.

The lateral force on vehicle due to wind has been considered as acting 1.50 m above the roadway and 2 m above the centre of the bottom chord. This force on the side of the vehicle tends to tip it over the leeward wheels, thus causing an increase of the load on these and a relief of the load on the windward wheels, which in turn produces an increase in the downward load on the leeward main truss, and a relief of the downward load on the windward main truss. The amount of increase in the downward load on the leeward truss will be

$$3 \text{ kN/m} \left(\frac{2}{6}\right) = 1 \text{ kN/m}.$$

The effect of the readjustment of the downward load may be taken as not significant in the top chord and web system of the leeward truss or in any member of the windward truss. Its effect however is significant in the bottom chord and end posts of the leeward truss. The total effect of the lateral forces on the end posts then is as follows :

Portal frame

Axial force = −21 kN; B.M. = 27 kNm; S.F. = 9 kN.

Overturning,

$$(1)(6)\left(\frac{5}{2}\right)\left(\frac{10}{6}\right) = -25 \text{ kN}.$$

Total axial force = −46 kN; B.M. = 27 kNm; S.F. = 9 kN

Total effect of lateral force on the leeward main truss bottom chord member, say on L_2L_3, is as follows

As member of lateral truss (Refer Fig. 15.35)

Axial force =+127.495

Portal frame effect=+12.60

Overturning effect

$$\frac{1.0}{2}\ \frac{(12)(24)}{6} = +24.0$$

Total : 164.095 kN

15.13 Effect of Longitudinal Force in Bridges

The most important longitudinal forces are the forces due to tractive effort and braking effect resulting from starting or stopping trains or other loads moving on a structure. The stresses resulting from them are not severe in ordinary truss bridges but may become important in long span structures and are always important in case of steel trestles and piers of bridges.

The forces on truss members from longitudinal forces even under the most unfavourable circumstances will not be large in an ordinary bridge. They are often neglected, even though it is desirable that proper investigation of their effects be made.

In investigating the importance of forces from longitudinal forces, the maximum possible resulting forces in a particular member should not be added to its maximum live load forces, unless the two occur for the same position of the loads. Sometimes, however, it is difficult to decide whether possible combinations of loading are sufficiently probable to justify considering them as occurring simultaneously.

Problems

15.1 Compute the bar forces in various members of the Portal Frame shown in Figs. P-15.1 and 15.2. Any reasonable and necessary assumptions may be made.

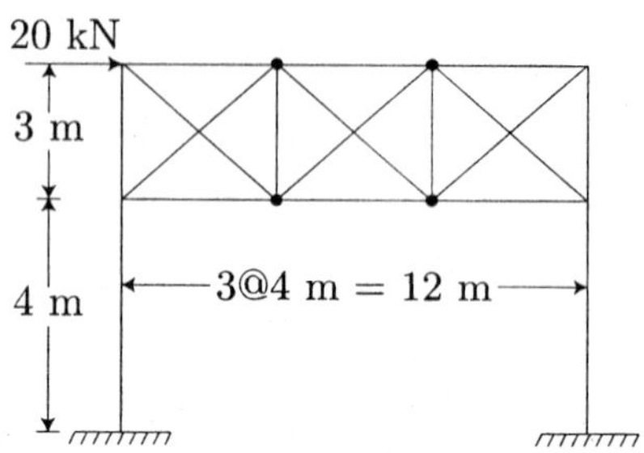

Fig. P-15.1

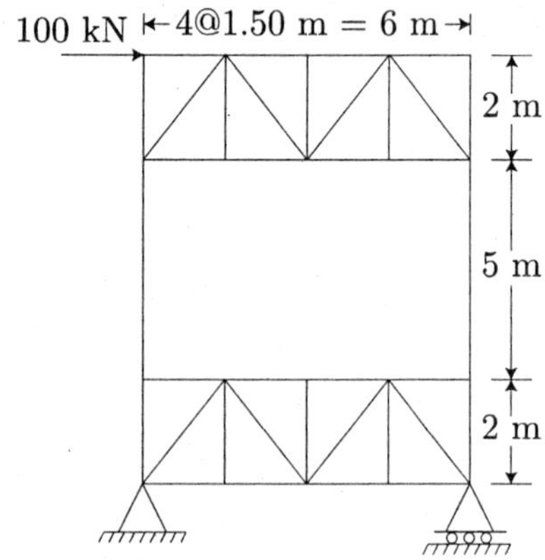

Fig. P-15.2

15.2 Analyse the bent in Fig. 15.1 assuming the column bases at A and B as hinged.

15.3 Analyse the bent in Fig. 15.13 except that columns AC and BD are 6 m and 10 m long respectively. The point of inflection may be assumed midway between knee brace and the base for each column. Assume that shear taken by the windward column is 20% and that on the leeward column is 80% of the total shear force.

15.4 A building bent has three equal bays of 6 m each and three stories of 4 m each. The columns may be considered fixed at their base. For each girder the dead load is 8 kN/m and the live load is 6 kN/m. Determine :

(a) the maximum positive and negative girder moments occurring in the bent.

(b) the maximum shear force in the girder.

(c) the maximum axial force in the exterior and interior columns.

(d) the maximum moment in the exterior and interior columns.

15.5 Analyse the bents in Problem 15.1 if it be subjected to a wind load of magnitude 2 kN/m^2. Assume bents to be spaced 5 m c/c.

Chapter 16

Matrix Method of Analysis

16.1 Introduction

Matrix Methods are based on the concept of replacing the actual indeterminate structure by a mathematical model made up from structural elements having known elastic and other properties which can be expressed in matrix form. Such methods represent the most powerful tool in structural analysis.

Matrix structural analysis programmes for digital computers which can be used to analyse and design general type of structures are nowadays available. In the matrix methods, the structure is initially idealised into discrete elements having the form of displacement or stress distribution. The complete solution is obtained by combining these displacements or stress distributions in a manner so as to satisfy the force equilibrium and compatibility of displacements at the junction of these elements. Formulation of the analysis in matrix algebra is ideally suited for subsequent solution on the digital computer.

16.2 Matrix Algebra

A matrix is an array of elements arranged in rows and columns which as a whole obey certain algebraic rules. The rows of a matrix are labelled with positive consecutive integers from top to bottom, starting with 1 at the top. Likewise, the columns of a matrix are labelled with consecutive integers from left to right starting with 1 at the left. A matrix element carries as the first subscript its row number and as the second subscript its column number. The elements of a matrix are displayed within brackets. For example, a rectangular matrix may be represented as

$$[A] = \begin{bmatrix} a_{11} & a_{12} & a_{13} \cdots a_{1_n} \\ a_{21} & a_{22} & a_{23} \cdots a_{2_n} \\ a_{31} & a_{32} & a_{33} \cdots a_{3_n} \\ \cdots & \cdots & \cdots\cdots \\ a_{m1} & a_{m2} & a_{m3} \cdots a_{mn} \end{bmatrix}. \tag{16.1}$$

The number of rows in a matrix is called the row order of the matrix and the number of columns is called the column order. Matrix $[A]$ in Eq. (16.1) may be referred to as a matrix of row order m and column order n or by an m by n matrix.

When there is only one row in a matrix, it is called a column matrix or simply a column. A column matrix of order $m \times 1$ may be expressed as

$$[A] = \begin{bmatrix} a_{11} \\ a_{21} \\ \vdots \\ a_{m1} \end{bmatrix} . \tag{16.2}$$

If a matrix has only one row (i.e., $m = 1$), it is known as a row matrix. A row matrix of order $1 \times n$ may be expressed as

$$[A] = [a_{11} \quad a_{12} \cdots a_{1n}] \; . \tag{16.3}$$

When the corresponding columns and rows of a matrix are interchanged, the resulting matrix is called the transpose of the original matrix. For example, the transpose of the matrix $[A]$ in Eq. (16.1) denoted by $[A^T]$ of the form

$$[A^T] = \begin{bmatrix} a_{11} & a_{21} & \cdots\cdots\cdots & a_{m1} \\ a_{12} & a_{22} & \cdots\cdots\cdots & a_{m2} \\ a_{13} & \cdots\cdots & \cdots\cdots\cdots & \cdots \\ \vdots & \vdots & & \vdots \\ a_{1n} & a_{2n} & & a_{nn} \end{bmatrix} \tag{16.4}$$

Transpose of a column matrix is a row matrix and that of a row matrix is a column matrix.

If all elements of a matrix are zeroes, it is called a zero matrix or null matrix. When the number of rows and columns of a matrix are the same, i.e., $m = n$, it is called a square matrix. The diagonal line established by the matrix elements with repeated subscripts such as a_{11}, a_{22}, etc. is called the main diagonal. Matrix elements on the main diagonal are called diagonal elements and the remaining elements are called off diagonal elements. A square matrix in which all the off diagonal elements are zero is called a diagonal matrix.

A diagonal matrix in which all the diagonal elements are equal to unity is called an identity matrix or a unit matrix. Identity matrices are shown by letter I and denoted by

$$[I] = \begin{bmatrix} 1 & 0 & 0\cdots\cdots0 \\ 0 & 1 & 0\cdots\cdots0 \\ 0 & 0 & 1\cdots\cdots0 \\ \cdots & \cdots & \cdots\cdots\cdots \\ 0 & 0 & 0\cdots\cdots1 \end{bmatrix} \tag{16.5}$$

The (i, j) elements may be denoted by the Kronecker delta δ_{ij} rather than i_{ij}. For the elements of I, it may be noted that,

$$\delta_{ij} = 0, \quad \text{for } i \neq j \quad \text{and} \quad \delta_{ii} = \delta_{jj} = 1.$$

A diagonal matrix in which the diagonal elements are equal to the same scalar is called a scalar matrix.

A symmetrical matrix is a square matrix whose elements are symmetrical about its main diagonal. Thus, for a symmetrical matrix, $a_{ij} = a_{ji}$. A symmetrical matrix remains unchanged if its columns and rows are interchanged, whence $[A] = -[A^T]$.

A skew matrix is a square matrix where elements are such that $a_{ij} = -a_{ji}$ for $i \neq j$ and the principal diagonal elements may have any value provided that they are not all zero, i.e., a_{ii} not all zero.

If the elements on the principal diagonal of a skew matrix are all zero, then the matrix is said to be skew symmetric or anti-symmetric and the transpose of the matrix is equal to the negative of the original matrix. Thus for the skew-symmetric matrix,

$$a_{ij} = -a_{ji} \text{ for } i \neq j, a_{ii} = a_{ij} = 0 \quad \text{and} \quad [A] = -[A^T].$$

From the definition it may be observed that all diagonal matrix are both symmetric and skew. Any square matrix can be expressed as the sum of a symmetric and anti-symmetric matrix.

In engineering analysis, it is not unusual to encounter square matrices having a large number of zero elements. Such a matrix with zero elements above and to the right of the principal diagonal, i.e. $a_{ij} = 0$ for $i < j$, is called a lower triangular matrix. Similarly, in case zero elements are below and to the left of the principal diagonal, i.e., $a_ij = 0$, for $i > j$, the matrix is called an upper triangular matrix.

Matrix of the form

$$[A] = \begin{bmatrix} a_{11} & 0 & 0 \cdots\cdots\cdots 0 \\ a_{21} & a_{22} & 0 \cdots\cdots\cdots 0 \\ a_{31} & a_{32} & a_{33} \cdots\cdots 0 \\ \cdots & \cdots & \cdots\cdots\cdots\cdots \\ a_{n1} & a_{n2} & a_{n3} \cdots\cdots a_{nn} \end{bmatrix} \tag{16.6a}$$

is a lower triangular matrix. An upper triangular matrix has the form,

$$[A] = \begin{bmatrix} a_{11} & a_{12} & a_{13} \cdots a_{1n} \\ 0 & a_{22} & a_{23} \cdots a_{2n} \\ 0 & 0 & a_{33} \cdots a_{3n} \\ 0 & 0 & 0 \cdots a_{nn} \end{bmatrix} \tag{16.6b}$$

A band matrix has zero elements everywhere except along a band or strip that runs diagonally through the matrix, usually (but not necessarily) being centred about the principal diagonal. Thus the following matrix is an example of a band matrix.

$$[A] = \begin{bmatrix} a_{11} & a_{12} & 0 & 0 & 0 & 0 \\ a_{12} & a_{22} & a_{23} & 0 & 0 & 0 \\ 0 & a_{32} & a_{33} & a_{34} & 0 & 0 \\ 0 & 0 & a_{43} & a_{44} & a_{45} & 0 \\ 0 & 0 & 0 & a_{54} & a_{55} & a_{56} \\ 0 & 0 & 0 & 0 & a_{65} & a_{66} \end{bmatrix} \tag{16.7}$$

The band width in the above illustration is three and the band is centered on the diagonal. This matrix is said to be tri-diagonal. In general, the width of the band may be of any value up to a number that would include the entire matrix within the band. It may be noted that a diagonal matrix is also a band matrix with bandwidth equal to one.

In carrying out matrix operation, it is sometimes convenient to partition them into submatrices and then deal with them separately. The matrix elements of a partitioned matrix are called submatrices. The pattern of the submatrices are indicated as shown below :

$$[A] = \begin{bmatrix} a_{11} & a_{12} & a_{13} & \vdots & a_{14} & a_{15} \\ a_{21} & a_{22} & a_{23} & \vdots & a_{24} & a_{25} \\ a_{31} & a_{32} & a_{33} & \vdots & a_{34} & a_{35} \\ \cdots & \cdots & \cdots & \cdots & \cdots & \cdots \\ a_{41} & a_{42} & a_{43} & \vdots & a_{44} & a_{45} \\ a_{51} & a_{52} & a_{53} & \vdots & a_{54} & a_{55} \end{bmatrix} = \begin{bmatrix} [A_{11}] & \vdots & [A_{12}] \\ \cdots & \cdots & \cdots \\ [A_{21}] & \vdots & [A_{22}] \end{bmatrix}, \tag{16.8}$$

where $[A_{11}] = \begin{bmatrix} a_{11} & a_{12} & a_{13} \\ a_{21} & a_{22} & a_{23} \\ a_{31} & a_{32} & a_{33} \end{bmatrix}$; $[A_{12}] = \begin{bmatrix} a_{14} & a_{15} \\ a_{24} & a_{25} \\ a_{34} & a_{35} \end{bmatrix}$

$[A_{21}] = \begin{bmatrix} a_{41} & a_{42} & a_{43} \\ a_{51} & a_{52} & a_{53} \end{bmatrix}$ and $[A_{22}] = \begin{bmatrix} a_{44} & a_{45} \\ a_{54} & a_{55} \end{bmatrix}$.

In Eq. (16.8), $[A_{11}], [A_{12}], [A_{21}]$ and $[A_{22}]$ are known as the submatrices. The original matrix in the partitioned form is called a partitioned matrix. The submatrices may be treated as if they are elements of the partitioned matrix.

16.3 Determinants

A determinant is an important scalar and it is related only to square arrays. It is useful for solution of simultaneous equations. A determinant of order n can be expressed by the equation,

$$\text{Det}[A] = |A| = \begin{vmatrix} a_{11} & a_{12} \cdots a_{1n} \\ a_{21} & a_{22} \cdots a_{2n} \\ \cdots & \cdots \\ a_{i1} & a_{i2} \cdots a_{ij} a_{in} \\ a_{n1} & a_{n2} \cdots a_{nn} \end{vmatrix}. \tag{16.9}$$

The determinant of order $(n-1)$ obtained by erasing the ith row and jth column is known as the minor of element a_{ij} denoted by M_{ij}. The cofactor of element a_{ij} is defined as the product of $(-1)^{i+j}$ and the minor of element a_{ij}.

$$\text{Thus } M_{ij} = \begin{vmatrix} a_{11} \cdots a_{1j} & \cdots a_{1n} \\ \cdots\cdots & \cdots\cdots \\ a_{i1} \cdots a_{ij} & \cdots a_{in} \\ \hline \cdots\cdots & \cdots\cdots \\ a_{n1} \cdots a_{nj} & \cdots a_{nn} \end{vmatrix} \tag{16.10a}$$

and cofactor $C_{ij} = (-1)^{i+j} M_{ij}$, (16.10b)

where M_{ij} = Minor of element a_{ij}.

Eq. (16.10b) indicates that the cofactor is equal to the minor with a positive or negative sign attached to it depending on whether the sum $(i+j)$ is even or odd. Hence, the cofactor is also known as the signed minor.

Every determinant has a definite numerical value. The value of the determinant may be computed by using Laplace expansion, whence

$$|A| = a_{i1}C_{i1} + a_{i2}C_{i2} + \cdot + a_{in}C_{in}$$
$$= \sum_{j=1}^{j=n} a_{ij}C_{ij}. \tag{16.11}$$

In Eq. (16.11), the value of the determinant has been computed by multiplying each element of the ith row by its cofactor and taking the sum of the products. Alternatively, the value of the determinant may be computed by multiplying each element of the jth column by its cofactor and taking the sum of the products. Thus,

$$|A| = a_{1j}C_{1j} + a_{2j}C_{2j} + \cdot + a_{nj}C_{nj}$$
$$= \sum_{i=j}^{i=n} a_{ij}C_{ij}. \tag{16.12}$$

A square matrix is known as non-singular, if its determinant is non-zero. If the determinant is zero, the matrix is called singular.

For a diagonal matrix, the determinant is given by

$$\text{Det}(\text{diagonal } x_i) = x_1 \cdot x_2 \cdot x_3 \cdot x_n \tag{16.13}$$

The definition given in Eq. (16.13) is also useful in computing the determinant of lower and upper triangular matrices as in Eqs. (16.6a) and (16.6b). In such cases,

$$\text{Det}(\text{Lower triangular matrix}) = a_{11} \cdot a_{22} \cdot a_{33} \cdots a_{nn} \tag{16.14a}$$

and $$\text{Det}(\text{Upper triangular matrix}) = a_{11}a_{22}a_{33} \cdots a_{nn}. \tag{16.14b}$$

From the basic definition of the determinant given in Eqs. (16.11) and (16.12), the following properties can be easily verified.

(i) The determinant of a matrix and its transpose are equal, i.e., Det$[A]$=Det$[A^T]$.

(ii) When any two rows or two columns are interchanged, the sign of the determinant is changed.

(iii) If one row or column can be generated by a linear combination of the rows or columns, the determinant is zero. It follows from this that, if two rows or columns are identical, the determinant is zero.

(iv) If the elements of a row (or column) of $[A]$ are all zero, then Det $[A] = 0$.

(v) If all the elements of a row (or column) of a matrix A be multiplied by a constant, the determinant is given by constant (Det $[A]$).

(vi) For a matrix obtained by adding any multiple of a row (or column) of matrix $[A]$ to any other row (or columns), the determinant will be same Det $[A]$.

(vii) The sum of the products of the elements in any row i with the corresponding cofactors of another row m is zero.

Example 16.1. *Compute the value of the determinant for the matrix.*

$$[A] = \begin{bmatrix} 8 & 4 & 2 \\ 3 & 1 & 4 \\ 6 & 4 & 5 \end{bmatrix}$$

Solution : Selecting the first column for Laplace expansion,

$$\begin{aligned} |A| &= 8(-1)^{1+1}\begin{vmatrix} 1 & 4 \\ 4 & 5 \end{vmatrix} + 3(-1)^{1+2}\begin{vmatrix} 4 & 2 \\ 4 & 5 \end{vmatrix} + 6(-1)^{1+3}\begin{vmatrix} 4 & 2 \\ 1 & 4 \end{vmatrix} \\ &= 8(5-16) - 3(20-8) + 6(16-2) \\ &= -88 - 36 + 84 = -40. \end{aligned}$$

Selecting the first row for Laplace transformation,

$$\begin{aligned} |A| &= 8(-1)^{1+1}\begin{vmatrix} 1 & 4 \\ 4 & 5 \end{vmatrix} + 4(-1)^{1+2}\begin{vmatrix} 3 & 4 \\ 6 & 5 \end{vmatrix} + 2(-1)^{1+3}\begin{vmatrix} 3 & 1 \\ 6 & 4 \end{vmatrix} \\ &= 8(5-16) - 4(15-24) + 2(12-6) \\ &= -88 + 36 + 12 = -40. \end{aligned}$$

16.4 Properties of Matrices

The following properties of the matrices are useful in solution of matrices.

(a) Matrix Equation. Two matrices are equal to each other, only if the elements in the corresponding positions are equal. For example, if

$$[A] = [B], \tag{16.15a}$$

this implies that $a_{ij} = b_{ij}$ for $i = 1, 2, \ldots, m$

$$j = 1, 2, \ldots, n \tag{16.15b}$$

In order that two matrices are equal to each other their row and column orders must be equal.

(b) Matrix Addition and Subtraction. Addition and substraction are possible only for matrices of the same order. Thus, if $[A]$ and $[B]$ denote two matrices both of row order m and column order n, the sum of these matrices, which may be denoted by $A+B$ or $B+A$ is also an m by n matrix. If $[C]$ denotes the sum matrix, then

$$[C] = [A] + [B] = [B] + [A]. \tag{16.16a}$$

The definition of matrix addition can thus be stated as

$$\begin{aligned} c_{ij} = a_{ij} + b_{ij}, \quad & i = 1, 2, \ldots, m \\ & j = 1, 2, \ldots n, \end{aligned} \tag{16.16b}$$

Since the definition is the same for either $[B] + [A]$ or $[A] + [B]$, matrix addition is said to be commulative. If more than two matrices are to be added, the sum can be obtained by adding any two and then adding the remaining matrices, one by one in any sequence. It is seen that the matrix addition is also associative, since

$$([A] + [B]) + [D] = [A] + ([B] + [D]). \tag{16.17}$$

Matrix subtraction is a special case of matrix addition, since

$$[E] = [A] - [B] = [A] + (-1)[B].$$

From Eqs. (16.16a) and (16.16b), it is seen that,

$$c_{ij} = a_{ij} - b_{ij}. \tag{16.18}$$

(c) Matrix Multiplication. A matrix may be multiplied by a constant (scalar) or by another matrix.

If $[A]$ denotes an m by n matrix and s denotes a scalar, the product of A by s, which may be indicated by sA is also an m by n matrix. If B denotes the product matrix, then

$$[B] = s[A]. \tag{16.19}$$

The definition of a matrix multiplied by a scalar can be stated as

$$b_{ij} = sa_{ij},\ i = 1, 2, \ldots, m \text{ and } j = 1, 2, \ldots, n$$

This means that to multiply a matrix by a scalar amounts to multiplying every element of a matrix by the same scalar.

If an identity matrix is multiplied by a scalar, the resulting matrix is a scalar matrix.

A matrix $[C]$ is known as the product of two matrices $[A]$ and $[B]$, if the element c_{ij} belonging to the ith row and jth column of matrix $[C]$ is obtained by the summation of the term by term product of the elements of the ith row of matrix $[A]$ and jth column of matrix $[B]$. The multiplication of the two matrices may be expressed by the equation,

$$\begin{bmatrix} a_{11} \cdots\cdots a_{1n} \\ \cdots\cdots\cdots\cdots \\ a_{i1} \cdots\cdots a_{in} \\ \cdots\cdots\cdots\cdots \\ a_{m1} \quad a_{mn} \end{bmatrix} \begin{bmatrix} b_{11} \cdots & b_{1j} \cdots & b_{1p} \\ \cdots\cdots & \cdots\cdots & \cdots \\ b_{n1} \cdots & b_{nj} \cdots & b_{np} \end{bmatrix} = \begin{bmatrix} C_{11} & C_{1j} & C_{1p} \\ \cdots\cdots & \cdots\cdots & \cdots\cdots \\ C_{i1} & C_{ij} & C_{ip} \\ \cdots\cdots & \cdots\cdots & \cdots\cdots \\ C_{m1} & C_{mj} & C_{mp} \end{bmatrix} \tag{16.20}$$

In this case, matrix $[A]$ is post multiplied by matrix $[B]$ or matrix $[B]$ is premultiplied by matrix $[A]$.

Eq. (16.20) may be written in a compact form

$$[A][B] = [C], \text{ implying}$$

$$C_{ij} = \sum_{k=1}^{k=n} a_{ik}b'_{kj}, \text{ where } i = 1, 2, \ldots, m, j = 1, 2, \ldots, p. \tag{16.21}$$

It is observed from these definitions that the row order of the product matrix is equal to that of the multiplier (the first matrix) and the column order of the product matrix is equal to that of the multiplicant (the second matrix). The definition also implies that the column order of the first matrix must equal the row order of the second matrix. If p is not equal to m, the premultiplication of $[A]$ by $[B]$, i.e., the product of $[B][A]$ is not possible. Even if p is equal to m, when $[B][A]$ is conformable, from Eq. (16.21), it is evident that, in general,

$$[B][A] \neq [A][B]. \tag{16.22}$$

In other words, matrix multiplication is not commutative.

The multiplication of partitioned matrices may be carried out in the usual manner, if each sub-matrix is treated as an element of the partitioned matrix. Considering two partitioned matrices $[A]$ and $[B]$ given by the equations,

$$[A] = \begin{vmatrix} [A_{11}] & \vdots & [A_{12}] \\ \cdots & \cdots & \cdots \\ [A_{21}] & \vdots & [A_{22}] \end{vmatrix} \text{ and } B = \begin{vmatrix} [B_{11}] & \vdots & [B_{12}] \\ \cdots & \cdots & \cdots \\ [B_{21}] & \vdots & [B_{22}] \end{vmatrix}, \tag{16.23a}$$

where, $[A_{11}], [A_{12}], [A_{21}]$ and $[A_{22}]$ are the submatrices of the partitioned matrix $[A]$ and $[B_{11}], [B_{12}], [B_{21}]$ and $[B_{22}]$ are the submatrices of partitioned matrix $[B]$, the product of the partitioned matrices $[A]$ and $[B]$ is given by the equation,

$$[A][B] = [C] = \begin{vmatrix} [C_{11}] & : & [C_{12}] \\ \cdots & \cdots & \cdots \\ [C_{21}] & : & [C_{22}] \end{vmatrix} \tag{16.23b}$$

In Eq. (16.23b), $[C_{11}], [C_{12}, [C_{21}]$ and $[C_{22}]$ are the submatrices of product matrix C and are given by the equation,

$$\left.\begin{aligned} [C_{11}] &= [A_{11}][B_{11}] + [A_{12}][B_{21}] \\ [C_{12}] &= [A_{11}][B_{12}] + [A_{12}][B_{22}] \\ [C_{21}] &= [A_{21}][B_{11}] + [A_{22}][B_{21}] \\ \text{and} \quad [C_{22}] &= [A_{21}][B_{12}] + [A_{22}][B_{22}] \end{aligned}\right] \tag{16.23c}$$

The following points are evident from the multiplication of the two matrices :

(i) For multiplication of the matrices to be possible, the number of columns in the first matrix must be equal to the number of rows in the second matrix.

(ii) If a matrix of order $m \times n$ is post-multiplied by a matrix of order $n \times p$, the order of the product is $m \times p$.

(iii) Multiplication of matrices is not necessarily commutative, i.e.,$[A][B] \neq [B][A]$.

(iv) The associative and distributive laws are applicable to matrix multiplication provided the sequence of matrices is strictly maintained, i.e.,

$$\begin{aligned} &[A][B][C] = [A][BC] = [AB][C] \\ \text{and} \quad &[D]([E] + [F]) = [D][E] + [D][F]. \end{aligned} \tag{16.24}$$

(v) Determinant of a matrix obtained by multiplication of two matrices is the same as the multiplication of the determinants of the two matrices. Thus,

$$\text{Det}([A][B]) = (\text{Det}[A])(\text{Det}[B]). \tag{16.25}$$

(vi) Post-multiplying or Pre-multiplying $[A]$ by the identity matrix of order n does not change its value. Thus,

$$[A][I] = [A] \quad \text{and} \quad [I][A] = [A].$$

(vii) Post-multiplying $[A]$ by a diagonal matrix of order n amounts to multiplying the columns of $[A]$ by the corresponding diagonal elements of the diagonal matrix. Thus,

$$[A](\text{diagonal matrix } d_i) = [d_1 a_1 \cdots d_n a_n].$$

(viii) Transpose of the multiplication $[AB]$ of two matrices A and B, is the same as the multiplication of transpose of $[B]$ and $[A]$, i.e., $[B^T][A^T]$.

$$\left.\begin{aligned} \text{Thus} \qquad ([A][B])^T &= [B^T][A^T] \\ ([A][B][C])^T &= [C^T][B^T][A^T] \end{aligned}\right] \tag{16.26}$$

(ix) Multiplication of any two partitioned matrices is possible, only if

(a) The two matrices before partitioning are conformable.

(b) Treating each submatrix as an element in the partitioned matrices, the partitioned matrices are comformable.

(c) The submatrices which are to be multiplied with each other are also conformable.

(d) Transpose of a partitioned matrix is formed as indicated.

$$[A]^T = \begin{bmatrix} [A_{11}]^T & [A_{12}]^T \\ [A_{21}]^T & [A_{22}]^T \end{bmatrix} \tag{16.27}$$

(d) Matrix Inversion. This operation is applicable only to nonsingular square matrices. The inverse of an nth order nonsingular square matrix $[A]$ is denoted by $[A]^{-1}$, which is also a unique nth order non singular square matrix. The property associated with inversion of matrices is

$$[A][A]^{-1} = [A]^{-1}[A] = [I], \tag{16.28}$$

where $[I]$ is the nth order identity matrix.

Eq. (16.28) shows that a matrix and its inverse are commutative.

There is an analogy between the reciprocal of a number in scalar algebra and the inverse of matrix. From the relationship in matrix algebra given by

$$[A][B] = [I], \tag{16.29}$$

it follows that $[B]$ is the inverse of $[A]$, i.e.,

$$[B] = [A]^{-1}. \tag{16.29a}$$

It is also true that $[A]$ is the inverse of $[B]$.

i.e., $[A] = [B]^{-1}$ (16.29b)

Matrix inversion can be used in solving simultaneous equations. Considering a set of equations of the form,

$$[A][X] = [D] \tag{16.30}$$

in which $[A]$ is a square matrix of order $(n \times n)$ and $[X]$ is a column matrix of the unknown and $[D]$ is a column matrix of constant terms, the order for both $[X]$ and $[D$ being $(n \times 1)$, if both sides of the equation be pre-multiplied by $[A]^{-1}$, the result is

$$[A]^{-1}[A][X] = [A]^{-1}[D], \text{ from which } [I][X] = [A]^{-1}[D], i.e., [X] = [A]^{-1}[D]. \tag{16.31}$$

The last expression gives a convenient way of representing the solution of a set of simultaneous equations and it shows also that finding the inverse of a matrix is essentially the same as solving the equations.

Matrix inversion may be made by various methods. Four such methods considered suitable are

(i) Adjoint method

(ii) Method based on Submatrices

(iii) Factorisation Method (Choleski Method)

(iv) Linear Transformation Method (Gauss-Jordan Method).

(i) Adjoint Method : The method may be expressed as under :

$$[A]^{-1} = \frac{\text{Adj } [A]}{\text{Det } [A]} = \frac{1}{\text{Det } [A]}(\text{Cofactor}[A])^T, \tag{16.32}$$

where Adjoint of matrix $[A]$ denoted by the symbol Adj $[A]$ is obtained by transposing matrix $[A]$ and then replacing each element of the transposed matrix by its cofactor.

From Eq. (16.32), it is evident that the inverse of a matrix can be determined, only if it is a square matrix, as only a square matrix has its determinant and that the inverse of a matrix exists, only if its determinant is non-singluar.

The inverse of a matrix is unique and the inverse of the transpose of a matrix is equal to the transpose of the inverse of the matrix.

i.e.,
$$([A]^T)^{-1} = ([A]^{-1})^T. \tag{16.33}$$

The inverse of the product of two matrices is equal to the product of the inverse of two matrices in the reverse order, i.e.,

$$([A][B])^{-1} = ([B]^{-1})([A]^{-1}). \tag{16.34}$$

The inverse of a symmetrical matrix is also a symmetrical matrix, that of a triangular matrix is a similar triangular matrix.

For a triangular matrix $[A]$ the elements on the main diagonal of $[A]^{-1}$ are the reciprocals of the elements on the main diagonal of $[A]$.

Any matrix when multiplied by its transpose results in a symmetrical matrix.

(ii) Method Based on Submatrices : If matrix $[B]$ is the inverse of matrix $[A]$, then

$$[A][B] = [I].$$

Partitioning the matrices $[A]$ and $[B]$ into four submatrices the following equation is obtained.

$$\begin{bmatrix} [A_{11}] & \vdots & [A_{12}] \\ \cdots & \cdots & \cdots \\ [A_{21}] & \vdots & [A_{22}] \end{bmatrix} \begin{bmatrix} [B_{11}] & \vdots & [B_{12}] \\ \cdots & \cdots & \cdots \\ [B_{21}] & : & [B_{22}] \end{bmatrix} = \begin{bmatrix} [I] & 0 \\ 0 & [I] \end{bmatrix}. \tag{16.35a}$$

Comparing the corresponding terms on both sides of Eq. (16.35a),

$$\left.\begin{aligned} &[A_{11}][B_{11}] + [A_{12}][B_{21}] = [I] \\ &[A_{11}][B_{12}] + [A_{12}][B_{22}] = [0] \\ &[A_{21}][B_{11}] + [A_{22}][B_{21}] = [0] \\ \text{and}\quad &[A_{21}][B_{12}] + [A_{22}][B_{22}] = [I]. \end{aligned}\right] \tag{16.35b}$$

Solveing Eq. (16.35b),

$$\begin{aligned} [A]^{-1} &= \begin{bmatrix} [B_{11}] & [B_{12}] \\ [B_{21}] & [B_{22}] \end{bmatrix} \\ &= \begin{bmatrix} ([A_{11}]^{-1} + [A_{11}]^{-1}[A_{12}]\beta[A_{21}][A_{11}]^{-1}), & -([A_{11}]^{-1}[A_{12}]\beta) \\ (-\beta[A_{21}][A_{11}]^{-1}), & \beta \end{bmatrix} \end{aligned} \tag{16.36}$$

where $\beta = ([A_{22}] - [A_{21}][A_{11}]^{-1}[A_{12}]^{-1}$.

Alternately, it can be shown that

$$\left.\begin{aligned} &[B_{11}] = ([A_{11}] - [A_{12}][A_{22}]^{-1}[A_{21}])^{-1} \\ &[B_{12}] = -[B_{11}] - [A_{12}][A_{22}]^{-1} \\ &[B_{21}] = -[A_{22}]^{-1}[A_{21}][B_{11}] \\ \text{and}\quad &[B_{22}] = [A_{22}]^{-1} - [B_{21}][A_{12}][A_{22}]^{-1}. \end{aligned}\right] \tag{16.36a}$$

(iii) Factorisation Method (Choleski Method) : In this method, the matrix to be inversed is pre-multiplied by its transpose. Any matrix can be expressed as multiplication of two matrices, one being a lower diagonal matrix denoted by $[L]$, the other being an upper diagonal matrix denoted by $[U]$. Thus,

$$[A] = [L][U]. \tag{16.37a}$$

Post-multiplying by $[U]^{-1}$,

$$[A][U]^{-1} = [L][U][U]^{-1} = [L].$$

Premultiplication by $[A]^{-1}$ gives

$$[A]^{-1}[A][U]^{-1} = [A]^{-1}[L],$$

whence $[U]^{-1} = [A]^{-1}[L] = [R][L]$ (16.37b)

and $[R] = [A]^{-1}$. (16.37c)

For symmetric matrix $[U] = [L]^T$, whence

$$[A] = [L][L]^T, \tag{16.38a}$$

where
$$[L] = \begin{bmatrix} l_{11} & 0 & 0 \cdots 0 \\ l_{21} & l_{22} & 0 \cdots 0 \\ \cdots & \cdots & \cdots \\ l_{n1} & l_{n2} & l_{n3} \cdots l_{nn} \end{bmatrix} \tag{16.38b}$$

and
$$[U] = [L]^T = \begin{bmatrix} l_{11} & l_{21} & \cdots & l_{n1} \\ 0 & l_{22} & \cdots & l_{n2} \\ \cdots & \cdots & \cdots & \cdots \\ 0 & 0 & & l_{nn} \end{bmatrix} \tag{16.38c}$$

Inverting $[A]^{-1} = [L^T]^{-1}[L]^{-1}$.

Also $\quad [A]^{-1}[L] = [L^T]^{-1}[L]^{-1}[L] = [L^T]^{-1}$

$$= \begin{bmatrix} \frac{1}{l_{11}} & ?? & ?? & \cdots\cdots & ?? \\ 0 & \frac{1}{l_{22}} & ?? & \cdots\cdots & ?? \\ \cdots & \cdots & \cdots & \cdots & \cdots \\ 0 & 0 & 0 & & \frac{1}{l_{nn}} \end{bmatrix}. \tag{16.38d}$$

If $[A]^{-1} = [B]$, then

$$\begin{bmatrix} b_{11} & b_{12} \cdots b_{1n} \\ b_{21} & b_{22} \cdots b_{2n} \\ \cdots & \cdots \\ b_{n1} & b_{n2} \cdots b_{nn} \end{bmatrix} \begin{bmatrix} l_{11} & 0 & 0 \\ l_{21} & l_{22} & 0 \\ \cdots & \cdots & \cdots \\ l_{n1} & l_{n2} \cdots l_{nn} \end{bmatrix} = \begin{bmatrix} \frac{1}{l_{11}} & ?? & ?? \cdots ?? \\ 0 & \frac{1}{l_{22}} & ?? \cdots ?? \\ \cdots & \cdots & \cdots \\ 0 & 0 & 0 \cdots \frac{1}{l_{nn}} \end{bmatrix}. \tag{16.38e}$$

Elements of matrix $[B]$ may be obtained by comparing the corresponding elements on both sides of Eq. (16.38c).

For unsymmetric matrix for the given matrix, lower matrix is taken as

$$[L] = \begin{bmatrix} 1 & 0 & 0 \cdots 0 \\ l_{21} & 1 & 0 \cdots 0 \\ \cdots & \cdots & \cdots \\ l_{n1} & l_{n2} & l_{n3} \cdots l_{nn} \end{bmatrix}$$

and upper matrix taken as

$$[U] = \begin{bmatrix} U_{11} & U_{12} \cdots U_{1n} \\ 0 & U_{22} \cdots U_{2n} \\ \cdots & \cdots \\ 0 & \cdots U_{nn} \end{bmatrix}.$$

From $[L][U] = [A]$, with definite values of lower matrix, the upper matrix $[U]$ can be obtained.

From $[U][U]^{-1} = 1, [U]^{-1}$ can be calculated.

Inverse of matrix $[A]$, i.e., $[A]^{-1}$ is obtained from, $[A]^{-1}[L] = [U]^{-1}$.

(iv) Linear Transformation Method : This is one of the fastest methods of inversion of a matrix by a computer. This method involves the construction of transformation matrices which transfer by means of successive operations the given matrix into identity matrix. Once the transformation matrices have been properly obtained, the inverse of the matrix will be equal to their product. For example, if in case of a matrix $[A], [T_1], [T_2] \dots [T_n]$ are the transformation matrices such that when $[A]$ is premultiplied by them in the order shown below, the result is the identity matrix, then it follows that,

$$[T_n] \cdots [T_2][T_1][A] = [I]. \tag{16.39}$$

From Eq. (16.39), it follows that the inverse of $[A]$ is the product

$$[A]^{-1} = [T_n] \cdots [T_2][T_1]. \tag{16.40}$$

If the transformation matrices are pre-multipliers as in Eq. (16.39), the successive operations are made on the rows of $[A]$. However, it is equally feasible to perform the operations on the columns of $[A]$ by post-multiplication with apropriate transformation matrices.

The essential features of the method of successive transformations is the determination of the transformation matrices themselves. Considering a matrix denoted by

$$[A] = \begin{bmatrix} A_{11} & A_{12} \cdots A_{1n} \\ A_{21} & A_{22} \cdots A_{2n} \\ \cdots & \cdots \\ A_{n1} & A_{n2} \cdots A_{nn} \end{bmatrix}, \tag{16.41a}$$

the first transformation matrix $[T_1]$ is constructed, so that it will transform the first column of $[A]$ into the first column of the identity matrix. The result can be accomplished by dividing the first row of $[A]$ by A_{11} and then adding appropriate multiples of the first row to the other rows. Thus, $[T_1]$ can be taken as

$$[T_1] = \begin{bmatrix} \frac{1}{A_{11}} & 0 \cdots 0 \\ -\frac{A_{21}}{A_{11}} & 1 \cdots 0 \\ \cdots & \cdots \\ -\frac{A_{n1}}{A_{11}} & 0 \cdots 1 \end{bmatrix}. \tag{16.41b}$$

Then $$[T_1][A] = \begin{bmatrix} 1 & \frac{A_{12}}{A_{11}} \cdots\cdots\cdots \frac{A_{1n}}{A_{11}} \\ 0 & \left(A_{22} - \frac{A_{21}}{A_{11}} A_{12}\right) \cdots \left(A_{2n} - \frac{A_{21}}{A_{11}} A_{1n}\right) \\ \cdots & \cdots \\ 0 & \left(A_{n2} - \frac{A_{n1}}{A_{11}} A_{12}\right) \cdots \left(A_{nn} - \frac{A_{n1}}{A_{11}} A_{1n}\right) \end{bmatrix}. \tag{16.41c}$$

Using new symbols for the transformation matrix,

$$[T_1][A] = \begin{bmatrix} 1 & B_{12} \cdots B_{1n} \\ 0 & B_{22} \cdots B_{2n} \\ \cdots & \cdots \\ 0 & B_{n2} \cdots B_{nn} \end{bmatrix} = [B]. \tag{16.41d}$$

A Transformation matrix $[T_2]$, which changes the second columns of the matrix $[B]$ into the second column of the identity matrix is then formed, using element B_{22} as the first.

$$\text{Thus,} \quad [T_2] = \begin{bmatrix} 1 & -\frac{B_{12}}{B_{22}} \cdots 0 \\ 0 & \frac{1}{B_{22}} \cdots 0 \\ \cdots & \cdots\cdots \\ 0 & -\frac{B_{n2}}{B_{22}} \cdots 1 \end{bmatrix} \tag{16.41e}$$

Pre-multiplication of $[B]$ by $[T_2]$ gives a new matrix $[C]$, in which the first two columns correspond to the identity matrix.

$$\text{Thus } [T_2][B] = [T_2][T_1][A] = \begin{bmatrix} 1 & 0 \cdots C_{1n} \\ 0 & 1 \cdots C_{2n} \\ \cdots & \cdots\cdots \\ 0 & 0 \cdots C_{nn} \end{bmatrix} = [C]. \tag{16.41f}$$

The process of transforming columns is applied a total of n times in the manner illustrated till the product becomes the identity matrix,

$$[T_n] \cdots [T_2][T_1[A] = [I]. \tag{16.41g}$$

Thus, the inverse of $[A]$ is the product of all of the transformation matrices.

16.5 Simultaneous Equations

Simultaneous algebraic equations are encountered frequently in engineering problems. Solutions of such equations can be obtained by (a) Method of Elimination. (b) Cramer's Rule and (c) Inversion.

(a) Method of Elimination. One of the widely used methods for solving simultaneous equations is the method of elimination (also called Gaussian elimination). The method can be illustrated readily by solving Eq. (16.42a) as shown below :

$$\left.\begin{aligned} 2X_1 + 4X_2 + X_3 &= -22 \\ -X_1 + 3X_2 - 2X_3 &= -32 \\ 2X_1 - 3X_2 + 5X_3 &= 42 \end{aligned}\right]. \tag{16.42a}$$

Solving the first equation for X_1 in terms of X_2 and X_3 gives

$$X_1 = -2X_2 - 0.5X_3 - 11. \tag{16.42b}$$

When the expression is substituted into the last two of Eq. (16.42a), they become

$$\left.\begin{aligned} & 5X_2 - 15X_2 = -43 \\ \text{and} \quad & -7X_2 + 4X_3 = 64 \end{aligned}\right]. \tag{16.42c}$$

Next the first Eq. (16.42c) is solved for X_2 in terms of X_3,

$$X_2 = 0.3X_3 - 8.6. \tag{16.42d}$$

This expression for X_2 is substituted into the second of Eq. (16.42c) which in turn can be solved for X_3, giving $X_3 = 2$.

Substitution of this value for X_3 into the equation for X_2 in terms of X_3, Eq. (16.42d) makes it possible for obtaining X_2, whence

$$X_2 = -8.$$

Lastly, the values for X_2 and X_3 are substituted into the equation for X_1 (Eq. 16.42b) yielding the result, $X_2 = 4$.

The same general scheme of solution can be used when there are any number of simultaneous equations to be solved.

There are various modifications which have been made to the method of elimination. One of these is the Gauss-Jordan Method, according to which for solving Eq. (16.42a), the first equation is taken as

$$X_1 + 2X_2 + 0.5X_3 = -11. \tag{16.42e}$$

Eq. (16.42a) then is added to the second equation, giving the result,

$$5X_2 - 1.5X_3 = -43. \tag{16.42f}$$

Also two times of Eq. (16.42e) is subtracted from the third equation,

$$-7X_2 + 4X_3 = 64. \tag{16.42g}$$

Eqs. (16.42f) and (16.42g) are treated in a similar manner. The equations are then interchanged so that Eq. (16.42g) becomes the leading equation,

$$\left.\begin{aligned} -7X_2 + 4X_3 &= 64 \\ \text{and} \quad 5XD_2 - 1.5X_3 &= -43. \end{aligned}\right] \tag{16.42h}$$

The first of these equations is divided by -7 and multiplying by 5, giving

$$5X_2 - \frac{20}{7} X_3 = -\frac{320}{7}.$$

Deducting this equation from the second of the Eq. (16.42h),

$$\frac{19}{14}X_3 = \frac{38}{14}, \quad \text{whence } X_3 = 2.$$

By similar process, $X_2 = -8$ and $X_3 = 4$.

The method of elimination for solving simultaneous equations is similar to the method of successive transformations for determining the inverse of a matrix and the method of pivotal consideration for evaluating a determinant. All these methods are related to one another and may be considered to be variations of the method of elimination.

(b) Cramer's Rule. A method for solving simultaneous equation by the use of determinants known as Cramer's rule can also be used. Thus, considering the series of simultaneous equations,

$$\left.\begin{aligned} A_{11}x_1 + a_{12}x_2 + \cdots + a_{1n}x_n &= b_1 \\ a_{21}x_1 + a_{22}x_2 + \cdots + a_{2n}x_n &= b_2 \\ \cdots\cdots\cdots\cdots\cdots\cdots\cdots\cdots \\ a_{n1}x_1 + a_{n2}x_2 + \cdots + a_{nn}x_n &= b_n \end{aligned}\right] \tag{16.43}$$

the equation is given by the expression,

$$x = \frac{\text{Adj}[A]}{\text{Det}[A]}[B], \tag{16.44}$$

provided Det $[A] \neq 0$.

Value of unknown according to Cramer's rule is given by

$$x_1 = \frac{\begin{vmatrix} b_1 & a_{12} & \cdots & a_{1n} \\ b_2 & a_{22} & \cdots\cdots & a_{1n} \\ \cdots & \cdots & \cdots & \cdots \\ b_n & a_{n2} & \cdots\cdots & a_{nn} \end{vmatrix}}{\begin{vmatrix} a_{11} & a_{12} & \cdots\cdots & a_{1n} \\ a_{21} & a_{22} & \cdots\cdots & a_{2n} \\ \cdots & \cdots & \cdots & \cdots \\ a_{n1} & a_{n2} & \cdots\cdots & a_{nn} \end{vmatrix}}$$

$$x_2 = \frac{\begin{vmatrix} a_{11} & b_1 & \cdots\cdots & a_{1n} \\ a_{21} & b_2 & \cdots\cdots & a_{2n} \\ \cdots & \cdots & \cdots & \cdots \\ a_{n1} & b_n & \cdots\cdots & a_{nn} \end{vmatrix}}{\text{Det}[A]}$$

and $$x_n = \frac{\begin{vmatrix} a_{11} & a_{12} & \cdots\cdots & b_1 \\ a_{21} & a_{22} & \cdots\cdots & b_2 \\ \cdots & \cdots & \cdots & \cdots \\ a_{n1} & a_{n2} & \cdots\cdots & b_n \end{vmatrix}}{\text{Det}[A]}$$

Considering Eq. (16.42a), if Cramer's rule is applied to the solution of the three equations expression for x_1 becomes

$$x_1 = \frac{\begin{vmatrix} -22 & 4 & 1 \\ -32 & 3 & -2 \\ 42 & -3 & 5 \end{vmatrix}}{\begin{vmatrix} 2 & 4 & 1 \\ -1 & 3 & -2 \\ 2 & -3 & 5 \end{vmatrix}} = \frac{76}{19} = 4$$

$$x_2 = \frac{\begin{vmatrix} 2 & -22 & 1 \\ -1 & -32 & -2 \\ 2 & 42 & 5 \end{vmatrix}}{19} = -\frac{152}{19} = -8$$

and $$x_3 = \frac{\begin{vmatrix} 2 & 4 & -22 \\ -1 & 3 & -32 \\ 2 & -3 & 42 \end{vmatrix}}{19} = \frac{38}{19} = 2.$$

Cramer's rule is useful in hand calculations only if the determinants can be evaluated easily.

(c) Method of Inversion. A direct means of simultaneous equations can be obtained by using matrix inversion as illustrated in Eq. (16.31). Considering the series of simultaneous equations in Eq. (16.42a), matrix $[A]$ and $[D]$ are given by

$$[A] = \begin{bmatrix} 2 & 4 & 1 \\ -1 & 3 & -2 \\ 2 & -3 & 5 \end{bmatrix} \quad \text{and} \quad [D] = \begin{bmatrix} -22 \\ -32 \\ -42 \end{bmatrix}.$$

Inverse of A is given by

$$[A]^{-1} = \frac{1}{19}\begin{bmatrix} 9 & -23 & -11 \\ 1 & 8 & 3 \\ -3 & 14 & 10 \end{bmatrix}$$

and $$x = \begin{bmatrix} x_1 \\ x_2 \\ x_3 \end{bmatrix} = \frac{1}{19}\begin{bmatrix} 9 & -23 & -11 \\ 1 & 8 & 3 \\ -3 & 14 & 10 \end{bmatrix}\begin{bmatrix} -22 \\ -32 \\ 42 \end{bmatrix} = \begin{bmatrix} 4 \\ -8 \\ 2 \end{bmatrix},$$

whence $x_1 = 4,\ x_2 = -8$ and $x_3 = 2$.

(d) Method of Partitioning. A stiffness matrix relates displacements Δ at a number of coordinates to the forces applied at the same coordinates by the equation,

$$[S][\Delta] = [F]. \tag{16.45}$$

This can be written in partitioned form as

$$\begin{bmatrix} [S_{11}] & : & [S_{12}] \\ \cdots & & \cdots \\ [S_{21}] & : & [S_{22}] \end{bmatrix}\begin{bmatrix} [\Delta_1] \\ \cdots \\ [\Delta_2] \end{bmatrix} = \begin{bmatrix} [F_1] \\ \cdots \\ [F_2] \end{bmatrix}. \tag{16.46}$$

If the value of the matrix $[F]$ be zero at some of the co-ordinates, i.e., the displacements at those co-ordinates can take place freely, the stiffness matrix corresponding to the remaining co-ordinates can be derived from the partitioned matrix. In this case, considering that the equations below the horizontal dotted line relate forces $[F_2]$, assumed to be zero, to the displacements $[\Delta_2] \neq 0$, at the corresponding co-ordinates it gives

$$\left.\begin{aligned} &[S_{11}][\Delta_1] + [S_{12}][\Delta_2] = [F_1] \\ \text{and} \quad &[S_{21}][\Delta_1] + [S_{22}][\Delta_2] = [0] \end{aligned}\right] \tag{16.47a}$$

Solving $\left[[S_{11}] - [S_{12}][S_{22}]^{-1}[S_{21}]\right][\Delta_1] = [F_1],$ (16.47b)

which can be written in the form,

$$[S^*][\Delta_1] = [F_1],$$

where $[S^*]$ is a condensed stiffness matrix relating forces $[F_1]$ to displacements $[\Delta_1]$ and is given by

$$[S^*] = [S_{11}] - [S_{12}][S_{22}]^{-1}[S_21]. \tag{16.47c}$$

As an example, consider the stiffness matrix of the beam given by

$$[S] = \frac{4EI}{L}\left[\begin{array}{cc:cc} 4 & 0 & 2 & \frac{12}{L} \\ 0 & 2 & 1 & -\frac{6}{L} \\ \hdashline 2 & 1 & 6 & \frac{6}{L} \\ \frac{12}{L} & -\frac{6}{L} & \frac{6}{L} & \frac{72}{L^2} \end{array}\right].$$

Forces P_1, P_2, P_3 and P_4 and the corresponding displacements, $\Delta_1, \Delta_2, \Delta_3$ and Δ_4 are related as follows :

$$\begin{bmatrix} [S_{11}] & [S_{12}] \\ [S_{21}] & [S_{22}] \end{bmatrix} \begin{bmatrix} [\Delta_1] \\ [\Delta_2] \end{bmatrix} = \begin{bmatrix} [F_1] \\ [F_2] \end{bmatrix},$$

where $[S_{11}], [S_{12}]$, etc. refer to the partitioned matrix in $[S]$ and

$$[\Delta_1] = \begin{bmatrix} \Delta_1 \\ \Delta_2 \end{bmatrix}, \ [\Delta_2] = \begin{bmatrix} \Delta_3 \\ \Delta_4 \end{bmatrix}, \ [F_1] = \begin{bmatrix} P_1 \\ P_2 \end{bmatrix}$$

and $[F_2] = \begin{bmatrix} P_3 \\ P_4 \end{bmatrix}$.

Taking $[F_2] = 0$, the matrix $[S^*]$ relating the forces to the displacements along the co-ordinates indicated, such that,

$$[S*][\Delta_1] = [F_1] \text{ can be obtained.}$$

Example 16.2. *Compute the product AB for the matrices $[A]$ and $[B]$ given below :*

$$[A] = \begin{bmatrix} 8 & -2 & 0 \\ 4 & 10 & -6 \end{bmatrix} \quad \textit{and} \quad [B] = \begin{bmatrix} -6 \\ 6 \\ 2 \end{bmatrix}.$$

Solution : $AB = \begin{bmatrix} 8(-6) - 2(6) + 0(2) \\ 4(-6) + 10(6) - 6(2) \end{bmatrix} = \begin{bmatrix} -60 \\ 24 \end{bmatrix}.$

In this example, BA cannot be obtained since the number of columns of matrix $[B]$ is not equal to the number of rows of matrix $[A]$.

Example 16.3. *Calculate the products AB and BA for the following matrices.*

$$[A] = \begin{bmatrix} 2 & 0 & 4 \\ -2 & 2 & 6 \end{bmatrix} \quad \textit{and} \quad [B] = \begin{bmatrix} 4 & 2 \\ -4 & 0 \\ 8 & 10 \end{bmatrix}.$$

Solution : The results are

$$AB = \begin{bmatrix} 40 & 44 \\ 32 & 56 \end{bmatrix} \text{ and } BA = \begin{bmatrix} 4 & 4 & 28 \\ -8 & -0 & -16 \\ -4 & 20 & 92 \end{bmatrix}.$$

The matrices A and B are conformable for multiplication for both products AB and BA, but $AB \neq BA$.

Example 16.4. *Calculate the product AB and AC.*

$$[A] = \begin{bmatrix} 2 & 2 & 2 \\ 6 & 6 & 6 \end{bmatrix}, \quad [B] = \begin{bmatrix} -4 & 6 \\ 2 & -10 \\ 8 & 2 \end{bmatrix} \quad \textit{and} \quad [C] = \begin{bmatrix} 8 & 2 \\ -8 & -6 \\ 6 & 2 \end{bmatrix}.$$

Solution : Products are

$$AB = \begin{bmatrix} 12 & -4 \\ 36 & -12 \end{bmatrix} \text{ and } AC = \begin{bmatrix} 12 & -4 \\ 36 & -12 \end{bmatrix}.$$

The example shows that the relationship $AB = AC$ does not imply that $[B] = [C]$.

Example 16.5. *Calculate the products of the matrices [A] and [B] by partitioning as indicated.*

$$[A] = \begin{bmatrix} 4 & 10 & 6 & : & 2 & 12 \\ 2 & 4 & 8 & : & 6 & 4 \\ 6 & 0 & 4 & : & 14 & 1 \\ \cdots & \cdots & \cdots & \cdots & \cdots & \cdots \\ 2 & 6 & 0 & : & 4 & 14 \\ 4 & 10 & 2 & : & 8 & 6 \end{bmatrix} \quad \textit{and} \quad [B] = \begin{bmatrix} 4 \\ 6 \\ 2 \\ \cdots \\ 8 \\ 4 \end{bmatrix}$$

Solution : If $AB = \begin{bmatrix} [A_{11}] & : & [A_{12}] \\ \cdots & \cdots & \cdots \\ [A_{21}] & : & [A_{22}] \end{bmatrix} \begin{bmatrix} [B_{11}] \\ \cdots \\ [B_{21}] \end{bmatrix} = \begin{bmatrix} [C_{11}] \\ \cdots \\ [C_{21}] \end{bmatrix}$,

$$[C_{11}] = [A_{11}][B_{11}] + [A_{12}][B_{21}]$$

$$= \begin{bmatrix} 4 & 10 & 6 \\ 2 & 4 & 8 \\ 6 & 0 & 4 \end{bmatrix} \begin{bmatrix} 4 \\ 6 \\ 2 \end{bmatrix} + \begin{bmatrix} 2 & 12 \\ 6 & 4 \\ 14 & 1 \end{bmatrix} \begin{bmatrix} 8 \\ 4 \end{bmatrix}$$

$$= \begin{bmatrix} 152 \\ 112 \\ 152 \end{bmatrix}.$$

Also $[C_{21}] = [A_{21}][B_{11}] + [A_{22}][B_{21}]$

$$= \begin{bmatrix} 2 & 6 & 0 \\ 4 & 10 & 2 \end{bmatrix} \begin{bmatrix} 4 \\ 6 \\ 2 \end{bmatrix} + \begin{bmatrix} 4 & 14 \\ 8 & 6 \end{bmatrix} \begin{bmatrix} 8 \\ 4 \end{bmatrix} = \begin{bmatrix} 132 \\ 168 \end{bmatrix}.$$

Hence, product $AB = \begin{bmatrix} 152 \\ 112 \\ 152 \\ 132 \\ 168 \end{bmatrix}$.

Example 16.6. *Obtain the inverse of matrix $[A]$ by various methods.*

$$[A] = \begin{bmatrix} 60 & -20 & 0 \\ -20 & 30 & -10 \\ 0 & -10 & 10 \end{bmatrix}.$$

Solution : (a) Adjoint Method :

$$\text{Cofactor } C = \begin{bmatrix} 200 & 200 & 200 \\ 200 & 600 & 600 \\ 200 & 600 & 1400 \end{bmatrix},$$

$$\text{whence} \quad \text{Adj } A = C^T = \begin{bmatrix} 200 & 200 & 200 \\ 200 & 600 & 600 \\ 200 & 600 & 1400 \end{bmatrix}.$$

In this case, $C = C^T$, since the matrix is symmetrical.

Also $\quad \text{Det}[A] = 60(200) + (-20)(200) + 0(200) = 8000.$

The inverse of the matrix is

$$= \frac{1}{8000}\begin{bmatrix} 200 & 200 & 200 \\ 200 & 600 & 600 \\ 200 & 600 & 1400 \end{bmatrix} = \begin{bmatrix} \frac{1}{40} & \frac{1}{40} & \frac{1}{40} \\ \frac{1}{40} & \frac{3}{40} & \frac{3}{40} \\ \frac{1}{40} & \frac{3}{40} & \frac{7}{40} \end{bmatrix}$$

(b) Inversion by Partitioning :

$$[A] = \begin{bmatrix} A_{11} & \vdots & A_{12} \\ \cdots & \cdots & \cdots \\ A_{21} & \vdots & A_{22} \end{bmatrix} = \begin{bmatrix} 60 & -20 & \vdots & 0 \\ -20 & 30 & \vdots & -10 \\ \cdots & \cdots & \cdots & \cdots \\ 0 & -10 & \vdots & 10 \end{bmatrix}.$$

Using Eq. (16.36), with $[A]^{-1} = [B]$,

$$[B_{22}] = \beta = ([A_{22}] - [A_{21}][A_{11}]^{-1}[A_{12}])^{-1}$$

$$= \left\{ 10 - [0 - 10]\begin{bmatrix} 60 & -20 \\ -10 & 30 \end{bmatrix}^{-1}\begin{bmatrix} 0 \\ -10 \end{bmatrix} \right\}^{-1} = \frac{7}{40}.$$

$$\text{Also} \quad B_{12} = -\beta[A_{11}]^{-1}[A_{12}] = -\frac{7}{40}\begin{bmatrix} 60 & -20 \\ -20 & 30 \end{bmatrix}^{-1}\begin{bmatrix} 0 \\ -10 \end{bmatrix}$$

$$= \begin{bmatrix} \frac{1}{40} \\ \frac{3}{40} \end{bmatrix}$$

$$[B_{21}] = -\beta[A_{21}][A_{11}]^{-1} = -\frac{7}{40}[0 - 10]\begin{bmatrix} 60 & -20 \\ -20 & 30 \end{bmatrix}^{-1}$$

$$= \begin{bmatrix} \frac{1}{40} & \frac{3}{40} \end{bmatrix}$$

and $[B_{11}] = [A_{11}]^{-1} + [A_{11}]^{-1}[A_{12}]\beta[A_{21}][A_{11}]^{-1} = [A_{11}]^{-1} + [A_{11}]^{-1}[B_{21}]$

$$= \frac{1}{140}\begin{bmatrix} 3 & 2 \\ 2 & 6 \end{bmatrix} - \frac{1}{140}\begin{bmatrix} 3 & 2 \\ 2 & 6 \end{bmatrix}\begin{bmatrix} 0 \\ -10 \end{bmatrix}\begin{bmatrix} \frac{1}{40} & \frac{3}{40} \end{bmatrix} = \begin{bmatrix} \frac{1}{40} & \frac{1}{40} \\ \frac{1}{40} & \frac{3}{40} \end{bmatrix}.$$

$$[A]^{-1} = \begin{bmatrix} B_{11} & : & B_{12} \\ \cdots & \cdots & \cdots \\ B_{21} & : & B_{22} \end{bmatrix} = \begin{bmatrix} \frac{1}{40} & \frac{1}{40} & \frac{1}{40} \\ \frac{1}{40} & \frac{3}{40} & \frac{3}{40} \\ \frac{1}{40} & \frac{3}{40} & \frac{7}{40} \end{bmatrix}.$$

(c) Factorization Method :

$$[A] = \begin{bmatrix} 60 & -20 & 0 \\ -20 & 30 & -10 \\ 0 & -10 & 10 \end{bmatrix} = \begin{bmatrix} l_{11} & 0 & 0 \\ l_{21} & l_{22} & 0 \\ l_{31} & l_{32} & l_{33} \end{bmatrix}\begin{bmatrix} l_{11} & l_{21} & l_{31} \\ 0 & l_{22} & l_{32} \\ 0 & 0 & l_{33} \end{bmatrix}.$$

Multiplying ,

$$l_{11}^2 = 60;\ l_{11}l_{21} = -20;\ l_{11}l_{31} = 0$$

$$l_{31}^2 + l_{22}^2 = 30;\ l_{21}l_{31} + l_{22}l_{32} = -10$$

and $$l_{31}^2 + l_{32}^2 + l_{33}^2 = 10.$$

Solving, $l_{11} = \sqrt{60}; l_{21} = -\sqrt{\frac{20}{3}};\ l_{31} = 0;$

$$l_{22} = \sqrt{\frac{70}{3}};\ l_{32} = -\frac{30}{7} \text{ and } l_{33} = \frac{40}{7}.$$

From $$\begin{bmatrix} 60 & -20 & 0 \\ -20 & 30 & -10 \\ 0 & -10 & 10 \end{bmatrix}^{-1}\begin{bmatrix} l_{11} & 0 & 0 \\ l_{21} & l_{22} & 0 \\ l_{31} & l_{32} & l_{33} \end{bmatrix} = \begin{bmatrix} l_{11} & l_{21} & l_{31} \\ 0 & l_{22} & l_{32} \\ 0 & 0 & l_{33} \end{bmatrix}^{-1}$$

$$\begin{bmatrix} C_{11} & C_{12} & C_{13} \\ C_{21} & C_{22} & C_{23} \\ C_{31} & C_{32} & C_{33} \end{bmatrix}\begin{bmatrix} \sqrt{60} & 0 & 0 \\ -\frac{\sqrt{20}}{3} & \sqrt{\frac{70}{3}} & 0 \\ 0 & -\sqrt{\frac{30}{7}} & \sqrt{\frac{40}{7}} \end{bmatrix} = \begin{bmatrix} \frac{1}{\sqrt{60}} & ? & ? \\ 0 & \sqrt{\frac{3}{70}} & ? \\ 0 & 0 & \sqrt{\frac{7}{40}} \end{bmatrix}$$

whence $C_{33}\sqrt{\frac{40}{7}} = \sqrt{\frac{7}{40}};\ C_{33} = \frac{7}{40};\ C_{32}\sqrt{\frac{70}{3}} - C_{33}\sqrt{\frac{30}{7}} = 0;$

$$C_{31}\sqrt{60} - C_{32}\sqrt{\frac{20}{3}} + C_{33}(0) = 0;\ C_{22}\sqrt{\frac{70}{3}} - C_{32}\sqrt{\frac{30}{7}} = \sqrt{\frac{3}{70}}$$

$$C_{21}\sqrt{60} - C_{22}\sqrt{\frac{20}{3}} = 0 \text{ and } C_{11}\sqrt{60} - C_{21}\sqrt{\frac{20}{3}} = \frac{1}{\sqrt{60}}.$$

Solving $$[A]^{-1} = \begin{bmatrix} \frac{1}{40} & \frac{1}{40} & \frac{1}{40} \\ \frac{1}{40} & \frac{3}{40} & \frac{3}{40} \\ \frac{1}{40} & \frac{3}{40} & \frac{7}{40} \end{bmatrix}$$

(d) Linear Transformation Method :

$$[A] = \begin{bmatrix} 60 & -20 & 0 \\ -20 & 30 & -10 \\ 0 & -10 & 10 \end{bmatrix}$$

With $[T_1] = \begin{bmatrix} \frac{1}{60} & 0 & 0 \\ \frac{1}{3} & 1 & 0 \\ 0 & 0 & 1 \end{bmatrix}$

$$[T_1][A] = \begin{bmatrix} 1 & -\frac{1}{3} & 0 \\ 0 & \frac{70}{3} & -10 \\ 0 & -10 & 10 \end{bmatrix} = [B]$$

With $[T_2] \quad = \begin{bmatrix} 1 & \frac{1}{70} & 0 \\ 0 & \frac{3}{70} & 0 \\ 0 & \frac{3}{7} & 1 \end{bmatrix}$

$$[T_2][B] = \begin{bmatrix} 1 & 0 & -\frac{1}{7} \\ 0 & 1 & -\frac{3}{7} \\ 0 & 0 & \frac{40}{7} \end{bmatrix} = [C]$$

With $[T_3] = \begin{bmatrix} 1 & 0 & \frac{1}{40} \\ 0 & 1 & \frac{3}{40} \\ 0 & 0 & \frac{7}{40} \end{bmatrix}$

$$[A]^{-1} = [T_3][T_2][T_1] = \begin{bmatrix} 1 & 0 & \frac{1}{40} \\ 0 & 1 & \frac{3}{40} \\ 0 & 0 & \frac{7}{40} \end{bmatrix} \begin{bmatrix} 1 & \frac{1}{70} & 0 \\ 0 & \frac{3}{70} & 0 \\ 0 & \frac{3}{7} & 0 \end{bmatrix} \begin{bmatrix} \frac{1}{60} & 0 & 0 \\ \frac{1}{3} & 1 & 0 \\ 0 & 0 & 1 \end{bmatrix}$$

$$= \begin{bmatrix} \frac{1}{40} & \frac{1}{40} & \frac{1}{40} \\ \frac{1}{40} & \frac{3}{40} & \frac{3}{40} \\ \frac{1}{40} & \frac{3}{40} & \frac{7}{40} \end{bmatrix}.$$

Example 16.7. *Obtain the inverse of the unsymmetrical matrix by Factorisation Method.*

$$[A] = \begin{bmatrix} 4 & 2 & 1 \\ 3 & 7 & 2 \\ 2 & 5 & 5 \end{bmatrix}.$$

Solution : Matrix $[A]$ can be considered as

$$[L][U] = [A],$$

where forms of $[L]$ and $[U]$ are

$$\begin{bmatrix} 1 & 0 & 0 \\ L_{21} & 1 & 0 \\ L_{31} & L_{32} & 1 \end{bmatrix} \begin{bmatrix} U_{11} & U_{12} & U_{13} \\ 0 & U_{22} & U_{23} \\ 0 & 0 & U_{33} \end{bmatrix}.$$

With $[L] = \begin{bmatrix} 1 & 0 & 0 \\ 0.75 & 1 & 0 \\ 0.50 & 0.7273 & 1 \end{bmatrix}$, $[U] = \begin{bmatrix} 4 & 2 & 1 \\ 0 & 5.50 & 1.25 \\ 0 & 0 & 3.5908 \end{bmatrix}$

From $[U][U]^{-1} = I$

$$[U]^{-1} = \begin{bmatrix} 0.25 & -0.0909 & -0.0380 \\ 0 & 0.1818 & -0.0633 \\ 0 & 0 & 0.2785 \end{bmatrix}.$$

From $[A]^{-1}[L] = [U]^{-1}$, it follows that,

$$[A]^{-1} = \begin{bmatrix} 0.3165 & -0.0633 & -0.0380 \\ -0.1392 & 0.2278 & -0.0633 \\ 0.0126 & -0.2025 & 0.2785 \end{bmatrix}.$$

Example 16.8. *Solve the matrix equations by partitioning the matrix as shown.*

$$\left|\begin{array}{c} x_1 \quad -18 \\ x_2 \quad -6 \\ \cdots\cdots \\ -15 \\ -15 \end{array}\right| = \left|\begin{array}{cc:cc} 0.024 & -0.012 & 0 & -0.06 \\ -0.012 & 0.012 & 0.06 & 0.06 \\ \hdashline 0 & 0.06 & 0.80 & 0.20 \\ -0.06 & 0.06 & 0.20 & 0.40 \end{array}\right| \left|\begin{array}{c} -10x_1 \\ -25x_2 \\ \cdots \\ x_3 \\ x_4 \end{array}\right|$$

Solution : On the basis of partitioning, the given matrix equation may be split up into the following two equations :

$$\begin{bmatrix} 0.024 & -0.012 \\ -0.012 & 0.012 \end{bmatrix}\begin{bmatrix} -10x_1 \\ -25x_2 \end{bmatrix} + \begin{bmatrix} 0 & -0.06 \\ 0.06 & 0.06 \end{bmatrix}\begin{bmatrix} x_3 \\ x_4 \end{bmatrix} = \begin{bmatrix} x_1 & -18 \\ x_2 & -6 \end{bmatrix} \tag{a}$$

and
$$\begin{bmatrix} 0 & 0.06 \\ -0.06 & 0.06 \end{bmatrix}\begin{bmatrix} -10x_1 \\ -25x_2 \end{bmatrix} + \begin{bmatrix} 0.80 & 0.20 \\ 0.20 & 0.80 \end{bmatrix}\begin{bmatrix} x_3 \\ x_4 \end{bmatrix} = \begin{bmatrix} -15 \\ -15 \end{bmatrix}. \tag{b}$$

From Eq. (a), $\begin{bmatrix} x_3 \\ x_4 \end{bmatrix} = \begin{bmatrix} 0 & -0.06 \\ 0.06 & 0.06 \end{bmatrix}^{-1} \begin{bmatrix} 0.76x_1 & -0.30x_2 & +18 \\ 0.80x_1 & +0.30x_2 & -6 \end{bmatrix}.$

Solving, for x_3 an x_4 in terms of x_1, x_2 and substituting in Eqn. (b), it is solved for x_1 and x_2.

Solving, $x_1 = 17.9$, $x_2 = 4.03, x_3 = 1.26$

and $x_4 = -49.91$.

16.6 Type of Framed Structure

Framed structure can be freely divided into six categeries, viz., beams, plane trusses, space trusses, plane frames, grids and space frames. Each of these represents a class of structures having special characteristic. These types are illustrated in Fig. 16.1.

A beam [Fig. 16.1(a)] consists of a straight member having one or more points of supports such as points A, B and C. Forces applied to a beam are assumed to act in a plane which contains an axis of symmetry of the cross section of the beam. All external couples acting on the beam have their moment vectors normal to this plane and the beam deflects in the same plane and does not twist. Internal stress resultants may exist at any cross section of the beam and in the general case may include an axial force, a shear force and a bending couple.

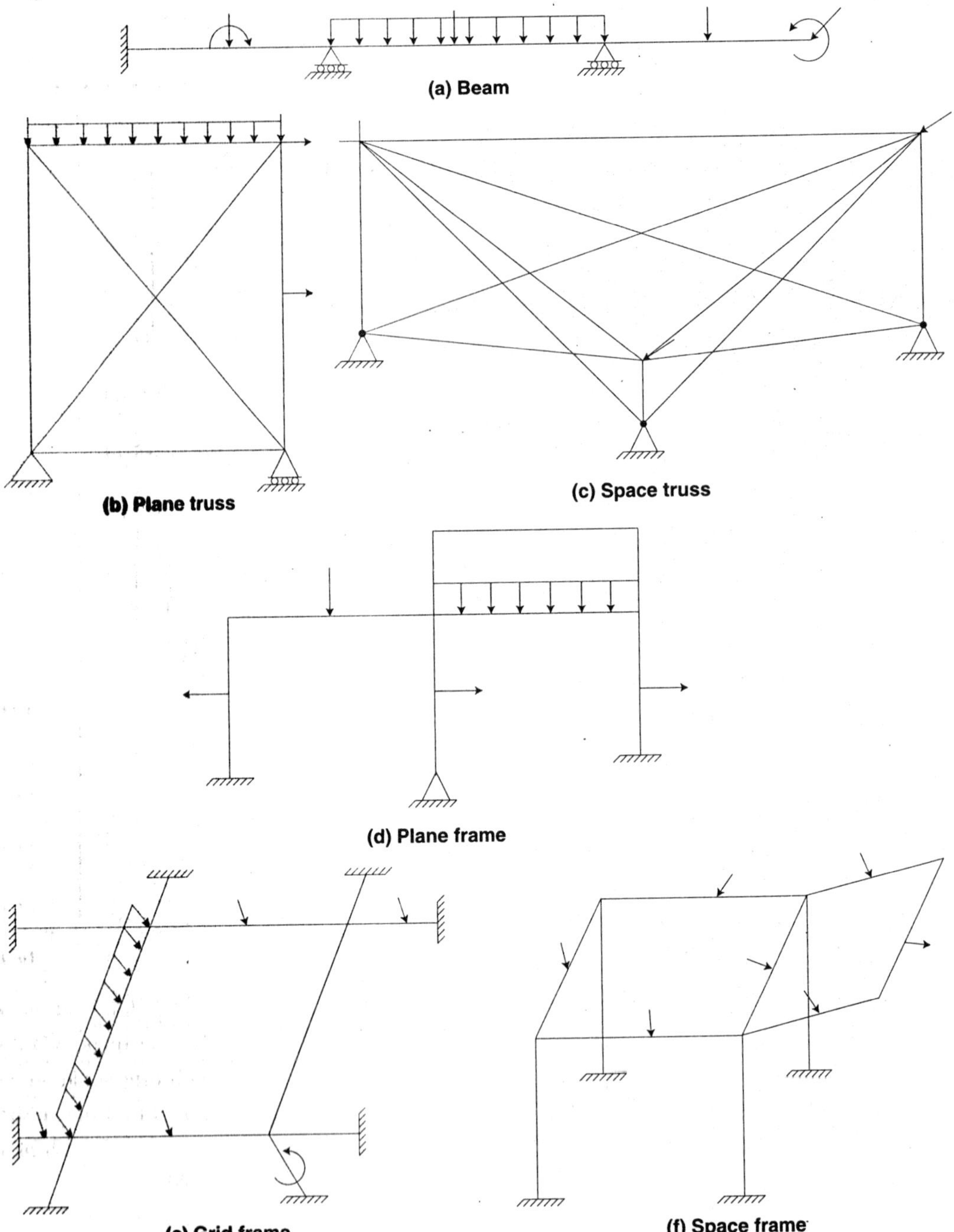

Fig. 16.1: Types of framed structures

A plane truss [Fig. 16.1(b)] is a system of members lying in a plane and interconnected at hinged joints. All forces act in the plane of the structure and external couples have their moment vectors normal to the plane. Loads may be concentrated force acting at the joints or on the members. For analysis, the latter force may be replaced by statically equivalent loads acting at the joints. Analysis of truss subjected to joint loads will result in axial forces only in members. In addition, there will be bending moment and shear forces in these members having loads which act directly on them.

A space truss [Fig. 16.1(c)] is similar to plane truss except that the members may have any direction in space. Forces acting may be in any directions, but couple acting must have its moment vector perpendicular to the axis of the member.

A plane frame [Fig. 16.1(d)] is composed of members lying in a single plane and having axis of symmetry in that plane as in case of a beam. Joints between members are rigid. Forces acting on a frame and the translations of the frame are in the plane of the structure. All couples acting on the frame have their moment vectors normal to the plane. Internal stress resultants may in general be a bending couple, a shear force and an axial force.

A grid [Fig. 16.1(e)] is a plane structure composed of continuous members, which either intersect or cross one another. Connections between members may be rigid or hinged. All forces are normal to the plane of the structure and all couples have their vectors in the plane of the grid. Loading may result in torsion as well as bending.

Space frames [Fig. 16.1(f)] are the most general type of framed structures. Members may carry internal axial forces, torsional couples, bending couples in both principal directions and shearing forces in both principal directions.

16.7 Structural Analysis

One of the objectives of any structural analysis is to determine various actions pertaining to the structure, such as reactions at the supports and internal stress resultant like bending moments, shear forces, etc. A correct solution for any of these quantities must satisfy all conditions of equilibrium, not only for the entire structure, but also for any part of the structure taken as a free body. If the resultant force vector and the resultant moment vector are both zero, the equations of static equilibrium obtained are as under :

$$\left.\begin{aligned} &\sum F_x = 0, \sum F_y = 0, \sum F_z = 0 \\ \text{and} \quad &\sum M_x = 0, \sum M_y = 0, \sum M_z = 0. \end{aligned}\right] \tag{16.48a}$$

In case of forces in the $xy-$ plane, the equations

$$\sum F_z = 0, \sum F_y = 0 \quad \text{and} \quad \sum M_z = 0 \tag{16.48b}$$

and these equations become the static equilibrium conditions for actions in the $xy-$ plane.

In the stiffness method of analysis, the basic equations to be solved are those which express the equilibrium conditions at the joints of the structure.

In addition to the static equilibrium conditions, it is necessary in any structural analysis that all conditions of compatibility be satisfied. These conditions refer to continuity of the displacements throughout the structure and are sometimes referred to as the conditions of geometry. As an example, compatibility conditions must be satisfied at all points of support, where it is necessary that the displacements of the structure be consistent with the support conditions. Compatibility conditions must also be satisfied at all points throughout the interior of a structure. In the flexibility method of analysis, the basic equations to be solved are equations which express the compatibility of the displacements.

16.8 Static and Kinematic Indeterminancies

The two types of indeterminancies, which must be considered in structural analysis depending on considerations of actions or displacements are the static and kinematic indeterminancies. When actions are the unknowns in the analysis as in flexibility method, then static indeterminancy must be considered. In this case, indeterminacy refers to an excess of unknown actions as compared to the number of equations of static equilibrium that are available.

In the Stiffness method of analysis, the displacements of the joints of the structure become the unknown quantities, and the second type of indeterminancy, known as kinematic indeterminancy becomes important. Joints in framed structures are located at all points where two or more members intersect, at points of supports and at free ends. When the structure is subjected to loads, each joint will undergo displacements in the form of translations and rotations depending on the configuration of the structure. In some cases, the joint displacement will be known because of restraint conditions imposed on the structure. At a fixed support, for instance, no displacement of any kind can occur. However, there will be other joint displacements which are not known in advance and which can be obtained only by making a complete analysis of the structure. These unknown joint displacements are the kinematically indeterminate quantities and are sometimes called kinematic redundants. Their number represents the degree of kinematic indeterminancy of the structure or the number of degrees of freedom for joint displacement.

Considering the beam in Fig. 16.2(a), it is seen that end A is fixed and cannot undergo any displacement. On the other hand, joint B has two degrees of freedom for joint displacement since it may translate in the horizontal direction and may rotate. Thus, the beam is kinematically indeterminant to the second degree and there are two unknown joint displacement to be calculated in a complete analysis of the beam. In many practical analysis, it would be permissible to neglect axial displacements of the beam. In such a case, joint B would have only one degree of freedom, i.e., rotation and the structure can be analysed as if it were kinematically indeterminate to the first degree.

Beam in Fig. 16.2(b) has no unknown joint displacements as the joints are fixed. The beam is, therefore, kinematically determinate.

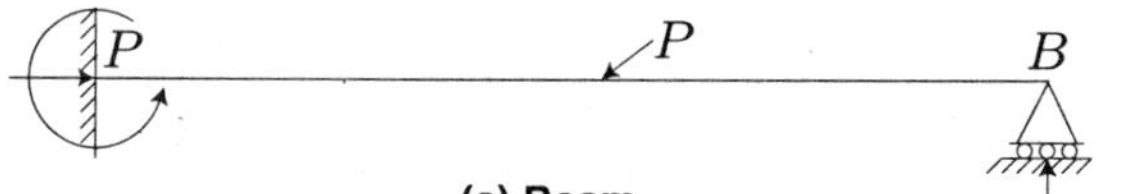

(a) Beam

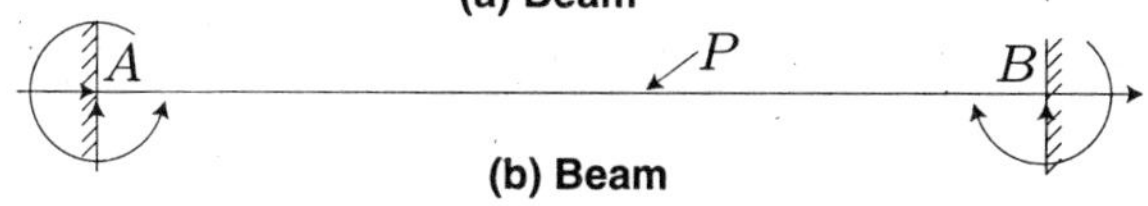

(b) Beam

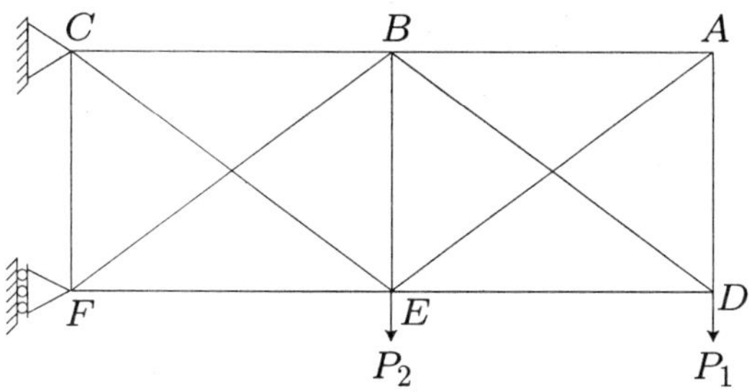

(c) Plane truss

Fig. 16.2: Statically indeterminate structures

In the plane truss in Fig. 16.2(c) joints A, B, D and E each, has two independent components of displacements, such as translations in two perpendicular directions and hence has two degrees of freedom. Joint C is restrained, while joint F has one degree of freedom. The truss thus has a total of nine degrees of freedom for joint translation and is kinematically indeterminate to the ninth degree.

In the rigid frame in Fig. 16.3(a), the supports at A, B and C being fixed, there cannot be any displacement at these joints. However, joints D, E and F each has three degrees of freedom, since each joint may undergo horizontal and vertical translations and a rotation. Thus the numbers of kinematic indeterminancies for this frame is nine. If the effects of axial deformations are not considered, the member of kinematic indeterminancy is reduced to four since there is no possibility of any vertical displacement of any of the joints and the horizontal translations of D, E and F are related.

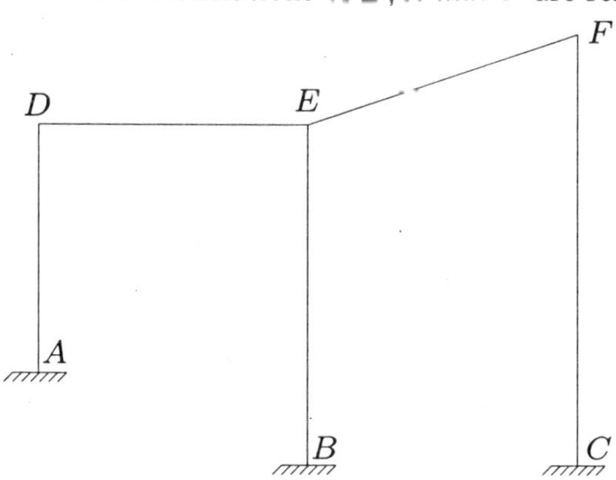

(a)

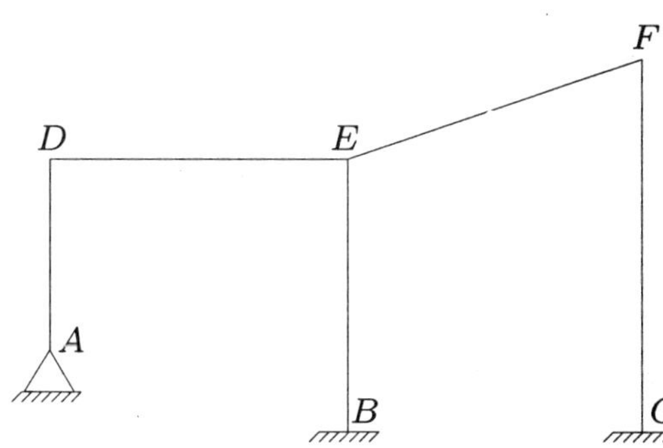

(b)

Fig. 16.3: Rigid frames

In case of the rigid frame in Fig. 16.3(b) the number of kinematic indeterminancies increases to ten and five respectively due to additional rotation at joint A.

Example 16.9. *Determine the degree of static and kinematic indeterminancies of the pin-jointed plane frame shown in Fig. 16.4.*

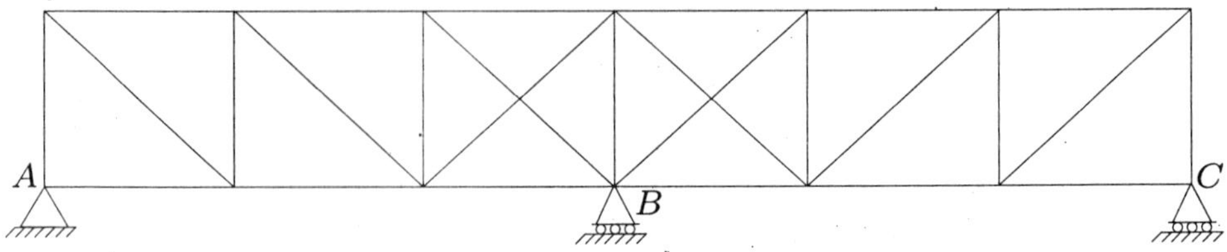

Fig. 16.4: For Example 16.9

Solution : The truss is stable.

Static indeterminancy

External indeterminancy = 1, Internal indeterminancy = 2 and Total indeterminancy = 3.

Kinematic Indeterminancy. Each joins except A, B and C has two displacements, one horizontal and the other vertical. Joint A has no displacement, whereas displacement at B and C each is unity.

Total Nos. of kinematic indeterminancy $= 11(2) + 1 + 1 = 24$.

Example 16.10. *Determine the degree of the static and kinematic indeterminancies for the space truss shown in Fig. 16.5.*

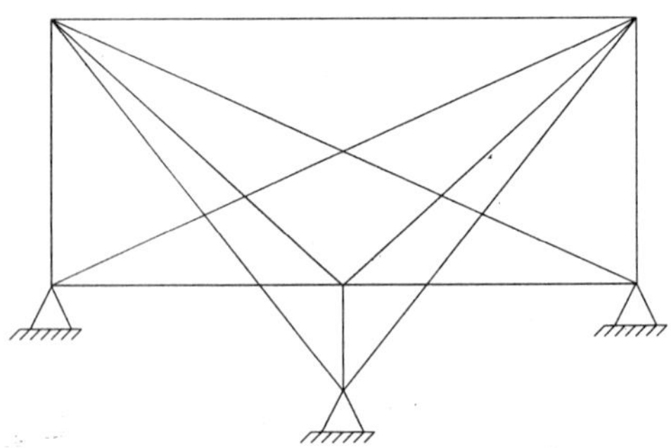

Fig. 16.5: For Example 16.10

Solution : Static Indeterminancy. For each joint, three equations of equilibrium as available. Thus there are 18 equations for six joints. Altogether there are 12 bar forces and 9 reactions to be found. The truss is, therefore, statically indeterminate to the third degree.

Kinematic Indeterminancy. Each of the joints D, E and F has three degrees of freedom and total number of kinematic indeterminancy is nine.

Example 16.11. *Determine the number of Static and Kinematic Indeterminancies for the rigid frame in Fig. 16.6.*

Solution : Static Indeterminancy. Cut on each of six beam can be replaced by three unknowns. The supports of B and C introduce two and one static indeterminancy respectively. Thus number of static indeterminancy is 15.

Kinematic Indeterminancy. Each joint except A, B and C provides three degree of freedom. Joint A has no degree of freedom, whereas degrees of freedom at B and C are two and one respectively. Thus total member of kinematic indeterminancy is 30.

If axial deformation be neglected, member of kinematic indeterminancy is reduced to 15.

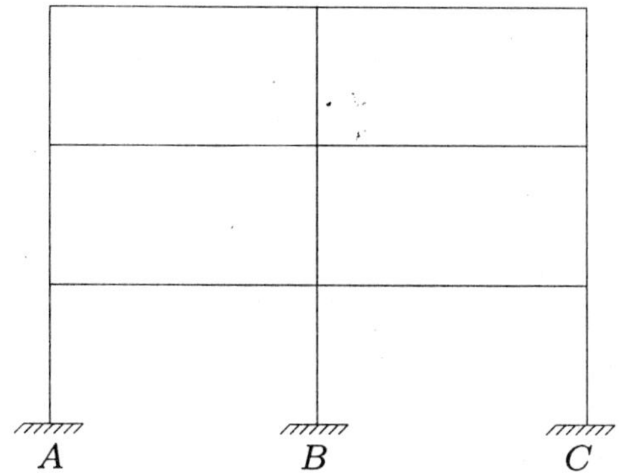

Fig. 16.6: For Example 16.11

Example 16.12. *Determine the Static and Kinematic indeterminancies for the rigid frame in Fig. 16.7.*

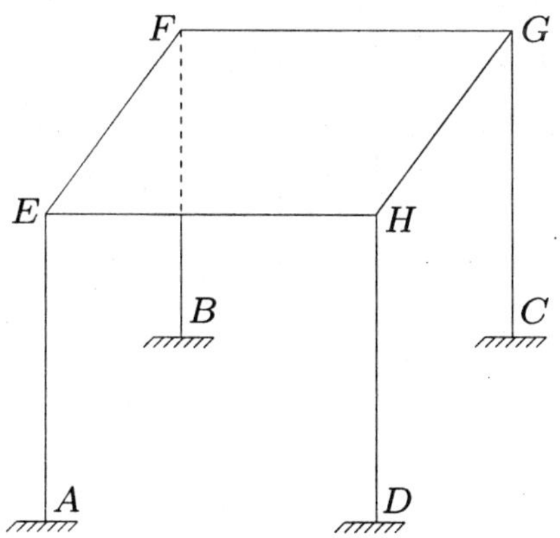

Fig. 16.7: For Example 16.12

Solution : Static Indeterminancy. Cuts in each of the members EF, EH, FG and GH can be replaced by six reactions, i.e., one axial force, two shear forces in vertical and horizontal directions, two couples and one twisting moments. Thus number of static indeterminancy is 24.

Kinematic Indeterminancy. Each joint E, F, G and H has six degrees of freedom, viz., three translations and three rotations : Thus number of kinematic indeterminancy is 24.

If axial deformations be neglected, number of indeterminancy is reduced to 16.

Example 16.13. *Determine the number of Static and Kinematic Indeterminancies for the rigid frame in Fig. 16.8.*

Solution : Static Indeterminancy. Proceeding as in Example 16.10, number of static indeterminancy$=21-(3+3)=15$.

Kinematic Indeterminancy. As in Example 16.10, number of Kinematic indeterminancy=9(3)+(3+3)=33.

However, if axial deformation be neglected number reduces to 17.

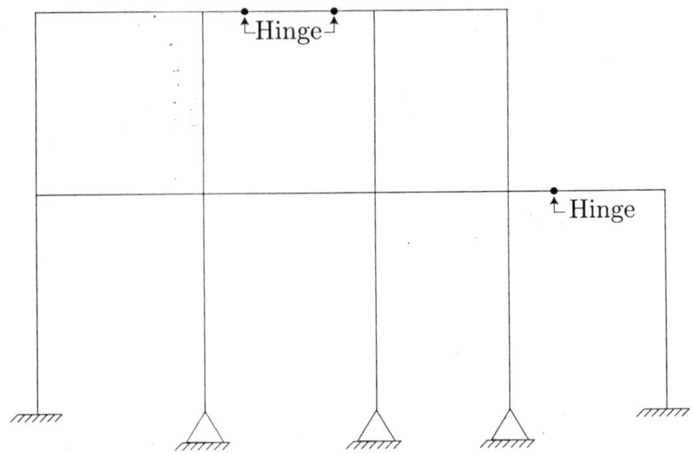

Fig. 16.8: For Example 16.13

Example 16.14. *Determine the number of static and Kinematic Indeterminancy for the grid in Figs. 16.9(a) and 16.9(b).*

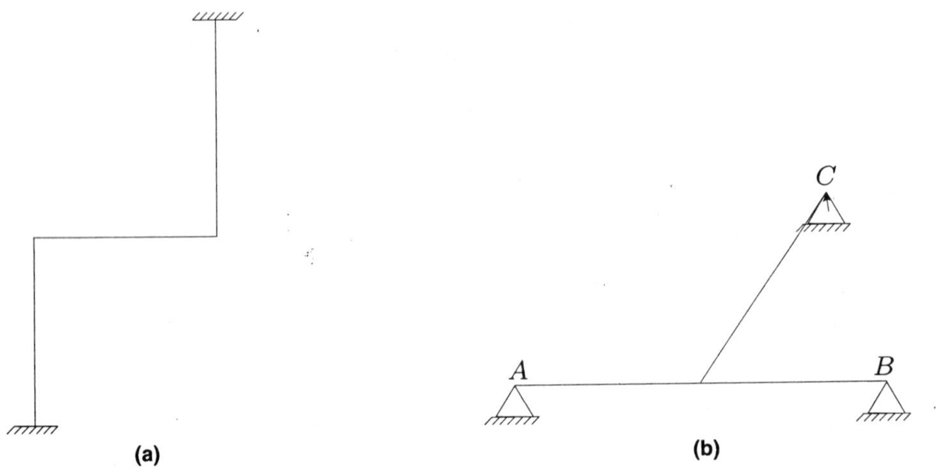

Fig. 16.9: For Example 16.14

Solution : (a) Static Indeterminancy. Total number of static indeterminancy is 3.

Kinematic Indeterminancy. Total number or Kinematic indeterminancy, considering joints=2(3)=6.

(b) Static Indeterminancy. The grid is statically determinate, as six reactions involving six conditions of statics are involved.

Kinematic Indeterminancy. Considering the joints, number of kinematic indeterminancy=3+3(2)=9.

16.9 Flexibility and Stiffness

Flexibility and Stiffness are important properties which characterise the response of a structure by means of the force displacement relationship.

Flexibility of a structure is the displacement caused by a unit force, the stiffness being the force required for units displacement. In case of a structure, there can be the following type of displacement :

(a) Axial displacement

(b) Transverse displacement

(c) Bending or Flexural displacement and

(d) Torsional displacement.

(a) Axial displacement. If an axial force P be applied in a member, the displacement Δ_a is given by

$$\Delta_a = \frac{PL}{AE},$$

where L=length of the member,

A=cross-sectional area of the member

and E=modulus of elasticity of the member.

It follows that

$$\text{axial flexibility} = \frac{L}{AE} \tag{16.49}$$

and

$$\text{axial stiffness} = \frac{AE}{L}. \tag{16.50}$$

The flexibility and stiffness with respect to axial displacements given by Eqs. (16.49) and (16.50) are of relevance in case of pin-jointed frames carrying only axial loads. In case of rigid jointed frames, axial displacements are small as compared to transverse displacements. It is, therefore, a common practice in analysis of rigid jointed frames to ignore the axial flexibility of the member. The members of the rigid frames are considered to be infinitely stiff with respect to axial displacements.

(b) Transverse displacement. Force P required for a deflection at one end of a member without any other displacement is given by

$$P = \frac{12Ei\Delta}{L^3}, \tag{16.51}$$

when far end is fixed,

and

$$P = \frac{3EI\Delta}{L^3}, \tag{16.52}$$

when far end is hinged.

From Eqs. (16.51) and (16.52).

$$\text{Transverse flexibility} = \frac{L^3}{12EI} \tag{16.53}$$

for far end fixed,

and

$$= \frac{L^3}{3EI} \tag{16.54}$$

for far end hinged.

Also Transverse or Translational stiffness $= \dfrac{12EI}{L^3}$ (16.55)

for far end fixed,

and $= \dfrac{3EI}{L^3}$ (16.56)

for far end hinged.

(c) Bending or Flexural displacement. Moment required for a rotation θ at an end of a member without any other displacement is given by

$$M = \frac{4EI\theta}{L}, \tag{16.57}$$

for far end fixed

$$\text{and} \quad M = \frac{3EI\theta}{L}, \tag{16.58}$$

for far end hinged.

From Eqs. (16.57) and (16.58),

flexural flexibility $= \dfrac{L}{4EI}$, (16.59)

for far end fixed,

and $= \dfrac{L}{3EI}$, (16.60)

for far end hinged.

Similarly, flexural or Rotational stiffness $= \dfrac{4EI}{L}$ (16.61a)

for far end fixed,

and $= \dfrac{3EI}{L}$, (16.61b)

for far end hinged.

(d) Torsional displacement or Twist. Angle of twist θ_t, due to torque is given by

$$\theta_t = \frac{TL}{GJ}, \tag{16.62}$$

where G = shear modulus of elasticity

and J = torsional constant.

From Eq. (16.63), it follows that,

Torsional flexibility $= \dfrac{L}{GJ}$

and Torsional stiffness $= \dfrac{GJ}{L}$.

16.10 Flexibility Method

A beam AB of constant EI fixed at A and propped at B, as shown in Fig. 16.10(a) is statically indeterminate to the first degree. Considering Reaction R_B as the redundant, the method of superposition can be used for analysis of the beam. Thus, considering a cantilever as the primary released structure, as in Fig. 16.10(b), the deflection at end B of the beam is given by

$$\Delta_{BE} = -\frac{wL^4}{8EI}.$$

The deflection at B due to unit load at B is given by

$$\delta_{BB} = \frac{L^3}{3EI}.$$

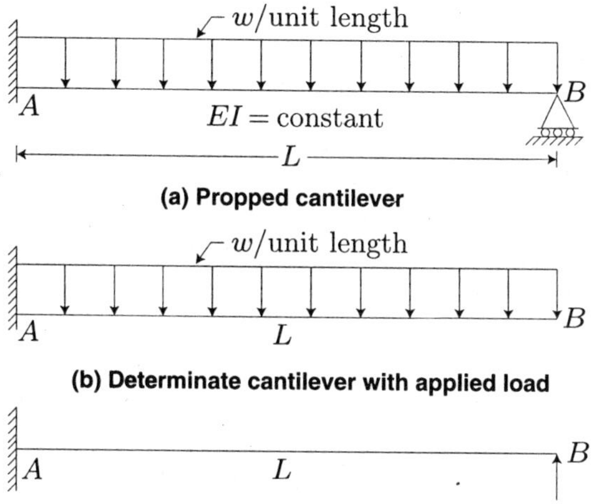

Fig. 16.10: Flexibility method

Using method of superposition, beam of Fig. 16.10(a) consists of beams shown in Figs. 16.10(b) and 16.10(c). Hence, from the compatibility equation, $\Delta_{BB} + R_b\delta_{BB} = 0$, it follows that,

$$-\frac{wL^4}{8EI} + R_B\frac{L^3}{3EI} = 0, \quad \text{whence} \quad R_B = \frac{3wL}{8}$$

The positive sign in the result indicates that the reaction R_B is in the upward direction as assumed.

An important part of the preceeding solution consists of writing the superposition equation, which expresses the geometrical fact that the beam does not deflect at the support B. Included in this equation are the effects of load and the redundant reaction. The displacement caused by the reaction has been conveniently expressed as the product of the reaction and the displacement caused by a unit value of the reaction. The latter is a flexibility influence coefficient, since it is the displacement due to unit value of the reaction.

This method is called flexibility method, since it is connected with flexibility influence coefficient. The method is also known as the force method and the compatibility method since force is considered as the redundant and the compatibility equation is used.

If the structure is statically indeterminate to more than one degree, the approach used in the preceeding Example has to be further organised and a more generalised notation must be introduced.

Fig. 16.11(a) shows a prismatic beam having two degrees of indeterminancy. A statically determinate released structure may be obtained by removing two redundant actions from the beam. Figs. 16.11(b), 16.11(c), 16.11(d) and 16.11(e) indicate four such possibilities.

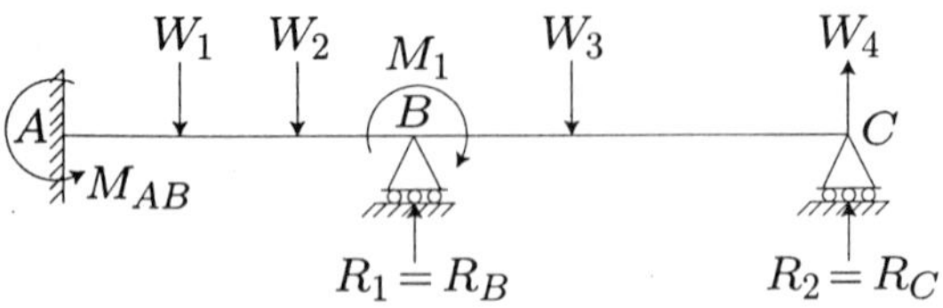

(a) Beam

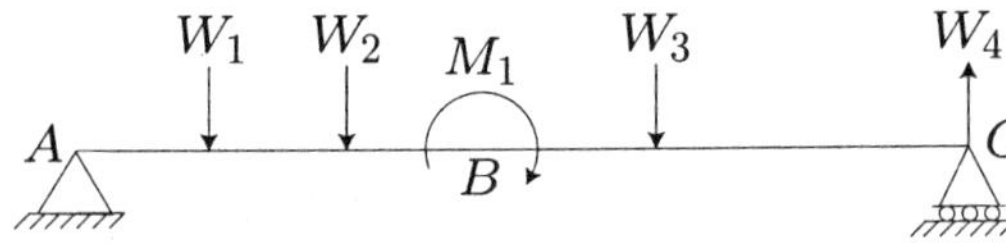

(b) Statically determinate released structure M_{AB} and R_B removed

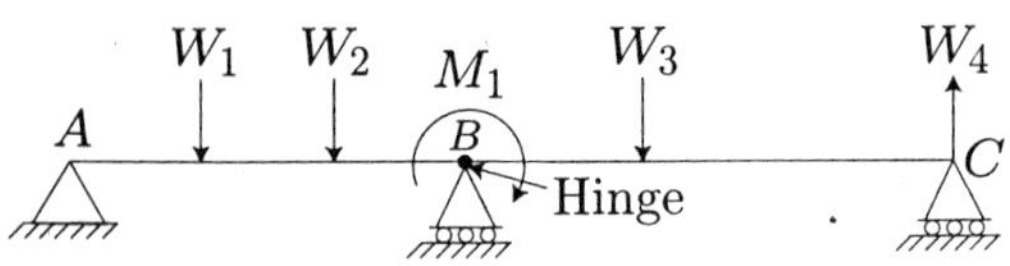

(c) Statically determinate released structure M_{AB} removed and hinge at B

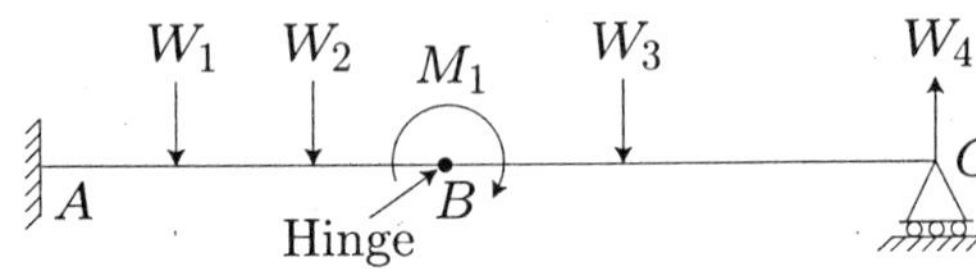

(d) Statically determinate released structure R_B removed and hinge at B

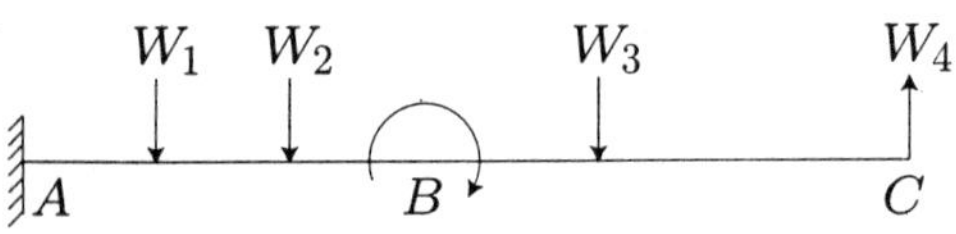

(e) Statically determinate released structure R_B and R_C removed

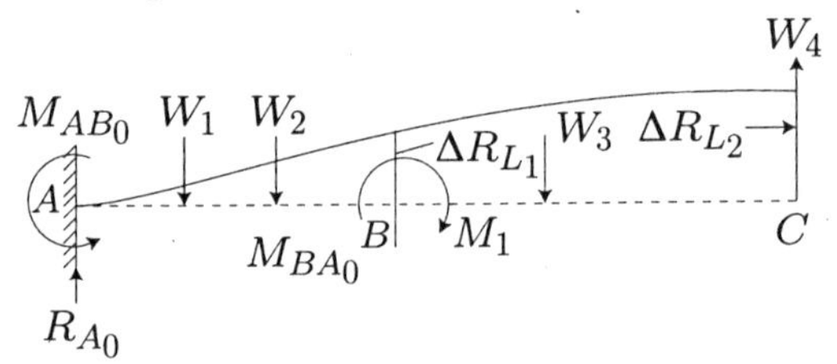

(f) Deflection of released primary structure due to external loading

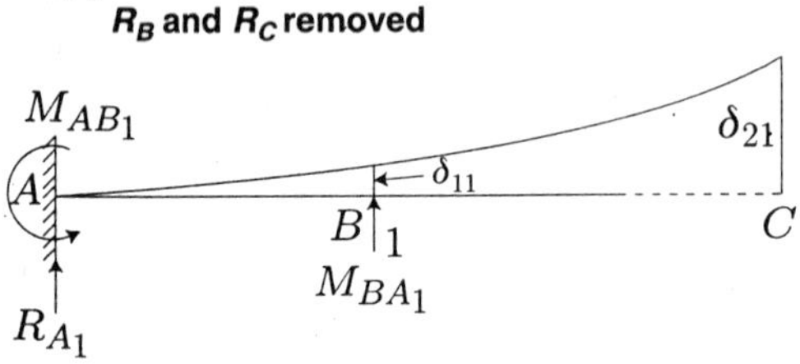

(g) Deflection of released primary structure due to unit load at B

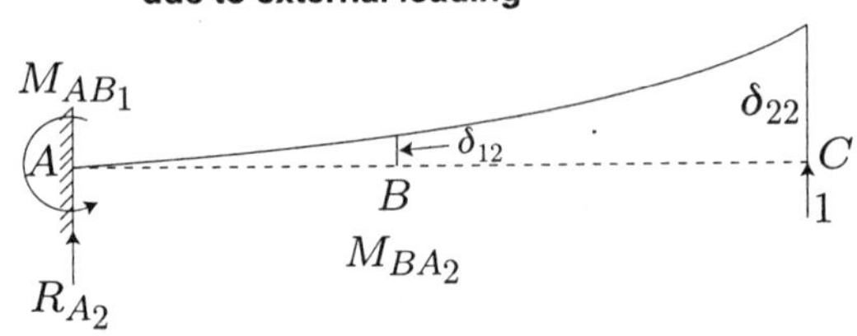

(h) Deflection of released structure due to unit load at C

Fig. 16.11: Development of flexibility method

The redundant actions which are selected for the analysis are denoted as R_1 and R_2 in Fig. 16.11(a). These actions are the reactive forces at joints B and C. The released primary structure acted upon by the various loads on the original beam is shown in Fig. 16.11(f). The loads produce displacement in the released primary structure and in particular, displacements corresponding to R_1 and R_2 are produced at joints B and C. Such displacements are denoted by Δ_{RL_1} and Δ_{RL_2} in the figure. In this notation Δ_{RL} is used to represent a displacements corresponding to a redundant P and caused by the loads on the structure. The numerical subscripts which follow the symbol denote the redundant to which the displacement corresponds. Thus, it is convenient to envisage Δ_{RL}, while the numbers following the symbol are the subscripts. The displacements Δ_{RL} are shown in their assumed positive direction, which is upward, since the redundants are assumed to be positive in the same direction.

In order to obtain the various flexibility influence coefficients which appear in the compatibility equations, unit values of the redundants $R_1 = 1$, as in Fig. 16.11(g), the flexibility coefficient δ_{11} is the displacement corresponding to R_1 due to a unit value of R_1, and the coefficient δ_{21} is the displacement corresponding to R_2 due to a unit value of R_1. For the condition $R_2 = 1$, shown in Fig. 16.11(h), δ_{12} is the displacement corresponding to R_1 due to a unit value of R_2, and δ_{22} is the displacement corresponding to R_2 due to a unit value of R_2.

The superposition equations expressing the conditions or compatibility at joints B and C of the actual beam may be written. Since the translational displacements at supports B and C are zero, the equations become

$$\left.\begin{aligned} &\Delta_{RL_1} + \delta_{11}R_1 + \delta_{12}R_2 = 0 \\ \text{and} \quad &\Delta_{RL_2} + \delta_{21}R_1 + \delta_{22}R_2 = 0. \end{aligned}\right] \tag{16.63}$$

The first of the equations represent the total displacement at B, which consists of three parts in case of the released structure : displacements due to loads, R_1 and R_2. Superposition of all the three displacements gives the total displacements which is zero in the case. This also applies to the second equation, i.e., at support C. The two equations can be solved simultaneously for R_1 and R_2 after which other functions in the beam can be obtained by statics.

The superposition equation can be written in a more general form. If there are support movements in the original beam corresponding to the redundants and denoted by Δ_{R_1} and Δ_{R_2} at supports B and C repectively, then equations are

$$\left.\begin{aligned} &\Delta_{R_1} = \Delta_{RL_1} + \delta_{11}R_1 + \delta_{12}R_2 \\ \text{and} \quad &\Delta_{R_2} = \Delta_{RL_2} + \delta_{21}R_1 + \delta_{22}R_2. \end{aligned}\right] \tag{16.64}$$

Eq. (16.64) can be written in the matrix form as

$$\Delta_R = \Delta_{RL} + \delta R, \tag{16.65}$$

in which Δ_R is the matrix of actual displacements corresponding to the redundants, Δ_{RL} is the matrix of displacements in the released structure corresponding to the redundant action R and due to loads and δ is the flexibility matrix for the released structure corresponding to the redundant action R. For the equations given above, these matrices are as under :

$$\Delta_R = \begin{bmatrix} \Delta_{R_1} \\ \Delta_{R_1} \end{bmatrix}; \ \Delta_{RL} = \begin{bmatrix} \Delta_{RL_1} \\ \Delta_{RL_2} \end{bmatrix}$$

$$\delta = \begin{bmatrix} \delta_{11} & \delta_{12} \\ \delta_{21} & \delta_{22} \end{bmatrix} \text{ and } R = \begin{bmatrix} R_1 \\ R_2 \end{bmatrix}.$$

These matrices have as many rows as there are unknown reactions. Thus, if n represents the number of unknown reactions, the order of flexibility matrix is $n \times n$, while Δ_R, Δ_{RL} and R are of order $n \times 1$.

Redundants R can be obtained by solving Eq. (16.65), whence

$$R = \delta^{-1}(\Delta_R - \Delta_{RL}), \tag{16.66}$$

in which δ^{-1} denotes the inverse of the flexibility matrix.

If it is required to include effect of displacements due to temperature, prestrain and support movements, the superposition equation can be written as

$$\Delta_R = \Delta_R S + \delta_R, \tag{16.67}$$

where $\Delta_{RS} = \Delta_{RL} + \Delta_{RT} + \Delta_{RP} + \Delta_{R_r}$. (16.68)

Δ_{RT} being the displacements due to temperature effects, Δ_{RP}, the displacements due to prestrain and Δ_{R_r} the restraint displacements in the released structures, corresponding to the redundants, R.

From Eq. (16.66), the result is

$$R = \delta^{-1}(\Delta_R - \Delta_{RS}). \tag{16.69}$$

The actions in the structure can be obtained using the method of superposition after the redundants can be obtained. Alternatively, their values can be obtained directly from the basic computations of flexibility method.

The redundant actions having been obtained, the remaining actions in the beam can be found from static equilibrium equations. Also the displacements at any point in the beam can now be obtained without difficulty in as much as all action can be considered as known. For example, one method of finding displacement is to subdivide the beam into two simple beams. Then each simple beam is acted upon by known moments at the ends as well as by the loads and the displacements can be calculated.

Alternatively, the actions and reactions can be obtained in matrix from considering characteristic matirix. Thus, referring to Figs. 16.11(f) to 16.11 (h), the various functions can be written in matrix form as under :

$$\begin{bmatrix} M_{AB} \\ R_A \\ M_{BA} \\ M_{CB} \end{bmatrix} = \begin{bmatrix} M_{AB_0} \\ R_{A_0} \\ M_{BA_0} \\ M_{CB_0} \end{bmatrix} + \begin{bmatrix} M_{AB_1} & M_{AB_2} \\ R_{A_1} & R_{A_2} \\ M_{BA_1} & M_{BA_2} \\ M_{CB_1} & M_{CB_2} \end{bmatrix} \begin{bmatrix} R_1 \\ R_2 \end{bmatrix}. \tag{16.70}$$

16.11 Stiffness Method

The Stiffness method differs from the flexibility method in the physical concepts which are involved, though the methods are similar in their mathematical formulation. In both the methods, the fundamental equations are derived by using the principle of superposition. In the flexibility method, the unknown quantities are redundant action but in the stiffness method, the unknown quantities are the joint-displacements in the structure. Therefore, in the stiffness method, the number of unknown to be calculated is the same as the degree of kinematic indeterminancy. The stiffness method involves extensive use of actions and displacements in members having fixed ends.

In order to illustrate the concepts of the stiffness method in the simplest form, the analysis of the propped cantilever as in Fig. 16.12(a) is considered. This beam has a fixed support at A, a roller support at B and is subjected to a uniform load of intensity w. The beam is kinematically indeterminate to the first degree, if axial deformations are neglected, the unknown joint displacement at point B, i.e., θ_B being the only unknown. After the rotation is obtained, the various functions throughout the beam can be determined.

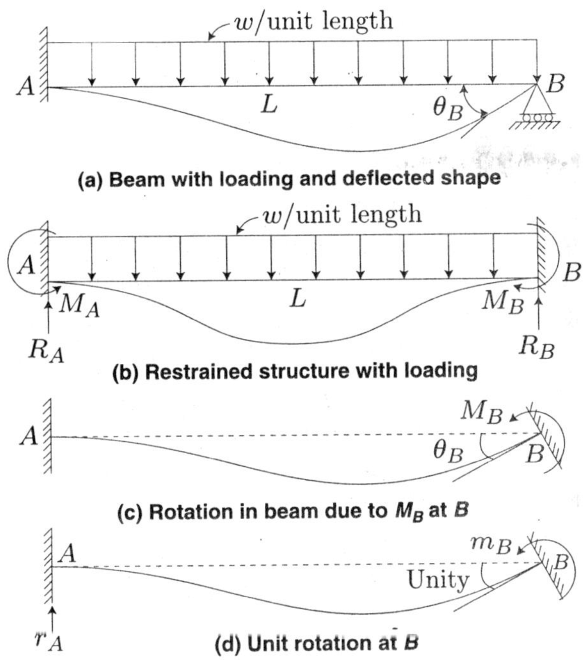

Fig. 16.12: Stiffness method

In the flexibility method, a statically determinate released structure is obtained by altering the actual structure in such a manner that the selected redundant actions are zero. The analogous operation in the stiffness method is to obtain a kinematically determinate structure by altering the actual structure in such a manner that all unknown displacements are zero. Since the unknown displacements are the translations and rotation of the joints, they can be made equal to zero by restraining the joints of the structure against displacements of any kind. The structure obtained by restraining all joints of the actual structure is called the restrained structure. For the beam in Fig. 16.12(a), the restrained structure is obtained by restraining joint B against rotation. The restrained structure is thus the fixed end beam shown in Fig. 16.12(b).

Due to the loads acting on the restrained beam in Fig. 16.12(b), fixed end moments $M_B = \frac{wL^2}{12}$ develop at support B. The couple M_B is an action corresponding to the rotation

θ_B, which is the unknown quantity in the analysis. As there cannot be any couple in the actual beam in Fig. 16.12(a), it is necessary to consider that the restraint beam is subjected to a couple equal and opposite to the couple M_B, as shown in Fig. 16.12(c). When the actions acting on the two beams in (b) and (c) are superimposed, they produce the actions on the actual beam. Thus, the analysis of the beam shown in Fig. 16.12(a) can be considered as the superposition of the analysis shown in Figs. 16.12(b) and 16.12(c). The rotation produced by the couple in Fig. 16.12(c) is equal to θ_B, the unknown rotation in the actual beam.

The relation between the moment M_B and the rotation θ_B in the beam of Fig. 16.12(c) is

$$M_B = \frac{4EI}{L}\theta_B, \tag{16.71}$$

where EI is the flxeural rigidity of the beam.

Equating the two expressions for the moment M_B, $\frac{wL^2}{12} = \frac{4EI}{L}\theta_B$, whence

$$\theta_B = \frac{wL^3}{48EI}.$$

As in case of flexibility method, it is convenient to consider the restrained structure under the effect of a unit value of the unknown rotation. It is more systematic to formulate the equation for the rotation as an equation of superposition and to use a consistent sign convention for all terms in the equation.

The effect of a unit value of the unknown rotation is shown in Fig. 16.12(d) where the restrained beam is acted on by a couple m_B which produces unit value of the rotation at B. Since the moment m_B is an action corresponding to the rotation θ_B and caused by a unit value of the rotation, while all other joint displacements are zero, it is recognised that m_B is given by stiffness coefficient for the restrained structure. The value of the couple m_B is given by

$$m_B = \frac{4EI}{L}.$$

Superposition equation has to be expressed with same sign convention. Considering the couples and rotations at joint B as positive when both are counter-clockwise (i.e., positive when joint moment is clockwise and rotation counter-clockwise) the fixed end moment M_B in Fig. 16.12(b) is negative, i.e., $M_B = -\frac{wL^2}{12}$.

Equation for the superposition of moments at support B now becomes

$$M_B + mB\theta_B = 0 \tag{16.72}$$

i.e., $$-\frac{wL^2}{12} + \frac{4EI}{L}\theta_B = 0.$$

Solving, $\theta_B = \frac{wL^3}{48EI}$.

The positive sign for the result means that the rotation is counter-clockwise.

Superposition as in Eq. (16.72) expresses the fact that the moment at B in the actual beam is zero. In the equation, moment caused by loads on the restrained structure and the moment caused by rotating the end B of the restrained structure are included. The two effects are summed algebrically using the same sign convention. The sign of the result gives the true direction of the displacement.

Eq. (16.72) may also be refered to as an equation of equilibrium and hence the method is also called 'equilibrium method'. The method is also called 'displacement method', since displacement are considered as redundants. As stiffness coefficient is involved in this method, it is also called 'stiffness method'.

For a structure kinematically indeterminate to more than one degree, a more organised approach to the solution, as well as a generalised notation have to be used. For this purpose a two span beam of constant EI shown in Fig. 16.13(a) will be analysed by stiffness method.

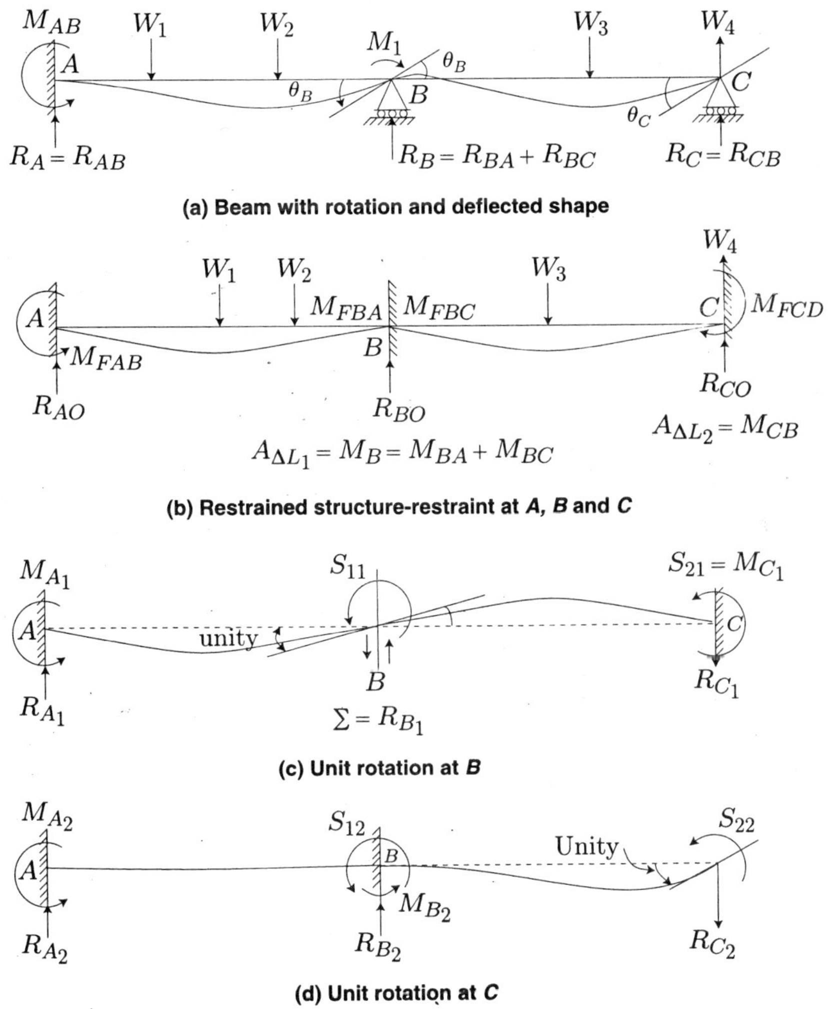

Fig. 16.13: Development of stiffness method

The beam is kinematically indeterminate to the second degree if axial deformations are neglected. The unknown rotations are θ_B and θ_C at joints B and C, and are considerd positive when anti-clockwise. These unknown displacements may be determined by solving equation of superposition for the actions at joints B and C.

The first step in the analysis consists of applying imaginary restraints at the joints to prevent all joint displacements. The restrained structure is shown in Fig. 16.13 and consists of two fixed end beams. The restrained structure is assumed to be acted upon by all of the loads except those which correspond to the unknown displacements. Only loads W_1, W_2, W_3 and W_4 are shown in Fig. 16.13(b). All loads which correspond to the unknown joint displacements such as couple M in this example are taken into account later. The moments $A_{\Delta L_1}$ and $A_{\Delta L_2}$ in Fig. 16.13(b) are the actions of the restraints, corresponding to θ_B (i.e., Δ_1) and θ_C (i.e., Δ_2) respectively and caused by loads acting on the structure. The restraint action $A_{\Delta L_1}$ is the sum of the reactive moment at B due to loads W_1 and W_2 acting on member AB and the reactive moment at B due to load W_3 and W_4 acting on member CB.

In order to obtain the stiffness coefficients at joints B and C, unit values of the unknown displacements θ_B (i.e., Δ_1) and θ_C (i.e., Δ_2) are induced separately in the restrained structure. A unit displacement corresponding to Δ_1 (i.e., θ_B) consists of a unit rotation of joint B as shown in Fig. 16.13(c). The displacement Δ_2 (i.e., θ_C) reamins equal to zero in this case. Thus the actions corresponding to Δ_1 and Δ_2 are the stiffness coefficients S_{11} and S_{21} respectively. These stiffnesses consist of the couples exerted by the restraints on the beam at joints B and C respectively. Calculation of these actions is not difficult when formulas for fixed end moments in beams are available.

The condition that Δ_2 (i.e., θ_C) is equal to unity, while Δ_1 (i.e.,θ_B) is equal to zero is shown in Fig. 16.13(d). In the figure stiffness S_{12} is the action corresponding to Δ_1 (i.e., θ_B), while the stiffness S_{22} is the action corresponding to Δ_2 (i.e., θ_C).

Two superposition equations expressing the conditions pertaining to the moments acting on the original structure [Fig. 16.13(a)] at joints B and C can be obtained. If $A_{\Delta 1}$ and $A_{\Delta 2}$ denote the actions in the actual structure corresponding to Δ_1 (i.e., θ_B) and Δ_2 (i.e., θ_C), the superposition equations are

$$\left.\begin{aligned} A_{\Delta 1} &= A_{\Delta L_1} + S_{11}\Delta_1 + S_{12}\Delta_2 \\ A_{\Delta 2} &= A_{\Delta L_2} + S_{21}\Delta_1 + S_{22}\Delta_2. \end{aligned}\right] \tag{16.73}$$

Eq. (16.73) can be expressed in matrix form, when

$$A_{\Delta 2} = A_{\Delta L_1} + S\Delta, \tag{16.74}$$

in which A_Δ represents the action in the original beam corresponding to the unknown joint displacements Δ, the vector $A_{\Delta L}$ represents actions in the restrained structure corresponding to the unknown joint displacements and caused by the loads, i.e., all loads except those corresponding to the unknown displacements and S is the stiffness corresponding to the unknown displacements. For the example, matrices,

$$A_\Delta = \begin{bmatrix} A_{\Delta 1} \\ A_{\Delta 2} \end{bmatrix}; A_{\Delta L} = \begin{bmatrix} A_{\Delta L_1} \\ A_{\Delta L_2} \end{bmatrix}$$

$$S = \begin{bmatrix} S_{11} & S_{12} \\ S_{21} & S_{22} \end{bmatrix} \text{ and } \Delta = \begin{bmatrix} \Delta_1 \\ \Delta_2 \end{bmatrix}.$$

These matrices have as many rows as there are unknown joint displacements. Thus, if n represents the number of unknown displacements the order of matrix S in $n \times n$, while $A_\Delta, A_{\Delta L}$ and Δ are of order $n \times 1$.

Subtracting $A_{\Delta L}$ from both sides of Eq. (16.73) gives

$$\Delta = S^{-1}[A_\Delta - A_{\Delta L}]. \tag{16.75}$$

The equations represents the solution for the displacements in matrix terms as the elements of A_Δ, $A_{\Delta L}$ and S are either known or may be obtained from the retained structure.

The member actions and reactions for the structure may be found after the joint displacements are known using slope deflection equations for prismatic sections which for a member AB gives

$$M_{AB} = M_{FAB} + \frac{2EI}{L}\left[2\theta_A + \theta_B - \frac{3\Delta}{L}\right]$$

and $$M_{BA} = M_{FBA} + \frac{2EI}{L}\left[\theta_A + 2\theta_B - \frac{3\Delta}{L}\right], \tag{16.76}$$

where M_{FAB} and M_{FBA} denote the fixed end moments at A and B respectively, θ_A and θ_B being the rotations at ends A and B of beam as obtained by stiffness method and Δ denotes the relative displacement of support B with respect to the supprt A.

Alternatively, the actions and reactions can be obtained in matrix form considering characteristic matrix.

Thus referring to Figs. 16.13(b) to 16.13(d), the various functions can be written in matrix form as

$$\begin{bmatrix} M_{AB} \\ R_A \\ M_{BA} \\ R_B \\ M_{CB} \\ R_C \end{bmatrix} = \begin{bmatrix} M_{FAB} \\ R_{AO} \\ M_{FBA} \\ R_{BO} \\ M_{FCB} \\ R_{CO} \end{bmatrix} + \begin{bmatrix} M_{A_1} & M_{A_2} \\ R_{A_1} & R_{A_2} \\ M_{B_1} & M_{B_2} \\ R_{B_1} & R_{B_2} \\ M_{C_1} & M_{C_2} \\ R_{C_1} & R_{C_2} \end{bmatrix} \begin{bmatrix} \theta_B \\ \theta_C \end{bmatrix} \tag{16.77}$$

Combining Eqs. (16.75) and (16.77) the complete solution for the various functions may be written in matrix form as

$$\begin{bmatrix} M_{AB} \\ R_A \\ M_{BA} \\ R_B \\ M_{CB} \\ R_C \end{bmatrix} = \begin{bmatrix} M_{FAB} \\ R_{AO} \\ M_{FBA} \\ R_{BO} \\ M_{FCB} \\ R_{CO} \end{bmatrix} + \begin{bmatrix} M_{A_1} & M_{A_2} \\ R_{A_1} & R_{A_2} \\ M_{B_1} & M_{B_2} \\ R_{B_1} & R_{B_2} \\ M_{C_1} & M_{C_2} \\ R_{C_1} & R_{C_2} \end{bmatrix} S^{-1}(A_\Delta - A_{\Delta L}). \tag{16.78}$$

The effects of temperature, prestrain and support displacements can be readily incorporated into the analysis of a structure by the stiffness method. It is necessary to calculate the actions $A_{\Delta L}$ corresponding to the unknown displacements. Such actions denoted by $A_{\Delta T}, A_{\Delta P}$ and $A_{\Delta R}$ for effects of temperature, prestrains and support displacements can be formed consistent with the symbol $A_{\Delta L}$.

The rotational stiffness of a rigid joint may be defined as the couple required at the joint for unit rotation without translation. Thus, considering a rigid joint O at which five members meet as shown in Fig. 16.14, the rotational stiffness S_{11} of the rigid joint at O is equal to the sum of the flexural stiffness of the members meeting at joint O. When a unit rotation is given at coordinate 1 without any rotation at other coordinates, 2 and 3 as shown in Fig. 16.14, the rotational stiffness of joint O is given by

$$S_{11} = \frac{4EI_{OA}}{L_{OA}} + \frac{4EI_{OB}}{L_{OB}} + \frac{4EI_{OC}}{L_{OC}} + \frac{4EI_{OD}}{L_{OD}} + \frac{4EI_{OE}}{L_{OE}}. \tag{16.79}$$

If rotation at coordinates 2 and 3 are permitted as in case of hinged supports at B and C, as shown in Fig. 16.14(c), the rotation stiffness of the joint O is given by

$$S_{11} = \frac{4EI_{OA}}{L_{OA}} + \frac{3EI_{OB}}{L_{OB}} + \frac{3EI_{OC}}{L_{OC}} + \frac{4EI_{OD}}{L_{OD}} + \frac{4EI_{OE}}{L_{OE}}. \tag{16.80}$$

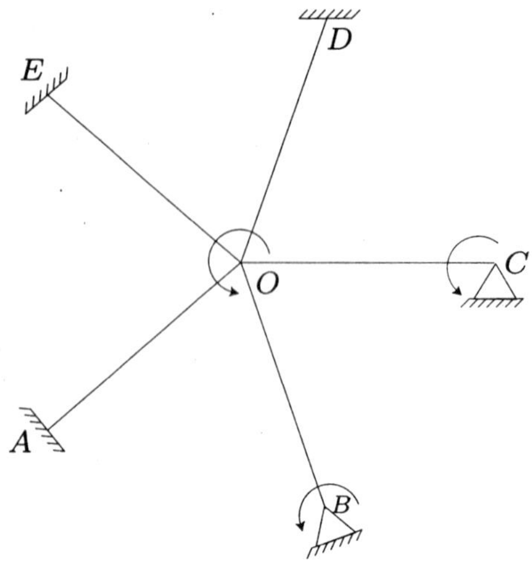

(a) Rigid joint at *O*

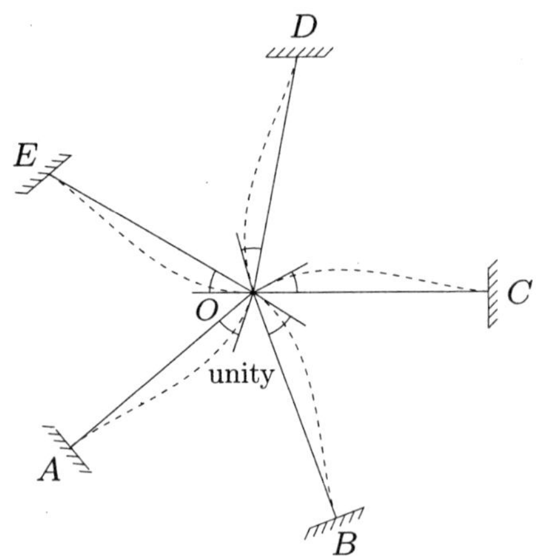

(b) Restrained at joints *B* and *C*

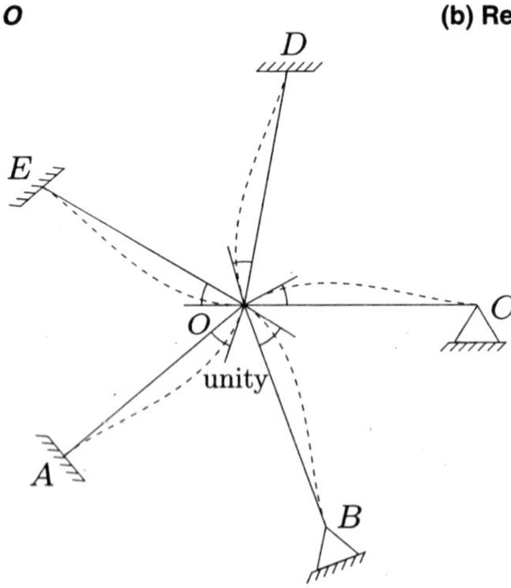

(c) Rotation in rigid joint *O*, for (a)

Fig. 16.14: Rotational stiffness for a rigid joint

The translational (sway) stiffness of a rigid joint in any direction is defined as the force required for a unit translation in the chosen direction without rotation of the joint. As the members are considered inextensible (neglecting effect of axial deformation), it is evident that the translational stiffness of a joint will have a finite value only, if translation of the joint is possible without change of the length of members meeting at the joint. If the translation of the joint in the chosen direction necessarily requires change in length of some of the members meeting at the joint, the translational stiffness of the joint in the chosen direction is to take that into consideration. Using the stiffness method, the computational works can be reduced, if partially restrained nature of the structure or symmetry or antisymmetry of the structures are taken into consideration. The procedure to be used in such cases are illustrated by Examples.

Example 16.15. *Analyse the fixed-end beam shown in Fig. 16.15(a) by flexibility matrix method, the directions of the displacements being as shown in the figure.*

Solution : In order to begin the analysis by the flexibility method, two redundant actions denoted by $R_1 = R_B$ and $R_2 = M_B$ are shown in Fig. 16.15(a). The redundant R_1 is considered positive in the upward direction and R_2, i.e., M_B is assumed positive in the clockwise direction.

The released structure which is a cantilever is shown in Fig. 16.15(b). The displacements Δ_{RL_1} and Δ_{RL_2} due to loads P and M as shown in Fig. 16.15(b) are as follows.

$$\Delta_{RL_1} = -\frac{5PL^3}{48EI} - \frac{3ML^2}{8EI} \quad \text{and} \quad \Delta_{RL_2} = \frac{PL^2}{8EI} + \frac{ML}{2EI}.$$

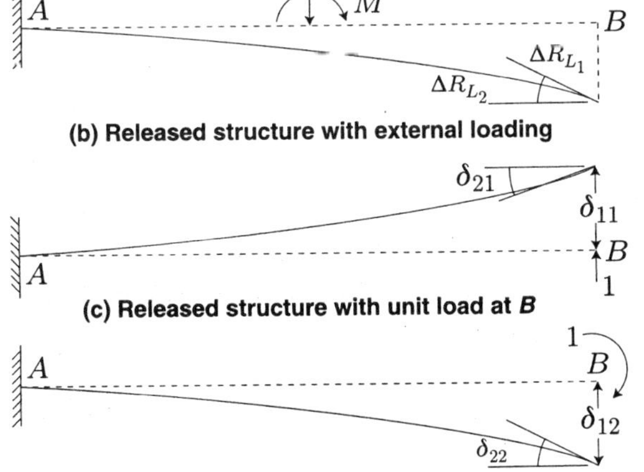

Fig. 16.15: For Example 16.15

Hence the matrix Δ_{RL} can be written as

$$\Delta_{RL} = \frac{L}{48EI}\begin{bmatrix}(-5PL^2 - 18ML)\\(6PL + 24M)\end{bmatrix}.$$

The flexibility influence coefficients are the displacements of the reduced structure due to unit values of R_1 and R_2 as shown in Figs. 16.15(c) and 16.15(d). These coefficients are

$$\delta_{11} = \frac{L^3}{3EI};\ \delta_{12} = \delta_{21} = -\frac{L^2}{2EI};\ \delta_{22} = \frac{L}{EI}.$$

The flexibility matrix δ then becomes

$$\delta = \frac{L}{6EI}\begin{bmatrix}2L^2 & -3L\\-3L & 6\end{bmatrix}.$$

Inverse of the flexibility matrix is

$$\delta^{-1} = \frac{2EI}{L^3}\begin{bmatrix}6 & 3L\\3L & 2L^2\end{bmatrix}.$$

Since displacements in the fixed end beams corresponding to R_1 and R_2 are both zero matrix, $\Delta_R = \begin{bmatrix}0\\0\end{bmatrix}$.

Substituting into Eq. (16.66),

$$\begin{bmatrix}R_1\\R_2\end{bmatrix} = \frac{2EI}{L^3}\begin{bmatrix}6 & 3L\\3L & 2L^2\end{bmatrix}\left\{\begin{bmatrix}0\\0\end{bmatrix} - \frac{L}{48EI}\begin{bmatrix}(-5PL^2 - 18ML)\\(6PL + 24M)\end{bmatrix}\right\}.$$

Solving, $R_1 = \frac{P}{2} + \frac{3M}{2L}$ and $R_2 = \frac{PL}{8} + \frac{M}{4}$.

Example 16.16. *Analyse the continuous beam as shown in Fig. 16.16(a) using flexibility method, the direction of the coordinates 1 and 2 being shown in figure.*

Solution : The released structure is shown in Fig. 16.16(b). The supports B and C have been removed so that the released structure is a cantilever. The displacements in the released structure at B and C due to external load can be calculated as

$$\Delta_{RL_1} = -\frac{47500}{EI} \text{ and } \Delta_{RL_2} = -\frac{128750}{EI},$$

minus sign indicating that the deflection is downward.

In the case, $\Delta R_1 = 0$ and $\Delta R_2 = 0$.

The flexibility matrix may be developed by applying unit force successively as shown in Figs. 16.16(c) and 16.16(d).Thus,

$$\delta_{11} = \frac{1000}{3EI};\ \delta_{12} = \delta_{21} = \frac{2500}{3EI} \text{ and } \delta_{22} = \frac{8000}{3EI}$$

Hence flexibility matrix is given by

$$\delta = \frac{1}{3EI}\begin{bmatrix}1000 & 2500\\2500 & 8000\end{bmatrix}.$$

Substituting in Eq. (16.66),

$$\begin{bmatrix} R_1 \\ R_2 \end{bmatrix} = 3EI \begin{bmatrix} 1000 & 2500 \\ 2500 & 8000 \end{bmatrix}^{-1} \cdot \left\{ \begin{bmatrix} 0 \\ 0 \end{bmatrix} - \begin{bmatrix} \dfrac{-47500}{EI} \\ \dfrac{-128750}{EI} \end{bmatrix} \right\}.$$

Solving, $R_1 = 99.65$ kN and $R_2 = 17.15$ kN.

With the redundants R_B and R_C known, the reaction components may be calculated by using equation of static equilibrium. The resulting B.M. diagram is shown in Fig. 16.16(e).

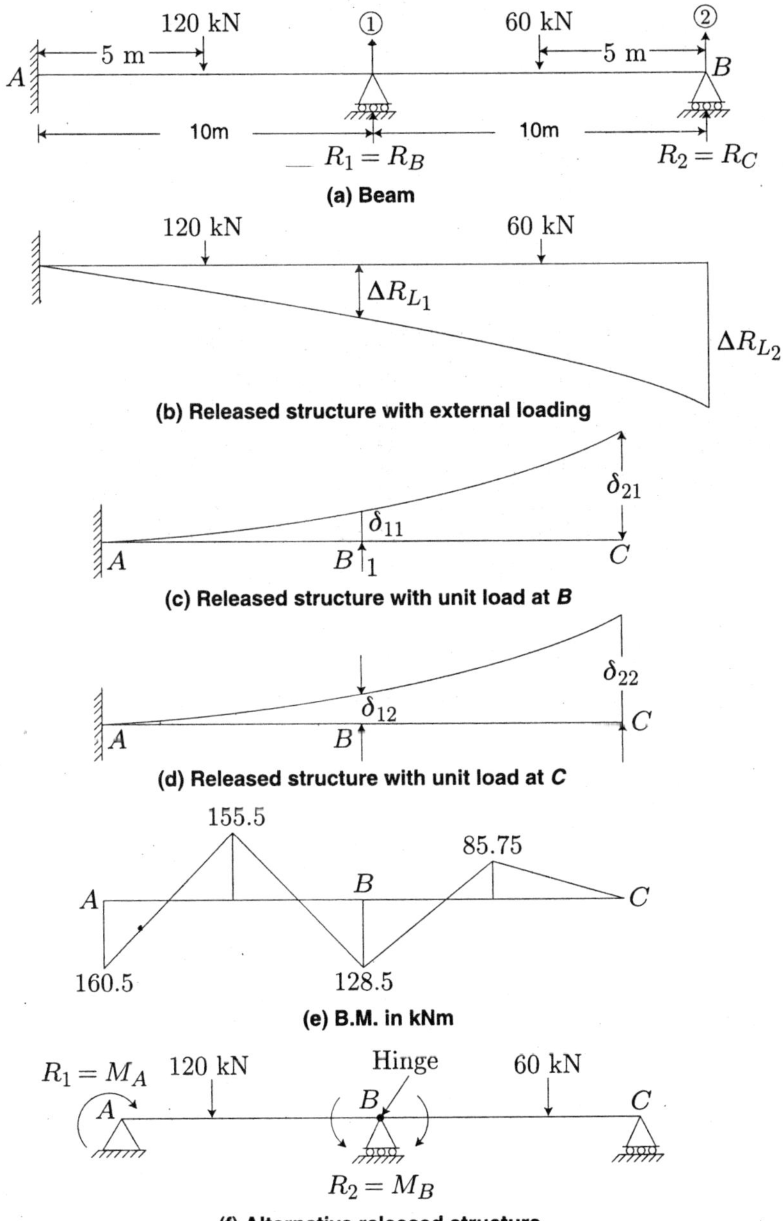

Fig. 16.16: For Example 16.16

The solution can also be done by using a released structure as shown in Fig. 16.16(f). In this case, the redundants are the moments at A and B, i.e., M_A and M_B respectively.

The displacements at A and B due to the external loading are the slope at A and B of the released structure. They are,

$$\Delta_{RL_1} = \theta_A = \frac{750}{EI} \quad \text{and} \quad \Delta_{RL_2} = \theta_{BA} + \theta_{BC} = \theta_B = \frac{1125}{EI}.$$

The flexibility matrix can be obtained by applying unit moment at A and B separately. Thus,

$$\delta_{11} = \frac{10}{3EI}, \delta_{12} = \delta_{21} = \frac{5}{3EI} \quad \text{and} \quad \delta_{22} = \frac{20}{3EI}.$$

Substituting in Eq. (16.66)

$$\begin{bmatrix} R_1 \\ R_2 \end{bmatrix} = 3EI \begin{bmatrix} 10 & 5 \\ 5 & 20 \end{bmatrix}^{-1} \left\{ \begin{bmatrix} 0 \\ 0 \end{bmatrix} - \begin{bmatrix} \frac{750}{EI} \\ \frac{1125}{EI} \end{bmatrix} \right\}.$$

Solving, $R_1 = -160.50 = M_A$ and $R_2 = -128.50 = M_B$.

Example 16.17. *Analyse the beam in Example 16.16, if the downward settlements of suppots at B and C be $\frac{1000}{EI}$ and $\frac{500}{EI}$ respectively.*

Solution : Two alternative methods of solution as in Example 16.16 will be given.

(i) The released structure is a cantilever as in Fig. 16.16(b).

As in Example 16.15.

$$\Delta_{RL_1} = -\frac{47500}{EI} \quad \text{and} \quad \Delta_{Rl_2} = -\frac{128750}{EI}.$$

Also flexibility matrix is given by

$$\delta = \frac{1}{3EI} \begin{bmatrix} 1000 & 2500 \\ 2500 & 8000 \end{bmatrix}.$$

In this case, $\Delta_{R_1} = -\dfrac{1000}{EI}$ and $\Delta_{R_2} = -\dfrac{500}{EI}$.

Substituting in Eq. (16.66),

$$\begin{bmatrix} R_1 \\ R_2 \end{bmatrix} = 3EI \begin{bmatrix} 1000 & 2500 \\ 2500 & 8000 \end{bmatrix}^{-1} \left\{ \begin{bmatrix} \frac{-1000}{EI} \\ \frac{-500}{EI} \end{bmatrix} - \begin{bmatrix} \frac{-47500}{EI} \\ \frac{-128750}{EI} \end{bmatrix} \right\}.$$

Solving, $R_1 = 88.07$ kN and $R_2 = 20.57$ kN. With these reactions at B and C known, the other reaction components may be calculated from static equilibrium. The bending moment diagram as obtained is shown in Fig. 16.17.

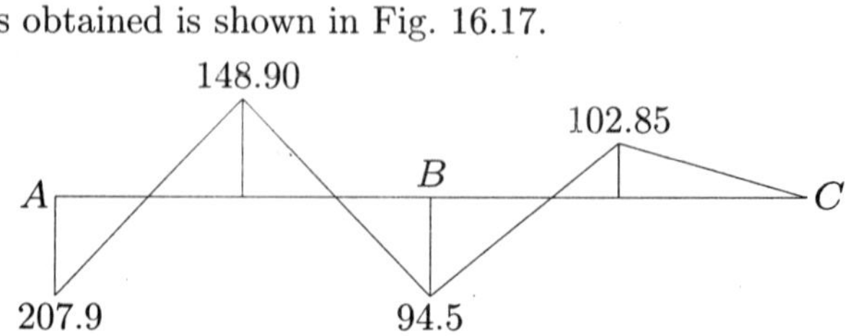

Fig. 16.17: B.M. in kNm for Example 16.17

(ii) The released structure chosen is as in Fig. 16.16(f). In this case, the rotation at A and B of the released structure are given by

$$\Delta_{RL_1} = \theta_A = \frac{750}{EI} + \frac{1000}{10(EI)} = \frac{850}{EI}$$

$$\Delta_{RL_2} = \theta_{BA} + \theta_{BC} = \frac{1125}{EI} - \frac{1000}{10(EI)} - \left(\frac{1000}{EI} - \frac{500}{EI}\right)\frac{1}{10} = \frac{975}{EI}.$$

In this case as in Example 16.15, flexibility matrix is given by

$$\delta = \frac{1}{3EI}\begin{bmatrix} 10 & 5 \\ 5 & 20 \end{bmatrix}.$$

Substituting in Eq. (16.66),

$$\begin{bmatrix} R_1 \\ R_2 \end{bmatrix} = 3EI\begin{bmatrix} 10 & 5 \\ 5 & 20 \end{bmatrix}^{-1}\left\{\begin{bmatrix} 0 \\ 0 \end{bmatrix} - \begin{bmatrix} \frac{850}{EI} \\ \frac{975}{EI} \end{bmatrix}\right\}.$$

Solving, $R_1 = -207.86$ kNm

and $R_2 = M_B = -94.29$ kNm.

Example 16.18. *Analyse the continuous beam shown in Fig. 16.18(a). The downward settlements of supports B and C are $\frac{3000}{EI}$ and $\frac{1500}{EI}$. The directions of coordinates 1 and 2 are shown in Fig. 16.18.*

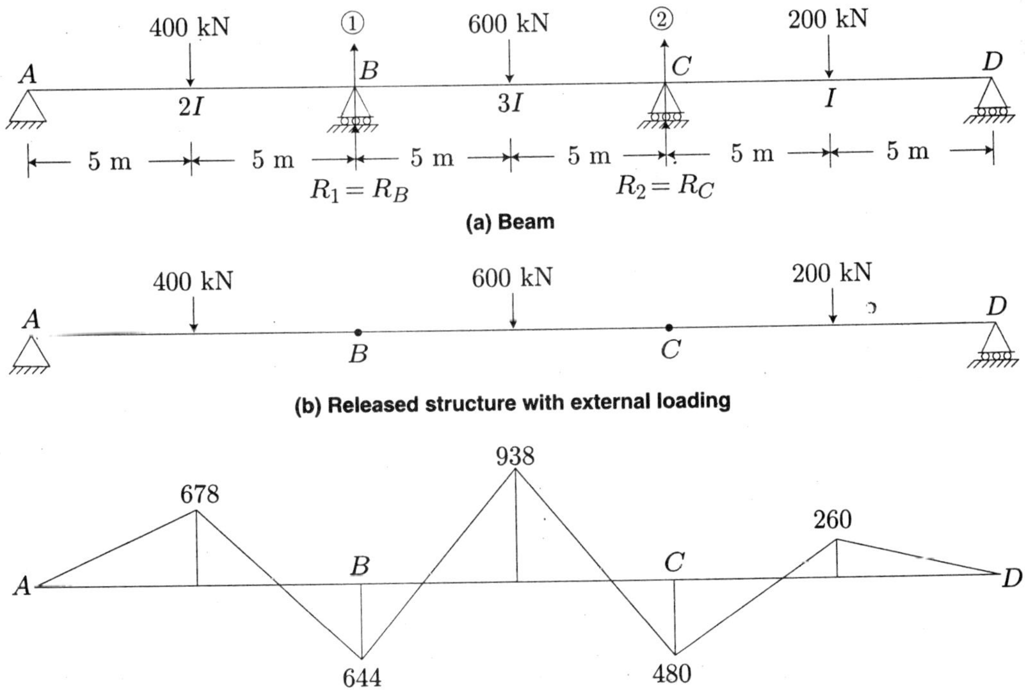

Fig. 16.18: For Example 16.18

Solution : The basic determinate structure may be obtained by considering the reactions at B and C as redundants. The released structure is then as shown in Fig. 16.18(b).

The displacements at B and C due to applied loads and elements of flexibility matrix may be computed using unit load method or Castigliano's Theorem.

They are found as below :

$$\Delta_{RL_1} = -\frac{200320}{EI}$$

$$\Delta_{RL_2} = -\frac{221880}{EI},$$

and $\delta_{11} = \frac{197.53}{EI}$, $\delta_{12} = \delta_{21} = \frac{191.36}{EI}$ and $\delta_{22} = \frac{253.09}{EI}$.

Also $\Delta_{R_1} = -\frac{3000}{EI}$ and $\Delta_{R_2} = -\frac{1500}{EI}$.

Substituting in Eq. (16.66),

$$\begin{bmatrix} R_1 \\ R_2 \end{bmatrix} = EI \begin{bmatrix} 197.53 & 191.36 \\ 191.36 & 253.09 \end{bmatrix}^{-1} \left\{ \begin{bmatrix} \frac{-3000}{EI} \\ \frac{-1500}{EI} \end{bmatrix} - \begin{bmatrix} \frac{-200320}{EI} \\ \frac{-221880}{EI} \end{bmatrix} \right\}.$$

Solving $R_1 = 580.8 \text{ kN} = R_B$ and $R_2 = 431.6 \text{ kN} = R_C$.

Knowing the reactions, the B.M. diagram may be obtained as shown in Fig. 16.18(c).

Example 16.19. *Analyse the beam shown in Fig. 16.19(a) by the force method. The beam rests on elastic supports at B and C, the flexibility of supports B and C being $\frac{10}{EI}$ and $\frac{25}{EI}$ respectively.*

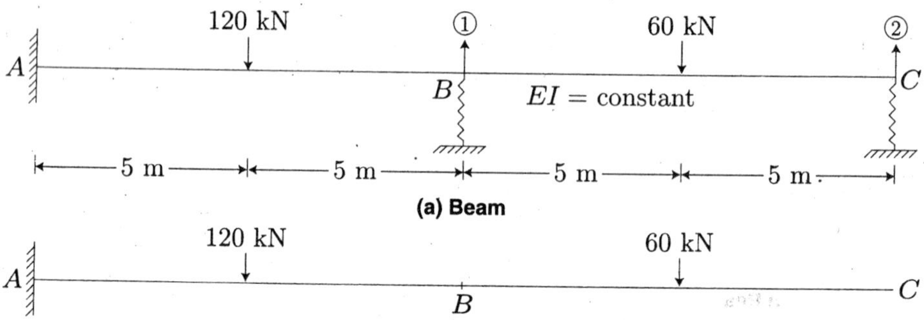

(a) Beam

(b) Released structure with external loading

Fig. 16.19: For Example 16.19

Solution : The released structure is shown in Fig. 15.19(b). As in Example 15.16, the displacements of B and C of the released structure are

$$\Delta_{RL_1} = -\frac{47500}{EI} \text{ and } \Delta_{RL_2} = -\frac{128750}{EI}.$$

As in Example 16.15,

Coefficients of matrices are given by

$$\delta_{11} = -\frac{1000}{3EI}, \delta_{12} = \delta_{21} = \frac{2500}{EI} \text{ and } \delta_{22} = \frac{8000}{3EI}.$$

Since flexibilities of support at B and C are $\frac{10}{EI}$ and $\frac{25}{EI}$ respectively, the modified flexibility matrix is given by

$$\frac{1}{EI}\begin{bmatrix} \left(\frac{1000}{3}+10\right) & \frac{2500}{3} \\ \frac{2500}{3} & \left(\frac{8000}{3}+25\right) \end{bmatrix}.$$

Substituting in Eq. (16.66) as before,

$$R_1 = 89.50 \text{ kN} = R_B \quad \text{and} \quad R_2 = 20.10 \text{ kN} = R_C.$$

The resulting reactions and B.M. can then be obtained by statics.

Example 16.20. *Analyse the Portal frame in Fig. 16.20(a) using force method, the direction of the coordinates being as shown.*

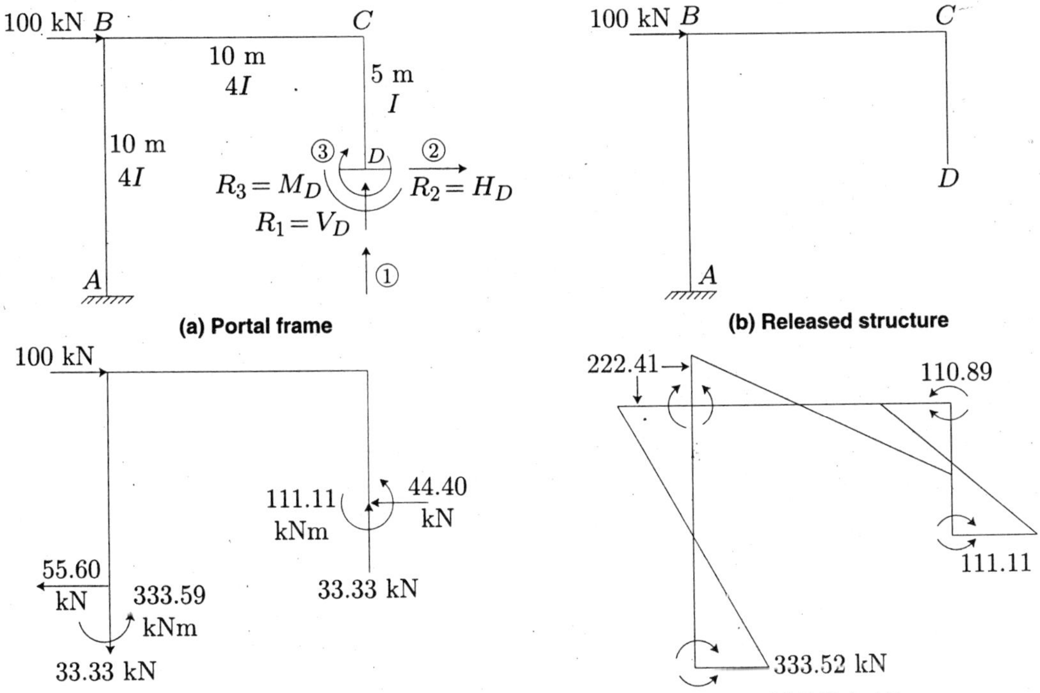

Fig. 16.20: For Example 16.20

Solution : The frame is statically indeterminate to the third degree. The released structure is shown in Fig. 16.20(b), considering the three reactions at D as redundants.

The displacements in the directions of the three redundants can be obtained using unit load method or any other suitable method. Thus,

$$\Delta_{RL_1} = -\frac{12500}{EI}$$

$$\Delta_{RL_2} = -\frac{6250}{3EI}$$

and $$\Delta_{RL_3} = -\frac{1250}{EI}.$$

Then coefficients of flexibility matrix can also be obtained similarly. Thus,

$$\delta_{11} = \frac{1000}{3EI}; \ \delta_{12} = \delta_{21} = \frac{125}{2EI}$$

$$\delta_{13} = \delta_{31} = -\frac{75}{2EI}; \ \delta_{22} = \frac{125}{2EI}$$

$$\delta_{23} = \delta_{32} = -\frac{25}{EI} \ \text{ and } \ \delta_{33} = \frac{10}{EI}.$$

The flexibility matrices can then be written as

$$\delta = \frac{1}{6EI}\begin{bmatrix} 2000 & 375 & -225 \\ 375 & 375 & -150 \\ -225 & -150 & 60 \end{bmatrix}.$$

As supports do not yield, substituting in Eq. (16.66),

$$\begin{bmatrix} R_1 \\ R_2 \\ R_3 \end{bmatrix} = 6EI\begin{bmatrix} 2000 & 375 & -225 \\ 375 & 375 & -150 \\ -225 & -150 & 60 \end{bmatrix}^{-1}\left\{\begin{bmatrix} 0 \\ 0 \\ 0 \end{bmatrix} - \begin{bmatrix} \frac{-12500}{EI} \\ \frac{-6250}{3EI} \\ \frac{1250}{EI} \end{bmatrix}\right\}.$$

Solving, $R_1 = V_D = 33.33$ kN

$R_2 = H_D = -44.40$ kN

and $R_3 = M_D = -111.11$ kNm.

Knowing these values, the B.M. diagram can be obtained from statics.

Example 16.21. *Analyse the portal frame in Example 16.20, if settlements of the supports at D be $\frac{178.6}{EI}$ to the right and $\frac{446.4}{EI}$ downwards.*

Solution : Values of Δ_{RL} and flexibility matrix are the same as in Example 16.20 Substituting in Eq. (16.66),

$$\begin{bmatrix} R_1 \\ R_2 \\ R_3 \end{bmatrix} = 6EI\begin{bmatrix} 2000 & 375 & -225 \\ 375 & 375 & -150 \\ -225 & -150 & 60 \end{bmatrix}^{-1}\left\{\begin{bmatrix} \frac{-446.4}{EI} \\ \frac{178.6}{EI} \\ 0 \end{bmatrix} - \begin{bmatrix} \frac{-12500}{EI} \\ \frac{-6250}{3EI} \\ \frac{1250}{EI} \end{bmatrix}\right\}.$$

Solving, $R_1 = 27.62 \text{ kN} = V_D$,

$R_2 = -65.40 \text{ kN} = H_D$

and $R_3 = -184.94 \text{ kNm} = M_D$.

Example 16.22. *Analyse the portal frame in Example 16.20, if the flexibilities of the support D for horizontal and vertical displacements be $\frac{10}{EI}$ and $\frac{20}{EI}$ respectively. Use force method.*

Solution : As in case of Example 16.20,

$$\Delta_{RL_1} = -\frac{12500}{EI} \quad \Delta_{RL_2} = \frac{-6250}{EI} \quad \text{and} \quad \Delta_{RL_3} = \frac{1250}{EI}.$$

Flexibility matrix is given by

$$\delta = \frac{1}{6EI}\begin{bmatrix} 2000 & 375 & -225 \\ 375 & 375 & -150 \\ -225 & -150 & 60 \end{bmatrix}.$$

Modified flexibility is given by

$$\delta_M = \frac{1}{6EI}\begin{bmatrix} 2120 & 375 & -225 \\ 375 & 435 & -150 \\ -225 & -150 & 60 \end{bmatrix}.$$

Substituting in Eq. (16.66),

$$\begin{bmatrix} R_1 \\ R_2 \\ R_3 \end{bmatrix} = 6EI\begin{bmatrix} 2120 & 375 & -225 \\ 375 & 435 & -150 \\ -225 & -150 & 60 \end{bmatrix}^{-1}\left\{\begin{bmatrix} 0 \\ 0 \\ 0 \end{bmatrix} - \begin{bmatrix} \frac{-12500}{EI} \\ \frac{-6250}{3EI} \\ \frac{1250}{EI} \end{bmatrix}\right\}.$$

Solving, $R_1 = 39.61$ kN $= V_D$; $R_2 = -19.62$ kN $= H_D$ and $R_3 = -12.58$ kNm $= M_D$.

Example 16.23. *Formulate the equation in matrix form for the rigid frame in Fig. 16.21(a) considering effect of both flexure and axial deformations. Use force method.*

Solution : The released structure with the released action, R_1, R_2 and R_3 is shown in Fig. 16.21(b). The displacement in the released structure corresponding to R_1 (i.e., V_B), R_2 (i.e., M_B) and R_3 (i.e., H_B) due to external load P as in Fig. 16.21(c) can be calculated by unit load method and their values are given by

$$\Delta_{RL_1} = -\frac{5PL^3}{48EI}, \quad \Delta_{RL_2} = -\frac{PL^2}{8EI} \quad \text{and} \quad \Delta_{RL_3} = 0.$$

The coefficients for flexibility matrix are given by

$$\delta_{11} = \frac{L^2}{3EI} + \frac{H}{EA}; \; \delta_{12} = \delta_{21} = \frac{L^2}{2EI}; \; \delta_{13} - \delta_{31} = 0$$

$$\delta_{22} = \frac{L}{EI} + \frac{H}{EI}; \; \delta_{23} = \delta_{32} = -\frac{H^2}{2EI} \quad \text{and} \quad \delta_{33} = \frac{L}{EA} + \frac{H^3}{3EI}.$$

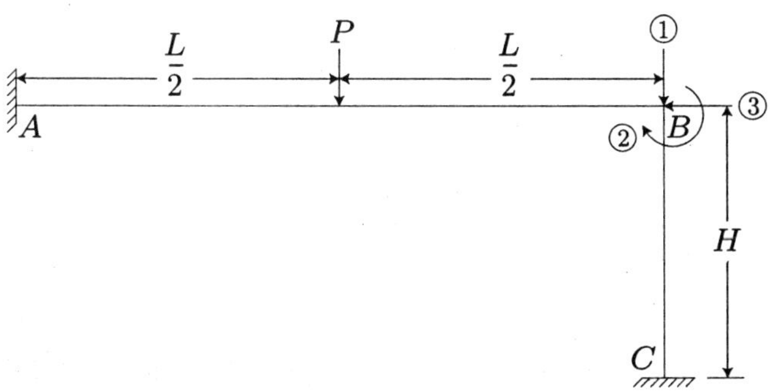

(a) Rigid frame

Fig. 16.21: For Example 16.23 (Contd.)

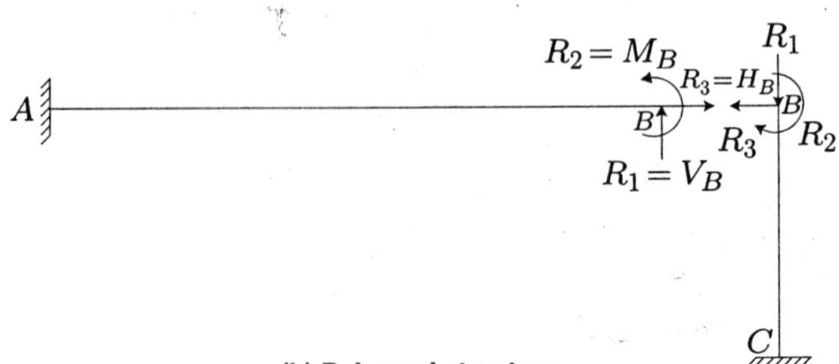

(b) Released structure

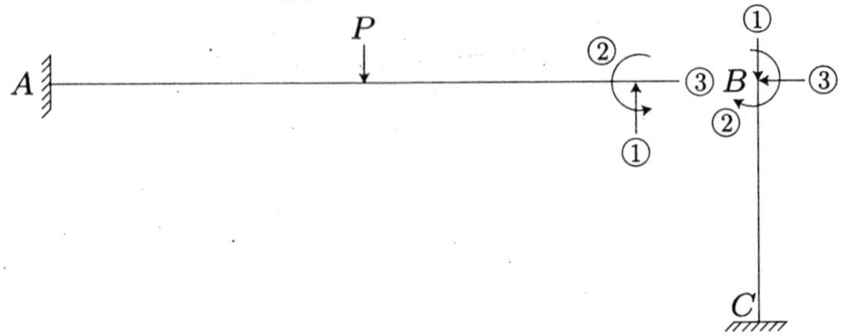

(c) Effect of external load on released structure

Fig. 16.21: For Example 16.23

The flexibility matrix can be written as

$$\delta = \begin{bmatrix} \left(\frac{L^3}{3EI} + \frac{H}{EA}\right) & \frac{L^2}{2EI} & 0 \\ \frac{L^2}{2EI} & \left(\frac{L}{EI} + \frac{H}{EI}\right) & -\frac{H^2}{2EI} \\ 0 & -\frac{H^2}{2EI} & \left(\frac{L}{EA} + \frac{H^3}{3EI}\right) \end{bmatrix}.$$

Substituting in Eq. (16.66),

$$\begin{bmatrix} R_1 \\ R_2 \\ R_3 \end{bmatrix} = \begin{bmatrix} \left(\frac{L63}{3EI} + \frac{H}{EA}\right) & \frac{L^2}{2EI} & 0 \\ \frac{L^2}{2EI} & \left(\frac{L}{EI} + \frac{H}{EI}\right) & -\frac{H^2}{2EI} \\ 0 & -\frac{H^2}{2EI} & \left(\frac{L}{EA} + \frac{H^3}{3EI}\right) \end{bmatrix}^{-1} \left\{ \begin{bmatrix} 0 \\ 0 \\ 0 \end{bmatrix} - \begin{bmatrix} \frac{5PL^3}{48EI} \\ -\frac{PL^2}{8EI} \\ 0 \end{bmatrix} \right\}.$$

Example 16.24. *Formulate the equation in matrix form for the grid structure shown in Fig. 16.22(a). Use flexibility matrix.*

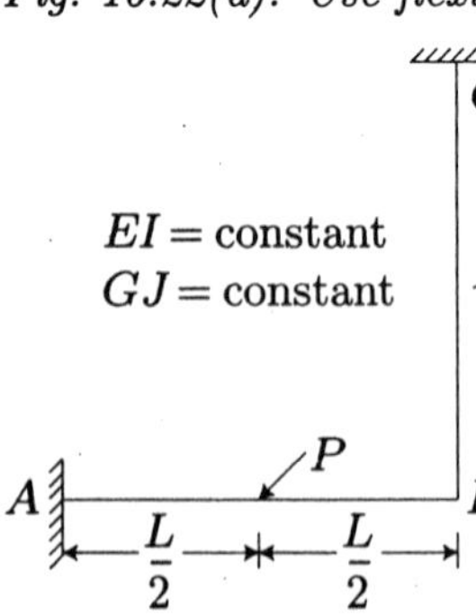

(a) Grid structure

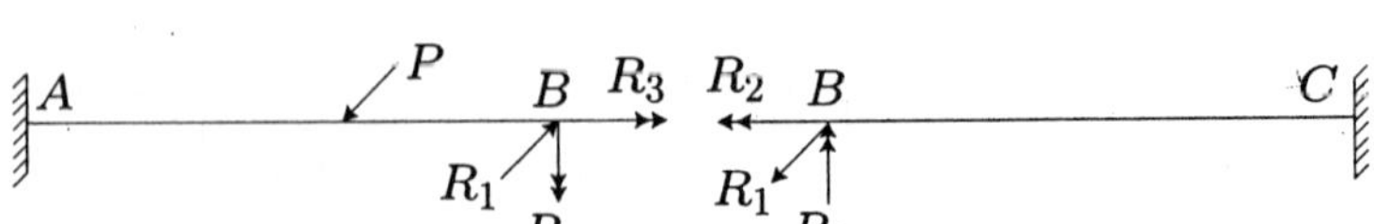

(b) Released structure

Fig. 16.22: For Example 16.24

Solution : The released structure with the reactions R_1, R_2 and R_3 are shown in Fig. 16.22(b). The displacement of B due to the application of the external load P are given by

$$\Delta_{RL_1} = -\frac{5PL^3}{48EI}, \; \Delta_{RL_2} = -\frac{PL^2}{8EI} \quad \text{and} \quad \Delta_{RL_3} = 0.$$

The flexibility matrix can be formed as

$$\delta = \begin{bmatrix} \frac{2L^3}{3EI} & -\frac{L^2}{2EI} & \frac{L^2}{2EI} \\ -\frac{L^2}{2EI} & \left(\frac{L}{EI}+\frac{L}{GJ}\right) & 0 \\ \frac{L^2}{2EI} & 0 & \left(\frac{L}{EI}+\frac{L}{GJ}\right) \end{bmatrix}$$

$$= \frac{L}{6EI}\begin{bmatrix} 4L^2 & -3L & 3L \\ -3L & 6(1+\alpha) & 0 \\ 3L & 0 & 6(1+\alpha) \end{bmatrix}, \text{ where } \alpha = \frac{EI}{GJ}.$$

Substituting in Eq. (16.66),

$$\begin{bmatrix} R_1 \\ R_2 \\ R_3 \end{bmatrix} = \frac{6EI}{L}\begin{bmatrix} 4L^2 & -3L & 3L \\ -3L & 6(1+\alpha) & 0 \\ 3L & 0 & 6(1+\alpha) \end{bmatrix}^{-1} \left\{ \begin{bmatrix} 0 \\ 0 \\ 0 \end{bmatrix} - \begin{bmatrix} -\frac{5PL^3}{48EI} \\ -\frac{PL^2}{8EI} \\ 0 \end{bmatrix} \right\}.$$

Solving, $R_1 = \frac{P}{8}\frac{2+5\alpha}{1+4\alpha}$

$$R_2 = \frac{3PL}{16}\frac{\alpha}{(1+\alpha)(1+4\alpha)}.$$

and $$R_3 = \frac{PL}{16}\frac{2+5\alpha}{(1+\alpha)(1+4\alpha)}.$$

Example 16.25. *Analyse the pin-jointed frame shown in Fig. 16.23(a) using force mehtod. The flexibility of each member is* 0.050 *mm/kN.*

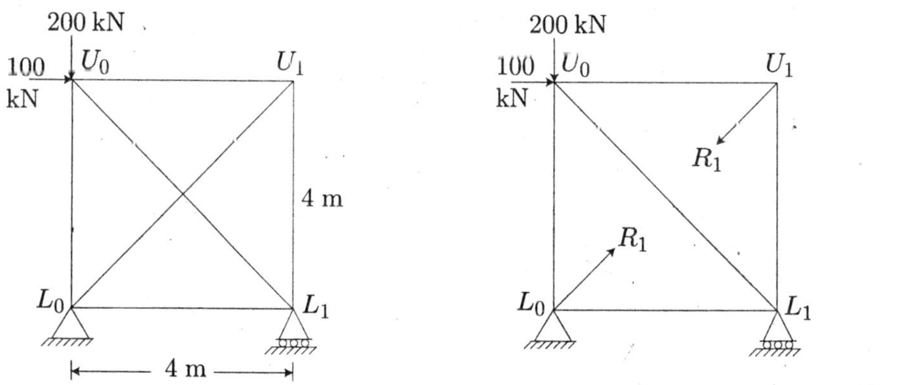

(a) Frame **(b) Released structure with indeterminate R_1**

Fig. 16.23: For Example 16.25

Solution : The degree of static indeterminancy of the frame is 1. The member L_0U_1 may be considered as indeterminate. Using unit load method, the displacement in the direction L_0U_1 is given by $\Delta_{RL_1} = -7.07$, including deformation of the redundant member.

Also $\delta_{11} = 0.20$.

Substituting in Eq. (16.66),

$$R_1 = [0.20]^{-1}\{[0] - [-7.07]\}, \quad \text{whence} \quad R_1 = 35.35 \text{ kN}.$$

Various forces in the member may then be obtained from statics.

Example 16.26. *Analyse the frame in Example 16.25, if the member L_0U_1 be too long by 2.5 mm.*

Solution : Using the same values of Δ_{RL_1} and δ_{11} as in Example 16.25, substitution in Eq. (16.64) gives

$$R_1 = [0.20]^{-1}\{[-2.5] - [-7.07]\}.$$

Solving, $R_1 = 22.85$ kN.

Example 16.27. *Analyse the frame shown in Fig. 16.24(a) using force method. Axial rigidity of the horizontal and vertical members is EA and for diagonal members is $2EA$.*

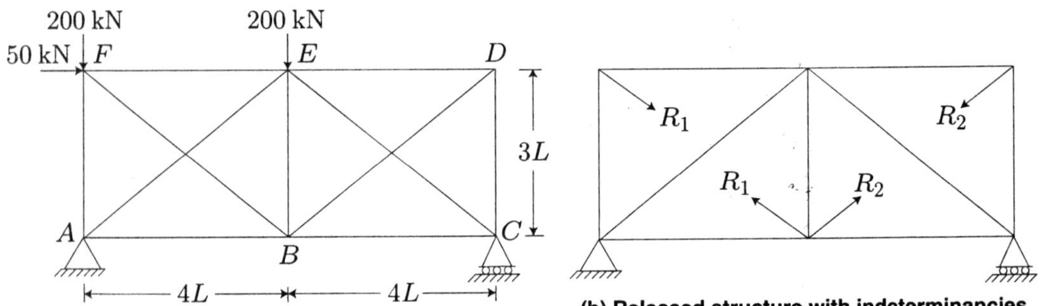

(a) Frame

(b) Released structure with indeterminancies R_1 and R_2

Fig. 16.24: For Example 16.27

Solution : The plane truss has two degrees of statical indeterminancies. The diagonals BF and BD are selected as redundants. These two bars are cut to make the structure statically determinate. The cut bars remain a part of the released structure and its deformations are to be included in the calculation of displacements.

Using unit load methods, displacements due to external directions in the direction of the redundants are given by

$$\Delta_{RL_1} = \frac{183L}{EA} \quad \text{and} \quad \Delta_{RL_2} = -\frac{579L}{EA}.$$

The coefficients of flexibility matrices are given by

$$\delta_{11} = \frac{8.64L}{EA}; \; \delta_{12} = \delta_{21} = \frac{1.08L}{EA} \quad \text{and} \quad \delta_{22} = \frac{8.64L}{EA}.$$

Substituting in Eq. (16.66),

$$\begin{bmatrix} R_1 \\ R_2 \end{bmatrix} = \frac{EA}{L}\begin{bmatrix} 8.64 & 1.08 \\ 1.08 & 8.64 \end{bmatrix}^{-1} \left\{ \begin{bmatrix} 0 \\ 0 \end{bmatrix} - \begin{bmatrix} \frac{183L}{EA} \\ -\frac{579L}{EA} \end{bmatrix} \right\}.$$

Solving, $R_1 = -30.03$ kN and $R_2 = 70.68$ kN.

If there had been lack of fit in the members by, say, S_1 and S_2, then Eq. (16.66) will give,

$$\begin{bmatrix} R_1 \\ R_2 \end{bmatrix} = \frac{EA}{L}\begin{bmatrix} 8.64 & 1.08 \\ 1.08 & 8.64 \end{bmatrix}^{-1}\left\{\begin{bmatrix} S_1 \\ S_2 \end{bmatrix} - \begin{bmatrix} \frac{183L}{EA} \\ -\frac{579L}{EA} \end{bmatrix}\right\}.$$

Example 16.28. *Analyse the continuous beam shown in Fig. 16.25(a) using stiffness matrix.*

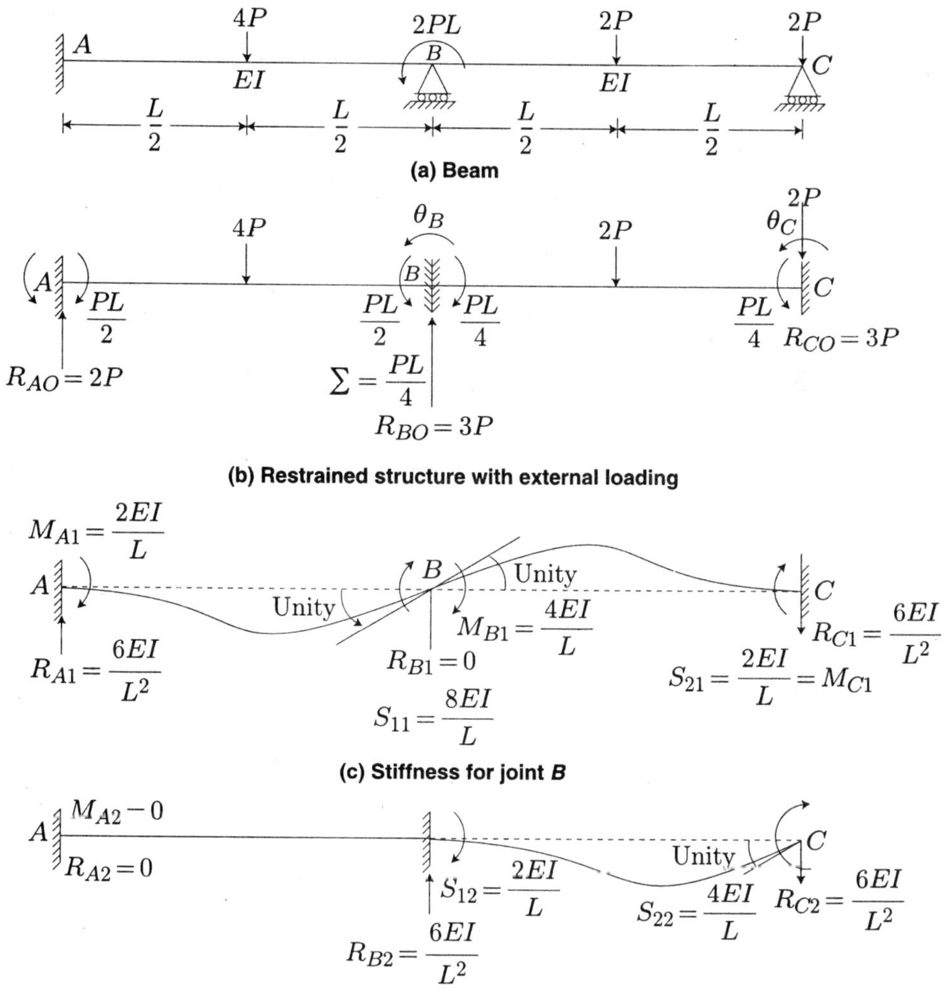

Fig. 16.25: For Example 16.28

Solution : The unknown joint displacements for the beam are the rotations at supports B and C denoted by Δ_1, i.e., θ_B and Δ_2, i.e., θ_C, respectively as shown in Fig. 16.25(b). The vector A_Δ of the actions corresponding to the unknown displacements are

$$A_{\Delta 1} = 2PL \quad \text{and} \quad A_{\Delta 2} = 0$$

i.e.,
$$A_\Delta = \begin{bmatrix} 2PL \\ 0 \end{bmatrix}.$$

The remaining loads on the beam are taken into account by considering them to act on the restrained structure shown in Fig. 16.25(b). The structure consists of fixed end beams and is obtained by preventing joints B and C from rotating. The actions $A_{\Delta L_1}$ and $A_{\Delta L_2}$ exerted on the beam by artificial restraints are the couples at supports B and C.

Each of the couples is an action corresponding to the displacements Δ_1 and Δ_2 caused by loads on the beam and can be obtained referring to Fig. 16.25(b). Thus,

$$A_{\Delta L_1} = -\frac{PL}{4} \quad \text{and} \quad A_{\Delta L_2} = -\frac{PL}{4},$$

whence $A_{\Delta L} = \dfrac{PL}{4}\begin{bmatrix} -1 \\ -1 \end{bmatrix}$.

The stiffness matrix consists of stiffness coefficients shown in Figs. 16.25(c) and 16.25(d). Each coefficient is a couple corresponding to one of the unknown displacements and due to a unit value of one of the displacements.

The task of calculating the joint stiffnesses S_{11} and S_{21} in Fig. 16.25(c) may now be performed through the use of member stiffnesses. When the beam is rotated through a unit angle at joint B, a moment equal to $\frac{4EI}{L}$ is developed at B because of the rotation of the end member AB. A moment equal to $\frac{4EI}{L}$ is also developed at B, because of the rotation of the end members BC. Thus the total moment at B, equal to S_{11} is

$$S_{11} = \frac{4EI}{L} + \frac{4EI}{L} = \frac{8EI}{L}.$$

The stiffness S_{21} is the moment developed at joint C when joint B is rotated through a unit angle. Since joint C is at the end of the member, the stiffness coefficient is

$$S_{21} = \frac{2EI}{L}.$$

Both S_{11} and S_{21} are positive since the joint members are clockwise. The stiffness coefficients S_{12} and S_{22} are shown in Fig. 16.25(d). The first of these is equal to $\frac{2EI}{L}$, since it is an action at the far end of the member BC, while the latter is equal to $\frac{4EI}{L}$, since it is at the near end of the member.

The stiffness matrix S can be formed from the stiffnesses coefficients and is given by

$$S = \begin{bmatrix} \frac{8EI}{L} & \frac{2EI}{L} \\ \frac{2EI}{L} & \frac{4EI}{L} \end{bmatrix} = \frac{EI}{L}\begin{bmatrix} 8 & 2 \\ 2 & 4 \end{bmatrix}.$$

In general, it will be found that stiffness elements on the principal diagonal are always composed of near end stiffnesses, while those of the diagonal may consist of either far end or near end stiffnesses.

Substituting in Eq. (16.73),

$$\begin{bmatrix} \theta_B \\ \theta_C \end{bmatrix} = \frac{L}{EI}\begin{bmatrix} 8 & 2 \\ 2 & 4 \end{bmatrix}^{-1} \left\{ \begin{bmatrix} 2PL \\ 0 \end{bmatrix} - \frac{PL}{4}\begin{bmatrix} -1 \\ -1 \end{bmatrix} \right\}.$$

Solving $\theta_B = \Delta_1 = \dfrac{17PL^2}{56EI}$ and $\theta_C = \Delta_2 = -\dfrac{5PL^2}{56EI}$.

Various moments may be obtained by slope deflection equation as given in Eq. (16.74).

$$M_{AB} = \frac{PL}{2} + \frac{2EI}{L}\left(0 + \frac{17PL^2}{56EI}\right)$$

$$= \frac{31}{28}PL$$

$$M_{BA} = -\frac{PL}{2} + \frac{2EI}{L}\left[0 + \frac{2(17PL^2)}{56EI}\right]$$

$$= \frac{5}{7}PL$$

$$M_{BC} = \frac{PL}{4} + \frac{2EI}{L}\left[\frac{2(17PL^2)}{56EI} - \frac{5PL^2}{56EI}\right]$$

$$= \frac{9}{7}PL$$

$$\sum M_B = \frac{5}{7}PL + \frac{9}{7}PL - 2PL = 0$$

$$M_{CB} = -\frac{PL}{4} + \frac{2EI}{L}\left[\frac{17PL^2}{56EI} - \frac{2(5PL^2)}{56EI}\right] = 0.$$

Various functions may also be obtained in matrix form as in Eq. (16.75). Referring to Figs. 16.25(b) to 16.25(d),

$$\begin{bmatrix} M_{AB} \\ R_A \\ M_{BA} \\ R_B \\ M_{CB} \\ R_C \end{bmatrix} = \begin{bmatrix} \frac{PL}{2} \\ +2P \\ -\frac{PL}{2} \\ +3P \\ -\frac{PL}{4} \\ +3P \end{bmatrix} + \begin{bmatrix} \frac{2EI}{L} & 0 \\ \frac{6EI}{L^2} & 0 \\ \frac{4EI}{L} & 0 \\ 0 & \frac{6EI}{L^2} \\ \frac{2EI}{L} & \frac{4EI}{L} \\ -\frac{6EI}{L^2} & \frac{6EI}{L^2} \end{bmatrix} \begin{bmatrix} \frac{17PL^2}{56EI} \\ -\frac{5PL^2}{56EI} \end{bmatrix}$$

Alternatively considering a partially restrained structures with kinematic indeterminancy only at B,

$$S = \frac{4EI}{L} + \frac{3EI}{L} = \frac{7EI}{L}.$$

In this case, $A_\Delta = 2PL$

and $$A_{\Delta L} = -\frac{PL}{2} + \frac{3PL}{8} = -\frac{PL}{8},$$

whence $$\theta_B = \frac{L}{7EI}\left\{2PL - \left(-\frac{PL}{8}\right)\right\} = \frac{17PL^2}{56EI}$$

$$M_{BA} = \frac{5}{7}PL$$

and $$M_{BC} = 2PL - \frac{5}{7}PL = \frac{9}{7}PL.$$

Example 16.29. *Analyse the continuous beam shown in Fig. 16.26(a) by displacement method.*

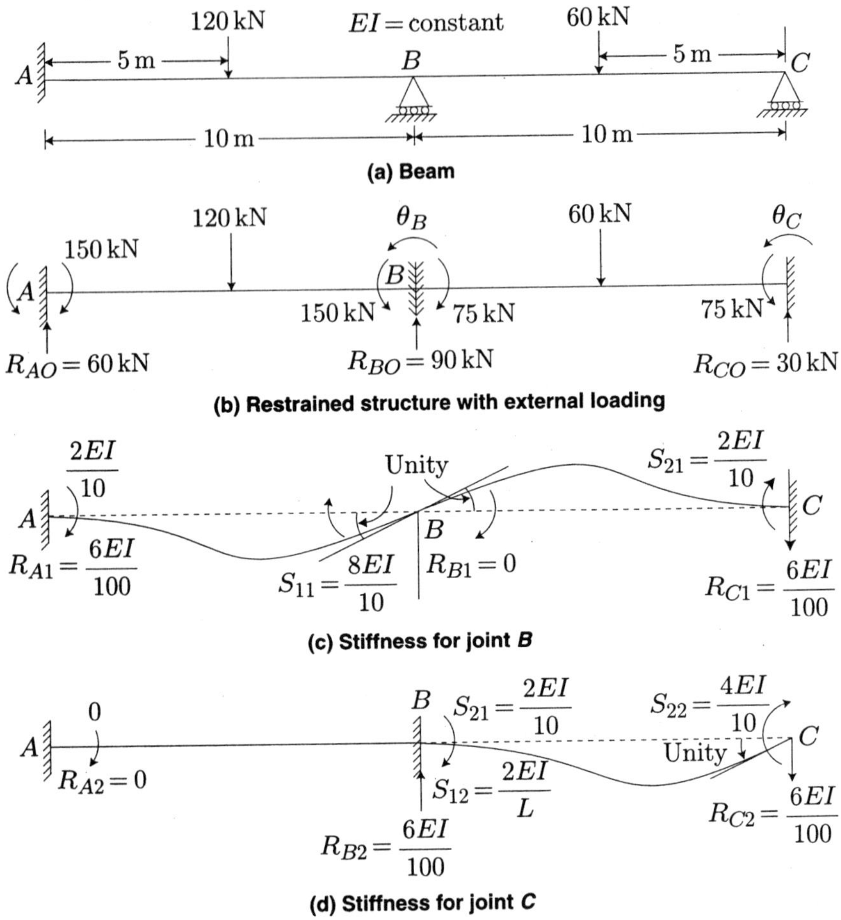

(a) Beam

(b) Restrained structure with external loading

(c) Stiffness for joint *B*

(d) Stiffness for joint *C*

Fig. 16.26: For Example 16.29

Solution : In this case, the only two unknown displacements are the rotations at B and C. Hence the degree of freedom is two. The restrained structure is shown in Fig. 16.26(b).

In this case, $A_\Delta = \begin{bmatrix} 0 \\ 0 \end{bmatrix}$ and $A_{\Delta L} = \begin{bmatrix} -75.0 \\ -75.0 \end{bmatrix}$

Stiffness matrix may be obtained from Figs. 16.26(c) and 16.26(d). Thus,

$$S = \begin{bmatrix} \frac{8EI}{10} & \frac{2EI}{10} \\ \frac{2EI}{10} & \frac{4EI}{10} \end{bmatrix} = EI \begin{bmatrix} 0.80 & 0.20 \\ 0.20 & 0.40 \end{bmatrix}.$$

Substituting in Eq. (16.73),

$$\begin{bmatrix} \theta_B \\ \theta_C \end{bmatrix} = \frac{1}{EI} \begin{bmatrix} 0.80 & 0.20 \\ 0.20 & 0.40 \end{bmatrix}^{-1} \left\{ \begin{bmatrix} 0 \\ 0 \end{bmatrix} - \begin{bmatrix} -75.00 \\ -75.00 \end{bmatrix} \right\}.$$

Solving, $\theta_B = \Delta_1 = \frac{53.57}{EI}$ and $\theta_C = \Delta_2 = \frac{160.72}{EI}$.

From slope deflection equation,

$$M_{AB} = 150 + \frac{2EI}{10}\left[0 + \frac{53.57}{EI}\right] = 160.70 \text{ kNm}$$

$$M_{BA} = -150 + \frac{2EI}{10}\left[\frac{2(53.57)}{EI} + 0\right] = -128.60 \text{ kNm}$$

$$M_{BC} = 75.0 + \frac{2EI}{10}\left[\frac{(160.72)}{EI} + \frac{2(53.57)}{EI}\right] = +128.60 \text{ kNm}$$

$$M_{CB} = -75.0 + \frac{2EI}{10}\left[\frac{2(160.72)}{EI} + \frac{53.57}{EI}\right] = 0.$$

Alternatively in matrix form

$$\begin{bmatrix} M_{AB} \\ R_A \\ M_{BA} \\ R_B \\ R_C \end{bmatrix} = \begin{bmatrix} 15 \\ 6 \\ -15 \\ 9 \\ 3 \end{bmatrix} + \begin{bmatrix} \frac{2EI}{10} & 0 \\ \frac{6EI}{100} & 0 \\ \frac{4EI}{10} & 0 \\ 0 & \frac{6EI}{100} \\ -\frac{6EI}{100} & -\frac{6EI}{100} \end{bmatrix} \begin{bmatrix} \frac{53.57}{EI} \\ \frac{160.72}{EI} \end{bmatrix}.$$

Alternatively, with a partially restrained structure considering indeterminancy at B only,

$$S = \frac{7EI}{10}; A_\Delta = 0 \text{ and } A_{\Delta L_1} = -150 + [75.0 + 37.5] = -37.50,$$

whence $\theta_B = \frac{10}{7EI}(37.5) = \frac{53.57}{EI}$

and $$M_{BA} = -150 + \frac{4EI}{10}\left(\frac{53.57}{EI}\right) = 128.572 \text{ kNm}$$

$$M_{BC} = +112.5 + \frac{3EI}{10}\left(\frac{53.57}{EI}\right) = 128.572 \text{ kNm}.$$

Example 16.30. *Analyse the beam in Example 16.29, if the downward settlements of supports at B and C are* $\frac{1000}{EI}$ *and* $\frac{500}{EI}$ *respectively.*

Solution : In this case, the fixed-end moments due to relative settlements are to be considered in calculating fixed end moments. Thus,

$$M_{FAB} = 150 + \frac{6EI}{(10)^2}\left(\frac{1000}{EI}\right) = 210 \text{ kNm}$$

$$M_{FBA} = -150 + 60 = -90 \text{ kNm}$$

$$M_{FBC} = +75.0 - \frac{6EI}{(10)^2}\left(\frac{500}{EI}\right) = +45 \text{ kNm}$$

$$M_{FCB} = -75 - 30 = -105 \text{ kNm}.$$

It follows that,

$$A_{\Delta L_1} = -90 + 45.0 = -45.0$$

and $$A_{\Delta L_2} = -105.0.$$

Taking stiffness matrix as in Example 16.29 and substituting in Eq. (16.75),

$$\begin{bmatrix} \theta_B \\ \theta_C \end{bmatrix} = \frac{1}{EI}\begin{bmatrix} 0.80 & 0.20 \\ 0.20 & 0.40 \end{bmatrix}^{-1}\left\{\begin{bmatrix} 0 \\ 0 \end{bmatrix} - \begin{bmatrix} 45.0 \\ -105.0 \end{bmatrix}\right\}.$$

Solving, $\theta_B = -\dfrac{10.72}{EI}$ and $\theta_C = \dfrac{267.85}{EI}$.

Resulting moments can be obtained as before,

$$M_{AB} = 208.0 \text{ kNm}; \; M_{BA} = 94.50 \text{ kNm}.$$

Alternatively, considering a partially restrained structure with kinematic indeterminancy at B only, i.e., θ_B only, $S = \dfrac{7EI}{10}$.

In this case, $A_{\Delta L} = -37.5 + 60 - 15.0 = 7.50$

and $$\theta_B = \frac{10}{7EI}[0 - (7.50)] = -\frac{10.72}{EI}.$$

$$M_{AB} = 208.0 \text{ kNm and } M_{BA} = 94.50 \text{ kNm}.$$

Example 16.31. *Analyse the continuous beam in Fig. 16.27 using displacement method considering EI as constant.*

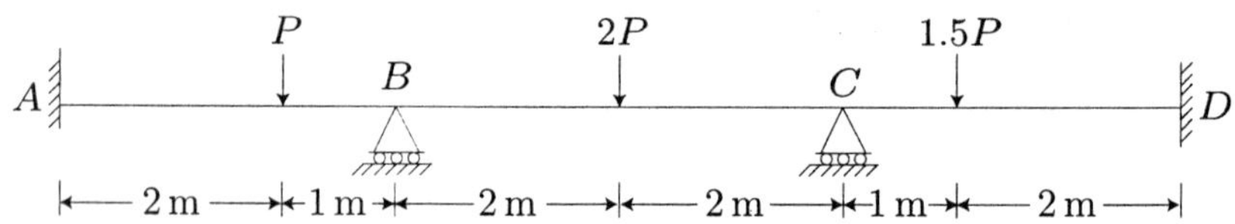

Fig. 16.27: For Example 16.31

Solution : In this case the unknown displacement components are the rotation at B and C and the degree of freedom is two.

Hence, $$A_\Delta = \begin{bmatrix} 0 \\ 0 \end{bmatrix}$$

$$M_{FAB} = \frac{P(2)(1)^2}{3^2} = 0.222P$$

$$M_{FBA} = -0.444P$$

$$M_{FBC} = \frac{2P(4)}{8} = P = -M_{FCB}$$

$$M_{FCD} = \frac{1.5P(1)(2)^2}{3^2} = 0.667P$$

$$M_{FDC} = -0.333P$$

$$A_{\Delta L} = \begin{bmatrix} 0.556P \\ -0.333P \end{bmatrix}.$$

For stiffness matrix in this case,

$$S_{11} = \frac{4EI}{3} + \frac{4EI}{4} = 2.33\ EI$$

$$S_{21} = \frac{2EI}{4} = 0.50\ EI = S_{12}$$

$$S_{22} = \frac{4EI}{3} + \frac{4EI}{4} = 2.33\ EI$$

$$S = EI \begin{bmatrix} 2.33 & 0.50 \\ 0.50 & 2.33 \end{bmatrix}.$$

Substituting in Eq. (16.75),

$$\begin{bmatrix} \theta_B \\ \theta_C \end{bmatrix} = \frac{1}{EI} \begin{bmatrix} 2.33 & 0.50 \\ 0.50 & 2.33 \end{bmatrix}^{-1} \left\{ \begin{bmatrix} 0 \\ 0 \end{bmatrix} - \begin{bmatrix} 0.556\ P \\ -0.333\ P \end{bmatrix} \right\}$$

Solving, $\theta_B = -\dfrac{0.28P}{EI}$ and $\theta_C = \dfrac{0.2031P}{EI}$.

Example 16.32. *Analyse continuous beam in Fig. 16.28 using displacement method.*

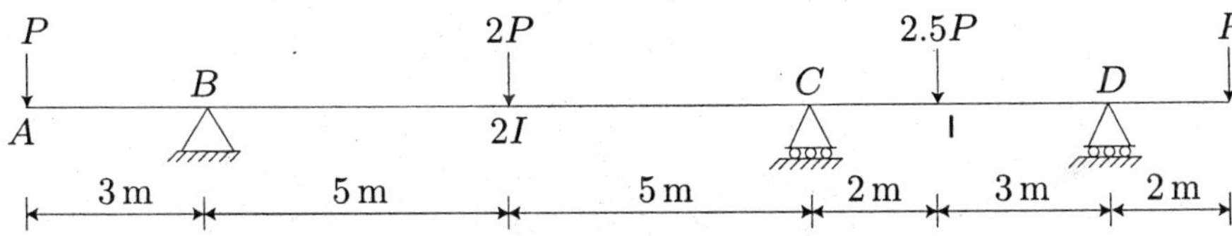

Fig. 16.28: For Example 16.32

Solution : In this case, the unknown displacements are the rotations at B, C and D and the degree of freedom is three.

Also $\quad A_\Delta = \begin{bmatrix} 0 \\ 0 \\ 0 \end{bmatrix}.$

Here $\quad A_{\Delta L_1} = -3P + 2.5P = -0.50P$

$$A_{\Delta L_2} = -2.5P + 1.80P = -0.70P$$

$$A_{\Delta L_3} = -1.20P + 2P = +0.80P.$$

For stiffness matrix,

$$S_{11} = \frac{8EI}{10};\ S_{12} = \frac{4EI}{10} = S_{21};\ S_{31} = 0 = S_{13}$$

$$S_{22} = \frac{8EI}{10} + \frac{4EI}{5} = \frac{16EI}{10};\ S_{32} = \frac{2EI}{5} = S_{22} \text{ and } S_{33} = \frac{4EI}{9},$$

whence $S = EI \begin{bmatrix} 0.80 & 0.40 & 0 \\ 0.40 & 1.60 & 0.40 \\ 0 & 0.40 & 0.80 \end{bmatrix}.$

Substituting in Eq. (16.75),

$$\begin{bmatrix} \theta_B \\ \theta_C \\ \theta_D \end{bmatrix} = \frac{1}{EI} \begin{bmatrix} 0.80 & 0.40 & 0 \\ 0.40 & 1.60 & 0.40 \\ 0 & 0.40 & 0.80 \end{bmatrix}^{-1} \left\{ \begin{bmatrix} 0 \\ 0 \\ 0 \end{bmatrix} - \begin{bmatrix} -0.50P \\ -0.70P \\ +0.80P \end{bmatrix} \right\}$$

Solving, $\theta_B = \dfrac{0.2708P}{EI}$

$$\theta_C = \frac{0.7083P}{EI}$$

and $\theta_D = -\dfrac{1.3542P}{EI}.$

Alternatively, using partially restrained structure as moments at B and D are known, permitting rotation at the supports B, and D, only θ_C is to be considered as the kinematic redundent. Then,

$$S = \frac{3}{4}\left[\frac{8EI}{10} + \frac{4EI}{5}\right] = \frac{12EI}{10}$$

$$A_\Delta = 0 \text{ and } A_{\Delta L} = -\left[2.5P - \left(\frac{3P - 2.5P}{2}\right)\right] + \left[1.80P - \left(\frac{2P - 1.20P}{2}\right)\right]$$

$$= -2.25P + 1.40P = -0.85P.$$

Solving, $\theta_C = \dfrac{10}{12EI}[0 - (-0.85P)] = \dfrac{0.7083}{EI}$

and $M_{CB} = -2.25P + \dfrac{3}{4}\dfrac{8EI}{10}\left(\dfrac{0.7083P}{EI}\right) = -1.826P.$

$$M_{CD} = +1.40P + \frac{3}{4}\frac{4EI}{5}\left(\frac{0.7083P}{EI}\right) = +1.826P.$$

Example 16.33. *Analyse the beam shown in Fig. 16.29(a) using displacement method, if supports B and C settled by $\frac{3000}{EI}$ and $\frac{1500}{EI}$ respectively.*

Solution : The degree of freedom is four and the rotation at A, B, C and D are the unknown displacements.

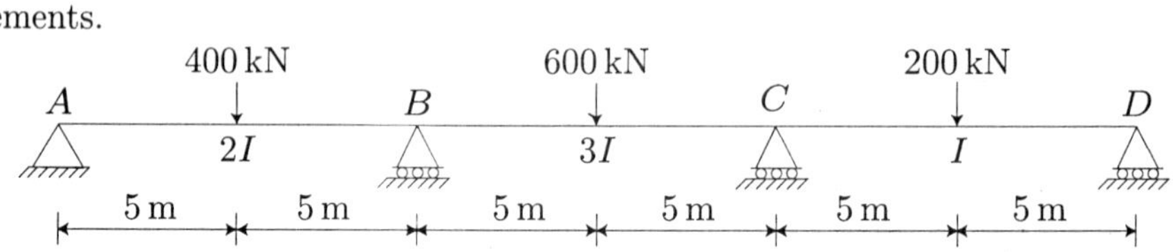

Fig. 16.29: For Example 16.33

The fixed-end moments due to relative settlements are to be considered in calculating fixed-end moments. Thus,

$$M_{FAB} = 500 + \frac{6E(2I)}{(10)^2}\frac{3000}{EI} = 860$$

$$M_{FBA} = -500 + 360 = -140$$

$$M_{FBC} = 750 - \frac{6E(3I)}{(10)^2}\frac{1500}{EI} = 480$$

$$M_{FCB} = -750 - 270 = -1020$$

$$M_{FCD} = 250 - \frac{6EI}{(10)^2}\frac{1500}{EI} = 160$$

$$M_{FDC} = -250 - 90 = -340.$$

From this it follows that

$$A_{\Delta L} = \begin{bmatrix} 860 \\ 340 \\ -860 \\ -340 \end{bmatrix}.$$

Stiffness matrix in this case may be obtained by giving unit displacement successively at A, B, C and D. Thus,

$$S = EI \begin{bmatrix} 0.80 & 0.40 & 0 & 0 \\ 0.40 & 2.0 & 0.60 & 0 \\ 0 & 0.60 & 1.60 & 0.20 \\ 0 & 0 & 0.20 & 0.40 \end{bmatrix}.$$

Substituting in Eq. (16.75),

$$\begin{bmatrix} \theta_A \\ \theta_B \\ \theta_C \\ \theta_D \end{bmatrix} = \frac{1}{EI} \begin{bmatrix} 0.80 & 0.40 & 0 & 0 \\ 0.40 & 2.0 & 0.60 & 0 \\ 0 & 0.60 & 1.60 & 0.20 \\ 0 & 0 & 0.20 & 0.40 \end{bmatrix}^{-1} \left\{ \begin{bmatrix} 0 \\ 0 \\ 0 \\ 0 \end{bmatrix} - \begin{bmatrix} 860 \\ 340 \\ -860 \\ -340 \end{bmatrix} \right\}$$

Solving, $\theta_A = -\dfrac{1015.2}{EI}$; $\theta_B = -\dfrac{118.6}{EI}$

$$\theta_C = -\frac{507.4}{EI} \quad \text{and} \quad \theta_D = -\frac{595.8}{EI}.$$

It follows that,

$$M_{AB} = 0 : M_{BA} = 644 \text{ kNm}$$

$$M_{BC} = +64.40 \text{ kNm}; \; M_{CB} = -480 \text{ kNm}$$

$$M_{CD} = 480 \text{ kNm} \quad \text{and} \quad M_{DC} = 0.$$

Example 16.34. *Analyse the beam shown in Fig. 16.30(a) by the displacement method. The beam rests on elastic supports at B and C, the flexibility of supports B and C being $\frac{10}{EI}$ and $\frac{25}{EI}$ respectively.*

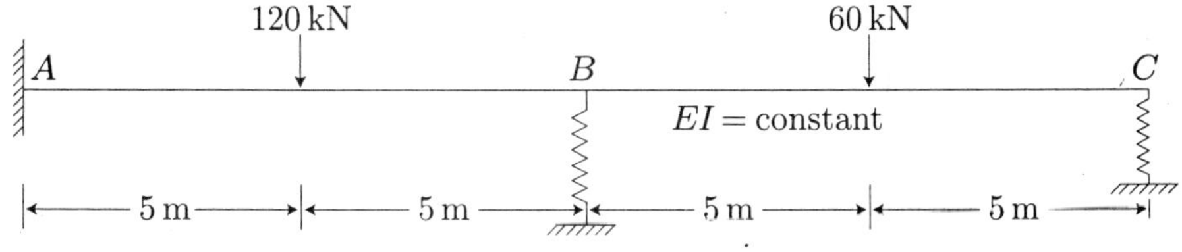

Fig. 16.30: For Example 16.34

Solution : In this case, the settlements Δ_B and Δ_C at B and C are unknown displacements over and above the rotation θ_B and θ_C at B and C. The degree of freedom is, therefore, four.

Considering θ_B, Δ_B, θ_C and Δ_C as Δ_1, Δ_2, Δ_3 and Δ_4 respectively,

$$A_{\Delta L} = \begin{bmatrix} -75 \\ 90 \\ -75 \\ 30 \end{bmatrix}.$$

Proceeding in the usual manner the stiffness matrix coefficients are

$$S_{11} = \frac{8EI}{10};\ S_{21} = 0 = S_{12};\ S_{31} = \frac{2EI}{10} = S_{13};\ S_{41} = +\frac{6EI}{(10)^3} = S_{14}$$

$$S_{22} = \frac{24EI}{(10)^3};\ S_{32} = -\frac{6EI}{(10)^3} = S_{23};\ S_{42} = -\frac{12EI}{(10)^3} = S_{24}$$

$$S_{33} = \frac{4EI}{10};\ S_{43} = \frac{6EI}{(10)^3} = S_{34};\ S_{44} = \frac{12EI}{(10)^3}.$$

Stiffness matrix then is given by

$$S = EI \begin{bmatrix} 0.80 & 0 & 0.20 & 0.06 \\ 0 & 0.024 & -0.06 & -0.012 \\ 0.20 & -0.06 & 0.40 & 0.06 \\ 0.60 & -0.012 & 0.06 & 0.012 \end{bmatrix}.$$

The stiffness of the elastic support at B and C are $\frac{EI}{10}$, i.e., $0.10EI$ and $\frac{EI}{25}$, i.e., $0.04EI$ respectively.

Modified stiffness matrix S_M is given by

$$S_M = EI \begin{bmatrix} 0.80 & 0 & 0.20 & 0.06 \\ 0 & (0.024 + 0.1) & -0.06 & -0.012 \\ 0.20 & 0.06 & 0.40 & 0.06 \\ 0.06 & -0.012 & 0.06 & (0.012 + 0.04) \end{bmatrix}$$

$$= EI \begin{bmatrix} 0.80 & 0 & 0.20 & 0.06 \\ 0 & 0.124 & -0.06 & -0.012 \\ 0.20 & -0.06 & 0.40 & 0.06 \\ 0.06 & -0.012 & 0.06 & 0.052 \end{bmatrix}.$$

Substituting in Eq. (16.75),

$$\begin{bmatrix} \theta_B \\ \Delta_B \\ \theta_C \\ \Delta_C \end{bmatrix} = \frac{1}{EI} \begin{bmatrix} 0.80 & 0 & 0.20 & 0.06 \\ 0 & 0.124 & -0.06 & -0.012 \\ 0.20 & -0.06 & 0.40 & 0.06 \\ 0.06 & -0.012 & 0.06 & 0.052 \end{bmatrix}^{-1} \left\{ \begin{bmatrix} 0 \\ 0 \\ 0 \\ 0 \end{bmatrix} - \begin{bmatrix} -75.0 \\ 90 \\ -75.0 \\ 30 \end{bmatrix} \right\},$$

$$\theta_B = -\frac{6.30}{EI}; \Delta_B = \frac{895}{EI}$$

$$\theta_C = \frac{249.5}{EI} \text{ and } \Delta_C = \frac{502.5}{EI}.$$

Reactions at B, i.e., $R_B = -0.10EI\left(\frac{895.0}{EI}\right) = -89.5\text{ kN}\uparrow$.

Also $\quad R_C = -0.04EI\left(\frac{502.5}{EI}\right) = -20.10\text{ kN}\uparrow$

Alternatively, the problem can be solved by partitioning the matrices. From the given data,

$$\Delta_B = -\frac{10}{EI}R_B \quad\text{and}\quad \Delta_C = -\frac{25}{EI}R_C.$$

Equation in matrix form may be written as

$$\begin{bmatrix} 0 \\ R_B \\ \cdots\cdots \\ 0 \\ R_C \end{bmatrix} \Bigg| \begin{bmatrix} -75.0 \\ 90 \\ \cdots\cdots \\ -75.0 \\ 30 \end{bmatrix} = EI\begin{bmatrix} 0.80 & 0 & : & 0.20 & 0.06 \\ 0 & 0.124 & : & -0.06 & -0.012 \\ \cdots & \cdots & \cdots & \cdots & \cdots \\ 0.20 & -0.06 & : & 0.40 & 0.06 \\ 0.06 & -0.012 & : & 0.06 & 0.002 \end{bmatrix}\begin{bmatrix} \theta_B \\ \frac{-10}{EI}R_B \\ \cdots\cdots \\ \theta_C \\ -\frac{25}{EI}R_C \end{bmatrix}.$$

Partitioning matrices

$$\begin{bmatrix} 75.0 \\ R_B - 90 \end{bmatrix} = EI\begin{bmatrix} 0.80 & 0 \\ 0 & 0.124 \end{bmatrix}\begin{bmatrix} \theta_B \\ -\frac{10}{EI}R_B \end{bmatrix} + EI\begin{bmatrix} 0.20 & 0.06 \\ -0.06 & -0.012 \end{bmatrix}\begin{bmatrix} \theta_C \\ -\frac{25}{EI}R_C \end{bmatrix}$$

and

$$\begin{bmatrix} 75.0 \\ R_C - 30 \end{bmatrix} = EI\begin{bmatrix} 0.20 & -0.06 \\ 0.06 & -0.012 \end{bmatrix}\begin{bmatrix} \theta_B \\ -\frac{10}{EI}R_B \end{bmatrix} + EI\begin{bmatrix} 0.40 & 0.06 \\ 0.06 & 0.052 \end{bmatrix}\begin{bmatrix} \theta_C \\ -\frac{25}{EI}R_C \end{bmatrix}.$$

Solving, $\quad R_B = -89.50\text{ kN} \quad\text{and}\quad R_B = -20.1\text{ kN}$

$$\theta_B = -\frac{6.3}{EI} \quad\text{and}\quad \theta_C = -\frac{249.5}{EI}.$$

Example 16.35. *Analyse the rigid frame in Fig. 16.31(a) using displacement method.*

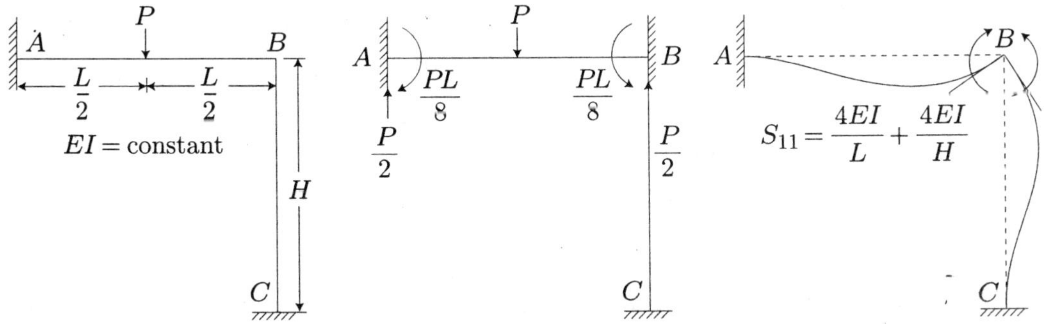

(a) Rigid frame **(b) Restraint structure** **(c) Stiffness at joint *B***

Fig. 16.31: For Example 16.35

Solution : In this case, the only unknown displacement is the rotation at B neglecting effect of axial deformation.

For the case, $A_\Delta = 0$ and $A_{\Delta L_1} = -\frac{PL}{8}$.

Stiffness matrix $S = S_{11} = \frac{4EI}{L} + \frac{4EI}{H} = 4EI\left[\frac{1}{L} + \frac{1}{H}\right].$

Substituting in Eq. (16.75),

$$\theta_B = \frac{1}{4EI}\left[\frac{1}{L}+\frac{1}{H}\right]^{-1}\left\{[0]-\left[-\frac{PL}{8}\right]\right\}.$$

Solving, $\theta_B = \dfrac{PL^2}{32EI}\left[\dfrac{H}{L+H}\right]$.

Using matrix form the various functions may be given by

$$\begin{bmatrix} M_{AB} \\ R_A \\ M_{BA} \\ M_{CB} \\ R_C \\ H_C \end{bmatrix} = \begin{bmatrix} \frac{PL}{8} \\ \frac{P}{2} \\ -\frac{PL}{8} \\ 0 \\ \frac{P}{2} \\ 0 \end{bmatrix} + \begin{bmatrix} \frac{2EI}{L} \\ \frac{6EI}{L^2} \\ \frac{4EI}{L} \\ \frac{2EI}{H} \\ -\frac{6EI}{L^2} \\ \frac{6EI}{H^2} \end{bmatrix}\left[\frac{PL^2}{32EI}\left(\frac{H}{L+H}\right)\right]$$

Example 16.36. *Analyse the portal frame in Fig. 16.32(a) using displacement method.*

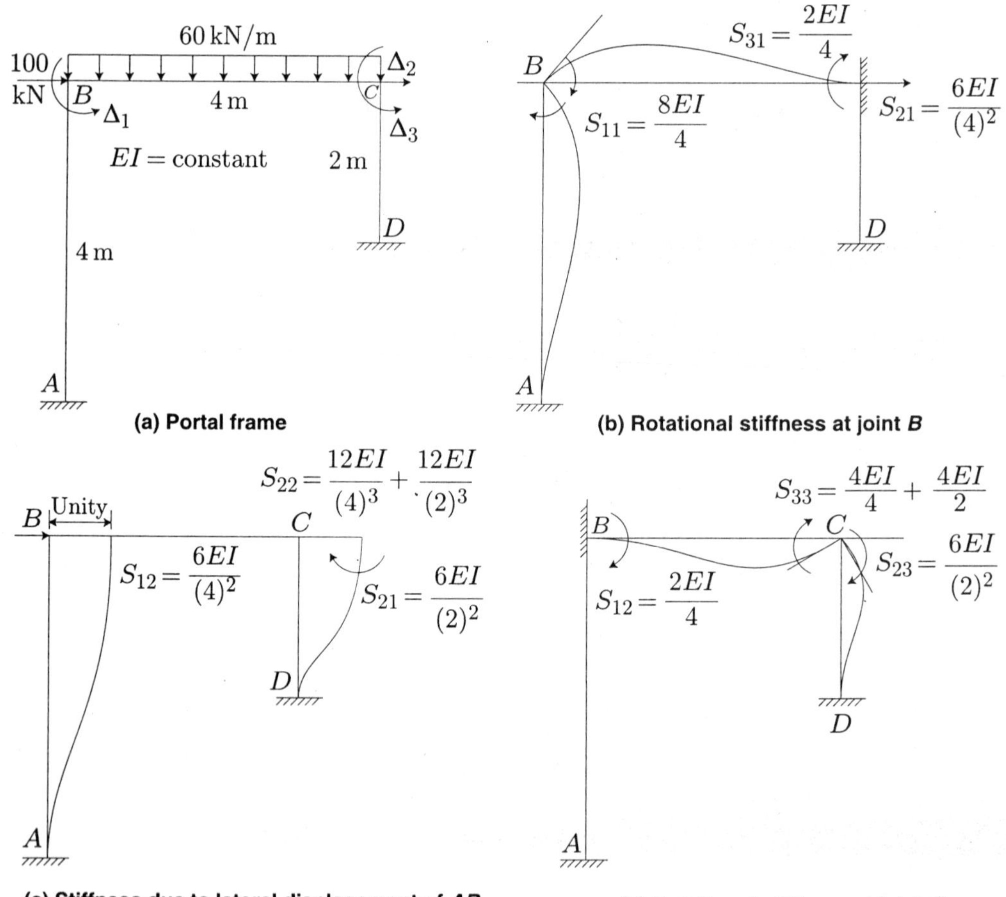

Fig. 16.32: For Example 16.36

Solution : As ends A and D are fixed, the rotations at B and C and the horizontal displacement at B are the three independent displacement components. The degree of freedom of the portal frame is three, neglecting effects of axial deformations.

Considering Δ_1 as θ_B, Δ_2 as Δ_{BH} and Δ_3 as θ_c, values of

$$A_\Delta = \begin{bmatrix} 0 \\ 100 \\ 0 \end{bmatrix}.$$

Also $$A_{\Delta L} = \begin{bmatrix} 80 \\ 0 \\ -80 \end{bmatrix}.$$

The stiffness matrix may be developed by giving a unit displacement successively according to Δ_1, Δ_2 and Δ_3. Thus

$$S_{11} = \frac{4EI}{4} + \frac{4EI}{4} = 2EI$$

$$S_{21} = \frac{6EI}{(4)^2} = \frac{3}{8}EI = S_{12}$$

$$S_{31} = \frac{2EI}{4} = \frac{EI}{2} = S_{13}$$

$$S_{22} = \frac{12EI}{(4)^3} + \frac{12EI}{(2)^3} = \frac{27}{16}EI$$

$$S_{32} = \frac{6EI}{(2)^2} = 1.5EI = S_{23}$$

$$S_{33} = \frac{4EI}{4} + \frac{4EI}{2} = 3EI.$$

Stiffness matrix then is given by

$$S = \frac{EI}{16}\begin{bmatrix} 32 & 6 & 8 \\ 6 & 27 & 24 \\ 8 & 24 & 48 \end{bmatrix}.$$

Substituting in Eq. (16.75),

$$\begin{bmatrix} \theta_B \\ \Delta_{BH} \\ \theta_C \end{bmatrix} = \frac{16}{EI}\begin{bmatrix} 32 & 6 & 8 \\ 6 & 27 & 24 \\ 8 & 24 & 48 \end{bmatrix}^{-1} \left\{ \begin{bmatrix} 0 \\ 100 \\ 0 \end{bmatrix} - \begin{bmatrix} 80 \\ 0 \\ -80 \end{bmatrix} \right\}.$$

Solving, $\theta_B = -\dfrac{533}{EI}; \Delta_{BH} = \dfrac{71.2}{EI}$ and $\theta_C = 0.$

Moments calculated are

$$M_{AB} = 0; M_{BA} = -26.6 = -M_{BC}$$

$$M_{CB} = -106.60 = -M_{CD} \text{ and } M_{DC} = 106.60.$$

Example 16.37. *Analyse the rigid frame shown in Fig. 16.33(a), using displacement method. The directions of the co-ordinates are shown in figure and a couple PL as shown at the joint B acts on the members.*

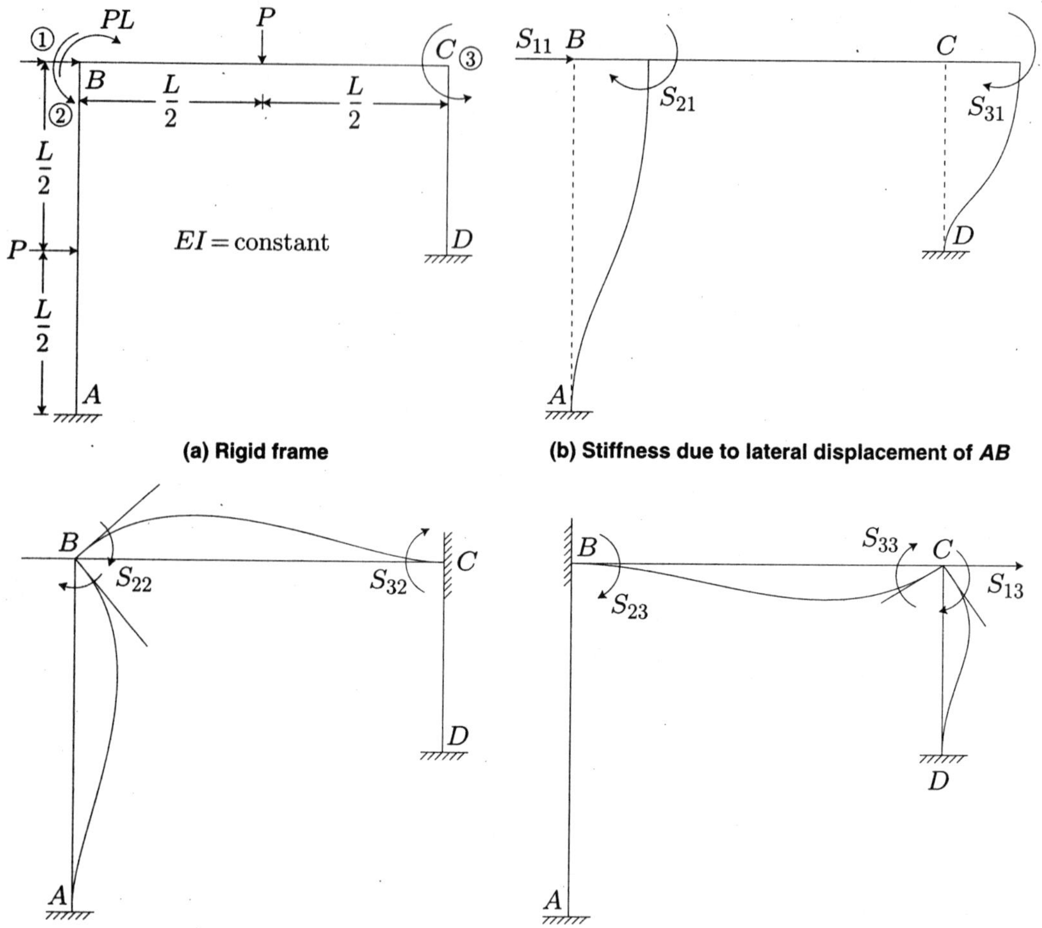

(a) Rigid frame

(b) Stiffness due to lateral displacement of *AB*

(c) Stiffness for Joint *B*

(d) Stiffness for Joint *C*

Fig. 16.33: For Example 16.37

Solution : In this case,

$$S_{11} = \frac{12EI}{L^3} + \frac{12EI}{\left(\frac{L}{2}\right)^3} = \frac{108EI}{L^3}$$

$$S_{21} = \frac{6EI}{L^2} = S_{12}$$

$$S_{31} = \frac{24EI}{L^2} = S_{13}$$

$$S_{22} = \frac{4EI}{L} + \frac{4EI}{L} = \frac{8EI}{L}$$

$$S_{32} = \frac{2EI}{L} = S_{23}$$

$$S_{33} = \frac{12EI}{L}.$$

Stiffness matrix is

$$S = \frac{EI}{L}\begin{bmatrix} \frac{108}{L^2} & \frac{6}{L} & \frac{24}{L} \\ \frac{6}{L} & 8 & 2 \\ \frac{24}{L} & 2 & 12 \end{bmatrix}.$$

In this case

$$A_\Delta = \begin{bmatrix} 0 \\ -PL \\ 0 \end{bmatrix} \quad \text{and} \quad A_{\Delta L} = \begin{bmatrix} -\frac{P}{2} \\ 0 \\ -\frac{PL}{8} \end{bmatrix}$$

whence solving, $\Delta_B = \Delta_1 = \dfrac{0.0087PL^3}{EI}$

$$\theta_B = \Delta_2 = -0.1355\frac{PL^2}{EI} \quad \text{and} \quad \theta_C = \Delta_3 = 0.0156\frac{PL^2}{EI}.$$

End moments can be calculated from

$$\begin{bmatrix} M_{AB} \\ M_{BA} \\ M_{BC} \\ M_{CB} \\ M_{CD} \\ M_{DC} \end{bmatrix} = PL\begin{bmatrix} 0.125 \\ -0.125 \\ 0.125 \\ -0.125 \\ 0 \\ 0 \end{bmatrix} + \frac{EI}{L}\begin{bmatrix} \frac{6}{L} & 2 & 0 \\ \frac{6}{L} & 4 & 0 \\ 0 & 4 & 2 \\ 0 & 2 & 4 \\ \frac{24}{L} & 0 & 8 \\ \frac{24}{L} & 0 & 4 \end{bmatrix} \frac{PL^2}{EI}\begin{bmatrix} 0.0087L \\ -0.1355 \\ 0.0156 \end{bmatrix} = PL\begin{bmatrix} -0.0938 \\ -0.6148 \\ -0.3852 \\ -0.3356 \\ +0.3336 \\ +0.2672 \end{bmatrix}.$$

Example 16.38. *Analyse the rigid frame shown in Fig. 16.34(a) using displacement method, if the settlements of supports at D be $-\frac{179.9}{EI}$ to the right and $\frac{446.4}{EI}$ downwards.*

Solution : In this case, the five degree of freedom are $\theta_B, \Delta_{BH}, \theta_C, \Delta_{DM}$ and Δ_{DV} neglecting effects of axial deformation.

Stiffness matrix corresponding to these freedom as shown in Figs. 15.34(b) to 15.34(f) :

$$S = EI\begin{bmatrix} 3.20 & 0.24 & 0.80 & 0 & 0.24 \\ 0.24 & 0.144 & 0.24 & -0.096 & 0 \\ 0.80 & 0.24 & 2.40 & -0.24 & 0.24 \\ 0 & -0.096 & -0.24 & 0.096 & 0 \\ 0.24 & 0 & 0.24 & 0 & 0.048 \end{bmatrix}.$$

In this case,

$$A_\Delta = \begin{bmatrix} 0 \\ 100 \\ 0 \\ P_H \\ P_V \end{bmatrix} \quad \text{and } A_{\Delta L} = \begin{bmatrix} 0 \\ 0 \\ 0 \\ 0 \\ 0 \end{bmatrix}.$$

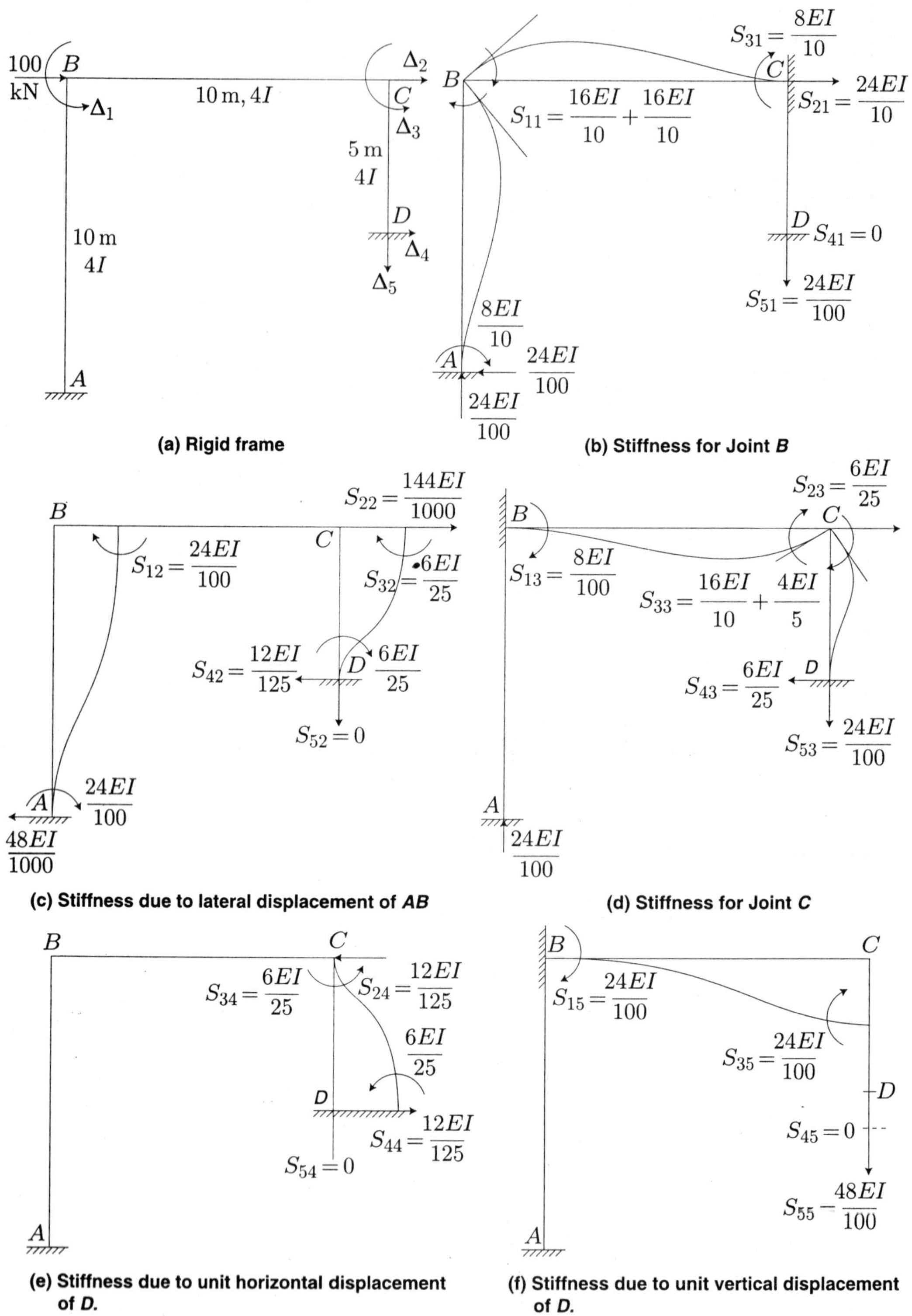

(a) Rigid frame

(b) Stiffness for Joint *B*

(c) Stiffness due to lateral displacement of *AB*

(d) Stiffness for Joint *C*

(e) Stiffness due to unit horizontal displacement of *D*.

(f) Stiffness due to unit vertical displacement of *D*.

Fig. 16.34: For Example 16.38

Substituting in Eq. (16.75),

$$\begin{bmatrix} \theta_B \\ \Delta_{BH} \\ \theta_C \\ -\frac{179.9}{EI} \\ \frac{446.4}{EI} \end{bmatrix} = \frac{1}{EI}\begin{bmatrix} 3.2 & 0.24 & 0.80 & 0 & 0.24 \\ 0.24 & 0.144 & 0.24 & -0.096 & 0 \\ 0.80 & 0.24 & 2.40 & -0.24 & 0.24 \\ 0 & -0.096 & -0.24 & 0.096 & 0 \\ 0.24 & 0 & 0.24 & 0 & 0.048 \end{bmatrix}^{-1} \left\{ \begin{bmatrix} 0 \\ 100 \\ 0 \\ P_H \\ P_V \end{bmatrix} - \begin{bmatrix} 0 \\ 0 \\ 0 \\ 0 \\ 0 \end{bmatrix} \right\}.$$

Rearranging terms,

$$\begin{bmatrix} 0 \\ 100 \\ 0 \\ \cdots\cdots \\ P_H \\ P_V \end{bmatrix} = EI \begin{bmatrix} 3.2 & 0.24 & 0.80 & : & 0 & 0.24 \\ 0.24 & 0.144 & 0.24 & : & -0.096 & 0 \\ 0.80 & 0.24 & 2.40 & : & -0.24 & 0.24 \\ \cdots & \cdots & \cdots & \cdots & \cdots & \cdots \\ 0 & -0.096 & -0.24 & : & 0.096 & 0 \\ 0.24 & 0 & 0.24 & : & 0 & 0.048 \end{bmatrix} \begin{bmatrix} \theta_B \\ \Delta_{BH} \\ \theta_C \\ \cdots\cdots \\ \frac{179.9}{EI} \\ \frac{446.4}{EI} \end{bmatrix}$$

Partitioning the matrices as indicated

$$\begin{bmatrix} 0 \\ 100 \\ 0 \end{bmatrix} = EI \begin{bmatrix} 3.20 & 0.24 & 0.80 \\ 0.24 & 0.144 & 0.24 \\ 0.80 & 0.24 & 2.40 \end{bmatrix} \begin{bmatrix} \theta_B \\ \Delta_{BH} \\ \theta_C \end{bmatrix} + \begin{bmatrix} 0 & 0.24 \\ -0.096 & 0 \\ -0.24 & 0.24 \end{bmatrix} \begin{bmatrix} 179.9 \\ 446.4 \end{bmatrix}$$

Solving $\theta_B = -\dfrac{92.38}{EI}$; $\Delta_{BH} = \dfrac{1157.3445}{EI}$ and $\theta_C = -\dfrac{111.59}{EI}$.

Also $$\begin{bmatrix} P_H \\ P_V \end{bmatrix} = EI \begin{bmatrix} 0 & -0.096 & -0.24 \\ 0.24 & 0 & 0.24 \end{bmatrix} \begin{bmatrix} \theta_B \\ \Delta_{BH} \\ \theta_C \end{bmatrix} + \begin{bmatrix} 0.096 & 0 \\ 0 & 0.048 \end{bmatrix} \begin{bmatrix} 179.9 \\ 446.4 \end{bmatrix}$$

Solving $P_H = -67.05$ kN $\longleftarrow$ and $P_V = -27.53$ kN $\uparrow$.

Using slope deflection equations,

$$M_{AB} = \frac{8EI}{10}\left[-\frac{92.38}{EI} - \frac{3}{10}\left(\frac{1157.3445}{EI}\right)\right] = 203.86 \text{ kNm}$$

$$M_{BA} = \frac{8EI}{10}\left[-\frac{2(92.38)}{EI} - \frac{3}{10}\left(0 - \frac{1157.3445}{EI}\right)\right] = 129.94 \text{ kNm} = -M_{BC}$$

$$M_{CB} = \frac{8EI}{10} - \left[\frac{92.38}{EI} - 2\frac{(111.59)}{EI} - \frac{3}{10}\left(0 - \frac{446.4}{EI}\right)\right] = -145.31 \text{ kNm}$$

$$M_{CD} = \frac{2EI}{5} - \left[\frac{2(111.59)}{EI} - \frac{3}{5}\left(\frac{179.9}{EI} - \frac{1157.3445}{EI}\right)\right] = 145.315 \text{ kNm.}$$

$$M_{DC} = \frac{2EI}{5}\left[-\frac{111.59}{EI} - \frac{3}{5}\left(\frac{179.9}{EI} - \frac{1157.3445}{EI}\right)\right] = 189.95 \text{ kNm.}$$

Alternatively, taking the unknown displacements as θ_B, Δ_{BH} and θ_C, the stiffness matrix developed is

$$S = EI \begin{bmatrix} 3.20 & 0.24 & 0.80 \\ 0.24 & 0.144 & 0.24 \\ 0.80 & 0.24 & 0.10 \end{bmatrix}$$

In this case, $A_\Delta = \begin{bmatrix} 0 \\ 100 \\ 0 \end{bmatrix}$.

For $A_{\Delta L}$ restraining force in the direction of θ_B, Δ_{BH} and θ_C at BC are to be calculated and taken into consideration.

Due to displacement of $\frac{179.9}{EI}$ horizontally at D to the right.

At B, Restraining forces = 0.

At C, Restraining Horizontal force $= -\frac{12EI}{(5)^3}\left(\frac{179.9}{EI}\right) = -\ 17.2704$ kNm $\longleftarrow$.

Restraining Moment $= -\frac{6EI}{(5)^2}\frac{179.9}{EI} = -43.176$ kNm.

Due to vertical displacement of $\frac{446.4}{EI}$ at D downwards.

At B and C, Restraining Moment $= \frac{6E(4I)}{(10)^2}\frac{446.4}{EI} = 107.136$ kNm.

Restraining horizontal force = 0.

Substituting in Eq. (16.75),

$$\begin{bmatrix} \theta_B \\ \Delta_{BH} \\ \theta_C \end{bmatrix} = \frac{1}{EI}\begin{bmatrix} 3.20 & 0.24 & 0.80 \\ 0.24 & 0.144 & 0.24 \\ 0.80 & 0.24 & 2.40 \end{bmatrix}^{-1}\left\{\begin{bmatrix} 0 \\ 10 \\ 0 \end{bmatrix} - \begin{bmatrix} 107.136 \\ -17.2704 \\ 107.136 - 43.176 \end{bmatrix}\right\}.$$

Solving, $\theta_B = -\frac{92.38}{EI}$

$\Delta_{BH} = \frac{1157.3445}{EI}$

and $\theta_C = -\frac{111.59}{EI}$.

Example 16.39. *Analyse the portal frame in Fig. 16.35 using displacement method. The flexibility of support D for horizontal and vertical displacements are $\frac{10}{EI}$ and $\frac{20}{EI}$ respectively.*

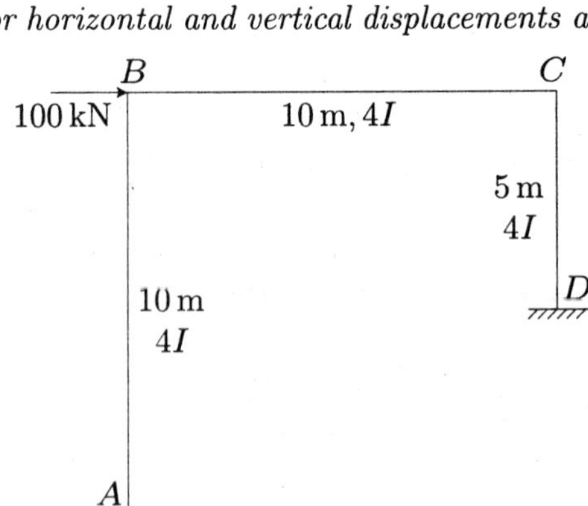

Fig. 16.35: For Example 16.39

Solution : As in Example 16.37, considerting the unknown displacements as $\theta_B, \Delta_{BH}, \theta_C$, Δ_{DH} and Δ_{DV}, the stiffness matrix is given by

$$S = EI\begin{bmatrix} 3.20 & 0.24 & 0.80 & 0 & 0.24 \\ 0.24 & 0.144 & 0.24 & -0.096 & 0 \\ 0.80 & 0.24 & 2.40 & -0.24 & 0.24 \\ 0 & -0.096 & -0.24 & 0.096 & 0 \\ 0.24 & 0 & 0.24 & 0 & 0.048 \end{bmatrix}.$$

Stiffness relative to the horizontal and vertical displacement at support D are $0.10EI$ and $0.50EI$ respectively.

The modified stiffness S_M is given by

$$S_M = \begin{bmatrix} 3.20 & 0.24 & 0.80 & 0 & 0.24 \\ 0.24 & 0.144 & 0.24 & -0.096 & 0 \\ 0.80 & 0.24 & 2.40 & -0.24 & 0.24 \\ 0 & -0.096 & -0.24 & \underline{0.196} & 0 \\ 0.24 & 0 & 0.24 & 0 & \underline{0.098} \end{bmatrix}$$

Substituting in Eq. (16.75),

$$\begin{bmatrix} \theta_B \\ \Delta_{BH} \\ \theta_C \\ \Delta_{DH} \\ \Delta_{DV} \end{bmatrix} = [S_M]^{-1}\left\{\begin{bmatrix} 0 \\ 100 \\ 0 \\ 0 \\ 0 \end{bmatrix} - \begin{bmatrix} 0 \\ 0 \\ 0 \\ 0 \\ 0 \end{bmatrix}\right\}.$$

Solving, $\theta_B = -\dfrac{126.25}{EI}$; $\Delta_{BH} = \dfrac{1463.04}{EI}$

$\theta_C = -\dfrac{100.36}{EI}$; $\Delta_{DH} = \dfrac{593.93}{EI}$

and $\Delta_{DV} = \dfrac{555}{EI}$.

From $\Delta_{DH} = -\dfrac{10}{EI}P_H$ and $\Delta_{DV} = -\dfrac{20}{EI}P_V$

$P_H = -\ 59.39$ kN $\longleftarrow$ and $P_V = -27.75$ kN $\uparrow$.

Alternatively, the solution can be done by partitioning the matrix.

From given data,

$$\Delta_{DH} = -\frac{10}{EI}P_H \quad \text{and} \quad \Delta_{DV} = -\frac{20}{EI}P_V.$$

In this case, $A_\Delta = \begin{bmatrix} 0 \\ 100 \\ 0 \\ P_H \\ P_V \end{bmatrix}$ and $A_{\Delta L} = \begin{bmatrix} 0 \\ 0 \\ 0 \\ 0 \\ 0 \end{bmatrix}$.

Substituting,

$$EI\begin{bmatrix} 3.20 & 0.24 & 0.80 & \vdots & 0 & 0.24 \\ 0.24 & 0.144 & 0.24 & \vdots & -0.096 & 0 \\ 0.80 & 0.24 & 2.40 & \vdots & -0.24 & 0.24 \\ \cdots & \cdots & \cdots & & \cdots & \cdots \\ 0 & -0.096 & -0.24 & \vdots & 0.096 & 0 \\ 0.24 & 0 & 0.24 & \vdots & 0 & 0.048 \end{bmatrix}\begin{bmatrix} \theta_B \\ \Delta_{BH} \\ \theta_C \\ \cdots \\ -\frac{10}{EI}P_H \\ \frac{20}{EI}P_V \end{bmatrix}$$

$$= \begin{bmatrix} 0 \\ 100 \\ 0 \\ \cdots \\ P_H \\ P_V \end{bmatrix} - \begin{bmatrix} 0 \\ 0 \\ 0 \\ \cdots \\ 0 \\ 0 \end{bmatrix}.$$

Partitioning the matrices as indicated by the dotted lines,

$$\begin{bmatrix} 0 \\ 100 \\ 0 \end{bmatrix} = EI\begin{bmatrix} 3.20 & 0.24 & 0.80 \\ 0.24 & 0.144 & 0.24 \\ 0.80 & 0.24 & 2.40 \end{bmatrix}\begin{bmatrix} \theta_B \\ \Delta_{BH} \\ \theta_C \end{bmatrix} + \begin{bmatrix} 0 & 0.24 \\ -0.096 & 0 \\ 0.24 & 0.24 \end{bmatrix}\begin{bmatrix} -10P_H \\ -20P_V \end{bmatrix}$$

and
$$\begin{bmatrix} P_H \\ P_V \end{bmatrix} = EI\begin{bmatrix} 0 & -0.096 & -0.24 \\ 0.24 & 0 & 0.24 \end{bmatrix}\begin{bmatrix} \theta_B \\ \Delta_{BH} \\ \theta_C \end{bmatrix} + \begin{bmatrix} 0.096 & 0 \\ 0 & 0.048 \end{bmatrix}\begin{bmatrix} -10P_H \\ -20P_V \end{bmatrix}.$$

Solving $\theta_B = -\dfrac{126.25}{EI}$; $\Delta_{BH} = \dfrac{1463.04}{EI}$

$\theta_C = -\dfrac{100.36}{EI}$; $P_H = -\ 59.39 \longleftarrow$ kN

and $P_V = -27.75$ kN $\uparrow$.

It follows that $\Delta_{DH} = \dfrac{593.93}{EI}$ and $\Delta_{DV} = \dfrac{555}{EI}$.

Example 16.40. *Analyse by displacement method the right angled bent in Fig. 16.36, if the support at A permits a clockwise rotation of* 0.001 radian. *Assume* $EI = 10000$ kNm^2.

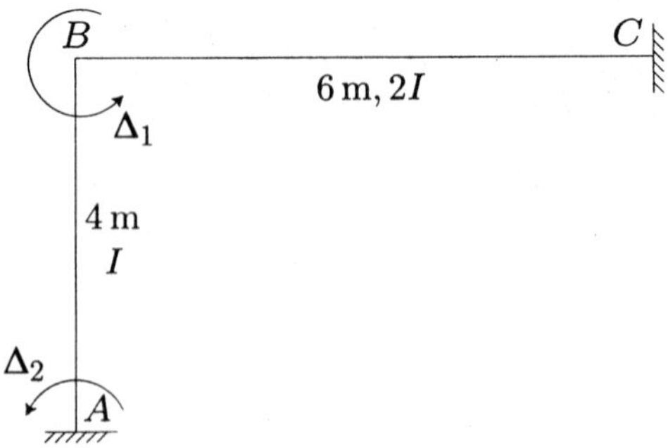

Fig. 16.36: For Example 16.40

Solution : In this case, the rotations at A and B, i.e., θ_B and θ_A are the unknown displacements. Of course, $\theta_A = -0.0010$ radians,

and $$M_{FAB} = -\frac{4EI}{4}(0.001).$$

Stiffness matrix is given by

$$S = EI\begin{bmatrix} \frac{14}{6} & \frac{1}{2} \\ \frac{1}{2} & 1 \end{bmatrix}.$$

Equation then becomes

$$EI\begin{bmatrix} \frac{14}{6} & \frac{1}{2} \\ \frac{1}{2} & 1 \end{bmatrix}\begin{bmatrix} \theta_B \\ -0.001 \end{bmatrix} = \begin{bmatrix} 0 \\ M_{AB} \end{bmatrix} - \begin{bmatrix} 0 \\ 0 \end{bmatrix}.$$

Solving $-0.001 = \frac{28}{25EI}(M_{AB})$ and $M_{AB} = -\frac{25EI}{28}(0.001) = -8.929$ kNm.

Also $$\theta_B = -\frac{6}{25EI}M_{AB} = \frac{0.006}{28}.$$

From slope deflection equations, $M_{CB} = \frac{4EI}{6}\left[\frac{0.006}{28}\right] = +1.43$ kNm.

Example 16.41. *Analyse the rigid frame in Fig. 16.37(a) considering effect of axial forces also. Formulate the equation in matrix from using stiffness method.*

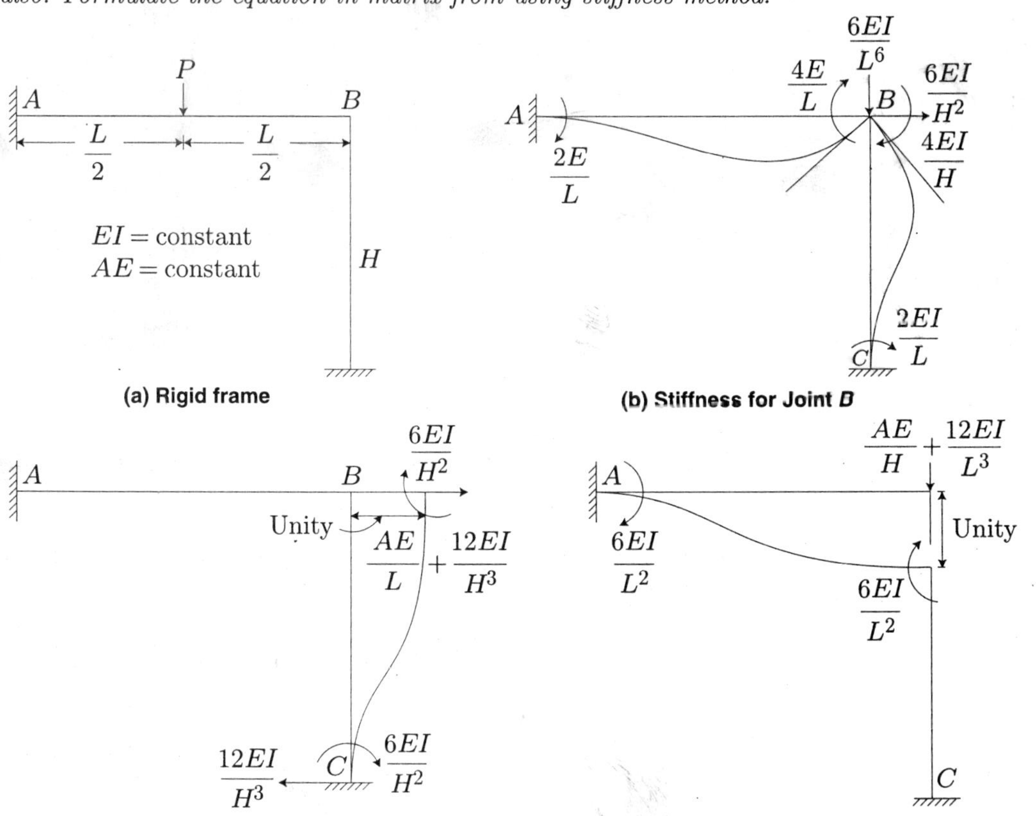

Fig. 16.37: For Example 16.41

Solution : In this case, the unknown displacements are θ_B, Δ_{BH} and Δ_{BV}. The stiffness matrix representing to Figs. 16.37(b) to 16.37(c) is given by

$$\begin{bmatrix} \left(\frac{4EI}{L}+\frac{4EI}{H}\right) & \frac{6EI}{H^2} & \frac{6EI}{L^2} \\ \frac{6EI}{H^2} & \left(\frac{12EI}{H^3}+\frac{EA}{L}\right) & 0 \\ \frac{6EI}{L^2} & 0 & \left(\frac{12EI}{L^3}+\frac{EA}{H}\right) \end{bmatrix}$$

$$A_{\Delta L} = \begin{bmatrix} -\frac{PL}{8} \\ 0 \\ -\frac{P}{L} \end{bmatrix} \text{ and } A_\Delta = \begin{bmatrix} 0 \\ 0 \\ 0 \end{bmatrix}.$$

Formulating,

$$\begin{bmatrix} \theta_B \\ \Delta_{BH} \\ \Delta_{BV} \end{bmatrix} = \begin{bmatrix} \left(\frac{4EI}{L}+\frac{4EI}{H}\right) & \frac{6EI}{H^2} & \frac{6EI}{L^2} \\ \frac{6EI}{H^2} & \left(\frac{12EI}{H^3}+\frac{EA}{L}\right) & 0 \\ \frac{6EI}{L^2} & 0 & \left(\frac{12EI}{L^3}+\frac{EA}{H}\right) \end{bmatrix}^{-1} \left\{ \begin{bmatrix} 0 \\ 0 \\ 0 \end{bmatrix} - \begin{bmatrix} -\frac{PL}{8} \\ 0 \\ -\frac{P}{L} \end{bmatrix} \right\}.$$

Example 16.42. *Determine the elements of stiffness matrix for the portal frame with reference to the coordinates shown in Fig. 16.38(a).*

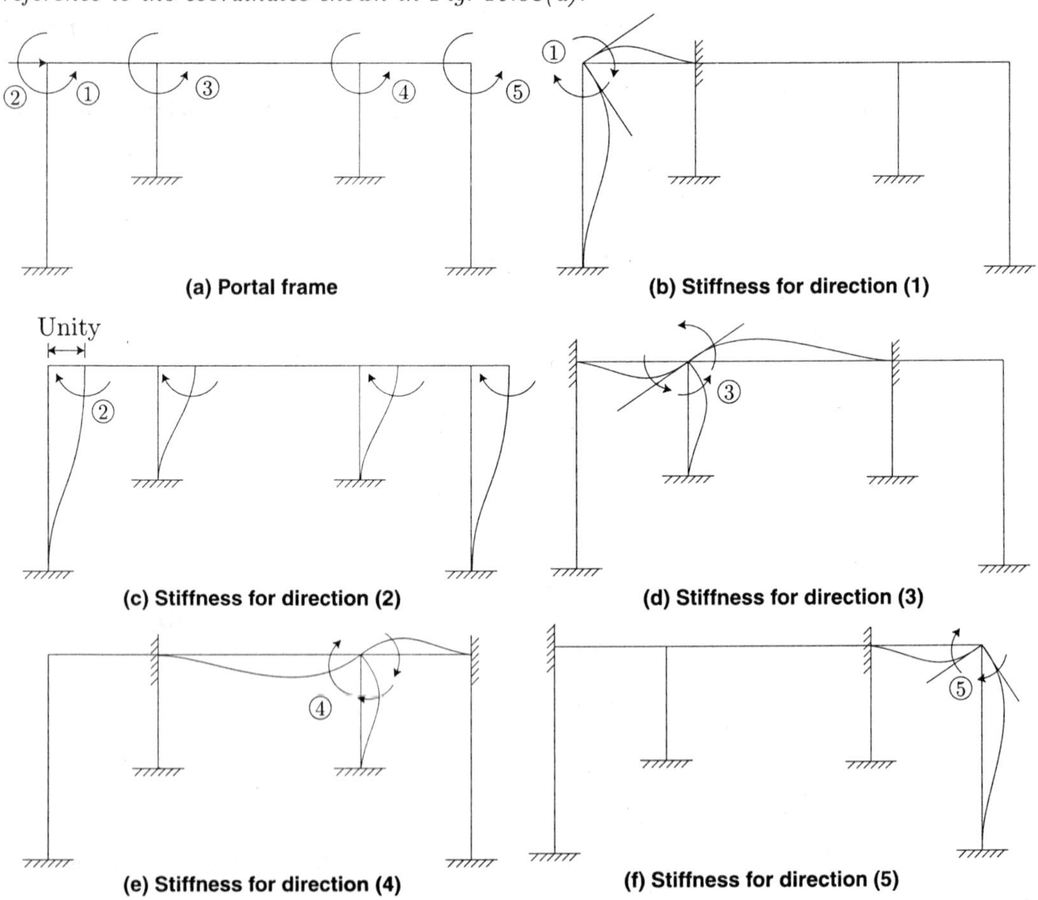

(a) Portal frame
(b) Stiffness for direction (1)
(c) Stiffness for direction (2)
(d) Stiffness for direction (3)
(e) Stiffness for direction (4)
(f) Stiffness for direction (5)

Fig. 16.38: For Example 16.42

Solution : The stiffness matrix can be developed by giving unit displacements successively at co-ordinates 1 to 5 as shown in Figs. 16.38(b) to 16.38(f).

Referring to these figures, then

$$S_{11} = \frac{4E(4I)}{10} + \frac{4E(I)}{5} = 2.40EI; \; S_{21} = \frac{6E(4I)}{(10)^2} = 0.24EI = S_{12}$$

$$S_{31} = \frac{2EI}{5} = 0.40EI = S_{13}; \; S_{41} = 0 = S_{14} = S_{51} = S_{15}$$

$$S_{22} = \frac{12E(4I)}{(10)^3} + \frac{12EI}{(5)^3} + \frac{12E(4I)}{(10)^3} = 0.288EI$$

$$S_{32} = \frac{6EI}{(5)^2} = 0.240EI = S_{23}; \; S_{42} = \frac{6EI}{(5)^2} = 0.240EI = S_{24}$$

$$S_{52} = \frac{6E(4I)}{(10)^2} = 0.240EI = S_{25}; \; S_{33} = \frac{4EI}{5} + \frac{4EI}{5} + \frac{4E(4I)}{10} = 3.20EI$$

$$S_{43} = \frac{2E(4I)}{10} = 0.80EI = S_{34}; \; S_{53} = 0 = S_{35}; \; S_{44} = 3.20EI$$

$$S_{54} = \frac{2EI}{5} = 0.40EI = S_{45}; \; S_{55} = 2.40EI.$$

Example 16.43. *Analyse the frame shown in Fig. 16.39(a) using displacement method. The co-ordinates may be taken as shown in figure.*

Solution : The stiffness matrix can be obtained from Figs. 16.39(b) to 16.39(d) and given by

$$S = \frac{EI}{L}\begin{bmatrix} 8 & 2 & \frac{3}{L} \\ 2 & 9 & \frac{4.875}{L} \\ \frac{3}{L} & \frac{4.875}{L} & \frac{48.938}{L^2} \end{bmatrix}.$$

In this case,

$$A_\Delta = \begin{bmatrix} 0 \\ 0 \\ 0 \end{bmatrix} \text{ and } A_{\Delta L} - \begin{bmatrix} 0.41PL \\ -0.6PL \\ -2.625P \end{bmatrix},$$

whence solving, $\theta_H = \Delta_1 = -\dfrac{0.0862PL^2}{EI}$

$$\theta_C = \Delta_2 = -\frac{0.0570PL^2}{EI}$$

$$\Delta_C = \Delta_3 = -\frac{0.0531PL^3}{EI}.$$

$$\begin{bmatrix} M_{AB} \\ M_{BA} \\ M_{BC} \\ M_{CB} \\ M_{CD} \\ M_{DC} \end{bmatrix} = \begin{bmatrix} \frac{PL}{12} \\ -\frac{PL}{12} \\ \frac{PL}{2} \\ -\frac{PL}{2} \\ -\frac{PL}{10} \\ \frac{PL}{10} \end{bmatrix} + EI\begin{bmatrix} \frac{2}{L} & 0 & \frac{7.50}{L^2} \\ \frac{4}{L} & 0 & \frac{7.5}{L^2} \\ \frac{4}{L} & \frac{2}{L} & \frac{4.5}{L^2} \\ \frac{2}{L} & \frac{4}{L} & \frac{4.5}{L^2} \\ 0 & \frac{5}{L} & \frac{9.375}{L^2} \\ 0 & \frac{5}{2L} & \frac{9.375}{L^2} \end{bmatrix} \frac{PL^2}{EI}\begin{bmatrix} -0.0862 \\ 0.570 \\ 0.0531L \end{bmatrix}.$$

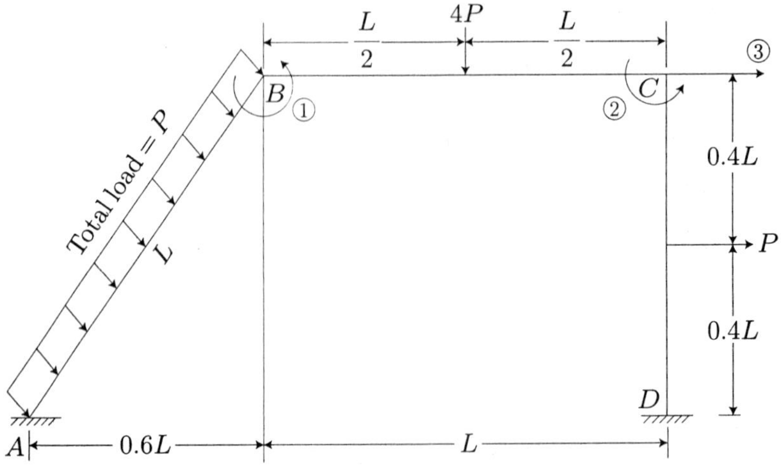

(a) Frame

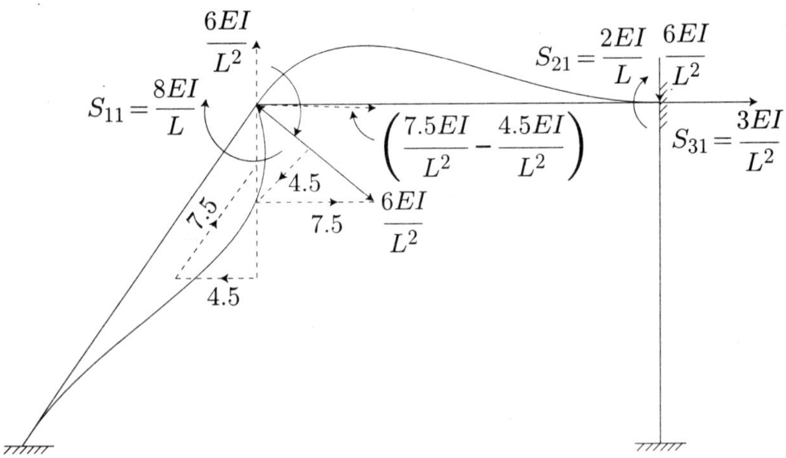

(b) Stiffness corresponding to (1)

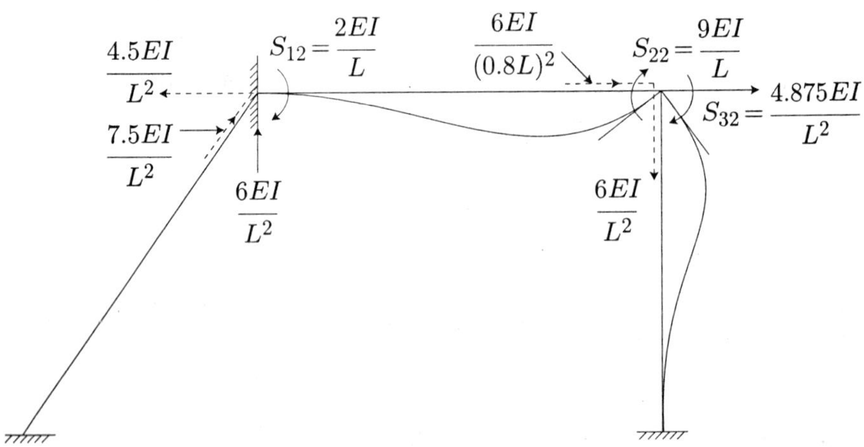

(c) Stiffness corresponding to (2)

Fig. 16.39: For Example 16.43 (Contd.)

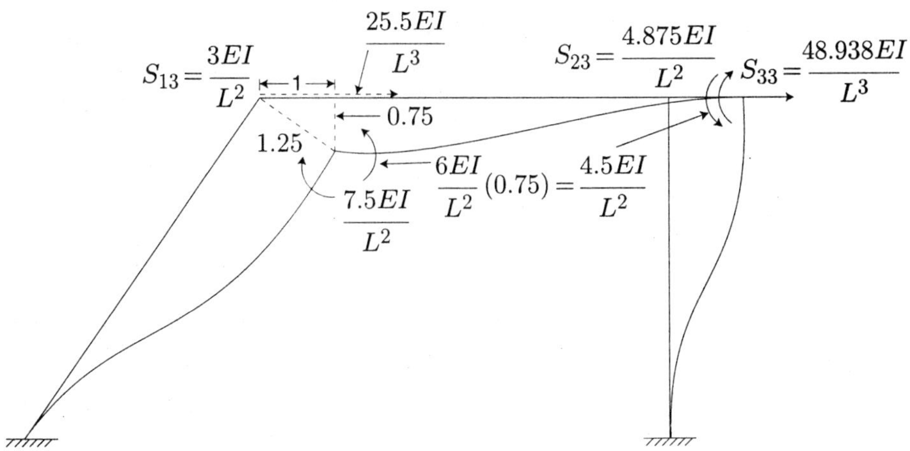

(d) Stiffness corresponding to (3)

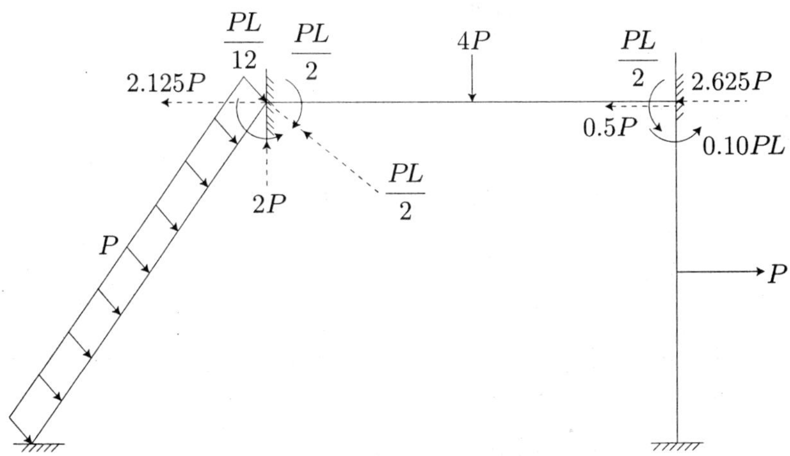

(e) Restrained structure

Fig. 16.39: For Example 16.43

Example 15.44. *Analyse the grid frame shown in Fig. 15.40(a), using stiffness method.*

Solution : In this case, the number of kinematic indeterminancy, viz., θ_{BA}, Δ_{BV} and θ_{BC} are shown in Fig. 16.40(b).

In this case, $A_\Delta = \begin{bmatrix} 0 \\ 0 \\ 0 \end{bmatrix}$ and $A_{\Delta L} = \begin{bmatrix} -\frac{PL}{8} \\ -\frac{P}{2} \\ 0 \end{bmatrix}$.

Referring to Figs. 16.40(c) to 16.40(e),

$$S_{11} = S'_{11} + S''_{11} = \frac{4EI}{L} + \frac{GJ}{L}; \; S_{12} = S'_{12} + S''_{12} = \frac{6EI}{L^2} + 0 = \frac{6EI}{L^2} = S_{21}$$

$$S_{13} = S'_{13} + S''_{13} = 0 + 0 = 0 = S_{31}; \; S_{21} = S'_{21} + S''_{21} = \frac{6EI}{L^2} + 0 = \frac{6EI}{L^2} = S_{12}$$

$$S_{22} = S'_{22} + S''_{22} = \frac{12EI}{L^3} + \frac{12EI}{L^3} = \frac{24EI}{L^3}$$

$$S_{23} = S'_{23} + S''_{23} = 0 - \frac{6EI}{L^2} = -\frac{6EI}{L^2} = S_{32}$$

$$S_{31} = S'_{31} + S''_{31} = 0 + 0 = 0 = S_{13}$$

$$S_{32} = S'_{32} + S''_{32} = 0 - \frac{6EI}{L^2} = -\frac{6EI}{L^2} = S_{23}$$

$$S_{33} = S'_{33} + S''_{33} = \frac{4EI}{L} + \frac{GJ}{L}.$$

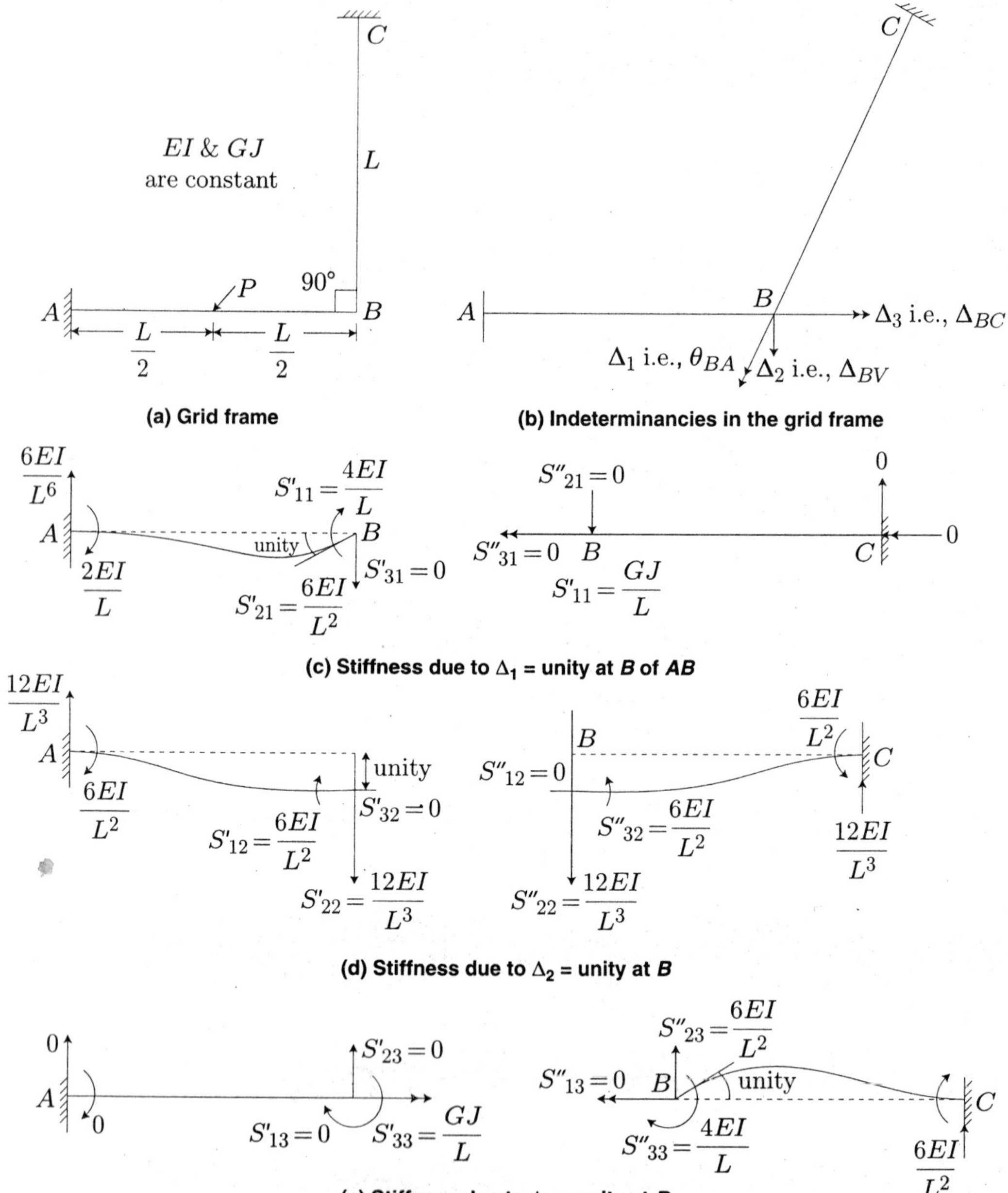

(a) Grid frame **(b) Indeterminancies in the grid frame**

(c) Stiffness due to Δ_1 = unity at *B* of *AB*

(d) Stiffness due to Δ_2 = unity at *B*

(e) Stiffness due to Δ_3 = unity at *B*

Fig. 16.40: For Example 16.44

With $\dfrac{GJ}{EI}$ as α,

$$S = \frac{EI}{L^3}\begin{bmatrix} (4+\alpha)L^2 & 6L & 0 \\ 6L & 24 & -6L \\ 0 & -6L & (4+\alpha)L^2 \end{bmatrix}.$$

Subsituting in Eq. (16.73),

$$\begin{bmatrix} \theta_{BA} \\ \Delta_{BV} \\ \theta_{BC} \end{bmatrix} = \frac{L^3}{6E}\begin{bmatrix} (4+\alpha)L^2 & 6L & 0 \\ 6L & 24 & -6L \\ 0 & -6L & (4+\alpha)L^2 \end{bmatrix}^{-1}\left\{\begin{bmatrix} 0 \\ 0 \\ 0 \end{bmatrix} - \begin{bmatrix} -\frac{PL}{8} \\ -\frac{P}{2} \\ 0 \end{bmatrix}\right\}.$$

Solving, $\theta_{BA} = -\dfrac{3PL^2}{16EI}\dfrac{1}{(1+\alpha)(4+\alpha)}$

$$\Delta_{BV} = \frac{PL^3}{96EI}\frac{5+2\alpha}{1+\alpha}$$

and $\theta_{BC} = \dfrac{PL^2}{16EI}\dfrac{5+2\alpha}{(1+\alpha)(4+\alpha)}.$

Substituting in matrix form,

$$\begin{bmatrix} M_{AB} \\ T_{AB} \\ R_A \\ M_{BA} \\ M_{BC} \\ M_{CB} \\ T_{CB} \\ R_C \end{bmatrix} = \begin{bmatrix} \frac{PL}{8} \\ 0 \\ -\frac{P}{2} \\ -\frac{PL}{8} \\ 0 \\ 0 \\ 0 \\ 0 \end{bmatrix} + \frac{EI}{L^3}\begin{bmatrix} 2L^2 & 6L & 0 \\ 0 & 0 & -\alpha L^2 \\ -6L & -12 & 0 \\ 4L^2 & 6L & 0 \\ 0 & -6L & 4L^2 \\ 0 & -6L & 2L^2 \\ -\alpha L^2 & 0 & 0 \\ 0 & -12 & 6L \end{bmatrix}\frac{PL^2}{EI}\begin{bmatrix} -\frac{3}{16}\frac{1}{(1+\alpha)(4+\alpha)} \\ \frac{5+2\alpha}{96(1+\alpha)}L \\ \frac{1}{16}\frac{5+2\alpha}{(1+\alpha)(4+\alpha)} \end{bmatrix}.$$

Example 16.45. *Analyse the grid frame in Example 16.43, if u.d.l. of magnitude w/unit length be placed on the frame.*

Solution : In this case, $A_\Delta = \begin{bmatrix} 0 \\ 0 \\ 0 \end{bmatrix}$ and $A_{\Delta L} = \begin{bmatrix} -\frac{wL^2}{12} \\ -wL \\ +\frac{wL^2}{12} \end{bmatrix}$

and $$\begin{bmatrix} \theta_{BA} \\ \Delta_{BV} \\ \theta_{BC} \end{bmatrix} = \frac{L^3}{EI}\begin{bmatrix} (4+\alpha)L^2 & 6L & 0 \\ 6L & 24 & -6L \\ 0 & -6L & (4+\alpha)L^2 \end{bmatrix}^{-1}\left\{\begin{bmatrix} 0 \\ 0 \\ 0 \end{bmatrix} - \begin{bmatrix} -\frac{wL^2}{12} \\ -wL \\ +\frac{wL^2}{12} \end{bmatrix}\right\}.$$

Solving $\theta_{BA} = -\dfrac{4wL^3}{24EI(1+\alpha)}$

$$\Delta_{BV} = \frac{wL^4(3+\alpha)}{24EI(1+\alpha)}$$

and $\theta_{BC} = +\dfrac{4wL^3}{24E(l+\alpha)}.$

Example 16.46 *Analyse the grid frame shown in Fig. 16.41 using stiffness method.*

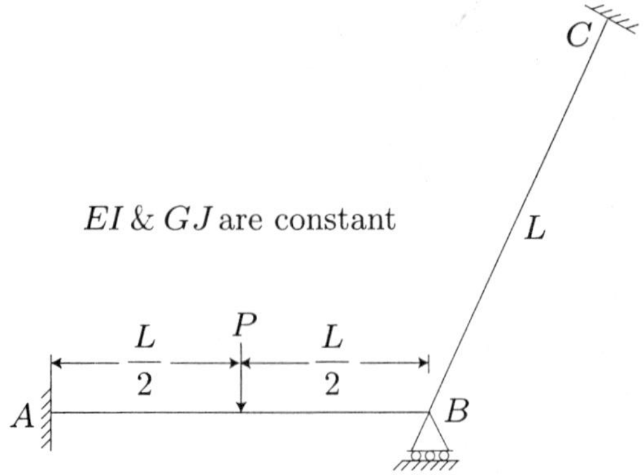

Fig. 16.41: For Example 16.46

Solution : In this case, there are two kinematic indeterminancies, i.e., θ_{BA} and θ_{BC}.

For the case, $A_\Delta = \begin{bmatrix} 0 \\ 0 \end{bmatrix}$ and $A_{\Delta L} = \begin{bmatrix} -\frac{PL}{8} \\ 0 \end{bmatrix}$.

The stiffness matrix is a diagonal matrix

and $$S = \frac{EI}{L^3}\begin{bmatrix} (4+\alpha)L^2 & 0 \\ 0 & (4+\alpha)L^2 \end{bmatrix}.$$

Substituting in Eq. (16.75),

$$\begin{bmatrix} \theta_{BA} \\ \theta_{BC} \end{bmatrix} = \frac{L^3}{EI}\begin{bmatrix} \frac{1}{(4+\alpha)L^2} & 0 \\ 0 & \frac{1}{(4+\alpha)L^2} \end{bmatrix}^{-1}\left\{\begin{bmatrix} 0 \\ 0 \end{bmatrix} - \begin{bmatrix} -\frac{PL}{8} \\ 0 \end{bmatrix}\right\}.$$

Solving, $\theta_{BA} = \dfrac{PL^2}{8EI(4+\alpha)}$ and $\theta_{BC} = 0$.

Example 16.47. *Analyse the grid frame shown in Fig. 16.42 using stiffness method.*

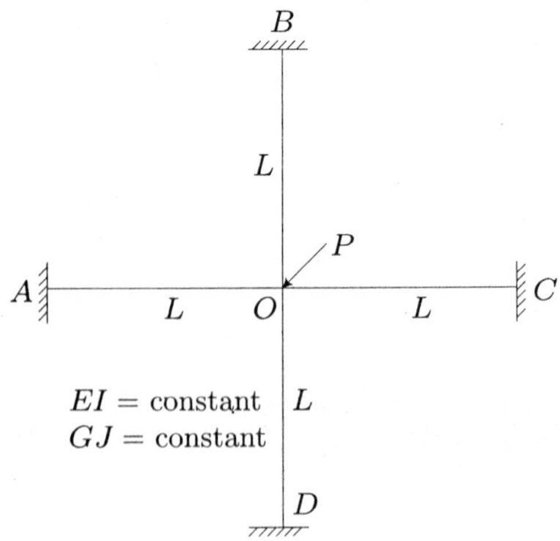

Fig. 16.42: For Example 16.47

Solution : In this case, the number of kinematic indeterminancy is three, i.e., $\theta_{OA}(\theta_{OC})$, $\theta_{OB}(\theta_{OD})$ and θ_{OV}.

$$A_\Delta = \begin{bmatrix} 0 \\ 0 \\ P \end{bmatrix} \text{ and } A_{\Delta L} = \begin{bmatrix} 0 \\ 0 \\ 0 \end{bmatrix}.$$

For stiffness matrix,

$$S_{11} = \frac{2(4EI)}{L} + 2\left(\frac{GJ}{L}\right) = \frac{EI}{L}[8 + 2\alpha] = S_{22}$$

$$S_{21} = 0 = S_{11};\ S_{31} = S_{13} = 0$$

$$S_{32} = 0 = S_{23};\ S_{33} = \frac{4(12EI)}{L^2} = \frac{48EI}{L^3}.$$

Substituting in Eq. (16.75),

$$\begin{bmatrix} \theta_{OA} \\ \theta_{OB} \\ \Delta_{OV} \end{bmatrix} = \frac{L^3}{EI}\begin{bmatrix} (8+2\alpha)L^2 & 0 & 0 \\ 0 & (8+2\alpha)L^2 & 0 \\ 0 & 0 & 48 \end{bmatrix}^{-1} \left\{\begin{bmatrix} 0 \\ 0 \\ P \end{bmatrix} - \begin{bmatrix} 0 \\ 0 \\ 0 \end{bmatrix}\right\},$$

whence
$$\begin{bmatrix} \theta_{OA} \\ \theta_{OB} \\ \Delta_{OV} \end{bmatrix} = \frac{L^3}{EI}\begin{bmatrix} \frac{1}{(8+2\alpha)L^2} & 0 & 0 \\ 0 & \frac{1}{(8+2\alpha)L^2} & 0 \\ 0 & 0 & \frac{1}{48} \end{bmatrix} \begin{bmatrix} 0 \\ 0 \\ P \end{bmatrix}.$$

Solving, $\theta_{OA} = \theta_{OB} = 0$ and $\Delta_{OV} = \dfrac{PL^3}{48EI}$.

Example 16.48. *Determine the elements of he stiffness matrix for the grid frame shown in Fig. 16.43.*

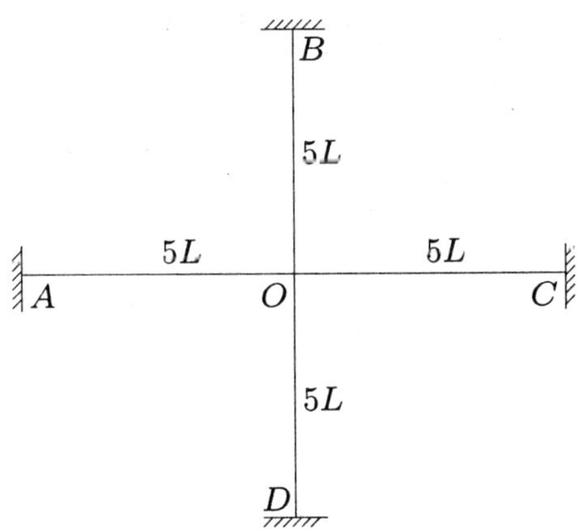

Fig. 16.43: For Example 16.48

Solution : Considering $\theta_{OA}(\theta_{OC}), \theta_{OB}(\theta_{OD})$ and Δ_{OV} as the kinematic indeterminancies,

$$S_{11} = 2\left(\frac{4EI}{5L}\right) + \frac{GJ}{5L} + \frac{GJ}{4L} = \frac{EI}{L}[1.60 + 0.45\alpha]$$

$$S_{21} = 0 = S_{12};\ S_{31} = 0 = S_{13}$$

$$S_{22} = \frac{4EI}{4L} + \frac{4EI}{5L} + \frac{2GJ}{5L} = \frac{EI}{L}[1.80 + 0.40\alpha]$$

$$S_{32} = \frac{6EI}{(4L)^2} - \frac{6EI}{(5L)^2} = 0.135\frac{EI}{L^2} = S_{23}$$

$$S_{33} = 2\frac{12EI}{(5L)^3} + 12EI\left[\frac{1}{(5L)^3} + \frac{1}{(4L)^3}\right] = 0.4755\frac{EI}{L^3}.$$

Stiffness matrix then is $S = \frac{EI}{L^3}\begin{bmatrix} (1.60 + 0.45\alpha)L^2 & 0 & 0 \\ 0 & (1.80 + 0.40\alpha)L^2 & 0.135L \\ 0 & 0.135L & 0.4755 \end{bmatrix}.$

Example 16.49. *Obtain the three reaction components vertical force, bending and twisting moments at end A of the horizontal grid shown in Fig. 16.44(a) due to a uniformly distributed load of intensity w/unit length acting on AC. All bars of the grid have the same cross section and* $\frac{GJ}{EI} = 0.50$.

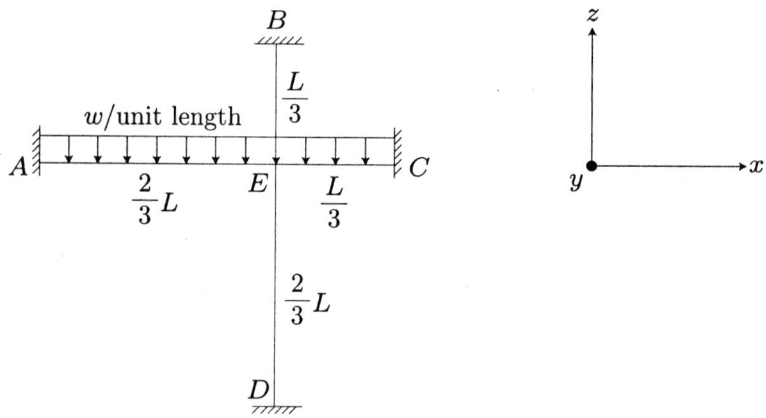

(a) Horizontal grid

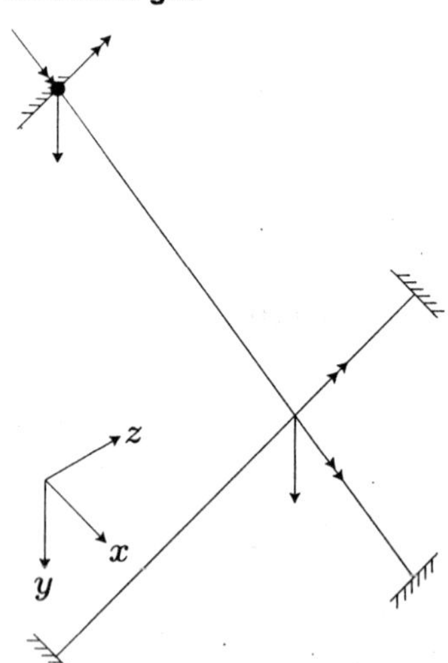

(b) Co-ordinates for the grid

Fig. 16.44: For Example 16.49

Solution : Referring to the co-ordinates shown in Fig. 16.44(b), stiffness matrix is given by

$$S = \frac{EI}{L}\begin{bmatrix} \frac{72.9}{L^2} & -\frac{40.5}{L} & \frac{40.5}{L} \\ -\frac{40.5}{L} & 20.25 & 0 \\ \frac{40.5}{L} & 0 & 20.25 \end{bmatrix}.$$

In this case, $A_\Delta = \begin{bmatrix} 0 \\ 0 \\ 0 \end{bmatrix}$ and $A_{\Delta L} = \begin{bmatrix} -\frac{wL}{2} \\ 0 \\ -\frac{wL^2}{36} \end{bmatrix}$.

Solving, $\Delta_E = \Delta_1 = 0.0010\dfrac{wL^4}{EI}$

$$\theta_{EB} = \Delta_2 = -\frac{0.0020wL^3}{EI}$$

and $\theta_{EA} = \Delta_3 = +\dfrac{0.0034wL^3}{EI}$.

Example 16.50. *The girder system shown in Fig. 16.45(a) has to support the load of a typical grid floor of a building. Longitudinal girders AB and CD carry a uniformly distributed load of* 20 kN/m *inclusive of its own weight. The uniformly distributed load on corss girders EF and GH inclusive of the self weight is* 18 kN/m. *If EI and GJ for all the girder be* 360, 000 kNm^2 *and* 120, 000 kNm^2 *respectively, analyse the grid by the displacement method. Use double symmetry of the structure to reduce the number of co-ordinates to 3. Develop the stiffness matrix with reference to co-ordinates 1,2 and 3.*

Solution : The coefficients for stiffness matrix may be obtained from unit displacements as shown in Figs. 16.45(b) and 16.45(c). From symmetry, in Fig. 16.45(b),

$$M_{BC} = M_{CB} = \frac{2EI}{L}$$

$$S_{11} = \frac{6(360,000)}{5} + \frac{120,000}{4} = 46.200(10)^4$$

$$S_{21} = 0 = S_{12}; S_{31} - 6(360,000)5^2 - 8.64(20)^4$$

$$S_{22} = \frac{6(360,000)}{4} + \frac{120,000}{5} = 56.400(10)^4$$

$$S_{32} = \frac{6(360,000)}{(4)^2} = 13.50(10)^4$$

$$S_{33} = \frac{12(360,000)}{(5)^3} + \frac{12(360,000)}{(4)^3} = 10.206(10)^4.$$

Stiffness matrix is given by

$$S = (10)^4\begin{bmatrix} 46.20 & 0 & 8.64 \\ 0 & 56.40 & 13.50 \\ 8.64 & 13.50 & 10.206 \end{bmatrix}.$$

In this case, $A_\Delta = \begin{bmatrix} 0 \\ 0 \\ 0 \end{bmatrix}$ and $A_{\Delta L} = \begin{bmatrix} 0 \\ 0 \\ -172.0 \end{bmatrix}$.

Substituting in Eq. (16.75),

$$\begin{bmatrix} \theta_{IA} \\ \theta_{IE} \\ \Delta_{IV} \end{bmatrix} = \frac{1}{10^3} \begin{bmatrix} 46.20 & 0 & 8.64 \\ 0 & 56.40 & 13.50 \\ 8.64 & 13.50 & 10.206 \end{bmatrix}^{-1} \left\{ \begin{bmatrix} 0 \\ 0 \\ 0 \end{bmatrix} - \begin{bmatrix} 0 \\ 0 \\ -172.0 \end{bmatrix} \right\}.$$

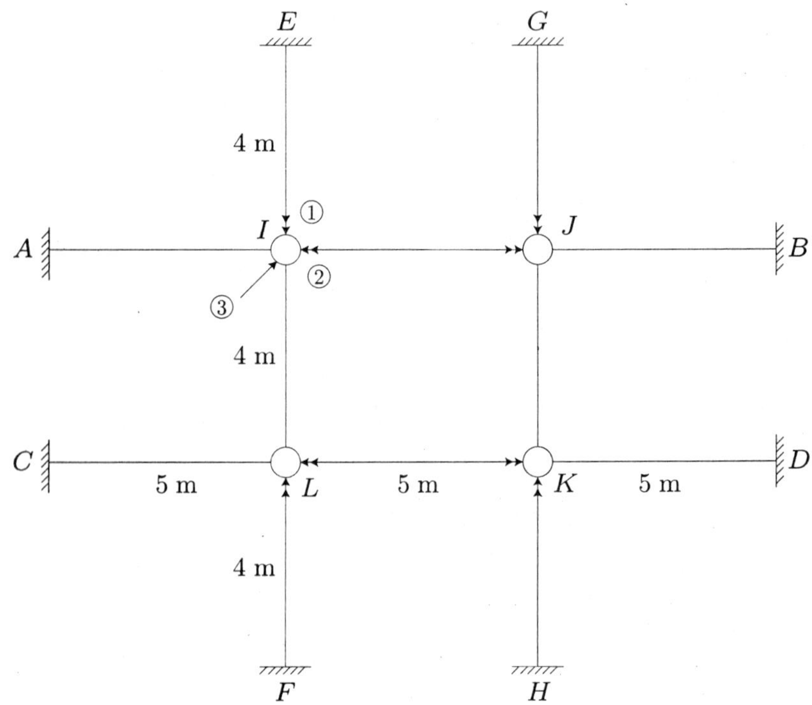

(a) Girder system

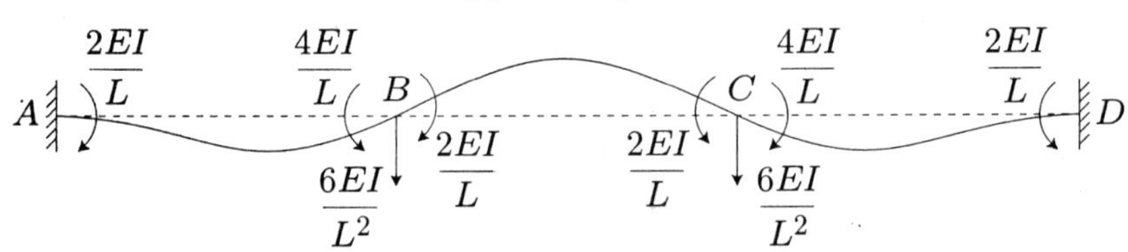

(b) Unit displacement in direction ①

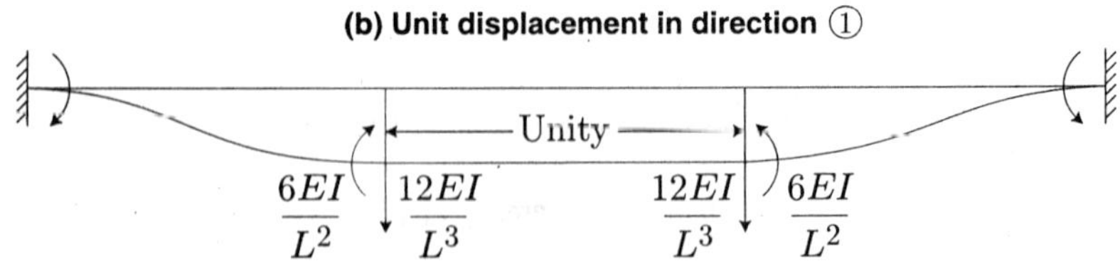

(c) Unit displacement in direction ②

Fig. 16.45: For Example 16.50

Example 16.51. *Analyse the continuous beam in Fig. 16.26(a) (Example 16.29) considering a partially restrained structure in which only the rotation at B is prevented.*

Solution : In this case, only θ_B is considered as the unknown displacement. For the partially restrained structure,

$$S_{11} = \frac{4EI}{10} + \frac{3EI}{10} = 0.70EI.$$

If spans AB and BC are considered fixed ended,

$$M_{FAB} = 150 \text{ kNm} = -M_{FAB}$$

$$M_{FBC} = 75 \text{ kNm} = -M_{FCB}.$$

With end C permanently free, the net moment at B

$$= -150 + 75 + 37.5 = 37.50 \text{ kNm}.$$

Hence, $A_{\Delta L} = -37.5$ and $A_\Delta = 0$.

Substituting in Eq. (16.75),

$$\theta_B = \frac{1}{0.70EI}\{[0] - [37.5]\} = +\frac{53.57}{EI}.$$

Using slope deflection equation,

$$M_{AB} = 150 + \frac{2EI}{10}\left(\frac{53.57}{EI}\right) = 160.71 \text{ kNm}$$

$$M_{BA} = -150 + \frac{2EI}{10}\left(\frac{2(53.57)}{EI}\right) = -128.37 \text{ kNm}.$$

Example 16.52. *Analyse the rigid frame in Fig. 16.46, using modified stiffness matrix with a partially restrained structure.*

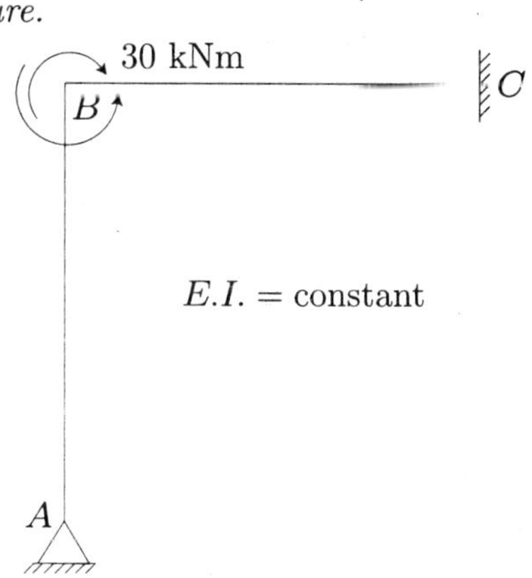

Fig. 16.46: For Example 16.52

Solution : In this case, the rotation at B, i.e., θ_B is considered as the only indeterminancy. Stiffness matrix is

$$S = \frac{7EI}{L}.$$

Solution is given by

$$\theta_B = \frac{L}{7EI}(-30) = -\frac{30L}{7EI}.$$

End moments are given by

$$\begin{bmatrix} M_{BA} \\ M_{BC} \\ M_{CB} \end{bmatrix} = \begin{bmatrix} 0 \\ 0 \\ 0 \end{bmatrix} + \begin{bmatrix} \frac{3EI}{L} \\ \frac{4EI}{L} \\ \frac{2EI}{L} \end{bmatrix} \begin{bmatrix} -\frac{30L}{7EI} \end{bmatrix} = \begin{bmatrix} -\frac{90}{7} \\ -\frac{120}{7} \\ -\frac{60}{7} \end{bmatrix}.$$

Example 16.53. *Analyse the rigid frame shown in Fig. 16.47 by stiffness matrix using modified stiffness considering antisymmetry.*

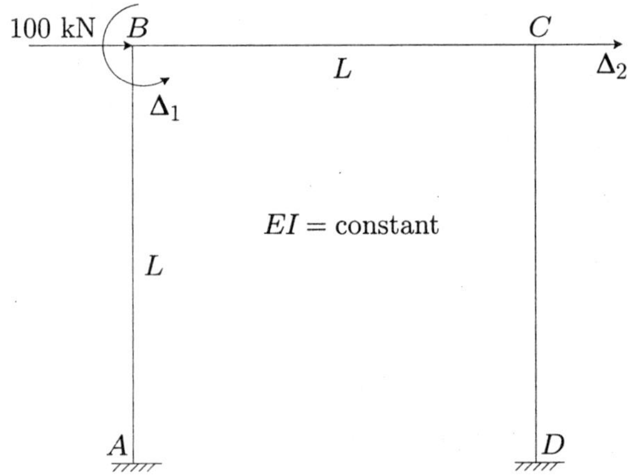

Fig. 16.47: For Example 16.53

Solution : In this case modified stiffness matrix considering antisymetry taking the coordinates as indicated is given by

$$S = \frac{EI}{L}\begin{bmatrix} 10 & \frac{6}{L} \\ \frac{6}{L} & \frac{12}{L^2} \end{bmatrix},$$

whence, as $A_\Delta = \begin{bmatrix} 0 \\ 50 \end{bmatrix}$ and $A_{\Delta L} = \begin{bmatrix} 0 \\ 0 \end{bmatrix}$,

it follows that,

$$\begin{bmatrix} \theta_B \\ \Delta_{BH} \end{bmatrix} = \frac{L^3}{84EI}\begin{bmatrix} \frac{12}{L^2} & -\frac{6}{L} \\ -\frac{6}{L} & 10 \end{bmatrix}\left\{\begin{bmatrix} 0 \\ 50 \end{bmatrix} - \begin{bmatrix} 0 \\ 0 \end{bmatrix}\right\}$$

$$\text{Solving, } \theta_B = -\frac{300}{L}\frac{L^3}{84EI} = -\frac{3L^2}{0.84EI}$$

$$\Delta_{BH} = \frac{500L^3}{84EI} = \frac{5L^3}{0.84EI}.$$

End moments are given by

$$\begin{bmatrix} M_{AB} \\ M_{BA} \\ M_{BC} \end{bmatrix} = \begin{bmatrix} 0 \\ 0 \\ 0 \end{bmatrix} + \begin{bmatrix} \frac{2EI}{L} & \frac{6EI}{L} \\ \frac{4EI}{L} & \frac{6EI}{L^2} \\ \frac{6EI}{L} & 0 \end{bmatrix}\begin{bmatrix} \frac{-3L^2}{0.84EI} \\ \frac{5L^3}{0.84EI} \end{bmatrix} = \begin{bmatrix} 28.57L \\ 21.43L \\ -21.43L \end{bmatrix}.$$

Example 16.54. *Analyse the rigid frame in Fig. 16.48 by displacement method considering symmetry.*

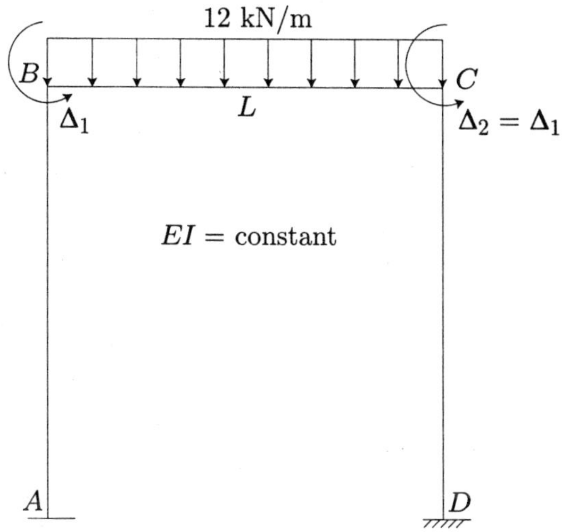

Fig. 16.48: For Example 16.54

Solution : The portal frame is symmetrical about the centre line of the horizontal member. Modified stiffness matrix considering symmetry is given by

$$S = \frac{4EI}{L} + \frac{2EI}{L} = \frac{6EI}{L} \text{ as}$$

rotation at B, i.e., $\theta_B = -$ rotation at $C = -\theta_C$.

Solution is given by

$$\theta_B = \frac{L}{6EI}\{[0] - [L^2]\} = -\frac{L^3}{6EI}.$$

End moments are given by

$$\begin{bmatrix} M_{AB} \\ M_{BA} \\ M_{BC} \end{bmatrix} = \begin{bmatrix} 0 \\ 0 \\ L^2 \end{bmatrix} + \begin{bmatrix} \frac{2EI}{L} \\ \frac{4EI}{L} \\ \frac{2EI}{L} \end{bmatrix} \left[-\frac{L^3}{6EI}\right] = \begin{bmatrix} -\frac{L^2}{3} \\ -\frac{2L^2}{3} \\ +\frac{2L^2}{3} \end{bmatrix}.$$

Example 16.55. *Obtain the stiffness matrix for the Truss shown in Fig. 16.49(a). Calculate the value for $\theta_1 = \theta_4' = 45°$ and $\theta_2 = \theta_3' = 60°$.*

Solution : The only joint which can be displaced in this case is the joint '0' which can have two translational displacements, one in the vertical and the other in the horizontal direction as indicated by (1) and (2) in Fig. 16.49(a).

To obtain the stiffness matrix, the displacements in the two directions are considered as shown in Figs. 16.49(b) and 16.49(c) respectively.

Length of the members are $\frac{L}{\sin\theta_1}$, $\frac{L}{\sin\theta_2}$, $\frac{L}{\sin\theta_3}$, and $\frac{L}{\sin\theta_4}$.

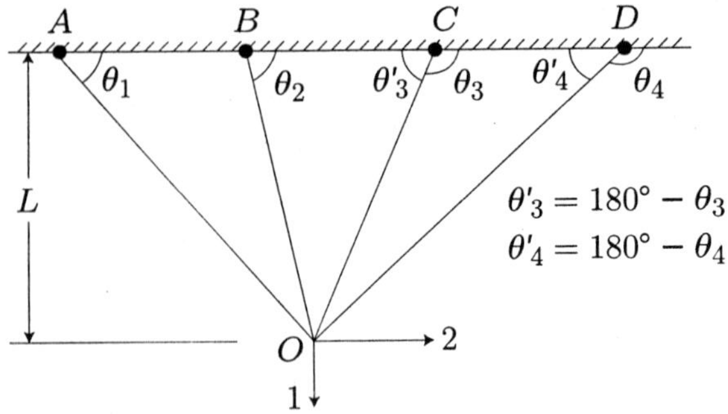

(a) Truss

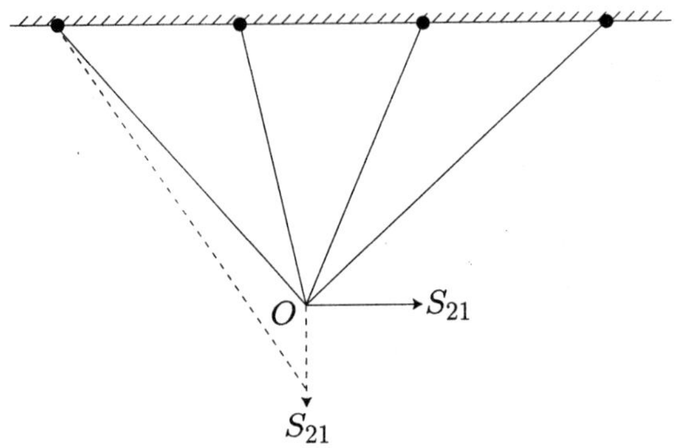

(b) Unit movement in vertical direction

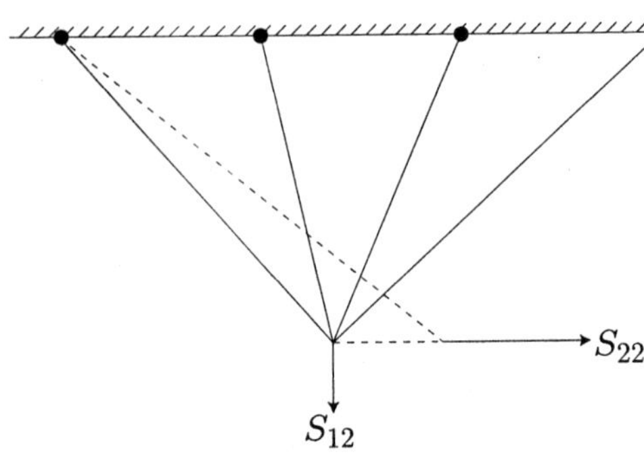

(c) Unit movement in horizontal direction

Fig. 16.49: For Example 16.55

Due to unit movement in the vertiacal direction of joint 0 members OA, OB, OC and OD elongate and tensile forces developed are given by

$$T_1 = \frac{A_1 E_1}{L} \sin^2 \theta_1 \text{ on member } OA,$$

$$T_2 = \frac{A_2 E_2}{L} \sin^2 \theta_2 \text{ on member } OB,$$

$$T_3 = \frac{A_3 E_3}{L} \sin^2 \theta_3 \text{ on member } OC,$$

and $$T_4 = \frac{A_4 E_4}{L} \sin^2 \theta_4 \text{ on member } OD.$$

Vertical and horizontal components of these tensile forces are

$$\frac{A_1E_1}{L}\sin^3\theta_1, \quad \text{and} \quad \frac{A_1E_1}{L}\sin^2\theta_1\cos\theta_1 \text{ on member } OA,$$

$$\frac{A_2E_2}{L}\sin^3\theta_2, \text{ and } \frac{A_2E_2}{L}\sin^2\theta_2\cos\theta_2 \text{ on member } OB,$$

$$\frac{A_3E_3}{L}\sin^3\theta_3, \text{ and } \frac{A_3E_3}{L}\sin^2\theta_3\cos\theta_3 \text{ on member } OC,$$

and
$$\frac{A_4E_4}{L}\sin^3\theta_4, \text{ and } \frac{A_4E_4}{L}\sin^2\theta_4\cos\theta_4 \text{ on member } OD,$$

To determine S_{11}, the contribution of the vertical components of the forces from all members meeting at that joint is to be taken into account. Thus,

$$\begin{aligned}&\frac{A_1E_1}{L}\sin^3\theta_1 + \frac{A_2E_2}{L}\sin^3\theta_2 + \frac{A_3E_3}{L}\sin^3\theta_3 + \frac{A_4E_4}{L}\sin^3\theta_4\\ &= \frac{A_1E_1}{L_1}\sin^2\theta_1 + \frac{A_2E_2}{L_2}\sin^2\theta_2 + \frac{A_3E_3}{L_3}\sin^2\theta_3 + \frac{A_4E_4}{L_4}\sin^2\theta_4\\ &= \sum\frac{AE}{L}\sin^2\theta.\end{aligned}$$

For AE constant, $S_{11} = \sum\frac{AE}{L}\sum\sin^2\theta.$

Also $\quad S_{21} = \sum\frac{AE}{L}\sin^2\theta\cos\theta = \frac{AE}{L}\sum\sin^2\theta\cos\theta = S_{12}.$

Similarly, $S_{22} = \sum\frac{AE}{L}\sin\theta\cos^2\theta = \frac{AE}{L}\sum\sin\theta\cos^2\theta$, for $AE =$ constant.

For the case, $\theta_1 = 45° = (180° - \theta_4)$ and $\theta_2 = 60° = (180° - \theta_3)$

and $\quad S_{11} = 2.000\frac{AE}{L}$

$$S_{12} = S_{21} = 0$$

and $\quad S_{22} = 1.140\frac{AE}{L}$

and stiffness matrix $= \frac{AE}{L}\begin{bmatrix} 2.000 & 0 \\ 0 & 1.140 \end{bmatrix}.$

Unknown displacements are given by

$$[\Delta] = S^{-1}\{[A_\Delta] - [A_{\Delta L}]\}.$$

Forces in the various member are given by

$$[F_M] = [A_{ML}] + \frac{AE}{L}\begin{bmatrix} \sin^2\theta_1 & \sin\theta_1\cos\theta_1 \\ \sin^2\theta_2 & \sin\theta_2\cos\theta_2 \\ \sin^2\theta_3 & \sin\theta_3\cos\theta_3 \\ \sin^2\theta_4 & \sin\theta_4\cos\theta_4 \end{bmatrix}\{S^{-1}[(A_\Delta) - (A_{\Delta L})]\}.$$

Example 16.56. *Analyse the Truss shown in Fig. 16.50 using displacement method if $AE =$ constant.*

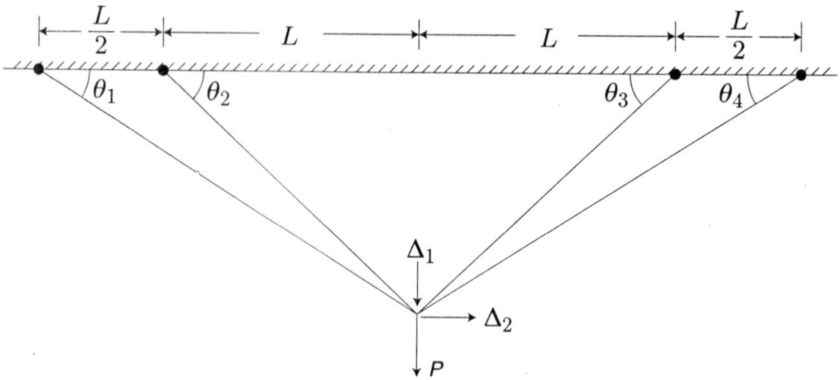

Fig. 16.50: For Example 16.56

Solution : In this case, $\theta_1 = \theta_4 = \tan^{-1}\dfrac{2}{3} = 33.69°$

and $\theta_2 = \theta_3 = \tan^{-1} 1 = 45°$.

Stiffness matrix is given by, using symmetry,

$$S = \frac{AE}{L}\begin{bmatrix} 1.048 & 0 \\ 0 & 1.475 \end{bmatrix} \quad \text{and} \quad S^{-1} = \frac{L}{AE}\begin{bmatrix} 0.954 & 0 \\ 0 & 0.678 \end{bmatrix}.$$

Solution is given by

$$\begin{bmatrix} \Delta_1 \\ \Delta_2 \end{bmatrix} = \frac{L}{AE}\begin{bmatrix} 0.954 & 0 \\ 0 & 0.678 \end{bmatrix}\left\{\begin{bmatrix} P \\ 0 \end{bmatrix} - \begin{bmatrix} 0 \\ 0 \end{bmatrix}\right\}$$

and $\Delta_1 = \dfrac{0.954PL}{AE}$

$\Delta_2 = 0.$

It is also evident from symmetry that $\Delta_2 = 0$.

Forces in the members are given by

$$\begin{bmatrix} F_1 \\ F_2 \\ F_3 \\ F_4 \end{bmatrix} = \begin{bmatrix} 0 \\ 0 \\ 0 \\ 0 \end{bmatrix} + \frac{AE}{L}\begin{bmatrix} \sin^2\theta_1 \\ \sin^2\theta_2 \\ \sin^2\theta_3 \\ \sin^2\theta_4 \end{bmatrix}\left[\frac{0.954PL}{AE}\right]$$

$$= \frac{AE}{L}\begin{bmatrix} 0.308 \\ 0.50 \\ 0.50 \\ 0.308 \end{bmatrix}\left[\frac{0.954PL}{AE}\right]$$

$$= P\begin{bmatrix} 0.294 \\ 0.477 \\ 0.477 \\ 0.294 \end{bmatrix}.$$

Example 16.57. *Analyse the truss shown in Fig. 16.49 if member OB in Example 16.55 be too short by 'e'.*

Solution : In this case, $S = \frac{AE}{L}\begin{bmatrix} 2.000 & 0 \\ 0 & 1.140 \end{bmatrix}$.

In order to bring the member OB to joint 0, a tensile force of magnitude $\frac{e(AE)\sin 60°}{L} = \frac{\sqrt{3}}{2} \cdot \frac{AEe}{L}$ is to be applied.

The components of the is force in the vertical and horizontal directions are $\frac{\sqrt{3}}{2}\frac{AEe}{L}\left(\frac{\sqrt{3}}{2}\right)$ and $\frac{\sqrt{3}}{2}\frac{AEe}{L}\left(\frac{1}{2}\right)$ respectively.

If forces applied in the truss by X_1 and X_2 in the vertical and horizontal directions, final equation is given by

$$\begin{bmatrix} X_1 \\ X_2 \end{bmatrix} = \frac{L}{AE}\begin{bmatrix} 0.500 & 0 \\ 0 & 0.877 \end{bmatrix}\left\{\begin{bmatrix} 0 \\ 0 \end{bmatrix} - \begin{bmatrix} \frac{\sqrt{3}}{2} \\ \frac{1}{2} \end{bmatrix}\frac{\sqrt{3}}{2}\frac{AEe}{L}\right\}.$$

Solving $\begin{bmatrix} X_1 \\ X_2 \end{bmatrix} = -\begin{bmatrix} 0.375 \\ 0.380 \end{bmatrix} e.$

Final forces in the members are given by

$$\begin{bmatrix} F_1 \\ F_2 \\ F_3 \\ F_4 \end{bmatrix} = \frac{AE}{L}\begin{bmatrix} \frac{1}{2} & \frac{1}{2} \\ \frac{3}{4} & \frac{\sqrt{3}}{4} \\ \frac{3}{4} & -\frac{\sqrt{3}}{4} \\ \frac{1}{2} & -\frac{1}{2} \end{bmatrix}\begin{bmatrix} -0.375e \\ -0.380e \end{bmatrix} + \frac{AE}{2L}\begin{bmatrix} 0 \\ \sqrt{3}e \\ 0 \\ 0 \end{bmatrix}$$

$$= \frac{AEe}{L}\begin{bmatrix} -0.3775 \\ 1.287 \\ -0.1160 \\ 0.0030 \end{bmatrix}$$

Example 16.58. *Analyse the pin-jointed truss shown in Fig. 16.51. The cross-sectional area of each member is* 2000 mm^2. *Assume* $E = 200$ *kN-mm* ^{2}.

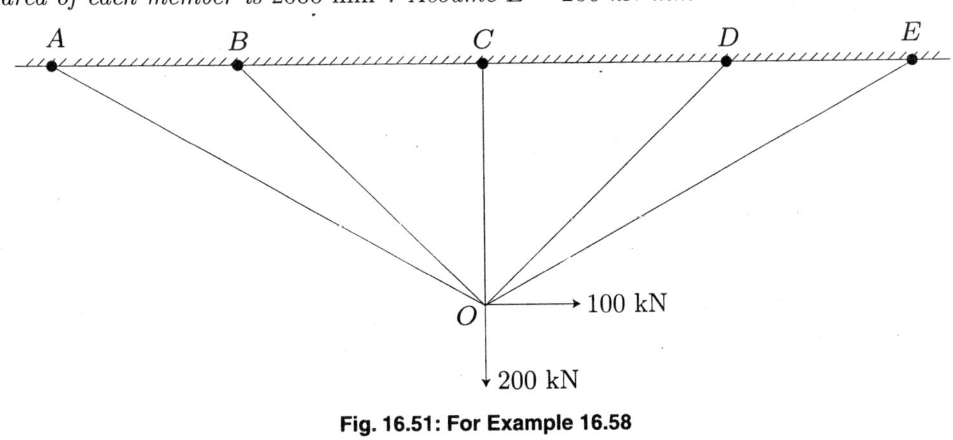

Fig. 16.51: For Example 16.58

Solution : The degree of freedom in this case is two as joint 0 can move in the horizontal and vertical directions and no moment is possible at joints A, B, C, D and E. Calculations of values required for stiffness matrix are given in tabular form :

Member	$\frac{AE}{L}$ (kN/mm)	θ	$\sin\theta$	$\cos\theta$	$\frac{AE}{L}\sin^2\theta$	$\frac{AE}{L}\cos^2\theta$	$\frac{AE}{L}\sin\theta\cos\theta$
OA	40	30°	$\frac{1}{2}$	$\frac{\sqrt{3}}{2}$	10	30	17.32
OB	56.57	45°	$\frac{1}{\sqrt{2}}$	$\frac{1}{\sqrt{2}}$	28.28	28.28	28.28
OC	80	90°	1	0	80.00	0	0
OD	69.28	120°	$\frac{\sqrt{3}}{2}$	$-\frac{1}{2}$	51.96	17.32	−30.00
OF	56.57	135°	$\frac{1}{\sqrt{2}}$	$-\frac{1}{\sqrt{2}}$	28.28	28.28	−28.28
$\sum$					198.52	103.88	−12.68

Solution given by

$$\begin{bmatrix} \Delta_1 \\ \Delta_2 \end{bmatrix} = \begin{bmatrix} 198.52 & -12.68 \\ -12.68 & 103.88 \end{bmatrix}^{-1} \begin{bmatrix} 200 \\ 100 \end{bmatrix}$$

$$= \frac{1}{20461.4752} \begin{bmatrix} 103.88 & 12.68 \\ 12.68 & 198.52 \end{bmatrix} \begin{bmatrix} 200 \\ 100 \end{bmatrix}$$

$$= \begin{bmatrix} 1.0773 \\ 1.0942 \end{bmatrix}$$

Example 16.59. *Analyse the pin-jointed truss shown in Fig. 16.52(a). The axial stiffness of each member is 40 kN/mm.*

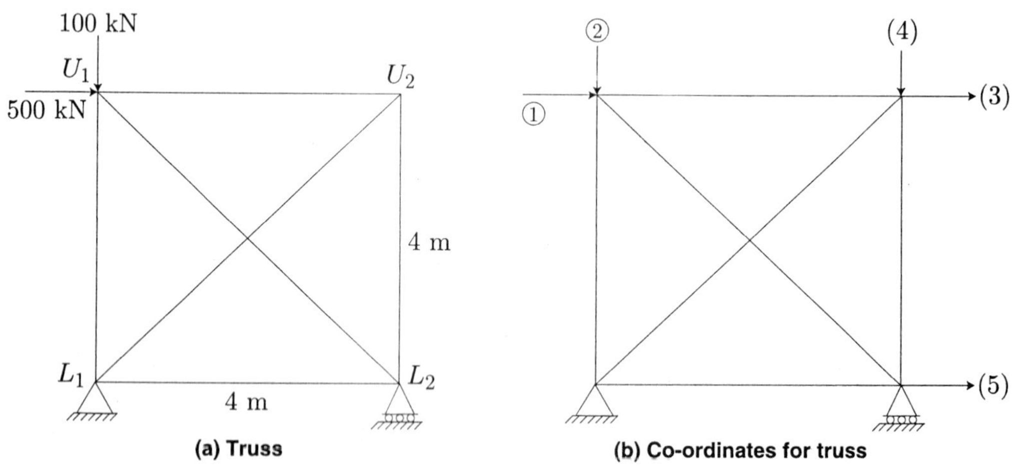

Fig. 16.52: For Example 16.59

Solution : In this case, each joint U_1 and U_2 has two degrees of freedom as they can move in the horizontal and vertical direction. Joint L_2 supported by a roller can move only in the horizontal directions. Joint L_1 supported by a hinge has zero degree of freedom as it cannot move in any direction. Degrees of freedom of the structure is five as shown in Fig. 16.52(b).

Calculations for evaluation of stiffness matrix are given in Tabular form.

Member	θ	$\cos\theta$	$\sin\theta$	$\cos^2\theta$	$\sin^2\theta$	$\sin\theta\cos\theta$
L_1L_2	0	1	0	1	0	0
L_2U_2	90°	0	1	0	1	0
U_1U_2	180°	−1.0	0	1	0	0
U_1L_2	270°	0	−1	0	1	0
L_1U_2	45°	$\frac{1}{\sqrt{2}}$	$\frac{1}{\sqrt{2}}$	$\frac{1}{2}$	$\frac{1}{2}$	$\frac{1}{2}$
L_2U_1	135°	$-\frac{1}{\sqrt{2}}$	$\frac{1}{\sqrt{2}}$	$\frac{1}{2}$	$\frac{1}{2}$	$-\frac{1}{2}$

Axial stiffness of the member $\dfrac{AE}{L} = 40$ kN/mm.

From the Table it follows that,

$$S_{11} = \sum \frac{AE}{L}\cos^2\theta, \text{ for members } U_1U_2, U_1L_1 \text{ and } U_1L_2$$

$$= 40(1 + 0 + 0.50) = 60.$$

$$S_{21} = S_{12} = \sum \frac{AE}{L}\sin\theta\cos\theta = 40(0 + 0 - 0.50) = -20$$

for members U_1U_2, U_1L_1 and U_1L_2.

Similarly, $S_{31} = S_{13} = -\sum \dfrac{AE}{L}\cos^2\theta$ for member U_1U_2 only $= -40(1) = -40$

$$S_{41} = S_{14} = -\sum \frac{AE}{L}\sin\theta\cos\theta \text{ for member } U_2L_2 = -40(0) = 0$$

$$S_{51} = S_{15} = -\sum \frac{AE}{L}\cos^2\theta \text{ for member } L_2U_1 = -40(0.50) = -20$$

$$S_{22} = \sum \frac{AE}{L}\sin^2\theta \text{ for member } U_1U_2, U_1L_1 \text{ and } U_1L_2 = 40(0 + 1 + 0.5) = 60$$

$$S_{32} = S_{23} = -\sum \frac{AE}{L}\sin\theta\cos\theta \text{ for member } U_2U_1 = -40(0) = 0$$

$$S_{42} = S_{24} = -\sum \frac{AE}{L}\sin^2\theta \text{ for member } U_2U_1 = 40(0) = 0$$

$$S_{52} = S_{25} = -\sum \frac{AE}{L}\sin\theta\cos\theta \text{ for member } L_2U_1 = -40(-0.5) = 20$$

$$S_{33} = \sum \frac{AE}{L}\cos^2\theta \text{ for member } U_2L_2, U_2U_1 \text{ and } U_2L_1 = 40(0 + 1 + 0.5) = 60$$

$$S_{43} = S_{34} = \sum \frac{AE}{L}\sin\theta\cos\theta, \text{ for member } U_2L_2, U_2U_1 \text{ and } U_2L_1$$

$$= 40(0 + 0 + 0.5) = 20$$

$$S_{53} = S_{35} = -\sum \frac{AE}{L}\cos^2\theta \text{ for member } L_2U_2 = -40(0) = 0$$

$$S_{44} = \sum \frac{AE}{L}\sin^2\theta \text{ for member } U_2L_2, U_2L_1 \text{ and } U_2U_1 = 40(1+0+0+0.5) = 60$$

$$S_{54} = S_{45} = -\sum \frac{AE}{L}\sin\theta\cos\theta \text{ for member } L_2U_2 = -40(0) = 0$$

$$S_{55} = \sum \frac{AE}{L}\cos^2\theta \text{ for member } L_2L_1, L_2U_2 \text{ and } L_2U_1$$
$$= 40(1 + 0 + 0.5) = 60.$$

Stiffness matrix then is given by

$$S = \begin{bmatrix} 60 & -20 & -40 & 0 & -20 \\ -20 & 60 & 0 & 0 & 20 \\ -40 & 0 & 60 & 20 & 0 \\ 0 & 0 & 20 & 60 & 0 \\ -20 & 20 & 0 & 0 & 60 \end{bmatrix}.$$

16.12 Conclusions

The suitability of the matrix method increases as the structure becomes larger. For elementary problems, the use of the matrix aproach may be time consuming. For a large structure with a high degree of indeterminancies, the matrix method provides a systematic and generalised solution particularly amenable to computer application. It is essential that the analysis of a large structure by the matrix method will almost invariably be carried out with the help of a digital computer. From this point of view, the displacement method is prferable for the analysis of large structure as all steps of the method can be executed by a computer more conveniently, as compared to those of the force method.

The selection of the released structure in the force method is not a simple problem. Usually several alternatives are available, each one having its own advantages. In general, it is difficult to select a released structure, which meets the triple requirements of minimum computational effort, accuracy and simplicity. On the other hand, the restrained structure in the displacement method is self-evident and follows automatically as soon as the indelpendent displacement components are recognised. From this consideration, the advantage of the displacement method is obvious.

In general, the development of the flexibility matrix in the force method is more time consuming and more diffcult than the development of stiffness matrix in the displacement method. The elements of the flexibility matrix represent the displacement components of the released structure. The computation of these displacements in case of many structures requires extensive calculations unless the released structure has been chosen to ensure a localised effect.

The flexibility matrix in the force method is not necessarily well conditined unless great care is exercised in the selection of the released structure and numbering the co-ordinates. Consequently, the inversion of the flexibility matrix which may lead to significant accuracy. On the other hand, the stiffness matrix in the displacement method is generally well conditioned if the co-ordinates are numbered properly.

The capacity of the computer may be a constraint in the analysis of a large structure. The size of the matrix may become an important consideration in the choice between the two methods. In the case of rigid jointed frames, the degree of static indeterminancy is usually higher than the degree of kinematic indeterminancy. For such frames the displacement method is, therefore, preferable. The degree of kinematic indeterminancy on the other hand is generally higher than the degree of static indeterminancy in case of pin-jointed frames. Consequently, the force method may appear to be preferable for the analysis of such frame. Wherever possible, advantage should be taken of the symmetry of the matrix and the band-width should be minimised through proper numbering of co-ordinates.This is more easily accomplilshed in case of the stiffness matirx.

The displacement method provides a complete analysis of the structure including displacements which may sometimes govern the design. In the case of the force method, the displacements are generally to be obtained from separate computations.

The suitability of the two methods for different types of structures and under different situations is enumerated below.

Points under consideration	Suitable Methods
1. Method of computation :	
(a) By Computer	Displacement Method
(b) By hand	
(i) High degree of static indeterminancy and low degree of kinematic indeterminancy.	Displacement Method
(ii) Low degree of static indeterminancy and high degree of kinematic indeterminancy.	Force Method
(iii) Low degree of both static and kinematic indeterminancies.	Both methods
(iv) High degree of both static and kinematic indeterminancies.	None
2. Choice of released or restrained structure.	Displacement Method.
3. Development of flexibility and stiffness matrices and their inversion.	Displacement Method.
(a) Computational effect (b) Accuracy (c) Simplicity	Displacement Method
4. Localised phenomenon banded and well conditioned matrix.	Displacement Method.
5. Size of Matrix	
(a) Continuous beams	Displacement Method
(b) Rigid jointed frames	Displacement Method
(c) Pin-jointed frame	Force Method
6. Completeness of analysis including displacements	Displacement Method

Problems

16.1 Define the static and kinematic indeterminancies and illustrate taking a multistoreyed building frame. How are these values modified with the introduction of internal hinges in some of the members? Illustrate.

16.2 Distinguish between the flexibility (Force) Method and Stiffness (Displacement) Method of matrix method for solution of indeterminate structures. Illustrate.

16.3 Discuss critically the merits and demerits of the two main methods of matrix analysis for different types of structures.

16.4 Analyse the beams shown in Fig. P-16.1 using Flexibility (Force) matrix method.

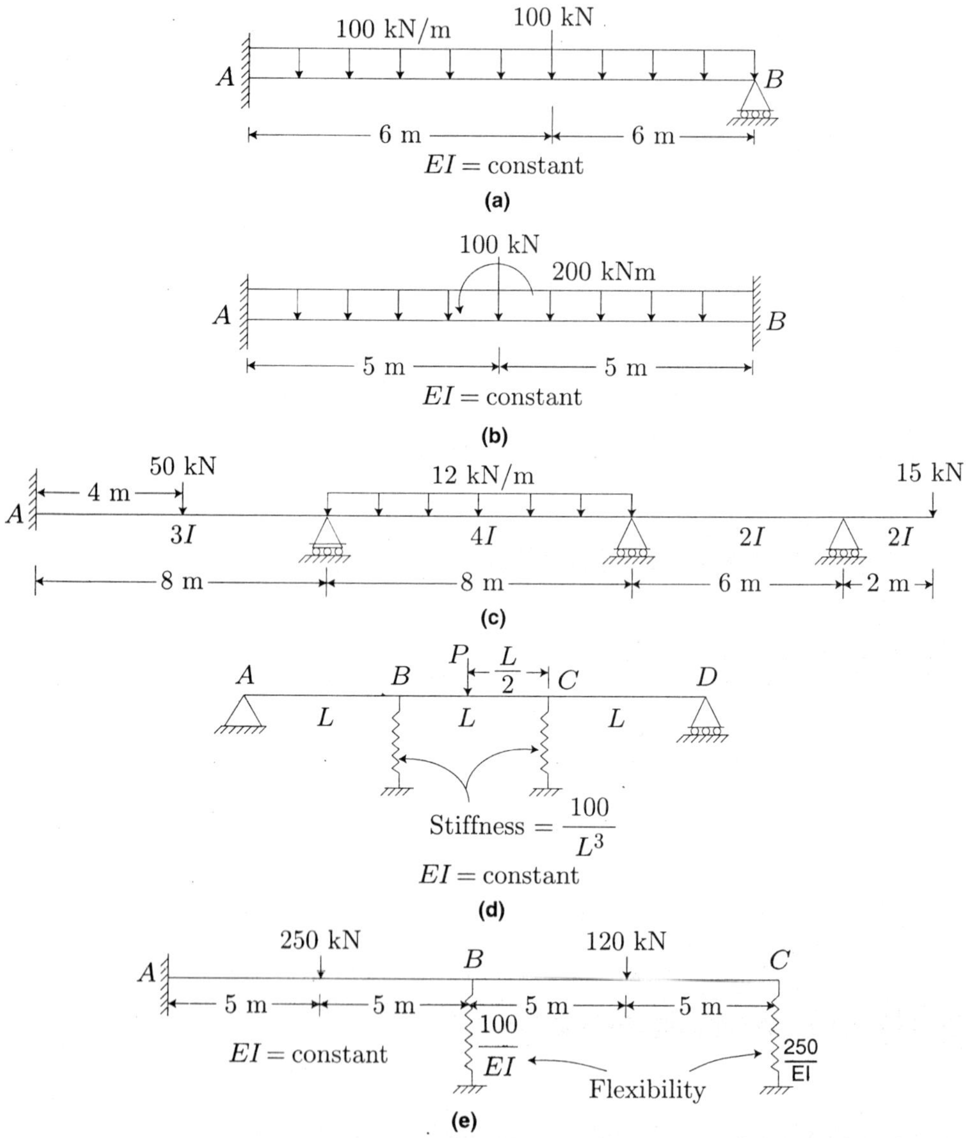

Fig. P-16.1

16.5 Develop the flexibility matrix corresponding to the coordinates indicated (or choosing suitable coordinates) in Fig. P-16.2.

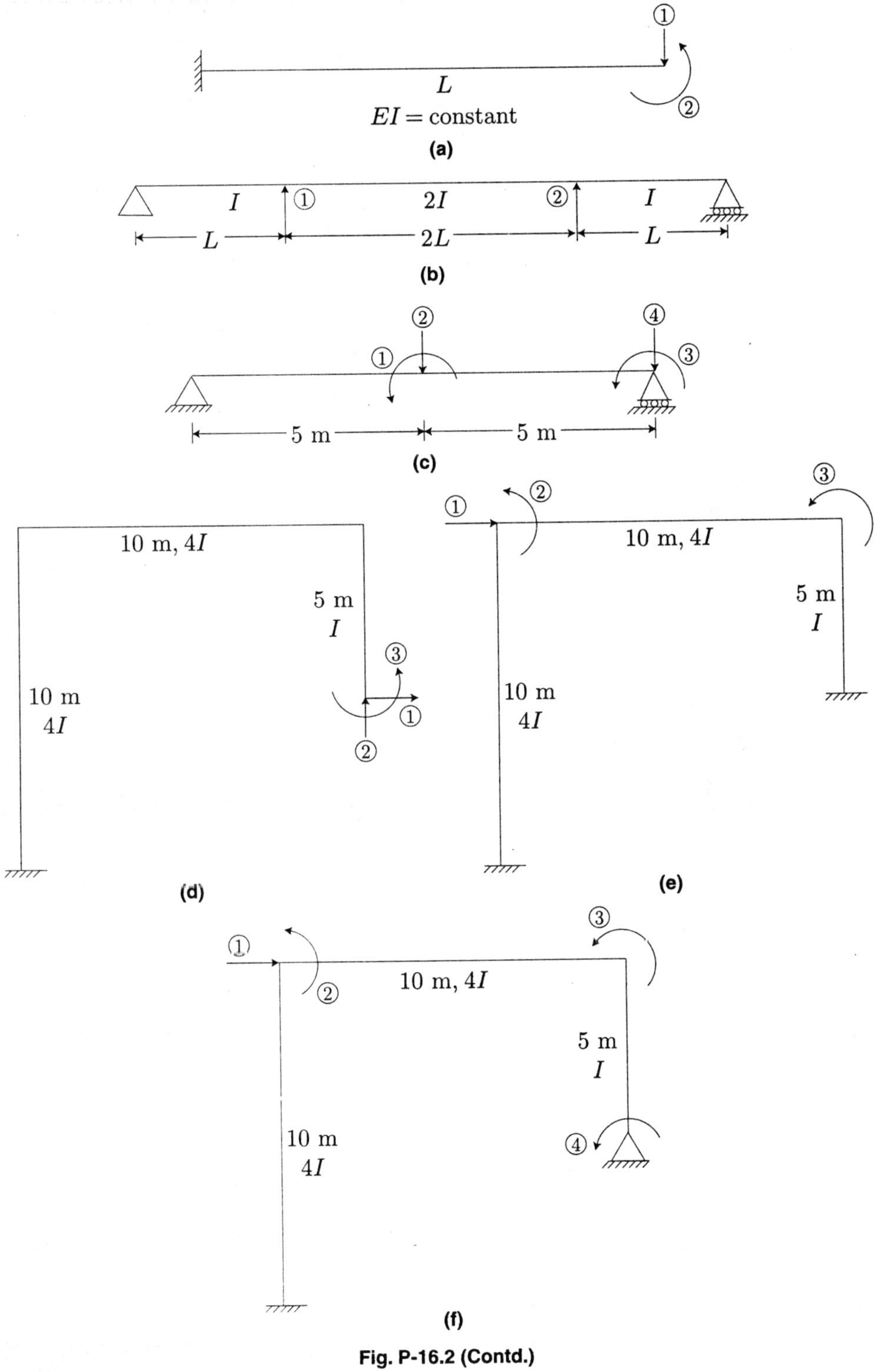

Fig. P-16.2 (Contd.)

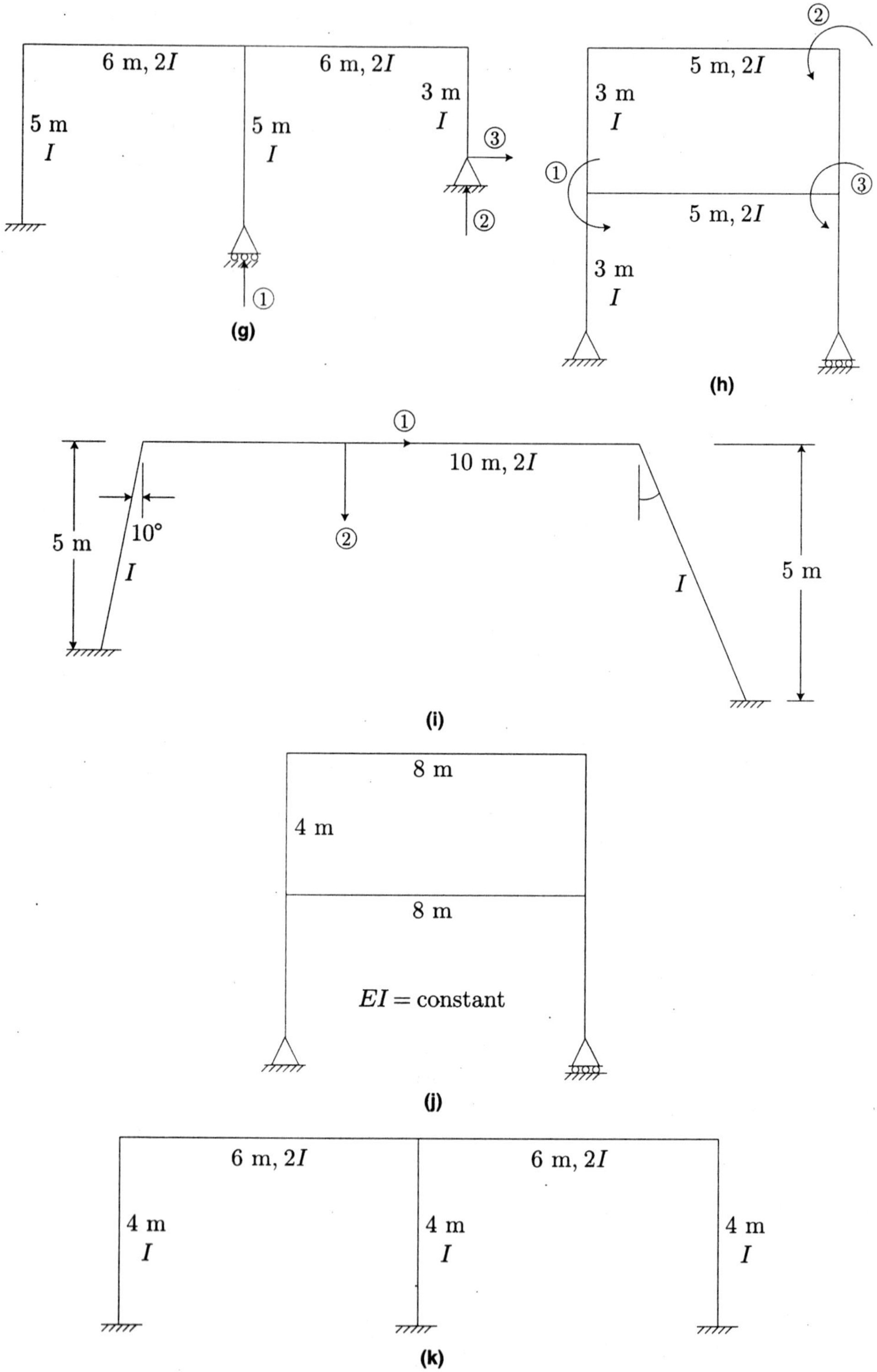
6 m, 2I
6 m, 2I
5 m
I
5 m
I
3 m
I
(g)
5 m, 2I
3 m
I
5 m, 2I
3 m
I
(h)
10 m, 2I
5 m
10°
I
I
5 m
(i)
8 m
4 m
8 m
EI = constant
(j)
6 m, 2I
6 m, 2I
4 m
I
4 m
I
4 m
I
(k)

Fig. P-16.2

16.6 The continuous beams of Fig. P-16.3 have an internal hinge at B. Analyse the beam by the force method and obtain the B.M. diagram.

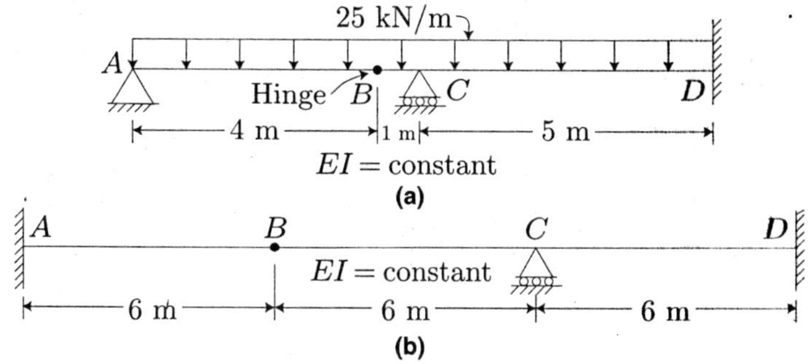

Fig. P-16.3

16.7 Using the co-ordinates shown in Fig. P-16.4, analyse the continuous beam by the force method for the following effects :

(a) A concentrated load of 250 kN at the centre of span AB, and a u.d.l. of 50 kN/m on BC.

(b) Downward displacement of $\frac{120}{EI}$ and $\frac{360}{EI}$ at A and B respectively.

(c) Clockwise rotation of $\frac{6}{EI}$ at A. Consider each effect separately.

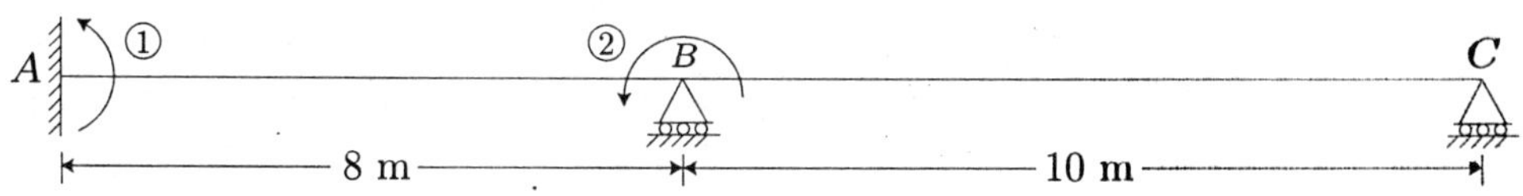

Fig. P-16.4

16.8 The continuous beams shown in Fig. P-16.5 have elastic supports. Analyse the beams by force method.

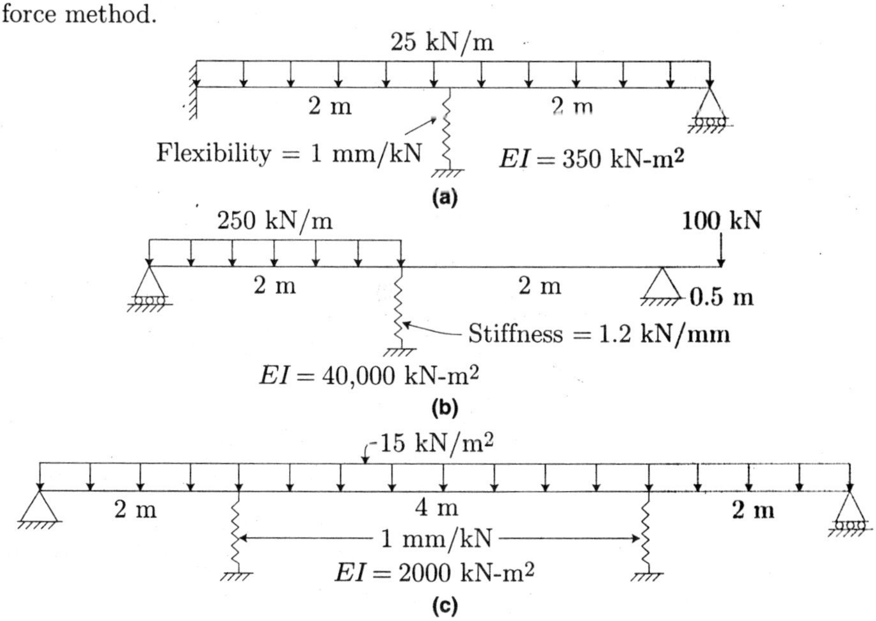

Fig. P-16.5

16.9 Analyse the beams in Fig. P-16.1(a) to P-16.1(c), if the end A rotates by unit radian due to a moment of 10,000 kNm, and the support at B deflects by 1 mm.

16.10 Analyse the rigid frame shown in Fig. P-16.6 using force method.

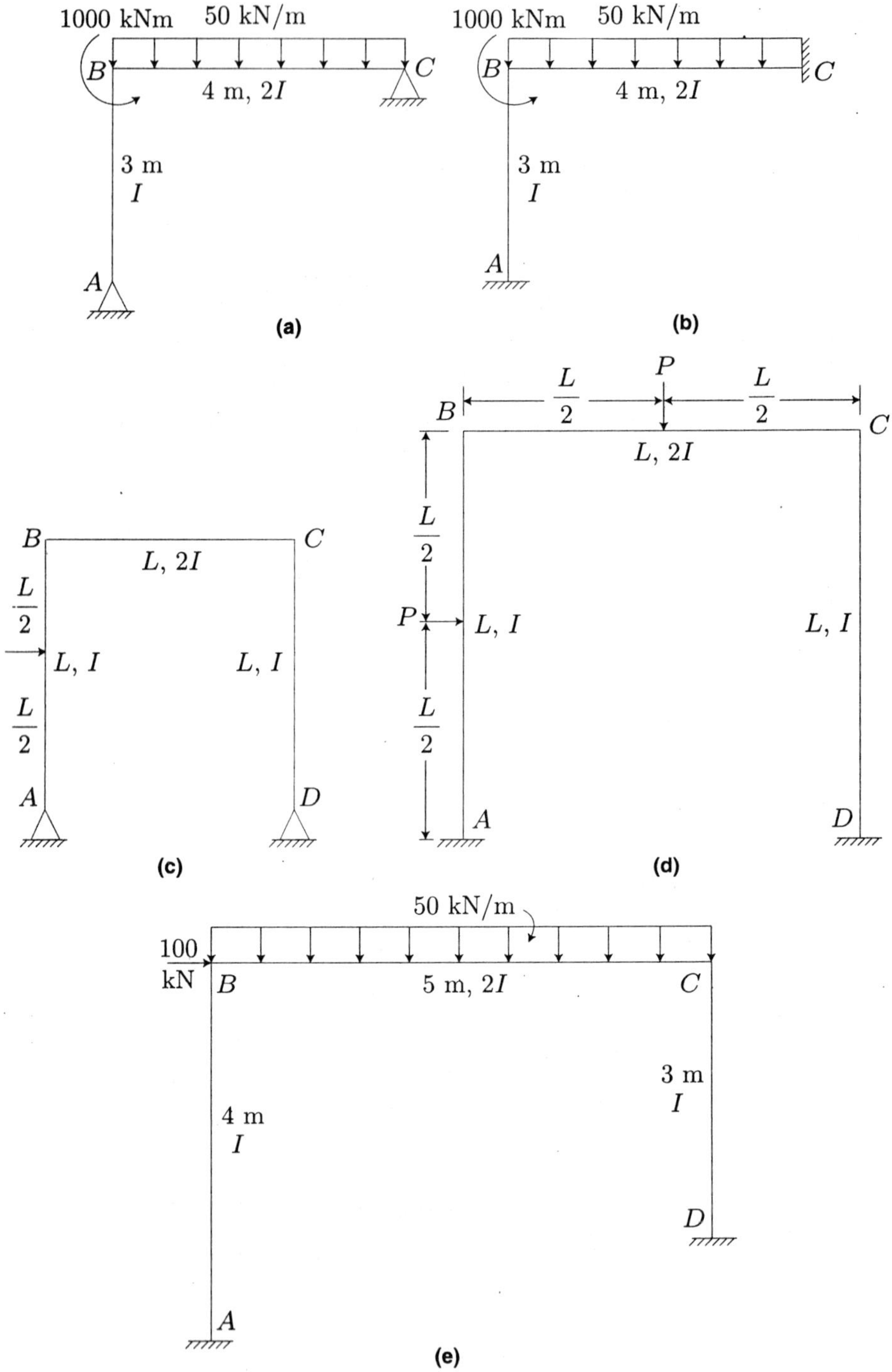

Fig. P-16.6 (Contd.)

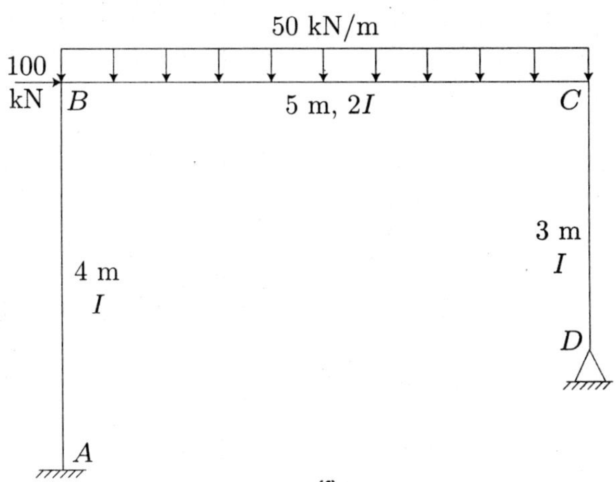

(f)

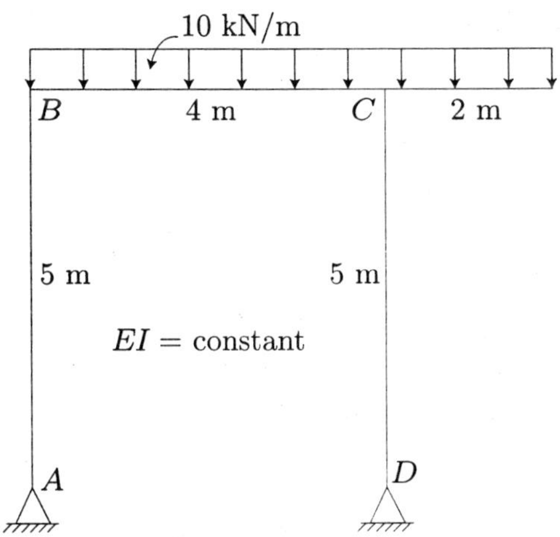

(g)

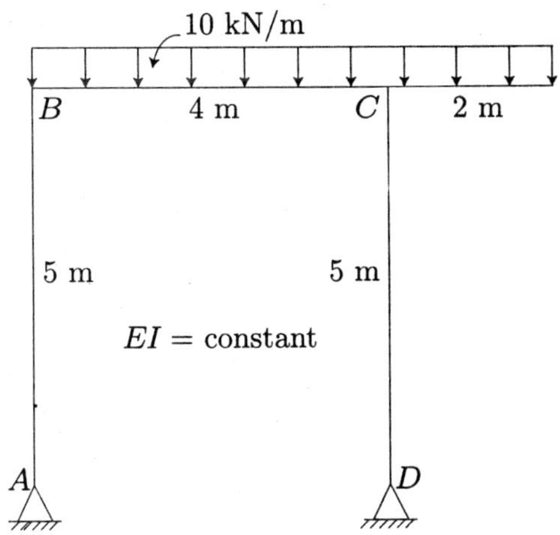

(h)

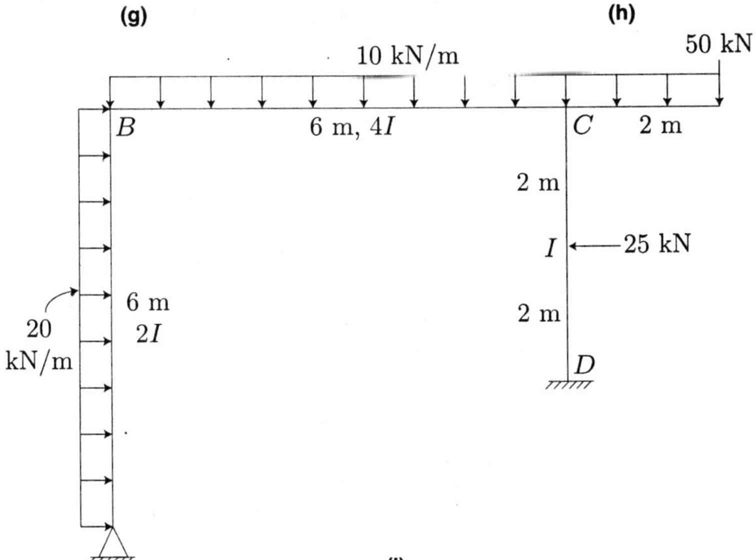

(i)

Fig. P-16.6 (Contd.)

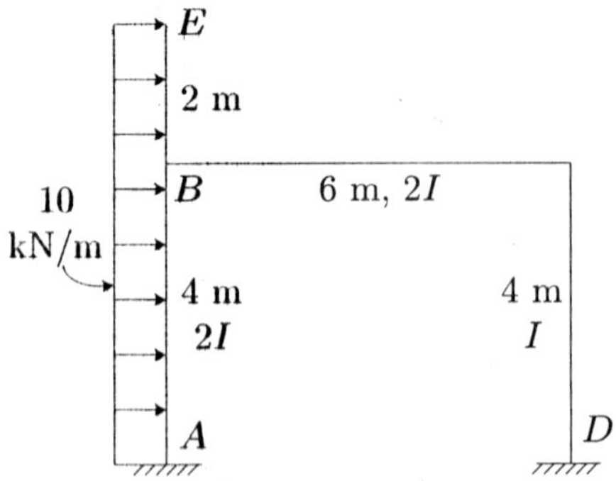

(j)

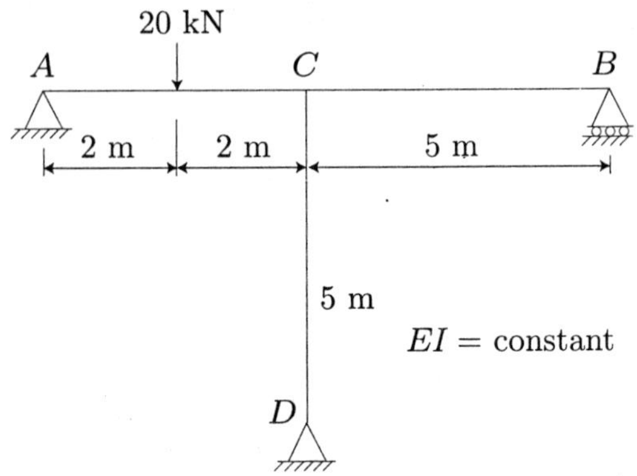

(k)

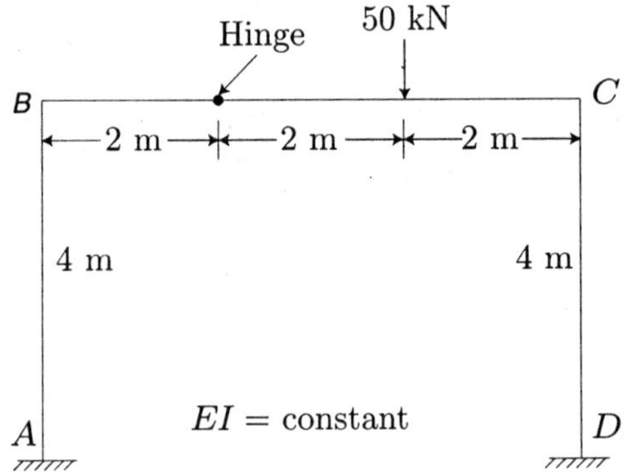

(l)

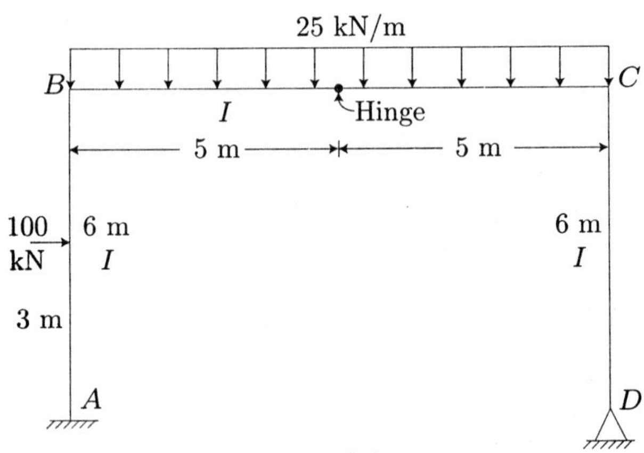

(m)

Fig. P-16.6 (Contd.)

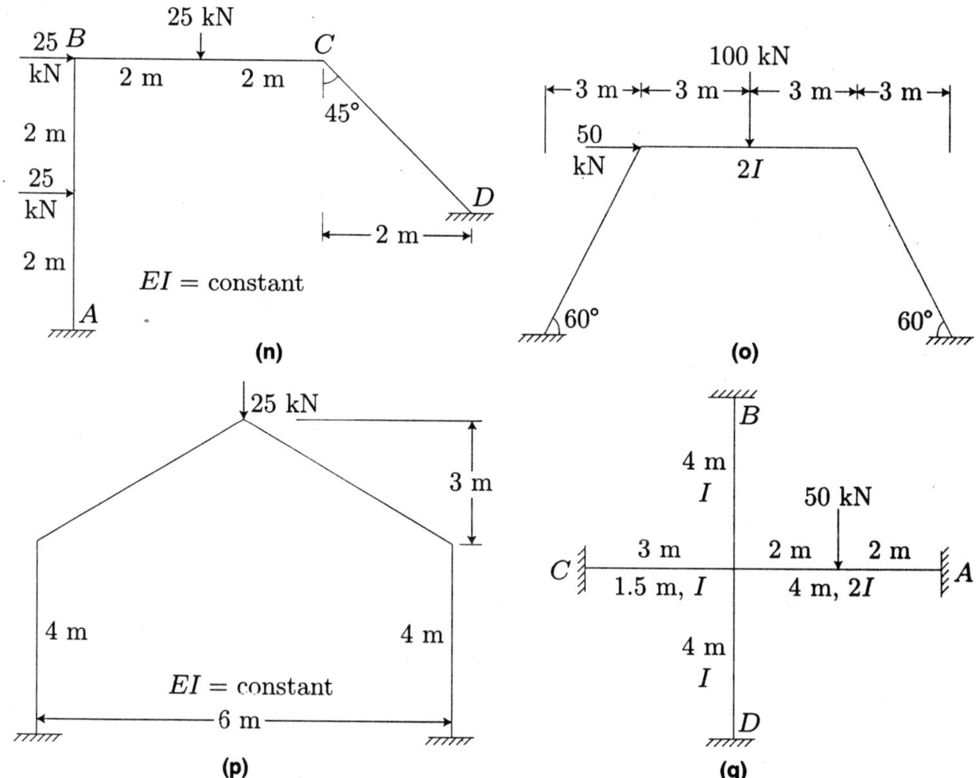

Fig. P-16.6

16.11 Analyse the right-angled bent in Fig. P-16.6(b), if support at A permits a clockwise rotation of 0.001 radian.

16.12 Analyse the rigid frame in Fig. P-16.5 if support D permits a clockwise rotation of 0.002 radian and a vertical displacement of 4 mm.

16.13 Analyse the pin-jointed trusses in Fig. P-16.7 using force method.

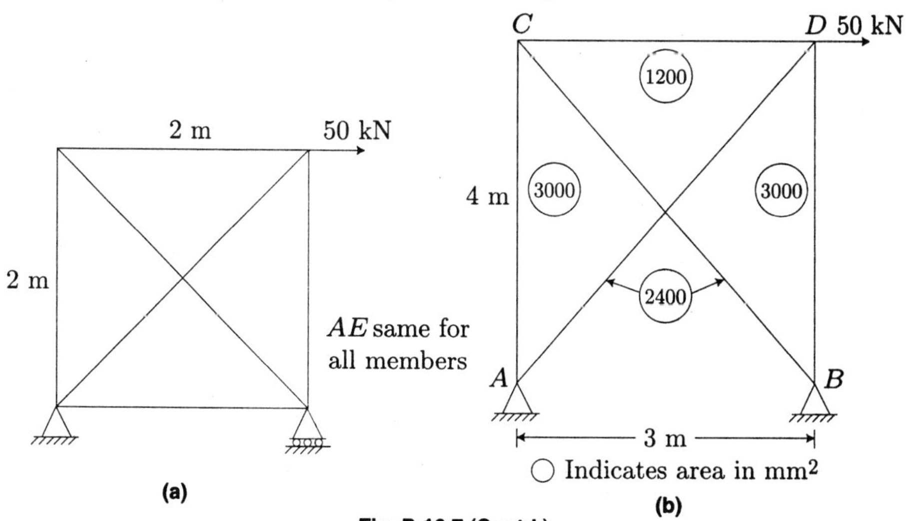

Fig. P-16.7 (Contd.)

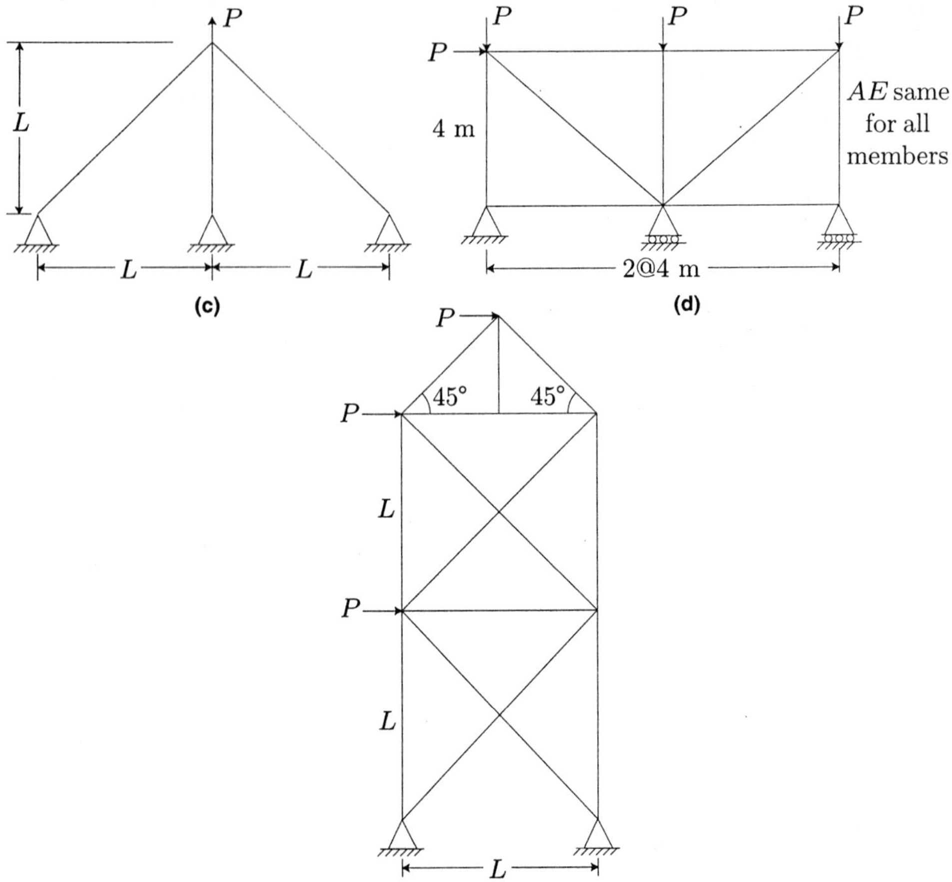

For horizontal and vertical members $2AE$
For others AE

(e)

Fig. P-16.7

16.14 Obtain the flexibility matrix for the simply supported beam of Fig. P-16.8 with reference to the co-ordinates shown. Also develop the stiffness matrix and verify that it is the reciprocal of the flexibility matrix. EI is constant unless stated in the figure.

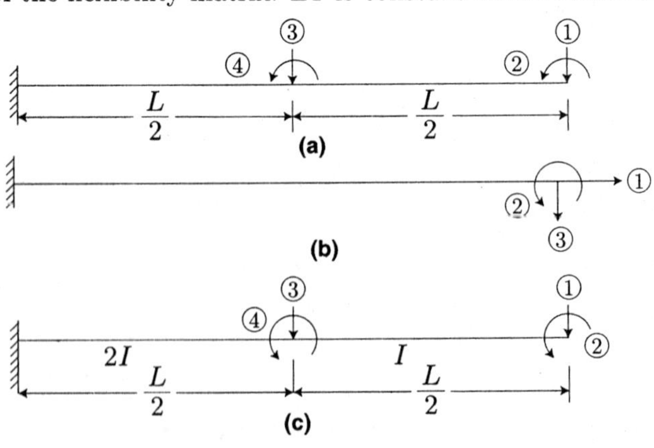

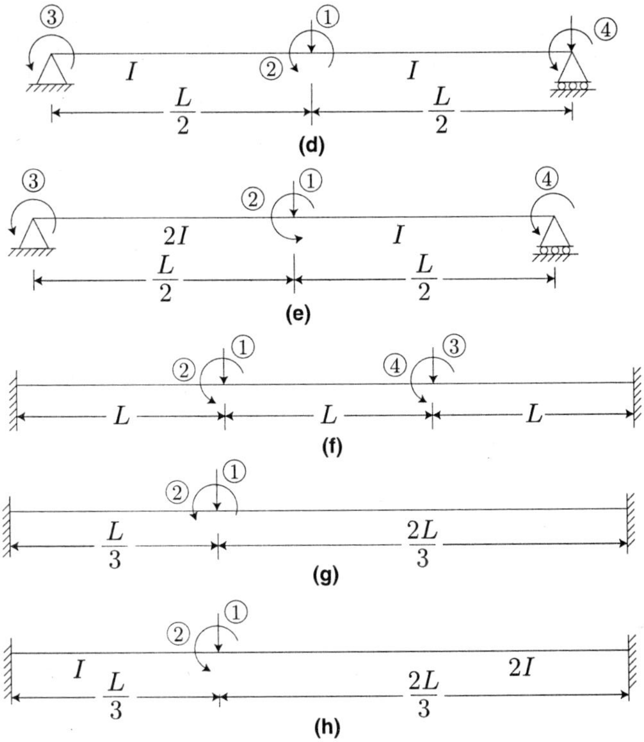

Fig. P-16.8

16.15 What are the types of flexibility and stiffness of a member with respect to the type of displacement. Obtain their values considering a prismatic member.

16.16 Distinguish between rotational and translational stiffnesses. Illustrate with suitable examples.

16.17 Develop the stiffness matrix for the members shown in Fig. P-16.9 choosing suitable co-ordinates wherever not shown. Consider EI and EA as constants unless otherwise mentioned.

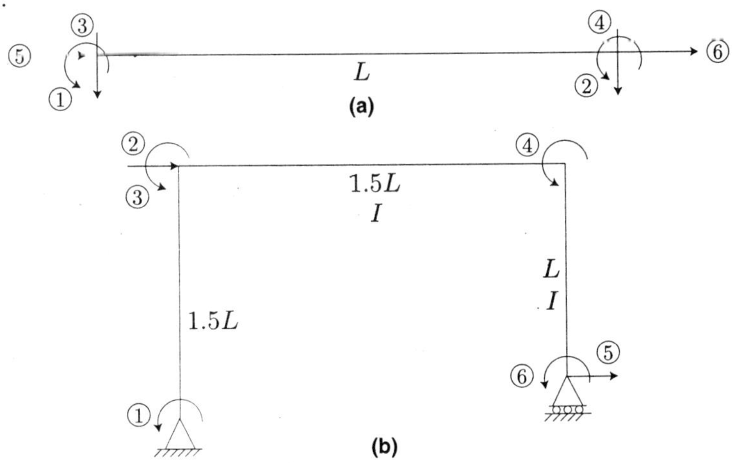

Fig. P-16.9 (Contd.)

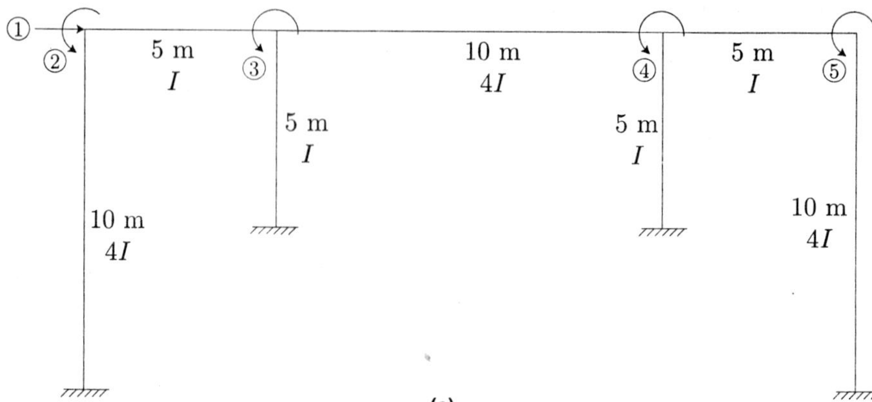

①
②
5 m
I
③
10 m
4I
④
5 m
I
⑤
5 m
I
5 m
I
10 m
4I
10 m
4I

(c)

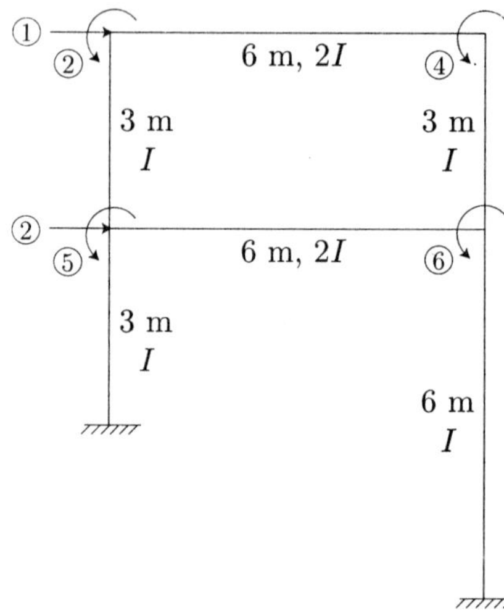

①
②
6 m, 2I
④
3 m
I
3 m
I
②
⑤
6 m, 2I
⑥
3 m
I
6 m
I

(d)

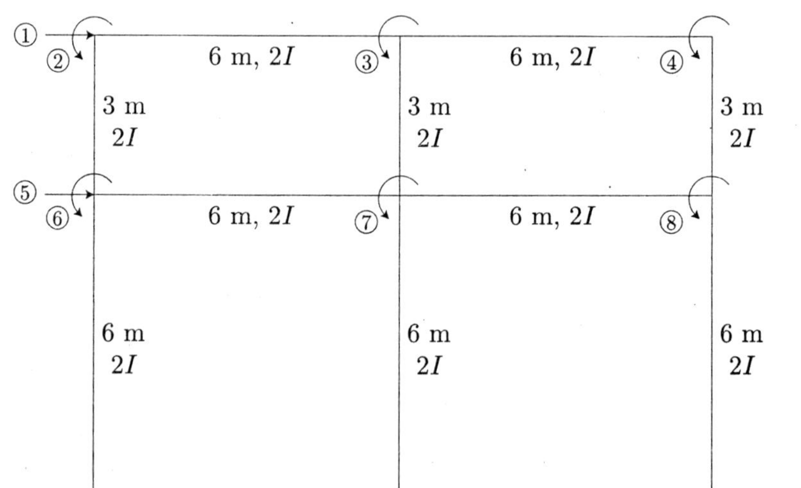

①
②
6 m, 2I
③
6 m, 2I
④
3 m
2I
3 m
2I
3 m
2I
⑤
⑥
6 m, 2I
⑦
6 m, 2I
⑧
6 m
2I
6 m
2I
6 m
2I

(e)

Fig. P-16.9

16.18 Develop the stiffness matrix for the frame in Fig. P-15.9(b) and P-15.9(c), if axial deformations of members are considered.

16.19 Solve problem 16.4, 16.6, 16.7, 16.8, 16.9, 16.10, 16.11, 16.12 and 16.13 using stiffness method :

(a) Neglecting effect of axial deformations in case of rigid frames.

(b) Considering effect of axial deformations in case of rigid frames.

16.20 Develop the flexibility matrix and stiffness matrix for the grid frame shown in Fig. P-16.10, choosing suitable co-ordinates. Assume EI and GJ as co-ordinates.

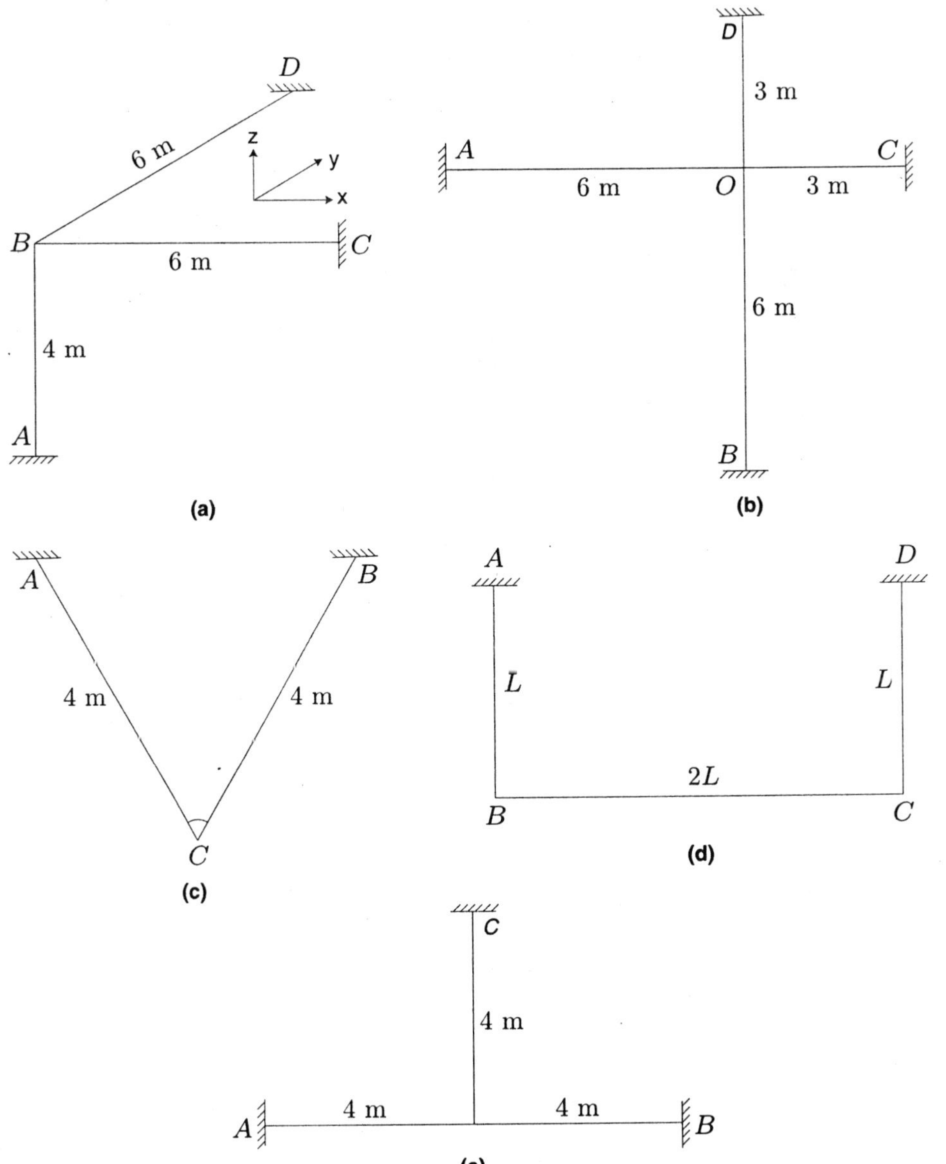

Fig. P-16.10 (Contd.)

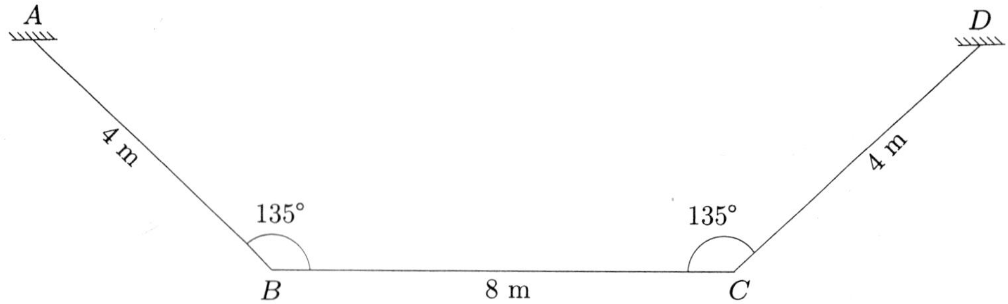

(f)

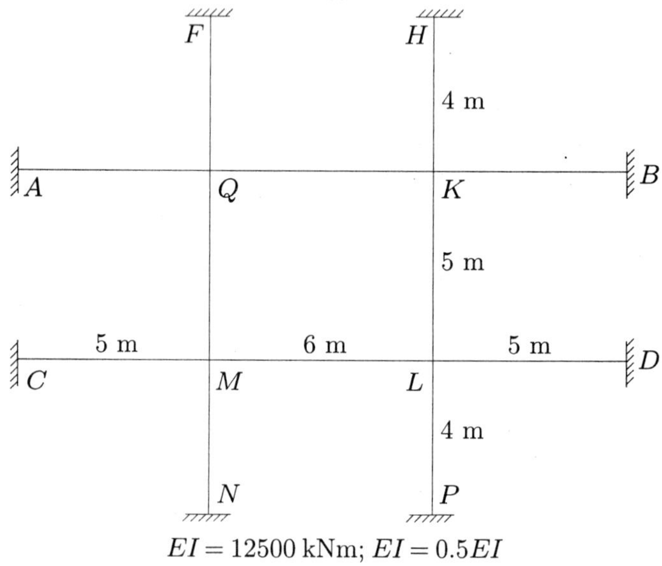

(g)

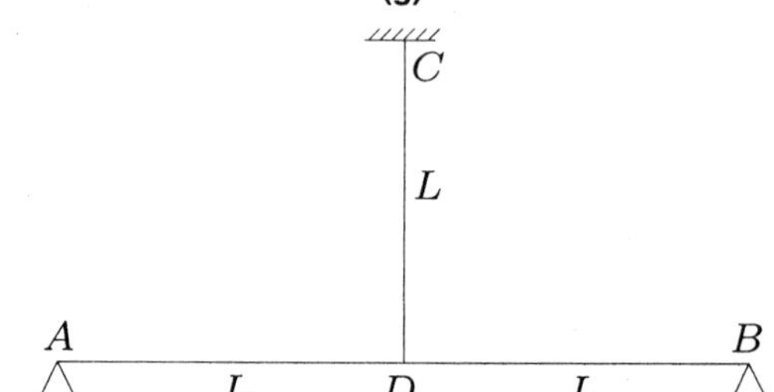

(h)

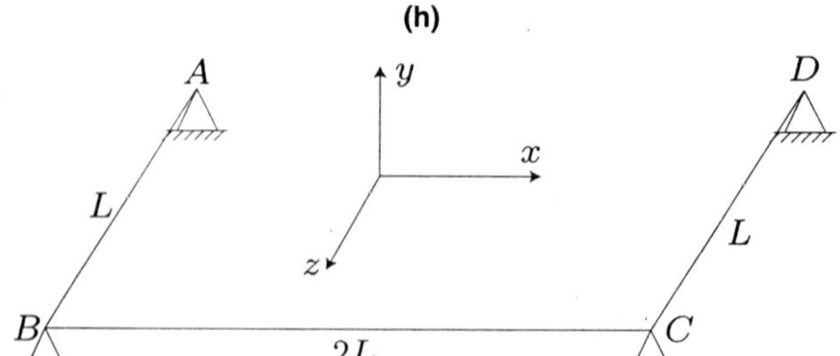

(i)

Fig. P-16.10 (Contd.)

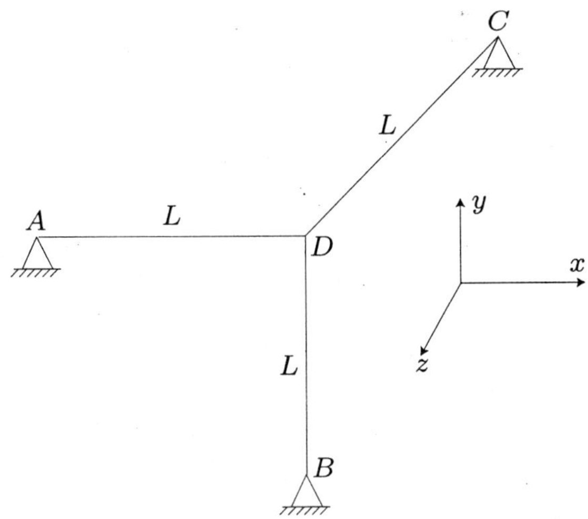

(j)

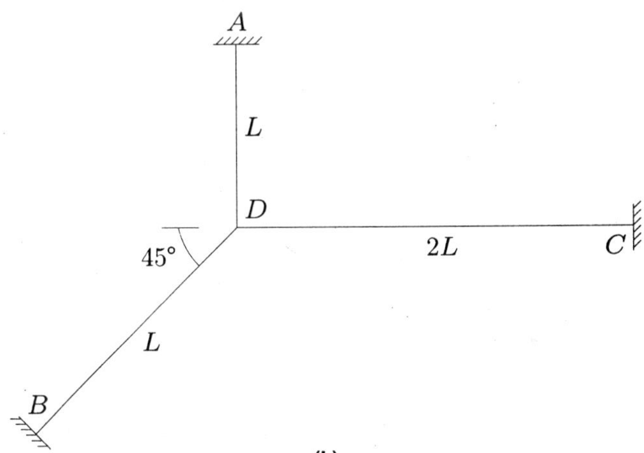

(k)

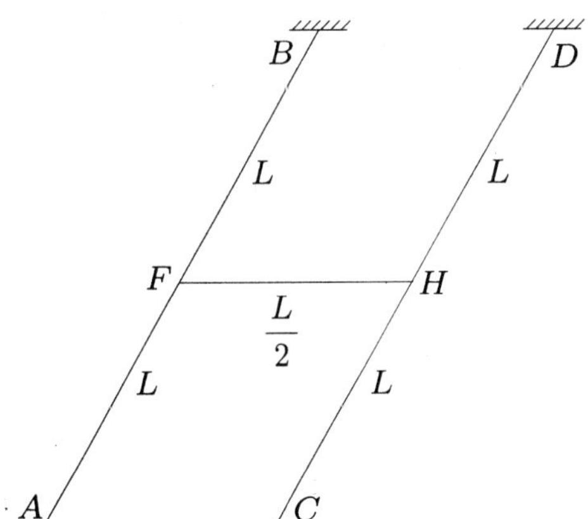

(l)

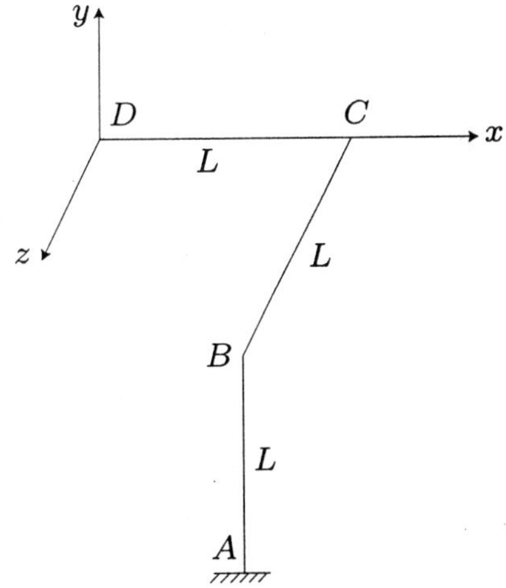

(m)

Fig. P-16.10

16.21 For the grids shown in Fig. P-16.11, analyse the frames. Consider EI and GJ as constants unless stated in the frames.

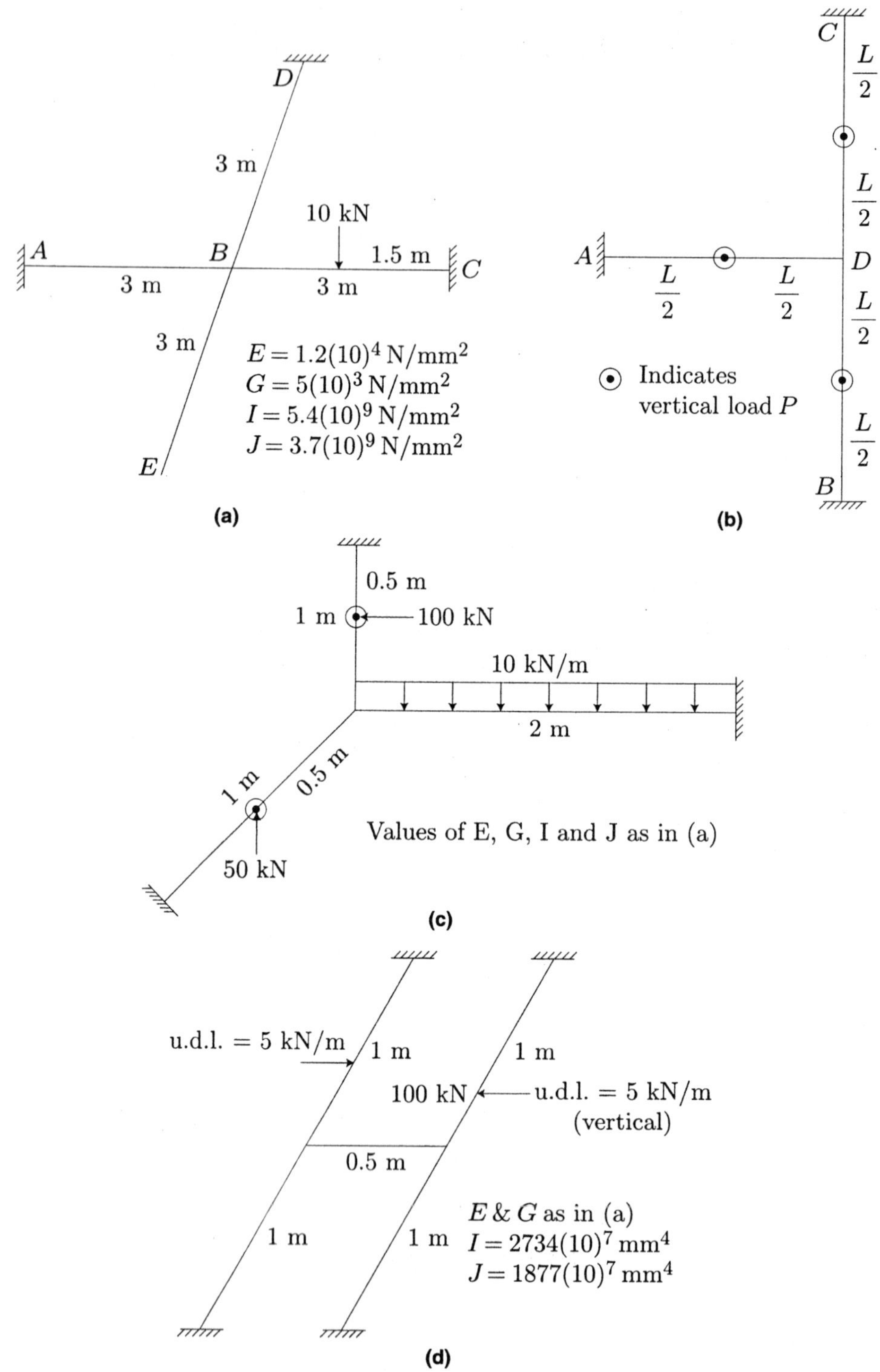

Fig. P-16.11

Chapter 17

Finite Difference Method, Relaxation Procedure and Finite Element Method

17.1 Introduction

The finite difference method is a very useful technique of applied mathematics which enables one to solve in a numerical manner many differential equations, which cannot be solved otherwise. Generally, the finite difference method substitutes for a given differential equation and boundary conditions, a set of simultaneous equations. With the use of the electronic computer and the consequent ability to solve simultaneous equations with large numbers of unknown very easily and quickly, the finite difference method has an additional advantage. It is now a recognised tool in all fields of engineering and is specially useful in the field of engineering elasticity.

The idea behind this method is to replace the governing differential equations and the equations defining the boundary conditions by the corresponding finite difference equations. This then reduces the problem to a set of simultaneous algebraic equations which can be solved without mathematical difficulty.

17.2 Outline of Finite Difference Method

For converting differential equations to finite difference equations some fundamental concepts regarding the finite difference approximations have to be mentioned. In the finite difference method, the difference (i.e. finite) quantities are substituted for differential quantities.

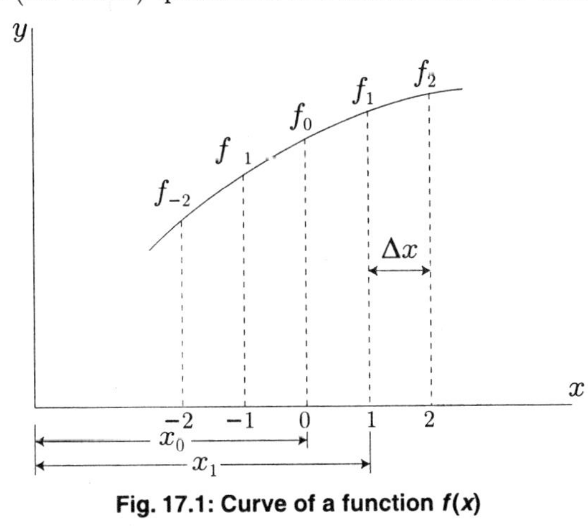

Fig. 17.1: Curve of a function *f*(*x*)

Fig. 17.1 shows the curve of a function $f(x)$ of the variable x. At some point 0, where $x = x_0$, the derivative $\frac{df}{dx}$ is defined as

$$\left(\frac{df}{dx}\right)_0 = \lim_{\Delta x \to 0}\left(\frac{\Delta f}{\Delta x}\right)_0 = \lim_{\Delta x \to 0}\frac{f_1 - f_0}{x_1 - x_0} = \lim_{h \to 0}\frac{f_1 - f_0}{h}, \tag{17.1}$$

where $\Delta x = h$ is the interval; Δf, the difference between the values of f at points 1 and 0; f_1 and f_0 being the values of f at the points 1 and 0 respectively. The subscript 0 of $\frac{df}{dx}$ indicate the point at which the derivative is taken. If h be small, $\frac{df}{dx}$ is approximately equal to $\frac{\Delta f}{\Delta x}$ and $\left(\frac{df}{dx}\right)_i \approx \frac{1}{h}(f_1 - f_0)$.

In general, at any point, x,

$$\frac{df}{dx} \approx \frac{\Delta f}{h} = \frac{1}{h}[f(x+h) - f(x)], \tag{17.2}$$

where $[f(x+h) - f(x)]$ is called the first difference of the function at x. In Eq. (17.2), the difference is given in terms of the values of the function at the point under consideration and at a point ahead of it. The difference in this form is called the forward difference.

In a similar manner, the derivative can be approximated by the backward difference, viz., at point 0,

$$\left(\frac{df}{dx}\right)_0 \approx \frac{1}{h}[f_0 - f_{-1}].$$

In general this gives

$$\frac{df}{dx} \approx \frac{1}{h}[f(x) - f(x-h)]. \tag{17.3}$$

In terms of central difference again in a similar manner,

$$\left(\frac{df}{dx}\right) \approx \frac{1}{2h}[f(x+h) - f(x-h)], \tag{17.4}$$

where interval of $\Delta x = 2h$ is considered.

Using the symbols $\overline{\Delta}$, Δ and $\underline{\Delta}$ to denote the forward, central and backward differences respectively, then

$$\left.\begin{aligned} \overline{\Delta} f &= [f(x+h) - f(x)] \\ \Delta f &= \frac{1}{2}[f(x+h) - f(x-h)] \\ \text{and} \quad \underline{\Delta} f &= [f(x) - f(x-h)]. \end{aligned}\right] \tag{17.5}$$

In a similar manner, the second derivative can be approximated by the second difference, whence

$$\frac{d^2 f}{dx^2} = \frac{d}{dx}\left(\frac{df}{dx}\right) \approx \frac{\Delta^2 f}{h^2}.$$

The symbols $\overline{\Delta}^2$, Δ^2 and $\underline{\Delta}^2$ may be taken as denoting the second forward, central and backward differences respectively. The second forward difference at point 0 can be obtained as

$$(\overline{\Delta}^2 f)_0 = \overline{\Delta}(\overline{\Delta} f) = \overline{\Delta}(f_1 - f_0) = (f_2 - f_1) - (f_1 - f_0) = f_2 - 2f_1 + f_0.$$

In general at any point x,

$$(\overline{\Delta}^2 f) = f(x+2h) - 2f(x+h) + f(x) \tag{17.6}$$

$$\text{and} \quad \frac{d^2 f}{dx^2} \approx \frac{f(x+2h) - 2f(x+h) + f(x)}{h^2}. \tag{17.6a}$$

Similarly, the second backward difference at any point x is given by

$$\underline{\Delta}^2 f = f(x) - 2f(x-h) + f(x-2h) \tag{17.7}$$

and $$\frac{d^2 f}{dx^2} \approx \frac{f(x) - 2f(x-h) + f(x-2h)}{h^2}. \tag{17.7a}$$

The second central difference at any point is given by

$$\Delta^2 f = \frac{f(x+2h) - 2f(x) + f(x-2h)}{4} \tag{17.8}$$

and $$\frac{d^2 f}{dx^2} \approx \frac{f(x+2h) - 2f(x) + f(x-2h)}{4h^2}. \tag{17.8a}$$

A better approximation can be obtained if the second central difference be defined in terms of the values of the function at one interval ahead of and behind it. This is accomplished by defining $\Delta^2 f$ as

$$(\Delta^2 f)_0 = \overline{\Delta}(\underline{\Delta} f)_0, \text{ or } = \underline{\Delta}(\overline{\Delta} f)_0.$$

From the first and the second one,

$$\begin{aligned}(\Delta^2 f)_0 = \overline{\Delta}(\underline{\Delta} f)_0 &= \overline{\Delta}(f_0 - f_{-1}) = (f_1 - f_0) - (f_0 - f_{-1})\\ &= f_1 - 2f_0 + f_{-1}\end{aligned}$$

and $$\begin{aligned}(\Delta^2 f)_0 = \underline{\Delta}(\overline{\Delta} f_0) &= \underline{\Delta}(f_1 - f_0) = (f_1 - f_0) - (f_0 - f_{-1})\\ &= f_1 - 2f_0 + f_{-1}.\end{aligned}$$

This indicates that the same result is obtained from both the cases. In general, at any point x, therefore,

$$\Delta^2 f = f(x+h) - 2f(x) + f(x-h) \tag{17.9}$$

and $$\frac{d^2 f}{dx^2} \approx \frac{f(x+h) - 2f(x) + f(x-h)}{h^2}. \tag{17.9a}$$

In general, the derivatives at any point can be better approximated by the corresponding central differences, as such differences involve points on both sides of the point involved along the curve thus giving better average values.

The higher order central differences can be easily obtained in a similar manner. The third central difference at a point x is given with $\Delta^2 f$ as the unit operator. Thus,

$$\begin{aligned}\Delta^3 f &= \Delta(\Delta^2 f) = \Delta[f(x+h) - 2f(x) + f(x-h)]\\ &= \frac{f(x+2h) - f(x)}{2} - 2\frac{f(x+h) - f(x-h)}{2} + \frac{f(x) - f(x-2h)}{2}\\ &= \frac{1}{2}[f(x+2h) - 2f(x+h) + 2f(x-h) - f(x-2h)].\end{aligned} \tag{17.10}$$

Similarly, $$\begin{aligned}\Delta^4 f &= \Delta^2(\Delta^2 f) = \Delta^2[f(x+h) - 2f(x) + f(x-h)]\\ &= f(x+2h) - 4f(x+h) + 6f(x) - 4f(x-h) + f(x-2h).\end{aligned} \tag{17.11}$$

It is, therefore, evident that the coefficient of the even ordered differences are those of the binomial expansion of $(a-b)^n$. For even differences, i.e., for even n, the nth central difference is given by

$$\Delta^n f = f\left(x+\frac{n}{2}h\right) - nf\left(x+\frac{n}{2}h-h\right) + \frac{n(n-1)}{2!}f\left(x+\frac{n}{2}h-2h\right)$$

$$\cdots\cdots\cdots$$

$$+(-1)^r\frac{n!}{r!(n-r)!}f\left(x+\frac{n}{2}h-rh\right)+\cdots+f\left(x-\frac{n}{2}h\right), \tag{17.12}$$

where r is any integer equal to or less than n.

For odd differentials, the corresponding difference may be obtained from

$$\Delta^{n+1}f = \Delta(\Delta^n f) = \frac{1}{2}[\Delta^n f(x+h) - \Delta^n f(x-h)], \tag{17.13}$$

where n is an even number and

$\Delta^n f(x+h)$ and $\Delta^n f(x-h)$ may be obtained from Eqn. (17.12).

Example 17.1. *Values of the function in case of an equation $\frac{d^2y}{dx^2} = f(x)$, at points 0, 1, 2, 3 and 4 at distance 3 units apart are 0, $-\frac{2}{9}$, $-\frac{2}{3}$, $-\frac{1}{3}$ and 0 respectively. Calculate the values of y at the points. Compare the values with those obtained when the function $f(x)$ is represented by a polymonial in the form $C_1x + C_2x^2 + C_3x^3$ and using values of $f(x)$, at $x = 3$, $x = 6$ and $x = 9$.*

Solution : $\dfrac{d^2y}{dx^2} = f(x)$.

Representing in the finite difference form,

$$\left(\frac{d^2y}{dx^2}\right)_n = \frac{y_{n+1} - 2y_n + y_{n-1}}{(\Delta x)^2} = f(x)_n,$$

at points 1, 2 and 3, relationships are

$$\frac{y_2 - 2y_1 + y_0}{(3)^2} = -\frac{2}{9}$$

$$\frac{y_3 - 2y_2 + y_1}{(3)^2} = -\frac{2}{3}$$

and $$\frac{y_4 - 2y_3 + y_2}{(3)^2} = -\frac{1}{3}.$$

It is known that, $y_0 = y_4 = 0$.

Solving, $y_1 = 5\dfrac{1}{4}$, $y_2 = 8\dfrac{1}{3}$ and $y_3 = 5\dfrac{3}{4}$.

Considering $f(x) = C_1x + C_2x^2 + C_3x^3$,

at $x = 3$, 6 and 9, equations are

$$3C_1 + 9C_2 + 27C_3 = -\frac{2}{9}$$

$$6C_1 + 36C_2 + 216C_3 = -\frac{2}{3}$$

and $\quad 9C_1 + 81C_2 + 729C_3 = -\dfrac{1}{3}.$

Solving, $C_1 = \dfrac{2}{27},\ C_2 = -\dfrac{11}{162}$ and $C_3 = \dfrac{1}{162}.$

From $\dfrac{d^2y}{dx^2} = \dfrac{2}{27}x - \dfrac{11}{162}x^2 + \dfrac{1}{162}x^3$, integrating twice and solving,

$$y_0 = 0,\ y_1 = 4.75,\ y_2 = 7.33 \quad \text{and} \quad y_3 = 4.50.$$

17.3 Finite Difference Equations

In the previous Section, finite difference approximations of the various derivatives were obtained considering that f is a function of x only. If f is a function of both x and y, it is the case of partial derivatives. Following the same reasoning as before and with a square grid, i.e., $\Delta x = \Delta y = h$, from definition of partial derivatives,

$$\frac{\partial f}{\partial x} \approx \frac{\Delta_x f}{2h} \quad \text{and} \quad \frac{\partial f}{\partial y} \approx \frac{\Delta_y f}{2h},$$

$$\frac{\partial^2 f}{\partial x^2} \approx \frac{\Delta_x^2 f}{h^2} \quad \text{and} \quad \frac{\partial^2 f}{\partial y^2} \approx \frac{\Delta_y^2 f}{h^2},$$

where the subscripts Δ_x and Δ_y denote the directions in which the differences are taken.

If the domain be divided into a square mesh with $\Delta_x = \Delta_y = h$, as shown in Fig. 17.2, the finite difference approximations at point 0 are given by

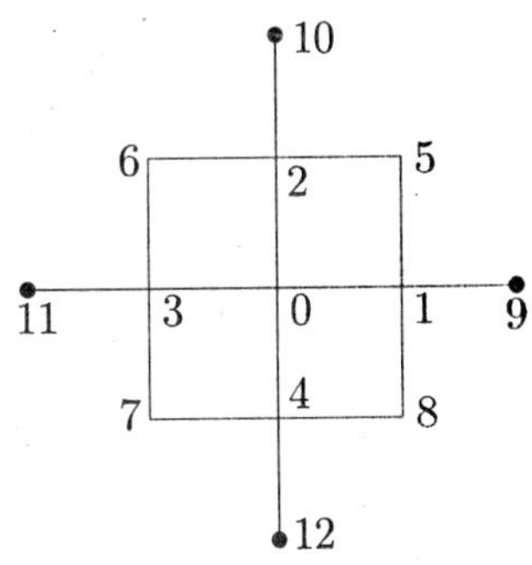

Fig. 17.2: Square mesh

$$\left(\frac{\partial f}{\partial x}\right)_0 \approx \frac{1}{2h}[f(x+h,y) - f(x-h,y)] = \frac{f_1 - f_3}{2h}.$$

$$\left(\frac{\partial f}{\partial y}\right)_0 \approx \frac{1}{2h}[f(x,y+h) - f(x,y-h)] = \frac{f_2 - f_4}{2h}.$$

$$\left(\frac{\partial^2 f}{\partial x^2}\right)_0 \approx \frac{1}{h_2}[f(x+h,y) - 2f(x,y) + f(x-h,y)]$$
$$= \frac{f_1 - 2f_0 + f_3}{h^2}. \tag{17.14a}$$

$$\left(\frac{\partial^2 f}{\partial y^2}\right)_0 \approx \frac{1}{h^2}[f(x,y+h) - 2f(x,y) + f(x,y-h)]$$
$$= \frac{f_2 - 2f_0 + f_4}{h^2}. \tag{17.14b}$$

Considering Laplace operator,

$$\nabla^2 f = \frac{\partial^2 f}{\partial x^2} + \frac{\partial^2 f}{\partial y^2} \approx \frac{f_1 + f_2 + f_3 + f_4 - 4f_0}{h^2}.$$

The finite difference approximation of

$$\left(\frac{\partial^2 f}{\partial x \partial y}\right)_0 = \left[\frac{\partial}{\partial x}\left(\frac{\partial f}{\partial y}\right)\right] \approx \frac{\left(\frac{\partial f}{\partial y}\right)_1 - \left(\frac{\partial f}{\partial y}\right)_3}{2h} \approx \frac{f_5 - f_6 + f_7 - f_8}{4h^2},$$

so that
$$\left(\frac{\partial^2 f}{\partial x \partial y}\right)_0 \approx \frac{f_5 - f_6 + f_7 - f_8}{4h^2}. \tag{17.15}$$

The finite difference approximation for the third and fourth derivatives are similarly given by

$$\left(\frac{\partial^3 f}{\partial x^3}\right)_0 = \frac{f_9 - 2f_1 + 2f_3 - f_{11}}{2h^3} \tag{17.16a}$$

$$\left(\frac{\partial^3 f}{\partial y^3}\right)_0 = \frac{f_{10} - 2f_2 + 2f_4 - f_{12}}{2h^3} \tag{17.16b}$$

$$\left(\frac{\partial^4 f}{\partial x^4}\right)_0 = \frac{6f_0 - 4f_1 - 4f_3 + f_9 + f_{11}}{h^4} \tag{17.17a}$$

$$\left(\frac{\partial^4 f}{\partial y^4}\right)_0 = \frac{6f_0 - 4f_2 - 4f_4 + f_{10} + f_{12}}{h^4} \tag{17.17b}$$

and
$$\left(\frac{\partial^4 f}{\partial x^2 \partial y^2}\right)_0 = \frac{4f_0 - 2f_1 - 2f_2 - 2f_3 - 2f_4 + f_5 + f_6 + f_7 + f_8}{h^4}. \tag{17.18}$$

If the domain be divided into a net with a inner points, n nos. of finite difference equations are obtained. Together with the boundary conditions, such equations can be solved for the values of the functions at every point of the domain into a system of algebraic equations which can easily be solved.

17.4 Higher-order Finite Difference Approximations

The accuracy of the finite difference approximation can be improved by taking a finer mesh. However, the amount of labour involved for solution of a problem increases considerably with the increase of the number of points. The difficulty can be avoided by a method proposed by F_{CX}, whereby the higher order finite difference approximation formulas are used instead of the standard first order formulas given in previous Sections. Even though such approach was criticised by Southwell, it nevertheless appears that in most cases this method will give a better solution with minimum increase in labour.

In order to obtain the higher order finite difference approximation formulas, it is necessary to express a curve connecting any set of points by a power series. The greater the number of known points, the greater will be the number of terms in the power series that can be found and the better will be the approximation to the actual curve. If the values of f are known at $x = x_0,\ x_1,\ x_2, \cdots$, Newton's formula for fitting such a curve is given by

$$f = a_0 + a_1(x - x_0) + a_2(x - x_0)(x - x_1) + a_3(x - x_0)(x - x_1)(x - x_2) + \cdots \tag{17.19}$$

If the points are given for equally spaced value of x, with an interval of h, then $x_1 = x_0 + h,\ x_2 = x_0 + 2h$, etc. Substituting the successive values of the co-ordinates in Eqn. (17.19), the coefficients $a_0,\ a_1,\ a_2, \cdots$ can be easily determined.

For $x = x_0$, taking $f = f_0$ from Eq. (17.19), $f_0 = a_0$. Newton's formula then becomes

$$f = f_0 + a_1(x - x_0) + a_2(x - x_0)(x - x_1) + a_3(x - x_0)(x - x_1)(x - x_2) + \cdots$$

For $x = x_1 = x_0 + h$, taking $f = f_1$, from Eq. (17.19),

$$f_1 = f_0 + a_1(h) \text{ and } a_1 = \frac{f_1 - f_0}{h} = \frac{\overline{\Delta} f_0}{h},$$

where $\overline{\Delta} f_0$ is the first forward difference at $x = x_0$.

Proceeding similarly values of $a_2,\ a_3, \cdots$ can be obtained and Newton's formula becomes

$$f = f_0 + \overline{\Delta} f_0 \left(\frac{x - x_0}{h}\right) + \frac{1}{2!}\overline{\Delta}^2 f_0 \left(\frac{x - x_0}{h}\right)\left(\frac{x - x_1}{h}\right)$$
$$+ \frac{1}{3!}\overline{\Delta}^3 f_0 \left(\frac{x - x_0}{h}\right)\left(\frac{x - x_1}{h}\right)\left(\frac{x - x_2}{h}\right)$$
$$+ \frac{1}{4!}\overline{\Delta}^4 f_0 \left(\frac{x - x_0}{h}\right)\left(\frac{x - x_1}{h}\right)\left(\frac{x - x_2}{h}\right)\left(\frac{x - x_3}{h}\right) + \cdots$$

With $\dfrac{x - x_0}{h} = u$,

$$\frac{x - x_1}{h} = \frac{(x - x_0) - (x_1 - x_0)}{h} = u - 1$$

$$\frac{x - x_2}{h} = u - 2 : \frac{x - x_3}{h} = u - 3 \text{ and so on.}$$

Newton's formula then becomes

$$f = f_0 + u\overline{\Delta} f_0 + \frac{u(u-1)}{2!}\overline{\Delta}^2 f_0 + \frac{u(u-1)(u-2)}{3!}\overline{\Delta}^3 f_0$$
$$+ \frac{u(u-1)(u-2)(u-3)}{4!}\overline{\Delta}^4 f_0 + \cdots \quad (17.20)$$

From Eq. (17.20), it is seen that the derivatives of f at the point $x = x_0$ can be expressed in terms of the finite differences at x_0 by a simple differentiation. Thus, as

$$u = \frac{x - x_0}{h} \quad \text{and} \quad \frac{du}{dx} = \frac{1}{h}, \quad \left(\frac{du}{dx}\right)_0 = \left(\frac{df}{dx}\right)_{x=x_0} = \left(\frac{df}{du}\frac{du}{dx}\right)_{u=0}$$

$$\left(\frac{df}{dx}\right)_0 = \frac{1}{h}\left[\overline{\Delta} f_0 + \frac{2u-1}{2!}\overline{\Delta}^2 f_0 + \frac{3u^2 - 6u + 2}{3!}\overline{\Delta}^3 f_0\right.$$
$$\left. + \frac{4u^3 - 18u^2 + 22u - 6}{4!}\Delta^4 f_0 + \cdots\right]_{u=0},$$

$$= \frac{1}{h}\left[\overline{\Delta} f_0 - \frac{1}{2}\overline{\Delta}^2 f_0 + \frac{1}{3}\overline{\Delta}^3 f_0 - \frac{1}{4}\overline{\Delta}^4 f_0 + \cdots\right] \quad (17.21)$$

and $$\left(\frac{d^2 f}{dx^2}\right)_0 = \frac{1}{h^2}\left[\overline{\Delta}^2 f_0 - \overline{\Delta}^3 f_0 + \frac{11}{12}\overline{\Delta}^4 f_0 - \cdots\right] \quad (17.22)$$

In a similar manner, the derivatives can be expressed in terms of the backward differences. Thus,

$$\left(\frac{df}{dx}\right)_0 = \frac{1}{h}\left[\underline{\Delta} f_0 + \frac{1}{2}\underline{\Delta}^2 f_0 + \frac{1}{3}\underline{\Delta}^3 f_0 + \frac{1}{4}\underline{\Delta}^4 f_0 + \cdots\right] \quad (17.23)$$

and $$\left(\frac{d^2 f}{dx^2}\right)_0 = \frac{1}{h^2}\left[\underline{\Delta}^2 f_0 + \underline{\Delta}^3 f_0 + \frac{11}{12}\underline{\Delta}^4 f_0 + \cdots\right]. \quad (17.24)$$

For expressing the derivatives in terms of the central differences, Newton's formula has to be transformed in terms of central differences. From definition,

$$\overline{\Delta} f_0 = f_1 - f_0 = \frac{1}{2}(f_1 - f_{-1}) + \frac{1}{2}(f_1 - 2f_0 + f_{-1}) = \Delta f_0 + \frac{1}{2}\Delta^2 f_0.$$

$$\overline{\Delta}^2 f_0 = \Delta^2 f_0 + \Delta^3 f_0 + \frac{1}{2}\Delta^4 f_0$$

$$\overline{\Delta}^3 f_0 = \Delta^3 f_0 + \frac{3}{2}\Delta^4 f_0 + \Delta^5 f_0 + \frac{1}{2}\Delta^6 f_0.$$

$$\overline{\Delta}^4 f_0 = \Delta^4 f_0 + 2\Delta^5 f_0 + 2\Delta^6 f_0 + \Delta^7 f_0 + \frac{1}{2}\Delta^8 f_0, \text{ and so on.}$$

Substituting these terms in Eq. (17.20) and collecting terms,

$$f = f_0 + u\Delta f_0 + \frac{u^2}{2!}\Delta^2 f_0 + \frac{u(u^2 - 1^2)}{3!}\Delta^3 f_0 + \frac{u^2(u^2 - 1^2)}{4!}\Delta^4 f_0 + \frac{u(u^2 - 1^2)(u^2 - 2^2)}{5!}\Delta^5 f_0 + \cdots \qquad (17.25)$$

Differentiating Eq. (17.25),

$$\frac{df}{dx} = \frac{1}{h}\left[\Delta f_0 + u\Delta^2 f_0 + \frac{3u^2 - 1}{6}\Delta^3 f_0 + \frac{4u^3 - 2u}{24}\Delta^4 f_0 + \frac{5u^4 - 15u^2 + 4}{120}\Delta^5 f_0 + \cdots\right]. \qquad (17.26)$$

At $x = x_0$, $u = 0$ and the derivative $\dfrac{df}{dx}$ at $x = x_0$, is given by

$$\left(\frac{df}{dx}\right)_0 = \frac{1}{h}\left[\Delta f_0 - \frac{1}{6}\Delta^3 f_0 + \frac{1}{30}\Delta^5 f_0 \cdots\right] \qquad (17.27a)$$

and
$$\left(\frac{d^2 f}{dx^2}\right)_0 = \frac{1}{h^2}\left[\Delta^2 f_0 - \frac{1}{12}\Delta^4 f_0 + \frac{1}{90}\Delta^6 f_0 - \cdots\right]. \qquad (17.27b)$$

Higher derivatives can be obtained in a similar manner.

Eqs. (17.21) to (17.24), (17.27) and (17.28) reduce to the standard finite difference formulas, if the higher order differences in these formulas are neglected. Comparing Eq. (17.27) with Eqs. (17.21) and (17.23), it is seen that in reducing to the standard first order formulas, the largest term neglected in Eq. (17.27) is $-\frac{1}{6}\Delta^3 f_0$, while those in Eqns. (17.21) and (17.23) are $-\frac{1}{2}\overline{\Delta}^2 f_0$ and $\frac{1}{2}\underline{\Delta}^2 f_0$ respectively. As $\frac{1}{6}\Delta^3 f_0$ will be smaller than both $\frac{1}{2}\overline{\Delta}^2 f_0$ and $\frac{1}{2}\underline{\Delta}^2 f_0$, the central difference formulas will in general give a better approximation than the forward or backward difference formulas. Similar conclusions can be drawn in the approximation of higher derivatives by the various higher differences.

In the above derivation, f is assumed to be a function of x only. If f is a function of both x and y, the partial derivatives can be easily obtained by replacing the symbol $\overline{\Delta}$, Δ or $\underline{\Delta}$ by $\overline{\Delta}_x$, Δ_x, $\underline{\Delta}_x$ or $\overline{\Delta}_y$, Δ_y, $\underline{\Delta}_y$ respectively. Thus in terms of central differences,

$$\left.\begin{aligned} \left(\frac{d^2 f}{dx^2}\right)_0 &\approx \frac{1}{h^2}\left[\Delta_x^2 f_0 - \frac{1}{12}\Delta_x^4 f_0 + \frac{1}{90}\Delta_x^6 f_0 - \cdots\right] \\ \left(\frac{d^2 f}{dy^2}\right)_0 &\approx \frac{1}{h_2}\left[\Delta_y^2 f_0 - \frac{1}{12}\Delta_y^4 f_0 + \frac{1}{90}\Delta_y^6 f_0 - \cdots\right]. \end{aligned}\right] \qquad (17.28)$$

Neglecting terms above $\Delta^4 f_0$, Eq. (17.28) can be represented as

$$\left(\frac{d^2f}{dx^2}\right) = \frac{1}{h^2}\left[f(x+h) - 2f(x) + f(x-h) - \frac{1}{12}\{f(x+2h) - 4f(x+h) + 6f(x) - 4f(x-h) + f(x-2h)\}\right]$$

$$= \frac{1}{12h^2}[-f(x+2h) + 16f(x+h) - 30f(x) + 16f(x-h) - f(x-2h)], \quad (17.29)$$

where values of $\Delta^2 f$ and $\Delta^4 f$ are as given in Eqs. (17.9) and (17.11).

17.5 Use of Finite Difference Method

Finite difference method can be used for solutions of various differential equations involved in (a) Obtaining deflections of beam (b) Solution of statically indeterminate structures (c) Calculating buckling loads for columns (d) Vibration of beam and (e) Solutions of plate and torsion and beam problems based on Theory of Elasticity.

Finite difference method provides a technique which can be applied directly to the most complicated cases of beams and loadings, including those problems which cannot be readily solved by the usual means. The finite difference formulation furthermore will permit us to set up the problem for simple electronic computer solution.

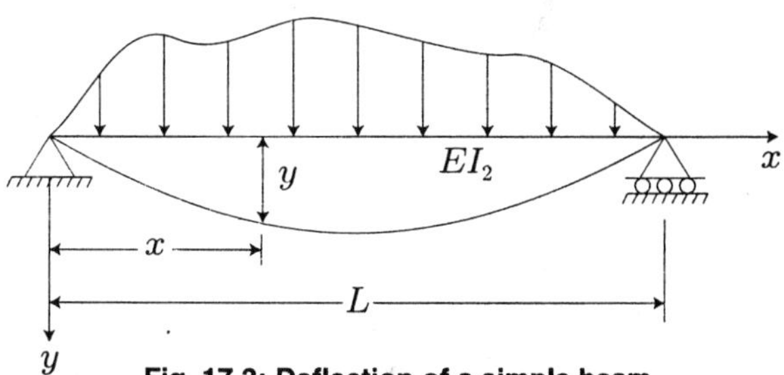

Fig. 17.3: Deflection of a simple beam

(a) Deflection of Beams. It is known that if the deflection of the beam is small, the basic equation following Bernoulli-Euler relation holds good. Thus considering Fig. 17.3,

$$\frac{d^2y}{dx^2} = -\frac{M_x}{EI_z}, \quad (17.30)$$

in which y is the deflection in the y direction (considered positive downward) of the neutral plane of the beam,

M_x is the bending moment at any point x of the beam, and EI_x is the beam stiffness, Eq. (17.30) can be represented in finite form notations as

$$y_{n+1} - 2y_n + y_{n-1} = -\frac{M_x}{EI_z}(\Delta_x)^2, \quad (17.31a)$$

following Eq. (17.8a) and neglecting higher terms.

If higher terms up to $\Delta^4 f$ are considered, Eq. (17.30) can be represented as

$$\frac{-y_{n+2} + 16y_{n+1} - 30y_n + 16y_{n-1} - y_{n-2}}{12} = -\left[\frac{M_x}{EI_z}\right](\Delta_x)^2. \quad (17.31b)$$

In Eqs. (17.31a) and (17.31b) y's are the deflections at points $(n+2)$, $(n+1)$, (n), $(n-1)$ and $(n-2)$.

Deflections of beams can also be obtained by finite difference solution of the basic relationship,

$$EI_z = \frac{d^4y}{dx^4} = w_x, \tag{17.32}$$

where w_x denotes the intensity of the load at x.

In finite difference from Eq. (17.32) can be represented as

$$\begin{aligned} y_{n+2} - 4y_{n+1} + 6y_n - 4y_{n-1} + y_{n-2} &= \frac{d^4y}{dx^4}(\Delta x)^4 \\ &= \frac{w_x}{EI_z}(\Delta x)^4, \end{aligned} \tag{17.33}$$

if higher order terms are neglected.

Example 17.2. *Determine the deflection at the quarter points and also at the centre of the beam for the loading shown in Fig. 17.4.*

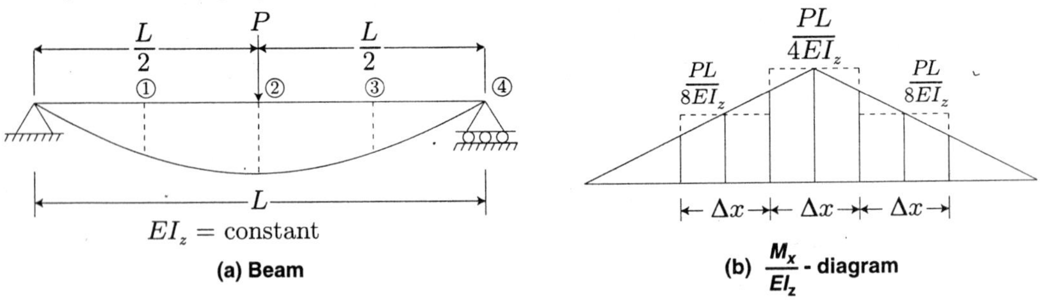

Fig. 17.4: For Example 17.2

Solution : The beam is assumed as divided into four equal sections as shown in Fig. 17.4, whence $\Delta x = \frac{L}{4}$. At each point 1, 2 and 3, the right-hand term of Eqs. (17.31a) and (17.31b) is just equal to the $\frac{M_x}{EI_z}$ at that point. Actually as $\frac{M_x}{EI_z}$ varies from point to point of the beam, a weighted value should be used.

Actual $\frac{M_x}{EI_z}$ curve denoted by full lines in Fig. 17.4(b) is replaced by an approximate curve as shown by the dotted lines. It is obvious that the smaller Δx is made, the more closely the approximation will approach the true curve. However, smaller Δx involves large computational works. A compromise has to be made between desired accuracy and the amount of permissible labour.

From Eq. (17.31a), at point 1, considering points 0, 1 and 2 and as $\Delta x = \frac{L}{4}$, the relationship comes to

$$y_2 - 2y_1 + y_0 = -\frac{PL}{EI_z}\left(\frac{L}{4}\right)^2. \tag{i}$$

Similarly at point 2, considering points 1, 2 and 3,

$$y_3 - 2y_2 + y_1 = -\frac{PL}{4EI_z}\left(\frac{L}{4}\right)^2. \tag{ii}$$

At point 3, considering points 2, 3 and 4,

$$y_4 - 2y_3 + y_2 = -\frac{PL}{EI_z}\left(\frac{L}{4}\right)^2. \tag{iii}$$

With $y_0 = y_4 = 0$ and solving Eqs. (i), (ii) and (iii),

$$y_1 = y_3 = \frac{PL^3}{64EI_z}, \text{ against } \frac{11}{768}\frac{PL^3}{EI_Z}$$

$$y_2 = \frac{9PL^3}{384EI_z}, \text{ against } \frac{PL^3}{48EI_Z}.$$

Better results for deflections can be obtained if higher order terms are taken into consideration. Thus, considering Eq. (17.31b), for point 1, considering points −1, 0, 1, 2 and 3,

$$-y_3 + 16y_2 - 30y_1 + 16y_0 - y_{-1} = -\frac{12PL}{8EI_z}\left(\frac{L}{4}\right)^2.$$

In this, y_{-1} is the deflection at a point −1, i.e., at a distance of $\frac{L}{4}$ on the left side of point 0 taken from the consideration that $y_{-1} = y_1 = y_3$. As $y_0 = 0$, the relationship for point 1 reduces to

$$16y_2 - 32y_1 = -\frac{12PL^3}{128EI_Z}. \quad \text{(iv)}$$

Similarly, for point 2, considering points 0, 1, 2, 3 and 4,

$$-y_4 + 16y_3 - 30y_2 + 16y_1 - y_0 = -\frac{12PL}{4EI_z}\left(\frac{L}{4}\right)^2.$$

As $y_0 = y_4 = 0$ and $y_1 = y_3$, relationship for point 2 reduces to

$$32y_1 - 30y_2 = -\frac{12PL^3}{64EI_z}. \quad \text{(v)}$$

Solving (iv) and (v),

$$y_2 = \frac{9PL^3}{448EI_z} \quad \text{and} \quad y_1 = \frac{93PL^3}{7168EI_z}.$$

Example 17.3. *Compute the deflection at quarter points and centre of a simply supported beam for the loading shown in Fig. 17.5.*

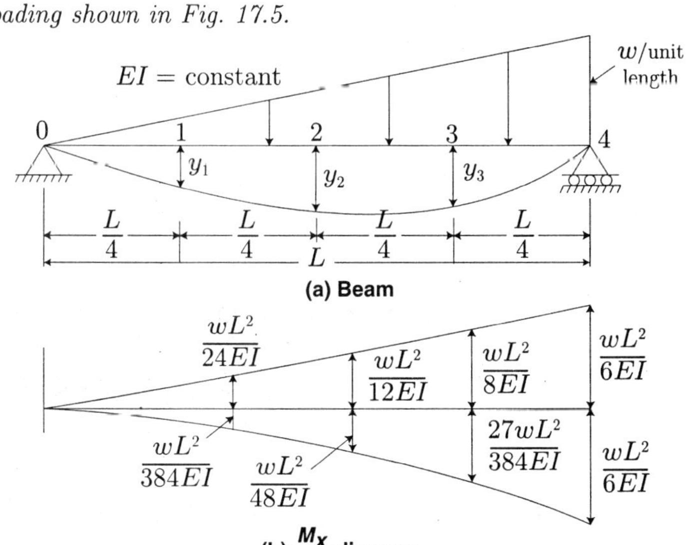

Fig. 17.5: For Example 17.3

Solution : The beam is divided into four equal spaces, so that $\Delta x = \frac{L}{4}$. The $\frac{M_x}{EI}$-diagram is drawn by parts as shown in Fig. 17.5(b).

Using finite difference relationship as per Eq. (17.31a),

at point 1, $y_2 - 2y_1 + y_0 = -\left[\frac{wL^2}{24EI} - \frac{wL^2}{384EI}\right]\frac{L^2}{16} = -\frac{15wL^2}{384EI}\left(\frac{L^2}{16}\right)$

at point 2, $y_3 - 2y_2 + y_1 = -\frac{24wL^2}{384EI}\left(\frac{L^2}{16}\right)$

and at point 3, $y_4 - 2y_3 + y_2 = -\frac{21wL^2}{384EI}\left(\frac{L^2}{16}\right)$.

As $y_0 = y_4 = 0$, solving

$$y_1 = \frac{28.5wL^4}{6144EI} \text{ against } \frac{27.15wL^4}{6144EI}$$

$$y_2 = \frac{42wL^4}{6144EI} \text{ against } \frac{39.92wL^4}{6144EI}$$

$$y_3 = \frac{31.5wL^4}{6144EI} \text{ against } \frac{29.60wL^4}{6144EI}.$$

Example 17.4. *Compute deflections at quarter points and centre of a simply supported beam for the loading shown in Fig. 17.6(a).*

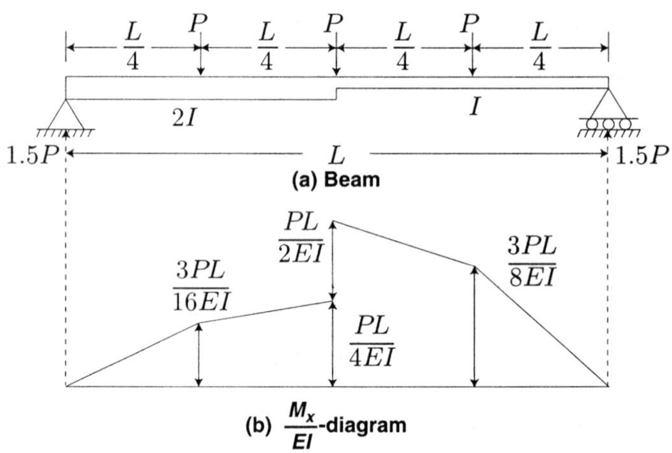

Fig. 17.6: For Example 17.4

Solution : The beam is divided into four equal sections, so that $\Delta x = \frac{L}{4}$. The $\frac{M_x}{EI}$-diagram is drawn by parts as shown in Fig. 17.6(b).

From Eq. (17.31a),

at point 1, $y_2 - 2y_1 + y_0 = -\left(\frac{3PL}{16EI}\right)\left(\frac{L^2}{16}\right)$

at point 2, $y_3 - 2y_2 + y_1 = -\left(\frac{PL}{4EI}\right)\left(\frac{L^2}{16}\right)$ or $-\left(\frac{PL}{2EI}\right)\left(\frac{L^2}{16}\right)$

at point 3, $y_4 - 2y_3 + y_2 = -\left(\frac{3PL}{8EI}\right)\left(\frac{L^2}{16}\right)$.

With $y_0 = y_4 = 0$, solving with the first value of right-hand side of equation for point 2,

$$y_2 = \frac{13PL^3}{512EI}, \text{ in place of } \frac{19PL^3}{512EI}.$$

With the second value of right-hand side of equation for point 2, $y_2 = \dfrac{17PL^3}{512EI}$.

Average value of $y_2 = \dfrac{15PL^3}{512EI}$.

Also $y_1 = \dfrac{10PL^3}{256EI}, \dfrac{14PL^3}{256EI}$ and $\dfrac{12PL^3}{256EI}$ (average).

Example 17.5. *Compute deflections of the cantilever beam shown in Fig. 17.7(a).*

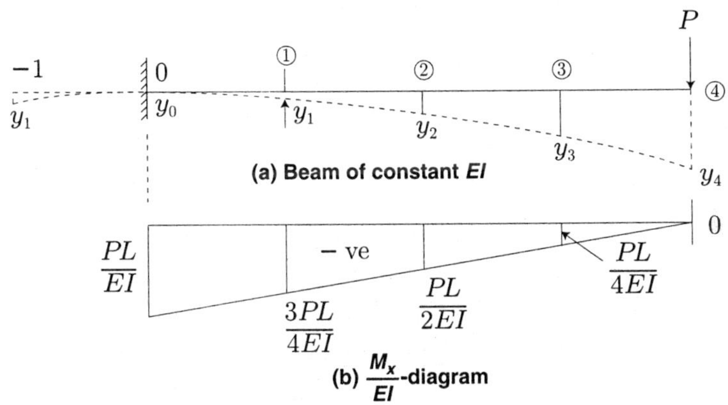

Fig. 17.7: For Example 17.5

Solution : The beam is divided into four equal sections, so that $\Delta x = \frac{L}{4}$. The $\frac{M_x}{EI}$-diagram is drawn by parts as shown in Fig. 17.7(b).

From Eq. (17.31a),

at point 0, $y_1 - 2y_0 + y_1 = +\dfrac{PL}{EI}\left(\dfrac{L}{4}\right)^2$.

Since $y_0 = 0$ and $y_{-1} = y_1$, as at point 0, slope is zero, it follows that, $y_1 = +\dfrac{PL^3}{32EI}$.

Considering point 1, $y_2 - 2y_1 + y_0 = \dfrac{3PL}{4EI}\left(\dfrac{L}{4}\right)^2$,

whence $y_2 = \dfrac{7PL^3}{64EI}$.

Considering point 2, $y_3 - 2y_2 + y_1 = \dfrac{PL}{2EI}\left(\dfrac{L}{4}\right)^2$,

whence $y_3 = \dfrac{7PL^3}{32EI}$.

Considering point 3, $y_4 - 2y_3 + y_2 = \dfrac{PL}{4EI}\left(\dfrac{L}{4}\right)^2$,

whence $y_4 = \dfrac{11PL^3}{32EI}$ in place of $\dfrac{PL^3}{3EI}$.

Example 17.6. *Compute deflections for the beam shown in Fig. 17.8 using equations*

$$EI\frac{d^4y}{dx^4} = w_x.$$

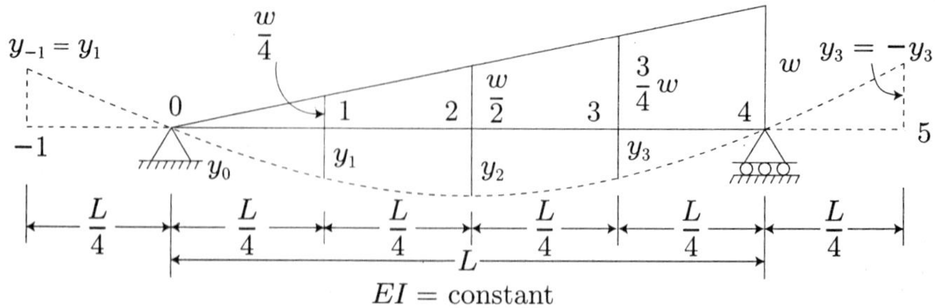

Fig. 17.8: For Example 17.6

Solution : $\left(\dfrac{d^4y}{dx^4}\right)_n = \dfrac{y_{n+2} - 4y_{n+1} + 6y_n - 4y_{n-1} + y_{n-2}}{(\Delta x)^4}$

The beam is divided into four equal sections, so that,

$$\Delta x = \frac{L}{4}.$$

From Fig. 14.8, considering slope at 0, it is given by

$$\frac{y_1 - y_0}{\Delta x} = \frac{y_0 - y_{-1}}{\Delta x}, \text{ whence } y_{-1} = -y_1.$$

Considering point 1,

$$y_2 - 4y_2 + 6y_1 - 4y_0 + y_{-1} = \frac{w}{4EI}\left(\frac{L}{4}\right)^4.$$

For point 2,

$$y_4 - 4y_3 + 6y_2 - 4y_1 + y_0 = \frac{w}{2EI}\left(\frac{L}{4}\right)^4.$$

For point 3,

$$y_5 - 4y_4 + 6y_3 - 4y_2 + y_1 = \frac{3w}{4EI}\left(\frac{L}{4}\right)^4.$$

As $y_0 = y_4 = 0$ and $y_{-1} = -y_1$ and $y_5 = -y_3$,

solution yields, $y_1 = \dfrac{28.5wL^4}{6144EI}$, in place of $\dfrac{27.15wL^4}{6144EI}$;

$$y_2 = \frac{42wL^4}{6144EI}, \text{ in place of } \frac{39.92wL^4}{6144EI};$$

and $y_3 = \dfrac{31.5wL^4}{6144EI}$, in place of $\dfrac{29.60wL^4}{6144EI}$.

Example 17.7. *Compute deflection for the overhung beam shown in Fig. 17.9 using equation*

$$d^2ydx^2 = -\frac{M_x}{EI}.$$

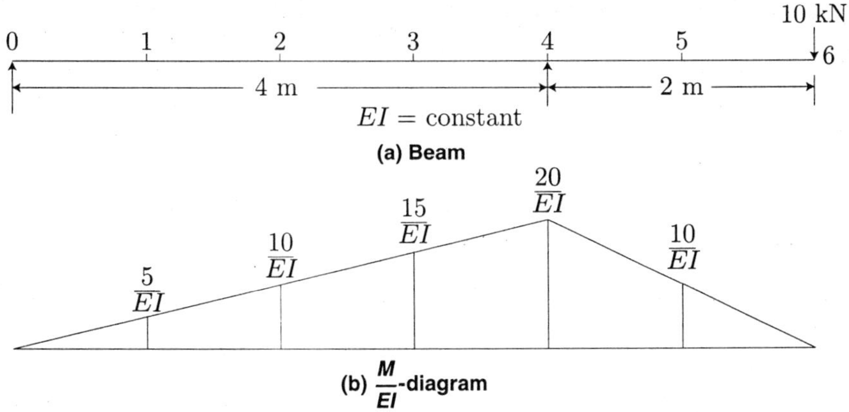

Fig. 17.9: For Example 17.7

Solution : In this case, use $\Delta x = 1$ m.

For point 1, $y_2 - 2y_1 + y_0 = \frac{5}{EI}(1)^2.$

For point 2, $y_3 - 2y_2 + y_1 = \frac{10}{EI}(1)^2.$

For point 3, $y_4 - 2y_3 + y_2 = \frac{15}{EI}(1)^2.$

For point 4, $y_5 - 2y_4 + y_3 = \frac{20}{EI}(1)^2.$

For point 5, $y_6 - 2y_5 + y_4 = \frac{10}{EI}(1)^2.$

As $y_0 = y_4 = 0$, there will be five unknowns namely, $y_1,\ y_2,\ y_3,\ y_5$ and y_6.

Solving, $y_6 = \frac{85}{EI}$, in place of $\frac{80}{EI}$.

(b) Solution of Statically Indeterminate Cases. Application of the finite difference method in the solution of statically indeterminate beams will be best understood by solution of Examples.

Example 17.8. *Determine the reactions for the beam and loading of Fig. 17.10.*

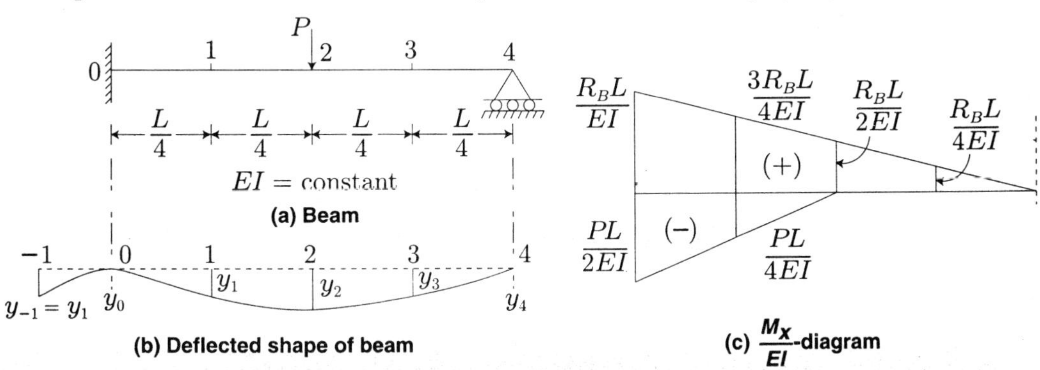

Fig. 17.10: For Example 17.8

Solution : The beam length is divided into four equal parts, so that $\Delta x = \frac{L}{4}$. R_B is considered as unknown. A built-in end physically is a means for preventing rotation of the end of the beam. At point 0, the slope is zero. In finite difference terminology, as shown in Fig. 17.10(b), this is entirely equivalent to the effect, which would be introduced if the beam were extended a distance beyond point 0 to a point -1, with the deflection at the point -1, equal to the deflection at point 1, i.e., $y_{-1} = y_1$, since at point 1, $\frac{dy}{dx} = 0$, whence

$$\frac{y_1 - y_{-1}}{2\Delta x} = 0, \quad \text{and} \quad y_1 = y_{-1}.$$

For point 0, $y_1 - 2y_0 + y_{-1} = -\left[\frac{R_B L}{EI} - \frac{PL}{2EI}\right]\left(\frac{L}{4}\right)^2.$

For point 1, $y_2 - 2y_1 + y_0 = -\left[\frac{3R_B L}{4EI} - \frac{PL}{4EI}\right]\left(\frac{L}{4}\right)^2.$

For point 2, $y_3 - 2y_2 + y_1 = -\left[\frac{R_B L}{2EI}\right]\left(\frac{L}{4}\right)^2.$

For point 3, $y_4 - 2y_3 + y_2 = -\left[\frac{R_B L}{4EI}\right]\left(\frac{L}{4}\right)^2.$

As $y_0 = y_4 = 0$, unknowns are y_1, y_2, y_3 and R_B which can be solved from four equations.

Solving, $R_B = \frac{7}{22}P$, in place of $\frac{5}{16}P$.

It is seen that the finite difference method gives a result which is within 1.8% of the value obtained by more accurate solution.

Example 17.9. *Determine the reactions for the beam and loading of Fig. 17.11.*

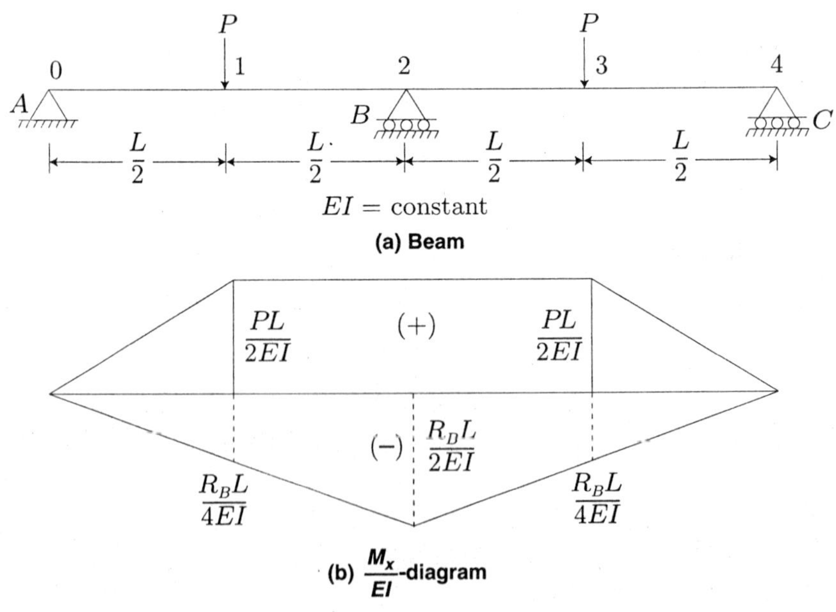

Fig. 17.11: For Example 17.9

Solution : In this case, R_B is considered as the redundant and Δx is taken as $\frac{L}{2}$

For point 1, $y_2 - 2y_1 + y_0 = -\left[\frac{PL}{2EI} - \frac{R_B L}{4EI}\right]\left(\frac{L}{2}\right)^2.$

For point 2, $y_3 - 2y_2 - y_1 = -\left[\frac{PL}{2EI} + \frac{R_B L}{2EI}\right]\left(\frac{L}{2}\right)^2.$

In this case, $y_0 = 0$ and $y_2 = 0$ and from symmetry $y_3 = y_1$.

Hence there are two unknowns y_1 and R_B. These can be solved from two equations, whence $R_B = \frac{4}{3}P$ in place of $\frac{11}{8}P$.

Thus value of R_B is within 3% of accurate value.

More accurate values can be obtained, if $\Delta x = \frac{L}{4}$ is chosen.

Example 17.10. *Analyse the propped cantilever in Fig. 17.12 by finite difference method, with $\Delta x = \frac{L}{4}$.*

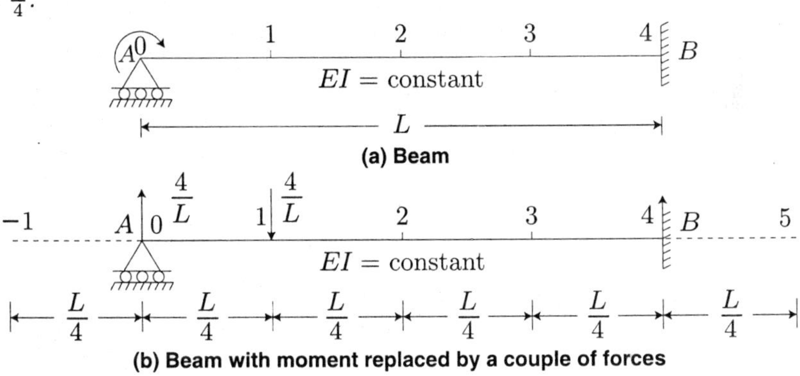

Fig. 17.12: For Example 17.10

Solution : Couple at A is replaced by equal and opposite forces as in Fig. 17.12(b). Conditions are $y_0 = y_4 = 0$ and $y_{-1} = -y_1$, and $y_5 = y_3$.

The basic equation to be used in this case is as in Eqn. (17.53), whence

$$\frac{EI}{\left(\frac{L}{4}\right)^4}\left[y_{n+2} - 4y_{n+1} + 6y_n - 4y_{n-1} + y_{n-3}\right] = w_x.$$

In order to substitute the nodal loading,

$$\frac{EI}{\left(\frac{L}{4}\right)^3}\left[y_{n+2} - 4y_{n+1} + 6y_n - 4y_{n-1} + y_{n-2}\right] = w_x\left(\frac{L}{4}\right).$$

Considering points 1, 2 and 3,

$$\frac{EI}{\left(\frac{L}{4}\right)^3}\left[5y_1 + 4y_2 - y_3\right] = \left(\frac{4}{L}\right)$$

$$\frac{EI}{\left(\frac{L}{4}\right)^3}\left[-4y_1 + 6y_2 - 4y_2\right] = 0$$

and $$\frac{EI}{\left(\frac{L}{4}\right)^3}\left[y_1 - 4y_2 + 7y_3\right] = 0$$

Solving, $y_1 = \dfrac{26L^2}{704EI}$; $y_2 = \dfrac{24L^2}{704EI}$ and $y_3 = \dfrac{10L^2}{704EI}$

and $\left(\dfrac{dy}{dx}\right)_{\text{at } A} = \dfrac{4}{2L}(2)\left(\dfrac{26L^2}{709EI}\right) = \dfrac{L}{6.82EI}$ in place of $\dfrac{L}{4EI}$.

(c) Buckling Loads. Critical bending buckling loads may be determined using the finite difference method which is particularly useful as it sets up a solution in a form specially applicable to electronic computer application. This is true also in case of column buckling problems particularly in case of complicated column problems, which cannot be solved exactly, but which can be approximately analysed using the finite difference formulation and electronic computers.

The basic equation

$$\frac{d^2y}{dx^2} = -\frac{M_x}{EI}, \text{ in this case reduces to, } \frac{d^2y}{dx^2} = -\frac{Py}{EI}, \tag{17.34}$$

whence $\dfrac{d^2y}{dx^2} + k^2y = 0$, where $k^2 = \dfrac{P}{EI}$.

Finite difference equation at any point n then is

$$\frac{y_{n+1} - 2y_n + y_{n-1}}{(\Delta x)^2} + k^2 y_n = 0,$$

whence $y_{n+1} - \left[k^2(\Delta x)^2 - 2\right] y_n + y_{n-1} = 0.$ (17.35)

Use of finite difference method for solution of the buckling problem will be best illustrated by means of Examples which follow.

Example 17.11. *Determine the critical load in case of the hinged column shown in Fig. 17.13. Solve taking (a)* $\Delta x = \frac{1}{2}$ *(b)* $\Delta x = \frac{L}{3}$ *(c)* $\Delta x = \frac{L}{4}$ *(d)* $\Delta x = \frac{L}{5}$ *(e)* $\Delta x = \frac{L}{6}$ *and (f)* $\Delta x = \frac{L}{7}$.

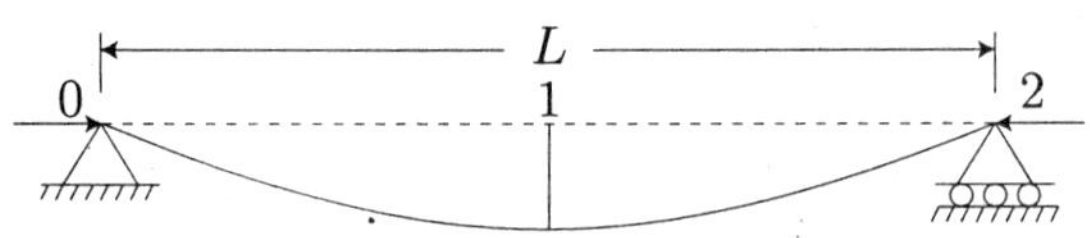

Fig. 17.13: For Example 17.11

Solution : (a) For $\Delta x = \dfrac{L}{2}$, i.e., $n = 2$,

for point 1, $y_2 + \left[k^2\left(\dfrac{L}{2}\right)^2 - 2\right] y_1 + y_0 = 0$

As $y_0 = y_2 = 0$, $k^2\dfrac{L^2}{4} - 2 = 0$,

whence $k^2 = \dfrac{8}{L^2}$, i.e., $P_{cr} = \dfrac{8EI}{L^2}$,

Compared with $\dfrac{\pi^2 EI}{L^2}$, i.e., $\dfrac{9.87EI}{L^2}$, error is -19%.

(b) For $\Delta x = \frac{L}{3}$, i.e., $n = 3$.

For point 1, $y_2 + \left[k^2\left(\frac{L}{3}\right)^2 - 2\right] y_1 + y_3 = 0.$

For point 2, $y_3 + \left[k^2\left(\frac{L}{3}\right)^2 - 2\right] y_2 + y_1 = 0.$

With $y_0 = y_3 = 0$ and $y_2 = y_1$, it follows that,

$$\left[k^2\left(\frac{L}{3}\right)^2 - 2 + 1\right] = 0, \quad \text{whence} \quad k^2 = \frac{9}{L^2}$$

and $\quad P_{cr} = \frac{9EI}{L^2}$, error being -9%.

(c) For $\Delta x = \frac{L}{4}$, i.e., $n = 4$.

For point 1, $y_2 + \left[k^2\left(\frac{L}{4}\right)^2 - 2\right] y_1 + y_0 = 0.$

For point 2, $y_3 + \left[k^2\left(\frac{L}{4}\right)^2 - 2\right] y_2 + y_1 = 0.$

With $y = 0$, and $y_1 = y_3$, it follows that,

$$y_2 + \left[k^2\left(\frac{L}{4}\right)^2 - 2\right] y_1 = 0 \quad \text{and} \quad \left[k^2\left(\frac{L}{4}\right)^2 - 2\right] y_2 + 2y_1 = 0.$$

Solution is given in matrix form when the determinant

$$\begin{vmatrix} 1 & \left[k^2\left(\frac{L}{4}\right)^2 - 2\right] \\ \left[k^2\left(\frac{L}{4}\right)^2 - 2\right] & 2 \end{vmatrix} = 0,$$

whence $\left[k^2\left(\frac{L}{4}\right)^2 - 2\right]^2 - 2 = 0.$

Solving, $k^2\frac{L}{16} = 2 - \sqrt{2}$,

whence $P_{cr} = \frac{9.3726EI}{L^2}$, error being -5%.

(d) For $\Delta x = \frac{L}{5}$, i.e., $n = 5$;

Points 1 and 4 are similar and points 2 and 3 are similar.

For point 1, $y_2 + \left[k^2\left(\frac{L}{5}\right)^2 - 2\right] y_1 + y_0 = 0.$

For point 2, $y_3 + \left[k^2\left(\frac{L}{5}\right)^2 - 2\right] y_2 + y_1 = 0.$

As $y_2 = y_3$ and $y_0 = 0$, it follows that,

$$y_2 + \left[k^2\left(\frac{L}{5}\right)^2 - 2\right] = 0 \quad \text{and} \quad \left[k^2\left(\frac{L}{5}\right)^2 - 1\right] y_2 + y_1 = 0.$$

For nontrivial solution,

$$\begin{vmatrix} 1 & \left[k^2\left(\frac{L}{5}\right)^2 - 2\right] \\ \left[k^2\left(\frac{L}{5}\right)^2 - 1\right] & 1 \end{vmatrix} = 0,$$

whence $\left[k^2\left(\frac{L}{5}\right)^2 - 2\right]\left[k^2\left(\frac{L}{5}\right)^2 - 1\right] - 1 = 0.$

Solving $P_{cr} = \frac{9.549EI}{L^2}$, error being -3.2%.

(e) For $\Delta x = \frac{L}{6}$, i.e., $n = 6$,

$$y_1 = y_5, \text{ and } y_2 = y_4.$$

For point 1, 2 and 3,

$$y_2 + \left[k^2\left(\frac{L}{6}\right)^2 - 2\right] y_1 + y_0 = 0$$

$$y_3 + \left[k^2\left(\frac{L}{6}\right)^2 - 2\right] y_2 + y_1 = 0$$

and
$$y_4 + \left[k^2\left(\frac{L}{6}\right)^2 - 2\right] y_3 + y_2 = 0.$$

Since $y_0 = 0$, and $y_2 = y_4$, it follows that, $\left[k^2\left(\frac{L}{6}\right)^2 - 2\right] y_1 + y_2 = 0$

$$y_1 + \left[k^2\left(\frac{L}{6}\right)^2 - 2\right] y_2 + y_3 = 0 \quad \text{and} \quad 2y_2 + \left[k^2\left(\frac{L}{6}\right)^2 - 2\right] y_3 = 0.$$

For nontrivial solution,

$$\begin{vmatrix} \left[k^2\left(\frac{L}{6}\right)^2 - 2\right] & 1 & 0 \\ 1 & \left[k^2\left(\frac{L}{6}\right)^2 - 2\right] & 1 \\ 0 & 2 & \left[k^2\left(\frac{L}{6}\right)^2 - 2\right] \end{vmatrix} = 0,$$

$$\text{i.e.,}\quad \left[k^2\left(\frac{L}{6}\right)^2 - 2\right]\left[\left\{k^2\left(\frac{L}{6}\right)^2 - 2\right\}^2 - 3\right] = 0,$$

whence $P_{cr} = \dfrac{9.646EI}{L^2}$, error being -2.30%.

(f) For $\Delta x = \dfrac{L}{7}$, $n = 7$

and $\quad y_1 = y_6,\ y_2 = y_5$ and $y_3 = y_4$.

As in the previous case, considering points 1, 2 and 3 and as $y_0 = 0$, it follows that,

$$\left[k^2\left(\frac{L}{7}\right)^2 - 2\right]y_1 + y_2 = 0$$

$$y_1 + \left[k^2\left(\frac{L}{7}\right)^2 - 2\right]y_2 + y_3 = 0$$

and

$$y_2 + \left[k^2\left(\frac{L}{7}\right)^2 - 1\right]y_3 = 0.$$

As before

$$\begin{vmatrix} \left[k^2\left(\frac{L}{7}\right)^2 - 2\right] & 1 & 0 \\ 1 & \left[k^2\left(\frac{L}{7}\right)^2 - 2\right] & 1 \\ 0 & 1 & \left[k^2\left(\frac{L}{7}\right)^2 - 1\right] \end{vmatrix} = 0$$

$$\text{whence}\quad \left[k^2\left(\frac{L}{7}\right)^2 - 1\right]^3 + \left[k^2\left(\frac{L}{7}\right)^2 - 2\right]^2 - 2\left[k^2\left(\frac{L}{7}\right)^2 - 2\right] - 1 = 0.$$

Solving for smallest k^2, $k^2\left(\dfrac{L}{7}\right)^2 - 2 = -1.80$

and $\quad P_{cr} = \dfrac{9.705EI}{L^2}$, error being -1.7%.

The above indicates that more accuracy is obtained with smaller Δx, i.e., larger n.

Example 17.12. *Solve Example 17.11 with higher order finite difference approximations with (a)* $\Delta x = \frac{L}{2}$, *(b)* $\Delta x = \frac{L}{3}$ *and (c)* $\Delta x = \frac{L}{4}$.

Solution : From Eq. (17.23) at any point n, it follows that,

$$-y_{n+2} + 16y_{n+1} + [12k^2(\Delta x)^2 - 30]y_n + 16_{n-1} - y_{n-2} = 0.$$

This difference equations involves the values of deflections one point on the left of support, i.e., point 0. From $\dfrac{d^2y}{dx^2} = 0$ at support,

$$y_1 - 2y_0 + y_{-1} = 0,\quad \text{whence}\quad y_1 = -y_{-1},\quad \text{since}\quad y_0 = 0.$$

Similarly, $y_1 = -y_3$.

(a) For $n = 2$, i.e., $\Delta x = \frac{L}{2}$.

Considering points 1, i.e., at $x = \frac{L}{2}$, as $y_{-1} = -y_1$ and $y_0 = y_2 = 0$,

$$y_1 + 0 + \left[12k^2\left(\frac{L}{2}\right)^2 - 30\right] y_1 + 0 + y_1 = 0,$$

whence $$12k^2\left(\frac{L}{2}\right)^2 - 28 = 0$$

and $$P_{cr} = \frac{9.33EI}{L^2}, \text{ error being } 5.5\%.$$

(b) For $n = 3$, i.e., $\Delta x = \frac{L}{3}$.

Considering point 1, i.e., at $x = \frac{L}{3}$,

$$y_1 + 0 + \left(12k^2\frac{L^2}{9} - 30\right) y_1 + 16y_1 + 0 = 0.$$

Solving $P_{cr} = \dfrac{9.75EI}{L^2}$, error being -1.21%.

(c) For $n = 4$, i.e., $\Delta x = \frac{L}{4}$.

Considering points 1 and 2, i.e., at $x = \frac{L}{4}$ and $\frac{L}{2}$

and as $y_{-1} = -y_1$ and $y_1 = y_3$ and $y_0 = y_4 = 0$.

$$-y_1 + 16y_2 + \left[12k^2\left(\frac{L}{4}\right)^2 - 30\right] y_1 + 0 + y_1 = 0$$

and $$0 + 16y_1 + \left[12k^2\left(\frac{L}{4}\right)^2 - 30\right] y_2 + 16y_1 + 0 = 0.$$

For non-trivial solution,

$$\begin{vmatrix} \left(\dfrac{12k^2L^2}{16} - 30\right) & 16 \\ 32 & \left(\dfrac{12k^2L^2}{16} - 30\right) \end{vmatrix} = 0.$$

Solving $P_{cr} = \dfrac{9.83EI}{L^2}$, error being -0.40%.

This indicates that with higher order approximation also, more accurate results are obtained.

Example 17.13. *Obtain the buckling loads of a simply supported column with variable moment of inertia given by,* $I_x = I_o\left[1 + \sin\frac{\pi x}{L}\right]$ *using* $\Delta x = \frac{L}{2}$.

Solution : In this case finite difference equation at any point n reduces to

$$EI_o\left[1 + \sin\frac{\pi x}{L}\right]\frac{y_{n+1} - 2y_n + y_{n-1}}{(\Delta x)^2} + P(y) = 0.$$

With $\Delta x = \frac{L}{2}$, considering point 1, i.e., at $x = \frac{L}{2}$,

$$y_2 + \left[\frac{k^2\left(\frac{L}{2}\right)^2}{-1+\sin\frac{\pi}{2}} - 2\right] y_1 + y_0 = 0,$$

where $k^2 = \frac{P_{cr}}{EI_0}$.

At $\quad x = 0,\;$ and $\;L,\; y_0 = y_2 = 0$

and $\quad P_{cr} = \frac{16EI_0}{L^2}$, error being 11.4%.

(d) Free Vibration of Beams. The differential equation for the lateral free vibration of beam is given by

$$\frac{d^2}{dx^2}\left[EI\frac{d^2y}{dx^2}\right] = p^2my, \tag{17.36}$$

where p = period of vibration

and $\quad m$ = mass per unit length.

For the special case, where the flexural rigidity EI is constant, Eqn. (14.36) may be written as

$$EI\frac{d^4y}{dx^4} - mp^2y = 0. \tag{17.37}$$

When expressed in finite difference form, Eq. (17.37) becomes for a point n,

$$\frac{EI}{(\Delta x)^4}\left[y_{n+2} - 4y_{n+1} + 6y_n - 4y_{n-1} + y_{n-2}\right] = p^2my_n. \tag{17.38}$$

Introducing the term $\lambda = \frac{p^2m}{EI}$, Eq. (17.38) becomes

$$y_{n+2} - 4y_{n+1} + 6y_n - 4y_{n-1} = \lambda(\Delta x)^4y_n,$$

which reduces to

$$y_{n+2} - 4y_{n+1} + [6 - \lambda(\Delta x)^4]y_n - 4y_{n-1} + y_{n-2} = 0. \tag{17.39}$$

Use of the equations will be illustrated by a suitable Example.

Example 17.14. *Analyse a propped cantilever beam as in Fig. 14.14 for free vibration.*

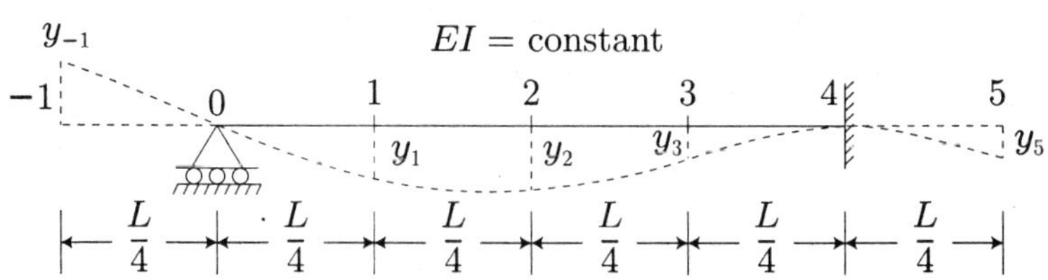

Fig. 17.14: For Example 17.14

Solution : The beam is divided into 4 equal parts, so that $\Delta x = \frac{L}{4}$. Eq. (17.39) is then applied to points 1, 2, 3. As in Example 17.10, $y_0 = y_4 = 0$, and $y_{-1} = -y_1$ and $y_5 = y_3$.

Eqns. then are

$$\left[5-\lambda\left(\frac{L}{4}\right)^4\right]y_1-4y_2+y_3=0$$

$$-4y_1+\left[6-\lambda\left(\frac{L}{4}\right)^4\right]y_2-4y_3=0$$

and $y_1-4y_2+\left[7-\lambda\left(\frac{L}{4}\right)^4\right]y_3=0.$

For non-trivial solution

$$\begin{vmatrix}\left[5-\lambda\left(\frac{L}{4}\right)^4\right] & -4 & 1\\ -4 & \left[6-y\left(\frac{L}{4}\right)^4\right] & -4\\ 1 & -4 & \left[7-\lambda\left(\frac{L}{4}\right)^4\right]\end{vmatrix}=0.$$

Solving the cubical equation as obtained from the determinant, the lowest root is given by

$$\lambda\left(\frac{L}{4}\right)^4=0.72,$$

whence $\dfrac{p^2m\left(\frac{L}{4}\right)^4}{EI}=0.72$

and $p=13.6\sqrt{\dfrac{EI}{mL^4}}$, in place of $15.9\sqrt{\dfrac{EI}{mL^4}}$.

(e) Plate Problems. The differential equation governing the small deflection of a thin plate with constant thickness under bending is given by

$$\frac{\partial^4 w}{\partial x^4}+2\frac{\partial^4 w}{\partial x^2\partial y^2}+\frac{\partial^4 w}{\partial y^4}=\frac{q}{D}, \tag{17.40}$$

where w refers to the deflection of middle surface of the plate, x and y being the two co-ordinate axis, q refers to the lateral load on the slab and D is the plate stiffness given by $\frac{Et^3}{12(1-\mu^2)}$, where 't' is the thickness of the plate and μ, the Poisson's ratio.

In terms of the operation $\nabla^2=\dfrac{\partial^2}{\partial x^2}+\dfrac{\partial^2}{dy^2}$.

Eq. (17.40) becomes

$$\nabla^2(\nabla^2 w)=\nabla^4 w=\left(\frac{\partial^2}{\partial x^2}+\frac{\partial^2}{\partial y^2}\right)\left(\frac{\partial^2 w}{\partial x^2}+\frac{\partial^2 w}{\partial y^2}\right),$$

whence $\nabla^4 w=\dfrac{q}{D}$. (17.40a)

It is interesting to note the similarity between form of Eq. (17.40a) and the fourth order differential equation of simple beam theory, i.e., $\frac{d^4w}{dx^4}=\frac{q}{EI}$.

The solution of a specific problem of bending of a simple rectangular plate requires finding a function w which satisfies Eq. (17.40) and also the boundary conditions of the specific problem. When w is known, moments, twists and shears can be calculated from

$$M_x = -D\left[\frac{\partial^2 w}{\partial x^2} + \mu\frac{\partial^2 w}{\partial y^2}\right] \tag{17.41a}$$

$$M_y = -D\left[\mu\frac{\partial^2 w}{\partial x^2} + \frac{\partial^2 w}{\partial y^2}\right] \tag{17.41b}$$

$$M_{xy} = -M_{yx} = D(1-\mu)\frac{\partial^4 w}{\partial x^2 \partial y^2}. \tag{17.41c}$$

Stresses can be found from

$$\sigma_x = -\frac{Ez}{1-\mu^2}\left(\frac{\partial^2 w}{\partial x^2} + \mu\frac{\partial^2 w}{\partial y^2}\right) \tag{17.42a}$$

$$\sigma = -\frac{Ez}{1-\mu^2}\left(\mu\frac{\partial^2 w}{\partial x^2} + \frac{\partial^2 w}{\partial y^2}\right) \tag{17.42b}$$

and $$\tau_{xy} = \frac{Ez}{1+\mu}\frac{\partial^2 w}{\partial x \partial y}, \tag{17.42c}$$

in which z is the distance of the point in question from the middle surface, positive Z being measured down from the middle surface.

Applications of the finite difference method to plate problems can be developed by first defining basic quantities and then introducing these in the differential equation $\nabla^4 w = \frac{q}{D}$. The result will be a series of linear equations which can be solved simultaneously to give the required solution. The rectangular plate is divided into strips and the load intensity at the corner of each strip is obtained as in Fig. 17.15(a).

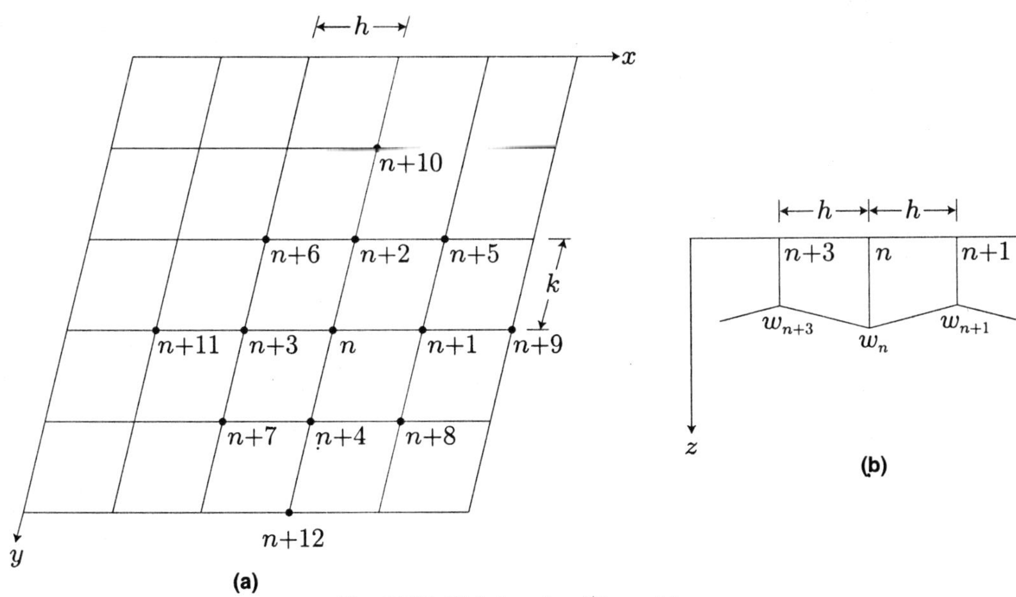

Fig. 17.15: Divisions in plate problem

In accordance with difference procedure, the first and second partial derivations of w for the point n is give by

$$\left.\frac{\partial w}{\partial x}\right|_{n,n+1} = \frac{w_{n+1} - w_n}{h}, \tag{17.43a}$$

as shown in Fig. 17.15(b).

The term $\left.\frac{\partial w}{\partial x}\right|_{n,n+1}$ denotes that it is the first partial derivation of w with respect to x taken at point n and written in terms of w_{n+1} and w_n.

The average value of $\frac{\partial w}{\partial x}$ is given by

$$\begin{aligned}\frac{\partial w_n}{\partial x} &= \frac{1}{2}\left[\frac{w_{n+1} - w_n}{h} + \frac{w_n - w_{n+3}}{h}\right]\\ &= \frac{w_{n+1} - w_{n+3}}{2h}.\end{aligned} \tag{17.43b}$$

In terms of w_n and w_{n+3}, $\frac{\partial w}{\partial x}$ can be expressed as

$$\left.\frac{\partial w}{\partial x}\right|_{n,n+3} = \frac{w_n - w_{n+3}}{h}.$$

It follows that.

$$\begin{aligned}\left.\frac{\partial^2 w}{\partial x^2}\right|_n &= \frac{\left.\frac{\partial w}{\partial x}\right|_{n,n+1} - \left.\frac{\partial w}{\partial x}\right|_{n,n+3}}{h}\\ &= \frac{w_{n+1} - 2w_n + w_{n+3}}{h^2}.\end{aligned} \tag{17.43c}$$

Similarly, $$\left.\frac{\partial^2 w}{\partial y^2}\right|_n = \frac{w_{n+2} - 2w_n + w_{n+4}}{k^2}. \tag{17.43d}$$

Then $$\nabla^2 w_n = \left.\frac{\partial^2 w}{\partial x^2}\right|_n + \left.\frac{\partial^2 w}{\partial y^2}\right|_n. \tag{17.43e}$$

For the case when $h = k$,

$$\nabla^2 w_l = \frac{1}{h^2}\left[w_{n+4} + w_{n+3} + w_{n+2} + w_{n+1} - 4w_n\right]. \tag{14.43f}$$

The deflection of a square simply supported plate can be obtained by using finite differences. The basic equations,

$$\nabla^2(\nabla^2 w) = \frac{q}{D}, \text{ can be written as}$$

$$\nabla^2 \phi = q, \tag{17.44a}$$

where $$\phi = (\nabla^2 w)D. \tag{17.44b}$$

If values of ϕ can be found for various points on the plate, w can be found by solving Eq. (17.44b), $\nabla^2 w = \frac{\phi}{D}$.

This approach can be used in cases where values of $\phi = 0$ at the boundaries. In cases of plates where $\phi \neq 0$ at the boundaries, it is necessary to set up finite difference expression for $\nabla^4 w$ in order to solve the plate problems.

In such cases, finite difference equation can be formulated from expressions given in Eqns. (17.15a) to (17.15c).

Use of finite difference method for plate problems is illustrated by Examples which follow.

Example 17.15. *Analyse a simply supported uniform square plate of Fig. 17.16 for uniformly distributed load by finite difference method.*

Solution : The square plate is divided into 16 equal parts with $h = \frac{L}{4}$. The deflections are w_1, w_2, w_3, w_4, considering the symmetry. Values of w at the boundary is zero. Also at boundary $\nabla^2 w = 0$.

From Eq. (17.44a), for points 1, 2 and 3 in Fig. 17.16,

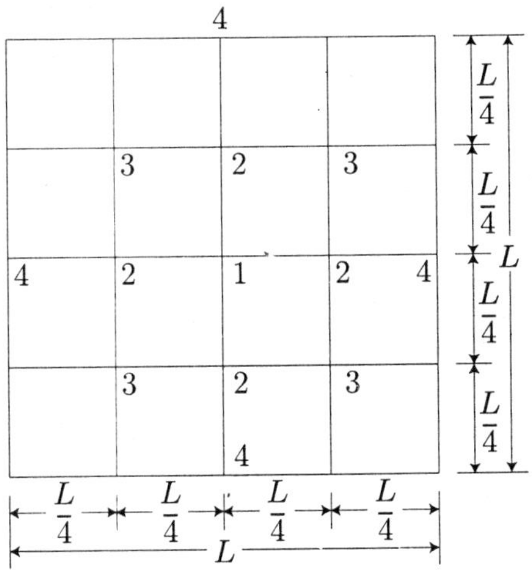

Fig. 17.16: For Example 17.15

$$-\phi_1 + \phi_2 = \frac{qL^2}{64}$$

$$\phi_1 + \phi_2 + 2\phi_3 - 4\phi_2 - \frac{qL^2}{16}$$

and $$2\phi_2 + 2\phi_4 - 4\phi_3 = \frac{qL^2}{16}.$$

As $\phi_4 = 0$, solving, $\phi_1 = -\frac{9}{128}qL^2$

$$\phi_2 = -\frac{7}{128}qL^2 \text{ and } \phi_3 = -\frac{3qL^2}{256}.$$

From the values of ϕ, values of w can be obtained solving Eq. (17.44b). Considering points 1, 2 and 3, and noting that $w_4 = 0$,

$$4w_2 - 4w_1 = \frac{\phi_1}{D}\left(\frac{L}{4}\right)^2 = -\frac{9}{128(16)D}qL^4$$

$$w_1 + 2w_3 - 4w_2 = -\frac{7}{7(128)D}qL^4$$

$$2w_2 - 4w_3 = -\frac{3}{128(16)D}qL^4.$$

Solving simultaneously,

$$w_1 = \frac{0.0045166qL^4}{D};\ w_2 = \frac{0.0028889qL^4}{D} \quad \text{and} \quad w_3 = \frac{0.0018107qL^4}{D}.$$

With $\mu = 0.30$,

$$M_{x1} = -\frac{16D}{L^2}[(w_2 - 2w_1 + w_2) + 0.3(w_2 - 2w_2 + w_3)]$$

$$= -\frac{16D}{L^2}(2)(13)(w_2 - w_1) = 0.0677qL^2.$$

Example 17.16. *Solve Example 14.15 using Eqns. (17.16) to (17.18).*

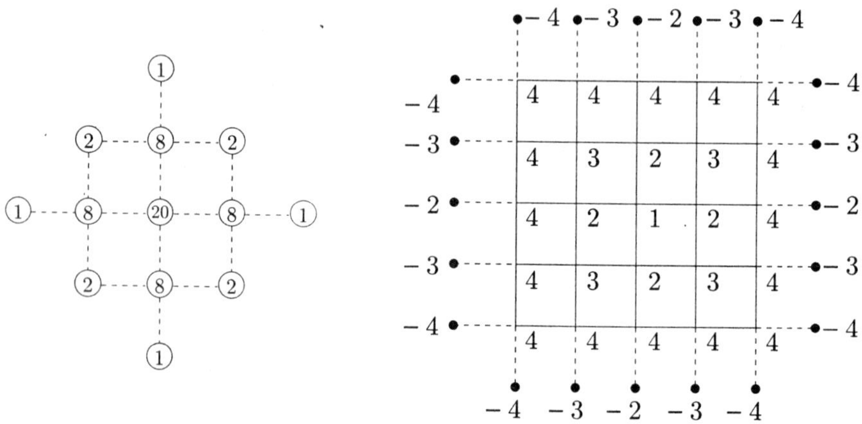

(a) Finite difference relationship (b) Difference nodes

Fig. 17.17: For Example 17.16

Solution : In this case, the finite difference at a point for the symmetrical case is as shown in Fig. 17.17(a). The different nodes are designated in Fig. 17.17(b), taking into account the symmetry of the problem. The simply supported end condition gives,

$$w_2 = -w_{-2} \quad \text{and} \quad w_3 = -w_{-3}.$$

Applying the finite difference relationship as in Fig. 17.17(a), considering points 1, 2 and 3.

$$20w_1 - 32w_2 + 8w_3 = \frac{qL^4}{256D}$$

$$-8w_1 + 24w_2 - 16w_3 = \frac{qL^4}{256D}$$

and $$2w_1 - 16w_2 + 20w_3 = \frac{qL^4}{256D}$$

Solving $w_1 = 0.00403\frac{qL^4}{D}$; $w_2 = 0.00293\frac{qL^4}{D}$ and $w_3 = 0.00214\frac{qL^4}{D}$

The B.M. at point 1 is given by,

$$M_{x1} = -\frac{16D}{L^2}[(w_2 - 2w_1 + w_2) + 0.3(w_2 - 2w_1 + w_2)] = 0.0458qL^2.$$

Example 17.17. *Analyse a square plate of sides L, with built-in ends and subjected to uniformly distributed load.*

Solution : For solution of the problem, the plate is divided into four segments as in Fig. 17.18. The boundary conditions for a built-in plate are zero deflection and zero slope. The two stage solution using the auxiliary function is not zero on the boundary. It is, therefore, necessary to set up a finite difference expression for $\nabla^4 w$ in order to solve this type of problem.

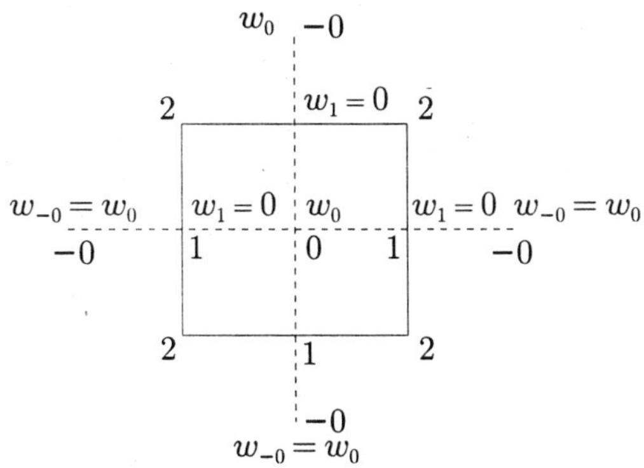

Fig. 17.18: For Example 17.17

$$\nabla^4 w_0 = \frac{1}{\left(\frac{L}{2}\right)^4}[w_{-0} + w_{-0} + w_{-0} + w_{-0} + 2(w_2 + w_2 + w_2 + w_2) - 8(w_1 + w_1 + w_1 + w_1) + 20w_0]$$

$$= \frac{24w_0}{\left(\frac{L}{2}\right)^4} = \frac{q}{D},$$

since $w_{-0} = w_0$ and $w_1 = 0$, and $w_2 = 0$, whence $w_0 = 00.0026\,\frac{qL^4}{D}$.

The error in the above answer is 118%. The accuracy can be improved by using small strip length with more segments. Thus instead of 4 sections, if the net work is made of 16 sections, the resulting solution will be in error by 42%. Better values can be obtained with finer segments.

Example 17.18. *Analyse the continuous plate shown in Fig. 17.19 using finite difference relationship if u.d.l. is placed on one half.*

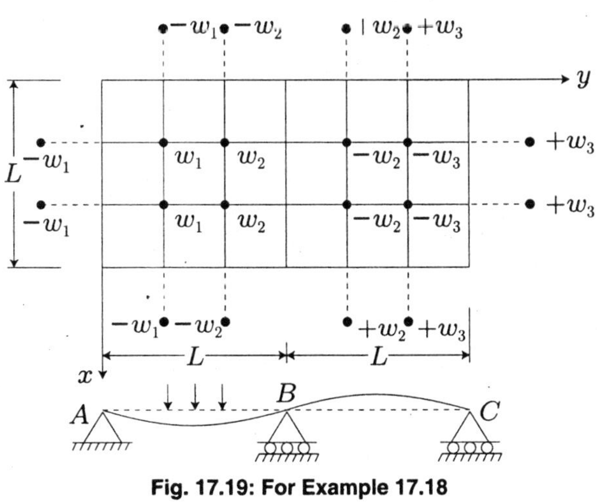

Fig. 17.19: For Example 17.18

Solution : The plate is divided into 18 equal segments. Considering points 1, 2 and 3 using operator as in Fig. 14.17(a),

$$-2w_1 + 2w_2 - 8w_2 - 8w_1 + 20w_1 = \frac{qL^4}{81D}$$

$$-2w_2 + 2w_1 - 8w_2 - 8w_1 + 20w_2 = \frac{qL^4}{81D}$$

and $$2w_3 - 2w_2 + 8w_2 + 8w_3 - 20w_3 = 0.$$

Solving, $w_1 = 1.50\frac{qL^4}{81D}$; $w_2 = 0.25\frac{qL^4}{81D}$ and $w_3 = 0.150\frac{qL^4}{81D}$.

(f) Beam Problems. The differential equation governing the behaviour of bending of a beam is given by the biharmonic equation,

$$\Delta^4\psi = 0, \tag{17.45a}$$

where $\nabla^2 = \left(\frac{\partial^2}{\partial x^2}\right) + \left(\frac{\partial^2}{\partial y^2}\right)$, called the Laplace or biharmonic operator and ψ is a stress function commonly known as Airy's stress function.

Eq. (17.45a) reduces to

$$\frac{\partial^4\psi}{\partial x^4} + 2\frac{\partial^4\psi}{\partial x^2\partial y^2} + \frac{\partial^4\psi}{\partial y^4} = 0. \tag{17.45b}$$

Eq. (14.45b) is similar to Eq. (14.40) in case of plate except that stress function ψ is involved in place of the deflection function w.

Solution of Eq. (17.45b) by finite difference method can be done in a similar way as in case of plates.

(g) Torsion Problem. The differential equation governing the case of Torsion is given by

$$\left.\begin{aligned}&\frac{\partial^2\psi}{\partial x^2} + \frac{\partial^2\psi}{\partial y^2} = -2 \text{ on domain } R\\ &\text{and} \quad \psi = 0 \text{ on boundary.}\end{aligned}\right] \tag{17.46}$$

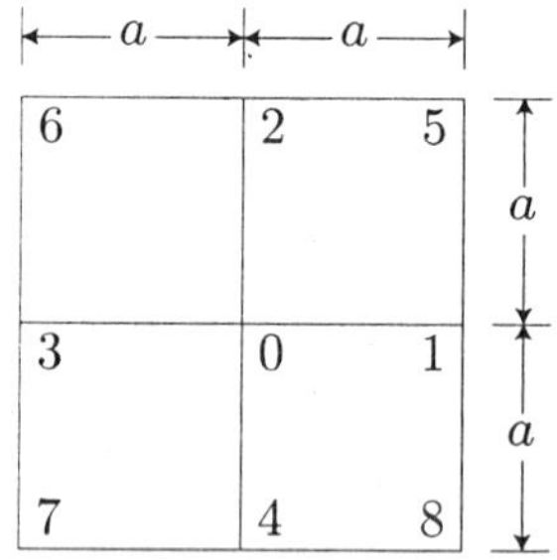

Fig. 17.20: Square section

Considering a square section and dividing the section into four square elements and numbering the nodal points as shown in Fig. 17.20, the finite difference equation corresponding of Eqn. (17.46) at point 0 is given by

$$\psi_1 + \psi_2 + \psi_3 + \psi_4 - 4\psi_0 = -2h^2. \tag{17.47}$$

From the boundary condition in Eq. (17.46),

$$\psi_i = 0, \; i = 1, 2, \cdots, 8. \tag{17.48}$$

Substituting Eq. (17.48) into Eq. (17.47) and with $h = a$,

$$-4\psi_0 = -2a^2, \quad \text{whence} \quad \psi_0 = 0.5a^2.$$

The derivation $\frac{\partial \psi}{\partial x}$ at the nodal point 1 can be approximated by the first backward difference, viz.,

$$\left(\frac{\partial \psi}{\partial x}\right)_1 \approx \frac{1}{h}(\psi_1 - \psi_0) = -0.5a.$$

The Torsion constant is given by

$$J = 2\int\int_R \psi dx dy,$$

where J is twice the volume under the ψ surface. Using Simpson's one-third rule for even numbers,

$$I_1 = \int_{x_o}^{x_a+nh} f dx = \frac{h}{3}\left[f_0 + 4f_1 + 2f_2 + 4f_3 + 2f_4 + \cdots + 2f_{n-2} + 4f_{n-1} + f_n\right],$$

from which the volume can be obtained as

$$J \approx 2\left[\frac{h^2}{9}(16\psi_0)\right] = 1.778a^4.$$

The maximum shear stress is then given by

$$\begin{aligned}\tau_{max} &= -\frac{T}{J}\frac{\partial \psi}{\partial x} \approx -\frac{T}{1.778a^4}(-0.5a) \\ &= 0.281\frac{T}{a^3}.\end{aligned} \tag{17.49}$$

The analytical solution gives the value of τ_{max} as 0.600 $\frac{T}{a^3}$. The approximate solution gives a value which is 53.2% low. With such a coarse mesh such a poor approximation is naturally to be expected.

To improve the approximation, the domain can be divided into sixteen parts with $h = \frac{a}{2}$ as in Fig. 17.21.

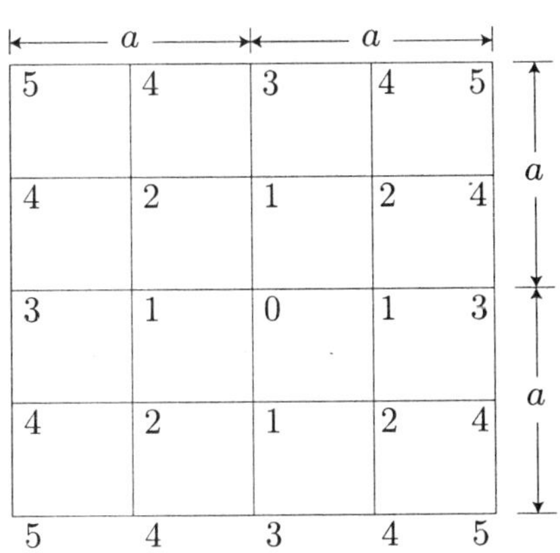

Fig. 17.21: Square section with large number of divisions

In this case,

$$\left.\begin{aligned} 4\psi_1 - 4\psi_2 &= -2\left(\frac{a}{2}\right)^2 \\ \psi_0 + 2\psi_2 + \psi_3 - 4\psi_4 &= -2\left(\frac{a}{2}\right)^2 \\ 2\psi_1 + 2\psi_4 - 4\psi_2 &= -2\left(\frac{a}{2}\right)^2. \end{aligned}\right] \tag{17.50}$$

On the boundary, $\psi_3 = \psi_4 = \psi_5 = 0$.

Solving, $\psi_0 = 0.5625a^2$; $\psi_1 = 0.4375a^2$ and $\psi_2 = 0.3438a^2$.

Then $\left(\dfrac{\partial \psi}{\partial x}\right)_3 \approx \dfrac{1}{h}(\psi_3 - \psi_1) = -0.8750a$

and $$J = 2\int\int_R \psi \, dxdy = 2\frac{h^2}{9}(4)[\psi_5 + 4\psi_4 + \psi_3 + 4\psi_4 + 16\psi_2 + 4\psi_1 + \psi_3 + 4\psi_1 + \psi_0] = 2.125a^4$$

and $$\tau_{max} = -\frac{T}{J}\left(\frac{\partial \psi}{\partial x}\right)_3 \approx 0.412\frac{T}{a^3}, \tag{17.51}$$

which is 31.3 per cent low.

Further better results can be obtained by reducing the size of the elements still further. Thus it can be shown that with $h = \frac{a}{4}$, $\tau_{max} = 0.498 \frac{T}{a^3}$, i.e., 22.1 per cent low.

However, improvements can be done by using higher order finite difference approximations also.

17.6 Relaxation Procedure

The first application of the Relaxation Method was made in the year 1935 by Sir Richard Southwell, then Professor of Engineering Science in the University of Oxford. The method had been developed gradually as a powerful computational technique for the solution of the practical engineering and physical problems.

The basic computational process underlying all applications of the relaxation method can be illustrated by employing them for the solution of a pair of linear algebraic equations.

Considering the simultaneous equations,

$$-6x + 2y + 234 = 0 \quad \text{and} \quad 2x - 4y + 102 = 0,$$

it is assumed that F_x and F_y are the residuals of the equations for assigned values of x and y.

Taking $-6x + 2y - 234 = F_x$ and $2x - 4y - 102 = F_y$,

values assigned to x and y and F_x and F_y can be calculated. Values of residuals are to be made zero gradually adopting a suitable scheme.

The operation Table for the scheme may be considered as follows :

	ΔF_x	ΔF_y
$x = 1$	-6	2
$y = 1$	2	-4

The computation is conveniently set out in tabular form in a relaxation table, each line of which corresponds to a single step of the process. The columns consist of one for each residual, showing the successive values taken by that residual and one showing the nature of the successive operations used. The typical relaxation table for the case is shown in Table 17.1.

Table 17.1. Relaxation Table

	F_x	F_y	
$x = y = 0$	234	102	$\Delta x = \frac{234}{6} = 39$
$\Delta x = 39$	0	180	$\Delta y = \frac{180}{4} = 45$
$\Delta y = 45$	90	0	$\Delta x = \frac{90}{6} = 15$
$\Delta x = 15$	0	30	$\Delta y = \frac{30}{4} \approx 7$
$\Delta y = 7$	14	2	$\Delta x = \frac{14}{6} \approx 2$
$\Delta x = 2$	2	6	$\Delta y = \frac{6}{4} \approx 1$
$\Delta y = 1$	4	2	$\Delta x = \frac{4}{6} \approx 1$
$\Delta x = 1$	-2	4	$\Delta y = \frac{4}{4} = 1$
$\Delta y = 1$	0	0	
$x = 57$ and $y = 54$	0	0	

If arithmetical mistakes are done F_x or F_y or both do not become zero. In such a case further steps are necessary for arriving at the correct solution.

Calculations can be speeded up by assuming any suitable initial values for x and y and then carrying on the Relaxation Table. For example, considering the equations considered assuming initial values of x and y as 55 and 48, respectively, the continuation of the table will be as shown in Table 17.2.

Table 17.2. Relaxation Table

	F_x	F_y	
$x = 55,\ y = 48$	0	20	$\Delta y = \frac{20}{4} = 5$
$\Delta y = 5$	10	0	$\Delta x = \frac{10}{6} \approx 2$
$\Delta x = 2$	-2	4	$\Delta y = \frac{4}{4} = 1$
$\Delta y = 1$	0	0	
$x = 57,\ y = 54$	0	0	

Various modifications on the basic rules of relaxation can be done for quicker convergence and solution of the equations. They are

(a) Over relaxation

(b) Block relaxation

and (c) Group relaxation.

They will be illustrated by considering the previous simultaneous equation.

(a) Over Relaxation. In case of over relaxation, the operation table, corresponding to the pair of equations is given as follows :

	ΔF_x	ΔF_y
$\Delta x = 1$	-6	2
$\Delta y = 1$	2	-4

It is evident from this Table that each operation when used to reduce one residual, automatically increases the other residual. Thus the operation $\Delta y = 1$ used to decrease F_y by 4, at the same time increases F_x by 2. This circumstance would of course be an advantage if the residuals are of the opposite signs, so that in order to bring them both to zero, one had to be decreased, while the other needed to be increased algebrically. This property is a hindrance to convergence, when as in this example, the initial residuals are both of the same sign (here positive), F_x being 234 and F_y beng 102. It is clear that the use of the basic rule will in the first step of the calculation reduce the value of F_x to zero, but increase the value of F_y. In the second step, F_y would be brought to zero but F_x increased to become again positive and so on. Much work would be saved if the effect of the second step could be anticipated before making the first. With the knowledge that in the second step F_x is automatically going to be increased as F_y is reduced, it would obviously be helpful if F_x could be negative and not merely zero, at the beginning of the second step. The basic rule as originally used in the first step has to be modified in order to achieve this. It is thus

Table 17.3. Over-Relaxation

	F_x	F_y	
$x = y = 0$	234	102	$\Delta x = \frac{234}{6} = 39$ Use $\Delta x = 60$ considering over-relaxation.
$\Delta x = 60$	-126	222	Use $\Delta y = \frac{222}{4} \approx 55$
$\Delta y = 55$	-16	2	$\Delta x = -\frac{16}{6} \approx -3$
$\Delta x = -3$	2	-4	$\Delta y = -\frac{4}{4} = -1$
$\Delta y = -1$	0	0	
$y = 57 : y = 54$	0	0	

better not just to reduce F_x to zero, but instead deliberately to go further and make F_x negative. This device is called over-relaxation. It is of course necessary to decide as to by how much it is appropriate to over-relax, i.e., to what extent F_x should be made negative. It depends on the equations being solved and a good division is largely a matter of experience. As a rough rule, it can be suggested that in the solution of a set of algebraic equations, the over-relaxing operation should be of a magnitude of about half as much against the operation which would be given by the basic rule of relaxation. The complete relaxation following over-relaxation pattern is shown in Table 17.3.

Situations in which over relaxation is desirable frequently arise in all applications of relaxation. It is useful whenever the available operations are such that in reducing one residual, the other residuals automatically increase and when at the same time the individual residuals are predominantly of the same sign.

(b) Block Relaxation. A second and even more important device involves the use of some new operations in the operation Table. This modification is known as block relaxation and it consists in the use of other operations than the basic unit operations. These new operations are called unit block operations and they entail the simultaneous application of increments to more than one of the unknown at the same time. In the case of the previous simultaneous equation only one block operation is possible, as there are only two unknown x and y. This unit block operations is $\Delta x = \Delta y = 1$. The changes to the residuals produced by applying it are easily calculated as shown in the operation Table shown below.

	F_x	F_y
$\Delta x = 1$	-6	2
$\Delta y = 1$	2	-4
$\Delta x = 1,\ \Delta y = 1$	-4	-2

The block operations shown above may be used at any stage in the relaxation process. The complete relaxation Table for the same using block relaxation is found to be as in Table 17.4.

Table 17.4. Block Relaxation

	F_x	F_y	
$x \doteq y = 0$	234	102	
$\Delta x = \Delta y = 56$	10	-10	$\Delta x = \Delta y = 56$
$\Delta y = -2$	6	-2	$\Delta y = -\dfrac{10}{4} \approx -2$
$\Delta x - 1$	0	0	$\Delta x = \dfrac{6}{6} = 1$
$x = 57;\ y = 54$	0	0	

The advantage of using block relaxation is quite evident from Table 17.4 as lesser number of steps is required for the complete solution.

(c) Group Relaxation. The method can be illustrated by considering the solution of a more complicated set of equations. A set for which the convergence is slower is required. Considering the three simultaneous equations,

$$\left.\begin{aligned} F_x &= -20x + 12y + 6z + 200 = 0 \\ F_y &= 12x - 20y + 10z + 200 = 0 \\ F_z &= 6x + 10y - 20z + 200 = 0, \end{aligned}\right] \tag{17.52}$$

the operation Table may be written as follows :

	ΔF_x	ΔF_y	ΔF_z
$\Delta x = 1$	-20	12	6
$\Delta y = 1$	12	-20	10
$\Delta z = 1$	6	10	-20

No block operations are included in this Table. The relaxation with the operation Table is shown in Table 17.5.

Table 17.5. Relaxation with Operation Table

	F_x	F_y	F_z	
$x = y = z = 0$	200	200	200	$\Delta x = \dfrac{200}{20} = 10$
$\Delta x = 10$	0	320	260	$\Delta y = \dfrac{320}{20} = 16$
$\Delta y = 16$	192	0	420	$\Delta z = \dfrac{420}{20} = 21$
$\Delta z = 21$	318	210	0	$\Delta x = \dfrac{318}{20} \approx 16$
$\Delta x = 16$	-2	402	96	(a)
$\Delta y = 20$	238	2	296	
$\Delta z = 15$	328	152	-4	
$\Delta x = 16$	8	344	92	(b)

The process is very slowly convergent and it would be practicable to carry on until the solution is reached with only the basic use of relaxation. For quicker convergence, a group operation Table can be used for reducing F_y. This new operation may be recorded as a new line in the operation Table shown below.

	ΔF_x	ΔF_y	ΔF_z
$\Delta x = 16, \Delta y = 20$ $\Delta z = 15$	10	-58	-4

A group operation may be used to any extent and to continue the relaxation process, the group operation should next be applied to such an extent as will cause F_y to be reduced to zero, while F_x and F_z are left comparatively unchanged. The necessary extent is clearly six times (i.e., approximately $\frac{344}{58}$) and the continuation of the relaxation Table gives

	F_x	F_y	F_z
	8	344	92
$\Delta x = 96,\ \Delta y = 120,\ \Delta z = 90$	68	-4	68

Values of residuals are recorded above. For complete convergence, the following Table may be made with the residuals obtained above.

	F_x	F_y	F_z
	68	-4	68
$\Delta z = 2$	80	16	28
$\Delta x = 2$	40	40	40
$x = 140,\ y = 156,\ z = 128$	40	40	40

The residuals obtained is $\frac{1}{5}$th of the values in the original Eq. (17.52), so that the multiplying factor is $\frac{5}{4}$, which means that the final solution to the Eq. (17.52) is given by $x = \frac{5}{4}(140) = 175$; $y = \frac{5}{4}(156) = 195$ and $z = \frac{5}{4}(128) = 160$.

17.7 Use of Relaxation Method for Differential Equations

The most widespread and perhaps the most important application of relaxation method is to obtain particular solutions of partial differential equations in two dimensions. However, the use of the method will be illustrated by considering a simple type of ordinary differential equations. The relaxation method is essentially a systematic procedure for solving finite differential equations.

Considering a beam of flexural rigidity EI, the transverse deflection can be related to the bending moment M along the beam by the equation,

$$\frac{d^2y}{dx^2} = -\frac{M}{EI}. \tag{17.53}$$

Expressing in finite difference notation, the differential equation becomes

$$y_{n+1} - 2y_n + y_{n-1} + \frac{M}{EI}(\Delta x)^2 = 0. \tag{17.54}$$

Instead of an operation table, a relaxation pattern is constructed from Eqn. (17.54), by means of which each residual F is calculated. It is obvious that the residual at any point of subdivisions has its value change by -2, if a unit increment is added to the value of y at that point and has its value changed by 1, if a unit increment is added to the value of y at either neighbouring point of subdivision. Hence, the complete effect of a unit increment applied to the value of y_0 at typical point is to change the residual at the two adjacent points by 1. This effect is illustrated in Fig. 17.22, where the changes of the residuals are recorded in circles in the same relative positions as would be occupied by any set of three typical points of subdivision actually marked out along the range of integration. The unit increment (+1) is also recorded in Fig. 17.22 in a position near the typical point where it is applied. This relaxation pattern represents a basic unit operation available for use in the solution of Eq. (17.53). Being typical of the effect produced by the application of a unit increment to y at any point of subdivision, it is in fact a diagrammatic representation of all the basic unit operations available.

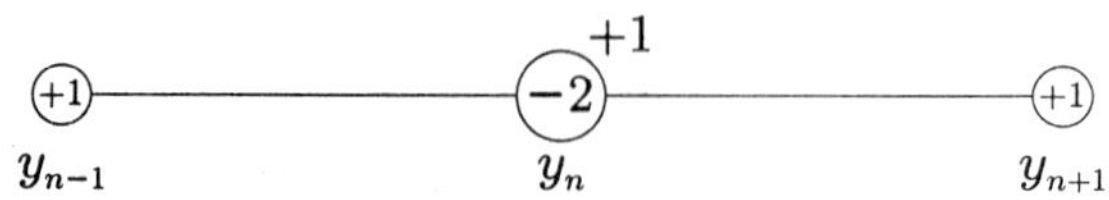

Fig. 17.22: Relaxation pattern

The actual process of liquidation of the residuals together with the consequent determination of the solution is carried out on a relaxation diagram instead of in a relaxation Table. Fig. 17.23 illustrates such a diagram, consisting of a diagrammatic representation of a range of integration, divided for example into five dimensions. At the end points of the range, values of y given as end conditions are recorded, which in this case are taken as zero, i.e., $y = 0$. Below each point of subdivision, the values of the quantity $\frac{M}{EI}(\Delta x)^2$ which in this case, are found to be 23, −29, 42 and −31 at the first, second, third and fourth points of subdivision. In Fig. 17.23, at each point of subdivision initially guessed values of y (above and to the left) and the corresponding initially calculated values of the residuals are recorded (above and to the right).

In this case all the initial values of y are taken as zero.

The liquidation process is recorded directly on this same diagram using the pattern in Fig. 17.22 and the complete relaxation pattern is shown in the diagram, the successive 1st, 2nd, 3rd, 4th, 5th and 6th steps being marked as (1), (2), (3), (4), (5) and (6) besides the values in the diagram. The largest residual is chosen as the first one. When liquidation of the residuals is completed, i.e., when they have all been made no larger in magnitude than 1, the solution is obtained by adding up the columns of increment of Δy, for each point separately. These totals are recorded in Fig. 17.23 at the top of their respective columns above a line. As a last step of the solution, the final values of the residuals at the top of the right-hand columns should be checked by a direct substitution of the totalled y values in Eqn. (17.54). If any error are then found, the relaxation must be continued until eventually, the residuals are correctly brought to zero.

Final value of y	0	11		0		18		− 6		0
					0(6)	(6)−1	0(6)		−1(6)	
							−2(5)	(5)−1	0(5)	
					1−(4)	(4)−2	−1(4)		−2(4)	
		11			3(2)		−5(3)	(3)−5	0(3)	
		(2) 11	1(2)		−8(1)	(1) 21	0(1)		−10(1)	
Assumed values of y	0	0	23	0	−29	0	42	0	−31	0
	y_0	y_1	Residual	y_2	Residual	y_3	Residual	y_4	Residual	y_5

Fig. 17.23: Relaxation procedure

If the basic relationship between load and y be considered, then

$$\frac{d^4y}{dx^4} = \frac{w}{EI}. \tag{17.55}$$

Relaxation pattern then follows from the finite difference notation given by

$$y_{n+2} - 4y_{n+1} + 6y_n - 4y_{n-1} + y_{n-2} + \frac{w}{EI}(\Delta x)^2 = 0 \tag{17.56}$$

and is shown in Fig. 17.24.

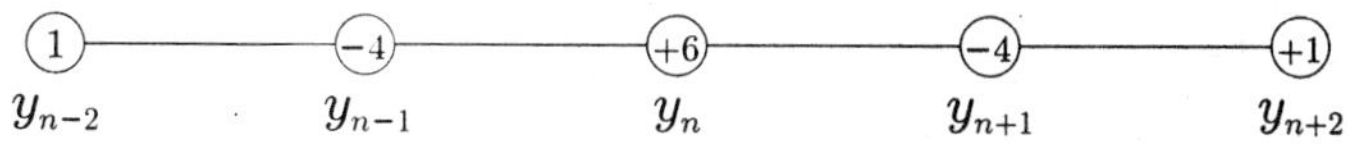

Fig. 17.24: Relaxation pattern

Example 17.19. *Obtain deflection of a simply supported beam as shown in Fig. 17.25, using relaxation procedure for solution of finite difference equations for* $\frac{d^2y}{dx^2} = -\frac{M}{EI}$. *Consider EI as constant.*

Solution : Finite difference equation is given by

$$y_{n+1} - 2y_n + y_{n-1} - \frac{M}{EI}(\Delta x)^2 = 0.$$

Dividing the beam into 4 equal parts so that $\Delta x = \frac{L}{4}$, as shown in Fig. 17.24, $y_0 = y_4 = 0$ from boundary conditions. Assume other values of deflections as

$$y_1 = y_2 = y_2 = 0.0020\frac{PL^3}{EI}.$$

For points 1 and 3,

$$\text{Residual} = \frac{PL^3}{EI}[0 - 2(0.0020) + 0.0020] - \frac{PL}{8}\left(\frac{L}{4}\right)^2\frac{1}{EI}$$

$$= -0.0098\frac{PL^3}{EI}.$$

For point 2,

$$\text{Residual} = \frac{PL^3}{EI}[0.0020 - 2(0.0020) + 0.0020] - \frac{PL}{4}\left(\frac{L}{4}\right)^2\frac{1}{EI}$$

$$= -0.0156\frac{PL^3}{EI}.$$

These residuals are recorded on the right hand side of the vertical lines drawn through the point as in Fig. 17.25(b). The other steps of calculations are shown in Fig. 17.25(b) starting from point 1.

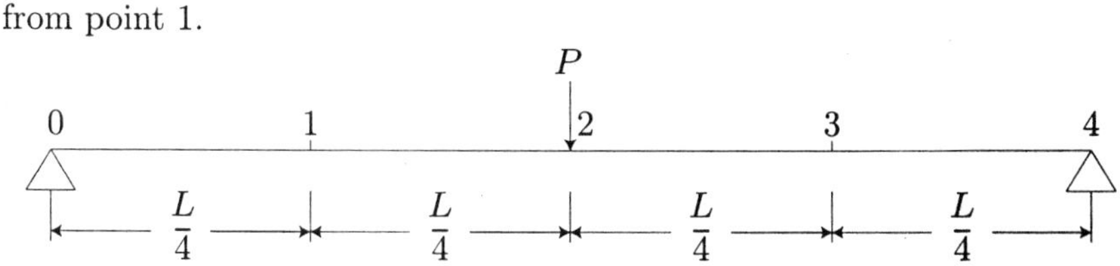

(a) Beam

Fig. 17.25: For Example 17.19 (Contd.)

Final $y/\left(\frac{PL}{EI}\right)^3 \to 0$	0 −0.01545	−0.0230	−0.0152	0
	(18) − 0.000025	0	−0.00025 (18)	
		−0.0005 (17) −0.0005	0	−0.0005 (17)
	(16) −0.0005	0	−0.0010 (16)	
			−0.0005 (15) −0.0005	0
		−0.0010 (14) −0.0010	0	−0.0009 (14)
	(13) −0.0010	+0.0001 13	−0.0020 (13)	
			−0.0010 (12) −0.0010	+0.0001
		−0.0019 (11) −0.0019	0	−0.0019 (11)
	(10) −0.0019	0	−0.0038 (10)	
			−0.0019 (9) −0.0019	0
		−0.0038 (8) −0.0038	0	−0.0038 (8)
	(7) −0.0038	0	−0.0076 (7)	
			−0.0038 (6) −0.0038	0
		−0.0076 (5) −0.0076	0	−0.0076 (5)
	(4) −0.0051	0	−0.0152 (4)	
			−0.0101 (3) −0.0100	0
		−0.0102 (2) −0.0102	−0.001	−0.002 (2)
	(1) −0.0049	0	−0.0205 (1)	
Assumed $y/\left(\frac{PL}{EI}\right)^3 \to 0$	0 +0.0020	−0.0098 +0.0020	−0.0156 +0.0020	−0.0098 0

(b) Deflection by relaxation procedure

Fig. 17.25: For Example 17.19

Example 17.20. *Obtain deflection for the beam shown in Fig. 14.26(a), using the basic relationship* $\frac{d^4y}{dx^4} = \frac{w}{EI}$ *with* $\Delta x = \frac{L}{4}$.

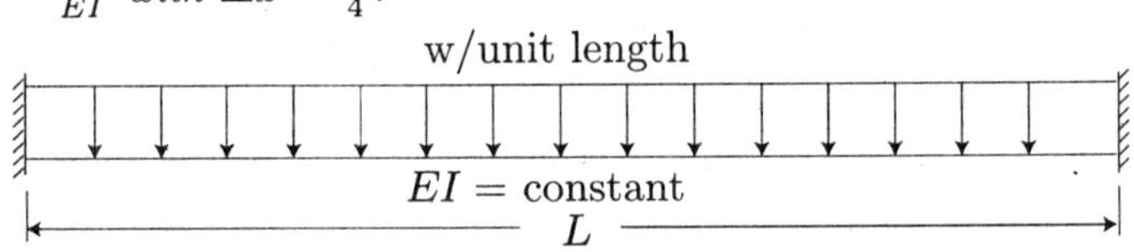

(a) Beam

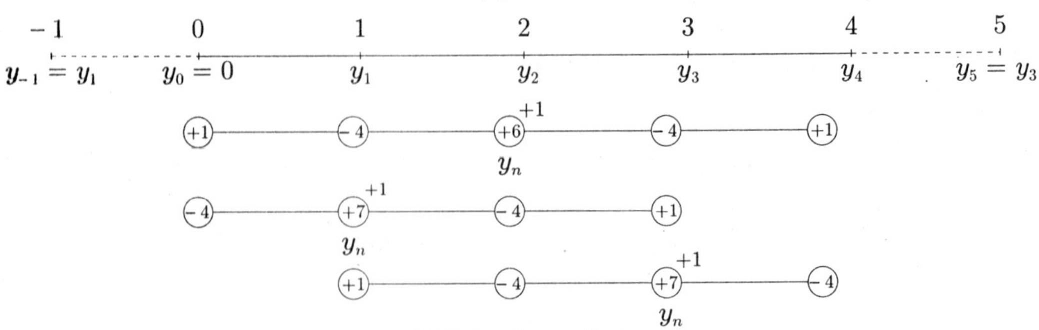

(b) Relaxation pattern

Final $y/\left(\frac{wL^4}{256EI}\right) \rightarrow 0$	+64	+102		+64			0
		+4 ⑩	−1	0		+4	
		0		+6 ⑨	2	0	
	⑧ − 4	−2		+14		−14	
		+26 ⑦	3	−2		−10	
		+38		−20 ⑥	42	2	
	⑤ + 44	−4		+148		−292	
		−312 ④	84	324		−336	
		+24		−180 ③	20	0	
	② + 24	+4		−100		−140	
		−164 ①	16	−4		−164	
Assumed $y/\frac{wL^4}{256Ex}$ 0	0	−100	0	−100	0	−100	0
y_0	y_1		y_2		y_3		y_4

(c) Relaxation procedure

Fig. 17.26: For Example 17.20

Solution : The finite difference relationship is given by

$$y_{n+2} - 4y_{n+1} + 6y_n - 4y_{n-1} + y_{n-2} + \frac{w}{EI}(\Delta x)^4 = 0.$$

The relaxation patterns for the cases with

$y_2 = y_n$, $y_1 = y_n$ and $y_3 = y_n$ are shown in Fig. 17.26(b).

As slope at ends is zero, $y_{-1} = y_1$ and $y_5 = y_3$.

The complete relaxation procedure is shown in Fig. 14.26(c), assuming initial values of all deflections as zero and taking the residuals for y_1, y_2 and y_3 as −100 with

$$\frac{100EI}{w(\Delta x)^4} \cdot \frac{w(\Delta x)^4}{EI}.$$

From boundary conditions $y_0 = y_4 = 0$.

17.8 Block Relaxation Patterns

Block patterns consider the simultaneous application of increments to y values at two or more adjacent points of subdivision. Unit block patterns are easily constructed and Fig. 17.27(a) represents the simplest possible block pattern for the case when the increment $\Delta y = 1$ is applied at only two points of subdivision for the relationship $\frac{d^2y}{dx^2} = -\frac{M}{EI}$. Fig. 17.27(b) shows the next larger block patterns, when the increment is applied at three points and Fig. 17.27(c) illustrates the general form of block patterns where an increment is applied at four number of points. Of course, once the form of the general block pattern in Fig. 17.27(c) has been realised as an obvious generalizations of the simplest patterns in Figs. 17.27(a) and 17.27(b), it becomes apparent that the general pattern is the only one which really need to be recorded as it includes both the first two cases as special cases, as well as in fact the point pattern in case of an increment applied at a single point only.

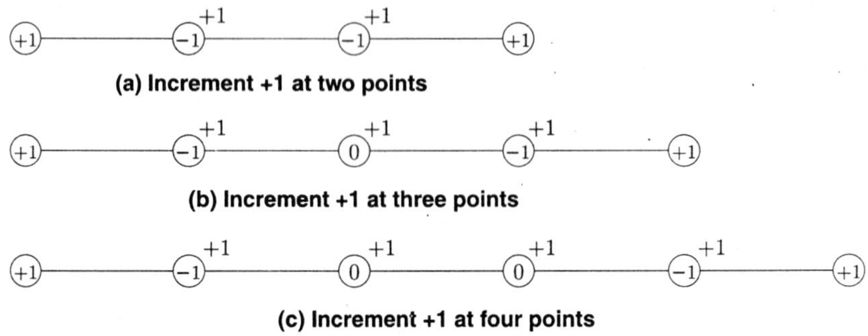

Fig. 17.27: Block relaxation patterns

Such patterns facilitate the relaxation process and their use is known as block relaxation in contrast to the use of single point patterns which is usually referred to as point relaxation. In most problems, the major portion of the liquidation of the residuals is achieved by block relaxation, with point relaxation reserved for making small local adjustments.

Fig. 17.28 illustrates the use of block relaxation in a particular case with $\Delta x = \frac{L}{9}$. The values of y at end points are taken as 0 and 100, and a reasonable first guess for the values of y, at the eight points of subdivision is taken as showing the values of the residuals being calculated and recorded. Liquidation of these residuals with use of only point relaxation even with some over relaxation, would be a lengthly process, mostly as all the residuals are of the same sign and because the point pattern merely redistributes the residual values from one point to the next and does not effect any change in their total, except when an increment is applied at either the first or last point of subdivision. Relaxation can be complete only when all the residual values have been gradually redistributed towards the two end points. The relaxation process must sweep all the initial residuals away from the centre of the range towards and finally onto the end points. This sweeping out of the residuals may be rapidly accomplished by the process of blocking outward. It consists in the application of a series of centrally situated and gradually more and more extensive operations. Such successive operations cover either even points like two, four, six, eight, etc. or odd points like one, three, five, seven, etc. In the case in Fig. 17.28, the number is even, so the first block operation to be applied extends over the central two points and should be of such a magnitude as to leave a zero total residual at these two points. The effect of any unit block operation as in Fig. 17.27 is to change the residuals at the two points at the ends of the block, each by -1, while 1 is added to the adjacent residuals at the two adjacent points on either side of the block. Thus, each unit block operation has the effect of reducing the total residuals at the points where it is applied by 2. The various steps in the block relaxation process are shown in Fig. 17.28. After the steps shown, point relaxation and small local block relaxation will complete the solution.

	311	−9 311	−133 311	68 311	7 311	−6 311	−68 311	133 311	9	4
		302 292	−133 292	68 292	7 292	−6 292	−68 292	133	320	3
			159 138	68 138	7 138	−6 138	−68	425		2
				206 31	7 31	−6	70			1
0	11	10 22	21 33	175 44	38 55	25 66	39 77	287 88	28 100	Assumed, $y/\left(\frac{M}{EI}\right)$

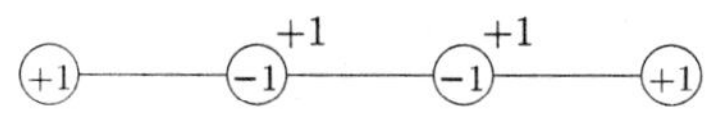

$$\frac{38 + 25}{2} \approx 31$$

1st Relaxation Pattern

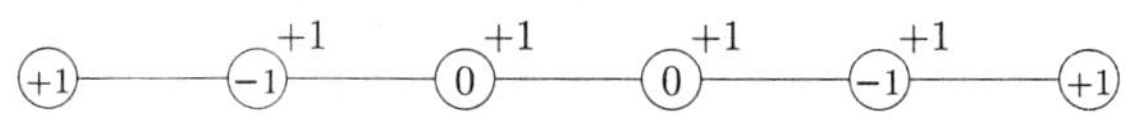

$$\frac{206 + 7 - 6 + 70}{2} \approx 138$$

2nd Relaxation Pattern

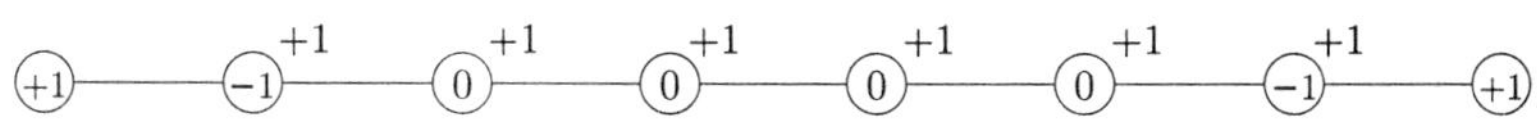

$$\frac{159 + 68 + 7 - 6 - 68 + 425}{2} \approx 292$$

3rd Relaxation Pattern

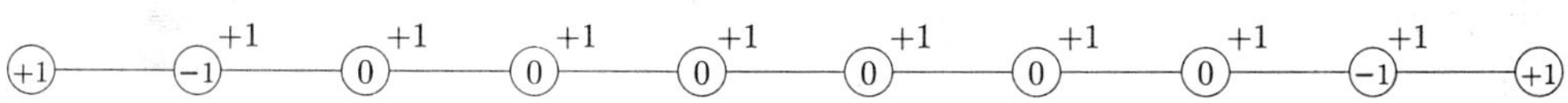

$$\frac{302 - 133 + 68 + 7 - 6 - 68 + 133 + 320}{2} \approx 311$$

4th Relaxation Pattern

$$\sum \text{Residuals} = -9 - 133 + 68 + 7 - 6 - 68 + 133 + 9 \approx 0.$$

Fig. 17.28: Block relaxation for a case with Δ*x* = *L*/9

The liquidation of a set of residuals can be regarded as completed when each individual residual has been made as small as possible. The criteria is necessary, but it is not quite sufficient and with the use of block patterns two additional criteria can be considered. The total of all the residuals should be brought to zero and residuals which are not individually exactly zero, should be well distributed along the range of integration.

17.9 Relaxation Nets

In two dimensions, relaxation solution to a differential equation consists of a finite number of values of the wanted quantity y at a number of points within the region of integration. Whereas in one dimension a range of integration was divided up by points of subdivision into a number of equal intervals, in two dimensions area of integration is subdivided by a uniform network and the required values are calculated at the node points of this network. It is geometrically possible to cover an area by various shapes of uniform meshes, viz. square, equilateral triangular or equilateral hexagonal as illustrated in Fig. 17.29. Such networks are known as Relaxation nets. The hexagonal one is never utilised. The square one is extensively used, although on rare occasions, a triangular net can be used. In all cases, the lines joining the node points are referred to as strings and the length denoted preferably by $\Delta x = \Delta y = h$.

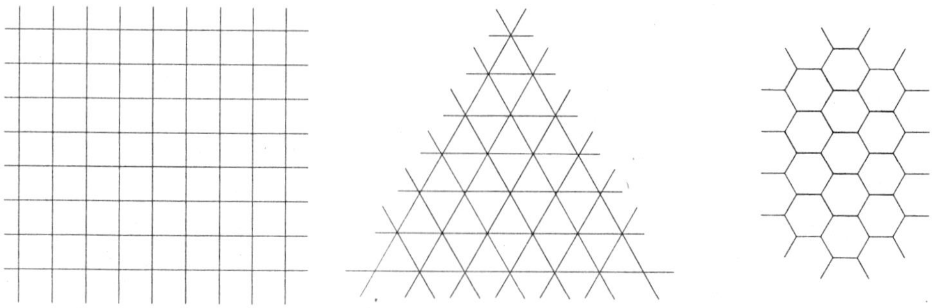

(a) Square mesh **(b) Equilateral triangle mesh** **(c) Equilateral hexagonal mesh**

Fig. 17.29: Various types of meshes

17.10 Use of Relaxation Nets

In two dimensions, a direct extension of Eq. (17.53) is provided by Poisson's equation,

$$\frac{\partial^2 w}{\partial x^2} + \frac{\partial^2 w}{\partial y^2} + w(x.y) = 0, \tag{17.57}$$

of which a special case is the important Laplace's equation,

$$\nabla^2 Z = 0, \tag{17.58}$$

where the operator ∇^2 stands for

$$\frac{\partial^2 w}{\partial x^2} + \frac{\partial^2 w}{\partial y^2}.$$

Laplace's equation governs a large number of physical quantities arising in almost every branch of engineering and applied mathematices.

For solution of any differential equation, it is replaced by a set of finite difference approximations relating to the values of the wanted quantity at the node points of a relaxation net. For Poisson's equation, this can be easily done, since it involves use of approximation of the second derivative as done previously. Fig. 17.30 illustrates the convention used in two dimensions to denote a typical node point 0 of the net and four surrounding node points 1, 2, 3 and 4 as previously.

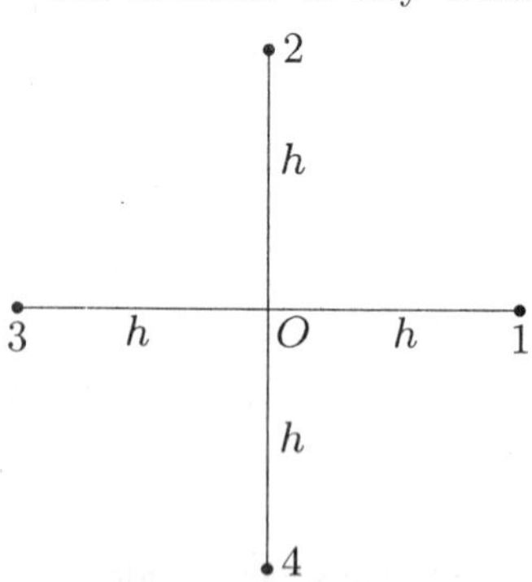

Fig. 17.30: Nodal points for two dimensional relaxation

Considering the finite difference approximation for a second derivative of z, at the point 0, the finite difference approximation for $\nabla^2 w$ at a typical node is obtained as

$$h^2\left(\nabla^2 w\right)_0 = w_1 + w_2 + w_3 + w_4 - 4w_0 = \sum w_1 - 4w_0,$$

from which the approximation to Eq. (17.57) at a typical node point 0 is given by

$$\sum w_1 - 4w_0 + h^2 w_0 = 0. \tag{17.59}$$

There is again an error in terms of h^4 in this equation and the effect of the error diminishes with the mesh size h. Eq. (17.59) must be satisfied by the wanted values of w at every group of five nodes. For any other values of w which in general might be assigned at the nodes of the net, Eq. (17.59) will not usually be true anywhere. Instead a non-zero residual given by

$$F_0 = \sum w_1 - 4w_0 + h^2 w_0 \tag{17.60}$$

will be calculable at each node, the process of relaxation being aimed at eventually reducing all these residuals to zero, by the application of increments Δw, so that automatically the w's acquire their wanted values simultaneously. The basic unit operation by means of which this process may be accomplished is contained in the relaxation pattern of Fig. 17.31, which shows diagramatically the changes produced in the values of the residuals, when a unit increment $\Delta w_0 = 1$ is added to the value of w at a typical mode of the net.

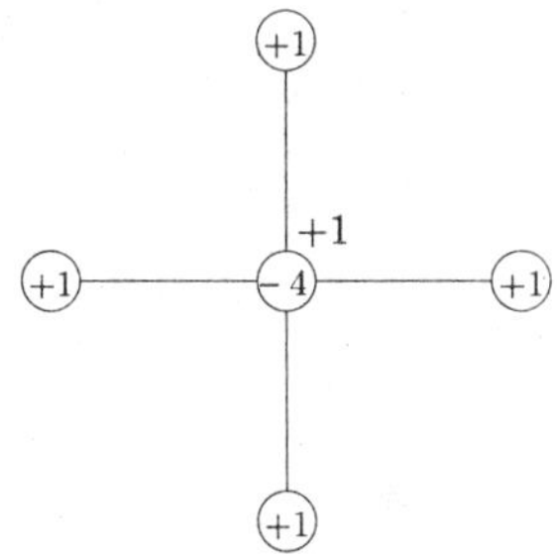

Fig. 17.31: Basic unit two dimensional relaxation

In order that a particular solution of Poisson's equation may be defined, certain conditions in terms of w must be specified to hold on the boundary of the area of integration. A single condition is usually specified at all points of the boundary and the most frequently encountered types of conditions are either that the values of w itself are known at the boundaries or the values of the normal gradiant $\frac{\partial w}{\partial x}$ or $\frac{\partial w}{\partial y}$ are given. Sometimes value of w may be given at some parts of the boundary and $\frac{\partial w}{\partial x}$ or $\frac{\partial w}{\partial y}$ may be given at other parts.

17.11 Block Relaxation and Lines of Symmetry

Use of block pattern is particularly suited to the solution of a Poisson type equation as the patterns may be built up, corresponding to unit increments applied to the w values at small blocks of nodes.

They may be relaxation in a line or for blocks. Fig. 17.32 shows five patterns, viz., for two, three and four point line relaxation operations and block relaxation operations for four and three nodes in a square. Each may be verified by superposing three and four patterns of Fig. 17.32. In both the patterns of Fig. 17.32, a dotted contour has been drawn surrounding the nodes at which the increment $\Delta w = 1$ is applied and which separates these nodes from the remaining nodes of the net. In both cases, it may be seen that the effect on the residuals of the application of the block increment is to transfer one unit of residual down each string of the net which crosses the bounding contour from the node at the end of each string inside the contour to the node at the end outside. The effect of a unit increment added to the w values at the nodes of a block of any size or shape will be similar. In order to construct a

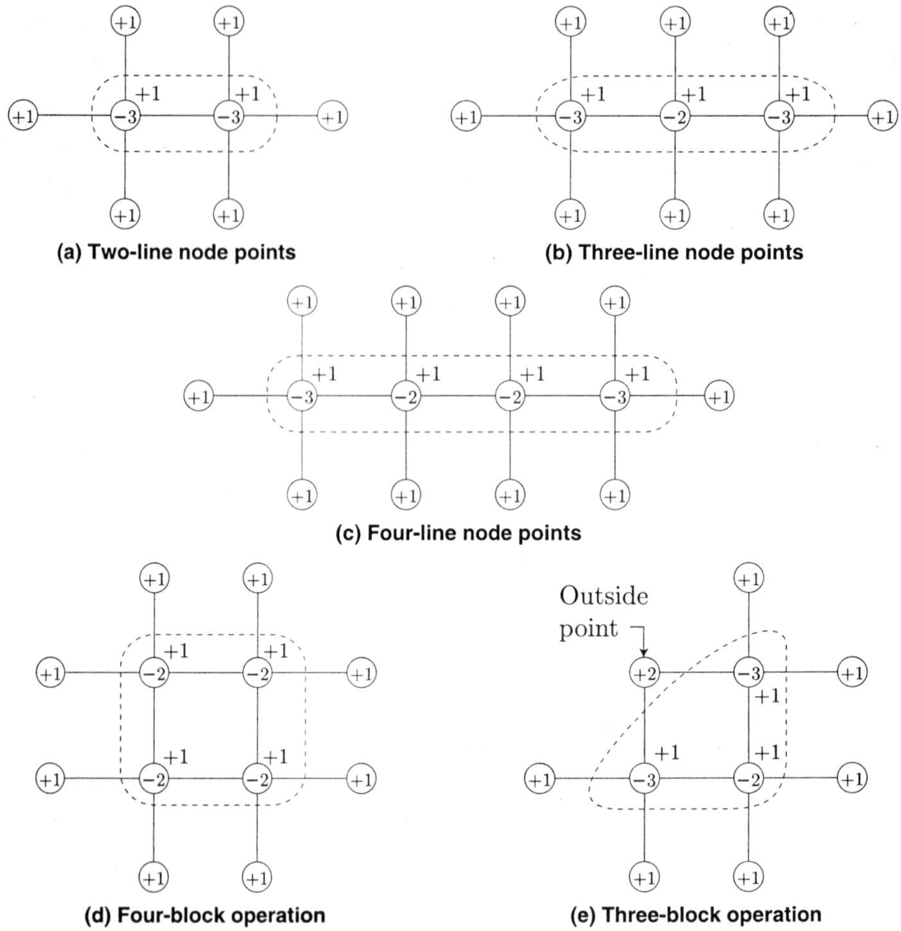

Fig. 17.32: Examples of line and block relaxation operators

unit block patterns it is necessary to imagine a contour to be drawn in, surrounding the nodes which make up the block and then to write changes, to the residuals of +1 at the outer ends of all the strings, which cross that contour and of −1 at their inner ends. Occasionally, two or three strings may have ends at a common node, either inside or outside the contour, the total changes to the residuals at such points being −2, +2, −3, or +3 as the case may be.

There may be nodes in a block from which no strings radiate across the bounding contour. In such a case, the residual at that node is unchanged by the application of the block operation although the increment is added to the w values at all the nodes which make up the block.

To reduce the total residual at any block of nodes to zero by use of single block operation covering the group of nodes, the total of the residuals is added up (say R). Considering N as the numbers of nodes in the net, block increment of magnitude $\frac{R}{N}$ would reduce the total residual by R, so that it becomes zero. In this way, the total residual over the whole of a relaxation diagram or over any part of it, can always be made to become zero in a single step.

The advantage of the line and block relaxation can easily be seen from Fig. 17.33. While the residuals at the points on the boundary of the block are altered, the residuals at points inside the block are not changed. Judicious use of block relaxation can prevent increase of residuals in certain cases, thus saving much time in obtaining a solution.

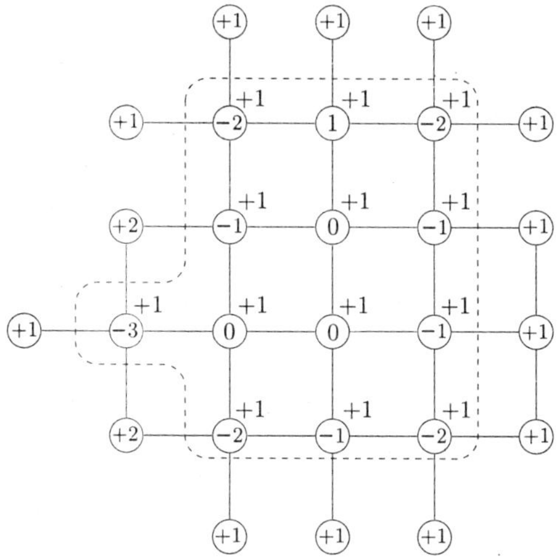

Fig. 17.33: Line and block relaxation

For any case having complete symmetry about some central line, the solution can be done on a diagram representing only one quarter of the complete field. In a similar way, the computation of the solution can be accomplished in a diagram representing only one half of the whole area in case of symmetry about one axes only. Fig. 17.34(a) illustrates a part of a net in the neighbourhood of a vertical central line. At a typical node 0 of a central line, a residual must be calculable, the centre line being not a real boundary of the area of integration, but only an apparent boundary of the working diagram. The usual formula given by Eq. (17.59) cannot be immediately used to find F_0, as on the diagram there, no w is recorded. However w_1 is not a fictitious value. It does exist, but is simply not recorded as due to symmetry it is identical to w_3. Eq. (17.39) can be written in the form,

$$F_0 = w_2 + 2w_3 + w_4 - 4w_0 + h^2 w_0. \qquad (17.61)$$

Corresponding modifications in the relaxation patterns in the neighbourhood of a central line can also be made. Figs. 17.33(b) and 17.33(c) illustrate the patterns for nodes 0 and 3 of the nets shown in Fig. 17.34(a).

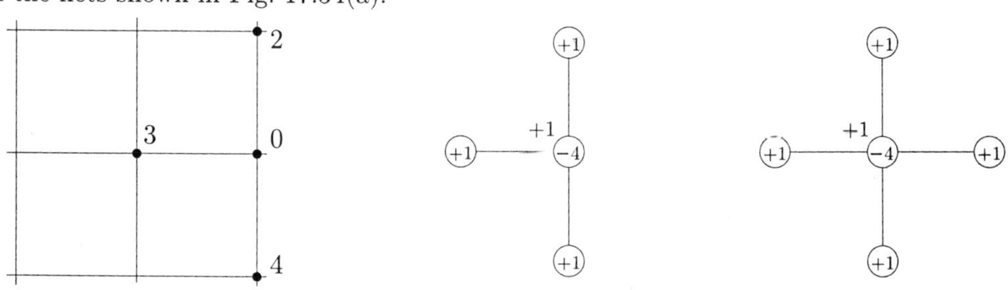

(a) Node on central vertical line **(b) Relaxation pattern for node 0** **(c) Relaxation pattern for node 3**

Fig. 17.34: Symmetry in relaxation operation

17.12 Finite Element Method

In all the previous chapters, the matrix methods are applied to one dimensional structures. For them one member of the structure is considered at a time. The choice of element was obvious – the member was the discrete element considered. However, for two dimensional structures like plates and shells, difficulty is faced regarding the choice of the discretization. In such cases, like conventional one dimensional structures which are considered as an assemblage of structural elements interconnected at their intersections, the continuum will be assumed to consist of imaginary divisions. Thus, the real continuum is divided into a finite number of discrete elements as shown in Fig. 17.35.

The continuum is considered as an assemblage of these 'finite elements'. Finite element analysis presented is the extension of techniques of analysis of one-dimensional structure to two-dimensional structures such as plates and shells, are equally applicable to three-dimensional structures as well.

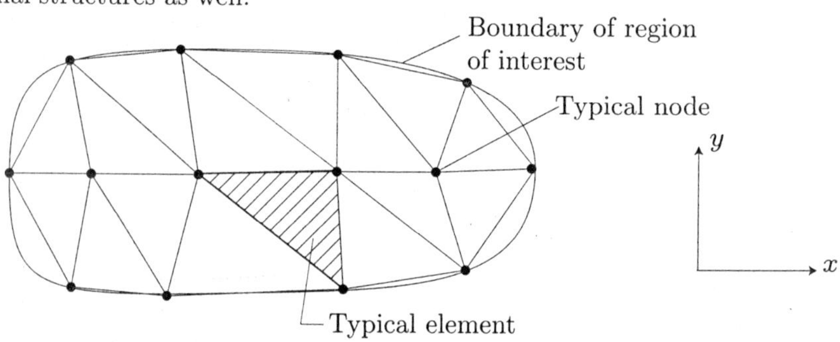

Fig. 17.35: A continuum divided into finite elements

In a continuum which is divided into a mesh, two adjacent elements placed side by side will have a common edge. Though it will be theoretically correct to satisfy the continuity requirements all along the edge, it has been found that this will mean a more complicated analysis. In order to make the analysis simpler, it is assumed that the elements are connected only at nodal points and it is only there that the continuity requirements are to be satisfied. Once the discretization is made, the analysis follows a rather set procedure. The stiffness matrix of the individual element need to be formulated. The forces actually distributed in the real structure are transformed to act at the nodal points. Total structure obtained as an assembly of individual elements can then be analysed on similar lines as that of one-dimensional structure. In the finite element analysis, the continuum is divided into a 'finite' number of elements, having 'finite' dimensions and reducing the continuum having infinite degrees of freedom to 'finite' degrees of unknowns.

The accuracy of solution in general increases with the number of elements taken. However, division of the continuum into more number of elements will result in increased computer cost. A question which will arise in the mind of the analyst is what should be the optimum number of divisions which should give a satisfactory solution. The answer given by the expert to this query is that (a) the required subdivision of the continuum should be based on past experience, and (b) if an analysis is attempted for the first time, then the convergence is to be tested based on varied mesh gradings and then suitable choice is to be made. For regions where higher concentration of stress is expected such as around openings, a finer mesh division is to be made as in Fig. 17.36.

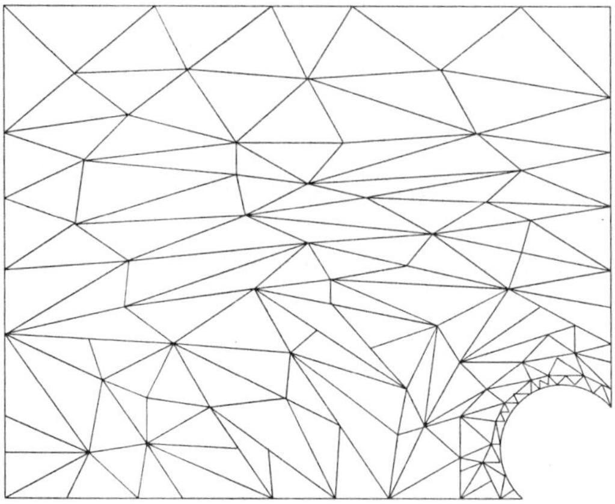

Fig. 17.36: A plate with a hole

Like one-dimensional structures, in finite element method the problem is formulated in two stages :

(a) the element formulation

(b) the system formulation.

The first stage involves the derivation of the element stiffness matrix (or element flexibility matrix) which yields a relationship between nodal point forces and nodal point displacements.

The next stage is the formulation of the stiffness and loads of entire structure. It is seen in matrix analysis of structures in the previous chapters that this stage is independent of the type of element used. Hence, the computer programme for this stage is same for any structure irrespective of the subdivision.

17.13 Advantages and Disadvantages of Finite Element Method

The main advantage of the finite element analysis can be put in one sentence – the physical problems which were so far intractable and complex for any closed-bound solution, can be analysed by this method. The advantages in relation to the complexity of the problem is stated below :

(a) The method can be efficiently applied to cater irregular geometry. It can take care of any type of boundary.

(b) Material anisotropy and inhomogeneity can be catered without much difficulty.

(c) Any type of loading can be handled.

One should not form the idea that the finite element method is the most efficient for the analysis of any type of structural engineering or physical problem. There are many types of problems where some other method of analysis may prove efficient than the finite element method. One of the main disadvantages of this method is the cost involved in the solution of the problem. For vibration and stability problem in many cases, the cost of analysis by finite element method may be prohibitive. It may, therefore, be a luxury to undertake vibration and stability analysis of simpler structures where application of even simpler computer methods such as finite strip or other semi-analytic methods will lead to much economic solution. But those methods will work within their own limitations and will not be as versatile as the finite element method.

17.14 Finite Element Method—The Displacement Approach

Of the various approaches prevalent for the analysis by the finite element method, one based on the assumed displacement pattern within an element is widely used. It is perhaps the simplest approach and has been widely used and successfully applied to a variety of stress analysis problems.

17.15 The Procedure

The following are the steps to be followed for carrying out the finite element analysis :

(a) The continuum is divided into a number of finite elements by imaginary lines or surfaces.

(b) The elements are assumed to be connected at their intersecting points referred to as nodal points. At each node, the unknown displacements are to be prescribed. They are dependent on the problem at hand. For example, in a plane stress problem the unknowns are two linear translations at each node. In the simpler model of the plate bending problem, the unknowns at a node are the deflection and two rotations. In ordinary cases, to every unknown displacement at a node there corresponds a nodal force which can be physically visualised. But it has already been discussed that the finite element method transcends the physical based reasoning. The problem may be identified in such a way that in addition to displacement which occurs at the nodes depending on the physical nature of the problem, certain other quantities such as strain, curvature, etc. may need to be specified in the element which, however, may not have a corresponding physical quantity in the generalised forces. The value of these quantities can be obtained from variational principles only.

(c) Displacement function is the starting point of the mathematical analysis. The displacement function represents the variation of the displacement within the element. This is conveniently represented by a polynomial expression.

(d) The polynomial representing the displacement functions contains arbitrary constraints. They are to be eliminated in such a way that the displacement function is related to its nodal displacements.

(e) Based on the problem at hand, the stresses and strains associated with an element are to be defined. The strains are to be related to the nodal displacements.

(f) Based on theory of elasticity relation, the stresses and strains are to be related. Stresses are finally expressed as a function of nodal displacements.

(g) By applying the appropriate variational principle, the stiffness matrix of the element is derived connecting the nodal forces to the nodal displacements.

(h) The element loading matrix is formulated.

(i) The overall stiffness matrix of the structure is generated as an assemblage of the stiffness matrix of the individual element.

(j) The overall loading matrix is generated as an assemblage of the loading matrix of the individual element.

(k) Boundary conditions of the problem are to be imposed.

(l) Unknown nodal displacements are obtained from the solution of simultaneous equations.

(m) Based on the known values of the nodal displacements, the stresses in any element can be obtained.

The element stiffness matrix formulation involves steps 2 to 6. For some elements such as isoparametric elements, the shape function instead of the displacement function is the starting point. Steps (a) to (m) presented above is typical of any finite element analysis.

Problems

17.1 Using Finite Difference Method, obtain the deflection and slope at points B and C of the beams shown in Fig. P-17.1.

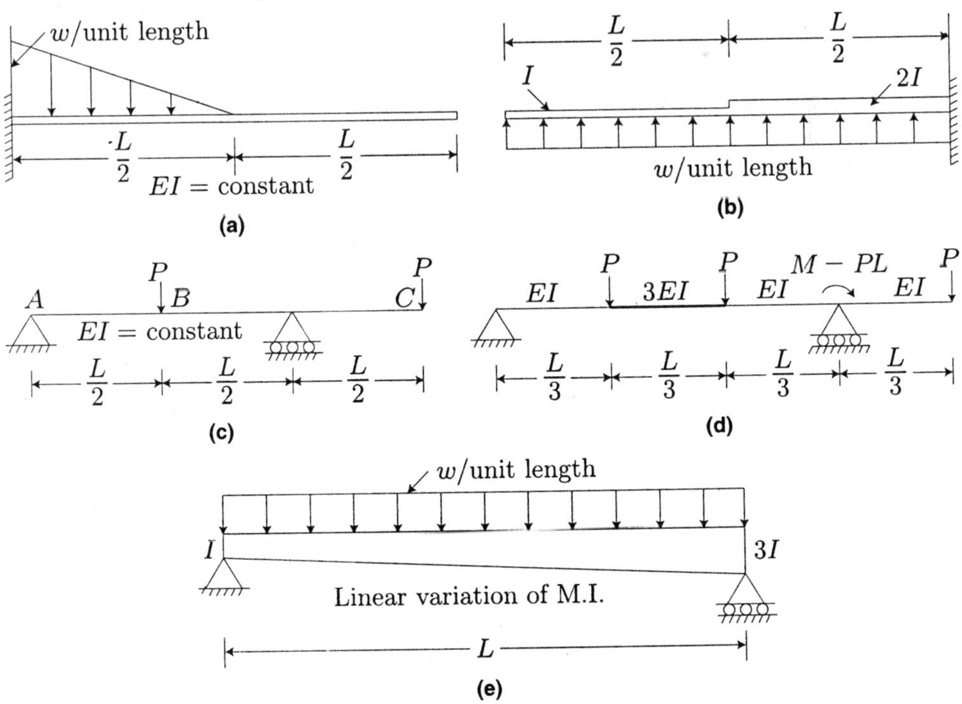

Fig. P-17.1

17.2 Analyse the statically indeterminate beams shown in Fig. P-17.2 using Finite Difference Method.

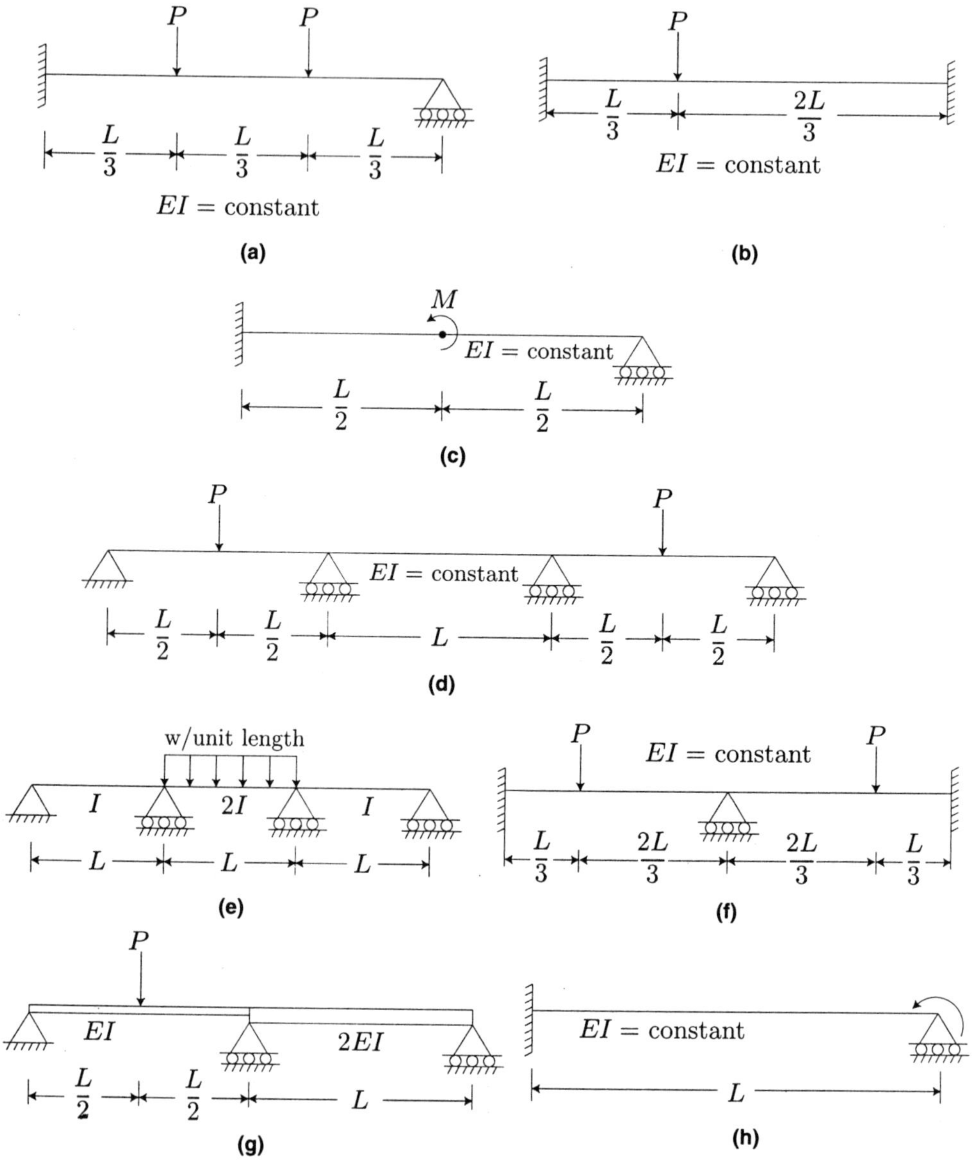

Fig. P-17.2

17.3 The moment of inertia of a column varies as

$$I_x = I_0\left(1 + \frac{2x}{L}\right) \text{ for } 0 \le x \le \frac{L}{2}$$

and $$= I_0\left(3 - \frac{2x}{L}\right) \text{ for } \frac{L}{2} \le x \le L.$$

Assuming $y = C_1 \sin\left(\frac{\pi x}{L}\right) + C_2 \sin\left(\frac{3\pi x}{L}\right)$, find the buckling loads for the column.

17.4 Determine the critical loads for the columns shown in Fig. P-17.3 using finite difference equations, if both ends are hinged.

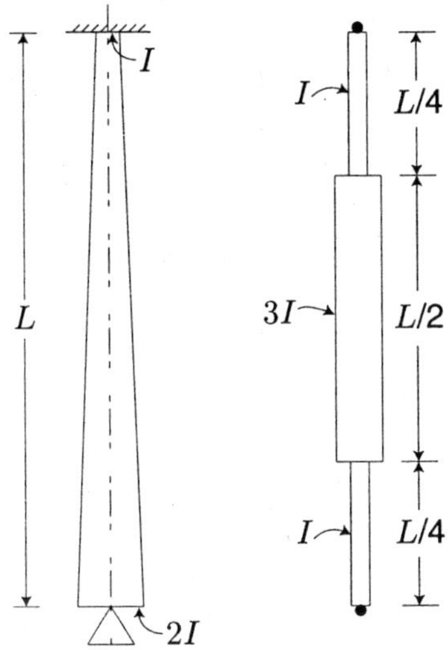

Fig. P-17.3

17.5 Using finite difference method determine the critical load of a column having one end fixed and the other end free. The column is of uniform cross section. Use equation

$$\frac{d^4y}{dx^4} + \frac{P}{EI}\frac{d^2y}{dx^2} = 0.$$

17.6 For a uniform square plate shown in Fig. P-17.4, determine deflection at each nodal point using finite difference method.

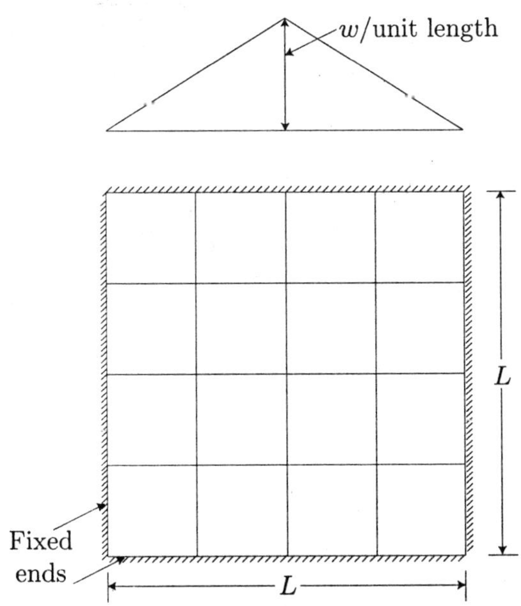

Fig. P-17.4

17.7 For the uniform plate shown in Fig. P-17.5, determine the maximum B.M. at midspan.

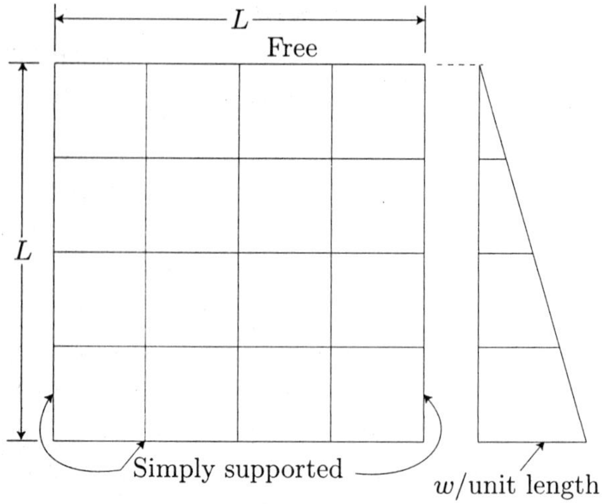

Fig. P-17.5

17.8 Find the solution to the equation $\frac{d^2w}{dx^2} + w = 0$, satisfying end conditions that $w = 0$ at both ends and dividing the range into fixed intervals. Consider residuals at the free points of subdivision as (a) 39, −66, 42 and −20 and (b) 39, 66, 42 and 20 from left to right.

17.9 Solve the equation $\frac{d^2w}{dx^2} + \sqrt{1+x^3} = 0$, over the range of $0 \leq x \leq 8$, when $w = 1$ at $x = 0$, and $\frac{dw}{dy} = -10$, at $x = 8$. Use successive intervals of $h = 4$, (for w_4), $h = 2$ (for w_2) and $h = 1$ (for w_1) and estimate correct value at $h = 8$ (for w_8) from

$$w_8 = \frac{w_1 w_4 - w_2^2}{w_1 + w_4 - 2w_2}.$$

17.10 A uniform beam is 3 m long and is simply supported at ends. The beam carries a u.d.l. of 20 kN/m along its whole length. The M.I. of the beam varies as in Fig. P-17.6.

Determine the maximum deflection of the beam using finite difference equation.

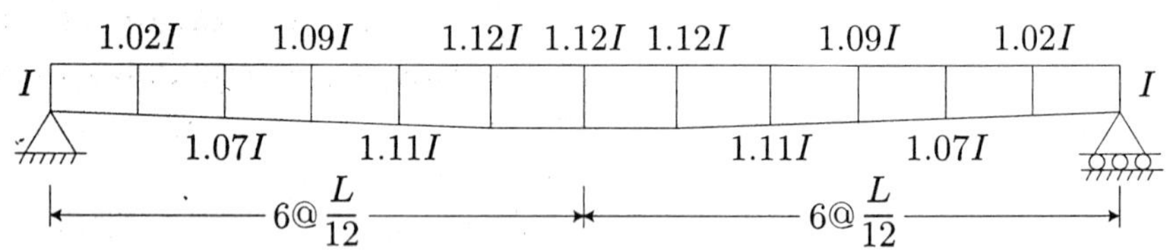

Fig. P-17.6

Chapter 18

Introduction to Structural Dynamics

18.1 Introduction

Civil engineering structures possessing mass and elasticity are capable of relative motion. In a structure usually called a system, the motion repeats itself after a given interval of time and is known as Vibration. Two types of vibration are **Free Vibration** and **Forced Vibration**.

(a) Free Vibration

The vibration persisting in a structure after the force causing the motion has been removed is known as **Free Vibration**. This type of vibration is maintained in the structure purely by the elastic forces present in the structure or gravity force acting on the structure. The former can be considered as **Elastic Type** and the latter as **Gravity Type** of vibration.

The vibration continues after the wind or earthquake has stopped in tall buildings, water tanks, antenna towers, cooling towers, chimneys (stacks), etc.

Free vibration is nothing but the periodic motion observed as the system is displaced from its static equilibrium. Free vibration is also called **Natural Vibration** or **Transient Vibration**.

(b) Forced Vibration

The vibration maintained in a system by a steady periodic force acting on the system is known as **Forced Vibration**.

In case of forced vibration, the system will vibrate at its own natural frequency as well as follow the frequency of the excitation force. In the presence of friction, motion not sustained by the excitation force will gradually die out. As a result, the system will vibrate at the frequency of the excitation force regardless of the initial conditions or the natural frequency of the system. That part of sustained vibration is called **Steady State Vibration** or **Response** of the system.

18.2 Single Degree Freedom System

Many systems can vibrate in more than one pattern and direction. If a system is constrained so that it can vibrate in only one mode or manner, or if only one independent coordinate is required to specify completely the geometric location of the masses of the system in x-y plane, then the system is known as **Single Degree of Freedom System**, commonly referred to as **SDOF** system.

The governing differential equation describing the motion is known as **Equation of Motion**, which for a single degree-of-freedom system can be obtained by the following methods :

(a) Simple Harmonic Motion.

(b) Newton's Law of Motion.

(c) Energy Method.

(d) Rayleigh's Method (A form of energy method).

(e) D'Alembert's Principle.

(a) Simple Harmonic Motion

If x be the rectilinear displacement, $\frac{dx}{dt} = \dot{x}$ is the velocity, by differentiating the same once again, $\frac{d^2x}{dt^2} = \ddot{x}$, which is the acceleration.

A particle in rectilinear motion is said to execute Simple Harmonic Motion, if its acceleration is always proportional to the distance of the particle from a fixed point on the path and is directed toward the fixed point, and can be represented by

$$\ddot{x} \propto (-x) \tag{18.1}$$

The fixed point is the equilibrium position. The minus (−) sign indicates that the particle is directed towards the fixed point which is opposite to the direction of motion. If the constant of proportionality be p^3, an unknown parameter, Eq. (18.1) can be written as,

$$\ddot{x} = -p^2 x,$$

hence $$\ddot{x} + p^2 x = 0, \tag{18.2}$$

which is known as Equation of Motion. This is a second order linear differential equation with constant coefficient.

(b) Newton's Law of Motion

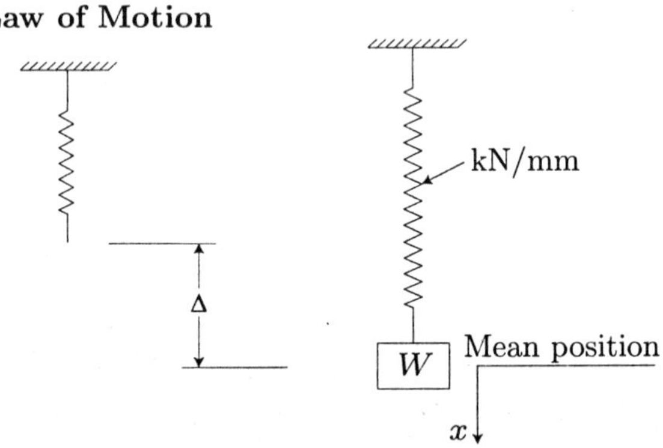

(a) Spring (b) Spring with load

Fig. 18.1: Massless spring

A massless spring shown in Fig. 18.1(a), is subjected to a load $W = mg$, where m is mass. Under the weight of the spring, the spring elongates to a distance of Δ. The spring factor designated by k is given by

$$k = \frac{W}{\Delta} \quad \text{and} \quad W = k\Delta.$$

The position occupied by the load after elongation having been taken placed is known as the **"Equilibrium Position"**, or **"Stable Position"** or **"Mean Position"** or **"Datum Position"**, or **"Reference Position"** or **"Neutral Position"**.

As the load is pulled a little down by some means and left out, it will be able to move up and down only. No lateral motion is possible in this system. Then only one co-ordinate will describe the motion of the load at any time. In other words, in the system, only one motion, i.e., up and down movement is possible. Such a system is known as spring mass system. The system is said to execute vibration or oscillation, in single degree of freedom.

From Newton's law of motion, the equation of motion can be derived. The laws states that $F = ma$.

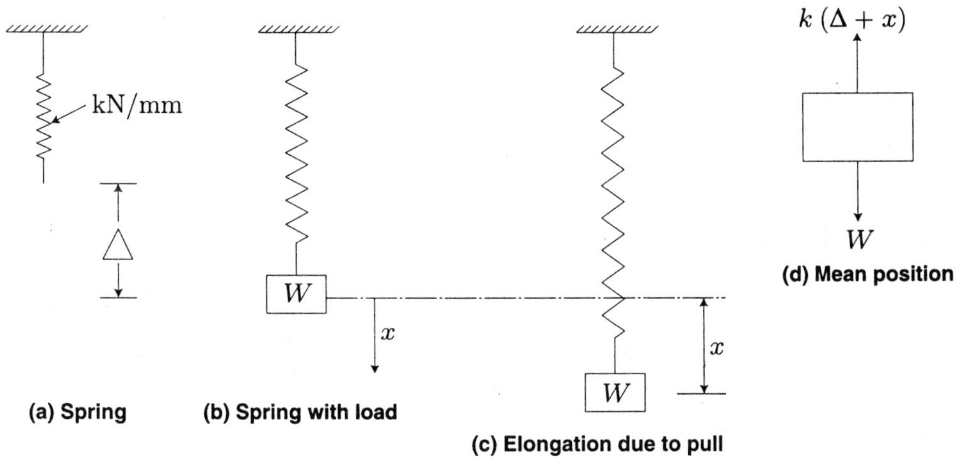

Fig. 18.2: Spring with load

Consider a spring of stiffness kN/mm as shown in Fig. 18.2(a). Under the load W, the spring elongates to a distance Δ and occupies the mean position as shown in Fig. 18.2(b). The +ve direction of x is shown in Fig. 18.2(b). From the mean position, the load W is pulled down a little and the pulling force is removed. The load W will continue to oscillate down which is known as Free Vibration. At any instant during vibration, the position of load ($W = mg$) is shown in Fig. 18.2(c). At that time, the inertia force acting in the positive direction of x is equal to $m\ddot{x}$. The other forces acting on the particle are shown in Fig. 18.2(d).

Force in the positive direction of $x = +W - k(\Delta + x)$. If $W = k\Delta$,

Force in x direction $= +W - W - kx = -kx$. According to Newton's law,

$$m\ddot{x} = -kx \quad \text{and} \quad m\ddot{x} + kx = 0. \tag{18.2a}$$

With $p^2 = \dfrac{k}{m}$, $\ddot{x} + p^2 x = 0$. (18.2)

The parameter 'p' is known as the **Natural Circular Frequency** or **Angular velocity** or **Vibrating Frequency** or sometimes simply as **Frequency**. It is measured in radians per second.

(c) Energy Method

For a conservative system, the total energy of the system is unchanged at all time. Since there is no friction or damping force, the total energy of the system consists of two parts, viz., **Potential Energy** (P.E.) and **Kinetic Energy** (K.E.) and P.E.+K.E. = constant = Total Energy (T.E.). It follows that time rate of change of energy must be zero,

whence $\frac{d}{dt}$ (K.E. + P.E.) = 0.

This resulting equation is the equation of motion of the system under consideration.

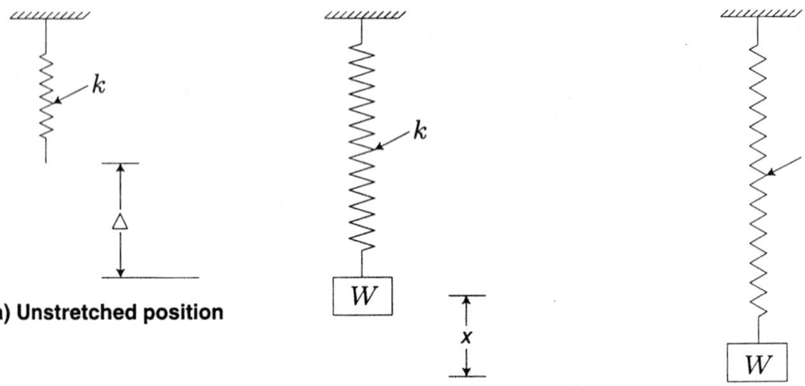

(a) Unstretched position (b) Equilibrium position (c) Position at any moment

Fig. 18.3

The equilibrium position is shown in Fig. 18.3(b). Due to external force, the motion of the weight is effected. The force causing the motion purely in vertical direction is removed. The resulting motion is known as Free Vibration. At any particular time t the position of the weight is shown in Fig. 18.3(a). The particle (weight) possesses two types of energies, viz., potential energy and kinetic energy.

$$\text{Kinetic Energy} = \frac{1}{2}mV^2 = \frac{1}{2}m\dot{x}^2 = \frac{1}{2}\frac{W}{g}\dot{x}^2.$$

Potential energy consists of two types of energies, one due to position and the other due to deformation of the spring.

The amount of strain energy stored in the system due to deformation from the position of equilibrium to the position of particle (weight) at any instant of time t is obtained as shown below :

$$\text{Strain Energy} = \text{Average Force (distance moved)}$$

$$= \left\{\frac{k[\Delta + x]}{2}\right\}[\Delta + x] - \frac{k\Delta}{2}(\Delta) = \frac{1}{2}kx^2 + k\Delta x.$$

Loss in potential energy $= -Wx$.

As $W = k\Delta$,

$$\text{Potential Energy} = \frac{1}{2}kx^2 + k\Delta x - Wx$$

$$= \frac{1}{2}kx^2 + Wx - Wx = \frac{1}{2}kx^2.$$

Total Energy = Potential Energy + Kinetic Energy

$$= \frac{1}{2}kx^2 + \frac{1}{2}mx^2 = \text{constant.}$$

Also, $\frac{d}{dt}[T.E.] = -kx\frac{dx}{dt} + m\dot{x}\frac{d^2x}{dt^2} = 0$

$$= kx + m\ddot{x} = 0.$$

With $p^2 = \frac{k}{m}$,

$\ddot{x} + p^2x = 0$, same as before.

(d) Rayleigh's Method

This method is based on the concept of the total energy. It is an inverse procedure. It is applicable only to single degree of freedom. At extreme positions,

$(\text{K.E.})_{\max} = (\text{P.E.})_{\max}$ = Total energy of the system. This is known as **Rayleigh's** method. The principle used in this method is to obtain the natural circular frequency. From this the equations of motion can be formulated,

(e) D'Alembert's Principle

The solution of problems involving the relationship between forces and accelerations was enunciated by **D'Alembert**. The principle enables the solution of a dynamics problem by the methods of statics. This is a remarkable simplification of the problem. Newton's law states that

$$F = ma.$$

If the above basic equation of motion of a particle is rewritten as $F - ma = 0$, it takes the form of an equation of motion of force equilibrium where the sum of a number of force terms equals zero. Thus, if a fictitious force equal to ma were applied to the accelerating particle in the direction opposite to the acceleration, the particle could then be considered to be in equilibrium under the action of the real force F and the fictitious force ma. The fictitious force ma is often called then **Force of Inertia**, and the artificial state of equilibrium is known as **Dynamic Equilibrium**, which is only a concept rather than a reality. The resultant of the actual forces is called the **Effective Force**, and the equal and opposite fictitious force of inertia is known as the **Reversed Effective Force**. The apparent transformation of a problem in dynamics to one in statics is known as **D'Alembert's Principle.**

In extending D'Alembert's principle to rigid-body motion it is merely necessary to add a fictitious force of inertia ma through the centre of mass in the direction opposite to the acceleration a of the mass center. This addition to the free body of diagram together with the real forces constitutes a system which obeys the force equilibrium equation.

Considering a body of mass m attached to a spring whose stiffness is k, the body is at rest, the spring being unstretched as in Fig. 18.4(a). The body is pulled by a force P to a distance x_0 which is known as initial amplitude. At time $t = 0$, the force P is removed. The body is in a state of **Free Vibration**.

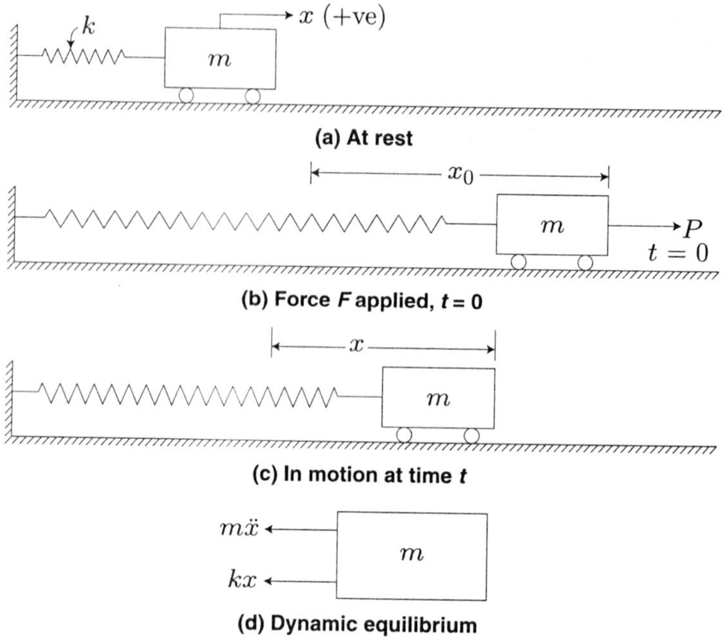

Fig. 18.4: Mass *m* attached to a spring

At any time t the body is away from the original position on the right side, the distance being x. The acceleration of the particle will be in the right direction. The only external force acting is the spring force equal to kx. The spring force will be acting towards left.

To bring the body to a state of artificial equilibrium known as Dynamic Equilibrium, all that is needed is to add the force of inertia mx in the direction opposite to the direction of motion in accordance with D'Alembert's principle.

The dynamic equilibrium of the body is as shown in Fig. 18.4(d).

Applying the Equilibrium condition $\sum F_x = 0$, to the body,

$$-m\ddot{x} = kx = 0.$$

The sign convention is that forces in the x direction (+ve direction) are positive. Since the forces are acting in the negative direction (−) sign is attached to each force,

$$m\ddot{x} + kx = 0. \tag{18.2a}$$

Taking $p^2 = \dfrac{k}{m}$,

$$\ddot{x} + p^2 x = 0. \tag{18.2}$$

Eq. (18.2) is the basic differential equation of motion obtained earlier by other methods.

A solution for the above problem is $x = C \cos pt$. This may be verified by differentiating twice and substituting in Eq. (18.2).

The time t is measured from the instant of release from rest at $x = x_0$. To evaluate the constant C in equation $x = C \cos pt$, the initial condition is $t = 0$, $x = x_0$. Substituting the same in $x = C \cos pt$,

$$C = x_0.$$

The solution is $x = x_0 \cos pt$. (18.3)

In Eq. (18.2), the angle pt changes by 2π as time changes by τ. This condition may be expressed as

$$p(t+\tau) - pt = 2\pi, \quad \text{i.e.,} \quad p\tau = 2\pi \quad \text{and} \quad \tau = \frac{2\pi}{p}.$$

τ is defined as the Natural Period of vibration. In other words, τ is defined as the time required to go through one cycle of free vibration. Eq. (18.3) is depicted in Fig. 18.5.

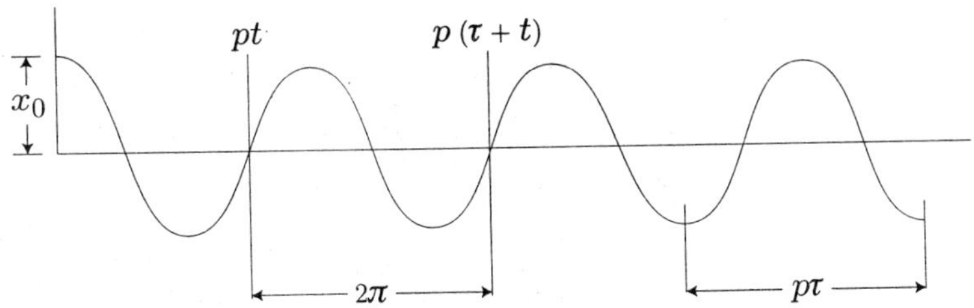

Fig. 18.5: Plot of Equation 18.3

It is obvious that

$$x_0 \cos pt = x_0 \cos(pt + p\tau).$$

From the above it follows that

$$p\tau = 2\pi \text{ and } \tau = \frac{2\pi}{p}. \qquad (18.4)$$

Also $p = \sqrt{\dfrac{k}{m}}$ and $k = \dfrac{W}{\delta_{st}}$.

Substituting the above in Eq. (18.3) and as $m = \dfrac{W}{g}$, $\tau = 2\pi\sqrt{\dfrac{\delta_{st}}{g}}$.

The time required for one cycle is τ seconds. In one second the number of cycles $= \frac{1}{\tau}$. This is designated as f and is called as Natural Frequency, which is given by $f = \frac{1}{\tau}$ cycles per second.

This is known as Hertz, i.e., Cps.

The term natural frequency is used to denote both the parameters 'p' and 'f' without distinction.

The parameter p is of considerable value around which the subject structural dynamics is built whereas the term 'f' is only of academic interest which will be useful at times.

Also $\tau = \dfrac{1}{f}$. (18.5)

Substituting, $\tau = \dfrac{1}{f} = \dfrac{2\pi}{p}$

and $p = 2\pi f$. (18.6)

Another form of solution is $x = A \sin pt$.

Differentiating the same twice and substituting in Eq. (18.2), it can be verified that this equation is a solution. However, the constant A is to be evaluated.

If the above vibration is described by counting time from the instant the body passes the origin with the velocity x_0, then the initial conditions are

$$x = 0 \quad \text{when} \quad t = 0.$$

Also when $t = 0$, $\dfrac{dx}{dt} = x_0 = Ap \cos pt.$ (18.7)

Now applying the initial conditions to Eq. (18.7),

$$\dot{x}_0 = Ap \quad \text{as } t = 0 \quad \text{and} \quad A = \frac{\dot{x}_0}{p}.$$

The equation of motion becomes

$$x = \frac{\dot{x}_0}{p} \sin pt, \tag{18.8}$$

which, of course, has the same period and frequency as the cosine expression.

Now Eqs. (18.3) and (18.8) are solutions for the differential equation. The general solution is

$$x = x_0 \cos pt + \frac{\dot{x}_0}{p} \sin pt. \tag{18.9}$$

Another solution is

$$x = a \cos(pt - \alpha).$$

By expanding,

$$x = a \cos pt \cos \alpha + a \sin pt \sin \alpha \tag{18.10}$$

and $x_0 = a \cos \alpha.$ (18.10a)

$$\frac{\dot{x}_0}{p} = a \sin \alpha. \tag{18.10b}$$

$$\because \quad \frac{a \sin \alpha}{a \cos \alpha} = \frac{\dot{x}_0}{p}\left(\frac{1}{x_0}\right)$$

$$\tan \alpha = \frac{\dot{x}_0}{p x_0}.$$

As $a^2 \sin^2 \alpha + \alpha^2 \cos^2 \alpha = x_0^2 + \dfrac{\dot{x}_0^2}{p^2}$, and $a = \sqrt{x_0^2 + \left(\dfrac{\dot{x}_0}{p}\right)}$.

Example 18.1. *An oscillating system with a natural frequency of 4.00 Hz starts with an initial amplitude of $x_0 = 15$ mm and an initial velocity of $\dot{x}_0 = 150$ mm per sec. Calculate all the vibratory parameters involved, the time taken to reach the first peak.*

Solution : (a) Natural frequency $p = 2\pi f = 2(\pi)(4) = 25.133$ rad/ sec

(b) Natural period $\tau = \dfrac{2\pi}{p} = \dfrac{1}{f} = \dfrac{1}{4} = 0.25$ sec

(c) Amplitude $a = \sqrt{x_0^2 + \left(\dfrac{\dot{x}_0}{p}\right)^2} = \sqrt{15^2 + \left(\dfrac{150}{25.133}\right)^2} = 16.144$ mm

(d) Velocity $|x| = ap = 16.144(25.133) = 405.747$ mm/ sec

(e) Acceleration $= |\ddot{x}| = ap^2 = 16.144(25.143^2) = 10197.64$ mm/ sec^2

(f) Phase angle $\alpha = \tan^{-1}\dfrac{\dot{x}_0}{px_0} = \tan^{-1}\dfrac{15}{(25.133)(15)}$ and $\alpha = 21.70°$

(g) First peak occurs when $\cos(pt - \alpha) = 1$ and $t = \dfrac{\alpha}{p} = 0.0150693$

(h) Displacement $x = a\cos(pt - \alpha) = 16.144\cos(25.133t - 0.3987)$

Example 18.2. *Show that the gravity force does not have any effect on the equation of motion if the spring is vertical.*

Solution : The various forces maintaining dynamic equilibrium of the body are shown in Fig. 18.6. From the equation of equilibrium in the direction of motion

$$-\frac{W}{g}\ddot{x} - k(x + \Delta) + W = 0,$$

where W is the gravity force acting downwards.

$$\Delta = \frac{W}{k} \text{ and } W = k\Delta$$

whence $-\dfrac{W}{g}\ddot{x} - kx - W + W = 0$

and $\dfrac{W}{g}\ddot{x} + kx = 0,$

i.e., $m\ddot{x} + kx = 0$, where $m = \dfrac{W}{g}$

$$\ddot{x} + p^2 x = 0.$$

Thus the above equation is free of the gravity force $W = mg$.

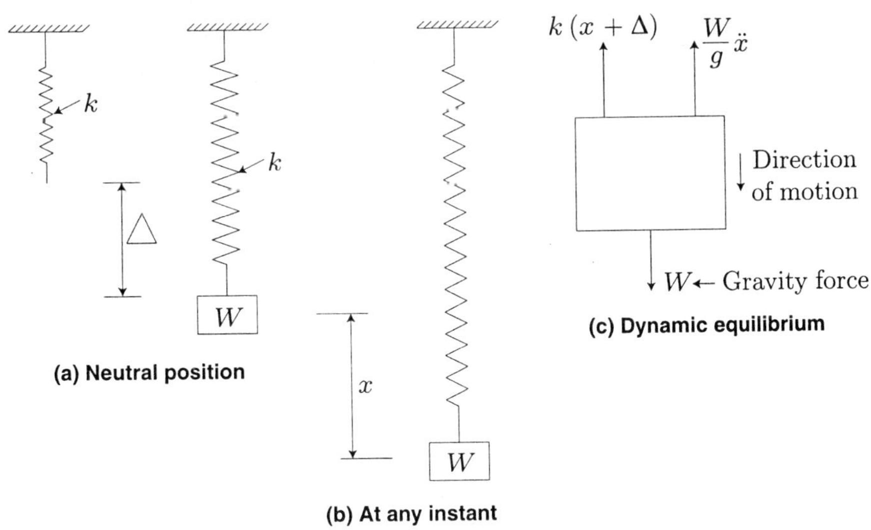

Fig. 18.6: For Example 18.2

The inference is that the gravity force need not be considered whenever the spring is vertical.

18.3 Types of Vibrations

Whatever may be the cause of vibration, a structure vibrates in one of the following four ways of deformation or a combination thereof. The types of vibrations are shown in Fig. (18.7).

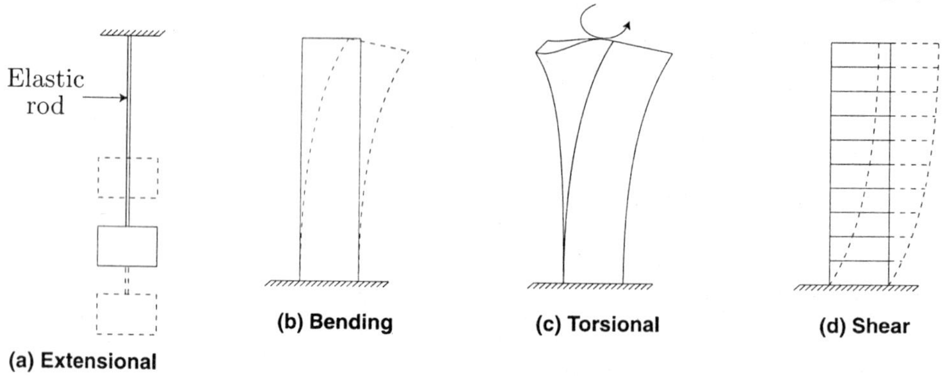

Fig. 18.7: Types of vibrations

(a) **Extensional Vibration.** In Case of machine foundation, the foundation is assumed to be exceedingly rigid. The rigid block resting on the soil vibrates up and down.

Another example is an elastic rod subjected to a load which vibrates up and down. The stiffness k associated with this type of vibration is known as "**Extensional Stiffness**".

(b) **Bending Vibration.** Examples are a tall water tank, a chimney and a tall building. Bending action is predominant in these structures during vibration. The stiffness connected with this type of vibration is known as **Bending or Flexural Stiffness.**

(c) **Torsional Vibration.** Twisting of a building occur during earthquake. The corresponding stiffness is known as **Torsional Stiffness.**

(d) **Shearing Vibration.** Example is horizontal vibration of tall building. In this type of building, the columns are connected by infinitely rigid beam at different levels. If the beams connecting the columns at different levels are not very stiff, this type of vibration is not possible.

Example 18.3. *If a vertical cable 2.54 m long has a cross-sectional area 322.58 mm^2 supports a weight of 45.36 kN, what is the natural period or natural frequency of the system?* $E = 2.1(10)5\ N/mm^2$; $g = 9810\ mm/\sec^2$.

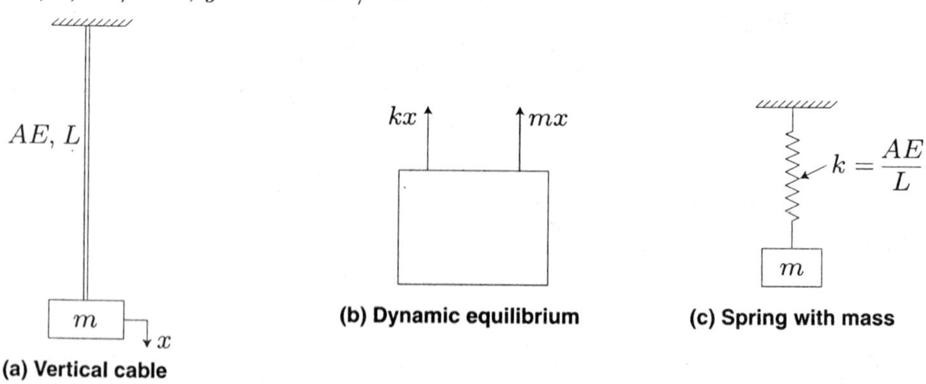

Fig. 18.8: For Example 18.3

Solution : The weight is attached to a cable whose extensional stiffness $k = \dfrac{P}{\Delta} = \dfrac{AE}{L}$.

The cable is assumed to be weightless and from the dynamic equilibrium,

$$-m\ddot{x} - kx = 0 \quad \text{and} \quad m\ddot{x} + kx = 0.$$

$$k = \frac{AE}{L}\frac{322.58(2.1)(10)^5}{2540} = 26670 \text{ N/mm}$$

$$p = \sqrt{\frac{k}{m}} \text{ and } \tau = \frac{2\pi}{p}$$

$$\tau = 2\pi\sqrt{\frac{W}{gk}} = 2\pi\sqrt{\frac{45360}{26670(9810)}} = 0.08273 \text{ sec}$$

$$f = \text{natural frequency} = \frac{1}{\tau} = \frac{1}{0.08273} = 12.0875 \text{ cps.}$$

Example 18.4. *A cantilever 2540 mm long and fixed at the base supports a concentrated weight of 4.536 kN. The mass of the vertical pole will be neglected. Find the natural period and natural frequency. Moment of inertia is 11712752 mm*^{4}, g = 9810 *mm/sec*^{2}, E = 2.1(10)^5 *N/mm*^{2}.

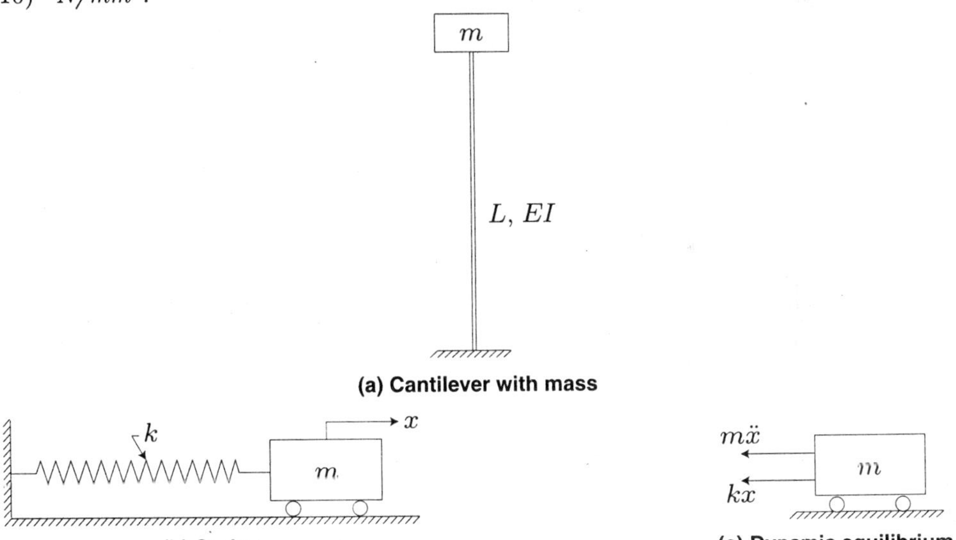

Fig. 18.9: For Example 18.4

Solution : The problem belongs to the type of single degree of freedom with single weight attached to the structure. To find out the flexural stiffness k, the weight is replaced by a force P in the direction of motion Δ.

$$\text{Flexural stiffness} = k = \frac{P}{\Delta} = \frac{P}{\frac{PL^3}{3EI}} = \frac{3EI}{L^3}.$$

From Fig. 18.9(c), $-m\ddot{x} - kx = 0$ and $m\ddot{x} + kx = 0$.

$$k = \frac{3EI}{l^3} = \frac{3(2.1)(10)^5(11712752)}{(2540)^3} = 450.296 \text{ N/mm}$$

$$\tau = \frac{2\pi}{p} = 2\pi\sqrt{\frac{m}{k}} = 2\pi\sqrt{\frac{(4536)}{(9810)(450.296)}} = 0.2014\,\text{sec.}$$

$$\text{Natural frequency} = \frac{1}{\tau} = \frac{1}{0.2014} = 4.965 \text{ cps.}$$

Example 18.5. *A simply supported beam of span L with flexural rigidity EI is carrying a weight W at the centre of the span. Compute the natural period and natural frequency.*

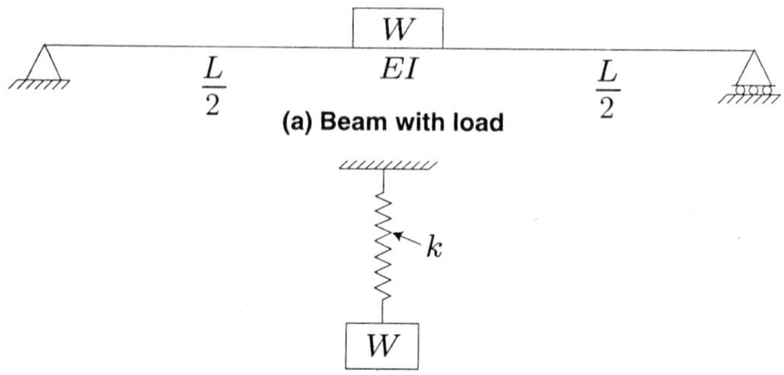

(a) Beam with load

(b) Spring mass system

Fig. 18.10: For Example 18.5

Solution : The equivalent spring mass system is shown in Fig. 18.10. Since it is a single degree of freedom with single weight attached, the flexural stiffness k is given by

$$k = \frac{P}{\Delta} = \frac{P}{\frac{PL^3}{48EI}} = \frac{48EI}{L^3}.$$

$$\text{Natural circular frequency, } p = \sqrt{\frac{k}{m}} = \sqrt{\frac{48EI}{L^3}\left(\frac{g}{W}\right)}$$

$$\text{Natural period } \tau = \frac{2\pi}{p} = 2\pi\sqrt{\frac{WL^3}{48EIg}}$$

$$\text{Natural frequency} = \frac{1}{\tau} \text{ cps.}$$

Example 18.6. *A mass suspended from a massless spring, vibrates vertically up and down between two positions 20 mm and 30 mm above the ground. During each second it reaches the top position 30 mm above ground twenty times. What are τ, f, p and x_o? If the weight of the spring is 50 N, find the stiffness of the spring when $g = 9800$ mm/sec^2.*

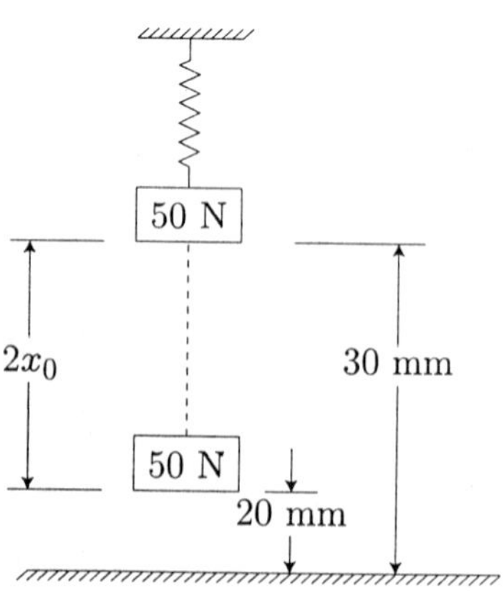

Fig. 18.11: For Example 18.6

Solution : It is seen, $2x_0 = 10$ mm, $x_0 = 5$ mm.

Number of cycles performed in one second = 20

and natural period, $\tau = \dfrac{1}{20}$.

As $\quad \tau = \dfrac{2\pi}{p} \quad$ and $\quad p = \dfrac{2\pi}{\tau} = \dfrac{2\pi}{0.05} = 40\pi.$

Angular velocity $p = 40\pi$ rad/ sec,

$$p = \sqrt{\frac{k}{m}} = \sqrt{\frac{k(9800)}{50}} \quad \text{and} \quad k = \frac{(40\pi)^2(50)}{9800} = 80.57 \text{ N/mm}.$$

18.4 Translational Stiffness k Using Moment Distribution and Reduced Stiffness Matrix Approach

In Examples 18.4 and 18.5, k was found using the definition of strength of materials. It was the flexural stiffness. The stiffness is related to the translational degree of freedom.

There are two methods available for determining the same. Even though for simple problem the procedure will appear to be tedious, the same will prove to be very elegant and neat procedure when applied to complicated problems, especially indeterminate structures. The two methods are :

(a) Moment Distribution Procedure and

(b) Reduced Stiffness Approach.

Example 18.7. *Find the translational stiffness for the cantilever by the moment distribution method and Reduced Stiffness Approach. The mass is attached at the free end.*

Solution :

(a) Moment Distribution Procedure

In the cantilever, the translation occurs at the mass point. Remove the mass temporarily and arrest the translation as shown in Fig. 18.12(b). The distribution factors for this structure is computed. A translation equal to unity at the arrested point is made as shown in Fig. 18.12(c). Due to this moments (anticlockwise) of magnitude $\frac{6EI}{L^2}$ are induced. Only the coefficient 6 is carried out from Fig. 18.12(c).

The moment distribution procedure is shown in Fig. 18.12(d). The FM (Final moment) is shown as −3. It is actually $-\frac{3EI}{L^2}$. From this it follows that the force induced on the right side in the downward direction which is the direction of the translation is nothing but Translational Stiffness $= \frac{3EI}{L^3}$.

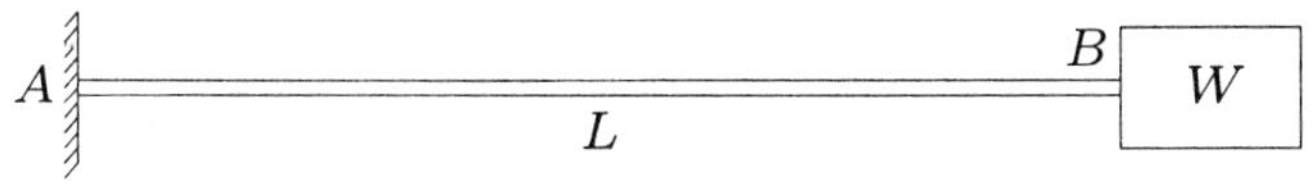

(a) Cantilever with mass at end

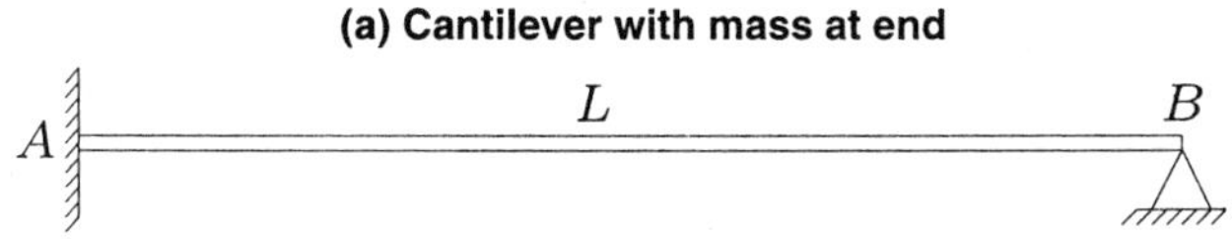

(b) Support at *B*

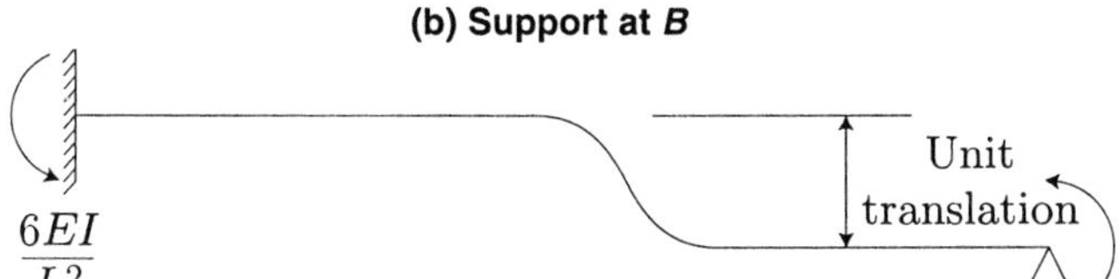

(c) Unit translation

Fig. 18.12: Moment distribution procedure (Contd.)

	0	1
FEM	−6	−6
DM	0	+6
CO	+3	0
DM	0	0
FM	−3	0

(d) Moment distribution

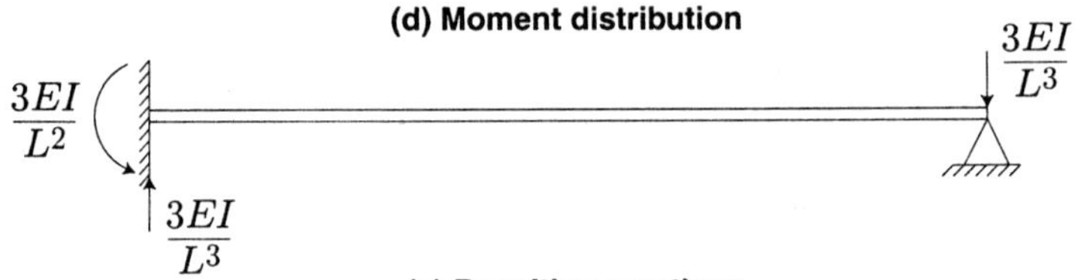

(e) Resulting reactions

Fig. 18.12: Moment distribution procedure

(b) Reduced Stiffness Matrix Approach

From the cantilever, the mass is removed temporarily. Ignoring the axial elongation, the total degree of freedom is equal to two; one translation and one rotation. They are indicated by means of co-ordinates in Fig. 18.13(b). For such co-ordinates either by element approach or physical approach the total Stiffness Matrix, k is given by

$$k = \left[\begin{array}{c|c} \frac{12EI}{L^3} & -\frac{6EI}{L^2} \\ \hline -\frac{6EI}{L^2} & +\frac{4EI}{L} \end{array}\right] = \left[\begin{array}{c|c} k_{11} & k_{12} \\ \hline k_{21} & k_{22} \end{array}\right]$$

The stiffness is associated with the translational degree of freedom only. Hence the above matrix can be partitioned as shown.

Also $k' = k_{11} - k_{12}k_{22}^{-1}k_{21}$

$$= \frac{12EI}{L^3} - \frac{6EI}{L^2}\left(\frac{L}{4EI}\right)\left(\frac{6EI}{L^2}\right) = \frac{12EI}{L^3} - \frac{9EI}{L^3} = +\frac{3EI}{L^3}.$$

Translational Stiffness Matrix $k = \frac{3EI}{L^3}$.

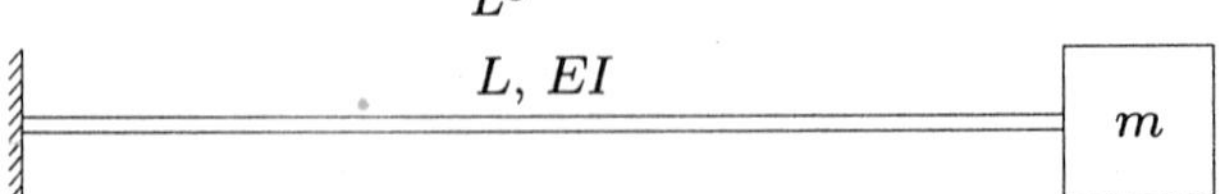

(a)

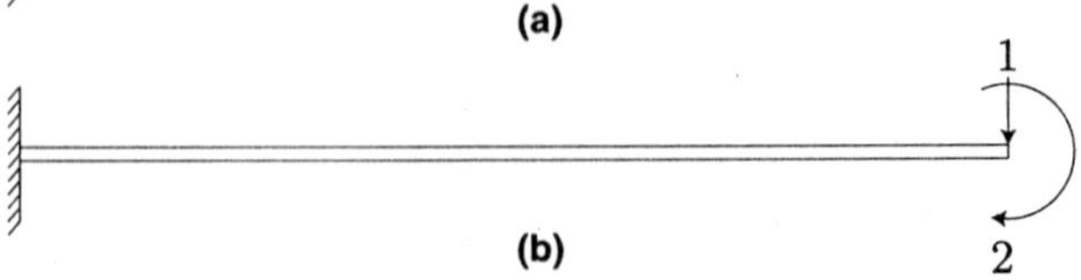

(b)

Fig. 18.13: Reduced stiffness matrix

Example 18.8. *Find the translational stiffness at the mass point shown in Fig. 18.14 by the moment distribution method and the reduced stiffness approach.*

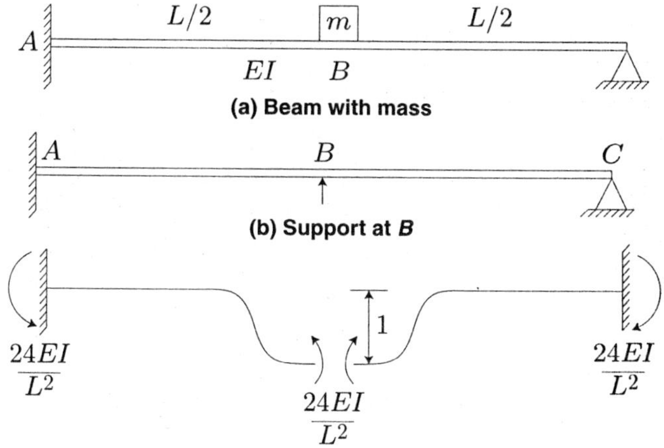

(a) Beam with mass

(b) Support at *B*

(c) Unit deflection at *B*

	A	B (AB)	B (BC)	C
DF	0	$\frac{4}{7}$	$\frac{3}{7}$	1
FEM	−24	−24	+24	+24
DM	0	0	0	−24
CO	0	0	−12	0
DM	0	$+\frac{48}{7}$	$+\frac{36}{7}$	0
CO	$+\frac{24}{7}$	0	0	0
DM	0	0	0	0
F.M.	$-\frac{144}{7}$	$-\frac{120}{7}$	$+\frac{120}{7}$	0
	$\frac{528}{7}$ ↑	$\frac{528}{7}$ ↓	↓ $\frac{240}{7}$	↑ $\frac{240}{7}$
		↓ $\frac{768EI}{7L^3}$		

(d) Moment distribution

Fig. 18.14: For Example 18.8

Solution :

Moment Distribution Method

According to structural dynamics theory, there is only one degree of freedom, i.e., translation at the mass point. Removing the mass m temporarily and arresting the translation as shown in Fig. 18.14(b), distribution factors have to be obtained as for

$$\text{for } AB : BC = \frac{4}{7} : \frac{3}{7} = 4 : 3.$$

A translation equal to unity in downward direction is made at the mass point. The moments induced are shown in Fig. 18.14(d).

The coefficients of the moments with proper signs are indicated in Fig. 18.14(d). The translation stiffness comes to $\frac{768EI}{7L^3}$.

Reduced Stiffness Approach

Ignoring axial elongation, the total number of degrees of freedom is equal to three. The total Stiffness Matrix can be formulated.

The total stiffness matrix for the co-ordinates is shown in Fig. 18.15(b).

$$k = \begin{bmatrix} \dfrac{192EI}{L^3} & 0 & \dfrac{24EI}{L^2} \\ 0 & \dfrac{16EI}{L} & \dfrac{4EI}{L} \\ \dfrac{EI}{L^2} & \dfrac{4EI}{L} & \dfrac{8EI}{L} \end{bmatrix}$$

Set EI = unity and L = unity for a while. This avoids the writing of EI and L, etc., every time.

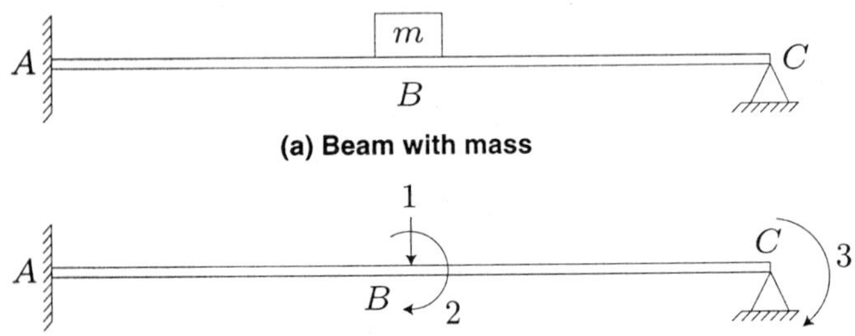

(a) Beam with mass

(b) Unit reactions at *B* and *C*

Fig. 18.15: Reduced stiffness approach

Partitioning the matrix the translations and rotations are separated.

$$k = \left[\begin{array}{c|cc} 192 & 0 & 24 \\ \hline 0 & 16 & 4 \\ 24 & 4 & 8 \end{array}\right] = \left[\begin{array}{c|c} k_{11} & k_{21} \\ \hline k_{12} & k_{22} \end{array}\right]$$

Now $k' = k_{11} - k_{12}\, k_{22}^{-1}\, k_{21}$

$$= 192 - [0.24] \begin{bmatrix} 8 & -4 \\ -4 & 16 \end{bmatrix} \begin{bmatrix} 0 \\ 24 \end{bmatrix} \left(\frac{1}{112}\right) \quad \text{as Det } K^{10} = 112$$

$$= \left[192 - \frac{576}{7}\right] = \frac{768}{7}.$$

The term $\dfrac{EI}{L^3}$ is always be associated with translational stiffness. Reintroducing the same,

$$k' = \frac{768EI}{L^3}.$$

Example 18.9. *Find the translational stiffness at the mass point shown in Fig. 18.16 by the moment distribution method and the reduced stiffness approach.*

Solution :

Moment Distribution Procedure

In the simply support beam, the translation occurs at the mass point. As per structural dynamics theory, it has got one degree of freedom. The mass is removed temporarily and the degree of freedom is arrested as shown in Fig. 18.16(b). The distribution factors are computed as for :

$$AB : BC = 0.50 : 0.50.$$

The moments with proper signs and the procedure are indicated in Fig. 18.16(d). Finally translation stiffness is shown in Fig. 18.16(d) as 6 in place of $\frac{6EI}{L^5}$.

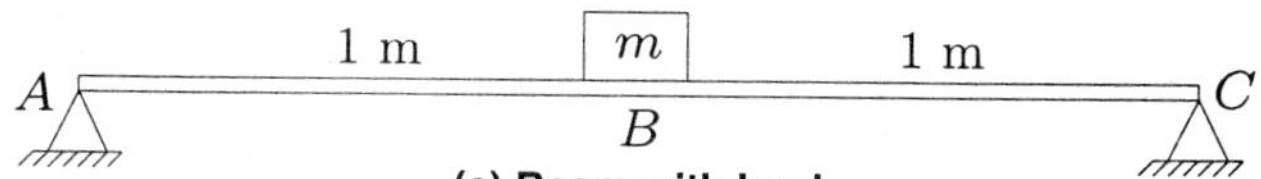

(a) Beam with load

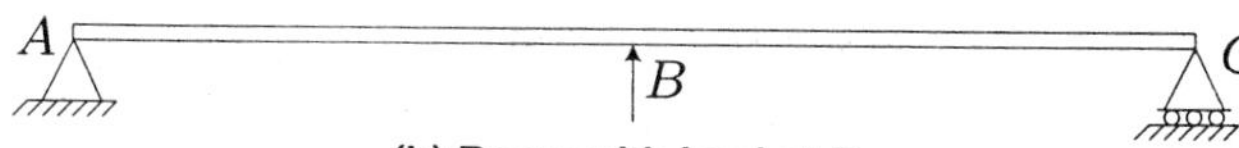

(b) Beam with load at *B*

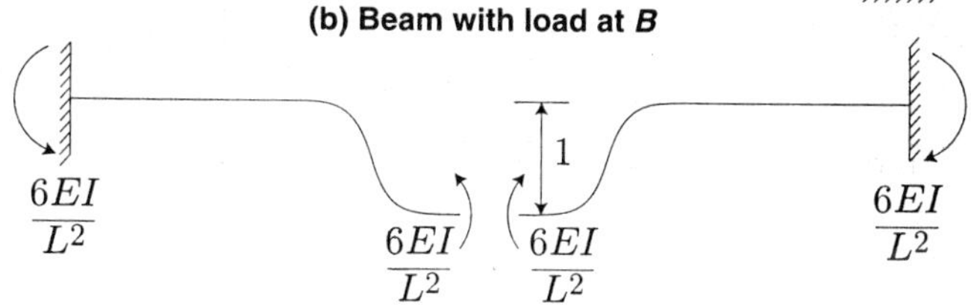

(c) Moment due to displacement at *B*

	A	B (BA)	B (BC)	C
	1	0.5	0.5	1
FEM	−6	−6	+6	+6
DM	+6	0	0	−6
CO	0	+3	−3	0
DM	0	0	0	0
FM	0	−3	+3	0
	↑3	3↓	↓3	3↑

↓6

(d) Moment distribution

Fig. 18.16: For Example 18.9

Example 18.10. *A pulley weighing 500 N is attached at the free end of a shaft which is fixed at the other end. The solid shaft has a diameter of 50 mm for a length of 1 m from the fixed end and diameter of 30 mm for the remaining length of 1 m. Calculate the natural frequency of transverse vibration* $E = 2.2(10)^5$ *N/mm*^{2} *and* $g = 9800$ mm/sec^2.

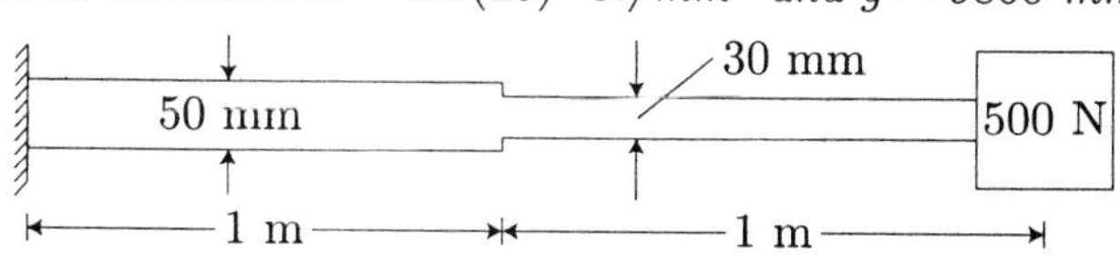

(a) Beam with load

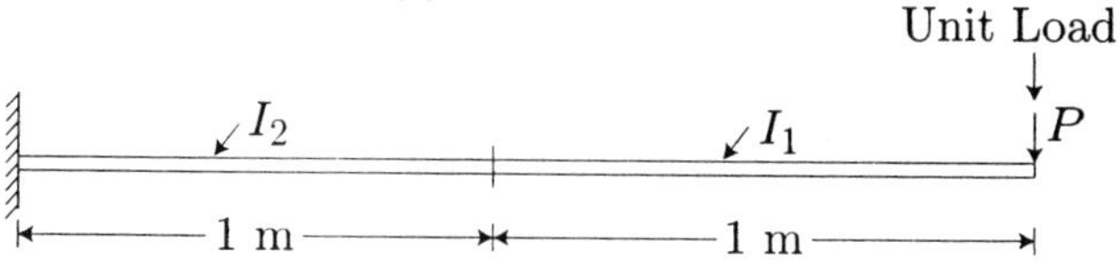

(a) Load at end

Fig. 18.17: For Example 18.10

Solution : The mass is replaced by a single load P as shown in Fig. 18.17(b). The unit load also is applied.

$$\Delta = \int_0^L \frac{(-Px)(-x)dx}{EI_1} + \int_L^{2L} \frac{(-Px)(-x)dx}{EI_2} = \frac{PL^2}{E}\left[\frac{1}{3I_1} + \frac{7}{3I_2}\right].$$

$$\text{Stiffness } k = \frac{P}{\Delta} \Rightarrow k = \frac{P}{\frac{PL^3}{3E}\left[\frac{1}{I_1} + \frac{7}{I_2}\right]} = \frac{1}{\frac{L^3}{3E}\left[\frac{1}{I_1} + \frac{7}{I_2}\right]}.$$

$$\text{Now } I_1 = \frac{\pi(30)^4}{64} = 39760.782 \text{ mm}^4$$

$$I_2 = \frac{\pi(50)^4}{64} = 306796.1576 \text{ mm}^4$$

and $k = 13.759$ N/mm.

$$\text{Natural period } \tau = \frac{2\pi}{p} = 2\pi\sqrt{\frac{W}{kg}}$$

$$\tau = 2\pi\sqrt{\frac{500}{13.759(9800)}} = 0.3826 \text{ sec.}$$

$$\text{Natural frequency } f = \frac{1}{\tau} = \frac{1}{0.3826} = 2.6137 \text{ cps.}$$

Example 18.11. *A simply supported beam of length l and rigidity EI supports a weight $3W_0$ acting at mid-span. At $t = 0$, one-third of the weight is suddenly removed. What will be the equation for the displacement at mid-span at any time thereafter?*

Solution : The weight $3W_0$ is attached at the centre of the simply supported beam. The translational stiffness k is a static parameter which is independent of the mass acting at the centre. the structure vibrates in single degree of freedom with single mass at the centre.

The translational stiffness $k = \frac{48EI}{L^3}$.

Now the natural circular frequency $p = \sqrt{\frac{k}{m}}$, where m is the mass present in the structure during vibration. By removing one-third of the weight W_0, the mass is set in vibration. The mass present during vibration is, therefore,

$$m = \frac{2W_0}{g}.$$

At $t = 0$, the structure is at rest under the load 3 W_0.

$$\therefore \quad x(t) = x(0) = x_0 = \frac{3W_0L^3}{48EI} = \frac{W_0L^3}{16EI}.$$

Velocity was zero at $t = 0$.

$\therefore$ the displacement equation $x = x_0 \cos pt$.

$$\therefore \quad x = \frac{W_0L^3}{16EI}\cos pt \quad \text{and} \quad p = \sqrt{\frac{k}{m}} = \sqrt{\frac{48EI(g)}{L^3(2W_0)}} = \sqrt{\frac{24EIg}{W_0L^3}}.$$

Example 18.12. *A weightless truss is subjected to a single mass m as shown in Fig. 18.18. The area of cross section is same for all members. Represent the structure as spring mass system, compute the natural period for the system in terms of m, L, A and E.*

Solution : Since there is only one mass, it can be represented as a single degree of freedom. For finding the stiffness, replace the mass by a load W and compute the deflection as indicated in the Table.

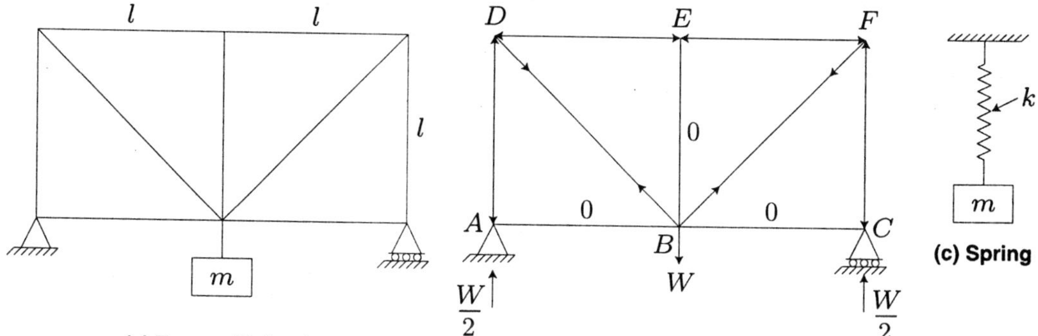

(a) Truss with load **(b) Forces in members** **(c) Spring**

Fig. 18.18: For Example 18.12

Member	Length	S	u	$\frac{SuL}{AE}(AE)$
AD	L	$-W/2$	$-1/2$	$WL/4$
DB	$\sqrt{2}L$	$W/\sqrt{2}$	$1/\sqrt{2}$	$WL/\sqrt{2}$
DE	L	$-W/2$	$-1/2$	$WL/4$
EF	L	$-W/2$	$-1/2$	$WL/4$
BF	$\sqrt{2}L$	$W/\sqrt{2}$	$1/\sqrt{2}$	$WL/\sqrt{2}$
FC	L	$-W/2$	$-1/2$	$WL/4$

$$\frac{\sum SuL}{AE} = [W{\cdot}L + 2WL/\sqrt{2}]\frac{1}{AE}$$

$$\Delta = WL[1+\sqrt{2}]\frac{1}{AE}$$

$$k = \frac{W}{\Delta} = \frac{AE}{(2.414)L}.$$

Natural circular frequency $p = \sqrt{\frac{k}{m}}$

$$= \sqrt{\frac{AE}{(2.414)L}\left(\frac{1}{m}\right)} \text{ rad/ sec.}$$

$$\tau = \text{ natural period } = \frac{2\pi}{p} \text{ sec.}$$

18.5 Equivalent Stiffness—Springs in Series—Springs in Parallel

(a) Springs in series :

Consider a system of springs with stiffnesses k_1, k_2 and k_3 attached in series as shown in Fig. 18.19(a). A load W is suspended on the system. The total deflection is Δ. Each spring gets elongated to a distance Δ_1, Δ_2 and Δ_3 as shown in Fig. 18.19(b).

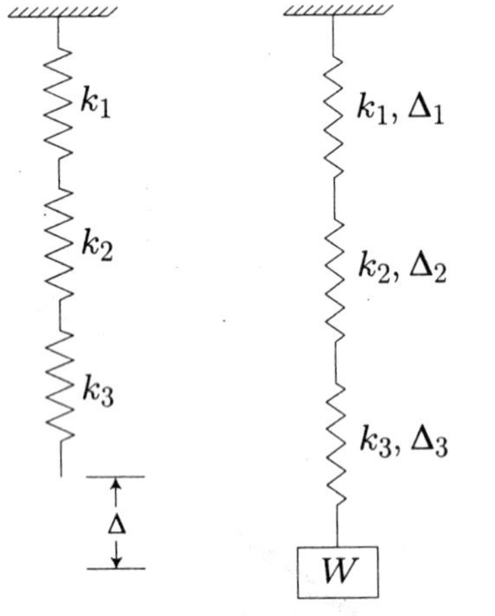

(a) Spring (b) Spring with load

Fig. 18.19: Springs series

$$\Delta = \Delta_1 + \Delta_2 + \Delta_3$$

$$\Delta_1 = \frac{W}{k_1}, \quad \Delta_2 = \frac{W}{k_2}$$

and $\Delta_3 = \dfrac{W}{k^3}$.

$$\text{Equivalent spring factor} = k_c = \frac{W}{\Delta} = \frac{W}{\frac{W}{k_1} + \frac{W}{k_2} + \frac{W}{k_3}}$$

$$= \frac{1}{\frac{1}{k_c}} = \frac{1}{\left(\frac{1}{k_1} + \frac{1}{k_2} + \frac{1}{k_3}\right)}$$

and
$$\frac{1}{k_0} = \frac{1}{k_1} + \frac{1}{k_2} + \frac{1}{k_3} + \cdots \tag{18.11}$$

(b) Springs in parallel :

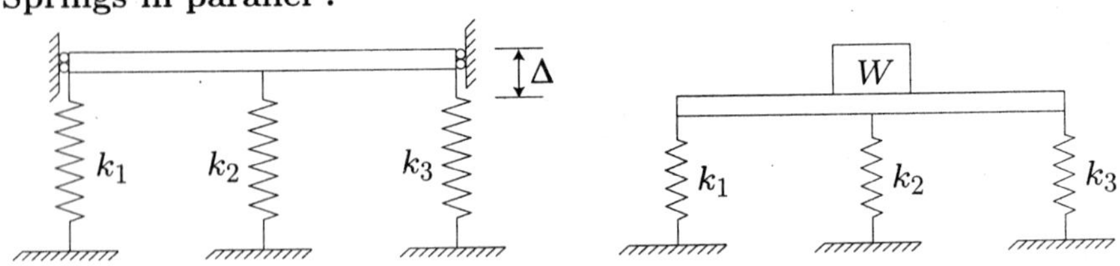

(a) Springs (b) Spring with load

Fig. 18.20: Springs in parallel

A system of springs arranged in parallel is shown in Fig. 18.20(a). When a load W is placed on the system, all the springs undergo the same deflection Δ. Since the deflection is same for all the springs, the load carried by each spring is $k_1\Delta$, $k_2\Delta$ and $k_3\Delta$.

$$W = W_1 + W_2 + W_3 = k_1\Delta + k_2\Delta + k_3\Delta$$

$$W = \Delta(k_1 + k_2 + k_3).$$

Now
$$\frac{W}{\Delta} = k_0 = (k_1 + k_2 + k_3), \tag{18.12}$$

where k_0 is equivalent stiffness.

Example 18.13. *A steel beam simply supported has a span of 500 mm. The cross section is rectangular 50 mm wide and 5 mm depth. This is connected at mid-span by means of a spring having a stiffness of 50 N/mm, and a load of 1500 N is applied at the other end of a spring.*

Determine the natural frequency of the system.

$$E = 2(10)^5 \text{ N/mm}^2; \quad \text{and} \quad g = 9800 \text{ mm/sec}^2$$

Solution : The beam acts as a spring whose stiffness is

$$k_1 = \frac{48EI}{L^3}.$$

Now $I = \frac{1}{12}(b)(d^3) = \frac{1}{12}(50)(5)^3 = 520.833 \text{ mm}^4$

$$k_1 = \frac{48(2)(10^5)(520.833)}{500^3} = 40 \text{ N/mm}.$$

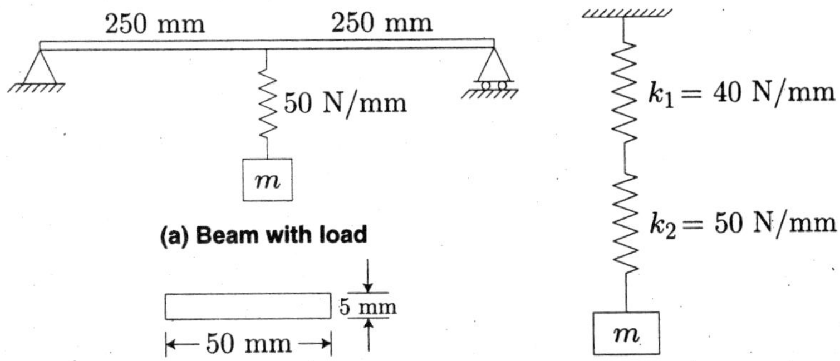

(a) Beam with load

(b) Section of beam

(c) Spring with load

Fig. 18.21: For Example 18.13

The two springs are in series as shown in Fig. 18.22

and $\frac{1}{k_e} = \frac{1}{40} + \frac{1}{50}$, whence $k_e = 22.22$ N/mm.

The system vibrates in a single degree of freedom with spring constant $k_e = 22.2$ N/mm.

$$p = \sqrt{\frac{k_e}{m}} = \sqrt{\frac{22.22(9800)}{1500}} = 12.0486 \text{ rad/ sec.}$$

$$\tau = \frac{2\pi}{p} = \frac{2\pi}{12.0486} = 0.5215 \text{ sec}$$

$$f = \frac{1}{\tau} = \frac{1}{0.5215} = 1.9176 \text{ cps.}$$

Example 18.14. *A mass is held by three springs as shown in Fig. 18.22. Assume the rolling friction on the floor is negligible, as are the inertial effects of the rollers. The spring constants are $k_1 = 6$ N/mm, $k_2 = 4$ N/mm, $k_3 = 2$ N/mm, $g = 9810$ mm/sec^2. Find p.*

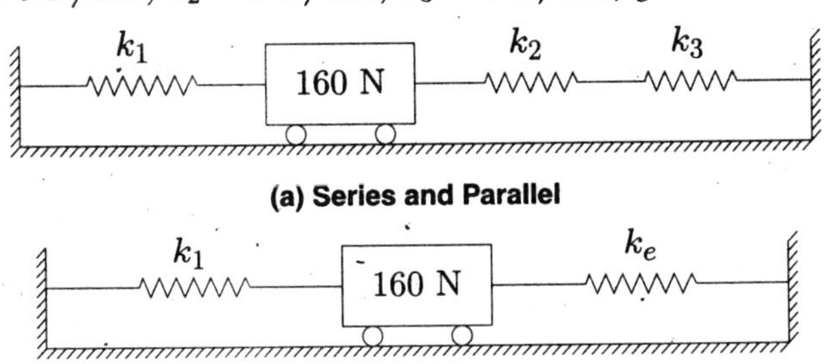

(a) Series and Parallel

(b) Parallel

Fig. 18.22: For Example 18.14

Solution : The right two springs are in series. The equivalent spring of the right acts in parallel with the left-side spring.

Springs in series :

$$\frac{1}{k_e} = \frac{1}{k_2} + \frac{1}{k_3} = \frac{1}{4} + \frac{1}{2}$$

$$k_e = \frac{4}{3}.$$

Springs in parallel :

$$k'_e = k_e + k_1 = \frac{4}{3} + 6 = \frac{22}{3} \text{ N/mm.}$$

and $p = \sqrt{\frac{k'_e}{m}} = \sqrt{\frac{22}{3}\left(\frac{9810}{160}\right)} = 21.2$ rad/second.

Example 18.15. *For small oscillations, what is the natural frequency of the system? Assume the rod to be infinitely rigid and massless.*

Solution : The spring is vertical. Hence the gravity force mg need not be considered. The only two forces acting on the system are the force of inertia $mb\ddot{\theta}$ and spring force $ka\theta$. Note the direction of the force of inertia is opposite to the direction of angular motion. The angle θ is very very small.

Taking moment of the forces about the fulcrum,

$$(mb\ddot{\theta})b + (ka\ \theta)a = 0$$

$$\ddot{\theta} + \frac{ka^2}{mb^2} = \theta = 0$$

and $p = \sqrt{\frac{ka^2}{mb^2}}.$

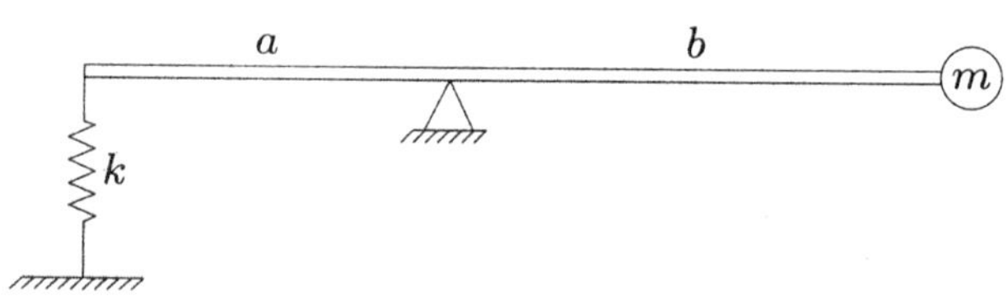

(a) Beam with spring

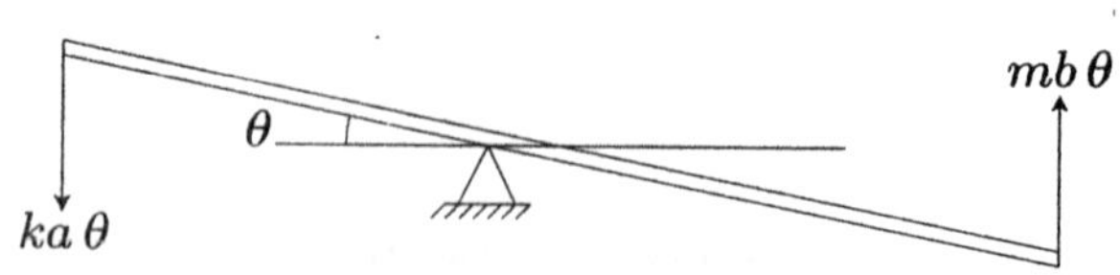

(b) Dynamic equilibrium

Fig. 18.23: For Example 18.15

Example 18.16. *A uniform bar of total weight W and length L is pivoted at a quarter.length point, which at the other quarter.length point a spring of stiffness k applies at 45° as shown. Find the natural frequency.*

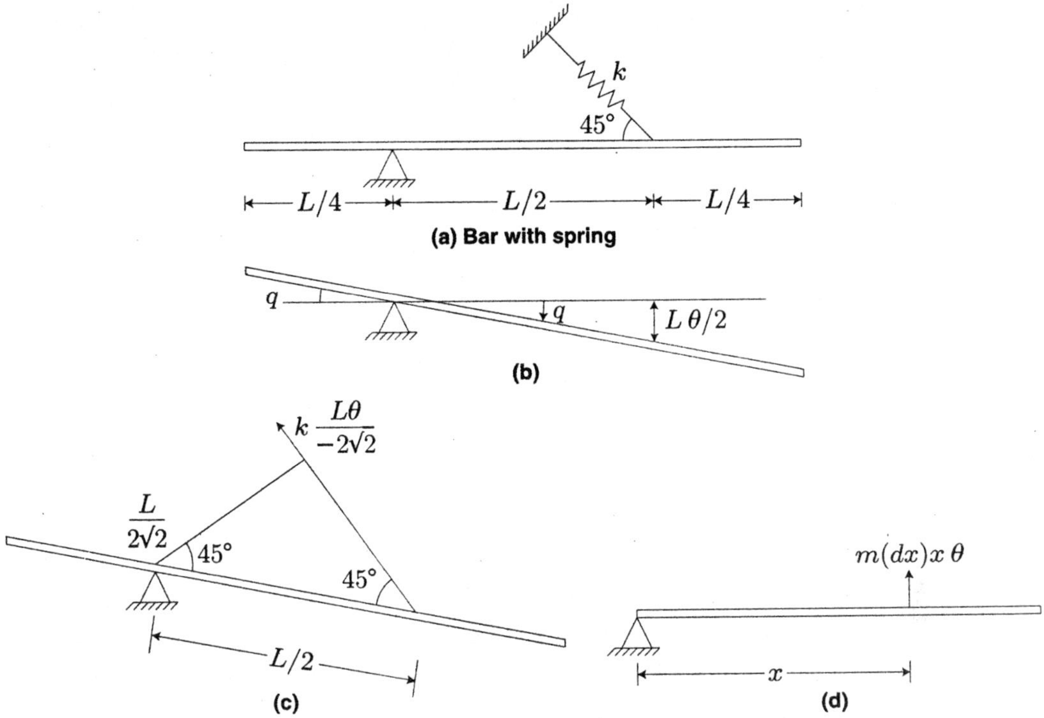

Fig. 18.24: For Example 18.16

Solution : Considering Fig. 18.24(d), the mass is m per unit run. The bar is given a small angular movement θ which is not shown in Fig. 18.24(d).

The moment of the small inertia force about the fulcrum is $= [m(dx)x\ddot{\theta}](x)$.

$$\text{Total moment} = \int_0^L m\ddot{\theta}x^2 dx = \frac{1}{3}mL^3\ddot{\theta}.$$

Now $\quad ml = \dfrac{W}{g}$, where W is the total weight of the rod.

$$\text{Total moment} = \frac{1}{3}\frac{W}{g}L^2\ddot{\theta} = I_0\ddot{\theta}, \quad \text{where} \quad I_0 = \frac{1}{3}\frac{W}{g}L^2.$$

Now from Fig. 18.24(c), taking moment about the fulcrum,

$$\ddot{\theta}\left[\left(\frac{W}{4g}\right)\frac{1}{3}\left(\frac{L}{4}\right)^2 + \left(\frac{3}{4}\frac{W}{g}\right)\frac{1}{3}\left(\frac{3L}{4}\right)^2\right] + k\frac{L}{2\sqrt{2}}\frac{L}{2\sqrt{2}}\theta = 0$$

whence $\dfrac{28}{192}\dfrac{WL^2}{g}\ddot{\theta} + \dfrac{kL^2}{8}\theta = 0$ and $\ddot{\theta} + \dfrac{g}{W}k\dfrac{192}{28(8)}\theta = 0.$

and $\quad p^2 = \dfrac{24}{28}\dfrac{ky}{W} = \dfrac{6}{7}\dfrac{kg}{W}.$

Example 18.17. *A weightless bar of length 21 is hinged in the middle. At its top end it carries a concentrated weight W and at its bottom end a concentrated weight 2W. Midway between the hinge and the bottom weight there is a spring k. Find the natural frequency of small oscillations in vertical plane.*

Solution : Since the motion is taking place in the horizontal direction, the gravity force will enter in the equilibrium equation. Again, the displacement angle θ is very very small. Taking moment of all the forces about the fulcrum,

$$\left(\frac{2W}{g}L\ddot{\theta}\right)L + \left(\frac{W}{g} - L\ddot{\theta}\right)L + \left(\frac{kL}{2}\theta\right)\frac{L}{2} + WL\theta = 0$$

$$\ddot{\theta} + \frac{g}{3W}\frac{k}{4}\theta + \frac{g}{3L}\theta = 0, \quad \text{and} \quad p^2 = \frac{gk}{12W} + \frac{g}{3L}.$$

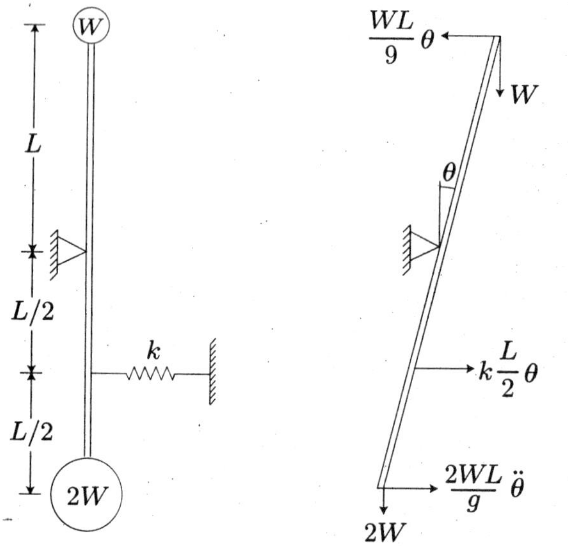

(a) Beam (b) Dynamic equilibrium

Fig. 18.25: For Example 18.17

Example 18.18. *A rigid uniform rod connected to a spring is as shown in Fig. 18.26. Total mass of the rod is m. Find the value of p.*

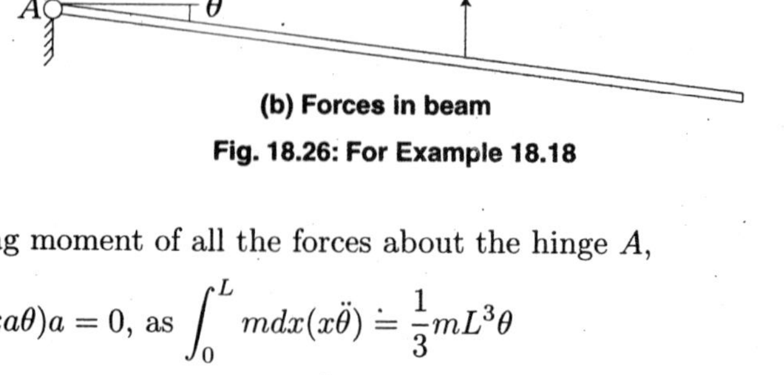

Fig. 18.26: For Example 18.18

Solution : Taking moment of all the forces about the hinge A,

$$\frac{1}{3}mL^2\ddot{\theta} + (ka\theta)a = 0, \text{ as } \int_0^L mdx(x\ddot{\theta}) \doteq \frac{1}{3}mL^3\theta$$

$$\ddot{\theta} + \frac{3ka^2}{mL^2} = \theta = 0 \quad \text{and} \quad p^2 = \frac{3ka^2}{mL^2}.$$

Example 18.19. *A weightless rigid bar of length $3L$ carries three distinct equal masses m. It is pivoted at A and carries two springs k. Find the natural frequency of small vibrations. The bar is horizontal so that gravity has no influence.*

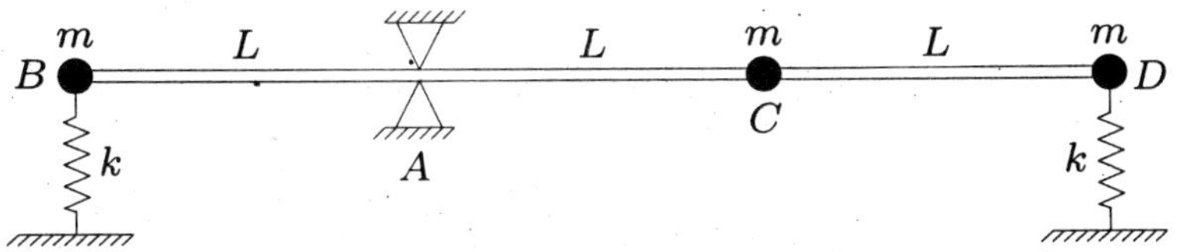

(a) Beam with hinge

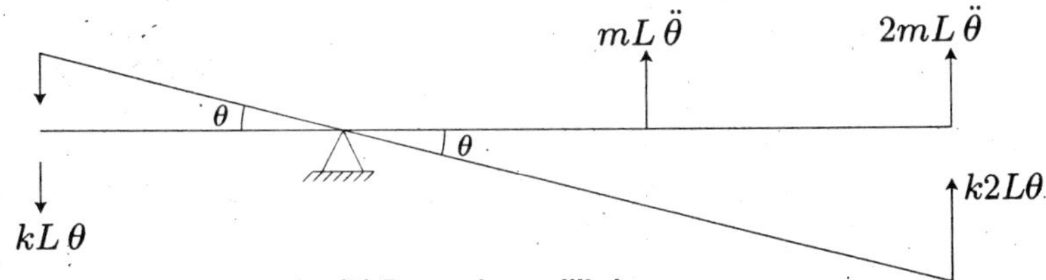

(b) Dynamic equilibrium

Fig. 18.27: For Example 18.19

Solution : Even though there are three masses, the vibration takes place in single degree of freedom as the bar is rigid.

As stated in the problem, the gravity forces mg need not be considered.

The dynamic equilibrium of the system is shown in Fig. 18.27(b).

Taking moment of the forces about the flucrum,

$$(mL\ddot{\theta})L + (mL\ddot{\theta})L + (2mL\ddot{\theta})2L + (k2L\theta)2L + (kL\theta)L = 0$$

$$6mL^2\ddot{\theta} + 5kL^2\theta = 0$$

$$\ddot{\theta} + \frac{5k}{6m}\theta = 0 \quad \text{and} \quad p^2 = \frac{5k}{6m}.$$

Example 18.20. *Compute the natural frequency of the system in Fig. 18.28. Assume the rod to be massless and infinitely rigid.*

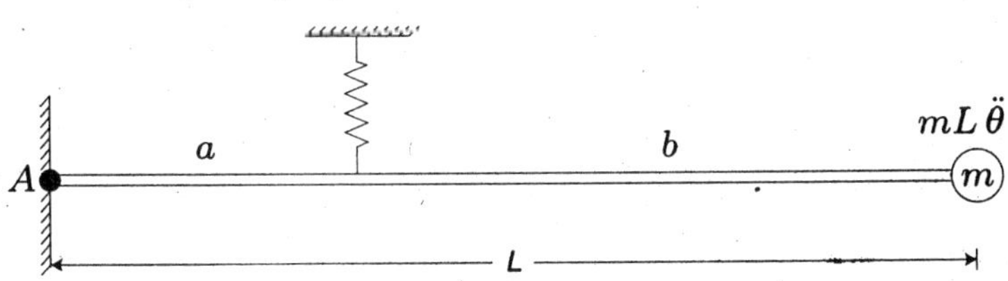

(a) Beam with spring

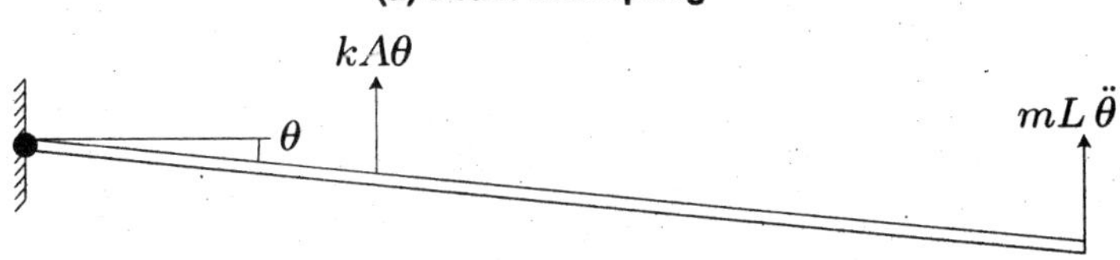

(b) Dynamic equilibrium

Fig. 18.28: For Example 18.20

Solution : The dynamic equilibrium is as shown in Fig. 18.27(b). Taking moment of the forces about the hinge,

$$(mL\ddot{\theta})L + (ka\,\theta)a = 0, \quad \text{whence,} \quad \ddot{\theta} + \frac{ka^2}{mL^2}\theta = 0,$$

and $\quad p^2 = \dfrac{ka^2}{mL^2}.$

Example 18.21. *A compound beam AE consists of two identical portions AC and CE hinged together at C and loaded with a mass m at the free end. The beam is elastic. Compute the natural frequency of the system.*

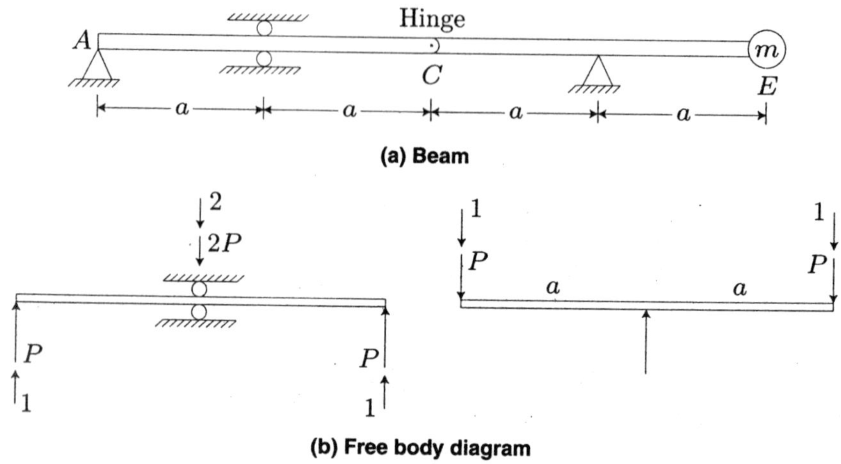

Fig. 18.29: For Example 18.21

Solution : The compound beam is a statically determinate structure. The mass m is replaced by a force P. The free body diagrams are shown in Fig. 18.29(b). The unit load is shown on the top of P.

From $\quad \Delta = \displaystyle\int \frac{Mm\,dx}{EI}$

$$\Delta = 4\int_0^a \frac{(px)(1x)dx}{EI}$$

$$\Delta = \frac{4pa^3}{3EI}.$$

Translational stiffness $k = \dfrac{P}{\Delta} = \dfrac{P}{\frac{4pa^3}{3EI}} = \dfrac{3EI}{4a^2}$

and $\quad p = \sqrt{\dfrac{k}{m}} = \sqrt{\dfrac{3EI}{4a^3m}}.$

Example 18.22. *A lamp post system is shown in Fig. 18.30. Assuming the structure to be massless and flexible, compute the natural frequency of the system. Assume SDOF in vertical direction.*

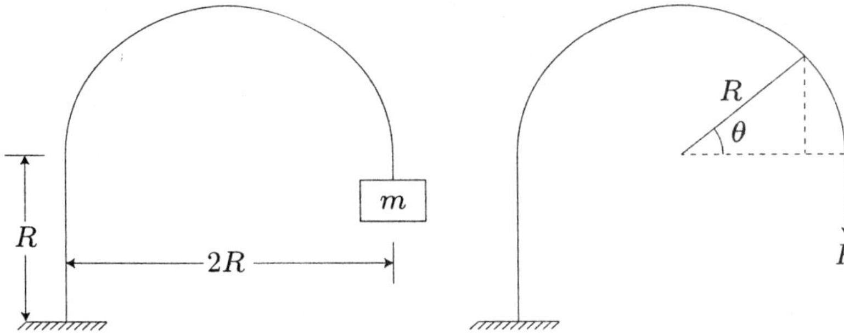

(a) Lampost **(b) Lampost with load *P***

Fig. 18.30: For Example 18.22

Solution : The system is a single degree of freedom. The mass m is replaced temporarily by a load P and the deflection is computed as follows :

For Semicircular segment

$$\Delta_1 = \frac{1}{EI}\int_0^{\pi}[P(R - R\cos\theta)][R - R\cos\theta]Rd\theta$$

$$\Delta_1 = \frac{3\pi}{2}\frac{PR^3}{EI}.$$

For Vertical Post

$$\Delta_2 = \int_0^{R}[2PR][2R]\frac{dx}{EI} = \frac{4PR^3}{EI}$$

$$\Delta = \Delta_1 + \Delta_2 = \frac{PR^3}{EI}\left[\frac{3\pi}{2} + 4\right] = 8.7124\frac{PR^3}{EI}.$$

and Translational Stiffness, $k = \dfrac{P}{\Delta} = \dfrac{EI}{8.7124R^3}$, whence

$$p\sqrt{\frac{k}{m}} = \sqrt{\frac{EI}{8.7124R^3m}}.$$

Example 18.23. *A massless rigid bar supports a mass m as shown in Fig. 18.31. Compute the natural circular frequency of the system.*

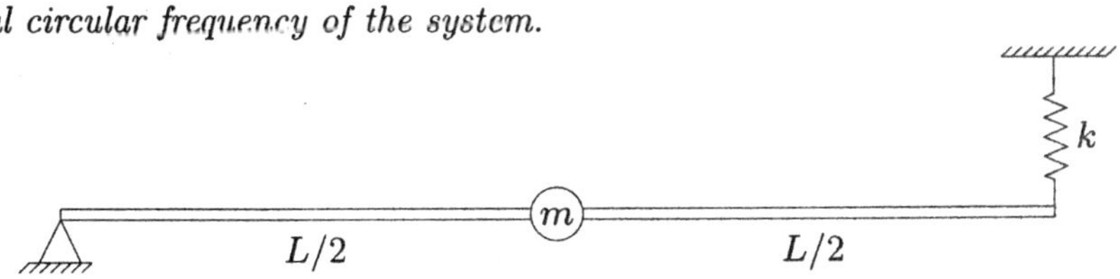

(a) Beam with mass

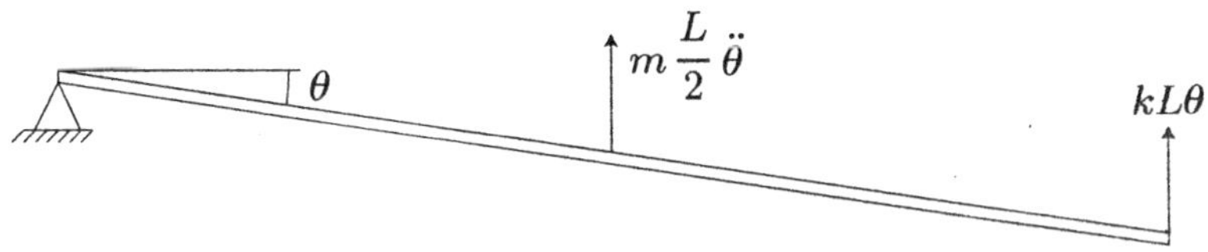

(b) Dynamic equilibrium

Fig. 18.31: For Example 18.23

Solution : The dynamic equilibrium is as shown in Fig. 18.31(b). Since the spring is vertical, the gravity force need not be considered.

Taking moment of the forces about the fulcrum,

$$\left(\frac{mL}{2}\ddot{\theta}\right)\frac{L}{2} + (kL\theta)L = 0 \quad \text{or,} \quad \ddot{\theta} + \frac{4k}{m}\theta = 0.$$

and $p^2 = \dfrac{4k}{m}$.

Example 18.24. *The uniform rigid massless rod is carrying a mass m as shown in Fig. 18.32. The rod is restrained to move vertically by both linear and torsional spring. Calculate the frequency of the vertical oscillation of the rod.*

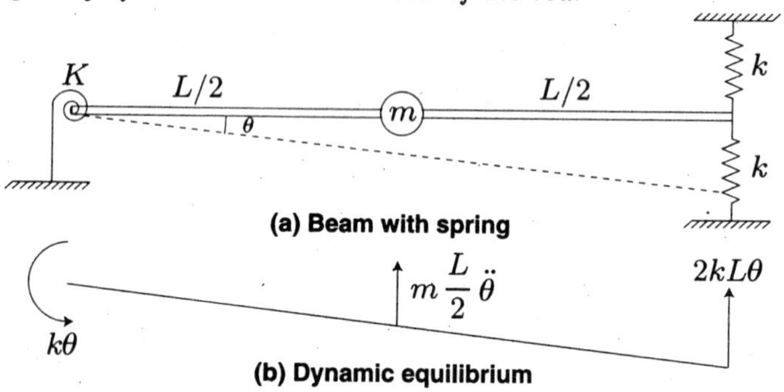

Fig. 18.32: For Example 18.24

Solution : The springs are connected in parallel. Therefore, $k_e = k + k = 2k$. The couple in the left spring will act in the direction opposite to the direction of motion whose magnitude is $K\theta$. The dynamic equilibrium is shown in Fig. 18.32(b).

Taking moment of all the forces about the left spring,

$$K\theta + \frac{mL}{2}\frac{L}{2}\ddot{\theta} + (2kL\theta)L = 0$$

$$\ddot{\theta} + \frac{4(2kL^2 + K)\theta}{mL^2} = 0 \quad \text{and} \quad p^2 = \frac{4}{mL^2}(2kL^2 + K).$$

Example 18.25. *The uniform stiff rod with mass m per unit run is to move vertically by both linear and torsional springs as shown in Fig. 18.33. Calculate the frequency of the vertical oscillation of the rod.*

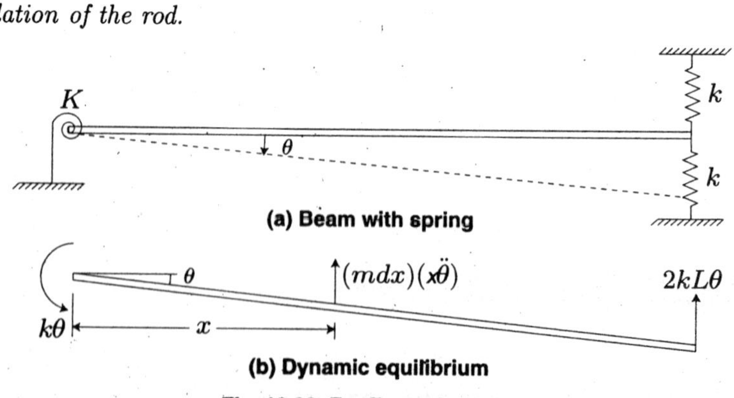

Fig. 18.33: For Example 18.25

Solution : The dynamic equilibrium is shown in Fig. 18.33(b). Taking moment of all the forces about the left end,

$$\int_0^L [(mdx)(x\ddot{\theta})][x] + [2kL\theta]L + K\theta = 0$$

whence $\frac{1}{3}mL^3\ddot{\theta} + 2kL\theta + K\theta = 0$

$$\ddot{\theta} + \frac{[2kL^2 + K]}{ml^3}\theta(3) \quad \text{and} \quad p^2 = \frac{3}{mL^2}[2kL^2 + K],$$

where m is the mass per unit run.

Example 18.26. *A simply supported beam of span L, is subjected to a single mass as shown is Fig. 18.34. Compute the natural frequency of the system.*

Solution : It is necessary to replace the load W by a force P in the direction of the motion and compute the spring constant k,

$$\Delta = \frac{1}{EI}\int_0^{2L/3}\left[\frac{P}{3}x\right]\left[\frac{x}{3}\right]dx + \frac{1}{2EI}\int_0^{L/3}\left[\frac{2}{3}Px\right]\left[\frac{2}{3}x\right]dx = \frac{10PL^3}{729EI}$$

$$k = \frac{P}{\Delta} = \frac{P}{\frac{10}{729}\frac{PL^3}{EI}} = \frac{729EI}{L^3(10)}, \quad \text{and} \quad p = \sqrt{\frac{k}{m}} = \sqrt{\frac{729EI(g)}{L^3(W)(10)}}.$$

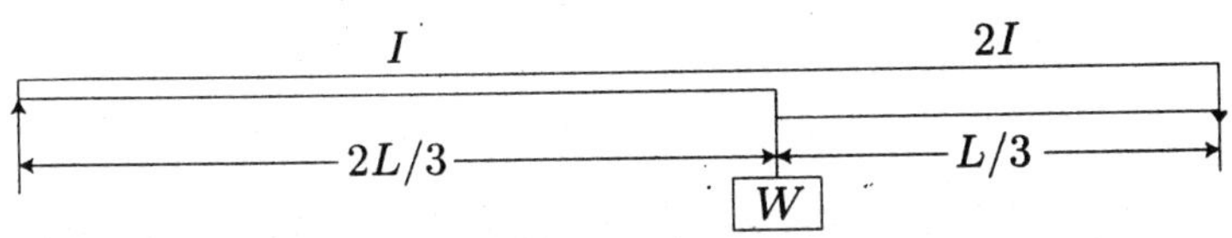

(a) Beam with load

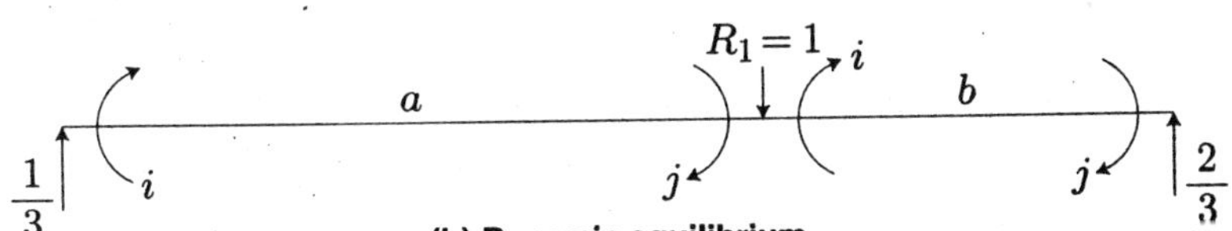

(b) Dynamic equilibrium

Fig. 18.34: For Example 18.26

Example 18.27. *The frame supports a weight W as shown in Fig. 18.35. Neglecting the weight of the frame, compute the natural period of horizontal vibration in the plane of the frame. Assume SDOF.*

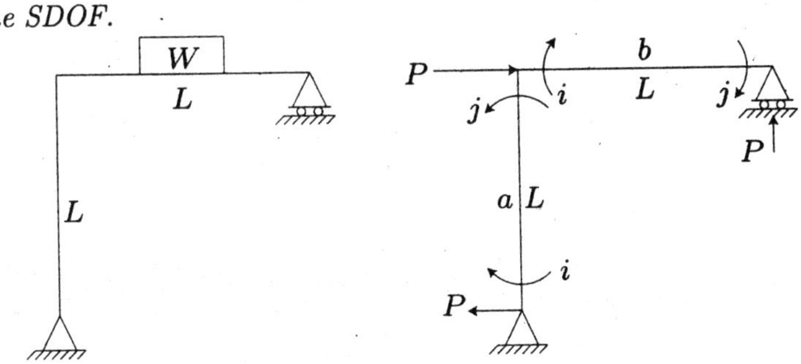

(a) Rigid frame (b) Dynamic equilibrium

Fig. 18.35: For Example 18.27

Solution : The weight W is removed temporarily and a force P is introduced as shown in Fig. 18.35(b).

$$R_1 = 1$$

Then
$$\underset{Q}{\begin{bmatrix} Q_i^a \\ Q_j^a \\ Q_i^b \\ Q_j^b \end{bmatrix}} = \underset{B_R}{\begin{bmatrix} 0 \\ -L \\ +L \\ 0 \end{bmatrix}} \quad \underset{R}{[R_1 = P]}$$

Then
$$f = \frac{L}{6EI}\begin{bmatrix} 2 & -1 & 0 & 0 \\ -1 & 2 & 0 & 0 \\ 0 & 0 & 2 & -1 \\ 0 & 0 & -1 & 2 \end{bmatrix}.$$

Now $F_{RR} = B_R{}^T f B_R$

$$F_{RR} = \begin{bmatrix} 0 & -L & +L & 0 \end{bmatrix} \frac{L}{6EI}\begin{bmatrix} 2 & -1 & 0 & 0 \\ -1 & 2 & 0 & 0 \\ 0 & 0 & 2 & -1 \\ 0 & 0 & -1 & 2 \end{bmatrix}\begin{bmatrix} 0 \\ -L \\ +L \\ 0 \end{bmatrix} = \frac{4L^2}{6EI}.$$

Now $r = F_{RR}\,R = \dfrac{4L^3}{6EI}(P) = \Delta$ and $k = \dfrac{P}{\Delta} = \dfrac{P}{\frac{4PL^3}{6EI}} = \dfrac{3EI}{2L^3}$

$$p = \sqrt{\frac{k}{m}} = \sqrt{\frac{3EI(g)}{2L^3(W)}} \quad \text{and} \quad \tau = \frac{2\pi}{p} = 2\pi\sqrt{\frac{2WL^3}{g(3I)}}.$$

Alternative Method of Computing *k*

$$\Delta = \int_0^L \frac{\overset{M}{(Px)}\ \overset{m}{(x)}\ dx}{EI} + \int_0^L \frac{\overset{M}{(Px)}\ \overset{m}{(x)}\ dx}{EI}.$$

$\therefore \quad \Delta = \dfrac{2}{3}\dfrac{PL^3}{EI}$. This is the same as obtained above.

Example 18.28. *The frame supports a weight W as shown in Fig. 18.36. Neglecting the weight of the frame, compute the natural period of horizontal vibration in the plane of the frame. Assume SDOF.*

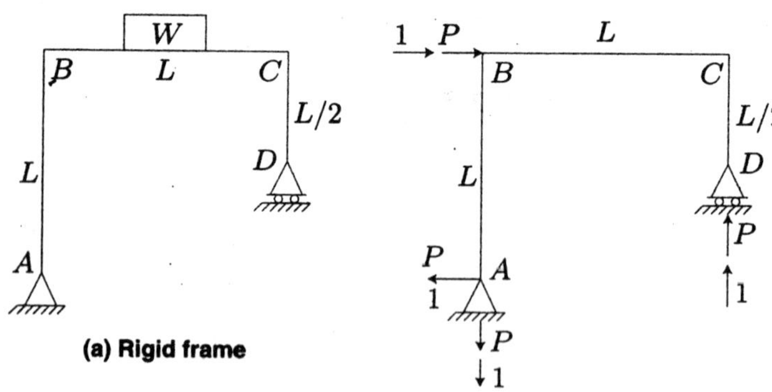

Fig. 18.36: For Example 18.28

Solution : To spring factor 'k' in the horizontal direction is to be found out.

To compute Δ using $\int Mm\frac{dx}{EI}$ method.

Portion DC:

$$\int 0(0)\left(\frac{dx}{EI}\right) = 0$$

Portion BC:

$$\int_0^L (Px)(1x)\frac{dx}{EI} = \frac{PL^3}{3EI}$$

Portion AB:

$$\int_0^L (Px)(1x)\frac{dx}{EI} = \frac{PL^3}{3EI}$$

$$\text{Total} = \Delta = \frac{PL^3}{3EI} + \frac{PL^3}{3EI} + 0 = \frac{2}{3}\frac{PL^3}{EI}$$

$$k = \frac{P}{\Delta} = \frac{3}{2}\frac{EI}{L^3}.$$

and $\quad \tau = 2\pi\sqrt{\frac{m}{k}} = 2\pi\sqrt{\frac{W}{g}\left(\frac{2L^3}{3EI}\right)}.$

Example 18.29. *The frame supports a weight W as shown in Fig. 18.37. Neglecting the weight of the frame, compute the natural period of vibration (vertical) in the plane of the frame. Assume SDOF.*

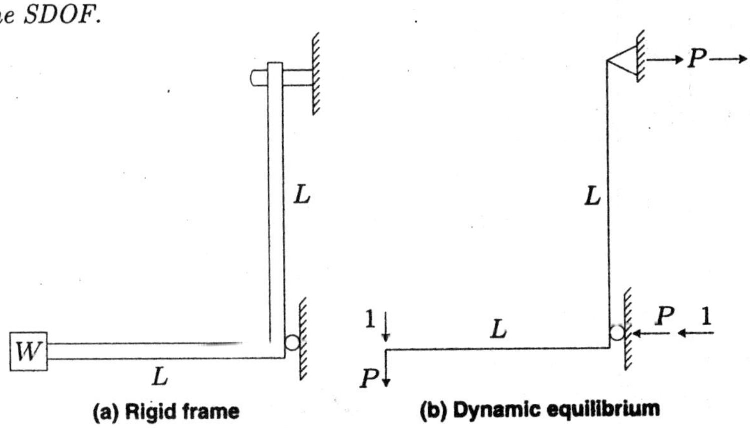

Fig. 18.37: For Example 18.29

Solution : For computing the spring factor in the vertical direction, use of $\int \frac{Mmdx}{EI}$ formula will be used.

Referring to Fig. 18.37(b),

$$\Delta = \int_0^L \underset{[Px]}{M}\ \underset{[x]}{m}\ \frac{dx}{EI} + \int_0^L \underset{[PL-Px]}{M}\ \underset{[L-x]}{m}\ \frac{dx}{EI} = \frac{2}{3}\frac{PL^2}{EI}$$

$$k = \frac{P}{\Delta} = \frac{3}{2}\frac{EI}{L^3} \quad \text{and} \quad \tau = 2\pi\sqrt{\frac{m}{k}} = 2\pi\sqrt{\frac{W}{g}\left(\frac{2L^3}{3EI}\right)}.$$

Example 18.30. *The frame supports a weight W as shown in Fig. 18.38. Neglecting the weight of the frame, compute the natural period of horizontal vibration in the plane of the frame. Assume SDOF.*

Solution : For computing the spring factor in the horizontal direction use of $\int \frac{Mm dx}{EI}$ formula will be used.

Portion AB: $\int_0^L \underset{(Px)}{M} \underset{(x)}{m} \frac{dx}{EI} = \frac{PL^3}{3EI}$

Portion DC: $\frac{1}{\sqrt{2}EI} \int_0^{\sqrt{2}L} \left(\frac{P}{2}x\right)\left(\frac{1}{2}x\right) ds$

As $\frac{x}{s} = \frac{1}{\sqrt{2}}, \; x = \frac{s}{\sqrt{2}},$

whence $\frac{1}{\sqrt{EI}} \int_0^{\sqrt{2}L} \left[\frac{P}{2}\frac{s}{\sqrt{2}}\right]\left[\frac{1}{2}\frac{s}{\sqrt{2}}\right] ds = \frac{PL^3}{12EI}.$

Portion BC (origin at B)

$$\int_0^l \left[\underset{PL - \frac{P}{2}x}{M}\right]\left[\underset{L - \frac{x}{2}}{m}\right] \frac{dx}{EI} = \frac{7}{12}\frac{PL^3}{EI}.$$

Total $\Delta = \left[\frac{1}{3} + \frac{1}{12} + \frac{7}{12}\right]\frac{PL^3}{EI} = \frac{PL^3}{EI}$

$$k = \frac{P}{\Delta} = \frac{EI}{L^3}.$$

and $\tau = 2\pi\sqrt{\frac{m}{k}} = 2\pi\sqrt{\frac{W}{g}\left(\frac{L^3}{EI}\right)}.$

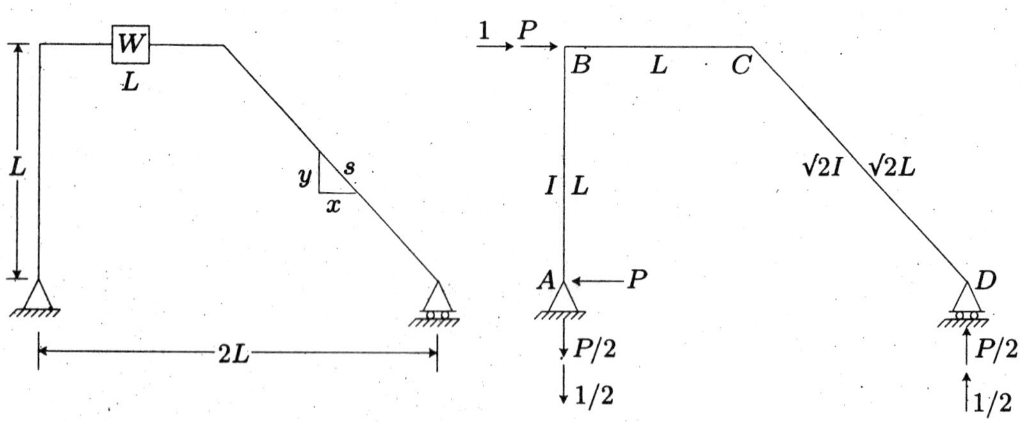

(a) Frame (b) Dynamic equilibrium

Fig. 18.38: For Example 18.30

Example 18.31. *A uniform beam of flexural rigidity EI supports a weight W as shown in Fig. 18.39. Neglecting the self-weight of the beam, compute the actual period of vertical vibration.*

Solution : Unit Load Method :

The weight W is replaced by a force P as shown in Fig. 18.39(b).

$$\Delta = \int_0^L \left[\frac{P}{2}x\right]\left[\frac{L}{2}x\right]\frac{dx}{EI} + \int_0^{L/2} [Px]^m [x] \frac{dx}{EI}$$

$$\Delta = \frac{PL^3}{8EI}.$$

and $k = \dfrac{P}{\Delta} = \dfrac{8EI}{L^3}.$

Natural period $= \tau = \dfrac{2\pi}{p} = 2\pi\sqrt{\dfrac{m}{k}} = 2\pi\sqrt{\dfrac{W}{g}\left(\dfrac{L^3}{8EI}\right)}.$

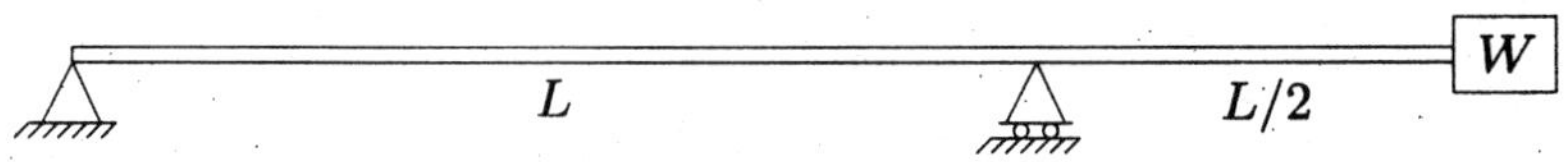

(a) Beam with load

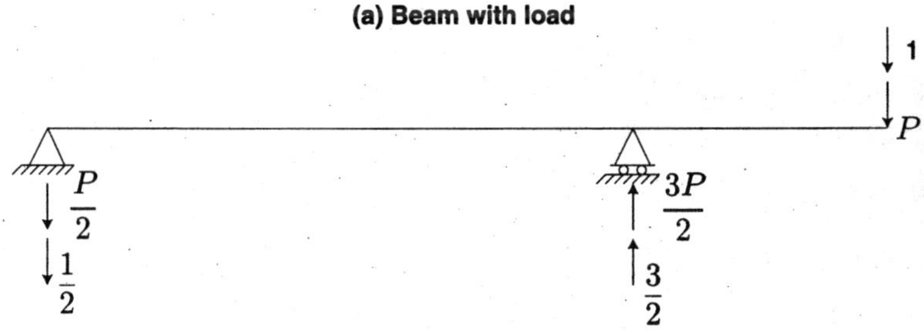

(b) Reactions

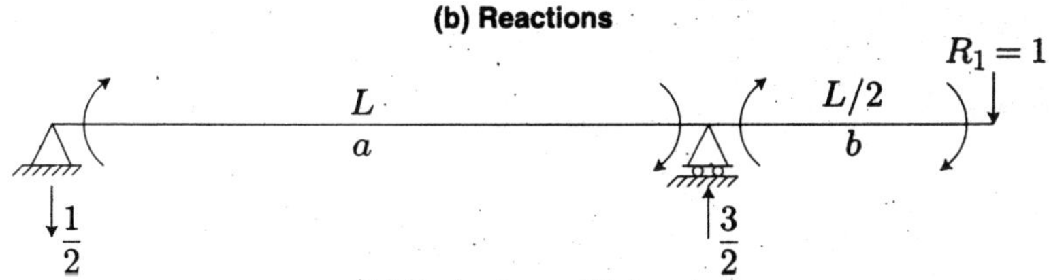

(c) Dynamic equilibrium

Fig. 18.39: For Example 18.31

Example 18.32. *A uniform frame carries a weight W as shown in Fig. 18.40. Compute the natural period of the system. Assume SDOF in vertical direction.*

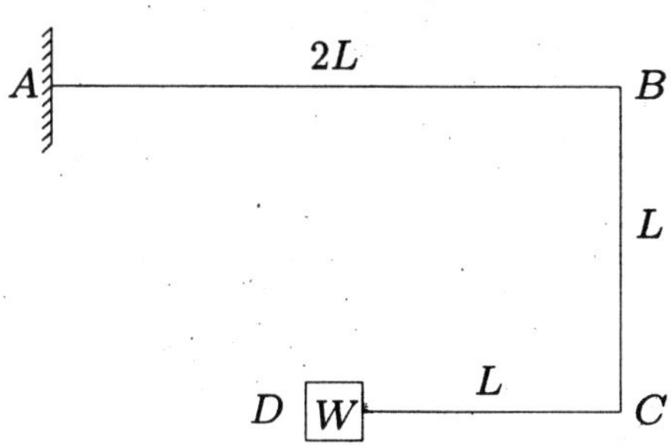

(a) Frame

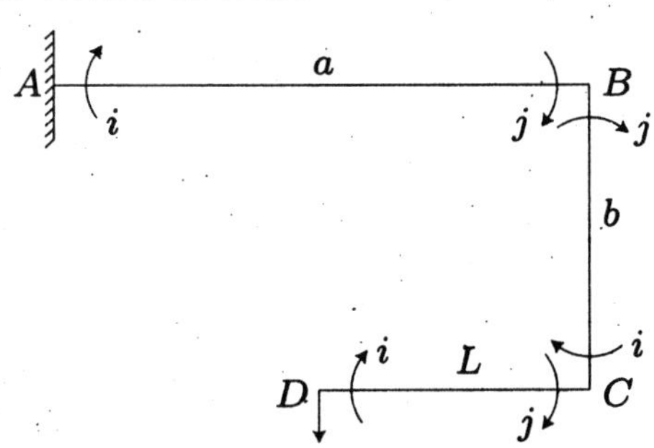

(b) Unit load at D

Fig. 18.40: For Example 18.32

Solution :

$$\begin{matrix} & R_1 = 1 & \\ \begin{bmatrix} Q_i^a \\ Q_j^a \\ Q_i^b \\ Q_j^b \\ Q_i^c \\ Q_j^c \end{bmatrix} = & \begin{bmatrix} 0 \\ +L \\ -L \\ +L \\ -L \\ -L \end{bmatrix} & [R_1 = P] \\ Q \qquad\quad & B_R & R \end{matrix}$$

Flexibility matrix "f" is given by

$$f = \frac{L}{6EI}\begin{bmatrix} 2 & -1 & 0 & 0 & 0 & 0 \\ -1 & 2 & 0 & 0 & 0 & 0 \\ 0 & 0 & 2 & -1 & 0 & 0 \\ 0 & 0 & -1 & 2 & 0 & 0 \\ 0 & 0 & 0 & 0 & 4 & -2 \\ 0 & 0 & 0 & 0 & -2 & 4 \end{bmatrix}.$$

The flexibility matrix F_{RR} can be obtained by the triple product $B_{R^T}\ f\ B_R$

$$F_{RR} = B_R{}^T\ f\ B_R = \frac{2L^3}{EI}$$

$$r = F_{RR}R = \Delta = \frac{2L^3}{EI}(P)$$

$$k = \text{Translational stiffness matrix} = \frac{P}{\Delta} = \frac{EI}{2L^3}$$

$$\tau = \frac{2\pi}{p} = 2\pi\sqrt{\frac{m}{k}} = 2\pi\sqrt{\frac{W}{g}\left(\frac{2L^3}{EI}\right)}.$$

Example 18.33. *The frame shown in Fig. 18.41 carries a weight W. Compute the natural period of the system. Assume SDOF in vertical direction.*

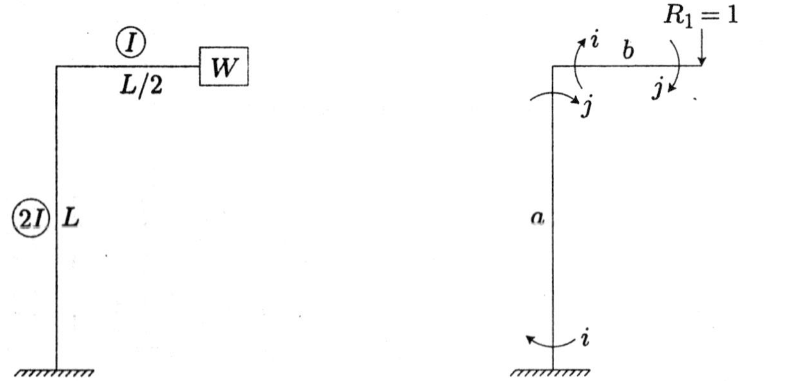

(a) Frame (b) Load in frames due to unit load

Fig. 18.41: For Example 18.33

Solution : To compute the deflection under the weight, replace the same by $R_1 = 1$.

$$\overset{R_1 = 1}{\begin{bmatrix} Q_i^a \\ Q_j^a \\ Q_i^b \\ Q_j^b \end{bmatrix} = \begin{bmatrix} -L/2 \\ +L/2 \\ -L/2 \\ 0 \end{bmatrix}} \; [R_1 = P]$$

$$Q \qquad B_R \qquad R$$

Flexibility matrix "f" is set up as follows :

$$f = \frac{L}{6EI}\begin{bmatrix} 1 & -1/2 & 0 & 0 \\ -1/2 & 1 & 0 & 0 \\ 0 & 0 & 1 & -1/2 \\ 0 & 0 & -1/2 & 1 \end{bmatrix}$$

$$F_{RR} = B_R{}^T \; f \; B_R = [-L/2 + L/2 - L/2.0]$$

$$\frac{L}{6EI}\begin{bmatrix} 1 & -0.5 & 0 & 0 \\ -0.5 & 1 & 0 & 0 \\ 0 & 0 & 1 & -0.5 \\ 0 & 0 & -0.5 & 1 \end{bmatrix}\begin{bmatrix} -L/2 \\ +L/2 \\ -L/2 \\ 0 \end{bmatrix} = \frac{L^3}{6EI}$$

$$r = \Delta = F_{RR}R = \frac{L^3}{6EI}(P) \quad \text{and} \quad k = \frac{P}{\Delta} = \frac{6EI}{L^3}.$$

Example 18.34. *A simply supported beam carries a weight W as shown in Fig. 18.42. Compute the natural period of the system.*

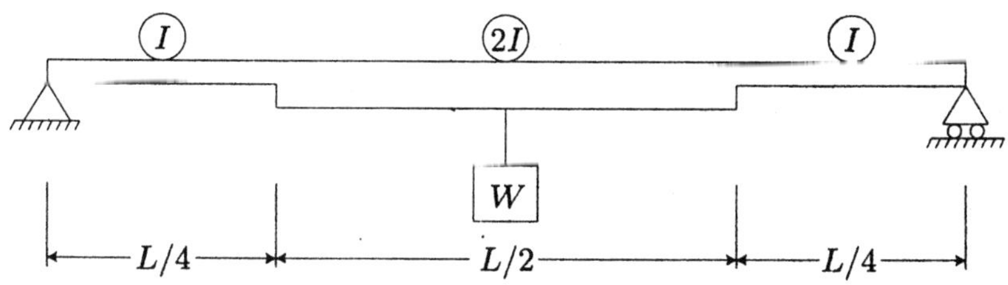

Fig. 18.42: For Example 18.34

Solution : A load P is applied in the direction of motion.

$$\Delta = 2\left[\frac{1}{EI}\int_0^{L/4}\left(\frac{P}{2}x\right)\left(\frac{x}{2}\right)dx + \frac{1}{2EI}\int_{L/4}^{L/2}\left(\frac{P}{2}x\right)\left(\frac{x}{2}\right)dx\right] = \frac{3}{256}\frac{pL^3}{EI}$$

and $$k = \frac{P}{\Delta} = \frac{256EI}{3L^3}$$

$$\tau = \frac{2\pi}{p} = 2\pi\sqrt{\frac{m}{p}} = 2\pi\sqrt{\frac{W}{g} \times \left(\frac{3L^3}{256EI}\right)}.$$

Example 18.35. *A propped cantilever is subjected to a weight W a shown in Fig. 18.43. Compute the natural period of the system.*

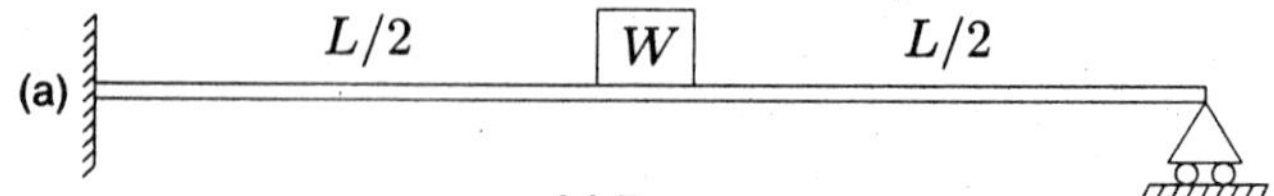

(a) Beam

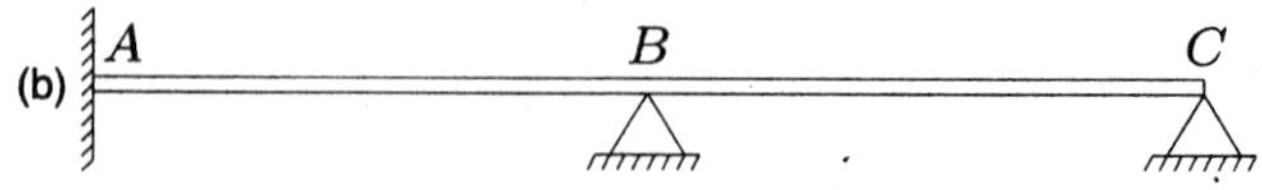

(b) Restrained structure

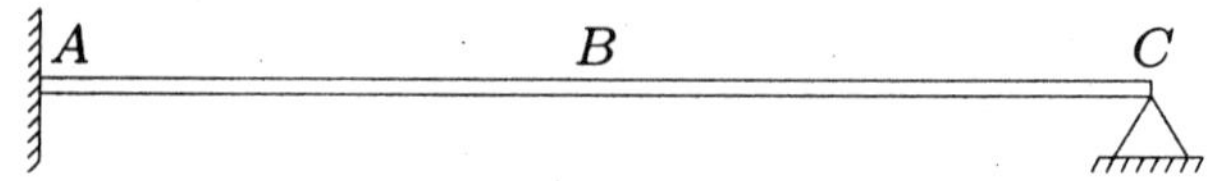

	A	B		C
		$\frac{4}{7}$	$\frac{3}{7}$	
FM	−24	−24	+24	+24
DM	0	0	0	−24
CO	0	0	−12	0
DM	0	$+\frac{48}{7}$	$+\frac{36}{7}$	0
CO	$+\frac{24}{7}$	0	0	0
DM	0	0	0	0
	$-\frac{144}{7}$	$-\frac{120}{7}$	$+\frac{120}{7}$	0
	$\uparrow \frac{528}{7}$	$\frac{528}{7} \downarrow$	$\downarrow \frac{240}{7}$	$\frac{240}{7} \uparrow$

$\downarrow k = \frac{768EI}{7L^3}$

(c) Moment distribution

Fig. 18.43: For Example 18.35

Solution : The structure is a statically indeterminate one. There is only one translational degree of freedom. For obtaining the translational stiffness, the best method would be moment distribution.

Distribution factors are based on restrained structure.

$$AB : BC = \frac{4}{7} : \frac{3}{7} = 4 : 3$$

Now unit displacement is impressed at B in the downward direction.

$$MF_{AB} = MF_{BA} = -6EI/(L/2)^2 = 24EI/L^2.$$

Similarly, $MF_{BC} = MF_{CB} = +24EI/L^2$.

Finding of k, the translational stiffness is shown in Fig. 18.43(c).

$$\tau = \frac{2\pi}{p} = 2\pi\sqrt{\frac{m}{k}} = 2\pi\sqrt{\frac{W}{g}\left(\frac{7L^3}{768EI}\right)}.$$

Example 18.36. *A propped cantilever is subjected to a weight W as shown in Fig. 18.44 Compute the natural period of vertical oscilation of the system.*

Solution : The structure is statically an indeterminate one. There is only one translational degree of freedom at the free end. Distribution Factors are based on the restrained structure.

$$AB : BC = 2 : 3$$

Now unit displacement is impressed at C in the downward direction.

$$MF_{BC} = MF_{CB} = -\frac{6EI(1)}{L^2} = -6 \text{ dropping}$$

for a while the quantity is EI/L^2.

The translational stiffness k is found out as shown in Fig. 18.44.

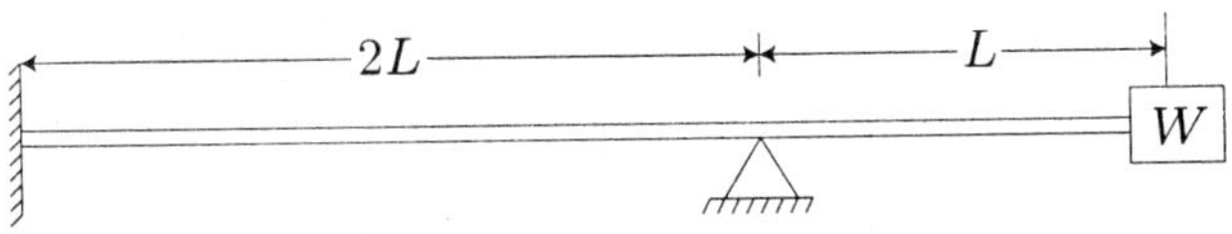

(a) Beam with load

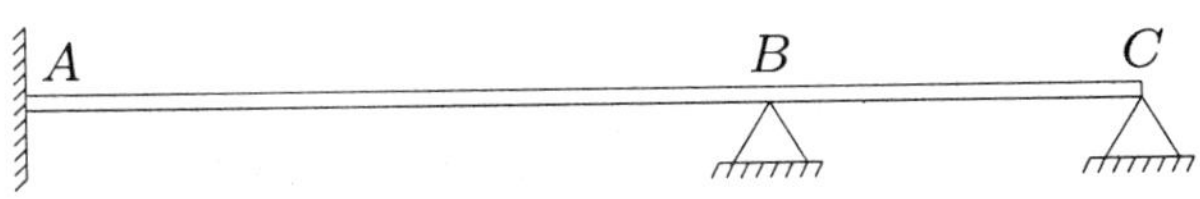

(b) Restrained structure

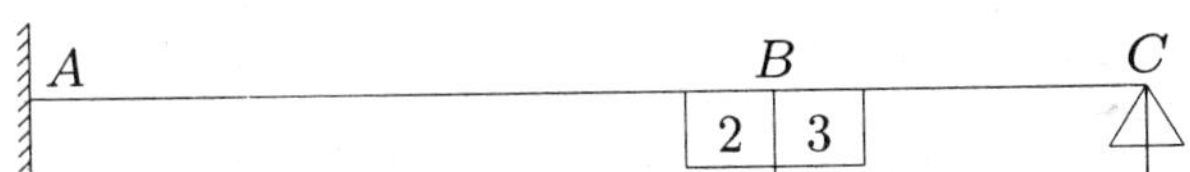

	A	B (AB)	B (BC)	C
FM	0	0	−6	+6
DM	0	$+\frac{12}{5}$	$+\frac{18}{5}$	−6
CO	$+\frac{6}{5}$	0	−3	0
DM	0	$-\frac{6}{5}$	$-\frac{9}{5}$	0
CO	$-\frac{3}{5}$	0	0	0
DM	0	0	0	0
	$+\frac{3}{5}$	$+\frac{6}{5}$	$-\frac{6}{5}$	0

$\downarrow \frac{9}{10}$ $\quad k = \frac{12EI}{10L^3}$ $\quad \frac{9}{10}\uparrow \ \uparrow\frac{12}{10}$ $\quad \frac{12}{10}\downarrow$

(c) Moment distribution

Fig. 18.44: For Example 18.36

$$k = \frac{12EI}{10L^3}$$

$$\text{Natural period } = \tau = \frac{2\pi}{p} = 2\pi\sqrt{\frac{m}{k}} = 2\pi\sqrt{\frac{W}{g}\left(\frac{10L^3}{12EI}\right)}.$$

Example 18.37. *A fixed beam is subjected to a weight W as shown in Fig. 18.45. Compute the natural period of the system.*

Solution : The corresponding stiffness may be obtained by moment distribution method.

D.F. : $AB : BC = 0.50 : 0.50$

Now unit displacement is impressed at B.

$$MF_{AB} = MF_{BA} = -6EI[1/(L/2)^2] = -24EI/L^2$$

$$MF_{BC} = MF_{CB} = +6E(2I)[1/L^2] = +12EI/L^2.$$

The quantity EI/L^2 is dropped for a while.

The translational stiffness k is found out as shown in Fig. 18.45.

$\therefore \quad k = 111EI/L^3.$

$$\text{Natural period } \tau = \frac{2\pi}{p} = 2\pi\sqrt{\frac{m}{k}} = 2\pi\sqrt{\frac{W}{g}\left(\frac{L^3}{111EI}\right)}.$$

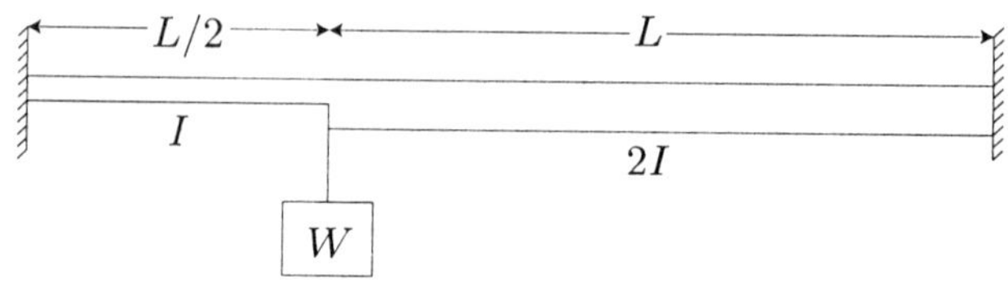

(a) Beam

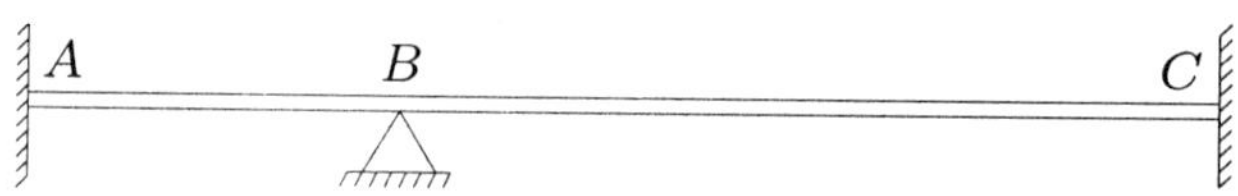

(b) Restrained structure

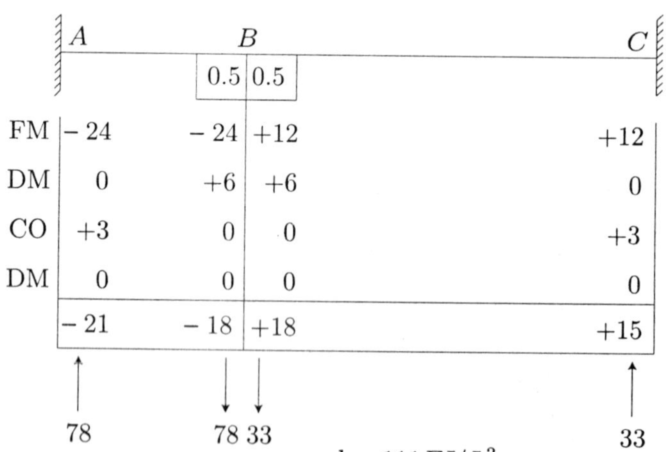

	A	B (0.5)	B (0.5)	C
FM	− 24	− 24	+12	+12
DM	0	+6	+6	0
CO	+3	0	0	+3
DM	0	0	0	0
	− 21	− 18	+18	+15

$k = 111EI/L^3$

(c) Moment distribution

Fig. 18.45: For Example 18.37

Example 18.38. *Compute the natural period of vertical oscillation of the system shown below in Fig. 18.46.*

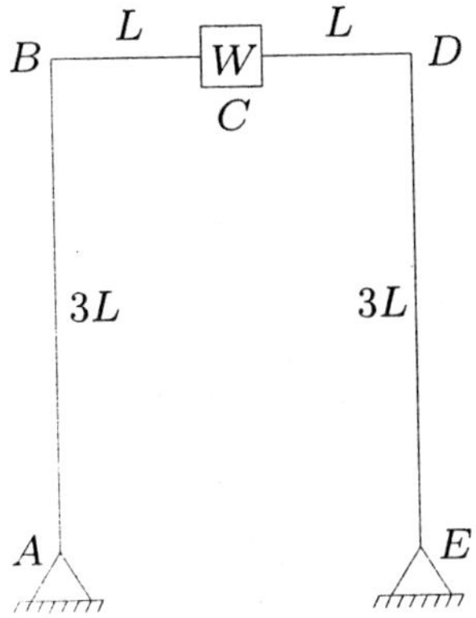

Fig. 18.46: For Example 18.38

Solution : This can be treated as a continuous beam as shown in Fig. 18.47.

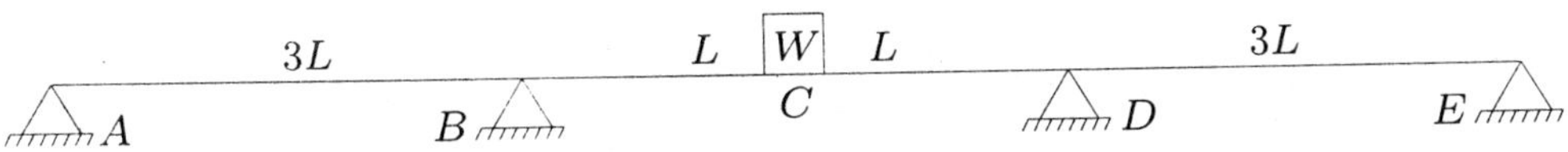

	A	B		C		D		E
	0	0.2	0.8	0.5	0.5	0.8	0.2	0
FM	0	0	− 6	− 6	+6	+6	0	0
DM	0	1.2	4.8	0	0	− 4.8	− 1.2	0
CO	0	0	0	+2.4	− 2.4	0	0	0
DM	0	0	0	0	0	0	0	0
		+1.2	− 1.2	− 3.6	+3.6	+1.2	− 1.2	0
				4.8↓	↓4.8			

$$k = \frac{9.6EI}{L^3}$$

Fig. 18.47: For Example 18.38

Distribution Factors :

$$AB : BC = 0.20 : 0.80$$

Distribution Factors :

$$BC : CD = 0.50 : 0.50$$

The translational stiffness k is obtained as shown in Fig. 18.47

$$k = \frac{9.6EI}{L^3}.$$

$$\text{Natural period } \tau = \frac{2\pi}{p} = 2\pi\sqrt{\frac{W}{g}\left(\frac{L^3}{9.6EI}\right)}.$$

Example 18.39. *Compute the natural period of horizontal vibration for the determinate frame shown in Fig. 18.48.*

Solution : The translational stiffness in the horizontal direction can be easily found using the Dummy Unit Load Method or Matrix Flexibility Method.

By Unit Load Method :

$$\text{For} \quad AB : \Delta_1 = \int_0^L \begin{pmatrix} M \\ Px \end{pmatrix} \begin{pmatrix} m \\ x \end{pmatrix} \frac{dx}{EI} = \frac{PL^3}{3EI}.$$

$$\text{For} \quad BC : \Delta_2 = \int_0^L (Px)(x) \frac{dx}{EI} = \frac{PL^3}{3EI}.$$

$$\Delta = \frac{PL^3}{3EI} + \frac{PL^3}{3EI} = \frac{2PL^3}{3EI} \quad \text{and} \quad k = \frac{P}{\Delta} = \frac{3}{2}\frac{EI}{L^3}.$$

$$\text{Natural period, } \tau = \frac{2\pi}{p} = 2\pi\sqrt{\frac{m}{k}} = 2\pi\sqrt{\frac{W}{g}\left(\frac{2L^3}{3EI}\right)}.$$

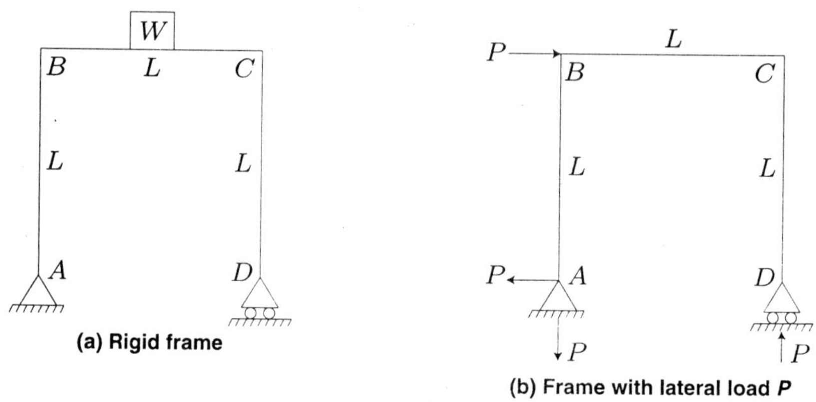

Fig. 18.48: For Example 18.39

Example 18.40. *Compute the natural period of vibration for the system shown in Fig. 18.49. Assume EI is equal to unity.*

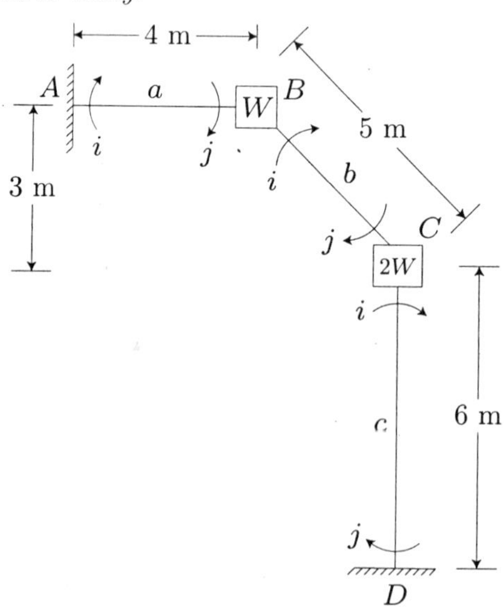

Fig. 18.49: For Example 18.40

The unit translation is applied at the bottom mass as shown in Fig. 18.50 in the right direction.

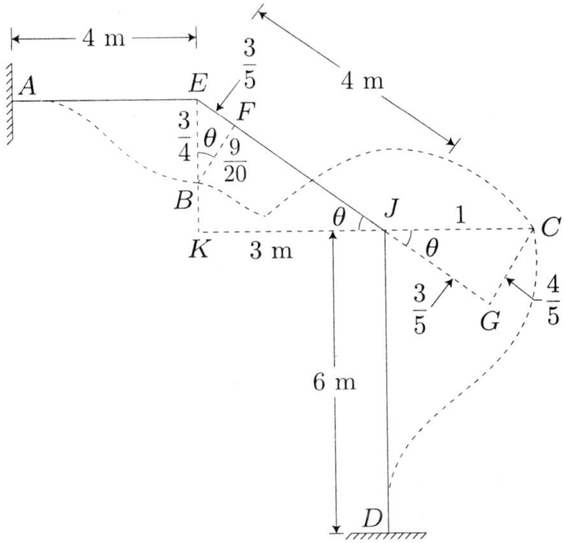

Fig. 18.50: For Example 18.40

From triangle EKJ, $\sin\theta = \dfrac{KE}{EJ} = \dfrac{4}{5}$.

Now $\quad CG = JC\sin\theta = 1\left(\dfrac{4}{5}\right) = \dfrac{4}{5}$

$$JG = JC\cos\theta = 1\left(\frac{3}{5}\right) = \frac{3}{5}.$$

Now $\quad EF = JG = \dfrac{3}{5}$, as there is no shortening.

$$BF = EF\cot\theta = \frac{3}{5}\left(\frac{3}{4}\right) = \frac{9}{20}$$

$$BE = BF\sec\theta = \frac{9}{20}\left(\frac{5}{3}\right) = \frac{3}{4}.$$

The perpendicular distance between B and C is $-\dfrac{9}{20} + \dfrac{4}{5} = \dfrac{5}{4}$.

The displacement transformation matrix A can be easily set up. If r_1 be the translation at C and r_2 and r_3 are rotations at B and C respectively in Fig. 18.49, then

$$\begin{array}{ccc} & r_1 = 1 \quad r_2 = 1 \quad r_3 = 1 & \\ \begin{bmatrix} q_i^a \\ q_j^a \\ q_i^b \\ q_j^b \\ q_i^c \\ q_j^c \end{bmatrix} = & \begin{bmatrix} -3/16 & 0 & 0 \\ -3/16 & +1 & 0 \\ +5/20 & +1 & 0 \\ +5/20 & 0 & +1 \\ -1/6 & 0 & +1 \\ -1/6 & 0 & 0 \end{bmatrix} & \begin{bmatrix} r_1 \\ r_2 \\ r_3 \end{bmatrix} \\ q & A & r \end{array}$$

The '−' (minus) sign is attached to the various coefficients because of the fact that angle measured from the chord to the tangent is anticlockwise.

The unassembled stiffness matrix 's' is formulated as

$$s = \begin{bmatrix} 4/4 & 2/4 & 0 & 0 & 0 & 0 \\ 2/4 & 4/4 & 0 & 0 & 0 & 0 \\ 0 & 0 & 4/5 & 2/5 & 0 & 0 \\ 0 & 0 & 2/5 & 4/5 & 0 & 0 \\ 0 & 0 & 0 & 0 & 4/6 & 2/6 \\ 0 & 0 & 0 & 0 & 2/6 & 4/6 \end{bmatrix}.$$

Matrix product $A^T\ s\ A$ will yield the total stiffness matrix, k, as follows :

$$k = \left[\begin{array}{c|cc} 0.31100 & 0.01875 & 0.13333 \\ \hline 0.01875 & 1.8000 & 0.40000 \\ 0.13333 & 0.40000 & 1.4666 \end{array}\right]$$

The translational stiffness k', will be obtained by suitably partitioning as shown above and applying the condensation formula given below :

$$k' = k_{11} - k_{12}k_{22}^{-1}k_{21}$$

$$k' = 0.3110 - [0.01875\ \ 0.1333]\ \frac{1}{(2.4788)}\begin{bmatrix} 1.4666 & -0.4 \\ -0.4 & 1.80 \end{bmatrix}\begin{bmatrix} 0.01875 \\ 0.13333 \end{bmatrix}$$

$$k^t = 0.31100 - 0.0123072 = 0.298693 \text{ units.}$$

The present system vibrates in single degree of freedom. In any mode, the masses reach their extreme position and cross the neutral equilibrium position simultaneously. The velocity of motion is maximum when the mass crosses the equilibrium position. The formula is $p^2 = \frac{k}{m}$ in the case of single degree of freedom.

Considering the bottom mass, if the motion be described by the equation

$$x_1 = a_1 \sin\ pt,$$

$$\frac{dx_1}{dt} = a_1 p \cos\ pt$$

$$\nu_{1\,\max} = a_1\ p.$$

Similarly, for the top mass,

$$x_2 = a_2 \sin\ pt,$$

$$\frac{dx_2}{dt} = a_2 p \cos\ pt$$

$$\nu_{2\,\max} = a_2 p.$$

If the equivalent mass be concentrated in the bottom, equating the kinetic energy

$$\frac{1}{2}\frac{W}{g}\left(\frac{3}{4}p\right)^2 + \frac{1}{2}\frac{2W}{g}(1p)^2 = \frac{1}{2}M_{eq}(1p)^2$$

$$M_{eq} = \left[\frac{9}{16} + 2\right]\frac{W}{g} = \frac{41}{16}\frac{W}{g}$$

and $$p^2 = \frac{k'}{M_{eq}} = \frac{0.298697}{41(W)}(g)(16) = 0.11656\,\frac{g}{W}.$$

Example 18.41. *What is the natural frequency of motion for block A for small oscillation? Consider the beam BC to have negligible mass and infinitely rigid and body A to be a particle. It is known that when body A is attached to the rod there is a static deflection of 0.25 m and the spring constant k_1 is 2000 N/m. Body A has a mass 12.5 N sec^2 and $g = 9.81$ m/sec^2 of 12.5 N sec^2 and $g = 9.81$ m/sec^2.*

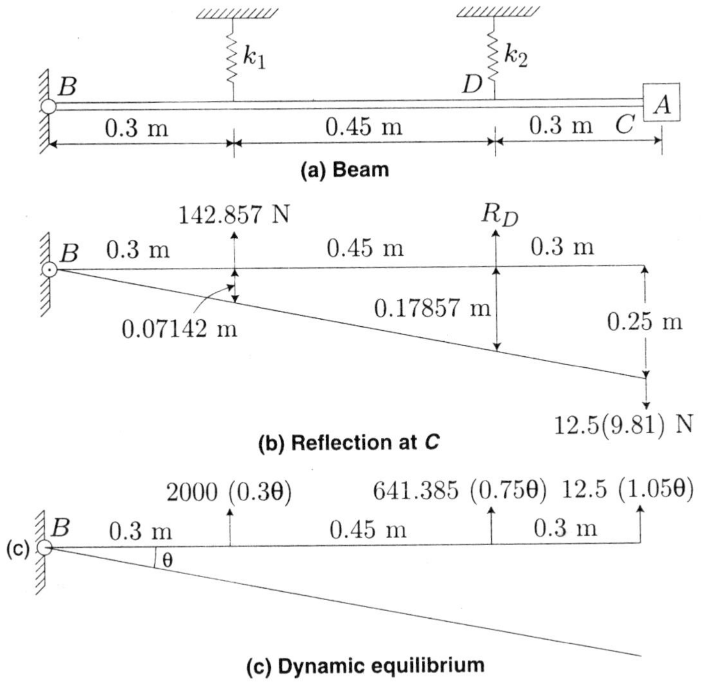

Fig. 18.51: For Example 18.41

Solution : To find k_2 :

Consider Fig. 18.51(b), when there is a downward deflection of 0.25 m at the free end, the deflection under k_1 is given by

$$\Delta = \frac{0.25}{1.05}(0.3) = 0.0714285 \text{ m}.$$

Force in the spring $= 2000(0.0714285) = 142.857$ N

If the force in spring at D be R_D, taking moment of forces about B,

$$R_D(0.75) = 12.5(9.81)(1.05) - 0.3(142.857)$$

$$R_D = 114.5322 \text{ N}.$$

The deflection at $D = 0.17857$ m

and stiffness $$k_3 = \frac{R_D}{\Delta} = \frac{114.5322}{0.17857} = 641.385 \text{ N/m}.$$

For a small angular displacement θ, the forces maintaining the dynamic equilibrium are shown in Fig. 18.51(c). Taking moment of the forces about B,

$$12.5(1.05)(1.05\ddot{\theta}) + 2000(0.3)(0.3\theta) + 641.385(0.75)(0.75\theta) = 0$$

$$13.7812\ddot{\theta} + 540.779\theta = 0 \quad \text{and} \quad p^2 = \frac{540.779}{13.7812} = 39.24.$$

and $p = 6.264$ rad/ sec.

Example 18.42. *Compute the natural period of the system shown in Fig. 18.52. The arm carrying the load is infinitely rigid. Assume SDOF in vertical direction.*

Solution : The load W is replaced by a force P and the components of forces acting on the system are shown in Fig. 18.52(b).

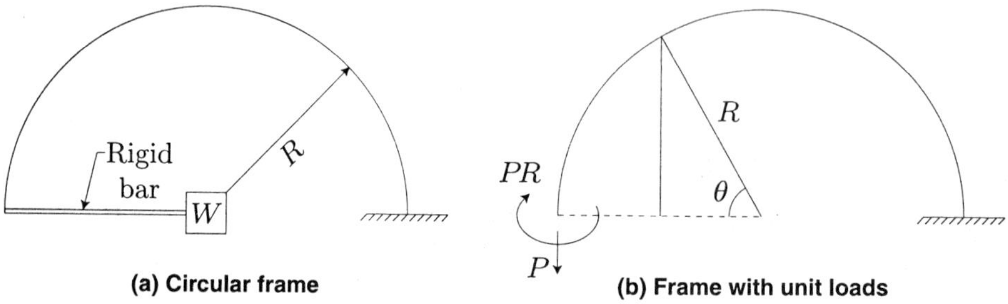

Fig. 18.52: For Example 18.42

$$\Delta = \int_0^{\pi} \left[\overset{M}{P(R - R\cos\theta) - PR} \right] \left[\overset{m}{(R - R\cos\theta) - R} \right] \frac{Rd\theta}{EI}$$

$$= \int_0^{\pi} \frac{PR^8}{EI} \cos^2\theta d\theta = \frac{\pi PR^3}{2EI}$$

$$k = \frac{P}{\Delta} = \frac{2EI}{\pi R^3}.$$

$$\text{and natural period, } \tau = \frac{2\pi}{p} = 2\pi\sqrt{\frac{m}{k}} = 2\pi\sqrt{\frac{W}{g}\left(\frac{\pi R^3}{2EI}\right)}.$$

Example 18.43. *Compute the natural period of the system shown in Fig. 18.53. EI is uniform throughout. Assume SDOF in vertical direction.*

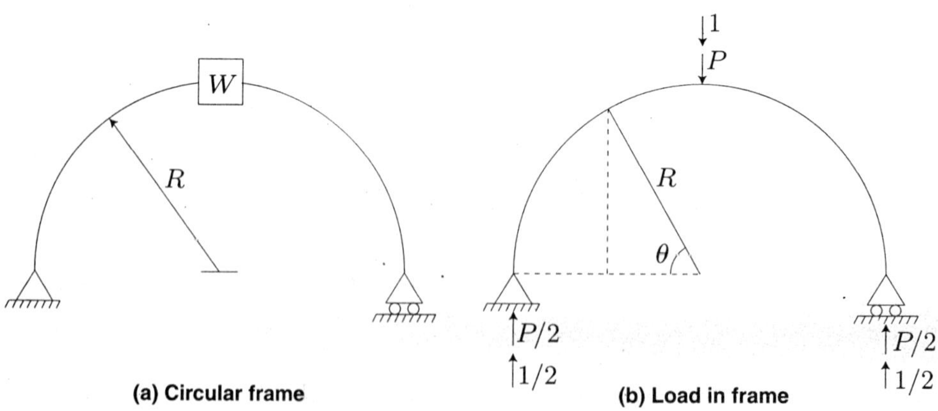

Fig. 18.53: For Example 18.43

Solution : The weight W is replaced by a force P as shown in Fig. 18.53(b).

$$\Delta = 2\int_0^{\pi/2}\left[\frac{P}{2}R(1-\cos\theta)\right]R(1-\cos\theta)\left(\frac{1}{2}\right)\frac{Rd\theta}{EI}$$

$$\Delta = 0.178\frac{PR^3}{EI}$$

$$k = \frac{P}{\Delta} = \frac{EI}{0.178R^3}.$$

Natural period $\tau = 2\pi\sqrt{\frac{m}{k}} = 2\pi\sqrt{\frac{W}{g}\left(\frac{0.178R^3}{EI}\right)}.$

Example 18.44. *Compute the natural period of vertical oscillation of the system shown below in Fig. 18.54. Assume SDOF in vertical direction.*

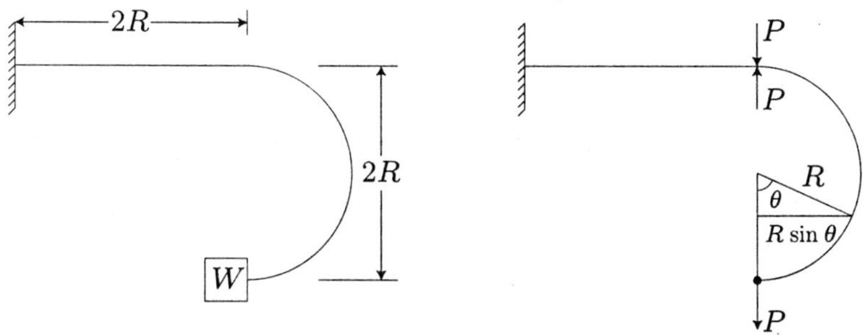

(a) Load in frame (b) Unit load in frame

Fig. 18.54: For Example 18.44

Solution : To compute the vertical deflection at the point of application of the load, replace the weight W by a force P as shown in Fig. 18.54(b).

$$\Delta_1(\text{circular portion}) = \int_0^{\pi}[PR\sin\theta][R\sin\theta]\frac{Rd\theta}{EI}$$

$$\Delta_1 = \frac{\pi PR^3}{2EI}$$

$$\Delta_2(\text{horizontal portion}) = \int_0^{2R}[Px][x]\frac{dx}{EI} = \frac{8PR^3}{3EI}$$

$$\Delta = \Delta_1 + \Delta_2 = \frac{PR^3}{EI}\left[\frac{\pi}{2}+\frac{8}{3}\right] = 4.237\frac{PR^3}{EI}.$$

Stiffness, $k = \frac{P}{\Delta} = \frac{EI}{4.237R^3}.$

Natural period, $\tau = 2\pi\sqrt{\frac{m}{k}} = 2\pi\sqrt{\frac{W}{g}\left(\frac{4.237R^3}{EI}\right)}.$

Example 18.45. *Compute the natural period of vertical oscillation of the system shown below in Fig. 18.55. Consider only flexural deformation of the system. Assume SDOF in vertical direction.*

Solution : The plan of the system is as shown in Fig. 18.55(b). For computing the vertical deflection, replace the weight W by a force P.

$$\Delta = \int_0^{\pi/2} [PR\sin\theta][R\sin\theta]\frac{Rd\theta}{EI} = \frac{\pi PR^3}{4EI}$$

$$k = \frac{P}{\Delta} = \frac{4EI}{\pi R^3}.$$

Natural period, $\tau = 2\pi\sqrt{\frac{m}{k}} = 2\pi\sqrt{\frac{W}{g}\left(\frac{\pi R^3}{4EI}\right)}$, as $\tau = \frac{2\pi}{p}$.

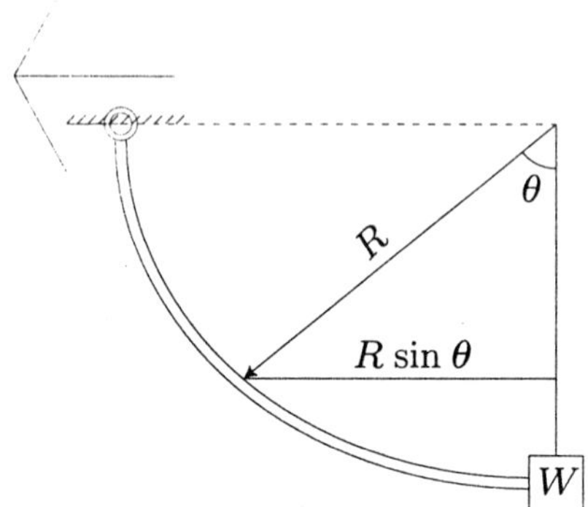

Curved Frame (Plain)

Fig. 18.55: For Example 18.45

Example 18.46. *Compute the natural period of horizontal oscillation of the system shown in Fig. 18.56. EI is uniform throughout. Assume SDOF in horizontal direction.*

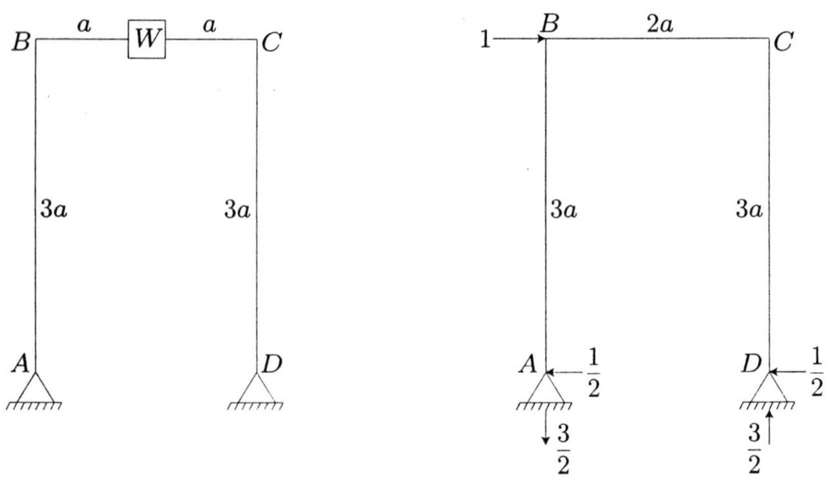

(a) Frame **(b) Unit load**

Fig. 18.56: For Example 18.46

Solution : For computing the horizontal deflection, replace the weight W by a horizontal force P as shown in Fig. 18.56(b). The system is a statically indeterminate onc. However, the horizontal thrust at A and D can be easily determined because of symmetry as $P/2$.

For AB, $\Delta_1 = \int_0^{3a} \left(\frac{P}{2}x\right)\left(\frac{1}{2}x\right)\frac{dx}{EI} = \frac{27}{12}\frac{Pa^3}{EI}$

For DC, $\Delta_2 = \Delta_1$

For BC, $\Delta_3 = \int_0^{2a} \left[\frac{3Pa}{2} - \frac{3Px}{2}\right]\left[\frac{3}{2}a - \frac{3}{2}x\right]\frac{dx}{EI} = \frac{18}{12}\frac{Pa^3}{EI}$.

Total $\Delta = \frac{Pa^3}{EI}\left[\frac{27}{12} + \frac{27}{12} + \frac{18}{12}\right] = \frac{6Pa^3}{EI}$.

Stiffness, $k = \frac{P}{\Delta} = \frac{EI}{6a^3}$.

Natural period, $\tau = 2\pi\sqrt{\frac{m}{k}} = 2\pi\sqrt{\frac{W}{g}\left(\frac{6a^3}{EI}\right)}$.

Example 18.47. *Compute the natural period of horizontal oscillation of the system shown in Fig. 18.57.*

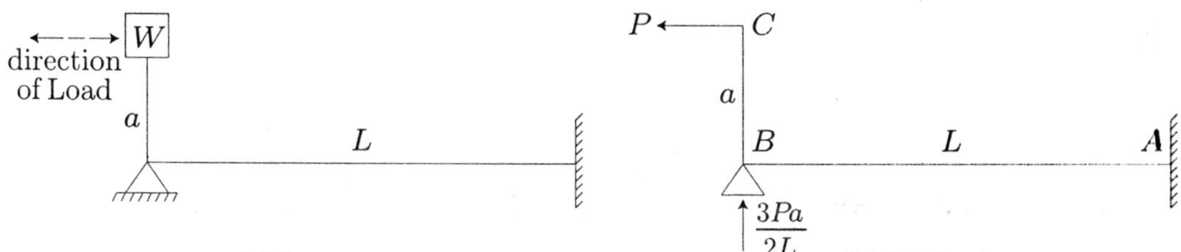

(a) Frame **(b) Unit load**

Fig. 18.57: For Example 18.47

Solution : For computing the horizontal deflection, the weight W is replaced by a force P as shown in Fig. 18.57(b). The system is statically indeterminate to first degree.

Deflection (horizontal)

For BC, $\Delta_1 = \frac{1}{EI}\int_0^a (Px)(x)dx = \frac{Pa^3}{3EI}$

For AB, $\Delta_2 = \frac{1}{EI}\int_0^L \left[Pa - \frac{3Pa}{2L}x\right]\left[a - \frac{3ax}{2L}\right]dx = \frac{Pa^2L}{4EI}$.

$$\Delta \doteq \Delta_1 + \Delta_2 = \frac{Pa^2}{12EI}[3L + 4a].$$

Example 18.48. *Compute the natural period of vertical oscillation of the system shown in Fig. 18.58, EI is uniform throughout. Assume SDOF in vertical direction.*

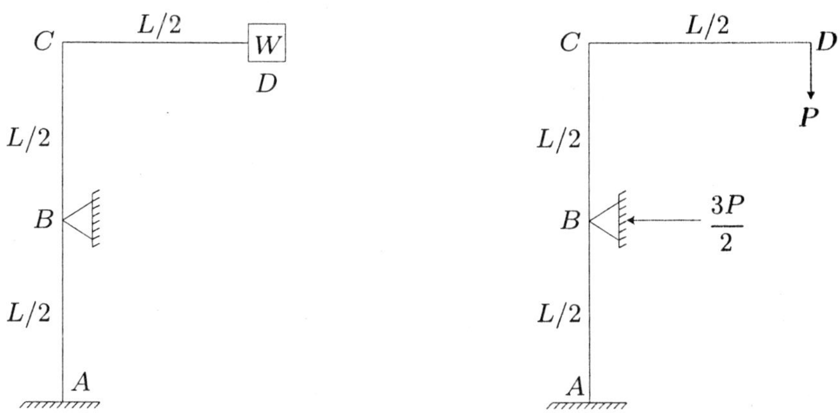

(a) Frame **(b) Load *P* at *D***

Fig. 18.58: For Example 18.48

Solution : The value of the redundant at B is found out by the standard procedures and is shown in Fig. 18.58(b). The vertical deflection is obtained as follows :

Portion CD :

$$\Delta_1 = \frac{1}{EI}\int_0^{L/2}\left[\begin{array}{c} M \\ Px \end{array}\right]\left[\begin{array}{c} m \\ x \end{array}\right] dx = \frac{PL^3}{24EI}$$

Portion BC : (origin at C)

$$\Delta_2 = \frac{1}{EI}\int_0^{L/2}\left[\frac{PL}{2}\right]\left[\frac{L}{2}\right] dx = \frac{PL^3}{8EI}$$

For Portion AB : (origin at B)

$$\Delta_3 = \frac{1}{EI}\int_0^{L/2}\left[\begin{array}{c} M \\ \frac{PL}{2} - \frac{3P}{2}x \end{array}\right]\left[\begin{array}{c} m \\ \frac{L}{2} - \frac{3}{2}x \end{array}\right] dx = \frac{3PL^3}{32EI}.$$

$$\therefore \quad \Delta = \frac{PL^3}{96EI}[9+12+4] = \frac{25PL^3}{96EI}$$

$$k = \frac{P}{\Delta} = \frac{96EI}{25L^3}.$$

and natural period, $\tau = 2\pi\sqrt{\frac{m}{k}} = 2\pi\sqrt{\frac{W}{g}\left(\frac{25L^3}{96EI}\right)}.$

18.6 Gravity Type of Free Vibration

There are systems in which the free vibration is maintained by gravitational force present in the system. However, this type of motion is rarely met with in civil engineering practice.

18.7 Rayleigh's Method

In applying the energy method to systems with distributed masses, such as the beam, the deflection configuration of the system is necessary for the evaluation of the kinetic and potential energies. Rayleigh showed that the fundamental frequency of such systems can be determined with good accuracy by assuming any reasonable deflection curve. If the true deflection curve of the vibrating system is assumed, the fundamental frequency found by this method will be the correct frequency. For any other curve, the frequency determined will be higher than the correct frequency. In general, the use of the static deflection curve of the elastic body results in a fairly accurate value of the frequency. If greater accuracy is desired, the approximate curve can be repeatedly improved.

Although Rayleigh's method is applied to systems of more than one degree of freedom, only the fundamental frequency can be found thereby. Also, the calculated frequency can never be lower than the true fundamental frequency of the system. In the examples to follow, the energy method is used to lump the distributed mass at a particular point where the stiffness is known. Thus in effect, the distributed system is reduced to an equivalent single degree of freedom system.

Example 18.49. *Fig. 18.59 shows a system with spring. It is desired to determine the effect of the mass of the spring on the vibration frequency.*

If the lower end of the spring has a maximum displacement of x_0, then any point a distance y from the fixed end can be assumed to have a displacement and velocity of

$$\frac{y}{L} x_0 \sin \omega t, \text{ and } \frac{y}{L} \omega x_0 \cos \omega t,$$

where L is the length of the spring at static equilibrium and $\omega = 2\pi f$.

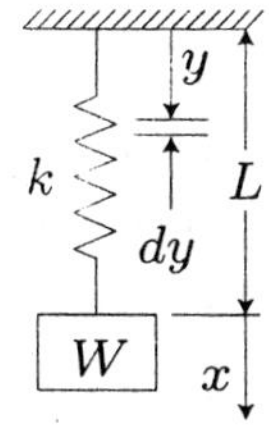

Fig. 18.59

The maximum kinetic energy of an element dy of the spring, a distance y from the fixed end is then equal to

$$dT'_{\max} = \frac{w}{2g} dy \left(\frac{y}{L} \omega x_0 \right)^2,$$

where w is the weight per unit length of the spring. Integrating, the maximum kinetic energy of the spring becomes

$$T'_{\max} = \frac{w}{2g} \left(\frac{\omega x_0}{L} \right)^2 \int_0^L y^2 dy = \frac{1}{2} \left(\frac{wL}{3g} \right) \omega^2 x_0^2.$$

Adding this to the kinetic energy of the rigid mass, the total kinetic energy of the system becomes

$$T_{\max} = \frac{1}{2} \left(\frac{W + \frac{1}{3} wL}{g} \right) \omega^2 x_0^2.$$

The potential energy of the system is equal to

$$U_{\max} = \int_0^{x_0} k(\Delta + x) dx - W x_0 = \frac{1}{2} k x_0^2.$$

Equating the two, the natural frequency of the system becomes

$$f = \frac{\omega}{2\pi} = \frac{1}{2\pi} \sqrt{\frac{kg}{W + \frac{1}{3} wL}}.$$

It is evident from this example that for a spring-mass system of Fig. 18.59, the effect of the mass of the spring can be taken into account by adding one-third its weight to the weight W of the rigid mass.

Example 18.50. *A cantilever beam is shown in Fig. 18.60. The system represents a continuous elastic body and hence possesses infinite number of degrees of freedom. However, by assuming the form of the deflection curve to be known, Rayleigh's method enables the system to be treated as one with a single degree of freedom.*

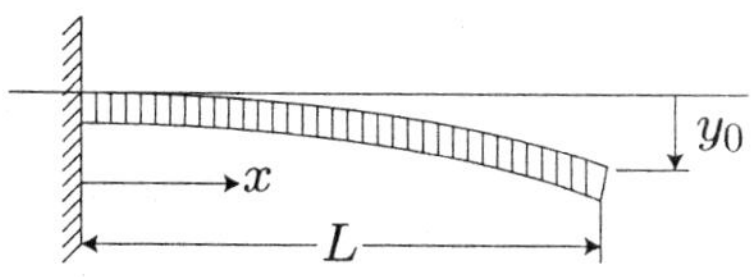

Fig. 18.60: For Example 18.50

Solution : The amplitude of the beam at any point x is given with sufficient accuracy by the statical deflection curve of a light cantilever beam with a concentrated load at the end. Writing this equation in the form,

$$y = \frac{1}{2y_0} \left[3 \left(\frac{x}{L} \right)^2 - \left(\frac{x}{L} \right)^3 \right],$$

where $y_0 = PL^3/3EI$ is the amplitude of the free end, the stiffness at the free end becomes $k = P/y_0 = 3EI/L^3$. The potential energy which is equal to the work done, is then

$$U_{\max} = \frac{1}{2}ky_0^2 = \frac{3EI}{2L^3}y_0^2,$$

which is justified only when statical deflection curve of the beam is used. The kinetic energy is next determined by integrating the product of mass and the square of the velocity over the length of the beam.

$$T_{\max} = \frac{w}{2g}\int_0^L (\omega y)^2 dx = \frac{w}{2g}\left(\frac{\omega y_0}{2}\right)\int^L \left[3\left(\frac{x}{L}\right)^2 - \left(\frac{x}{L}\right)^3\right] dx$$

$$= \frac{1}{2}\left(\frac{33wL}{140g}\right)\omega^2 y_0^2.$$

The above equation indicates that for the assumed deflection curve, the continuous beam of w N/m is equivalent in vibration characteristics to that of a weightless beam with a concentrated weight $(\frac{33}{140}wl)$ at the end.

Equating the two energies, the fundamental frequency of vibration in radians per second becomes

$$\omega = \sqrt{\frac{(3EI/L^3)g}{\frac{33}{140}wL}} = 3.56\sqrt{\frac{gEI}{wL^4}}.$$

Applying the result to the extreme case, where $W = 0$,

$$\delta_{st} = \frac{33}{140}wL\frac{L^3}{3EI}.$$

The corresponding period of vibration will be

$$\tau = 2\pi\sqrt{\frac{\delta_{st}}{g}} = \frac{2\pi}{3.567}\sqrt{\frac{wL^4}{EIg}}.$$

The exact solution for the same case is

$$\tau = \frac{2\pi}{3.515}\sqrt{\frac{wL^4}{EIg}}.$$

It is seen that the error of the approximate solution is about $1\frac{1}{2}\%$. For the case $W = 0$, a better approximation can be obtained. It is only necessary to assume that during the vibration the shape of the deflection curve of the beam is the same as the one produced by a uniformly distributed load. The deflection y_0 at any cross section distant c from the built-in section will then be given by the following equation,

$$y_0 = x_0\left[-\frac{1}{3} + \left(\frac{4c}{3L}\right) + \frac{1}{3}\left(1 - \frac{c}{L}\right)^4\right],$$

in which $x_0 = \dfrac{wL^4}{8EI}$

represents the deflection of the end of the cantilever.

The potential energy of bending will be

$$V = \frac{w}{2}\int_0^L y_0 dc = \frac{8}{5}\cdot\frac{EIx_0^2}{L^3}.$$

The kinetic energy of the vibrating beam is

$$T = \frac{1}{2}\int_0^L \frac{w}{g}\dot{y}^2 dc.$$

Taking $y = y_0 \cos pt$ and $(\dot{y})_{\max} = y_0 p$,

for determining p,

$$\frac{1}{2}\int_0^L \frac{w}{g}(y_0 p)^2 dc = \frac{8}{5}\frac{EIx_0^2}{L^3}.$$

Substituting for y_0 and performing the integration,

$$p = 3.530\sqrt{\frac{EIg}{wL^4}}.$$

The corresponding period of vibration is

$$\tau = \frac{2\pi}{3.530}\sqrt{\frac{wL^4}{EIg}}.$$

Comparing this result with the exact solution it can be concluded that in this case the error of the approximate solution is only about $\frac{1}{2}\%$.

As a second example, consider the case of vibration of a beam of uniform cross section loaded at the middle (see Fig. 18.61). If the weight wL of the beam being small in comparison with the load W. It can be assumed with sufficient accuracy that the deflection curve of the beam during vibration has the same shape as the statical deflection curve. Then, denoting by x the displacement of the load W during vibration, the displacement of any element wdc of the beam at a distance c from support will be

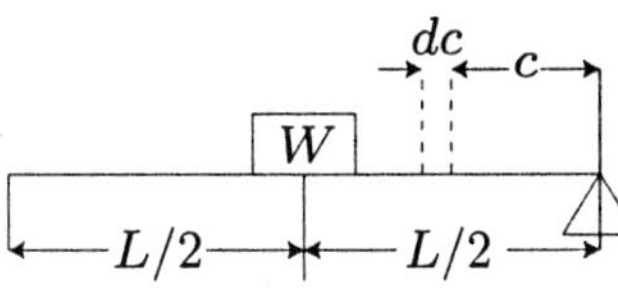

Fig. 18.61

$$x\frac{3cL^2 - 4c^3}{L^3}.$$

Kinetic Energy of the beam will be

$$2\int_0^{\frac{L}{2}} \frac{w}{2g}\left(\dot{x}\frac{3cL^2 - 4c^3}{L^3}\right)^2 dc = \frac{17}{35}wL\frac{\dot{x}^2}{2g}.$$

Kinetic Energy of vibrating beam must be added to the energy $W\dot{x}^2/2g$ of the load concentrated at the middle in order to estimate the effect of the weight of the beam on the period of vibration, i.e., the period of vibration will be the same as for a massless beam loaded at the middle by the load

$$W + (17/35)wL.$$

It must be noted that value of equivalent load thus obtained on the assumption that the weight of the beam is small in comparison to load W can be used in all practical cases. Even in the extreme case, when $W = 0$ and where the assumption is made that $(17/35)wL$ is concentrated at middle of the beam, the accuracy of the approximate method is sufficiently close for all practical cases. The deflection of the beam under the action of the load $(\frac{17}{35})wL$ applied at the middle is

$$\delta_{st} = \frac{17}{35}wL\frac{L^3}{48EI}.$$

With this the period of the natural vibration is

$$\tau = 2\pi\sqrt{\frac{\delta_{st}}{g}} = 0.632\sqrt{\frac{wL^4}{EIg}}.$$

The exact solution comes to

$$\tau = \frac{2}{\pi}\sqrt{\frac{wL^4}{EIg}} = 0.637\sqrt{\frac{wL^4}{EIg}}.$$

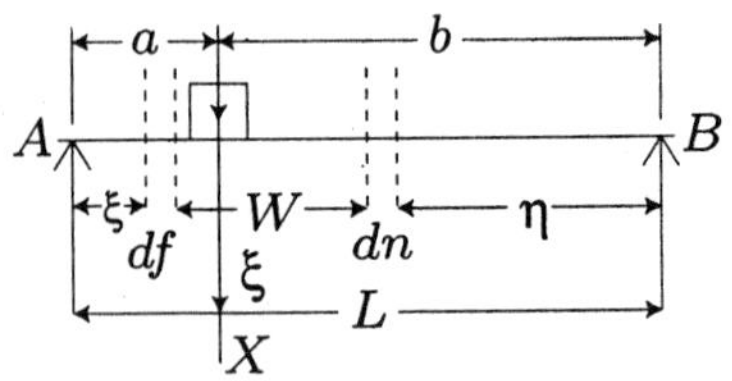

Fig. 18.62: For Example 18.51

Example 18.51. *Determine the frequency of natural vibrations of the load W supported by a beam AB, (Fig. 18.62), of constant cross section. (a) Assuming that the weight of the beam can be neglected; (b) taking the weight of the beam into consideration and using Rayleigh's method, obtain frequency of vibration.*

Solution : If a and b are the distances of the load from the ends of the beam the static deflection of the beam under the load is $\delta = Wa^2b^2/3LEI$. Taking for the spring constant the expression $k = 3LEI/a^2b^2$ and neglecting the mass of the beam the angular frequency of vibration is obtained from the equation of energy,

$$\frac{W}{2g}\dot{x}^2_{\max} = \frac{kx_0}{2}. \tag{a}$$

in which $x_{\max} = x_0 p$. Hence,

$$p = \sqrt{\frac{kg}{W}} = \sqrt{\frac{3LEIg}{Wa^2b^2}}.$$

To take the mass of the beam into account, the deflection curve of the beam under static action of the load W is considered. The deflection at any point of the left portion of the beam at the distance ξ from the support A is

$$x_1 = \frac{W\xi b}{6LEI}[a(L+b) - \xi^2].$$

For the deflection at any point to the right of the load W and at a distance η from the support B,

$$x_2 = \frac{Wa\eta}{6LEI}[b(L+a) - \eta^2].$$

Applying Rayleigh's method and assuming that during vibration the maximum velocity of any point of the left portion of the beam at a distance ξ from the support A is given by the equation

$$(\dot{x_1})_{\max} = \dot{x}_{\max}\frac{x_1}{\delta} = \dot{x}_{\max}\frac{\xi}{2a^2b}[a(L+b) - \xi^2],$$

in which $\dot{x}_{\max}$ is the maximum velocity of the load W, To take into account the mass of the left portion of the beam, to the left side of Eq. (a) the quantity to be added is,

$$\frac{w\dot{x}_{\max}^2}{2g}\int_0^a\left(\frac{x_1}{\delta}\right)^2 d\xi = \frac{w\dot{x}_{\max}^2}{2g}\int_0^a \frac{L^2}{4a^4b^2}[a(L+b)-\xi^2]^2 d\xi$$

$$= \dot{x}_{\max}^2\frac{wa}{2g}\left[\frac{1}{3}\frac{L^2}{b^2}+\frac{23}{105}\frac{a^2}{b^2}-\frac{8}{15}\frac{aL}{b^2}\right].$$

In the same manner, considering the right portion of the beam, add to the left side of above the expression

$$\frac{\dot{x}_{\max}^2 wb}{2g}\left[\frac{1}{12}\frac{(L+a)^2}{a^2}+\frac{1}{28}\frac{b^2}{a^2}-\frac{1}{10}\frac{b(L+a)}{a^2}\right].$$

The equation of energy becomes

$$\frac{(W+\alpha wa+\beta wb)}{2g}\dot{x}_{\max}^2 = \frac{kx_0^2}{2},$$

where α and β denote the quantities in the brackets of expressions (k) and (l) and we obtain for the angular frequency of vibration the following formula :

$$p=\sqrt{\frac{3LEIg}{(W+\alpha aw+\beta bw)a^2b^2}}.$$

Example 18.52. *Determine the frequency of the natural vertical vibrations of the load W supported by a frame hinged at A and B, Fig. 18.63(a),. assuming that the three members of the frame have the same length and the same cross section and the load is applied at the middle of the bar CD. In the calculation (a) neglect the mass of the frame; (b) consider the mass of the frame by using Rayleigh's method.*

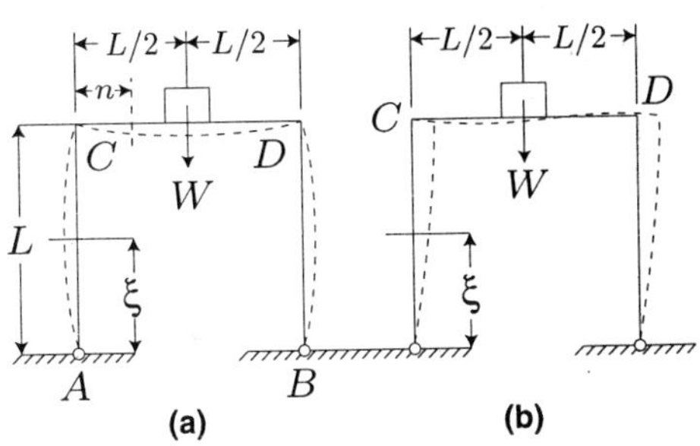

Fig. 18.63: For Example 18.52

Solution : By using the known formulas for deflections of beams, we find that the bending moments at the joints C and D are equal to $3WL/40$. The deflections of vertical bars at a distance ξ from the bottom is

$$x_1 = \frac{3WL^2\xi}{240EI}\left(1-\frac{\xi^2}{L^2}\right).$$

The deflections of the horizontal bar to the left of the load is

$$x_2 = \frac{W\eta}{48EI}(3L^2-4\eta^2)-\frac{3}{80}\frac{WL}{EI}\eta(L-\eta).$$

The deflections under the load W is

$$\delta=(x_3)_{\eta=\frac{L}{2}}=\frac{11}{960}\frac{WL^3}{EI}.$$

By neglecting the mass of the frame the angular frequency,

$$p=\sqrt{\frac{g}{\delta}}=\sqrt{\frac{960EIg}{11WL^3}}.$$

In calculating the effect of this mass on the frequency, denoting by $\dot{x}_{\max}$ the maximum velocity of the vibrating body W, the maximum velocity of any point of the vertical bars at a distance ξ from the bottom is

$$(\dot{x}_1)_{\max} = \dot{x}_{\max}\frac{x_1}{\delta} = \dot{x}_{\max}\frac{12}{11}\frac{\xi}{L}\left(1-\frac{\xi^2}{L^2}\right) \tag{a}$$

and the maximum velocity at any point of the left portion of the horizontal bar CD,

$$(\dot{x}_2)_{\max} = \dot{x}_{\max}\frac{x_2}{\delta} = \dot{x}_{\max}\left[\frac{20}{11}\frac{\eta}{L}\left(3-\frac{4\eta^2}{L^2}\right) - \frac{36}{11}\frac{\eta}{L}\left(1-\frac{\eta}{L}\right)\right]. \tag{b}$$

The kinetic energy of the frame which to be added to the kinetic energy of the load W is

$$2\int_0^L \frac{w\dot{x}^2_{\max}}{2g}\left(\frac{x_1}{\delta}\right)^2 d\xi + 2\int_0^{L/2}\frac{w\dot{x}^2_{\max}}{2g}\left(\frac{x_2}{\delta}\right)^2 d\eta.$$

Substituting for the ratios x_1/δ and x_2/δ their expressions from (a) and (b) and integrating, the additional kinetic energy can be represented in the following form

$$\frac{w\alpha L}{2g}(\dot{x})^2_{\max},$$

where α is a constant factor.

The equation for angular frequency of vibration now becomes

$$p = \sqrt{\frac{960EIg}{11(W+\alpha wL)L^3}}.$$

Example 18.53. *Determine the frequency of lateral vibrations of the frame shown in Fig. 18.63(b).*

Solution : The frequency of these vibrations, if the mass of the frame is neglected, can be calculated by taking into account the mass of the frame, in the bending of the frame. If x is the lateral displacement of the load W together with the horizontal bar CD, the horizontal displacement of any point of the vertical bars at a distance ξ from the bottom, from consideration of the bending of the frame, is

$$x_1 = x - \frac{x}{3}\left(1-\frac{\xi}{L}\right) - \frac{2}{3}x\left[\frac{3}{2}\left(1-\frac{\xi}{l}\right)^2 - \frac{1}{2}\left(1-\frac{\xi}{L}\right)^3\right].$$

The kinetic energy of the vertical bars is

$$2\int_0^L \frac{w\dot{x}_1^2}{2g}d\xi = \frac{\alpha wL}{g}\dot{x}^2,$$

where α is a constant factor which is obtained after substituting for x_1 its expression from (u) and integrating. In considering the kinetic energy of the horizontal bar, only the horizontal component $\dot{x}$ of the velocities of the particles of the bar are used. Then the total kinetic energy of the entire system is

$$\frac{W\dot{x}^2}{2g} + \frac{(1+2\alpha)wL\dot{x}^2}{2g},$$

and the frequency is obtained from

$$f = \frac{1}{2\pi}\sqrt{\frac{4EIg}{[W+(1+2\alpha)wL]L^3}}.$$

18.8 Damping

In reality a vibration without decrease in amplitude is never realized. The presence of damping forces causes the dissipation of energy that progressively reduces the amplitude of vibration and ultimately stops the motion when all energy initially stored in the system has been dissipated.

If the energy of a nonconservative system is to be maintained at a constant level, an external source must supply energy to the system at a rate equal to the rate of energy dissipation. In free vibration, as it is not possible to supply energy to the system, the vibration dies out in course of time.

In short, damping is a phenomenon in which the vibrational energy of the system is gradually reduced or the amplitude of vibration is slowly decreased. Eventually, the vibration of the system is completely eliminated and the system is brought to rest at neutral position.

18.9 Nature of Damping

Damping forces may depend on the vibrating system as well as on elements exterior to it. The nature of damping is usually of the following types :

(a) Internal Friction of Material

(b) Fluid action

(c) Magnetic Damping

(d) Dry Friction or Coulomb Damping

(e) Viscous Damping

(f) Structural or Solid Damping.

(a) Internal Friction of Material :

When a member is subjected to vibration, the unit stress-strain or load-strain diagram shows a hysteresis loop. The area of this loop represents the amount of energy transformed into heat per cycle due to internal friction in the material. Experiments show that this is a function of the material and the manner in which it is loaded, which may be tension, shear, or combination thereof.

(b) Fluid Actions :

Systems vibrate in an environment such as air, gas, oil, water steam, etc. The amount of damping depends on the frequency and amplitude, or the velocity, as well as on the size and shape of the vibrating member, the viscosity of the medium, and the clearance between the stationary and the vibrating members.

(c) Magnetic Damping :

Systems which vibrate in a magnetic field undergo a damping action as the masses cut the magnetic lines of force. Examples of this action occur in electric machinery such as motors, generators and galvanometers.

(d) Dry Friction :

This would apply to a system where the mass slides on a surface such that the resistance to motion is provided by simple friction. The magnitude of this friction force F_f is constant and depends only upon the coefficient of friction and the weight of the body.

(e) Viscous Damping :

This damping results when a system is made to vibrate under the control of fluid, usually through the dashpot. Shock absorbers, hydraulic dashpots, and sliding of a body on a lubricated surface are some examples where viscous damping may be encountered. This type is the most commonly assumed in dynamics of structures.

(f) Structural or Solid damping :

Structural damping is due to internal friction within the material or at connections between elements of a structural system. The resulting damping forces are a function of the strain or deflections, in the structure.

For purposes of analysis structural damping may be assumed to be of the viscous type; i.e., the damping force is opposite but proportional to the velocity. The damping force applied to a lumped mass, may be expressed as

$$\text{Damping force} = -c\dot{x},$$

where c is a numerical constant, and $\dot{x}$ is the velocity of the mass.

18.10 Viscous Damping

Fig. 18.64 represents a system that is subject to free vibrations with viscous damping.

It consists of a mass m suspended on a weightless spring of scale k and connected by means of a dashpot that provides viscous damping with a factor c. This dashpot opposes the motion of the mass. The force provided by the dash pot is equal to $c\dot{x}$ where c is known as damping factor or damping coefficient.

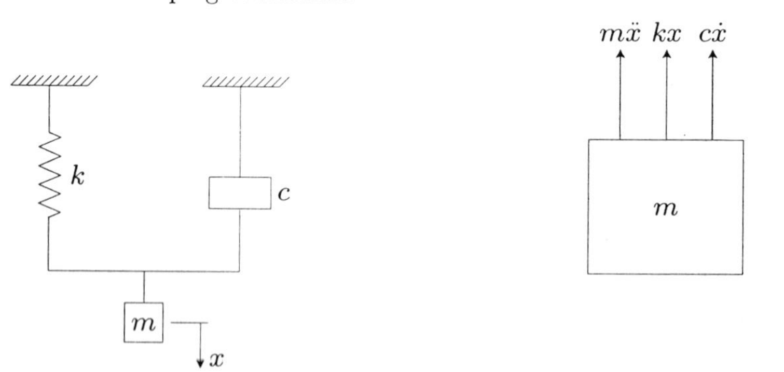

(a) System (b) Dynamic equilibrium

Fig. 18.64: Viscous damping

The dynamic equilibrium of the mass is as shown in Fig. 18.64(b) and the differential equation of motion is given by

$$m\ddot{x} + cx + kx = 0. \tag{18.13}$$

The term "**Viscous damping**" is usually associated with the expression $c\dot{x}$ since it represents fairly well the conditions of damping due to the viscosity of oil in a dashpot.

The solution for Eq. (18.13) cannot be found as simply as that of

$$m\ddot{x} + kx = 0.$$

If the function $x = e^{\lambda t}$ is considered where t is the time and λ an unknown constant, it is seen that upon differentiation the same function results, but multiplied by a constant.

If the solution is taken as $e^{\lambda t}$, whence

$$(m\lambda^2 + c\lambda + k)e^{\lambda t} = 0. \quad (18.14)$$

There are two values λ_1 and λ_2 that will make left side of Eq. (18.14) equal to zero.

Thus $\lambda_1 = \dfrac{-c + \sqrt{c^2 - 4mk}}{2m} = -\dfrac{c}{2m} + \sqrt{\dfrac{c^2}{4m^2} - \dfrac{4mk}{4m^2}}$

and $\lambda_2 = \dfrac{-c - \sqrt{c^2 - 4mk}}{2m}.$

If $p^2 = \dfrac{k}{m}$,

$$\lambda_1 = -\frac{c}{2m}\left(\frac{p}{p}\right) + \sqrt{\frac{p^2c^2}{p^2 4m} - p^2}.$$

With $\alpha = \dfrac{c}{2mp}$, where α is known as Damping Ratio,

$$\left.\begin{aligned} \text{then} \quad \lambda_1 &= +p[-\alpha + \sqrt{\alpha^2 - 1}] \\ \text{and} \quad \lambda_2 &= p[-\alpha - \sqrt{\alpha^2 - 1}]. \end{aligned}\right\} \quad (18.15)$$

Case (1) : $\alpha > 1$ for overdamping.

When $\alpha > 1$, both the roots λ_1 and λ_2 are real and negative. The solution is

$$x = Ae^{-\lambda_1 t} + Be^{-\lambda_2 t}. \quad (18.16)$$

The function is exponentially decaying function and is shown in Fig. 18.65.

The damping is so much that it causes the eventual dying out of oscillation. System possessing heavy damping is said to be Overdamped and the motion is non-oscillatory or aperiodic. The mass upon being released will simply tend to return to its static equilibrium position. However, the time taken to return back to the neutral position is exceedingly high.

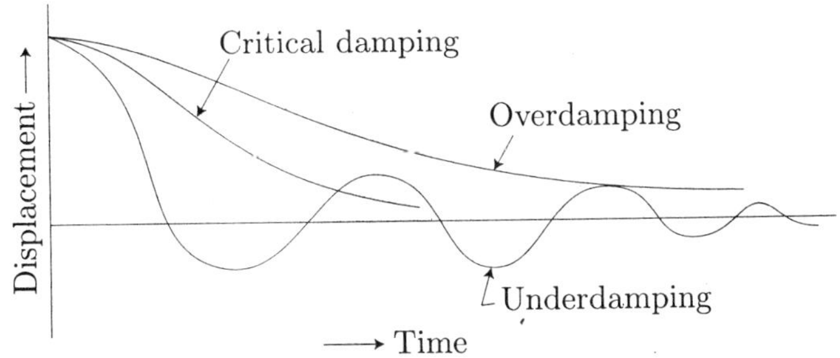

Fig. 18.65: Damping function for underdamping

Case (2) : $\alpha < 1$ for underdamping

When $\alpha < 1$, the quantity within the radical becomes negative.

$$\lambda_1 = p[-\alpha + i\sqrt{1-\alpha^2}]$$

$$\lambda_2 = p[-\alpha - i\sqrt{1-\alpha^2}]$$

and $x = Ae^{-p\alpha t}e^{(pi\sqrt{1-\alpha^2})t} + Be^{-\alpha pt}e^{(-pi\sqrt{1-\alpha^2})}t.$

If $p_d = p\sqrt{1-\alpha^2}$, which is known as Damped frequency,

$$x = e^{-p\alpha t}\left[Ae^{ip_dt} + Be^{-ip_dt}\right]. \tag{18.17a}$$

Also $x = e^{-p\alpha t}[C\cos p_dt + D\sin p_dt]$ (18.17b)

Underdamping causes the system to oscillate more slowly than without damping. In other words, the amplitude decreases with each subsequent cycle of oscillation and the frequency of oscillation with viscous damping is lower than the frequency of the system.

Case (3) : $\alpha = 1$ for critical damping

For $\alpha > 1$, the quantity under the radical $\sqrt{\alpha^2 - 1}$ becomes positive.

For $\alpha < 1$, this quantity becomes negative.

The transition from positive quantity to negative quantity occurs when $\alpha = 1$. The value becomes zero.

$$\text{With } \alpha = \frac{c}{2mp},$$

when $\alpha = 1$, the corresponding damping coefficient c is known as Critical Damping Coefficient and is represented by the letter c_c.

$$c_c = 2mp = 2m\sqrt{\frac{k}{m}} = 2\sqrt{km} \tag{18.18}$$

and $\alpha = \dfrac{c}{c_c}$. (18.18a)

Now when $\alpha = 1$, $\lambda_1 = -p$ and $\lambda_2 = -p$ from Eq. (18.15). The solution is

$$x = (C + Dt)e^{-pt}. \tag{18.19}$$

When both the roots are equal to s

$$y = Ae^{sx} + Be^{sx}.$$

Varying s by a small amount h,

$$y = Ae^{sx} + Be^{(s+h)x} \quad \text{and} \quad y = e^{sx}\left[A + Be^{hx}\right].$$

As $e^{hx} = 1 + \dfrac{(hx)}{1} + \dfrac{(hx)^2}{21} + \cdots$

and $y = e^{sx}[A + B(1 + hx)]$ neglecting higher order terms.

If $y = e^{sx}[(A+B) + Bhx]$,

$$(A + B) = C \text{ and } Bh = D.$$

$$y = e^{sx}[C + Dx]. \tag{18.20}$$

Eq. (18.19) is plotted in Fig. 18.65. This gives the minimum viscous damping that allows a displaced system to return to its initial position without oscillation. The vibration is completely eliminated and the time taken to come to the initial position is less than that of the overdamping situation.

18.11 Logarithmic Decrement

The logarithm of the ratio of the two successive amplitudes of the underdamped system is known as **Logarithmic Decrement**.

In the case of underdamped system, $x_1 = x_0 e^{-p\alpha t} \cos p_d t$.

After a time interval T, $x_2 = x_0 e^{-p\alpha(t+T)} \cos p_d(t+T)$.

As $\quad p_d T = 2\pi, \; \cos(p_d t + 2\pi) = \cos p_d t$

and $\quad \dfrac{x_1}{x_2} = \dfrac{(x_0 e^{-p\alpha t} \cos p_d t)}{(x_0 e^{-p\alpha t} e^{-p\alpha T} \cos p_d t)} = e^{p_\alpha T}$

$$\log_e \frac{x_1}{x_2} = \log_e e^{p\alpha T} = p\alpha T = \delta.$$

$\delta = p\alpha T$ is known as **Logarithmic Decrement.**

Now $\quad p\alpha T = p\dfrac{c}{2mp}T = \dfrac{cT}{2m}$, since

$$\alpha = \frac{c}{2mp} \quad \text{and} \quad \delta = \frac{cT}{2m}.$$

Also $\quad T = \dfrac{2\pi}{p_d} = \dfrac{2\pi}{p\sqrt{1-d^2}} \quad$ as $p_d = p\sqrt{1-d^2}$

and $\quad \delta = \dfrac{c2\pi}{2mp\sqrt{1-\alpha^2}}$. $\quad$ But $\alpha = \dfrac{c}{2mp}$,

whence $\delta = \dfrac{2\pi\alpha}{\sqrt{1 \quad \alpha^2}}$.

For very small α, $\delta = 2\pi\alpha$,

where $\quad \delta =$ logarithmic decrement

and $\quad \alpha = \dfrac{c}{2mp} =$ damping ratio.

Example 18.54. *The successive amplitudes from a free vibration test for a structure are 0.69, 0.362, 0.19 and 0.099 units respectively. Determine the damping ratio (assuming it to be very small) of the system considering each cycle separately and considering them all together.*

Solution 1 : (i) $\dfrac{x_0}{x_1} = e^{\delta_1} = \dfrac{0.69}{0.362} = 1.906$

$$\delta_1 = 0.6449 = 2\pi\alpha_1$$

and $\alpha_1 = 0.1026$.

(ii) $\dfrac{x_1}{x_2} = e^{\delta_2} = \dfrac{0.362}{0.19} = 1.905$

$\delta_2 = 0.6445 = 2\pi\alpha_2$

and $\alpha_2 = 0.10259.$

(iii) $\dfrac{x_2}{x_3} = e^{\delta_3} = \dfrac{0.19}{0.099} = 1.919$

$\delta_3 = 0.652 = 2\pi\alpha_3$

and $\alpha_3 = 0.10377.$

Solution 2 : $\dfrac{x_0}{x_3} = \dfrac{x_0}{x_1}\left(\dfrac{x_1}{x_2}\right)\left(\dfrac{x_2}{x_3}\right) = e^{\delta}(e^{\delta})(e^{\delta}) = e^{3\delta}$

$$e^{3\delta} = \frac{0.69}{0.099} = 6.97$$

$\therefore \quad \delta = 0.6472 = 2\pi\alpha$

$\alpha = 0.1030.$

From solution 1:

$$\text{Average value of } \alpha = \frac{0.1026 + 0.10259 + 0.10378}{3} = 0.10299.$$

Example 18.55. *The following data are given for a vibratory system with viscous damping :*

$$\text{Weight} = 50 \text{ N}$$

$$\text{Spring constant} = 15 \text{ N/mm}.$$

And the amplitude decreases to $\frac{1}{4}$th of the initial value after five consecutive cycles. Determine the damping coefficient.

Solution : $\dfrac{x_0}{x_5} = \dfrac{x_0}{x_1}\,\dfrac{x_1}{x_2}\,\dfrac{x_2}{x_3}\,\dfrac{x_3}{x_4}\,\dfrac{x_4}{x_5} = e^{\delta}\ e^{\delta}\ e^{\delta}\ e^{\delta}\ e^{\delta} = e^{5\delta}.$

As, $\quad x_5 = \dfrac{x_0}{4}, \quad \dfrac{x_0}{x_5} = 4 = e^{5\delta}$

$$\delta = 0.278 = \frac{2\pi\alpha}{\sqrt{1-\alpha^2}}$$

$\alpha = 0.0442097.$

Now $\quad c = 2\alpha\sqrt{km} = 2(0.0442)\sqrt{15\left(\dfrac{50}{9800}\right)} = 0.0244551\dfrac{\text{N/ sec}}{\text{mm}}.$

Example 18.56. *For a damping ratio $\alpha = 0.2$ what is*

(i) the damped and undamped frequency?

(ii) the logarithmic decrement?

(iii) the ratio of successive amplitudes?

Solution : (i) $P_d = p\sqrt{1-\alpha^2} = p\sqrt{1-(0.2)^2} = 0.9797958\,p \approx 0.98\,p.$

(ii) $\delta = \dfrac{2\pi\alpha}{\sqrt{1-\alpha^2}} = \dfrac{2\pi \times 0.2}{\sqrt{1-(0.2)^2}} = 1.2823.$

(iii) $\dfrac{x_0}{x_1} = e^{\delta} = e^{1.2823} = 3.6049.$

Example 18.57. *A rigid uniform beam of total mass m, and span L is hinged at one end; it is restrained by viscous damper of damping coefficient c at the free end, and by a spring (spring constant k) at a distance a from the hinged end. If the bar is moved in vertical plane, show that the critical damping coefficient is given by the expression :*

$$c = \frac{2}{\sqrt{3}}\frac{a}{L}\sqrt{km}.$$

Solution : The dynamic equilibrium of the system is shown in Fig. 18.66(b). Taking moment of all the forces about the hinge.

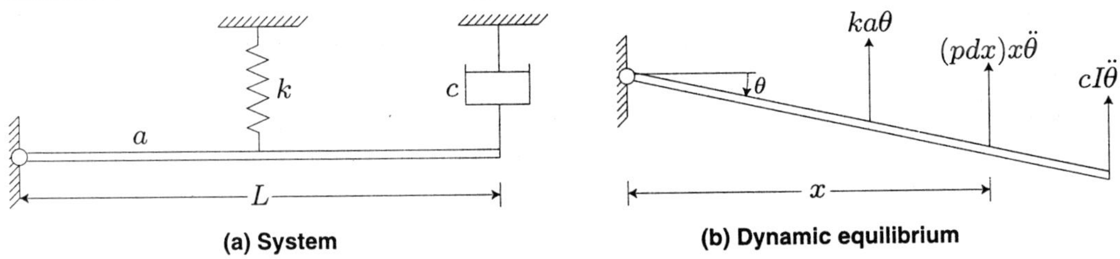

Fig. 18.66: For Example 18.57

$$\int_0^l [p(dx)x\ddot{\theta}]x + ka^2\theta + cl^2\theta = 0$$

$$\frac{1}{3}ml^2\ddot{\theta} + ka^2\theta + cl^2\theta = 0, \tag{1}$$

where $m = pL..$

Let
$$\left.\begin{aligned} \theta &= e^{\lambda t} \\ \dot{\theta} &= \lambda e^{\lambda t} \\ \ddot{\theta} &- \lambda^2 e^{\lambda t}. \end{aligned}\right\} \tag{2}$$

Substituting (2) in Eq. (1) and simplifying,

$$\lambda^2 + \frac{3cL^2}{m}\lambda + \frac{3ka^2}{mL^2} = 0.$$

Solving, $\lambda = \dfrac{-\frac{3c}{m} \pm \sqrt{\frac{9c^2}{m^2} - \frac{12ka^2}{mL^2}}}{2}.$

For critical damping, the quantity within the radical sign must be equal to zero,

whence $\dfrac{9c^2}{m^2} = \dfrac{12ka^2}{mL^2} = 0$ and $c = \dfrac{2}{\sqrt{3}}\dfrac{a}{L}\sqrt{mk}.$

Example 18.58. *Set up the differential equation for the motion of the system shown in Fig. 18.67. Determine from it the expression for the damped natural circular frequency of the* *oscillation for small displacements of the system, and also the critical damping coefficient of the dashpot. The weightless rigid rod is hinged at A.*

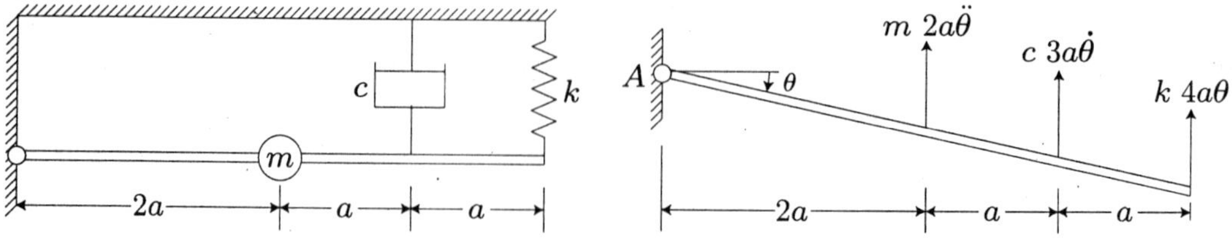

(a) System (b) Dynamic equilibrium

Fig. 18.67: For Example 18.58

Solution : The dynamic equilibrium of the system is shown in Fig. 18.67(b). Taking moment of the forces about the hinge,

$$4ma^2\ddot{\theta} + 9ca^2\dot{\theta} + 16ka^2\theta = 0$$

$$4m\ddot{\theta} + 9c^2\dot{\theta} + 16k\theta = 0. \tag{1}$$

If $\theta = e^{\lambda t}$

$\dot{\theta} = \lambda e^{\lambda t}$ (2)

and $\ddot{\theta} = \lambda^2 e^{\lambda t}$,

substituting Eq. (2) in Eq. (1),

$$4m\lambda^2 + 9c^2\lambda + 16k = 0 \quad \text{and} \quad \lambda = \frac{-9c \pm \sqrt{81c^2 - 256\ mk}}{8m}.$$

For critical damping, the quantity within the radical must be equal to zero,

whence $81c^2 = 256\ mk$ and $c_c = \dfrac{16}{9}\sqrt{mk}.$

For finding the undamped natural frequency, substitute $c = 0$ in Eq. (1), whence

$$4m\ddot{\theta} + 16k\theta = 0 \quad \text{and} \quad p^2 = \frac{16}{4}\frac{k}{m}$$

and undamped natural frequency, $p = 2\sqrt{\dfrac{k}{m}}$.

$$\alpha = \frac{c}{2mp}, \quad \text{when} \quad \alpha = 1$$

$$c_c = 2mp.$$

and
$$\alpha = \frac{c}{c_c} = \frac{c}{16}\left(\frac{9}{\sqrt{km}}\right)$$

$$\alpha^2 = \frac{81c^2}{256\ km}.$$

$$\therefore \quad p_d = p\sqrt{1-\alpha^2} = 2\sqrt{\frac{k}{m}}\sqrt{1 - \frac{81c^2}{256\ km}} = \sqrt{\frac{4k}{m} - \left(\frac{9c}{8m}\right)^2}.$$

Example 18.59. *Determine an equation for the natural frequency of damped vibration and the critical damping coefficient of the dashpot of the system shown in Fig. 18.68. The* *weightless rod is pivoted at point P and undergoes small displacements. Note that pendulum action does not exist since the spring is vertical.*

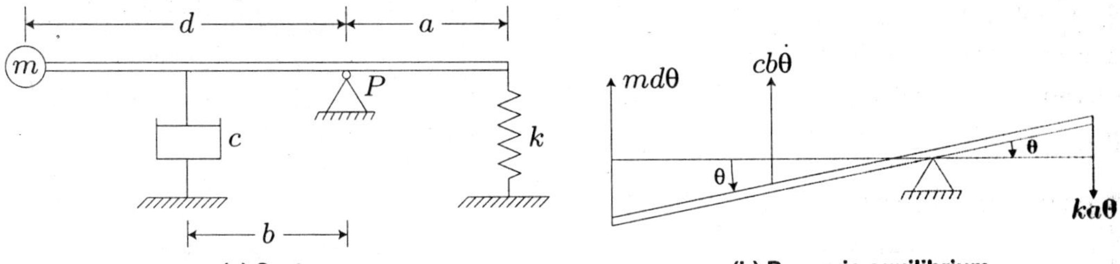

Fig. 18.68: For Example 18.59

Solution : The dynamic equilibrium of the system is shown in Fig. 18.68(b). **Taking** moment of the forces about the pivot,

$$(md\ddot{\theta})d + (cb\dot{\theta})b + (ka\theta)a = 0$$

$$\ddot{\theta}md^2t + \dot{\theta}cb^2 + \theta\ ka^2 = 0. \qquad \textbf{(1)}$$

If
$$\left.\begin{aligned} \theta &= e^{\lambda t} \\ \dot{\theta} &= \lambda e^{\lambda t} \\ \ddot{\theta} &= \lambda^2 e^{\lambda t}, \end{aligned}\right] \qquad \textbf{(2)}$$

substituting Eq. (2) in Eq. (1)

$$\lambda^2 md^2 + \lambda cb^2 + ka^2 = 0 \quad \text{and} \quad \lambda = \frac{-cb^2 + \sqrt{c^2b^4 - 4km\ a^2d^2}}{2md^2}.$$

For critical damping, the quantity within the bracket must be equal to zero.

$$c^2b^4 = 4km\ a^2d^2$$

$$c_c = \frac{2\text{ad}}{b^2}\sqrt{km}.$$

To find undamped circular frequency :

when $c \to 0$, Eq. (1) becomes $\ddot{\theta}md^2 + ka^2\theta = 0$

and $p^2 = \dfrac{ka^2}{md^2}$ and $p = \dfrac{a}{d}\sqrt{\dfrac{k}{m}}$.

Also $p_d = p\sqrt{1-\alpha^2}$,

where $\alpha = \dfrac{c}{c_c} = \dfrac{cb^2}{2ad\sqrt{km}}$

$$\therefore \quad \alpha^2 = \frac{c^2b^4}{4a^2d^2\ km}.$$

$\therefore$ p_d = damped circular frequency

$$= p\sqrt{1-\alpha^2} = \sqrt{p^2 - p^2\alpha^2}$$

$$= \sqrt{\frac{ka^2}{md^2} - \left(\frac{b^2c}{2md^2}\right)^2}.$$

18.12 Forced Vibration

In structural engineering practice, many problems arise in which forced vibrations occur due to applied disturbing force or motions as in case of machine foundations, floors mounted with rotating machinery, etc. If these disturbing forces or motions follow a periodic pattern, the forced vibration is termed the steady state.

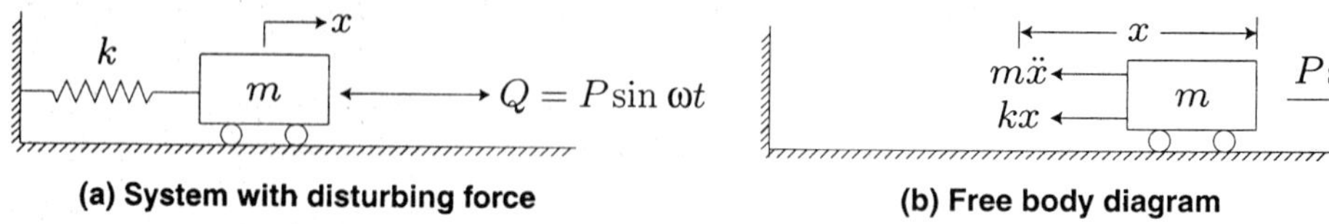

Fig. 18.69: Applied disturbing force

In Fig. 18.69, a single degree of freedom system is subjected to an applied disturbing force p sin ωt where ω is the impressed circular frequency or angular velocity of the system in radians per second. The periodic nature of the disturbing force is defined by the period of repetition τ_1 so that

$$\omega\tau_1 = 2\pi, \quad \text{whence} \quad \tau_1 = \frac{2\pi}{\omega}.$$

Frequency in cycles per second $= f_1 = \dfrac{1}{\tau_1} = \dfrac{\omega}{2\pi}$ and $\omega = 2\pi f_1$.

The dynamic equilibrium of the system at any instant during vibration is shown in Fig. 18.68(b), whence

$$m\ddot{x} + kx = P\sin\omega t \quad \text{and} \quad \ddot{x} + \frac{k}{m}x = \frac{P}{m}\sin\omega t.$$

If $\quad p^2 = \dfrac{k}{m}$ and $q = \dfrac{P}{m}$, $\ddot{x} + p^2x = q\sin\omega t$. (18.21)

The solution may be in the form, $x = C_3\sin\omega t$.

Substitution in Eq. (18.21) yields

$$-\omega^2C_3\sin\omega t + p^2C_3\sin\omega t = q\sin\omega t \quad \text{and} \quad C_3 = \frac{q}{(p^2-\omega^2)}.$$

The general solution is $x = C_1\cos\ pt + C_2\sin\ pt$. Combining the two, the complete solution is

$$x = C_1\cos\ pt + C_2\sin\ pt + \frac{q\sin\omega t}{(p^2-\omega^2)}. \tag{18.22}$$

The first two terms represent free vibration whose period is $\tau = \dfrac{2\pi}{p}$ and the last term represents a periodic forced vibration whose period $\tau_1 = \dfrac{2\pi}{\omega}$.

Omitting the first two terms, the response equation becomes

$$x = \frac{q\sin\omega t}{(p^2-\omega^2)}. \tag{18.23}$$

Substituting $q = \dfrac{P}{m}$ and manipulating further with $p^2 = \dfrac{k}{m}$,

$$x = \frac{P}{m}\frac{\sin\omega t}{p^2\left(1-\frac{\omega^2}{p^2}\right)} = \frac{P}{k}\frac{1}{\left(1-\frac{\omega^2}{p^2}\right)}\sin\omega t.$$

In the above equation, $\frac{P}{k}$ represents the statical deflection of spring if P were applied as a maximum static force. The term $\frac{1}{\left(1-\frac{\omega^2}{p^2}\right)}$ modifies the amplitude for dynamic effect. This factor is known as Magnification Factor.

Magnification Factor $= \frac{1}{\left(1-\frac{\omega^2}{p^2}\right)}$, which is plotted in Fig. 18.70.

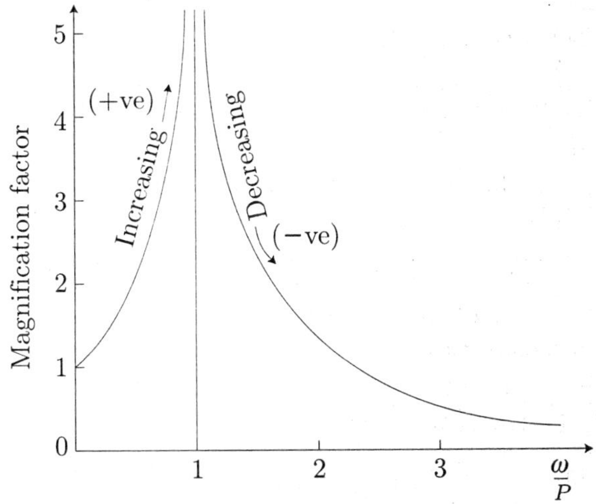

Fig. 18.70: Magnifaction factor

As $\frac{\omega}{p}$ varies up to unity, the magnification steadily increases. For small values of $\frac{\omega}{p}$, or when the natural period is small compared with the period of the disturbing force, the magnification factor is nearly unity. If $\omega = p$ or the two periods $(\tau = \tau_1)$ are equal, resonance occurs and a vibration of infinite amplitude would occur if there is no damping. For values of ω/p greater than unity, the magnification starts decreasing.

Example 18.60. *A machine weighing 10 kN is supported on its foundation by spring mounting. The piston on the machine moves up and down with a harmonic frequency of 10 cps. The piston has a weight of 500 N and a total stroke of 500 mm. Determine the max force transmitted to the foundation if the total 'k' of the springs is 5 N/mm.*

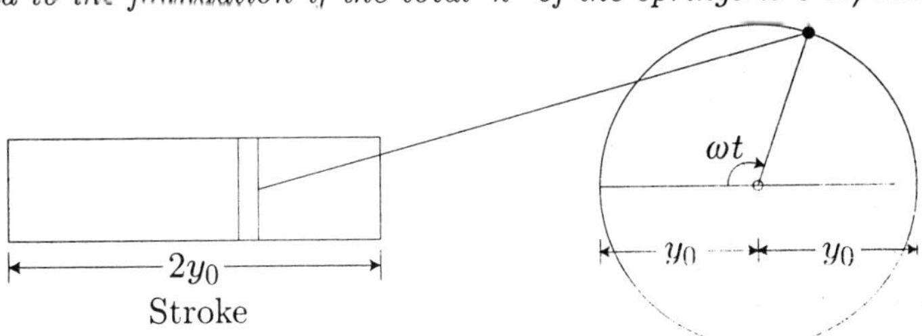

Fig. 18.71: For Example 18.60

Solution : Natural circular frequency of the machine foundation

$$p^2 = \frac{k}{m} = \frac{(5000)(9810)}{10000} = 4905 \text{ units.}$$

Impressed circular frequency $= \omega = 2\pi f_1$.

$$\omega^2 = (2\pi f_1)^2 = [2\pi(10)]^2 = 400\pi^2 \text{ units.}$$

The piston is undergoing harmonic motion and its vertical motion may be represented by.

$$y = y_0 \sin \omega t = \frac{500}{2} \sin \omega t = 250 \sin \omega t$$

$$\ddot{y} = 250\omega^2 \sin \omega t.$$

Maximum acceleration, $\ddot{y}_{max} = 250\omega^2$.

Force exerted by the piston = mass of the piston (acceleration)

$$P = m\ddot{y} = \frac{500}{9810}(400)(\pi^2)(250) = 50303.79 \text{ N}.$$

The periodic disturbing force $= 50303.79 \sin \omega t = P \sin \omega t$.

The vertical displacement of the machine on its supporting spring may be obtained by the equation,

$$x = \frac{P}{k} \frac{1}{\left(1 - \frac{\omega^2}{p^2}\right)}$$

and exerting force $= P \frac{1}{\left(1 - \frac{\omega^2}{p^2}\right)} = \frac{50303.79}{\left[1 - \frac{400(\pi^2)}{4905}\right]} = 257784$ N.

18.13 Earthquake Motion

It is very essential that the structures are designed and built to resist earthquakes. During earthquake, the structure is set in motion and is subject to forces due to its own mass and acceleration. The structure usually undergoes vertical and horizontal motion. Any structure is designed to resist vertical forces. In general, vertical motion during earthquake is not a major problem. It is the horizontal motion which causes maximum damage to the structure due to induction of what is known as *Base Shear.*

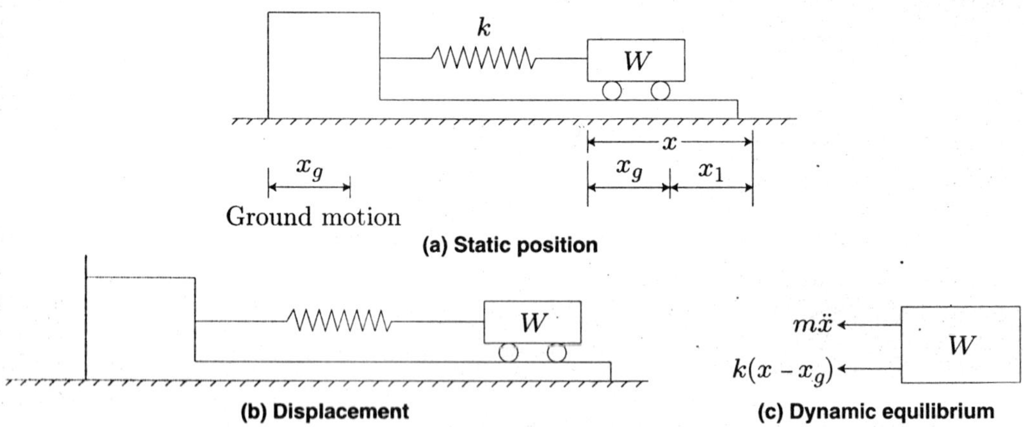

Fig. 18.72: System with load

Fig. 18.72(a) shows the static position due to a weight W connected to a rigid frame by a spring with a spring factor of k. The frame is firmly attached to the ground. Due to the ground motion, the weight W gets displaced to a distance x. This displacement consists of two translations. One due to the rigid body displacement of the frame x_0 and the other x_1 due to the elongation of the spring.

$$\text{Therefore}\ , x = x_g + x_1.$$

The acceleration felt by the weight is $\ddot{x}$ and the elongation suffered by the spring is x_1. In other words, the spring changes length by the difference between x and x_g. The difference $x - x_g$ is the relative displacement. Now $m = \frac{W}{g}$. The dynamic equilibrium of the system is shown in Fig. 18.72(c).

It follow that

$$-m\ddot{x} - k(x - x_g) = 0 \quad \text{and} \quad m\ddot{x} + kx = kx_g.$$

whence $\ddot{x} + \dfrac{k}{m}x = \dfrac{k}{m}x_g$

with $\dfrac{k}{m} = p^2$ and $\ddot{x} + p^2x = p^2x_g.$ (18.24)

The solution is $x = C_1 \cos\ pt + C_2 \sin\ pt+$ particular solution dependent upon a form of ground motion.

The first two terms represent free vibration which will die only in course of time. The ground motion is assumed to be a periodic steady state motion as $x_g = G \sin \omega t$, whence G is the maximum amplitude of the ground motion and ω is the circular frequency related to the period of the ground motion. The horizontal ground motion is assumed to be steady and sinusoidal with a stated maximum acceleration of 0.05 g to 1.00 g. The period of ground motion is assumed to be anywhere around 1 to 2 seconds. Eq. (18.24) becomes

$$\ddot{x} + p^2x = p^2\ G \sin \omega t. \tag{18.24a}$$

If the weight was at rest at time $t = 0$, its motion may be described by taking $x = A \sin \omega t$, where A is a constant. The solution of the Eq. (18.24a) is assumed to be

$$x = A \sin \omega t. \tag{18.24b}$$

Substituting $-\omega^2 A \sin \omega t + p^2 A \sin \omega t = p^2 G \sin \omega t$.

Solving, $A = \dfrac{p^2G}{(p^2 - \omega^2)}$.

Solution then is $x = \dfrac{p^2G}{(p^2 - \omega^2)} \sin \omega t.$ (18.24c)

Eq. (18.24c) is the equation of motion for a steady and repetitive condition of ground movement with free vibrations damped out.

Displacement of spring is $x_1 = x - x_g.$ (18.25)

With $x_g = G \sin \omega t.$

$$x_1 = \left[\left(\frac{p^2}{(p^2 - \omega^2)}\right) - 1\right] G \sin \omega t.$$

The maximum value of x_1 occurs when $\sin \omega t = 1$,

whence $x_{1\,\max} = \left[\left\{\dfrac{p^2}{(p^2 - \omega^2)}\right\} - 1\right] G.$

With the knowledge of p, ω and G, the relative displacement of the spring x_1 can be determined. Then

$$\text{Base shear} = \text{stiffness (displacement)} = kx_1.$$

Example 18.61. *An elevated water tank weighs 2500 kN when filled with water, the centroid of the mass being at a height of 10 metres from the ground. The horizontal spring factor for the tank is $k = 9$ kN/mm.*

Estimate the maximum shear generated at the base of the water tank by the horizontal inertia force. Assume the ground motion has a period 1.5 sec and a maximum acceleration of 0.05 g.

Solution : The natural circular frequency of the system $p = \sqrt{\dfrac{k}{m}} = \sqrt{\dfrac{9000}{250000}(9810)}$ $= 5.9427$.

The steady state circular frequency of the ground motion is

$$\omega = \frac{2\pi}{\tau_1} = \frac{2\pi}{1.5} = 4.1888 \text{ rad/sec.}$$

As $x_g = G \sin \omega t$,

$$\ddot{x}_g = \omega^2 G \sin \omega t,$$

when $\sin \omega t = 1$, $\ddot{x}_g$ becomes maximum.

and $(\ddot{x}_g)_{\max} = \omega^2 G$

$$G = \frac{\ddot{x}_g}{\omega^2} = \frac{0.05\ g}{\omega^2}$$

$$G = \frac{0.05(9810)}{(4.1888)^2} = 27.955 \text{ mm.}$$

To determine the shear generated at the base of tower by the horizontal inertia force acting through the centre of mass the relative displacement of the top of tower with respect to the base (x_1) is needed.

This relative deflection when multiplied by the spring factor will equal the base shear.

$$(x_1)_{\max} = \left[\left(\frac{p^2}{(p^2 - \omega^2)}\right) - 1\right] G = \left[\frac{(5.9427)^2}{(5.9427)^2 - (4.1888)^2} - 1\right](27.955)$$

$$= 27.6032 \text{ mm.}$$

Base shear $= k(x_1) = 9000(27.6032) = 248428.80$ N.

18.14 Shear Building

In concrete multistorey buildings, part of the slab constitutes the beam portion. This beam portion possesses considerable rigidity when compared with the columns. For practical purposes, the rigidity of the beam may be considered infinite. Consequently, during displacement due to wind or earthquake, the joints rotate very little and can be neglected. The displacement is purely translational. Therefore, all the columns in the same floor translate the same amount. Such an arrangement can be considered as springs in parallel. A schematic view of the shear building is shown in Fig. 18.73.

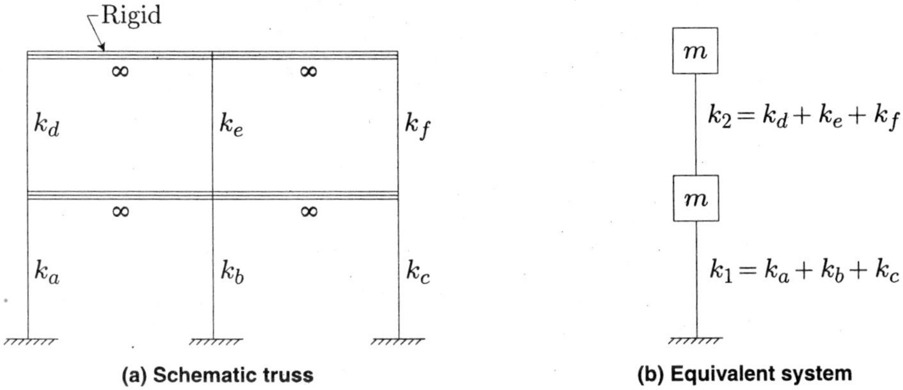

Fig. 18.73: Shear building

The equivalent system is shown in Fig. 18.73(b).

Example 18.62. *A shear building is shown in Fig. 18.74. Set up the translational stiffness and flexibility matrices.*

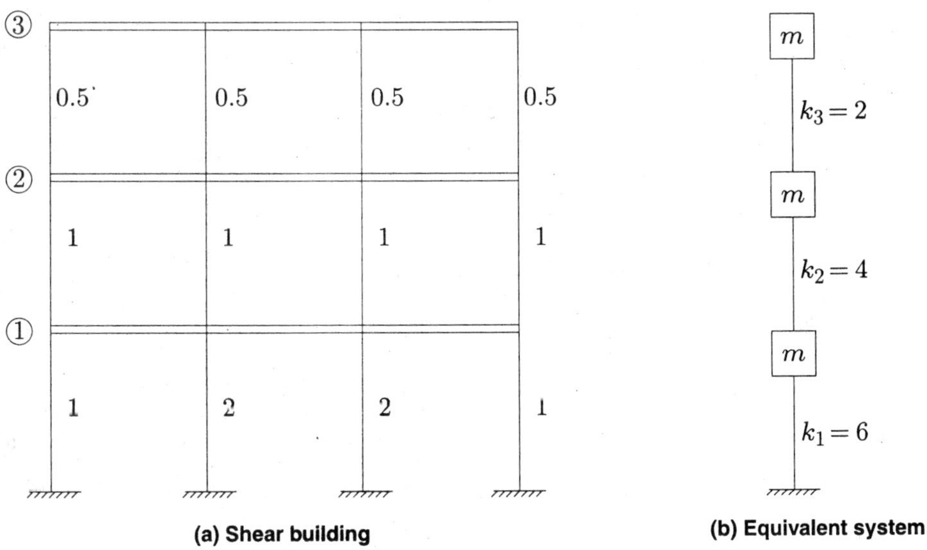

Fig. 18.74: For Example 18.62

Solution : (a) Flexibility Matrix :

Unit load is applied at stations 1, 2 and 3 successively as in Fig. 18.75.

$$\delta_{11} = \frac{1}{k_1} = \frac{1}{6}, \ \delta_{21} = \frac{1}{6}, \ \delta_{31} = \frac{1}{6}, \quad \text{and} \quad \delta_{12} = \frac{1}{k_1} = \frac{1}{6}$$

Also $\delta_{22} = \dfrac{1}{k_1} + \dfrac{1}{k_2} = \dfrac{1}{6} + \dfrac{1}{4} = \dfrac{10}{24}$

$$\delta_{32} = \frac{1}{k_1} + \frac{1}{k_2} = \frac{10}{24} \quad \text{and} \quad \delta_{13} = \frac{1}{k_1} = \frac{1}{6}$$

$$\delta_{23} = \frac{1}{k_1} + \frac{1}{k_2} = \frac{1}{6} + \frac{1}{4} = \frac{10}{24}$$

$$\delta_{33} = \frac{1}{k_1} + \frac{1}{k_2} + \frac{1}{k_3} = \frac{1}{6} + \frac{1}{4} + \frac{1}{2} = \frac{22}{24}.$$

Collecting all the terms

$$\delta = \begin{bmatrix} 1/6 & 1/6 & 1/6 \\ 1/6 & 10/24 & 10/24 \\ 1/6 & 10/24 & 22/24 \end{bmatrix}.$$

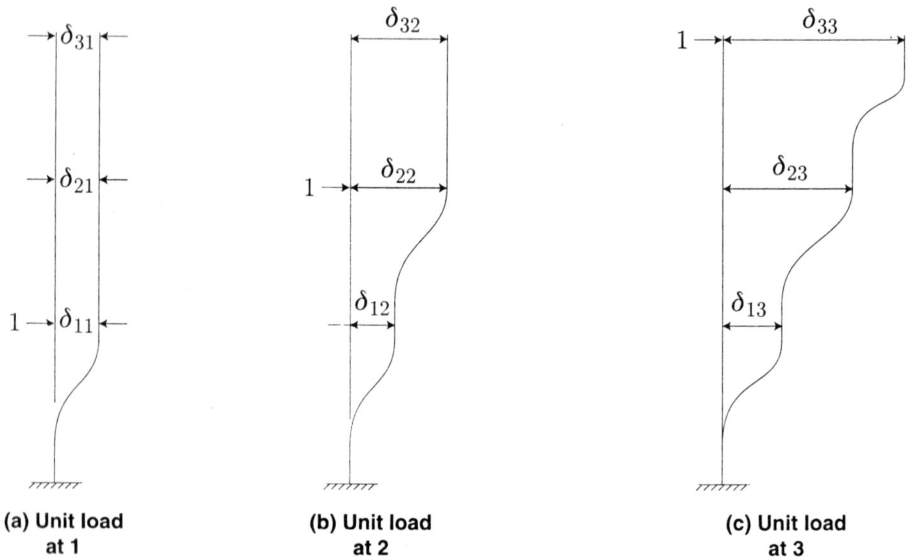

Fig. 18.75: Flexibility constants

(b) Stiffness Matrix :

Unit displacement is effected at stations 1, 2 and 3 one at a time successively and the stiffness influence coefficients are computed as shown in Fig. 18.76.

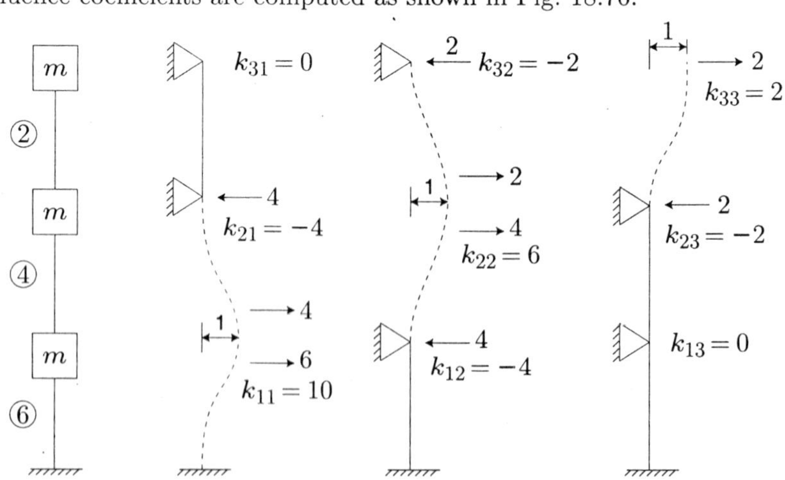

Fig. 18.76: Stiffness coefficients

Collecting all the terms,

$$k = \begin{bmatrix} 10 & -4 & 0 \\ -4 & 6 & -2 \\ 0 & -2 & 2 \end{bmatrix}.$$

By multiplying k and δ, $k\delta = I$.

$$\delta k = \begin{bmatrix} 1/6 & 1/6 & 1/6 \\ 1/6 & 10/24 & 10/24 \\ 1/6 & 10/24 & 22/24 \end{bmatrix} \begin{bmatrix} 10 & -4 & 0 \\ -4 & 6 & -2 \\ 0 & -2 & 2 \end{bmatrix} = \begin{bmatrix} 1 & 0 & 0 \\ 0 & 1 & 0 \\ 0 & 0 & 1 \end{bmatrix}.$$

Example 18.63. *Determine the equation of motion of the mass m for the system shown in Fig. 18.77. Assume, the horizontal bar is rigid and is of negligible mass.*

$$k_{31} = 0$$

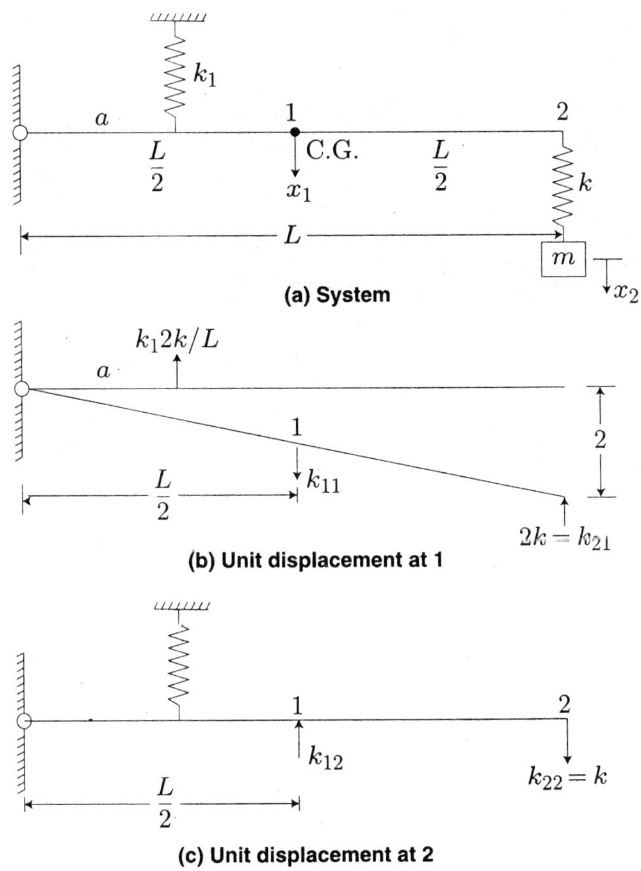

Fig. 18.77: For Example 18.63

Solution : The system possesses two translational degrees of freedom.

Considering Fig. 18.77(b), unit displacement is impressed at co-ordinate 1 arresting the co-ordinate at 2. Due to this the spring at co-ordinate 2 gets compressed by 2 units. The force developed in the spring is $2k$. The force in the left spring is $k_1\ 2a/L$.

For computing k_{11}, moment of the forces about the hinge is taken.

$$k_{11}\left(\frac{L}{2}\right) = 2kL + \frac{2ak_1a}{L}$$

$$k_{11} = 4k + \frac{4k_1a^2}{L^2}$$

and $k_{21} = -2k$.

At co-ordinate 2, a unit displacement is impressed arresting co-ordinate 1. The force in the left spring will be zero. Taking moment about the hinge,

$$k_{12}\left(\frac{L}{2}\right) = kL$$

$k_{12} = -2k$ (The minus sign is taken since the force acts opposite to the co-ordinate)

$k_{22} = k$.

Substituting these values in $m\ddot{x} + kx = 0$,

$$\begin{bmatrix} 0 & 0 \\ 0 & m \end{bmatrix}\begin{bmatrix} \ddot{x}_1 \\ \ddot{x}_2 \end{bmatrix} + \begin{bmatrix} \left(4k + \frac{4k_1a^2}{L^2}\right) & -2k \\ -2k & k \end{bmatrix}\begin{bmatrix} x_1 \\ x_2 \end{bmatrix} = \begin{bmatrix} 0 \\ 0 \end{bmatrix}.$$

Expanding the matrix,

$$\left(4k + 4k_1\frac{a^2}{L^2}\right)x_1 - 2kx_2 = 0 \quad \text{and} \quad x_1 = \frac{2kx_2}{\left(4k + 4k_1\frac{a^2}{L^2}\right)}.$$

The second equation is $m\ddot{x}_2 - 2kx_1 + kx_2 = 0$.

Substituting for x_1

$$m\ddot{x}_2 - \frac{4k^2}{\left(4k + 4k_1\frac{a^2}{L^2}\right)}x_2 + kx_2 = 0.$$

Simplifying, $mx_2 + \dfrac{kk_1}{(k + k_1)}x_2 = 0.$

A number of particular cases can be investigated depending on values of L, a, k and k_1.

18.15 Fundamental Frequency

In a system within degrees of freedom, there are n natural frequencies. The lowest of them is known as the **Fundamental Frequency**. All these frequencies contribute to the response of the system. In fact, when dealing with many degrees of freedom, analysis is often based on the first few modes alone, while the higher modes are neglected. The fundamental frequency contributes to a very great extent and is of greatest interest in structural dynamics problems.

There are many methods available for computing the fundamental frequency. The most important ones are :

(a) Rayleigh's Method

(b) Stodala's Method

(c) Dunkerley's Method.

18.16 Rayleigh's Method (Discrete System)

This method is also known as Static Deflection Method for multi-mass system. By invoking the energy principle a formula can be derived for the fundamental frequency.

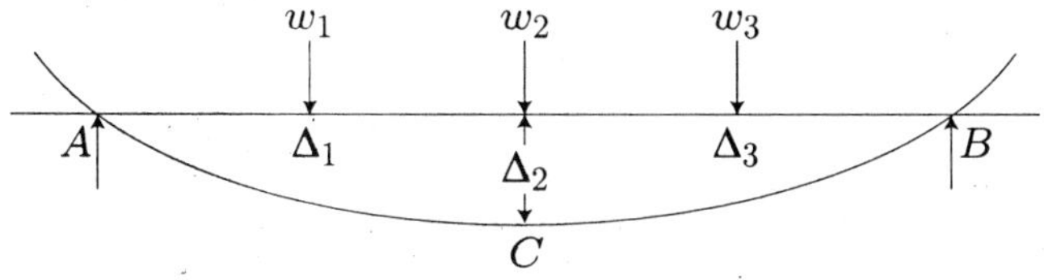

Fig. 18.78: Beam with loads

For a set of concentrated loads, W_1, W_2, and W_3, acting on a beam which is simply supported, ACB is the static deflection curve, i.e., Δ_1, Δ_2, and Δ_3 are the deflections under the loads. Hence Δ_1, Δ_2, and Δ_3 are the amplitudes of motion. At the extreme position, the velocity is zero and the energy is purely potential (strain energy).

Potential energy maximum $= V_{\max} = \frac{1}{2}W_1\Delta_1 + \frac{1}{2}W_2\Delta_2 + \frac{1}{2}W_3\Delta_3.$

The velocity is maximum, when the masses cross the neutral equilibrium position. When the potential energy is zero, the kinetic energy is maximum.

$$(\dot{x}_1)_{\max} = |\Delta_1 p|$$

$$(\dot{x}_2)_{\max} = |\Delta_2 p|$$

$$(\dot{x}_3)_{\max} = |\Delta_3 p|$$

Kinetic energy maximum, K.E.$_{\max} = \frac{1}{2}mv^2$

$$(K.E.)_{\max} = \frac{1}{2}\,\frac{W_1}{g}\,(\Delta_1 p)^2 + \frac{1}{2}\,\frac{W_2}{g}\,(\Delta_2 p)^2 + \frac{1}{2}\,\frac{W_3}{g}\,(\Delta_3 p)^2.$$

In a conservative system, where no energy is lost or gained, $V_{\max} = T_{\max}$

$$\text{and} \quad p^2 = \frac{g[W_1\Delta_1 + W_2\Delta_2 + W_3\Delta_3]}{[W_1\Delta_1^2 + W_2\Delta_2^2 + W_3\Delta_3^2]} = \frac{g\sum_1^n W_i\Delta_i}{\sum_1^n W_i\Delta_i^2}. \tag{18.26}$$

$$\text{Period of vibration, } \tau = \frac{2\pi}{p} = 2\pi\sqrt{\frac{\sum_1^n W_i\Delta_i^2}{g\sum_1^n W_i\Delta_i}}. \tag{18.27}$$

Example 18.64. *A solid shaft of diameter with two discs is represented by a simply supported beam carrying two weights as shown in Fig. 18.79. The corresponding deflections are shown in the Figure. Determine the fundamental frequency of the system. Assume that the links are weightless and are infinitely rigid in the axial direction..*

Solution : Substituting the various values in the formula

$$p^2 = \frac{(9800)[600(0.168) + 1000(0.272)](10)^2}{[600(0.168)^2 + 1000(0.272)^2]10^3} = 40183.72519$$

and $p = 200.4588$ radians per second.

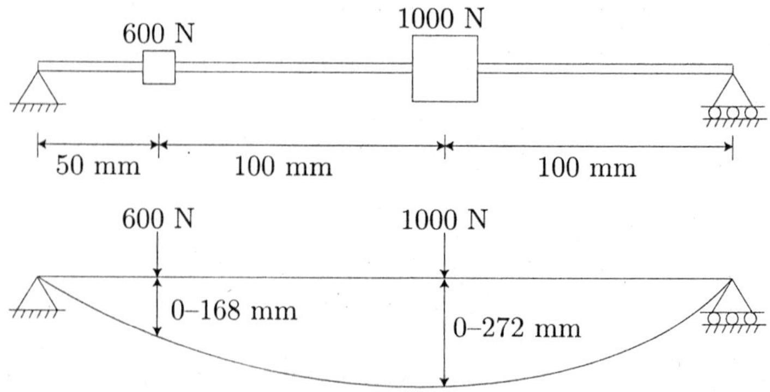

Fig. 18.79: For Example 18.64

Example 18.65. *The flexibility matrix and the load matrix of a vibrating system are given below. Compute the fundamental frequency using the Rayleigh method.*

$$F = \begin{bmatrix} 0.40 & 0.40 & 0.32 \\ 0.40 & 0.48 & 0.40 \\ 0.32 & 0.40 & 0.40 \end{bmatrix} \text{mm}$$

$$R = \begin{bmatrix} 60 \\ 100 \\ 80 \end{bmatrix} \text{kN}$$

Solution : To compute the deflections,

$$r = FR \quad \text{and} \quad \Delta = FW,$$

where $r = \Delta$, and $R = W$.

$$\begin{bmatrix} \Delta_1 \\ \Delta_2 \\ \Delta_2 \end{bmatrix} = \begin{bmatrix} 0.40 & 0.40 & 0.32 \\ 0.40 & 0.48 & 0.40 \\ 0.32 & 0.40 & 0.40 \end{bmatrix} \begin{bmatrix} 60 \\ 100 \\ 80 \end{bmatrix} = \begin{bmatrix} 89.60 \\ 104.00 \\ 91.20 \end{bmatrix}$$

Substituting in the formula, $p^2 = 101.452$ and $p = 10.072$ rad/second.

Example 18.66. *A simply supported beam of span L carries two equal loads W as shown in Fig. 18.80. Using the physical approach for computing the deflections, obtain the value of the fundamental frequency by the Rayleigh method.*

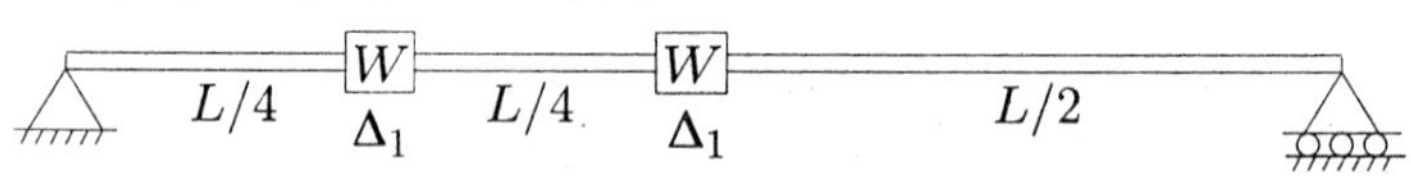

(a) Beam with load

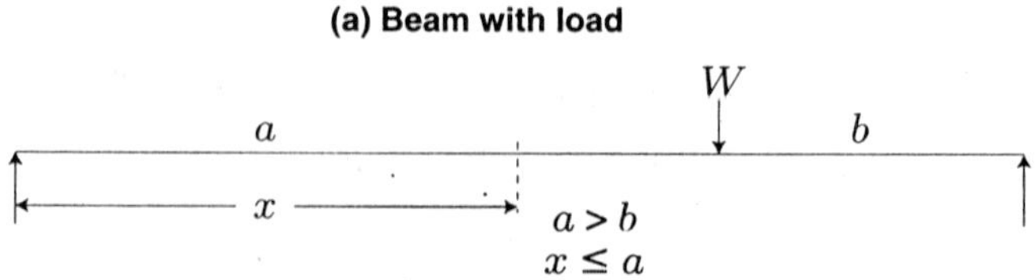

(b)

Fig. 18.80: For Example 18.66

Solution : The following equation for computing deflection for any section x-x lying in portion a (where $a > b$) is given as follows :

$$\Delta = \frac{Wbx(L^2 - b^2 - x^2)}{6EIL} \text{ for } a > b;\ x \le a.$$

Due to central load W, $x = L/4$: $b = L/2$.

$$\Delta_1 = \frac{W\left(\frac{L}{2}\right)\left(\frac{L}{4}\right)\left[L^2 - \left(\frac{L}{2}\right)^2 - \left(\frac{L}{4}\right)^2\right]}{6IEL} = \frac{11}{768}\frac{WL^3}{EI}.$$

Due to non-central load (under the load)

$$\Delta_1 = \frac{Wa^2b^2}{3EIL} = \frac{W\left(\frac{L}{4}\right)^2\left(\frac{3}{4}L\right)^2}{3EIL} = \frac{3}{256}\frac{WL^3}{EI}.$$

$$\text{Total } \Delta_1 = \frac{WL^3}{EI}\left(\frac{11}{768} + \frac{3}{256}\right) = \frac{20}{768}\frac{WL^3}{EI}.$$

$$\text{Also } \Delta_2 = \frac{WL^2}{EI}\left[\frac{11}{768} + \frac{1}{48}\right] = \frac{27WL^2}{768EI}.$$

Substituting Δ_1 and Δ_2 in the frequency formula,

$$p^2 = \frac{EIg}{WL^3}\left[\frac{\left(\frac{27}{768}\right) + \left(\frac{20}{768}\right)}{\left(\frac{27}{768}\right)^2 + \left(\frac{20}{768}\right)^2}\right]$$

$$p^2 = \frac{EIg}{WL^3}(31.97).$$

Example 18.67. *simply supported beam carries two weights W at 1/3 d span as shown in Fig. 18.81. Assuming the links to be weightless and rigid in axial direction, compute the fundamental frequency using the Rayleigh method.*

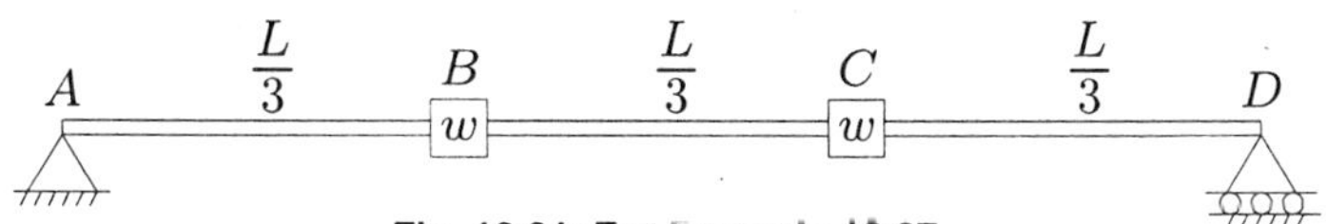

Fig. 18.81: For Example 18.67

Solution : Δ_1 and Δ_2 can be obtained as in the previous example or from the Flexibility Matrix whose formulation is shown using the element approach. The flexibility matrix is as follows :

$$F = \frac{L^3}{486}\begin{bmatrix} 8 & 7 \\ 7 & 8 \end{bmatrix}\left(\frac{1}{EI}\right)$$

$$\text{Load matrix } R = \begin{bmatrix} W \\ W \end{bmatrix};\ r = FR \text{ and } \Delta = FW.$$

$$\text{Then } \Delta_1 = \frac{15}{486}\frac{WL^3}{EI} \text{ and } \Delta_2 = \frac{15WL^3}{486EI}.$$

Substituting the same in the frequency formula,

$$p^2 = \frac{g[W\Delta + W\Delta]}{[W\Delta^2 + W\Delta^2]} = \frac{g}{\Delta};\ p^2 = \frac{EIg}{WL^3}\left(\frac{486}{15}\right).$$

Example 18.68. *An overhang beam is subjected to two weights W, as shown in Fig. 18.82. The links are weightless and inextensible. Compute the fundamental frequency using the Rayleigh method.*

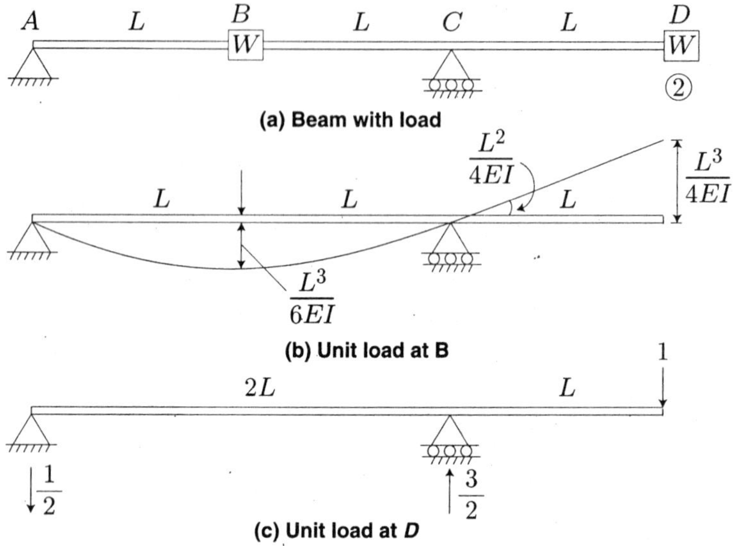

Fig. 18.82: For Example 18.68

Solution : The flexibility matrix can be formulated using the element approach or physical approach. Or, even without using the flexibility matrix, Δ_1 and Δ_2 can be computed from the strength of materials approach directly. Here the physical approach will be used.

From Fig. 18.82, by inspection,

$$\delta_{11} = \frac{L^3}{6EI}$$

$$\delta_{21} = -\frac{L^3}{4EI} = \delta_{12}.$$

From Fig. 18.82(c),

$$\delta_{22} = \left[\int_0^{2L} \left(\frac{x}{2}\right)\left(\frac{x}{2}\right) dx + \int_0^{L} (x)(x) dx\right] \frac{1}{EI} = \frac{L^3}{EI}.$$

The flexibility matrix becomes

$$F = \frac{L^3}{EI}\begin{bmatrix} 1/6 & -1/4 \\ -1/4 & 1 \end{bmatrix}$$

$$r = FR; \ \Delta = FW$$

and $$\begin{bmatrix} \Delta_1 \\ \Delta_2 \end{bmatrix} = \frac{L^3}{EI}\begin{bmatrix} 1/6 & -1/4 \\ -1/4 & 1 \end{bmatrix}\begin{bmatrix} -W \\ +W \end{bmatrix} = \begin{bmatrix} -10W/24 \\ 30W/24 \end{bmatrix}\frac{L^3}{EI}.$$

Substituting Δ_1 and Δ_2 in the frequency formula,

$$p^2 = \frac{EIg\left[+\frac{10}{24} + \frac{30}{24}\right]}{WL^3\left[\left(\frac{10}{24}\right)^2 + \left(\frac{30}{24}\right)^2\right]} = \frac{0.96EIg}{WL^3}.$$

Example 18.69. *The two storey frame shown in Fig. 18.83 carries two weights concentrated in the middle of each girder. The links are weightless and are inextensible. Using the Portal method of analysis, compute the fundamental frequency. EI is constant throughout.*

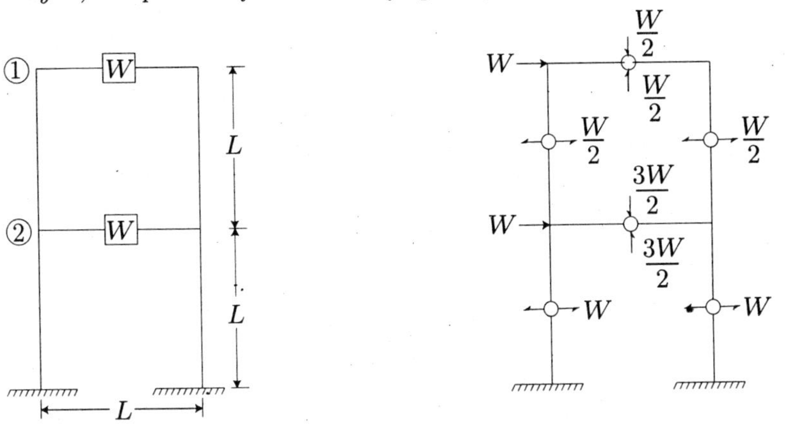

(a) Frame (b) Portal analysis

Fig. 18.83: For Example 18.69

Solution : The oscillation takes place in the horizontal direction. For finding the deflections Δ_1 and Δ_2 in the horizontal direction, the method of virtual work will be used. Unit loads are applied at stations (1) and (2) separately and the portal analysis are carried out as shown in Fig. 18.84.

$$\Delta_1 = \frac{6}{EI}\int_0^{L/2}\left(\frac{W}{2}x\right)\left(\frac{1}{2}x\right)dx + \frac{4}{EI}\int_0^{L/2}(Wx)\left(\frac{x}{2}\right)dx + \frac{2}{EI}\int_0^{L/2}\left(\frac{3}{2}Wx\right)(x)dx$$

$$= \frac{13WL^3}{48EI}.$$

$$\Delta_2 = \frac{4}{EI}\int_0^{L/2}(Wx)\left(\frac{1}{2}x\right)dx + \frac{2}{EI}\int_0^{L/2}\left(\frac{3W}{2}x\right)\left(\frac{1}{2}x\right)dx = \frac{7}{48}\frac{WL^3}{EI}$$

and $$p^2 = \frac{EIg}{WL^3}\frac{\left[\frac{13}{48}+\frac{7}{48}\right]}{\left[\left(\frac{13}{48}\right)^2+\left(\frac{7}{48}\right)^2\right]} = 4.4037\frac{EIg}{WL^2}.$$

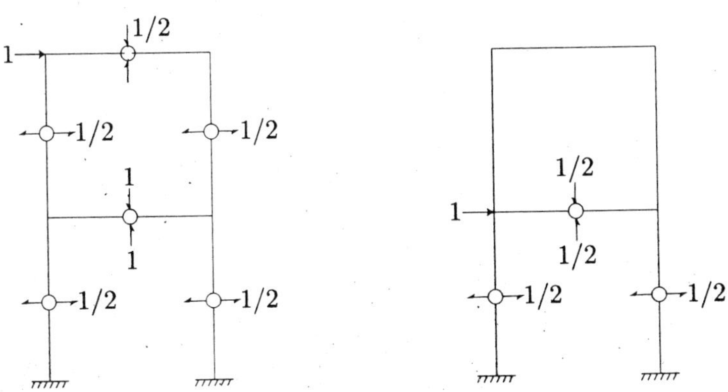

(a) Unit load at 1 (b) Unit load at 2

Fig. 18.84: Fo Example 18.69

Example 18.70. *A shear building is shown in Fig. 18.85 with k values ringed. The top floor carries a weight of 2W and the other two floors have weights equal to 3W. Compute the fundamental frequency using the Rayleigh method.*

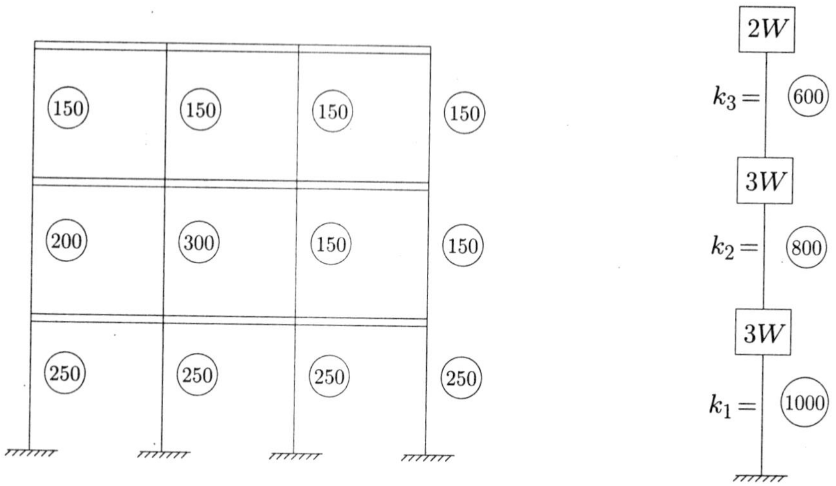

(a) Shear building

(b) Equivalent structure

Fig. 18.85: For Example 18.70

Solution : The equivalent structure is shown in Fig. 18.85(b). The flexibility influence coefficients are

$$\delta_{11} = \frac{1}{k_1} = \frac{1}{1000}, \delta_{21} = \frac{1}{1000} \cdot \delta_{31} = \frac{1}{1000}$$

$$\delta_{12} = \frac{1}{k_1} = \frac{1}{1000}, \delta_{22} = \frac{1}{k_1} + \frac{1}{k_2} = \frac{1}{1000} + \frac{1}{800} = 0.00225.$$

$$\delta_{32} = 0.00225$$

$$\delta_{13} = \frac{1}{k_1} = \frac{1}{1000}$$

$$\delta_{23} = \frac{1}{1000} + \frac{1}{800} = 0.00225$$

$$\delta_{33} = \frac{1}{1000} + \frac{1}{800} + \frac{1}{600} = 0.0039166$$

$$\tau = FR$$

$$\Delta = FW$$

$$\begin{bmatrix} \Delta_1 \\ \Delta_2 \\ \Delta_3 \end{bmatrix} = \begin{bmatrix} 0.001 & 0.001 & 0.001 \\ 0.001 & 0.00225 & 0.00225 \\ 0.001 & 0.00225 & 0.0039166 \end{bmatrix} \begin{bmatrix} 3 & W \\ 3 & W \\ 2 & W \end{bmatrix} = \frac{1}{1000} \begin{bmatrix} 8.00 \\ 12.25 \\ 17.58 \end{bmatrix}.$$

Substituting Δ_1, Δ_2 and Δ_3 in the frequency formula,

$$p^2 = \frac{1000g[3(8) + 3(12.25) + 2(17.58)]}{W[3(8)^2 + 3(12.25)^2 + 2(17.58)^2]} = 76.10\frac{g}{W}$$

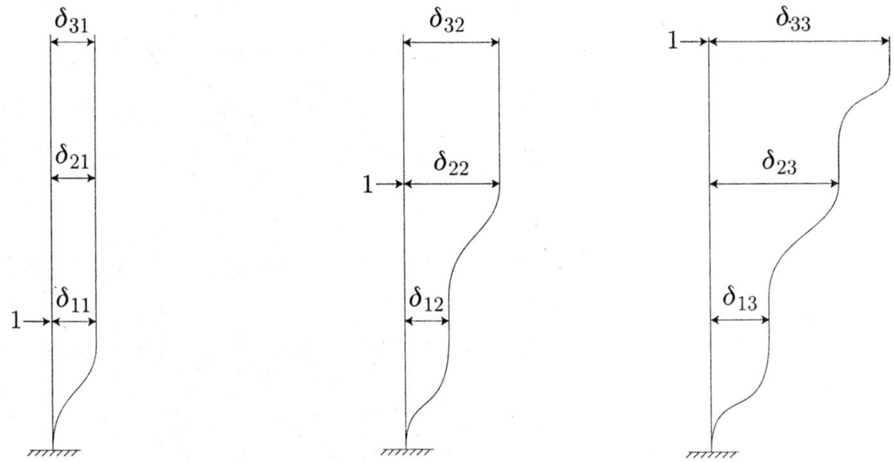

(a) Unit load at 1 (b) Unit load at 2 (c) Unit load at 3

Fig. 18.86: For Example 18.70

18.17 Rayleigh's Method for Distributed Mass System

When the system is at its extremum, the kinetic energy is zero and the potential energy (strain energy) is maximum.

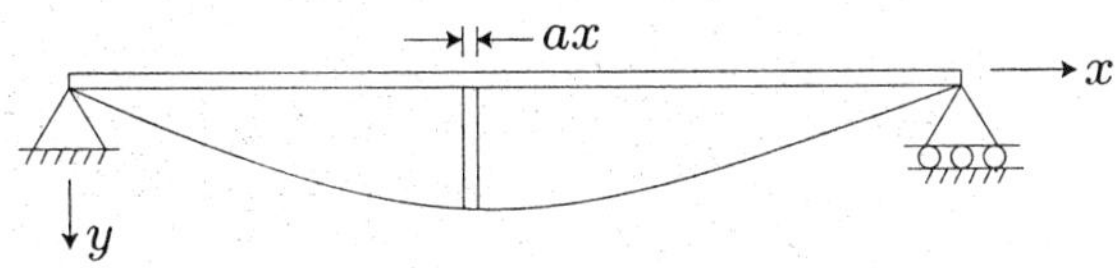

Fig. 18.87: Distributed load

$$\text{Total Strain Energy} = \int_0^L \frac{1}{2} M^2 \frac{dx}{EI}.$$

Also $\quad M = EI\dfrac{d^2y}{dx^2}$

and $\quad \text{T.E.}_{\max} = \displaystyle\int_0^L \frac{1}{2}\left(EI\frac{d^2y}{dx^2}\right)^2 \left(\frac{dx}{EI}\right) = \int_0^L \frac{EI}{2}\left(\frac{d^2y}{dx^2}\right)^2 dx.$

Now Kinetic Energy is maximum when the particle crosses the neutral position $= \dfrac{m}{2} p^2 \displaystyle\int_0^L y^2 dx,$

$$\text{whence} \quad p^2 = \frac{g\int_0^L EI\left(\frac{d^2y}{dx^2}\right)^2 dx}{w\int_0^L y^2 dx}. \tag{18.28}$$

Example 18.71. *A simply supported beam of span 6 metres is subjected to a dynamic force as shown in Fig. 18.88(b). The beam carries a weight of* 45 kN *at the centre. The various other data are as follows :*

Section modulus, $Z = 1500(10)^3$ mm^3

Maximum value of $F_1 = 90$ kN

Rise time, $t_r = 0.075\,\text{sec}$

$E = 2.1(105)\ \text{N/mm}^2$

$I = 33400(10)^4\ \text{mm}^4$

Dynamic Load Factor $(\text{D.L.F.})_{\max}$ *given by*

$\frac{tr}{\tau}$	$(\text{DLF})_{\max}$
0.6935	*1.32*
1.0000	*1.00*
0.5000	*1.62*

Considering the weight of the beam to be negligible, compute the maximum dynamic bending stress.

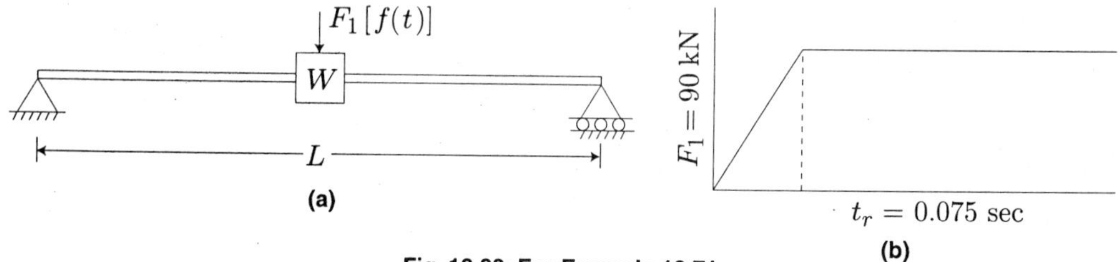

Fig. 18.88: For Example 18.71

Solution : $k = \dfrac{48EI}{L^3} = \dfrac{48(2.1)(10)^5(33400)(10)^4}{(6000)^3} = 15586.67\ \text{N/mm}$

$$p = \sqrt{\frac{k}{m}} = \sqrt{\frac{9800(15586.67)}{45000}} = 58.26$$

$$\tau = \frac{2\pi}{p} = \frac{2\pi}{58.26} = 0.1078\,\text{sec.}$$

Now $\dfrac{L_r}{\tau} = \dfrac{0.075}{0.1078} = 0.6957$

From the table $(\text{DLF})_{\max} = 1.318$.

The maximum dynamic stress equals the static stress due to the 90 kN force multiplied by $(\text{DLF})_{\max}$.

$$\text{max stress} = \sigma_{\max} = (\text{DLF})_{\max}(\sigma_s t) = (\text{DLF})_{\max}\left(\frac{F_2 L}{4}\right)\left(\frac{1}{Z}\right) = 118.617\ \text{N/mm}$$

Example 18.72. *An elevated water tank weighs 2500 kN when filled with water, the centroid of the mass being at a height of 10 m from the ground. The horizontal spring factor for the tank is* $k = 9000$ *N/mm. The tank is subjected to a dynamic load with arise time* L_r *of* 0.075 sec *and a maximum value of* F_1 *of 100 kN. The table below gives the various values of* $(DLF)_{\max}$. *Compute the BM at the base of the tank, if* $g = 9800$ *mm/sec*^{2}.

Dynamic Load Factor (DLF) given by

$\frac{t_r}{\tau}$	$(\text{DLF})_{\max}$
0.0001	*1.98*
0.0709	*1.59*
1.00	*0.00*

Solution : $p = \sqrt{\dfrac{k}{m}} = 5.940 \text{ rad/ sec}$

$$\tau = \frac{2\pi}{p} = \frac{2\pi}{5.940} = 1.058 \text{ sec}$$

$$\frac{t_r}{\tau} = \frac{0.075}{1.058} = 0.0709$$

From the Table, $(\text{DLF})_{\text{max}} = 1.59$

B.M. in the bottom = (100)(10)(1.59) = 1590 kNm.

Example 18.73. *A circular water tank carries a weight of water equal to 212 kN at a height of 10 metres from the ground. The horizontal spring factor is equal to 4140 N per mm. It is subjected to dynamic load with a rise time of* 0.075 sec *and maximum force* $F_1 = 100$ *kN. Compute the Bending Moment due to dynamic load at the bottom.* $(g = 9800 \text{ N/mm}^2)$

Dynamic Load Factor (D.L.F.) is given by,

$\frac{t_r}{\tau}$	$(\text{DLF})_{\text{max}}$
0.165	*1.34*
1.000	*1.00*
0.500	*1.62*

Solution : $p = \sqrt{\dfrac{k}{m}} = 13.83 \text{ rad/ sec}$

$$\tau = \frac{2\pi}{p} = 0.454 \text{ sec}$$

$$\frac{t_r}{\tau} = \frac{0.075}{0.454} = 0.165$$

$$(\text{D.L.F.})_{\text{max}} = 1.34.$$

B.M. in the bottom = 100(1.34)(10) = 1340 kNm.

Example 18.74. *An elevated water tank weighs 6600 kN when full and 2600 kN when empty. The centroid of the mass is 19 m from the bottom. The horizontal spring factor for the column is 30000 N/mm. The tank is subjected to a dynamic wind load with a rise of time* $t_r = 0.1$ sec *and a maximum value of 150 kN. Compute the B.M. at the base of the tank* $g = 9800 \text{ mm/ sec}^2$.

$\frac{t_r}{\tau}$	$(\text{DLF})_{\text{max}}$
0.10623	*1.95*
0.1692	*1.92*

Solution :(a) When the tank is full :

$$p = \sqrt{\frac{k}{m}} = 6.674 \text{ rad/ sec}$$

$$\tau = \frac{2\pi}{p} = \frac{2\pi}{6.674} = 0.9414 \text{ sec}$$

$$\frac{t_r}{\tau} = \frac{0.1}{0.9414} = 0.10622$$

$$(\text{D.L.F.})_{\max} = 1.95$$

$$(\text{B.M.})_{\max} = 5557.50 \text{ kNm.}$$

(b) When the tank is empty :

$$p = 10.63 \text{ rad/ sec}$$

$$\tau = \frac{2\pi}{p} = \frac{2\pi}{10.63} = 0.5909 \text{ sec}$$

$$\frac{t_r}{\tau} = \frac{0.1}{0.5909} = 0.1692$$

$$(\text{DLF})_{\max} = 1.92$$

$$(\text{B.M.})_{\max} = 1.92(150)(19) = 5472 \text{ kNm.}$$

Example 18.75. *Compute the horizontal spring factor of the water tower shown in Fig. 18.89. The horizontal braces in each level possess infinite rigidity. The columns are of 400 mm square. The inclined braces are 20 mm diameter steel rods.* $E_c = 3(10)^4$ *N/mm*^{2} *and* $E_s = 2.10(10)^4$ *N/mm*^{2}.

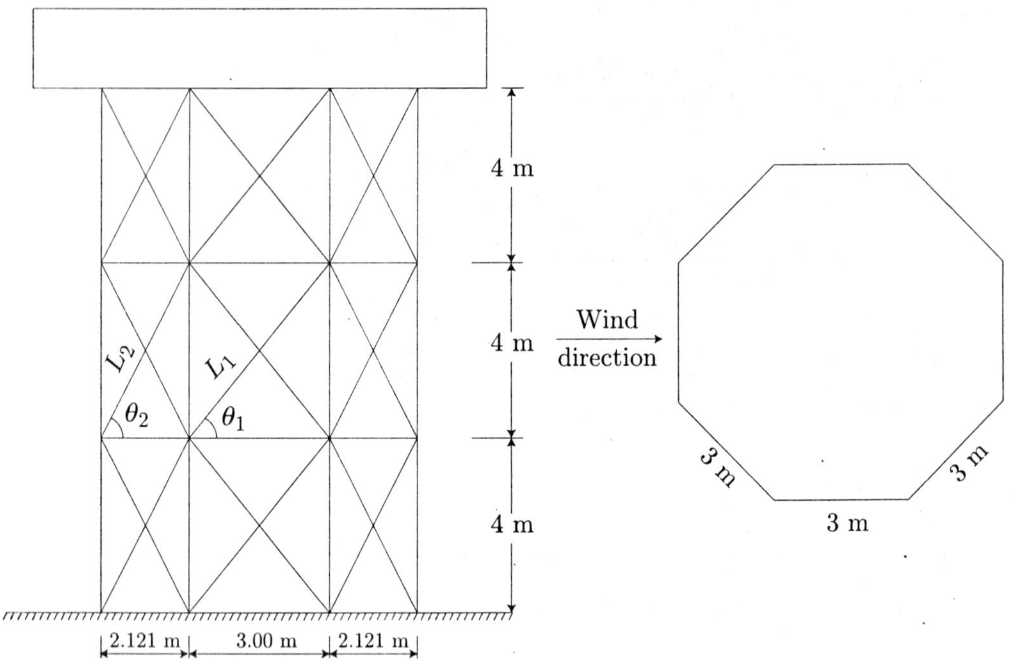

Fig. 18.89: For Example 18.75

Solution : Before proceeding with the computation, consider a derivation for stiffness for inclined member.

Derivation. Let AB be the inclined member with angle θ to the horizontal whose axial stiffness is $\frac{AE}{L}$. A unit displacement BC in the horizontal direction is made as shown in Fig. 18.90. Its component along AB direction is $\cos\theta$.

Therefore, stiffness along AB direction when unit displacement was impressed in the horizontal direction $= \frac{AE}{L}\cos\theta$. Resolving the same in the horizontal direction $= \left(\frac{AE}{L}\cos\theta\right) \times \cos\theta$,

$$(k)_{\text{horizontal}} = \frac{AE}{L}\cos^2\theta,$$

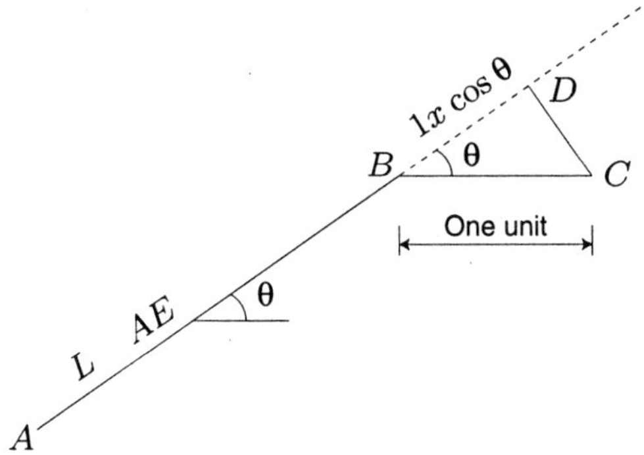

Fig. 18.90: Unit displacement in *BC*

Solution : Since the horizontal bracing possesses infinite rigidity, the tower acts as a shear building.

Considering any one bay, there are eight columns which are in parallel. Of the inclined braces only the tension bracing will contribute to the stiffness.

$$k = 8\left(12\frac{EI_c}{h^3}\right) + 4\left(\frac{AE}{L^2}\cos^2\theta_2\right) + 2\left(\frac{AE}{L_1}\cos^2\theta_1\right)$$

$$I_c = \frac{1}{12}(400)^4 = 21.33533(10)^8 \text{ mm}^4$$

$$\frac{12EI}{h^3} = \frac{12(3)(10)^4(2133333333)}{(4000)^3} = 12000 \text{ N/mm}$$

$$L_2 = 4.5275 \text{ m}$$

$$\cos\theta_2 = 0.4684$$

$$\frac{AE}{L_2}\cos^2\theta_2 = \frac{(2.1)(10)^5(0.2193)}{4527.5}\left(\frac{\pi}{4}\right)(20)^2 = 3195.58 \text{ N/mm.}$$

Now $L_1 = 5000$ mm

$$\cos\theta_1 = \frac{3}{5} = 0.60$$

$$\frac{AE}{L_1}\cos^2\theta_1 = \frac{2.1(10)^5(0.36)}{5000}\frac{\pi}{4}(20)^2 = 4750 \text{ N/mm}$$

$$k = 118282.32 \text{ N/mm.}$$

In shear building, the springs are connected in series in vertical direction

and $$\frac{1}{k_e} = \frac{1}{k} + \frac{1}{k} + \frac{1}{k} = \frac{3}{k},$$

whence $$k_e = \frac{k}{3} = \frac{118282.32}{3} = 39427.94 \text{ N/mm.}$$

Problems

18.1 What portion of the uniformly distributed weight of the simply supported beam in Fig. 18.61 should be added to the weight W at the middle if, during lateral vibration, the beam is assumed to have the configuration of a half sine wave instead of a static deflection curve?

18.2 If the beam in Fig. 18.61 has both ends built-in instead of simply supported, what portion of its total weight should be added to the weight W at the mid-span in calculating the natural period of lateral vibration? Assume a configuration of the beam during vibration similar to a static deflection curve under the load W.

18.3 Repeat the preceding problem, assuming that during vibration the beam maintains a configuration identical in shape with a full wave of a cosine curve.

With the left end of the beam as an origin, the equation of the deflection curve may be taken as,

$$y = \frac{\Delta}{2}\left(1 - \cos\frac{2\pi x}{L}\right),$$

where Δ is the displacement under the load W.

18.4 A 50 N weight attached to a light spring elongates it by 8 mm. Determine the natural frequency of the system.

18.5 What weight attached to the spring of Prob. 18.4 would result in a natural frequency of 1.67 cycles per second.

18.6 In a spring-mass system k_1, m has a natural frequency of f_1. If a second spring k_2 is added in series with the first spring, the natural frequency is lowered to $\frac{1}{2}f_1$. Determine k_2 in terms of k_1.

18.7 A 50 N weight, attached to the lower end of a spring whose upper end is fixed, vibrates with a natural period of 0.45 sec. Determine the natural period when a 25 N weight is attached to the mid-point of the same spring with the upper and lower ends fixed.

18.8 An unknown weight W N attached to the end of an unknown spring k has a natural frequency of 94 c.p.m. When a 5 N weight is added to W, the natural frequency is lowered to 76.7 c.p.m. Determine the unknown weight W N and the spring constant k.

18.9 Determine the natural frequency of vibration of the system shown in Fig. P-18.1. Assume the bar AB to be rigid and weightless with C as the mid-point.

18.10 Determine the natural frequency of vibration of the system shown in Fig. P-18.2. Assume the bar AB to be rigid and weightless.

18.11 Determine the natural frequency of vibration of a weight W attached to the centre of a beam of length L as shown in Fig. P-18.3. The beam is simply supported, its weight and lateral stiffness being w N and EI N-mm^2 respectively. Assume the deflection to be equal to the state curve for concentrated load

$$y = y_{\max}\left[3\left(\frac{x}{L}\right) - 4\left(\frac{x}{L}\right)^3\right], \quad 0 < x < L/2,$$

where $y_{\max} = PL^3/48EI$.

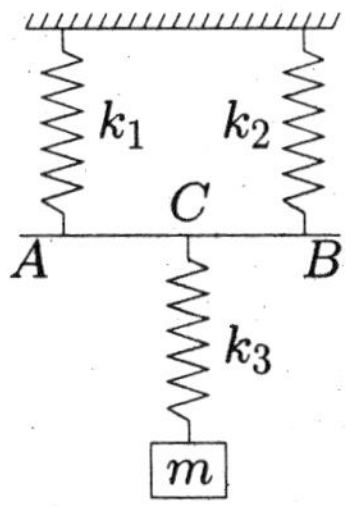

Fig. P-18.1

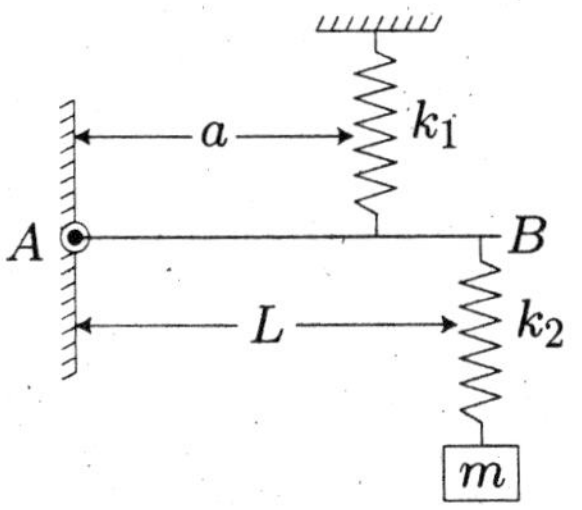

Fig. P-18.2

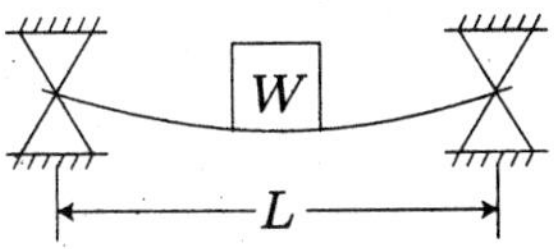

Fig. P-18.3

18.12 If in Prob. 18.11 $W = 75$ N, and the beam is steel of dimensions 1.6 m × 75 mm × 10 mm, determine the numerical value of the natural frequency. ($E = 2.1(10)^5$ N/mm^2).

18.13 A steel cantilever beam 1.90 mm long has a rectangular cross section 25 × 13 mm with the 25 mm side in the vertical direction. Determine the natural frequency for oscillation in the vertical and horizontal planes.

18.14 The free end of the beam of Prob. 18.13 is given an initial displacement 6 mm down and 12 mm to the left. Determine the path described by a point on the end face.

18.15 What portion of the uniformly distributed weight of the beam *ABC* in Fig. P-18.4 should be added to the weight *W* at the free end in calculating the natural frequency of lateral vibration? Assume a static deflection curve.

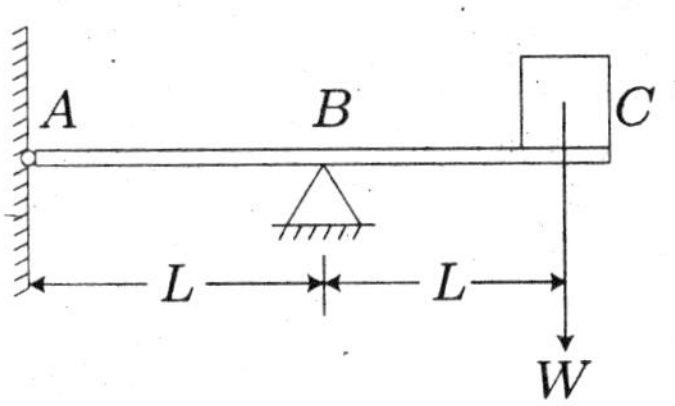

Fig. P-18.4

18.16 The following data are given for a vibrating system with viscous damping

$$W = 100 \text{ N};\ k = 30 \text{ N/mm};\ C = 0.12 \text{ N/mm per sec.}$$

Determine the logarithmic decrement and ratio of any two successive amplitude.

18.17 Show that the logarithmic decrement is also given by, $\frac{1}{n} ln(\frac{x_0}{x_n})$, where x_n represents the amplitude after n cycles have elasped.

Appendix

Statically Indeterminate Frames (Predominant Values)

Table 1 : Rigid Bents

Frame Constants

$$\alpha = \frac{I_2}{I_1} \cdot \frac{h}{L},\ \beta = \frac{b}{L},\ \gamma = \frac{a}{h},\ N_1 = \alpha + 1$$

Loading	Bending Moment Diagram	Formulas	
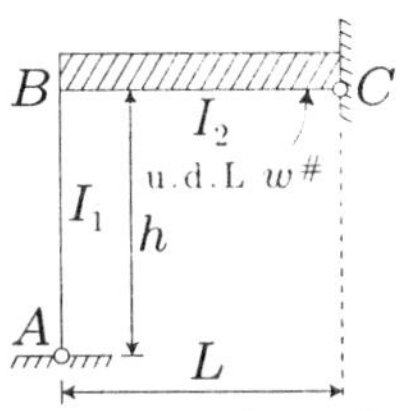	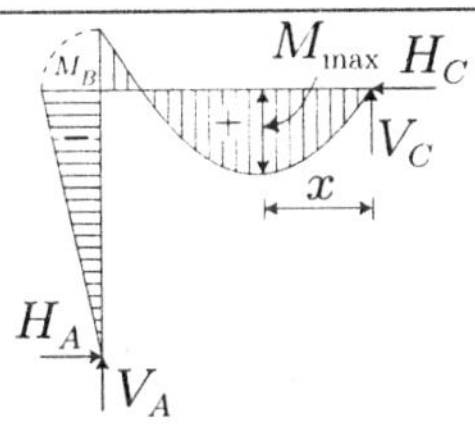	$M_B = -\frac{wL^2}{8N_1}$ $H_A = H_C = -\frac{M_B}{h}$ $V_A = \frac{wL}{2} - \frac{M_B}{L}$ $V_C = \frac{wL}{2} + \frac{M_B}{L}$	$x = \frac{V_C}{w}$ $M_{max} = V_C x - \frac{wx^2}{2}$
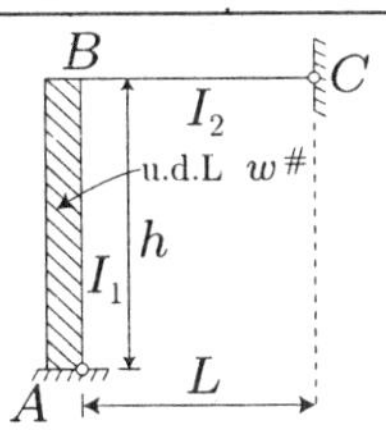	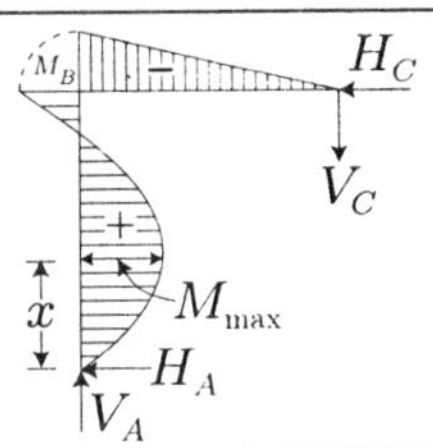	$M_B = -\frac{wh^2\alpha}{8N_1}$ $H_A = -\frac{wh}{2} - \frac{M_B}{h}$ $H_C = \frac{wh}{2} - \frac{M_B}{h}$ $V_A = -V_C = -\frac{M_B}{L}$	$x = \frac{H_A}{w}$ $M_{max} = H_A x - \frac{wx^2}{2}$
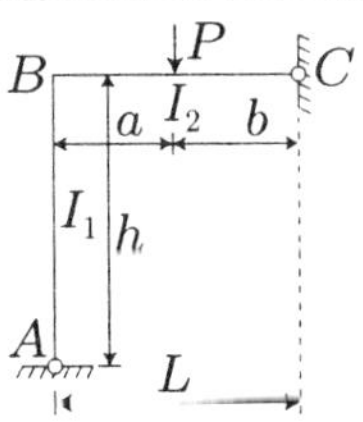	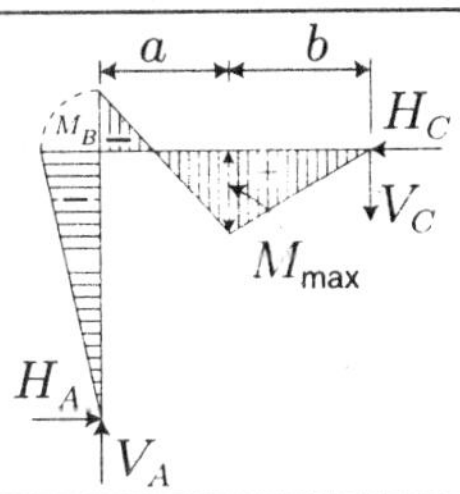	$M_B = -\frac{Pa\beta(1+\beta)}{2N_1}$ $H_A = H_C = -\frac{M_B}{h}$ $V_A = \frac{Pb - M_B}{L}$ $V_C = P - V_A$	$M_{max} = V_C \cdot b$
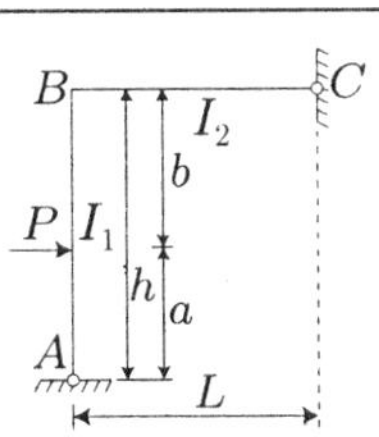	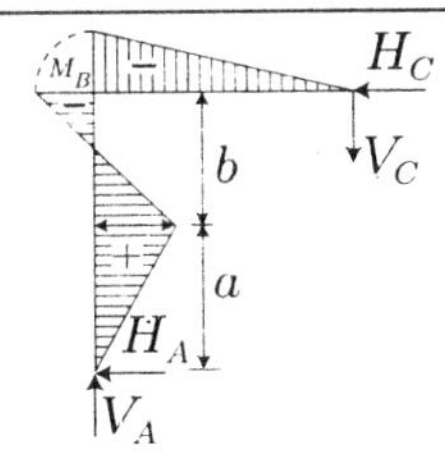	$M_B = -\frac{Pb\gamma(1+\gamma)\alpha}{2N_1}$ $H_A = -(P - H_C)$ $H_C = \frac{Pa - M_B}{h}$ $V_A = -V_C = -\frac{M_B}{L}$	$M_{max} = H_A \cdot a$

Contd.

*Adapted from *Rahmenformeln* by Kleinlogel

Table 1 (contd.)

Frame Constants

$$\alpha = \frac{I_2}{I_1}\cdot\frac{h}{L},\ \beta = \frac{b}{L},\ \gamma = \frac{a}{h},\ \epsilon = \frac{b}{h},\ \gamma + \epsilon = 1,\ N_1 = \alpha + 1,\ N_2 = 3\alpha + 4\ N_3 = 4\alpha + 3$$

Loading	Bending Moment Diagram	Formulas	
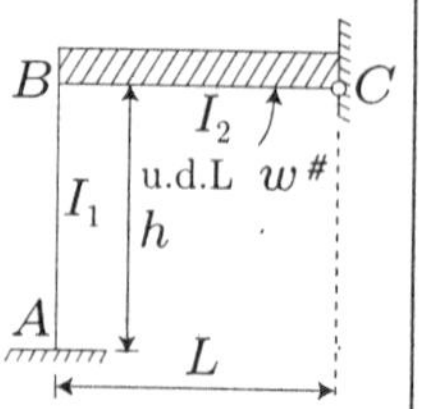	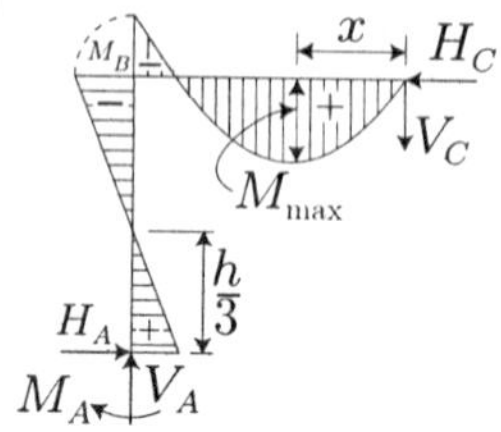	$M_B = -\frac{wL^2}{2N_2}$ $M_A = -\frac{M_B}{2}$ $V_A = \frac{wL}{2} - \frac{M_B}{L}$ $V_C = \frac{wL}{2} + \frac{M_B}{L}$ $N_2 = 3\alpha + 4$	$H_A = H_C = \frac{3M_A}{h}$ $x = \frac{V_C}{a}$ $M_{max} = \frac{V_C \cdot x}{2}$
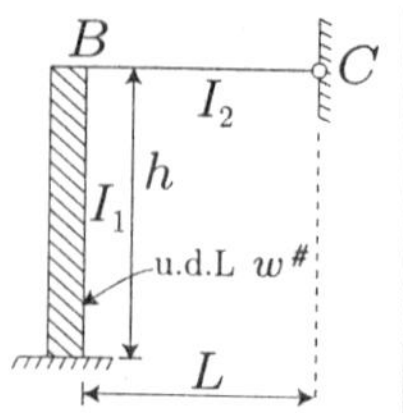	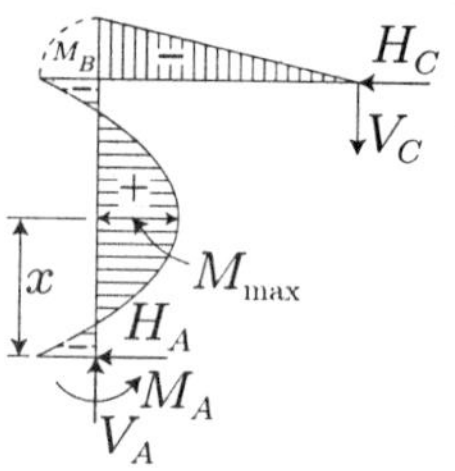	$M_B = -\frac{wh^2\alpha}{4N_2}$ $M_A = -\frac{wh^2(\alpha+2)}{4N_2}$ $V_A = -V_C = -\frac{M_B}{L}$ $H_A = H_C - wh$ $N_2 = 3\alpha + 4$	$H_C = \frac{wh}{2} - \frac{M_A - M_B}{h}$ $x = \frac{H_A}{w}$ $M_{max} = -M_A - \frac{H_A^2}{2w}$
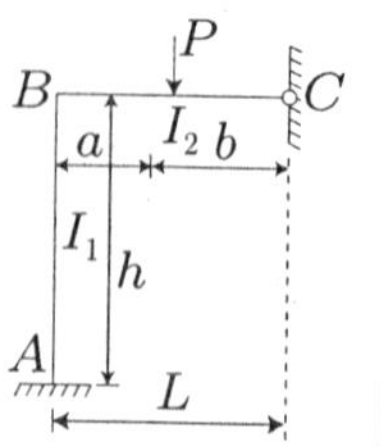	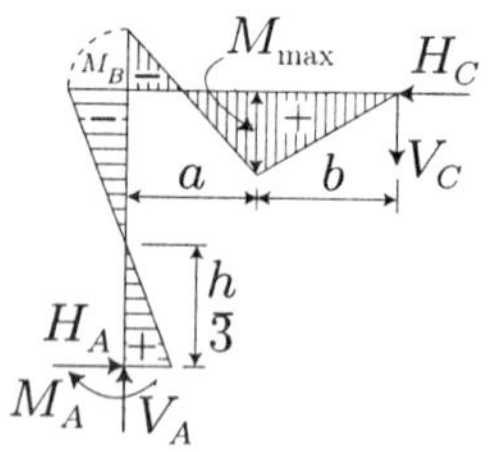	$M_A = \frac{Pa\beta(1+\beta)}{N_2}$ $M_B = -2M_A$ $V_A = \frac{Pb - M_B}{L}$ $V_C = P - V_A$ $N_2 = 3\alpha + 4$	$H_A = H_C = \frac{3M_A}{h}$ $M_{max} = (Pa + M_B)\beta$
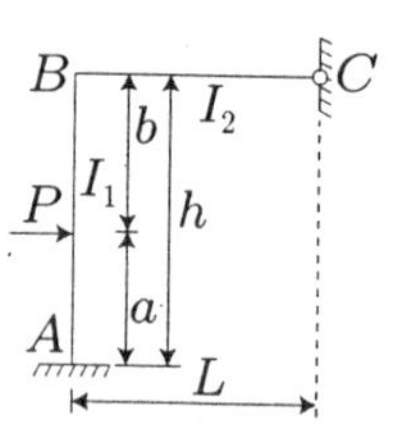	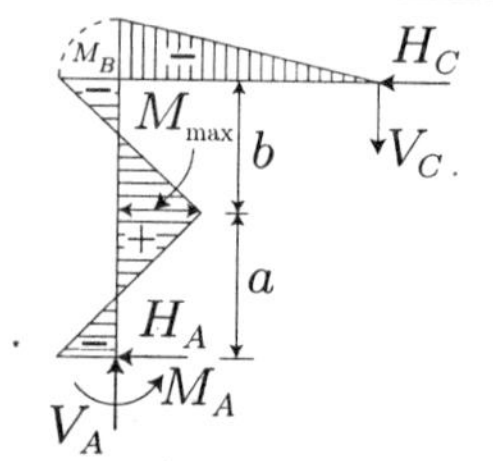	$M_A = -\frac{Pab}{h}\cdot\frac{3\epsilon\alpha + 2(1-\epsilon)}{N_2}$ $M_B = -\frac{Pab}{h}\cdot\frac{3\gamma\alpha}{N_2}$ $V_A = -V_C = -\frac{M_B}{L}$ $H_A = H_C - P$ $N_2 - N_2 = 3\alpha + 4$	$H_C = \frac{Pa - M_B + M_A}{h}$ $M_{max} = \frac{Pab}{h} + \epsilon M_A$ $+ \gamma M_B$
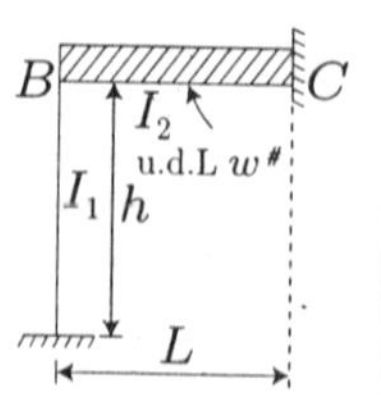	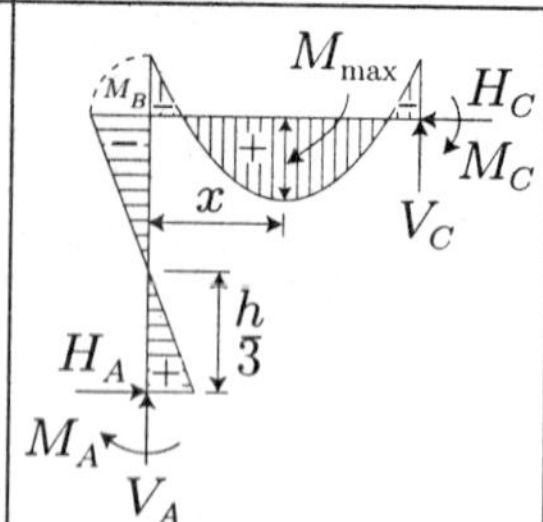	$M_B = -\frac{wL^2}{12N_1}$ $M_A = -\frac{M_B}{2}$ $M_C = -\frac{wL^2(3\alpha+2)}{24N_1}$ $V_A = \frac{wL}{2} - \frac{M_B - M_C}{L}$ $N_2 = \alpha + 1$	$V_C = \frac{wL}{2}\frac{M_B - M_C}{L}$ $H_A = M_C\frac{3M_A}{h}$ $x = \frac{V_A}{w}$ $M_{max} = \frac{V_A x}{2} + M_B$

Contd.

*Adapted from *Rahmenformeln* by Kleinlogel

Table 1 (contd.)

Frame Constants

$$\alpha = \frac{I_2}{I_1}\frac{h}{L},\ \beta = \frac{b}{L},\ \delta = \frac{a}{L},\ \beta + \delta = 1,\ N_1 = \alpha + 1,\ N_2 = 3\alpha + 4\ N_3 = 4\alpha + 3$$

Loading	Bending Moment Diagram	Formulas	
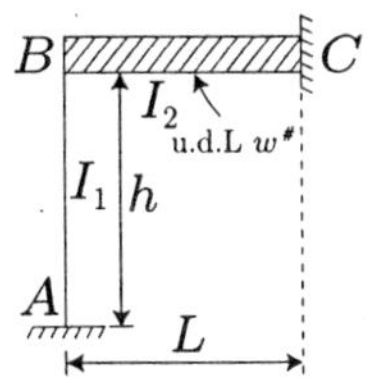	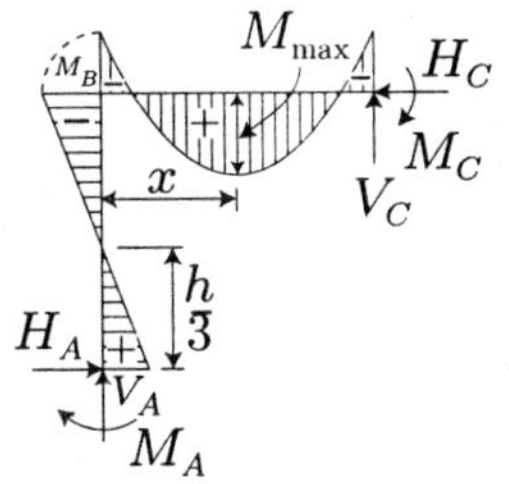	$M_B = -\frac{wL^2}{12N_1}$ $M_A = -\frac{M_B}{2}$ $M_C = -\frac{wL^2(3\alpha+2)}{24N_1}$ $V_A = \frac{wL}{2} - \frac{M_B-M_C}{L}$	$V_C = \frac{wL}{2}\frac{M_B-M_C}{L}$ $H_A = M_C\frac{3M_A}{h}$ $x = \frac{V_A}{w}$ $M_{max} = \frac{V_A x}{2} + M_B$
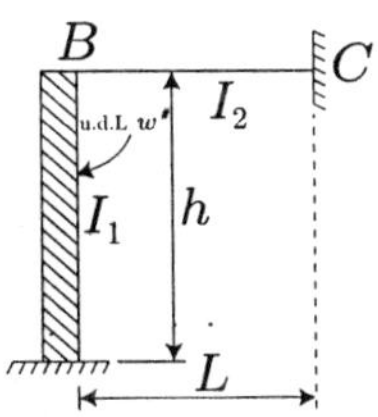	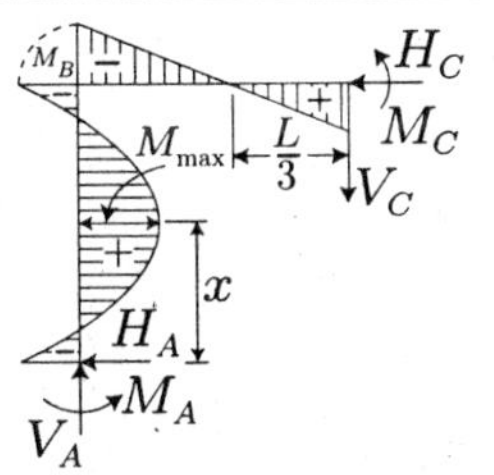	$M_B = -\frac{wh^2\alpha}{12N_1}$ $M_A = -\frac{wh^2(2\alpha+3)}{24N_1}$ $M_C = -\frac{M_B}{2}$ $V_A = -V_C = \frac{3M_C}{L}$	$H_A = -(wh - H_C)$ $H_C = \frac{wh}{2} + \frac{M_A-M_B}{h}$ $x = \frac{H_A}{w}$ $M_{max} = -M_A - \frac{H_A^2}{2w}$
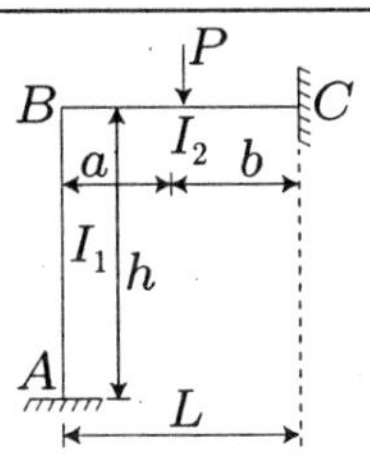	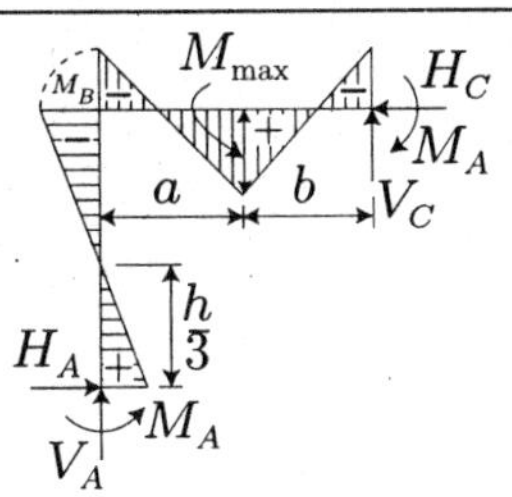	$M_B = -\frac{Pab}{L}\cdot\frac{\beta}{N_1}$ $M_A = -\frac{M_B}{2}$ $M_C = -\frac{Pab}{L}\cdot\frac{(1+\delta)=\alpha+2\delta}{2N_1}$ $V_A = \frac{Pb+M_C-M_B}{L}$	$V_C = -V_A + P$ $H_A = H_C = \frac{3M_A}{h}$ $M_{max} = \frac{Pab}{L} + \beta M_B$ $+ \delta M_C$
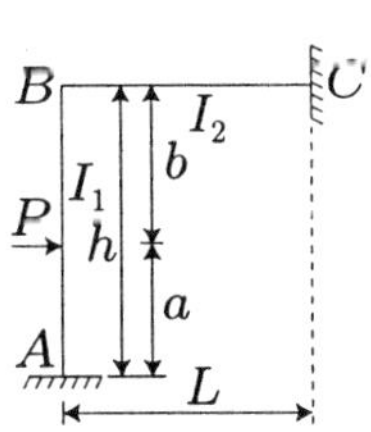	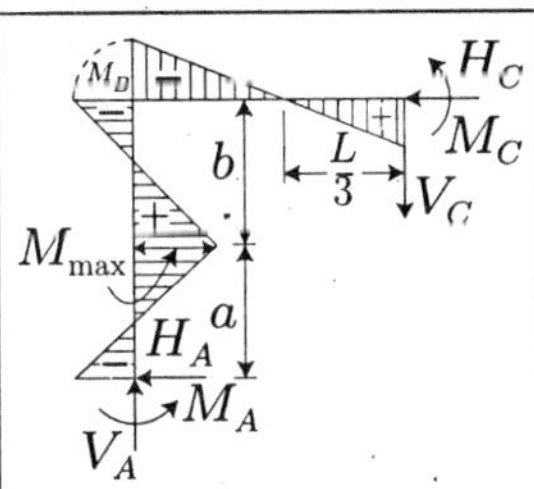	$M_B = -\frac{Pab}{h}\frac{\delta\alpha}{N_1}$ $M_A = -\frac{Pab}{h}\cdot\frac{1+\beta+2\alpha\beta}{2N_1}$ $M_C = -\frac{M_B}{2}$ $V_A = -V_C = \frac{3M_C}{L}$	$H_A = +H_C - P$ $H_C = \frac{Pa+M_B-M_A}{h}$ $M_{max} = \frac{Pab}{h} + \beta M_A$ $+ \delta M_B$
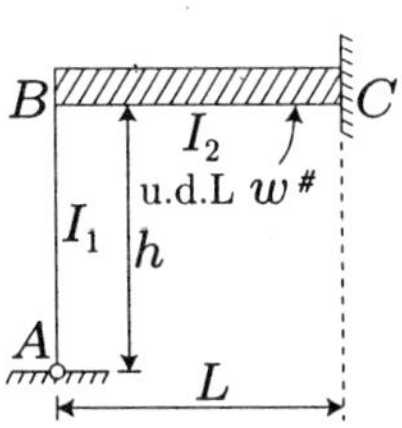	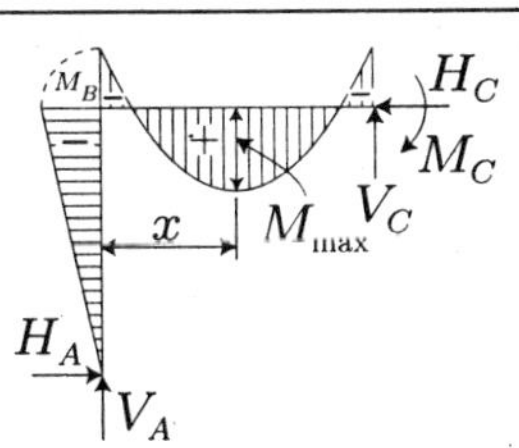	$M_B = -\frac{wL^2}{4N_3}$, $M_C = -wL^2\frac{(2\alpha+1)}{4N_3}$ $H_A = H_C = -\frac{M_B}{h}$ $V_A = \frac{wL}{2} - \frac{M_B-M_C}{L}$	$V_C = \frac{wL}{2} + \frac{M_B-M_C}{L}$ $x = \frac{V_A}{w}$ $M_{max} = \frac{V_A\cdot x}{2} + M_B$

Contd.

Table 1 (contd.)

Frame Constants

$$\alpha = \frac{I_2}{I_1} \cdot \frac{h}{L},\ \beta = \frac{b}{L},\ \delta = \frac{a}{L},\ \gamma = \frac{a}{h},\ \epsilon = \frac{b}{h}.\quad \begin{matrix} \gamma + \epsilon = 1 \\ \beta + \delta = 1 \end{matrix},\ N_3 = 4\alpha + 3$$

Loading	Bending Moment Diagram	Formulas	
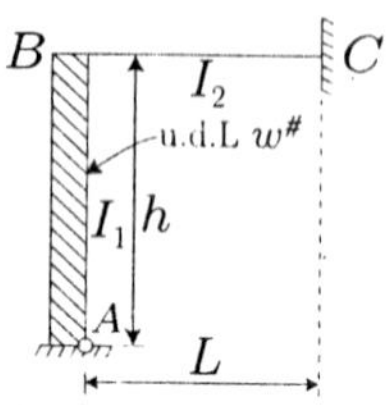	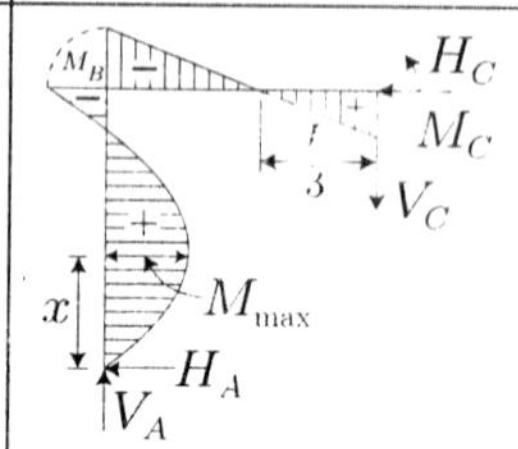	$M_B = -\frac{wh^2\alpha}{2N_3}$, $M_C = -\frac{M_B}{2}$ $H_A = -\frac{wh}{2} - \frac{M_B}{h}$ $H_C = \frac{wh}{2} - \frac{M_B}{h}$	$V_A = -V_C = \frac{3M_C}{L}$ $x = \frac{H_A}{w}$ $M_{\max} = H_A \cdot x - \frac{wx^2}{2}$
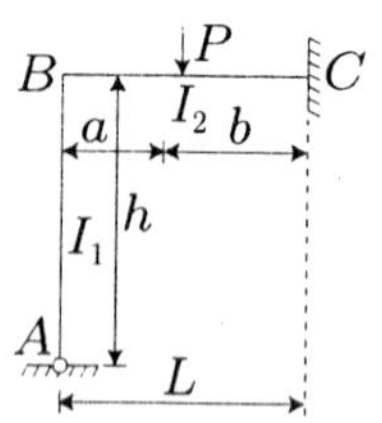	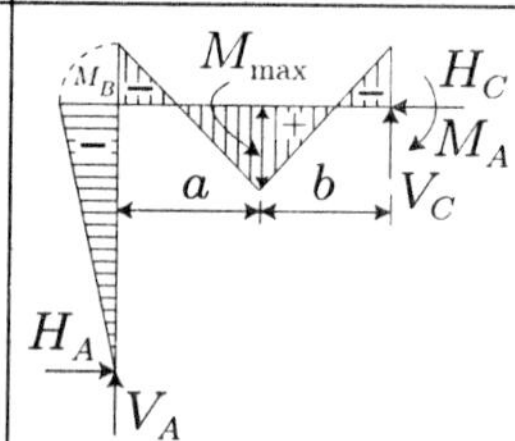	$M_B = -\frac{Pab}{L} \cdot \frac{3\beta}{N_3}$ $M_C = -\frac{Pab}{L} \cdot \frac{2\alpha(1+\beta)+3\delta}{N_3}$ $H_A = H_C = -\frac{M_B}{h}$ $V_A = \frac{Pb - M_B + M_C}{L}$	$V_C = P - V_A$ $M_{\max} = \frac{Pab}{L} + \beta M_B + \delta M_C$
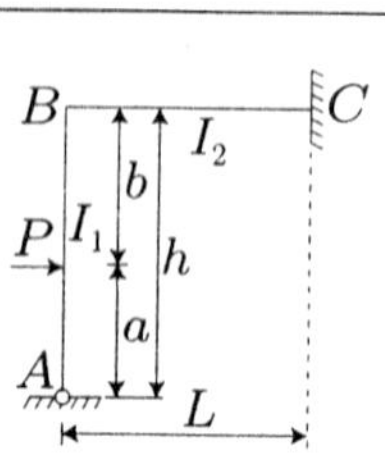	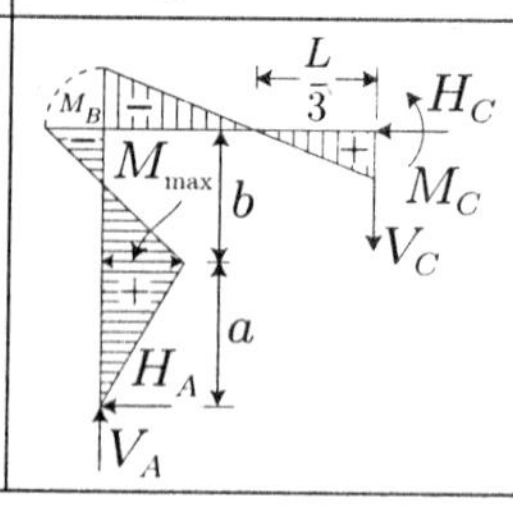	$M_C = -\frac{Pa\epsilon\alpha(1+\gamma)}{N_3}$ $M_B = -2M_c$ $H_A = H_C - P$ $H_C = \frac{Pa - M_B}{h}$	$V_A = -V_C = \frac{3M_C}{L}$ $M_{\max} = (Pb + M_B)\gamma$

*Adapted from *Rahmenformeln* by Kleinlogel

Table 2 : Portal Frames

Frame Constants

$$\alpha = \frac{I_2}{I_1} \cdot \frac{h}{L},\ \beta = \frac{b}{L},\ \delta = \frac{a}{L}.\ \gamma = \frac{a}{h},\ \epsilon = \frac{b}{h},\ \gamma + \epsilon = 1,\ \delta + \beta = 1,\ N = 2\alpha + 3$$

Loading	Bending Moment Diagram	Formulas	
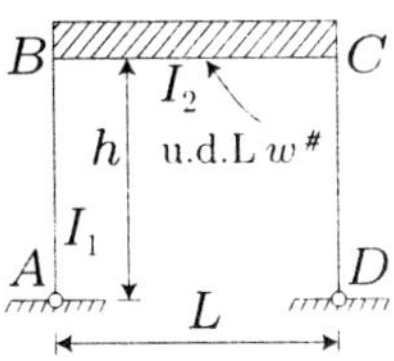	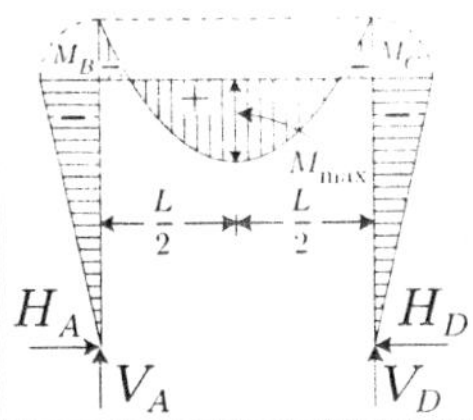	$M_B = M_C = -\frac{wL^2}{4N}$ $V_A = V_D = \frac{wL}{2}$ $H_A = H_D = -\frac{M_B}{h}$	$M_{max} = \frac{wL^2}{8} + M_B$
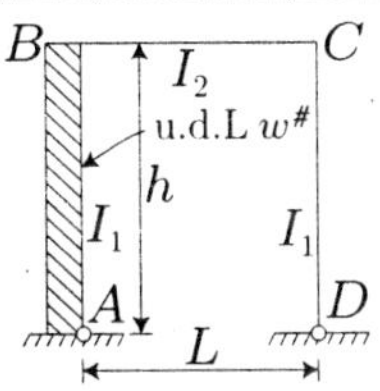	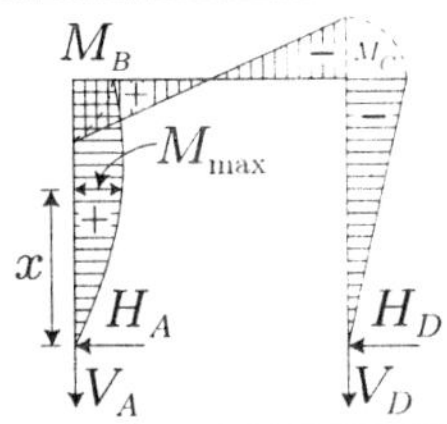	$M_B = \frac{wh^2}{4}\left[\frac{-\alpha}{2N} + 1\right]$ $M_C = \frac{wh^2}{4}\left[\frac{-\alpha}{2N} - 1\right]$ $V_A = -V_D = -\frac{wh^2}{2L}$ $H_A = -wh + H_D$	$H_D = -\frac{M_C}{h}$ $x = \frac{H_A}{w}$ $M_{max} = \frac{H_A \cdot x}{2}$
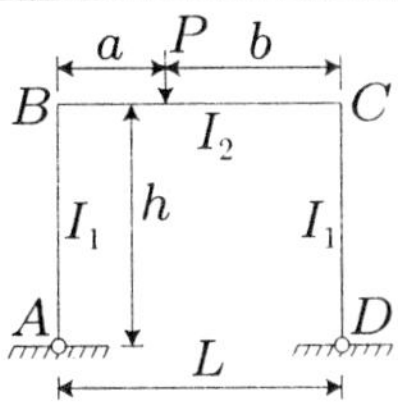	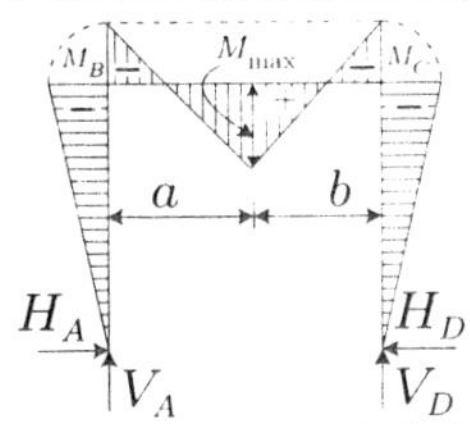	$M_B = M_C = -\frac{Pab}{L} \cdot \frac{3}{2N}$ $V_A = \frac{Pb}{L}$ $V_D = \frac{Pa}{L}$ $H_A = H_D = -\frac{M_B}{A}$	$M_{max} = \frac{Pab}{L} + M_B$
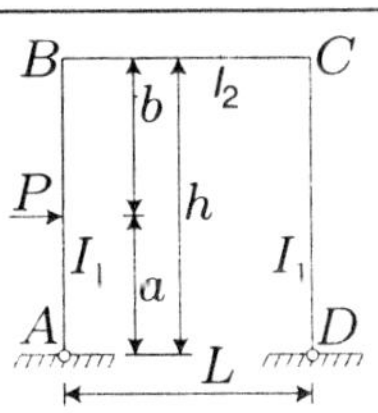	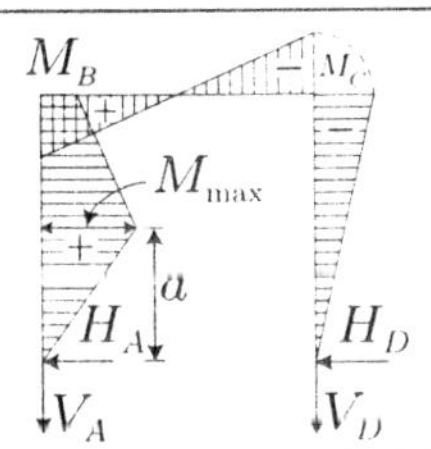	$M_B = \frac{Pa}{2}\left[-\frac{(1+\gamma)\epsilon\alpha}{N} + 1\right]$ $M_C = \frac{Pa}{2}\left[-\frac{(1+\gamma)\epsilon\alpha}{N} - 1\right]$ $V_A = -V_D = -\frac{Pa}{L}$ $H_D = -\frac{M_C}{h}$	$H_A = -(P - H_D)$ $M_{max} = \gamma(P \cdot b + M_B)$
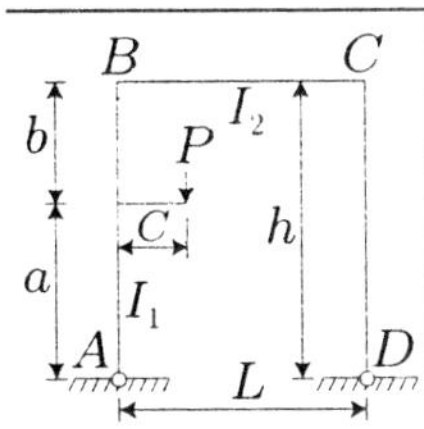	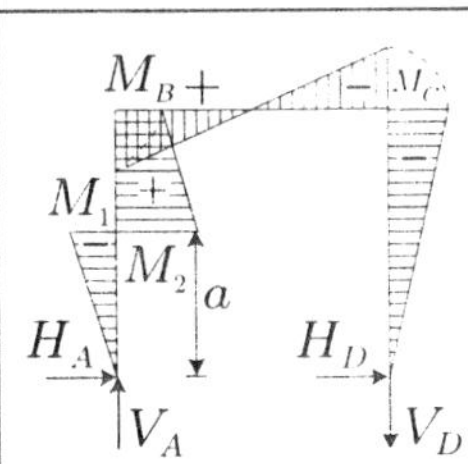	$M_B = \frac{P_c}{2}\left[\frac{(3\gamma^2-1)\alpha}{N} + 1\right]$ $M_C = \frac{P_c}{2}\left[\frac{(3\gamma^2-1)\alpha}{N} - 1\right]$ $M_1 = -H_A \cdot a$ $M_2 = P_c - H_A \cdot a$ $V_D = \frac{P_c}{L}$.	$V_A = P - V_D$ $H_A = H_D = -$ [illegible]/h

Contd.

*Adapted from *Rahmenformeln* by Kleinlogel

Table 2 (contd.)

Frame Constants

$$\alpha = \frac{I_2}{I_1}\cdot\frac{h}{L},\ \gamma = \frac{a}{h},\ \delta = \frac{a}{L},\ \beta = \frac{b}{L},\ \epsilon = \frac{b}{h},\quad \delta+\beta = 1,\ \gamma+\epsilon = 1 \quad N = 2\alpha+3,\ N_1 = \alpha+2,\ N_2 = 6\alpha+1$$

Loading	Bending Moment Diagram	Formulas	
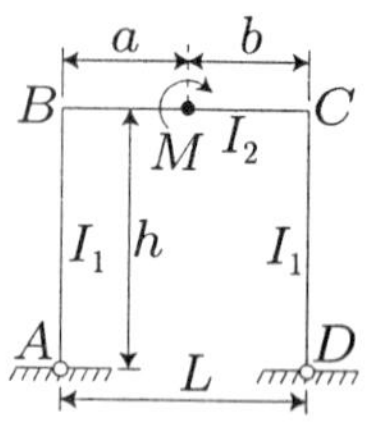	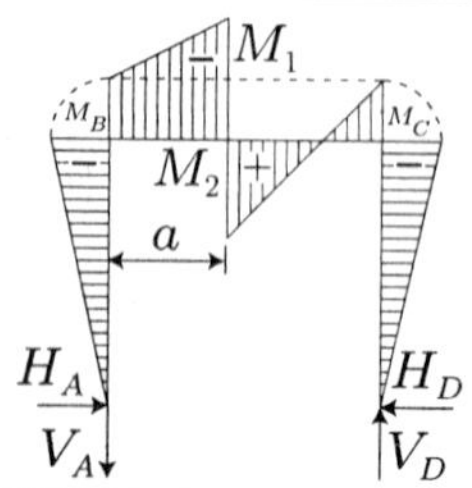	$M_B = M_C = -\frac{3M(\beta-\delta)}{2N}$ $M_1 = M_B - \delta M$ $M_2 = M_C + \beta M$ $H_A = H_D = -\frac{M_B}{h}$ $V_A = -V_C = -\frac{M}{L}$	
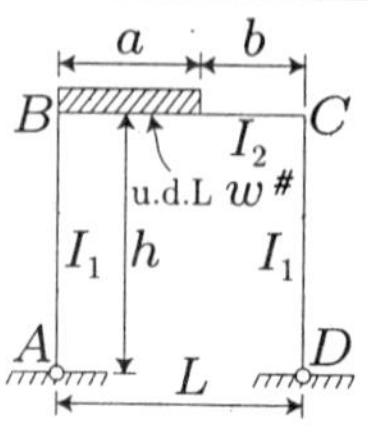	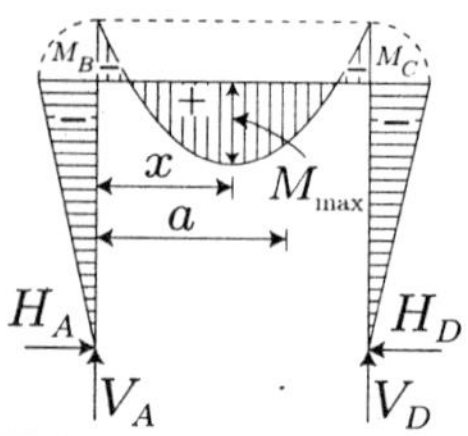	$M_B = M_C = -\frac{wa^2}{4N}(1+2\beta)$ $H_A = H_D = -\frac{M_B}{h}$ $V_A = wa - V_D$ $V_D = \frac{wa^2}{2L}$	$x = \frac{V_A}{w}$ $M_{max} = -H_A\cdot h + \frac{V_A x}{2}$
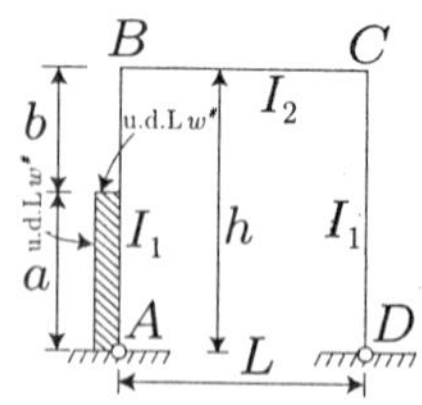	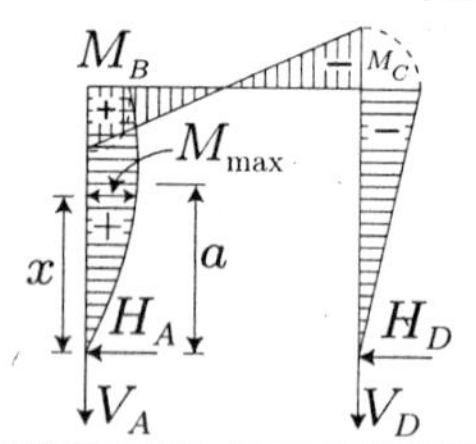	$M_B = \frac{wa^2}{4}\left[-\frac{(2-\gamma^2)\alpha}{2N}+1\right]$ $M_C = \frac{wa^2}{4}\left[-\frac{(2-\gamma^2)\alpha}{2N}-1\right]$ $V_A = V_D = \frac{wa^2}{2L}$ $H_D = -\frac{M_C}{h}$	$H_A = H_D - wa$ $x = \frac{H_A}{w}$ $M_{max} = \frac{H_A\cdot x}{2}$
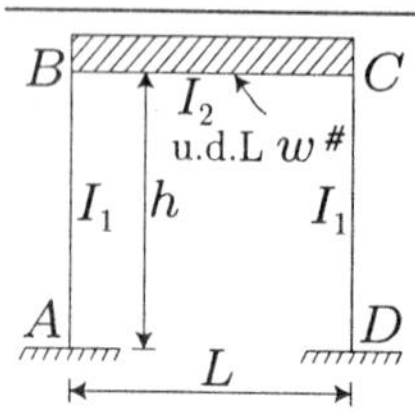	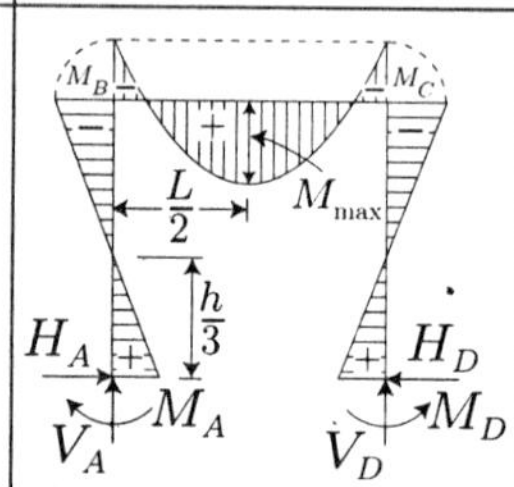	$M_A = M_D = +\frac{wL^2}{12N_1}$ $M_B = M_C = -2M_A$ $V_A = V_D = \frac{wL}{2}$ $H_A = H_D = \frac{3M_A}{h}$	$M_{max} = \frac{wL^2}{8} + M_B$
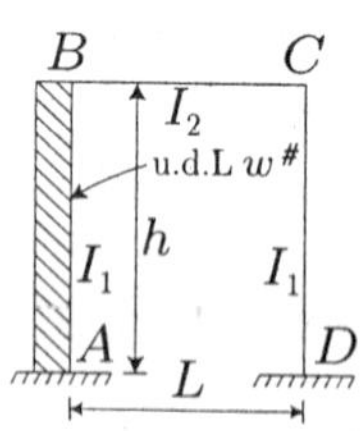	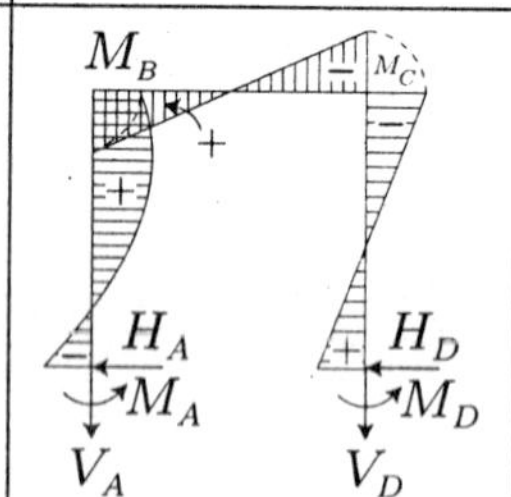	$M_A = \frac{wh^2}{4}\left[-\frac{\alpha+3}{6N_1}-\frac{4\alpha+1}{N_2}\right]$ $M_D = \frac{wh^2}{4}\left[-\frac{\alpha+3}{6N_1}+\frac{4\alpha+1}{N_2}\right]$ $M_B = \frac{wh^2}{4}\left[-\frac{\alpha}{6N_1}+\frac{2\alpha}{N_2}\right]$ $M_C = \frac{wh^2}{4}\left[-\frac{\alpha}{6N_1}-\frac{2\alpha}{N_2}\right]$ $H_A = H_D - wh$	$H_D = \frac{wh(2\alpha+3)}{8N_1}$ $V_A = -V_D = -\frac{wh^2\alpha}{LN_2}$

Contd.

*Adapted from *Rahmenformeln* by Kleinlogel

Table 2 (contd.)

Frame Constants

$$\alpha = \frac{I_2}{I_1}\cdot\frac{h}{L},\ \delta = \frac{a}{L},\ \beta = \frac{b}{L},\ \gamma = \frac{a}{h},\ \epsilon = \frac{b}{h},\quad \delta+\beta = 1,\ \gamma+\epsilon = 1,\quad N = 2\alpha+3,\ N_1 = \alpha+2,\ N_2 = 6\alpha+1$$

Loading	Bending Moment Diagram	Formulas	
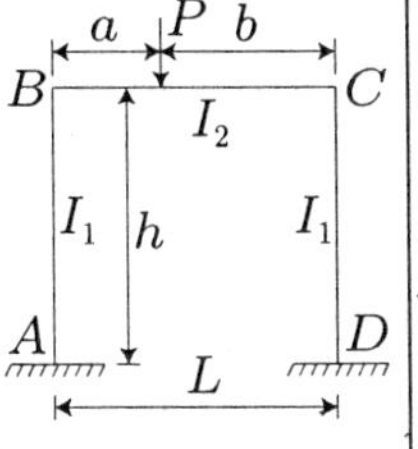	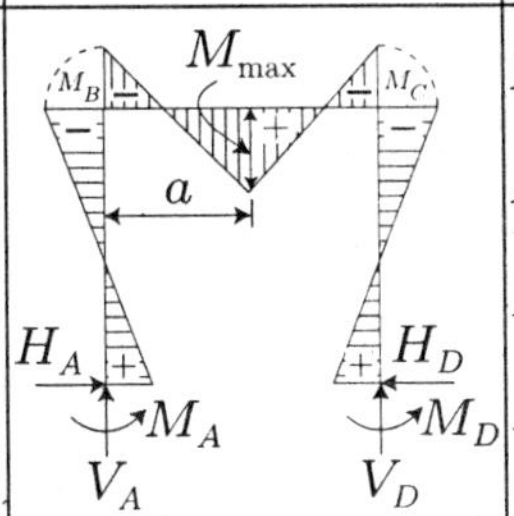	$M_A = \frac{Pab}{L}\left[\frac{1}{2N_1} - \frac{\beta-\delta}{2N_2}\right]$ $M_D = \frac{Pab}{L}\left[\frac{1}{2N_1} + \frac{\beta-\delta}{2N_2}\right]$ $M_B = -\frac{Pab}{L}\left[\frac{1}{N_1} + \frac{\beta-\delta}{2N_2}\right]$ $M_C = -\frac{Pab}{L}\left[\frac{1}{N_1} - \frac{\beta-\delta}{2N_2}\right]$ $M_{\max} = \frac{Pab}{L} + \beta M_B + \delta M_C$	$H_A = H_D = \frac{3Pab}{2LhN_1}$ $V_D = P - V_A$ $V_A = P_\beta\left[\frac{1+\delta(\beta-\delta)}{N_2}\right]$
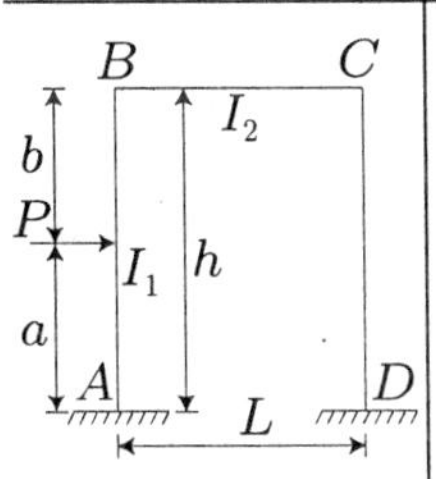	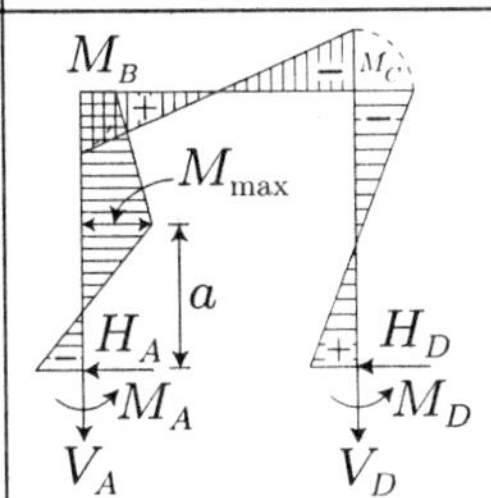	$C_1 = \frac{Pab}{h}\frac{1+\epsilon+\epsilon\alpha}{2N_1}$ $C_2 = \frac{Pab}{h}\cdot\frac{\gamma\alpha}{2N_1}$ $C_3 = \frac{3Pa\gamma\alpha}{2N_2}$ $V_A = -V_D = -\frac{2C_3}{L}$ $M_{\max} = \frac{Pab}{h} + \epsilon M_A + \gamma M_B$	$M_A = -C_1 - \left(\frac{Pa}{2} - C_3\right)$ $M_D = -C_1 + \left(\frac{Pa}{2} - C_3\right)$ $M_B = -C_2 + C_3$ $M_C = -C_2 - C_3$ $H_D = \frac{Pa}{2h} - \frac{C_1 - C_2}{h}$ $H_A = -P + H_D$
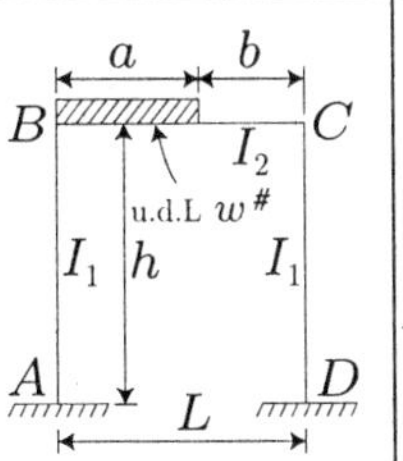	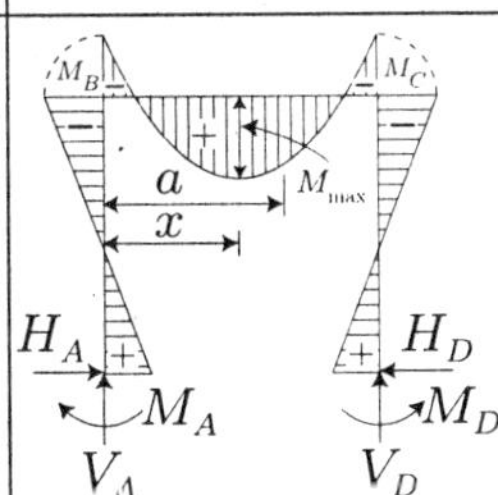	$M_A = +\frac{wa^2}{2}\left[\frac{1+2\beta}{6N_1} - \frac{\beta^2}{2N_2}\right]$ $M_D = +\frac{wa^2}{2}\left[\frac{1+2\beta}{6N_1} + \frac{\beta^2}{2N_2}\right]$ $M_B = -\frac{wa^2}{2}\left[\frac{1+2\beta}{3N_1} + \frac{\beta^2}{2N_2}\right]$ $M_C = -\frac{wa^2}{2}\left[\frac{1+2\beta}{3N_1} - \frac{\beta^2}{2N_2}\right]$ $V_D = \frac{wa^2}{2L}\left(1 - \frac{\beta^2}{N_2}\right)$ $H_A = H_D = \frac{wa^2(1+2\beta)}{4hN_1}$	$V_A = wa - V_D$ $x = \frac{V_A}{2}$ $M_{\max} = M_A - H_A\cdot h + \frac{V_A\cdot x}{2}$
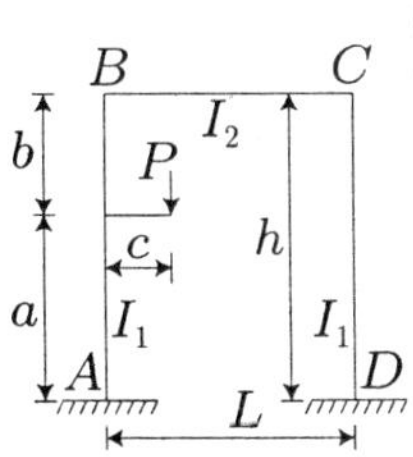	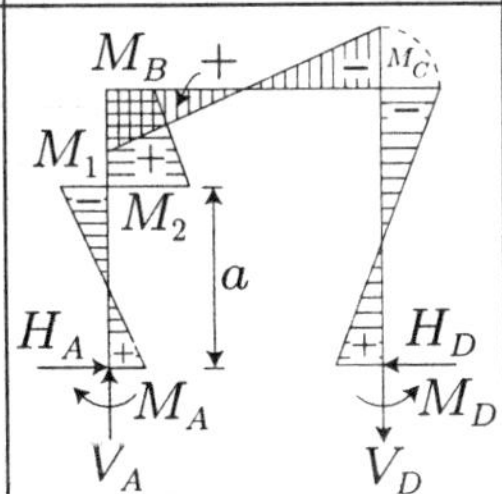	$C_1 = \frac{P_c}{2N_1}[1 + 2\epsilon\alpha - 3\epsilon^2(\alpha+1)]$ $C_2 = \frac{P_c\alpha\gamma(3\gamma-2)}{2N_1}$ $C_3 = \frac{3P_c\alpha\gamma}{N_2}$ $M_A = +C_1 - \left(\frac{P_c}{2} - C_3\right)$ $M_D = +C_1 + \left(\frac{P_c}{2} - C_3\right)$ $M_1 = M_A - H_A\cdot a,\ M_2 = M_B + H_D\cdot b$	$M_B = +C_2 + C_3$ $M_C = +C_2 - C_3$ $H_A = H_D = \frac{P_c}{2h} + \frac{C_1 - C_2}{h}$ $V_D = \frac{2C_3}{L}$, $V_A = P - V_D$

Contd.

Table 2 (contd.)

Frame Constants

$$\alpha = \frac{I_2}{I_1}\cdot\frac{h}{L},\ \delta = \frac{a}{L},\ \beta = \frac{b}{L},\ \gamma = \frac{a}{h},\ \epsilon = \frac{b}{h},\quad \gamma+\epsilon = 1,\ \delta+\beta = 1,\quad N = 2\alpha+3,\ N_1 = \alpha+2,\ N_2 = 6\alpha+1$$

Loading	Bending Moment Diagram	Formulas	
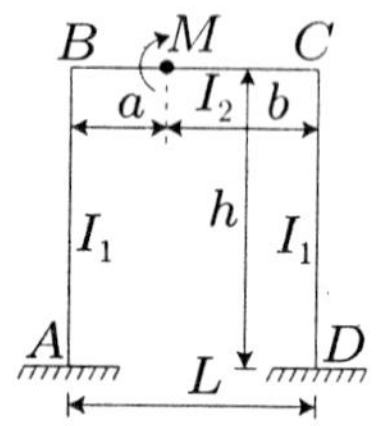	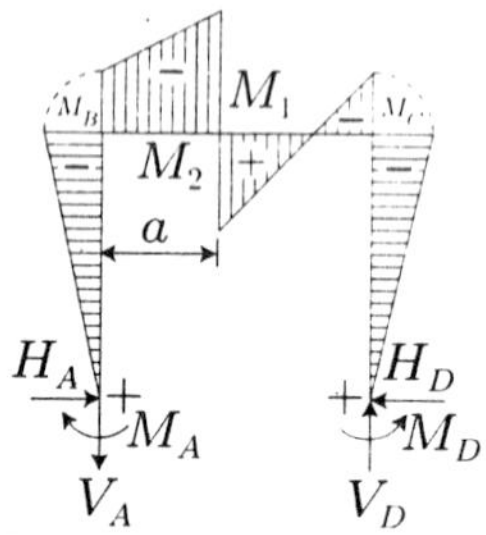	$M_A = +M\left[\frac{\beta-\delta}{2N_1} - \frac{1-6\delta\beta}{2N_2}\right]$ $M_D = +M\left[\frac{\beta-\delta}{2N_1} + \frac{1-6\delta\beta}{2N_2}\right]$ $M_B = +M\left[-\frac{\beta-\delta}{N_1} - \frac{1-6\delta\beta}{2N_2}\right]$ $M_C = +M\left[-\frac{\beta-\delta}{N_1} + \frac{1-6\delta\beta}{2N_2}\right]$ $H_A = H_D = \frac{3M(\beta-\delta)}{2N_1}$	$V_A = -V_D = \frac{6M(\alpha+\delta\beta)}{LN_2}$ $M_1 = M_B + V_A\cdot a$ $M_2 = M_C + V_D\cdot b$
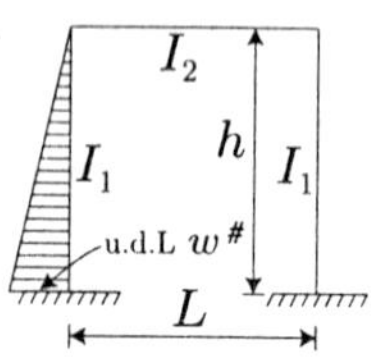	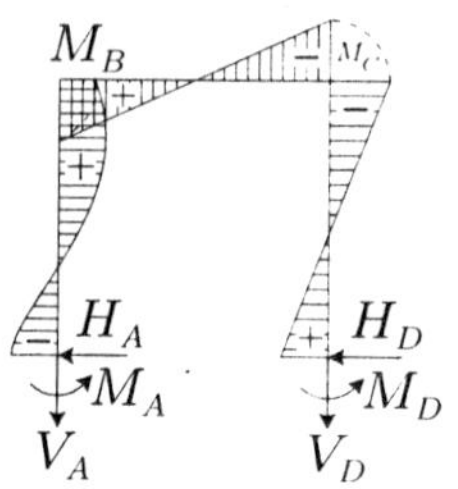	$M_A = \frac{wh^2}{24}\left[-\frac{3\alpha+8}{5N_1} - \frac{9\alpha+2}{N_2}\right]$ $M_D = \frac{wh^2}{24}\left[-\frac{3\alpha+8}{5N_1} + \frac{9\alpha+2}{N_2}\right]$ $M_B = \frac{wh^2}{24}\left[-\frac{2\alpha}{5N_1} + \frac{3\alpha}{N_2}\right]$ $M_C = \frac{wh^2}{24}\left[-\frac{2\alpha}{5N_1} - \frac{3\alpha}{N_2}\right]$ $H_D = \frac{wh}{40}\left[\frac{3\alpha+4}{N_1}\right]$ $H_A = -\left(\frac{wh}{2} - H_D\right)$	$V_A = -V_D = -\frac{wh^2\alpha}{4LN_2}$

Table 3 : Portal Frames with Inclined Members

Frame Constants

$$\alpha = \frac{I_2}{I_1}\cdot\frac{s_1}{b},\ \alpha_1 = \frac{I_2}{I_3}\frac{s_2}{b},\ A = 3+\alpha+\alpha_1$$

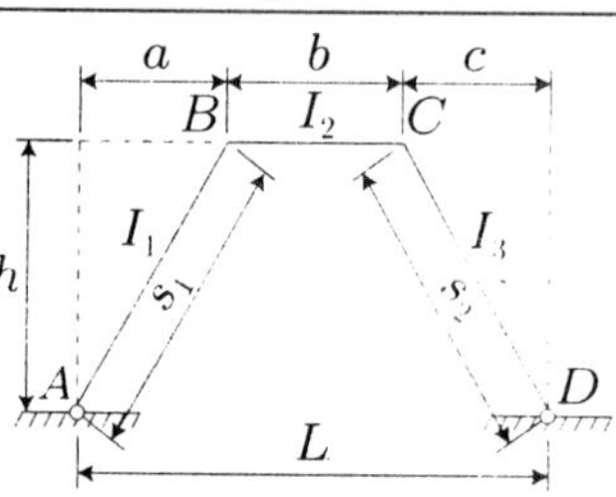

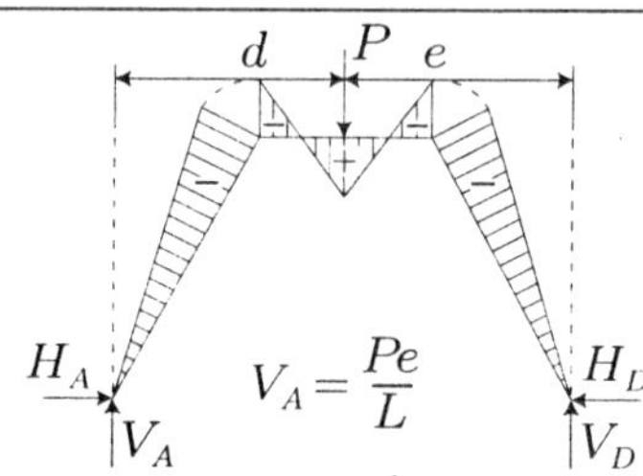

$$V_A = \frac{Pe}{L}$$

$$H_A = \frac{P}{2hL}\,\frac{\frac{3}{b}[e(dL-a^2)-dc^2]+2ae\alpha+2cd\alpha_1}{A}$$

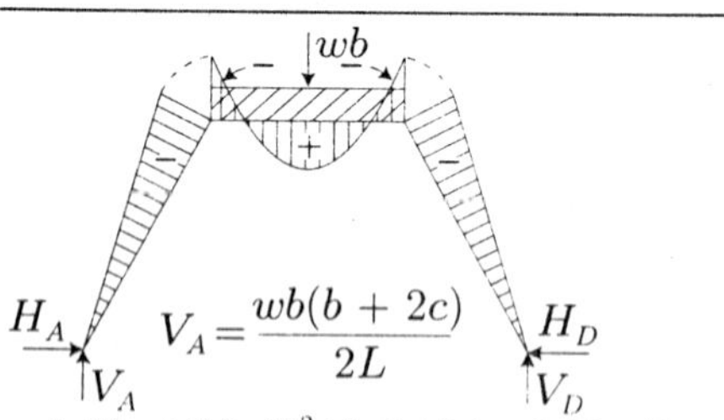

$$V_A = \frac{wb(b+2c)}{2L}$$

$$H_A = \frac{wb}{4hL}\,\frac{12ac+4bL-3b^2+2a(b+2c)\alpha+2c(2a+b)\alpha_1}{A}$$

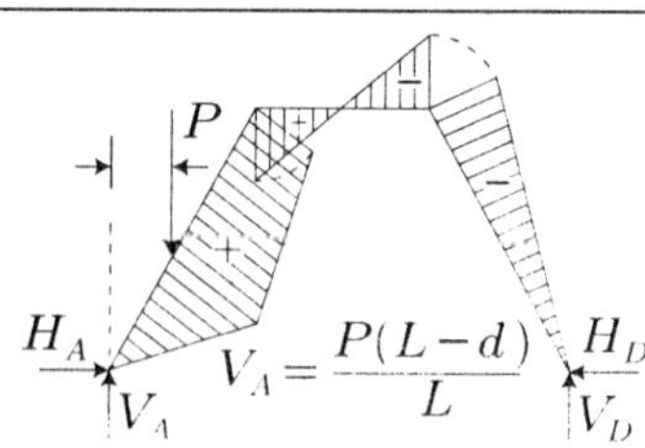

$$V_A = \frac{P(L-d)}{L}$$

$$H_A = \frac{Pd}{2Lh}\,\frac{3(b+2c)+(3L-L\left(\frac{d}{a}\right)^2-2a)\alpha+2c\alpha_1}{A}$$

Contd.

Table 3 (contd.)

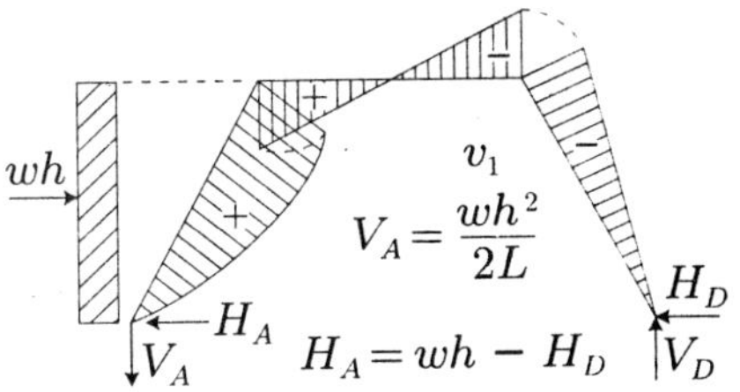

$$V_A = \frac{wh^2}{2L} \qquad H_A = wh - H_D$$

$$H_D = \frac{wh}{8L}\,\frac{6(b+2c)+(5L-4a)\alpha+4c\alpha_1}{A}$$

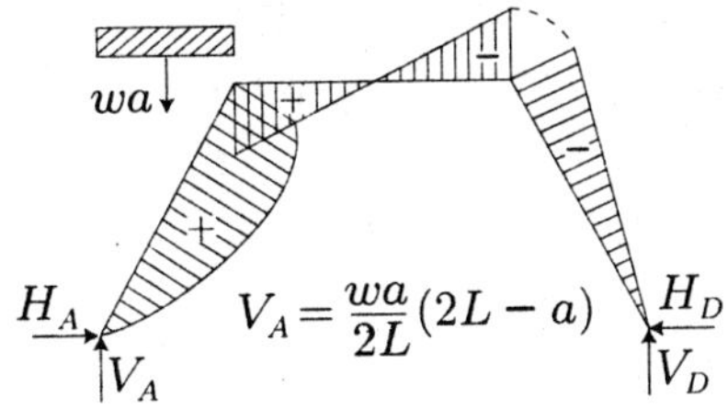

$$V_A = \frac{wa}{2L}(2L-a)$$

$$H = \frac{wa^2}{8Lh}\,\frac{6(b+2c)+(5L-4a)\alpha+4c\alpha_1}{A}$$

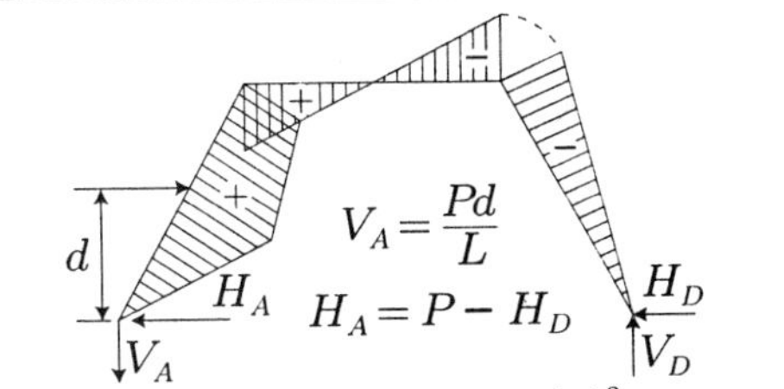

$$V_A = \frac{Pd}{L} \qquad H_A = P - H_D$$

$$H_D = \frac{Pd}{2hL}\,\frac{3(b+2c)+(3L-2a-L\left(\frac{d}{h}\right)^2)\alpha+2c\alpha_1}{A}$$

For this frame let $a = 0$ in the above formulas

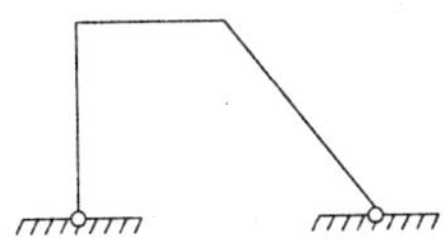

For this frame let $c = 0$ in the above formulas

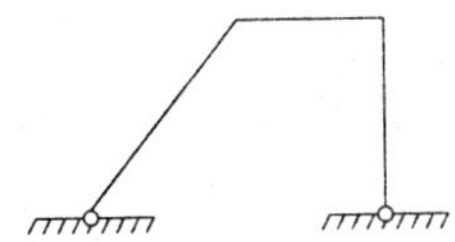

*Adapted from *Rahmenformeln* by Kleinlogel

Table 4 : Portal Frames with Inclined Beam

Frame Constants

$$n = \frac{h_1}{h},\quad \alpha = \frac{I_2}{I_1}\cdot\frac{h}{s},\quad A = \alpha(1+n^3)+1+n+n^2$$

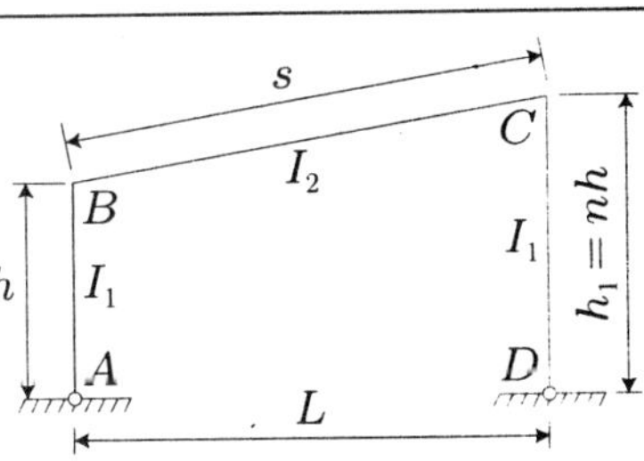

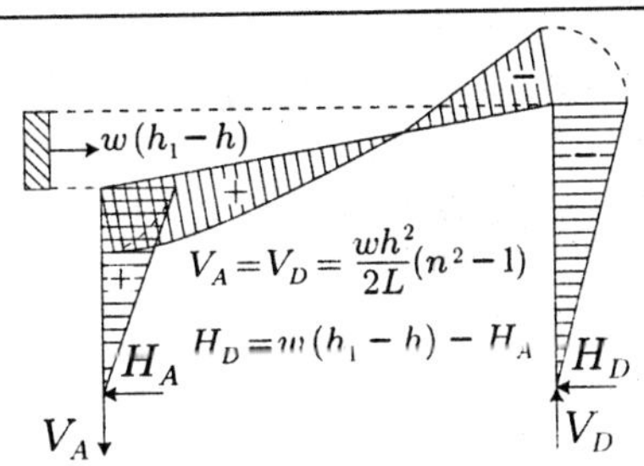

$$V_A = V_D = \frac{wh^2}{2L}(n^2-1)$$

$$H_D = w(h_1 - h) - H_A$$

$$H_A = \frac{wh}{24}\,\frac{(n-1)(24\alpha n^3+3+12n+21n^2)}{A}$$

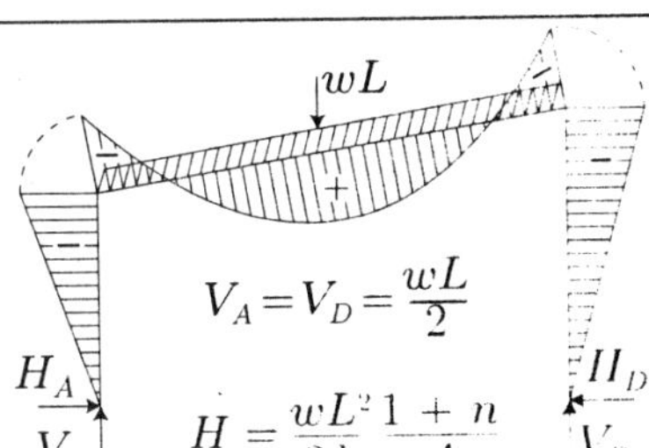

$$V_A = V_D = \frac{wL}{2}$$

$$H = \frac{wL^2}{3h}\,\frac{1+n}{A}$$

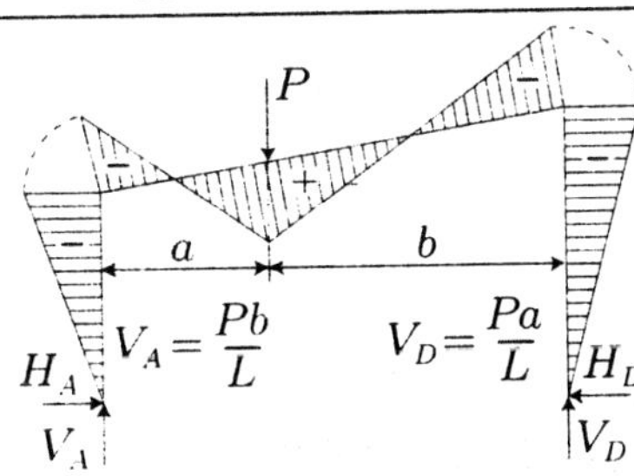

$$V_A = \frac{Pb}{L} \qquad V_D = \frac{Pa}{L}$$

$$H_A = H_D = \frac{Pab}{2L^2h}\,\frac{L(2+n)+an-a}{A}$$

Contd.

Table 4 (contd.)

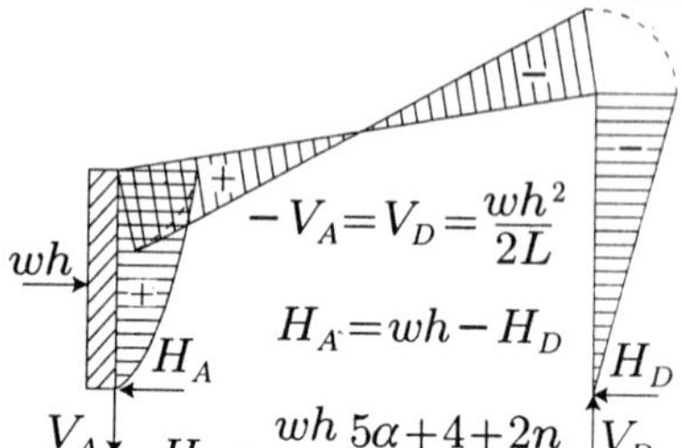 $-V_A = V_D = \frac{wh^2}{2L}$; $H_A = wh - H_D$; $H_D = \frac{wh}{8}\,\frac{5\alpha+4+2n}{A}$	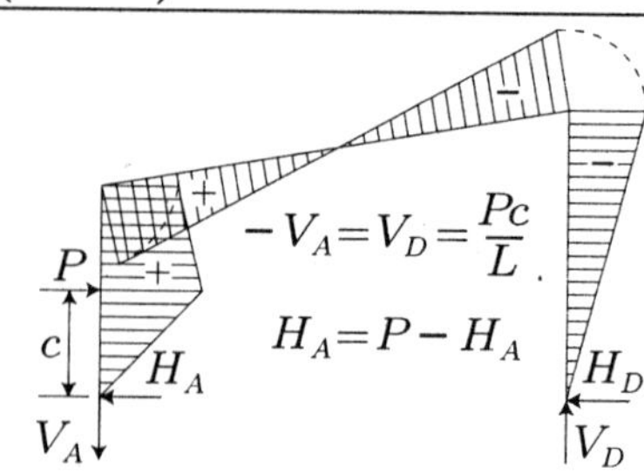 $-V_A = V_D = \frac{Pc}{L}$; $H_A = P - H_A$; $H_A = \frac{P}{h^3}\,\frac{\alpha[h^3(1+n^3)+\frac{c}{2}(c^2-3h^2)]+h^3(1+n+n^2)-\frac{ch^2}{2}(2+n)}{A}$
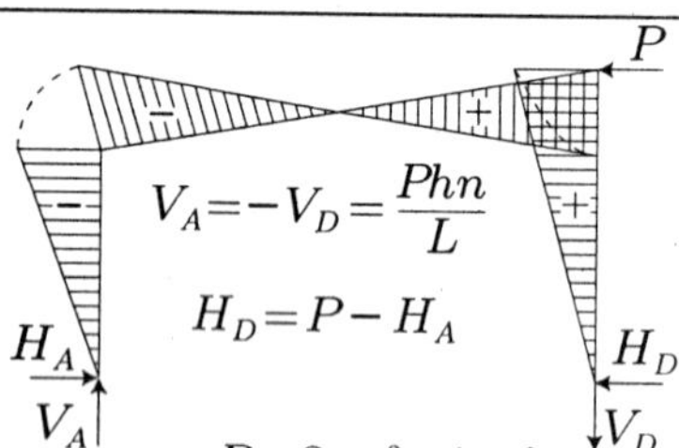 $V_A = -V_D = \frac{Phn}{L}$; $H_D = P - H_A$; $H_A = \frac{Pn}{2}\cdot\frac{2\alpha n^2+1+2n}{A}$	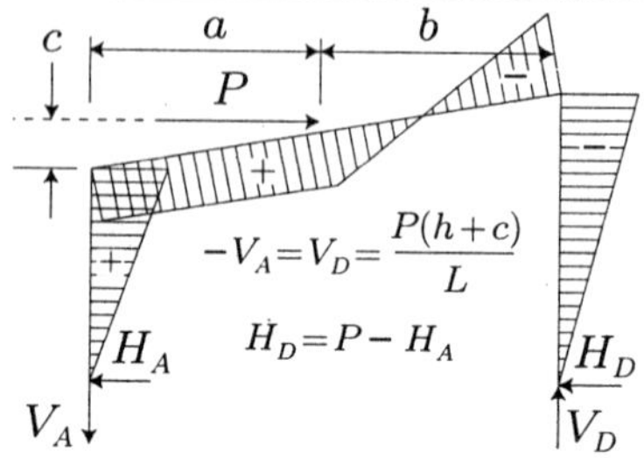 $-V_A = V_D = \frac{P(h+c)}{L}$; $H_D = P - H_A$; $H_A = \frac{P}{2h}\,\frac{\left[2\alpha n^3 h+(c+h)(1+2n)-\frac{bh}{L}(1+n-2n^2)-\frac{bc}{L}\left(2+n+\frac{c}{h}\right)\right]}{A}$

*Adapted from *Rahmenformeln* by Kleinlogel

Table 5 : Gable Frames

Frame Constants	
$\alpha = \frac{I_2}{I_1}\frac{h}{s},\; A = h^2(3+\alpha) + f(3h+f)$	
Frame dimensions 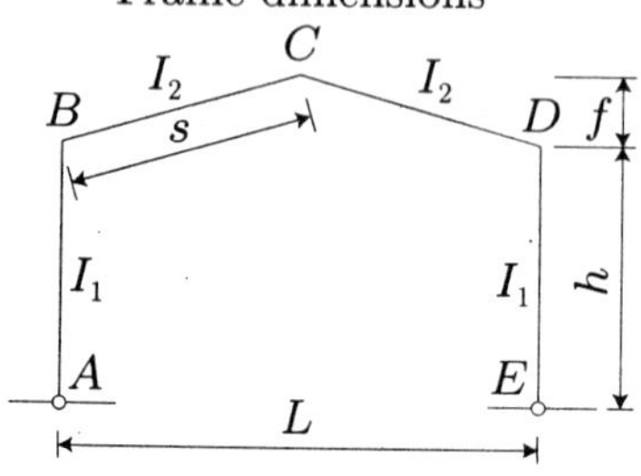	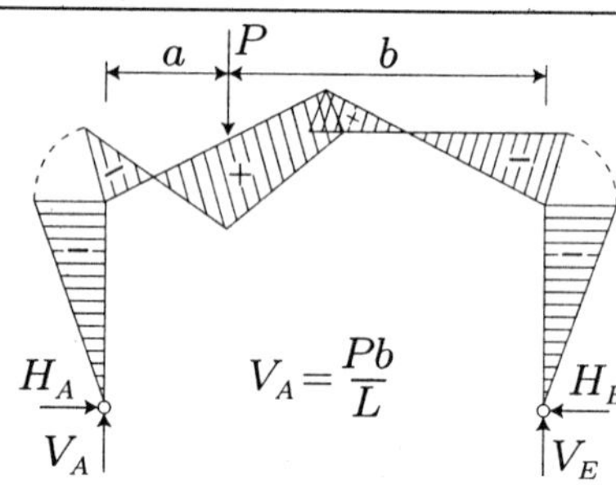 $V_A = \frac{Pb}{L}$; $H_A = \frac{Pa}{4L^2A}[6bLh + f(3L^2-4a^2)]$
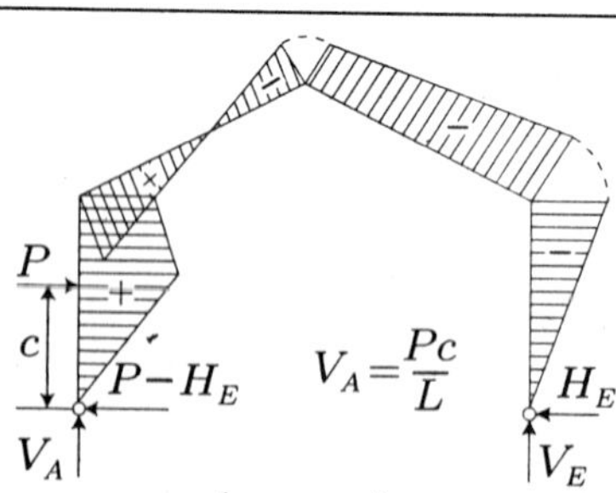 $V_A = \frac{Pc}{L}$; $H_E = \frac{Pc}{4A}\left[\alpha\left(3h-\frac{c^2}{h}\right)+3(2h+f)\right]$	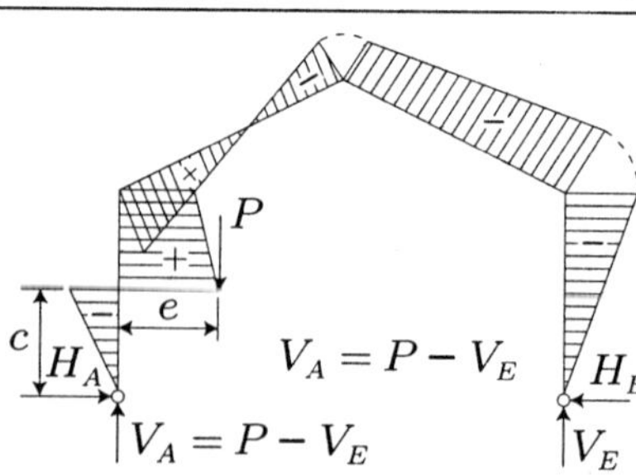 $V_A = P - V_E$; $V_A = P - V_E$; $H_E = \frac{3Pe}{4hA}[\alpha(h^2-c^2)+h(2h+f)]$

Contd.

Table 5 (contd.)

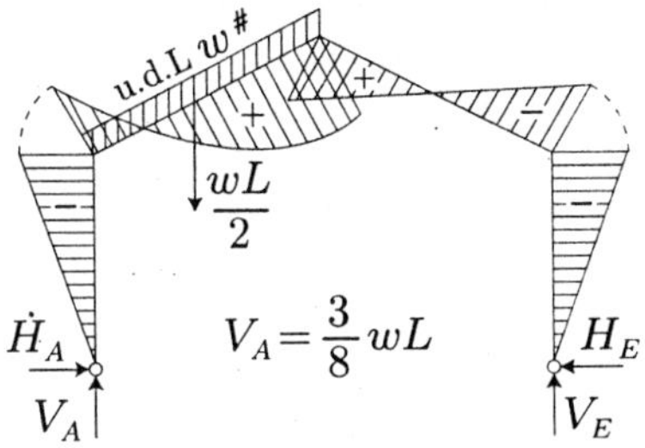 $V_A = \frac{3}{8}wL$ $H = \frac{wL^2}{64A}[8h + 5f]$	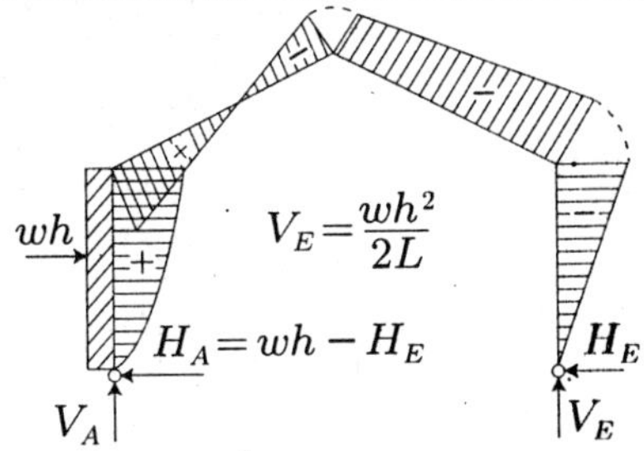 $V_E = \frac{wh^2}{2L}$ $H_A = wh - H_E$ $H_E = \frac{wh^2}{16A}[5h\alpha + 6(2h + f)]$
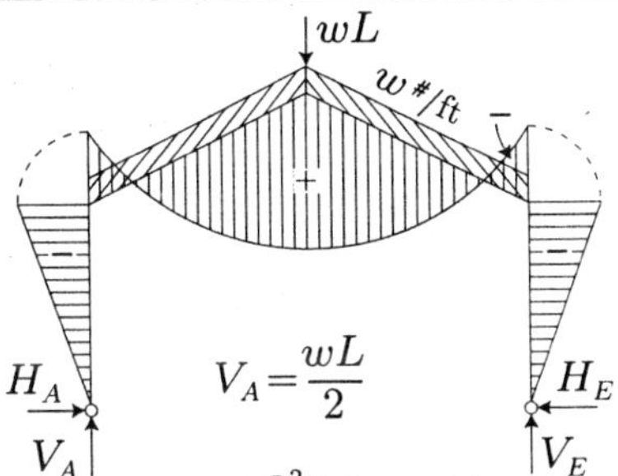 $V_A = \frac{wL}{2}$ $H = \frac{wL^2}{32A}[8h + 5f]$	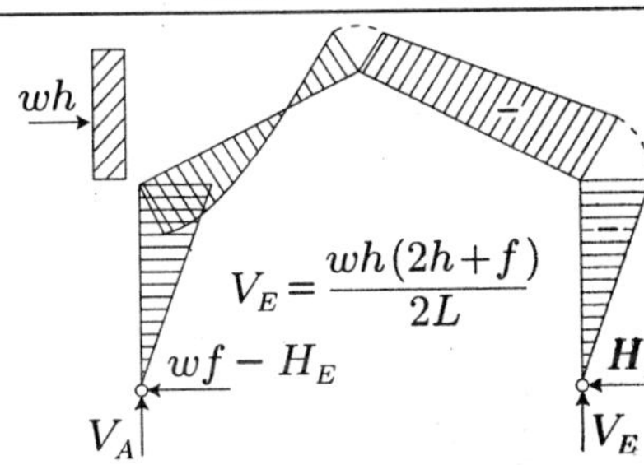 $V_E = \frac{wh(2h+f)}{2L}$ $wf - H_E$ $H_E = \frac{wf}{16A}[8h^2(3 + \alpha) + 5f(4h + f)]$
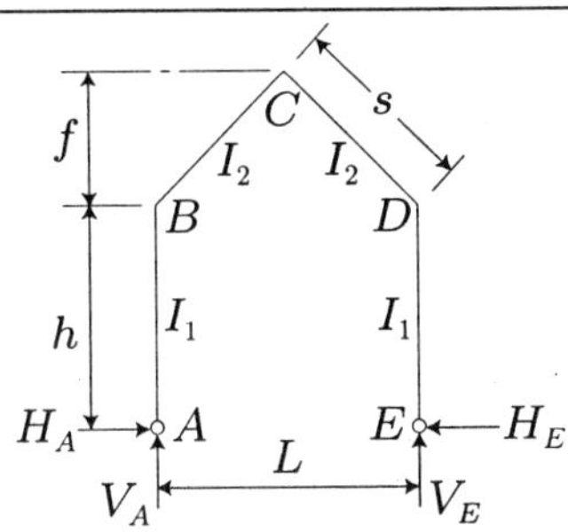	**FOR HINGED BASE** Coefficients $K = \frac{I_2}{I_1}\frac{h}{s};\ \phi = \frac{f}{h};\ m = 1 + \phi$ $B = 2(K + 1) + m;\ C = 1 + 2m,\ N = \boldsymbol{B + mC}$ Moment Sign Convention $M(+ve) \rightarrow$ Inside Tension
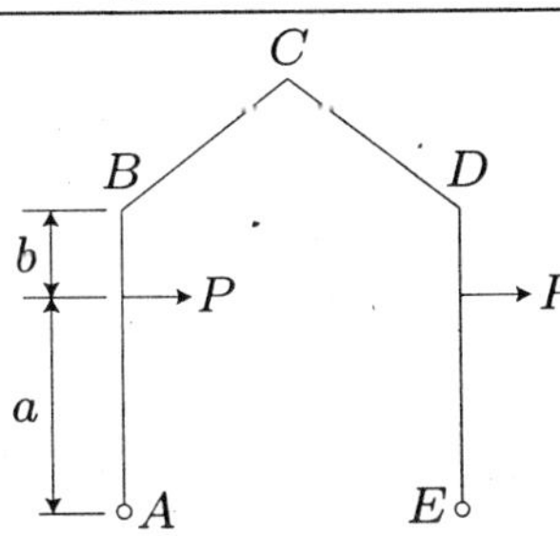	$M_B = -M_D = P_a$ $M_C = 0$ $H_A = -H_E = -P$ $V_A = -V_E = -\frac{2Pa}{L}$ Moments at Loads $= \pm Pa$
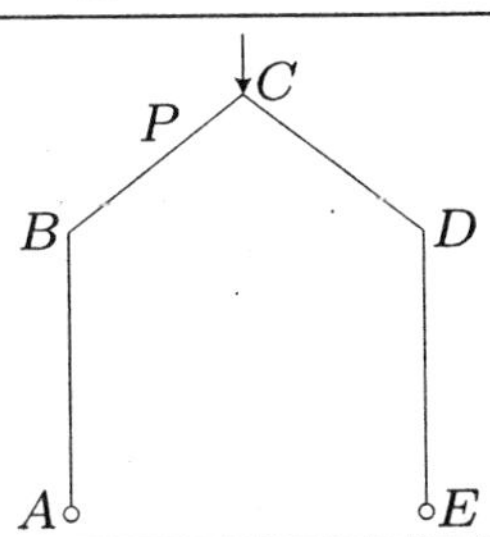	$M_B = M_D = -\frac{PL}{4}\cdot\frac{C}{N}$ $M_C = +\frac{PL}{4}\cdot\frac{B}{N}$ $V_A = V_E = \frac{P}{2},\ H_A = H_E = -\frac{M_B}{h}$

Contd.

Table 5 (contd.)

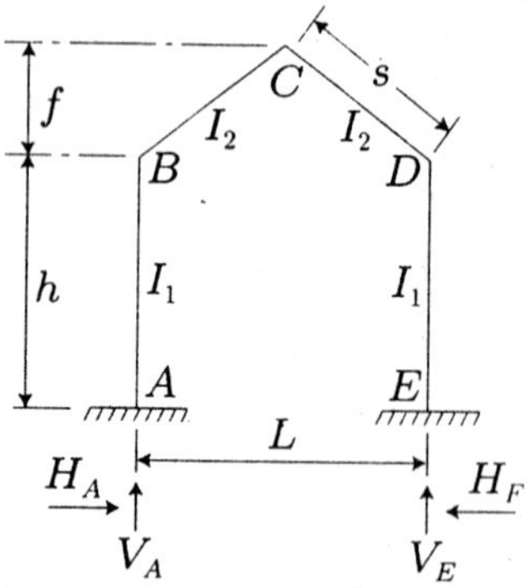

For Fixed Base

Coefficients :

$$K = \frac{I_2}{I_1} \cdot \frac{h}{s},\ \phi = \frac{f}{h},\ m = 1 + \phi,\ B = 3K + 2$$

$$C = 1 + 2m,\ K_1 = 2(K + 1 + m + m^2)$$

$$K_2 = 2(K + \phi^2)$$

$$R = \phi C - K;\ N_1 = K_1 K_2 - R^2,\ N_2 = 3K + B$$

Moment Sign Convention

(*i*) + *ve* moment – Inside tension

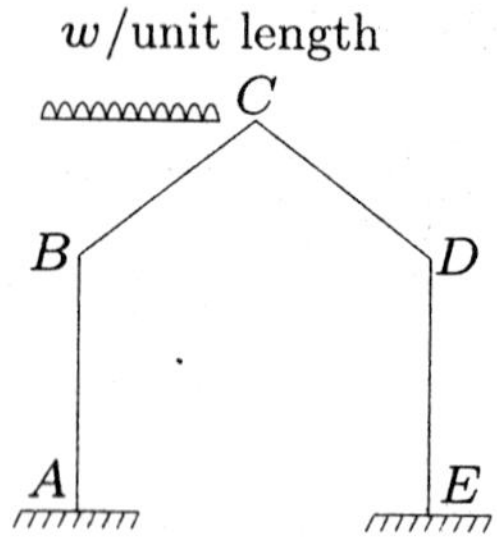

Constants :

$$X_1 = \frac{wL^2}{32} \cdot \frac{K(8 + 15\phi) + \phi(6 - \phi)}{N_1}$$

$$X_2 = \frac{wL^2}{32} \cdot \frac{K(16 + 15\phi) + \phi^2}{N_1}$$

$$X_3 = \frac{wL^2}{32 N_2}$$

$$M_A = +X_1 - X_3;\ M_E = +X_1 + X_3$$

$$M_C = \frac{wL^2}{16} - \phi X_1 - m X_2,\ M_B = -X_2 - X_3;$$

$$H_A = H_E = \frac{X_1 + X_2}{h}$$

$$M_D = -X_2 + X_3;\ V_E = \frac{wl}{8} - \frac{2X_3}{L};\ V_A = \frac{wl}{2} - V_E$$

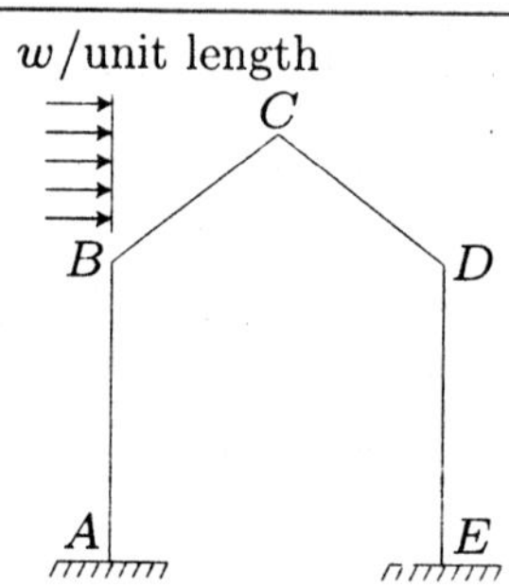

Constants :

$$X_1 = \frac{wf^2}{8} \cdot \frac{K(9\phi + 4) + \phi(6 + \phi)}{N_1}$$

$$X_2 = \frac{wf^2}{8} \cdot \frac{K(8 + 9\phi) - \phi^2}{N_1}$$

$$X_3 = \frac{wfh}{8} \cdot \frac{4B + \phi}{N_2};$$

$$M_A = -X_1 - X_3,\ M_B = +X_2 + \left(\frac{wfh}{2} - X_3\right)$$

$$M_E = -X_1 + X_3;\ M_D = +X_2 - \left(\frac{wfh}{2} - X_3\right)$$

$$M_C = -\frac{wf^2}{4} + \phi X_1 + m X_2;$$

$$V_A = -V_E = -\frac{wfh(2 + \phi)}{2L} + \frac{2X_3}{L}$$

$$H_E = \frac{wf}{2} - \frac{X_1 + X_2}{h};\ H_A = -(wf - H_E)$$

Contd.

Table 5 (contd.)

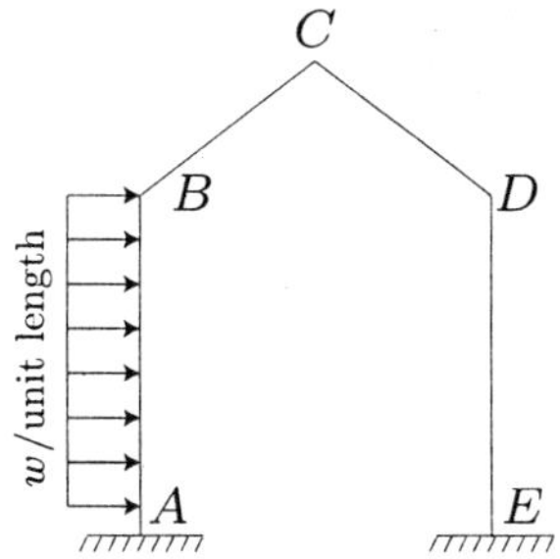 	**Constants :** $X_1 = \frac{wh^2}{8} \cdot \frac{K(K+6) + K\phi(15+16\phi) + 6\phi^2}{N_1}$ $X_2 = \frac{wh^2}{8} \cdot \frac{K(9\phi + 8\phi^2 - K)}{N_1}$; $X_3 = \frac{wh^2(2K+1)}{2N_2}$ $M_A = -X_1 - X_3$; $M_B = +X_2 + \left(\frac{wh^2}{4} - X_3\right)$ $M_E = -X_1 + X_3$, $M_D = +X_2 - \left(\frac{wh^2}{4} - X_3\right)$ $M_C = -\frac{wfh}{4} + \phi X_1 + mX_2$; $V_A = -V_E = -\frac{wh^2}{2L} + \frac{2X_3}{L}$ $H_E = \frac{wh}{4} - \frac{X_1 + X_2}{h}$; $H_A = -(wh - H_E)$
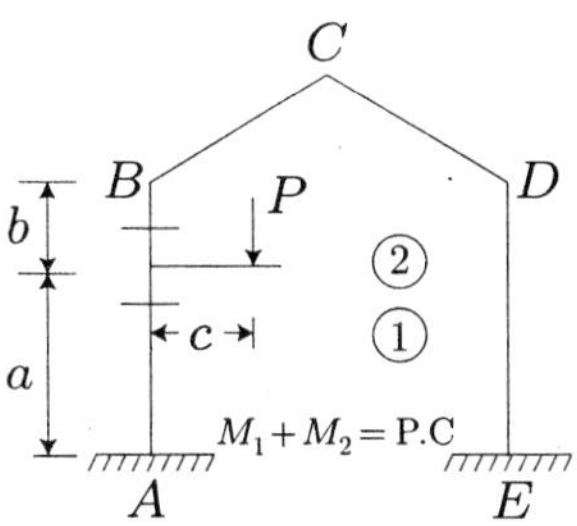 M_1 = Moment (outside) M_2 = Moment (inside)	**Constants :** $a_1 = \frac{a}{h}$, $b_1 = \frac{b}{h}$, $Y_1 = Pc[2\phi^2 - (1 - 3b_1^2)K]$ $Y_2 = Pc[\phi - (3a_1^2 - 1)K]$; $X_1 = \frac{Y_1K_1 - Y_2R}{2N_1}$ $X_2 = \frac{Y_2K_2 - Y_1R}{2N_1}$; $X_3 = \frac{PC}{2} \cdot \frac{B - 3(a_1 - b_1)K}{N_2}$ $M_A = -X_1 - X_3$; $M_B = +X_2 + \left(\frac{PC}{2} - X_3\right)$ $M_E = -X_1 + X_3$; $M_D = +X_2 - \left(\frac{PC}{2} - X_3\right)$ $M_C = -\frac{\phi PC}{2} + \phi X_1 + mX_2$, $M_1 = M_A - H_A a$ $M_2 = M_B + H_E b$; $V_E = \frac{Pc - 2X_3}{L}$ $V_A = P - V_E$; $H_A = H_E = \frac{PC}{2h} - \frac{X_1 + X_2}{h}$
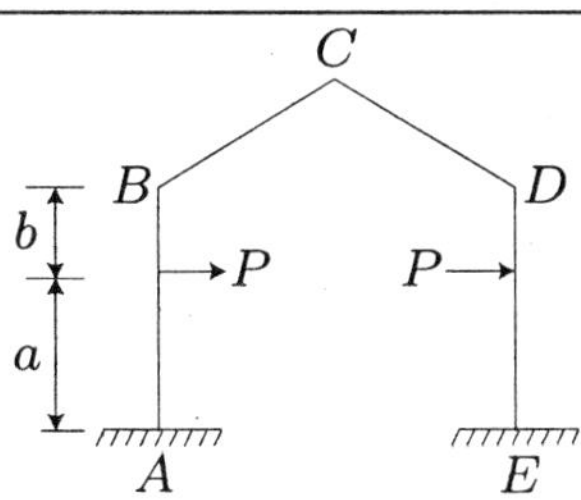 	**Constants :** $X_1 = \frac{P \cdot a(B + 3b_1 + K)}{N_2}$ $M_A = -M_E = -X_1$; $M_B = -M_D = P \cdot a - X_1$ $M_C = 0$; $V_A = -V_E = +2\left[\frac{P \cdot a - X_1}{L}\right]$ $H_A = -H_E = -P$

Selected References

1. Wilbur, J.B., Norris, C.B., and Utku, S., *Elementary Structural Analysis*, McGraw Hill, New York, 1976.

2. Mtheson, A.L., *Hyperstatic Structures*, Vols. I and II, Butterworth's Scientific Publications, London, 1959.

3. Cross, H., and Morgan, N.D., *Continuous Frames of Reinforced Concrete*, John Wiley & Sons Inc., New York, 1932.

4. Ghali, A. and Neville, A.M., *Structural Analysis*, Chapman and Hill, London, New York, 1978.

5. Gere, J.M., *Moment Distribution*, D. Van Nostrand Co., Inc., New York, 1963.

6. Lightfoot, E., *Moment Distribution*, Span, London, 1961.

7. Gere, J.M. and Weaver (Jr.), W., *Analysis of Framed Structures*, D. Van Nostrand Co., Inc., New York, 1965.

8. Wright, J., *Distribution Method of Analysis*, Thames and Hudson, London, 1963.

9. Borg, S.F. and Gennaro, *Advanced Structural Analysis*, Van Nostrand Co., Inc., New York, 1959.

10. Mough, L.C. *Statically Indeterminate Structures*, John Wiley and Sons, New York, 1964.

11. Wang, C.K., *Statically Indeterminance Structures*, McGraw Hill Book Co., New York.

12. Kinney, J.S., *Indeterminate Structural Analysis*, Addison-Wiley Publishing Co., Mass.

13. Borg, S.F., *Fundamentals of Engineering Elasticity*, D. Van Nostrand Co., Inc., New York, 1962.

14. Wang, C., *Applied Elasticity*, McGraw Hill, New York.

15. Parcel, J.L. and Moorman, R.B.B., *Analysis of Statically Indeterminate Structures*, John Wiley and Sons, New York, 1955.

16. Gere, J.M. and Weaver (Jr.), W., *Matrix Algebra for Engineers*, D. Van Nostrand Co., Inc., New York.

17. Pippard, A.J.S. and Baker, J.F., *The Analysis of Engineering Structures*, Edward Arnold and Co., London, 1953.

18. Allen, D.W.G., *Relaxation Methods in Engineering and Science*, Mc.Graw Hill Book Co., Inc., New York, 1954.

19. Morice, P.B., *Matrices for Structural Analysis*, E. & F.N. Span, London, 1962.

20. Wang, C.K., *Intermediate Structure Analysis*, International Student Ed., Auckland, McGraw Hill, New Delhi, 1983.

21. Charlton, T.M., *Model Analysis of Structures*, E. & F.N., Span, London, 1954.

22. Shedd and Vawter, *Theory of Simple Structure*, John Wiley and Sons, New York.

23. Amenakes, Anthony E., *Classical Structural Analysis*, McGraw Hill Book Company, 1988.

24. Roy, S.K. and Chakrabarty, S., *Fundamentals of Structural Analysis*, S. Chand and Co. Ltd., New Delhi.

Index